Lecture Notes in Mathematics 1996

Thomas Lorenz

Mutational Analysis

A Joint Framework for Cauchy Problems In and Beyond Vector Spaces

Springer

Thomas Lorenz
Goethe University Frankfurt am Main
Institute of Mathematics
Robert-Mayer-Str. 10
60325 Frankfurt am Main
Germany
lorenz@math.uni-frankfurt.de

ISBN: 978-3-642-12470-9 e-ISBN: 978-3-642-12471-6
DOI: 10.1007/978-3-642-12471-6
Springer Heidelberg Dordrecht London New York

Lecture Notes in Mathematics ISSN print edition: 0075-8434
ISSN electronic edition: 1617-9692

Library of Congress Control Number: 2010928326

Mathematics Subject Classification (2000): 34A60, 34G10, 35K20, 49J53, 60H20, 93B03

Cover design: SPi Publisher Services

Printed on acid-free paper

springer.com

Preface

Differential problems should not be restricted to vector spaces in general.

The Main Goal of This Book

Ordinary differential equations play a central role in science. Newton's Second Law of Motion relating force, mass and acceleration is a very famous and old example formulated via derivatives. The theory of ordinary differential equations was extended from the finite-dimensional Euclidean space to (possibly infinite-dimensional) Banach spaces in the course of the twentieth century. These so-called *evolution equations* are based on strongly continuous semigroups.
For many applications, however, it is difficult to specify a suitable normed vector space. Shapes, for example, do not have an obvious linear structure if we dispense with any a priori assumptions about regularity and thus, we would like to describe them merely as compact subsets of the Euclidean space.

Hence, this book generalizes the classical theory of ordinary differential equations beyond the borders of vector spaces. It focuses on the well-posed Cauchy problem in any finite time interval.

In other words, states are evolving in a set (not necessarily a vector space) and, they determine their own evolution according to a given "rule" concerning their current "rate of change" — a form of feedback (possibly even with finite delay). In particular, the examples here do not have to be gradient systems in metric spaces.

The Driving Force of Generalization: Solutions via Euler Method

The step-by-step extension starts in metric spaces and ends up in nonempty sets that are merely supplied with suitable families of distance functions (not necessarily symmetric or satisfying the triangle inequality).
Solutions to the abstract Cauchy problem are usually constructed by means of the Euler method and so the key question for each step of conceptual generalization is: Which aspect of the a priori given structures can be still weakened so that the Euler method does not fail ?

Diverse Examples Have Always Given Directions ... Towards a Joint Framework.

In the 1990s, Jean-Pierre Aubin suggested what he called *mutational equations* and applied them to systems of ordinary differential equations and time-dependent compact subsets of $\mathbb{R}^N$ (equipped with the popular Pompeiu-Hausdorff metric). They are the starting point of this monograph.
Further examples, however, reveal that Aubin's a priori assumptions (about the additional structure of the metric space) are quite restrictive indeed. There is no obvious way for applying the original theory to semilinear evolution equations.

Our basic strategy to generalize mutational equations is simple: Consider several diverse examples successively and, whenever it does not fit in the respective mutational framework, then find some extension for overcoming this obstacle.

Mutational Analysis is definitely not just to establish another abstract term of solution though. Hence, it is an important step to check for each example individually whether there are relations to some more popular meaning (like classical, strong, weak or mild solution).

Here are some of the examples under consideration in this book:

- Feedback evolutions of nonempty compact subsets of $\mathbb{R}^N$
  Application to image segmentation
- Birth-and-growth processes of random closed sets (not necessarily convex)
- Semilinear evolution equations in arbitrary Banach spaces
- Nonlocal parabolic differential equations in noncylindrical domains
- Nonlinear transport equations for Radon measures on $\mathbb{R}^N$
- Structured population model with Radon measures on $\mathbb{R}_0^+$
- Stochastic ordinary differential equations with nonlocal sample dependence

In particular, these examples can now be coupled in systems immediately – due to the *joint* framework of Mutational Analysis. This possibility provides new tools for modelling in future.

The Structure of This Extended Book ... for the Sake of the Reader

This monograph is written as a synthesis of two aims: first, the reader should have quick access to the results of individual interest and second, all mathematical conclusions are presented in detail so that they are sufficiently comprehensible.

Each chapter is elaborated in a quite self-contained way so that the reader has the opportunity to select freely according to the examples of personal interest. Hence some arguments typical for mutational analysis might appear rather frequently, but they are always adapted to the respective framework. Moreover, the proofs are usually collected at the end of each subsection so that they can be skipped easily if wanted. References to results elsewhere in the monograph are usually supplied with page numbers. Each example contains a table that summarizes the choice of basic sets, distances etc. and indicates where to find the main results.

The introductory Chapter 0 summarizes the essential notions and motivates the generalizations in this book. Many of the subsequent conclusions have their origins in §§ 1.1 – 1.6 and so these subsections facilitate understanding the modifications later.

Experience has already taught that such a monograph cannot be written free from any errors or mistakes. I would like to apologize in advance and hope that the gist of both the approach and examples is clear. Comments are very welcome.

Heidelberg and Frankfurt, January 2010 Thomas Lorenz

Acknowledgments

This monograph would not have been completed if I had not benefited from the harmony and the support in my vicinity. Both the scientific and the private aspect are closely related in this context.

Prof. Willi Jäger has been my academic teacher since my very first semester at Heidelberg University. Infected by the "virus" of analysis, I followed his courses, full of insights into mathematical relations. As a part of his scientific support, he drew my attention to set-valued maps quite early and gave me the opportunity to gain experience very autonomously. I would like to express my deep gratitude to Prof. Jäger.
Moreover, I am deeply indebted to Prof. Jean–Pierre Aubin and Hélène Frankowska. Their mathematical influence on me started quite early — as a consequence of their monographs. During three stays at CREA of Ecole Polytechnique in Paris, I benefited from collaborating with them and meeting several colleagues sharing my mathematical interests partly.
Furthermore, I would like to thank all my friends, collaborators and colleagues for the inspiring discussions and observations over time. This list (in alphabetical order) is neither complete nor a representative sample, of course: Zvi Artstein, Robert Baier, Bruno Becker, Hans Belzer, Christel Brüschke, Eva Crück, Roland Dinkel, Herbert-Werner Diskut, Tzanko Donchev, Matthias Gerdts, Piotr Gwiazda, Peter E. Kloeden, Roger Kömpf, Stephan Luckhaus, Anna Marciniak-Czochra, Reinhard Mohr, Jerzy Motyl, José Alberto Murillo Hernández, Janosch Rieger, Ina Scheid, Ursula Schmitt, Roland Schnaubelt, Oliver Schnürer, Jens Starke, Angela Stevens, Martha Stocker, Christina Surulescu, Manfred Taufertshöfer, Friedrich Tomi, Edelgard Weiß-Böhme, Kurt Wolber.

Heidelberg University and, in particular, its Interdisciplinary Center for Scientific Computing (IWR) were extraordinary home institutions for me until my habilitation (for venia lengendi in mathematics) in 2009. In addition, some results of this monograph were elaborated as parts of projects or during research stays funded by

- German Research Foundation DFG (SFB 359 and LO 273)
- Hausdorff Research Institute for Mathematics in Bonn (spring 2008)
- Research Training Network "Evolution Equations for Deterministic and Stochastic Systems" (HPRN–CT–2002–00281) of the European Community
- Minerva Foundation for scientific cooperation between Germany and Israel.

Finally, I would like to express my deep gratitude to my family.
My parents have always supported me and have provided a harmonic environment so that I have been able to concentrate on my studies. I surely would not have reached my current situation without them as a permanent pillar.
Meanwhile my wife Irina Surovtsova has been at my side for several years. I have always trusted her to give me good advice and so she has often enabled me to overcome obstacles — both in everyday life and in science. I am optimistic that together we can cope with the challenges that Daniel, Michael and the "other aspects" of life provide for us.

TL

Contents

Chapter 0
Introduction

Think beyond vector spaces !

0.1 Diverse Evolutions Come Together Under the Same Roof

Many applications consist of diverse components and thus, their mathematical description as functions often starts with long preliminaries (like restrictive assumptions about regularity).
However, *shapes and images are basically sets, not even smooth* (Aubin [10]). This observation leads to the question how to specify models in which both real- or vector-valued functions and shapes are involved. The components usually depend on time and have a huge amount of influence over each other. Consider e.g.

→ A bacterial colony is growing in a nonhomogeneous nutrient broth. For the bacteria, both speed and direction of expansion depend on the nutrient concentration close to the boundary in particular. On the other hand, the nutrient concentration is changing due to consumption and diffusion. (Further applications of set-valued flows in biological modeling are sketched in [57].)

→ A chemical reaction in a liquid is endothermic and depends strongly on the dissolved catalyst. However, this catalyst is forming crystals due to temperature decreasing.

→ In image segmentation, a computer is to detect the region belonging to one and the same object. An example of a so-called region growing method (presented here in § 1.10) is based on constructing time-dependent compact segments so that an error functional is decreasing in the course of time. So far, smoothing effects on the image within the current segment are not taken into account. Basically speaking, it is an example how to extend Lyapunov methods to shape optimization. Further examples can be found in [58, 71].

→ In dynamic economic theory, the results of control theory form the mathematical basis for important conclusions (e.g. [11]). Coalitions of economic agents, technological progress and social effects due to migration, however, have an important impact on the dynamic process that is difficult to quantify by vector-valued functions. Thus, some parameters ought to be described as sets of permissible values and, these subsets might depend on current and former states.

Our goal consists in a joint framework for Cauchy problems of maybe completely different types. In particular, examples of evolving shapes motivate the substantial aspect that we dispense with any (additional) linear structure whenever possible. In other words, the key question here is how to extend ordinary differential equations beyond vector spaces.

T. Lorenz, *Mutational Analysis: A Joint Framework for Cauchy Problems In and Beyond Vector Spaces*, Lecture Notes in Mathematics 1996,
DOI 10.1007/978-3-642-12471-6_1, © Springer-Verlag Berlin Heidelberg 2010

Why We Need a "Nonvectorial" Approach to Evolving Subsets of $\mathbb{R}^N$

In regard to time-dependent subsets of the Euclidean space $\mathbb{R}^N$, several formulations in vector spaces have already been suggested and, they have proved to be very useful. Each of these "detours" via a vector space, however, has conceptual constraints for analytical (but not geometric) reasons. This observation strengthens our interest in describing shape dynamics on the basis of distances (not vectors).

Osher and Sethian, for example, devised new numerical algorithms for fronts propagating with curvature-dependent speed in 1988 [149]. Describing these fronts as level sets of a real-valued auxiliary function leads to equations of motion which resemble Hamilton-Jacobi equations with parabolic right-hand sides. As an essential advantage, their numerical methods can handle topological merging and breaking naturally.
Meanwhile this level set approach has a solid analytical base in the form of viscosity solutions introduced by Crandall and Lions (see e.g. [51, 52], [43, 44], [24, 175]). The viscosity approach, however, has two constraints due to the parabolic maximum principle as its conceptual starting point:

(1.) All these geometric evolutions have to obey the so-called *inclusion principle*, i.e., whenever an initial set contains another initial subset, this inclusion is always preserved while evolving.
De Giorgi even suggested to use this inclusion principle for constructing subsolutions and supersolutions whose values are sets with nonsmooth boundaries — similarly to Perron's method for elliptic partial differential equations [54], [28, 29]. Cardaliaguet extended this notion to set evolutions depending on their nonlocal properties [36, 37, 38]. However, there is no obvious way how to apply these concepts to the easy example that the normal velocity at the boundary is $\frac{1}{1 + \text{set diameter}} > 0$.

(2.) There is no popular theory for the existence of viscosity solutions to *systems* so far.

Replacing viscosity solutions by weak (distributional) solutions to the equations of motion, we always have to neglect any influence of subsets with measure 0.
The distance from a given subset might provide a suitable alternative to the characteristic function of this set, but in general, the distance is just Lipschitz continuous. The choice of the function space is directly related to the regularity of the topological boundary. Delfour and Zolésio pointed out that the *oriented distance function* is often a more appropriate way to characterize a closed subset $K \subset \mathbb{R}^N$, i.e.

$$\mathbb{R}^N \longrightarrow \mathbb{R}, \quad x \longmapsto \begin{cases} \ \ \operatorname{dist}(x, K) \stackrel{\text{Def.}}{=} \inf\{|x-y| \,:\, y \in K\} & \text{if } x \in \mathbb{R}^N \setminus K \\ -\operatorname{dist}(x, \partial K) & \text{if } x \in K. \end{cases}$$

If its restriction to a neighborhood of the topological boundary ∂K belongs to the Sobolev space $W^{2,p}_{\text{loc}}$ with $p > N$, for example, then the well-known embedding theorem of Sobolev implies immediately that the set K is of class $C^{1,\alpha}$ [55, § 5.6.3].

0.2 Some Introductory Examples

0.2.1 A Region Growing Method of Image Segmentation

An important problem of computer vision is the detection of image segments which belong to the same object. Meanwhile many concepts have been developed to find their boundaries on grey-scale images. We mention only few earlier approaches for clarifying the differing aspects here.

The early methods use real-valued "detectors" to check if a point belongs to the contours or not. These criteria mostly depend on large changes of the grey values that are reflected by their gradients. For finding the segments of the same objects, the detected points have to be combined to boundaries, but the algorithms of each step lose more information about the image and, errors can hardly be corrected. For avoiding this weakness, other methods are based on approximations that are improved in some sense while time is increasing. Active contour models (*snakes*) belong to the popular examples that have been implemented efficiently (e.g. [39, 98, 185]). Restricted to two dimensions, they describe each contour as a Jordan curve that is deformed for minimizing some energy functional. These curves are to approximate the solution of a variational problem while time is increasing.

Many algorithms of image segmentation rely on analytical concepts that use a priori assumptions about regularity. Snakes (in their classical form), for example, are described as Jordan curves that are even twice continuously differentiable. Therefore edges can be found only in some smoothed shapes. Furthermore it is impossible to change the topological properties of the resulting segment. Meanwhile there have been several suggestions to overcome such weaknesses. Level set methods represent probably the most popular approach [148, 172]. Many of these ideas follow former directions and develop abstract generalizations which are to bridge the gaps. Level set methods, for example, use viscosity solutions of (generalized) Hamilton-Jacobi equations as mentioned before.

On the Way to an Approach (Just) by Means of Set-Valued Analysis

Our goal here is a (hopefully rather simple) region growing method – just on the basis of evolving compact subsets of $\mathbb{R}^N$, i.e. in comparison with many preceding approaches, there are:

- no a priori restrictions on the regularity of final contours and
- no parameterization of boundaries while expanding.

Indeed, searching for the (connected) image segment of an object, the basic graphical notion is only to decide which points belong to the segment. If we omit any additional conditions on regularity we want to detect a compact subset of $\mathbb{R}^N$ and so, the approximations depending on time are described as a set-valued map which associates each time $t \in [0,T[$ with a nonempty compact subset $K(t) \subset \mathbb{R}^N$:

$$K(\cdot) : [0,T[\longrightarrow \mathscr{K}(\mathbb{R}^N).$$

For quantifying the "quality" of the approximations we need a real-valued functional of compact subsets of $\mathbb{R}^N$. We prefer regarding it as "measurement of error" to interpreting it as "energy". The variance of grey values $G|_M$ (restricted to a subset $M \subset \mathbb{R}^N$ with positive Lebesgue measure), for example, gives a quantitative impression of their oscillation in M. More generally, we consider

$$\begin{aligned} \Phi : \mathscr{K}(\mathbb{R}^N) &\longrightarrow \mathbb{R}, \\ M &\longmapsto \psi\Big(\mathscr{L}^N(M),\ \int_M G\,dx,\ \int_M G^2\,dx\Big) \end{aligned}$$

with a function $\psi \in C_c^2(]0,\infty[\times\mathbb{R}^2,\mathbb{R})$. The composition

$$\Phi \circ K :\ [0,T[\ \longrightarrow \mathbb{R}$$

is a usual real-valued function which ought to decrease for improving the approximation $K(t) \subset \mathbb{R}^N$ in the course of time.

Finally, the aim of a *region growing* method (in a stricter sense) can be formulated as the following mathematical problem:

Given:	function of grey values $G \in C_c^0(\mathbb{R}^N)$, $N \geq 2$ error functional $\Phi : \mathscr{K}(\mathbb{R}^N) \longrightarrow \mathbb{R}$ s.t. $\Phi(M) = \psi\Big(\mathscr{L}^N(M),\ \int_M G\,dx,\ \int_M G^2\,dx\Big)$ with some $\psi \in C_c^2(]0,\infty[\times\mathbb{R}^2,\mathbb{R})$, initial set $K_0 \in \mathscr{K}(\mathbb{R}^N)$.
Wanted:	$K(\cdot) : [0,T[\ \longrightarrow \mathscr{K}(\mathbb{R}^N)\quad (T \in]0,\infty])$: (i) $K(0) = K_0$ (ii) $K(s) \subset K(t)$ whenever $s \leq t$ (iii) $K(\cdot)$ continuous w.r.t. Hausdorff metric (iv) $\Phi \circ K(\cdot) : [0,T[\ \longrightarrow \mathbb{R}$ nonincreasing (v) $M := \bigcup_{0\leq t<T} K(t)$ is "critical" w.r.t. Φ

The term of a "critical" set in $\mathbb{R}^N$ remains to be specified precisely. Intuitively we are looking for a (not necessarily closed) set $M \subset \mathbb{R}^N$ which cannot be "improved" in an obvious way by decreasing $\Phi \circ K(\cdot)$ and thus, M is the final candidate for the wanted image segment surrounding K_0.

The ansatz for $K(t)$ is based on the notion of prescribing the *speed* of set expansion (but not the direction of the corresponding *velocity*). We can easily avoid restrictions of regularity if this speed function is not specified just on the boundary $\partial K(t)$, but on the whole space $\mathbb{R}^N$. Then for a function $c : [0,T[\times\mathbb{R}^N \longrightarrow [0,\infty[$ given, the initial compact set $K_0 \in \mathscr{K}(\mathbb{R}^N)$ is deformed to

$$K(t) := \big\{x(t) \in \mathbb{R}^N \mid \exists\, x(\cdot) \in W^{1,1}([0,t],\mathbb{R}^N) : \; x(0) \in K_0, \\ |x'(s)| \le c(s, x(s)) \;\text{ for } \mathscr{L}^1\text{-almost every } s \in [0,t]\,\big\}.$$

In other words, this is the *reachable set* of K_0 and the differential inclusion

$$x'(\cdot) \in \mathbb{B}_{c(\cdot,x(\cdot))}(0) \quad \text{a.e. in } [0,t].$$

Here $\mathbb{B}_{c(s,x(s))}(0) \subset \mathbb{R}^N$ denotes the closed ball with center at 0 and radius $c(s, x(s))$. The key criterion for constructing $c(\cdot,\cdot)$ is that the real-valued composition

$$[0,T[\longrightarrow \mathbb{R}, \qquad t \longmapsto \Phi(K(t)) = \psi\Big(\mathscr{L}^N(K(t)), \int_{K(t)} G\,dx, \int_{K(t)} G^2\,dx\Big)$$

should be decreasing. Reynolds Transport Theorem for differential inclusions (in § A.6 on page 476 ff.) provides sufficient conditions on $c(\cdot,\cdot)$ such that each time-dependent argument

$$[0,T[\longrightarrow \mathbb{R}, \qquad t \longmapsto \int_{K(t)} G^k\,dx \qquad\qquad (k = 0,1,2)$$

is absolutely continuous with the (weak) derivative

$$\tfrac{d}{dt} \int_{K(t)} G^k\, dx \;=\; \int_{\partial K(t)} G(x)^k\; c(t,x)\; d\mathscr{H}^{N-1}x$$

Here $\mathscr{H}^{N-1}$ denotes the $(N-1)$-dimensional Hausdorff measure in $\mathbb{R}^N$. Now the chain rule for absolutely continuous functions provides the weak derivative of the relevant composition

$$\tfrac{d}{dt}\, \Phi(K(t)) \;=\; \int_{\partial K(t)} \varphi(x, K(t)) \cdot c(t,x)\; d\mathscr{H}^{N-1}x$$

with the coefficient function

$$\varphi(z,M) := \sum_{k=0}^{2} \partial_{k+1}\psi\Big(\mathscr{L}^N(M), \int_M G\,dx, \int_M G^2\,dx\Big) \cdot G(z)^k.$$

The basic idea of solving the segmentation problem is quite easy: The composition $\Phi \circ K(\cdot)$ is nonincreasing if the integrand of its (weak) derivative is nonpositive, i.e. $\varphi(x, K(t)) \cdot c(t,x) \le 0$ for all $t \in [0,T[$, $x \in \partial K(t)$. As a consequence we get the following criterion of the construction of $c(\cdot,\cdot)$: for all $t \in [0,T[$, $x \in \partial K(t)$,

$$\varphi(x, K(t)) > 0 \implies c(t,x) = 0.$$

Roughly speaking, the sign of $\varphi(\cdot,K(t))$ ought to be locally "stable" because Reynolds Transport Theorem (in § A.6) supposes $c(\cdot,\cdot)$ to be continuous with respect to space (at least). In this context we benefit from the assumption $G \in C^0_c(\mathbb{R}^N)$ for the first time:

Lemma 1. *Let $K_0 \in \mathscr{K}(\mathbb{R}^N)$ and $x \in \partial K$ satisfy $\varphi(x, K_0) < 0$.*
Then there exist both an autonomous Lipschitz continuous function $c : \mathbb{R}^N \longrightarrow [0,\infty[$ and a time period $\Delta > 0$ such that

$$[0,\Delta] \longrightarrow \mathbb{R}, \quad t \longmapsto \Phi\big(K(t)\big)$$

is strictly decreasing with $K(t) := \vartheta_{\mathbb{B}_{c(\cdot)}(0)}(t, K_0) \subset \mathbb{R}^N$ denoting the reachable set of K_0 and the differential inclusion $x' \in \mathbb{B}_{c(x(\cdot))}(0)$ (a.e.) at time $t \geq 0$.

The details are presented in § 1.10 (on page 79 ff.) below. In the proof of this lemma, the sign of $\varphi(x, K_0)$ plays a decisive role. Its application fails if $\varphi(\cdot, K_0)$ is nonnegative on the boundary of K_0 and so, we coin the term of a "critical set" w.r.t. Φ in the following way:

Definition 2. A set $M \subset \mathbb{R}^N$ with $\Phi(M) < \infty$ is called *critical* with respect to the integral shape functional Φ if all boundary points $z \in \partial M$ satisfy $\varphi(z, M) \geq 0$.

In subsequent section 1.10.2 (on page 83 ff.), we present a method how to construct the speed function $c(\cdot,\cdot) : [0,T[\times \mathbb{R}^N \longrightarrow [0,\infty[$ piecewise with respect to time and Lipschitz continuous with respect to space such that the curve of corresponding reachable sets $K(\cdot) : [0,T[\longrightarrow \mathscr{K}(\mathbb{R}^N)$, $t \longmapsto \vartheta_{\mathbb{B}_{c(\cdot,\cdot)}(0)}(t, K_0)$ solves the continuous segmentation problem (formulated on page 4).

Some Examples by Means of a Simple Computer Algorithm

For computer images we start with two central aspects: The smallest suitable unit of a computer image is one pixel or one voxel respectively. Moreover the grey values within each of these units are constant.
Hence we intend a combination: On the one hand we use the preceding concept of continuous deformation for decreasing the error functional. On the other hand we want to restrict ourselves finally to the decision whether or not a pixel (or a voxel) belongs to the next approximating set K_{n+1}.

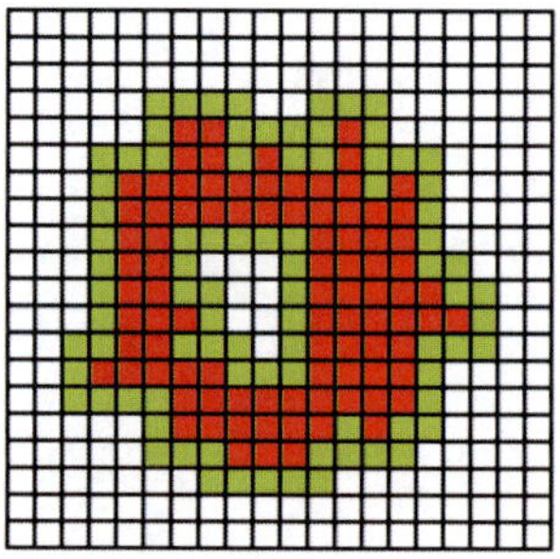

This combination is based on an explicit ansatz (of the speed function) that describes the expansion to a neighboring pixel. If the error functional is guaranteed to stay decreasing during this deformation then the pixel P is admitted to the next set K_{n+1}. A condition sufficient for decreasing Φ is

$$\varphi(x, K_n) < \delta$$

with any point x in the tested neighboring pixel P and some "appropriately small" $\delta > 0$. The final computer algorithm (discussed in § 1.10 on page 79 ff.) is based on the variance of grey values in combination with Lebesgue measure and thus, $\varphi(x, K_n)$ is a quadratic polynomial in $G(x)$. Its sign check can be executed more quickly than the explicit comparison of $\Phi(K_n)$ and $\Phi(K_n \cup P)$. This algorithm is quite simple indeed, but shows remarkable results. More details are in § 1.10.3 (on page 90 ff.).

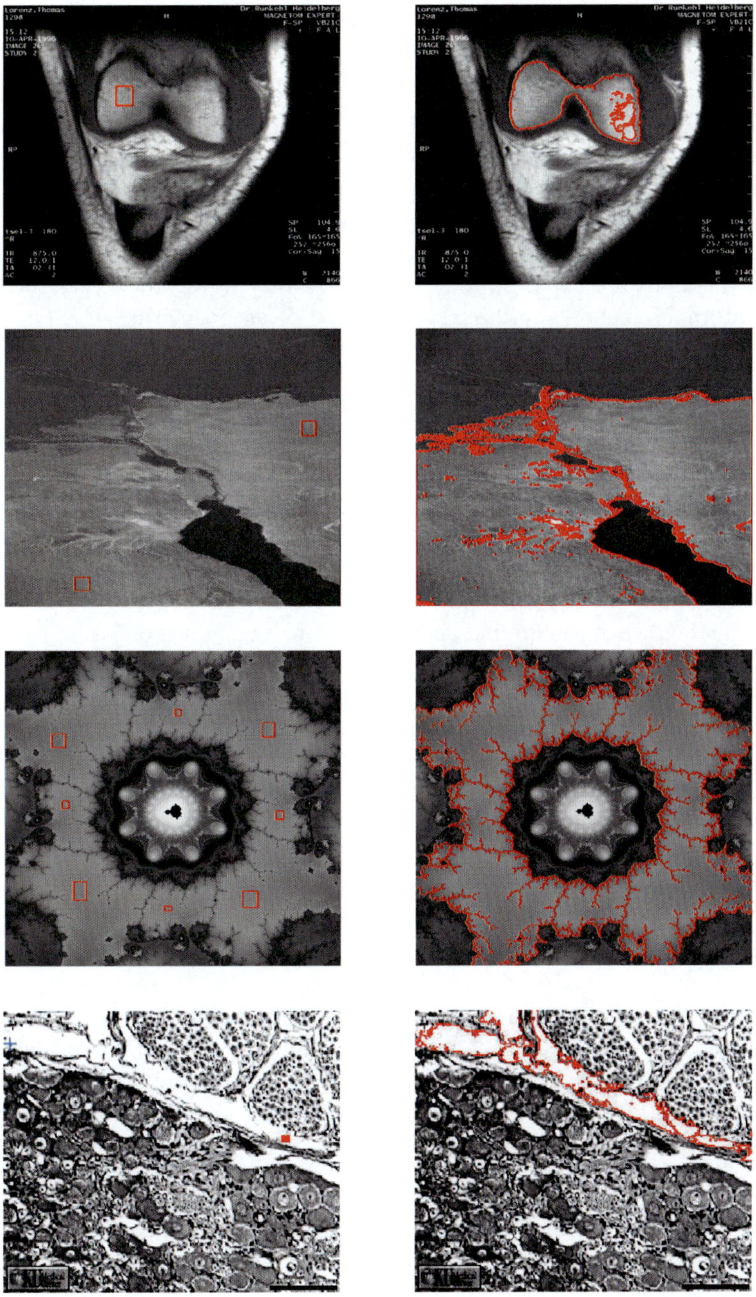

Fig. 0.1 Some examples of the discrete region growing method based on a simple sign check. *Left:* The initial set K_0 with its boundary marked brightly. *Right:* The resulting set [131]. (a) MR of right human knee. (b) Satellite image of Suez Canal. (c) Fractal sets give an impression of the precision at the boundary. (d) Nerve cells (dorsal root ganglion). The anisotropic expansion in direction to a given pixel (marked by a small cross in the top left-hand corner) is succeeded by the standard isotropic sign checks

0.2.2 Image Smoothing via Anisotropic Diffusion

Image segmentation is just one of the many problems in digital image processing. Smoothing images, for example, usually has top priority because the quality of all real photographs is reduced by some form of noise and so, image pre-processing is regarded as indispensable. The key challenge is to smooth the image in such a way that the essential information about edges and contours is not lost.

Suitable methods are expected to smooth adaptively in regard to the "relevance" of the information shown in the respective part of an image. In 1987, Perona and Malik [156] implemented this notion in a PDE based algorithm, which has become a pioneer in the expanding field of nonlinear diffusion filters. The basic notion is to evolve the grey-scale function $u_0 : \mathbb{R}^N \longrightarrow \mathbb{R}$ of the original image according to the nonlinear parabolic differential equation

$$\begin{cases} \partial_t u - \operatorname{div}_x \big(g(|\nabla u|)\, \nabla u\big) &= 0 \\ u(0,\cdot) &= u_0 \end{cases}$$

The diffusion coefficient $g(|\nabla u|)$ depends on the spatial gradient of the smoothed grey-scale function $u = u(t,x)$ via a given cut-off function $g \geq 0$. If $|\nabla u(t,x)|$ is quite large then we expect to find an edge close to x and thus, diffusion is reduced, possibly even "switched off" completely due to $g(|\nabla u|) = 0$. Otherwise, i.e. if $|\nabla u(t,x)|$ rather small, the algorithm is to smooth the grey values close to x by means of diffusion. Perona and Malik suggested the ansatz

$$g(s) \stackrel{\text{Def.}}{=} \lambda_0 \, \frac{\lambda^2}{s^2 + \lambda^2}$$

with adjustable parameters $\lambda, \lambda_0 > 0$ [156]. This implementation makes a quite simple and straightforward impression indeed, but it has some mathematical weaknesses: The underlying initial value problem is not well-posed in general. Indeed, the diffusion process may become instable as Weickert explains in more details in his monograph [183, § 1.3.1].

Many suggestions have been made for overcoming this weakness of the Perona-Malik model. From the mathematical point of view, the approach of Catté, Lions, Morel and Coll [41] is very elegant. Their key idea is to regularize the spatial gradient on which the diffusion coefficient depends. The standard tool of convolution proves to be useful and motivates the following modification

$$\begin{cases} \partial_t u - \operatorname{div}_x \big(g(|u * \nabla G_\sigma|)\, \nabla u\big) &= 0 \qquad \text{in }]0,T[\times \Omega \\ \nabla u \cdot \nu_{\partial\Omega} &= 0 \qquad \text{on }]0,T] \times \partial\Omega \\ u(0,\cdot) &= u_0 \qquad \text{in } \Omega \subset \mathbb{R}^N \end{cases}$$

where $G_\sigma(\cdot)$ is a smoothing kernel, which can be regarded as a "low pass filter". The Gaussian function

$$G_\sigma(\cdot) : \mathbb{R}^N \longrightarrow [0,\infty[, \qquad x \longmapsto \frac{1}{(2\,\pi\,\sigma)^{N/2}} \cdot e^{-\frac{1}{2\sigma}|x|^2}$$

is probably the most popular choice in this context.

Strictly speaking, this diffusion coefficient takes spatially nonlocal properties of the function $u(t,\cdot) : \Omega \longrightarrow \mathbb{R}$ into consideration due to the convolution. In contrast to the Perona-Malik model, the modified initial-boundary value problem is well-posed as Catté et. al conclude from Schauder fixed point theorem and classical results about parabolic differential equations [41].

Weickert suggests in his monograph [183] how to improve this concept: The *scalar* function g in the diffusion coefficient is replaced by a matrix-valued function $A : \mathbb{R}^{N\times N} \longrightarrow \mathrm{Sym}(\mathbb{R}^N) \subset \mathbb{R}^{N\times N}$. This form of diffusion can take the direction of the smoothed spatial gradient into consideration (instead of its absolute value only) and, it is called *anisotropic*.
The continuous dependence of the solution on the initial image has significant practical impact as it ensures "stability" with respect to perturbations of the original image. This is of importance when considering stereo image pairs, spatio-temporal image sequences or slices from medical CT or MRI sequences because similar images remain similar to each other after filtering [183, § 2.3, Remark].
According to Weickert, continuous diffusion filtering uses the unique distributional solution $u \in C^0([0,T], L^2(\Omega)) \cap L^2([0,T], W^{1,2}(\Omega))$ to the initial-boundary value problem with Neumann boundary conditions

$$\begin{cases} \partial_t u \;-\; \mathrm{div}_x \big(A\big(G_\rho * ((u * \nabla G_\sigma) \otimes (u * \nabla G_\sigma))\big)\; \nabla u\big) & = \; 0 \;\text{ in }]0,T[\times \Omega \\ \qquad\qquad\qquad\qquad\qquad\qquad\qquad \nabla u \cdot \nu_{\partial\Omega} & = \; 0 \;\text{ on }]0,T] \times \partial\Omega \\ \qquad\qquad\qquad\qquad\qquad\qquad\qquad u(0,\cdot) & = \; u_0 \text{ in } \Omega \subset \mathbb{R}^N \end{cases}$$

From the mathematical point of view, this anisotropic diffusion approach and the preceding image segmentation algorithm in § 0.2.1 (on page 3 ff.) belong to completely different fields, i.e. parabolic differential equations and set-valued analysis.

The challenging question is now whether these two methods can be handled as components of one system in a joint analytical framework so that we can apply them simultaneously: The set-valued approach approximates the image segment of a wanted object by means of a compact-valued curve $[0,T[\longrightarrow \mathscr{K}(\mathbb{R}^N),\ t \longmapsto K(t)$ and, the respective compact set $K(t) \subset \mathbb{R}^N$ has an explicit influence on the diffusion coefficient

$$\partial_t u \;-\; \mathrm{div}_x \big(\widetilde{A}\big(K(t),\ G_\rho * ((u * \nabla G_\sigma) \otimes (u * \nabla G_\sigma))\big)\; \nabla u\big) \;=\; 0 \qquad \text{in }]0,T[\times \Omega$$

Vice versa, the smoothed grey-scale function $u(t,\cdot) : \Omega \longrightarrow \mathbb{R}$ should have the opportunity to influence the dynamic evolution of $K(t)$ at each time t.

Such a joint framework for dynamic processes is the main goal of this monograph. It is to specify sufficient conditions for well-posed Cauchy problems both *in* and *beyond* vector spaces in finite time intervals. Evolution equations represent a form of generalizing ordinary differential equations from the Euclidean space to Banach spaces. The mutational framework here is to make the step beyond vector spaces.

In subsequent section 3.9 (on page 278 ff.), we apply this general framework to so-called *conormal* parabolic problems in cylindrical domains, which cover the continuous diffusion filtering of Weickert as a special case. One of the main results in this example is the following statement quoting Theorem 3.103 (on page 281 f.):

Theorem 3. *Let $\Omega \subset \mathbb{R}^N$ be a bounded domain with C^2 boundary and $T, \lambda > 0$. Assume the coefficient functions*

$$\begin{aligned}
\mathbf{A}: &\ [0,T] \times \big(W^{1,2}(\Omega), \|\cdot\|_{L^2(\Omega)}\big) \longrightarrow W^{1,\infty}(\Omega, \operatorname{Sym}(\mathbb{R}^{N\times N}))\\
\mathbf{b}: &\ [0,T] \times \big(W^{1,2}(\Omega), \|\cdot\|_{L^2(\Omega)}\big) \longrightarrow W^{1,\infty}(\Omega, \mathbb{R}^N)\\
\mathbf{c}: &\ [0,T] \times \big(W^{1,2}(\Omega), \|\cdot\|_{L^2(\Omega)}\big) \longrightarrow L^\infty(\Omega, \mathbb{R}^N)\\
b_0: &\ [0,T] \times \big(W^{1,2}(\Omega), \|\cdot\|_{L^2(\Omega)}\big) \longrightarrow L^\infty(\partial\Omega)\\
c_0: &\ [0,T] \times \big(W^{1,2}(\Omega), \|\cdot\|_{L^2(\Omega)}\big) \longrightarrow L^\infty(\Omega, \mathbb{R})\\
\mathbf{f}: &\ [0,T] \times \big(W^{1,2}(\Omega), \|\cdot\|_{L^2(\Omega)}\big) \longrightarrow W^{1,2}(\Omega, \mathbb{R}^N)\\
g: &\ [0,T] \times \big(W^{1,2}(\Omega), \|\cdot\|_{L^2(\Omega)}\big) \longrightarrow L^2(\Omega, \mathbb{R})\\
\psi: &\ [0,T] \times \big(W^{1,2}(\Omega), \|\cdot\|_{L^2(\Omega)}\big) \longrightarrow L^2(\partial\Omega)
\end{aligned}$$

to be uniformly bounded w.r.t. the respective Sobolev norms of their values,
continuous w.r.t. the L^∞ or L^2 norm of their values respectively and
satisfy the condition of uniform parabolicity $\xi^T \cdot \mathbf{A}(t,v)(x)\,\xi \geq \lambda\,|\xi|^2$
for all $t \in [0,T]$, $v \in W^{1,2}(\Omega)$, $x \in \Omega$, $\xi \in \mathbb{R}^N$.

Then for every initial $u_0 \in W^{1,2}(\Omega)$, there exists a weak solution $u(\cdot) : [0,T] \longrightarrow W^{1,2}(\Omega)$ to the nonlinear conormal problem

$$\begin{cases}
-\frac{\partial u}{\partial t} + \operatorname{div}_x\big(\mathbf{A}|_{(t,u(t,\cdot))} \cdot \nabla u\big) + \operatorname{div}_x\big(\mathbf{b}|_{(t,u(t,\cdot))} \cdot u\big) & \\
\quad + \ \mathbf{c}|_{(t,u(t,\cdot))} \cdot \nabla u + c_0|_{(t,u(t,\cdot))}\, u = \operatorname{div}_x \mathbf{f}|_{(t,u(t,\cdot))} + g|_{(t,u(t,\cdot))} & \\
\big(\mathbf{A}|_{(t,u(t,\cdot))}\, \nabla u + \mathbf{b}|_{(t,u(t,\cdot))}\, u - \mathbf{f}|_{(t,u(t,\cdot))}\big) \cdot \nu_\Omega & \\
\quad + \ b_0|_{(t,u(t,\cdot))}\, u = \psi|_{(t,u(t,\cdot))} & \text{on } \partial\Omega\\
u(0,\cdot) = u_0 & \text{in } \Omega
\end{cases}$$

in the sense of Definition 3.101 (on page 279) with the composed coefficient functions

$$\begin{aligned}
\check{\mathbf{A}} &:= \mathbf{A}(\cdot,u(\cdot)) : [0,T] \times \Omega \longrightarrow \mathbb{R}^{N\times N}, \quad (t,x) \longmapsto \mathbf{A}(t,\, u(t,\,\cdot))\,(x)\\
\check{\mathbf{b}} &:= \mathbf{b}(\cdot,u(\cdot)) : [0,T] \times \Omega \longrightarrow \mathbb{R}^N, \quad (t,x) \longmapsto \mathbf{b}(t,\, u(t,\,\cdot))\,(x) \qquad \text{etc.}
\end{aligned}$$

In fact, $u \in C^0\big([0,T], L^2(\Omega)\big) \cap L^2\big([0,T], W^{1,2}(\Omega)\big)$, $\partial_t u \in L^2\big([0,T], W^{1,2}(\Omega)^*\big)$.

Lipschitz continuity of the coefficients $\mathbf{A}, \mathbf{b}, b_0, \mathbf{c}, c_0, \mathbf{f}, g$ (w.r.t. the L^∞ or L^2 norms of their values respectively) implies uniqueness of this weak solution and its continuous dependence on the given data. As a consequence, the existence result can then be extended to initial functions $u_0 \in L^2(\Omega)$ approximatively.

0.2.3 A Stochastic Differential Game without Precisely Known Realizations of Opponents

Many processes with aspects of uncertainty are modelled by means of stochastic differential equations. Now consider the situation that two players play a stochastic game with each other. Their states are described by real-valued stochastic processes $(X_t)_{0 \le t \le T}$, $(Y_t)_{0 \le t \le T}$ respectively and, the strategies might take control parameters $u, v \in \mathbb{R}^N$ into consideration. Then the underlying system of stochastic differential equations usually has the form

$$\begin{cases} dX_t = f_1(t, X_t, Y_t, u_t)\ dt + f_2(t, X_t, Y_t, u_t)\ dW_t, & u_t \in U \subset \mathbb{R}^N \\ dY_t = g_1(t, X_t, Y_t, v_t)\ dt + g_2(t, X_t, Y_t, v_t)\ dW_t, & v_t \in V \subset \mathbb{R}^N \end{cases}$$

where $(W_t)_{t \ge 0}$ is a standard scalar Wiener process. We do not discuss any examples of payoff functions in detail, but we are interested in the following question instead:

How to modify this stochastic control problem if each player knows all differential rules, but not the precise current value of the opponent's realization ?

A potential way out is to "estimate" the current value of the opponent's state and, the expected value is a quite simple candidate that takes all possible states of the opponent into consideration. Then the stochastic control problem has the form

$$\begin{cases} dX_t = f_1(t, X_t, \quad \mathbb{E}(Y_t),\ u_t)\ dt + f_2(t, X_t, \quad \mathbb{E}(Y_t), u_t)\ dW_t, & u_t \in U \subset \mathbb{R}^N \\ dY_t = g_1(t, \mathbb{E}(X_t),\ Y_t, \quad v_t)\ dt + g_2(t, \mathbb{E}(X_t),\ Y_t, \quad v_t)\ dW_t, & v_t \in V \subset \mathbb{R}^N. \end{cases}$$

It differs from the counterpart above in the dependence of the right-hand side on *nonlocal* properties (with respect to the probability space). Some additional interest in the confidence of the estimator can be quantified by means of the second moments.

Lacking precise information about current realizations is just one example motivating the following type of stochastic ordinary differential equations

$$dX_t = f_1\big(t, X_t, \mathbb{E}(\varphi_1(X_t)), \mathbb{E}(|X_t|^2)\big)\ dt + f_2\big(t, X_t, \mathbb{E}(\varphi_2(X_t)), \mathbb{E}(|X_t|^2)\big)\ dW_t$$

with sufficiently smooth functions $f_1, f_2, \varphi_1, \varphi_2$. The right-hand side does not depend only on the current realization of the random variable X_t, but also on some properties which are nonlocal with respect to the probability space (not with respect to time). Hence this differential problem differs from what is usually called "stochastic functional differential equations" (as e.g. in [135, 187]) because we do not focus on pathwise dependence.

Strong solutions can be constructed via the Euler method and, this type of stochastic differential equations will be handled in the mutational framework in section 3.6 (on page 231 ff.) in detail. The general mutational approach gives us the advantage that we can solve systems of these nonlocal stochastic differential equations and any other preceding example (like parabolic conormal problems) immediately, i.e. without further preliminary work for the respective system.

0.3 Extending the Traditional Horizon: Evolution Equations Beyond Vector Spaces

In fact, we regard nonlocal set evolutions just as a motivating example.
When introducing mutational equations in metric spaces, Aubin's key motivation was to extend ordinary differential equations to compact subsets of the Euclidean space. It should provide, for example, the framework for control problems

$$\begin{cases} x'(t) \;=\; f\big(t,\, x(t),\, u(t)\big) \;\in\; \mathbb{R}^N \\ u(t) \;\in\; U(t) \qquad\qquad\quad\;\; \subset\; \mathbb{R}^M \end{cases}$$

whose compact control set $U(t) \subset \mathbb{R}^M$ had the opportunity to evolve according to the current state $x(t)$ and itself (i.e. $U(t)$).
This approach of mutations has a much larger potential though. Indeed, the main goal here is a common analytical framework for continuous Cauchy problems *within and beyond* the traditional borders of vector spaces.

Whenever a dynamical system proves to fit in this framework, the mutational theory immediately opens the door to existence results about *systems* with other suitable components – no matter whether their mathematical origins are completely different. A nonlocal geometric evolution can be combined, for example, with an ordinary differential equation and a semilinear evolution equation. This is the main advantage of mutational equations – in comparison to more popular concepts like viscosity solutions and thus, all our generalizations here are to preserve this feature. It is to lay the foundations of future results about free boundary problems.

If a component does not fit in this framework, however, it might serve as motivation for generalizing the mutational theory and weakening the conditions in its definitions.

This interaction between the general mutational framework – without the linear structure of vector spaces – and diverse examples of dynamical systems facilitates a better understanding of very popular results in functional analysis. How can weak sequential compactness, for example, be defined in a metric space without linear structure (and thus, without linear functionals)?

0.3.1 Aubin's Initial Notion: Regard Affine Linear Maps Just as a Special Type of "Elementary Deformations" (Alias Transitions)

Roughly speaking, the starting point consists in extending the term "velocity" from vector spaces to metric spaces. Then the basic idea of first-order approximation leads to a definition of derivative for curves in a metric space and step by step, we can apply the same notions as for ordinary differential equations.

First let us focus on velocities of curves $[0,T] \longrightarrow \mathbb{R}^N$.
A vector $v \in \mathbb{R}^N$ represents the velocity of the curve $x(\cdot) : [0,T] \longrightarrow \mathbb{R}^N$ at time $t \in [0,T[$ if it is the limit of difference quotients:

$$v = \lim_{h \to 0} \frac{x(t+h) - x(t)}{h}.$$

Such a difference quotient is difficult to specify in metric spaces and thus, we use an equivalent condition which became very popular in connection with functions in Banach spaces. Indeed, $v \in \mathbb{R}^N$ represents the velocity of $x(\cdot) : [0,T] \longrightarrow \mathbb{R}^N$ at time $t \in [0,T[$ if it provides a first-order approximation in the following sense:

$$\lim_{h \to 0} \tfrac{1}{h} \cdot \bigl| x(t+h) - \bigl(x(t) + h\,v\bigr) \bigr| = 0. \qquad (*)$$

This condition is reflecting a quantitative comparison between the curve of interest $x(t+\cdot)$ and the affine linear map $h \longmapsto x(t) + h\,v$ for $h \longrightarrow 0$. Such a comparison can also be formulated in a metric space as soon as we have specified a counterpart of the affine linear map.

From a more conceptual point of view, each vector $v \in \mathbb{R}^N$ determines an affine linear map of *two* variables, namely

$$[0,\infty[\times \mathbb{R}^N \longrightarrow \mathbb{R}^N, \qquad (h,\, x) \longmapsto x + h\,v.$$

The first argument h can be interpreted as time whereas the second argument $x \in \mathbb{R}^N$ has the geometric meaning of an initial point in the Euclidean space $\mathbb{R}^N$. After the period $h \geq 0$, it is moved to the end point $x + h\,v \in \mathbb{R}^N$.
Moreover, the asymptotic features leading to time derivatives require comparisons only for short periods. Thus, for the sake of simplicity, let us always choose $h \in [0,1]$ instead of $h \in [0,\infty[$.

Passing the traditional borders of vector spaces, we are free to skip the affine linear structure of this auxiliary map. In a metric space (E,d), a function

$$\vartheta : [0,1] \times E \longrightarrow E, \qquad (h,\, x) \longmapsto \vartheta(h,x)$$

is to play the role of such an affine linear map instead. ϑ determines to which point $\vartheta(h,x) \in E$ any initial point $x \in E$ is moved at time $h \in [0,1]$ and thus, it can be regarded as a kind of "elementary deformation" of E.

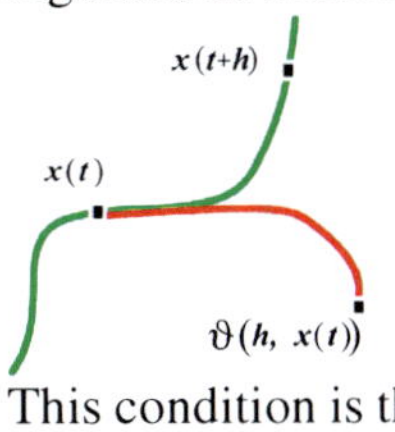

Such a function ϑ represents the time derivative of a curve $x(\cdot) : [0,T] \longrightarrow E$ at time $t \in [0,T[$ if it provides a first-order approximation in the following sense:

$$\lim_{h \downarrow 0} \tfrac{1}{h} \cdot d\bigl(x(t+h),\ \vartheta(h,x(t))\bigr) = 0. \qquad (**)$$

This condition is the (almost) exact analogue of preceding statement $(*)$ as we have merely restricted the limit to $h > 0$ tending to 0. Strictly speaking, it is the precise counterpart of the right-hand Dini derivative of a curve in a vector space like $\mathbb{R}^N$.

Of course, there might be more than just one of these "elementary deformations" $\vartheta : [0,1] \times E \longrightarrow E$ satisfying the characterizing condition $(**)$ at time $t \in [0,T[$. Following the proposal of Aubin in [10], we first specify the class $\Theta(E,d)$ of such functions $[0,1] \times E \longrightarrow E$ appropriate for the metric space (E,d) under consideration and then, the set of *all* functions $\vartheta \in \Theta(E,d)$ satisfying this condition $(**)$ is called *mutation* of the curve $x(\cdot) : [0,T] \longrightarrow E$ at time $t \in [0,T[$:

$$\overset{\circ}{x}(t) := \big\{ \vartheta \in \Theta(E,d) \;\big|\; \lim_{h \downarrow 0} \; \tfrac{1}{h} \cdot d\big(\vartheta(h, x(t)),\; x(t+h)\big) = 0 \big\}.$$

Here the mutation plays the role of the time derivative, but it may consist of more than one function in $\Theta(E,d)$. There is no obvious additional advantage of boiling it down to single elements by means of equivalent classes and thus, we use these sets.

Finally, the step to differential equations in a metric space (E,d) is rather small and based on the notion of feedback.
Indeed, we prescribe such an "elementary deformation" $\vartheta : [0,1] \times E \longrightarrow E$ for each state $y \in E$ and at time $t \in [0,T]$ by means of a function $E \times [0,T] \longrightarrow \Theta(E,d)$. Then the wanted continuous *solution* $x : [0,T] \longrightarrow E$ to the corresponding *mutational equation* is expected to obey the underlying law of first-order approximations $(**)$ — at Lebesgue-almost every time $t \in [0,T]$ at least.

Constructing a differential calculus for curves in a metric space (E,d) can only succeed if these "elementary deformations" $[0,1] \times E \longrightarrow E$ are sufficiently regular with respect to both arguments. In this context, Aubin introduced a set of four conditions on a so-called *transition* $\vartheta : [0,1] \times E \longrightarrow E$. His rather local formulations in [10] (quoted in Definition 1.1 on page 32) imply the following typical features:

(1.) $\vartheta(0,\cdot) = \mathbb{I}\mathrm{d}_E$,

(2.) ϑ has the semigroup property for any $x \in E$, $h_1, h_2 \geq 0$ with $h_1 + h_2 \leq 1$, i.e.
$$\vartheta\big(h_2,\; \vartheta(h_1,x)\big) = \vartheta(h_1 + h_2,\; x),$$

(3.) there exists $\alpha(\vartheta) < \infty$ such that for every $h \in [0,1]$ and $x, y \in E$,
$$d\big(\vartheta(h,x),\; \vartheta(h,y)\big) \leq d(x,\, y) \cdot e^{\alpha(\vartheta) \cdot h},$$

(4.) there exists $\beta(\vartheta) < \infty$ such that for every $h_1, h_2 \in [0,1]$ and $x \in E$,
$$d\big(\vartheta(h_1,x),\; \vartheta(h_2,x)\big) \leq \beta(\vartheta) \cdot |h_2 - h_1|.$$

In other words, transitions are restrictions of semidynamical systems on (E,d) which are ω-contractive w.r.t. state and uniformly Lipschitz continuous w.r.t. time. They prove to be appropriate for extending classical results like the existence theorems of Cauchy-Lipschitz and Nagumo from ordinary differential equations in $\mathbb{R}^N$ to the so-called *mutational equations* in a metric space (E,d). Aubin's concept is presented in more detail in Chapter 1.
His typical geometric examples are so-called *morphological equations*: The set $\mathcal{K}(\mathbb{R}^N)$ of nonempty compact subsets of $\mathbb{R}^N$ is supplied with the classical Pompeiu-Hausdorff metric $d\!l$ and, transitions are induced by reachable sets of differential inclusions (with bounded and Lipschitz continuous right-hand side).

0.3.2 Mutational Analysis as an "Adaptive Black Box" for Initial Value Problems

Let us now discuss in more detail how to solve initial value problems by means of *mutational analysis*.

The first step consists in specifying the mathematical environment of the problem under consideration. Basically, we choose a set $E \neq \emptyset$, a metric $d : E \times E \longrightarrow \mathbb{R}$ and a suitable set of transitions $[0,1] \times E \longrightarrow E$, denoted by $\Theta(E,d)$.

The transitions are usually induced by simpler problems in the same environment, e.g. on the basis of fixing the coefficients or considering the corresponding linear problem (instead of the full nonlinear one). It is essential to verify the characterizing properties of transitions for the respective choice on E, i.e. in particular, the appropriate continuity with respect to initial state and time.

For constructing wanted solutions approximatively, the two most popular concepts in analysis are compactness and completeness. Comparing the classical theorem of Peano (about ordinary differential equations in $\mathbb{R}^N$) with Cauchy-Lipschitz Theorem reveals that compactness usually opens the door to existence theorems under weaker assumptions of continuity. Thus, we mostly intend to verify a form of sequential compactness for the respective mathematical environment (rather than completeness).

These are the main "ingredients" of mutational analysis.
Indeed, the full problem under consideration is determined by a "feedback" function

$$f : E \times [0,T] \longrightarrow \Theta(E,d)$$

and, the theorems in mutational analysis specify sufficient conditions on f such that for every initial element $x_0 \in E$, there exists a Lipschitz continuous curve $x(\cdot) : [0,T] \longrightarrow (E,d)$ with $x(0) = x_0$ such that at $\mathscr{L}^1$-almost every time $t \in [0,T[$,

$$\overset{\circ}{x}(t) \ni f\big(x(t), t\big)$$

i.e.,
$$\lim_{h \downarrow 0} \ \tfrac{1}{h} \cdot d\big(x(t+h),\ f(x(t),t)\,(h, x(t))\big) \ = \ 0.$$

This result corresponds to Peano's Theorem about ordinary differential equations in $\mathbb{R}^N$ and, its proof is based on Euler approximations evaluating transitions successively in equidistant partitions of $[0,T]$. Moreover, mutational analysis provides sufficient conditions on f for uniqueness of solutions in bounded time intervals and their continuous dependence on data. Last, but not least, we can also handle initial value problems with state constraints leading to the counterpart of Nagumo's Theorem.
Strictly speaking, however, all these results deal with curves $x(\cdot) : [0,T] \longrightarrow E$ in some abstract set $E \neq \emptyset$ — with some supplementary properties in regard to first-order approximations via transitions.
If we stopped here, mutational analysis would hardly provide new insights in more traditional fields like partial differential equations.

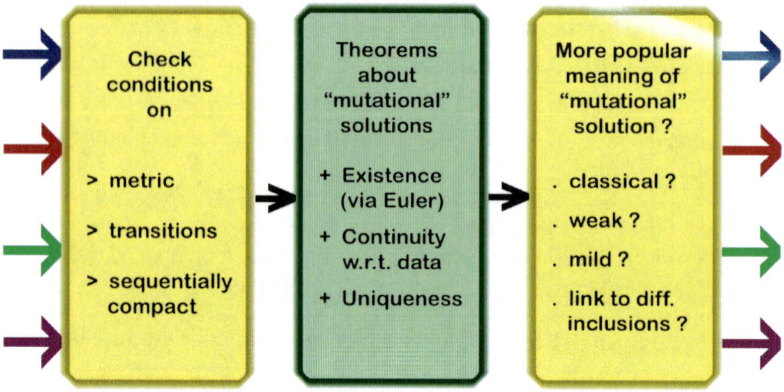

For this reason, the last step of our method focuses on respective links between such a solution to the mutational equation $\overset{\circ}{x}(\cdot) \ni f\big(x(\cdot), \cdot\big)$ and a popular concept of solution (whenever possible).

Such a connection strongly depends on the type of considered problem, of course. In regard to partial differential equations, for example, it might lead to classical, strong or weak solutions. Alternatively, for evolution equations, we can often prove a relation to mild solutions and, some set evolutions in $(\mathscr{K}(\mathbb{R}^N), dl)$ are characterized as reachable sets of nonautonomous differential inclusions (whose coefficients depend on the wanted curve in $\mathscr{K}(\mathbb{R}^N)$). § 0.3.3 sketches three conceptual approaches how to establish a link with other solution concepts.

As a precipitate result of this summary, mutational analysis might be regarded as "just" some complicated formalism providing a very long list of features sufficient for the convergence of Euler approximation in a mathematical environment without linear structure.

This evaluation, however, ignores an essential advantage of the mutational framework which we have already mentioned in a preceding subsection:

Mutational analysis can handle systems in regard to existence and stability.

As soon as an example fulfills the conditions on distance, transitions etc., we are immediately free to apply the existence results about systems of mutational equations and couple this example with any other one fitting in this mutational framework. Nonlocal set evolutions in $\mathbb{R}^N$, for example, can be combined with nonlinear transport equations for Radon measures.

This flexibility in regard to systems makes mutational analysis very attractive.

Whenever an example does not fit in the mutational framework, it might serve as motivation for generalizing mutational analysis. In particular, several examples of Cauchy problems have demonstrated that Aubin's four conditions on transitions are quite restrictive for deriving significantly more benefit from this concept. Thus it is our goal to adapt them step by step — motivated by diverse examples (see § 0.3.4).

0.3.3 *The Initial Problem Decomposition and the Final Link to More Popular Meanings of Abstract Solutions*

Mutational analysis is to provide useful tools for solving dynamical problems that are continuous in time and have a possibly very complex structure. The main idea is always to split the full problem up into a class of simpler problems and "feedback". In particular, the general mutational framework clarifies the main features of such decomposition that are relevant to the existence of solutions by means of Euler method. Then conclusions about the uniqueness of solutions and their continuous dependence on data are usually drawn via Gronwall's inequality.

Hence, the initial decomposition of the problem plays a key role, but there is no global recipe (so far). Each of the simpler initial value problems induces a transition on the basic set and should be more than just "well-posed". Indeed, we will use an appropriate form of Lipschitz continuity with respect to time and initial state — for ensuring the convergence of Euler approximations later on.
Most of the subsequent examples in Banach spaces have in common that fixing the coefficients usually leads to *linear* problems which prove to satisfy the required conditions on transitions.

Furthermore, the benefit which the mutational framework can bring to a complicated Cauchy problem depends very much on the question if a solution to such an abstract mutational equation can be interpreted in a more popular way. Does it prove to be a classical, strong, weak or mild solution, for example?
Obviously, the answer has to be given for each application separately, but there are three useful approaches to proving such a connection between the original full problem denoted here as $z' = F(z,t)$ (just for a moment) and the related mutational equation $\overset{\circ}{x}(t) \ni f(x(t),t)$ (after the problem decomposition):

(1.) The direct check:

Verify that the asymptotic comparison (with the evolution along a transition) for vanishing time always implies the relevant features of solutions in an alternative sense. This check is often quite intricate though.

(2.) Use existing solutions $z(\cdot)$ to the nonautonomous, but "simplified" problem $z' = F(x(t),t)$ for comparison:

Assume that a solution $x(\cdot)$ to the mutational equation $\overset{\circ}{x} \ni f(x,t)$ is known. Prove that a solution $z(\cdot)$ to the nonautonomous problem $z' = F(x(t),t)$, whose right-hand side depends only on t, but not on the wanted state z, starts at $x(0)$ and solves the mutational equation $\overset{\circ}{z}(t) \ni f(x(t),t)$. Then uniqueness of solutions to this last simple mutational equation can imply $z(\cdot) \equiv x(\cdot)$.

(3.) Both solution criteria are stable w.r.t. the same type of convergence:

Every transition ϑ provides a joint solution $y(\cdot) = \vartheta(\cdot, y_0)$ of both a simple original problem and $\overset{\circ}{y}(\cdot) \ni \vartheta$ – by definition. In the mutational framework, each piecewise Euler approximation solves a "perturbed" mutational equation and, the Convergence Theorems specify the type of convergence w.r.t. which the solution property is preserved (e.g. pointwise convergence). Check if the solution property of the full problem is also preserved by the same type of convergence. Then the (initially piecewise) equivalence of the solution criteria for Euler approximation holds for the limit curve.

0.3.4 The New Steps of Generalization in Mutational Analysis

Step (A) Linear Examples in Vector Spaces Exclude Uniform Parameters of Transitions

The affine linear maps $[0,1] \times \mathbb{R}^N \longrightarrow \mathbb{R}^N,\ (h,x) \longmapsto x + h\,v$ with fixed vectors $v \in \mathbb{R}^N$ are the first and probably simplest examples of transitions on the Euclidean space $\mathbb{R}^N$. Obviously, each of them is Lipschitz continuous with respect to both arguments and thus fulfills Aubin's conditions on transitions.
This situation changes, however, if the transitions are based on the unique solutions to *linear* initial value problems. In connection with a nonlinear continuity equation

$$\partial_t\, u \;+\; \mathrm{div}_x\, \big(h(u)\, u\big) \;=\; 0 \qquad\qquad \text{in } [0,T] \times \mathbb{R}^N,$$

for example, the linear Cauchy problem with a fixed coefficient function b

$$\begin{cases} \partial_t\, u \;+\; \mathrm{div}_x\, \big(b\, u\big) \;=\; 0 & \text{in } [0,h] \times \mathbb{R}^N \\ \qquad\qquad u(0,\cdot) \;=\; u_0 & \text{in } \mathbb{R}^N \end{cases}$$

provides an obvious ansatz for a transition $(h, u_0) \longmapsto u(h,\cdot)$ on the corresponding function space, but Aubin's conditions on transitions reveal obstacles due to linearity immediately: The family of curves $h \longmapsto u(h,\cdot)$ for all permissible initial functions $u_0 : \mathbb{R}^N \longrightarrow \mathbb{R}^N$ can hardly be expected to be Lipschitz continuous with a globally bounded Lipschitz constant. How to choose the parameter of continuity $\beta(\vartheta)$ then?

Whenever a parameter cannot be chosen globally, *local* bounds might be recommendable to check instead. This is our first step for generalizing Aubin's mutational framework.
In particular, we need a criterion for which subsets of permissible states each transition should have uniform parameters of continuity (denoted by $\alpha(\vartheta), \beta(\vartheta)$ above). Another glance at the linear examples in vector spaces motivates us to specify counterparts of the norm. Such an "absolute value" reflects the properties of a single state whereas a metric usually compares two elements.
In addition to a metric space (E,d), any function $\lfloor\cdot\rfloor : E \longrightarrow [0,\infty[$ is now given at the very beginning of the (new) mutational framework and, a transition ϑ : $[0,1] \times E \longrightarrow E$ on the tuple $(E, d, \lfloor\cdot\rfloor)$ is supposed to have the following features:

(1.) $\vartheta(0,\cdot) = \mathbb{I}\mathrm{d}_E$,

(2.) ϑ has the semigroup property for any $x \in E$, $h_1, h_2 \geq 0$ with $h_1 + h_2 \leq 1$, i.e.
$$\vartheta\big(h_2,\ \vartheta(h_1,x)\big) = \vartheta(h_1+h_2,\ x),$$

(3.') for every $R > 0$, there exists $\alpha(\vartheta;R) < \infty$ such that for every $h \in [0,1]$ and $x,y \in E$ with $\lfloor x \rfloor \leq R$ and $\lfloor y \rfloor \leq R$,
$$d\big(\vartheta(h,x),\ \vartheta(h,y)\big) \leq d(x,y) \cdot e^{\alpha(\vartheta;R)\cdot h},$$

(4.') for every $R > 0$, there exists $\beta(\vartheta;R) < \infty$ such that for every $h_1, h_2 \in [0,1]$ and $x \in E$ with $\lfloor x \rfloor \leq R$,
$$d\big(\vartheta(h_1,x),\ \vartheta(h_2,x)\big) \leq \beta(\vartheta;R) \cdot |h_2 - h_1|.$$

This list of conditions has to be extended though. Indeed, the concatenation of transitions leads to curves $x(\cdot) : [0,T] \longrightarrow E$ for any period $T > 1$ and, they will be used for solving mutational equations later on. Thus we are obliged to keep the "absolute value" $\lfloor x(\cdot) \rfloor : [0,T] \longrightarrow [0,\infty[$ under control so that the propagation of initial errors can be estimated properly. Each transition $\vartheta : [0,1] \times E \longrightarrow E$ is expected to fulfill a growth condition whose structure is preserved by concatenation:

(5.) there exists $\gamma(\vartheta) < \infty$ such that for every $h \in [0,1]$ and $x \in E$,
$$\lfloor \vartheta(h,x) \rfloor \leq \big(\lfloor x \rfloor + \gamma(\vartheta)\ h\big) \cdot e^{\gamma(\vartheta)\cdot h}.$$

Now the modified "machinery" of mutational analysis is ready to start again and, Euler method together with suitable compactness assumptions ensure the existence of solutions to the Cauchy problem in Chapter 2. One of the consequences is the following theorem presented in § 2.5.3. It deals with the nonlinear transport equation for finite real-valued Radon measures on $\mathbb{R}^N$ whose set is denoted by $\mathcal{M}(\mathbb{R}^N)$.

Theorem 4 (Existence of solution to nonlinear transport equation).
For $\mathbf{f} = (\mathbf{f}_1, f_2) : \mathcal{M}(\mathbb{R}^N) \times [0,T] \longrightarrow W^{1,\infty}(\mathbb{R}^N,\mathbb{R}^N) \times W^{1,\infty}(\mathbb{R}^N,\mathbb{R})$ *suppose*

(i) $\sup_{\mu,t} \big(\|\mathbf{f}_1(\mu,t)\|_{W^{1,\infty}} + \|f_2(\mu,t)\|_{W^{1,\infty}}\big) < \infty$,

(ii) $\mathbf{f}$ *is continuous in the following sense: For* $\mathcal{L}^1$*-almost every* $t \in [0,T]$ *and any sequences* $(t_m)_m$, $(\mu_m)_m$ *in* $[0,T]$, $\mathcal{M}(\mathbb{R}^N)$ *respectively with* $t_m \longrightarrow t$, $\mu_m \longrightarrow \mu$ *narrowly for* $m \longrightarrow \infty$ *and* $\sup_m |\mu_m|(\mathbb{R}^N) < \infty$, *it fulfills*
$$\mathbf{f}(\mu_m,t_m) \longrightarrow \mathbf{f}(\mu,t) \quad \text{in } L^\infty(\mathbb{R}^N,\mathbb{R}^N) \times L^\infty(\mathbb{R}^N,\mathbb{R}) \quad \text{for } m \longrightarrow \infty.$$

Then for every initial Radon measure $\mu_0 \in \mathcal{M}(\mathbb{R}^N)$, *there exists a narrowly continuous distributional solution to the nonlinear transport equation*
$$\partial_t\, \mu_t + \mathrm{div}_x\big(\mathbf{f}_1(\mu_t,t)\,\mu_t\big) = f_2(\mu_t,t)\ \mu_t \qquad \text{in } \mathbb{R}^N \times\,]0,T[$$
in the sense that
$$\int_{\mathbb{R}^N} \varphi\ d\mu_t - \int_{\mathbb{R}^N} \varphi\ d\mu_0 = \int_0^t \int_{\mathbb{R}^N} \Big(\nabla\varphi(x)\cdot \mathbf{f}_1(\mu_s,s)(x) + f_2(\mu_s,s)(x)\Big)\ d\mu_s(x)\ ds$$
for every $t \in [0,T]$ *and any test function* $\varphi \in C_c^\infty(\mathbb{R}^N,\mathbb{R})$.

Mutational Equations on Function Spaces Are "Functional Equations"

The recent example about the nonlinear transport equation reflects a typical feature
of mutational equations on function spaces: Each function (like a Radon measure
here) comes into play as one single element of a basic set E and, the function $f(\cdot,\cdot)$
on the right-hand side of the mutational equation

$$\overset{\circ}{x}(t) \;\ni\; f\big(x(t),t\big)$$

is relating each state in E and time in $[0,T]$ to a transition on $(E,d,\lfloor\cdot\rfloor)$.
In connection with a function space for E, this relation can take nonlocal properties
of the functions $u \in E$ into consideration immediately, but on the other hand, the
hypotheses about the continuity of f might exclude pointwise composition of these
functions.
Due to this structural consequence of $f : E \times [0,T] \longrightarrow \Theta(E,d,\lfloor\cdot\rfloor)$ as given data,
most examples of mutational equations on a function space belong to the field of
functional differential equations.

Step (B)   Admit More Than One Distance Function on the Basic Set E

Compactness often plays the basic role for concluding the existence of a solution
from an approximative sequence. It is very restrictive, however, if a vector space is
supplied with a norm because its closed unit ball is compact if and only if the space
is finite-dimensional. This observation has already aroused the frequent interest in
the weak topology on Banach spaces. Indeed, the weak sequential compactness of
the closed unit ball is equivalent to its reflexivity.

The short excursion to linear functional analysis motivates us to provide simple
access to the mutational framework for the weak topology on metric vector spaces.
Our suggestion is to replace the metric $d : E \times E \longrightarrow [0,\infty[$ by a family $(d_j)_{j \in \mathscr{I}}$
of distance functions $E \times E \longrightarrow [0,\infty[$.   It is an excellent opportunity to weaken
the conditions on each distance function d_j, $j \in \mathscr{I}$. The example induced by linear
functionals on a metric vector space makes clear that d_j does not have to be positive
definite. In this next step of generalization, we assume each $d_j : E \times E \longrightarrow [0,\infty[$
to be reflexive, symmetric and to satisfy the triangle inequality. These three properties characterize a so-called *pseudo-metric* on E.
Similarly, a family $(\lfloor\cdot\rfloor_j)_{j \in \mathscr{I}}$ of functions $E \longrightarrow [0,\infty[$ substitutes for $\lfloor\cdot\rfloor$ indicating
the "absolute value" of states in E. All conditions on transitions and solutions are
then formulated or verified for each d_j, $j \in \mathscr{I}$, simultaneously and hence, this extension does not have any significant influence on the proofs. It is also implemented
in Chapter 2.

How to Compare the Evolution of Two Initial States along Two Transitions: The Key Inequality About Error Propagation

We are still lacking tools how to compare the evolution of two initial states $x, y \in E$ along two (possibly different) transitions ϑ, τ on $(E, (d_j)_{j\in\mathscr{I}}, (\lfloor\cdot\rfloor_j)_{j\in\mathscr{I}})$. Indeed, the only inequality about error propagation so far deals with a single transition ϑ and states that the initial error may grow at most exponentially:

$$d_j\big(\vartheta(h,x),\ \vartheta(h,y)\big) \ \le\ d_j\big(x,\ y\big)\cdot e^{\alpha_j(\vartheta;R)\ h}$$

for every $h \in [0,1]$ and $x, y \in E$ with $\lfloor x\rfloor_j,\ \lfloor y\rfloor_j \le R$.

In other words, the qualitative influence of initial error has already been clarified. Now we focus on the effect of two transitions ϑ, τ on one and the same initial state $x \in E$. The curves $\vartheta(\cdot,x),\ \tau(\cdot,x) : [0,1] \longrightarrow E$ are both continuous with respect to each d_j $(j \in \mathscr{I})$ by definition and thus,

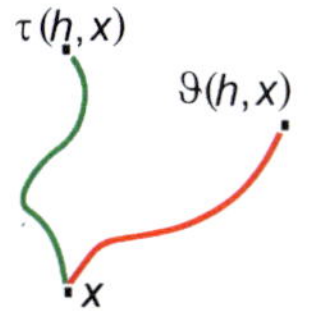

$$d_j\big(\vartheta(h,x),\ \tau(h,x)\big) \longrightarrow 0 \qquad\qquad \text{for } h \downarrow 0.$$

The first-order features of this time-dependent distance might be more informative and hence, Aubin suggested

$$\sup_{x\in E}\ \limsup_{h\downarrow 0}\ \tfrac{1}{h}\cdot d\big(\vartheta(h,x),\ \tau(h,y)\big)$$

as distance between two transitions ϑ, τ on a metric space (E,d). It is always finite because the triangle inequality of the metric d reveals the upper bound $\beta(\vartheta)+\beta(\tau)$. Now our two recent steps of generalization lead to the following counterpart for transitions ϑ, τ on the tuple $(E, (d_j)_{j\in\mathscr{I}}, (\lfloor\cdot\rfloor_j)_{j\in\mathscr{I}})$

$$D_j(\vartheta,\tau;\,r) \ :=\ \sup_{x\in E:\ \lfloor x\rfloor_j\le r}\ \limsup_{h\downarrow 0}\ \tfrac{1}{h}\cdot d_j\big(\vartheta(h,x),\ \tau(h,x)\big)\ <\ \infty$$

for any radius $r \ge 0$ and index $j \in \mathscr{I}$. (If $\{x\in E\,|\,\lfloor x\rfloor_j \le r\} = \emptyset$, set $D_j(\,\cdot\,,\cdot\,;r) := 0$.) If d_j is a pseudo-metric on E, then $D_j(\cdot,\cdot;r)$ proves to be a pseudo-metric on the set of transitions for each $r \ge 0$.

This supplementary information about transitions is based on *local* features because it takes only joint initial states and short periods into consideration. Now we need to bridge the gap to curves $[0,1] \longrightarrow E$ with possibly different initial points and, Gronwall's inequality plays the essential role for this step to estimates in $[0,1]$. Indeed, the distance function $\varphi_j : [0,1] \longrightarrow [0,\infty[,\ \ h \longmapsto d_j\big(\vartheta(h,x),\ \tau(h,y)\big)$ is continuous and, the triangle inequality of d_j ensures at every time $t \in [0,1[$

$$\limsup_{h\downarrow 0}\ \frac{\varphi(t+h)-\varphi(t)}{h}\ \le\ \alpha_j(\vartheta;R_j)\cdot\varphi(t)\ +\ D_j(\vartheta,\tau;R_j)$$

with a sufficiently large radius $R_j > 0$ depending only on $\lfloor x\rfloor_j,\ \lfloor y\rfloor_j,\ \gamma_j(\vartheta),\ \gamma_j(\tau)$. Then Gronwall's inequality provides directly the "global" estimate at any time $h \le 1$

$$d_j\big(\vartheta(h,x),\ \tau(h,y)\big)\ \le\ \big(d_j(x,\ y)\ +\ h\cdot D_j(\vartheta,\tau;R_j)\big)\cdot e^{\alpha_j(\vartheta;R_j)\,h}.$$

Such a step from an upper Dini derivative to an upper bound in a compact time interval is typical for mutational analysis and, it usually results from some modification of Gronwall's Lemma. (Hence, we present several extensions in Appendix A.1.)

Furthermore, this general inequality of error propagation has a quite intuitive structure on its right-hand side. Indeed, the initial distance $d_j(x, y)$ can be regarded a term of order 0 (w.r.t. h) whereas the transitions ϑ, τ contribute to the "term of first order", i.e. $h \cdot D_j(\vartheta, \tau; R_j)$. Both of them are free to increase at most exponentially. This form of influence is quite similar to Taylor expansions in vector spaces.

Step (C) Separate Families of Distances for Regularity in State and Time

Ordinary differential equations in the Euclidean space were extended to Banach spaces in a very successful way a long time ago. Nowadays, the result is known as *evolution equations* and, its conceptual starting points are strongly continuous semigroups $(S(t))_{t\geq 0}$ on a fixed Banach space X and their respective generators A. This historic background justifies our attempt to deal with evolution equations

$$z'(t) \;=\; A\,z(t) + f(z,t)$$

in the mutational framework. It does not necessarily provide new results about mild solutions, but it opens the door to coupling evolution equations with other examples (like nonlocal set evolutions or nonlinear transport equations) immediately.

Strong continuity, however, causes difficulties. Indeed, the variation of constants formula motivates the following ansatz for a transition

$$\tau_v : [0,1] \times X \longrightarrow X, \quad (h,x) \longmapsto \tau_v(h,x) := S(h)\,x + \int_0^h S(h-s)\; v\; ds$$

with an arbitrarily fixed vector v in the Banach space X. If the semigroup $(S(t))_{t\geq 0}$ is assumed to be ω-contractive, then it is easy to verify that initial errors with respect to norm can grow at most exponentially, i.e. for any $x, y \in X$ and $h \in [0,1]$,

$$\big\|\tau_v(h,x) - \tau_v(h,y)\big\|_X \;\leq\; \|x-y\|_X \cdot e^{\omega\, h}.$$

In regard to potential transitions on $(X, \|\cdot\|_X, \|\cdot\|_X)$, the continuity with respect to time is an obstacle: All curves $\tau_v(\cdot,x) : [0,1] \longrightarrow X$ with x in the unit ball of X are expected to be uniformly Lipschitz continuous and, this condition is likely to fail whenever the dimension of X is infinite. The situation is much easier in the following estimate, for example,

$$\big\|\tau_v(h,x) - S(h)\,x\big\|_X \;\leq\; \int_0^h \|S(h-s)\; v\|_X\; ds \;\leq\; h\, e^{\omega\, h}\; \|v\|_X,$$

but then it is probably more difficult to verify a counterpart of the exponentially growing initial error and to provide a link to mild solutions in the end.

Our proposal to overcome this difficulty in the general mutational framework is to use separate families $(d_j)_{j\in\mathcal{I}}, (e_j)_{j\in\mathcal{I}}$ of distance functions $E \times E \longrightarrow [0,\infty[$ for the regularity with respect to state and time (if it is advantageous). Then a transition $\vartheta : [0,1] \times E \longrightarrow E$ on $(E, (d_j)_{j\in\mathcal{I}}, (e_j)_{j\in\mathcal{I}}, (\lfloor\cdot\rfloor_j)_{j\in\mathcal{I}})$ is expected to satisfy

$$\begin{cases} d_j\big(\vartheta(h,\ x),\ \vartheta(h,\ y)\big) \;\leq\; d_j(x,y) \cdot e^{\alpha_j(\vartheta;r)\, h} \\ e_j\big(\vartheta(h_1,x),\ \vartheta(h_2,x)\big) \;\leq\; \beta_j(\vartheta;r)\; |h_1 - h_2| \end{cases}$$

for all $r \geq 0$, $j \in \mathcal{I}$, $h, h_1, h_2 \in [0,1]$ and $x, y \in E$ with $\lfloor x\rfloor_j, \lfloor y\rfloor_j \leq r$.

In fact, $(e_j)_{j\in\mathscr{I}}$ is supposed to represent the same "topology" as $(d_j)_{j\in\mathscr{I}}$ in the sense that every sequence $(x_n)_{n\in\mathbb{N}}$ tends to $x \in E$ with respect to each $e_j (j \in \mathscr{I})$ if and only if it converges to x with respect to each d_i $(i \in \mathscr{I})$. We adhere to distance functions for specifying continuity in time mainly because we need equi-continuity of Euler approximations for the continuity of their limit function.

Separate distance functions of the same "topology" for the regularity in state and time have proved to be a good starting point for handling semilinear evolution equations with ω-contractive semigroups by means of mutational equations. More details are discussed in § 3.10.
These results are then used for some initial-boundary value problems with second-order parabolic differential equations in noncylindrical domains — without assuming any transformation to a reference domain (§ 3.11).

Step (D) How the ω-Contractivity of Transitions Can Become Dispensable

So far, all transitions ϑ (even in their generalized form) are obliged to let the initial distance grow at most exponentially in time, i.e. for each $r > 0$ and $j \in \mathscr{I}$, there exists a parameter $\alpha_j(\vartheta;r) > 0$ with

$$d_j\big(\vartheta(h,x),\ \vartheta(h,y)\big) \ \le\ d_j(x,y)\ \cdot\ e^{\alpha_j(\vartheta;r)\,h}$$

for all $h \in [0,1]$ and $x, y \in E$ with $\lfloor x \rfloor_j, \lfloor y \rfloor_j \le r$. In the terminology of semidynamical systems, ϑ is said to be (locally) *ω-contractive* with respect to each d_j $(j \in \mathscr{I})$.

In the example of semilinear evolution equations (in § 3.10), however, this condition implies a significant restriction: the underlying strongly continuous semigroup $(S(t))_{t\in\ge 0}$ is assumed to be ω-contractive so that the variation of constants formula ensures the corresponding property of $\tau_v(h,x) := S(h)\,x + \int_0^h S(h-s)\,v\,ds$ $(v\in X)$.

Mutational analysis would be much more helpful indeed if we could draw essentially the same conclusions about existence of solutions from the weaker hypothesis

$$d_j\big(\vartheta(h,x),\ \vartheta(h,y)\big) \ \le\ C_j\ \cdot\ d_j(x,y)\ \cdot\ e^{\alpha_j(\vartheta;r)\,h}$$

with a constant C_j possibly larger than 1, but then the standard piecewise approach will fail to estimate the maximal distance between two Euler approximations in $[0,T]$ when the partition is refined (and the number of subintervals tends to ∞).

This dilemma can be overcome if we *suppose* suitable inequalities for the maximal distance between any Euler approximations in arbitrary time intervals $[0,T]$.
The main idea presented in § 3.4 is motivated by the well-known step from Hille-Yosida Theorem (about contractive C^0 semigroups) to the Theorem of Feller, Miyadera and Phillips (about arbitrary C^0 semigroups) (e.g. [76, Theorem II.3.8]).

Indeed, we construct a family of auxiliary distance functions $\check{d}_j$ ($j \in \mathcal{I}$) which are "equivalent" to d_j respectively such that ϑ is ω-contractive with respect to each $\check{d}_j$. In contrast to the semigroup theory in Banach spaces, we cannot use the linear resolvent operator here, but the construction of $\check{d}_j$ is based on the figurative question how "far away from each other" two states can come along any (nonequidistant) Euler curves after we subtract the potential influence of transitions on the distance.

The so-called *candidates* for transitions fulfill all required properties except for ω-contractivity, but the results of § 3.4 fill this gap via an auxiliary family $(\check{d}_j)_{j\in\mathcal{I}}$.

Step (E) Less Restrictive Conditions on Distance Functions d_j, e_j ($j \in \mathcal{I}$): Continuity Assumptions Instead of Triangle Inequality

Examples with stochastic differential equations are quite difficult to consider in the mutational framework up to now. Let us take a glance at real-valued solutions $(X_t)_{0\le t\le T}$ to the stochastic initial value problem

$$\begin{cases} dX_t = a(t,X_t)\,dt + b(t,X_t)\,dW_t \\ X_0 \quad \text{given} \end{cases}$$

with a fixed Wiener process $W = (W_t)_{t\ge 0}$ on a complete probability space $(\Omega, \mathcal{A}, P)$. Under suitable assumptions about the coefficients $a,b : [0,T]\times\mathbb{R} \longrightarrow \mathbb{R}$, a pathwise unique strong solution $(X_t)_{0\le t\le T}$ is known to exist and, the following estimates hold with constants C_1, C_2, C_3 depending only on $a(\cdot), b(\cdot), T$

$$\begin{aligned} \mathbb{E}\big(|X_t|^2\big) &\le \quad \big(\mathbb{E}\big(|X_0|^2\big) + C_2\,t\big)\; e^{C_1 t}, \\ \mathbb{E}\big(|X_t - X_0|^2\big) &\le C_3\,\big(\mathbb{E}\big(|X_0|^2\big) + 1\big) \quad e^{C_1 t}\cdot t\,. \end{aligned}$$

If we regard these solutions as possible candidates for transitions, then the first inequality provides a suitable upper bound of growth. The second inequality indicates Lipschitz continuity with respect to time – exactly in the form we usually want it, but the estimate considers the square deviation which does not satisfy the triangle inequality in general.

This observation exemplifies that the triangle inequality of pseudo-metrics on the one hand and the familiar types of distance estimates like

$$\begin{cases} d_j\big(\vartheta(h,\ x),\ \ \vartheta(h,\ y)\big) \ \le\ d_j(x,y)\cdot e^{\alpha_j(\vartheta;R_j)\,h} \\ e_j\big(\vartheta(h_1,x),\ \ \vartheta(h_2,x)\big) \ \le\ \beta_j(\vartheta;R_j)\,|h_1-h_2| \\ d_j\big(\vartheta(h,\ x),\ \ \tau(h,\ y)\big) \ \le\ \big(d_j(x,y) + h\cdot D_j(\vartheta,\tau;R_j)\big)\cdot e^{\alpha_j(\vartheta;R_j)\,h} \end{cases}$$

on the other hand might exclude each other. Now we have to make a decision which aspect to preserve in the mutational framework.

We prefer the key inequality of error propagation to the triangle inequality.

The main goal of mutational analysis is to extend the familiar results about ordinary differential equations beyond the traditional borders of vector spaces. Meanwhile we have even left metric spaces by means of the tuples $\big(E, (d_j)_{j\in\mathcal{I}}, (e_j)_{j\in\mathcal{I}}, \lfloor\cdot\rfloor_j)_{j\in\mathcal{I}}\big)$, but the key inequality of error propagation for transitions

$$d_j\big(\vartheta(h,x),\ \tau(h,y)\big) \ \le\ \big(d_j(x,y) + h\cdot D_j(\vartheta,\tau;R_j)\big)\cdot e^{\alpha_j(\vartheta;R_j)\,h}$$

still reflects the notion of first-order approximation.

The triangle inequality has become a very popular condition on distance functions and, it seems to be indispensable in many standard textbook about topology and calculus as it is one of the defining conditions on metrics. A closer look at its role in proofs reveals that it mostly serves a single purpose: verifying continuity. In particular, the triangle inequality guarantees that the metric on a set is continuous with respect to its topology.

In regard to the mutational framework, our new suggestion is to ensure the "continuity" of each distance function d_j, e_j $(j\in\mathcal{I})$ by means of explicit hypotheses about converging sequences in E (instead of the triangle inequality). If, for example, sequences $(x_n)_{n\in\mathbb{N}}$ and $(y_n)_{n\in\mathbb{N}}$ satisfy

$$\begin{cases} d_j(x_n,x) & \longrightarrow\ 0\\ d_j(y_n,y) & \longrightarrow\ 0\end{cases}$$

for $n\longrightarrow\infty$ and each $j\in\mathcal{I}$, then we expect for every index $i\in\mathcal{I}$ quite intuitively

$$d_i(x,y) \ =\ \lim_{n\to\infty} d_i(x_n,y_n).$$

At the beginning of Chapter 3, we list a few conditions on $d_j, e_j, \lfloor\cdot\rfloor_j$ $(j\in\mathcal{I})$ which admit all steps on the way to the main results of mutational analysis. As a special consequence of this step, we obtain the existence of strong solutions to a class of stochastic functional differential equations (in § 3.6) like

$$dX_t \ =\ h_1\big(t,\ \mathbb{E}(|X_t|),\ \mathbb{E}(|X_t|^2)\big)\cdot h_2(X_t)\ dt\ +\ b(t)\ dW_t.$$

Step (F) How to Extend the Weak Topology Beyond Normed Vector Spaces

Many of our subsequent results about the existence of solutions to examples are based on the counterpart of Peano's Theorem in the mutational framework. It states that continuity of the right-hand side and an appropriate form of sequential compactness always guarantee the existence of a solution to the given mutational equation. Hence, sequential compactness forms the basis for many existence results below — on the one hand.

On the other hand, evolution equations in an arbitrary Banach space exemplify that the norm of a vector space is frequently the most obvious choice for (at least) one of the distance functions d_j, e_i.

Norm compactness of the unit ball in a vector space, however, implies necessarily finite dimensions.

The weak topology is the typical way out of this conflict: The (norm-) closed unit ball in a reflexive Banach space is known to be weakly compact. In contrast to step (B), this observation encourages us now to generalize the concept of weak sequential *compactness* to the tuple $(E,(d_j)_{j\in\mathcal{I}},\ (e_j)_{j\in\mathcal{I}},(\lfloor\cdot\rfloor_j)_{j\in\mathcal{I}})$, but we are lacking any linear functionals on a set E in general.

Thus, we suggest starting from another connection between norm and weak topology of a real vector space X (rather than from linear functionals on X). A popular characterization of the norm concludes from the Theorem of Hahn-Banach

$$\|x\|_X \;=\; \sup\big\{y'(x)\;\big|\;y' : (X,\|\cdot\|_X) \longrightarrow \mathbb{R} \text{ linear, continuous, } \|y'\|_{\mathrm{Lin}(X,\mathbb{R})} \le 1\big\}.$$

As a first consequence, we become aware (again) that the substantial difference between weak and norm convergence of a sequence in X results from switching limit and supremum. The linear features of the functionals y' on X are of rather subordinate importance here.
Secondly, the basic structure of this characterization can be extended to abstract sets easily: The distance between two points is *represented as supremum* of further distance functions.

Now we apply this notion to the tuple $(E,\ (d_j)_{j\in\mathcal{I}},\ (e_j)_{j\in\mathcal{I}},\ (\lfloor\cdot\rfloor_j)_{j\in\mathcal{I}})$. The distance functions d_j, e_j $(j\in\mathcal{I})$ continue their role for transitions and solutions, but in addition, we assume distance functions $d_{j,\kappa}, e_{j,\kappa} : E\times E \longrightarrow [0,\infty[$ (with a further index set $\mathcal{J}\neq\emptyset$) such that for each index $j\in\mathcal{I}$,

$$d_j \;=\; \sup_{\kappa\in\mathcal{J}} d_{j,\kappa}, \qquad e_j \;=\; \sup_{\kappa\in\mathcal{J}} e_{j,\kappa}.$$

Then a sequence $(x_n)_{n\in\mathbb{N}}$ in E is said to converge "weakly" to an element $x\in E$ if for every $j\in\mathcal{I}$ and $\kappa\in\mathcal{J}$,

$$\lim_{n\to\infty} d_{j,\kappa}(x_n,x) \;=\; 0.$$

The families $(d_{j,\kappa})_{j\in\mathcal{I},\kappa\in\mathcal{J}}$ and $(e_{j,\kappa})_{j\in\mathcal{I},\kappa\in\mathcal{J}}$ do not have to consist of pseudometrics, but they are expected to specify the same "topology" on E again. Thus, we usually suppose the corresponding list of hypotheses as for $(d_j)_{j\in\mathcal{I}},\ (e_j)_{j\in\mathcal{I}}$.
In § 3.3.6, we clarify which forms of "weak" sequential compactness and "weak" continuity (of the right-hand side of mutational equations) are sufficient for extending Peano's Theorem about the existence of solutions.
These general results are applied to the nonlinear continuity equation, for example,

$$\begin{cases} \frac{d}{dt}\,\mu \,+\, \mathrm{div}_x\,(\mathbf{f}(\mu,\cdot)\,\mu) = 0 & \text{in } \mathbb{R}^N\times\,]0,T[\\ \mu(0) = \rho_0\,\mathscr{L}^N \in \mathbb{L}^{\infty\cap 1}(\mathbb{R}^N) & \end{cases}$$

with a given functional relationship in the form of

$$\mathbf{f}\colon\ \mathbb{L}^{\infty\cap 1}(\mathbb{R}^N)\times[0,T] \longrightarrow \mathrm{BV}_{\mathrm{loc}}(\mathbb{R}^N,\mathbb{R}^N)\cap L^\infty(\mathbb{R}^N,\mathbb{R}^N)$$

in § 3.8. Here the distributional solutions $\mu(\cdot) : [0,T] \longrightarrow \mathbb{L}^{\infty\cap 1}(\mathbb{R}^N)$ have their values in $\mathbb{L}^{\infty\cap 1}(\mathbb{R}^N) := \big\{\rho\,\mathscr{L}^N \;\big|\; \rho\in L^1(\mathbb{R}^N)\cap L^\infty(\mathbb{R}^N),\ \rho\ge 0\big\}$ and are constructed by means of Prokhorov's Compactness Theorem.

Step (G) Less Restrictive Conditions on Distance Functions $d_j, e_j \ (j \in \mathscr{I})$: Dispense with Symmetry

The evolution of compact subsets of the Euclidean space $\mathbb{R}^N$ might depend explicitly on their topological boundary and, we would like to take such an influence into consideration — still without making any a priori assumptions about regularity. Even simple examples, however, indicate obstacles in the current mutational framework.

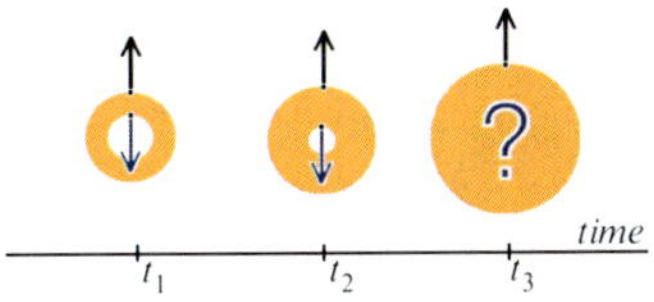

Consider just an annulus expanding isotropically at a constant speed 1. After a finite period, the "hole" in the center of the annulus disappears suddenly. Hence, the topological boundary of the expanding annulus does not evolve continuously (in the sense of Painlevé-Kuratowski).

The classical Pompeiu-Hausdorff distance between the boundaries of such an annulus $K \subset \mathbb{R}^N$ and its expanding counterpart $\mathbb{B}_t(K) \subset \mathbb{R}^N$ does not have to be continuous with respect to time t and thus, it is unsuitable for comparing topological boundaries in regard to transitions.
In search of an alternative pseudo-metric, we realize that some topological components of $\partial\mathbb{B}_t(K)$ might "disappear" while time t is increasing, but each boundary point of $\partial\mathbb{B}_t(K)$ has close counterparts at earlier sets $\partial\mathbb{B}_s(K)$ (with $s < t$). Indeed,

$$\operatorname{dist}\big(\partial\mathbb{B}_t(K),\ \partial\mathbb{B}_s(K)\big) \quad \leq \quad t - s$$

for all $0 \leq s \leq t$, but a corresponding estimate does not have to hold for $0 \leq t < s$. In other words, we find properties similar to some requirements for transitions if we compare only *later* sets with *earlier* sets (in regard to their topological boundaries), but not vice versa.

For this reason, we aim at a mutational framework for a tuple $(E, (d_j)_{j \in \mathscr{I}}, (e_j)_{j \in \mathscr{I}}, (\lfloor\cdot\rfloor_j)_{j \in \mathscr{I}})$ without assuming symmetry of d_j and e_j $(j \in \mathscr{I})$. Broadly speaking, the first argument of each distance usually refers to the earlier state whereas the second argument is the later element (in Chapter 4).

The same geometric example also demonstrates an analytical obstacle which we have to overcome after dispensing with symmetry. Indeed, consider a further initial set $K' \subset \mathbb{R}^N$. Of course, the preceding inequality still holds for $t \longmapsto \partial\mathbb{B}_t(K')$, but the distance of $\partial\mathbb{B}_t(K)$ from the other boundary $\partial\mathbb{B}_t(K')$ at the same time t, i.e.

$$[0,\infty[\longrightarrow [0,\infty[, \qquad t \longmapsto \operatorname{dist}\big(\partial\mathbb{B}_t(K),\ \partial\mathbb{B}_t(K')\big),$$

might be discontinuous. As a general consequence for mutational equations, we have to ensure (at least) lower semicontinuity of some time-dependent distances which had always been continuous before so that the adapted program of mutational analysis still works.

Step (H) Distribution-Like Solutions to Mutational Equations

Examples with compact subsets of $\mathbb{R}^N$ evolving according to their topological boundaries are still difficult to handle in the mutational framework though. Indeed, an additional challenge is closely related to the regularity of transitions with respect to state (and its continuity parameter $\alpha_j(\vartheta; r) < \infty$).

It is an essential feature of transitions that the initial distance between two states may grow at most exponentially while evolving along one and the same transition.
Although this condition does not require continuity of distances with respect to time, the boundaries of two time-dependent compact sets and their normals might not satisfy it whenever one of the boundaries is not continuous with respect to time.

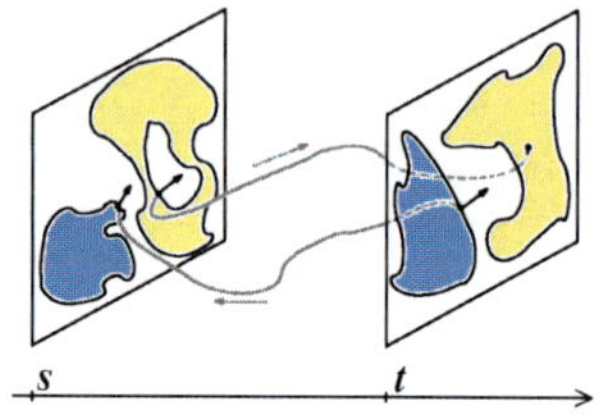

With regard to the geometric situation sketched in the figure on the right, there is no general rule for compact sets when the next topological component of the boundary disappears, i.e., when the distance from another boundary might be discontinuous for the next time.

This obstacle can be overcome in the mutational framework if we introduce a less restrictive concept of transition and solution.
In the theory of partial differential equations, similar difficulties have already led to distributions and distributional solutions, but their defining property, i.e. partial integration with smooth functions, requires more mathematical structure than a set $E \neq \emptyset$ provides in general. For this reason, we suggest a more general interpretation of the step from classical to distributional derivatives:

> *Select an essential property in the "classical" theory and demand to preserve it (only) for all elements of a given fixed "test set" – instead of the whole "basic set".*

Usually this important feature is the rule of partial integration and, it is preserved for smooth test functions with compact support (or Schwartz functions).
In the mutational framework, the inequality of error propagation plays a central role and specifies in which sense transitions represent first-order approximations:

$$d_j\big(\vartheta(h,x),\ \tau(h,y)\big) \ \le\ \big(d_j(x,y) + h \cdot D_j(\vartheta,\tau;R_j)\big) \cdot e^{\alpha_j(\vartheta;R_j)\,h}$$

with the radius $R_j > 0$ just depending on $\max\{\lfloor x \rfloor_j,\ \lfloor y \rfloor_j\},\ \gamma_j(\vartheta),\ \gamma_j(\tau) < \infty$. At time $t \in [0,T]$, a curve $x(\cdot) : [0,T] \longrightarrow E$ has the "same properties up to first order" as a transition τ (in a generalized sense) if essentially the same *asymptotic* inequalities of error propagation hold for $\tau(\cdot, x(t))$, $x(t+\cdot)$ and $h \downarrow 0$:

$$d_j\big(\vartheta(h,z),\ \tau(h,x(t))\big) \ \le\ \big(d_j(z,x(t)) + h \cdot D_j(\vartheta,\tau;R_j)\big) \cdot e^{\alpha_j(\vartheta;R_j)\,h}$$

$$d_j\big(\vartheta(h,z),\ x(t+h)\big) \ \le\ \big(d_j(z,x(t)) + h \cdot D_j(\vartheta,\tau;R_j)\big) \cdot e^{\alpha_j(\vartheta;R_j)\,h} + o(h).$$

Strictly speaking, the latter inequality "in an asymptotic sense for $h \downarrow 0$" means

$$\limsup_{h \downarrow 0} \tfrac{1}{h} \cdot \Big(d_j\big(\vartheta(h,z),\, x(t+h)\big) \;-\; d_j(z,\, x(t)) \cdot e^{\alpha_j(\vartheta;R_j)\, h} \Big) \;\le\; D_j(\vartheta,\tau;R_j). \quad (\lozenge)$$

In Aubin's original theory of mutational equations, this condition being satisfied by all elements $z \in E$ and all transitions ϑ proves to be equivalent to $\tau \in \overset{\circ}{x}(t)$ and thus, it characterizes the mutation of $x(\cdot)$ at time t. All our steps of generalizations before have not changed this situation. (In fact, we have even preferred the error inequality of transitions to the triangle inequality of distances in step (E).)
For the step to distribution-like mutations, we are now free to fix a nonempty "test set" $\mathscr{D}$ arbitrarily and to demand the property $(\lozenge)$ for all elements $z \in \mathscr{D}$ (instead of E) and all transitions ϑ. This feature is central to the generalized definition of $\tau \in \overset{\circ}{x}(t)$. Motivated by the finite element methods of Petrov-Galerkin, we avoid the assumption $\mathscr{D} \subset E$ deliberately.
More details about this step are presented in Chapter 4. Afterwards this most general theory of mutational equations so far is applied to two examples with compact subsets of $\mathbb{R}^N$ evolving according to their graphs of limiting normal cones.

0.4 Mutational Inclusions

In Chapter 5, mutational inclusions are introduced. Correspondingly to differential inclusions in $\mathbb{R}^N$, they are based on the idea that more than one transition can be admitted at each element and time. For this purpose, the single-valued function $f : E \times [0,T] \longrightarrow \Theta$ (on the right-hand side of the mutational equation) is replaced by a set-valued map $\mathscr{F} : E \times [0,T] \rightsquigarrow \Theta$ and, we are looking for a continuous curve $x(\cdot) : [0,T] \longrightarrow E$ such that at $\mathscr{L}^1$-almost every time, a transition $\vartheta \in \mathscr{F}(x(t),t) \subset \Theta$ also belongs to the mutation $\overset{\circ}{x}(t)$.

Dispensing with state constraints in § 5.1, we prove a selection principle generalizing the Theorem of Antosiewicz-Cellina. For technical reasons, however, both the basic set E and the transition set Θ are supposed to be separable metric spaces. Then continuity of $\mathscr{F}$ and a suitable form of sequential compactness in E are sufficient for existence of solutions in Theorem 5.4 (on page 388).

Inclusions with state constraints are discussed (only) for morphological transitions on compact subsets of $\mathbb{R}^N$ because we need more compactness properties for measurable curves in the transition set. A quite general viability theorem is presented and proven in § 5.2. Finally, § 5.3 deals with applications to control problems for nonlocal set evolutions. It is remarkable that these control equations with state constraints have the *states* in a metric space (and not only the controls).

Chapter 1
Extending Ordinary Differential Equations to Metric Spaces: Aubin's Suggestion

This chapter is devoted to Aubin's original concept of *mutational equations* introduced in the early 1990s. They provide an interesting extension of ordinary differential equations to a metric space (instead of the classical Euclidean space $\mathbb{R}^N$).
The main challenge to which Aubin suggested an interesting answer is how to dispense with any linear structure of the basic set while following the popular track of ordinary differential equations up to solutions to the initial value problem.

1.1 The Key for Avoiding (Affine) Linear Structures: Transitions

For extending ordinary differential equations beyond the traditional borders of vector spaces, we start with a given metric space (E,d) as suitable mathematical environment. Independently from dispensing with any linear structure of the basic set, we still need a quantitative tool for investigating the asymptotic features of the relationship between time-dependent states.
Roughly speaking, the starting point now consists in extending elementary terms like "velocity" (in the sense of time derivative of a curve) from vector spaces to the given metric space (E,d).

Considering a curve $x(\cdot) : [0,T] \longrightarrow \mathbb{R}^N$ in the Euclidean space $\mathbb{R}^N$, its derivative $x'(t)$ at time $t \in [0,T[$ is usually defined as limit of difference quotients, i.e.

$$x'(t) \;=\; \lim_{h \to 0} \; \frac{x(t+h) \,-\, x(t)}{h} \, .$$

This definition, however, cannot be extended to a metric space in an obvious way – due to lacking differences. Hence, we consider the alternative characterization which is based on affine linear approximation of first order. Indeed, a vector $v \in \mathbb{R}^N$ represents the time derivative of $x(\cdot)$ at time $t \in [0,T[$ if and only if there exists a residual function $w(\cdot)$ with $\lim_{h \to 0} \frac{1}{h} \cdot w(h) = 0$ such that

T. Lorenz, *Mutational Analysis: A Joint Framework for Cauchy Problems In and Beyond Vector Spaces*, Lecture Notes in Mathematics 1996,
DOI 10.1007/978-3-642-12471-6_2,

$$x(t+h) = x(t) + h \cdot v + w(h)$$

is satisfied for every $h \in \mathbb{R}$ sufficiently close to 0. The equivalent formulation

$$\lim_{h \to 0} \tfrac{1}{h} \left| x(t+h) - \left(x(t) + h \cdot v \right) \right| = 0$$

motivates how this classical notion might be extended to a metric space. Indeed, we now compare the asymptotic features of the curve $h \longmapsto x(t+h)$ to the affine linear map $h \longmapsto x(t) + h \cdot v$ with respect to the Euclidean metric $|\cdot|$.
For dispensing with any aspects of affine linearity in a moment, we focus on the continuous map

$$[0,\infty[\times \mathbb{R}^N \longrightarrow \mathbb{R}^N, \quad (h,z) \longmapsto z + h \cdot v$$

for a fixed vector $v \in \mathbb{R}^N$ of direction. Geometrically speaking, it indicates the final point $z + h \cdot v$ to which the initial point z is moved at time h and, it serves as a kind of "elementary deformation" of the Euclidean space $\mathbb{R}^N$ for approximating the curve $x(t + \cdot)$ up to first order.
For avoiding any linear structure of the basic set, Aubin suggested to consider such maps of time and state as counterparts of affine linear maps in vector spaces, i.e. in the given metric space (E,d), a continuous map

$$\vartheta : [0,1] \times E \longrightarrow E, \quad (h,z) \longmapsto \vartheta(h,z)$$

is to play the role of (not necessarily affine linear) "deformations" in a fixed direction. It specifies the point $\vartheta(h,z) \in E$ to which each initial point $z \in E$ is moved at time $h \in [0,1]$. Such a map ϑ can be interpreted as first-order approximation of a curve $x(\cdot) : [0,T[\longrightarrow E$ at time $t \in [0,T[$ if it satisfies

$$\lim_{h \to 0} \tfrac{1}{h} \cdot d\big(x(t+h),\ \vartheta(h, x(t))\big) = 0.$$

This is a characterization corresponding to time derivative, but completely free of any affine linear structure indeed.
Obviously, such a homotopy-like map ϑ can serve as starting point for a differential calculus in (E,d) only if it satisfies appropriate continuity conditions. Aubin introduced the term of "transition" in the following way:

Definition 1. Let (E,d) be a metric space. A map $\vartheta : [0,1] \times E \longrightarrow E$ is called *transition* on (E,d) if it satisfies the following four conditions:

1.) for every $x \in E$: $\quad \vartheta(0,x) = x$

2.) for every $x \in E,\ t \in [0,1[$: $\quad \lim_{h \downarrow 0} \tfrac{1}{h} \cdot d\big(\vartheta(t+h,x),\ \vartheta(h, \vartheta(t,x))\big) = 0$

3.) $$\alpha(\vartheta) := \sup_{\substack{x,y \in E \\ x \neq y}} \ \limsup_{h \downarrow 0} \ \max\left\{0,\ \frac{d(\vartheta(h,x),\ \vartheta(h,y)) - d(x,y)}{h \cdot d(x,y)}\right\} < \infty$$

4.) $$\beta(\vartheta) := \sup_{x \in E} \ \limsup_{h \downarrow 0} \ \frac{d(x,\ \vartheta(h,x))}{h} < \infty$$

Condition (1.) guarantees that the second argument x of ϑ represents the initial point at time $t = 0$. Moreover condition (2.) can be regarded as a weakened form of the semigroup property. Due to Gronwall's Lemma, it even implies that ϑ satisfies the semigroup condition

$$\vartheta(t+h,x) = \vartheta(h, \vartheta(t,x))$$

for every element $x \in E$ and time $t,h \in [0,1]$ with $t+h \leq 1$ (as we will verify in subsequent Corollary 22).
Finally the parameters $\alpha(\vartheta), \beta(\vartheta) < \infty$ guarantee the continuity of ϑ with respect to both arguments. In particular, condition (4.) implies the uniform Lipschitz continuity of ϑ with respect to time:

$$d\big(\vartheta(s,x), \vartheta(t,x)\big) \leq \beta(\vartheta) \cdot |t-s|$$

for all times $s,t \in [0,1]$ and initial elements $x \in E$ (as subsequent Lemma 8 shows in detail). Due to Condition (3.), the distance of initial points can grow at most exponentially with respect to time (as we will verify in subsequent Proposition 7):

$$d\big(\vartheta(h,x), \vartheta(h,y)\big) \leq d(x,y) \cdot e^{\alpha(\vartheta)\,h}$$

for all $h \in [0,1]$ and $x,y \in E$. In terms of semigroups or dynamical systems, ϑ is said to be *ω-contractive*.

Example 2. The most popular transitions on the Euclidean space $(\mathbb{R}^N, |\cdot|)$ are induced by the affine linear functions

$$\vartheta_v : [0,1] \times \mathbb{R}^N \longrightarrow \mathbb{R}^N, \quad (h,x) \longmapsto x + h \cdot v$$

in any fixed direction $v \in \mathbb{R}^N$. Then, $\alpha(\vartheta_v) = 0$ and $\beta(\vartheta_v) = |v|$.

Example 3. The constant velocity $v \in \mathbb{R}^N$ of translation in $\mathbb{R}^N$ is now replaced by a vector field, i.e. for a given bounded Lipschitz function $f : \mathbb{R}^N \longrightarrow \mathbb{R}^N$, every initial point $x_0 \in \mathbb{R}^N$ is moving along the unique solution $x(\cdot) : [0,\infty[\longrightarrow \mathbb{R}^N$ to the ordinary differential equation $x'(t) = f(x(t))$.
Hence, $\vartheta_f(t,x_0) := x(t)$ with the unique solution $x(\cdot) \in C^1([0,t],\mathbb{R}^N)$ to the initial value problem

$$\begin{cases} x'(t) = f(x(t)), \\ x(0) = x_0. \end{cases}$$

The classical Theorem of Cauchy-Lipschitz about ordinary differential equations can be regarded as a special case of Filippov's Theorem A.6 about differential inclusions and, it implies that $\vartheta_f : [0,1] \times \mathbb{R}^N \longrightarrow \mathbb{R}^N$ satisfies the four conditions on transitions with $\alpha(\vartheta_f) \leq \operatorname{Lip} f$ and $\beta(\vartheta_f) \leq \|f\|_{\sup}$.
$\mathrm{BLip}(\mathbb{R}^N,\mathbb{R}^N)$ consists of all bounded Lipschitz continuous functions $\mathbb{R}^N \longrightarrow \mathbb{R}^N$.

Example 4. Leaving now the familiar field of points in $\mathbb{R}^N$, we consider compact subsets of the Euclidean space $\mathbb{R}^N$ (instead of single state vectors).
$\mathscr{K}(\mathbb{R}^N)$ denotes the set of all nonempty compact subsets of $\mathbb{R}^N$. Subsets of $\mathbb{R}^N$, however, do not have any obvious linear structure, but $\mathscr{K}(\mathbb{R}^N)$ is usually supplied

with a very useful metric: The so-called *Pompeiu-Hausdorff distance* between two sets $K_1, K_2 \in \mathscr{K}(\mathbb{R}^N)$ is defined as

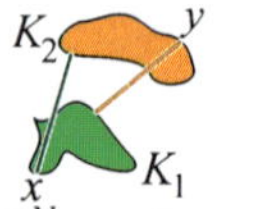

$$dl(K_1, K_2) := \max \Big\{ \sup_{x \in K_1} \operatorname{dist}(x, K_2), \quad \sup_{y \in K_2} \operatorname{dist}(y, K_1) \Big\}.$$

Correspondingly to the preceding Example 3, suppose $f : \mathbb{R}^N \longrightarrow \mathbb{R}^N$ to be a bounded and Lipschitz vector field. Now the initial points $x_0 \in \mathbb{R}^N$ are replaced by initial sets $K_0 \in \mathscr{K}(\mathbb{R}^N)$ and, we focus on *all* points that can be reached by a solution $x(\cdot)$ of $x'(\cdot) = f(x(\cdot))$ starting in K_0, i.e.

$$\begin{aligned} \vartheta_f : [0,1] \times \mathscr{K}(\mathbb{R}^N) &\longrightarrow \mathscr{K}(\mathbb{R}^N) \\ (t,\ K_0) &\longmapsto \big\{ x(t) \;\big|\; \text{there exists } x(\cdot) \in C^1([0,t], \mathbb{R}^N) : \\ &\qquad\qquad x'(\cdot) = f(x(\cdot)),\ x(0) \in K_0 \big\}. \end{aligned}$$

$\vartheta_f(t, K_0)$ is called *reachable set* of the vector field f and the initial set K_0 at time t. It provides an approach how to "deform" any compact subset of $\mathbb{R}^N$ – without any regularity assumptions about the set or its topological boundary. In fact, these set evolutions belong to the basic tools of the so-called velocity method (alias speed method) and have led Céa, Delfour, Sokolowski, Zolésio and others to excellent results about shape optimization (e.g. [42, 55, 56, 174, 190]).
The classical Theorem of Cauchy-Lipschitz about ordinary differential equations provides estimates that are even uniform with respect to the initial point and thus, the same conclusions as in Example 3 ensure that ϑ_f is a transition on $(\mathscr{K}(\mathbb{R}^N), dl)$ with $\alpha(\vartheta_f) \leq \operatorname{Lip} f$, $\beta(\vartheta_f) \leq \|f\|_{\sup}$ (see subsequent Example 54 for details).

Reachable sets of Lipschitz vector fields, however, are always reversible in time. Indeed, every reachable set $\vartheta_f(t, K_0) \subset \mathbb{R}^N$ can be deformed to the initial set K_0 by means of the flow along $-f$, i.e.

$$\vartheta_{-f}\big(t,\ \vartheta_f(t, K_0)\big) = K_0$$

for every set $K_0 \in \mathscr{K}(\mathbb{R}^N)$. This results directly from the uniqueness of solutions $x(\cdot) :]-\infty, \infty[\longrightarrow \mathbb{R}^N$ to the initial value problem

$$\begin{cases} x'(t) = f(x(t)), \\ x(0) = x_0. \end{cases}$$

Example 5. The class of set evolutions described as reachable set can be extended very easily if we admit more than one velocity at each point of the Euclidean space. Thus, the bounded and Lipschitz vector fields $f : \mathbb{R}^N \longrightarrow \mathbb{R}^N$ mentioned in Example 4 are now replaced by set-valued maps $F : \mathbb{R}^N \rightsquigarrow \mathbb{R}^N$ whose values are nonempty compact subsets of $\mathbb{R}^N$ and, we consider the flow along the differential inclusion $x'(\cdot) \in F(x(\cdot))$ (Lebesgue-almost everywhere) instead of the ordinary differential equation $x'(\cdot) = f(x(\cdot))$.
The *reachable set* $\vartheta_F(t, K_0) \subset \mathbb{R}^N$ of the initial set $K_0 \in \mathscr{K}(\mathbb{R}^N)$ and the set-valued map $F : \mathbb{R}^N \rightsquigarrow \mathbb{R}^N$ at time $t \geq 0$ consists of all points that can be attained at time t via an absolutely continuous solution $x(\cdot)$ of $x'(\cdot) \in F(x(\cdot))$ a.e. starting in K_0. If $F : \mathbb{R}^N \rightsquigarrow \mathbb{R}^N$ is bounded and Lipschitz continuous with nonempty compact values, then Filippov's Theorem A.6 implies that

$$\begin{aligned}\vartheta_F : [0,1] \times \mathscr{K}(\mathbb{R}^N) &\longrightarrow \mathscr{K}(\mathbb{R}^N)\\ (t,\ K_0) &\longmapsto \big\{x(t) \;\big|\; \text{there exists } x(\cdot) \in W^{1,1}([0,t],\mathbb{R}^N): \\ &\qquad\qquad x'(\cdot) \in F(x(\cdot))\ \mathscr{L}^1\text{-a.e. in } [0,t],\ x(0) \in K_0\big\}\end{aligned}$$

is a transition on $(\mathscr{K}(\mathbb{R}^N), d\!\!l)$ with $\alpha(\vartheta_F) \leq \operatorname{Lip} F$ and $\beta(\vartheta_F) \leq \sup_{x\in\mathbb{R}^N} \sup_{v\in F(x)} |v|$.
Aubin called it *morphological transition* and used it in most of his examples about set evolutions. It will be discussed in more detail in § 1.9.2 (on page 60 ff.) below.

Let us now return to a metric space (E,d) and some nonempty set $\Theta(E,d)$ of transitions in the (very general) sense of Definition 1.
The "flow" along these transitions can form the basis for differential calculus (considering curves in E) only if we have an opportunity to "compare" the evolution of two arbitrary initial states along two different transitions. For this reason, Aubin suggested a distance between transitions:

Definition 6. Let (E,d) be a metric space and $\Theta(E,d)$ be a nonempty set of transitions on (E,d). For any $\vartheta, \tau \in \Theta(E,d)$, define

$$D(\vartheta, \tau) := \sup_{x\in E} \limsup_{h\downarrow 0} \tfrac{1}{h} \cdot d\big(\vartheta(h,x),\ \tau(h,x)\big).$$

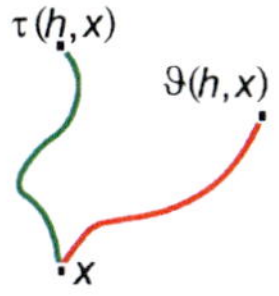

The basic idea of $D(\vartheta,\tau)$ is to compare the two curves $\vartheta(\cdot,x),\ \tau(\cdot,x) : [0,1] \longrightarrow E$ with the same initial point $x \in E$ for $h \downarrow 0$. As each of these curves is continuous, their joint initial point always implies $d\big(\vartheta(h,x),\ \tau(h,x)\big) \longrightarrow 0$ for $h \downarrow 0$. Thus we consider its asymptotic properties of first order – represented by the factor $\frac{1}{h}$ in Definition 6.
The parameters of continuity $\beta(\vartheta), \beta(\tau)$ (specified in Definition 1) guarantee that $D(\vartheta,\tau)$ is always finite. Indeed, due to the triangle inequality of the metric d,

$$D(\vartheta,\tau) \leq \sup_{x\in E} \limsup_{h\downarrow 0} \tfrac{1}{h} \cdot \Big(d\big(\vartheta(h,x),\ x\big) + d\big(x,\ \tau(h,x)\big)\Big) \leq \beta(\vartheta) + \beta(\tau).$$

Furthermore, $D : \Theta(E,d) \times \Theta(E,d) \longrightarrow [0,\infty[$ is symmetric and always satisfies the triangle inequality, i.e. for any transitions $\vartheta_1, \vartheta_2, \tau$ on (E,d),

$$D(\vartheta_1, \vartheta_2) \leq D(\vartheta_1, \tau) + D(\tau, \vartheta_2).$$

$D(\cdot,\cdot)$ is not a metric on $\Theta(E,d)$, though, because it does not have to be positive definite, i.e. $D(\vartheta,\tau) = 0$ does not imply $\vartheta \equiv \tau$ in general. Indeed, $D(\vartheta,\tau)$ focuses on the transitions ϑ, τ merely for $h \downarrow 0$.
Now all tools are available for comparing two initial states in E while evolving along two different transitions respectively:

Proposition 7. *Let (E,d) be a metric space and $\Theta(E,d)$ be a nonempty set of transitions on (E,d). For any transitions $\vartheta, \tau \in \Theta(E,d)$ and elements $x, y \in E$, the following estimate is satisfied at each time $h \in [0,1[$*

$$d\big(\vartheta(h,x),\ \tau(h,y)\big) \leq \big(d(x,y) + h \cdot D(\vartheta,\tau)\big) \cdot e^{\alpha(\vartheta)\,h}.$$

The subdifferential version of Gronwall's Lemma (Proposition A.2) is the key tool for concluding global estimates from local information. In this regard, the proof of Proposition 7 exemplifies the basic technique for most of our subsequent results:

Lemma 8. *For every transition ϑ on a metric space (E,d) and initial point $x \in E$, the curve $\vartheta(\cdot,x) : [0,1[\longrightarrow E$ is $\beta(\vartheta)$-Lipschitz continuous.*

Proof. Choose $x \in E$ and $\varepsilon > 0$ arbitrarily. Due to conditions (2.),(4.) of Definition 1, i.e.

$$\begin{cases} \beta(\vartheta) \stackrel{\text{Def.}}{=} \sup\limits_{y\in E} \limsup\limits_{h\downarrow 0} \frac{1}{h} \cdot d(y, \vartheta(h,y)) < \infty \\ \lim\limits_{h\downarrow 0} \frac{1}{h} \cdot d\big(\vartheta(h, \vartheta(t,x)), \vartheta(t+h, x)\big) = 0 \end{cases}$$

we obtain for each $t \in [0,1[$ that some sufficiently small $\delta_t \in]0, 1-t[$ satisfies

$$\tfrac{1}{h} \cdot d\big(\vartheta(t,x), \vartheta(t+h,x)\big) \le \beta(\vartheta) + \varepsilon \qquad \text{for all } h \in]0, \delta_t].$$

For any $0 \le s_1 \le s_2 \le 1-\varepsilon$ given, covering $[s_1,s_2]$ with (at most countably many) subintervals $[t, t+\delta_t]$ (with $t \in [s_1,s_2[$) and the triangle inequality of d imply

$$d\big(\vartheta(s_1,x), \vartheta(s_2,x)\big) \le (\beta(\vartheta)+\varepsilon) \cdot (s_2 - s_1).$$

As $\varepsilon > 0$ was chosen arbitrarily, $\vartheta(\cdot,x)$ is $\beta(\vartheta)$-Lipschitz continuous in $[0,1[$. □

Proof (of Proposition 7). The auxiliary function

$$\psi : [0,1[\longrightarrow [0,\infty[, \quad h \longmapsto d\big(\vartheta(h,x), \tau(h,y)\big)$$

is Lipschitz continuous due to Lemma 8 and the triangle inequality of d. Moreover it satisfies for every $t \in [0,1[$

$$\begin{aligned}
&\limsup_{h\downarrow 0} \frac{\psi(t+h) - \psi(t)}{h} \\
&\quad = \limsup_{h\downarrow 0} \tfrac{1}{h} \cdot \Big(d\big(\vartheta(t+h, x), \tau(t+h, y)\big) - d\big(\vartheta(t, x), \tau(t, y)\big)\Big) \\
&\quad \le \limsup_{h\downarrow 0} \tfrac{1}{h} \cdot \Big(d\big(\vartheta(t+h, x), \vartheta(h, \vartheta(t,x))\big) \\
&\qquad\qquad + d\big(\vartheta(h, \vartheta(t,x)), \vartheta(h, \tau(t,y))\big) - d\big(\vartheta(t, x), \tau(t, y)\big) \\
&\qquad\qquad + d\big(\vartheta(h, \tau(t,y)), \tau(h, \tau(t,y))\big) \\
&\qquad\qquad + d\big(\tau(h, \tau(t,y)), \tau(t+h, y)\big)\Big) \\
&\quad \le 0 + \alpha(\vartheta) \cdot \psi(t) + D(\vartheta,\tau) + 0.
\end{aligned}$$

Finally, the Gronwall estimate in Proposition A.2 implies for each $h \in [0,1[$

$$\psi(h) \le \psi(0)\, e^{\alpha(\vartheta)\,h} + D(\vartheta,\tau)\, \frac{e^{\alpha(\vartheta)\,h} - 1}{\alpha(\vartheta)}. \qquad \square$$

Remark 9. The same arguments lead to the inequality for any $t_1,t_2 \in [0,1[$

$$d\big(\vartheta(t_1+h,x), \tau(t_2+h,y)\big) \le \Big(d\big(\vartheta(t_1,x), \tau(t_2,y)\big) + h \cdot D(\vartheta,\tau)\Big) \cdot e^{\alpha(\vartheta)\,h}.$$

1.2 The Mutation as Counterpart of Time Derivative

Consider a curve $x(\cdot) : [0,T] \longrightarrow E$ in a metric space (E,d). A transition ϑ on (E,d) can be regarded as (generalized) time derivative of $x(\cdot)$ at time $t \in [0,T[$ if the comparison with $x(t+\cdot)$ reveals an approximation of first order in the following sense:

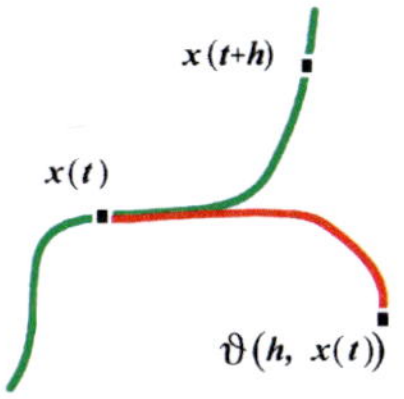

$$\lim_{h \downarrow 0} \frac{1}{h} \cdot d\big(\vartheta(h, x(t)),\ x(t+h)\big) = 0.$$

In general this asymptotic condition may be satisfied by more than one transition since only the properties for $h \downarrow 0$ are taken into consideration. Aubin suggested to introduce a new term for the set of *all* these transitions – rather than considering the underlying equivalent classes of transitions because the latter do not provide additional mathematical insight:

Definition 10. Let $\Theta(E,d)$ be a nonempty set of transitions on a metric space (E,d) and, $x(\cdot) : [0,T] \longrightarrow E$ denotes a curve. For $t \in [0,T[$, the set

$$\overset{\circ}{x}(t) := \Big\{\vartheta \in \Theta(E,d) \ \Big|\ \lim_{h \downarrow 0} \frac{1}{h} \cdot d\big(\vartheta(h, x(t)),\ x(t+h)\big) = 0\Big\}$$

is called *mutation* of $x(\cdot)$ at time t.

Remark 11. For every transition ϑ on (E,d) and initial element $x_0 \in E$, the curve $x_{x_0}(\cdot) := \vartheta(\cdot, x_0) : [0,1] \longrightarrow E$ has ϑ in its mutation at each time $t \in [0,1[$:

$$\vartheta \in \overset{\circ}{x}_{x_0}(t)$$

for every $t \in [0,1[$. This results directly from condition (2.) in Definition 1.

In regard to real-valued functions, the classical concepts of derivative and integral are closely related by the fundamental theorem of calculus. Similarly, we can also start with a curve of transitions and look for an appropriate curve in the metric space:

Definition 12. Let $\Theta(E,d)$ be a nonempty set of transitions on a metric space (E,d) and, $\vartheta(\cdot) : [0,T] \longrightarrow \Theta(E,d)$ denotes a curve of transitions.
A curve $x(\cdot) : [0,T] \longrightarrow E$ is called *primitive* of $\vartheta(\cdot)$ if $x(\cdot)$ is Lipschitz continuous with respect to d and satisfies for Lebesgue-almost every $t \in [0,T]$

$$\vartheta(t) \in \overset{\circ}{x}(t)$$

i.e. $$\lim_{h \downarrow 0} \frac{1}{h} \cdot d\big(\vartheta(t)\,(h, x(t)),\ x(t+h)\big) = 0 \qquad \text{for a.e. } t \in [0,T].$$

Lemma 8 and Remark 11 imply that constructing a primitive of $\vartheta(\cdot) : [0,T] \longrightarrow \Theta(E,d)$ with given initial element $x_0 \in E$ is particularly easy if $\vartheta(\cdot)$ is piecewise constant with $\sup_t \beta(\vartheta(t)) < \infty$.

1.3 Feedback Leads to Mutational Equations

Ordinary differential equations are based on the notion that the derivative of the wanted solution is prescribed by a given function of the current state. This form of feedback can be extended to curves in a metric space (E,d) and their mutations. Aubin introduced the following definition:

Definition 13. Let $\Theta(E,d)$ be a nonempty set of transitions on a metric space (E,d). Furthermore, a single-valued function $f : E \times [0,T] \longrightarrow \Theta(E,d)$ is given. A curve $x(\cdot) : [0,T] \longrightarrow E$ is called *solution* to the mutational equation

$$\overset{\circ}{x}(\cdot) \ni f(x(\cdot), \cdot)$$

if $x(\cdot)$ is primitive of the composition $f(x(\cdot),\cdot) : [0,T] \longrightarrow \Theta(E,d)$ in the sense of Definition 12, i.e. $x(\cdot)$ is Lipschitz continuous with respect to d and satisfies

$$\lim_{h \downarrow 0} \frac{1}{h} \cdot d\big(f(x(t),t)\,(h,\, x(t)),\ x(t+h)\big) = 0$$

for Lebesgue-almost every $t \in [0,T]$.

Remark 14. At first glance, the symbol $\ni$ here seems to be contradictory to the term "equation". The mutation $\overset{\circ}{x}(t)$, however, is defined as *subset* of all transitions in $\Theta(E,d)$ providing a first-order approximation of $x(t+\cdot)$ (Definition 10). The transition on the "right-hand side" $f(x(t),t) \in \Theta(E,d)$ is required to be one of its elements at Lebesgue-almost every time t.
Example 2 lays the foundations for applying this framework to Lipschitz continuous solutions to ordinary differential equations in $\mathbb{R}^N$. In this special case, the mutation of a Lipschitz continuous curve $x : [0,T] \longrightarrow \mathbb{R}^N$ consists of just one vector at almost every time – as a consequence of Rademacher's Theorem.
In general, however, the mutation $\overset{\circ}{x}(t)$ might consists of more than one transition.

Adapting the classical arguments about ordinary differential equations, the next step is now to solve initial value problems with mutational equations. As mentioned at the end of § 1.2, a primitive of piecewise constant functions is easy to construct and this opens the door to applying Euler method in the mutational framework. Aubin has already presented the following counterpart of Cauchy-Lipschitz Theorem about existence and uniqueness of solutions to the initial value problem:

Theorem 15 (Aubin's adaptation of Cauchy-Lipschitz Theorem).
Let (E,d) be a metric space in which all closed bounded balls are compact. $\Theta(E,d)$ denotes a nonempty set of transitions on (E,d).
Let $f : E \longrightarrow \Theta(E,d)$ be a λ-Lipschitz continuous function, i.e.

$$D(f(y),\ f(z)) \le \lambda \cdot d(y,z) \qquad \textit{for any } y,z \in E.$$

Furthermore assume $\widehat{\alpha} := \sup_{z \in E} \alpha(f(z)) < \infty.$

Fix an element $x_0 \in E$ and a curve $y(\cdot) : [0,T] \longrightarrow E$ with $\overset{\circ}{y}(t) \neq \emptyset$ for all $t \in [0,T]$.

Then there exists a unique solution $x(\cdot) : [0,T] \longrightarrow E$ *to the initial value problem*

$$\begin{cases} \overset{\circ}{x}(\cdot) \ni f\big(x(\cdot)\big) \\ x(0) = x_0 \end{cases}$$

In addition, it satisfies the following inequality for all $t \in [0,T]$

$$\begin{aligned} d\big(x(t),\ y(t)\big) \ \le\ & d\big(x_0,\ y(0)\big) \cdot e^{(\widehat{\alpha}+\lambda)\,t} \\ & + \int_0^t e^{(\widehat{\alpha}+\lambda)\,(t-s)} \cdot \inf_{\vartheta \,\in\, \overset{\circ}{y}(s)} D\big(f(y(s)),\ \vartheta\big)\ ds. \end{aligned}$$

In particular, this theorem implies for autonomous mutational equations with Lipschitz continuous right-hand side that solutions depend continuously on the initial element and the transition function (on the right-hand side). Here $D(\cdot,\cdot)$ is usually the distance function used for transitions on (E,d).

The second important result that Aubin extended from ordinary differential equations to mutational equations is Nagumo's Theorem. It provides sufficient and necessary conditions on initial value problems with state constraints.
In addition to the mutational equation, a nonempty subset $\mathscr{V} \subset E$ is given for specifying the state constraints and, we want to ensure that each element of $\mathscr{V}$ is the initial point of *at least* one solution "viable in $\mathscr{V}$" (i.e. with all its values in $\mathscr{V}$).
Similarly to the classical form of Nagumo's Theorem about ordinary differential equations, the "tangential" properties of the (generalized) directions come into play.
Aubin introduced the following counterpart of Bouligand's contingent cone:

Definition 16. Let $\Theta(E,d) \neq \emptyset$ be a set of transitions on a metric space (E,d). Fix a nonempty set $\mathscr{V} \subset E$ and an element $x \in E$.

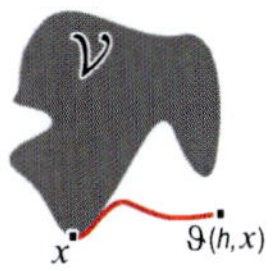

$$\mathscr{T}_{\mathscr{V}}(x) := \Big\{\vartheta \in \Theta(E,d) \ \Big|\ \liminf_{h \downarrow 0} \tfrac{1}{h} \cdot \mathrm{dist}\big(\vartheta(h,x),\ \mathscr{V}\big) = 0\Big\}$$

is called the *contingent transition set* of $\mathscr{V}$ at x.

Remark 17. The transitions in $\mathscr{T}_{\mathscr{V}}(x) \subset \Theta(E,d)$ are specified by means of the distances of elements from $\mathscr{V} \subset E$. By definition,

$$\mathrm{dist}\big(\vartheta(h,x),\ \mathscr{V}\big) \overset{\text{Def.}}{=} \inf_{z \in \mathscr{V}} d\big(\vartheta(h,x),\ z\big).$$

Example 18. For the affine linear transitions on $\mathbb{R}^N$ introduced in Example 2, i.e.

$$\vartheta_v : [0,1] \times \mathbb{R}^N \longrightarrow \mathbb{R}^N, \quad (h,x) \longmapsto x + h \cdot v \qquad (\text{with } v \in \mathbb{R}^N),$$

we can identify the contingent transition set of $V \subset \mathbb{R}^N$ at $x \in V$ directly with

$$\mathscr{T}_V(x) \cong \Big\{v \in \mathbb{R}^N \ \Big|\ \liminf_{h \downarrow 0} \tfrac{1}{h} \cdot \mathrm{dist}\big(x + h \cdot v,\ V\big) = 0\Big\}$$

and, the latter set is the contingent cone of Bouligand (see Definition 63 on page 68). In general, such an immediate link cannot be expected for the morphological transitions on $(\mathscr{K}(\mathbb{R}^N), d\!l)$ in Example 5.

Theorem 19 (Aubin's adaptation of Nagumo's Theorem).
Let $\Theta(E,d)$ be a nonempty set of transitions on a metric space (E,d). Assume that all closed bounded balls in (E,d) are compact.
Suppose $f : (E,d) \longrightarrow (\Theta(E,d), D)$ to be continuous with

$$\sup_{z\in E} \alpha(f(z)) < \infty, \qquad \sup_{z\in E} \beta(f(z)) < \infty.$$

Then the following two statements are equivalent for any closed subset $\mathscr{V} \subset E$:

1. *Every element $x_0 \in \mathscr{V}$ is the initial point of at least one solution $x : [0,1] \longrightarrow E$ to the mutational equation*

$$\overset{\circ}{x}(\cdot) \ni f(x(\cdot))$$

   with $x(t) \in \mathscr{V}$ for all $t \in [0,1]$.

2. *$\mathscr{V} \subset E$ is a viability domain of f in the sense that $f(z) \in \mathscr{T}_{\mathscr{V}}(z)$ for every $z \in \mathscr{V}$.*

1.4 Proofs for Existence and Uniqueness of Solutions without State Constraints

In the previous section, some of Aubin's results about existence and uniqueness of solutions are quoted. They exemplify the analogies between mutational equations and ordinary differential equations. but they are restricted to *autonomous* mutational equations.
Now we prove these analogies for *nonautonomous* mutational equations in more detail. The proofs presented here, however, differ from their counterparts in Aubin's monograph because we follow another track which will be generalized successively in the subsequent chapters.

The following result about existence corresponds to Peano's Theorem about ordinary differential equations, i.e. continuity of the "right-hand side" implies existence of a solution:

Theorem 20 (Peano's Theorem for nonautonomous mutational equations).
Let (E,d) be a metric space in which all closed bounded balls are compact and, $\Theta(E,d)$ denotes a nonempty set of transitions on (E,d).
Assume $f : (E,d) \times [0,T] \longrightarrow (\Theta(E,d),D)$ to be continuous with

$$\sup_{\substack{z\in E\\ 0\le t\le T}} \alpha(f(z,t)) < \infty, \qquad \sup_{\substack{z\in E\\ 0\le t\le T}} \beta(f(z,t)) < \infty.$$

Then for every initial element $x_0 \in E$, there exists a solution $x(\cdot) : [0,T] \longrightarrow E$ to the mutational equation

$$\overset{\circ}{x}(\cdot) \ni f(x(\cdot), \cdot)$$

with $x(0) = x_0$.

The proof (presented at the end of this section) is based on Euler's method in combination with Arzelà-Ascoli Theorem A.82 about compactness of continuous functions. In particular, we have to verify the solution property of the limit function for a convergent subsequence of Euler approximations. This is based on comparing two solutions to mutational equations:

Proposition 21. *Assume for* $f, g : E \times [0,T] \longrightarrow \Theta(E,d)$ *and* $x, y : [0,T] \longrightarrow E$ *that* $x(\cdot)$ *is a solution to the mutational equation* $\overset{\circ}{x}(\cdot) \ni f(x(\cdot),\cdot)$ *and* $y(\cdot)$ *is a solution to the mutational equation* $\overset{\circ}{y}(\cdot) \ni g(y(\cdot),\cdot)$.
Furthermore, let $\widehat{\alpha} > 0$ *and* $\varphi \in C^0([0,T])$ *satisfy for* $\mathscr{L}^1$*-almost every* $t \in [0,T]$

$$\begin{cases} \alpha(g(y(t),t)) \le \widehat{\alpha} \\ D(f(x(t),t),\, g(y(t),t)) \le \varphi(t). \end{cases}$$

Then, $\quad d(x(t), y(t)) \le \left(d(x(0),y(0)) + \int_0^t \varphi(s)\, e^{-\widehat{\alpha} s} ds\right) e^{\widehat{\alpha} t} \quad$ *for any* $t \in [0,T]$.

Similarly to the estimate comparing two transitions in Proposition 7, this upper bound results from generalized Gronwall's Lemma (Proposition A.2) as we will verify at the end of this section. It lays the basis for three important conclusions: Firstly, we can now verify easily that all transitions have the semigroup property in the following sense:

Corollary 22 (Semigroup property of transitions).
Every transition ϑ *on a metric space* (E,d) *satisfies*

$$\vartheta\big(h,\, \vartheta(t,x)\big) = \vartheta(t+h,\, x)$$

for any $x \in E$ *and* $t, h \in [0,1]$ *with* $t + h \le 1$.

Indeed, both $[0, 1-t] \longrightarrow E,\ h \longmapsto \vartheta(h,\, \vartheta(t,x))$ and $h \longmapsto \vartheta(t+h, x)$ solve the mutational equation $\overset{\circ}{x}(\cdot) \ni \vartheta$ according to Remark 11 (on page 37) and share the initial element at time $h = 0$. Essentially the same arguments provide the uniqueness of primitives as second result:

Corollary 23 (Uniqueness of primitives).
Let $\vartheta(\cdot) : [0,T] \longrightarrow \Theta(E,d)$ *satisfy* $\sup_{t \in [0,T]} \alpha(\vartheta(t)) < \infty$.
If $x(\cdot), y(\cdot) : [0,T] \longrightarrow E$ *are primitives of* $\vartheta(\cdot)$ *with* $x(0) = y(0)$, *then* $x(\cdot) \equiv y(\cdot)$.

Finally Proposition 21 even guarantees that the solutions depend on the initial data and the "right-hand side" in a continuous way — under the additional assumption that the "right-hand side" of a mutational equation is Lipschitz continuous.

Proposition 24 (Continuity w.r.t. initial data and the right-hand side).
Let $\Theta(E,d)$ be a nonempty set of transitions on a metric space (E,d).
For $f : E \times [0,T] \longrightarrow \Theta(E,d)$ suppose $\widehat{\alpha} := \sup_{z,t} \alpha(f(z,t)) < \infty$ and that there exists $\lambda > 0$ such that $f(\cdot,t) : (E,d) \longrightarrow (\Theta(E,d), D)$ is λ-Lipschitz continuous for $\mathscr{L}^1$-almost every $t \in [0,T]$.
Let $g : E \times [0,T] \longrightarrow \Theta(E,d)$ fulfill $\sup\limits_{z,s} D(f(z,s), g(z,s)) < \infty$,

Then every solutions $x(\cdot), y(\cdot) : [0,T] \longrightarrow E$ to the mutational equations

$$\overset{\circ}{x}(\cdot) \ni f(x(\cdot),\cdot) \qquad \overset{\circ}{y}(\cdot) \ni g(y(\cdot),\cdot)$$

satisfy the following inequality for every $t \in [0,T]$

$$d(x(t), y(t)) \le \big(d(x(0),y(0)) + t \cdot \sup_{z,s} D(f(z,s),g(z,s))\big)\, e^{(\widehat{\alpha}+\lambda)\,t}.$$

The combination of Theorem 20 and Proposition 24 implies directly Aubin's adaptation of Cauchy-Lipschitz Theorem formulated in Theorem 15 (on page 38). Let us now prove the three main results of this section:

Proof (of Theorem 20). This existence proof is based on Euler approximations $x_n(\cdot) : [0,T] \longrightarrow E$ ($n \in \mathbb{N}$ with $2^n > T$) together with Arzelà-Ascoli Theorem A.82 in metric spaces. Indeed, for each $n \in \mathbb{N}$ with $2^n > T$, set

$$\begin{aligned} h_n &:= \tfrac{T}{2^n}, \qquad t_n^j := j\, h_n && \text{for } j = 0 \dots 2^n,\\ x_n(0) &:= x_0, \\ x_n(t) &:= f(x_n(t_n^j), t_n^j)\big(t - t_n^j,\ x_n(t_n^j)\big) && \text{for } t \in \,]t_n^j, t_n^{j+1}],\ j < 2^n. \end{aligned}$$

According to Remark 11,

$$\overset{\circ}{x}_n(t) \ni f(x_n(t_n^j), t_n^j)$$

for every $t \in [t_n^j, t_n^{j+1}[$ with $j \in \{0, 1 \dots 2^n - 1\}$.
Due to Lemma 8 and the piecewise construction of each $x_n(\cdot)$, the constant $\widehat{\beta} := \sup_{z,s} \beta(f(z,s)) < \infty$ is a uniform Lipschitz constant of every curve $x_n(\cdot)$. Moreover, the set of all values $\{x_n(t) \mid n \in \mathbb{N},\ t \in [0,T], 2^n > T\}$ is contained in the ball $B := \{y \in E \mid d(x_0,y) \le \widehat{\beta}\, T\}$ which is compact with respect to d by assumption.
The Arzelà-Ascoli Theorem states that $\{x_n(\cdot) \mid n \in \mathbb{N}, 2^n > T\} \subset C^0([0,T],B)$ is precompact with respect to uniform convergence and therefore, there exists a subsequence $\big(x_{n_j}(\cdot)\big)_{j\in\mathbb{N}}$ converging uniformly to a function $x(\cdot) \in C^0([0,T],B)$.

Finally, we verify that $x(\cdot)$ solves the mutational equation $\overset{\circ}{x}(\cdot) \ni f(x(\cdot),\cdot)$. Indeed, $x(\cdot)$ is $\widehat{\beta}$-Lipschitz continuous with respect to d by virtue of its construction. Furthermore, using the notation $\delta_n := \sup_{[0,T]} d(x_n(\cdot), x(\cdot))$, we conclude from Proposition 21 that for any $t \in [0,T[$, $h \in [0, T-t[$ and $n \in \mathbb{N}$ with $2^n > T$

$$\begin{aligned} d\,(f(x(t),t)\,(h,\,x(t)),\ \ x(t+h)) \\ &\le\ d\big(f(x(t),t)\,(h,x(t)),\ x_n(t+h)\big) + d\,(x_n(t+h),\ x(t+h)) \\ &\le\ \Big(\delta_n + h \cdot \sup_{\substack{-h_n \le s \le h \\ y:\, d(y,x(t+s)) \le \delta_n}} D\big(f(x(t),t),\ f(y,\,t+s)\big)\Big)\, e^{\widehat{\alpha}\,h} + \delta_n \end{aligned}$$

with $\widehat{\alpha} \overset{\text{Def.}}{=} \sup_{z,s} \alpha(f(z,s)) < \infty$.

Due to the continuity of f with respect to D, the limit for $n \longrightarrow \infty$ implies that

$$d\big(f(x(t),t)\,(h,x(t)),\ x(t+h)\big)\ \le\ h \cdot \sup_{0 \le s \le h} D\big(f(x(t),t),\ f(x(t+s),t+s)\big)\ e^{\widehat{\alpha}\,h}$$

and thus,

$$\limsup_{h \downarrow 0}\ \tfrac{1}{h} \cdot d\big(f(x(t),t)\,(h,x(t)),\ x(t+h)\big)\ \le\ 0. \qquad \square$$

Remark 25. This proof reveals that the continuity of $f : E \times [0,T] \longrightarrow \Theta(E,d)$ implies the first-order approximation at even *every* time $t \in [0,T[$ (and not just at Lebesgue-almost every time as Definition 13 demands).

Proof (of Proposition 21). Similarly to the proof of Proposition 7 comparing two transitions, we consider the auxiliary function

$$\psi : [0,T] \longrightarrow [0,\infty[, \quad t \longmapsto d\big(x(t),\ y(t)\big).$$

It is Lipschitz continuous because any solutions $x(\cdot), y(\cdot)$ to mutational equations

$$\overset{\circ}{x}(\cdot) \ni f(x(\cdot),\cdot), \qquad \overset{\circ}{y}(\cdot) \ni g(y(\cdot),\cdot)$$

are Lipschitz continuous due to Definition 13.

Furthermore, we obtain for Lebesgue-almost every $t \in [0,T[$

$$\begin{aligned} \limsup_{h \downarrow 0}\ \tfrac{1}{h} \cdot d\big(x(t+h),\ & f(x(t),t)\,(h,x(t))\big) &&= 0 \\ \limsup_{h \downarrow 0}\ \tfrac{1}{h} \cdot d\big(f(x(t),t)\,(h,x(t)),\ & g(y(t),t)\,(h,x(t))\big) &&\le D\big(f(x(t),t),\ g(y(t),t)\big) \\ \limsup_{h \downarrow 0}\ \tfrac{1}{h} \cdot d\big(g(y(t),t)\,(h,y(t)),\ & y(t+h)\big) &&= 0 \end{aligned}$$

due to Definition 6 and Definition 13. For estimating $\psi(t+h)$, we now use

$$\limsup_{h \downarrow 0}\ \tfrac{1}{h} \cdot \Big(d\big(g(y(t),t)\,(h,x(t)),\ g(y(t),t)\,(h,y(t))\big)\ -\ \psi(t)\Big)\ \le\ \widehat{\alpha} \cdot \psi(t)$$

and conclude from the triangle inequality of d

$$\begin{aligned} \limsup_{h \downarrow 0}\ \frac{\psi(t+h)\ -\ \psi(t)}{h}\ &\le\ \widehat{\alpha} \cdot \psi(t)\ +\ D\big(f(x(t),t),\ g(y(t),t)\big) \\ &\le\ \widehat{\alpha} \cdot \psi(t)\ +\ \varphi(t) \end{aligned}$$

at Lebesgue-almost every time $t \in [0,T[$. Finally the claimed estimate results from generalized Gronwall's Lemma (Proposition A.2 on page 440). $\square$

Proof (of Proposition 24). Assuming $f : E \times [0,T] \longrightarrow \Theta(E,d)$ to be λ-Lipschitz continuous in the first argument with $\widehat{\alpha} := \sup_{z,t} \alpha(f(z,t)) < \infty$, we obtain for any solutions $x(\cdot), y(\cdot)$ to the mutational equations $\overset{\circ}{x}(\cdot) \ni f(x(\cdot),\cdot)$, $\overset{\circ}{y}(\cdot) \ni g(y(\cdot),\cdot)$ the following inequality at $\mathscr{L}^1$-almost every time $t \in [0,T]$

$$\begin{aligned} D\big(f(x(t),t),\, g(y(t),t)\big) &\le D\big(f(x(t),t),\, f(y(t),t)\big) + D\big(f(y(t),t),\, g(y(t),t)\big) \\ &\le \lambda \cdot d(x(t),y(t)) \qquad\quad + \sup_{z,s} D\big(f(z,s), g(z,s)\big). \end{aligned}$$

Proposition 21 implies for the Lipschitz continuous auxiliary function

$$\psi : [0,T] \longrightarrow [0,\infty[, \quad t \longmapsto d\big(x(t),y(t)\big)$$

the implicit integral inequality

$$\psi(t) \le \Big(\psi(0) + \int_0^t \big(\lambda \cdot \psi(s) + \sup D(f(\cdot,\cdot),g(\cdot,\cdot))\big)\, e^{-\widehat{\alpha}s} ds\Big)\, e^{\widehat{\alpha}t}$$

at every time $t \in [0,T]$. Finally the integral version of Gronwall's Lemma (Proposition A.1 on page 439) bridges the last gap and provides the claimed explicit estimate. □

1.5 An Essential Advantage of Mutational Equations: Solutions to Systems

Roughly speaking, mutational equations provide a joint framework for diverse time-dependent systems whose evolutions are determined by a form of generalized differential equation – without requiring any linear structure.
In regard to applications, it is of particular interest that we can consider more than one mutational equation simultaneously. The analytical origin of the individual components (like set evolutions in $(\mathscr{K}(\mathbb{R}^N), d\!l)$) does not really matter as long as each component satisfies the conditions on transitions. This opens the door for coupling nonlocal set evolutions with an ordinary differential equation, for example.

The main basis for considering systems of mutational equations is the following counterpart of Peano's Theorem and thus, all the generalizations of mutational equations in subsequent chapters are to ensure that the same existence result about systems holds in the extended framework.

Theorem 26 (Peano's Theorem for systems of mutational equations).
Let $(E_1,d_1), (E_2,d_2)$ be metric spaces in which all closed bounded balls are compact. $\Theta(E_1,d_1)$ and $\Theta(E_2,d_2)$ denote nonempty sets of transitions on (E_1,d_1) and (E_2,d_2) respectively. Assume

$$\begin{aligned} f_1 &: (E_1,d_1) \times (E_2,d_2) \times [0,T] \longrightarrow (\Theta(E_1,d_1), D_1) \\ f_2 &: (E_1,d_1) \times (E_2,d_2) \times [0,T] \longrightarrow (\Theta(E_2,d_2), D_2) \end{aligned}$$

to be continuous with

$$\sup_{z_1,z_2,t} \{\alpha(f_1(z_1,z_2,t)),\ \alpha(f_2(z_1,z_2,t))\} < \infty,$$
$$\sup_{z_1,z_2,t} \{\beta(f_1(z_1,z_2,t)),\ \beta(f_2(z_1,z_2,t))\} < \infty.$$

Then for every elements $x_0 \in E_1, y_0 \in E_2$, there exist solutions $x(\cdot) : [0,T] \longrightarrow E_1$, $y(\cdot) : [0,T] \longrightarrow E_2$ to the two mutational equations

$$\begin{cases} \overset{\circ}{x}(\cdot) \ni f_1(x(\cdot), y(\cdot), \cdot) \\ \overset{\circ}{y}(\cdot) \ni f_2(x(\cdot), y(\cdot), \cdot) \end{cases}$$

with $x(0) = x_0$ and $y(0) = y_0$.

In this mutational framework, such an existence result is an immediate consequence of the following relationship between transitions on two separate metric spaces and on their product space:

Lemma 27 (Product of transitions and mutations).
Let (E_1,d_1) and (E_2,d_2) be metric spaces. $\Theta(E_1,d_1)$ and $\Theta(E_2,d_2)$ denote nonempty sets of transitions on (E_1,d_1) and (E_2,d_2) respectively. The product space $E := E_1 \times E_2$ is supplied with the metric

$$\begin{array}{rccl} d_+ : & E \times E & \longrightarrow & [0,\infty[, \\ & ((x_1,x_2),\ (y_1,y_2)) & \longmapsto & d_1(x_1,y_1) + d_2(x_2,x_2). \end{array}$$

1. *For every $\vartheta_1 \in \Theta(E_1,d_1)$ and $\vartheta_2 \in \Theta(E_2,d_2)$, the tuple*
$$\begin{array}{rcl} \vartheta := (\vartheta_1, \vartheta_2) : [0,1] \times (E_1 \times E_2) & \longrightarrow & E_1 \times E_2, \\ (h,\ (x_1,x_2)) & \longmapsto & (\vartheta_1(h,x_1),\ \vartheta_2(h,x_2)) \end{array}$$
is a transition on $(E_1 \times E_2,\ d_+)$ with
$$\begin{cases} \alpha(\vartheta) \le \max\{\alpha(\vartheta_1),\ \alpha(\vartheta_2)\} \\ \beta(\vartheta) \le \max\{\beta(\vartheta_1),\ \beta(\vartheta_2)\} \\ D_+((\vartheta_1,\vartheta_2),\ (\tau_1,\tau_2)) \le D_1(\vartheta_1,\tau_1) + D_2(\vartheta_2,\tau_2). \end{cases}$$

2. *Let the product space $E \overset{\text{Def.}}{=} E_1 \times E_2$ be now supplied with the transitions in $\Theta(E,d_+) := \Theta(E_1,d_1) \times \Theta(E_2,d_2)$. For arbitrary curves $x_1(\cdot) : [0,T] \longrightarrow E_1$ and $x_2(\cdot) : [0,T] \longrightarrow E_2$ set $x(\cdot) := (x_1(\cdot),x_2(\cdot)) : [0,T] \longrightarrow E$.*

   Then $\vartheta = (\vartheta_1,\vartheta_2) \in \Theta(E,d_+)$ belongs to the mutation $\overset{\circ}{x}(t)$ if and only if $\vartheta_1 \in \overset{\circ}{x}_1(t)$ and $\vartheta_2 \in \overset{\circ}{x}_2(t)$.

Proof (of Lemma 27) results directly from the definitions and the essential estimate of Proposition 7 (on page 35) and thus, we dispense with its details. Obviously, not every transition on $(E_1 \times E_2,\ d_+)$ is necessarily induced by a tuple of two "decoupled" transitions on the components as in Lemma 27 (1.).

The close relationship between the mutation of a tuple and the product of the componentwise mutations cannot be extended to all subsequent generalizations of mutational equations. For this reason, we present an alternative (and simple) proof of Theorem 26 whose basic notion will be reused later on.

Proof (of Theorem 26). Correspondingly to the proof of Theorem 20 (page 42), we use Euler approximations for each component. Arzelà-Ascoli Theorem A.82 applied to the corresponding curves $[0,T] \longrightarrow E_1 \times E_2$ provides a subsequence such that each component has a continuous limit curve in E_1 and E_2 respectively. Finally we verify the solution property for each component separately.
Indeed, for each $n \in \mathbb{N}$ with $2^n > T$, set

$$\begin{aligned} h_n &:= \tfrac{T}{2^n}, \qquad t_n^j := j\,h_n \qquad \text{for } j = 0 \dots 2^n, \\ x_n(0) &:= x_0, \\ y_n(0) &:= y_0, \\ x_n(t) &:= f_1(x_n(t_n^j),\, y_n(t_n^j),\, t_n^j)\,\big(t - t_n^j,\; x_n(t_n^j)\big) \\ y_n(t) &:= f_2(x_n(t_n^j),\, y_n(t_n^j),\, t_n^j)\,\big(t - t_n^j,\; y_n(t_n^j)\big) \qquad \text{for } t \in\,]t_n^j, t_n^{j+1}],\ j < 2^n. \end{aligned}$$

According to Remark 11,

$$\begin{aligned} \overset{\circ}{x}_n(t) &\ni f_1(x_n(t_n^j),\, y_n(t_n^j)),\, t_n^j) \\ \overset{\circ}{y}_n(t) &\ni f_2(x_n(t_n^j),\, y_n(t_n^j)),\, t_n^j). \end{aligned}$$

for every $t \in [t_n^j, t_n^{j+1}[$ with $j \in \{0, 1 \dots 2^n - 1\}$.
Due to Lemma 8 and the piecewise construction of each $x_n(\cdot), y_n(\cdot)$, the constant

$$\widehat{\beta} := \sup_{z_1, z_2, s} \big\{\beta(f_1(z_1, z_2, s)),\ \beta(f_2(z_1, z_2, s))\big\} < \infty$$

is a joint Lipschitz constant of all curves $x_n(\cdot) : [0,T] \longrightarrow E_1$, $y_n(\cdot) : [0,T] \longrightarrow E_2$ $(2^n > T)$. As a consequence, the sets of all values

$$\begin{aligned} \{x_n(t) \,|\, n \in \mathbb{N},\ 2^n > T,\ t \in [0,T]\} &\subset E_1, \\ \{y_n(t) \,|\, n \in \mathbb{N},\ 2^n > T,\ t \in [0,T]\} &\subset E_2 \end{aligned}$$

are contained in closed balls of radius $\widehat{\beta} \cdot T$ respectively. Considering now the sequence of Lipschitz continuous curves

$$(x_n, y_n) : [0,T] \longrightarrow (E_1 \times E_2, d_1 + d_2)$$

the Arzelà-Ascoli Theorem guarantees a subsequence $\big(x_{n_j}(\cdot), y_{n_j}(\cdot)\big)_{j \in \mathbb{N}}$ converging uniformly to a continuous curve $(x(\cdot), y(\cdot)) : [0,T] \longrightarrow E_1 \times E_2$.

Finally, we verify that $x(\cdot)$ solves the mutational equation $\overset{\circ}{x}(\cdot) \ni f_1(x(\cdot), y(\cdot), \cdot)$. The corresponding proof for $y(\cdot)$ is based on exactly the same steps.
Indeed, $x(\cdot)$ is $\widehat{\beta}$-Lipschitz continuous with respect to d_1 by virtue of its construction. Now we focus on the nonautonomous mutational equation in (E_1, d_1) with

$$(E_1, d_1) \times [0,T] \longrightarrow \Theta(E_1, d_1), \qquad (z_1, t) \longmapsto f_1\big(z_1,\, y(t),\, t\big)$$

on its right-hand side.

Using the notations $\widehat{\alpha}_1 := \sup_{z_1,z_2,s} \alpha(f_1(z_1,z_2,s)) < \infty$ and

$$\delta_n^1 := \sup_{[0,T]} d_1(x_n(\cdot),x(\cdot)), \qquad \delta_n^2 := \sup_{[0,T]} d_2(y_n(\cdot),y(\cdot)),$$

Proposition 21 implies for any $t \in [0,T[$, $h \in [0,T-t[$ and $n \in \mathbb{N}$

$$\begin{aligned} & d_1\big(f_1(x(t),\,y(t),\,t)\,(h,\,x(t)),\;\; x(t+h)\big) \\ &\quad \le\; d_1\big(f_1(x(t),\,y(t),\,t)\,(h,x(t)),\;\; x_n(t+h)\big) + d_1\,(x_n(t{+}h),\,x(t{+}h)) \\ &\quad \le\; \Big(\delta_n^1 + h\cdot \sup_{\substack{-h_n\le s\le h \\ z_1:\,d_1(z_1,x(t+s))\le\delta_n^1 \\ z_2:\,d_2(z_2,y(t+s))\le\delta_n^2}} D_1\,(f_1(x(t),\,y(t),\,t),\;\; f_1(z_1,z_2,t+s))\Big)\, e^{\widehat{\alpha}_1\,h} + \delta_n^1. \end{aligned}$$

Due to the continuity of f_1 with respect to D_1, the limit for $n \longrightarrow \infty$ reveals

$$\begin{aligned} & d_1\big(f_1(x(t),\,y(t),\,t)\,(h,\,x(t)),\;\; x(t+h)\big) \\ &\qquad \le\; h\cdot \sup_{0\le s\le h} D_1\big(f_1(x(t),\,y(t),\,t),\;\; f_1(x(t+s),\,y(t+s),\,t+s)\big)\, e^{\widehat{\alpha}_1\,h} \end{aligned}$$

at every time $t \in [0,T[$ and thus,

$$\limsup_{h\downarrow 0}\; \tfrac{1}{h}\cdot d_1\big(f_1(x(t),y(t),t)\,(h,x(t)),\;\; x(t+h)\big) \;\le\; 0.$$ □

1.6 Proof for Existence of Solutions Under State Constraints

Theorem 19 (on page 40) specifies Aubin's adaptation of Nagumo's Theorem to mutational equations with state constraint. In this section, we give a slightly modified proof that the viability condition is sufficient:

Proposition 28.
Let $\Theta(E,d)$ be a nonempty set of transitions on a metric space (E,d). Assume that all closed bounded balls in (E,d) are compact.
Suppose $f : (E,d) \longrightarrow (\Theta(E,d),\, D)$ to be continuous with

$$\widehat{\alpha} := \sup_{z\in E} \alpha(f(z)) < \infty, \qquad \widehat{\beta} := \sup_{z\in E} \beta(f(z)) < \infty.$$

For the nonempty closed subset $\mathscr{V} \subset E$ assume the following viability condition:

$$f(z) \in \mathscr{T}_{\mathscr{V}}(z) \qquad \text{for every } z \in \mathscr{V}.$$

Then every $x_0 \in \mathscr{V}$ is the initial point of at least one solution $x : [0,1] \longrightarrow E$ to the mutational equation

$$\overset{\circ}{x}(\cdot) \;\ni\; f(x(\cdot))$$

with $x(t) \in \mathscr{V}$ for all $t \in [0,1]$.

For proving this proposition, the first step consists in constructing approximative solutions satisfying a weakened form of state constraints:

Lemma 29 (Aubin's construction of approximative solutions).
Choose any $\varepsilon > 0$. Under the assumptions of Proposition 28, there exists a $\widehat{\beta}$-Lipschitz continuous function $x_\varepsilon(\cdot) : [0,1] \longrightarrow E$ satisfying with $R_\varepsilon := \varepsilon\ e^{\widehat{\alpha}}$

(a) $x_\varepsilon(0) = x_0$,

(b) $\operatorname{dist}\big(x_\varepsilon(t),\ \mathscr{V}\big) \le R_\varepsilon$ *for all* $t \in [0,1]$,

(c) $\overset{\circ}{x}_\varepsilon(t) \cap \big\{ f(z) \,\big|\, z \in E : d(z,\ x_\varepsilon(t)) \le R_\varepsilon \big\} \ne \emptyset$ *for all* $t \in [0,1[$.

Considering a sequence of these approximative solutions $(x_{1/n}(\cdot))_{n\in\mathbb{N}}$, Arzelà-Ascoli Theorem A.82 provides a subsequence $(x_{1/n_j}(\cdot))_{j\in\mathbb{N}}$ that converges uniformly to a Lipschitz continuous curve $x(\cdot) : [0,T] \longrightarrow E$. Moreover, $x(\cdot)$ has all its values in the closed set of constraints $\mathscr{V} \subset E$.
Finally we have to verify that $x(\cdot)$ solves the mutational equation $\overset{\circ}{x}(\cdot) \ni f\big(x(\cdot)\big)$. This is a consequence of the following general result:

Theorem 30 (Convergence of solutions to mutational equations).
Let $\Theta(E,d)$ be a nonempty set of transitions on a metric space (E,d). Consider $f, f_m : E \times [0,T] \longrightarrow \Theta(E,d)$ and $x, x_m : [0,T] \longrightarrow E$ for each $m \in \mathbb{N}$ and, suppose the following properties:

1. *for each $m \in \mathbb{N}$, $x_m(\cdot)$ is a solution to the mutational equation $\overset{\circ}{x}_m(\cdot) \ni f_m(x_m(\cdot), \cdot)$*
2. $\widehat{\beta} := \sup_{m\in\mathbb{N}} \operatorname{Lip} x_m(\cdot) < \infty$
3. $\widehat{\alpha} := \sup_{m\in\mathbb{N}} \sup_{\substack{z\in E \\ 0\le t\le T}} \big\{\alpha(f_m(z,t)),\ \alpha(f(z,t))\big\} < \infty$
4. *for Lebesgue-almost every $t \in [0,T]$, any $y \in E$ and all sequences $t_m \to t$, $y_m \to y$ in $[0,T], E$ respectively:* $\lim_{m\to\infty} D\big(f_m(y,t),\ f_m(y_m,t_m)\big) = 0$
5. *for Lebesgue-almost every $t \in [0,T]$:* $\lim_{m\to\infty} D\big(f(x(t),t),\ f_m(x(t),t)\big) = 0$
6. *for each $t \in [0,T]$:* $\lim_{m\to\infty} d\big(x(t),\ x_m(t)\big) = 0.$

Then $x(\cdot)$ is a solution to the mutational equation $\overset{\circ}{x}(\cdot) \ni f(x(\cdot), \cdot)$.

Proof (of Lemma 29). For $\varepsilon > 0$ fixed, let $\mathscr{A}_\varepsilon(x_0)$ denote the set of all tuples $(T_x,\ x(\cdot))$ consisting of some $T_x \in [0,1]$ and a $\widehat{\beta}$-Lipschitz continuous function $x(\cdot) : [0,T_x] \longrightarrow (E,d)$ such that

(a) $x(0) = x_0$,

(b') 1.) $\operatorname{dist}\big(x(T_x),\ \mathscr{V}\big) \le r_\varepsilon(T_x)$ with $r_\varepsilon(t) := \varepsilon\, e^{\widehat{\alpha}t}\, t$,
2.) $\operatorname{dist}\big(x(t),\ \mathscr{V}\big) \le R_\varepsilon$ for all $t \in [0,T_x]$,

(c) $\overset{\circ}{x}(t) \cap \big\{ f(z) \,\big|\, z \in E : d(z,\ x(t)) \le R_\varepsilon \big\} \ne \emptyset$ for all $t \in [0,T_x[$.

Obviously, $\mathscr{A}_\varepsilon(x_0)$ is not empty since it contains $(0,\, x(\cdot) \equiv x_0)$. Moreover, an order relation $\preceq$ on $\mathscr{A}_\varepsilon(x_0)$ is specified by

$$(T_x,\, x(\cdot)) \,\preceq\, (T_y,\, y(\cdot)) \quad :\Longleftrightarrow \quad T_x \le T_y \text{ and } x = y\big|_{[0,T_x]}.$$

Thus, Zorn's Lemma provides a maximal element $\big(T,\, x_\varepsilon(\cdot)\big) \in \mathscr{A}_\varepsilon(x_0)$.
As all considered functions with values in E have been supposed to be $\widehat{\beta}$-Lipschitz continuous, $x_\varepsilon(\cdot)$ is also $\widehat{\beta}$-Lipschitz continuous in $[0,T[$. In particular, $x_\varepsilon(\cdot)$ can always be extended to the closed interval $[0,T] \subset [0,1]$ in a unique way.

Assuming $T < 1$ for a moment, we obtain a contradiction if $x_\varepsilon(\cdot)$ can be extended to a larger interval $[0,T+\delta] \subset [0,1]$ $(\delta > 0)$ preserving conditions (b'), (c).
Since closed bounded balls of (E,d) are compact, the closed set $\mathscr{V}$ contains an element $z \in E$ with $d(x_\varepsilon(T),z) = \operatorname{dist}(x_\varepsilon(T),\, \mathscr{V}) \le r_\varepsilon(T)$ and, the assumed viability condition states

$$f(z) \in \mathscr{T}_{\mathscr{V}}(z) \subset \Theta(E,d).$$

Due to Definition 16 of the contingent transition set $\mathscr{T}_{\mathscr{V}}(z)$, there is a sequence $h_m \downarrow 0$ in $]0,1-T[$ such that

$$\operatorname{dist}\big(f(z)\,(h_m,z),\, \mathscr{V}\big) \le \varepsilon\, h_m \qquad \text{for all } m \in \mathbb{N}.$$

Now set for each $t \in [T,\, T+h_1]$

$$x_\varepsilon(t) := f(z)\,\big(t-T,\, x_\varepsilon(T)\big).$$

Obviously, Remark 11 implies $f(z) \in \overset{\circ}{x}_\varepsilon(t)$ for all $t \in [T,\, T+h_1[$. Moreover, Lemma 8 leads to

$$\begin{aligned} d\big(x_\varepsilon(t),\, z\big) &\le d\big(f(z)\,(t-T, x_\varepsilon(T)),\; x_\varepsilon(T)\big) + d\big(x_\varepsilon(T),\; z\big) \\ &\le \widehat{\beta}\cdot(t-T) + \varepsilon\; e^{\widehat{\alpha}T}\; T \\ &\le R_\varepsilon \end{aligned}$$

for every $t \in [T,\, T+\delta[$ with $\delta := \min\big\{h_1,\; \varepsilon\, e^{\widehat{\alpha}}\, \frac{1-T}{1+\widehat{\beta}}\big\}$, i.e. conditions (b') (2.) and (c) hold in the interval $[T, T+\delta]$.
For any index $m \in \mathbb{N}$ with $h_m < \delta$, we conclude from Proposition 7

$$\begin{aligned} \operatorname{dist}\big(x_\varepsilon(T+h_m),\, \mathscr{V}\big) &\le d\big(f(z)\,(h_m, x_\varepsilon(T)),\; f(z)\,(h_m,z)\big) + \operatorname{dist}\big(f(z)\,(h_m,z),\; \mathscr{V}\big) \\ &\le d\big(x_\varepsilon(T),\, z\big) \cdot e^{\widehat{\alpha}\, h_m} + \varepsilon \cdot h_m \\ &\le \varepsilon\; e^{\widehat{\alpha}T}\; T \quad \cdot e^{\widehat{\alpha}\, h_m} + \varepsilon\cdot h_m \\ &\le r_\varepsilon(T+h_m), \end{aligned}$$

i.e. condition (b')(1.) is also satisfied at time $t = T + h_m$ with any large $m \in \mathbb{N}$.
Finally, $x_\varepsilon(\cdot)\big|_{[0,\,T+h_m]}$ provides the wanted contradiction and thus, $T = 1$.

□

Proof (of Convergence Theorem 30). The limit curve $x(\cdot) : [0,T] \longrightarrow E$ is $\widehat{\beta}$-Lipschitz continuous due to assumption (6.) and the $\widehat{\beta}$-Lipschitz continuity of each $x_m(\cdot)$, $m \in \mathbb{N}$. (This is an easy consequence of the triangle inequality of d.) Choose $t \in [0,T[$ and $h \in [0,T-t[$ arbitrarily. Proposition 21 (comparing solutions to mutational equations on page 41) implies

$$\begin{aligned} & d\big(f(x(t),t)\,(h,\, x(t)),\ x(t+h)\big) \\ & \quad \le\ d\big(f(x(t),t)\,(h,\, x(t)),\ x_m(t+h)\big) \qquad\qquad + d\big(x_m(t+h),\, x(t+h)\big) \\ & \quad \le\ \big(d\big(x(t),\, x_m(t)\big) \,+\, h \cdot \Delta(t,t+h,m)\big)\, e^{\widehat{\alpha}\, h} + d\big(x_m(t+h),\, x(t+h)\big) \end{aligned}$$

with the abbreviation $\Delta(t,t+h,m) := \sup\limits_{t \le s \le t+h} D\big(f(x(t),t),\ f_m(x_m(s),s)\big)$.
As mentioned after Definition 6 (on page 35), $D(\cdot,\cdot)$ satisfies the triangle inequality and thus,

$$\Delta(t,t+h,m) \le D\big(f(x(t),t),\ f_m(x(t),t)\big) + \sup_{t \le s \le t+h} D\big(f_m(x(t),t),\ f_m(x_m(s),s)\big).$$

Considering now the limits for $m \longrightarrow \infty$ (with fixed t,h), we conclude from assumption (5.) for Lebesgue-almost every $t \in [0,T[$ and any $h \in [0,T-t[$

$$d\big(f(x(t),t)\,(h,\, x(t)),\ x(t+h)\big) \le h\, e^{\widehat{\alpha}\, h} \cdot \limsup_{m \to \infty}\ \sup_{t \le s \le t+h} D\big(f_m(x(t),t),\ f_m(x_m(s),s)\big).$$

Finally $x(\cdot)$ is a solution to the mutational equation $\overset{\circ}{x}(\cdot) \ni f(x(\cdot),\,\cdot)$ if we can verify the following asymptotic condition for Lebesgue-almost every $t \in [0,T]$

$$\limsup_{h \downarrow 0}\ \limsup_{m \to \infty}\ \sup_{t \le s \le t+h} D\big(f_m(x(t),t),\ f_m(x_m(s),s)\big) = 0.$$

If this last condition was not correct (at time t), we could find some $\varepsilon > 0$ and sequences $(m_j)_{j \in \mathbb{N}}$, $(s_j)_{j \in \mathbb{N}}$ satisfying for each $j \in \mathbb{N}$

$$t \le s_j \le t + \tfrac{1}{j}, \qquad D\big(f_{m_j}(x(t),t),\ f_{m_j}(x_{m_j}(s_j),s_j)\big) \ge \varepsilon > 0$$

and this would induce a contradiction to assumption (4.) at $\mathscr{L}^1$-a.e. time t.

□

Remark 31. Lemma 27 lays the foundations for extending Proposition 28 to systems of mutational equations and a joint set of constraints in the product space. Some examples with compact subsets of $\mathbb{R}^N$ are given in subsequent section 1.9.6 (on page 74 ff.).

1.7 Some Elementary Properties of the Contingent Transition Set

In Definition 16 (on page 39), the contingent transition set of a nonempty set $\mathscr{V} \subset E$ at an element $x \in \mathscr{V}$ was introduced as counterpart of Bouligand's contingent cone:

$$\mathscr{T}_{\mathscr{V}}(x) \stackrel{\text{Def.}}{=} \big\{\vartheta \in \Theta(E,d) \;\big|\; \liminf_{h\downarrow 0} \tfrac{1}{h} \cdot \operatorname{dist}\big(\vartheta(h,x),\, \mathscr{V}\big) \;=\; 0\big\}.$$

It has proved to be useful in connection with Nagumo's Theorem 19 about solutions to mutational equations with state constraints (on page 40).
Now we summarize some properties of the contingent transition set. Most of them result directly from the definition or can be verified in exactly the same way as their counterparts about Bouligand's contingent cone of subsets in $\mathbb{R}^N$ (see e.g. [19, § 4.1], [162]).

Lemma 32. *Let $\Theta(E,d) \neq \emptyset$ be a set of transitions on a metric space (E,d). $\vartheta \in \Theta(E,d)$ belongs to the contingent transition set of $\mathscr{V} \subset E$ at $x \in \mathscr{V}$ if and only if there exist sequences $(h_n)_{n\in\mathbb{N}}$, $(y_n)_{n\in\mathbb{N}}$ in $]0,1[$ and $\mathscr{V}$ respectively satisfying*

$$h_n \longrightarrow 0, \qquad \tfrac{1}{h_n} \cdot d\big(\vartheta(h_n,x),\, y_n\big) \longrightarrow 0 \qquad\qquad \text{for } n \longrightarrow \infty.$$

□

Proposition 33. *Let $\Theta(E,d) \neq \emptyset$ be a set of transitions on a metric space (E,d). $\mathscr{V}_1, \mathscr{V}_2, \mathscr{V}_3 \ldots$ denote nonempty closed subsets of E. Then,*

(a) $\mathscr{T}_{\mathscr{V}_1 \cup \mathscr{V}_2 \cup \ldots}(x) \quad \supset \bigcup_{k\in\mathbb{N}:\, x\in\mathscr{V}_k} \mathscr{T}_{\mathscr{V}_k}(x)$ *for any* $x \in \bigcup_{k\in\mathbb{N}} \mathscr{V}_k$.

(b) $\mathscr{T}_{\mathscr{V}_1 \cup \mathscr{V}_2 \cup \ldots \cup \mathscr{V}_j}(x) \quad = \bigcup_{k\in\{1\ldots j\}:\, x\in\mathscr{V}_k} \mathscr{T}_{\mathscr{V}_k}(x)$ *for any* $j \in \mathbb{N}$, $x \in \mathscr{V}_1 \cup \ldots \cup \mathscr{V}_j$.

(c) $\mathscr{T}_{\mathscr{V}_1 \cap \mathscr{V}_2 \cap \ldots}(x) \quad \subset \bigcap_{k\in\mathbb{N}} \mathscr{T}_{\mathscr{V}_k}(x)$ *for any* $x \in \mathscr{V}_1 \cap \mathscr{V}_2 \cap \ldots \cap \mathscr{V}_j$.

□

Considering the contingent transition set of an intersection (as in statement (c)), there is still an "inner" approximation lacking, i.e. a subset of $\mathscr{T}_{\mathscr{V}_1 \cap \mathscr{V}_2 \cap \ldots}(x)$ in (separate) terms of $\mathscr{V}_1, \mathscr{V}_2 \ldots \subset E$. For this purpose, we introduce the counterpart of the tangent cone in the sense of Dubovitsky-Miliutin:

Definition 34. Let $\Theta(E,d)$ be a nonempty set of transitions on a metric space (E,d). Fix a nonempty set $\mathscr{V} \subset E$ and an element $x \in E$.

$$\mathscr{T}^{DM}_{\mathscr{V}}(x) := \big\{\vartheta \in \Theta(E,d) \;\big|\; \exists\, \varepsilon, \rho \in]0,1[\;\; \forall\, h \in]0,\varepsilon] : \;\; \mathbb{B}_{\rho h}\big(\vartheta(h,x)\big) \;\subset\; \mathscr{V}\big\}$$

is called *Dubovitsky-Miliutin transition set* of $\mathscr{V}$ at x.

Remark 35. For a boundary point x of a nonempty set $V \subset \mathbb{R}^N$, the tangent cone in the sense of Dubovitsky-Miliutin is usually defined as

$$T_V^{DM}(x) := \{v \in \mathbb{R}^N \mid \exists\, \varepsilon, \rho > 0: \ x +]0,\varepsilon] \cdot \mathbb{B}_\rho(v) \subset V\}$$

(see e.g. [14, Definition 4.3.1]). Adapting such a tangent cone to transitions on a metric space should be done rather carefully. Indeed, not all elements of E close to $\vartheta(h,x)$ have to be values of a transition close to ϑ and thus in general,

$$\mathbb{B}_\rho(\vartheta(h,x)) \not\subset \{\tau(s,y) \in E \mid \tau \in \Theta(E,d),\ s \in [0,1],\ y \in \mathbb{B}_r(x)\}.$$

for fixed $h \in]0,1]$, $x \in E$ and even arbitrarily small radii $\rho, r > 0$. The Euclidean space $\mathbb{R}^N$, supplied with affine linear transitions of Example 2, distinguishes from many other metric examples in regard to this form of local surjectivity.

Lemma 36. *Let $\Theta(E,d) \neq \emptyset$ be a set of transitions on a metric space (E,d). Suppose x to belong to the topological boundary of a nonempty closed set $\mathscr{V} \subset E$. Then,* $\mathscr{T}_{\mathscr{V}}^{DM}(x) = \Theta(E,d) \setminus \mathscr{T}_{E\setminus\mathscr{V}}(x)$.

Proof is an immediate consequence of Definition 16 and 34.

Proposition 37. *Let $\Theta(E,d) \neq \emptyset$ be a set of transitions on a metric space (E,d). $\mathscr{V}_1, \mathscr{V}_2 \dots \mathscr{V}_j$ denote nonempty closed subsets of E. Then,*

$$\bigcup_{k \in \{1\dots j\}} \Big(\mathscr{T}_{\mathscr{V}_k}(x) \cap \bigcap_{l \neq k} \mathscr{T}_{\mathscr{V}_l}^{DM}(x)\Big) \subset \mathscr{T}_{\mathscr{V}_1 \cap \dots \cap \mathscr{V}_j}(x)$$

for every element $x \in \mathscr{V}_1 \cap \mathscr{V}_2 \cap \dots \cap \mathscr{V}_j \subset E$.

Proof. Choose any element $x \in \mathscr{V}_1 \cap \mathscr{V}_2 \cap \dots \cap \mathscr{V}_j$ and transition $\vartheta \in \mathscr{T}_{\mathscr{V}_1}(x) \cap \mathscr{T}_{\mathscr{V}_2}^{DM}(x) \cap \dots \cap \mathscr{T}_{\mathscr{V}_j}^{DM}(x)$. As a consequence of Definition 34 for each set $\mathscr{V}_k$ ($k \in \{2 \dots j\}$), there exist $\varepsilon, \rho \in]0,1[$ such that for all $h \in]0,\varepsilon]$,

$$\mathbb{B}_{\rho h}(\vartheta(h,x)) \subset \mathscr{V}_2 \cap \mathscr{V}_3 \cap \dots \cap \mathscr{V}_j.$$

Due to $\vartheta \in \mathscr{T}_{\mathscr{V}_1}(x)$, there is a sequence $(h_n)_{n\in\mathbb{N}}$ in $]0,\varepsilon[$ tending to 0 and satisfying

$$\operatorname{dist}(\vartheta(h_n,x), \mathscr{V}_1) < \tfrac{\rho}{n}\, h_n \qquad \text{for all } n \in \mathbb{N}.$$

For each $n \in \mathbb{N}$, we can choose an element

$$y_n \in \mathscr{V}_1 \cap \mathbb{B}_{\frac{\rho h_n}{n}}(\vartheta(h_n,x)) \subset \mathscr{V}_1 \cap \mathscr{V}_2 \cap \dots \cap \mathscr{V}_j$$

and thus, $\vartheta \in \mathscr{T}_{\mathscr{V}_1 \cap \dots \cap \mathscr{V}_j}(x)$. □

1.8 Example: Ordinary Differential Equations in $\mathbb{R}^N$

Mutational equations are motivated by the goal of extending ordinary differential equations to metric spaces. For this reason, we are obliged to verify that ordinary differential equations fit in the mutational framework as an example.
This example reflects an essential point of mutational analysis. Indeed, the results of previous sections provide sufficient conditions for the existence of a "generalized" solution (namely to a mutational equation in the sense of Definition 13). Whenever we apply this general framework to a classical type of dynamical problem (such as ordinary differential equations here), we have to investigate the link with a classical concept of solution. This can be done for each example individually and, the results prove to be of particular interest when applying them to separate components of a system of mutational equations as explained in § 1.5.

For linking ordinary differential equations and mutational equations on $(\mathbb{R}^N, |\cdot|)$, we consider the maps of Example 2 (on page 33)

$$\vartheta_v : [0,1] \times \mathbb{R}^N \longrightarrow \mathbb{R}^N, \quad (h,x) \longmapsto x + h \cdot v$$

for each vector $v \in \mathbb{R}^N$ and summarize some obvious properties in regard to Definitions 1 and 6:

Lemma 38. *For each vector $v \in \mathbb{R}^N$, the affine linear map*

$$\vartheta_v : [0,1] \times \mathbb{R}^N \longrightarrow \mathbb{R}^N, \quad (h,x) \longmapsto x + h \cdot v$$

is a transition on the Euclidean space $(\mathbb{R}^N, |\cdot|)$ with

$$\begin{aligned} \alpha(\vartheta_v) &= 0, \\ \beta(\vartheta_v) &= |v|, \\ D(\vartheta_v, \vartheta_w) &= |v - w|. \end{aligned}$$

□

Basic set	$E := \mathbb{R}^N$
Distance	Euclidean distance $d : \mathbb{R}^N \times \mathbb{R}^N \longrightarrow \mathbb{R}, \ (x,y) \longmapsto \|x-y\|$
Transition	For each vector $v \in \mathbb{R}^N$, $\vartheta_v : [0,1] \times \mathbb{R}^N \longrightarrow \mathbb{R}^N, \ (h,x) \longmapsto x + h \cdot v$
Compactness	Closed bounded balls are compact due to Heine-Borel Theorem.
Mutational solutions	Lipschitz continuous solutions to ordinary differential equations
List of main results formulated in § 1.8	Classical version of Cauchy-Lipschitz Theorem: Corollary 41 Classical version of Nagumo Theorem: Corollary 42 Classical version of Peano Theorem: Corollary 43 Continuous dependence on data: Corollary 44

Table 1.1 Brief summary in mutational terms: Ordinary differential equations in $\mathbb{R}^N$

For the sake of simplicity, we identify this transition $\vartheta_v : [0,1] \times \mathbb{R}^N \longrightarrow \mathbb{R}^N$ on the Euclidean space $(\mathbb{R}^N, |\cdot|)$ with its directional vector $v \in \mathbb{R}^N$: $\quad \Theta(\mathbb{R}^N, |\cdot|) \cong \mathbb{R}^N$.

Proposition 39. *Let $f : \mathbb{R}^N \times [0,T] \longrightarrow \mathbb{R}^N$ be given.*
A curve $x(\cdot) : [0,T] \longrightarrow \mathbb{R}^N$ is a solution to the mutational equation

$$\overset{\circ}{x}(\cdot) \ni f\big(x(\cdot), \cdot\big)$$

if and only if $x(\cdot)$ is Lipschitz continuous and its weak derivative $x' \in L^\infty([0,T], \mathbb{R}^N)$ satisfies

$$x'(t) = f\big(x(t), t\big)$$

at Lebesgue-almost every time $t \in [0,T]$.

This proposition, whose proof is postponed to the end of this section, implies several well-known results about ordinary differential equations – now, however, as consequences of the theorems in § 1.3 – § 1.6. This is based on the Heine-Borel theorem ensuring that all closed bounded sets of the Euclidean space $\mathbb{R}^N$ are compact.

Corollary 40. *Let $f : \mathbb{R}^N \times [0,T] \longrightarrow \mathbb{R}^N$ be continuous.*
A curve $x(\cdot) : [0,T] \longrightarrow \mathbb{R}^N$ is a solution to the mutational equation

$$\overset{\circ}{x}(\cdot) \ni f\big(x(\cdot), \cdot\big)$$

if and only if $x(\cdot)$ is continuously differentiable and its derivative $x'(\cdot)$ satisfies

$$x'(t) = f\big(x(t), t\big)$$

at every time $t \in [0,T]$. □

Corollary 41 (Cauchy-Lipschitz: Classical version for ODEs).
Let $f : \mathbb{R}^N \longrightarrow \mathbb{R}^N$ be λ-Lipschitz continuous. Fix $x_0 \in \mathbb{R}^N$ and $y(\cdot) \in C^1([0,T], \mathbb{R}^N)$. Then there exists a unique continuously differentiable solution $x(\cdot) : [0,T] \longrightarrow \mathbb{R}^N$ to the initial value problem

$$\begin{cases} x'(\cdot) = f(x(\cdot)) \\ x(0) = x_0. \end{cases}$$

In addition, it satisfies the following inequality for all $t \in [0,T]$

$$|x(t) - y(t)| \le |x_0 - y(0)|\, e^{\lambda t} + \int_0^t e^{\lambda (t-s)} \, \big|f(y(s)) - y'(s)\big| \, ds.$$

Proof results directly from Theorem 15 (on page 38) with $\widehat{\alpha} := \sup \alpha(f(\cdot)) = 0$.

Corollary 42 (Nagumo: Classical version for autonomous ODE).
Suppose $f : \mathbb{R}^N \longrightarrow \mathbb{R}^N$ to be continuous and bounded. Then the following two statements are equivalent for any closed nonempty subset $V \subset \mathbb{R}^N$:

1. *Every state $x_0 \in V$ is the initial point of at least one solution $x(\cdot) : [0,1] \longrightarrow \mathbb{R}^N$ to the ordinary differential equation*
$$x'(\cdot) = f\big(x(\cdot)\big)$$
with all its values in V.

2. *$V \subset \mathbb{R}^N$ is a viability domain of f in the sense that for every $z \in V$, the vector $f(z) \in \mathbb{R}^N$ belongs to Bouligand's contingent cone of $V \subset \mathbb{R}^N$ at z, i.e.*
$$\liminf_{h \downarrow 0} \tfrac{1}{h} \cdot \text{dist}\big(z + h \cdot f(z),\, V\big) = 0.$$

Proof is an immediate consequence of Theorem 19 (on page 40) due to the remarks (about contingent cones) mentioned in Example 18.

Corollary 43 (Peano: Classical version for nonautonomous ODE).
Suppose $f : \mathbb{R}^N \times [0,T] \longrightarrow \mathbb{R}^N$ to be continuous and bounded.
Then for every initial state $x_0 \in \mathbb{R}^N$, there exists a solution $x(\cdot) : [0,T] \longrightarrow \mathbb{R}^N$ to the ordinary differential equation
$$x'(\cdot) = f\big(x(\cdot),\, \cdot\big)$$
with $x(0) = x_0$.

Proof results from Theorem 20 (on page 40).

Corollary 44 (Continuity w.r.t. initial data and the right-hand side).
Suppose $f : \mathbb{R}^N \times [0,T] \longrightarrow \mathbb{R}^N$ to be λ-Lipschitz continuous in the first argument. Let $g : \mathbb{R}^N \times [0,T] \longrightarrow \mathbb{R}^N$ be continuous with $\Delta := \sup_{z,s} \big|f(z,s) - g(z,s)\big| < \infty$.

Then every continuously differentiable solutions $x(\cdot), y(\cdot) : [0,T] \longrightarrow \mathbb{R}^N$ to the ordinary differential equations
$$\begin{cases} x'(\cdot) = f\big(x(\cdot),\, \cdot\big) \\ y'(\cdot) = g\big(y(\cdot),\, \cdot\big) \end{cases}$$
satisfy the following inequality for every $t \in [0,T]$
$$|x(t) - y(t)| \leq \big(|x(0) - y(0)| + \Delta \cdot t\big)\; e^{\lambda t}.$$

Proof is an obvious conclusion from Proposition 24 (on page 42).

Proof (of Proposition 39). The key tool is Rademacher's Theorem stating that every Lipschitz continuous function $h : \mathbb{R}^M \longrightarrow \mathbb{R}^N$ is differentiable at Lebesgue-almost every point of its domain (see e.g. [162]). In particular, the weak derivative of h coincides with its Fréchet derivative Lebesgue-almost everywhere in $\mathbb{R}^M$.

"$\Longleftarrow$" Obviously, every Lipschitz continuous curve $x(\cdot) : [0,T] \longrightarrow \mathbb{R}^N$ with $x'(t) = f(x(t), t)$ at Lebesgue-almost every time $t \in [0,T]$ fulfills

$$\lim_{h \downarrow 0} \tfrac{1}{h} \left| x(t+h) - \left(x(t) + h \cdot f(x(t),t)\right) \right| = 0$$

for Lebesgue-almost every $t \in [0,T]$ and thus, $x(\cdot)$ solves the mutational equation $\overset{\circ}{x}(\cdot) \ni f(x(\cdot), \cdot)$ in the sense of Definition 13 (on page 38).

"$\Longrightarrow$" Let $x(\cdot) : [0,T] \longrightarrow \mathbb{R}^N$ be a solution to the mutational equation $\overset{\circ}{x}(\cdot) \ni f(x(\cdot), \cdot)$. According to Definition 13, $x(\cdot)$ is Lipschitz continuous and satisfies

$$0 = \lim_{h \downarrow 0} \tfrac{1}{h} \left| x(t+h) - \left(x(t) + h \cdot f(x(t),t)\right) \right| = \lim_{h \downarrow 0} \left| \tfrac{x(t+h) - x(t)}{h} - f(x(t),t) \right|$$

for Lebesgue-almost every $t \in [0,T]$. Rademacher's Theorem ensures the differentiability of $x(\cdot)$ Lebesgue-almost everywhere in $[0,T]$ and thus, the one-sided differential quotient even reflects the time derivative, i.e. $x'(\cdot) = f(x(\cdot),\cdot)$ a.e. in $[0,T]$. □

1.9 Example: Morphological Equations for Compact Sets in $\mathbb{R}^N$

$\mathscr{K}(\mathbb{R}^N)$ consists of all nonempty compact subsets of the Euclidean space $\mathbb{R}^N$.
It is no obvious linear structure though. To be more precise, Minkowski suggested a very popular definition of the sum, i.e.

$$K_1 + K_2 \stackrel{\text{Def.}}{=} \{x+y \mid x \in K_1, y \in K_2\} \subset \mathbb{R}^N$$

for $K_1, K_2 \in \mathscr{K}(\mathbb{R}^N)$. This addition has the obvious neutral element $\{0\} \subset \mathbb{R}^N$, but it is not invertible in general, i.e. for any given $K_1 \in \mathscr{K}(\mathbb{R}^N)$, the equation $K_1 + K_2 = \{0\}$ does not always have a solution $K_2 \in \mathscr{K}(\mathbb{R}^N)$.
$\mathscr{K}(\mathbb{R}^N)$ can be supplied with a metric instead:

1.9.1 The Pompeiu-Hausdorff Distance $d\!l$

Definition 45. The *Pompeiu-Hausdorff excesses* between two nonempty subsets $K_1, K_2 \subset \mathbb{R}^N$ are defined as

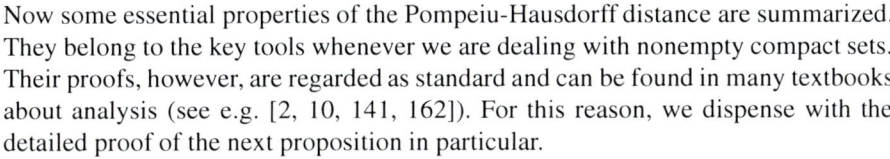

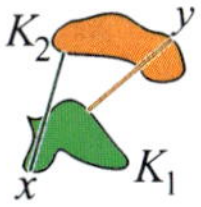

$$e^{\subset}(K_1, K_2) := \sup_{x \in K_1} \operatorname{dist}(x,\ K_2),$$
$$e^{\supset}(K_1, K_2) := \sup_{y \in K_2} \operatorname{dist}(y,\ K_1) \in [0, \infty]$$

Their maximum is called the *Pompeiu-Hausdorff distance* $d\!l(K_1, K_2)$.

Now some essential properties of the Pompeiu-Hausdorff distance are summarized. They belong to the key tools whenever we are dealing with nonempty compact sets. Their proofs, however, are regarded as standard and can be found in many textbooks about analysis (see e.g. [2, 10, 141, 162]). For this reason, we dispense with the detailed proof of the next proposition in particular.

Proposition 46. *The Pompeiu-Hausdorff distance $d\!l$ is a metric on $\mathscr{K}(\mathbb{R}^N)$ and has the equivalent characterizations for any $K_1, K_2 \in \mathscr{K}(\mathbb{R}^N)$*

$$\begin{aligned} d\!l(K_1, K_2) &= \sup_{z \in \mathbb{R}^N} \left|\operatorname{dist}(z, K_1) - \operatorname{dist}(z, K_2)\right| \\ &= \inf \left\{\rho > 0 \mid K_1 \subset K_2 + \rho\,\mathbb{B} \text{ and } K_2 \subset K_1 + \rho\,\mathbb{B}\right\} \end{aligned}$$

with the standard abbreviation $\mathbb{B}$ for the closed unit ball in $\mathbb{R}^N$

$$\mathbb{B} := \mathbb{B}_1(0) \stackrel{\text{Def.}}{=} \{x \in \mathbb{R}^N \mid |x| \le 1\}.$$

Moreover, the metric space $(\mathscr{K}(\mathbb{R}^N), d\!l)$ is locally compact in the following sense:

Proposition 47. *In the metric space $(\mathscr{K}(\mathbb{R}^N), d\!l)$, every closed bounded ball*

$$\mathbb{B}_R^{d}(K) := \{K' \in \mathscr{K}(\mathbb{R}^N) \mid d\!l(K', K) \le R\}$$

with any center $K \in \mathscr{K}(\mathbb{R}^N)$ and arbitrary radius $R \ge 0$ is compact.

Basic set	$E := \mathscr{K}(\mathbb{R}^N)$ the set of nonempty compact subsets of the Euclidean space $\mathbb{R}^N$
Distance	Pompeiu-Hausdorff metric $d\!\!l : \mathscr{K}(\mathbb{R}^N) \times \mathscr{K}(\mathbb{R}^N) \longrightarrow \mathbb{R}$, $d\!\!l(K_1, K_2) := \max \Big\{ \sup_{x \in K_1} \mathrm{dist}(x, K_2), \quad \sup_{y \in K_2} \mathrm{dist}(y, K_1) \Big\}$
Transition	For each $F \in \mathrm{LIP}(\mathbb{R}^N, \mathbb{R}^N)$, i.e. bounded and Lipschitz continuous set-valued map $F : \mathbb{R}^N \rightsquigarrow \mathbb{R}^N$ with compact values, define $\vartheta_F : [0,1] \times \mathscr{K}(\mathbb{R}^N) \longrightarrow \mathscr{K}(\mathbb{R}^N)$ by means of reachable sets of the autonomous differential inclusion $x'(\cdot) \in F(x(\cdot))$ a.e.: $\vartheta_F(t, K_0) := \big\{ x(t) \mid \text{there exists } x(\cdot) \in W^{1,1}([0,t], \mathbb{R}^N) : x'(\cdot) \in F(x(\cdot)) \ \mathscr{L}^1\text{-a.e. in } [0,t], \ x(0) \in K_0 \big\}.$
Compactness	Closed bounded balls in $(\mathscr{K}(\mathbb{R}^N), d\!\!l)$ are compact: Proposition 47 (page 57)
Mutational solutions	Reachable sets of a nonautonomous differential inclusion whose set-valued right-hand side is determined via feedback: Proposition 57 (page 64), Proposition 70 (page 74)
List of main results formulated in § 1.9	Existence due to compactness (Peano): Proposition 71 (page 75) Cauchy-Lipschitz Theorem: Proposition 72 (page 75) Continuity w.r.t. data: Proposition 73 (page 75) Existence under state constraints (Nagumo): Proposition 74
Key tools	Filippov's Theorem A.6 about differential inclusions (page 443) Integral funnel equation for reachable sets of nonautonomous differential inclusions: Proposition A.13 (page 447)

Table 1.2 Brief summary of the example in § 1.9 in mutational terms:
Morphological equations for compact sets in $\mathbb{R}^N$

Proof. Choose any set $K \in \mathscr{K}(\mathbb{R}^N)$, radius $R \geq 0$ and any sequence $(K_n)_{n\in\mathbb{N}}$ in $\mathscr{K}(\mathbb{R}^N)$ satisfying $dl(K_n, K) \leq R$ for all $n \in \mathbb{N}$.
Now we prove that some subsequence $(K_{n_j})_{j\in\mathbb{N}}$ is convergent with respect to the Pompeiu-Hausdorff distance. Then $\mathbb{B}^{dl}_R(K)$ is sequentially compact with respect to dl and (as in every metric space) this is equivalent to the property that every open cover of $\mathbb{B}^{dl}_R(K) \subset \mathscr{K}(\mathbb{R}^N)$ has a finite subcover (see e.g. [170, Chapter 12]).
Using the abbreviation $\mathbb{B}_{R+1}(K) \stackrel{\text{Def.}}{=} \{x \in \mathbb{R}^N \mid \text{dist}(x,K) \leq R+1\}$, set for each $n \in \mathbb{N}$

$$\delta_n : \mathbb{B}_{R+1}(K) \longrightarrow [0,\infty[, \quad z \longmapsto \text{dist}(z, K_n).$$

Obviously each function $\delta_n(\cdot)$ is 1-Lipschitz continuous and has the uniform bound

$$\delta_n(\cdot) \leq \text{diam}\, K + 2\,(R+1).$$

Arzelà-Ascoli Theorem A.82 implies that a subsequence $(\delta_{n_j})_{j\in\mathbb{N}}$ converges uniformly to a continuous function $\delta : \mathbb{B}_{R+1}(K) \longrightarrow [0,\infty[$. In particular, $\delta(\cdot)$ is also 1-Lipschitz continuous.

Then $K_\infty := \{x \in \mathbb{B}_{R+1}(K) \mid \delta(x) = 0\}$ is the limit of $(K_{n_j})_{j\in\mathbb{N}}$ with respect to dl. Indeed, K_∞ is closed because $\delta(\cdot)$ is continuous. Furthermore, K_∞ is nonempty since any sequence $(x_{n_j})_{j\in\mathbb{N}}$ with $x_{n_j} \in K_{n_j} = \delta_{n_j}^{-1}(\{0\})$ for each $j \in \mathbb{N}$ is contained in the compact subset $\mathbb{B}_R(K) \subset \mathbb{R}^N$ and thus, it has an accumulation point $x \in \mathbb{B}_R(K)$. The uniform convergence of the 1-Lipschitz functions $\delta_{n_j}(\cdot)$ implies $\delta(x) = 0$, i.e. $x \in K_\infty$. Hence, $K_\infty \in \mathscr{K}(\mathbb{R}^N)$.
Moreover, $\delta(z) \leq \text{dist}(z, K_\infty)$ holds for every vector $z \in \mathbb{B}_{R+1}(K) \subset \mathbb{R}^N$ because for every element $x \in K_\infty$, we conclude from the 1-Lipschitz continuity of $\delta(\cdot)$

$$\delta(z) = \delta(z) - \delta(x) \leq |z - x|.$$

For proving the opposite inequality $\delta(z) \geq \text{dist}(z, K_\infty)$ with arbitrary $z \in \mathbb{B}_{R+1}(K)$, we can restrict our considerations to any element $z \in \mathbb{B}_{R+1}(K)$ with $\text{dist}(z, K_\infty) > 0$. In particular, $z \notin K_\infty$. Choose any positive $r < \text{dist}(z, K_\infty)$. Then every point $y \in \mathbb{B}_r(z)$ does not belong to K_∞ either, i.e. $\delta(y) > 0$. Due to the continuity of $\delta(\cdot)$, we even have $\mu := \inf_{\mathbb{B}_r(z)} \delta(\cdot) > 0$. For all $j \in \mathbb{N}$ sufficiently large,

$$\sup_{x \in \mathbb{B}_{R+1}(K)} \left|\delta_{n_j}(x) - \delta(x)\right| < \tfrac{\mu}{2}.$$

and thus, all $y \in \mathbb{B}_r(z)$ satisfy $\delta_{n_j}(y) > \delta(y) - \frac{\mu}{2} > 0$. We have just verified $\mathbb{B}_r(z) \cap K_{n_j} = \emptyset$ for all large indices $j \in \mathbb{N}$. As a consequence,

$$\delta(z) = \lim_{j\to\infty} \delta_{n_j}(z) = \lim_{j\to\infty} \text{dist}(z, K_{n_j}) \geq r$$

with any positive $r < \text{dist}(z, K_\infty)$. Finally, $\delta(z) \geq \text{dist}(z, K_\infty)$ for any $z \in \mathbb{B}_{R+1}(K)$. The resulting equality $\delta(\cdot) = \text{dist}(\cdot, K_\infty)$ in $\mathbb{B}_{R+1}(K) \subset \mathbb{R}^N$ opens the door to proving the convergence of $(K_{n_j})_{j\in\mathbb{N}}$ with respect to dl :

$$\begin{aligned} dl(K_{n_j}, K_\infty) &= \max\Big\{ \sup_{x\in K_{n_j}} \delta(x), \ \sup_{y \in K_\infty} \delta_{n_j}(y) \Big\} \\ &\leq \sup_{z\in\mathbb{B}_{R+1}(K)} \left|\delta(z) - \delta_{n_j}(z)\right| \quad \stackrel{j\to\infty}{\longrightarrow} 0. \end{aligned}$$

□

1.9.2 Morphological Transitions on $(\mathscr{K}(\mathbb{R}^N), d\!l)$

As mentioned briefly in Example 5 (on page 34), differential inclusions can serve as a tool for specifying "deformations" of compact subsets of $\mathbb{R}^N$. The so-called reachable set of such a differential inclusion at time $t \geq 0$ consists of all points $x(t)$ that can be reached by an absolutely continuous solution $x(\cdot) : [0,t] \longrightarrow \mathbb{R}^N$ (to this differential inclusion) starting in the given set. This notion is not necessarily restricted to autonomous differential inclusions, of course.

Definition 48. Let $F : \mathbb{R}^N \leadsto \mathbb{R}^N$ be a set-valued map. Then the set

$$\begin{aligned}\vartheta_F(t,\ K_0) := \big\{x(t) \;\big|\; &\text{there exists } x(\cdot) \in W^{1,1}([0,t],\mathbb{R}^N) : \\ &x'(\cdot) \in F(x(\cdot)) \ \mathscr{L}^1\text{-a.e. in } [0,t],\ \ x(0) \in K_0\big\}\end{aligned}$$

is called *reachable set* of the initial set $K_0 \in \mathscr{K}(\mathbb{R}^N)$ and the map F at time $t \geq 0$. Correspondingly for any set-valued map $\widetilde{F} : [0,T] \times \mathbb{R}^N \leadsto \mathbb{R}^N$, we define the *reachable set* of $K_0 \in \mathscr{K}(\mathbb{R}^N)$ and the map $\widetilde{F}$ at time $t \in [0,T]$ as

$$\begin{aligned}\vartheta_{\widetilde{F}}(t,\ K_0) := \big\{x(t) \;\big|\; &\text{there exists } x(\cdot) \in W^{1,1}([0,t],\mathbb{R}^N) : \\ &x'(\cdot) \in \widetilde{F}(\cdot,x(\cdot)) \ \mathscr{L}^1\text{-a.e. in } [0,t],\ \ x(0) \in K_0\big\}.\end{aligned}$$

Filippov's Theorem A.6 about solutions to differential inclusions provides the key tool for investigating compact reachable sets of Lipschitz continuous set-valued maps with nonempty compact values. It motivates the following abbreviation introduced by Aubin:

Definition 49. $\mathrm{LIP}(\mathbb{R}^N,\mathbb{R}^N)$ consists of all set-valued maps $F : \mathbb{R}^N \leadsto \mathbb{R}^N$ satisfying the following two conditions:

1.) F has nonempty compact values that are uniformly bounded in $\mathbb{R}^N$,
2.) F is Lipschitz continuous with respect to the Pompeiu-Hausdorff distance $d\!l$.

Furthermore define for any maps $F, G \in \mathrm{LIP}(\mathbb{R}^N,\mathbb{R}^N)$

$$\begin{aligned}\|F\|_\infty &:= \sup_{x\in\mathbb{R}^N}\ \sup_{v\in F(x)} |v|, \\ d\!l_\infty(F,G) &:= \sup_{x\in\mathbb{R}^N} d\!l\big(F(x),G(x)\big).\end{aligned}$$

Proposition 50. *For any initial sets $K_1, K_2 \in \mathscr{K}(\mathbb{R}^N)$ and set-valued maps $F, G \in \mathrm{LIP}(\mathbb{R}^N,\mathbb{R}^N)$ with $\Lambda := \max\{\mathrm{Lip}\,F,\ \mathrm{Lip}\,G\}$, the reachable sets $\vartheta_F(t,K_1)$, $\vartheta_G(t,K_2)$ are closed subsets of $\mathbb{R}^N$ and, the Pompeiu-Hausdorff distance between the reachable sets at time $t \geq 0$ satisfies*

$$d\!l\big(\vartheta_F(t,K_1),\ \vartheta_G(t,K_2)\big) \;\leq\; \big(d\!l(K_1,K_2) + t\cdot d\!l_\infty(F,G)\big)\cdot e^{\Lambda\,t}.$$

Proof. $\vartheta_F(t,K_1), \vartheta_G(t,K_2) \subset \mathbb{R}^N$ are closed due to Filippov's Theorem A.6 (on page 443). Due to the symmetry of $d\!\!l$, it is sufficient to prove for every $x_1 \in \vartheta_F(t,K_1)$

$$\operatorname{dist}\big(x_1,\ \vartheta_G(t,K_2)\big) \ \le\ \big(d\!\!l(K_1,K_2) + t\cdot d\!\!l_\infty(F,G)\big)\cdot e^{\Lambda\, t}.$$

According to Definition 48, there exists a solution $x(\cdot) \in W^{1,1}([0,t],\mathbb{R}^N)$ to the differential inclusion $x'(\cdot) \in F(x(\cdot))$ ($\mathscr{L}^1$-almost everywhere in $[0,t]$) satisfying

$$x(0) \in K_1, \quad x(t) = x_1.$$

Choose now any point $y_0 \in K_2$ with $|x(0) - y_0| = \operatorname{dist}(x(0),\ K_2) \le d\!\!l(K_1,K_2)$. Filippov's Theorem A.6 guarantees a solution $y(\cdot) \in W^{1,1}([0,t],\mathbb{R}^N)$ to the differential inclusion $y'(\cdot) \in G(y(\cdot))$ a.e. in $[0,t]$ satisfying in addition

$$\begin{aligned}\big|y(t) - x(t)\big| \ &\le\ \big|y_0 - x(0)\big|\, e^{\Lambda t} \ +\ \int_0^t e^{\Lambda\cdot(t-s)} \operatorname{dist}\big(x'(s),\ G(x(s))\big)\, ds\\ &\le\ d\!\!l(K_1,K_2)\ e^{\Lambda t} \ +\ t\ \ e^{\Lambda t}\quad d\!\!l_\infty(F,G)\end{aligned}$$

In particular, $y(t) \in \vartheta_G(t,K_2)$ and thus, $\operatorname{dist}\big(x_1,\ \vartheta_G(t,K_2)\big) \le |x(t) - y(t)|$. □

This proof of Proposition 50 reveals that the same estimate holds for any Lipschitz continuous set-valued maps with nonempty compact values. The uniform bound of their set values, in particular, is not required for applying Filippov's Theorem here. It is used for the Lipschitz continuity with respect to time instead:

Lemma 51. *For any initial set $K \in \mathscr{K}(\mathbb{R}^N)$ and map $F \in \mathrm{LIP}(\mathbb{R}^N,\mathbb{R}^N)$, the reachable set $\vartheta_F(\cdot,K) : [0,\infty[\rightsquigarrow \mathbb{R}^N$ is Lipschitz continuous with respect to $d\!\!l$, i.e.*

$$d\!\!l\big(\vartheta_F(s,K),\ \vartheta_F(t,K)\big) \ \le\ \|F\|_\infty \cdot |s-t| \qquad \text{for any } s,t \ge 0.$$

Proof results directly from Definition 48 because every absolutely continuous solution $x(\cdot)$ of $x'(\cdot) \in F(x(\cdot))$ is even $\|F\|_\infty$-Lipschitz continuous. □

Lemma 52. *For any initial set $K \in \mathscr{K}(\mathbb{R}^N)$ and map $F \in \mathrm{LIP}(\mathbb{R}^N,\mathbb{R}^N)$, the reachable set $\vartheta_F(\cdot,K) : [0,\infty[\longrightarrow \big(\mathscr{K}(\mathbb{R}^N), d\!\!l\big)$ has the semigroup property in the following sense*

$$\vartheta_F\big(h,\ \vartheta_F(t,K)\big) \ =\ \vartheta_F(t+h,\ K) \qquad \text{for any } t,h \ge 0.$$

Proof is an immediate consequence of Definition 48 and the following concatenation properties of solutions to differential inclusions: Let $x_1(\cdot) \in W^{1,1}([0,t],\mathbb{R}^N)$ and $x_2(\cdot) \in W^{1,1}([0,h],\mathbb{R}^N)$ be solutions to the autonomous differential inclusion $x_j' \in F(x_j)$ a.e. with $x_1(t) = x_2(0)$. Then

$$[0,\ t+h] \ \longrightarrow\ \mathbb{R}^N, \qquad s \longmapsto \begin{cases} x_1(s) & \text{for } 0 \le s \le t\\ x_2(s-t) & \text{for } t \le s \le t+h\end{cases}$$

is an absolutely continuous solution of $x' \in F(x)$ a.e. (and vice versa). □

Now we have collected all the analytical tools for verifying that reachable sets of maps in $\mathrm{LIP}(\mathbb{R}^N, \mathbb{R}^N)$ induce transitions on $(\mathscr{K}(\mathbb{R}^N), d\!l)$. Aubin called them *morphological transition* and used them in most of his examples about evolving sets.

Proposition 53. *For every set-valued map $F \in \mathrm{LIP}(\mathbb{R}^N, \mathbb{R}^N)$,*

$$\begin{aligned} \vartheta_F : [0,1] \times \mathscr{K}(\mathbb{R}^N) &\longrightarrow \mathscr{K}(\mathbb{R}^N) \\ (t,\ K) &\longmapsto \vartheta_F(t,K) \end{aligned}$$

is a transition on $(\mathscr{K}(\mathbb{R}^N), d\!l)$ with

$$\begin{aligned} \alpha(\vartheta_F) &\le \mathrm{Lip}\, F, \\ \beta(\vartheta_F) &\le \|F\|_\infty, \\ D(\vartheta_F, \vartheta_G) &\le d\!l_\infty(F,G). \end{aligned}$$

Proof. Obviously, $\vartheta_F(0,K) = K$ for every initial set $K \in \mathscr{K}(\mathbb{R}^N)$. According to Proposition 50 and Lemma 51, the reachable set $\vartheta_F(t,K) \subset \mathbb{R}^N$ is closed and bounded for every $K \in \mathscr{K}(\mathbb{R}^N)$ and $t \ge 0$. Thus, $\vartheta_F(t,K)$ is compact due to Heine-Borel Theorem, i.e. $\vartheta_F(t,K) \in \mathscr{K}(\mathbb{R}^N)$.
Moreover Lemma 52 implies condition (2.) on transitions (in Definition 1 on page 32), i.e. for every set $K \in \mathscr{K}(\mathbb{R}^N)$ and time $t \in [0,1[$

$$\lim_{h \downarrow 0} \tfrac{1}{h} \cdot d\!l\big(\vartheta_F(t+h,\, K),\ \vartheta_F(h,\, \vartheta_F(t,K))\big) = 0.$$

The estimate in Proposition 50 (applied to $G := F$) guarantees

$$\begin{aligned} \alpha(\vartheta_F) &\overset{\text{Def.}}{=} \sup_{\substack{K_1,K_2 \in \mathscr{K}(\mathbb{R}^N) \\ K_1 \ne K_2}} \limsup_{h \downarrow 0} \max\left\{0,\ \frac{d\!l(\vartheta_F(h,K_1),\ \vartheta_F(h,K_2)) - d\!l(K_1,K_2)}{h \cdot d\!l(K_1,K_2)}\right\} \\ &\le \limsup_{h \downarrow 0} \frac{e^{\mathrm{Lip}\, F \cdot h} - 1}{h} \\ &= \mathrm{Lip}\, F. \end{aligned}$$

Due to Lemma 51, we obtain

$$\beta(\vartheta_F) \overset{\text{Def.}}{=} \sup_{K \in \mathscr{K}(\mathbb{R}^N)} \limsup_{h \downarrow 0} \tfrac{1}{h} \cdot d\!l\big(K,\ \vartheta_F(h,K)\big) \le \|F\|_\infty.$$

Finally, Proposition 50 lays also the basis for estimating $D(\vartheta_F, \vartheta_G)$ (in the sense of Definition 6) for arbitrary maps $F, G \in \mathrm{LIP}(\mathbb{R}^N, \mathbb{R}^N)$ and $\Lambda := \max\{\mathrm{Lip}\, F,\ \mathrm{Lip}\, G\}$

$$\begin{aligned} D(\vartheta_F, \vartheta_G) &\overset{\text{Def.}}{=} \sup_{K \in \mathscr{K}(\mathbb{R}^N)} \limsup_{h \downarrow 0} \frac{1}{h} \cdot d\!l\big(\vartheta_F(h,K),\ \vartheta_G(h,K)\big) \\ &\le \limsup_{h \downarrow 0} d\!l_\infty(F,G) \cdot e^{\Lambda h} \\ &= d\!l_\infty(F,G). \end{aligned}$$

□

Example 54. In Example 4 (on page 33), we have already mentioned the flow of compact subsets along a bounded Lipschitz continuous vector field $f : \mathbb{R}^N \longrightarrow \mathbb{R}^N$. This type of set deformations lays the basis for the so-called *velocity method* used in approaches to shape optimization by Céa, Delfour, Sokolowski, Zolésio and others. Now the flow along such a vector field proves to be a special case of morphological transitions. Indeed, we just consider a single-valued map f in $\mathrm{LIP}(\mathbb{R}^N, \mathbb{R}^N)$.
As an immediate consequence of Proposition 53, the corresponding reachable set $\vartheta_f(\cdot,\cdot)$ induces a transition on $(\mathscr{K}(\mathbb{R}^N), d\!l)$ with

$$\begin{aligned} \alpha(\vartheta_f) &\leq \mathrm{Lip}\, f, \\ \beta(\vartheta_f) &\leq \|f\|_{\sup}, \\ D(\vartheta_f, \vartheta_g) &\leq \|f-g\|_{\sup} \end{aligned}$$

for any bounded and Lipschitz continuous vector fields $f, g : \mathbb{R}^N \longrightarrow \mathbb{R}^N$.

Example 55.
Considering a fixed compact convex neighborhood $C \subset \mathbb{R}^N$ of the origin, we find a further special case of morphological transitions: the so-called *morphological dilation*, that became very popular in image processing, for example, due to publications of Matheron and Serra:

Each reachable set of the differential inclusion $x'(\cdot) \in C$ (with constant convex right-hand side) coincides with a Minkowski sum in the following sense

$$\vartheta_C(h,K) \;=\; K + h\,C \;\stackrel{\text{Def.}}{=}\; \{x + h\,v \mid x \in K,\ v \in C\}$$

for every initial set $K \in \mathscr{K}(\mathbb{R}^N)$ and at any time $h \geq 0$. Indeed, $K + h\,C \subset \vartheta_C(h,K)$ results from the obvious statement that for each $x \in K$ and $v \in C$, the curve

$$y(\cdot) : [0,h] \longrightarrow \mathbb{R}^N, \qquad s \longmapsto x + s\,v$$

solves the differential inclusion $y'(\cdot) \in C$. In regard to the opposite inclusion $\vartheta_C(h,K) \subset K + h\,C$, choose $z \in \vartheta_C(h,K)$ arbitrarily. It is related to an initial point $x \in K$ and a Lebesgue-integrable function $u(\cdot) : [0,h] \longrightarrow \mathbb{R}^N$ with

$$z = x + \int_0^h u(s)\,ds, \qquad u(t) \in C \qquad \text{for every } t \in [0,h].$$

Now the convexity of the closed set $C \subset \mathbb{R}^N$ implies $\frac{1}{h} \cdot \int_0^h u(s)\,ds \in \overline{\mathrm{co}}\; C = C$ and thus, $z \in x + h\,C$.

In Serra's framework of "mathematical morphology" [171], the fixed set $C \subset \mathbb{R}^N$ is usually called *structural element* (of the corresponding morphological operations like dilation). In a figurative sense, every reachable set $\vartheta_F(h,K) \subset \mathbb{R}^N$ of an initial set $K \in \mathscr{K}(\mathbb{R}^N)$ and a set-valued map $F \in \mathrm{LIP}(\mathbb{R}^N, \mathbb{R}^N)$ can be interpreted as a generalized dilation of K with the structural element depending on space, namely $F = F(x)$. This was (probably) Aubin's motivation for seizing the term "morphological" in connection with these transitions on $(\mathscr{K}(\mathbb{R}^N), d\!l)$.

1.9.3 Morphological Primitives as Reachable Sets

Each morphological transition is induced by a set-valued map in $\text{LIP}(\mathbb{R}^N, \mathbb{R}^N)$ by definition. For the sake of simplicity, we sometimes identify the morphological transition ϑ_F on $(\mathscr{K}(\mathbb{R}^N), d\!l)$ with its corresponding map $F \in \text{LIP}(\mathbb{R}^N, \mathbb{R}^N)$ representing the right-hand side of the autonomous differential inclusion.

Definition 56. A curve $[0,T] \longrightarrow \mathscr{K}(\mathbb{R}^N)$ is usually called *tube* in $\mathbb{R}^N$.

According to Definition 10 (on page 37), the (morphological) mutation of a tube $K(\cdot)$ at time t consists of all morphological transitions providing a first-order approximation of $K(t+\cdot)$ with respect to $d\!l$. Identifying now morphological transitions with the respective set-valued maps in $\text{LIP}(\mathbb{R}^N, \mathbb{R}^N)$, we obtain

$$\overset{\circ}{K}(t) = \big\{F \in \text{LIP}(\mathbb{R}^N, \mathbb{R}^N) \;\big|\; \lim_{h \downarrow 0} \tfrac{1}{h} \cdot d\!l\big(\vartheta_F(h, K(t)),\; K(t+h)\big) = 0\big\}.$$

Each tube $K(\cdot) : [0,T] \rightsquigarrow \mathbb{R}^N$ induces a set-valued map $\overset{\circ}{K}: [0,T] \rightsquigarrow \text{LIP}(\mathbb{R}^N, \mathbb{R}^N)$ whose values might be empty.

Primitives are linked to this relation in the opposite direction: Now a curve of morphological transitions is given, i.e.

$$\mathscr{F} : [0,T] \longrightarrow \text{LIP}(\mathbb{R}^N, \mathbb{R}^N).$$

According to Definition 12, a tube $K(\cdot) : [0,T] \rightsquigarrow \mathbb{R}^N$ is a (morphological) primitive of $\mathscr{F}(\cdot)$ if and only if $K(\cdot)$ is Lipschitz continuous with respect to $d\!l$ and satisfies at Lebesgue-almost every time $t \in [0,T]$:

$$\mathscr{F}(t) \in \overset{\circ}{K}(t)$$

or, equivalently,
$$\lim_{h \downarrow 0} \tfrac{1}{h} \cdot d\!l\big(\vartheta_{\mathscr{F}(t)}(h, K(t)),\; K(t+h)\big) = 0.$$

This is a differential criterion – in a figurative sense. The following proposition is an equivalent "integral" characterization of primitives using reachable sets of nonautonomous differential inclusions:

Proposition 57. *Suppose $\mathscr{F} : [0,T] \longrightarrow \big(\text{LIP}(\mathbb{R}^N, \mathbb{R}^N), d\!l_\infty\big)$ to be Lebesgue-measurable with $\sup_{t \in [0,T]} \big(\|\mathscr{F}(t)\|_\infty + \text{Lip}\, \mathscr{F}(t)\big) < \infty$ and define the set-valued map*

$$\widehat{F} : \; [0,T] \times \mathbb{R}^N \rightsquigarrow \mathbb{R}^N, \qquad (t,x) \mapsto \mathscr{F}(t)(x).$$

A tube $K : [0,T] \rightsquigarrow \mathbb{R}^N$ is a morphological primitive of $\mathscr{F}(\cdot)$ if and only if at every time $t \in [0,T]$, its value $K(t) \subset \mathbb{R}^N$ coincides with the reachable set of the nonautonomous differential inclusion $x' \in \widehat{F}(\cdot, x)$ a.e. (in the sense of Definition 48), i.e.

$$K(t) = \vartheta_{\widehat{F}}\big(t,\, K(0)\big).$$

Proof results directly from the uniqueness of primitives (Corollary 23 on page 41) and the following lemma about reachable sets:

Lemma 58. *Suppose* $\mathscr{F} : [0,T] \longrightarrow \big(\mathrm{LIP}(\mathbb{R}^N,\mathbb{R}^N), d\!l_\infty\big)$ *to be* $\mathscr{L}^1$*-measurable with* $C := \sup_{t\in[0,T]} \big(\|\mathscr{F}(t)\|_\infty + \mathrm{Lip}\,\mathscr{F}(t)\big) < \infty$ *and define the set-valued map*

$$\widehat{F}: \ [0,T]\times\mathbb{R}^N \leadsto \mathbb{R}^N, \qquad (t,x) \mapsto \mathscr{F}(t)(x).$$

Then for every initial set $K_0 \in \mathscr{K}(\mathbb{R}^N)$, *the reachable set of the nonautonomous differential inclusion* $x' \in \widehat{F}(\cdot,x)$ *a.e.*

$$\vartheta_{\widehat{F}}(\cdot,K_0) : \ [0,T] \longrightarrow \mathscr{K}(\mathbb{R}^N)$$

is a primitive of $\mathscr{F}(\cdot)$.

Proof. $\vartheta_{\widehat{F}}(\cdot, K_0) : [0,T] \longrightarrow \big(\mathscr{K}(\mathbb{R}^N), d\!l\big)$ is C-Lipschitz continuous because the bound $C<\infty$ of $\mathscr{F}(\cdot)$ implies $|v| \le C$ for all $t\in[0,T]$, $x\in\mathbb{R}^N$ and $v\in\widehat{F}(t,x)$.

Denote the pointwise convex hull of $\widehat{F}$ as $G: [0,T]\times\mathbb{R}^N \leadsto \mathbb{R}^N$, $(t,x)\mapsto \overline{\mathrm{co}}\ \widehat{F}(t,x)$. Then for Lebesgue-almost every $t\in[0,T]$, the set-valued map $G(t,\cdot):\mathbb{R}^N \leadsto \mathbb{R}^N$ is C-Lipschitz continuous with nonempty compact convex values and $\|G(t,\cdot)\|_\infty \le C$. For every $x\in\mathbb{R}^N$, the map $G(\cdot,x) : [0,T] \leadsto \mathbb{R}^N$ is measurable.
Furthermore Relaxation Theorem A.19 of Filippov-Ważewski (on page 453) implies

$$\vartheta_{\widehat{F}(t+\cdot,\cdot)}(h,K) \;=\; \vartheta_{G(t+\cdot,\cdot)}(h,K)$$

for every initial set $K\in\mathscr{K}(\mathbb{R}^N)$ and any $t,h\in[0,T]$ with $t+h\le T$.

According to Proposition A.13 (on page 447), there exists a set $J\subset[0,T]$ of full Lebesgue measure (i.e. $\mathscr{L}^1([0,T]\setminus J)=0$) such that at every time $t\in J$ and for any set $K_t\in\mathscr{K}(\mathbb{R}^N)$,

$$\tfrac{1}{h}\cdot d\!l\Big(\vartheta_{G(t+\cdot,\cdot)}(h,\,K_t),\ \ \bigcup_{x\in K_t}\big(x+h\cdot G(t,x)\big)\Big) \longrightarrow 0 \qquad \text{for } h\downarrow 0.$$

Applying the same Proposition A.13 to the autonomous differential inclusion with $G(t,\cdot):\mathbb{R}^N\leadsto\mathbb{R}^N$ and arbitrary $t\in[0,T]$, we obtain

$$\tfrac{1}{h}\cdot d\!l\Big(\vartheta_{G(t,\cdot)}(h,\,K_t),\ \ \bigcup_{x\in K_t}\big(x+h\cdot G(t,x)\big)\Big) \longrightarrow 0 \qquad \text{for } h\downarrow 0.$$

The triangle inequality of $d\!l$ implies for every $t\in J$ and $K_t\in\mathscr{K}(\mathbb{R}^N)$

$$\tfrac{1}{h}\cdot d\!l\Big(\vartheta_{G(t+\cdot,\cdot)}(h,\,K_t),\ \ \vartheta_{G(t,\cdot)}(h,\,K_t)\Big) \longrightarrow 0 \qquad \text{for } h\downarrow 0,$$

i.e. for $K_t := \vartheta_{\widehat{F}}(t,K_0)\in\mathscr{K}(\mathbb{R}^N)$ with an arbitrary initial set $K_0\in\mathscr{K}(\mathbb{R}^N)$:

$$\tfrac{1}{h}\cdot d\!l\Big(\vartheta_{\widehat{F}}(t+h,\,K_0),\ \ \vartheta_{\mathscr{F}(t)}\big(h,\ \vartheta_{\widehat{F}}(t,K_0)\big)\Big) \longrightarrow 0 \qquad \text{for } h\downarrow 0.$$

Thus, $\mathscr{F}(t)\in\mathrm{LIP}(\mathbb{R}^N,\mathbb{R}^N)$ belongs to the morphological mutation of $\vartheta_{\widehat{F}}(\cdot,K_0)$ at every time $t\in J$. □

1.9.4 Some Examples of Morphological Primitives

Proposition 57 (on page 64) has just provided an equivalent characterization of morphological primitives by means of reachable sets. This property can be very useful as the following tubes exemplify:

Example 59. For a Lipschitz continuous function $g : [0,T] \longrightarrow \mathbb{R}^N$, we consider the set-valued map (with just one element in each value)

$$K : [0,T] \rightsquigarrow \mathbb{R}^N, \quad t \mapsto \{g(t)\}.$$

Due to Rademacher's Theorem, there is a set $J \subset [0,T]$ of full Lebesgue measure (i.e. $\mathscr{L}^1([0,T] \setminus J) = 0$) such that $g(\cdot)$ is differentiable at every time $t \in J \subset [0,T]$. Now we can easily specify an element F_t of the mutation $\overset{\circ}{K}(t) \subset \mathrm{LIP}(\mathbb{R}^N, \mathbb{R}^N)$ for every $t \in J$: Choose *any* set-valued map $F_t \in \mathrm{LIP}(\mathbb{R}^N, \mathbb{R}^N)$ with

$$F_t(\cdot) \equiv \{g'(t)\} \subset \mathbb{R}^N$$

in some neighborhood $U_t \subset \mathbb{R}^N$ of $g(t)$. Indeed, the differentiability of $g(\cdot)$ at $t \in J$ implies for $h \downarrow 0$

$$\tfrac{1}{h} \cdot dl\big(K(t+h),\ \vartheta_{F_t}(h, K(t))\big) \;=\; \tfrac{1}{h} \cdot \big|g(t+h) - \big(g(t) + h \cdot g'(t)\big)\big| \;\longrightarrow\; 0.$$

Hence, $K(\cdot)$ is a primitive of any curve $F : [0,T] \longrightarrow \mathrm{LIP}(\mathbb{R}^N, \mathbb{R}^N)$, $t \longmapsto F_t$ with this feature close to $g(\cdot)$.

Example 60. Let $A : [0,T] \longrightarrow \mathbb{R}^{N \times N}$ be a continuous map of real matrices and $K_0 \in \mathscr{K}(\mathbb{R}^N)$. We focus on the morphological primitive $K(\cdot) : [0,T] \rightsquigarrow \mathbb{R}^N$ of

$$[0,T] \longrightarrow \mathrm{LIP}(\mathbb{R}^N, \mathbb{R}^N), \quad t \longmapsto A(t)\, \mathbb{I}\mathrm{d}_{\mathbb{R}^N}$$

with $K(0) = K_0$. Due to Proposition 57, $K(t) = \vartheta_{A(\cdot)\, \mathbb{I}\mathrm{d}_{\mathbb{R}^N}}(t, K_0)$. For simplifying this reachable set, let $\Phi(\cdot) : [0,T] \longrightarrow \mathbb{R}^{N \times N}$ denote the unique matrix-valued solution to the initial value problem

$$\begin{cases} \Phi'(t) = A(t)\, \Phi(t) & \text{for every } t \in [0,T] \\ \Phi(0) = \mathbb{I}\mathrm{d}_{\mathbb{R}^{N \times N}} \end{cases}$$

and the theory of linear differential equations implies immediately $K(t) = \Phi(t)\, K_0$ for every $t \in [0,T]$.

Example 61. Similarly to the preceding Example 60, let $A, B : [0,T] \longrightarrow \mathbb{R}^{N \times N}$ be two continuous maps of real matrices, $U \in \mathscr{K}(\mathbb{R}^N)$ convex and $K_0 \in \mathscr{K}(\mathbb{R}^N)$ given. Now we use Proposition 57 for determining the morphological primitive $K(\cdot)$ of

$$[0,T] \longrightarrow \mathrm{LIP}(\mathbb{R}^N, \mathbb{R}^N), \quad t \longmapsto A(t)\, \mathbb{I}\mathrm{d}_{\mathbb{R}^N} + B(t)\, U$$

with $K(0) = K_0$.

Using again the fundamental matrix $\Phi(\cdot) : [0,T] \longrightarrow \mathbb{R}^{N\times N}$ related to $A(\cdot)$, the well-known variation of constants formula implies for every $t \in [0,T]$

$$K(t) \;=\; \vartheta_{A(\cdot)\,\mathbb{I}\mathrm{d}_{\mathbb{R}^N}+B(\cdot)\,U}(t,K_0) \;=\; \Phi(t)\,K_0 \;+\; \int_0^t \Phi(t)\Phi(s)^{-1}\;B(s)\;U\;ds$$

with the set integral at the end to be understood in the sense of Aumann.

Example 62. The product of primitives is always a primitive of the product – in the following sense: For any two curves $F_1(\cdot), F_2(\cdot) : [0,T] \longrightarrow \mathrm{LIP}(\mathbb{R}^N,\mathbb{R}^N)$, let $K_j(\cdot) : [0,T] \longrightarrow \mathscr{K}(\mathbb{R}^N)$ denote a morphological primitive of $F_j(\cdot)$ for $j = 1,2$ respectively. Then

$$K_1 \times K_2 : [0,T] \longrightarrow \mathscr{K}(\mathbb{R}^N \times \mathbb{R}^N), \quad t \longmapsto K_1(t) \times K_2(t) \subset \mathbb{R}^N \times \mathbb{R}^N = \mathbb{R}^{2N}$$

is a morphological primitive of

$$F_1 \times F_2 : [0,T] \longrightarrow \mathrm{LIP}(\mathbb{R}^N \times \mathbb{R}^N,\; \mathbb{R}^N \times \mathbb{R}^N)$$

with $\quad (F_1 \times F_2)(t) :\; \mathbb{R}^N \times \mathbb{R}^N \rightsquigarrow \mathbb{R}^N \times \mathbb{R}^N, \quad (z_1,z_2) \mapsto F_1(z_1) \times F_2(z_2).$

Indeed, this property results from the representation of morphological primitives as reachable sets according to Proposition 57.
This example shows once more that mutations have useful features in regard to cartesian products. Essentially the same statement about primitives holds even for the product of metric spaces (and their transitions respectively) as we can conclude from the results of § 1.5 (and the proof of Theorem 26 on page 46, in particular).

1.9.5 Some Examples of Contingent Transition Sets

Considering mutational equations under state constraints, the contingent transition set plays an essential role. It was introduced in Definition 16 (on page 39) and, Nagumo's Theorem 19 (on page 40) uses it for conditions being sufficient and necessary for the existence of solutions under state constraints.

Now we consider the contingent transition set of a nonempty subset $\mathscr{V} \subset \mathscr{K}(\mathbb{R}^N)$. Using the morphological transitions on the metric space $(\mathscr{K}(\mathbb{R}^N), d\!\!l)$, its definition at $K \in \mathscr{V}$ can be reformulated as

$$\mathscr{T}_{\mathscr{V}}(K) \;\stackrel{\text{Def.}}{=}\; \big\{F \in \mathrm{LIP}(\mathbb{R}^N,\mathbb{R}^N) \;\big|\; \liminf_{h \downarrow 0} \tfrac{1}{h} \cdot \mathrm{dist}\big(\vartheta_F(h,K),\, \mathscr{V}\big) \;=\; 0\big\}$$

with $$\mathrm{dist}\big(\vartheta_F(h,K),\, \mathscr{V}\big) \;\stackrel{\text{Def.}}{=}\; \inf_{S \in \mathscr{V}} d\!\!l\big(\vartheta_F(h,K),\, S\big).$$

Corollary A.21 of Filippov-Ważewski Relaxation Theorem A.19 (on page 453) implies $\vartheta_F(t,K) = \vartheta_{\overline{\mathrm{co}}\,F}(t,K)$ for any $K \in \mathscr{K}(\mathbb{R}^N)$, $F \in \mathrm{LIP}(\mathbb{R}^N,\mathbb{R}^N)$ and $t \geq 0$ and thus, we can restrict our search for criteria to convex-valued maps in $\mathrm{LIP}(\mathbb{R}^N,\mathbb{R}^N)$.

Definition 63. Let $K \subset \mathbb{R}^N$ be a nonempty closed subset and $x \in K$.

$$T_K(x) \;:=\; \big\{v \in \mathbb{R}^N \;\big|\; \liminf_{h \downarrow 0} \tfrac{1}{h} \cdot \operatorname{dist}(x + h\,v,\, K) \;=\; 0\big\}$$

is called *contingent cone* to K at x (in the sense of Bouligand).

The *Clarke tangent cone* or *circatangent cone* $T_K^C(x)$ is defined (equivalently) by

$$\begin{aligned} T_K^C(x) &:= \operatorname{Liminf}_{\substack{h \downarrow 0,\\ y \xrightarrow[K]{} x}} \frac{K-y}{h} \\ &= \Big\{v \in X \;\Big|\; \forall\ h_n \downarrow 0,\ y_n \to x \text{ with } y_n \in K:\ \operatorname{dist}\big(v,\, \tfrac{K-y_n}{h_n}\big) \stackrel{n\to\infty}{\longrightarrow} 0\Big\} \\ &= \Big\{v \in X \;\Big|\; \forall\ h_n \downarrow 0,\ y_n \to x \text{ with } y_n \in K:\ \frac{\operatorname{dist}(y_n + h_n \cdot v,\, K)}{h_n} \stackrel{n\to\infty}{\longrightarrow} 0\Big\}. \end{aligned}$$

Example 64. For a fixed nonempty closed subset $M \subset \mathbb{R}^N$, define

$$\mathscr{V}_{\subset M} := \big\{K \in \mathscr{K}(\mathbb{R}^N) \;\big|\; K \subset M\big\}.$$

Following the arguments of Anne Gorre [89], we can characterize the contingent transition setl $\mathscr{T}_{\mathscr{V}_{\subset M}}(K) \subset \operatorname{LIP}(\mathbb{R}^N, \mathbb{R}^N)$ for each $K \in \mathscr{V}_{\subset M}$:

$$\mathscr{T}_{\mathscr{V}_{\subset M}}(K) \;=\; \big\{F \in \operatorname{LIP}(\mathbb{R}^N, \mathbb{R}^N) \;\big|\; \forall\, x \in K: \overline{\operatorname{co}}\, F(x) \subset T_M(x)\big\}.$$

For proving "$\subset$" choose any set-valued map $F \in \mathscr{T}_{\mathscr{V}_{\subset M}}(K) \subset \operatorname{LIP}(\mathbb{R}^N, \mathbb{R}^N)$. Then the definition of $\mathscr{T}_{\mathscr{V}_{\subset M}}(K)$ provides two sequences $(h_n)_{n\in\mathbb{N}}$, $(K_n)_{n\in\mathbb{N}}$ in $]0,1[$ and $\mathscr{V}_{\subset M} \subset \mathscr{K}(\mathbb{R}^N)$ respectively satisfying for each $n \in \mathbb{N}$

$$h_n \leq \tfrac{1}{n}, \qquad \tfrac{1}{h_n} \cdot d\!\!l\big(\vartheta_F(h_n, K),\, K_n\big) \;\leq\; \tfrac{1}{n}.$$

For each point $x \in K$ and velocity $v \in \overline{\operatorname{co}}\, F(x)$, we have to verify $v \in T_M(x)$. Due to Filippov's Theorem A.6, there exists a solution $x(\cdot) \in W^{1,1}([0,T], \mathbb{R}^N)$ to the differential inclusion $x'(\cdot) \in \overline{\operatorname{co}}\, F(x(\cdot))$ a.e. with $x(0) = x$ and the additional property that $x(\cdot)$ is differentiable at $t = 0$ with $x'(0) = v$ (e.g. [14, Corollary 5.3.2]). For each $n \in \mathbb{N}$, select $y_n \in K_n \subset M$ with

$$\big|x(h_n) - y_n\big| \;=\; \operatorname{dist}(x(h_n),\, K_n) \;\leq\; d\!\!l\big(\vartheta_{\overline{\operatorname{co}}\,F}(h_n, K), K_n\big) \;=\; d\!\!l\big(\vartheta_F(h_n, K), K_n\big) \;\leq\; \tfrac{h_n}{n}.$$

Then, we obtain

$$\begin{aligned} \tfrac{1}{h_n} \cdot \operatorname{dist}\big(x + h_n\, v,\, M\big) \;&\leq\; \tfrac{1}{h_n} \cdot \big|x + h_n\, v - x(h_n)\big| \;+\; \tfrac{1}{h_n} \cdot \big|x(h_n) \;-\; y_n\big| \\ &\leq\; \Big|v \;-\; \frac{x(h_n) - x}{h_n}\Big| \qquad\quad +\; \tfrac{1}{n} \qquad\qquad \longrightarrow 0 \end{aligned}$$

for $n \longrightarrow \infty$, i.e. $v \in T_M(x)$.

In regard to the inclusion "$\supset$", let $F \in \operatorname{LIP}(\mathbb{R}^N, \mathbb{R}^N)$ satisfy $\overline{\operatorname{co}}\, F(x) \subset T_M(x)$ for every $x \in K$. The Invariance Theorem about differential inclusions (Proposition A.8 on page 445) ensures that *every* solution $x(\cdot) \in W^{1,1}([0,1], \mathbb{R}^N)$ of $x'(\cdot) \in \overline{\operatorname{co}}\, F(x(\cdot))$ with $x(0) \in K$ has all its values in $M \subset \mathbb{R}^N$ and thus,

$$\vartheta_F(h, K) \;\subset\; \vartheta_{\overline{\operatorname{co}}\,F}(h, K) \;\subset\; M$$

for every $h \in [0,1]$. In particular, $\operatorname{dist}\big(\vartheta_F(h,K),\, \mathscr{V}_{\subset M}\big) \;=\; 0$ for all $h \in [0,1]$, i.e. $F \in \mathscr{T}_{\mathscr{V}_{\subset M}}(K)$. This completes the proof of the preceding characterization of the contingent transition set $\mathscr{T}_{\mathscr{V}_{\subset M}}(K)$ for any nonempty closed subset $M \subset \mathbb{R}^N$.

This Example 64 focuses on a subset $\mathscr{V}_{\subset M}$ of the metric space $\big(\mathscr{K}(\mathbb{R}^N), d\!l\big)$ prescribing a condition on just *one* compact set. Mutational equations, however, have the important advantage that many existence results can be extended to systems as explained in § 1.5 (on page 44). For this reason, we consider now some examples with tuples of two or even three compact sets.

Strictly speaking, the product $\mathscr{K}(\mathbb{R}^N)^2 := \mathscr{K}(\mathbb{R}^N) \times \mathscr{K}(\mathbb{R}^N)$ is supplied with the metric

$$\begin{aligned} d\!l_2 : \big(\mathscr{K}(\mathbb{R}^N) \times \mathscr{K}(\mathbb{R}^N)\big) \times \big(\mathscr{K}(\mathbb{R}^N) \times \mathscr{K}(\mathbb{R}^N)\big) &\longrightarrow [0,\infty[, \\ \big((K_1,K_2),\ \ (L_1,L_2)\big) &\longmapsto d\!l(K_1,L_1) + d\!l(K_2,L_2) \end{aligned}$$

and, the product of maps in $\mathrm{LIP}(\mathbb{R}^N \times \mathbb{R}^N, \mathbb{R}^N)$ serve as transitions, i.e. for any tuple $(F,G) \in \mathrm{LIP}(\mathbb{R}^N \times \mathbb{R}^N, \mathbb{R}^N) \times \mathrm{LIP}(\mathbb{R}^N \times \mathbb{R}^N, \mathbb{R}^N)$ define

$$\begin{aligned} \vartheta_{(F,G)} : [0,1] \times \mathscr{K}(\mathbb{R}^N)^2 &\longrightarrow \mathscr{K}(\mathbb{R}^N)^2 \\ \big(h,\ \ (K_1,K_2)\big) &\longmapsto \big\{(x(h),\, y(h)) \;\big|\; \exists\, x(\cdot), y(\cdot) \in W^{1,1}([0,h], \mathbb{R}^N) : \\ &\qquad\qquad x(0) \in K_1, \quad y(0) \in K_2, \\ &\qquad\qquad x' \in F(x,y),\ \ y' \in G(x,y) \text{ a.e.}\big\} \end{aligned}$$

Indeed, the transition properties of $\vartheta_{(F,G)}(\cdot,\cdot)$ result from Filippov's Theorem about differential inclusions for the same reasons as Proposition 53 (on page 62).
Similarly to Example 64, Anna Gorre has already used the so-called paratingent cones (of Bouligand) and characterized the contingent transition sets of

$$\mathscr{V}_{\cap} := \big\{(K,L) \in \mathscr{K}(\mathbb{R}^N)^2 \;\big|\; K \cap L \neq \emptyset\big\} :$$

Definition 65. Let $K, L \subset \mathbb{R}^N$ be nonempty closed subsets and $x \in K \cap L$.

$$P_L^K(x) \;:=\; \Big\{v \in \mathbb{R}^N \;\Big|\; \liminf_{\substack{h \downarrow 0 \\ y \to x\ (y \in K)}} \tfrac{1}{h} \cdot \mathrm{dist}(y + h\,v,\, L) \;=\; 0\Big\}$$

is called *Bouligand paratingent cone* to L relative to K at x.
Furthermore, the *adjacent cone* to K at x (in the sense of Bouligand) is defined as

$$T_K^\flat(x) \;:=\; \Big\{v \in \mathbb{R}^N \;\Big|\; \lim_{h \downarrow 0} \tfrac{1}{h} \cdot \mathrm{dist}(x + h\,v,\, K) \;=\; 0\Big\}.$$

Proposition 66 (Gorre [10, Theorem 4.2.4], [90]**).**

$$\mathscr{V}_{\cap} := \big\{(K,L) \in \mathscr{K}(\mathbb{R}^N)^2 \;\big|\; K \cap L \neq \emptyset\big\}$$

is a closed subset of $\big(\mathscr{K}(\mathbb{R}^N)^2, d\!l_2\big)$.
For any tuples $(K,L) \in \mathscr{V}_{\cap}$ *and* $(F,G) \in \mathrm{LIP}(\mathbb{R}^N \times \mathbb{R}^N, \mathbb{R}^N)^2$, *the following two statements are equivalent:*

1. (F,G) *belongs to the contingent transition set of* $\mathscr{V}_{\cap}$ *at* (K,L).
2. *There exists* $x \in K \cap L \subset \mathbb{R}^N$ *with* $\big(\overline{\mathrm{co}}\, F(x,x) - \overline{\mathrm{co}}\, G(x,x)\big) \cap P_L^K(x) \neq \emptyset$.

For the similar characterization related to

$$\mathscr{V}_{\subset} := \{(K,L) \in \mathscr{K}(\mathbb{R}^N)^2 \mid K \subset L\},$$

we prefer the simpler transitions on $\mathscr{K}(\mathbb{R}^N)^2$ that are induced by two *decoupled* differential inclusions and thus specified by tuples in $\mathrm{LIP}(\mathbb{R}^N,\mathbb{R}^N) \times \mathrm{LIP}(\mathbb{R}^N,\mathbb{R}^N)$.

Proposition 67 (Gorre). *$\mathscr{V}_{\subset}$ is closed in $\big(\mathscr{K}(\mathbb{R}^N)^2, d\!\!l_2\big)$.*
Whenever $(F,G) \in \mathrm{LIP}(\mathbb{R}^N,\mathbb{R}^N)^2$ belongs to the contingent transition set of $\mathscr{V}_{\subset}$ at (K,L), then

$$\overline{\mathrm{co}}\, F(x) \subset \overline{\mathrm{co}}\, G(x) + T_L(x).$$

holds for every $x \in K$.
For every $(K,L) \in \mathscr{V}_{\subset}$, the tuple $(F,G) \in \mathrm{LIP}(\mathbb{R}^N,\mathbb{R}^N)^2$ belongs the contingent transition set of $\mathscr{V}_{\subset}$ at (K,L) if every $x \in K$ satisfies the inclusion

$$\overline{\mathrm{co}}\, F(x) \subset \overline{\mathrm{co}}\, G(x) + T_L^C(x).$$

This implication is a special case of the following statement considering tuples of three compact sets. Strictly speaking, $\mathscr{K}(\mathbb{R}^N)^3 \stackrel{\text{Def.}}{=} \mathscr{K}(\mathbb{R}^N) \times \mathscr{K}(\mathbb{R}^N) \times \mathscr{K}(\mathbb{R}^N)$ is now supplied with the distance

$$\begin{array}{rcl} d\!\!l_3: \quad \mathscr{K}(\mathbb{R}^N)^3 \times \mathscr{K}(\mathbb{R}^N)^3 & \longrightarrow & [0,\infty[, \\ \big((K_1,K_2,K_3),\ (L_1,L_2,L_3)\big) & \longmapsto & d\!\!l(K_1,L_1) + d\!\!l(K_2,L_2) + d\!\!l(K_3,L_3) \end{array}$$

and, tuples of three morphological transitions serve as transitions on the metric space $(\mathscr{K}(\mathbb{R}^N)^3, d\!\!l_3)$ – following the notion of Lemma 27 (on page 45). This is equivalent to considering reachable sets of three decoupled differential inclusions. The original form of the next proposition also goes back to Anne Gorre [90] and concerns contingent transition sets as formulated in [10, Theorem 4.2.8]. Here, however, we prefer a slightly stronger assumption in statement (2.) so that we can specify sufficient conditions for the "adjacent transition set". This feature will be useful for verifying Proposition 69 below. Proofs are again postponed to the end of this section.

Proposition 68 (Gorre). *The subset*

$$\mathscr{V}_{\subset\cap} := \{(K,L,M) \in \mathscr{K}(\mathbb{R}^N)^3 \mid K \subset L \cap M\}$$

is closed in the metric space $\big(\mathscr{K}(\mathbb{R}^N)^3,\ d\!\!l_3\big)$. Furthermore,

1. *If $(F,G,H) \in \mathrm{LIP}(\mathbb{R}^N,\mathbb{R}^N)^3$ belongs to the contingent transition set of $\mathscr{V}_{\subset\cap}$ at $(K,L,M) \in \mathscr{V}_{\subset\cap}$ then*
$$\overline{\mathrm{co}}\, F(z) + T_K^\flat(z) \subset \big(\overline{\mathrm{co}}\, G(z) + T_L(z)\big) \cap \big(\overline{\mathrm{co}}\, H(z) + T_M(z)\big) \quad \text{for every } z \in K.$$

2. *If $(F,G,H) \in \mathrm{LIP}(\mathbb{R}^N,\mathbb{R}^N)^3$ satisfies*
$$\overline{\mathrm{co}}\, F(z) \subset \big(\overline{\mathrm{co}}\, G(z) + T_L^C(z)\big) \cap \big(\overline{\mathrm{co}}\, H(z) + T_M^C(z)\big) \quad \text{for every } z \in K$$
then (F,G,H) even fulfills $\lim_{h\downarrow 0} \frac{1}{h} \cdot \mathrm{dist}\big(\vartheta_{(F,G,H)}(h,\ (K,L,M)),\ \mathscr{V}_{\subset\cap}\big) = 0$ and so, (F,G,H) belongs to the contingent transition set of $\mathscr{V}_{\subset\cap}$ at $(K,L,M) \in \mathscr{V}_{\subset\cap}$.

Finally we extend this list of Gorre's earlier results by considering a further set of constraints in detail:

$$\mathscr{V}_{\cap,\cup\subset} := \left\{(K,L,M) \in \mathscr{K}(\mathbb{R}^N)^3 \;\middle|\; K\cap L \neq \emptyset,\; K\cup L \subset M\right\}$$

Proposition 69. *The subset $\mathscr{V}_{\cap,\cup\subset} \subset \mathscr{K}(\mathbb{R}^N)^3$ is closed with respect to $d\!l_3$. Moreover,*

1. *If $(F,G,H) \in \mathrm{LIP}(\mathbb{R}^N,\mathbb{R}^N)^3$ belongs to the contingent transition set of $\mathscr{V}_{\cap,\cup\subset}$ at $(K,L,M) \in \mathscr{V}_{\cap,\cup\subset}$ then*
$$\begin{cases} \emptyset \neq \left(\overline{\mathrm{co}}\, F(x) - \overline{\mathrm{co}}\, G(x)\right) \cap P_L^K(x) & \text{for some } x \in K\cap L,\\ \overline{\mathrm{co}}\, F(z) \subset \overline{\mathrm{co}}\, H(z) + T_M(z) & \text{for every } z \in K,\\ \overline{\mathrm{co}}\, G(z) \subset \overline{\mathrm{co}}\, H(z) + T_M(z) & \text{for every } z \in L. \end{cases}$$
2. *If $(F,G,H) \in \mathrm{LIP}(\mathbb{R}^N,\mathbb{R}^N)^3$ satisfies*
$$\begin{cases} \emptyset \neq \left(\overline{\mathrm{co}}\, F(x) - \overline{\mathrm{co}}\, G(x)\right) \cap P_L^K(x) & \text{for some } x \in K\cap L,\\ \overline{\mathrm{co}}\, F(z) \subset \overline{\mathrm{co}}\, H(z) + T_M^C(z) & \text{for every } z \in K,\\ \overline{\mathrm{co}}\, G(z) \subset \overline{\mathrm{co}}\, H(z) + T_M^C(z) & \text{for every } z \in L, \end{cases}$$
then (F,G,H) belongs to the contingent transition set of $\mathscr{V}_{\cap,\cup\subset}$ at $(K,L,M) \in \mathscr{V}_{\cap,\cup\subset}$.

Proof (of Proposition 68). First, we verify that $\mathscr{V}_{\subset\cap}$ is closed in the metric space $\left(\mathscr{K}(\mathbb{R}^N)^3,\, d\!l_3\right)$. Let $\left((K_n,L_n,M_n)\right)_{n\in\mathbb{N}}$ be any sequence in $\mathscr{V}_{\subset\cap}$ that converges to a tuple $(K,L,M) \in \mathscr{K}(\mathbb{R}^N)^3$ with respect to $d\!l_3$. Then for every $\varepsilon > 0$, there exists an index $n_\varepsilon \in \mathbb{N}$ with $K \subset \mathbb{B}_\varepsilon(K_{n_\varepsilon})$, $L_{n_\varepsilon} \subset \mathbb{B}_\varepsilon(L)$, $M_{n_\varepsilon} \subset \mathbb{B}_\varepsilon(M)$ and thus,

$$K \subset \mathbb{B}_\varepsilon(K_{n_\varepsilon}) \subset \mathbb{B}_\varepsilon(L_{n_\varepsilon} \cap M_{n_\varepsilon}) \subset \mathbb{B}_\varepsilon\big(\mathbb{B}_\varepsilon(L) \cap \mathbb{B}_\varepsilon(M)\big).$$

As K,L,M are nonempty compact subsets of $\mathbb{R}^N$, the limit for $\varepsilon \downarrow 0$ provides $K \subset L\cap M$ (indirectly), i.e. $(K,L,M) \in \mathscr{V}_{\subset\cap}$.

(1.) Assume that $(F,G,H) \in \mathrm{LIP}(\mathbb{R}^N,\mathbb{R}^N)^3$ belongs to the contingent transition set of $\mathscr{V}_{\subset\cap}$ at $(K,L,M) \in \mathscr{V}_{\subset\cap}$. Without loss of generality, all values of F,G,H are supposed to be convex in addition. Now fix any $z \in K$, $v \in F(z)$ and $u \in T_K^\flat(z)$. We show that $v+u \in \big(G(z)+T_L(z)\big) \cap \big(H(z)+T_M(z)\big)$.
According to Definition 16 of contingent transition sets, there exist sequences $(h_n)_{n\in\mathbb{N}}$, $(\varepsilon_n)_{n\in\mathbb{N}}$, $\left((K_n,L_n,M_n)\right)_{n\in\mathbb{N}}$ in $\mathbb{R}_0^+$ and $\mathscr{V}_{\subset\cap}$ respectively with $\varepsilon_n + h_n \leq \frac{1}{n}$,

$$d\!l\left(\vartheta_F(h_n,K),\, K_n\right) + d\!l\left(\vartheta_G(h_n,L),\, L_n\right) + d\!l\left(\vartheta_H(h_n,M),\, M_n\right) \leq \varepsilon_n\, h_n$$

for all $n \in \mathbb{N}$. Moreover, there is a sequence $(u_n)_{n\in\mathbb{N}}$ converging to $u \in T_K^\flat(z)$ such that $z + h_n\, u_n \in K$ holds for every $n \in \mathbb{N}$. Filippov's Theorem A.6 ensures a solution $z_n(\cdot) \in W^{1,1}([0,1],\mathbb{R}^N)$ of $z_n'(\cdot) \in F(z_n(\cdot))$ $\mathscr{L}^1$-a.e. with $z_n(0) = z + h_n\, u_n \in K$ and

$$\left|z_n(t) - (z + h_n\, u_n + t\, v)\right| \leq \tfrac{|v|}{\mathrm{Lip}\, F}\left(e^{\mathrm{Lip}\, F\cdot t} - 1 - \mathrm{Lip}\, F\cdot t\right) \leq \mathrm{const}(F)\cdot t^2.$$

Then, we obtain

$$\begin{aligned} z_n(h_n) \in \vartheta_F(h_n,K) &\subset K_n + \varepsilon_n\, h_n\, \mathbb{B} \\ &\subset L_n \cap M_n + \varepsilon_n\, h_n\, \mathbb{B} \\ &\subset \big(\mathbb{B}_{\varepsilon_n h_n}\big(\vartheta_G(h_n,L)\big) \cap \mathbb{B}_{\varepsilon_n h_n}\big(\vartheta_H(h_n,M)\big)\big) + \varepsilon_n\, h_n\, \mathbb{B} \end{aligned}$$

for every $n \in \mathbb{N}$ and, we can select $x_n \in L,\ y_n \in M$ with

$$z_n(h_n) \in \big(\mathbb{B}_{\varepsilon_n h_n}\big(\vartheta_G(h_n,x_n)\big) \cap \mathbb{B}_{\varepsilon_n h_n}\big(\vartheta_H(h_n,y_n)\big)\big) + \varepsilon_n\, h_n\, \mathbb{B}.$$

In particular, $(x_n)_{n\in\mathbb{N}}$ and $(y_n)_{n\in\mathbb{N}}$ have the limit $z \in K$ in common with $\big(z_n(h_n)\big)_{n\in\mathbb{N}}$. Due to $\varepsilon_n \downarrow 0$, Lemma A.14 (on page 448) implies

$$\begin{cases} \operatorname{Limsup}_{n\to\infty} \ \frac{z_n(h_n) - x_n}{h_n} \subset G(z), \\ \operatorname{Limsup}_{n\to\infty} \ \frac{z_n(h_n) - y_n}{h_n} \subset H(z). \end{cases}$$

Choosing subsequences for any accumulation points $\widehat{g} := \lim\limits_{k\to\infty} \frac{z_{n_k}(h_{n_k}) - x_{n_k}}{h_{n_k}} \in G(z)$ and $\widehat{h} := \lim\limits_{k\to\infty} \frac{z_{n_k}(h_{n_k}) - y_{n_k}}{h_{n_k}} \in H(z)$, the following limits exist

$$\begin{cases} \lim\limits_{k\to\infty} \frac{x_{n_k} - z}{h_{n_k}} = \lim\limits_{k\to\infty} \left(\frac{x_{n_k} - z_{n_k}(h_{n_k})}{h_{n_k}} + \frac{z_{n_k}(h_{n_k}) - z}{h_{n_k}} \right) = -\widehat{g} + u + v \in T_L(z), \\ \lim\limits_{k\to\infty} \frac{y_{n_k} - z}{h_{n_k}} = \lim\limits_{k\to\infty} \left(\frac{y_{n_k} - z_{n_k}(h_{n_k})}{h_{n_k}} + \frac{z_{n_k}(h_{n_k}) - z}{h_{n_k}} \right) = -\widehat{h} + u + v \in T_M(z). \end{cases}$$

(2.) Now suppose $(F,G,H) \in \mathrm{LIP}(\mathbb{R}^N,\mathbb{R}^N)^3$ to have convex values with

$$F(z) \subset \big(G(z) + T_L^C(z)\big) \cap \big(H(z) + T_M^C(z)\big) \qquad \text{for every } z \in K \subset L \cap M.$$

Set $\gamma := \max\big\{\|F\|_\infty + \operatorname{Lip} F,\ \|G\|_\infty + \operatorname{Lip} G,\ \|H\|_\infty + \operatorname{Lip} H\big\}$ as an abbreviation. Fix $\varepsilon > 0$ arbitrarily. Lemmas A.14 and A.17 (on pages 448, 450 respectively) state that for every $x \in K$ there exists $\rho = \rho(\varepsilon,x) \in\,]0,\varepsilon[$ such that

$$\begin{cases} \vartheta_F(h,y) \subset y + h\cdot F(x) + \varepsilon\, h\, \mathbb{B} \\ y + h\cdot G(y) + h\cdot \big(T_L^C(x)\cap \mathbb{B}_{2\gamma}\big) \subset \vartheta_G(h,L) \quad + \varepsilon\, h\, \mathbb{B} \\ y + h\cdot H(y) + h\cdot \big(T_M^C(x)\cap \mathbb{B}_{2\gamma}\big) \subset \vartheta_H(h,M) \quad + \varepsilon\, h\, \mathbb{B} \end{cases}$$

for all $y \in K \cap \mathbb{B}_\rho(x)$ and $h \in [0,\rho]$. Now finitely many points $x_1 \ldots x_k \in K$ suffice for an open cover of set $K \in \mathscr{K}(\mathbb{R}^N)$:

$$K \subset \bigcup_{j=1\ldots k} \mathbb{B}_{\rho(\varepsilon,x_j)}(x_j)^\circ.$$

For each $y \in K$, let $j(y) \in \{1 \ldots k\}$ denote an index with $\big|y - x_{j(y)}\big| < \rho(\varepsilon,x_j) < \varepsilon$. Then we obtain for all positive $h < \min\big\{\rho(\varepsilon,x_j)\,\big|\, 1 \le j \le k\big\} < \varepsilon$

$$\begin{aligned} &\operatorname{dist}\big(\vartheta_F(h,K),\ \vartheta_G(h,L)\big) \\ &\quad \le \operatorname{dist}\Big(\bigcup_{y\in K} \big(y + h\cdot F(x_{j(y)})\big), \vartheta_G(h,L)\Big) + \varepsilon\, h \\ &\quad \le \operatorname{dist}\Big(\bigcup_{y\in K} \big(y + h\cdot (G(x_{j(y)}) + (T_L^C(x_{j(y)})\cap \mathbb{B}_{2\gamma}))\big),\ \vartheta_G(h,L)\Big) + \varepsilon\, h\ , \end{aligned}$$

$$\begin{aligned}
&\mathrm{dist}\big(\vartheta_F(h,K),\ \vartheta_G(h,L)\big)\\
&\quad \le \mathrm{dist}\Big(\bigcup_{y\in K}\Big(y+h\cdot\big(G(y)\ +\ (T_L^C(x_{j(y)})\cap\mathbb{B}_{2\gamma})\big)\Big),\ \vartheta_G(h,L)\Big)+\ (\gamma+1)\,\varepsilon\,h\\
&\quad \le \varepsilon\,h+(\gamma+1)\,\varepsilon\,h.
\end{aligned}$$

The corresponding inequality for H and M implies for all $h>0$ sufficiently small

$$\vartheta_F(h,K)\ \subset\ \mathbb{B}_{(\gamma+2)\,\varepsilon\,h}\big(\vartheta_G(h,L)\big)\ \cap\ \mathbb{B}_{(\gamma+2)\,\varepsilon\,h}\big(\vartheta_H(h,M)\big)$$

and thus,

$$\lim_{h\downarrow 0}\ \tfrac{1}{h}\cdot\ \mathrm{dist}\,\big(\vartheta_{(F,G,H)}\big(h,\ (K,L,M)\big),\ \mathcal{V}_{\subset\cap}\big)\ =\ 0. \qquad \square$$

Proof (of Proposition 69). The set $\mathcal{V}_{\cap,\cup\subset}\subset\mathcal{K}(\mathbb{R}^N)^3$ can be regarded as an intersection of three sets similar to the types investigated by Gorre:

$$\begin{aligned}
\mathcal{V}_{\cap,\cup\subset}\ =\ &\big(\{(K,L)\in\mathcal{K}(\mathbb{R}^N)^2\ \big|\ K\cap L\neq\emptyset\}\times\mathcal{K}(\mathbb{R}^N)\big)\\
&\cap\ \big(\mathcal{K}(\mathbb{R}^N)\times\{(L,M)\in\mathcal{K}(\mathbb{R}^N)^2\ \big|\ L\subset M\}\big)\\
&\cap\ \big\{(K,L,M)\in\mathcal{K}(\mathbb{R}^N)^3\ \big|\ K\subset M,\quad L\in\mathcal{K}(\mathbb{R}^N)\ \text{arbitrary}\big\}
\end{aligned}$$

As each of these three sets is closed w.r.t. $d\!l_3$, so is their intersection $\mathcal{V}_{\cap,\cup\subset}$.

(1.) According to Proposition 33 (c) (on page 51), the contingent transition set of an intersection is contained in the intersection of the contingent transition sets. Statement (1.) thus results from Gorre's characterizations in Proposition 66 (just with the restricted class of transitions in $\mathrm{LIP}(\mathbb{R}^N,\mathbb{R}^N)^2$ instead of $\mathrm{LIP}(\mathbb{R}^N\times\mathbb{R}^N,\mathbb{R}^N)^2$) and Proposition 67 respectively.

(2.) As a consequence of Proposition 66, the tuple $(F,G)\in\mathrm{LIP}(\mathbb{R}^N,\mathbb{R}^N)^2$ is contingent to $\mathcal{V}_\cap$ at (K,L). Hence there exist sequences $(h_n)_{n\in\mathbb{N}}$, $\big((K_n,L_n)\big)_{n\in\mathbb{N}}$ in $]0,1[$ and $\mathcal{K}(\mathbb{R}^N)^2$ respectively satisfying for all $n\in\mathbb{N}$

$$h_n\ \le\ \tfrac{1}{n},\quad K_n\cap L_n\neq\emptyset,\quad d\!l\big(\vartheta_F(h_n,K),\ K_n\big)\ +\ d\!l\big(\vartheta_G(h_n,L),\ L_n\big)\ \le\ \tfrac{h_n}{n}.$$

This implies $\mathbb{B}_{\frac{h_n}{n}}\big(\vartheta_F(h_n,K)\big)\cap\mathbb{B}_{\frac{h_n}{n}}\big(\vartheta_G(h_n,L)\big)\neq\emptyset$ for every $n\in\mathbb{N}$.
In the proof of Proposition 68, we have just concluded from

$$\begin{cases} F(z)\subset H(z)\ +\ T_M^C(z) & \text{for every } z\in K,\\ G(z)\subset H(z)\ +\ T_M^C(z) & \text{for every } z\in L\end{cases}$$

that for every $\varepsilon>0$, all sufficiently small $h>0$ fulfill

$$\begin{cases} \vartheta_F(h,K)\ \subset\ \mathbb{B}_{\varepsilon\,h}\big(\vartheta_H(h,M)\big)\\ \vartheta_G(h,L)\ \subset\ \mathbb{B}_{\varepsilon\,h}\big(\vartheta_H(h,M)\big)\end{cases}$$

Hence, the inclusion

$$\mathbb{B}_{\frac{h_n}{n}}\big(\vartheta_F(h_n,K)\big)\cup\mathbb{B}_{\frac{h_n}{n}}\big(\vartheta_G(h_n,L)\big)\ \subset\ \mathbb{B}_{(\varepsilon+\frac{1}{n})\,h_n}\big(\vartheta_H(h_n,M)\big)$$

holds for all large $n\in\mathbb{N}$ depending on $\varepsilon>0$. An appropriate subsequence for $\varepsilon\downarrow 0$ clarifies that (F,G,H) belongs to the contingent transition set of $\mathcal{V}_{\cap,\cup\subset}$ at $(K,L,M)\in\mathcal{V}_{\cap,\cup\subset}$. $\square$

1.9.6 Solutions to Morphological Equations

Now we apply the rather general results about mutational equations to the metric space $(\mathscr{K}(\mathbb{R}^N), d\!\!l)$ and the morphological transitions (represented by the set-valued maps in $\text{LIP}(\mathbb{R}^N, \mathbb{R}^N)$).
Let $\mathscr{F} : \mathscr{K}(\mathbb{R}^N) \times [0,T] \longrightarrow \text{LIP}(\mathbb{R}^N, \mathbb{R}^N)$ be given. According to Definition 13 (on page 38), a compact-valued tube $K(\cdot) : [0,T] \leadsto \mathbb{R}^N$ is a solution to the so-called morphological equation

$$\overset{\circ}{K}(\cdot) \ni \mathscr{F}(K(\cdot), \cdot)$$

if (and only if) $K(\cdot)$ is a morphological primitive of the composition

$$\mathscr{F}(K(\cdot), \cdot) : [0,T] \longrightarrow \text{LIP}(\mathbb{R}^N, \mathbb{R}^N),$$

i.e. $K(\cdot)$ is Lipschitz continuous with respect to $d\!\!l$ and satisfies

$$\lim_{h \downarrow 0} \tfrac{1}{h} \cdot d\!\!l\big(\vartheta_{\mathscr{F}(K(t),t)}(h, K(t)),\ \ K(t+h)\big) = 0$$

at Lebesgue-almost every time $t \in [0,T]$.
Proposition 57 (on page 64) has already provided an equivalent characterization of morphological primitives:

Proposition 70 (Solutions to morphological equations as reachable sets). *Suppose* $\mathscr{F} : \big(\mathscr{K}(\mathbb{R}^N), d\!\!l\big) \times [0,T] \longrightarrow \big(\text{LIP}(\mathbb{R}^N, \mathbb{R}^N), d\!\!l_\infty\big)$ *to be a Carathéodory function (i.e. here continuous with respect to the first argument and measurable with respect to time) satisfying*

$$\sup_{\substack{M \in \mathscr{K}(\mathbb{R}^N) \\ t \in [0,T]}} \big(\|\mathscr{F}(M,t)\|_\infty + \text{Lip}\,\mathscr{F}(M,t)\big) < \infty.$$

Then a continuous tube $K : [0,T] \leadsto \mathbb{R}^N$ *is a solution to the morphological equation*

$$\overset{\circ}{K}(\cdot) \ni \mathscr{F}(K(\cdot), \cdot)$$

if and only if at every time $t \in [0,T]$, *the set* $K(t) \subset \mathbb{R}^N$ *coincides with the reachable set of the initial set* $K(0) \subset \mathbb{R}^N$ *and the nonautonomous differential inclusion*

$$x'(\cdot) \in \mathscr{F}\big(K(\cdot), \cdot\big)\,(x(\cdot)).$$

Proof. Suppose the tube $K(\cdot) : [0,T] \leadsto \mathbb{R}^N$ to be continuous. As a consequence of the Carathéodory property of $\mathscr{F}(\cdot,\cdot)$, the composition

$$\mathscr{F}(K(\cdot), \cdot) : [0,T] \longrightarrow \text{LIP}(\mathbb{R}^N, \mathbb{R}^N)$$

is always measurable and thus, we can conclude the claimed equivalence directly from Proposition 57. □

First we focus on the initial value problem of morphological equations *without* state constraints:

Proposition 71 (Peano's Theorem for morphological equations).
Suppose $\mathscr{F} : \big(\mathscr{K}(\mathbb{R}^N), d\!\!l\big) \times [0,T] \longrightarrow \big(\mathrm{LIP}(\mathbb{R}^N,\mathbb{R}^N), d\!\!l_\infty\big)$ *to be continuous*

$$\sup_{M \in \mathscr{K}(\mathbb{R}^N),\, t \in [0,T]} \big(\|\mathscr{F}(M,t)\|_\infty + \mathrm{Lip}\,\mathscr{F}(M,t)\big) < \infty.$$

Then for every initial set $K_0 \in \mathscr{K}(\mathbb{R}^N)$, *there exists a solution* $K : [0,T] \rightsquigarrow \mathbb{R}^N$ *to the morphological equation*

$$\overset{\circ}{K}(\cdot) \ni \mathscr{F}\big(K(\cdot), \cdot\big)$$

with $K(0) = K_0$.

Proof results directly from Theorem 20 (on page 40) in combination with Proposition 47 (on page 57) and Proposition 53 (on page 62). □

Proposition 72 (Cauchy-Lipschitz Theorem for morphological equations).
Suppose the continuous function $\mathscr{F} : \big(\mathscr{K}(\mathbb{R}^N), d\!\!l\big) \times [0,T] \longrightarrow \big(\mathrm{LIP}(\mathbb{R}^N,\mathbb{R}^N), d\!\!l_\infty\big)$ *to be Lipschitz continuous in the first argument with*

$$\sup_{M \in \mathscr{K}(\mathbb{R}^N),\, t \in [0,T]} \big(\|\mathscr{F}(M,t)\|_\infty + \mathrm{Lip}\,\mathscr{F}(M,t)\big) < \infty.$$

Then for every initial set $K_0 \in \mathscr{K}(\mathbb{R}^N)$, *there exists a unique solution* $K : [0,T] \rightsquigarrow \mathbb{R}^N$ *to the morphological equation*

$$\overset{\circ}{K}(\cdot) \ni \mathscr{F}\big(K(\cdot), \cdot\big)$$

with $K(0) = K_0$.

Proof. The existence of a solution results from preceding Proposition 71 and, Proposition 24 (on page 42) implies uniqueness. □

Proposition 73 (Continuity w.r.t. initial data and the right-hand side).
Suppose $\mathscr{F} : \big(\mathscr{K}(\mathbb{R}^N), d\!\!l\big) \times [0,T] \longrightarrow \big(\mathrm{LIP}(\mathbb{R}^N,\mathbb{R}^N), d\!\!l_\infty\big)$ *to be* λ*-Lipschitz continuous in the first argument with*

$$\widehat{\alpha} := \sup_{M \in \mathscr{K}(\mathbb{R}^N),\, t \in [0,T]} \mathrm{Lip}\,\mathscr{F}(M,t) < \infty.$$

For $\mathscr{G} : \mathscr{K}(\mathbb{R}^N) \times [0,T] \longrightarrow \mathrm{LIP}(\mathbb{R}^N,\mathbb{R}^N)$ *assume* $\sup_{M,t} d\!\!l_\infty\big(\mathscr{F}(M,t), \mathscr{G}(M,t)\big) < \infty$.

Then every solutions $K_1(\cdot), K_2(\cdot) : [0,T] \rightsquigarrow \mathbb{R}^N$ *to the morphological equations*

$$\begin{cases} \overset{\circ}{K}_1(\cdot) \ni \mathscr{F}\big(K_1(\cdot), \cdot\big) \\ \overset{\circ}{K}_2(\cdot) \ni \mathscr{G}\big(K_2(\cdot), \cdot\big) \end{cases}$$

satisfy the following inequality for every $t \in [0,T]$

$$d\!\!l\big(K_1(t), K_2(t)\big) \le \Big(d\!\!l\big(K_1(0), K_2(0)\big) + t \cdot \sup_{M,s} d\!\!l_\infty\big(\mathscr{F}(M,s), \mathscr{G}(M,s)\big)\Big)\, e^{(\lambda+\widehat{\alpha})\,t}.$$

Proof is also a consequence of Proposition 24 in combination with Proposition 53 (about morphological transitions). □

Now we consider the initial value problem *with* state constraints and apply Nagumo's Theorem 19 (on page 40) to morphological transitions on $(\mathscr{K}(\mathbb{R}^N), d\!l)$:

Proposition 74 (Nagumo's Theorem for morphological equations). *Suppose* $\mathscr{F} : (\mathscr{K}(\mathbb{R}^N), d\!l) \longrightarrow (\mathrm{LIP}(\mathbb{R}^N, \mathbb{R}^N), d\!l_\infty)$ *to be continuous with*

$$\sup_{M \in \mathscr{K}(\mathbb{R}^N)} \big(\|\mathscr{F}(M)\|_\infty + \mathrm{Lip}\,\mathscr{F}(M)\big) < \infty.$$

Then the following statements are equivalent for any closed subset $\mathscr{V} \subset \mathscr{K}(\mathbb{R}^N)$ *:*

1. *Every set* $K_0 \in \mathscr{V}$ *is the initial set of at least one solution* $K : [0,1] \longrightarrow \mathscr{K}(\mathbb{R}^N)$ *to the morphological equation* $\overset{\circ}{K}(\cdot) \ni \mathscr{F}\big(K(\cdot)\big)$ *with* $K(t) \in \mathscr{V}$ *for all* $t \in [0,1]$.
2. $\mathscr{V} \subset \mathscr{K}(\mathbb{R}^N)$ *is a viability domain of* $\mathscr{F}$ *in the sense that* $\mathscr{F}(M) \in \mathscr{T}_{\mathscr{V}}(M)$ *for every* $M \in \mathscr{V}$. □

Corollary 75. *Suppose* $\mathscr{F} : (\mathscr{K}(\mathbb{R}^N), d\!l) \longrightarrow (\mathrm{LIP}(\mathbb{R}^N, \mathbb{R}^N), d\!l_\infty)$ *to be continuous with*

$$\sup_{M \in \mathscr{K}(\mathbb{R}^N)} \big(\|\mathscr{F}(M)\|_\infty + \mathrm{Lip}\,\mathscr{F}(M)\big) < \infty.$$

Let $M \subset \mathbb{R}^N$ *be a nonempty closed set satisfying* $\overline{\mathrm{co}}\,\mathscr{F}(K)(x) \subset T_M(x) \subset \mathbb{R}^N$ *for every nonempty compact subset* $K \subset M$ *and element* $x \in K$.

Then for any compact initial set $K_0 \subset M$, *there exists a solution* $K(\cdot) : [0,1] \rightsquigarrow \mathbb{R}^N$ *to the morphological equation* $\overset{\circ}{K}(\cdot) \ni \mathscr{F}\big(K(\cdot)\big)$ *with* $K(0) = K_0$ *and* $K(t) \subset M$ *for all* $t \in [0,1]$.

Proof results from Proposition 74 and Example 64 (on page 68). □

As mentioned briefly in Remark 31, the existence of viable solutions can also be guaranteed for systems of morphological equations. Now Propositions 66 and 68 respectively imply the following statements (as Aubin has already concluded in [10, §§ 4.3.2, 4.3.3]):

Corollary 76. *Suppose* $\mathscr{F}, \mathscr{G} : (\mathscr{K}(\mathbb{R}^N)^2, d\!l_2) \longrightarrow (\mathrm{LIP}(\mathbb{R}^N, \mathbb{R}^N), d\!l_\infty)$ *to be continuous with*

$$\begin{cases} \sup\limits_{M_1, M_2 \in \mathscr{K}(\mathbb{R}^N)} \big(\|\mathscr{F}(M_1, M_2)\|_\infty + \mathrm{Lip}\,\mathscr{F}(M_1, M_2)\big) < \infty, \\ \sup\limits_{M_1, M_2 \in \mathscr{K}(\mathbb{R}^N)} \big(\|\mathscr{G}(M_1, M_2)\|_\infty + \mathrm{Lip}\,\mathscr{G}(M_1, M_2)\big) < \infty, \end{cases}$$

Assume for any sets $M_1,\ M_2 \in \mathscr{K}(\mathbb{R}^N)$ *with* $M_1 \cap M_2 \neq \emptyset$

$$\big(\overline{\mathrm{co}}\,\mathscr{F}(M_1, M_2)(x) - \overline{\mathrm{co}}\,\mathscr{G}(M_1, M_2)(x)\big) \cap P_{M_2}^{M_1}(x) \neq \emptyset \quad \textit{for some } x \in M_1 \cap M_2.$$

Then for any sets $K_0, L_0 \in \mathscr{K}(\mathbb{R}^N)$ with $K_0 \cap L_0 \neq \emptyset$, there exist solutions $K(\cdot), L(\cdot) : [0,1] \rightsquigarrow \mathbb{R}^N$ to the morphological equations

$$\begin{cases} \overset{\circ}{K}(\cdot) \ni \mathscr{F}\big(K(\cdot),\, L(\cdot)\big) \\ \overset{\circ}{L}(\cdot) \ni \mathscr{G}\big(K(\cdot),\, L(\cdot)\big) \end{cases}$$

with $K(0) = K_0$, $L(0) = L_0$ and $K(t) \cap L(t) \neq \emptyset$ for all $t \in [0,1]$. □

Corollary 77.
Suppose $\mathscr{F}, \mathscr{G}, \mathscr{H} : (\mathscr{K}(\mathbb{R}^N)^3, dl_3) \longrightarrow (\mathrm{LIP}(\mathbb{R}^N, \mathbb{R}^N), dl_\infty)$ to be continuous with

$$\begin{cases} \sup\limits_{\widetilde{M} \in \mathscr{K}(\mathbb{R}^N)^3} \big(\|\mathscr{F}(\widetilde{M})\|_\infty + \mathrm{Lip}\, \mathscr{F}(\widetilde{M})\big) < \infty, \\ \sup\limits_{\widetilde{M} \in \mathscr{K}(\mathbb{R}^N)^3} \big(\|\mathscr{G}(\widetilde{M})\|_\infty + \mathrm{Lip}\, \mathscr{G}(\widetilde{M})\big) < \infty, \\ \sup\limits_{\widetilde{M} \in \mathscr{K}(\mathbb{R}^N)^3} \big(\|\mathscr{H}(\widetilde{M})\|_\infty + \mathrm{Lip}\, \mathscr{H}(\widetilde{M})\big) < \infty, \end{cases}$$

Assume for any $\widetilde{M} = (M_1, M_2, M_3) \in \mathscr{K}(\mathbb{R}^N)^3$ with $M_1 \subset M_2 \cap M_3$ and every $x \in M_1$

$$\overline{\mathrm{co}}\ \mathscr{F}(\widetilde{M})(x) \subset \big(\overline{\mathrm{co}}\, \mathscr{G}(\widetilde{M})(x) + T^C_{M_2}(x)\big) \cap \big(\overline{\mathrm{co}}\, \mathscr{H}(\widetilde{M})(x) + T^C_{M_3}(x)\big)$$

Then for any sets $K_0, L_0, M_0 \in \mathscr{K}(\mathbb{R}^N)$ with $K_0 \subset L_0 \cap M_0$, there exist solutions $K(\cdot), L(\cdot), M(\cdot) : [0,1] \rightsquigarrow \mathbb{R}^N$ to the morphological equations

$$\begin{cases} \overset{\circ}{K}(\cdot) \ni \mathscr{F}\big(K(\cdot),\, L(\cdot),\, M(\cdot)\big) \\ \overset{\circ}{L}(\cdot) \ni \mathscr{G}\big(K(\cdot),\, L(\cdot),\, M(\cdot)\big) \\ \overset{\circ}{M}(\cdot) \ni \mathscr{H}\big(K(\cdot),\, L(\cdot),\, M(\cdot)\big) \end{cases}$$

with $K(0) = K_0$, $L(0) = L_0$, $M(0) = M_0$ and $K(t) \subset L(t) \cap M(t)$ for all $t \in [0,1]$. □

Finally we extend this list of conclusions here on the basis of Proposition 69 (2.):

Corollary 78. *Suppose $\mathscr{F}, \mathscr{G}, \mathscr{H} : (\mathscr{K}(\mathbb{R}^N)^3, dl_3) \longrightarrow (\mathrm{LIP}(\mathbb{R}^N, \mathbb{R}^N), dl_\infty)$ to be continuous with*

$$\begin{cases} \sup\limits_{\widetilde{M} \in \mathscr{K}(\mathbb{R}^N)^3} \big(\|\mathscr{F}(\widetilde{M})\|_\infty + \mathrm{Lip}\, \mathscr{F}(\widetilde{M})\big) < \infty, \\ \sup\limits_{\widetilde{M} \in \mathscr{K}(\mathbb{R}^N)^3} \big(\|\mathscr{G}(\widetilde{M})\|_\infty + \mathrm{Lip}\, \mathscr{G}(\widetilde{M})\big) < \infty, \\ \sup\limits_{\widetilde{M} \in \mathscr{K}(\mathbb{R}^N)^3} \big(\|\mathscr{H}(\widetilde{M})\|_\infty + \mathrm{Lip}\, \mathscr{H}(\widetilde{M})\big) < \infty, \end{cases}$$

Assume for any $\widetilde{M} = (M_1, M_2, M_3) \in \mathscr{K}(\mathbb{R}^N)^3$ with $M_1 \cap M_2 \neq \emptyset$ and $M_1 \cup M_2 \subset M_3$

$$\begin{cases} \emptyset \neq \big(\overline{\mathrm{co}}\ \mathscr{F}(\widetilde{M})(x) - \overline{\mathrm{co}}\, \mathscr{G}(\widetilde{M})(x)\big) \cap P^{M_1}_{M_2}(x) & \text{for some } x \in M_1 \cap M_2, \\ \overline{\mathrm{co}}\, \mathscr{F}(\widetilde{M})(z) \subset \overline{\mathrm{co}}\, \mathscr{H}(\widetilde{M})(z) + T^C_{M_3}(z) & \text{for every } z \in M_1, \\ \overline{\mathrm{co}}\ \mathscr{G}(\widetilde{M})(z) \subset \overline{\mathrm{co}}\, \mathscr{H}(\widetilde{M})(z) + T^C_{M_3}(z) & \text{for every } z \in M_2. \end{cases}$$

Then for any sets $K_0, L_0, M_0 \in \mathscr{K}(\mathbb{R}^N)$ *with* $K_0 \cap L_0 \neq \emptyset$ *and* $K_0 \cup L_0 \subset M_0$, *there exist solutions* $K(\cdot), L(\cdot), M(\cdot) : [0,1] \rightsquigarrow \mathbb{R}^N$ *to the morphological equations*

$$\begin{cases} \overset{\circ}{K}(\cdot) \ni \mathscr{F}\big(K(\cdot),\, L(\cdot),\, M(\cdot)\big) \\ \overset{\circ}{L}(\cdot) \ni \mathscr{G}\big(K(\cdot),\, L(\cdot),\, M(\cdot)\big) \\ \overset{\circ}{M}(\cdot) \ni \mathscr{H}\big(K(\cdot),\, L(\cdot),\, M(\cdot)\big) \end{cases}$$

with $K(0) = K_0$, $L(0) = L_0$, $M(0) = M_0$ *and* $K(t) \cap L(t) \neq \emptyset$, $K(t) \cup L(t) \subset M(t)$ *for all* $t \in [0,1]$.

□

1.10 Example: Morphological Transitions for Image Segmentation

In the introductory section 0.2.1 (on page 3 ff.), we have already made a concrete suggestion how to formulate the popular problem of image segmentation by means of set-valued analysis. Demongeot and Leitner were the first to suggest mutational equations as tools for image segmentation in [58]. Their proposal is mathematically rather vague though and so it has served as motivation for the rigorous approach in [131] presented now.

A grey-scale image in $\mathbb{R}^N$ ($N \geq 2$) is given as a spatial function of real grey values $G : \mathbb{R}^N \longrightarrow \mathbb{R}$. In the search of an image segment which belongs to one and the same object, the basic idea is to make a given initial compact set $K_0 \subset \mathbb{R}^N$ expand such that the composition with a specified "error functional"

$$\Phi : \mathscr{K}(\mathbb{R}^N) \longrightarrow \mathbb{R}, \qquad M \longmapsto \Phi(M) = \psi\Big(\mathscr{L}^N(M),\ \int_M G\,dx,\ \int_M G^2\,dx\Big)$$

is a decreasing function of time. The subsequent examples of computer images use essentially the variation with a form of penalty term $\alpha \cdot \mathscr{L}^N(M)$ ($\alpha > 0$):

$$\begin{aligned}\Phi(M) &:= \tfrac{1}{\mathscr{L}^N(M)} \cdot \int_M \Big(G(x) \quad - \tfrac{1}{\mathscr{L}^N(M)} \cdot \quad \int_M G(y)\,dy\Big)^2 dx \;-\; \alpha \cdot \mathscr{L}^N(M) \\ &= \tfrac{1}{\mathscr{L}^N(M)} \cdot \int_M \quad G(x)^2\,dx \;-\; \tfrac{1}{\mathscr{L}^N(M)} \cdot \Big(\int_M G(y)\,dy\Big)^2 \quad - \alpha \cdot \mathscr{L}^N(M).\end{aligned}$$

This set evolution is to be continued as long as possible, i.e. the final set $M \subset \mathbb{R}^N$ should not admit any strict decrease in Φ via the same approach of expansion. In other words, we aim for a descent method in $(\mathscr{K}(\mathbb{R}^N), d\!\!l)$ with respect to Φ, but due to potential difficulties with regularity, we do not insist on the *steepest* descent "direction".

Given:	function of grey values $G \in C_c^0(\mathbb{R}^N)$, $N \geq 2$ error functional $\Phi : \mathscr{K}(\mathbb{R}^N) \longrightarrow \mathbb{R}$ s.t. $\Phi(M) = \psi\Big(\mathscr{L}^N(M),\ \int_M G\,dx,\ \int_M G^2\,dx\Big)$ with some $\psi \in C_c^2(]0,\infty[\times \mathbb{R}^2, \mathbb{R})$, initial set $K_0 \in \mathscr{K}(\mathbb{R}^N)$.
Wanted:	$K(\cdot) : [0,T[\longrightarrow \mathscr{K}(\mathbb{R}^N) \quad (T \in]0,\infty])$: (i) $K(0) = K_0$ (ii) $K(s) \subset K(t)$ whenever $s \leq t$ (iii) $K(\cdot)$ continuous w.r.t. Hausdorff metric (iv) $\Phi \circ K(\cdot) : [0,T[\longrightarrow \mathbb{R}$ nonincreasing (v) $M := \bigcup_{0 \leq t < T} K(t)$ is "critical" w.r.t. Φ

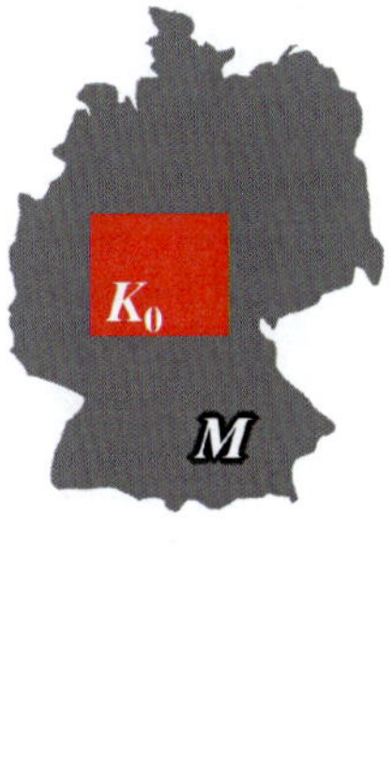

1.10.1 Analytical Tools of the Continuous Segmentation Problem

The central aspect of constructing $K(t)$ is how the Lebesgue integral depends on the current set. For the flow along smooth vector fields, the following version of so-called Reynolds Transport Theorem is well-known (see e.g. [55, § 8.3]):

Proposition 79. *Suppose* $w \in C^1(\mathbb{R}^N, \mathbb{R}^N)$. *For a nonempty compact set* $K_0 \subset \mathbb{R}^N$, *let* $K(t) = \vartheta_w(t, K_0)$ *contain all points* $x(t)$ *of solutions* $x(\cdot) \in C^1([0,t], \mathbb{R}^N)$ *of* $x' = w(x),\ x(0) \in K_0$.

Then for every $\Psi \in C^1(\mathbb{R} \times \mathbb{R}^N)$, *the function* $\mathbb{I}_w : t \longmapsto \int_{K(t)} \Psi(t,x)\, dx$ *fulfills*

$$\frac{d^+}{dt^+}\, \mathbb{I}_w(0) \ \stackrel{\text{Def.}}{=}\ \lim_{t \downarrow 0} \frac{\mathbb{I}_w(t) - \mathbb{I}_w(0)}{t} \ = \ \int_{K_0} \Big(\partial_t \Psi(0,x) \ + \ \operatorname{div}\big(\Psi(0,x)\, w(x)\big)\Big)\ dx.$$

If, in addition, K_0 *satisfies the assumptions of Gauss' Integral Theorem then*

$$\frac{d^+}{dt^+}\, \mathbb{I}_w(0) \ = \ \int_{K_0} \partial_t \Psi(0,x)\ dx \ + \ \int_{\partial K_0} \Psi(0,x)\, w(x) \cdot \nu_{K_0}(x)\ d\sigma_x$$

with the exterior unit normal ν_{K_0} *to* K_0.

Our construction of $K(t)$, however, prefers reachable sets of differential *inclusions* $x' \in \widetilde{F}(\cdot, x)$ (instead of differential equations) because this ansatz covers a more general class of possibly irreversible set deformations.

This theorem cannot be generalized immediately to the Lebesgue measure of reachable sets $\vartheta_{\widetilde{F}}(\cdot, K)$, but it gives a hint about the form of its derivative at time t: Naturally K is replaced by $\vartheta_{\widetilde{F}}(t, K)$ and all velocities of $\widetilde{F}(t,x) \subset \mathbb{R}^N$ have to be considered (instead of $w(x)$). Moreover the unit normal vector $\nu_K(x)$ might not be defined uniquely. If ∂K is not sufficiently smooth there may be more than one vector satisfying a normal condition instead, and all of them have to be taken into consideration. We use the following definition of tangent and normal vectors respectively.

Definition 80. Let $V \subset \mathbb{R}^N$ be a nonempty subset and x belong to the closure of V. $T_V(x)$ denotes Bouligand's contingent cone, i.e.

$$T_V(x) \ \stackrel{\text{Def.}}{=}\ \Big\{ w \in \mathbb{R}^N \ \Big|\ \liminf_{h \downarrow 0}\ \tfrac{1}{h} \cdot \operatorname{dist}(x + h\,w,\ V) \ = 0 \Big\}.$$

Bouligand's normal cone $N_V^B(x)$ is defined as its negative polar cone, i.e.

$$N_K^B(x) := T_V(x)^- \ \stackrel{\text{Def.}}{=}\ \big\{ p \in \mathbb{R}^N \,\big|\ p \cdot w \le 0 \quad \text{for all } \ w \in T_V(x) \big\}.$$

As a further abbreviation, set ${}^\flat N_V^B(x) \ := \ N_V^B(x) \cap \mathbb{B} \ = \ \big\{ w \in N_V^B(x) \,\big|\, |w| \le 1 \big\}$.

According to Proposition 79, the derivative of the Lebesgue measure $\mathscr{L}^N(\vartheta_w(\cdot, K))$ (at time $t = 0$) is the surface integral of the normal component of velocity $w(x)$.

In other words, the graphical interpretation regards it as resulting from the question how much the initial set K_0 is deformed in normal direction at each boundary point $x \in \partial K_0$. For the generalization to the reachable set $\vartheta_{\widetilde{F}}(t, K_0)$ this question has to be modified slightly:

How much is the current set deformed *at most*
by the normal components of all velocities ?

As a consequence, we expect the integrand $w(x) \cdot \nu_K(x)$ to be replaced by

$$\sup \left\{ w \cdot p \,\middle|\, w \in \widetilde{F}(t,x),\ p \in {}^\flat N^B_{\vartheta_{\widetilde{F}}(t,K_0)}(x) \right\} \;=\; \sup \left(\widetilde{F}(t,x) \cdot {}^\flat N^B_{\vartheta_{\widetilde{F}}(t,K_0)}(x) \right).$$

Finally the boundary of $\vartheta_{\widetilde{F}}(t, K_0)$ does not have to be sufficiently smooth for the complete definition of the surface integral. This problem of regularity is avoided by the $(N-1)$-dimensional Hausdorff integral. Under the assumptions of Proposition 79 the latter provides an equivalent result:

$$\tfrac{d}{dt}\, \mathscr{L}^N(\vartheta_w(t,K))\Big|_{t=0} \;=\; \int_{\partial K} w(x) \cdot \nu_K(x)\ d\mathscr{H}^{N-1}x,$$

but the important difference is that the Hausdorff measure $\mathscr{H}^{N-1}$ is defined for any subset of $\mathbb{R}^N$ (if the value ∞ is tolerated). Combining all these ideas, we presume that if the derivative of $\mathscr{L}^N(\vartheta_{\widetilde{F}}(t, K_0))$ exists it has the form

$$\int_{\partial\, \vartheta_{\widetilde{F}}(t,K_0)} \sup \left(\widetilde{F}(t,x) \cdot {}^\flat N^B_{\vartheta_{\widetilde{F}}(t,K_0)}(x) \right)\ d\mathscr{H}^{N-1}x.$$

The precise result is the following statement of Theorem A.54 (on page 476):

Proposition 81. *Assume* $N \geq 2$. *Let* $\rho_{\widetilde{F}}, \mu_{\widetilde{F}} > 0$, $v_{\widetilde{F}} \in C^0([0,T] \times \mathbb{R}^N, \mathbb{R}^N)$ *and* $\widetilde{F} : [0,T] \times \mathbb{R}^N \leadsto \mathbb{R}^N$ *be a Carathéodory map with compact convex values and*

$$\mathbb{B}_{\rho_{\widetilde{F}}}(v_{\widetilde{F}}(t,x)) \;\subset\; \widetilde{F}(t,x) \;\subset\; \mu_{\widetilde{F}}\,(1+|x|) \cdot \mathbb{B}$$

for every $(t,x) \in [0,T] \times \mathbb{R}^N$. *Furthermore assume* $K_0 \in \mathscr{K}(\mathbb{R}^N)$, $\psi \in C^0(\mathbb{R}^N)$.

Then the Lebesgue integral of ψ *over the reachable set* $\vartheta_{\widetilde{F}}(t, K_0)$

$$\mathbb{I}_{\widetilde{F}} :\ [0,T] \longrightarrow \mathbb{R}, \qquad t \longmapsto \int_{\vartheta_{\widetilde{F}}(t,K_0)} \psi(x)\, dx$$

is absolutely continuous and has the weak derivative

$$\tfrac{d}{dt}\, \mathbb{I}_{\widetilde{F}}(t) \;=\; \int_{\partial\, \vartheta_{\widetilde{F}}(t,K_0)} \psi(x) \cdot \sup \left(\widetilde{F}(t,x) \cdot {}^\flat N^B_{\vartheta_{\widetilde{F}}(t,K_0)}(x) \right)\ d\mathscr{H}^{N-1}x.$$

Remark 82. The complete proof (presented in [127]) provides some additional regularity properties of the boundary $\partial \vartheta_{\widetilde{F}}(t, K_0)$ at $\mathscr{L}^1$-almost every time $t \in [0,T]$ in terms of geometric measure theory: First, $\vartheta_{\widetilde{F}}(t, K_0)$ has locally finite perimeter. Second, $\partial \vartheta_{\widetilde{F}}(t, K_0)$ can be represented as level set of a Lipschitz continuous function and thus is countably $(\mathscr{H}^{N-1}, N-1)$-rectifiable at $\mathscr{L}^1$-almost every time t.

In regard to image segmentation, Proposition 81 has an essential disadvantage though. The set-valued map $\widetilde{F} : [0,T] \times \mathbb{R}^N \rightsquigarrow \mathbb{R}^N$ is assumed to have all values with nonempty interior. Hence, the reachable sets always undergo a form of translated expansion in *every* boundary point. It is difficult to stop near the boundary of the wanted segment.
This obstacle motivates us to prefer the following consequence as basis for solving the continuous segmentation problem. It results from Proposition 81 by means of vanishing neighborhoods of K_0 and monotone convergence of integrals [127, § 5], but this proof is restricted to *autonomous* differential inclusions with Lipschitz continuous right-hand side.

Proposition 83. *Assume $N \geq 2$. Let $F : \mathbb{R}^N \rightsquigarrow \mathbb{R}^N$ be a Lipschitz continuous map with compact convex values satisfying $0 \in F(x)^\circ$ or $F(x) = \{0\}$ for every $x \in \mathbb{R}^N$. Furthermore assume $K_0 \in \mathscr{K}(\mathbb{R}^N)$, $\psi \in C^0(\mathbb{R}^N)$.*

Then, $\mathbb{I}_F : [0,T] \longrightarrow \mathbb{R}, \; t \longmapsto \int_{\vartheta_F(t,K_0)} \psi(x)\, dx$ *is absolutely continuous with*

$$\tfrac{d}{dt}\, \mathbb{I}_F(t) \;=\; \int_{\partial\, \vartheta_F(t,K_0)} \psi(x) \,\cdot\, \sup\, \big(F(x) \cdot {}^\flat N^B_{\vartheta_F(t,K_0)}(x)\big)\; d\mathscr{H}^{N-1}x.$$

Now we need an ansatz for the region growing method in image segmentation. As explained in § 0.2.1 (on page 5), the reachable sets of K_0 and

$$F : \mathbb{R}^N \rightsquigarrow \mathbb{R}^N, \quad x \mapsto \mathbb{B}_{c(x)}(0) \overset{\text{Def.}}{=} \big\{ w \in \mathbb{R}^N \;\big|\; |w| \leq c(x) \big\}$$

reflect the notion that we prescribe the speed of propagation, but not the directions of the admitted velocity vectors. The assumptions of the preceding proposition are satisfied if $c : \mathbb{R}^N \longrightarrow [0,\infty[$ is Lipschitz continuous. If $c(\cdot)$ is bounded in addition, this ansatz is an example of a morphological transition in the sense of Example 5 (on page 34 f.).
Due to the rotational symmetry of the balls $\mathbb{B}_{c(\cdot)}(0)$ (with space-dependent radius), the weak derivative of $\mathbb{I}_{c(\cdot)\,\mathbb{B}}(t)$ does not depend on the normal cones explicitly

$$\tfrac{d}{dt}\, \mathbb{I}_{c(\cdot)\,\mathbb{B}}(t) \;=\; \int_{\partial\, \vartheta_{c(\cdot)\,\mathbb{B}}(t,K_0)} \psi(x) \cdot c(x)\;\; d\mathscr{H}^{N-1}x$$

and so, the composition with the given shape functional

$$\Phi : \mathscr{K}(\mathbb{R}^N) \longrightarrow \mathbb{R}, \quad M \longmapsto \Phi(M) = \psi\Big(\mathscr{L}^N(M),\; \int_M G\, dx,\; \int_M G^2\, dx\Big)$$

has the weak derivative

$$\tfrac{d}{dt}\, \Phi\big(\vartheta_{c(\cdot)\,\mathbb{B}}(t,K_0)\big) \;=\; \int_{\partial\, \vartheta_{c(\cdot)\,\mathbb{B}}(t,K_0)} \varphi(x,\, \vartheta_{c(\cdot)\,\mathbb{B}}(t,K_0)) \cdot c(x)\;\; d\mathscr{H}^{N-1}x$$

with the coefficient function, which is a quadratic polynomial in $G(x)$,

$$\varphi(z,M) \;:=\; \sum_{k=0}^{2}\; \partial_{k+1}\psi\Big(\mathscr{L}^N(M),\; \int_M G\, dx,\; \int_M G^2\, dx\Big) \cdot G(z)^k.$$

1.10.2 Solving the Continuous Segmentation Problem

For starting deforming the initial set $K_0 \in \mathcal{K}(\mathbb{R}^N)$, we are to construct a bounded Lipschitz continuous speed function $c : \mathbb{R}^N \longrightarrow [0,\infty[$ such that the real-valued composition $t \longmapsto \Phi\big(\vartheta_{c(\cdot)\,\mathbb{B}}(t,K_0)\big)$ is decreasing – for a *short* period (at least).

The region growing approach in this section has a rather constructive character and does not apply the general results about continuous mutational equations explicitly – due to lacking regularity of Φ (in a broad sense).
Indeed, we cannot expect to find the proper answer for all times immediately since the analytical basis in Proposition 83 implies the restriction to reachable sets of *autonomous* differential inclusions. For this reason, we present a construction which makes the sets evolve piecewise in time.
In more conceptual words, for each compact subset of $\mathbb{R}^N$, we "select" a morphological transition (represented by a bounded Lipschitz continuous speed function) which induces a descent direction with respect to Φ, but the regularity of φ does not guarantee the continuity of this selection principle in an obvious way. The final goal of converging to a "critical" set makes the selection even more difficult.

The sign of $\varphi(\,\cdot\,,K_0)$ is to play the decisive role where speed $c(\cdot)$ is equal to 0. In particular, it is sufficient to demand $c(x) = 0$ whenever $\varphi(x,K_0) > 0$. This connection can serve as a starting point for constructing some Lipschitz continuous $c(\cdot)$ only if we can rely on the continuity of $(t,x) \longmapsto \varphi(x,\,\vartheta_{c(\cdot)\,\mathbb{B}}(t,K_0))$ (at least). The assumption $G \in C^0_c(\mathbb{R}^N)$ proves here to be useful:

Lemma 84. *Assume $N \geq 2$. Let $F : \mathbb{R}^N \rightsquigarrow \mathbb{R}^N$ be a Lipschitz continuous map with compact convex values satisfying $0 \in F(x)^\circ$ or $F(x) = \{0\}$ for every $x \in \mathbb{R}^N$. Furthermore assume $K_0 \in \mathcal{K}(\mathbb{R}^N)$.*

Then, $\quad [0,\infty[\, \times \mathbb{R}^N \longrightarrow \mathbb{R}, \quad (t,x) \longmapsto \varphi(x,\,\vartheta_F(t,K_0)) \quad$ *is continuous.*

The continuity of $\varphi(\,\cdot\,,\,\vartheta_F(\cdot,K_0))$ gives an opportunity to reduce the error functional Φ by means of "local" expansion:

Lemma 85. *Let $K_0 \in \mathcal{K}(\mathbb{R}^N)$ and $x \in \partial K_0$ satisfy $\varphi(x,K_0) < 0$.*
Then there exist both a bounded Lipschitz continuous function $c : \mathbb{R}^N \longrightarrow [0,\infty[$ and some $\Delta > 0$ such that the composition

$$[0,\,\Delta] \longrightarrow \mathbb{R}, \quad t \longmapsto \Phi(\vartheta_{c(\cdot)\,\mathbb{B}}(t,K_0))$$

is strictly decreasing.

A strictly increasing sequence $(t_n)_{n\in\mathbb{N}}$ in $[0,\infty[$ is to describe the partition of the time axis with respect to which the speed functions $c_n : \mathbb{R}^N \longrightarrow [0,\infty[$ $(n \in \mathbb{N})$ are defined constant in time.

The expansion property of $\vartheta_{c_n(\cdot)\mathbb{B}}(\cdot, K_0)$ has an important consequence: It is impossible to correct mistakes during the image segmentation. Roughly speaking, whenever the reachable set $\vartheta_{c_n(\cdot)\mathbb{B}}(t, K_0)$ has passed the wanted contour on the image there is no opportunity to return. For this reason we prefer a criterion of the form

$$\varphi(x, \vartheta_{\widetilde{F}}(t, K_0)) < -\delta$$

(with a threshold $\delta > 0$ decreasing) to a simple check of the sign. This slight modification has the additional advantage that the points with larger absolute values of φ are taken into consideration first. In the following let $(\delta_m)_{m\in\mathbb{N}}$ denote a sequence in $]0,1]$ converging monotonically to 0.

The two preceding notions may be combined to an algorithm providing a solution of the continuous segmentation problem (on page 79) explicitly. In addition to the sequence $(t_n)_{n\in\mathbb{N}}$, we introduce a monotone sequence $(n_m)_{m\in\mathbb{N}}$ of indices such that the deformation during the intervals

$$[t_{n_m}, t_{n_m+1}], \quad [t_{n_m+1}, t_{n_m+2}] \quad \ldots \quad [t_{n_{m+1}-1}, t_{n_{m+1}}]$$

serves the supplementary purpose that

$$\varphi(x, \vartheta_{\widetilde{F}}(t_{n_{m+1}}, K_0)) \geq -2\,\delta_{m+1}$$

for all $x \in \partial\,\vartheta_{\widetilde{F}}(t_{n_{m+1}}, K_0) \cap \mathbb{B}_{1/\delta_{m+1}}(0)$. This is to guarantee

$$\varphi(x, M) \geq 0$$

for all boundary points x of the final set $M := \bigcup_t \vartheta_{\widetilde{F}}(t, K_0) \subset \mathbb{R}^N$, i.e. M is critical w.r.t. Φ in the sense of Definition 0.2 (on page 6). We do not insist on the resulting set M being closed (and thus compact) because $\varphi(\cdot,\cdot)$ is continuous with respect to both arguments, but $\varphi(\cdot, M) = \varphi(\cdot, \overline{M})$ can be ensured only by means of additional assumptions like $\mathscr{L}^N(\partial M) = 0$.
Starting with $n_0 := 0,\ t_0 := 0$ the inductive algorithm is shown in Fig. 1.1 below. It has four properties whose short (but merely technical) proofs are given in the end of this section.

Proposition 86. *The iterative algorithm in Fig. 1.1 has the following properties:*

(i) $\tau_j > 0$ *for every* $j \in \mathbb{N}$ *(if we set* $\inf \emptyset \stackrel{\text{Def.}}{=} \infty$*)*

(ii) n_{m+1} *is finite for each* $m \in \mathbb{N}$*, i.e. starting with index* $j = n_m$ *and time* t_{n_m}*, the condition* $\varphi(x, K_{j+1}) \geq -2\,\delta_{m+1}$ *for all* $x \in \partial K_{j+1} \cap \mathbb{B}_{1/\delta_m}$ *is satisfied after (at most) finitely many iterations w.r.t.* j*.*

(iii) $\vartheta_{\widetilde{F}}(\cdot, K_0) : [0, T[\longrightarrow \mathscr{K}(\mathbb{R}^N)$ *is expanding, i.e. the inclusion* $\vartheta_{\widetilde{F}}(s, K_0) \subset \vartheta_{\widetilde{F}}(t, K_0)$ *holds whenever* $s \leq t$*.*

(iv) $M := \bigcup_t \vartheta_{\widetilde{F}}(t, K_0) \subset \mathbb{R}^N$ *is critical w.r.t.* Φ *in the sense of Definition 0.2, i.e.* $\varphi(\cdot, M) \geq 0$ *on* ∂M*.*

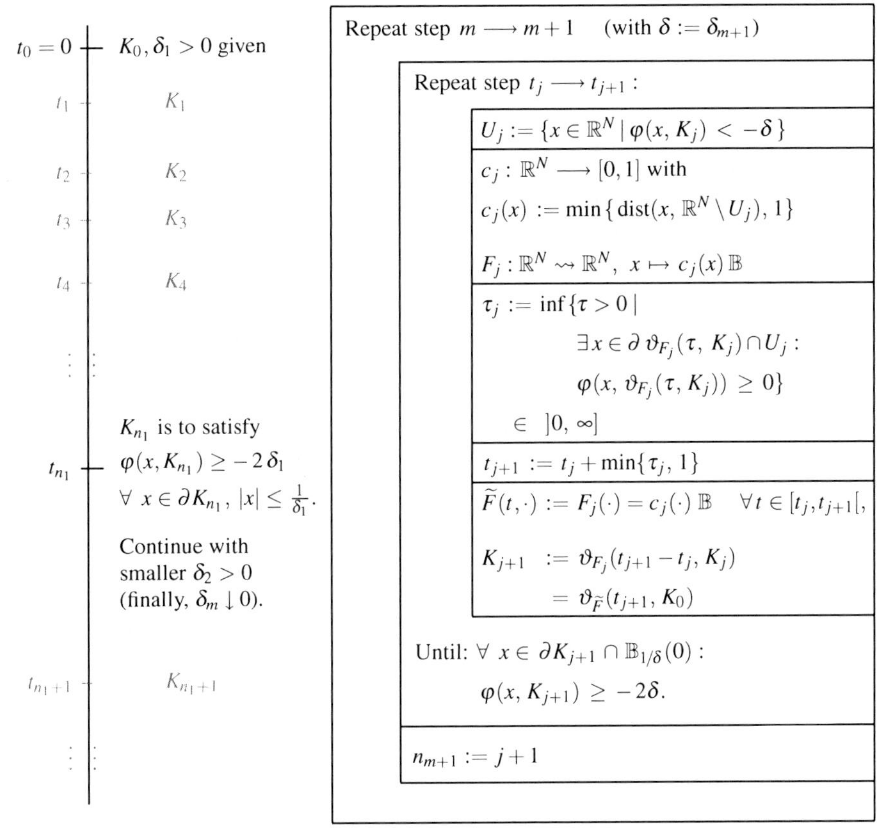

Fig. 1.1 Iterative algorithm for the solution of the continuous segmentation problem

Proof (of Lemma 84 on page 83). By definition,

$$\begin{aligned} \varphi : \mathbb{R}^N \times \mathscr{K}(\mathbb{R}^N) &\longrightarrow \mathbb{R}, \\ (z, M) &\longmapsto \sum_{k=0}^{2} \partial_{k+1}\psi\Big(\mathscr{L}^N(M),\ \int_M G\,dx,\ \int_M G^2\,dx\Big) \cdot G(z)^k \end{aligned}$$

is a quadratic polynomial in $G(z)$ and thus, $\psi \in C_c^2(]0,\infty[\times\mathbb{R}^2, \mathbb{R})$ implies the equi-continuity of $\varphi(\cdot, M) : \mathbb{R}^N \longrightarrow \mathbb{R}$ in a neighborhood of each tuple (z, M) with $\mathscr{L}^N(M) > 0$. Moreover, the Lebesgue integral over the reachable set $\vartheta_F(t, K_0)$ depends continuously on time t due to Proposition 83 (on page 82) and $G \in C_c^0(\mathbb{R}^N)$. These two features guarantee the claimed continuity of the composition

$$[0,\infty[\times\mathbb{R}^N \longrightarrow \mathbb{R}, \quad (t,x) \longmapsto \varphi(x,\ \vartheta_F(t, K_0)).$$

□

Proof (of Lemma 85 on page 83). Due to the continuity of $\varphi(\cdot, K_0)$, there exists a bounded neighborhood $V \subset \mathbb{R}^N$ of x with $\varphi(\cdot, K_0) \le \frac{\varphi(x,K_0)}{2} < 0$ in V. The distance from its complement induces the bounded Lipschitz function

$$c(\cdot) : \mathbb{R}^N \longrightarrow [0,1], \quad z \longmapsto \min\big\{\mathrm{dist}(z,\ \mathbb{R}^N \setminus V),\ 1\big\}.$$

Then the corresponding composition $[0,\infty[\, \times \mathbb{R}^N \longrightarrow \mathbb{R},\ (t,z) \longmapsto \varphi(z,\ \vartheta_{c(\cdot)\mathbb{B}}(t, K_0))$ is continuous due to Lemma 84 and in particular, it is uniformly continuous in the compact set $[0,1] \times \mathbb{B}_1(K_0) \subset \mathbb{R} \times \mathbb{R}^N$. Hence there exists some $\Delta \in\,]0,1]$ satisfying

$$\begin{cases} \vartheta_{c(\cdot)\mathbb{B}}(\Delta,\ K_0) \ \subset \ \mathbb{B}_1(K_0), \\ \varphi(z,\ \vartheta_{c(\cdot)\mathbb{B}}(t,\ \ K_0)) \ < \ 0 \qquad\qquad \text{for all } t \in [0,\Delta],\ z \in V \cap \mathbb{B}_1(K_0). \end{cases}$$

Then $\Phi\big(\vartheta_{c(\cdot)\mathbb{B}}(\cdot, K_0)\big)$ is strictly decreasing in $[0,\Delta]$ as a consequence of Proposition 83 (on page 82). □

The (Elementary, But Complete) Proof of Proposition 86 (on Page 84)

For the sake of simplicity we do not present the detailed induction with respect to m and suppose $\delta > 0$ to be fixed instead. Moreover the inductive definition of $U_j \ldots K_j$ is continued for any $j \in \mathbb{N}$ (with the same δ):

Initialize $\hat{K}_0 \in \mathscr{K}(\mathbb{R}^N),\ \ \mathscr{L}^N(\hat{K}_0) > 0,$
$\hat{t}_0 \ge 0$ and $\delta > 0$ (fixed).

Repeat step $j \longrightarrow j+1$:

- $\hat{U}_j := \big\{x \in \mathbb{R}^N \,\big|\, \varphi(x,\ \hat{K}_j) \ < \ -\delta\big\}$
- $\hat{c}_j : \mathbb{R}^N \longrightarrow [0,1]$ with
  $\hat{c}_j(x) := \min\big\{\mathrm{dist}(x,\ \mathbb{R}^N \setminus \hat{U}_j),\ 1\big\}$
  $\hat{F}_j : \mathbb{R}^N \rightsquigarrow \mathbb{R}^N,\ \ x \mapsto \hat{c}_j(x)\,\mathbb{B}$
- $\hat{\tau}_j := \inf\{\tau > 0 \mid$
  $\exists x \in \partial\, \vartheta_{\hat{F}_j}(\tau,\ \hat{K}_j) \cap \hat{U}_j :$
  $\varphi(x,\ \vartheta_{\hat{F}_j}(\tau,\ \hat{K}_j)) \ge 0\}$
  $\in\ ]0,\ \infty]$
- $\hat{t}_{j+1} := \hat{t}_j + \min\{\hat{\tau}_j,\ 1\}$
- $\hat{K}_{j+1} := \vartheta_{\hat{F}_j}(\hat{t}_{j+1} - \hat{t}_j,\ K_j)$

Then Proposition 86 is a consequence of the following claims:

(i') $\hat{\tau}_j > 0$ for every $j \in \mathbb{N}$,

(ii') $0 < \mathscr{L}^N(\hat{K}_j) \leq \max\{\mathrm{const}(\mathrm{supp}\ \psi),\ \mathscr{L}^N(\hat{K}_0)\}$ for every $j \in \mathbb{N}$,

(iii') there exists $j_0 \in \mathbb{N}$ such that $\varphi(x, \hat{K}_{j_0}) \geq -2\,\delta$ for all $x \in \partial \hat{K}_{j_0} \cap \mathbb{B}_{1/\delta}$.

The return to the complete iterative algorithm in Fig. 1.1 is based on aborting this (modified) construction at index j_0 and continuing with a smaller value of $\delta > 0$. Statement (ii') results directly from the assumption (on page 79) that the support of ψ is compact.

Lemma 87. $\hat{\tau}_j > 0$ *for every* $j \in \mathbb{N}$.

Proof. Due to Lemma 84, $[0,\infty[\, \times \mathbb{R}^N \longrightarrow \mathbb{R},\ (\tau,x) \longmapsto \varphi(x,\ \vartheta_{\hat{F}_j}(\tau,\ \hat{K}_j))$ is continuous and so, it is uniformly continuous on the compact set $[0,1] \times \mathbb{B}_1(\hat{K}_j)$. Hence there exists some $\tau_0 \in\,]0,1[$ such that all $x \in \mathbb{B}_1(\hat{K}_j)$ and $\tau \in\,]0,\tau_0]$ fulfill

$$\left|\varphi(x,\ \vartheta_{\hat{F}_j}(\tau,\ \hat{K}_j)) - \varphi(x,\hat{K}_j)\right| < \tfrac{1}{2}\,\delta.$$

$\tau \leq \tau_0 < 1$ and $\hat{c}_j(\cdot) \leq 1$ imply $\vartheta_{\hat{F}_j}(\tau,\hat{K}_j) \subset \mathbb{B}_1(\hat{K}_j)$ and thus every boundary point $x \in \partial\, \vartheta_{\hat{F}_j}(\tau,\ \hat{K}_j) \cap \hat{U}_j \subset \mathbb{B}_1(\hat{K}_j) \cap \hat{U}_j$ satisfies

$$\varphi(x,\ \vartheta_{\hat{F}_j}(\tau,\ \hat{K}_j)) < \varphi(x,\hat{K}_j) + \tfrac{1}{2}\,\delta \leq -\tfrac{1}{2}\,\delta,$$

i.e. $\hat{\tau}_j \geq \tau_0 > 0$. □

In regard to statement (iii'), the next lemma provides a connection between the changes of $\varphi(x,\ \vartheta_F(\cdot,K))$ and $\mathscr{L}^N(\vartheta_F(\cdot,K))$. It is the only context in which the second derivatives of ψ are used. (The assumption $\psi \in C^1_c(]0,\infty[\times\mathbb{R}^2)$ is enough for the remaining conclusions.)

Lemma 88. *Let* $F : \mathbb{R}^N \rightsquigarrow \mathbb{R}^N$ *be a Lipschitz continuous map with compact convex values satisfying* $0 \in F(x)^\circ$ *or* $F(x) = \{0\}$ *for every* $x \in \mathbb{R}^N$ *(as in Proposition 83). Moreover suppose that* $G : \mathbb{R}^N \longrightarrow \mathbb{R}$ *is bounded and continuous,* $K \in \mathscr{K}(\mathbb{R}^N)$ *with* $\mathscr{L}^N(K) \geq \lambda > 0$ *and* $\psi \in C^2_c(]0,\infty[\, \times \mathbb{R}^2, \mathbb{R})$.

Then, for any $t \geq 0$ *and* $x \in \mathbb{R}^N$,

$$\left|\varphi(x,\ \vartheta_F(t,K)) - \varphi(x,K)\right| \leq \mathrm{const}(\psi, \|G\|_\infty, \lambda) \cdot \big(\mathscr{L}^N(\vartheta_F(t,K)) - \mathscr{L}^N(K)\big).$$

Proof. Since all values of F contain 0 the reachable set $\vartheta_F(t,K)$ is expanding and hence $\mathscr{L}^N(\vartheta_F(t,K)) \geq \mathscr{L}^N(K) \geq \lambda$. Proposition 83 and $\psi \in C^2$ imply the absolute continuity of $t \mapsto \varphi(x,\ \vartheta_F(t,K))$ for each $x \in \mathbb{R}^N$. The (weak) derivative is

$$\sum_{j=0}^{2} \sum_{k=0}^{2} \int_{\partial\,\vartheta_F(t,K)} \partial_{j+1}\partial_{k+1}\psi \cdot G(y)^j\, G(x)^k \cdot \sup\left(F(y) \cdot {}^\flat N^B_{\vartheta_F(t,K)}(y)\right)\, d\mathscr{H}^{N-1}y$$

and, its absolute value has the upper bound

$$\left| \tfrac{d}{dt}\, \varphi(x, \vartheta_F(t,K)) \right|$$

$$\leq \ \text{const}\big(\|D^2\, \psi\|_{\sup},\, \|G\|_\infty\big) \cdot \int_{\partial\, \vartheta_F(t,K)} \sup\Big(F(y) \cdot {}^b N^B_{\vartheta_F(t,K)}(y) \Big) \ d\mathcal{H}^{N-1}y$$

$$= \ \text{const}\big(\|D^2\, \psi\|_{\sup},\, \|G\|_\infty\big) \cdot \tfrac{d}{dt}\, \mathcal{L}^N(\vartheta_F(t,K)).$$

$\square$

Lemma 89. $\hat{\tau}_j = \infty$ *for almost every* $j \in \mathbb{N}$ *and thus,* $\hat{t}_j \longrightarrow \infty$ *for* $j \longrightarrow \infty$.

Proof. If $\hat{\tau}_j < \infty$, we obtain a lower estimate of $\mathcal{L}^N(\hat{K}_{j+1}) - \mathcal{L}^N(\hat{K}_j) > 0$:
For any $\varepsilon > 0$ there exist some $\tau \in [\hat{\tau}_j, \hat{\tau}_j + \varepsilon[$ and a point x satisfying

$$\begin{cases} x \in \partial\, \vartheta_{\hat{F}_j}(\tau, \hat{K}_j) \ \cap\ \hat{U}_j \\ \varphi(x, \vartheta_{\hat{F}_j}(\tau, \hat{K}_j)) \ \geq\ 0 \end{cases}$$

The definition of $\hat{U}_j$ implies $\varphi(x, \vartheta_{\hat{F}_j}(\tau, \hat{K}_j)) - \varphi(x, \hat{K}_j) \geq \delta$. Now preceding Lemma 88 leads to an estimate for the change of Lebesgue measure

$$\mathcal{L}^N(\vartheta_{\hat{F}_j}(\tau, \hat{K}_j)) - \mathcal{L}^N(\hat{K}_j) \ \geq\ c\,\delta.$$

with a constant $c = c(\psi, G, \mathcal{L}^N(K_0)) > 0$. Due to Levi's Theorem of Monotone Convergence, $\varepsilon \longrightarrow 0$ provides $\mathcal{L}^N(\hat{K}_{j+1}) - \mathcal{L}^N(\hat{K}_j) \geq c\,\delta$.
Hence the situation $\hat{\tau}_j < \infty$ can arise only for a finite number of indices $j \in \mathbb{N}$ until $\mathcal{L}^N(\hat{K}_j)$ leaves the bounded set $\text{proj}_1(\text{supp }\psi) \subset \mathbb{R}$ (and then $\hat{U}_j = \emptyset$, $\hat{\tau}_j = \infty$ do not change any longer while j is increasing).

$\square$

Lemma 90. *Let* $\hat{K}_\infty := \bigcup_j \hat{K}_j$. *Then,* $\varphi(\cdot, \hat{K}_j) \overset{j \to \infty}{\longrightarrow} \varphi(\cdot, \hat{K}_\infty)$ *locally uniformly.*

Proof. Due to statement (ii'), the Lebesgue measures of $\hat{K}_j$, $j \in \mathbb{N}$, are uniformly bounded. As the support of ψ is assumed to be compact, for all $x \in \mathbb{R}^N$, the argument of $\partial_{k+1}\psi$ at $\varphi(x, \hat{K}_j)$, i.e. $\Big(\mathcal{L}^N(\hat{K}_j), \int_{\hat{K}_j} G\,dz, \int_{\hat{K}_j} G^2\,dz \Big)$, is contained in a compact subset of $]0, \infty[\times \mathbb{R}^2$ (independent of j). Hence the partial derivatives $\partial_{k+1}\psi\Big(\mathcal{L}^N(\hat{K}_j), \int_{\hat{K}_j} G\,dz, \int_{\hat{K}_j} G^2\,dz \Big)$ are uniformly bounded for all $j \in \mathbb{N}$.

As a consequence, the functions $\varphi(\cdot, \hat{K}_j) = \sum_{k=0}^2 \partial_{k+1}\psi(\mathcal{L}^N(\hat{K}_j) \ldots) \cdot G(\cdot)^k$, $j \in \mathbb{N}$, are equi-continuous in any compact subset of $\mathbb{R}^N$.
Moreover $\psi \in C^2$ and Lebesgue's Theorem of Dominated Convergence lead to the pointwise convergence $\varphi(x, \hat{K}_j) \longrightarrow \varphi(x, \hat{K}_\infty)$ for every $x \in \mathbb{R}^N$ and so finally, these properties provide the locally uniform convergence. $\square$

Definition 91 ([19, Definition 1.1.1], [162, Definition 4.1]). The *upper limit*, alias *outer limit*, of a sequence $(M_n)_{n\in\mathbb{N}}$ of subsets in a metric space (X,d)

$$\mathrm{Limsup}_{n\to\infty} M_n := \left\{x \in X \;\middle|\; \liminf_{n\to\infty} \operatorname{dist}(x, M_n) = 0\right\}$$

consists of all cluster points of sequences $(x_n)_{n\in\mathbb{N}}$ with $x_n \in M_n$.

Lemma 92. $\varphi(x, \hat{K}_\infty) \geq -\delta$ *for every point* $x \in \mathrm{Limsup}_{j\to\infty} \partial \hat{K}_j$.

Proof. Suppose $\varphi(x, \hat{K}_\infty) < -\delta$ for $x \in \mathrm{Limsup}\, \partial \hat{K}_j$. Due to Lemma 90 and the continuity of $\varphi(\cdot, \hat{K}_\infty)$, there exist a radius $\Delta \in]0,1]$ and an index $k \in \mathbb{N}$ with

$$\varphi(z, \hat{K}_j) < -\delta \qquad \text{for all } z \in \mathbb{B}_{3\Delta}(x) \text{ and } j \geq k.$$

Hence the construction implies $\mathbb{B}_{3\Delta}(x) \subset \hat{U}_j$ and $\hat{c}_j(z) \geq \Delta$ for all $z \in \mathbb{B}_{2\Delta}(x)$, $j \geq k$. According to Lemma 89, there are some $j \geq k$ and $y \in \hat{K}_j$ with $\hat{\tau}_j = \infty$, $|x - y| < \frac{\Delta}{2}$. Then we conclude for each $\tau \in [0,1] = [0, \hat{t}_{j+1} - \hat{t}_j]$

$$\begin{aligned}
\mathbb{B}_{\Delta\cdot\tau}(y) \cap \mathbb{B}_\Delta(x) &= \vartheta_{\Delta\cdot\mathbb{B}}(\tau, y) \cap \mathbb{B}_\Delta(x) \\
&\subset \vartheta_{\hat{F}_j}(\tau, y) \\
&\subset \vartheta_{\hat{F}_j}(\tau, \hat{K}_j) \\
&\subset \vartheta_{\hat{F}_j}(\hat{t}_{j+1} - \hat{t}_j, \hat{K}_j) \quad = \hat{K}_{j+1},
\end{aligned}$$

i.e. $\mathbb{B}_{\Delta/2}(x) \subset \mathbb{B}_\Delta(y) \cap \mathbb{B}_\Delta(x) \subset \hat{K}_{j+1}$ and the expanding property of $(\hat{K}_n)$ implies $\mathbb{B}_{\Delta/2}(x) \subset \hat{K}_n$ for any $n \geq j+1$, but this contradicts $x \in \mathrm{Limsup}_{n\to\infty} \partial \hat{K}_n$. □

In regard to Proposition 86, the same conclusion implies for the whole iterative algorithm that $\varphi(x, M) \geq -\delta$ holds for every boundary point $x \in \partial M \subset \mathrm{Limsup}_{n\to\infty} \partial K_n$ and any threshold $\delta > 0$, i.e. M is "critical" with respect to Φ.
Now we have the technical preparations at our disposal for proving statement (iii'):

Lemma 93. *There exists an index* $j_0 \in \mathbb{N}$ *such that every boundary point* $x \in \partial \hat{K}_{j_0} \cap \mathbb{B}_{1/\delta}(0)$ *satisfies* $\varphi(x, \hat{K}_{j_0}) \geq -2\delta$.

Proof. It is based on the continuity of $\varphi(\cdot, \hat{K}_\infty)$ and the locally uniform convergence of $\varphi(\cdot, \hat{K}_j)$. Indeed there is an open neighborhood V of the compact set $\left(\mathrm{Limsup}_{j\to\infty} \partial \hat{K}_j\right) \cap \mathbb{B}_{1/\delta}$ with

$$\varphi(\cdot, \hat{K}_\infty) \geq -\tfrac{3}{2}\,\delta \qquad \text{in } V.$$

Lemma 90 provides an index $k \in \mathbb{N}$ with $\varphi(\cdot, \hat{K}_j) \geq -2\delta$ in $\overline{V \cap \mathbb{B}_{1/\delta}}$ for all $j \geq k$. The final link $\partial \hat{K}_j \cap \mathbb{B}_{1/\delta} \subset V \cap \mathbb{B}_{1/\delta}$ for every large $j \in \mathbb{N}$ results from a simple indirect conclusion. □

1.10.3 The Application to Computer Images

1.10.3.1 Applying the Continuous Approach to Images

For computer images we start with two central aspects: The smallest suitable unit of a computer image is one pixel or one voxel respectively. Moreover the grey values within each of these units are constant.
Hence we intend a combination: On the one hand, we apply the preceding concept of continuous set deformation (along differential inclusions) for decreasing the error functional Φ. On the other hand, we want to restrict ourselves just to the decision whether or not a pixel (or a voxel) belongs to the approximating set.
This combination is based on an explicit ansatz (of the speed function) that describes the expansion to a single neighboring pixel. If the error functional is surely nonincreasing during this deformation then the pixel is admitted to the next set.
Before deriving a sufficient condition (on the neighboring pixel) we would like to point out that it does not result from discretizing but from applying the continuous deformation presented for the continuous segmentation problem in § 1.10.2. The main advantage of the computer algorithm here is its simplicity in regard to calculations. Its implementation is essentially based on the successive evaluation of quadratic polynomials of the grey value – in spite of all the set-valued theory before.

Let K_n denote the finite union of pixels representing the current set at the nth step. Furthermore suppose P to be a closed neighboring pixel, i.e. $P \cap K_n^\circ = \emptyset$ and $P \cap K_n$ contains (at least) a side of ∂P.
The continuous expansion of K_n to $K_n \cup P$ is described as reachable set of K_n and $F_n(\cdot) = \mathbb{B}_{c_n(\cdot)}(0) : \mathbb{R}^N \rightsquigarrow \mathbb{R}^N$ with

$$c_n(\cdot) := \operatorname{dist}(\,\cdot\,,\ \mathbb{R}^N \setminus (K_n \cup P)),$$

for example. In more graphical words, $\vartheta_{F_n}(\cdot, K_n)$ starts expanding at the common sides of $\partial K_n \cap \partial P$ and then keeps "filling" the interior P° while time is increasing. Thus,

$$K_n \cup P^\circ \;=\; \bigcup_{t \geq 0} \vartheta_{F_n}(t, K_n),$$

$$\lim_{t \longrightarrow \infty} \varphi(x,\ \vartheta_{F_n}(t, K_n)) \;=\; \varphi(x,\ K_n \cup P^\circ) \;=\; \varphi(x,\ K_n \cup P)$$

since $(K_n \cup P) \setminus (K_n \cup P^\circ) \subset \partial P$ has Lebesgue measure 0.
A condition sufficient to make the real-valued composition $\Phi \circ \vartheta_{F_n}(\cdot, K_n) : [0, \infty[\longrightarrow \mathbb{R}$ decrease is

$$\varphi(x,\ \vartheta_{F_n}(t, K_n)) < 0 \qquad \forall\, x \in P^\circ,\, t > 0. \qquad (*)$$

Here $\varphi(x,\ \vartheta_{F_n}(t, K_n))$ does not dependent on $x \in P^\circ$ because the grey values are constant within each pixel.

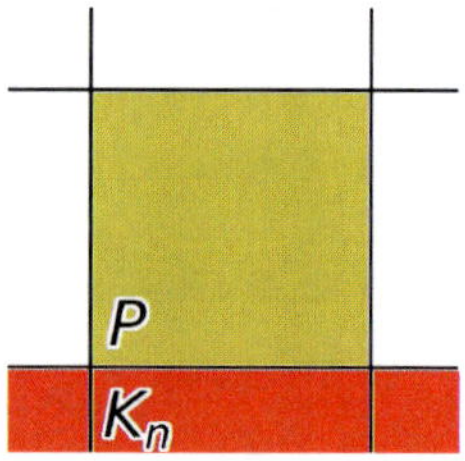

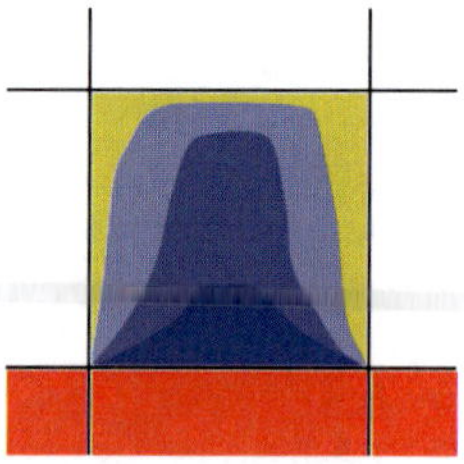

The maximal change of $t \longmapsto \varphi(x,\ \vartheta_{F_n}(t, K_n))$ can be estimated by the increase of Lebesgue measure due to Lemma 88 (on page 87):

$$\begin{aligned}\left|\varphi(x,\ \vartheta_{F_n}(t, K_n)) - \varphi(x, K_n)\right| \;&\leq\; \operatorname{const}(\psi, \|G\|_\infty) \cdot \left(\mathscr{L}^N(\vartheta_{F_n}(t, K_n)) - \mathscr{L}^N(K_n)\right) \\ &\leq\; \operatorname{const}(\psi, \|G\|_\infty) \cdot \mathscr{L}^N(P).\end{aligned}$$

It leads to a sufficient condition depending only on K_n (and not on $\vartheta_{F_n}(t, K_n), t > 0$): Using the abbreviation $\delta := \text{const}(\psi, \|G\|_\infty) \cdot \mathscr{L}^N(P) > 0$ the condition

$$\varphi(\,\cdot\,, K_n) \; < \; -\,\delta \qquad \text{in } P^\circ \qquad\qquad (**)$$

guarantees the preceding property $(*)$ and thus, $\Phi \circ \vartheta_{F_n}(\cdot, K_n)$ is decreasing in $[0, \infty[$.

Obviously condition $(**)$ has the same form as the inequality used for solving the continuous segmentation problem in § 1.10.2, but we want to point out an important difference between these two concepts:
For solving the continuous segmentation problem iteratively (Fig. 1.1 on page 85), the parameter δ is given at each step and it determines both the speed $c_j(\cdot)$ and the next time interval (i.e. τ_j). Now for computer images, we start with an explicit ansatz of the radius $c_n(\cdot)$ which implies the time interval $[0, \infty[$ for expanding to $K_n \cup P^\circ$. This provides an adequate definition of δ. Both methods have Proposition 83 (on page 82) as analytical basis in common.

1.10.3.2 A (Very) Simple Implementation for Computer Images

Now condition $(**)$ underlies an implementation for image segmentation. The final simplicity of this algorithm is to justify dispensing with analytical accuracy for the first time and following heuristic arguments at three steps.

Firstly, condition $(**)$ is sufficient to prevent that

$$\varphi(x, \vartheta_{\tilde{F}_n}(t, K_n)) > 0 \qquad\qquad \text{at some } x \in \partial\, \vartheta_{\tilde{F}_n}(t, K_n) \cap P^\circ$$

while expanding to a neighboring pixel P. It might be much too strong. For weakening it gradually we replace the fixed value of $\delta(\psi, \|G\|_\infty, \mathscr{L}^N(P))$ by a decreasing sequence until reaching a given threshold δ_{end}.
Similarly to the algorithm in § 1.10.2, this notion has the advantage that boundary pixels with larger absolute values of $\varphi(\cdot, K_n)$ are taken into consideration first.

Strictly speaking, it is necessary to update K_n and $\varphi(\cdot, K_n)$ after each neighboring pixel P fulfilling condition $(**)$. But higher accuracy requires much higher costs of calculation. This observation leads to the second step based on rather heuristic reasons: Applying $\varphi(\cdot, K_n)$ to *all* neighboring pixels of K_n is easier to perform since

$$\varphi(x, K_n) \; = \; \sum_{k=0}^{2} \, \partial_{k+1}\psi\big(\mathscr{L}^N(K_n),\; \int_{K_n} G\, dy,\; \int_{K_n} G^2\, dy\big) \cdot G(x)^k$$

is a quadratic polynomial in the grey value $G(x)$ whose coefficients are calculated only once. This slightly weakened concept improves the speed of the algorithm and makes it feasible even for large images. Moreover it prevents the dependence on the order in which the neighboring pixels are checked (and possibly included), i.e. this modification preserves the isotropic character of updating.

These two steps motivate the following simple version of implementation:

Let $\delta > 0$. Suppose that K_n (still) denotes the finite union of pixels representing the current approximating set. For each neighboring pixel P of K_n, condition $(**)$ is checked. We define

$$K_{n+1} := K_n \cup \bigcup_{\substack{\text{neighbor } P \text{ of } K_n \\ P \text{ satisfies } (**)}} \overline{P}$$

This step is repeated until $K_{n+1} = K_n$. Then we continue with a smaller value of $\delta > 0$, e.g. $\frac{\delta}{2}$. Finally the algorithm ends if δ falls below a given threshold $\delta_{end} > 0$.

The symmetry properties of the implementation motivate the last step: We consider all closed pixels P satisfying $\emptyset \neq K_n \cap P \subset \partial P$, i.e. we dispense with the stronger condition that P and K_n have at least one side of ∂P in common.
This step goes beyond the continuous framework in § 1.10.3.1 because there is no point of ∂K_n at which the speed $c_n(\cdot)$ is positive. But it may keep simple shapes invariant while expanding. Consider a black rectangle on a white background, for example. Starting with a small rectangle inside, this modification guarantees that the approximating sets are also rectangles at any time. The stronger restriction (on neighboring pixels) instead leads to rhombi temporarily while expanding.

Hence the final implementation is founded on the inductive definition

$$K_{n+1} := K_n \cup \bigcup_{\substack{\text{Pixel } P:\, \emptyset \neq K_n \cap \overline{P} \subset \partial P \\ P \text{ satisfies } (**)}} \overline{P}$$

This step is again repeated until $K_{n+1} = K_n$. Then we replace $\delta > 0$ by a smaller value, e.g. $\frac{\delta}{2}$. The algorithm finishes if δ falls below a given positive threshold δ_{end}.

Fig. 1.2 Example: Picture of a 300-year-old globe. *Left:* The original image. *Middle:* The initial set consists of black pixels that are detected by global thresholds. *Right:* The binary result

The detection of uniform segments (of the image) may be realized by the error functional

$$\begin{aligned}\Phi(M) &:= \text{Variance}(G|_M) && - \alpha\,\mathscr{L}^N(M)\\ &= \frac{1}{\mathscr{L}^N(M)}\int_M G^2\,dx - \frac{1}{\mathscr{L}^N(M)^2}\Big(\int_M G\,dx\Big)^2 && - \alpha\,\mathscr{L}^N(M)\end{aligned}$$

for all measurable sets $M \subset \mathbb{R}^N$, $\mathscr{L}^N(1\,\text{pixel}) < \mathscr{L}^N(M) < \infty$ with a weight parameter $\alpha \geq 0$. Roughly speaking, the additional term $-\alpha\,\mathscr{L}^N(M)$ reduces "sensitivity" in search of uniform segments. It has the consequence that expansion continues despite increasing oscillations if their increase is small (in comparison with the change of Lebesgue measure). Then,

$$\varphi(x,M) = \frac{1}{\mathscr{L}^N(M)}\,G(x)^2 - \frac{2\int_M G\,dy}{\mathscr{L}^N(M)^2}\,G(x) - \frac{\int_M G^2\,dy}{\mathscr{L}^N(M)^2} + \frac{2\,(\int_M G\,dy)^2}{\mathscr{L}^N(M)^3} - \alpha.$$

Hence the implementation depends principally on merely two parameters. K_0 and the difference between α, δ_{end} determine where the algorithm starts and when it stops. They have to be adapted to the given image (somehow – as usual).

Such a simple algorithm cannot be free from weaknesses of course. In particular the resulting set depends largely on the parameters because they determine its sensitivity to oscillations of grey values and so, this error functional is not really suited to ignoring strong noise.
Moreover the choice of the initial segment K_0 has an influence on which connected components can be found and on which boundary pixels are detected first. Indeed, the earlier a pixel of the *wanted* contour is neighbor of K_n the more often it is checked later and, whenever it is included in K_n the computer starts searching for next contour outside.

Despite this weakness, the computer implementation leads to quite good results for strongly differing images (see also Fig. 0.1 on page 7, for example) and, it has the important advantage of rather simple computations. Furthermore it does not change the original image and thus facilitates combining it with other methods of image processing. Fig. 1.2 (on page 92) illustrates this extension as its binary result can help us to remove noise from the original picture by different means later.

1.10.3.3 Extensions of the Implementation

There are several aspects how to extend the computer algorithm. Finally we summarize such possibilities that have the analytical basis (presented in §1.10.2) in common. They exemplify that the preceding concept of set-valued maps provides the mathematical background of several variants of region growing methods.

Although the algorithm is mostly illustrated with 2-dimensional images, it is independent of dimension N and the computer implementation may be extended directly to 3-dimensional images. Then, of course, neighboring pixels in the third dimension are also taken into consideration during each step.

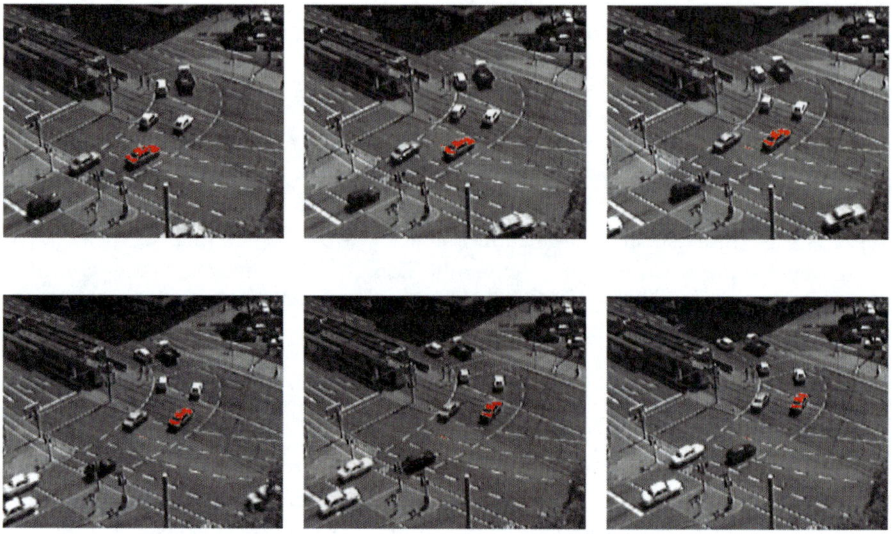

Fig. 1.3 Extract of 50 images with slow changes. One car is marked in the first image and then the discrete algorithm detects it in the subsequent images while passing the crossroads

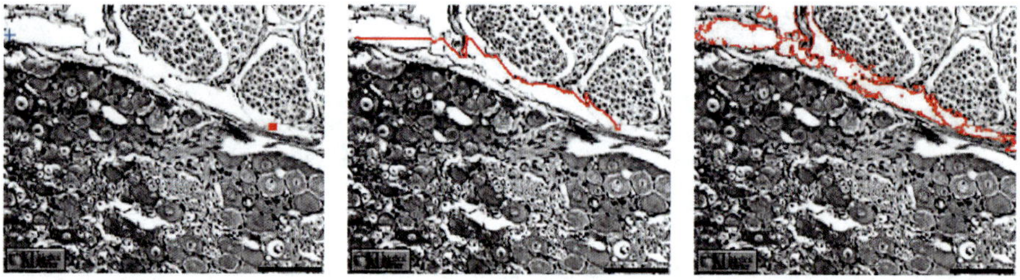

Fig. 1.4 Example of the anisotropic version: Nerve cells (dorsal root ganglion) *Left:* Initial set (small rectangle on the right-hand side) and target pixel P_0 (marked by a small cross in the top left-hand corner). Then the neighboring pixels are checked according to their distances from P_0. *Middle:* The approximating set when reaching P_0. Now the implementation continues with the isotropic version discussed in § 1.10.3.2 *Right:* Final segment

Moreover this approach can even be applied to image sequences. Then we consider time τ (of the image sequence) as additional dimension and use coordinates $\hat{x} = (x, \tau) \in \mathbb{R}^N \times \mathbb{R}$. The last component of $\hat{x}$, however, is not related with the parameter t of set expansion (during segmentation). For example, a sequence of 2-dimensional images is regarded as one image of dimension 3 (see Fig. 1.3).
In comparison with previous results, the only difference concerns the permitted directions of expansion. Information about segmentation should not be transported to the past, i.e. the expanding sets do not result from trajectories in the opposite τ direction. Thus the values of $F : \mathbb{R}^{N+1} \rightsquigarrow \mathbb{R}^{N+1}$ are contained in $\mathbb{R}^N \times [0, \infty[$. Considering the ansatz of $F(\hat{x})$, the closed balls $\mathbb{B}_{c(\hat{x})}(0) \subset \mathbb{R}^{N+1}$ are replaced by semi-balls

$$\mathbb{B}_{c(\hat{x})}(0)^+ := \mathbb{B}_{c(\hat{x})}(0) \cap (\mathbb{R}^N \times [0, \infty[ ).$$

In morphological words, we use the new "structuring element" $\mathbb{B}_1(0)^+$ instead of $\mathbb{B}_1(0) \subset \mathbb{R}^{N+1}$ while its size still depends on $\hat{x}$ (according to the respective speed function $c(\cdot)$) as mentioned in Example 55 (on page 63).
Then we can still apply the arguments proving fundamental Proposition 83 (on page 82) and use these results for image sequences. Furthermore the computer implementation considers only neighboring pixels which refer to the same or to a later point of time τ.

Up to now the algorithms presented here have the isotropic property in common, but they can be enhanced easily with respect to anisotropic expansion.
A slight modification of the computer algorithm concerns the directions of expansion. Condition $(**)$ (on page 91) provides a sufficient criterion whether a neighboring pixel is included in the next approximating set. There is no restriction on the order of checking though. In fact, we can first sort neighboring pixels (of K_n) by their distances from a given pixel P_0, for example. The modified algorithm then tries to detect a connection with P_0. After reaching P_0, we return to the isotropic version and, this combination of methods is more sensitive to considering given structures. An example is presented in Fig. 1.4.

Last but not least, the analytical results in § 1.10.2 can also be applied to more general types of error functionals $\Phi(\cdot)$ and set constraints. The first modification is based on transforming the grey values $G(\cdot)$, i.e. we use shape functionals of the form

$$\Phi(M) := \psi\Big(\int_M \rho_1(x, G(x))\, dx, \ldots \int_M \rho_m(x, G(x))\, dx\Big)$$

for measurable sets $M \subset \mathbb{R}^N$, $0 < \mathscr{L}^N(M) < \infty$ with functions $\psi : \mathbb{R}^m \to \mathbb{R}$, $\rho_1 \ldots \rho_m : \mathbb{R}^N \times \mathbb{R} \to \mathbb{R}$ sufficiently smooth. As a consequence the compositions $\rho_j(\cdot, G(\cdot))$ replace the grey values $G(\cdot)$ at preceding conclusions.
Weight functions are typical examples. In particular they comprise excluding parts of the initial set K_0 from calculations, i.e. they admit considering

$$\Phi(M) := \psi\Big(\int_{M\setminus E} \rho_1(\cdot, G)\, dx, \ \ldots \int_{M\setminus E} \rho_m(\cdot, G)\, dx\Big)$$

with some $E \subset K_0^\circ$. This extension is useful for starting with a large segment K_0 but taking into account only regions along ∂K_0. Moreover the integrand may depend on derivatives of $G(\cdot)$. If G is not differentiable then convolution provides well-known methods of smoothing (see e.g. [39, 41, 95, 172]).
In comparison with standard level set methods, the segmentation problem in terms of time-dependent subsets of $\mathbb{R}^N$ has the essential advantage that we can formulate nonlocal set constraints (on the maximal diameter, for example) more easily.

1.11 Example: Modified Morphological Equations for Compact Sets in $\mathbb{R}^N$ via Bounded One-Sided Lipschitz Continuous Maps

Reachable sets of differential inclusions can serve as transitions on $(\mathscr{K}(\mathbb{R}^N), d\!l)$ only if they are stable with respect to initial set and the right-hand side of the inclusion. For this reason, we have considered Lipschitz continuous maps with uniformly bounded compact values so far.
In [9, Remark 5.2], Artstein poses the question which other assumptions (alternative to classical Lipschitz continuity) might guarantee such an estimate of stability as in Proposition 50 (on page 60) here. Donchev and Farkhi suggest an answer in [69] introducing the so-called one-sided Lipschitz continuity (with respect to space). Their existence theorem (quoted in subsequent Theorem A.63 on page 480) provides an estimate of the distance between a given curve and the wanted solution being very similar to the inequality of Filippov. Some key aspects of their nonautonomous differential inclusions are summarized in Appendix A.7 (on page 480 f.).

In this section, we use this type of set-valued maps as right-hand side of autonomous differential inclusions so that their reachable sets induce more general transitions on $(\mathscr{K}(\mathbb{R}^N), d\!l)$. In regard to Theorem A.63 applied to *autonomous* differential inclusions, we introduce similarly to Definition 49 (on page 60):

Definition 94. OSLIP$(\mathbb{R}^N, \mathbb{R}^N)$ consists of all set-valued maps $F : \mathbb{R}^N \leadsto \mathbb{R}^N$ satisfying the following three conditions:

1. F has nonempty compact convex values that are uniformly bounded in $\mathbb{R}^N$,
2. F is upper semicontinuous,
3. F is one-sided Lipschitz continuous, i.e. there is a constant $L \in \mathbb{R}$ such that for every $x, y \in \mathbb{R}^N$ and $v \in F(x)$, there exists some $w \in F(y)$ satisfying
$$\langle x - y,\, v - w \rangle \;\leq\; L\,|x - y|^2.$$

The smallest constant $L \in \mathbb{R}$ with this property is usually abbreviated as Lip F.

Remark 95. Every map $F \in \text{LIP}(\mathbb{R}^N, \mathbb{R}^N)$ with convex values is contained in OSLIP$(\mathbb{R}^N, \mathbb{R}^N)$. Set-valued maps in OSLIP$(\mathbb{R}^N, \mathbb{R}^N)$, however, do not have to be continuous in general, just consider the example (in addition to Remark A.62)

$$\mathbb{R} \leadsto \mathbb{R}, \quad x \mapsto \begin{cases} -1 & \text{for } x > 0 \\ [-1, 1] & \text{for } x = 0 \\ 1 & \text{for } x < 0 \end{cases}$$

Proposition 96. *For any sets $K_1, K_2 \in \mathscr{K}(\mathbb{R}^N)$ and maps $F, G \in \text{OSLIP}(\mathbb{R}^N, \mathbb{R}^N)$ with $\Lambda := \max\{\text{Lip } F, \text{ Lip } G\} \in \mathbb{R}$, the reachable sets $\vartheta_F(t, K_1)$, $\vartheta_G(t, K_2)$ are closed subsets of $\mathbb{R}^N$ and, the Pompeiu-Hausdorff distance between the reachable sets at time $t \geq 0$ satisfies*

$$d\!l\big(\vartheta_F(t, K_1),\, \vartheta_G(t, K_2)\big) \;\leq\; \big(d\!l(K_1, K_2) + t \cdot d\!l_\infty(F, G)\big) \cdot e^{\Lambda\, t}.$$

Basic set	$E := \mathscr{K}(\mathbb{R}^N)$ the set of nonempty compact subsets of the Euclidean space $\mathbb{R}^N$
Distance	Pompeiu-Hausdorff metric $d\!l : \mathscr{K}(\mathbb{R}^N) \times \mathscr{K}(\mathbb{R}^N) \longrightarrow \mathbb{R}$, $d\!l(K_1, K_2) := \max\left\{ \sup_{x \in K_1} \operatorname{dist}(x, K_2), \quad \sup_{y \in K_2} \operatorname{dist}(y, K_1) \right\}$
Transition	For each $F \in \text{OSLIP}(\mathbb{R}^N, \mathbb{R}^N)$, i.e. bounded, upper semicontinuous and one-sided Lipschitz continuous set-valued map $F : \mathbb{R}^N \rightsquigarrow \mathbb{R}^N$ with compact convex values (Definition 94), define $\vartheta_F : [0,1] \times \mathscr{K}(\mathbb{R}^N) \longrightarrow \mathscr{K}(\mathbb{R}^N)$ by means of reachable sets of the autonomous differential inclusion $x'(\cdot) \in F(x(\cdot))$ a.e.: $\vartheta_F(t, K_0) := \big\{ x(t) \mid \text{there exists } x(\cdot) \in W^{1,1}([0,t], \mathbb{R}^N) : x'(\cdot) \in F(x(\cdot)) \ \mathscr{L}^1\text{-a.e. in } [0,t], \ x(0) \in K_0 \big\}.$
Compactness	Closed bounded balls in $(\mathscr{K}(\mathbb{R}^N), d\!l)$ are compact: Proposition 47 (page 57)
Mutational solutions	Reachable sets of a nonautonomous differential inclusion whose set-valued right-hand side is determined via feedback – if the transitions are induced by additionally continuous maps, i.e. each $F \in \text{COSLIP}(\mathbb{R}^N, \mathbb{R}^N)$ (Definition 103): Proposition 105, Corollary 106 (page 101)
List of main results formulated in § 1.11	Existence due to compactness (Peano): Proposition 99 (page 98) Cauchy-Lipschitz Theorem: Proposition 100 (page 99) Continuity w.r.t. data: Proposition 101 (page 99) Existence under state constraints (Nagumo): Proposition 102
Key tools	Filippov-like Theorem A.63 of Donchev and Farkhi [69] about differential inclusions with one-sided Lipschitz continuous right-hand side (page 480) Integral funnel equation for reachable sets of nonautonomous differential inclusions: Proposition A.13 (page 447)

Table 1.3 Brief summary of the example in § 1.11 in mutational terms:
Modified morphological equations for compact sets in $\mathbb{R}^N$ via bounded one-sided Lipschitz maps

Proof follows from Theorem A.63 (on page 480) in exactly the same way as Proposition 50 about morphological transitions in $\mathrm{LIP}(\mathbb{R}^N,\mathbb{R}^N)$ resulted from Filippov's Theorem A.6 (see page 61 for details). □

Obviously, $[0,\infty[\longrightarrow \big(\mathscr{K}(\mathbb{R}^N), d\!\!l\big),\ t\longmapsto \vartheta_F(t,K_0)$ is $\|F\|_\infty$-Lipschitz continuous for every $F\in\mathrm{OSLIP}(\mathbb{R}^N,\mathbb{R}^N)$ and, the semigroup property of reachable sets still holds (as in Lemma 52 on page 61). The same conclusions as for morphological transitions in § 1.9.2 (on page 60 ff.) now lead to

Proposition 97. *For every set-valued map $F\in\mathrm{OSLIP}(\mathbb{R}^N,\mathbb{R}^N)$,*

$$\begin{aligned}\vartheta_F : [0,1]\times\mathscr{K}(\mathbb{R}^N) &\longrightarrow \mathscr{K}(\mathbb{R}^N)\\ (t,\ K) &\longmapsto \vartheta_F(t,K)\end{aligned}$$

with $\vartheta_F(t,K)\subset\mathbb{R}^N$ denoting the reachable set of the initial set $K\in\mathscr{K}(\mathbb{R}^N)$ and the differential inclusion $x'\in F(x)$ a.e. at time t is a transition on $(\mathscr{K}(\mathbb{R}^N), d\!\!l)$ with

$$\begin{aligned}\alpha(\vartheta_F) &\le \max\{0,\ \mathrm{Lip}\,F\},\\ \beta(\vartheta_F) &\le \|F\|_\infty,\\ D(\vartheta_F,\vartheta_G) &\le d\!\!l_\infty(F,G).\end{aligned}$$ □

Remark 98. We prefer excluding negative values of the transition parameter $\alpha(\vartheta_F)$ because Gronwall's estimate (in form of Proposition A.2 on page 440) often serves as key analytic tool, but does not cover exponential decrease here.

The next step consists in existence of solutions to initial value problems without state constraints:

Proposition 99 (Peano's Theorem for modified morphological equations). *Suppose $\mathscr{F}:\big(\mathscr{K}(\mathbb{R}^N), d\!\!l\big)\times[0,T]\longrightarrow\big(\mathrm{OSLIP}(\mathbb{R}^N,\mathbb{R}^N), d\!\!l_\infty\big)$ to be continuous and*

$$\sup_{\substack{M\in\mathscr{K}(\mathbb{R}^N)\\ t\in[0,T]}}\big(\|\mathscr{F}(M,t)\|_\infty+\max\{0,\ \mathrm{Lip}\,\mathscr{F}(M,t)\}\big)<\infty.$$

Then for every initial set $K_0\in\mathscr{K}(\mathbb{R}^N)$, there exists a solution $K:[0,T]\leadsto\mathbb{R}^N$ to the modified morphological equation

$$\overset{\circ}{K}(\cdot)\ni\mathscr{F}\big(K(\cdot),\cdot\big)$$

with $K(0)=K_0$, i.e. $K(\cdot)$ is Lipschitz continuous with respect to $d\!\!l$ and satisfies for $\mathscr{L}^1$-almost every $t\in[0,T]$

$$\lim_{h\downarrow 0}\ \tfrac{1}{h}\cdot d\!\!l\big(\vartheta_{\mathscr{F}(K(t),t)}(h,\ K(t)),\ \ K(t+h)\big)=0$$

Proof results directly from Theorem 20 (on page 40) in combination with Proposition 47 (on page 57) and Proposition 97. □

Proposition 100 (Cauchy-Lipschitz for modified morphological equations). *Suppose the continuous function* $\mathscr{F} : \mathscr{K}(\mathbb{R}^N) \times [0,T] \longrightarrow \big(\mathrm{OSLIP}(\mathbb{R}^N,\mathbb{R}^N), d\!l_\infty\big)$ *to be Lipschitz continuous in the first argument with*

$$\sup_{\substack{M \in \mathscr{K}(\mathbb{R}^N)\\ t\in[0,T]}} \big(\|\mathscr{F}(M,t)\|_\infty + \max\{0,\ \mathrm{Lip}\,\mathscr{F}(M,t)\}\big) < \infty.$$

Then for each initial set $K_0 \in \mathscr{K}(\mathbb{R}^N)$, *there exists a unique solution* $K : [0,T] \leadsto \mathbb{R}^N$ *to the modified morphological equation*

$$\overset{\circ}{K}(\cdot) \ni \mathscr{F}\big(K(\cdot), \cdot\big)$$

with $K(0) = K_0$.

Proof. The existence of a solution results from preceding Proposition 99 and, Proposition 24 (on page 42) implies uniqueness. □

Proposition 101 (Continuity w.r.t. initial data and the right-hand side). *Suppose* $\mathscr{F} : \big(\mathscr{K}(\mathbb{R}^N), d\!l\big) \times [0,T] \longrightarrow \big(\mathrm{OSLIP}(\mathbb{R}^N,\mathbb{R}^N), d\!l_\infty\big)$ *to be* λ*-Lipschitz continuous in the first argument with*

$$\widehat{\alpha} := \sup_{\substack{M \in \mathscr{K}(\mathbb{R}^N)\\ t\in[0,T]}} \max\{0,\ \mathrm{Lip}\,\mathscr{F}(M,t)\} < \infty.$$

For $\mathscr{G} : \mathscr{K}(\mathbb{R}^N) \times [0,T] \longrightarrow \mathrm{OSLIP}(\mathbb{R}^N,\mathbb{R}^N)$ *assume*

$$\sup_{M,t} d\!l_\infty\big(\mathscr{F}(M,t),\ \mathscr{G}(M,t)\big) < \infty.$$

Any solutions $K_1(\cdot), K_2(\cdot) : [0,T] \leadsto \mathbb{R}^N$ *to the modified morphological equations*

$$\begin{cases} \overset{\circ}{K}_1(\cdot) \ni \mathscr{F}\big(K_1(\cdot), \cdot\big) \\ \overset{\circ}{K}_2(\cdot) \ni \mathscr{G}\big(K_2(\cdot), \cdot\big) \end{cases}$$

satisfy the following inequality for every $t \in [0,T]$

$$d\!l\big(K_1(t), K_2(t)\big) \le \Big(d\!l\big(K_1(0), K_2(0)\big) + t \cdot \sup_{M,s} d\!l_\infty\big(\mathscr{F}(M,s),\ \mathscr{G}(M,s)\big)\Big)\ e^{(\lambda+\widehat{\alpha})\,t}.$$

Proof is also a consequence of Proposition 24 in combination with Proposition 97. □

Furthermore, the existence of solutions *with* state constraints is again guaranteed by a consequence of Nagumo's general Theorem 19 (on page 40):

Proposition 102 (Nagumo's Theorem for modified morphological equations). *Suppose* $\mathscr{F} : \big(\mathscr{K}(\mathbb{R}^N), d\!l\big) \longrightarrow \big(\mathrm{OSLIP}(\mathbb{R}^N,\mathbb{R}^N),\ d\!l_\infty\big)$ *to be continuous with*

$$\sup_{M \in \mathscr{K}(\mathbb{R}^N)} \big(\|\mathscr{F}(M)\|_\infty + \max\{0,\ \mathrm{Lip}\,\mathscr{F}(M)\}\big) < \infty.$$

Then the following statements are equivalent for any closed subset $\mathscr{V} \subset \mathscr{K}(\mathbb{R}^N)$:

1. *Every set* $K_0 \in \mathscr{V}$ *is the initial set of at least one solution* $K : [0,1] \longrightarrow \mathscr{K}(\mathbb{R}^N)$ *to the modified morphological equation* $\overset{\circ}{K}(\cdot) \ni \mathscr{F}\big(K(\cdot)\big)$ *with* $K(t) \in \mathscr{V}$ *for all* $t \in [0,1]$.

2. $\mathscr{V} \subset \mathscr{K}(\mathbb{R}^N)$ *is a viability domain of* $\mathscr{F}$ *in the sense that*
$$\mathscr{F}(M) \in \mathscr{T}_{\mathscr{V}}(M) \subset \mathrm{OSLIP}(\mathbb{R}^N, \mathbb{R}^N) \qquad \textit{for every } M \in \mathscr{V}.$$
□

This, however, seems to be the critical point at which the obvious analogies to the morphological equations discussed in § 1.9 (on page 57 ff.) end.
In particular, Proposition 70 (on page 74) specifies the close link between any solution of a morphological equation and reachable sets of a suitable nonautonomous differential inclusion. Its counterpart for modified morphological equations can be formulated here only under additional assumptions about the continuity of each value $\mathscr{F}(M,t) \in \mathrm{OSLIP}(\mathbb{R}^N, \mathbb{R}^N)$.
This results from the following feature: Replacing the Lipschitz continuity of § 1.9 by the one-sided Lipschitz continuity (in combination with upper semicontinuity) implies an essential gap that is also pointed out in Remark A.64 (on page 481). Indeed, every map $F \in \mathrm{OSLIP}(\mathbb{R}^N, \mathbb{R}^N)$ satisfies the assumptions of Theorem A.63, but not every point $x_0 \in \mathbb{R}^N$ and vector $v_0 \in F(x_0)$ has to be related to a solution $x(\cdot) \in W^{1,1}([0,T], \mathbb{R}^N)$ of $x'(\cdot) \in F(x(\cdot))$ satisfying $x(0) = x_0$ and
$$\lim_{h \downarrow 0} \tfrac{1}{h} \cdot \big(x(h) - x(0)\big) \;=\; v_0.$$

Definition 103. $\mathrm{COSLIP}(\mathbb{R}^N, \mathbb{R}^N)$ consists of all maps in $\mathrm{OSLIP}(\mathbb{R}^N, \mathbb{R}^N)$ that are continuous in addition, i.e. every set-valued map $F : \mathbb{R}^N \rightsquigarrow \mathbb{R}^N$ satisfying

1. F has nonempty compact convex values that are uniformly bounded in $\mathbb{R}^N$,
2. F is continuous,
3. F is one-sided Lipschitz continuous, i.e. there is a constant $L \in \mathbb{R}$ such that for every $x, y \in \mathbb{R}^N$ and $v \in F(x)$, there exists some $w \in F(y)$ satisfying
$$\langle x - y,\, v - w \rangle \;\le\; L\, |x - y|^2.$$

Lemma 104. *Let* $\mathscr{F} : [0,T] \longrightarrow \big(\mathrm{COSLIP}(\mathbb{R}^N, \mathbb{R}^N), dl_\infty\big)$ *be* $\mathscr{L}^1$*-measurable with* $\sup\limits_{t \in [0,T]} \big(\|\mathscr{F}(t)\|_\infty + \max\{0,\, \mathrm{Lip}\, \mathscr{F}(t)\}\big) < \infty$ *and define the set-valued map*
$$\widehat{F} :\; [0,T] \times \mathbb{R}^N \rightsquigarrow \mathbb{R}^N, \qquad (t,x) \mapsto \mathscr{F}(t)(x).$$
Then for every set $K_0 \in \mathscr{K}(\mathbb{R}^N)$, *the reachable set* $\vartheta_{\widehat{F}}(\cdot, K_0) : [0,T] \longrightarrow \mathscr{K}(\mathbb{R}^N)$ *of the nonautonomous differential inclusion* $x' \in \widehat{F}(\cdot, x)$ *a.e. is a modified morphological primitive of* $\mathscr{F}(\cdot)$.

Proof results from Proposition A.13 (on page 447) in exactly the same way as Lemma 58 (on page 65). Indeed, continuity of the set-valued maps with respect to space (and not Lipschitz continuity) is assumed for proving the integral funnel equation in Proposition A.13. □

As a direct consequence of the uniqueness of primitives (Corollary 23 on page 41), we obtain the counterpart of Proposition 57 (on page 64) and can characterize these modified morphological primitives as reachable sets of nonautonomous differential inclusions:

Proposition 105 (Modified morphological primitives as reachable sets). *Suppose* $\mathscr{F} : [0,T] \longrightarrow \big(\mathrm{COSLIP}(\mathbb{R}^N,\mathbb{R}^N), d\!l_\infty\big)$ *to be Lebesgue-measurable with* $\sup_{t\in[0,T]} \big(\|\mathscr{F}(t)\|_\infty + \max\{0,\, \mathrm{Lip}\,\mathscr{F}(t)\}\big) < \infty$ *and define the set-valued map*

$$\widehat{F}:\; [0,T]\times\mathbb{R}^N \rightsquigarrow \mathbb{R}^N, \qquad (t,x) \mapsto \mathscr{F}(t)(x).$$

A tube $K : [0,T] \rightsquigarrow \mathbb{R}^N$ *is a modified morphological primitive of* $\mathscr{F}(\cdot)$ *if and only at every time* $t \in [0,T]$*, its value* $K(t) \subset \mathbb{R}^N$ *coincides with the reachable set of the nonautonomous differential inclusion* $x' \in \widehat{F}(\cdot,x)$ *a.e.*

$$K(t) \;=\; \vartheta_{\widehat{F}}\big(t,\, K(0)\big).$$

Corollary 106 (Solutions to modified morpholog. equations as reachable sets). *Let* $\mathscr{F} : \big(\mathscr{K}(\mathbb{R}^N), d\!l\big) \times [0,T] \longrightarrow \big(\mathrm{COSLIP}(\mathbb{R}^N,\mathbb{R}^N), d\!l_\infty\big)$ *be a Carathéodory function (i.e. here continuous with respect to the first argument and measurable with respect to time) satisfying*

$$\sup_{\substack{M\in\mathscr{K}(\mathbb{R}^N)\\ t\in[0,T]}} \big(\|\mathscr{F}(M,t)\|_\infty + \max\,\{0,\, \mathrm{Lip}\,\mathscr{F}(M,t)\}\big) < \infty.$$

Then a continuous tube $K : [0,T] \rightsquigarrow \mathbb{R}^N$ *is a solution to the modified morphological equation*

$$\overset{\circ}{K}(\cdot) \;\ni\; \mathscr{F}\big(K(\cdot),\cdot\big)$$

if and only if at every time $t \in [0,T]$*, the set* $K(t) \subset \mathbb{R}^N$ *coincides with the reachable set of the initial set* $K(0) \subset \mathbb{R}^N$ *and the nonautonomous differential inclusion*

$$x'(\cdot) \;\in\; \mathscr{F}\big(K(\cdot),\cdot\big)\,(x(\cdot)).$$

Chapter 2
Adapting Mutational Equations to Examples in Vector Spaces: Local Parameters of Continuity

The notion of transitions instead of affine linear maps in a given direction has proved to be very powerful. Aubin's definition of transition (Definition 1.1), however, is too restrictive.

Indeed, many examples in vector spaces share the feature that the Lipschitz constant of $t \longmapsto \vartheta(t,x)$ cannot be bounded uniformly for all initial states x. In this chapter we will study several examples in which the transitions are based on solutions to linear problems in vector spaces. Doubling the initial state implies doubling the transition value and thus doubling the Lipschitz constant with respect to time.

The main goal of the subsequent chapters is to weaken the conditions on transitions and solutions in the mutational framework such that Euler method still provides existence of (generalized) solutions.

In this chapter, we implement two additional aspects in the recently introduced terms: Firstly, we use an analog of the absolute value in the metric space (E,d). Indeed, $\lfloor \cdot \rfloor : E \longrightarrow [0, \infty[$ is just to specify the "absolute magnitude" of each element in E, but does not have to satisfy structural conditions such as homogeneity or triangle inequality. In contrast to a metric, $\lfloor \cdot \rfloor$ does not serve the comparison of two elements in E, but the continuity parameters $\alpha(\vartheta), \beta(\vartheta)$ will be assumed to be uniform in all "balls" $\{x \in E \mid \lfloor x \rfloor \leq r\}$ with positive "radius" $r > 0$. The proofs do not change substantially if we impose appropriate bounds on the growth of $\lfloor \vartheta(\cdot, x) \rfloor$ for each initial element $x \in E$.

Secondly, we admit more than just one distance function on E simultaneously. A family $(d_j)_{j \in \mathscr{I}}$ of pseudo-metrics on E (i.e. reflexive, symmetric and satisfying the triangle inequality, but not necessarily positive definite) replaces the metric d always used in Chapter 1. The weak topology of a Banach space, for example, is much easier to describe by means of many linear forms than by just a single metric and, the suitable choice of linear forms will prove to be very helpful for semilinear evolution equations discussed in subsequent § 2.4.

In a word, these extensions of the mutational framework do not require significant improvements of the proofs in comparison with the preceding chapter. They share the basic notion with later generalizations: For implementing additional "degrees of freedom", we focus on the question which parameter may depend on which others.

T. Lorenz, *Mutational Analysis: A Joint Framework for Cauchy Problems In and Beyond Vector Spaces*, Lecture Notes in Mathematics 1996, DOI 10.1007/978-3-642-12471-6_3, © Springer-Verlag Berlin Heidelberg 2010

2.1 The Topological Environment of This Chapter

E always denotes a nonempty set, but we do not restrict our considerations to a metric space (E,d) as in Chapter 1.

Definition 1. Let E be a nonempty set. A function $d : E \times E \longrightarrow [0,\infty[$ is called *pseudo-metric* on E if it satisfies the following conditions:

1. d is reflexive, i.e. for all $x \in E$: $d(x, x) = 0$,
2. d is symmetric, i.e. for all $x, y \in E$: $d(x, y) = d(y, x)$
3. d satisfies the triangle inequality, i.e. for all x, y, z: $d(x, z) \leq d(x, y) + d(y, z)$.

In particular, a pseudo-metric d on E does not have to be positive definite, i.e. $d(x,y) = 0$ does not always imply $x = y$.

General assumptions for Chapter 2. E is a nonempty set and, $\mathscr{I} \neq \emptyset$ denotes an index set. For each index $j \in \mathscr{I}$, $d_j : E \times E \longrightarrow [0,\infty[$ is a pseudo-metric on E and, $\lfloor\cdot\rfloor_j : E \longrightarrow [0,\infty[$ is a given function that is lower semicontinuous with respect to the topology of $(d_i)_{i\in\mathscr{I}}$, i.e. strictly speaking,

$$\lfloor x \rfloor_j \;\leq\; \liminf_{n\longrightarrow\infty} \; \lfloor x_n \rfloor_j$$

for any $x \in E$ and sequence $(x_n)_{n\in\mathbb{N}}$ in E with $d_i(x_n,x) \overset{n\to\infty}{\longrightarrow} 0$ and $\sup_n \lfloor x_n \rfloor_i < \infty$ for each $i \in \mathscr{I}$.

Now the main goal of this chapter is to extend the mutational framework from a metric space to the tuple $\big(E, (d_j)_{j\in\mathscr{I}}, (\lfloor\cdot\rfloor_j)_{j\in\mathscr{I}}\big)$. Several examples in vector spaces like semilinear evolution equations and nonlinear transport equations will follow.

2.2 Specifying Transitions and Mutation on $\big(E, (d_j)_{j\in\mathscr{I}}, (\lfloor\cdot\rfloor_j)_{j\in\mathscr{I}}\big)$

Definition 2. $\vartheta : [0,1] \times E \longrightarrow E$ is called *transition* on $\big(E, (d_j)_{j\in\mathscr{I}}, (\lfloor\cdot\rfloor_j)_{j\in\mathscr{I}}\big)$ if it satisfies the following conditions for each $j \in \mathscr{I}$:

1.) for every $x \in E$: $\quad \vartheta(0,x) = x$

2.) for every $x \in E$, $t \in [0,1[$: $\quad \lim_{h\downarrow 0} \frac{1}{h} \cdot d_j\big(\vartheta(t+h,x),\ \vartheta(h,\vartheta(t,x))\big) = 0$

3.) there exists $\alpha_j(\vartheta;\cdot) : [0,\infty[\longrightarrow [0,\infty[$ such that for any $x, y \in E$ with
$$\lfloor x \rfloor_j \leq r,\ \lfloor y \rfloor_j \leq r: \ \limsup_{h\downarrow 0} \frac{d_j(\vartheta(h,x),\ \vartheta(h,y)) - d_j(x,y)}{h} \leq \alpha_j(\vartheta;r)\cdot d_j(x,y)$$

4.) there exists $\beta_j(\vartheta;\cdot) : [0,\infty[\longrightarrow [0,\infty[$ such that for any $s,t \in [0,1]$ and $x \in E$ with $\lfloor x \rfloor_j \leq r$: $\quad d_j\big(\vartheta(s,x),\ \vartheta(t,x)\big) \leq \beta_j(\vartheta;r)\cdot|t-s|$

5.) there exists $\gamma_j(\vartheta) \in [0,\infty[$ such that for any $t \in [0,1]$ and $x \in E$:
$$\lfloor \vartheta(t,x) \rfloor_j \;\leq\; \big(\lfloor x \rfloor_j + \gamma_j(\vartheta)\ t\big)\cdot e^{\gamma_j(\vartheta)\,t}$$

Remark 3. In particular, this definition covers the special case of a transition $\vartheta : [0,1] \times E \longrightarrow E$ on a metric space (E,d) in the sense of Definition 1.1 (on page 32). Indeed, set $\mathscr{I} = \{0\}$, $d_0 := d$ and $\lfloor\cdot\rfloor_0 := 0$. Then $\alpha(\vartheta;\cdot)$ and $\beta(\vartheta;\cdot)$ can be chosen constant for each transition ϑ on (E,d). $\gamma_0(\vartheta)$ is defined as 0 arbitrarily.

Now the continuity parameters of a transition are fixed for each "ball" $\{x \in E \mid \lfloor x \rfloor_j \le r\}$ $(r > 0, j \in \mathscr{I})$. This does not cause analytical difficulties since condition (5.) provides a suitable a priori bound of $\lfloor \vartheta(t,x) \rfloor_j$ for $t \in [0,1]$. The choice of its structure is rather arbitrary (we admit), but it covers many examples and, the following lemma lays the foundations inductively for extending many results of Chapter 1 to transitions on $(E, (d_j)_{j\in\mathscr{I}}, (\lfloor\cdot\rfloor_j)_{j\in\mathscr{I}})$.

Lemma 4. *Let $\vartheta_1 \ldots \vartheta_K$ be finitely many transitions on $(E, (d_j)_{j\in\mathscr{I}}, (\lfloor\cdot\rfloor_j)_{j\in\mathscr{I}})$ with*

$$\widehat{\gamma}_j := \sup_{k\in\{1\ldots K\}} \gamma_j(\vartheta_k) < \infty \qquad \textit{for some } j \in \mathscr{I}.$$

For any $x_0 \in E$ and $0 = t_0 < t_1 < \ldots < t_K$ with $\sup_k t_k - t_{k-1} \le 1$ define the curve $x(\cdot) : [0, t_K] \longrightarrow E$ piecewise as $x(0) := x_0$ and

$$x(t) := \vartheta_k\big(t - t_{k-1},\, x(t_{k-1})\big) \qquad \textit{for } t \in \,]t_{k-1}, t_k],\ k \in \{1 \ldots K\}.$$

Then, $\lfloor x(t) \rfloor_j \le \big(\lfloor x_0 \rfloor_j + \widehat{\gamma}_j \cdot t\big) \cdot e^{\widehat{\gamma}_j \cdot t}$ *at every time $t \in [0, t_K]$.*

Proof is given via induction with respect to k : The claim is obvious at time $t_0 = 0$. Assuming this estimate at time t_{k-1}, we conclude for each $t \in \,]t_{k-1}, t_k]$

$$\begin{aligned}
\lfloor x(t) \rfloor_j &= \big\lfloor \vartheta_k\big(t - t_{k-1},\, x(t_{k-1})\big) \big\rfloor_j \\
&\le \big(\lfloor x(t_{k-1}) \rfloor_j + \widehat{\gamma}_j \cdot (t - t_{k-1})\big) \cdot e^{\widehat{\gamma}_j \cdot (t - t_{k-1})} \\
&\le \big(\big(\lfloor x_0 \rfloor_j + \widehat{\gamma}_j \cdot t_{k-1}\big) \cdot e^{\widehat{\gamma}_j\, t_{k-1}} + \widehat{\gamma}_j \cdot (t - t_{k-1})\big) \cdot e^{\widehat{\gamma}_j \cdot (t - t_{k-1})} \\
&\le \big(\lfloor x_0 \rfloor_j + \widehat{\gamma}_j \cdot t\big) \cdot e^{\widehat{\gamma}_j \cdot t}.
\end{aligned}$$

□

The next step is to implement this locally uniform aspect of parameters in the distance between transitions. Following the basic idea of Definition 1.6 (on page 35), we introduce

Definition 5. $\Theta(E, (d_j)_{j\in\mathscr{I}}, (\lfloor\cdot\rfloor_j)_{j\in\mathscr{I}})$ denotes a nonempty set of transitions on $(E, (d_j)_{j\in\mathscr{I}}, (\lfloor\cdot\rfloor_j)_{j\in\mathscr{I}})$ satisfying additionally

$$D_j(\vartheta, \tau; r) := \sup_{x\in E:\, \lfloor x \rfloor_j \le r} \ \limsup_{h\downarrow 0} \ \tfrac{1}{h} \cdot d_j\big(\vartheta(h,x),\ \tau(h,x)\big) < \infty$$

for any $\vartheta, \tau \in \Theta(E, (d_j)_{j\in\mathscr{I}}, (\lfloor\cdot\rfloor_j)_{j\in\mathscr{I}})$ and $r \ge 0, j \in \mathscr{I}$. (If $\{x \in E \mid \lfloor x \rfloor_j \le r\} = \emptyset$, set $D_j(\cdot,\cdot; r) := 0$.)

For each $r \geq 0$, the distance function

$$D_j(\cdot,\cdot;r):\ \Theta\big(E,(d_j)_{j\in\mathscr{I}},(\lfloor\cdot\rfloor_j)_{j\in\mathscr{I}}\big)\times\Theta\big(E,(d_j)_{j\in\mathscr{I}},(\lfloor\cdot\rfloor_j)_{j\in\mathscr{I}}\big)\ \longrightarrow\ [0,\infty[$$

is reflexive, symmetric and satisfies the triangle inequality and thus, $D_j(\cdot,\cdot;r)$ is a pseudo-metric on the transition set $\Theta\big(E,(d_j)_{j\in\mathscr{I}},(\lfloor\cdot\rfloor_j)_{j\in\mathscr{I}}\big)$.
Similarly to Proposition 1.7 (on page 35), we can now compare the evolution of two states in E along two different transitions:

Proposition 6. *Let $\vartheta,\tau\in\Theta\big(E,(d_j)_{j\in\mathscr{I}},(\lfloor\cdot\rfloor_j)_{j\in\mathscr{I}}\big)$ be arbitrary, $r\geq 0, j\in\mathscr{I}$. Then for any elements $x,y\in E$ with $\lfloor x\rfloor_j\leq r,\ \lfloor y\rfloor_j\leq r$ and times $t_1,t_2\in[0,1[$, the following estimate is satisfied at each time $h\in[0,1[$ with $\max\{t_1+h,\ t_2+h\}\leq 1$*

$$d_j\big(\vartheta(t_1+h,x),\ \tau(t_2+h,y)\big)\ \leq\ \big(d_j\big(\vartheta(t_1,x),\ \tau(t_2,y)\big)+h\cdot D_j(\vartheta,\tau;R_j)\big)\cdot e^{\alpha_j(\vartheta;R_j)\,h}$$

with $R_j\ :=\ \big(r+\max\{\gamma_j(\vartheta),\ \gamma_j(\tau)\}\big)\cdot e^{\max\{\gamma_j(\vartheta),\ \gamma_j(\tau)\}}$.

Proof results from Gronwall's inequality (in Proposition A.2 on page 440) applied to the auxiliary function

$$\psi_j : h\ \longmapsto\ d_j\big(\vartheta(t_1+h,x),\ \tau(t_2+h,y)\big)$$

in exactly the same way as the proof of Proposition 1.7 (on page 36) because condition (5.) of Definition 2 ensures for each $h\in[0,1]$

$$\begin{cases}\lfloor\vartheta(h,x)\rfloor_j\ \leq\ R_j\\ \lfloor\tau(h,y)\rfloor_j\ \leq\ R_j\end{cases}$$
□

As in § 1.2 (on page 37), the notion of first-order approximation leads to the so-called mutation of a curve – as counterpart of its time derivative:

Definition 7. Let $x(\cdot):[0,T]\longrightarrow E$ be a function. The set

$$\overset{\circ}{x}(t) := \Big\{\vartheta\in\Theta\big(E,(d_j)_{j\in\mathscr{I}},(\lfloor\cdot\rfloor_j)_{j\in\mathscr{I}}\big)\ \Big|\ \forall\, j\in\mathscr{I}:\ \lim_{h\downarrow 0}\ \tfrac{1}{h}\cdot d_j\big(\vartheta(h,x(t)),\ x(t+h)\big)\ =\ 0\Big\}$$

is called *mutation* of $x(\cdot)$ at time $t\in[0,T[$ in $\big(E,(d_j)_{j\in\mathscr{I}},(\lfloor\cdot\rfloor_j)_{j\in\mathscr{I}}\big)$.

Remark 8. Remark 1.11 (on page 37) also holds for transitions on the tuple $\big(E,(d_j)_{j\in\mathscr{I}},(\lfloor\cdot\rfloor_j)_{j\in\mathscr{I}}\big)$: For every transition $\vartheta\in\Theta\big(E,(d_j)_{j\in\mathscr{I}},(\lfloor\cdot\rfloor_j)_{j\in\mathscr{I}}\big)$ and initial element $x_0\in E$, the curve $x_{x_0}(\cdot):=\vartheta(\cdot,x_0):[0,1]\longrightarrow E$ has ϑ in its mutation at each time $t\in[0,1[$:

$$\vartheta\in\ \overset{\circ}{x}_{x_0}(t).$$

This results directly from condition (2.) in Definition 2 and, it lays the basis for constructing solutions by means of Euler method in the next section.

2.3 Solutions to Mutational Equations

Now we focus on solving dynamical problems with feedback: For a given function relating each state in E and time to a transition on $\big(E, (d_j)_{j\in\mathscr{I}}, (\lfloor\cdot\rfloor_j)_{j\in\mathscr{I}}\big)$, we are looking for a curve in E whose mutation obeys this "law" at almost every time. In comparison with Definition 1.13 (on page 38) for a metric space, however, the families $(d_j)_{j\in\mathscr{I}}$, $(\lfloor\cdot\rfloor_j)_{j\in\mathscr{I}}$ should be taken into consideration appropriately:

Definition 9. A single-valued function $f : E\times[0,T] \longrightarrow \Theta\big(E, (d_j)_{j\in\mathscr{I}}, (\lfloor\cdot\rfloor_j)_{j\in\mathscr{I}}\big)$ is given. $x(\cdot) : [0,T] \longrightarrow E$ is called a *solution* to the mutational equation

$$\overset{\circ}{x}(\cdot) \ni f(x(\cdot), \cdot)$$

in $\big(E, (d_j)_{j\in\mathscr{I}}, (\lfloor\cdot\rfloor_j)_{j\in\mathscr{I}}\big)$ if it satisfies the following conditions for each $j\in\mathscr{I}$:

1.) $x(\cdot)$ is continuous with respect to d_j

2.) for $\mathscr{L}^1$-almost every $t\in[0,T[$: $\lim\limits_{h\downarrow 0} \; \frac{1}{h}\cdot d_j\big(f(x(t),t)\,(h, x(t)),\; x(t+h)\big) = 0$

3.) $\sup\limits_{t\in[0,T]} \; \lfloor x(t)\rfloor_j < \infty$.

A global bound of the continuity parameter $\beta_j(\,\cdot\,; R)$ implies that each solution is even (locally) Lipschitz continuous with respect to d_j.

Lemma 10. *For $f : E\times[0,T] \longrightarrow \Theta\big(E, (d_j)_{j\in\mathscr{I}}, (\lfloor\cdot\rfloor_j)_{j\in\mathscr{I}}\big)$ let $x(\cdot) : [0,T] \longrightarrow E$ be a solution to the mutational equation $\overset{\circ}{x}(\cdot) \ni f(x(\cdot),\cdot)$ such that some $j\in\mathscr{I}$ and $L_j, R_j\in\mathbb{R}$ satisfy for all $t\in[0,T]$*

$$\begin{cases} \lfloor x(t)\rfloor_j \le R_j \\ \beta_j\big(f(x(t),t);\, R_j\big) \le L_j. \end{cases}$$

Then $x(\cdot)$ is L_j-Lipschitz continuous with respect to d_j.

Proof. Fix $s\in[0,T[$ arbitrarily. Then the auxiliary function

$$\psi_j : [s,T] \longrightarrow \mathbb{R}, \quad t \longmapsto d_j\big(x(s),\, x(t)\big)$$

is continuous due to Definition 9 (1.) and, it satisfies for $\mathscr{L}^1$-almost every $t\in[s,T]$

$$\begin{aligned}
\limsup_{h\downarrow 0} \frac{\psi_j(t+h)-\psi_j(t)}{h} &\le \limsup_{h\downarrow 0} \frac{1}{h}\cdot \; d_j\big(x(t),\, x(t+h)\big) \\
&\le \limsup_{h\downarrow 0} \frac{1}{h}\cdot \Big(d_j\big(x(t),\, f(x(t),t)\,(h,x(t))\big) \\
&\qquad\qquad + d_j\big(f(x(t),t)\,(h,x(t)),\; x(t+h)\big)\Big) \\
&\le L_j + 0.
\end{aligned}$$

Finally $\psi_j(t) \le L_j\cdot(t-s)$ for all $t\in[s,T]$ results from Gronwall's inequality (Proposition A.2 on page 440). □

2.3.1 Continuity with Respect to Initial States and Right-Hand Side

The continuity of solutions with respect to given data plays a key role for solving mutational equations by explicit methods such as Euler algorithm. For this reason, we now extend Proposition 1.21 (on page 41) and Proposition 1.24 (on page 42) to mutational equations in $\big(E, (d_j)_{j\in\mathscr{I}}, (\lfloor\cdot\rfloor_j)_{j\in\mathscr{I}}\big)$:

Proposition 11. *Assume for* $f, g : E \times [0,T] \longrightarrow \Theta\big(E, (d_j)_{j\in\mathscr{I}}, (\lfloor\cdot\rfloor_j)_{j\in\mathscr{I}}\big)$ *and* $x, y : [0,T] \longrightarrow E$ *that* $x(\cdot)$ *is a solution to the mutational equation* $\overset{\circ}{x}(\cdot) \ni f(x(\cdot), \cdot)$ *and* $y(\cdot)$ *is a solution to the mutational equation* $\overset{\circ}{y}(\cdot) \ni g(y(\cdot), \cdot)$.
For some $j \in \mathscr{I}$, *let* $\widehat{\alpha}_j, R_j > 0$ *and* $\varphi_j \in C^0([0,T])$ *satisfy for almost every* $t \in [0,T]$

$$\begin{cases} \lfloor x(t)\rfloor_j,\ \lfloor y(t)\rfloor_j \ \le\ R_j \\ \alpha_j\,(g(y(t),t);\, R_j) \ \le\ \widehat{\alpha}_j \\ D_j\,(f(x(t),t),\ g(y(t),t);\, R_j) \ \le\ \varphi_j(t). \end{cases}$$

Then, $d_j(x(t), y(t)) \le \big(d_j(x(0), y(0)) + \int_0^t \varphi_j(s)\, e^{-\widehat{\alpha}_j\cdot s} ds\big)\, e^{\widehat{\alpha}_j\cdot t}$ *for any* $t \in [0,T]$.

By means of monotone approximation in the sense of Daniell-Lebesgue, this estimate can be extended to Lebesgue-integrable functions $\varphi_j : [0,T] \longrightarrow [0,\infty[$ easily. Assuming one of the functions on the right-hand side to be Lipschitz continuous in addition simplifies the comparison between two solutions w.r.t. a pseudo-metric d_j:

Corollary 12. *For some* $j \in \mathscr{I}$ *and each* $r > 0$, *suppose* $f : E \times [0,T] \longrightarrow \Theta\big(E, (d_i)_{i\in\mathscr{I}}, (\lfloor\cdot\rfloor_i)_{i\in\mathscr{I}}\big)$ *to satisfy* $\widehat{\alpha}_{j,r} := \sup_{z,t}\ \alpha_j(f(z,t);\, r) < \infty$ *and to fulfill with a constant* $\lambda_{j,r} > 0$ *that for* $\mathscr{L}^1$*-almost every* $t \in [0,T]$,

$$f(\cdot,t) : (E, d_j) \longrightarrow \big(\Theta\big(E, (d_i)_{i\in\mathscr{I}}, (\lfloor\cdot\rfloor_i)_{i\in\mathscr{I}}\big),\ D_j(\cdot\,,\cdot\,; r)\big)$$

is $\lambda_{j,r}$*-Lipschitz continuous. For* $g : E \times [0,T] \longrightarrow \Theta\big(E, (d_i)_{i\in\mathscr{I}}, (\lfloor\cdot\rfloor_i)_{i\in\mathscr{I}}\big)$ *assume*

$$\sup_{z,s}\ D(f(z,s),\, g(z,s);\, r) < \infty \qquad \textit{for each } r > 0.$$

Then every solutions $x(\cdot), y(\cdot) : [0,T] \longrightarrow E$ *to the mutational equations*

$$\overset{\circ}{x}(\cdot) \ni f(x(\cdot),\cdot) \qquad \overset{\circ}{y}(\cdot) \ni g(y(\cdot),\cdot)$$

satisfy the following inequality for every $t \in [0,T]$

$$d_j(x(t), y(t)) \le \big(d_j(x(0), y(0)) + t\cdot \sup_{z,\,s} D_j(f(z,s), g(z,s));\, R_j)\big)\ e^{(\widehat{\alpha}_{j,R_j} + \lambda_{j,R_j})\,t}$$

with $R_j := \sup_{t\in[0,T]} \big\{\lfloor x(t)\rfloor_j,\ \lfloor y(t)\rfloor_j\big\} < \infty$.

Proof (of Proposition 11). As in the proof of Proposition 1.21 (on page 43), we consider the auxiliary function

$$\psi_j : [0,T] \longrightarrow [0,\infty[, \quad t \longmapsto d_j\big(x(t),\, y(t)\big).$$

It is continuous because any solutions $x(\cdot)$, $y(\cdot)$ to mutational equations are continuous with respect to d_j due to Definition 9.
Furthermore, we obtain for Lebesgue-almost every $t \in [0,T[$

$$\begin{aligned}
&\limsup_{h\downarrow 0} \tfrac{1}{h}\cdot d_j\big(x(t+h), \quad f(x(t),t)\,(h,x(t))\big) = 0\\
&\limsup_{h\downarrow 0} \tfrac{1}{h}\cdot d_j\big(f(x(t),t)\,(h,x(t)),\ g(y(t),t)\,(h,x(t))\big) \le D_j\big(f(x(t),t),\, g(y(t),t); R_j\big)\\
&\limsup_{h\downarrow 0} \tfrac{1}{h}\cdot d_j\big(g(y(t),t)\,(h,y(t)),\ y(t+h)\big) = 0
\end{aligned}$$

due to Definition 5 and Definition 9. For estimating $\psi_j(t+h)$, we conclude from the assumed bound of $\alpha_j(g(y(t),t); R_j)$, i.e.

$$\limsup_{h\downarrow 0} \tfrac{1}{h}\cdot \Big(d_j\big(g(y(t),t)\,(h,x(t)),\ g(y(t),t)\,(h,y(t))\big) - \psi_j(t)\Big) \le \widehat{\alpha}_j\cdot\psi_j(t),$$

and the triangle inequality of d_j

$$\begin{aligned}
\limsup_{h\downarrow 0} \frac{\psi_j(t+h) - \psi_j(t)}{h} &\le \widehat{\alpha}_j\cdot\psi_j(t) + D_j\big(f(x(t),t),\, g(y(t),t);\, R_j\big)\\
&\le \widehat{\alpha}_j\cdot\psi_j(t) + \varphi_j(t)
\end{aligned}$$

at Lebesgue-almost every time $t \in [0,T[$. Finally the claimed estimate results from generalized Gronwall's Lemma (Proposition A.2 on page 440). □

Proof (of Corollary 12). It results from Proposition 11 in exactly the same way as Proposition 1.24 was concluded from Proposition 1.21 (on page 44). □

2.3.2 *Limits of Pointwise Converging Solutions: Convergence Theorem*

In preceding Proposition 11, the continuity of solutions (with respect to initial data and right-hand side) is based on the assumption that two solutions are given. Hence this result can hardly be used as a tool for proving an existence theorem.

Now we consider a sequence of solutions instead. If it converges with respect to the topology of $(d_j)_{j\in\mathscr{I}}$ then the limit function might be a solution to a mutational equation. The following theorem extends Convergence Theorem 1.30 (on page 48) and specifies the details.
It is worth pointing out briefly that we do not require *uniform* convergence of the sequence with respect to each d_j, $j \in \mathscr{I}$, but just pointwise convergence of subsequences, which can even depend on time. Moreover, perturbations of the right-hand sides are also taken into consideration. This aspect will be very helpful for the Euler approximations used in subsequent § 2.3.3.

Theorem 13 (Convergence of solutions to mutational equations).
For each $j \in \mathcal{I}$, suppose the following properties of

$$\begin{array}{rrl} f_n, f: E \times [0,T] & \longrightarrow & \Theta\big(E, (d_i)_{i\in\mathcal{I}}, (\lfloor\cdot\rfloor_i)_{i\in\mathcal{I}}\big) \\ x_n, x: \quad [0,T] & \longrightarrow & E: \end{array} \qquad (n \in \mathbb{N})$$

1.) $R_j := \sup_{n,t} \lfloor x_n(t)\rfloor_j < \infty,$

$\widehat{\alpha}_j := \sup_{n,t,y} \alpha_j\big(f_n(y,t); R_j\big) < \infty,$

$\widehat{\beta}_j := \sup_{n} \operatorname{Lip}\big(x_n(\cdot) : [0,T] \longrightarrow (E,d_j)\big) < \infty,$

2.) $\overset{\circ}{x}_n(\cdot) \ni f_n(x_n(\cdot), \cdot)$ *(in the sense of Definition 9 on page 107) for every* $n \in \mathbb{N}$,

3.) $\lim_{n\to\infty} D_j\big(f_n(x(t),t),\ f_n(y_n,t_n);\ R_j\big) = 0$ *for $\mathcal{L}^1$-almost every $t \in [0,T]$ and any sequences $(t_n)_{n\in\mathbb{N}}$, $(y_n)_{n\in\mathbb{N}}$ in $[t,T]$ and E respectively satisfying* $\lim_{n\to\infty} t_n = t$ *and* $\lim_{n\to\infty} d_i\big(x(t), y_n\big) = 0$, $\sup_{n\in\mathbb{N}} \lfloor y_n\rfloor_i \le R_i$ *for each $i \in \mathcal{I}$,*

4.) for Lebesgue-almost every $t \in [0,T]$ and any $\widetilde{t} \in [0,T[$, there exists a sequence $n_m \nearrow \infty$ of indices (possibly depending on $t, \widetilde{t}, j$) that satisfies for $m \longrightarrow \infty$ and each $i \in \mathcal{I}$

$$\begin{cases} (i) & D_j\big(f(x(t),t),\ f_{n_m}(x(t),t);\ R_j\big) \longrightarrow 0 \\ (ii) & d_i\big(x(t),\ x_{n_m}(t)\big) \longrightarrow 0 \\ (iii) & d_j\big(x(\widetilde{t}),\ x_{n_m}(\widetilde{t})\big) \longrightarrow 0 \end{cases}$$

Then, $x(\cdot)$ is a solution to the mutational equation $\overset{\circ}{x}(\cdot) \ni f(x(\cdot), \cdot)$ in $[0,T[$.

Proof. Choose the index $j \in \mathcal{I}$ arbitrarily. Then $x(\cdot) : [0,T] \longrightarrow (E, d_j)$ is $\widehat{\beta}_j$-Lipschitz continuous. Indeed, for Lebesgue-almost every $t \in [0,T]$ and any $\widetilde{t} \in [0,T]$, assumption (4.) provides a subsequence $\big(x_{n_m}(\cdot)\big)_{m\in\mathbb{N}}$ satisfying

$$\begin{cases} d_j\big(x(t),\ x_{n_m}(t)\big) \longrightarrow 0 \\ d_j\big(x(\widetilde{t}),\ x_{n_m}(\widetilde{t})\big) \longrightarrow 0 \end{cases} \qquad \text{for } m \longrightarrow \infty.$$

The uniform $\widehat{\beta}_j$-Lipschitz continuity of $x_n(\cdot), n \in \mathbb{N}$, and the properties of d_j imply

$$\begin{array}{rlll} d_j\big(x(t),\ x(\widetilde{t})\big) & \le d_j\big(x(t), x_{n_m}(t)\big) & + d_j\big(x_{n_m}(t), x_{n_m}(\widetilde{t})\big) & + d_j\big(x_{n_m}(\widetilde{t}), x(\widetilde{t})\big) \\ & \le d_j\big(x(t), x_{n_m}(t)\big) & + \widehat{\beta}_j\, |\widetilde{t} - t| & + d_j\big(x_{n_m}(\widetilde{t}), x(\widetilde{t})\big) \\ & \longrightarrow 0 & + \widehat{\beta}_j\, |\widetilde{t} - t| & + 0 \qquad \text{for } m \longrightarrow \infty. \end{array}$$

This Lipschitz inequality even holds for *any* $t \in [0,T]$ due to the triangle inequality of d_j. Moreover the general hypothesis about lower semicontinuity of $\lfloor\cdot\rfloor_j$ ensures

$$\lfloor x(\widetilde{t})\rfloor_j \le \liminf_{m\to\infty} \lfloor x_{n_m}(\widetilde{t})\rfloor_j \le R_j.$$

Finally we verify the solution property

$$\lim_{h\downarrow 0} \tfrac{1}{h} \cdot d_j\big(f(x(t),t)\,(h, x(t)),\ x(t+h)\big) = 0$$

for Lebesgue-almost every $t \in [0,T[$. Indeed, for Lebesgue-almost every $t \in [0,T[$ and any $h \in]0, T-t[$, assumption (4.) guarantees a subsequence $\big(x_{n_m}(\cdot)\big)_{m\in\mathbb{N}}$ satisfying for each $i \in \mathcal{I}$ and $m \longrightarrow \infty$

$$\begin{cases} D_j\big(f(x(t),t),\ f_{n_m}(x(t),t);\ R_j\big) & \longrightarrow 0 \\ d_i\big(x(t),\ \ \ x_{n_m}(t)\big) & \longrightarrow 0 \\ d_j\big(x(t+h),\ x_{n_m}(t+h)\big) & \longrightarrow 0 \end{cases}$$

We conclude from Proposition 6 (on page 106) and Proposition 11 (on page 108) respectively

$$\begin{aligned} & d_j\big(f(x(t),t)\,(h,\,x(t)),\ \ x(t+h)\big) \\ & \quad \le\ \ d_j\big(f(x(t),t)\ \ (h,\,x(t)),\ \ f_{n_m}(x(t),t)\,(h,\,x(t))\big) \\ & \quad\quad +\, d_j\big(f_{n_m}(x(t),t)\,(h,\,x(t)),\ \ x_{n_m}(t+h)\big) \\ & \quad\quad +\, d_j\big(x_{n_m}(t+h),\ \ x(t+h)\big) \\ & \quad \le\ \ h\ e^{\widehat{\alpha}_j\,h}\cdot D_j\big(f(x(t),t),\ \ f_{n_m}(x(t),t);\ R_j\big) \\ & \quad\quad +\, d_j\big(x(t),\,x_{n_m}(t)\big)\ e^{\widehat{\alpha}_j\,h} + h\,e^{\widehat{\alpha}_j\,h}\cdot \sup_{t\le s\le t+h} D_j\big(f_{n_m}(x(t),t),f_{n_m}(x_{n_m}(s),s);\ R_j\big) \\ & \quad\quad +\, d_j\big(x_{n_m}(t+h),\ \ x(t+h)\big)\,. \end{aligned}$$

Now $m \longrightarrow \infty$ leads to the inequality

$$\begin{aligned} & d_j\big(f(x(t),t)\,(h,\,x(t)),\ x(t+h)\big) \\ & \quad \le\ h\,e^{\widehat{\alpha}_j\,h}\cdot \limsup_{m\to\infty}\ \sup_{[t,t+h]} D_j\big(f_{n_m}(x(t),t),\ f_{n_m}(x_{n_m}(\cdot),\cdot);\ R_j\big). \end{aligned}$$

For completing the proof, it is sufficient to verify

$$\limsup_{h\downarrow 0}\ \limsup_{m\to\infty}\ \sup_{[t,t+h]} D_j\big(f_{n_m}(x(t),t),\ f_{n_m}(x_{n_m}(\cdot),\cdot);\ R_j\big) \ = 0$$

for Lebesgue-almost every $t \in [0,T[$ and *any* subsequence $\big(x_{n_m}(\cdot)\big)_{m\in\mathbb{N}}$ satisfying

$$d_i\big(x(t),\,x_{n_m}(t)\big) \longrightarrow 0 \qquad \text{for } m \longrightarrow \infty \text{ and each } i \in \mathscr{I}.$$

Indeed, if this limit superior was positive then we could select some $\varepsilon > 0$ and sequences $(h_l)_{l\in\mathbb{N}}$, $(m_l)_{l\in\mathbb{N}}$, $(s_l)_{l\in\mathbb{N}}$ such that

$$\begin{cases} D_j\big(f_{n_{m_l}}(x(t),t),\ \ f_{n_{m_l}}(x_{n_{m_l}}(t+s_l),\,t+s_l);\ R_j\big) \ \ge\ \varepsilon \\ 0 \le s_l \le h_l \le \frac{1}{l}, \qquad m_l \ge l \end{cases} \qquad \text{for all } l \in \mathbb{N}.$$

The consequence

$$d_i\big(x(t),\,x_{n_{m_l}}(t+s_l)\big)\ \le\ d_i\big(x(t),\,x_{n_{m_l}}(t)\big)\ +\ \widehat{\beta}_i\,s_l\ \stackrel{l\to\infty}{\longrightarrow}\ 0$$

for each $i \in \mathscr{I}$ would lead to a contradiction to equi-continuity assumption (3.) at Lebesgue-almost every time $t \in [0,T[$. □

Remark 14. The continuity assumptions about $(x_n(\cdot))_{n\in\mathbb{N}}$ can be weakened easily. Supposing for each index $j \in \mathscr{I}$ that the sequence $(x_n(\cdot))_{n\in\mathbb{N}}$ is equi-continuous with respect to d_j (instead of uniformly $\widehat{\beta}_j$-Lipschitz continuous) admits the same conclusions and thus, the limit function $x(\cdot)$ is also a solution of $\overset{\circ}{x}(\cdot) \ni f(x(\cdot),\cdot)$ in the sense of Definition 9.

2.3.3 *Existence for Mutational Equations without State Constraints*

Whenever equations are solved constructively, two principles usually bridge the gap between approximations and the wanted solution: completeness or compactness. In fact, both principles guarantee the existence of a limit, but compactness refers to any sequence and focuses on a suitable subsequence whereas the concept of completeness is restricted to Cauchy sequences. In metric spaces, compactness usually implies completeness.
For the tuple $\big(E, (d_j)_{j\in\mathscr{I}}, (\lfloor\cdot\rfloor_j)_{j\in\mathscr{I}}\big)$, however, we usually prefer compactness as analytical basis for constructing solutions to mutational equations because a family $(d_j)_{j\in\mathscr{I}}$ of pseudo-metrics is admitted (and we have not even supposed the index set $\mathscr{I} \neq \emptyset$ to be at most countable).

Specifying a suitable form of sequential compactness in $\big(E, (d_j)_{j\in\mathscr{I}}, (\lfloor\cdot\rfloor_j)_{j\in\mathscr{I}}\big)$ plays an essential role in the mutational framework. Indeed, Aubin's initial concept (as presented in Chapter 1) considers metric spaces in which all closed bounded balls are assumed to be compact. Now we have more than just one distance function and thus, the classical equivalence of compactness (with regard to covers) and sequential compactness well-known in metric spaces might fail in this environment.

Our main goal is to construct solutions by means of Euler method and thus, the piecewise Euler approximations using transitions should provide a convergent subsequence. For this reason, we introduce the following version of compactness:

Definition 15 (Euler compact).
The tuple $\big(E,\, (d_j)_{j\in\mathscr{I}},\, (\lfloor\cdot\rfloor_j)_{j\in\mathscr{I}},\, \Theta\big(E, (d_i)_{i\in\mathscr{I}}, (\lfloor\cdot\rfloor_i)_{i\in\mathscr{I}}\big)\big)$ is called *Euler compact* if it satisfies the following condition for any initial element $x_0 \in E$, time $T \in]0,\infty[$ and bounds $\widehat{\alpha}_j, \widehat{\beta}_j, \widehat{\gamma}_j > 0$ $(j \in \mathscr{I})$:
Let $\mathscr{N} = \mathscr{N}(x_0, T, (\widehat{\alpha}_j, \widehat{\beta}_j, \widehat{\gamma}_j)_{j\in\mathscr{I}})$ denote the (possibly empty) subset of all curves $y(\cdot) : [0,T] \longrightarrow E$ constructed in the following piecewise way: Choosing an arbitrary equidistant partition $0 = t_0 < t_1 < \ldots < t_n = T$ of $[0,T]$ (with $n > T$) and transitions $\vartheta_1 \ldots \vartheta_n \in \Theta\big(E, (d_i)_{i\in\mathscr{I}}, (\lfloor\cdot\rfloor_i)_{i\in\mathscr{I}}\big)$ with

$$\begin{cases} \sup_k \;\; \gamma_j(\vartheta_k) & \leq \widehat{\gamma}_j \\ \sup_k \;\; \alpha_j\big(\vartheta_k;\; (\lfloor x_0\rfloor_j + \widehat{\gamma}_j\, T)\, e^{\widehat{\gamma}_j\, T}\big) & \leq \widehat{\alpha}_j \\ \sup_k \;\; \beta_j\big(\vartheta_k;\; (\lfloor x_0\rfloor_j + \widehat{\gamma}_j\, T)\, e^{\widehat{\gamma}_j\, T}\big) & \leq \widehat{\beta}_j \end{cases}$$

for each index $j \in \mathscr{I}$, define $y(\cdot) : [0,T] \longrightarrow E$ as

$$y(0) := x_0, \qquad y(t) := \vartheta_k\,(t - t_{k-1},\; y(t_{k-1})) \quad \text{for } t \in]t_{k-1}, t_k],\; k = 1, 2 \ldots n.$$

Then for each $t \in [0,T]$, every sequence $(z_n)_{n\in\mathbb{N}}$ in $\{y(t) \,|\, y(\cdot) \in \mathscr{N}\} \subset E$ has a subsequence $(z_{n_m})_{m\in\mathbb{N}}$ converging to an element $z \in E$ with respect to each pseudo-metric d_j $(j \in \mathscr{I})$.

Remark 16. Euler compactness weakens the condition that all bounded closed balls are compact – in the following sense: The family $(d_j)_{j\in\mathcal{I}}$ of pseudo-metrics induces a topology of the nonempty set E. If every " generalized ball" in E

$$\{y \in E \mid \forall\, j \in \mathcal{I}:\ d_j(x_0,y) \le r_j,\ \lfloor y \rfloor \le R_j\}$$

with arbitrary "center" $x_0 \in E$ and bounds $r_j, R_j \in]0,\infty[$ $(j \in \mathcal{I})$ is sequentially compact, then $\big(E, (d_j)_{j\in\mathcal{I}}, (\lfloor\cdot\rfloor_j)_{j\in\mathcal{I}}, \Theta\big(E,(d_i)_{i\in\mathcal{I}},(\lfloor\cdot\rfloor_i)_{i\in\mathcal{I}}\big)\big)$ is Euler compact. Indeed, fixing the parameters $x_0, T, (\widehat{\alpha}_j, \widehat{\beta}_j, \widehat{\gamma}_j)_{j\in\mathcal{I}}$ arbitrarily, every curve $y(\cdot)$: $[0,T] \longrightarrow E$ in $\mathcal{N} = \mathcal{N}(x_0, T, (\widehat{\alpha}_j, \widehat{\beta}_j, \widehat{\gamma}_j)_{j\in\mathcal{I}})$ satisfies

$$\lfloor y(t) \rfloor_j \ \le\ (\lfloor x_0 \rfloor_j + \widehat{\gamma}_j\, T)\ e^{\widehat{\gamma}_j\, T}$$

for each $t \in [0,T]$ and $j \in \mathcal{I}$ according to Lemma 4 (on page 105). Furthermore, condition (4.) of Definition 2 (about transitions) and the triangle inequality of d_j guarantee for each index $j \in \mathcal{I}$ that $y(\cdot) : [0,T] \longrightarrow (E,d_j)$ is $\widehat{\beta}_j$-Lipschitz continuous and thus,

$$d_j\big(x_0,\, y(t)\big) \ \le\ \widehat{\beta}_j\ T$$

for every $t \in [0,T]$. Hence the set of all values $\{y(t) \mid y(\cdot) \in \mathcal{N},\ t \in [0,T]\} \subset E$ is contained in such a "generalized ball".

The bound on the parameter α_j is not used explicitly, but it weakens the conditions of Euler compactness. Indeed, subsequent Theorem 18 about existence assumes such a bound anyway and thus, the Euler approximations are based on transitions with uniform bounds on all their parameters $\alpha_j, \beta_j, \gamma_j$.

In a word, Euler compactness ensures the existence of a convergent subsequence for each point of time separately. This even implies the existence of one and the same subsequence converging at every time. Specifying this conclusion in the following lemma, we realize a counterpart of Arzelà-Ascoli Theorem A.82 (on page 491) – now, however, in the tuple $\big(E, (d_j)_{j\in\mathcal{I}}, (\lfloor\cdot\rfloor_j)_{j\in\mathcal{I}}\big)$.

Lemma 17 (Uniform sequential compactness due to Euler compactness). *Assume* $\big(E, (d_j)_{j\in\mathcal{I}}, (\lfloor\cdot\rfloor_j)_{j\in\mathcal{I}}, \Theta\big(E,(d_i)_{i\in\mathcal{I}},(\lfloor\cdot\rfloor_i)_{i\in\mathcal{I}}\big)\big)$ *to be Euler compact. Using the notation of Definition 15, choose initial element* $x_0 \in E$*, time* $T \in]0,\infty[$ *and bounds* $\widehat{\alpha}_j, \widehat{\beta}_j, \widehat{\gamma}_j > 0$ $(j \in \mathcal{I})$ *arbitrarily.*

For every sequence $(y_n(\cdot))_{n\in\mathbb{N}}$ *of curves* $[0,T] \longrightarrow E$ *in* $\mathcal{N}\big(x_0, T, (\widehat{\alpha}_j, \widehat{\beta}_j, \widehat{\gamma}_j)_{j\in\mathcal{I}}\big)$*, there exists a subsequence* $(y_{n_m}(\cdot))_{m\in\mathbb{N}}$ *and a function* $y(\cdot) : [0,T] \longrightarrow E$ *such that for every* $j \in \mathcal{I}$,

$$\sup_{t\in[0,T]}\ d_j\big(y_{n_m}(t),\, y(t)\big) \ \longrightarrow\ 0 \qquad \text{for } m \longrightarrow \infty.$$

Furthermore if $(y_n(t_0))_{n\in\mathbb{N}}$ *is constant for some* $t_0 \in [0,T]$ *then* $y(\cdot)$ *can be chosen with the additional property* $y(t_0) = y_n(t_0)$.

The last statement does not result directly from the convergence because the set E supplied with the topology of $(d_j)_{j\in\mathcal{I}}$ does not have to be a Hausdorff space. The proof is postponed to the end of this section. As a consequence, we obtain the extension of Peano's Theorem 1.20 (on page 40) to the tuple $\big(E,\,(d_j)_j,\,(\lfloor\cdot\rfloor_j)_j\big)$ and its transitions.

Theorem 18 (Peano's Theorem for nonautonomous mutational equations). *Suppose* $\big(E,\,(d_j)_{j\in\mathcal{I}},\,(\lfloor\cdot\rfloor_j)_{j\in\mathcal{I}},\,\Theta\big(E,(d_i)_{i\in\mathcal{I}},(\lfloor\cdot\rfloor_i)_{i\in\mathcal{I}}\big)\big)$ *to be Euler compact. Assume for* $f : E\times[0,T] \longrightarrow \Theta\big(E,(d_i)_{i\in\mathcal{I}},(\lfloor\cdot\rfloor_i)_{i\in\mathcal{I}}\big)$ *and each* $j\in\mathcal{I}$, $R>0$,

1.) $\sup_{z,t}\ \alpha_j(f(z,t);\,R) < \infty$,

2.) $\sup_{z,t}\ \beta_j(f(z,t);\,R) < \infty$,

3.) $\sup_{z,t}\ \gamma_j(f(z,t)) < \infty$,

4.) $\lim_{n\to\infty} D_j\big(f(z_n,t_n),\,f(z,t);\,R\big) = 0$ *for $\mathcal{L}^1$-almost every $t\in[0,T]$ and any sequences $(t_n)_{n\in\mathbb{N}}$ in $[0,T]$ and $(z_n)_{n\in\mathbb{N}}$ in E satisfying* $\lim_{n\to\infty} t_n = t$ *and* $\lim_{n\to\infty} d_i(z_n,z) = 0$, $\sup_{n\in\mathbb{N}} \lfloor z_n\rfloor_i < \infty$ *for every* $i\in\mathcal{I}$.

Then for every initial element $x_0\in E$, there exists a solution $x(\cdot) : [0,T] \longrightarrow E$ to the mutational equation

$$\overset{\circ}{x}(\cdot) \ni f\big(x(\cdot),\cdot\big)$$

in the tuple $\big(E,(d_j)_{j\in\mathcal{I}},(\lfloor\cdot\rfloor_j)_{j\in\mathcal{I}}\big)$ *with* $x(0)=x_0$.

Proof (of Lemma 17). Fixing the parameters $x_0, T, (\widehat{\alpha}_j,\widehat{\beta}_j,\widehat{\gamma}_j)_{j\in\mathcal{I}}$ arbitrarily, we can assume the set $\mathcal{N} = \mathcal{N}(x_0,T,(\widehat{\alpha}_j,\widehat{\beta}_j,\widehat{\gamma}_j)_{j\in\mathcal{I}})$ to be nonempty (since otherwise the claim is trivial).
Let $(y_n(\cdot))_{n\in\mathbb{N}}$ be any sequence of functions $[0,T] \longrightarrow E$ in $\mathcal{N}$. Then for every $j\in\mathcal{I}$ and $n\in\mathbb{N}$, the curve $y_n : [0,T] \longrightarrow (E,d_j)$ is $\widehat{\beta}_j$-Lipschitz continuous due to condition (4.) of Definition 2 (about transitions) and the triangle inequality of d_j.

For each $t\in[0,T]$, the assumption of Euler compactness ensures a subsequence of $\big(y_n(t)\big)_{n\in\mathbb{N}}$ converging with respect to each d_j. Cantor's diagonal construction provides a subsequence $\big(y_{n_m}(\cdot)\big)_{m\in\mathbb{N}}$ of functions $[0,T] \longrightarrow E$ with the additional property that at every *rational* time $t\in[0,T]$, an element $y(t)\in E$ satisfies

$$d_j\big(y_{n_m}(t),\,y(t)\big) \longrightarrow 0 \qquad\qquad \text{for } m\longrightarrow\infty$$

and each $j\in\mathcal{I}$ since the subset $\mathbb{Q}\cap[0,T]$ of rational numbers in $[0,T]$ is countable.

Now we consider any $t\in[0,T]\setminus\mathbb{Q}$. Due to Euler compactness, there exists a subsequence $\big(y_{n_{m_l}}(t)\big)_{l\in\mathbb{N}}$ converging to an element $y(t)\in E$ with respect to each d_j (but maybe depending on t).
Then we even obtain $d_j\big(y_{n_m}(t),\,y(t)\big) \longrightarrow 0$ for $m\longrightarrow\infty$ and each $j\in\mathcal{I}$. Indeed, the triangle inequality of d_j and the $\widehat{\beta}_j$-Lipschitz continuity of each $y_n(\cdot)$, $n\in\mathbb{N}$, imply for every $s\in[0,T]\cap\mathbb{Q}$ and $l,m\in\mathbb{N}$

$$\begin{aligned} d_j\big(y_{n_m}(t),\, y(t)\big) &\le d_j\big(y_{n_m}(t),\, y_{n_m}(s)\big) + d_j\big(y_{n_m}(s),\, y_{n_{m_l}}(s)\big) \\ &\quad + d_j\big(y_{n_{m_l}}(s),\, y_{n_{m_l}}(t)\big) + d_j\big(y_{n_{m_l}}(t),\, y(t)\big) \\ &\le \widehat{\beta}_j\ |t-s| + d_j\big(y_{n_m}(s),\, y_{n_{m_l}}(s)\big) \\ &\quad + \widehat{\beta}_j\ |t-s| + d_j\big(y_{n_{m_l}}(t),\, y(t)\big)\,. \end{aligned}$$

$l \longrightarrow \infty$ leads to the following inequality for every $m \in \mathbb{N}$, $s \in [0,T] \cap \mathbb{Q}$ and $j \in \mathscr{I}$

$$d_j\big(y_{n_m}(t),\, y(t)\big) \ \le\ 2\,\widehat{\beta}_j\ |t-s| \ +\ d_j\big(y_{n_m}(s),\, y(s)\big)$$

and thus, $\displaystyle \limsup_{m\longrightarrow\infty}\ d_j\big(y_{n_m}(t),\, y(t)\big) \ \le \inf_{s\in[0,T]\cap\mathbb{Q}} 2\,\widehat{\beta}_j\ |t-s| \ +\ 0 \ =\ 0.$

Finally pointwise convergence of $\big(y_{n_m}(\cdot)\big)_{m\in\mathbb{N}}$ to $y(\cdot) : [0,T] \longrightarrow E$ and the $\widehat{\beta}_j$-Lipschitz continuity of each $y_{n_m}(\cdot) : [0,T] \longrightarrow (E,d_j)$, $m \in \mathbb{N}$, imply uniform convergence with respect to d_j in the compact interval $[0,T]$ for each index $j \in \mathscr{I}$. □

Proof (of Theorem 18). It is based on Euler approximations $x_n(\cdot) : [0,T] \longrightarrow E$ $(n \in \mathbb{N})$ on equidistant partitions of $[0,T]$. Indeed, for each $n \in \mathbb{N}$ with $2^n > T$, set

$$\begin{array}{llll} h_n & := \frac{T}{2^n}, \quad t_n^k := k\,h_n & & \text{for } k = 0 \ldots 2^n, \\ x_n(0) & := x_0, & & \\ x_n(t) & := f(x_n(t_n^k), t_n^k)(t - t_n^k,\ x_n(t_n^k)) & & \text{for } t \in\,]t_n^k, t_n^{k+1}],\ k < 2^n. \end{array}$$

Using the abbreviation $\widehat{\gamma}_j := \sup_{z,t}\ \gamma_j(f(z,t)) < \infty$, Lemma 4 (on page 105) ensures

$$\lfloor x_n(t) \rfloor_j \ \le\ \big(\lfloor x_0 \rfloor_j + \widehat{\gamma}_j\, T\big) \cdot e^{\widehat{\gamma}_j\, T} \ =:\ R_j$$

for every $t \in [0,T]$, $n \in \mathbb{N}$ (with $2^n > T$) and each $j \in \mathscr{I}$.
Due to Euler compactness and assumptions (1.)–(3.), Lemma 17 provides a subsequence $\big(x_{n_m}(\cdot)\big)_{m\in\mathbb{N}}$ and a function $x(\cdot) : [0,T] \longrightarrow E$ with $x(0) = x_0$ and

$$\sup_{t\in[0,T]}\ d_j\big(x_{n_m}(t),\, x(t)\big) \ \longrightarrow\ 0 \qquad \text{for } m \longrightarrow \infty$$

and each $j \in \mathscr{I}$.

Finally we conclude from Convergence Theorem 13 (on page 110) that $x(\cdot)$ is a solution to the mutational equation

$$\overset{\circ}{x}(\cdot) \ \ni\ f\big(x(\cdot), \cdot\big)$$

in the sense of Definition 9 (on page 107). Indeed, as a consequence of Remark 8 (on page 106), each Euler approximation $x_n(\cdot) : [0,T] \longrightarrow E$, $n \in \mathbb{N}$, is a solution to the mutational equation

$$\overset{\circ}{x}_n(\cdot) \ \ni\ f_n\big(x_n(\cdot), \cdot\big)$$

with the auxiliary function $f_n : E \times [0,T[\longrightarrow \Theta\big(E, (d_j)_{j\in\mathscr{I}}, (\lfloor\cdot\rfloor_j)_{j\in\mathscr{I}}\big)$ that is defined in a piecewise way: $f_n(y,t) \ :=\ f\big(x_n(t_n^k),\, t_n^k\big)$ for $t \in [t_n^k, t_n^{k+1}[$, $k < 2^n$.

At Lebesgue-almost every time $t \in [0,T]$, assumption (4.) about the continuity of f implies indirectly

$$\begin{aligned} D_j\big(f(x(t),t),\quad f_{n_m}(x(t),t);\ R_j\big) \ &\le \sup_{s:\,|s-t|\le h_{n_m}} D_j\big(f(x(t),t),\quad f(x_{n_m}(s),s);\ R_j\big) \\ &\longrightarrow 0 \qquad\qquad \text{for } m \longrightarrow 0, \end{aligned}$$

$$\begin{aligned} D_j\big(f_{n_m}(x(t),t),\ f_{n_m}(y_m,t_m);\ R_j\big) \ &\le \sup_{\substack{s:\,|s-t|\ \le h_{n_m} \\ \widetilde{s}:\,|\widetilde{s}-t_m|\le h_{n_m}}} D_j\big(f(x_{n_m}(s),s),\ f(x_{n_m}(\widetilde{s}),\widetilde{s});\ R_j\big) \\ &\longrightarrow 0 \qquad\qquad \text{for } m \longrightarrow 0 \end{aligned}$$

for each $j \in \mathscr{I}$ and any sequences $(t_m)_{m\in\mathbb{N}}$, $(y_m)_{m\in\mathbb{N}}$ in $[0,T]$, E respectively with $t_m \longrightarrow t$. (A similar indirect conclusion has already been drawn at the end of the proof of Convergence Theorem 13 on page 111.)
Thus, all hypotheses of Convergence Theorem 13 are satisfied by the subsequence $(x_{n_m}(\cdot))_{m\in\mathbb{N}}$ of Euler approximations and $x(\cdot)$. As a consequence, $x(\cdot)$ is a solution to the mutational equation $\overset{\circ}{x}(\cdot) \ni f\big(x(\cdot),\cdot\big)$. □

Remark 19. A pointwise Cauchy sequence $(x_n(\cdot))_{n\in\mathbb{N}}$ of solutions concluded from Theorem 18 always converges to a curve $x(\cdot) : [0,T] \longrightarrow E$ in a pointwise way. This form of completeness results from Euler compactness of $(E,d,\Theta(E,d,\lfloor\cdot\rfloor))$ and thus, it does not require assuming completeness of (E,d) additionally.
(This observation will be used in § 5.1.3.)

2.3.4 Convergence Theorem and Existence for Systems

The preceding results about convergence and existence of solutions can be extended to systems of finitely many mutational equations in a rather obvious way, but this is an important feature of the mutational framework as we have already pointed out in § 1.5 (on page 44 ff.).

Now a (possibly infinite) family $(d_j)_{j\in\mathscr{I}}$ of pseudo-metrics should be taken into consideration – instead of a single metric as in Chapter 1.
For this reason, we cannot use the same arguments as in Lemma 1.27 (on page 45) and supply a product $E_1 \times E_2$ simply with the sum of distance functions. In particular, the equivalence about componentwise mutations in Lemma 1.27 (2.) might lack a suitable counterpart for products of tuples $\big(E,(d_i)_{i\in\mathscr{I}},(\lfloor\cdot\rfloor_i)_{i\in\mathscr{I}}\big)$.

We prefer an alternative notion that has already been used for proving Peano's Theorem 1.26 for systems in metric spaces (on page 46 f.): The wanted mutational properties are verified for each component separately while the other components are regarded as additional time-dependent parameters. For proving existence of a joint solution to the system in particular, we again rely on Euler approximations for the system and select suitable subsequences successively according to Euler compactness in each component.
The assumptions, however, are now doubling ...

Theorem 20 (Convergence of solutions to systems of mutational equations). *Let the tuples* $\big(E_1, (d_j^1)_{j\in\mathscr{I}_1}, (\lfloor\cdot\rfloor_j^1)_{j\in\mathscr{I}_1}\big)$ *and* $\big(E_2, (d_j^2)_{j\in\mathscr{I}_2}, (\lfloor\cdot\rfloor_j^2)_{j\in\mathscr{I}_2}\big)$ *satisfy the general assumptions of this chapter (on page 104).* $\Theta\big(E_1, (d_j^1)_{j\in\mathscr{I}_1}, (\lfloor\cdot\rfloor_j^1)_{j\in\mathscr{I}_1}\big)$ *and* $\Theta\big(E_2, (d_j^2)_{j\in\mathscr{I}_2}, (\lfloor\cdot\rfloor_j^2)_{j\in\mathscr{I}_2}\big)$ *respectively denote nonempty sets of transitions as in Definition 5 (on page 105).*

For each $j_1 \in \mathscr{I}_1, j_2 \in \mathscr{I}_2$*, suppose the following properties of*

$$\begin{array}{llll}
f_n^1,\ f^1 : E_1 \times E_2 \times [0,T] & \longrightarrow \Theta\big(E_1, (d_i^1)_{i\in\mathscr{I}_1}, (\lfloor\cdot\rfloor_i^1)_{i\in\mathscr{I}_1}\big) & & (n\in\mathbb{N})\\
f_n^2,\ f^2 : E_1 \times E_2 \times [0,T] & \longrightarrow \Theta\big(E_2, (d_i^2)_{i\in\mathscr{I}_2}, (\lfloor\cdot\rfloor_i^2)_{i\in\mathscr{I}_2}\big) & & (n\in\mathbb{N})\\
x_n^1,\ x^1 : \qquad [0,T] & \longrightarrow E_1 : & & \\
x_n^2,\ x^2 : \qquad [0,T] & \longrightarrow E_2 : & &
\end{array}$$

1.)
$$\begin{aligned}
R_{j_1}^1 &:= \sup_{n,t} \lfloor x_n^1(t)\rfloor_{j_1}^1 < \infty, & \widehat{\alpha}_{j_1}^1 &:= \sup_{n,t,y^1,y^2} \alpha_{j_1}^1\big(f_n^1(y^1,y^2,t);\, R_{j_1}^1\big) < \infty,\\
R_{j_2}^2 &:= \sup_{n,t} \lfloor x_n^2(t)\rfloor_{j_2}^2 < \infty, & \widehat{\alpha}_{j_2}^2 &:= \sup_{n,t,y^1,y^2} \alpha_{j_2}^2\big(f_n^2(y^1,y^2,t);\, R_{j_2}^2\big) < \infty,\\
\widehat{\beta}_{j_1}^1 &:= \sup_n \ \mathrm{Lip}\,\big(x_n^1(\cdot) : [0,T] \longrightarrow (E, d_{j_1}^1)\big) < \infty, & &\\
\widehat{\beta}_{j_2}^2 &:= \sup_n \ \mathrm{Lip}\,\big(x_n^2(\cdot) : [0,T] \longrightarrow (E, d_{j_2}^2)\big) < \infty, & &
\end{aligned}$$

2.)
$$\begin{aligned}
&\overset{\circ}{x}{}_n^1(\cdot) \ni f_n^1(x_n^1(\cdot), x_n^2(\cdot), \cdot)\\
&\overset{\circ}{x}{}_n^2(\cdot) \ni f_n^2(x_n^1(\cdot), x_n^2(\cdot), \cdot) \quad \text{(in the sense of Definition 9) for every } n\in\mathbb{N},
\end{aligned}$$

3.)
$$\begin{aligned}
&\lim_{n\to\infty} D_{j_1}^1\big(f_n^1(x^1(t), x^2(t), t),\ f_n^1(y_n^1, y_n^2, t_n);\ R_{j_1}^1\big) = 0\\
&\lim_{n\to\infty} D_{j_2}^2\big(f_n^2(x^1(t), x^2(t), t),\ f_n^2(y_n^1, y_n^2, t_n);\ R_{j_2}^2\big) = 0
\end{aligned}$$
for $\mathscr{L}^1$*-almost every* $t\in[0,T]$ *and any sequences* $(t_n)_{n\in\mathbb{N}},\ (y_n^1)_{n\in\mathbb{N}},\ (y_n^2)_{n\in\mathbb{N}}$ *in* $[t,T]$, E_1 *and* E_2 *respectively satisfying*
$$\begin{aligned}
\lim_{n\to\infty} t_n = t \quad \text{and} \quad &\lim_{n\to\infty} d_i^1\big(x^1(t), y_n^1\big) = 0,\ \sup_{n\in\mathbb{N}} \lfloor y_n^1\rfloor_i^1 \le R_i^1 \quad \text{for each } i\in\mathscr{I}_1,\\
&\lim_{n\to\infty} d_i^2\big(x^2(t), y_n^2\big) = 0,\ \sup_{n\in\mathbb{N}} \lfloor y_n^2\rfloor_i^2 \le R_i^2 \quad \text{for each } i\in\mathscr{I}_2,
\end{aligned}$$

4.) *for Lebesgue-almost every* $t\in[0,T]$ *and any* $\widetilde{t}\in[0,T[$, *there exists a sequence* $n_m \nearrow \infty$ *of indices (possibly depending on* $t, \widetilde{t}, j_1, j_2$*) that satisfies for* $m\longrightarrow\infty$ *and each* $i_1\in\mathscr{I}_1, i_2\in\mathscr{I}_2$
$$\left\{\begin{array}{lll}
(i) & D_{j_1}^1\big(f^1(x^1(t), x^2(t), t),\ f_{n_m}^1(x^1(t), x^2(t), t);\ R_{j_1}^1\big) \longrightarrow 0 & \\
 & D_{j_2}^2\big(f^2(x^1(t), x^2(t), t),\ f_{n_m}^2(x^1(t), x^2(t), t);\ R_{j_2}^2\big) \longrightarrow 0 & \\
(ii) & d_{i_1}^1\big(x^1(t),\ x_{n_m}^1(t)\big) \longrightarrow 0, & d_{i_2}^2\big(x^2(t),\ x_{n_m}^2(t)\big) \longrightarrow 0\\
(iii) & d_{j_1}^1\big(x^1(\widetilde{t}),\ x_{n_m}^1(\widetilde{t})\big) \longrightarrow 0, & d_{j_2}^2\big(x^2(\widetilde{t}),\ x_{n_m}^2(\widetilde{t})\big) \longrightarrow 0
\end{array}\right.$$

Then, $x^1(\cdot)$ *and* $x^2(\cdot)$ *are solutions to the mutational equations*
$$\overset{\circ}{x}{}^1(\cdot) \ni f^1\big(x^1(\cdot), x^2(\cdot), \cdot\big), \qquad \overset{\circ}{x}{}^2(\cdot) \ni f^2\big(x^1(\cdot), x^2(\cdot), \cdot\big).$$

Theorem 21 (Peano's Theorem for systems of mutational equations).
Suppose the tuples $\big(E_1, (d_j^1)_{j\in\mathscr{I}_1}, (\lfloor\cdot\rfloor_j^1)_{j\in\mathscr{I}_1}, \ \Theta\big(E_1,(d_i^1)_{i\in\mathscr{I}_1},(\lfloor\cdot\rfloor_i^1)_{i\in\mathscr{I}_1}\big)\big)$ *and* $\big(E_2, (d_j^2)_{j\in\mathscr{I}_2}, (\lfloor\cdot\rfloor_j^2)_{j\in\mathscr{I}_2}, \ \Theta\big(E_2,(d_i^2)_{i\in\mathscr{I}_2},(\lfloor\cdot\rfloor_i^2)_{i\in\mathscr{I}_2}\big)\big)$ *to be Euler compact.*
Assume for

$$\begin{aligned} f^1 &: E_1 \times E_2 \times [0,T] \longrightarrow \Theta\big(E_1,(d_i^1)_{i\in\mathscr{I}_1},(\lfloor\cdot\rfloor_i^1)_{i\in\mathscr{I}_1}\big) \\ f^2 &: E_1 \times E_2 \times [0,T] \longrightarrow \Theta\big(E_2,(d_i^2)_{i\in\mathscr{I}_2},(\lfloor\cdot\rfloor_i^2)_{i\in\mathscr{I}_2}\big) \end{aligned}$$

and each $j_1 \in \mathscr{I}_1$, $j_2 \in \mathscr{I}_2$, $R > 0$:

1.) $\sup\limits_{z^1,z^2,t} \alpha_{j_1}^1(f^1(z^1,z^2,t);\,R) < \infty, \qquad \sup\limits_{z^1,z^2,t} \alpha_{j_2}^2(f^2(z^1,z^2,t);\,R) < \infty,$

2.) $\sup\limits_{z^1,z^2,t} \beta_{j_1}^1(f^1(z^1,z^2,t);\,R) < \infty, \qquad \sup\limits_{z^1,z^2,t} \beta_{j_2}^2(f^2(z^1,z^2,t);\,R) < \infty,$

3.) $\sup\limits_{z^1,z^2,t} \gamma_{j_1}^1(f^1(z^1,z^2,t)) < \infty, \qquad \sup\limits_{z^1,z^2,t} \gamma_{j_2}^2(f^2(z^1,z^2,t)) < \infty,$

4.) $\lim\limits_{n\to\infty} D_{j_1}^1\big(f^1(z_n^1,z_n^2,t_n),\ f^1(z^1,z^2,t);\,R\big) = 0$
$\lim\limits_{n\to\infty} D_{j_2}^2\big(f^2(z_n^1,z_n^2,t_n),\ f^2(z^1,z^2,t);\,R\big) = 0$
for $\mathscr{L}^1$*-almost every* $t \in [0,T]$ *and any sequences* $(t_n)_{n\in\mathbb{N}}$, $(z_n^1)_{n\in\mathbb{N}}$ $(z_n^2)_{n\in\mathbb{N}}$ *in* $[0,T], E_1, E_2$ *respectively satisfying*

$$\begin{aligned} \lim_{n\to\infty} t_n = t \quad \text{and} \quad &\lim_{n\to\infty} d_i^1(z^1,z_n^1) = 0, \ \sup_{n\in\mathbb{N}} \lfloor z_n^1\rfloor_i^1 < \infty \quad \text{for each } i \in \mathscr{I}_1, \\ &\lim_{n\to\infty} d_i^2(z^2,z_n^2) = 0, \ \sup_{n\in\mathbb{N}} \lfloor z_n^2\rfloor_i^2 < \infty \quad \text{for each } i \in \mathscr{I}_2, \end{aligned}$$

Then for any elements $x_0^1 \in E_1, x_0^2 \in E_2$, *there exist solutions* $x^1(\cdot) : [0,T] \longrightarrow E_1$ *and* $x^2(\cdot) : [0,T] \longrightarrow E_2$ *to the mutational equations*

$$\begin{cases} \overset{\circ}{x}{}^1(\cdot) \ni f^1\big(x^1(\cdot),\, x^2(\cdot),\, \cdot\big) \\ \overset{\circ}{x}{}^2(\cdot) \ni f^2\big(x^1(\cdot),\, x^2(\cdot),\, \cdot\big) \end{cases}$$

with $x^1(0) = x_0^1$, $x^2(0) = x_0^2$.

The proofs do not really provide new analytical aspects in comparison with the proofs of Theorem 13 (on page 110 f.) and Theorem 18 (on page 115 f.) respectively. Thus, we verify only Convergence Theorem 20 in detail and, the formulation is deliberately analogous to § 2.3.2:

Proof (of Theorem 20). Due to the symmetry with respect to $x^1(\cdot)$ and $x^2(\cdot)$, we can restrict ourselves to the solution properties of $x^1(\cdot)$.

For each index $j_1 \in \mathscr{I}_1$, the function $x^1(\cdot) : [0,T] \longrightarrow (E, d_{j_1}^1)$ is $\widehat{\beta}_{j_1}^1$-Lipschitz continuous. Indeed, for Lebesgue-almost every $t \in [0,T]$ and any $\widetilde{t} \in [0,T]$, assumption (4.) provides a subsequence $\big(x_{n_m}^1(\cdot)\big)_{m\in\mathbb{N}}$ with

$$\begin{cases} d_{j_1}^1\big(x^1(t),\, x_{n_m}^1(\widetilde{t})\big) \longrightarrow 0 \\ d_{j_1}^1\big(x^1(\widetilde{t}),\, x_{n_m}^1(\widetilde{t})\big) \longrightarrow 0 \end{cases} \qquad \text{for } m \longrightarrow \infty.$$

Now the uniform $\widehat{\beta}^1_{j_1}$-Lipschitz continuity of $x^1_n(\cdot), n \in \mathbb{N}$, implies

$$\begin{aligned} d^1_{j_1}\big(x^1(t),\, x^1(\widetilde{t})\big) \;&\le\; d^1_{j_1}\big(x^1(t),\, x^1_{n_m}(t)\big) + d^1_{j_1}\big(x^1_{n_m}(t),\, x^1_{n_m}(\widetilde{t})\big) + d^1_{j_1}\big(x^1_{n_m}(\widetilde{t}),\, x^1(\widetilde{t})\big) \\ &\le\; d^1_{j_1}\big(x^1(t),\, x^1_{n_m}(t)\big) + \widehat{\beta}^1_{j_1}\, |\widetilde{t}-t| \qquad\qquad + d^1_{j_1}\big(x^1_{n_m}(\widetilde{t}),\, x^1(\widetilde{t})\big) \\ &\longrightarrow\; 0 \qquad\qquad\qquad\quad + \widehat{\beta}^1_{j_1}\, |\widetilde{t}-t| \qquad\qquad + 0 \qquad \text{for } m \to \infty. \end{aligned}$$

This Lipschitz inequality can be easily extended to *all* $t \in [0,T]$ by means of the triangle inequality of $d^1_{j_1}$. Moreover the general hypothesis about lower semicontinuity of $\lfloor \cdot \rfloor^1_{j_1}$ ensures

$$\lfloor x^1(\widetilde{t}) \rfloor^1_{j_1} \;\le\; \liminf_{m \to \infty}\; \lfloor x^1_{n_m}(\widetilde{t}) \rfloor^1_{j_1} \;\le\; R^1_{j_1}.$$

Finally we focus on the feature of first-order approximation

$$\lim_{h \downarrow 0}\; \tfrac{1}{h} \cdot d^1_{j_1}\big(f^1(x^1(t),x^2(t),t)\,(h,\, x^1(t)),\;\, x^1(t+h)\big) \;=\; 0$$

at Lebesgue-almost every time $t \in [0,T[$. Indeed, for Lebesgue-almost every $t \in [0,T[$ and any $h \in]0,\, T-t[$, assumption (4.) provides a sequence $n_m \nearrow \infty$ of indices satisfying for each $i_1 \in \mathscr{I}_1$, $i_2 \in \mathscr{I}_2$ and $m \longrightarrow \infty$

$$\begin{cases} D^1_{j_1}\big(f^1(x^1(t), x^2(t), t),\;\, f^1_{n_m}(x^1(t), x^2(t), t);\;\, R^1_{j_1}\big) \;\longrightarrow\; 0 \\ d^1_{i_1}\big(x^1(t), \qquad x^1_{n_m}(t)\big) \qquad \longrightarrow 0 \\ d^2_{i_2}\big(x^1(t), \qquad x^1_{n_m}(t)\big) \qquad \longrightarrow 0 \\ d^1_{j_1}\big(x^1(t+h),\, x^1_{n_m}(t+h)\big) \;\longrightarrow 0. \end{cases}$$

We conclude from Proposition 11 (on page 108)

$$\begin{aligned} & d^1_{j_1}\big(f^1(x^1(t),x^2(t),t)\,(h,\, x^1(t)),\;\, x^1(t+h)\big) \\ &\le\quad d^1_{j_1}\big(f^1(x^1(t),x^2(t),t)\;\;(h,\, x^1(t)),\;\, f^1_{n_m}(x^1(t),x^2(t),t)\,(h,\, x^1(t))\big) \\ &\quad + d^1_{j_1}\big(f^1_{n_m}(x^1(t),x^2(t),t)\,(h,\, x^1(t)),\;\, x^1_{n_m}(t+h)\big) \\ &\quad + d^1_{j_1}\big(x^1_{n_m}(t+h),\;\, x^1(t+h)\big) \\ &\le\quad h\; e^{\widehat{\alpha}^1_{j_1}\, h} \cdot D^1_{j_1}\big(f^1(x^1(t),x^2(t),t),\;\, f^1_{n_m}(x^1(t),x^2(t),t);\;\, R_j\big) \\ &\quad + d^1_{j_1}\big(x^1(t),\, x^1_{n_m}(t)\big)\; e^{\widehat{\alpha}^1_{j_1}\, h} \\ &\quad + h\; e^{\widehat{\alpha}^1_{j_1}\, h} \cdot \sup_{[t,\,t+h]} D^1_{j_1}\big(f^1_{n_m}(x^1(t),x^2(t),t),\;\, f^1_{n_m}(x^1_{n_m}(\cdot),x^2_{n_m}(\cdot),\cdot);\;\, R_j\big) \\ &\quad + d^1_{j_1}\big(x^1_{n_m}(t+h),\;\, x^1(t+h)\big). \end{aligned}$$

Now $m \longrightarrow \infty$ leads to the inequality

$$\begin{aligned} & d^1_{j_1}\big(f^1(x^1(t),x^2(t),t)\,(h,\, x^1(t)),\;\, x^1(t+h)\big) \\ &\le h\, e^{\widehat{\alpha}^1_{j_1}\, h} \cdot \limsup_{m \to \infty}\; \sup_{[t,\,t+h]} D^1_{j_1}\big(f^1_{n_m}(x^1(t),x^2(t),t),\;\, f^1_{n_m}(x^1_{n_m}(\cdot),x^2_{n_m}(\cdot),\cdot);\;\, R_j\big). \end{aligned}$$

For completing the proof, it is sufficient to verify

$$0 = \limsup_{h\downarrow 0} \limsup_{m\to\infty} \sup_{[t,t+h]} D^1_{j_1}\left(f^1_{n_m}(x^1(t),x^2(t),t),\ f^1_{n_m}(x^1_{n_m}(\cdot),x^2_{n_m}(\cdot),\cdot);\ R_j\right)$$

for Lebesgue-almost every $t \in [0,T[$ and *any* subsequence $\left(x_{n_m}(\cdot)\right)_{m\in\mathbb{N}}$ satisfying

$$\begin{cases} d^1_{i_1}\left(x^1(t),\ x^1_{n_m}(t)\right) \stackrel{m\to\infty}{\longrightarrow} 0 & \text{for each } i_1 \in \mathscr{I}_1, \\ d^2_{i_2}\left(x^2(t),\ x^2_{n_m}(t)\right) \stackrel{m\to\infty}{\longrightarrow} 0 & \text{for each } i_2 \in \mathscr{I}_2. \end{cases}$$

Indeed, if this limit superior was positive then we could select some $\varepsilon > 0$ and sequences $(h_l)_{l\in\mathbb{N}}$, $(m_l)_{l\in\mathbb{N}}$, $(s_l)_{l\in\mathbb{N}}$ such that for every $l \in \mathbb{N}$,

$$\begin{cases} D^1_{j_1}\left(f^1_{n_{m_l}}(x^1(t),x^2(t),t),\ f^1_{n_{m_l}}(x^1_{n_{m_l}}(t+s_l),x^2_{n_{m_l}}(t+s_l),t+s_l);\ R_j\right) \ge \varepsilon \\ 0 \le s_l \le h_l \le \frac{1}{l}, \qquad m_l \ge l. \end{cases}$$

The consequence

$$\begin{cases} d^1_{i_1}\left(x^1(t),\ x^1_{n_{m_l}}(t+s_l)\right) \le d^1_{i_1}\left(x^1(t),\ x^1_{n_{m_l}}(t)\right) + \widehat{\beta}^1_{i_1}\, s_l \stackrel{l\to\infty}{\longrightarrow} 0 \\ d^2_{i_2}\left(x^2(t),\ x^2_{n_{m_l}}(t+s_l)\right) \le d^2_{i_2}\left(x^2(t),\ x^2_{n_{m_l}}(t)\right) + \widehat{\beta}^2_{i_2}\, s_l \stackrel{l\to\infty}{\longrightarrow} 0 \end{cases}$$

for any indices $i_1 \in \mathscr{I}_1$ and $i_2 \in \mathscr{I}_2$ would lead to a contradiction to equi-continuity assumption (3.) at Lebesgue-almost every time $t \in [0,T[$. □

2.3.5 *Existence for Mutational Equations with Delay*

Euler method in combination with Euler compactness proves to be useful indeed. Essentially the same approximations also provide solutions to mutational equations with delay. Pichard and Gautier formulated and proved their existence for Aubin's form of mutational equations in a metric space [158]. Now we present the counterpart for the tuple $\left(E,\ (d_j)_{j\in\mathscr{I}},\ (\lfloor\cdot\rfloor_j)_{j\in\mathscr{I}}\right)$. First we have to specify the type of functions that are admitted as argument in the delay equation:

Definition 22. Let $I \subset \mathbb{R}$ be a nonempty interval.
$\mathrm{BLip}\left(I,\ E;\ (d_j)_{j\in\mathscr{I}},\ (\lfloor\cdot\rfloor_j)_{j\in\mathscr{I}}\right)$ denotes the set of all functions $y(\cdot): I \longrightarrow E$ satisfying the following conditions for each index $j \in \mathscr{I}$:

1.) $y(\cdot): I \longrightarrow E$ is Lipschitz continuous with respect to d_j
2.) $\sup_{t\in I} \lfloor y(t)\rfloor_j < \infty$

Proposition 23 (Existence of solutions to mutational equations with delay). *Suppose* $\left(E,\ (d_j)_{j\in\mathscr{I}},\ (\lfloor\cdot\rfloor_j)_{j\in\mathscr{I}},\ \Theta\left(E,(d_i)_{i\in\mathscr{I}},(\lfloor\cdot\rfloor_i)_{i\in\mathscr{I}}\right)\right)$ *to be Euler compact. Moreover assume for some fixed* $\tau \ge 0$, *the function*

$$f: \mathrm{BLip}\left([-\tau,0],\ E;\ (d_j)_{j\in\mathscr{I}},\ (\lfloor\cdot\rfloor_j)_{j\in\mathscr{I}}\right) \times [0,T] \longrightarrow \Theta\left(E,(d_i)_{i\in\mathscr{I}},(\lfloor\cdot\rfloor_i)_{i\in\mathscr{I}}\right)$$

and each $j \in \mathscr{I}$, $R > 0$:

1.) $\sup_{z(\cdot),t} \alpha_j(f(z(\cdot),t);R) < \infty$,

2.) $\sup_{z(\cdot),t} \beta_j(f(z(\cdot),t);R) < \infty$,

3.) $\sup_{z(\cdot),t} \gamma_j(f(z(\cdot),t)) < \infty$,

4.) $\lim_{n\to\infty} D_j\big(f(z_n(\cdot),t_n),\ f(z(\cdot),t);R\big) = 0$ *for $\mathscr{L}^1$-almost every $t \in [0,T]$ and any sequences $(z_n(\cdot))_{n\in\mathbb{N}}$, $(t_n)_{n\in\mathbb{N}}$ in* $\mathrm{BLip}\big([-\tau,0],E;(d_j)_{j\in\mathscr{I}},(\lfloor\cdot\rfloor_j)_{j\in\mathscr{I}}\big)$ *and $[0,T]$ respectively satisfying*

$$\lim_{n\to\infty} t_n = t \quad \text{and} \quad \lim_{n\to\infty} \sup_{s\in[-\tau,0]} d_i\big(z_n(s),z(s)\big) = 0,$$
$$\sup_{n\in\mathbb{N}} \sup_{s\in[-\tau,0]} \lfloor z_n(s)\rfloor_i < \infty \quad \text{for every } i \in \mathscr{I}.$$

For every function $x_0(\cdot) \in \mathrm{BLip}\big([-\tau,0],E;(d_j)_{j\in\mathscr{I}},(\lfloor\cdot\rfloor_j)_{j\in\mathscr{I}}\big)$, there exists a curve $x(\cdot):[-\tau,T] \longrightarrow E$ with the following properties:

(i) $x(\cdot) \in \mathrm{BLip}\big([-\tau,T],E;(d_j)_{j\in\mathscr{I}},(\lfloor\cdot\rfloor_j)_{j\in\mathscr{I}}\big)$,

(ii) for $\mathscr{L}^1$-almost every $t \in [0,T]$, $f\big(x(t+\cdot)\big|_{[-\tau,0]},t\big)$ belongs to $\overset{\circ}{x}(t)$,

(iii) $x(\cdot)\big|_{[-\tau,0]} = x_0(\cdot)$.

In particular, the restriction $x(\cdot)\big|_{[0,T]}$ is a solution to the mutational equation

$$\overset{\circ}{x}(t) \ni f\big(x(t+\cdot)\big|_{[-\tau,0]},t\big)$$

in the sense of Definition 9 (on page 107).

Proof. Similarly to the proof of Peano's Theorem 18 (on page 115 f.), we construct a sequence of Euler approximations on equidistant partitions of $[0,T]$. The (only) new aspect is due to the appropriate restrictions as argument of $f(\cdot,t)$. For every $n \in \mathbb{N}$ with $2^n > T$, set

$$\begin{aligned} h_n &:= \tfrac{T}{2^n}, \qquad t_n^k := k\,h_n \qquad\qquad \text{for } k = 0\ldots 2^n,\\ x_n(\cdot)\big|_{[-\tau,0]} &:= x_0,\\ x_n(t) &:= f\big(x_n(t_n^k+\cdot)\big|_{[-\tau,0]},t_n^k\big)\big(t-t_n^k,\ x_n(t_n^k)\big) \quad \text{for } t \in\,]t_n^k,t_n^{k+1}],\ k < 2^n. \end{aligned}$$

With $\widehat{\gamma}_j := \sup \gamma_j(f(\cdot,\cdot)) < \infty$, Lemma 4 (on page 105) again provides a uniform bound for every $t \in [0,T]$, $n \in \mathbb{N}$ (with $2^n > T$) and each $j \in \mathscr{I}$:

$$\lfloor x_n(t)\rfloor_j \le \big(\lfloor x_0(0)\rfloor_j + \widehat{\gamma}_j\,T\big)\cdot e^{\widehat{\gamma}_j\,T} =: R_j.$$

Thus, exactly as in the proof of Peano's Theorem 18, we conclude from Euler compactness and assumptions (1.)–(3.) that a subsequence $\big(x_{n_m}(\cdot)\big)_{m\in\mathbb{N}}$ converges to a function $x(\cdot):[0,T] \longrightarrow E$ in the sense that

$$\sup_{t\in[0,T]} d_j\big(x_{n_m}(t),x(t)\big) \longrightarrow 0 \qquad \text{for } m \longrightarrow \infty$$

and each index $j \in \mathscr{I}$. In particular, $x(0) = x_0(0)$ due to Lemma 17.

For every $t \in [0,T]$, the estimate $\lfloor x(t) \rfloor_j \le R_j$ results from the general assumption about $\lfloor\cdot\rfloor_j$ (on page 104) and, $x(\cdot) : [0,T] \longrightarrow (E,d_j)$ is also $\widehat{\beta}_j$-Lipschitz continuous with $\widehat{\beta}_j := \sup \beta(f(\cdot,\cdot)) < \infty$. Defining $x(\cdot)\big|_{[-\tau,0]} := x_0(\cdot)$, we obtain

$$x(\cdot) \in \mathrm{BLip}\big([-\tau,T],\, E;\, (d_j)_{j\in\mathscr{I}},\, (\lfloor\cdot\rfloor_j)_{j\in\mathscr{I}}\big).$$

Finally it is again the conclusion of Convergence Theorem 13 (on page 110) that

$$\lim_{h\downarrow 0} \tfrac{1}{h}\cdot d_j\Big(f\big(x(t+\cdot)\big|_{[-\tau,0]},\, t\big)\,(h,\, x(t)),\;\; x(t+h)\Big) \;=\; 0$$

holds for arbitrarily fixed $j \in \mathscr{I}$ and $\mathscr{L}^1$-almost every $t \in [0,T]$. Indeed, each Euler approximation $x_n(\cdot) : [0,T] \longrightarrow E$, $n \in \mathbb{N}$, can be regarded as a solution of

$$\overset{\circ}{x}_n(t) \;\ni\; f_n\big(x_n(t+\cdot)\big|_{[-\tau,0]},\, t\big)$$

with the auxiliary function

$$\begin{aligned} f_n : \mathrm{BLip}\big([-\tau,0],\, E;\, (d_j)_{j\in\mathscr{I}},\, (\lfloor\cdot\rfloor_j)_{j\in\mathscr{I}}\big) \times [0,T] &\longrightarrow \Theta\big(E, (d_j)_{j\in\mathscr{I}}, (\lfloor\cdot\rfloor_j)_{j\in\mathscr{I}}\big),\\ f_n\big(y(\cdot),\, t\big) \;:=\; f\big(x_n(\cdot)\big|_{[t_n^k-\tau,\, t_n^k]},\, t_n^k\big) &\quad \text{for any } y(\cdot) \text{ and } t \in [t_n^k, t_n^{k+1}[,\ k < 2^n. \end{aligned}$$

Fix index $j \in \mathscr{I}$ arbitrarily. At $\mathscr{L}^1$-almost every time $t \in [0,T]$, assumption (4.) has two indirect consequences. First,

$$\begin{aligned} & D_j\big(f(x(t+\cdot)|_{[-\tau,0]},t),\ f_{n_m}(x(t+\cdot)|_{[-\tau,0]},t);\ R_j\big)\\ \le \sup_{s:\,|s-t|\le h_{n_m}} & D_j\big(f(x(t+\cdot)|_{[-\tau,0]},t),\ f(x_{n_m}(s+\cdot)|_{[-\tau,0]},s);\ R_j\big) \ \overset{m\to\infty}{\longrightarrow}\ 0, \end{aligned}$$

because for any index $i \in \mathscr{I}$ and $s,t \in [0,T]$,

$$\begin{aligned} \sup_{[-\tau,0]} d_i\big(x(t+\cdot),\ x_{n_m}(s+\cdot)\big) \;&\le\; \sup_{[-\tau,0]} d_i\big(x(t+\cdot),\ x_{n_m}(t+\cdot)\big) \;+\; \widehat{\beta}_i\,|s-t|\\ &\overset{m\to\infty}{\longrightarrow}\ 0 \;+\; \widehat{\beta}_i\,|s-t|. \end{aligned}$$

Second, we obtain for any sequences $(t_m)_{m\in\mathbb{N}}$ in $[0,T]$ tending to t and $(y_m(\cdot))_{m\in\mathbb{N}}$ in $\mathrm{BLip}\big([-\tau,0],\, E;\, (d_j)_{j\in\mathscr{I}},\, (\lfloor\cdot\rfloor_j)_{j\in\mathscr{I}}\big)$

$$\begin{aligned} & D_j\big(f_{n_m}(x(\cdot)|_{[t-\tau,t]},t),\ \ f_{n_m}(y_m(\cdot),t_m);\qquad R_j\big)\\ \le \sup_{\substack{s:\,|s-t|\ \le h_{n_m}\\ \widetilde{s}:\,|\widetilde{s}-t_m|\le h_{n_m}}} & D_j\big(f(x_{n_m}(\cdot)|_{[s-\tau,s]},s),\ f(x_{n_m}(\cdot)|_{[\widetilde{s}-\tau,\widetilde{s}]},\widetilde{s});\ R_j\big) \ \overset{m\to\infty}{\longrightarrow}\ 0. \end{aligned}$$

Finally we can now draw exactly the same conclusions as in the proof of Convergence Theorem 13 (on page 110 ff.) – considering, however, $x(\cdot)$ and the subsequence $(x_{n_m}(\cdot))_{m\in\mathbb{N}}$ of Euler approximations restricted to $[0,T]$. As a consequence,

$$\lim_{h\downarrow 0} \tfrac{1}{h}\cdot d_j\Big(f\big(x(t+\cdot)\big|_{[-\tau,0]},\, t\big)\,(h,\, x(t)),\;\; x(t+h)\Big) \;=\; 0$$

is satisfied for arbitrarily fixed index $j \in \mathscr{I}$ and at $\mathscr{L}^1$-a.e. time $t \in [0,T]$. □

2.3.6 Existence Under State Constraints for Finite Index Set $\mathscr{I}$

If the index set $\mathscr{I} \neq \emptyset$ consists of at most finitely many elements, then we even can restrict our considerations to a single index (i.e. $\mathscr{I} = \{0\}$). Indeed, all conditions on transitions and solutions respectively are then satisfied by

$$\begin{array}{rcl} d_0 := \max_{j \in \mathscr{I}} d_j : & E \times E & \longrightarrow [0, \infty[, \\ \lfloor \cdot \rfloor_0 := \max_{j \in \mathscr{I}} \lfloor \cdot \rfloor_j : & E & \longrightarrow [0, \infty[. \end{array}$$

Even in this special case, the recent mutational framework is more general than its counterpart in Chapter 1 because the parameters α, β of transitions and the distance between transitions require merely "local" bounds, i.e. in every "generalized ball" $\{x \in E \mid \lfloor x \rfloor_0 \leq r\}$ with arbitrary $r > 0$.
This additional feature, however, does not have any significant consequences for verifying the existence of solutions with state constraints. Now Proposition 1.28 (on page 47) has the following counterpart:

Proposition 24 (Existence of solutions under state constraints for $\mathscr{I} = \{0\}$). *In addition to $\mathscr{I} = \{0\}$, let (E, d_0) be a metric space and assume that for every $r_1, r_2 > 0$ and $x_0 \in E$, the (possibly empty) set $\{x \in E \mid d_0(x_0, x) \leq r_1,\ \lfloor x \rfloor_0 \leq r_2\}$ is sequentially compact. For each $r > 0$, suppose*

$$f : (E, d_0) \longrightarrow \big(\Theta\big(E, d_0, \lfloor \cdot \rfloor_0\big),\ D_0(\cdot, \cdot; r)\big)$$

to be continuous with

$$\begin{array}{rcl} \widehat{\alpha}(r) &:=& \sup_{z \in E}\ \alpha_0(f(z); r) < \infty, \\ \widehat{\beta}(r) &:=& \sup_{z \in E}\ \beta_0(f(z); r) < \infty, \\ \widehat{\gamma} &:=& \sup_{z \in E}\ \gamma_0(f(z)) \quad < \infty. \end{array}$$

Let the nonempty closed subset $\mathscr{V} \subset (E, d_0)$ satisfy the following viability condition (with the contingent transition set as specified in Definition 1.16 on page 39) :

$$f(z) \in \mathscr{T}_{\mathscr{V}}(z) \qquad \text{for every } z \in \mathscr{V},$$

i.e.

$$\liminf_{h \downarrow 0} \tfrac{1}{h} \cdot \inf_{y \in \mathscr{V}} d_0\big(f(z)(h, z),\ y\big) = 0 \qquad \text{for every } z \in \mathscr{V}.$$

Then every $x_0 \in \mathscr{V}$ is the initial point of at least one solution $x : [0, 1] \longrightarrow E$ to the mutational equation

$$\overset{\circ}{x}(\cdot) \ni f\big(x(\cdot)\big)$$

with $x(t) \in \mathscr{V}$ for all $t \in [0, 1]$.

The proof follows exactly the arguments of Proposition 1.28 and is based on the approximative solutions in subsequent Lemma 25 in combination with Arzelà-Ascoli Theorem A.82 and Convergence Theorem 13 (on page 110).

Lemma 25 (Constructing approximative solutions).
Choose any $\varepsilon > 0$. Under the assumptions of Proposition 24, there always exists a $\widehat{\beta}$-Lipschitz continuous function $x_\varepsilon(\cdot) : [0,1] \longrightarrow (E, d_0)$ satisfying

(a) $x_\varepsilon(0) = x_0,$

(b) $\operatorname{dist}\big(x_\varepsilon(t),\ \mathscr{V}\big) \le \varepsilon\ e^{\widehat{\alpha}}$ *for all* $t \in [0,1]$,

(c) $\overset{\circ}{x}_\varepsilon(t) \cap \big\{ f(z) \,\big|\, z \in E : d_0(z, x_\varepsilon(t)) \le \varepsilon\ e^{\widehat{\alpha}} \big\} \ne \emptyset$ *for all* $t \in [0,1[$,

(d) $\lfloor x_\varepsilon(t) \rfloor_0 \le \big(\lfloor x_0 \rfloor_0 + \widehat{\gamma}\, t \big)\ e^{\widehat{\gamma}\, t}$ *for all* $t \in [0,1]$.

This lemma differs from Aubin's metric counterpart in Lemma 1.29 (on page 48) merely in property (d). Following the proving arguments (on page 48 f.), however, this upper bound of $\lfloor x(t) \rfloor_0$ can be implemented easily due to Lemma 4 (on page 105). Now we dispense with further details verifying Lemma 25 and Proposition 24.

The analogy to Lemma 1.29 and its proof is a reason for restricting our considerations in this subsection to a finite index set $\mathscr{I}$. Indeed, the indirect arguments for Lemma 1.29 consider several points of time $T + h_m$, $m \in \mathbb{N}$, with a sequence $h_m \downarrow 0$ related to

$$\liminf_{h \downarrow 0} \ \tfrac{1}{h} \cdot \inf_{y \in \mathscr{V}} d_0\big(f(z)\,(h,z),\ y\big) \ = \ 0$$

for some $z \in \mathscr{V}$. Such a sequence should be chosen appropriately "uniformly" for all indices $j \in \mathscr{I}$ if more than one distance function d_j comes into play.

2.4 Example: Semilinear Evolution Equations in Reflexive Banach Spaces

In this example, we consider semilinear evolution equations

$$\tfrac{d}{dt}\, u(t) \;=\; A\, u(t) \,+\, f\big(u(t), t\big)$$

with a fixed generator A of a C^0 semigroup on a Banach space X. The goal is to specify sufficient conditions on X, its topology and the generator A so that initial value problems can be solved in the mutational framework.
Solutions to the corresponding mutational equations prove to be weak solutions. A proposition of John Ball [20] implies that they are even mild solutions. Considering these results separately, they have already been well-known, but the essential advantage of their fitting in the mutational framework is that we are free to combine these evolution equations with any other example in systems. This opens the door to coupling, for example, a reaction-diffusion equation (on the whole Euclidean space) with a modified morphological equation for compact subsets (in the sense of § 1.11). Such a result about existence for systems is formulated in Proposition 37 (on page 131) below.

Assumptions for § 2.4.

(1.) $(X, \|\cdot\|)$ is a separable reflexive Banach space.

(2.) The linear operator A generates a C^0 semigroup $(S(t))_{t\geq 0}$ on X.

(3.) The C^0 semigroup $(S(t))_{t\geq 0}$ of linear operators on X is ω-contractive, i.e. there is some $\omega > 0$ such that $\|S(t)\,x\| \leq e^{\omega\, t}\, \|x\|$ for all $x \in X$, $t \geq 0$.

(4.) The dual operator A' of A has a family of unit eigenvectors $\{v'_j\}_{j\in\mathscr{I}}$ spanning the dual space X'. λ_j denotes the eigenvalue of A' related to v'_j for each $j \in \mathscr{I}$.

(5.) For each index $j \in \mathscr{I}$, set $d_j : X \times X \longrightarrow [0,\infty[,\ (x,y) \longmapsto |\langle x-y,\, v'_j\rangle|$ and $\lfloor\cdot\rfloor_j := \|\cdot\|$.

Among these five assumptions, condition (4.) is probably the most restrictive one: The eigenvectors of A' are spanning the dual space X'. First we specify two classes of operators fulfilling this condition with an even countable family of eigenvectors. In particular, the separability of the dual space X' implies that X is also separable [188, Chapter V, Appendix § 4].

Example 26. Consider a normal compact operator $A : H \longrightarrow H$ on a separable Hilbert space H generating a C^0 semigroup $(S(t))_{t\geq 0}$.
Then there exists a countable orthonormal system $(e_i)_{i\in\widehat{\mathscr{I}}}$ of eigenvectors of A with $H = \ker A \oplus \overline{\sum_{i\in\widehat{\mathscr{I}}} \mathbb{R}\, e_i}$ [184, Theorem VI.3.2]. Since H is separable, $(e_i)_{i\in\widehat{\mathscr{I}}}$ induces a countable orthonormal basis $(e_i)_{i\in\mathscr{I}}$ of H with $A\, e_i = 0$ for all $i \in \mathscr{I} \setminus \widehat{\mathscr{I}}$.
In fact, each e_i $(i \in \mathscr{I})$ is also eigenvector of the dual operator A' as A is normal [184, Lemma VI.3.1]. Hence, assumption (3.) of this section is satisfied. Symmetric integral operators of Hilbert-Schmidt type provide typical examples of this class.

Basic set	$E := X$ a separable reflexive Banach space
Distances	Let $\big(S(t)\big)_{t\geq 0}$ be a strongly continuous semigroup of linear operators on X with its generator A. The dual operator A' of A is supposed to have a family of unit eigenvectors $\{v'_j\}_{j\in\mathcal{J}}$ spanning the dual space X'. $d_j : X \times X \longrightarrow [0, \infty[,\ (x,y) \longmapsto \lvert\langle x-y,\ v'_j\rangle\rvert$
Absolute values	$\lfloor\cdot\rfloor_j := \lVert\cdot\rVert$ norm of X
Transition	For each $v \in X$, set $\vartheta_v : [0,1] \times X \longrightarrow X,\quad (h,x) \longmapsto S(h)\,x + \int_0^h S(h-s)\ v\ ds,$ i.e. variation of constants formula for $u'(t) = A\,u(t) + v,\ u(0) = x$
Compactness	All norm-bounded and w.r.t. $(d_j)_{j\in\mathcal{J}}$ closed balls are sequentially compact due to Alaoglu's Theorem: Lemma 34 (page 129)
Mutational solutions	Mild solutions to semilinear evolution equations
List of main results formulated in § 2.4	Existence due to compactness (Peano): Theorem 33 (page 129) Existence for systems with modified morphological equations: Proposition 37 (page 131)
Key tool	Weak solutions to linear evolution equations are mild solutions (Ball [20], Lemma 36)

Table 2.1 Brief summary of the example in § 2.4 in mutational terms:
Semilinear evolution equations in reflexive Banach spaces

Example 27. Another example is the generator $A : D_A \longrightarrow H$ $(D_A \subset H)$ of a C^0 semigroup $(S(t))_{t \geq 0}$ on a Hilbert space H under the assumption that the resolvent $R(\lambda_0, A) := (\lambda_0 \cdot \mathrm{Id}_H - A)^{-1} : H \longrightarrow H$ is compact and normal for some λ_0.
For the same reasons as before, there exists a countable orthonormal system $(e_i)_{i \in \mathscr{I}}$ of eigenvectors of $R(\lambda_0, A)$ satisfying $H = \ker R(\lambda_0, A) \oplus \overline{\sum_{i \in \mathscr{I}} \mathbb{R}\, e_i} = \overline{\sum_{i \in \mathscr{I}} \mathbb{R}\, e_i}$.
$R(\lambda_0, A)\, e_i = \mu_i \cdot e_i$ implies $\mu_i \neq 0$ and that e_i is eigenvector of A corresponding to the eigenvalue $\lambda_0 - \frac{1}{\mu_i}$ since $(\lambda_0 - A)\, e_i = (\lambda_0 - A) \cdot \frac{1}{\mu_i}\, R(\lambda_0, A)\, e_i = \frac{1}{\mu_i}\, e_i$.
This example opens the door to considering strongly elliptic differential operators in divergence form with smooth (time-independent) coefficients.

The variation of constants formula motivates the following choice of candidates for transitions on $\big(X, (d_j)_{j \in \mathscr{I}}, (\|\cdot\|)_{j \in \mathscr{I}}\big)$.

Definition 28. For each $v \in X$, the function $\tau_v : [0,1] \times X \longrightarrow X$ is defined as mild solution to the initial value problem $\frac{d}{dt}\, u(t) = A\, u(t) + v$, $u(0) = x \in X$, i.e.

$$\tau_v(h, x) \;:=\; S(h)\, x + \int_0^h S(h-s)\, v\; ds.$$

Proposition 29. *For each vector $v \in X$ fixed, the function $\tau_v : [0,1] \times X \longrightarrow X$ has the following properties for every $j \in \mathscr{I}$, $x, y, w \in X$ and $t, h \in [0,1]$ with $t + h \leq 1$*

(*1.*) $\tau_v(0, x) \;=\; x$

(*2.*) $\tau_v(t+h, x) \;=\; \tau_v\big(h, \tau_v(t, x)\big)$

(*3.*) $\limsup_{h \downarrow 0} \frac{1}{h} \big(d_j(\tau_v(h,x), \tau_v(h,y)) - d_j(x,y)\big) \;\leq\; |\lambda_j|\; d_j(x,y)$

(*4.*) $d_j\big(x, \tau_v(h,x)\big) \;\leq\; \big(\|x\| + \|v\|\big)\quad e^{|\lambda_j|}\;\; h$

(*5.*) $\|\tau_v(h,x)\| \qquad \leq\; \big(\|x\| + \|v\|\, h\big)\; e^{\omega\, h}$

(*6.*) $\limsup_{h \downarrow 0} \frac{1}{h} \cdot d_j\big(\tau_v(h,x), \tau_w(h,x)\big) \;\leq\; d_j(v, w)$.

For preparing the proof, we summarize the essential tools about C^0 semigroups. Subsequent Lemma 30 bridges the gap between the linear semigroup operators and their dual counterparts. It is one of the reasons for assuming X to be reflexive. Afterwards Lemma 31 implies that each vector v'_j $(j \in \mathscr{I})$ is a eigenvector of every dual operator $S(t)'$ $(t \geq 0)$ belonging to the eigenvalue $e^{\lambda_j \cdot t}$.

Lemma 30 ([76, Proposition I.5.14], [154, Corollary 1.10.6]).
Let $(S(t))_{t \geq 0}$ be a C^0 semigroup on a reflexive Banach space with generator A. Then the dual operators $S(t)'$ $(t \geq 0)$ provide a C^0 semigroup on the dual space and its generator is the dual operator A'.

Lemma 31 ([76, Corollary IV.3.8]). *The eigenspaces of the generator A and of the C^0 semigroup operators $S(t)$ $(t \geq 0)$, respectively, fulfill for every $\mu \in \mathbb{C}$*

$$\ker\,(\mu - A) \;=\; \bigcap_{t \geq 0} \ker\,\big(e^{\mu\, t} - S(t)\big).$$

Proof (of Proposition 29). Statements (1.) and (2.) result directly from the semigroup property of $(S(t))_{t\geq 0}$.

(3.) For every $x, y \in X$, $h \in [0,1]$ and $j \in \mathscr{I}$, we obtain

$$\begin{aligned} d_j\big(\tau_v(h,x),\ \tau_v(h,y)\big) - d_j(x,y) &\leq |\langle x-y,\ (S(h)' - \mathbb{I}\mathrm{d}_{X'})\, v'_j\rangle| \\ \limsup_{h\downarrow 0}\ \tfrac{1}{h}\ \big(d_j\big(\tau_v(h,x),\ \tau_v(h,y)\big) - d_j(x,y)\big) &\leq |\langle x-y,\ A'\, v'_j\rangle| \\ &\leq |\lambda_j| \cdot |\langle x-y,\ v'_j\rangle|. \end{aligned}$$

(4.) Each $v'_j \in X'$ is unit eigenvector of A' related to eigenvalue λ_j by assumption. Thus, Lemma 31 implies for every $x \in X$, $h \in [0,1]$ and $j \in \mathscr{I}$

$$\begin{aligned} d_j\big(x,\ \tau_v(h,x)\big) &= \Big|\Big\langle (S(h) - \mathbb{I}\mathrm{d}_X)\, x + \int_0^h S(h-s)\, v\, ds,\ v'_j\Big\rangle\Big| \\ &\leq \Big|\big\langle x,\ (S(h)' - \mathbb{I}\mathrm{d}_{X'})\, v'_j\big\rangle\Big| + \Big|\Big\langle v,\ \int_0^h S(h-s)'\, v'_j\ ds\Big\rangle\Big| \\ &\leq \|x\|\ \ (e^{|\lambda_j|\, h} - 1)\, \|v'_j\| + \|v\|\ \Big\|\int_0^h e^{\lambda_j\,(h-s)}\, v'_j\ ds\Big\| \\ &\leq \big(\|x\| + \|v\|\big)\ e^{|\lambda_j|\, h}\ h. \end{aligned}$$

(5.) $(S(t))_{t\geq 0}$ is ω-contractive with $\omega > 0$. Thus, for every $x \in X$, $h \in [0,1]$

$$\begin{aligned} \|\tau_v(h,x)\| &\leq \Big\|S(h)\, x + \int_0^h S(h-s)\ v\ ds\Big\| \\ &\leq e^{\omega\, h}\, \|x\| + \int_0^h e^{\omega\,(h-s)}\ ds \cdot \|v\| \\ &\leq e^{\omega\, h}\, \|x\| + \frac{e^{\omega\, h} - 1}{\omega}\ \|v\|. \end{aligned}$$

(6.) For arbitrary vectors $v, w \in X$, the functions $\tau_v, \tau_w : [0,1] \times X \longrightarrow X$ satisfy for every $x \in X$ and $h \in [0,1]$

$$\begin{aligned} d_j\big(\tau_v(h,x),\ \tau_w(h,x)\big) &= \Big|\Big\langle \int_0^h S(h-s)\ (v-w)\ ds,\ v'_j\Big\rangle\Big| \\ &= \Big|\Big\langle v-w,\ \int_0^h S(h-s)'\, v'_j\ ds\Big\rangle\Big| \\ &= \Big|\Big\langle v-w,\ \int_0^h e^{\lambda_j\cdot(h-s)}\ v'_j\ ds\Big\rangle\Big| \\ &\leq \big|\big\langle v-w,\ v'_j\big\rangle\big|\ e^{|\lambda_j|\, h}\ h \\ \limsup_{h\downarrow 0}\ \tfrac{1}{h}\cdot d_j\big(\tau_v(h,x),\ \tau_w(h,x)\big) &\leq d_j(v,w). \end{aligned}$$

□

Corollary 32. *For each $v \in X$, the function $\tau_v : [0,1] \times X \longrightarrow X$ specified in Definition 28 is a transition on $\big(X,\ (d_j)_{j\in\mathscr{I}},\ (\|\cdot\|)_{j\in\mathscr{I}}\big)$ in the sense of Definition 2 (on page 104) with*

$$\begin{aligned} \alpha_j(\tau_v; r) &:= |\lambda_j| \\ \beta_j(\tau_v; r) &:= (r + 2\, \|v\|)\ e^{\omega + |\lambda_j|} \\ \gamma_j(\tau_v) &:= \max\big\{\|v\|,\ \omega\big\} \\ D_j(\tau_v,\ \tau_w;\ r) &\leq d_j(v,w). \end{aligned}$$

Theorem 33 (Existence of mild solutions to semilinear evolution equations). *In addition to the general assumptions of § 2.4, suppose for* $f : X \times [0,T] \longrightarrow X$

(*i*) $\sup_{x,t} \|f(x,t)\| < \infty$,

(*ii*) *f is weakly continuous in the following sense: For* $\mathscr{L}^1$*-almost every* $t \in [0,T]$ *and any sequences* $(t_m)_m$, $(y_m)_m$ *in* $[0,T]$, X *respectively with* $t_m \longrightarrow t$ *and* $y_m \longrightarrow y$ *weakly in X for* $m \longrightarrow \infty$, *it fulfills*

$$f(y_m, t_m) \longrightarrow f(y,t) \quad \text{weakly in } X \qquad \text{for } m \longrightarrow \infty.$$

Then for every initial vector $x_0 \in X$, *there exists a solution* $x(\cdot) : [0,T] \longrightarrow X$ *to the mutational equation*

$$\overset{\circ}{x}(\cdot) \ni \tau_{f(x(\cdot),\,\cdot)}$$

on the tuple $\big(X, (d_j)_{j\in\mathscr{I}}, (\|\cdot\|)_{j\in\mathscr{I}}\big)$ *with* $x(0) = x_0$.
Furthermore every solution $x(\cdot) : [0,T] \longrightarrow X$ *to this mutational equation is a mild solution to the semilinear evolution equation*

$$\tfrac{d}{dt}\, x(t) = A\, x(t) + f(x(t),t).$$

The proof results from Peano's Theorem 18 (on page 114) and the following three lemmas:

Lemma 34. (*1.*) *A sequence* $(y_m)_{m\in\mathbb{N}}$ *in X converges to y weakly in X if and only if* $\sup_m \|y_m\| < \infty$ *and* $\lim_{m\to\infty} d_j(y_m, y) = 0$ *for each index* $j \in \mathscr{I}$.
(2.) *Every ball* $\{y \in X \mid \|y\| \le r\}$ *with arbitrary radius* $r \ge 0$ *is sequentially compact w.r.t. the topology of* $(d_j)_{j\in\mathscr{I}}$. *Hence* $\big(X, (d_j)_j, (\|\cdot\|)_j\big)$ *is Euler compact.*

Lemma 35. *Under the assumptions of Theorem 33, any solution* $x(\cdot) : [0,T] \longrightarrow X$ *to the mutational equation*

$$\overset{\circ}{x}(\cdot) \ni \tau_{f(x(\cdot),\,\cdot)}$$

on the tuple $\big(X, (d_j)_{j\in\mathscr{I}}, (\|\cdot\|)_{j\in\mathscr{I}}\big)$ *has the following properties for every* $v' \in X'$:

(*1.*) $[0,T] \longrightarrow \mathbb{R},\ t \longmapsto \langle f(x(t),t),\, v'\rangle$ *is continuous at* $\mathscr{L}^1$*-almost every time t,*

(*2.*) $f(x(\cdot),\cdot) \in L^\infty([0,T],X)$,

(*3.*) $[0,T] \longrightarrow \mathbb{R},\ t \longmapsto \langle x(t), v'\rangle$ *is absolutely continuous for every* $v' \in D(A') \subset X'$ *and* $\tfrac{d}{dt}\langle x(t), v'\rangle = \langle x(t), A'v'\rangle + \langle f(x(t),t),\, v'\rangle$.

Lemma 36 (Ball [20]). *Let A be a densely defined closed linear operator on a real or complex Banach space Y and* $g \in L^1([0,T],Y)$.
There exists for each $x_0 \in Y$ *a unique weak solution* $u(\cdot)$ *of*

$$\begin{cases} \tfrac{d}{dt}\, u(t) = A\, u(t) + g(t) \ \text{on }]0,T] \\ u(0) = x_0 \end{cases}$$

i.e. for every $v' \in D(A') \subset Y'$, $\langle u(\cdot), v'\rangle \in W^{1,1}([0,T])$ *and*

$$\tfrac{d}{dt}\langle u(t), v'\rangle = \langle u(t), A'v'\rangle + \langle g(t), v'\rangle \qquad \text{for almost all } t,$$

if and only if A is the generator of a strongly continuous semigroup $(S(t))_{t\ge 0}$, *and in this case,* $u(t)$ *is given by* $u(t) = S(t)\,x_0 + \int_0^t S(t-s)\, g(s)\, ds.$

Proof (of Lemma 34). Statement (1.) is a standard result of linear functional analysis since $(v'_j)_{j\in\mathscr{I}}$ spans X' by assumption (see e.g. [188, § V.1, Theorem 3]). The sequential compactness (of closed norm balls) in statement (2.) results from Alaoglu's Theorem due to the reflexivity of X. Finally we obtain Euler compactness as a consequence of Remark 16 (on page 113). □

Proof (of Lemma 35). (1.) According to Definition 9 (on page 107), every solution $x(\cdot) : [0,T] \longrightarrow X$ to the mutational equation

$$\overset{\circ}{x}(\cdot) \ni \tau_{f(x(\cdot),\cdot)}$$

on the tuple $\big(X, (d_j)_{j\in\mathscr{I}}, (\|\cdot\|)_{j\in\mathscr{I}}\big)$ satisfies $\sup_t \|x(t)\| < \infty$ and is continuous with respect to each pseudo-metric d_j, $j \in \mathscr{I}$. Due to preceding Lemma 34, $x(\cdot) : [0,T] \longrightarrow X$ is weakly continuous. For each linear form $v' \in X'$, assumption (ii) of Theorem 33 guarantees the continuity of the composition

$$[0,T] \longrightarrow \mathbb{R}, \quad t \longmapsto \langle f(x(t),t),\, v'\rangle$$

at $\mathscr{L}^1$-almost every time $t \in [0,T]$.

(2.) Statement (1.) and the uniform bound

$$\sup_{t\in[0,T]} \; |\langle f(x(t),t),\, v'\rangle| \;\le\; \|f\|_{L^\infty}\, \|v'\|_{X'} \;<\; \infty$$

imply the weak Lebesgue measurability of $f(x(\cdot),\cdot)$. Banach space X is separable by assumption and thus, $f(x(\cdot),\cdot) : [0,T] \longrightarrow X$ is (strongly) Lebesgue-measurable due to the Theorem of Pettis (stated and proved in [188, § V.4], for example).

(3.) Choose any index $j \in \mathscr{I}$. At $\mathscr{L}^1$-almost every time $t \in [0,T]$, $x(\cdot)$ satisfies

$$\begin{aligned} 0 \;&=\; \lim_{h\downarrow 0} \tfrac{1}{h}\cdot d_j\big(\tau_{f(x(t),t)}(h, x(t)),\; x(t+h)\big) \\ &=\; \lim_{h\downarrow 0} \tfrac{1}{h}\cdot \big|\langle \tau_{f(x(t),t)}(h, x(t)) - x(t),\; v'_j\rangle \;-\; \langle x(t+h) - x(t),\; v'_j\rangle\big| \end{aligned}$$

Due to Definition 28 (on page 127), we obtain for $\mathscr{L}^1$-almost every $t \in [0,T]$

$$\lim_{h\downarrow 0} \; \tfrac{1}{h}\, \langle x(t+h) - x(t),\; v'_j\rangle \;=\; \langle x(t), A'v'_j\rangle + \langle f(x(t),t),\; v'_j\rangle$$

and, the right-hand side is $\mathscr{L}^1$-integrable with respect to t. These two properties ensure that $[0,T] \longrightarrow \mathbb{R}$, $t \longmapsto \langle x(t),\, v'_j\rangle$ is absolutely continuous for every $j \in \mathscr{I}$. The corresponding integral equation

$$\langle x(t),\, v'_j\rangle \;-\; \langle x(0),\, v'_j\rangle \;=\; \int_0^t \big(\langle x(s),\, A'v'_j\rangle + \langle f(x(s),s),\; v'_j\rangle\big)\, ds$$

with arbitrary $t \in [0,T]$ can be extended to every linear form $v' \in D(A') \subset X'$ since $(v'_j)_{j\in\mathscr{I}}$ spans the dual space X'. Hence, $[0,T] \longrightarrow \mathbb{R}$, $t \longmapsto \langle x(t),\, v'\rangle$ is absolutely continuous for every $v' \in D(A') \subset X'$ and satisfies

$$\tfrac{d}{dt}\, \langle x(t),\, v'\,\rangle \;=\; \langle x(t),\, A'\,v'\,\rangle \;+\; \langle\, f(x(t),t),\; v'\,\rangle.$$

□

Proposition 37 (Existence of solutions to a system with semilinear evolution equation and modified morphological equation).
In addition to the general assumptions of § 2.4, suppose for

$$f : X \times \mathscr{K}(\mathbb{R}^N) \times [0,T] \longrightarrow X,$$
$$\mathscr{G} : X \times \mathscr{K}(\mathbb{R}^N) \times [0,T] \longrightarrow \text{OSLIP}(\mathbb{R}^N, \mathbb{R}^N)$$

(i) $\sup_{x,M,t} \big(\|f(x,M,t)\|_X + \|\mathscr{G}(x,M,t)\|_\infty + \max\{0, \ \text{Lip}\, \mathscr{G}(x,M,t)\} \big) < \infty.$

(ii) *f and $\mathscr{G}$ are continuous in the following sense:*

$$\left\{ \begin{array}{rcl} f(y_n, M_n, t_n) - f(y, M, t) & \longrightarrow & 0 \ \text{weakly in } X \\ d\!l_\infty\big(\mathscr{G}(y_n, M_n, t_n), \ \mathscr{G}(y, M, t)\big) & \longrightarrow & 0 \end{array} \right. \qquad \text{for } n \longrightarrow \infty$$

holds for $\mathscr{L}^1$-almost every $t \in [0,T]$ and any sequences $(t_n)_{n\in\mathbb{N}}$, $(M_n)_{n\in\mathbb{N}}$ and $(y_n)_{n\in\mathbb{N}}$ in $[0,T], \mathscr{K}(\mathbb{R}^N), X$ respectively satisfying $t_n \longrightarrow t$, $d\!l(M_n, M) \longrightarrow 0$ and $y_n \longrightarrow y$ weakly in X for $n \longrightarrow \infty$.

Then for every initial vector $x_0 \in X$ and set $K_0 \in \mathscr{K}(\mathbb{R}^N)$, there exist solutions $x(\cdot) : [0,T] \longrightarrow X$, $K(\cdot) : [0,T] \longrightarrow \mathscr{K}(\mathbb{R}^N)$ to the system of mutational equations

$$\left\{ \begin{array}{rcl} \overset{\circ}{x}(\cdot) & \ni & \tau_{f(x(\cdot), K(\cdot), \cdot)} \\ \overset{\circ}{K}(\cdot) & \ni & \mathscr{G}\big(x(\cdot), K(\cdot), \cdot\big) \end{array} \right.$$

with $x(0) = x_0$ and $K(0) = K_0$. In particular,

(1.) *$x(\cdot) : [0,T] \longrightarrow X$ is a mild solution to the evolution equation*

$$\tfrac{d}{dt}\, x(t) = A\, x(t) + f(x(t), K(t), t).$$

(2.) *$K(\cdot)$ is Lipschitz continuous w.r.t. $d\!l$ and satisfies for $\mathscr{L}^1$-almost every t*

$$\lim_{h \downarrow 0} \ \tfrac{1}{h} \cdot d\!l\big(\vartheta_{\mathscr{G}(x(t), K(t), t)}(h, K(t)), \ K(t+h)\big) = 0.$$

(3.) *If, in addition, the set-valued map $\mathscr{G}(x(t), K(t), t) : \mathbb{R}^N \rightsquigarrow \mathbb{R}^N$ is continuous for each $t \in [0,T]$, then the set $K(t) \subset \mathbb{R}^N$ coincides with the reachable set $\vartheta_{\mathscr{G}(x(\cdot), K(\cdot), \cdot)}(t, K_0)$ of the nonautonomous differential inclusion*

$$y'(\cdot) \in \mathscr{G}\big(x(\cdot), K(\cdot), \cdot\big)\,(y(\cdot))$$

at every time $t \in [0,T]$.

Proof. It results from Peano's Theorem 21 about systems of mutational equations (on page 118), Theorem 33 about mild solutions (on page 129) and Proposition 1.97 in combination with Corollary 1.106 about modified morphological equations (on pages 98, 101). □

2.5 Example: Nonlinear Transport Equations for Radon Measures on $\mathbb{R}^N$

Now the focus of interest is the Cauchy problem of the nonlinear transport equation

$$\tfrac{d}{dt}\,\mu + \operatorname{div}_x\big(f(\mu,\cdot)\;\mu\big) \;=\; g(\mu,\cdot)\,\mu \qquad (\text{in } \mathbb{R}^N\times\,]0,T[)$$

together with its distributional solutions $\mu(\cdot) : [0,T] \longrightarrow \mathscr{M}(\mathbb{R}^N)$ whose values are Radon measures on the whole Euclidean space $\mathbb{R}^N$. The coefficients $f(\mu,t)$, $g(\mu,t)$ are assumed to be uniformly bounded and Lipschitz continuous vector fields on $\mathbb{R}^N$. Considering them as an example of the mutational framework here, we specify some sufficient conditions on the coefficients $f(\cdot,\cdot)$, $g(\cdot,\cdot)$ for existence, uniqueness of distributional solutions and even their continuous dependence on data.
In particular, this nonlinear transport equation takes nonlocal dependence into consideration because the arguments of the coefficient functions $f(\cdot,t)$ and $g(\cdot,t)$ are not restricted to *local* properties of measures, but consider the Radon measures on whole $\mathbb{R}^N$. This example provides some technical preparation for the structural population model in § 2.6 below.

2.5.1 *The* $W^{1,\infty}$ *Dual Metric* $\rho_{\mathscr{M}}$ *on Radon Measures* $\mathscr{M}(\mathbb{R}^N)$

For implementing these transport equations in the mutational framework, we first specify the basic set and an appropriate metric.

Definition 38. $C^0_c(\mathbb{R}^N)$ denotes the space of continuous functions $\mathbb{R}^N \longrightarrow \mathbb{R}$ with compact support and $C^0_0(\mathbb{R}^N)$ its closure with respect to the supremum norm, respectively.
Furthermore, $\mathscr{M}(\mathbb{R}^N)$ consists of all finite real-valued Radon measures on $\mathbb{R}^N$, i.e., it is the dual space of $\big(C^0_0(\mathbb{R}^N),\, \|\cdot\|_\infty\big)$ (due to Riesz theorem [5, Remark 1.57]).
$\mathscr{M}^+(\mathbb{R}^N)$ denotes the subset of *nonnegative* measures $\mu \in \mathscr{M}(\mathbb{R}^N)$, i.e. $\mu(\cdot) \geq 0$.

The weak* topology on $\mathscr{M}(\mathbb{R}^N)$ is a rather obvious choice. There is, however, a very useful alternative which proves to be equivalent if we restrict our considerations to subsets of Radon measures which are "concentrated not too far away from each other".

Definition 39. A sequence $(\mu_n)_{n\in\mathbb{N}}$ in $\mathscr{M}(\mathbb{R}^N)$ is said to *converge narrowly* to $\mu \in \mathscr{M}(\mathbb{R}^N)$ if for every bounded continuous function $\varphi : \mathbb{R}^N \longrightarrow \mathbb{R}$,

$$\lim_{n\to\infty} \int_{\mathbb{R}^N} \varphi\; d\mu_n \;=\; \int_{\mathbb{R}^N} \varphi\; d\mu.$$

Definition 40. A nonempty subset $\mathscr{V} \subset \mathscr{M}(\mathbb{R}^N)$ is called *tight* if for every $\varepsilon > 0$, there exists a compact set $K_\varepsilon \subset \mathbb{R}^N$ such that the total variations of all $\mu \in \mathscr{V}$ satisfy

$$\sup_{\mu\in\mathscr{V}} |\mu|(\mathbb{R}^N \setminus K_\varepsilon) \;<\; \varepsilon.$$

Basic set	$E := \mathcal{M}(\mathbb{R}^N) = \big(C_0^0(\mathbb{R}^N)\big)'$ the space of all finite real-valued Radon measures on $\mathbb{R}^N$
Distances Absolute values	$W^{1,\infty}$ dual metric $\rho_{\mathcal{M}} : \mathcal{M}(\mathbb{R}^N) \times \mathcal{M}(\mathbb{R}^N) \longrightarrow [0,\infty[$ $\rho_{\mathcal{M}}(\mu,\, \nu) := \sup\Big\{ \int_{\mathbb{R}^N} \psi \; d(\mu-\nu) \;\Big|\; \psi \in C^1(\mathbb{R}^N),\; \|\psi\|_\infty, \|\nabla\psi\|_\infty \leq 1 \Big\}$ (in tight subsets, it is equivalent to the narrow topology, Prop. 43) $\lfloor \cdot \rfloor := |\cdot|(\mathbb{R}^N)$ total variation of the Radon measure
Transition	For each $\mathbf{b} \in W^{1,\infty}(\mathbb{R}^N, \mathbb{R}^N)$ and $c \in W^{1,\infty}(\mathbb{R}^N, \mathbb{R})$, define $\vartheta_{\mathcal{M}(\mathbb{R}^N),\mathbf{b},c} : [0,1] \times \mathcal{M}(\mathbb{R}^N) \longrightarrow \mathcal{M}(\mathbb{R}^N), \quad (t, \nu_0) \longmapsto \mu_t$ as the distributional solution to the linear autonomous problem $\begin{cases} \partial_t \mu_t + \operatorname{div}_x(\mathbf{b}\,\mu_t) = c\,\mu_t & \text{in } [0,T] \\ \mu_0 = \nu_0 \end{cases}$
Compactness	Euler compactness since the flow $\mathbf{X_b}$ along $\mathbf{b}$ implies tightness: Lemma 52 (page 142)
Mutational solutions	Narrowly continuous distributional solution to nonlinear transport equations $\frac{d}{dt}\,\mu + \operatorname{div}_x\big(f(\mu,\cdot)\,\mu\big) = g(\mu,\cdot)\,\mu$ in $\mathbb{R}^N \times]0,T[$
List of main results formulated in § 2.5	Existence due to compactness (Peano): Theorem 53 (page 142) Uniqueness due to Lipschitz continuity: Theorem 54 (page 143)
Key tools	Explicit solutions to autonomous linear transport equations: Proposition 46 (page 137) Uniqueness of solutions to nonautonomous linear equations: Proposition 47 (page 138)

Table 2.2 Brief summary of the example in § 2.5 in mutational terms:
Nonlinear transport equations for Radon measures on $\mathbb{R}^N$

Remark 41. (1.) On every tight subset of $\mathcal{M}(\mathbb{R}^N)$, the narrow topology is equivalent to the weak* topology (with respect to $\mathcal{M}(\mathbb{R}^N) = C_0^0(\mathbb{R}^N)'$).

(2.) Tightness is just one of the many concepts which are often introduced (merely) for probability measures or positive Radon measures (see e.g. [3, 4, 6]). Many results also hold in $\mathcal{M}(\mathbb{R}^N)$ by considering the total variation (if necessary). Here we want to dispense with any global restrictions in regard to sign or total variation of Radon measures. and so, we cannot simply use any Wasserstein metric (for probability measures) in particular.

(3.) A nonempty subset $\mathcal{V} \subset \mathcal{M}(\mathbb{R}^N)$ is tight if and only if there is a function $\Psi : \mathbb{R}^N \longrightarrow [0,\infty]$ whose sublevel set $\{x \in \mathbb{R}^N \mid \Psi(x) \le c\}$ is compact for every $c \in [0,\infty[$ and which satisfies

$$\sup_{\mu \in \mathcal{V}} \int_{\mathbb{R}^N} \Psi(x) \; d|\mu|(x) \; < \; \infty$$

[6, Remark 5.1.5]. In regard to total variation $|\mu|$, the last condition is equivalent to

$$\sup_{\mu \in \mathcal{V}} \; \sup_{\substack{\phi \in C^0(\mathbb{R}^N): \\ |\phi| \le \Psi}} \int_{\mathbb{R}^N} \phi(x) \; d\mu(x) \; < \; \infty.$$

The topology of narrow convergence on $\mathcal{M}(\mathbb{R}^N)$ is metrizable on tight subsets with uniformly bounded total variation:

Definition 42.

$$\begin{aligned} \mathcal{M}(\mathbb{R}^N) \times \mathcal{M}(\mathbb{R}^N) &\longrightarrow [0,\infty[\\ (\mu,\nu) &\longmapsto \sup\left\{ \int_{\mathbb{R}^N} \psi \; d(\mu-\nu) \;\middle|\; \psi \in C^1(\mathbb{R}^N),\; \|\psi\|_\infty, \|\nabla\psi\|_\infty \le 1 \right\} \end{aligned}$$

is called $W^{1,\infty}$ *dual metric* $\rho_{\mathcal{M}}$ on $\mathcal{M}(\mathbb{R}^N)$.

Proposition 43. (1.) *For every* $\lambda > 0$ *and* $\mu, \nu \in \mathcal{M}(\mathbb{R}^N)$,

$$\begin{aligned} \rho_{\mathcal{M}}(\mu,\nu) &= \sup\left\{ \tfrac{1}{\lambda} \int_{\mathbb{R}^N} \varphi \; d(\mu-\nu) \;\middle|\; \varphi \in C_c^\infty(\mathbb{R}^N), \quad \|\varphi\|_\infty \le \lambda,\; \|\nabla\varphi\|_\infty \le \lambda \right\} \\ &= \sup\left\{ \tfrac{1}{\lambda} \int_{\mathbb{R}^N} \varphi \; d(\mu-\nu) \;\middle|\; \varphi \in W^{1,\infty}(\mathbb{R}^N),\; \|\varphi\|_\infty \le \lambda,\; \|\nabla\varphi\|_\infty \le \lambda \right\} \\ &= \|\mu-\nu\|_{(W^{1,\infty})'} \end{aligned}$$

(2.) *For any tight sequence* $(\mu_n)_{n\in\mathbb{N}}$ *and* μ *in* $\mathcal{M}(\mathbb{R}^N)$, *the following equivalence holds*

$$\left\{ \begin{array}{lcl} \lim_{n\to\infty} & \rho_{\mathcal{M}}(\mu_n,\mu) & = \; 0 \\ \sup_{n\in\mathbb{N}} & |\mu_n|(\mathbb{R}^N) & < \; \infty \end{array} \right\} \iff \mu_n \longrightarrow \mu \;\; \textit{weak*} \quad \textit{for } n \longrightarrow \infty$$

$$\iff \mu_n \longrightarrow \mu \;\; \textit{narrowly} \;\; \textit{for } n \longrightarrow \infty$$

(3.) *For any* $r > 0$, *the set* $\{\mu \in \mathcal{M}(\mathbb{R}^N) \mid |\mu|(\mathbb{R}^N) \le r\}$ *is complete w.r.t.* $\rho_{\mathcal{M}}$.

(4.) *Every tight set* $\mathcal{V} \subset \mathcal{M}(\mathbb{R}^N)$ *with* $\sup_{\mu\in\mathcal{V}} |\mu|(\mathbb{R}^N) < \infty$ *is relatively compact with respect to* $\rho_{\mathcal{M}}$.

Proof. (1.) Considering the restrictions to an arbitrarily fixed compact subset of $\mathbb{R}^N$, each function in $W^{1,\infty}(\mathbb{R}^N)$ can be approximated by elements of $C_c^\infty(\mathbb{R}^N) \subset C^1(\mathbb{R}^N) \cap W^{1,\infty}(\mathbb{R}^N)$ with respect to supremum norm. This implies the equivalent characterizations of $\rho_{\mathscr{M}}(\mu,\nu)$ claimed here.

(2.) The equivalence of narrow and weak* convergence results from the assumption of tightness according to Remark 41 (1.).
Now let $(\mu_n)_{n\in\mathbb{N}}$ be any sequence in $\mathscr{M}(\mathbb{R}^N)$ and $\mu \in \mathscr{M}(\mathbb{R}^N)$ satisfying

$$\lim_{n\longrightarrow\infty} \rho_{\mathscr{M}}(\mu_n,\mu) = 0, \qquad \sup_{n\in\mathbb{N}} |\mu_n|(\mathbb{R}^N) < \infty.$$

In particular, $\int_{\mathbb{R}^N} \varphi \, d\mu_n \longrightarrow \int_{\mathbb{R}^N} \varphi \, d\mu$ for $n \longrightarrow \infty$ and every $\varphi \in W^{1,\infty}(\mathbb{R}^N)$.

We obtain $\int_{\mathbb{R}^N} \varphi \, d\mu_n \longrightarrow \int_{\mathbb{R}^N} \varphi \, d\mu$ for $n \longrightarrow \infty$ and every $\varphi \in C_0^0(\mathbb{R}^N)$

since $W^{1,\infty}(\mathbb{R}^N) \cap C_0^0(\mathbb{R}^N)$ is dense in $(C_0^0(\mathbb{R}^N), \|\cdot\|_\infty)$ and the total variations of $(\mu_n)_{n\in\mathbb{N}}$ are bounded. Thus, the sequence $(\mu_n)_{n\in\mathbb{N}}$ converges also weakly* in $\mathscr{M}(\mathbb{R}^N) = C_0^0(\mathbb{R}^N)'$.

Finally, assume the tight sequence $(\mu_n)_{n\in\mathbb{N}}$ in $\mathscr{M}(\mathbb{R}^N)$ to converge weakly* to $\mu \in \mathscr{M}(\mathbb{R}^N)$. Then $C := \sup_{n\in\mathbb{N}} |\mu_n|(\mathbb{R}^N) < \infty$ due to the uniform boundedness theorem and, $|\mu|(\mathbb{R}^N) \leq \liminf_{n\longrightarrow\infty} |\mu_n|(\mathbb{R}^N) \leq C$. We still have to prove for $n \longrightarrow \infty$

$$\sup \Big\{ \int_{\mathbb{R}^N} \varphi \; d(\mu_n - \mu) \;\Big|\; \varphi \in C_c^\infty(\mathbb{R}^N),\ \|\varphi\|_\infty \leq 1,\ \|\nabla\varphi\|_\infty \leq 1 \Big\} \longrightarrow 0.$$

Choose $\varepsilon > 0$ arbitrarily. Then there exists a sufficiently large radius $R > 0$ with

$$\sup_{n\in\mathbb{N}} |\mu_n|(\mathbb{R}^N \setminus \mathbb{B}_R(0)) + |\mu|(\mathbb{R}^N \setminus \mathbb{B}_R(0)) \leq \varepsilon$$

since $\{\mu_n \,|\, n \in \mathbb{N}\}$ is tight. Due to Arzelà-Ascoli Theorem A.82,

$$\big\{\varphi \in C_c^\infty(\mathbb{B}_{R+1}(0)) \;\big|\; \|\varphi\|_\infty \leq 1,\ \|\nabla\varphi\|_\infty \leq 1\big\}$$

is relatively compact in $\big(C^0(\mathbb{B}_{R+1}(0)), \|\cdot\|_\infty\big)$. Hence, there always exist finitely many functions $\widetilde{\varphi}_1 \ldots \widetilde{\varphi}_{k_\varepsilon} \in C_c^\infty(\mathbb{R}^N)$ with support in $\mathbb{B}_{R+1}(0)$ and $\|\widetilde{\varphi}_i\|_\infty \leq 1$, $\|\nabla\widetilde{\varphi}_i\|_\infty \leq 1$ such that

$$\big\{\varphi \in C_c^\infty(\mathbb{B}_{R+1}(0)) \;\big|\; \|\varphi\|_\infty \leq 1, \|\nabla\varphi\|_\infty \leq 1\big\} \subset \bigcup_{i=1\ldots k_\varepsilon} \big\{\varphi \;\big|\; \|\varphi - \widetilde{\varphi}_i|_{\mathbb{B}_{R+1}(0)}\|_\infty \leq \varepsilon\big\}.$$

This implies

$$\begin{aligned}
&\sup\Big\{ \int_{\mathbb{R}^N} \varphi \, d(\mu_n - \mu) \;\Big|\; \varphi \in C_c^\infty(\mathbb{R}^N),\ \|\varphi\|_\infty \leq 1,\ \|\nabla\varphi\|_\infty \leq 1\Big\} \\
&\quad\leq \sup\Big\{ \int_{\mathbb{B}_R(0)} \varphi \, d(\mu_n - \mu) \;\Big|\; \varphi \in C_c^\infty(\mathbb{R}^N),\ \|\varphi\|_\infty \leq 1,\ \|\nabla\varphi\|_\infty \leq 1\Big\} + \varepsilon \\
&\quad\leq \sup\Big\{ \int_{\mathbb{B}_R(0)} \widetilde{\varphi}_i \; d(\mu_n - \mu) \;\Big|\; 1 \leq i \leq k_\varepsilon\Big\} + 2C\,\varepsilon + \varepsilon \\
&\quad\leq \varepsilon + 2C\,\varepsilon + \varepsilon
\end{aligned}$$

for all $n \in \mathbb{N}$ sufficiently large (merely depending on ε) since $\mu_n \longrightarrow \mu$ weakly*.

(3.) Let $(\mu_n)_{n\in\mathbb{N}}$ be a $\rho_{\mathcal{M}}$-Cauchy sequence satisfying $\sup_{n\in\mathbb{N}} |\mu_n|(\mathbb{R}^N) \le r < \infty$. The arguments proving the first part "$\Rightarrow$" of statement (2.) imply that $(\mu_n)_{n\in\mathbb{N}}$ is Cauchy sequence with respect to the weak* topology of $\mathcal{M}(\mathbb{R}^N)$. There is the unique measure $\mu \in \mathcal{M}(\mathbb{R}^N)$ as weak* limit of $(\mu_n)_{n\in\mathbb{N}}$ due to [5, Theorem 1.59]. In particular, $|\mu|(\mathbb{R}^N) \le \liminf\limits_{n\to\infty} |\mu_n|(\mathbb{R}^N) \le r$.
We still have to verify $\rho_{\mathcal{M}}(\mu_n,\mu) \longrightarrow 0$ for $n \longrightarrow \infty$. Indeed for arbitrary $\varepsilon > 0$, there exists $n_\varepsilon \in \mathbb{N}$ such that for all $m,n \ge n_\varepsilon$

$$\rho_{\mathcal{M}}(\mu_m,\mu_n) \overset{\text{Def.}}{=} \sup\left\{ \int_{\mathbb{R}^N} \varphi \; d(\mu_m - \mu_n) \,\middle|\, \varphi \in C_c^\infty(\mathbb{R}^N),\ \|\varphi\|_\infty,\ \|\nabla\varphi\|_\infty \le 1 \right\} \le \varepsilon.$$

Due to the weak* convergence of $(\mu_n)_{n\in\mathbb{N}}$ to μ in $\mathcal{M}(\mathbb{R}^N) = \big(C_0^0(\mathbb{R}^N),\ \|\cdot\|_\infty\big)'$, the limit for $n \longrightarrow \infty$ reveals for every $m \ge n_\varepsilon$

$$\begin{aligned}\rho_{\mathcal{M}}(\mu_m,\ \mu) &\overset{\text{Def.}}{=} \sup\left\{ \int_{\mathbb{R}^N} \varphi \, d(\mu_m - \mu) \,\middle|\, \varphi \in C_c^\infty(\mathbb{R}^N),\ \|\varphi\|_\infty,\ \|\nabla\varphi\|_\infty \le 1 \right\} \\ &\le \sup\left\{ \lim_{n\to\infty} \int_{\mathbb{R}^N} \varphi \, d(\mu_m - \mu_n) \,\middle|\, \varphi \in C_c^\infty(\mathbb{R}^N),\ \|\varphi\|_\infty,\ \|\nabla\varphi\|_\infty \le 1 \right\} \\ &\le \varepsilon.\end{aligned}$$

(4.) Due to the assumption of tightness, the relative compactness of $\mathcal{V}$ with respect to $\rho_{\mathcal{M}}$ results from its weak* compactness in $\mathcal{M}(\mathbb{R}^N) = C_0^0(\mathbb{R}^N)'$ and, the latter is ensured by the Banach-Alaoglu Theorem.
(Alternatively, the so-called *Prokhorov Theorem* states that bounded and tight subsets of *positive* Radon measures are sequentially relatively compact with respect to narrow convergence [3, 6, 169]. Finally the claim about signed Radon measures here can also be concluded from this compactness statement by means of Jordan decompositions.) □

2.5.2 Linear Transport Equations Induce Transitions on $\mathcal{M}(\mathbb{R}^N)$

Among the transport equations for Radon measures, the linear one is much simpler to solve, of course. Indeed, the method of characteristics even provides an explicit solution to the initial value problem:

Let $\mathbf{b} : \mathbb{R}^N \longrightarrow \mathbb{R}^N$, $c : \mathbb{R}^N \longrightarrow \mathbb{R}$ be bounded and Lipschitz continuous. For given $\nu_0 \in \mathcal{M}(\mathbb{R}^N)$, the linear problem here focuses on a measure-valued distributional solution $\mu : [0,T] \longrightarrow \mathcal{M}(\mathbb{R}^N)$, $t \longmapsto \mu_t$ of

$$\begin{cases} \partial_t \mu_t + \operatorname{div}_x(\mathbf{b}\,\mu_t) = c\,\mu_t & \text{in } [0,T] \\ \mu_0 = \nu_0 \end{cases}$$

in the sense that

$$\int_{\mathbb{R}^N} \varphi(x)\; d\mu_t(x) - \int_{\mathbb{R}^N} \varphi(x)\; d\nu_0(x) \;=\; \int_0^t \int_{\mathbb{R}^N} \Big(\nabla\varphi(x)\cdot\mathbf{b}(x) + c(x)\Big)\; d\mu_s(x)\; ds$$

for every $t \in [0,T]$ and any test function $\varphi \in C_c^\infty(\mathbb{R}^N,\mathbb{R})$.

Definition 44. $\mathbf{X_b} : [0,T] \times \mathbb{R}^N \longrightarrow \mathbb{R}^N$ is induced by the flow along $\mathbf{b}$, i.e. $\mathbf{X_b}(\cdot,x_0) : [0,T] \longrightarrow \mathbb{R}^N$ is the continuously differentiable solution to the Cauchy problem

$$\begin{cases} \frac{d}{dt}\, x(t) = \mathbf{b}(x(t)) & \text{in } [0,T], \\ \quad x(0) = x_0. \end{cases}$$

As a well-known result about ordinary differential equations, solutions to Cauchy problems are continuously differentiable with respect to initial data and right-hand side if the vector field (on the right-hand side) is continuously differentiable and, the following estimates result from the corresponding integral equations and Gronwall's Lemma (see e.g. [92, Chapter V], [93, Chapter 17], [181, § 13]).

Lemma 45. *For any vector fields* $\mathbf{b}, \widetilde{\mathbf{b}} \in C^1(\mathbb{R}^N,\mathbb{R}^N) \cap W^{1,\infty}(\mathbb{R}^N,\mathbb{R}^N)$, *the solution maps* $\mathbf{X_b}, \mathbf{X}_{\widetilde{\mathbf{b}}} : [0,T] \times \mathbb{R}^N \longrightarrow \mathbb{R}^N$ *are continuously differentiable with*

$$\begin{aligned} \operatorname{Lip} \mathbf{X_b}(t,\cdot) \;&\le\; e^{\operatorname{Lip} \mathbf{b} \cdot t}, \\ \|\mathbf{X_b}(t,\cdot) - \mathbf{X}_{\widetilde{\mathbf{b}}}(t,\cdot)\|_\infty \;&\le\; \|\mathbf{b} - \widetilde{\mathbf{b}}\|_\infty \cdot t\, e^{t \cdot \operatorname{Lip} \widetilde{\mathbf{b}}}. \end{aligned}$$

Proposition 46. *For any* $\mathbf{b} \in W^{1,\infty}(\mathbb{R}^N,\mathbb{R}^N)$, $c \in W^{1,\infty}(\mathbb{R}^N,\mathbb{R})$ *and initial measure* $\nu_0 \in \mathscr{M}(\mathbb{R}^N)$, *a solution* $\mu : [0,T] \longrightarrow \mathscr{M}(\mathbb{R}^N), t \longmapsto \mu_t$ *to the linear problem*

$$\begin{cases} \partial_t\, \mu_t \;+\; \operatorname{div}_x(\mathbf{b}\, \mu_t) \;=\; c\, \mu_t & \text{in } [0,T] \\ \qquad\qquad\qquad\quad \mu_0 \;=\; \nu_0 \end{cases}$$

(in the distributional sense) is given by

$$\int_{\mathbb{R}^N} \varphi\, d\mu_t \;=\; \int_{\mathbb{R}^N} \varphi(\mathbf{X_b}(t,x)) \cdot \exp\Big(\int_0^t c(\mathbf{X_b}(s,x))\, ds\Big)\, d\nu_0(x)$$

for all $\varphi \in C_c^1(\mathbb{R}^N)$.

Proof. First, we verify that the right-hand side provides a distributional solution to the linear problem with the initial measure ν_0. In fact, it is absolutely continuous with respect to t because for any subinterval $[s,t] \subset [0,T]$,

$$\begin{aligned} &\Big| \int_{\mathbb{R}^N} \varphi\, d\mu_t \;-\; \int_{\mathbb{R}^N} \varphi\, d\mu_s \Big| \\ &= \Big| \int_{\mathbb{R}^N} \Big(\varphi(\mathbf{X_b}(t,x)) \cdot e^{\int_0^t c(\mathbf{X_b}(r,x))\, dr} - \varphi(\mathbf{X_b}(s,x)) \cdot e^{\int_0^s c(\mathbf{X_b}(r,x))\, dr} \Big) \qquad d\mu_0(x) \Big| \\ &\le \int_{\mathbb{R}^N} \Big(\Big| \big[\varphi(\mathbf{X_b}(\sigma,x))\big]_{\sigma=s}^{\sigma=t} \Big|\, e^{t\, \|c\|_\infty} + |\varphi(\mathbf{X_b}(s,x))|\, \Big[e^{\int_0^\sigma c(\mathbf{X_b}(r,x))\, dr} \Big]_{\sigma=s}^{\sigma=t} \Big)\, d|\mu_0(x)| \\ &\le \Big(\|\nabla \varphi\|_\infty\, \|\mathbf{b}\|_\infty\, (t-s)\, e^{t\, \|c\|_\infty} + \|\varphi\|_\infty \qquad e^{t\, \|c\|_\infty}\, \|c\|_\infty\, (t-s) \Big)\, |\mu_0|(\mathbb{R}^N) \end{aligned}$$

At $\mathscr{L}^1$-almost every time $t \in [0,T]$, we conclude from the chain rule for weak derivatives

$$\begin{aligned}
&\tfrac{d}{dt}\int_{\mathbb{R}^N}\Big(\varphi(\mathbf{X_b}(t,x))\cdot\exp\Big(\int_0^t c(\mathbf{X_b}(s,x))\,ds\Big)\Big)\,d\nu_0(x)\\
&=\int_{\mathbb{R}^N}\Big(\nabla\varphi(\mathbf{X_b}(t,x))\cdot\mathbf{b}(\mathbf{X_b}(t,x))+\varphi(\mathbf{X_b}(t,x))\,c(\mathbf{X_b}(t,x))\Big)\,e^{\int_0^t c(\mathbf{X_b}(r,x))\,dr}\,d\nu_0\\
&=\int_{\mathbb{R}^N}\Big(\nabla\varphi(y)\cdot\mathbf{b}(y)+\varphi(y)\,c(y)\Big)d\mu_t(y).
\end{aligned}$$ □

This solution is already well-known and usually denoted in the form of a push-forward. Furthermore, it is unique because solutions to the nonautonomous linear transport equation fulfill the following comparison principle (see also [3, 6, 64]):

Proposition 47 (Maniglia [134, Lemma 3.5, Proposition 3.6]**).**
Let $v : t \longmapsto v_t$ *be a Borel vector field in* $L^1\big([0,T];W^{1,\infty}(\mathbb{R}^N,\mathbb{R}^N)\big)$ *and* $c(\cdot,\cdot)$ *a Borel bounded and locally Lipschitz continuous (w.r.t. the space variable) scalar function in* $]0,T[\times\mathbb{R}^N$.

(1.) For each probability measure $\widehat{\mu}_0$ *on RN (i.e. positive measure* $\widehat{\mu}_0 \in \mathcal{M}(\mathbb{R}^N)$ *with* $\widehat{\mu}_0(\mathbb{R}^N)=1$*), there exists a unique narrowly continuous* $\mu : [0,T] \longrightarrow \mathcal{M}(\mathbb{R}^N)$, $t \longmapsto \mu_t$ *solving the initial value problem (in the distributional sense)*

$$\partial_t\,\mu_t + \operatorname{div}_x(v_t\,\mu_t) = c_t\,\mu_t \quad \text{in }]0,T[\times\mathbb{R}^N, \qquad \mu_0=\widehat{\mu}_0.$$

(2.) The comparison principle holds in the following sense: Let $\sigma : t \longmapsto \sigma_t$ *be a narrowly continuous family of (possibly signed) measures solving*

$$\partial_t\,\sigma_t + \operatorname{div}_x(v_t\,\sigma_t) = c_t\,\sigma_t \qquad \text{in }]0,T[\times\mathbb{R}^N$$

with $\sigma_0 \le 0$ *and*

$$\begin{aligned}
&\int_0^T\int_{\mathbb{R}^N}\Big(|v_t(x)| + |c_t(x)|\Big)\quad d|\sigma_t|(x)\;dt < \infty\\
&\int_0^T\quad\Big(|\sigma_t|(B) + \sup_B|v_t| + \operatorname{Lip} v_t|_B\Big)\;dt < \infty\\
&\int_0^T\quad\Big(|\sigma_t|(B) + \sup_B|c_t| + \operatorname{Lip} c_t|_B\Big)\;dt < \infty
\end{aligned}$$

for any bounded closed set $B\subset\mathbb{R}^N$. *Then,* $\sigma_t \le 0$ *for any* $t\in[0,T[$.

Now the solutions to the linear problem lay the basis for transitions on $\mathcal{M}(\mathbb{R}^N)$:

Definition 48. For each $\mathbf{b}\in W^{1,\infty}(\mathbb{R}^N,\mathbb{R}^N)$ and $c\in W^{1,\infty}(\mathbb{R}^N,\mathbb{R})$, define

$$\vartheta_{\mathcal{M}(\mathbb{R}^N),\,\mathbf{b},\,c} : [0,1]\times\mathcal{M}(\mathbb{R}^N) \longrightarrow \mathcal{M}(\mathbb{R}^N), \quad (t,\mu_0)\longmapsto \mu_t$$

with $\mu : [0,T] \longrightarrow \mathcal{M}(\mathbb{R}^N)$, $t\longmapsto\mu_t$ denoting the unique solution of

$$\partial_t\,\mu_t + \operatorname{div}_x(\mathbf{b}\,\mu_t) = c\,\mu_t \quad \text{in } [0,T]$$

(in the distributional sense) as specified in Proposition 46.

Lemma 49. *For any* $\mathbf{b}, \widetilde{\mathbf{b}} \in C^1(\mathbb{R}^N, \mathbb{R}^N) \cap W^{1,\infty}(\mathbb{R}^N, \mathbb{R}^N)$ *and* $c, \widetilde{c} \in W^{1,\infty}(\mathbb{R}^N, \mathbb{R})$, *the measure-valued maps*

$$\vartheta_{\mathcal{M}(\mathbb{R}^N), \mathbf{b}, c}, \ \vartheta_{\mathcal{M}(\mathbb{R}^N), \widetilde{\mathbf{b}}, \widetilde{c}} : \ [0,1] \times \mathcal{M}(\mathbb{R}^N) \longrightarrow \mathcal{M}(\mathbb{R}^N)$$

fulfill for any $\mu_0, \nu_0 \in \mathcal{M}(\mathbb{R}^N)$ *and* $t, h \in [0,1]$ *with* $t + h \leq 1$

(a) $\vartheta_{\mathcal{M}(\mathbb{R}^N), \mathbf{b}, c}(0, \mu_0) = \mu_0$

(b) $\vartheta_{\mathcal{M}(\mathbb{R}^N), \mathbf{b}, c}\big(h, \vartheta_{\mathcal{M}(\mathbb{R}^N), \mathbf{b}, c}(t, \mu_0)\big) = \vartheta_{\mathcal{M}(\mathbb{R}^N), \mathbf{b}, c}(t+h, \mu_0)$

(c) $\big|\vartheta_{\mathcal{M}(\mathbb{R}^N), \mathbf{b}, c}(h, \mu_0)\big|(\mathbb{R}^N) \leq e^{\|c\|_\infty h} \cdot |\mu_0|(\mathbb{R}^N)$

(d) $\rho_{\mathcal{M}}\big(\vartheta_{\mathcal{M}(\mathbb{R}^N), \mathbf{b}, c}(t, \mu_0), \ \vartheta_{\mathcal{M}(\mathbb{R}^N), \mathbf{b}, c}(t+h, \mu_0)\big) \leq h\,\big(\|\mathbf{b}\|_\infty + \|c\|_\infty\big)\, e^{\|c\|_\infty} \cdot |\mu_0|(\mathbb{R}^N)$

(e) $\rho_{\mathcal{M}}\big(\vartheta_{\mathcal{M}(\mathbb{R}^N), \mathbf{b}, c}(h, \mu_0), \ \vartheta_{\mathcal{M}(\mathbb{R}^N), \mathbf{b}, c}(h, \nu_0)\big) \leq \rho_{\mathcal{M}}(\mu_0\ \nu_0)\, e^{(\mathrm{Lip}\ \mathbf{b} + \|c\|_{W^{1,\infty}})\,h}$

(f) $\rho_{\mathcal{M}}\big(\vartheta_{\mathcal{M}(\mathbb{R}^N), \mathbf{b}, c}(h, \mu_0), \ \vartheta_{\mathcal{M}(\mathbb{R}^N), \widetilde{\mathbf{b}}, \widetilde{c}}(h, \mu_0)\big) \leq$
$\leq \big(\|\mathbf{b} - \widetilde{\mathbf{b}}\|_\infty\, e^{h\|\nabla c\|_\infty} + \|c - \widetilde{c}\|_\infty\big)\ h\ e^{h \cdot (\mathrm{Lip}\ \mathbf{b} + \max\{\|c\|_\infty, \|\widetilde{c}\|_\infty\})} \cdot \big|\mu_0\big|(\mathbb{R}^N)$

The proof in detail is postponed to the end of this section.

Remark 50. Assuming $\mathbf{b}, \widetilde{\mathbf{b}} \in C^1(\mathbb{R}^N, \mathbb{R}^N)$ in addition to $\mathbf{b}, \widetilde{\mathbf{b}} \in W^{1,\infty}(\mathbb{R}^N, \mathbb{R}^N)$ serves the single purpose that we can use the estimates of preceding Lemma 45 for the comparisons specified in Lemma 49.
The additional regularity of $\mathbf{b}, \widetilde{\mathbf{b}}$ does not have any influence on the inequalities though. Indeed, for each $h \in [0,1]$ and $\mu_0 \in \mathcal{M}(\mathbb{R}^N)$, the map

$$(\mathbf{b}, c) \longmapsto \vartheta_{\mathcal{M}(\mathbb{R}^N), \mathbf{b}, c}(h, \mu_0)$$

is continuous with respect to the L^∞ norm according to statement (f). For this reason, we can extend all statements in Lemma 49 to arbitrary $\mathbf{b}, \widetilde{\mathbf{b}} \in W^{1,\infty}(\mathbb{R}^N, \mathbb{R}^N)$ because $C^1(\mathbb{R}^N, \mathbb{R}^N) \cap W^{1,\infty}(\mathbb{R}^N, \mathbb{R}^N)$ is dense in $W^{1,\infty}(\mathbb{R}^N, \mathbb{R}^N)$ with respect to the L^∞ norm and, bounded subsets of $\mathcal{M}(\mathbb{R}^N)$ are complete w.r.t. $\rho_{\mathcal{M}}$ as specified in Proposition 43 (3.) (on page 134).

Definition 2 (on page 104) and Definition 5 (on page 105) lead directly to

Proposition 51. *For every* $\mathbf{b} \in W^{1,\infty}(\mathbb{R}^N, \mathbb{R}^N)$ *and* $c \in W^{1,\infty}(\mathbb{R}^N, \mathbb{R})$,

$$\vartheta_{\mathcal{M}(\mathbb{R}^N), \mathbf{b}, c} : \ [0,1] \times \mathcal{M}(\mathbb{R}^N) \longrightarrow \mathcal{M}(\mathbb{R}^N)$$

is a transition on $\big(\mathcal{M}(\mathbb{R}^N), \ \rho_{\mathcal{M}}, \ |\cdot|(\mathbb{R}^N)\big)$ *with*

$$\begin{aligned}
\alpha(\vartheta_{\mathcal{M}(\mathbb{R}^N), \mathbf{b}, c}; r) &:= \mathrm{Lip}\, \mathbf{b} + \|c\|_{W^{1,\infty}} \\
\beta(\vartheta_{\mathcal{M}(\mathbb{R}^N), \mathbf{b}, c}; r) &:= \big(\|\mathbf{b}\|_\infty + \|c\|_\infty\big)\, e^{\|c\|_\infty}\ r \\
\gamma(\vartheta_{\mathcal{M}(\mathbb{R}^N), \mathbf{b}, c}) &:= \|c\|_\infty \\
D(\vartheta_{\mathcal{M}(\mathbb{R}^N), \mathbf{b}, c}, \ \vartheta_{\mathcal{M}(\mathbb{R}^N), \widetilde{\mathbf{b}}, \widetilde{c}}; r) &\leq \big(\|\mathbf{b} - \widetilde{\mathbf{b}}\|_\infty + \|c - \widetilde{c}\|_\infty\big)\ r
\end{aligned}$$

From now on, the set of these transitions on $\big(\mathcal{M}(\mathbb{R}^N), \ \rho_{\mathcal{M}}, \ |\cdot|(\mathbb{R}^N)\big)$ *is abbreviated as* $\Theta\big(\mathcal{M}(\mathbb{R}^N), \ \rho_{\mathcal{M}}, \ |\cdot|(\mathbb{R}^N)\big)$.

Proof (of Lemma 49). Statements (a) and (b) result directly from the explicit formula in Proposition 46 (on page 137) and the semigroup property of the flow $\mathbf{X_b}(\cdot,\cdot)$

$$\mathbf{X_b}\big(h,\ \mathbf{X_b}(t,x)\big) \;=\; \mathbf{X_b}(t+h,\ x)$$

for all $x \in \mathbb{R}^N$ and $t,h \geq 0$.

(c) The total variation of any measure $\mu \in \mathscr{M}(\mathbb{R}^N)$ in open set $A \subset \mathbb{R}^N$ is

$$|\mu|(A) \;=\; \sup\Big\{ \int_{\mathbb{R}^N} \varphi \, d\mu \;\Big|\; \varphi \in C^0_c(A),\ \|\varphi\|_\infty \leq 1 \Big\}$$

according to [5, Proposition 1.47]. Thus, we conclude from Proposition 46 for every $\mu_0 \in \mathscr{M}(\mathbb{R}^N)$ and $h \in [0,1]$

$$\begin{aligned}
&\big|\vartheta_{\mathscr{M}(\mathbb{R}^N),\mathbf{b},c}(h,\ \mu_0)\big|(\mathbb{R}^N) \\
&= \sup\Big\{ \int_{\mathbb{R}^N} \varphi \; d\vartheta_{\mathscr{M}(\mathbb{R}^N),\mathbf{b},c}(h,\ \mu_0) \;\Big|\; \varphi \in C^0_c(\mathbb{R}^N),\ \|\varphi\|_\infty \leq 1 \Big\} \\
&= \sup\Big\{ \int_{\mathbb{R}^N} \varphi(\mathbf{X_b}(t,x)) \cdot e^{\int_0^h c(\mathbf{X_b}(s,x))\,ds} \; d\mu_0(x) \;\Big|\; \varphi \in C^0_c(\mathbb{R}^N),\ \|\varphi\|_\infty \leq 1 \Big\} \\
&\leq e^{\|c\|_\infty h} \cdot \sup\Big\{ \int_{\mathbb{R}^N} |\varphi(\mathbf{X_b}(t,x))| \quad d|\mu_0|(x) \;\Big|\; \varphi \in C^0_c(\mathbb{R}^N),\ \|\varphi\|_\infty \leq 1 \Big\} \\
&\leq e^{\|c\|_\infty h} \cdot |\mu_0|(\mathbb{R}^N).
\end{aligned}$$

(d) Let $\varphi \in C^\infty_c(\mathbb{R}^N)$ be an arbitrary function with $\|\varphi\|_\infty \leq 1$, $\|\nabla\varphi\|_\infty \leq 1$. Due to Proposition 46 again, we obtain for every $\mu_0 \in \mathscr{M}(\mathbb{R}^N)$ and $t,h \in [0,1]$ with $t+h \leq 1$

$$\begin{aligned}
&\int_{\mathbb{R}^N} \varphi \; d\big(\vartheta_{\mathscr{M}(\mathbb{R}^N),\mathbf{b},c}(t+h,\ \mu_0) - \vartheta_{\mathscr{M}(\mathbb{R}^N),\mathbf{b},c}(t,\ \mu_0)\big) \\
&= \int_t^{t+h} \tfrac{d}{ds} \int_{\mathbb{R}^N} \varphi(y) \; d\vartheta_{\mathscr{M}(\mathbb{R}^N),\mathbf{b},c}(s,\ \mu_0)(y) \qquad ds \\
&= \int_t^{t+h} \int_{\mathbb{R}^N} \big(\nabla\varphi(y)\cdot \mathbf{b}(y) \;+\; \varphi(y)\,c(y)\big) \; d\vartheta_{\mathscr{M}(\mathbb{R}^N),\mathbf{b},c}(s,\ \mu_0)(y) \qquad ds \\
&\leq \int_t^{t+h} \big(\|\nabla\varphi\|_\infty\,\|\mathbf{b}\|_\infty + \|\varphi\|_\infty\,\|c\|_\infty\big)\, \big|\vartheta_{\mathscr{M}(\mathbb{R}^N),\mathbf{b},c}(s,\ \mu_0)\big|(\mathbb{R}^N) \; ds \\
&\leq h \cdot \big(\|\mathbf{b}\|_\infty + \|c\|_\infty\big) \qquad e^{\|c\|_\infty} \; |\mu_0|(\mathbb{R}^N)
\end{aligned}$$

as a consequence of statement (c). The supremum with respect to all these functions φ leads to claim (d) about $\rho_{\mathscr{M}}\big(\vartheta_{\mathscr{M}(\mathbb{R}^N),\mathbf{b},c}(t,\ \mu_0),\ \vartheta_{\mathscr{M}(\mathbb{R}^N),\mathbf{b},c}(t+h,\ \mu_0)\big)$.

(e) Let $\varphi \in C^\infty_c(\mathbb{R}^N)$ again denote any function with $\|\varphi\|_\infty \leq 1$, $\|\nabla\varphi\|_\infty \leq 1$. Then, any measures $\mu_0, \nu_0 \in \mathscr{M}(\mathbb{R}^N)$ satisfy at every time $h \in [0,1]$

$$\begin{aligned}
&\int_{\mathbb{R}^N} \varphi \; d\big(\vartheta_{\mathscr{M}(\mathbb{R}^N),\mathbf{b},c}(h,\ \mu_0) - \vartheta_{\mathscr{M}(\mathbb{R}^N),\mathbf{b},c}(h,\ \nu_0)\big) \\
&= \int_{\mathbb{R}^N} \varphi(\mathbf{X_b}(h,x)) \cdot \exp\Big(\int_0^h c(\mathbf{X_b}(s,x))\,ds\Big) \; d(\mu_0 - \nu_0)(x) \\
&\leq e^{(\mathrm{Lip}\ \mathbf{b} + \|c\|_{W^{1,\infty}})\,h} \cdot \rho_{\mathscr{M}}(\mu_0, \nu_0)
\end{aligned}$$

Indeed, the last estimate results from Proposition 43 (1.) (on page 134) because the composition

$$\psi_h : \mathbb{R}^N \longrightarrow \mathbb{R}^N, \quad x \longmapsto \varphi(\mathbf{X_b}(h,x)) \cdot \exp\Big(\int_0^h c(\mathbf{X_b}(s,x))\, ds\Big)$$

is continuously differentiable with compact support and, Lemma 45 (on page 137) implies

$$\begin{aligned}
\|\psi_h\|_\infty &\le \|\varphi\|_\infty\ e^{\|c\|_\infty h} \le e^{\|c\|_\infty h} \\
\|\nabla\psi_h\|_\infty &\le e^{\|c\|_\infty h} \Big(\|\nabla\varphi\|_\infty\ \|\nabla\mathbf{X_b}(h,\cdot)\|_\infty + \|\varphi\|_\infty \cdot \int_0^h \|\nabla c\|_\infty\ \|\nabla\mathbf{X_b}(s,\cdot)\|_\infty\, ds\Big) \\
&\le e^{\|c\|_\infty h} \Big(e^{\mathrm{Lip}\ \mathbf{b}\cdot h} + h\ \|\nabla c\|_\infty e^{\mathrm{Lip}\ \mathbf{b}\cdot h}\Big) \\
&\le e^{(\mathrm{Lip}\ \mathbf{b} + \|c\|_\infty)\, h}\ \big(1 + h\ \|\nabla c\|_\infty\big) \\
&\le e^{(\mathrm{Lip}\ \mathbf{b} + \|c\|_\infty)\, h}\quad e^{h\ \|\nabla c\|_\infty} \\
&= e^{(\mathrm{Lip}\ \mathbf{b} + \|c\|_{W^{1,\infty}})\, h}.
\end{aligned}$$

The supremum with respect to all $\varphi \in C_c^\infty(\mathbb{R}^N)$ satisfying $\|\varphi\|_\infty \le 1,\ \|\nabla\varphi\|_\infty \le 1$ leads to

$$\rho_{\mathscr{M}}\big(\vartheta_{\mathscr{M}(\mathbb{R}^N),\mathbf{b},c}(h,\ \mu_0),\ \vartheta_{\mathscr{M}(\mathbb{R}^N),\mathbf{b},c}(h,\ \nu_0)\big)\ \le\ e^{(\mathrm{Lip}\ \mathbf{b} + \|c\|_{W^{1,\infty}})\, h}\ \rho_{\mathscr{M}}(\mu_0, \nu_0).$$

(f) For estimating $\rho_{\mathscr{M}}\big(\vartheta_{\mathscr{M}(\mathbb{R}^N),\mathbf{b},c}(h,\ \mu_0),\ \vartheta_{\mathscr{M}(\mathbb{R}^N),\widetilde{\mathbf{b}},\widetilde{c}}(h,\ \mu_0)\big)$ with any $\mu_0 \in \mathscr{M}(\mathbb{R}^N)$ and $h \in [0,1]$, we again choose an arbitrary function $\varphi \in C_c^\infty(\mathbb{R}^N)$ with $\|\varphi\|_\infty \le 1,\ \|\nabla\varphi\|_\infty \le 1$ and consider now an appropriate convex combination $\psi : [0,1]\times[0,1]\times\mathbb{R}^N \longrightarrow \mathbb{R}^N$:

$$\psi(\lambda,h,x) :=\ \varphi\big(\lambda\ \mathbf{X_b}(h,x) + (1-\lambda)\ \mathbf{X}_{\widetilde{\mathbf{b}}}(h,x)\big)\ \cdot\ e^{\int_0^h \lambda\cdot c(\mathbf{X_b}(r,x))\, +\, (1-\lambda)\cdot\widetilde{c}(\mathbf{X}_{\widetilde{\mathbf{b}}}(r,x))\, dr}.$$

Obviously, ψ is continuously differentiable and, Lemma 45 (on page 137) ensures

$$\begin{aligned}
\Big\|\frac{\partial}{\partial\lambda}\ \psi(\lambda,h,\cdot)\Big\|_\infty &\le \|\nabla\varphi\|_\infty\ \big\|\mathbf{X_b}(h,\cdot) - \mathbf{X}_{\widetilde{\mathbf{b}}}(h,\cdot)\big\|_\infty \cdot e^{h\cdot\max\{\|c\|_\infty,\|\widetilde{c}\|_\infty\}} \\
&\quad + \|\varphi\|_\infty \cdot \int_0^h \big\|c(\mathbf{X_b}(r,\cdot)) - \widetilde{c}(\mathbf{X}_{\widetilde{\mathbf{b}}}(r,\cdot))\big\|_\infty\, dr\ e^{h\cdot\max\{\|c\|_\infty,\|\widetilde{c}\|_\infty\}} \\
&\le \|\mathbf{b} - \widetilde{\mathbf{b}}\|_\infty\ h\ e^{h\cdot\mathrm{Lip}\ \mathbf{b}} \cdot e^{h\cdot\max\{\|c\|_\infty,\|\widetilde{c}\|_\infty\}} \\
&\quad + h\ \big(\|c - \widetilde{c}\|_\infty + \|\nabla c\|_\infty \|\mathbf{b} - \widetilde{\mathbf{b}}\|_\infty\ h\, e^{h\cdot\mathrm{Lip}\ \mathbf{b}}\big)\ e^{h\cdot\max\{\|c\|_\infty,\|\widetilde{c}\|_\infty\}} \\
&\le \big(\|\mathbf{b} - \widetilde{\mathbf{b}}\|_\infty\, e^{h\|\nabla c\|_\infty} + \|c - \widetilde{c}\|_\infty\big)\ h\ e^{h\cdot(\mathrm{Lip}\ \mathbf{b} + \max\{\|c\|_\infty,\|\widetilde{c}\|_\infty\})}
\end{aligned}$$

Hence we obtain

$$\begin{aligned}
&\int_{\mathbb{R}^N} \varphi\ d\big(\vartheta_{\mathscr{M}(\mathbb{R}^N),\mathbf{b},c}(h,\ \mu_0)\ -\ \vartheta_{\mathscr{M}(\mathbb{R}^N),\widetilde{\mathbf{b}},\widetilde{c}}(h,\ \mu_0)\big) \\
&\quad = \int_{\mathbb{R}^N} \big(\psi(1,h,x)\ -\ \psi(0,h,x)\big)\ d\mu_0(x) \\
&\quad = \int_{\mathbb{R}^N} \int_0^1 \tfrac{\partial}{\partial\lambda}\ \psi(\lambda,h,x)\ d\lambda \quad d\mu_0(x) \\
&\quad \le \qquad \big\|\tfrac{\partial}{\partial\lambda}\ \psi(\lambda,h,\cdot)\big\|_\infty \qquad |\mu_0|(\mathbb{R}^N) \\
&\quad \le\ \big(\|\mathbf{b} - \widetilde{\mathbf{b}}\|_\infty\, e^{h\|\nabla c\|_\infty} + \|c - \widetilde{c}\|_\infty\big)\ h\ e^{h\cdot(\mathrm{Lip}\ \mathbf{b} + \max\{\|c\|_\infty,\|\widetilde{c}\|_\infty\})}\ \ |\mu_0|(\mathbb{R}^N). \quad \square
\end{aligned}$$

2.5.3 Conclusions About Nonlinear Transport Equations

Now we exploit the preparations and draw some conclusions about the nonlinear transport equation of Radon measures – in the mutational framework. Here Euler compactness plays the role of a key ingredient to existence, but its slightly technical proof is postponed to the end of this section (on page 144).

Lemma 52. *The tuple* $\big(\mathscr{M}(\mathbb{R}^N),\ \rho_{\mathscr{M}},\ |\cdot|(\mathbb{R}^N),\ \Theta\big(\mathscr{M}(\mathbb{R}^N),\ \rho_{\mathscr{M}},\ |\cdot|(\mathbb{R}^N)\big)\big)$ *is Euler compact (in the sense of Definition 15 on page 112), i.e.*

choose $\mu_0 \in \mathscr{M}(\mathbb{R}^N)$, $T > 0$, $R > 0$ *arbitrarily and let* $\mathscr{N} = \mathscr{N}(\mu_0, T, R)$ *denote the subset of all curves* $\mu(\cdot) : [0,T] \longrightarrow \mathscr{M}(\mathbb{R}^N)$ *constructed in the following piece-wise way: Choosing an arbitrary equidistant partition* $0 = t_0 < t_1 < \ldots < t_n = T$ *of* $[0,T]$ *(with* $n > T$*) and* $\mathbf{b}_1 \ldots \mathbf{b}_n \in W^{1,\infty}(\mathbb{R}^N,\mathbb{R}^N)$, $c_1 \ldots c_n \in W^{1,\infty}(\mathbb{R}^N,\mathbb{R})$ *with*

$$\max\ \big\{ \|\mathbf{b}_k\|_{W^{1,\infty}},\ \|c_k\|_{W^{1,\infty}} \ \big|\ 1 \le k \le n \big\} \ \le\ R,$$

define $\mu(\cdot) : [0,T] \longrightarrow E,\ t \longmapsto \mu_t$ *as*

$$\mu_t := \vartheta_{\mathscr{M}(\mathbb{R}^N),\,\mathbf{b}_k,\,c_k}\big(t - t_{k-1},\ \mu_{t_{k-1}}\big) \qquad \text{for } t \in\,]t_{k-1}, t_k],\ k = 1,2\ldots n.$$

Then at each time $t \in [0,T]$, *the set* $\{\mu_t \,|\, \mu(\cdot) \in \mathscr{N}\} \subset \mathscr{M}(\mathbb{R}^N)$ *is relatively sequentially compact with respect to the* $W^{1,\infty}$ *dual metric* $\rho_{\mathscr{M}}$.

Furthermore, the set of all measure values of $\mathscr{N}(\mu_0, T, R)$*, i.e.*

$$\big\{\mu_t \ \big|\ t \in [0,T],\ \mu(\cdot) \in \mathscr{N}\big\} \subset \mathscr{M}(\mathbb{R}^N),$$

is tight.

Theorem 53 (Existence of solution to nonlinear transport equation).
For $\mathbf{f} = (\mathbf{f}_1, f_2) : \mathscr{M}(\mathbb{R}^N) \times [0,T] \longrightarrow W^{1,\infty}(\mathbb{R}^N,\mathbb{R}^N) \times W^{1,\infty}(\mathbb{R}^N,\mathbb{R})$ *suppose*

(*i*) $\sup_{\mu,t}\ \big(\|\mathbf{f}_1(\mu,t)\|_{W^{1,\infty}} + \|f_2(\mu,t)\|_{W^{1,\infty}}\big) < \infty,$

(*ii*) $\mathbf{f}$ *is continuous in the following sense: For* $\mathscr{L}^1$*-almost every* $t \in [0,T]$ *and any sequences* $(t_m)_m$, $(\mu_m)_m$ *in* $[0,T]$, $\mathscr{M}(\mathbb{R}^N)$ *respectively with* $t_m \longrightarrow t$, $\rho_{\mathscr{M}}(\mu_m,\mu) \longrightarrow 0$ *for* $m \longrightarrow \infty$ *and* $\sup_m |\mu_m|(\mathbb{R}^N) < \infty$, *it fulfills*

$$\mathbf{f}(\mu_m, t_m) \longrightarrow \mathbf{f}(\mu, t) \ \text{ in } L^\infty(\mathbb{R}^N,\mathbb{R}^N) \times L^\infty(\mathbb{R}^N,\mathbb{R}) \quad \text{for } m \longrightarrow \infty.$$

Then for every initial Radon measure $\mu_0 \in \mathscr{M}(\mathbb{R}^N)$, *there exists a solution* $\mu(\cdot) : [0,T] \longrightarrow \mathscr{M}(\mathbb{R}^N)$ *to the mutational equation*

$$\overset{\circ}{\mu}(\cdot) \ni \vartheta_{\mathscr{M}(\mathbb{R}^N),\,\mathbf{f}_1(\mu(\cdot),\cdot),\,f_2(\mu(\cdot),\cdot)}$$

on the tuple $\big(\mathscr{M}(\mathbb{R}^N),\ \rho_{\mathscr{M}},\ |\cdot|(\mathbb{R}^N)\big)$ *with* $\mu(0) = \mu_0$ *and, all its values in* $\mathscr{M}(\mathbb{R}^N)$ *are tight.*

Furthermore every solution $\mu(\cdot) : [0,T] \longrightarrow \mathscr{M}(\mathbb{R}^N)$ (to this mutational equation) with tight values in $\mathscr{M}(\mathbb{R}^N)$ is a narrowly continuous distributional solution to the nonlinear transport equation

$$\partial_t \, \mu_t \; + \; \mathrm{div}_x \, (\mathbf{f}_1(\mu_t,t) \, \mu_t) \; = \; f_2(\mu_t,t) \, \mu_t \qquad \text{in } \mathbb{R}^N \times \,]0,T[$$

in the sense that

$$\int_{\mathbb{R}^N} \varphi \; d\mu_t - \int_{\mathbb{R}^N} \varphi \; d\mu_0 = \int_0^t \int_{\mathbb{R}^N} \Big(\nabla\varphi(x) \cdot \mathbf{f}_1(\mu_s,s)(x) + f_2(\mu_s,s)(x) \Big) \, d\mu_s(x) \; ds$$

for every $t \in [0,T]$ and any test function $\varphi \in C_c^\infty(\mathbb{R}^N, \mathbb{R})$.

Corollary 12 (on page 108) provides sufficient conditions for the uniqueness of solutions to mutational equations. Moreover, the comparison principle in Proposition 47 (2.) (on page 138) implies uniqueness for the *linear* (but) *nonautonomous* transport equation. The combination of these two results leads to uniqueness of measure-valued solutions to the nonlinear transport equation:

Theorem 54 (Uniqueness of solution to nonlinear transport equation).
For $\mathbf{f} = (\mathbf{f}_1, f_2) : \mathscr{M}(\mathbb{R}^N) \times [0,T] \longrightarrow W^{1,\infty}(\mathbb{R}^N, \mathbb{R}^N) \times W^{1,\infty}(\mathbb{R}^N, \mathbb{R})$ suppose

(i) $\sup_{\mu,t} \; \big(\|\mathbf{f}_1(\mu,t)\|_{W^{1,\infty}} + \|f_2(\mu,t)\|_{W^{1,\infty}} \big) \; < \; \infty,$

(ii) $\mathbf{f}$ *is Lipschitz continuous with respect to state in the following sense: There exists a constant $\lambda > 0$ such that for $\mathscr{L}^1$-almost every $t \in [0,T]$ and every $\mu_0, \mu_1 \in \mathscr{M}(\mathbb{R}^N)$,*

$$\|\mathbf{f}(\mu_0,t) - \mathbf{f}(\mu_1,t)\|_\infty \; \le \; \lambda \cdot \rho_{\mathscr{M}}(\mu_0, \, \mu_1).$$

Then for every initial $\mu_0 \in \mathscr{M}(\mathbb{R}^N)$, the solution $\mu(\cdot) : [0,T] \longrightarrow \mathscr{M}(\mathbb{R}^N)$ to the mutational equation

$$\overset{\circ}{\mu}(\cdot) \; \ni \; \vartheta_{\mathscr{M}(\mathbb{R}^N), \, \mathbf{f}_1(\mu(\cdot),\cdot), \, f_2(\mu(\cdot),\cdot)}$$

on the tuple $\big(\mathscr{M}(\mathbb{R}^N), \, \rho_{\mathscr{M}}, \, |\cdot|(\mathbb{R}^N)\big)$ with $\mu(0) = \mu_0$ is unique.

In particular, the distributional solution $\mu(\cdot) : [0,T] \longrightarrow \mathscr{M}(\mathbb{R}^N), \; t \longmapsto \mu_t$ to the nonlinear transport equation

$$\partial_t \, \mu_t \; + \; \mathrm{div}_x \, (\mathbf{f}_1(\mu_t,t) \, \mu_t) \; = \; f_2(\mu_t,t) \, \mu_t \qquad \text{in } \mathbb{R}^N \times \,]0,T[$$

being continuous with respect to $\rho_{\mathscr{M}}$, having initial Radon measure $\mu_0 \in \mathscr{M}(\mathbb{R}^N)$ at time $t = 0$ and satisfying $\sup_{t \in [0,T]} \, |\mu_t|(\mathbb{R}^N) < \infty$ is unique.

Remark 55. The two preceding theorems exemplify how to benefit from the mutational framework appropriately. Indeed, the results of § 2.3 (on page 107 ff.) cover a generalized type of solutions, namely to mutational equations. Theorem 53 reveals the connection to the more popular concept of distributional solutions.

On this basis, the results of § 2.3 lead to further statements about measure-valued distributional solutions to nonlinear transport equations with delay or in systems with other examples of mutational equations. We are not going to formulate them in detail here.

Proof (of Lemma 52). In regard to Definition 15 (on page 112), choose $\mu_0 \in \mathscr{M}(\mathbb{R}^N)$, $T > 0$ and $R > 0$ arbitrarily and let $\mathscr{N} = \mathscr{N}(\mu_0, T, R)$ denote the subset of all curves $\mu(\cdot) : [0,T] \longrightarrow \mathscr{M}(\mathbb{R}^N)$ constructed in the following piecewise way: Choosing an arbitrary equidistant partition $0 = t_0 < t_1 < \ldots < t_n = T$ of $[0,T]$ (with $n > T$) and $\mathbf{b}_1 \ldots \mathbf{b}_n \in W^{1,\infty}(\mathbb{R}^N, \mathbb{R}^N)$, $c_1 \ldots c_n \in W^{1,\infty}(\mathbb{R}^N, \mathbb{R})$ with

$$\max \left\{ \|\mathbf{b}_k\|_{W^{1,\infty}},\ \|c_k\|_{W^{1,\infty}} \;\middle|\; 1 \le k \le n \right\} \le R,$$

define $\mu(\cdot) : [0,T] \longrightarrow \mathscr{M}(\mathbb{R}^N)$, $t \longmapsto \mu_t$ as

$$\mu_t := \vartheta_{\mathscr{M}(\mathbb{R}^N), \mathbf{b}_k, c_k}\left(t - t_{k-1},\ \mu_{t_{k-1}}\right) \qquad \text{for } t \in]t_{k-1}, t_k],\ k = 1, 2 \ldots n.$$

Then we have to verify at each time $t \in [0,T]$: The set $\{\mu_t \,|\, \mu(\cdot) \in \mathscr{N}\} \subset \mathscr{M}(\mathbb{R}^N)$ is relatively sequentially compact with respect to the $W^{1,\infty}$ dual metric $\rho_{\mathscr{M}}$.

As a consequence of Lemma 49 (c) (on page 139), the total variation $|\nu|(\mathbb{R}^N)$ is uniformly bounded for all measures $\nu \in \{\mu_t \,|\, t \in [0,T],\ \mu(\cdot) \in \mathscr{N}\} \subset \mathscr{M}(\mathbb{R}^N)$:

$$|\nu|(\mathbb{R}^N) \le e^{RT}\, |\mu_0|(\mathbb{R}^N).$$

Thus, due to Proposition 43 (4.) (on page 134), it suffices to prove that this set $\{\mu_t \,|\, t \in [0,T],\ \mu(\cdot) \in \mathscr{N}\} \subset \mathscr{M}(\mathbb{R}^N)$ is tight.

For every $\varepsilon > 0$, there exists a compact subset $K_\varepsilon \subset \mathbb{R}^N$ with $|\mu_0|(\mathbb{R}^N \setminus K_\varepsilon) < \varepsilon$. Then,

$$\left|\mu_t\right|\left(\mathbb{R}^N \setminus \mathbb{B}_{RT}(K_\varepsilon)\right) \le \left|\mu_t\right|\left(\mathbb{R}^N \setminus \mathbb{B}_{Rt}(K_\varepsilon)\right) < \varepsilon\, e^{Rt} \le \varepsilon\, e^{RT}$$

holds for all $t \in [0,T]$ and $\mu(\cdot) \in \mathscr{N}(\mu_0, T, R)$.
Indeed, we consider the underlying equidistant partition $0 = t_0 < t_1 < \ldots < t_n = T$ of $[0,T]$ and $\mathbf{b}_1 \ldots \mathbf{b}_n \in W^{1,\infty}(\mathbb{R}^N, \mathbb{R}^N)$, $c_1 \ldots c_n \in W^{1,\infty}(\mathbb{R}^N, \mathbb{R})$ with

$$\mu_t = \vartheta_{\mathscr{M}(\mathbb{R}^N), \mathbf{b}_{k+1}, c_{k+1}}\left(t - t_k,\ \mu_{t_k}\right) \qquad \text{for } t \in]t_k, t_{k+1}],\ k = 0, 1 \ldots n-1.$$

Then, we obtain for each $t \in]t_k, t_{k+1}]$ – via induction with respect to k

$$\begin{aligned}
&\left|\mu_t\right|\left(\mathbb{R}^N \setminus \mathbb{B}_{Rt}(K_\varepsilon)\right) \\
&= \sup\left\{ \int_{\mathbb{R}^N} \varphi \; d\,\vartheta_{\mathscr{M}(\mathbb{R}^N), \mathbf{b}_{k+1}, c_{k+1}}(t - t_k,\ \mu_{t_k}) \;\middle|\; \varphi \in C_c^0(\mathbb{R}^N \setminus \mathbb{B}_{Rt}(K_\varepsilon)),\ \|\varphi\|_\infty \le 1 \right\} \\
&\le \sup\left\{ \int_{\mathbb{R}^N} \widetilde{\varphi}\big|_{(\mathbf{X}_{\mathbf{b}_{k+1}}(t - t_k, x))}\, d\mu_{t_k}(x)\;\; e^{(t-t_k)\,R} \;\middle|\; \widetilde{\varphi} \in C_c^0(\mathbb{R}^N \setminus \mathbb{B}_{Rt}(K_\varepsilon)),\ \|\widetilde{\varphi}\|_\infty \le 1 \right\} \\
&\le \sup\left\{ \int_{\mathbb{R}^N} \psi(y)\;\; d\mu_{t_k}(y) \qquad e^{(t-t_k)\,R} \;\middle|\; \psi \in C_c^0(\mathbb{R}^N \setminus \mathbb{B}_{Rt_k}(K_\varepsilon)),\ \|\psi\|_\infty \le 1 \right\} \\
&= e^{(t-t_k)\,R}\;\; \left|\mu_{t_k}\right|\left(\mathbb{R}^N \setminus \mathbb{B}_{Rt_k}(K_\varepsilon)\right).
\end{aligned}$$

□

Proof (of Theorem 53). The existence of a solution $\mu(\cdot) : [0,T] \longrightarrow \mathscr{M}(\mathbb{R}^N)$ to the mutational equation results directly from Peano's Theorem 18 (on page 114) and Proposition 51 (on page 139). Its proof is based on Euler approximations in combination with Lemma 52 (as presented on page 115 f.).
In addition, with $R > 0$ denoting the bound of assumption (i), Lemma 52 states that the values of all Euler approximations in $\mathscr{N}(\mu_0, T, R)$,

$$\big\{ \nu_t \,\big|\, t \in [0,T],\ \nu(\cdot) \in \mathscr{N}(\mu_0,T,R) \big\} \subset \mathscr{M}(\mathbb{R}^N),$$

are tight. Thus for every $\varepsilon > 0$, there exists a compact set $K_\varepsilon \subset \mathbb{R}^N$ satisfying

$$|\nu_t|(\mathbb{R}^N \setminus K_\varepsilon) < \varepsilon \qquad \text{for all } t \in [0,T] \text{ and } \nu(\cdot) \in \mathscr{N}(\mu_0,T,R).$$

Since the solution $\mu(\cdot) : t \longmapsto \mu_t$ is constructed as $\rho_{\mathscr{M}}$-limit of Euler approximations, each measure μ_t is weak* limit of a sequence in $\big\{ \nu_t \,\big|\, \nu(\cdot) \in \mathscr{N}(\mu_0,T,R) \big\}$ due to Proposition 43 (2.) and, the lower semicontinuity of total variation implies $|\mu_t|(\mathbb{R}^N \setminus K_\varepsilon) < \varepsilon$. Hence, $\{\mu_t \mid t \in [0,T]\} \subset \mathscr{M}(\mathbb{R}^N)$ is tight.

Now we provide the claimed link to distributional solutions.
Let $\mu(\cdot) : [0,T] \longrightarrow \mathscr{M}(\mathbb{R}^N)$, $t \longmapsto \mu_t$ be a solution to the mutational equation

$$\overset{\circ}{\mu}(\cdot) \ni \vartheta_{\mathscr{M}(\mathbb{R}^N),\, \mathbf{f}_1(\mu(\cdot),\cdot),\, f_2(\mu(\cdot),\cdot)}$$

with tight values in $\mathscr{M}(\mathbb{R}^N)$. In particular, $\mu(\cdot)$ is continuous w.r.t. $\rho_{\mathscr{M}}$ and, $R := 1 + \sup_{t\in[0,T]} |\mu_t|(\mathbb{R}^N) < \infty$. Due to Proposition 43 (2.) (on page 134), $\mu(\cdot)$ is narrowly continuous.
There exists a $\mathscr{L}^1$-measurable subset $A \subset [0,T]$ such that $\mathscr{L}^1([0,T] \setminus A) = 0$,

$$\lim_{h\downarrow 0} \tfrac{1}{h} \cdot \rho_{\mathscr{M}} \big(\mu_{t+h},\ \vartheta_{\mathscr{M}(\mathbb{R}^N),\, \mathbf{f}_1(\mu_t,t),\, f_2(\mu_t,t)}(h,\, \mu_t) \big) = 0$$

for every $t \in A$ and that assumption (ii) about the continuity of $\mathbf{f}$ is satisfied at every time $t \in A$. Choosing the test function $\varphi \in C_c^\infty(\mathbb{R}^N, \mathbb{R})$ arbitrarily, we obtain

$$\lim_{h\downarrow 0} \tfrac{1}{h} \cdot \int_{\mathbb{R}^N} \varphi \; d\big(\mu_{t+h} - \vartheta_{\mathscr{M}(\mathbb{R}^N),\, \mathbf{f}_1(\mu_t,t),\, f_2(\mu_t,t)}(h,\, \mu_t) \big) = 0$$

for each $t \in A$. The auxiliary function $\psi : [0,T] \longrightarrow \mathbb{R}$, $t \longmapsto \int_{\mathbb{R}^N} \varphi \, d\mu_t$ is continuous due to the $\rho_{\mathscr{M}}$-continuity of $\mu(\cdot)$ and, it fulfills at every time $t \in A \subset [0,T]$

$$\begin{aligned}
\lim_{h\downarrow 0} \tfrac{\psi(t+h) - \psi(t)}{h} &= \lim_{h\downarrow 0} \tfrac{1}{h} \int_{\mathbb{R}^N} \varphi \;\, d\big(\vartheta_{\mathscr{M}(\mathbb{R}^N),\, \mathbf{f}_1(\mu_t,t),\, f_2(\mu_t,t)}(h,\, \mu_t) - \mu_t \big) \\
&= \lim_{h\downarrow 0} \tfrac{1}{h} \int_{\mathbb{R}^N} \Big(\varphi(\mathbf{X}_{\mathbf{f}_1(\mu_t,t)}(h,x)) \cdot e^{\int_0^h f_2(\mu_t,t)\left(\mathbf{X}_{\mathbf{f}_1(\mu_t,t)}(s,x)\right)\, ds} \\
&\qquad\qquad - \varphi(x) \Big) \, d\mu_t(x) \\
&= \int_{\mathbb{R}^N} \Big(\nabla\varphi(x) \cdot \mathbf{f}_1(\mu_t,t)(x) + \varphi(x) \;\; f_2(\mu_t,t)(x) \Big) \, d\mu_t(x).
\end{aligned}$$

In particular, the last integral on the right-hand side is continuous with respect to t for each $t \in A$. Thus, $\psi : [0,T] \longrightarrow \mathbb{R}$ is even absolutely continuous and, its weak derivative is

$$\tfrac{d}{dt}\, \psi(t) = \int_{\mathbb{R}^N} \Big(\nabla\varphi(x) \cdot \mathbf{f}_1(\mu_t,t)(x) + \varphi(x) \;\; f_2(\mu_t,t)(x) \Big) \, d\mu_t(x)$$

for $\mathscr{L}^1$-almost every $t \in [0,T]$. As a consequence, $\mu(\cdot)$ is a distributional solution of

$$\partial_t \mu_t + \operatorname{div}_x(\mathbf{f}_1(\mu_t,t)\,\mu_t) = f_2(\mu_t,t)\,\mu_t \qquad \text{in } \mathbb{R}^N \times]0,T[$$

□

Proof (of Theorem 54). Lipschitz continuity of **f** with respect to state implies uniqueness of solutions to mutational equations according to Corollary 12 (on page 108).

Now let $\mu(\cdot) : [0,T] \longrightarrow \mathscr{M}(\mathbb{R}^N),\ t \longmapsto \mu_t$ be a distributional solution of

$$\partial_t \mu_t + \operatorname{div}_x(\mathbf{f}_1(\mu_t,t)\ \mu_t) = f_2(\mu_t,t)\,\mu_t \qquad \text{in } \mathbb{R}^N \times]0,T[$$

that is continuous with respect to $\rho_{\mathscr{M}}$ and satisfies $\sup_{t\in[0,T]} |\mu_t|(\mathbb{R}^N) < \infty$. Then we can show that $\mu(\cdot)$ is a solution to the mutational equation

$$\overset{\circ}{\mu}(\cdot) \ni \vartheta_{\mathscr{M}(\mathbb{R}^N),\,\mathbf{f}_1(\mu(\cdot),\cdot),\,f_2(\mu(\cdot),\cdot)}$$

on the tuple $(\mathscr{M}(\mathbb{R}^N),\ \rho_{\mathscr{M}},\ |\cdot|(\mathbb{R}^N))$ and thus, it is uniquely determined by $\mu_0 \in \mathscr{M}(\mathbb{R}^N)$. Indeed, the composition

$$\mathbf{g} :\ [0,T] \longrightarrow W^{1,\infty}(\mathbb{R}^N,\mathbb{R}^N) \times W^{1,\infty}(\mathbb{R}^N,\mathbb{R}), \quad t \longmapsto (\mathbf{f}_1(\mu_t,t),\ f_2(\mu_t,t))$$

is continuous with respect to the L^∞ norm Lebesgue-almost everywhere in $[0,T]$. Theorem 53 (on page 142) guarantees a solution $\nu(\cdot) : [0,T] \longrightarrow \mathscr{M}(\mathbb{R}^N),\ t \longmapsto \nu_t$ to the mutational equation

$$\overset{\circ}{\nu}(\cdot) \ni \vartheta_{\mathscr{M}(\mathbb{R}^N),\,\mathbf{g}_1(\cdot),\,g_2(\cdot)}$$

on the tuple $(\mathscr{M}(\mathbb{R}^N),\ \rho_{\mathscr{M}},\ |\cdot|(\mathbb{R}^N))$ with $\nu_0 = \mu_0$ and, it is a distributional solutions to the nonautonomous linear transport equation

$$\partial_t \nu_t + \operatorname{div}_x(\mathbf{g}_1(t)\,\nu_t) = g_2(t)\,\nu_t \qquad \text{in } \mathbb{R}^N \times]0,T[.$$

Finally the comparison principle in Proposition 47 (2.) (on page 138) implies

$$\nu(\cdot) \equiv \mu(\cdot).$$

□

2.6 Example: A Structured Population Model with Radon Measures over $\mathbb{R}_0^+ = [0,\infty[$

Now we focus on measure-valued solutions to a nonlocal first-order hyperbolic problem on $\mathbb{R}_0^+ \stackrel{\text{Def.}}{=} [0,\infty[$ describing a physiologically structured population:

$$\begin{cases} \partial_t\, \mu_t \;+\; \partial_x\big(F_2(\mu_t,t)\,\mu_t\big) & = \;F_3(\mu_t,t)\;\mu_t, & \text{in } \mathbb{R}_0^+ \times [0,T] \\ F_2(\mu_t,t)(0)\;\mu_t(0) & = \;\displaystyle\int_{\mathbb{R}_0^+} F_1(\mu_t,t)(x)\;d\mu_t(x), & \text{in }]0,T] \\ \mu_0 & = \;\nu_0, \end{cases}$$

Avoiding structural restrictions on its coefficients, we specify continuity assumptions sufficient for global existence of distributional solutions, whose values are tight finite Radon measures on $\mathbb{R}_0^+$, and their continuous dependence on the given data. These results can be easily extended to systems describing more than one species because this problem is considered in the mutational framework.

2.6.1 Introduction

A Joint Framework for Both Continuous and Discrete Distributions: Radon Measures

Global existence and stability of solutions to structured population models were established for states defined in Banach space L^1 [94, 182]. In this case it was possible to prove strong continuity and structural stability of solutions. However, it is often necessary to describe populations in which the initial distribution of the individuals is concentrated with respect to the structure, i.e., it is not absolutely continuous with respect to the Lebesgue measure.
In these cases it is relevant to consider initial data in the space of Radon measures as proposed in [137]. It covers both finite measures of the Euclidean space being absolutely continuous with respect to Lebesgue measure and all Dirac measures that are suitable for describing discrete distributions.
For linear age-dependent population dynamics, a qualitative theory using semigroup methods and spectral analysis has been laid out in [137]. The follow-up work [59] is devoted to constructing nonlinear models. Some analytical results concerning the existence of solutions are given in [60]. All results there about continuous dependence of solutions on time and initial state are based on the weak$*$ topology of Radon measures. Moreover, there exist even simple counterexamples indicating that continuous dependence, either with respect to time or to initial state, cannot be expected in the strong (dual) topology in general [60].

In this section, we use the $W^{1,\infty}$ dual metric on $\mathcal{M}(\mathbb{R}_0^+)$ as introduced in Definition 42 (on page 134). It metrizes both weakly* and narrow topology on each tight subset of Radon measures with uniformly bounded total variation according to Proposition 43.

Furthermore bounded Lipschitz continuous test functions have proved to be particularly useful for investigating continuity properties of solutions to the linear subproblems here in § 2.6.2.
In general, using a dual norm can be interpreted in regard to modelling biological processes. The basic notion of weak* topology is to compare *features* of two linear forms individually. Considering the dual space of any topological vector space, the features of interest result from the *effect* of a linear form *on each vector separately.* Here we use Radon measures μ, ν on $\mathbb{R}_0^+$ in combination with bounded Lipschitz continuous functions $\varphi : \mathbb{R}_0^+ \longrightarrow \mathbb{R}$. Then $\varphi(x)$ indicates the relevance of each structural state $x \in \mathbb{R}_0^+$ and, the integral $\int_{\mathbb{R}_0^+} \varphi(x)\, d(\mu - \nu)(x)$ reflects how much μ and ν differ from each other in regard to this weight function φ.
Restricting to bounded Lipschitz continuous functions instead of any real-valued function vanishing at infinity, however, is based on our interest *only* in those weight functions $\varphi : \mathbb{R}_0^+ \longrightarrow \mathbb{R}$ being *not too sensitive* with respect to structural state. For modelling biological systems, it is not recommended to take features into consideration which are extremely sensitive with respect to the structure parameter.

The Nonlinear Model of Physiologically Structured Population

The structured population models considered in [94, 182] focus on solutions $u(\cdot,t) \in L^1(\mathbb{R}_0^+)$ to first-order hyperbolic problems of the general form

$$\begin{aligned} \partial_t\, u(x,t) + \partial_x \big(F_2(u(\cdot,t), x,t)\, u(x,t)\big) &= F_3(u(\cdot,t),x,t)\, u(x,t) && \text{in } \mathbb{R}_0^+ \times [0,T], \\ F_2(u(\cdot,t), 0,t)\, u(0,t) &= \int_{\mathbb{R}_0^+} F_1(u(\cdot,t), x,t)\, u(x,t)\, dx && \text{in }]0,T], \\ u(x,0) &= u_0(x) && \text{in } \mathbb{R}_0^+. \end{aligned}$$

Here x denotes the state of individuals (for example, the size, level of neoplastic transformation, stage of differentiation) and $u(x,t)$ the density of individuals being in state $x \in \mathbb{R}_0^+$ at time t. By $F_3(u,x,t)$ we denote a function describing the individual's rate of evolution, such as growth or death rate. $F_2(u,x,t)$ describes the rate of the dynamics of the structure, i.e., the dynamics of the transformation of individual state. The boundary term describes influx of new individuals to state $x = 0$. Finally, u_0 denotes initial population density.
In the special case of the so-called Gurtin-MacCumy model, the coefficient functions F_j depend on the integral $\int_{\mathbb{R}_0^+} u(x,t)\, dx$ [182, § 1.3] and, additional weight functions were taken into consideration later (e.g. [60]).

In this section, we investigate existence of measure-valued solutions $\mu_t \in \mathscr{M}(\mathbb{R}_0^+)$ to the corresponding nonlinear equations

$$\left\{ \begin{aligned} \partial_t\, \mu_t + \partial_x \big(F_2(\mu_t,t)\, \mu_t\big) &= F_3(\mu_t,t)\, \mu_t && \text{in } \mathbb{R}_0^+ \times [0,T] \\ F_2(\mu_t,t)(0)\, \mu_t(0) &= \int_{\mathbb{R}_0^+} F_1(\mu_t,t)(x)\; d\mu_t(x) && \text{in }]0,T] \\ \mu_0 &= \nu_0 \end{aligned} \right. \tag{2.1}$$

and their dependence on both the initial measure $\nu_0 \in \mathscr{M}(\mathbb{R}_0^+)$ and three coefficient functions $F_1, F_2, F_3 : \mathscr{M}(\mathbb{R}_0^+) \times [0,T] \longrightarrow W^{1,\infty}(\mathbb{R}_0^+)$.
In particular, there are no structural assumptions about the coefficients F_j such as linearity with respect to the measure μ_t. Furthermore, the partial differential equation and the boundary condition on $]0,T]$ are nonlocal because the coefficients depend on the whole measures as elements of the space $\mathscr{M}(\mathbb{R}_0^+)$ – and not on their local properties in $\mathbb{R}_0^+$.

Problem (2.1) is interpreted in a distributional sense: The wanted solutions are weakly* continuous curves $\mu : [0,T] \longrightarrow \mathscr{M}(\mathbb{R}_0^+) = C_0^0(\mathbb{R}_0^+)'$ satisfying the problem in a distributional sense, i.e. in duality with all test functions in $C_c^\infty(\mathbb{R}_0^+ \times [0,T])$. The additional assumption $F_1(\cdot) \geq 0$ guarantees that positivity of initial measure ν_0 is preserved by the solution μ_t constructed here. This feature is of particular interest for modelling population dynamics. The main results of this section are:

Theorem 56 (Existence of solutions to nonlinear structured population model). *Suppose that* $\mathbf{F} : \mathscr{M}(\mathbb{R}_0^+) \times [0,T] \longrightarrow \big\{(a,b,c) \in W^{1,\infty}(\mathbb{R}_0^+)^3 \mid b(0) > 0\big\}$ *satisfies*

(*i*) $\sup_{t\in[0,T]} \sup_{\nu\in\mathscr{M}(\mathbb{R}_0^+)} \|\mathbf{F}(\nu,t)\|_{W^{1,\infty}} < \infty.$

(*ii*) $\mathbf{F} : (\mathscr{M}(\mathbb{R}_0^+), narrow) \times [0,T] \longrightarrow \big(W^{1,\infty}(\mathbb{R}_0^+)^3, \|\cdot\|_\infty\big)$ *is continuous.*

Then, for any initial measure $\nu_0 \in \mathscr{M}(\mathbb{R}_0^+)$, *there exists a narrowly continuous distributional solution* $\mu : [0,T] \longrightarrow \mathscr{M}(\mathbb{R}_0^+)$ *to the nonlinear population model (2.1) with* $\mu(0) = \nu_0$.
If, in addition, $\nu_0 \in \mathscr{M}^+(\mathbb{R}_0^+)$ *and* $F_1(\nu,t)(\cdot) \geq 0$ *for every* $\nu \in \mathscr{M}^+(\mathbb{R}_0^+), t \in [0,T]$, *then the solution* $\mu(\cdot)$ *has values in* $\mathscr{M}^+(\mathbb{R}_0^+)$.

Theorem 57 (Lipschitz contin. dependence of distributional solutions on data). *Assume that for* $\mathbf{F}, \mathbf{G} : \mathscr{M}(\mathbb{R}_0^+) \times [0,T] \longrightarrow \big\{(a,b,c) \in W^{1,\infty}(\mathbb{R}_0^+)^3 \,|\, b(0) > 0\big\}$,

(*i*) $M_F := \sup_{t\in[0,T]} \sup_{\mu\in\mathscr{M}(\mathbb{R}_0^+)} \|\mathbf{F}(\mu,t)\|_{W^{1,\infty}(\mathbb{R}_0^+)^3} < \infty,$

$M_G := \sup_{t\in[0,T]} \sup_{\mu\in\mathscr{M}(\mathbb{R}_0^+)} \|\mathbf{G}(\mu,t)\|_{W^{1,\infty}(\mathbb{R}_0^+)^3} < \infty,$

(*ii*) *for any* $R > 0$, *there are a constant* $L_R > 0$ *and a modulus of continuity* $\omega_R(\cdot)$ *with* $\|\mathbf{F}(\mu,s) - \mathbf{F}(\nu,t)\|_{L^\infty(\mathbb{R}_0^+)} \leq L_R \cdot \rho_{\mathscr{M}}(\mu,\nu) + \omega_R(|t-s|)$ *for all* $\mu,\nu \in \mathscr{M}(\mathbb{R}_0^+)$ *with* $|\mu|(\mathbb{R}_0^+), |\nu|(\mathbb{R}_0^+) \leq R$.

(*iii*) $\mathbf{G} : (\mathscr{M}(\mathbb{R}_0^+), \rho_{\mathscr{M}}) \times [0,T] \longrightarrow (W^{1,\infty}(\mathbb{R}_0^+)^3, \|\cdot\|_\infty)$ *is continuous.*

Let $\mu, \nu : [0,T] \longrightarrow \mathscr{M}(\mathbb{R}_0^+)$ *denote* $\rho_{\mathscr{M}}$*-continuous distributional solutions to the nonlinear population model (2.1) for the coefficients* $\mathbf{F}(\cdot), \mathbf{G}(\cdot)$ *respectively such that* $\sup_t |\mu_t|(\mathbb{R}_0^+) < \infty$, $\sup_t |\nu_t|(\mathbb{R}_0^+) < \infty$ *and all their values are tight in* $\mathscr{M}(\mathbb{R}_0^+)$.

Then there is $C = C(M_F, M_G, |\mu_0|(\mathbb{R}_0^+), |\nu_0|(\mathbb{R}_0^+)) \in [0,\infty[$ *such that for all* $t \in [0,T]$,

$$\rho_{\mathscr{M}}(\mu_t,\nu_t) \leq \Big(\rho_{\mathscr{M}}(\mu_0,\nu_0) + C\,t \cdot \sup_{\mathscr{M}(\mathbb{R}_0^+)\times[0,T]} \|\mathbf{F}(\cdot,\cdot) - \mathbf{G}(\cdot,\cdot)\|_\infty\Big)\, e^{C\,t}.$$

Comparison with Earlier Results of Diekmann and Getto

Model (2.1) is a generic formulation of a nonlinear single-species model with a one-dimensional structure. The model was considered by Diekmann and Getto [60] in a case where the functions F_i depend on the population density via weighted integrals $\int \gamma_i(x) d\mu_t$. Diekmann and Getto proved the global existence of solutions and their continuous dependence on time and initial state in the weak* topology of $\mathcal{M}(\mathbb{R}_0^+)$. The results were formulated under the assumptions of Lipschitz continuity of functions F_1, F_2 and F_3 and the global Lipschitz property of the output function γ_i. For solving the fully nonlinear problem, Diekmann and Getto applied the so-called method of interaction variables. The method consists of replacing the dependence on the measure μ incorporated in F_1, F_2 and F_3 by input $I(t)$ at time t, and splitting the nonlinear problem (2.1) into a nonautonomous linear problem coupled with a fixed point problem. Indeed, their linear problem is determined by parameter function $I(\cdot)$ of time and, it is solved by extending the concept of semigroup.
The feedback law relates the parameter function $I(\cdot)$ to the wanted solution and thus provides a fixed point problem equivalent to the original nonlinear problem. Appropriate assumptions about the coefficients lay the basis for applying Banach's contraction principle.

In this section, we investigate the nonlinear problem (2.1) in the mutational framework. Similarly to § 2.5 about the nonlinear transport equation, the transitions on $\big(\mathcal{M}(\mathbb{R}_0^+), \rho_{\mathcal{M}}, |\cdot|(\mathbb{R}_0^+)\big)$ are induced by the underlying linear problem, i.e.

$$\begin{cases} \partial_t \mu_t \; + \; \partial_x (b\, \mu_t) = c\, \mu_t, & \text{in } \mathbb{R}_0^+ \times [0,T], \\ b(0)\; \mu_t(0) = \displaystyle\int_{\mathbb{R}_0^+} a\, d\mu_t, & \text{in }]0,T], \\ \mu_0 = \nu_0. \end{cases} \tag{2.2}$$

with $a(\cdot), b(\cdot), c(\cdot) \in W^{1,\infty}(\mathbb{R}_0^+)$ and $b(0) > 0$.
The key estimates for this linear problem are obtained using the concepts of duality theory applied to transport equations similarly in [64]. In subsequent § 2.6.2, the smooth solution to a dual partial differential equation provides an integral representation of a measure-valued solution $\mu : [0,T] \longrightarrow \mathcal{M}(\mathbb{R}_0^+)$ to equation (2.2). In particular, this solution exists and depends continuously on the initial measure ν_0 and on the coefficients $a(\cdot)$, $b(\cdot)$ and $c(\cdot)$.

In comparison to the approach of Diekmann et al. [59, 60], the connection with the nonlinear problem (2.1) is not based on the contraction principle, but on Euler compactness in the mutational framework.
It has the advantage that existence of weak solutions to the nonlinear population model (2.1) does not require Lipschitz continuity of the coefficients $F_1(\cdot,t)$, $F_2(\cdot,t)$, $F_3(\cdot,t)$, but merely continuity. In addition, assuming Lipschitz continuity of the model coefficients $F_1(\cdot,t)$, $F_2(\cdot,t)$, $F_3(\cdot,t)$ ensures uniqueness of the weak solution.

Basic set	$E := \mathscr{M}(\mathbb{R}_0^+)$ the space of all finite real-valued Radon measures on $\mathbb{R}_0^+$
Distances	$W^{1,\infty}$ dual metric $\rho_{\mathscr{M}} : \mathscr{M}(\mathbb{R}_0^+) \times \mathscr{M}(\mathbb{R}_0^+) \longrightarrow [0,\infty[$ $\rho_{\mathscr{M}}(\mu, \nu) := \sup\Big\{ \int_{\mathbb{R}_0^+} \psi \; d(\mu-\nu) \;\Big\vert\; \psi \in C^1(\mathbb{R}_0^+),\; \Vert\psi\Vert_\infty, \Vert\nabla\psi\Vert_\infty \leq 1 \Big\}$ (in tight subsets, it is equivalent to the narrow topology, Prop. 43)
Absolute values	$\lfloor\cdot\rfloor := \vert\cdot\vert(\mathbb{R}_0^+)$ total variation of the Radon measure
Transition	For each $a(\cdot), b(\cdot), c(\cdot) \in W^{1,\infty}(\mathbb{R}_0^+)$ with $b(0) > 0$, define $\vartheta_{a,b,c} : [0,1] \times \mathscr{M}(\mathbb{R}_0^+) \longrightarrow \mathscr{M}(\mathbb{R}_0^+),\; (h, \nu_0) \longmapsto \mu_h$ as the narrowly continuous distributional solution to the linear autonomous problem $\begin{cases} \partial_t \mu_t + \partial_x(b\,\mu_t) = c\,\mu_t & \text{in } \mathbb{R}_0^+ \times [0,1], \\ b(0)\,\mu_t(0) = \int_{\mathbb{R}_0^+} a\,d\mu_t & \text{in }]0,1], \\ \mu_0 = \nu_0 \end{cases}$
Compactness	Euler compactness: Lemma 69 (page 165)
Mutational solutions	Narrowly continuous distributional solution to the nonlinear structured population model $\begin{cases} \partial_t\,\mu_t + \partial_x(F_2(\mu_t,t)\,\mu_t) = F_3(\mu_t,t)\,\mu_t & \text{in } \mathbb{R}_0^+ \times [0,T] \\ F_2(\mu_t,t)(0)\,\mu_t(0) = \int_{\mathbb{R}_0^+} F_1(\mu_t,t)(x)\;d\mu_t(x) & \text{in }]0,T] \\ \mu_0 = \nu_0 \end{cases}$ with tight values in $\mathscr{M}(\mathbb{R}_0^+)$
List of main results formulated in § 2.6	Existence due to compactness (Peano): Corollary 71 (page 166) Continuous dependence on data: Proposition 73 (page 167) Extension to models with delay: Remark 74
Key tools	The measure-valued solution to the linear problem is represented by means of a dual partial differential equation: Definition 61 (page 153), Proposition 63 (page 154) The dual PDE problem is equivalent to an integral equation, which is an inhomogeneous Volterra equation of second type at the initial point of time $t = 0$: Lemma 62 (page 153) The additional assumption $a(\cdot) \geq 0$ about $\vartheta_{a,b,c}$ preserves positivity of Radon measures: Corollary 64 (page 155)

Table 2.3 Brief summary of the example in § 2.6 in mutational terms:
A structured population model with Radon measures over $\mathbb{R}_0^+ = [0,\infty[$

2.6.2 The Linear Population Model

Now we consider the linear structured population model

$$\begin{cases} \partial_t \mu_t + \partial_x (b\, \mu_t) = c\, \mu_t & \text{in } \mathbb{R}_0^+ \times [0,T], \\ b(0)\, \mu_t(0) = \int_{\mathbb{R}_0^+} a\, d\mu_t & \text{in }]0,T], \\ \mu_0 = \nu_0, \end{cases} \tag{2.3}$$

where $a,b,c : \mathbb{R}_0^+ \longrightarrow \mathbb{R}$ are bounded and Lipschitz continuous functions with $b(0) > 0$ and, $\nu_0 \in \mathscr{M}(\mathbb{R}_0^+)$ is a given initial Radon measure.
Similarly to § 2.5.2 (about linear transport equations for Radon measures on $\mathbb{R}^N$), we first assume $b(\cdot) \in C^1(\mathbb{R}_0^+)$ in addition and then extend the subsequent estimates to $b(\cdot) \in W^{1,\infty}(\mathbb{R}_0^+)$ by means of L^∞ continuity (correspondingly to Remark 50 on page 139). All proofs of the following results about problem (2.3) are collected at the end of this subsection.

The Statements

Formal integration by parts motivates how to define a weak solution $[0,T] \longrightarrow \mathscr{M}(\mathbb{R}_0^+)$ to the linear problem (2.3).

Definition 58. $\mu : [0,T] \longrightarrow \mathscr{M}(\mathbb{R}_0^+)$, $t \longmapsto \mu_t$ is called a *weak solution* to problem (2.3) if μ is narrowly continuous with respect to time and, for all test functions $\varphi \in C^1(\mathbb{R}_0^+ \times [0,T]) \cap W^{1,\infty}(\mathbb{R}_0^+ \times [0,T])$,

$$\begin{aligned} &\int_{\mathbb{R}_0^+} \varphi(x,T)\, d\mu_T(x) \;-\; \int_{\mathbb{R}_0^+} \varphi(x,0)\, d\nu_0(x) \\ &= \int_0^T \int_{\mathbb{R}_0^+} \partial_t \varphi(x,t)\, d\mu_t(x)\, dt + \int_0^T \int_{\mathbb{R}_0^+} \Big(\partial_x \varphi(x,t)\, b(x) + \varphi(x,t)\, c(x) \Big)\, d\mu_t(x)\, dt \\ &\qquad + \int_0^T \varphi(0,t) \int_{\mathbb{R}_0^+} a(x)\, d\mu_t(x)\, dt. \end{aligned}$$

Now the key point is an implicit characterization of the solution to the linear problem (2.3) by an integral equation exploiting the notion of characteristics. This solution is derived for any initial finite Radon measure $\nu_0 \in \mathscr{M}(\mathbb{R}_0^+)$ and coefficient $b(\cdot) \in C^1(\mathbb{R}_0^+) \cap W^{1,\infty}(\mathbb{R}_0^+)$ with $b(0) > 0$.
Motivated by the application to population dynamics, we then specify a sufficient condition on $a(\cdot)$ for preserving nonnegativity of measures, namely $a(\cdot) \geq 0$. The corresponding solution map can easily be extended to less regular coefficients $b(\cdot) \in W^{1,\infty}(\mathbb{R}_0^+)$ as specified in Corollary 66 below (on page 156).

Remark 59. Adapting Definition 44 (on page 137), each function $b \in W^{1,\infty}(\mathbb{R}_0^+, \mathbb{R})$ induces the flow $X_b : [0,T] \times \mathbb{R}_0^+ \longrightarrow \mathbb{R}$ in the following sense: For any initial point $x_0 \in \mathbb{R}_0^+$, the curve $X_b(\cdot, x_0) : [0,T] \longrightarrow \mathbb{R}_0^+$ is the continuously differentiable solution to the Cauchy problem

$$\begin{cases} \frac{d}{dt}\, x(t) = b(x(t)), & \text{in } [0,T], \\ x(0) = x_0 \in \mathbb{R}_0^+. \end{cases}$$

The additional property $b(0) > 0$ ensures that all values of X_b are in $\mathbb{R}_0^+$.

The local assumptions $b \in C^1(\mathbb{R}_0^+) \cap W^{1,\infty}(\mathbb{R}_0^+)$, $b(0) > 0$ and Gronwall's Lemma imply continuous differentiability of solutions to ordinary differential equations with respect to parameters and initial data [92, 93, 181]. We summarize in the counterpart of Lemma 45 (on page 137):

Lemma 60. *If $b \in C^1(\mathbb{R}_0^+) \cap W^{1,\infty}(\mathbb{R}_0^+)$ and $b(0) > 0$, then $X_b : [0,T] \times \mathbb{R}_0^+ \longrightarrow \mathbb{R}_0^+$ is continuously differentiable with*

(*i*) $\|\partial_x X_b(t,\cdot)\|_\infty \;\leq\; e^{\|\partial_x b\|_\infty t}$,

(*ii*) $\operatorname{Lip} \partial_x X_b(\cdot,x) \;\leq\; \|\partial_x b\|_\infty \; e^{\|\partial_x b\|_\infty T}$,

(*iii*) $\|X_b(t,\cdot) - X_{\widetilde{b}}(t,\cdot)\|_\infty \;\leq\; \|b - \widetilde{b}\|_\infty \; t \; e^{\|\partial_x \widetilde{b}\|_\infty t}$ *for any* $\widetilde{b} \in W^{1,\infty}(\mathbb{R}_0^+)$, $\widetilde{b}(0) > 0$.

For every weak solution $\mu : [0,T] \longrightarrow \mathscr{M}(\mathbb{R}_0^+)$, integration by parts provides a characterization using a dual problem in the form of a partial differential equation:

Definition 61. Let $\psi \in C^1(\mathbb{R}_0^+) \cap W^{1,\infty}(\mathbb{R}_0^+)$. We call $\varphi_{t,\psi} \in C^1(\mathbb{R}_0^+ \times [0,t])$ the *solution to the dual problem* related to $\psi(\cdot)$ and t if it satisfies

$$\begin{cases} \partial_\tau\, \varphi_{t,\psi} + b(x)\, \partial_x\, \varphi_{t,\psi} + c(x)\, \varphi_{t,\psi} + a(x)\, \varphi_{t,\psi}(0,\tau) &= 0 \quad \text{in } \mathbb{R}_0^+ \times [0,t], \\ \varphi_{t,\psi}(\cdot,t) &= \psi \quad \text{in } \mathbb{R}_0^+. \end{cases} \tag{2.4}$$

The formulation of the dual problem is particularly useful as tool for proving existence of weak solutions. Knowing the solution to the dual problem, the solution to the linear problem (2.3) is given by the integral formula explicitly stated in Proposition 63. First we collect the properties of the dual problem though.

Lemma 62. *Let $a, b, c \in W^{1,\infty}(\mathbb{R}_0^+)$ and $b \in C^1(\mathbb{R}_0^+)$, $b(0) > 0$. For any function $\psi \in C^1(\mathbb{R}_0^+) \cap W^{1,\infty}(\mathbb{R}_0^+)$ and time $t \in]0,T]$, the solution $\varphi := \varphi_{t,\psi}$ to the related dual problem (2.4) is unique and, its equivalent characterization is given by the integral equation*

$$\begin{aligned} \varphi(x,\tau) \;=\; & \psi\big(X_b(t-\tau,x)\big) \cdot e^{\int_\tau^t c(X_b(r-\tau,x))\, dr} \\ & + \int_\tau^t a\big(X_b(s-\tau,x)\big)\; \varphi(0,s)\; e^{\int_\tau^s c(X_b(r-\tau,x))\, dr}\, ds. \end{aligned} \tag{2.5}$$

Moreover, for any $t > 0$ and $\psi \in C^1(\mathbb{R}_0^+) \cap W^{1,\infty}(\mathbb{R}_0^+)$ fixed, the following holds

(i) $\varphi(0,\cdot) : [0,t] \longrightarrow \mathbb{R}$ *is a bounded and continuously differentiable solution to the following inhomogeneous Volterra equation of second type*

$$\begin{aligned}\varphi(0,\tau) &= \psi(X_b(t-\tau,0))\ \ e^{\int_\tau^t c(X_b(r-\tau,0))\,dr} \\ &\quad + \int_\tau^t a(X_b(s-\tau,0))\ \ \varphi(0,s)\ \ e^{\int_\tau^s c(X_b(r-\tau,0))\,dr}\ ds \qquad (2.6)\end{aligned}$$

with
$$\begin{aligned}\|\varphi(0,\cdot)\|_\infty &\le \sup_{z \le \|b\|_\infty t} |\psi(z)| \cdot (1 + \|a\|_\infty\, t)\, e^{(\|a\|_\infty + \|c\|_\infty)\, t}, \\ \|\partial_\tau \varphi(0,\cdot)\|_\infty &\le \mathrm{const}(\|a\|_{W^{1,\infty}},\ \|b\|_\infty,\ \|c\|_{W^{1,\infty}}) \cdot \max\{\|\psi\|_\infty, \|\partial_x \psi\|_\infty\} \cdot \\ &\quad\ e^{2\,(\|a\|_\infty + \|c\|_\infty)\, t}\ \ (1+t).\end{aligned}$$

(ii) $\varphi(x,\cdot) : [0,t] \longrightarrow \mathbb{R}$ *is continuously differentiable for each $x \in \mathbb{R}_0^+$ with*

$$\begin{aligned}\|\partial_\tau \varphi(x,\cdot)\|_\infty \le\ &\mathrm{const}(\|a\|_{W^{1,\infty}},\ \|b\|_\infty,\ \|c\|_{W^{1,\infty}}) \cdot \max\{\|\psi\|_\infty, \|\partial_x \psi\|_\infty\} \\ &e^{2\,(\|a\|_\infty + \|c\|_\infty)\, t}\ \ (1+t).\end{aligned}$$

(iii) $\varphi(\cdot,\tau) : \mathbb{R}_0^+ \longrightarrow \mathbb{R}$ *is continuously differentiable for every $\tau \in [0,t]$ and satisfies*

$$\begin{aligned}\|\varphi(\cdot,\tau)\|_\infty &\le \|\psi\|_\infty\, e^{2\,(\|a\|_\infty + \|c\|_\infty)\, t}, \\ \|\partial_x \varphi(\cdot,\tau)\|_\infty &\le \max\{\|\partial_x \psi\|_\infty, 1\}\ \ e^{\max\{\|\psi\|_\infty, 1\}\, 3(\|a\|_{W^{1,\infty}} + \|\partial_x b\|_\infty + \|c\|_{W^{1,\infty}})\, t}.\end{aligned}$$

(iv) *For every $t > 0$ and $\psi \in C^1(\mathbb{R}_0^+) \cap W^{1,\infty}(\mathbb{R}_0^+)$, there exists a continuously differentiable solution $\varphi : \mathbb{R}_0^+ \times [0,t] \longrightarrow \mathbb{R}$ to integral equation (2.5). It is unique and has the regularity properties stated in parts (ii) and (iii).*

(v) *If additionally $\psi \in C^2(\mathbb{R}_0^+) \cap W^{2,\infty}(\mathbb{R}_0^+)$, then $\partial_x \varphi(x,\cdot) : [0,t] \longrightarrow \mathbb{R}$ is Lipschitz continuous and, its Lipschitz constant has an upper bound depending only on $\|a\|_{W^{1,\infty}}$, $\|b\|_{W^{1,\infty}}$, $\|c\|_{W^{1,\infty}}$, $\|\psi\|_{W^{2,\infty}}$ and, in particular, on t in an increasing way.*

Proposition 63. *Let $\varphi_{t,\psi} \in C^1(\mathbb{R}_0^+ \times [0,t])$ denote the solution to the dual problem (2.4) or equivalently, the integral equation (2.5) for any $t > 0$ and $\psi \in C^1(\mathbb{R}_0^+) \cap W^{1,\infty}(\mathbb{R}_0^+)$. For any Radon measure $\mu_0 \in \mathscr{M}(\mathbb{R}_0^+)$, let $\mu : [0,T] \longrightarrow \mathscr{M}(\mathbb{R}_0^+)$, $t \longmapsto \mu_t$ be given by*

$$\int_{\mathbb{R}_0^+} \psi(x)\, d\mu_t(x) = \int_{\mathbb{R}_0^+} \varphi_{t,\psi}(x,0)\, d\mu_0(x). \qquad (2.7)$$

Then

(i) *μ satisfies the following form of the semigroup property for every $0 \le s \le t \le T$ and $\psi \in C^1(\mathbb{R}_0^+) \cap W^{1,\infty}(\mathbb{R}_0^+)$:*

$$\int_{\mathbb{R}_0^+} \psi(x)\, d\mu_t(x) = \int_{\mathbb{R}_0^+} \varphi_{t,\psi}(x,s)\, d\mu_s(x). \qquad (2.8)$$

(ii) $t \longmapsto \int_{\mathbb{R}_0^+} \psi \, d\mu_t$ *is Lipschitz continuous for every* $\psi \in C^1(\mathbb{R}_0^+) \cap W^{1,\infty}(\mathbb{R}_0^+)$ *with Lipschitz constant* $\leq \ \mathrm{const}\big(\|a\|_{W^{1,\infty}}, \|b\|_\infty, \|c\|_{W^{1,\infty}}, T\big) \cdot \|\psi\|_{W^{1,\infty}} \ |\mu_0|(\mathbb{R}_0^+)$.
Furthermore, $|\mu_t|(\mathbb{R}_0^+) \ \leq \ e^{2(\|a\|_\infty + \|c\|_\infty)t} \cdot |\mu_0|(\mathbb{R}_0^+)$.

(iii) μ *is a weak solution to the linear problem (2.3) (in the sense of Definition 58).*

(iv) For any $\phi \in C^0(\mathbb{R}_0^+)$ *such that* $\operatorname{supp}\phi \subset [\|b\|_\infty t, \infty[$, *the following estimate holds with* $\widetilde{\phi}(x) := \sup_{z \leq x} \phi(z)$:

$$\int_{\mathbb{R}_0^+} \widetilde{\phi}(x + \|b\|_\infty t) \ d|\mu_0|(x) \ \geq \ e^{-\|c\|_\infty t} \int_{\mathbb{R}_0^+} \phi(x) \ d\mu_t(x).$$

We can also exploit the preceding properties to demonstrate that nonnegativity of finite Radon measures is preserved.

Corollary 64. *Under the additional hypothesis that* $a(\cdot) \geq 0$, *all values of the weak solution* $\mu : [0,T] \longrightarrow \mathscr{M}(\mathbb{R}_0^+)$ *presented in Proposition 63 are nonnegative Radon measures for every nonnegative initial measure* $\mu_0 \in \mathscr{M}^+(\mathbb{R}_0^+)$.

The preceding results provide more information than just the existence of solutions. Using the construction of Proposition 63, we obtain a continuous solution map for the linear problem (2.3). Furthermore, these solutions depend continuously on the coefficients $a(\cdot)$, $b(\cdot)$, $c(\cdot)$.

Proposition 65.
Let $a(\cdot),\ c(\cdot) \in W^{1,\infty}(\mathbb{R}_0^+)$ *and* $b(\cdot) \in C^1(\mathbb{R}_0^+) \cap W^{1,\infty}(\mathbb{R}_0^+)$ *satisfy* $b(0) > 0$. *The weak solutions to the linear problem (2.3), characterized in Proposition 63, induce a map*

$$\vartheta_{a,b,c} : [0,1] \times \mathscr{M}(\mathbb{R}_0^+) \longrightarrow \mathscr{M}(\mathbb{R}_0^+), \quad (t,\mu_0) \longmapsto \mu_t$$

satisfying for any $\mu_0, \nu_0 \in \mathscr{M}(\mathbb{R}_0^+)$, $t,h \in [0,1]$, $\widetilde{a}, \widetilde{c} \in W^{1,\infty}(\mathbb{R}_0^+)$, $\widetilde{b} \in C^1(\mathbb{R}_0^+) \cap W^{1,\infty}(\mathbb{R}_0^+)$ *with* $t+h \leq 1$, $\widetilde{b}(0) > 0$:

(i) $\vartheta_{a,b,c}(0,\cdot) = \mathrm{Id}_{\mathscr{M}(\mathbb{R}_0^+)}$
(ii) $\vartheta_{a,b,c}(h,\ \vartheta_{a,b,c}(t,\mu_0)) = \vartheta_{a,b,c}(t+h,\mu_0)$
(iii) $\big|\vartheta_{a,b,c}(h,\mu_0)\big|(\mathbb{R}_0^+) \ \leq \ |\mu_0|(\mathbb{R}_0^+) \cdot e^{2(\|a\|_\infty + \|c\|_\infty)h}$
(iv) $\rho_{\mathscr{M}}\big(\vartheta_{a,b,c}(t,\mu_0),\ \vartheta_{a,b,c}(t+h,\mu_0)\big) \leq \ h \cdot C(\|a\|_{W^{1,\infty}}, \|b\|_\infty, \|c\|_{W^{1,\infty}}) \cdot |\mu_0|(\mathbb{R}_0^+)$
(v) $\rho_{\mathscr{M}}\big(\vartheta_{a,b,c}(h,\mu_0),\ \vartheta_{a,b,c}(h,\nu_0)\big) \ \leq \ \rho_{\mathscr{M}}(\mu_0,\nu_0) \cdot e^{3(\|a\|_{W^{1,\infty}} + \|\partial_x b\|_\infty + \|c\|_{W^{1,\infty}})h}$
(vi) $\rho_{\mathscr{M}}\big(\vartheta_{a,b,c}(h,\mu_0),\ \vartheta_{\widetilde{a},\widetilde{b},\widetilde{c}}(h,\mu_0)\big) \ \leq \ h\,\big\|(a,b,c) - (\widetilde{a},\widetilde{b},\widetilde{c})\big\|_\infty \ \widehat{C} \ |\mu_0|(\mathbb{R}_0^+)$
with a constant $\widehat{C} = \widehat{C}(\|a\|_{W^{1,\infty}},\ \|\widetilde{a}\|_{W^{1,\infty}},\ \|b\|_{W^{1,\infty}},\ \|\widetilde{b}\|_{W^{1,\infty}},\ \|c\|_{W^{1,\infty}},\ \|\widetilde{c}\|_{W^{1,\infty}})$

(vii) If additionally $a(\cdot) \geq 0$, *then* $\vartheta_{a,b,c}([0,1], \mathscr{M}^+(\mathbb{R}_0^+)) \subset \mathscr{M}^+(\mathbb{R}_0^+)$.

The additional hypothesis $b(\cdot) \in C^1(\mathbb{R}_0^+)$ is dispensable – similarly to Remark 50 about the linear transport equation in $\mathcal{M}(\mathbb{R}^N)$ (on page 139):

Corollary 66. *For any functions $a(\cdot), b(\cdot), c(\cdot) \in W^{1,\infty}(\mathbb{R}_0^+)$ satisfying $b(0) > 0$, a map $\vartheta_{a,b,c} : [0,1] \times \mathcal{M}(\mathbb{R}_0^+) \longrightarrow \mathcal{M}(\mathbb{R}_0^+)$ can be constructed in such a way that $\vartheta_{a,b,c}(\cdot, \mu_0)$ is a weak solution to the linear problem (2.3) for each $\mu_0 \in \mathcal{M}(\mathbb{R}_0^+)$ and the statements (i)–(vii) of Proposition 65 hold for all $\mu_0, \nu_0 \in \mathcal{M}(\mathbb{R}_0^+)$, $t, h \in [0,1]$, $\widetilde{a}, \widetilde{b}, \widetilde{c} \in W^{1,\infty}(\mathbb{R}_0^+)$ with $t + h \leq 1$, $\widetilde{b}(0) > 0$.*

In terms of the mutational framework, we have obtained the following statement as main result of § 2.6.2:

Corollary 67 (Transitions due to linear problem (2.3)).
For arbitrary functions $a(\cdot), b(\cdot), c(\cdot) \in W^{1,\infty}(\mathbb{R}_0^+)$ satisfying $b(0) > 0$, the corresponding solution map of linear problem (2.3)

$$\vartheta_{a,b,c} : [0,1] \times \mathcal{M}(\mathbb{R}_0^+) \longrightarrow \mathcal{M}(\mathbb{R}_0^+)$$

is a transition on $\big(\mathcal{M}(\mathbb{R}_0^+),\ \rho_{\mathcal{M}},\ |\cdot|(\mathbb{R}_0^+)\big)$ with

$$\begin{array}{rcl} \alpha(\vartheta_{a,b,c};\ r) & := & 3\,(\|a\|_{W^{1,\infty}} + \|\partial_x b\|_\infty + \|c\|_{W^{1,\infty}}) \\ \beta(\vartheta_{a,b,c};\ r) & := & C(\|a\|_{W^{1,\infty}}, \|b\|_\infty, \|c\|_{W^{1,\infty}}) \cdot r \\ \gamma(\vartheta_{a,b,c}) & := & 2\,(\|a\|_\infty + \|c\|_\infty) \\ D(\vartheta_{a,b,c},\ \vartheta_{\widetilde{a},\widetilde{b},\widetilde{c}};\ r) & \leq & \big\|(a,b,c) - (\widetilde{a},\widetilde{b},\widetilde{c})\big\|_\infty \cdot \widehat{C}\ r \end{array}$$

From now on, the set of these transitions on $\big(\mathcal{M}(\mathbb{R}_0^+),\ \rho_{\mathcal{M}},\ |\cdot|(\mathbb{R}_0^+)\big)$ is abbreviated as $\Theta\big(\mathcal{M}(\mathbb{R}_0^+),\ \rho_{\mathcal{M}},\ |\cdot|(\mathbb{R}_0^+)\big)$.

The Proofs About the Linear Population Model

Proof (of Lemma 62 on page 153).
We start with the proof of integral characterization (2.5). Fix $t > 0$ arbitrarily. For any $\widetilde{b} \in C^1(\mathbb{R}_0^+) \cap W^{1,\infty}(\mathbb{R}_0^+)$, $\widetilde{c} \in W^{1,\infty}(\mathbb{R}_0^+)$ and $\widetilde{f} \in W^{1,\infty}(\mathbb{R}_0^+ \times [0,t])$ with $\widetilde{b}(0) < 0$ and every $\psi \in C^1(\mathbb{R}_0^+)$, the semilinear initial value problem

$$\begin{cases} \partial_\tau \xi(x,\tau) + \widetilde{b}(x)\,\partial_x \xi(x,\tau) + \widetilde{c}(x)\,\xi(x,\tau) + \widetilde{f}(x,\tau) = 0 & \text{in } \mathbb{R}_0^+ \times [0,t] \\ \xi(\cdot,0) = \psi & \text{in } \mathbb{R}_0^+ \end{cases}$$

has a unique solution $\xi \in C^1(\mathbb{R}_0^+ \times [0,t])$ given explicitly by

$$\begin{aligned} \xi(x,\tau) = {} & \psi\big(X_{-\widetilde{b}}(\tau,x)\big) \cdot e^{-\int_0^\tau \widetilde{c}(X_{-\widetilde{b}}(\tau-r,x))\,dr} \\ & - \int_0^\tau \widetilde{f}\big(X_{-\widetilde{b}}(\tau-s,x),\ s\big) \cdot e^{-\int_s^\tau \widetilde{c}(X_{-\widetilde{b}}(\tau-r,x))\,dr}\,ds. \end{aligned}$$

This explicit representation of $\xi(x,\tau)$ results from the classical method of characteristics. It was presented by Conway [49] for the corresponding problem in $\mathbb{R}^n$ instead of $\mathbb{R}_0^+$. Since $\widetilde{b}(0) < 0$, i.e., $\mathbb{R}_0^+$ is invariant under the characteristic flow of $-\widetilde{b}(\cdot)$, the expression obtained in [49] can be restricted to $\mathbb{R}_0^+$.

Substituting $\varphi(x,\tau) := \xi(x,t-\tau)$ yields the solution to the corresponding partial differential equation with an end-time condition and the coefficients $b(\cdot)$ and $c(\cdot)$ satisfying $b(0) > 0$. Indeed, let $t > 0$, $b \in C^1(\mathbb{R}_0^+) \cap W^{1,\infty}(\mathbb{R}_0^+)$, $c \in W^{1,\infty}(\mathbb{R}_0^+)$ and $f \in W^{1,\infty}(\mathbb{R}_0^+ \times [0,t])$ be arbitrary with $b(0) > 0$. For any function $\psi \in C^1(\mathbb{R}_0^+)$, the semilinear partial differential equation

$$\begin{cases} \partial_\tau \varphi(x,\tau) + b(x)\,\partial_x \varphi(x,\tau) + c(x)\,\varphi(x,\tau) + f(x,\tau) = 0 & \text{in } \mathbb{R}_0^+ \times [0,t], \\ \varphi(\cdot,t) = \psi & \text{in } \mathbb{R}_0^+, \end{cases}$$

has a unique solution $\varphi \in C^1(\mathbb{R}_0^+ \times [0,t])$ explicitly given by

$$\begin{aligned} \varphi(x,\tau) = {} & \psi\big(X_b(t-\tau,x)\big) \cdot e^{\int_\tau^t c(X_b(r-\tau,x))\,dr} \\ & + \int_\tau^t f\big(X_b(s-\tau,x),\, s\big) \cdot e^{\int_\tau^s c(X_b(r-\tau,x))\,dr}\, ds. \end{aligned}$$

Applying this result to $f(x,\tau) = a(x)\,\varphi(0,\tau)$, we obtain the equivalence between equations (2.4) and (2.5) for every function $\varphi \in C^1(\mathbb{R}_0^+ \times [0,t])$ (with Lipschitz continuous $\varphi(0,\cdot) : [0,t] \longrightarrow \mathbb{R}$).

Now we proceed with the proof of the statements (i)–(v) of Lemma 62:

(i) Volterra equation (2.6) results directly from equation (2.5) by setting $x = 0$. The upper bound of $|\varphi(0,\cdot)|$, restricted to $[0,t]$, is a consequence of

$$|\varphi(0,\tau)|\; e^{\|c\|_\infty \tau} \;\le\; \sup_{z \le \|b\|_\infty t} |\psi(z)|\; e^{\|c\|_\infty t} + \|a\|_\infty \int_\tau^t |\varphi(0,s)|\; e^{\|c\|_\infty s} ds$$

and Gronwall's Lemma (Proposition A.1 on page 439).
Moreover, the right-hand side of Volterra equation (2.6) is continuously differentiable with respect to τ and thus, $\varphi(0,\cdot) \in C^1([0,t])$. The product rule reveals that at every time $\tau \in [0,t]$

$$\begin{aligned} & \left|\tfrac{d}{d\tau}\, \varphi(0,\tau)\right| \\ \le {} & e^{\|c\|_\infty (t-\tau)} \Big(\|\partial_x \psi\|_\infty \cdot \|b\|_\infty + \|\psi\|_\infty \Big(\|c\|_\infty + (t-\tau) \cdot \|\partial_x c\|_\infty \cdot \|b\|_\infty \Big) \Big) \\ & + e^{\|c\|_\infty (t-\tau)} \Big(\|a\|_\infty\, \|\varphi(0,\cdot)\|_\infty + (t-\tau) \cdot \Big(\|\partial_x a\|_\infty \cdot \|b\|_\infty\, \|\varphi(0,\cdot)\|_\infty \\ & \qquad + \|a\|_\infty\, \|\varphi(0,\cdot)\|_\infty \Big(\|c\|_\infty + t \cdot \|\partial_x c\|_\infty\, \|b\|_\infty \Big) \Big) \Big). \end{aligned}$$

(ii) For arbitrarily fixed $x \in \mathbb{R}_0^+$, $\varphi(x,\cdot) : [0,t] \longrightarrow \mathbb{R}$ is continuously differentiable since it satisfies the integral equation (2.5) and $\varphi(0,\cdot)$ is continuous. The upper bound of the derivative $\|\partial_\tau \varphi(x,\cdot)\|_\infty$ results from considerations similar to the conclusions concerning $\sup |\partial_\tau \varphi(0,\cdot)|$ in statement (i).

(iii) The upper bound of $\|\varphi(\cdot,\tau)\|_\infty$ results directly from the integral equation (2.5) and property (i):

$$\begin{aligned}
\|\varphi(\cdot,\tau)\|_\infty &\le \|\psi\|_\infty \Big(e^{\|c\|_\infty t} + \int_0^t \|a\|_\infty \cdot (1+\|a\|_\infty\, s)\; e^{(\|a\|_\infty + \|c\|_\infty)\cdot s} \cdot e^{\|c\|_\infty s}\; ds \Big) \\
&\le \|\psi\|_\infty \Big(e^{\|c\|_\infty t} + \|a\|_\infty \int_0^t (1 + (\|a\|_\infty + 2\,\|c\|_\infty) s)\; e^{(\|a\|_\infty + 2\,\|c\|_\infty)\cdot s}\; ds \Big) \\
&= \|\psi\|_\infty \Big(e^{\|c\|_\infty t} + \|a\|_\infty\; t\; e^{(\|a\|_\infty + 2\,\|c\|_\infty)\cdot t} \Big) \\
&\le \|\psi\|_\infty\, e^{(\|a\|_\infty + 2\,\|c\|_\infty)\cdot t} \; \big(1 + \|a\|_\infty\, t\big) \\
&\le \|\psi\|_\infty\, e^{(2\,\|a\|_\infty + 2\,\|c\|_\infty)\cdot t}.
\end{aligned}$$

The last inequality results from $1+s \le e^s$ for all $s \ge 0$. The form of the right-hand side of integral equation (2.5) ensures that $\varphi(\cdot,\tau) : \mathbb{R}_0^+ \longrightarrow \mathbb{R}$ is continuously differentiable for every $\tau \in [0,t]$. Furthermore, for every $x \in \mathbb{R}_0^+$, the chain rule and Lemma 60 (on page 153) imply

$$\begin{aligned}
&\Big| \tfrac{\partial}{\partial x}\, \varphi(x,\tau) \Big| \;\cdot\; e^{\|c\|_\infty (\tau - t)} \\
&\le\; \|\partial_x \psi\|_\infty \cdot \|\partial_x X_b(t-\tau,\cdot)\|_\infty \qquad + \|\psi\|_\infty \int_\tau^t \|\partial_x c\|_\infty \cdot \|\partial_x X_b(r-\tau,\cdot)\|_\infty\, dr \\
&\quad + \int_\tau^t \Big(\|\partial_x a\|_\infty \cdot \|\partial_x X_b(s-\tau,\cdot)\|_\infty + \|a\|_\infty \int_\tau^s \|\partial_x c\|_\infty \cdot \|\partial_x X_b(r-\tau,\cdot)\|_\infty\, dr \Big) \\
&\qquad\qquad |\varphi(0,s)|\; ds,
\end{aligned}$$

and thus due to property (i),

$$\begin{aligned}
\|\partial_x \varphi\|_\infty \;&\le\; \|\partial_x \psi\|_\infty e^{(\|\partial_x b\|_\infty + \|c\|_\infty) t} + \|\psi\|_\infty\, \|\partial_x c\|_\infty\, e^{(\|\partial_x b\|_\infty + \|c\|_\infty)\, t}\; t \\
&\quad + \|\psi\|_\infty\, e^{(2\,\|a\|_\infty + \|\partial_x b\|_\infty + 2\,\|c\|_\infty) t} \Big(\|\partial_x a\|_\infty t + \|a\|_\infty \|\partial_x c\|_\infty\, \tfrac{t^2}{2} \Big) \\
&\le\; \max\{\|\partial_x \psi\|_\infty, 1\}\;\; e^{(2\,\|a\|_\infty + \|\partial_x b\|_\infty + 2\,\|c\|_\infty)\; t} \\
&\quad\; \Big(1 + \|\psi\|_\infty\, (\|\partial_x c\|_\infty + \|\partial_x a\|_\infty)\; t \;+\; \|\psi\|_\infty\, \|a\|_\infty\, \|\partial_x c\|_\infty\, \tfrac{t^2}{2} \Big) \\
&\le\; \max\{\|\partial_x \psi\|_\infty, 1\} \;\cdot\; e^{\max\{\|\psi\|_\infty, 1\} \cdot 3\, (\|a\|_{W^{1,\infty}} + \|\partial_x b\|_\infty + \|c\|_{W^{1,\infty}})\; t}.
\end{aligned}$$

(iv) Volterra equation (2.6) has a unique continuous solution, since the integrand is Lipschitz continuous with respect to $\varphi(0,s)$ [173, 181]. It induces directly the unique continuously differentiable solution to equation (2.5) and thus equivalently to dual problem (2.4).

(v) This feature results from differentiating equation (2.5) with respect to x. Indeed, due to Lemma 60 (on page 153), the functions $[0,T] \longrightarrow \mathbb{R},\; t \longmapsto \partial_x X_b(t,x)$ are uniformly Lipschitz continuous for all $x \in \mathbb{R}_0^+$. □

Proof (of Proposition 63 on page 154).
(i) Choose arbitrary $0 \le s < t \le T$ and $\psi \in C^1(\mathbb{R}_0^+) \cap W^{1,\infty}(\mathbb{R}_0^+)$.
Let $\xi \in C^1(\mathbb{R}_0^+ \times [0,s])$ denote a solution to the semilinear differential equation

$$\begin{aligned}
\partial_\tau \xi + b(x)\, \partial_x \xi \;+\; c(x)\, \xi + a(x)\, \xi(0,\tau) &= 0 \qquad &&\text{in } \mathbb{R}_0^+ \times [0,s], \\
\xi(\cdot,s) &= \varphi_{t,\psi}(\cdot,s) &&\text{in } \mathbb{R}_0^+,
\end{aligned}$$

or (as an equivalent formulation) to the integral equation for $(x,\tau) \in \mathbb{R}_0^+ \times [0,s]$

$$\begin{aligned}\xi(x,\tau) &= \varphi_{t,\psi}\big(X_b(s-\tau,x),\, s\big) \cdot e^{\int_\tau^s c(X_b(r-\tau,x))\,dr} \\ &\quad + \int_\tau^s a(X_b(\sigma-\tau,x))\ \xi(0,\sigma)\ e^{\int_\tau^\sigma c(X_b(r-\tau,x))\,dr}\ d\sigma.\end{aligned}$$

According to Lemma 62 (iv), such a solution exists and is unique since $\varphi_{t,\psi}(\cdot,s)$ is continuously differentiable and bounded in $W^{1,\infty}(\mathbb{R}_0^+)$. Thus, $\xi \equiv \varphi_{t,\psi}(\cdot,\cdot)|_{\mathbb{R}_0^+\times[0,s]}$ and, using the duality formula (2.7), we conclude that

$$\begin{aligned}\int_{\mathbb{R}_0^+} \psi(x)\ d\mu_t(x) &= \int_{\mathbb{R}_0^+} \varphi_{t,\psi}(x,0)\ d\mu_0(x) \\ &= \int_{\mathbb{R}_0^+} \xi(x,0)\quad d\mu_0(x) \;=\; \int_{\mathbb{R}_0^+} \varphi_{t,\psi}(x,s)\ d\mu_s(x).\end{aligned}$$

(ii) The total variation of μ_t can be characterized as a supremum [5, Proposition 1.47]. Therefore, due to Lemma 62 (iii),

$$\begin{aligned}|\mu_t|(\mathbb{R}_0^+) &= \sup\Big\{\int_{\mathbb{R}_0^+} u(x)\quad d\,\mu_t(x)\ \Big|\ u \in C_c^0(\mathbb{R}_0^+),\ \|u\|_\infty \le 1\Big\} \\ &= \sup\Big\{\int_{\mathbb{R}_0^+} u(x)\quad d\,\mu_t(x)\ \Big|\ u \in C_c^1(\mathbb{R}_0^+),\ \|u\|_\infty \le 1\Big\} \\ &\overset{(2.7)}{=} \sup\Big\{\int_{\mathbb{R}_0^+} \varphi_{t,u}(x,0)\ d\,\mu_0(x)\ \Big|\ u \in C_c^1(\mathbb{R}_0^+),\ \|u\|_\infty \le 1\Big\} \\ &\le \sup\Big\{\|\varphi_{t,u}(\cdot,0)\|_\infty\quad |\mu_0|(\mathbb{R}_0^+)\ \Big|\ u \in C_c^1(\mathbb{R}_0^+),\ \|u\|_\infty \le 1\Big\} \\ &\le \qquad e^{2(\|a\|_\infty+\|c\|_\infty)\cdot t}\quad |\mu_0|(\mathbb{R}_0^+).\end{aligned}$$

For arbitrary $0 \le s < t \le T$ and $\psi \in W^{1,\infty}(\mathbb{R}_0^+) \cap C^1(\mathbb{R}_0^+)$, we obtain

$$\begin{aligned}\Big|\int_{\mathbb{R}_0^+} \psi\,d\mu_t - \int_{\mathbb{R}_0^+} \psi\,d\mu_s\Big| &= \Big|\int_{\mathbb{R}_0^+} \varphi_{t,\psi}(x,s)\ d\mu_s(x) \;-\; \int_{\mathbb{R}_0^+} \varphi_{t,\psi}(x,t)\ d\mu_s(x)\Big| \\ &\le \int_{\mathbb{R}_0^+} \Big|\varphi_{t,\psi}(x,s) - \varphi_{t,\psi}(x,t)\Big|\ d\,|\mu_s|(x) \\ &\le \quad (t-s)\ \|\partial_\tau \varphi_{t,\psi}\|_\infty \qquad |\mu_s|(\mathbb{R}_0^+).\end{aligned}$$

Lemma 62 (ii) implies Lipschitz continuity due to $\psi \in W^{1,\infty}(\mathbb{R}_0^+)$.

(iii) First we focus on autonomous functions $\psi \in C^2(\mathbb{R}_0^+) \cap W^{2,\infty}(\mathbb{R}_0^+)$ and prove

$$\lim_{h \downarrow 0} \tfrac{1}{h} \cdot \Big(\int_{\mathbb{R}_0^+} \psi\ d\mu_t \;-\; \int_{\mathbb{R}_0^+} \psi\ d\mu_{t-h}\Big) \;=\; \int_{\mathbb{R}_0^+} \Big(b \cdot \partial_x \psi + c\,\psi + a\,\psi(0)\Big)\,d\mu_t$$

for any $t \in]0,T]$. Indeed, statement (i) implies for any $0 < h \le t \le T$

$$\tfrac{1}{h} \cdot \Big(\int_{\mathbb{R}_0^+} \psi\ d\mu_t \;-\; \int_{\mathbb{R}_0^+} \psi\ d\mu_{t-h}\Big) \;=\; \int_{\mathbb{R}_0^+} \tfrac{\varphi_{t,\psi}(x,t-h) \,-\, \psi(x)}{h}\ d\mu_{t-h}(x).$$

In particular, Lemma 62 (ii) and (v) provide upper bounds for the $W^{1,\infty}$ norm of $\mathbb{R}_0^+ \longrightarrow \mathbb{R},\ x \longmapsto \frac{\varphi_{t,\psi}(x,t-h) \,-\, \psi(x)}{h}$ which depend on $\|\psi\|_{W^{2,\infty}}$, but not on t,h:

$$\begin{aligned}\Big\|\tfrac{\varphi_{t,\psi}(\cdot,t-h) \,-\, \psi(\cdot)}{h}\Big\|_\infty &\le \mathrm{const}(\|a\|_{W^{1,\infty}},\ \|b\|_\infty,\quad \|c\|_{W^{1,\infty}},\ T) \cdot \|\psi\|_{W^{1,\infty}}, \\ \Big\|\tfrac{\partial_x \varphi_{t,\psi}(\cdot,t-h) \,-\, \partial_x \psi(\cdot)}{h}\Big\|_\infty &\le \mathrm{const}(\|a\|_{W^{1,\infty}},\ \|b\|_{W^{1,\infty}},\ \|c\|_{W^{1,\infty}},\ T,\ \|\psi\|_{W^{2,\infty}}).\end{aligned}$$

Hence property (ii) provides a constant $C(\|a\|_{W^{1,\infty}}, \|b\|_{W^{1,\infty}}, \|c\|_{W^{1,\infty}}, \|\psi\|_{W^{2,\infty}}, T)$ such that for every $h \in]0,t]$,

$$\left|\tfrac{1}{h}\Big(\int_{\mathbb{R}_0^+} \psi\, d\mu_t - \int_{\mathbb{R}_0^+} \psi\, d\mu_{t-h}\Big) - \int_{\mathbb{R}_0^+} \tfrac{\varphi_{t,\psi}(x,t-h) - \psi(x)}{h}\, d\mu_t(x)\right| \;\le\; C \cdot h \cdot |\mu_0|(\mathbb{R}_0^+).$$

In regard to the limit for $h \downarrow 0$, we conclude from $\varphi_{t,\psi} \in C^1(\mathbb{R}_0^+ \times [0,t])$ solving the dual problem (2.4)

$$\begin{aligned}\lim_{h\downarrow 0} \tfrac{1}{h} \cdot \Big(\int_{\mathbb{R}_0^+} \psi\, d\mu_t \;-\; \int_{\mathbb{R}_0^+} \psi\, d\mu_{t-h}\Big) &= \lim_{h\downarrow 0} \int_{\mathbb{R}_0^+} \tfrac{\varphi_{t,\psi}(x,t-h) - \psi(x)}{h}\, d\mu_t(x)\\ &= \int_{\mathbb{R}_0^+} \Big(b \cdot \partial_x \psi \,+\, c\,\psi \,+\, a\,\psi(0)\Big)\, d\mu_t.\end{aligned}$$

Finally we will provide the missing link to weak solutions to the linear problem (2.3) in the sense of Definition 58 (on page 152). Indeed, for any smooth test function $\varphi \in C_c^\infty(\mathbb{R}_0^+ \times [0,T])$, the auxiliary function

$$\zeta : [0,T] \times [0,T] \longrightarrow \mathbb{R}, \quad (s,t) \longmapsto \int_{\mathbb{R}_0^+} \varphi(x,t)\; d\mu_s(x)$$

has continuous partial derivatives

$$\begin{aligned}\tfrac{\partial}{\partial s}\, \zeta(s,t) &= \int_{\mathbb{R}_0^+} \Big(b \cdot \partial_x \varphi(\cdot,t) + c\,\varphi(\cdot,t) + a\,\varphi(0,t)\Big)\; d\mu_s\\ \tfrac{\partial}{\partial t}\, \zeta(s,t) &= \int_{\mathbb{R}_0^+} \partial_t \varphi(x,t)\; d\mu_s(x).\end{aligned}$$

Hence, $\zeta(\cdot,\cdot) \in C^1([0,T] \times [0,T])$. Due to the chain rule, the function $[0,T] \longrightarrow \mathbb{R}$, $t \longmapsto \zeta(t,t)$ is continuously differentiable with

$$\tfrac{d}{dt}\, \zeta(t,t) \;=\; \int_{\mathbb{R}_0^+} \Big(b \cdot \partial_x \varphi(\cdot,t) + c\,\varphi(\cdot,t) + a\,\varphi(0,t)\Big)\; d\mu_t \;+\; \int_{\mathbb{R}_0^+} \partial_t \varphi(\cdot,t)\; d\mu_t.$$

Thus, $\mu(\cdot)$ satisfies the integral condition on weak solutions for all smooth test functions $\varphi \in C_c^\infty(\mathbb{R}_0^+ \times [0,T])$. This property is easy to extend to all test functions $\varphi \in C^1(\mathbb{R}_0^+ \times [0,T]) \cap W^{1,\infty}(\mathbb{R}_0^+ \times [0,T])$ by means of continuity with respect to the $W^{1,\infty}$ norm.

(iv) $\operatorname{supp} \phi \subset \big[\|b\|_\infty t, \infty\big[$ implies $\|\varphi_{t,\phi}(0,\cdot)\|_\infty = 0$ due to Lemma 62 (i). Hence the integral equation (2.5) for $\varphi_{t,\phi}$ simplifies to

$$\varphi_{t,\phi}(x,\tau) \;=\; \phi\,(X_b(t-\tau,x))\; e^{\int_\tau^t c(X_b(r-\tau,x))dr}$$

for all $x \in \mathbb{R}_0^+$ and $\tau \in [0,t]$. Finally, we conclude for $\widetilde{\phi}(x) := \sup_{z \le x} \phi(z)$

$$\begin{aligned}e^{\|c\|_\infty t} \int_{\mathbb{R}_0^+} \widetilde{\phi}(x + t\,\|b\|_\infty)\; d|\mu_0|(x) &\ge \int_{\mathbb{R}_0^+} \widetilde{\phi}(X_b(t,x))\; e^{\int_0^t c(X_b(r,x))\,dr}\; d|\mu_0|(x)\\ &\ge \int_{\mathbb{R}_0^+} \phi(X_b(t,x))\; e^{\int_0^t c(X_b(r,x))\,dr}\; d\mu_0(x)\\ &= \int_{\mathbb{R}_0^+} \varphi_{t,\phi}(x,0)\; d\mu_0(x) = \int_{\mathbb{R}_0^+} \phi(x)\; d\mu_t(x).\end{aligned}$$

□

Proof (of Corollary 64 on page 155). The construction of μ_t via equation (2.7) implies that nonnegativity of measures is preserved if we can ensure that

$$\psi(\cdot) \geq 0 \quad \Longrightarrow \quad \varphi_{t,\psi}(\cdot,0) \geq 0.$$

Setting $x = 0$ in the integral characterization (2.5) of $\varphi_{t,\psi}$ leads to the Volterra equation (2.6) for $\varphi_{t,\psi}(0,\cdot)$. In particular, supposing $\psi(\cdot) \geq 0$ implies

$$\varphi_{t,\psi}(0,\tau) \;\geq\; \int_\tau^t a(X_b(s-\tau,0))\;\varphi_{t,\psi}(0,s)\;e^{\int_\tau^s c(X_b(r-\tau,0))\,dr}\,ds.$$

The additional hypothesis $a(\cdot) \geq 0$ guarantees for all $\tau \in [0,t]$

$$\begin{aligned}
&\max\left\{0,\,-\varphi_{t,\psi}(0,\tau)\right\} \\
&\leq \max\left\{0,\,-\int_\tau^t a\big(X_b(s-\tau,0)\big)\;\varphi_{t,\psi}(0,s)\;e^{\int_\tau^s c(X_b(r-\tau,0))dr}\;ds\right\} \\
&\leq \int_\tau^t a\big(X_b(s-\tau,0)\big)\;\max\left\{0,\,-\varphi_{t,\psi}(0,s)\right\}\;e^{\int_\tau^s c(X_b(r-\tau,0))\,dr}ds.
\end{aligned}$$

and, we conclude from Gronwall's Lemma (Proposition A.1 on page 439) that $\varphi_{t,\psi}(\cdot,t) = \psi(\cdot) \geq 0$ implies $\max\left\{0,\,-\varphi_{t,\psi}(0,\cdot)\right\} \equiv 0$, i.e. $\varphi_{t,\psi}(0,\cdot) \geq 0$. □

The next lemma is very useful for proving Proposition 65 (vi) afterwards because it provides a link between two solutions to the dual problems for different coefficient functions $a(\cdot), b(\cdot), c(\cdot)$ and $\widetilde{a}(\cdot), \widetilde{b}(\cdot), \widetilde{c}(\cdot)$ respectively. Appropriate convex combinations lay the foundations:

Lemma 68. *Suppose $a,\ \widetilde{a},\ c,\ \widetilde{c} \in W^{1,\infty}(\mathbb{R}_0^+)$, $b,\ \widetilde{b} \in C^1(\mathbb{R}_0^+) \cap W^{1,\infty}(\mathbb{R}_0^+)$ with $b(0) > 0$ and $\widetilde{b}(0) > 0$. Fixing $t \in]0,1]$, $\lambda \in [0,1]$ and $\psi \in C^1(\mathbb{R}_0^+) \cap W^{1,\infty}(\mathbb{R}_0^+)$ arbitrarily, let $\varphi^\lambda \in C^0(\mathbb{R}_0^+ \times [0,t])$ satisfy the integral equation*

$$\begin{aligned}
\varphi^\lambda(x,\tau) \;=\; &\psi\Big|_{\left(\lambda\,X_b(t-\tau,x)+(1-\lambda)\,X_{\widetilde{b}}(t-\tau,x)\right)}\;e^{\int_\tau^t\left(\lambda\,c(X_b(r-\tau,x))+(1-\lambda)\,\widetilde{c}(X_{\widetilde{b}}(r-\tau,x))\right)dr} \\
&+\int_\tau^t \left(\lambda\,a(X_b(s-\tau,x)) + (1-\lambda)\,\widetilde{a}\big(X_{\widetilde{b}}(s-\tau,x)\big)\right)\cdot\varphi^\lambda(0,s)\,\cdot \\
&\quad\times e^{\int_\tau^s\left(\lambda\,c(X_b(r-\tau,x))+(1-\lambda)\,\widetilde{c}(X_{\widetilde{b}}(r-\tau,x))\right)dr}\;ds. \qquad (2.9)
\end{aligned}$$

Then, $\lambda \longmapsto \varphi^\lambda(x,\tau)$ is continuously differentiable for every $x \in \mathbb{R}_0^+$ and $\tau \in [0,t]$ and there is a constant $C = C(\|a\|_{W^{1,\infty}}, \|\widetilde{a}\|_{W^{1,\infty}}, \|b\|_{W^{1,\infty}}, \|\widetilde{b}\|_{W^{1,\infty}}, \|c\|_{W^{1,\infty}}, \|\widetilde{c}\|_{W^{1,\infty}})$ such that

$$\begin{aligned}
\left|\frac{\partial}{\partial\lambda}\varphi^\lambda(x,\tau)\right| \;\leq\; &C\cdot\max\{\|\psi\|_\infty, \|\partial_x\psi\|_\infty, 1\}\cdot(t-\tau)\;e^{C\,(t-\tau)}\,\cdot \\
&\times\left(\|a-\widetilde{a}\|_\infty + \|b-\widetilde{b}\|_\infty + \|c-\widetilde{c}\|_\infty\right).
\end{aligned}$$

Proof (of Lemma 68). Similarly to Lemma 62 (on page 153 f.),

$$[0,t] \longrightarrow \mathbb{R}, \quad \tau \longmapsto \varphi^\lambda(0,\tau)$$

is a bounded and Lipschitz continuous solution to the following inhomogeneous Volterra equation of the second type

$$\begin{aligned}\varphi^\lambda(0,\tau) \;=\; & \psi\Big|_{\left(\lambda\, X_b(t-\tau,0)+(1-\lambda)\, X_{\widetilde{b}}(t-\tau,0)\right)} \; e^{\int_\tau^t \left(\lambda\, c(X_b(r-\tau,0))+(1-\lambda)\,\widetilde{c}(X_{\widetilde{b}}(r-\tau,0))\right)dr} \\ & + \int_\tau^t \left(\lambda\, a(X_b(s-\tau,0)) + (1-\lambda)\;\widetilde{a}\left(X_{\widetilde{b}}(s-\tau,0)\right)\right) \cdot \varphi^\lambda(0,s)\;\cdot \\ & \qquad \times\, e^{\int_\tau^s \left(\lambda\, c(X_b(r-\tau,0))+(1-\lambda)\,\widetilde{c}(X_{\widetilde{b}}(r-\tau,0))\right)dr}\; ds.\end{aligned}$$

The bounds on the L^∞ norm and the Lipschitz constant mentioned in Lemma 62 (i) can be adapted by considering $\max\{\|a\|_{W^{1,\infty}}, \|\widetilde{a}\|_{W^{1,\infty}}\}$ instead of $\|a\|_{W^{1,\infty}}$ and so forth.

Furthermore, $\varphi^\lambda(0,\tau)$ depends on λ in a continuously differentiable way [181, § 13] and, using the abbreviations $\widehat{a} := \max\{\|a\|_\infty,\, \|\widetilde{a}\|_\infty\}$, $\widehat{c} := \max\{\|c\|_\infty, \|\widetilde{c}\|_\infty\}$,

$$\begin{aligned}& \left|\tfrac{\partial}{\partial\lambda}\, \varphi^\lambda(0,\tau)\right|\; e^{-\widehat{c}\cdot(t-\tau)} \\ \le\; & \Big(\|\partial_x\psi\|_\infty \qquad \cdot \left|X_b(t-\tau,0) - X_{\widetilde{b}}(t-\tau,0)\right| \\ & \quad + \|\psi\|_\infty \qquad \cdot (t-\tau)\;\left(\|c-\widetilde{c}\|_\infty + \|\partial_x c\|_\infty \cdot \sup_{[\tau,t]} \left|X_b|_{(\cdot-\tau,0)} - X_{\widetilde{b}}|_{(\cdot-\tau,0)}\right|\right)\Big) \\ & + \int_\tau^t \Big(|\varphi^\lambda(0,s)|\;\left(\|a-\widetilde{a}\|_\infty + \|\partial_x a\|_\infty \cdot \left|X_b(s-\tau,0) - X_{\widetilde{b}}(s-\tau,0)\right|\right) \\ & \quad + |\partial_\lambda\, \varphi^\lambda(0,s)|\;\widehat{a} \\ & \quad + |\varphi^\lambda(0,s)|\widehat{a}\cdot(s-\tau)\left(\|c-\widetilde{c}\|_\infty + \|\partial_x c\|_\infty \sup_{[\tau,s]} |X_b|_{(\cdot-\tau,0)} - X_{\widetilde{b}}|_{(\cdot-\tau,0)}|\right)\Big)\, ds.\end{aligned}$$

Lemma 60 (on page 153) provides the estimate

$$\|X_b(s,\cdot) - X_{\widetilde{b}}(s,\cdot)\|_\infty \;\le\; \|b-\widetilde{b}\|_\infty \cdot s\;\; e^{\|\partial_x b\|_\infty s}$$

for all $s \ge 0$ and thus, Gronwall's Lemma implies the bound

$$\begin{aligned}\left|\tfrac{\partial}{\partial\lambda}\, \varphi^\lambda(0,\tau)\right| \;\le\; & C_0 \cdot \max\left\{\|\psi\|_\infty,\; \|\partial_x\psi\|_\infty,\; 1\right\} \cdot (t-\tau)\; e^{C_0\,(t-\tau)} \\ & \times \left(\|a-\widetilde{a}\|_\infty + \|b-\widetilde{b}\|_\infty + \|c-\widetilde{c}\|_\infty\right)\end{aligned}$$

with a constant $C_0 = C_0(\|a\|_{W^{1,\infty}},\, \|\widetilde{a}\|_{W^{1,\infty}},\, \|b\|_{W^{1,\infty}},\, \|\widetilde{b}\|_{W^{1,\infty}},\, \|c\|_{W^{1,\infty}},\, \|\widetilde{c}\|_{W^{1,\infty}})$.
Integral equation (2.9) ensures that $\varphi^\lambda(x,\tau)$ is continuously differentiable with respect to the parameter λ. Similarly to the preceding estimate of $\left|\tfrac{\partial}{\partial\lambda}\, \varphi^\lambda(0,\tau)\right|$, the differentiation of equation (2.9) yields for all $x \in \mathbb{R}_0^+$, $\tau \in [0,t]$

$$\begin{aligned}\left|\tfrac{\partial}{\partial\lambda}\varphi^\lambda(x,\tau)\right| \;\le\; & C\cdot \max\{\|\psi\|_\infty, \|\partial_x\psi\|_\infty, 1\} \cdot (t-\tau)\;\; e^{C\,(t-\tau)}\,. \\ & \times \left(\|a-\widetilde{a}\|_\infty + \|b-\widetilde{b}\|_\infty + \|c-\widetilde{c}\|_\infty\right).\end{aligned}$$

with a constant $C = C(\|a\|_{W^{1,\infty}},\, \|\widetilde{a}\|_{W^{1,\infty}},\, \|b\|_{W^{1,\infty}},\, \|\widetilde{b}\|_{W^{1,\infty}},\, \|c\|_{W^{1,\infty}},\, \|\widetilde{c}\|_{W^{1,\infty}})$. □

Proof (of Proposition 65 on page 155). (i) It is a consequence of equation (2.7) in Proposition 63 (on page 154).

(ii) It results from equation (2.8) in Proposition 63 (i), which can be written in the form

$$\int_{\mathbb{R}_0^+} \psi(x)\ d\mu_{t+h}(x) = \int_{\mathbb{R}_0^+} \varphi_{t+h,\psi}(x,t)\ d\mu_t(x) = \int_{\mathbb{R}_0^+} \varphi_{h,\psi}(x,0)\ d\mu_t(x).$$

for every $\psi \in C^1(\mathbb{R}_0^+) \cap W^{1,\infty}(\mathbb{R}_0^+)$. In particular, $\varphi_{t+h,\psi}(\cdot,t) \equiv \varphi_{h,\psi}(\cdot,0)$ results from partial differential equation (2.4) characterizing $\varphi_{h,\psi}$ since all its coefficients are autonomous.

(iii) It has already been verified in Proposition 63 (ii).

(iv) It results directly from Proposition 63 (ii) and the definition of $\rho_{\mathscr{M}}(\cdot,\cdot)$:

$$\begin{aligned}
&\rho_{\mathscr{M}}\big(\vartheta_{a,b,c}(t,\mu_0), \vartheta_{a,b,c}(t+h,\mu_0)\big)\\
&\quad= \sup\Big\{\int_{\mathbb{R}_0^+} \psi\ d\big(\vartheta_{a,b,c}(t+h,\mu_0) - \vartheta_{a,b,c}(t,\mu_0)\big)\ \Big|\\
&\qquad\qquad \psi \in C^1(\mathbb{R}_0^+),\ \|\psi\|_\infty \le 1,\ \|\partial_x\psi\|_\infty \le 1\Big\}\\
&\quad\le \operatorname{const}\big(\|a\|_{W^{1,\infty}},\ \|b\|_\infty,\ \|c\|_{W^{1,\infty}}\big)\cdot|\mu_0|(\mathbb{R}_0^+)\cdot h.
\end{aligned}$$

(v) Choose any $\psi \in C^1(\mathbb{R}_0^+)$ with $\|\psi\|_\infty \le 1$ and $\|\partial_x\psi\|_\infty \le 1$. Employing the notation of Proposition 63, we obtain

$$\int_{\mathbb{R}_0^+} \psi\ d\Big(\vartheta_{a,b,c}(h,\mu_0) - \vartheta_{a,b,c}(h,\nu_0)\Big) = \int_{\mathbb{R}_0^+} \varphi_{h,\psi}(x,0)\ d(\mu_0-\nu_0)(x),$$

and, due to Lemma 62 (iii), $x \longmapsto \varphi_{h,\psi}(x,t)$ is continuously differentiable with

$$\begin{aligned}
\|\varphi_{h,\psi}(\cdot,t)\|_\infty &\le e^{2(\|a\|_\infty+\|c\|_\infty)\,h},\\
\|\partial_x\,\varphi_{h,\psi}(\cdot,t)\|_\infty &\le e^{3(\|a\|_{W^{1,\infty}}+\|\partial_x b\|_\infty+\|c\|_{W^{1,\infty}})\,h}.
\end{aligned}$$

Now Proposition 43 (i) about the $W^{1,\infty}$ dual metric $\rho_{\mathscr{M}}(\cdot,\cdot)$ (on page 134) implies

$$\begin{aligned}
&\int_{\mathbb{R}_0^+} \varphi_{h,\psi}(\cdot,0)\ d(\mu_0-\nu_0)\\
&\quad\le \rho_{\mathscr{M}}(\mu_0,\nu_0)\ \max\big\{e^{2(\|a\|_\infty+\|c\|_\infty)\,h},\ e^{3(\|a\|_{W^{1,\infty}}+\|\partial_x b\|_\infty+\|c\|_{W^{1,\infty}})\,h}\big\}\\
&\quad\le \rho_{\mathscr{M}}(\mu_0,\nu_0)\ e^{3(\|a\|_{W^{1,\infty}}+\|\partial_x b\|_\infty+\|c\|_{W^{1,\infty}})\,h}
\end{aligned}$$

and thus,

$$\rho_{\mathscr{M}}\big(\vartheta_{a,b,c}(h,\mu_0),\ \vartheta_{a,b,c}(h,\nu_0)\big) \le \rho_{\mathscr{M}}(\mu_0,\nu_0)\cdot e^{3(\|a\|_{W^{1,\infty}}+\|\partial_x b\|_\infty+\|c\|_{W^{1,\infty}})\,h}.$$

(vi) It is based on the estimate in Lemma 68 (on page 161) and therefore it uses notation $\varphi^\lambda(\cdot,\cdot)$ for some arbitrary $\psi \in C^1(\mathbb{R}_0^+)$ with $\|\psi\|_\infty \le 1$, $\|\partial_x\psi\|_\infty \le 1$ (see equation (2.9)). Indeed, Proposition 63 (on page 154) implies that for every $\mu_0 \in \mathscr{M}(\mathbb{R}_0^+)$ and $t \in [0,1]$

$$\begin{aligned}\int_{\mathbb{R}_0^+} \psi \, d\Big(\vartheta_{a,b,c}(t,\mu_0) - \vartheta_{\widetilde{a},\widetilde{b},\widetilde{c}}(t,\mu_0)\Big) &= \int_{\mathbb{R}_0^+} \big(\varphi^1(x,0) - \varphi^0(x,0)\big)\, d\mu_0(x) \\ &= \int_{\mathbb{R}_0^+} \int_0^1 \tfrac{\partial}{\partial\lambda}\varphi^\lambda(x,0)\; d\lambda \quad d\mu_0(x).\end{aligned}$$

Lemma 68 guarantees that for every $x \in \mathbb{R}_0^+$

$$\Big|\tfrac{\partial}{\partial\lambda}\,\varphi^\lambda(x,0)\Big| \;\le\; C\cdot t\; e^{Ct}\cdot\big(\|a-\widetilde{a}\|_\infty + \|b-\widetilde{b}\|_\infty + \|c-\widetilde{c}\|_\infty\big),$$

with a constant $C = C(\|a\|_{W^{1,\infty}},\, \|\widetilde{a}\|_{W^{1,\infty}},\, \|b\|_{W^{1,\infty}},\, \|\widetilde{b}\|_{W^{1,\infty}},\, \|c\|_{W^{1,\infty}},\, \|\widetilde{c}\|_{W^{1,\infty}})$.
Now we obtain uniformly for all $\psi \in C^1(\mathbb{R}_0^+)$ with $\|\psi\|_\infty \le 1,\ \|\partial_x\psi\|_\infty \le 1$

$$\begin{aligned}\int_{\mathbb{R}_0^+} \psi\; d\big(\vartheta_{a,b,c}(t,\mu_0) - \vartheta_{\widetilde{a},\widetilde{b},\widetilde{c}}(t,\mu_0)\big) \;\le\; & C\cdot t\; e^{Ct}\cdot|\mu_0|(\mathbb{R}_0^+)\,\cdot \\ & \big(\|a-\widetilde{a}\|_\infty + \|b-\widetilde{b}\|_\infty + \|c-\widetilde{c}\|_\infty\big).\end{aligned}$$

(vii) If additionally $a(\cdot) \ge 0$, then nonnegative initial measures lead to solutions with nonnegative values in $\mathscr{M}(\mathbb{R}_0^+)$ according to Corollary 64 (on page 155). □

Proof (of Corollary 66 on page 156).
The solution map $\vartheta_{a,b,c} : [0,1]\times\mathscr{M}(\mathbb{R}_0^+) \longrightarrow \mathscr{M}(\mathbb{R}_0^+)$ is continuous with respect to the coefficients $\big(a(\cdot), b(\cdot), c(\cdot)\big)$. In particular, Proposition 65 (vi) (on page 155) indicates that the distance between two solutions to the problem with the same initial data but a different coefficient $b(\cdot)$ can be estimated by the L^∞ norm of the difference in the values of b.
Therefore, we can extend our obtained results to the problems with coefficients $b(\cdot) \in W^{1,\infty}(\mathbb{R}_0^+)\setminus C^1(\mathbb{R}_0^+)$. Indeed, $C^1(\mathbb{R}_0^+)\cap W^{1,\infty}(\mathbb{R}_0^+)$ is dense in $W^{1,\infty}(\mathbb{R}_0^+)$ with respect to the L^∞ norm and thus, any $b(\cdot)\in W^{1,\infty}(\mathbb{R}_0^+)$ can be approximated by a sequence $\big(b^n(\cdot)\big)_{n\in\mathbb{N}}$ in $C^1(\mathbb{R}_0^+)\cap W^{1,\infty}(\mathbb{R}_0^+)$ converging to $b(\cdot)$ in $L^\infty(\mathbb{R}_0^+)$.
According to Proposition 43 (3.) (on page 134), the subset of Radon measures $\big\{\mu\in\mathscr{M}(\mathbb{R}_0^+)\,\big|\;|\mu|(\mathbb{R}_0^+)\le r\big\}$ (with arbitrary $r>0$) is complete with respect to the $W^{1,\infty}$ dual metric $\rho_{\mathscr{M}}$ and, the sequence of solutions $\vartheta_{a,b^n,c}(t,\mu_0)$, $n\in\mathbb{N}$, has uniformly bounded variation due to Proposition 65 (iii) (on page 155). The Cauchy sequence $\big(\vartheta_{a,b^n,c}(t,\mu_0)\big)_{n\in\mathbb{N}}$ has a limit $\vartheta_{a,b,c}(t,\mu_0)\in\mathscr{M}(\mathbb{R}_0^+)$.
As a consequence, we can extend Proposition 65 to coefficients $b(\cdot)\in W^{1,\infty}(\mathbb{R}_0^+)$ with $b(0)>0$. □

2.6.3 Conclusions About the Full Nonlinear Population Model

As main result of § 2.6.2, the linear population model (2.3) provides transitions $\vartheta_{a,b,c}(\cdot,\cdot)$ on the tuple $\big(\mathscr{M}(\mathbb{R}_0^+),\ \rho_{\mathscr{M}},\ |\cdot|(\mathbb{R}_0^+)\big)$ and, Corollary 67 (on page 156) specifies the underlying parameters of continuity.

Now we pass to the nonlinear problem

$$\begin{cases} \partial_t \mu_t \;+\; \partial_x (F_2(\mu_t,t)\,\mu_t) \;=\; F_3(\mu_t,t)\,\mu_t & \text{in } \mathbb{R}_0^+ \times [0,T] \\ F_2(\mu_t,t)(0)\,\mu_t(0) \;=\; \int_{\mathbb{R}_0^+} F_1(\mu_t,t)(x)\,d\mu_t(x) & \text{in }]0,T] \\ \mu_0 \;=\; \nu_0 & \end{cases} \tag{2.10}$$

with $F : \mathscr{M}(\mathbb{R}_0^+) \times [0,T] \longrightarrow \{(a,b,c) \in W^{1,\infty}(\mathbb{R}_0^+)^3 \mid b(0) > 0\}$ and $\nu_0 \in \mathscr{M}(\mathbb{R}_0^+)$ given.
Due to Definition 58 (on page 152), $\mu : [0,T] \longrightarrow \mathscr{M}(\mathbb{R}_0^+),\ t \longmapsto \mu_t$ is regarded as a weak solution to this nonlinear problem (2.10) if it is narrowly continuous and satisfies for every test function $\varphi \in C^1(\mathbb{R}_0^+ \times [0,T]) \cap W^{1,\infty}(\mathbb{R}_0^+ \times [0,T])$

$$\begin{aligned} &\int_{\mathbb{R}_0^+} \varphi(x,T)\,d\mu_T(x) - \int_{\mathbb{R}_0^+} \varphi(x,0)\,d\nu_0(x) \\ &= \int_0^T \int_{\mathbb{R}_0^+} \Big(\partial_t \varphi(x,t) + \partial_x \varphi(x,t) \cdot F_2(\mu_t,t)(x) + \varphi(x,t) \cdot F_3(\mu_t,t)(x)\Big)\,d\mu_t(x)\,dt \\ &\quad + \int_0^T \varphi(0,t) \cdot \int_{\mathbb{R}_0^+} F_1(\mu_t,t)(x)\,d\mu_t(x)\,dt. \end{aligned}$$

Mutational equations (presented in § 2.3) serve as tools for proving existence, stability and uniqueness of weak measure-valued solutions to problem (2.10). In particular, we have to focus again on the relationship between solutions to the mutational equation in $\big(\mathscr{M}(\mathbb{R}_0^+),\ \rho_{\mathscr{M}},\ |\cdot|(\mathbb{R}_0^+)\big)$ and weak solutions to the nonlinear problem (2.10) (in the sense of distributions).
Let us formulate the main results of this section before giving all proofs in detail:

Lemma 69. *The tuple* $\big(\mathscr{M}(\mathbb{R}_0^+),\ \rho_{\mathscr{M}},\ |\cdot|(\mathbb{R}_0^+),\ \Theta\big(\mathscr{M}(\mathbb{R}_0^+),\ \rho_{\mathscr{M}},\ |\cdot|(\mathbb{R}_0^+)\big)\big)$ *is Euler compact in the sense of Definition 15 (on page 112) :*

For any initial measure $\mu_0 \in \mathscr{M}(\mathbb{R}_0^+)$, *time* $T \in]0,\infty[$ *and bound* $M > 0$, *let* $\mathscr{N} = \mathscr{N}(\mu_0,T,M)$ *denote the set of all measure-valued functions* $\mu : [0,T] \longrightarrow \mathscr{M}(\mathbb{R}_0^+)$ *constructed in the following piecewise way: For any finite equidistant partition* $0 = t_0 < t_1 < \ldots < t_n = T$ *of* $[0,T]$ *and* n *tuples* $\{(a_j^n, b_j^n, c_j^n)\}_{j=1}^n \subset W^{1,\infty}(\mathbb{R}_0^+)^3$ *with* $b_j^n(0) > 0,\quad \|a_j^n\|_{W^{1,\infty}} + \|b_j^n\|_{W^{1,\infty}} + \|c_j^n\|_{W^{1,\infty}} \le M$ *for each* $j = 1 \ldots n$ *define* $\mu :]0,T] \longrightarrow \mathscr{M}(\mathbb{R}_0^+),\ t \longmapsto \mu_t$ *by*

$$\mu_t := \vartheta_{a_j^n, b_j^n, c_j^n}\big(t - t_{j-1},\ \mu_{t_{j-1}}\big) \qquad \text{for } t \in]t_{j-1}, t_j],\ j = 1 \ldots n.$$

Then for each $t \in [0,T]$, *the union of all images* $\{\mu_t \mid \mu \in \mathscr{N}\} \subset \mathscr{M}(\mathbb{R}_0^+)$ *is tight and relatively compact in the metric space* $(\mathscr{M}(\mathbb{R}_0^+), \rho_{\mathscr{M}})$.

Proposition 70 (Solutions to the underlying mutational equation).
Suppose that $\mathbf{F} : \mathscr{M}(\mathbb{R}_0^+) \times [0,T] \longrightarrow \{(a,b,c) \in W^{1,\infty}(\mathbb{R}_0^+)^3 \mid b(0) > 0\}$ *satisfies*

(i) $\sup_{t \in [0,T]} \sup_{\nu \in \mathscr{M}(\mathbb{R}_0^+)} \|\mathbf{F}(\nu,t)\|_{W^{1,\infty}} < \infty.$

(ii) $\mathbf{F} : (\mathscr{M}(\mathbb{R}_0^+), \rho_{\mathscr{M}}) \times [0,T] \longrightarrow \big(W^{1,\infty}(\mathbb{R}_0^+)^3, \|\cdot\|_\infty\big)$ *is continuous.*

Then, for any initial Radon measure $\nu_0 \in \mathscr{M}(\mathbb{R}_0^+)$, *there exists a solution* $\mu : [0,T] \longrightarrow \mathscr{M}(\mathbb{R}_0^+)$, $t \longmapsto \mu_t$ *to the mutational equation*

$$\overset{\circ}{\mu}_t \ni \vartheta_{\mathbf{F}(\mu_t,t)}$$

in $\big(\mathscr{M}(\mathbb{R}_0^+),\ \rho_{\mathscr{M}},\ |\cdot|(\mathbb{R}_0^+)\big)$ *with* $\mu_0 = \nu_0$ *and tight values in* $\mathscr{M}(\mathbb{R}_0^+)$, *i.e.*

(*a*) $\mu(\cdot)$ *is continuous with respect to* $\rho_{\mathscr{M}}$,

(*b*) $\lim_{h\downarrow 0} \frac{1}{h} \cdot \rho_{\mathscr{M}}\big(\vartheta_{F_1(\mu_t,t),\,F_2(\mu_t,t),\,F_3(\mu_t,t)}(h,\mu_t),\ \mu_{t+h}\big) = 0$ *for* $\mathscr{L}^1$*-a.e.* $t \in [0,T[$,

(*c*) $\sup_{0\le t<T} |\mu_t|(\mathbb{R}_0^+) < \infty$.

If, in addition, $\nu_0 \in \mathscr{M}^+(\mathbb{R}_0^+)$ *and* $F_1(\nu,t)(\cdot) \ge 0$ *for every* $\nu \in \mathscr{M}^+(\mathbb{R}_0^+)$, $t \in [0,T]$, *then this solution* $\mu(\cdot)$ *has values in* $\mathscr{M}^+(\mathbb{R}_0^+)$.

Furthermore every solution $\mu : [0,T] \longrightarrow \mathscr{M}(\mathbb{R}_0^+)$, $t \longmapsto \mu_t$ *to this mutational equation with tight values in* $\mathscr{M}(\mathbb{R}_0^+)$ *is a narrowly continuous weak solution to nonlinear population model (2.10).*

The continuity conditions on $\mathbf{F} : \mathscr{M}(\mathbb{R}_0^+) \times [0,T] \longrightarrow W^{1,\infty}(\mathbb{R}_0^+)^3$ can be formulated for the narrow topology on $\mathscr{M}(\mathbb{R}_0^+)$ and, we obtain Theorem 56 (on page 149) as a corollary:

Corollary 71 (Existence of solutions to nonlinear structured population model). *Suppose that* $\mathbf{F} : \mathscr{M}(\mathbb{R}_0^+) \times [0,T] \longrightarrow \big\{(a,b,c) \in W^{1,\infty}(\mathbb{R}_0^+)^3 \mid b(0) > 0\big\}$ *satisfies*

(*i*) $\sup_{t\in[0,T]} \sup_{\nu\in\mathscr{M}(\mathbb{R}_0^+)} \|\mathbf{F}(\nu,t)\|_{W^{1,\infty}} < \infty$.

(*ii*) $\mathbf{F} : (\mathscr{M}(\mathbb{R}_0^+), \textit{narrow}) \times [0,T] \longrightarrow \big(W^{1,\infty}(\mathbb{R}_0^+)^3, \|\cdot\|_\infty\big)$ *is continuous.*

Then, for any initial measure $\nu_0 \in \mathscr{M}(\mathbb{R}_0^+)$, *there exists a narrowly continuous weak solution* $\mu : [0,T] \longrightarrow \mathscr{M}(\mathbb{R}_0^+)$ *to the nonlinear population model (2.10) with* $\mu(0) = \nu_0$.
If, in addition, $\nu_0 \in \mathscr{M}^+(\mathbb{R}_0^+)$ *and* $F_1(\nu,t)(\cdot) \ge 0$ *for every* $\nu \in \mathscr{M}^+(\mathbb{R}_0^+)$, $t \in [0,T]$, *then the solution* $\mu(\cdot)$ *has values in* $\mathscr{M}^+(\mathbb{R}_0^+)$.

Lipschitz continuity of the coefficient function $\mathbf{F}$ with respect to state measures implies the opposite inclusion, i.e. every weak solution to population model (2.10) is also a solution to the corresponding mutational equation.

Proposition 72 (Weak solutions solve the mutational equation). *Suppose that* $\mathbf{F} : \mathscr{M}(\mathbb{R}_0^+) \times [0,T] \longrightarrow \big\{(a,b,c) \in W^{1,\infty}(\mathbb{R}_0^+)^3 \mid b(0) > 0\big\}$ *satisfies*

(*i*) $\sup_{t\in[0,T]} \sup_{\nu\in\mathscr{M}(\mathbb{R}_0^+)} \|\mathbf{F}(\nu,t)\|_{W^{1,\infty}} < \infty$.

(*ii*) $\mathbf{F} : (\mathscr{M}(\mathbb{R}_0^+), \rho_{\mathscr{M}}) \times [0,T] \longrightarrow \big(W^{1,\infty}(\mathbb{R}_0^+)^3, \|\cdot\|_\infty\big)$ *is Lipschitz continuous.*

Then every narrowly continuous weak solution $\mu : [0,T] \longrightarrow \mathscr{M}(\mathbb{R}_0^+)$, $t \longmapsto \mu_t$ *to the nonlinear population model (2.10) with tight values and* $\sup_t |\mu_t|(\mathbb{R}_0^+) < \infty$ *is a solution to the mutational equation* $\overset{\circ}{\mu}_t \ni \vartheta_{\mathbf{F}(\mu_t,t)}$ *in* $\big(\mathscr{M}(\mathbb{R}_0^+),\ \rho_{\mathscr{M}},\ |\cdot|(\mathbb{R}_0^+)\big)$.

We conclude uniqueness of weak solutions and their continuous dependence on data directly from the more general Proposition 11 (on page 108) and Gronwall's inequality (Proposition A.2 on page 440). As a consequence, we obtain the estimate stated already in Theorem 57 (on page 149):

Proposition 73 (Lipschitz continuous dependence of weak solutions on data). *Assume that for* $\mathbf{F},\mathbf{G} : \mathcal{M}(\mathbb{R}_0^+) \times [0,T] \longrightarrow \{(a,b,c) \in W^{1,\infty}(\mathbb{R}_0^+)^3 \mid b(0) > 0\}$,

(i) $M_F := \sup_{t\in[0,T]} \sup_{\mu\in\mathcal{M}(\mathbb{R}_0^+)} \|\mathbf{F}(\mu,t)\|_{W^{1,\infty}} < \infty$,
$M_G := \sup_{t\in[0,T]} \sup_{\mu\in\mathcal{M}(\mathbb{R}_0^+)} \|\mathbf{G}(\mu,t)\|_{W^{1,\infty}} < \infty$,

(ii) *for any* $R > 0$, *there are a constant* $L_R > 0$ *and a modulus of continuity* $\omega_R(\cdot)$ *with* $\|\mathbf{F}(\mu,s) - \mathbf{F}(\nu,t)\|_\infty \le L_R \cdot \rho_{\mathcal{M}}(\mu,\nu) + \omega_R(|t-s|)$ *for all* $\mu,\nu \in \mathcal{M}(\mathbb{R}_0^+)$ *with* $|\mu|(\mathbb{R}_0^+), |\nu|(\mathbb{R}_0^+) \le R$.

(iii) $\mathbf{G} : (\mathcal{M}(\mathbb{R}_0^+), \rho_{\mathcal{M}}) \times [0,T] \longrightarrow (W^{1,\infty}(\mathbb{R}_0^+)^3, \|\cdot\|_\infty)$ *is continuous.*

Let $\mu,\ \nu : [0,T] \longrightarrow \mathcal{M}(\mathbb{R}_0^+)$ *denote* $\rho_{\mathcal{M}}$*-continuous distributional solutions to the nonlinear population model (2.10) for the coefficients* $\mathbf{F}(\cdot), \mathbf{G}(\cdot)$ *respectively such that* $\sup_t |\mu_t|(\mathbb{R}_0^+) < \infty$, $\sup_t |\nu_t|(\mathbb{R}_0^+) < \infty$ *and all their values are tight in* $\mathcal{M}(\mathbb{R}_0^+)$.

Then there is $C = C(M_F, M_G, |\mu_0|(\mathbb{R}_0^+), |\nu_0|(\mathbb{R}_0^+)) \in [0,\infty[$ *such that for all* $t \in [0,T]$,

$$\rho_{\mathcal{M}}(\mu_t,\nu_t) \le \left(\rho_{\mathcal{M}}(\mu_0,\nu_0) + C\,t \cdot \sup \|\mathbf{F}(\cdot,\cdot) - \mathbf{G}(\cdot,\cdot)\|_\infty\right) e^{C\,t}.$$

Remark 74. Furthermore, Lemma 69 and Proposition 70 lay the foundations for applying the mutational tools to a nonlinear population model *with delay*:

$$\begin{cases} \partial_t \mu_t + \partial_x \big(G_2(\mu|_{[t-\tau,t]}, t)\,\mu_t\big) = G_3(\mu|_{[t-\tau,t]}, t)\,\mu_t & \text{in } \mathbb{R}_0^+ \times [0,T] \\ G_2(\mu|_{[t-\tau,t]}, t)(0)\ \ \mu_t(0) = \int_{\mathbb{R}_0^+} G_1((\mu|_{[t-\tau,t]}, t)(x)\, d\mu_t(x) & \text{in }]0,T] \\ \mu|_{[-\tau,0]} = \nu_0 \end{cases}$$

with given initial data $\nu_0 \in \mathrm{BLip}\big([-\tau,0],\ \mathcal{M}(\mathbb{R}_0^+),\ \rho_{\mathcal{M}},\ |\cdot|(\mathbb{R}_0^+)\big)$ and

$$\mathbf{G} : \mathrm{BLip}\big([-\tau,0],\ \mathcal{M}(\mathbb{R}_0^+),\ \rho_{\mathcal{M}},\ |\cdot|(\mathbb{R}_0^+)\big) \times [0,T] \longrightarrow \big\{(a,b,c) \in W^{1,\infty}(\mathbb{R}_0^+)^3 \mid b(0) > 0\big\}$$

for a fixed time interval $[-\tau,0] \neq \emptyset$ (BLip is introduced in Definition 22 on page 120). Indeed, $\rho_{\mathcal{M}}$-continuous weak solutions are guaranteed by Proposition 23.

The Proofs About the Nonlinear Population Model

Proof (of Lemma 69 on page 165). Every subset of $\mathcal{M}(\mathbb{R}_0^+)$ with exactly one Radon measure is tight, of course. Therefore, Remark 41 (3.) (on page 134) provides a nondecreasing continuous function $\Psi_0 : \mathbb{R}_0^+ \longrightarrow \mathbb{R}_0^+$ with $\lim_{x\to\infty} \Psi_0(x) = \infty$ such that

$$\int_{\mathbb{R}_0^+} \Psi_0\ d|\mu_0| < \infty.$$

Setting $\bar{x} := M\,T \geq \sup_{j \in \{1...n\}} \|b_j^n\|_\infty\, T$, let us define $\psi_T : \mathbb{R}_0^+ \longrightarrow \mathbb{R}$ as

$$\psi_T(x) := \begin{cases} 0 & \text{for } x \leq \bar{x}, \\ \Psi_0(x - \bar{x}) & \text{for } x > \bar{x}. \end{cases}$$

Obviously, ψ_T is continuous, nondecreasing and thus nonnegative, but unbounded. Considering any measure-valued function $\mu(\cdot) \in \mathscr{N}$, Proposition 63 (iv) implies a uniform integral bound for any function $\phi_T \in C^0(\mathbb{R}_0^+)$ satisfying $|\phi_T| \leq \psi_T$ and for each time $t \in [0,T]$:

$$\int_{\mathbb{R}_0^+} \phi_T \; d\mu_t \;\leq\; e^{\|c\|_\infty T} \int_{\mathbb{R}_0^+} \psi_T(\cdot + \bar{x}) \; d|\mu_0| \;\leq\; e^{\|c\|_\infty T} \int_{\mathbb{R}_0^+} \Psi_0 \; d|\mu_0| \;<\; \infty$$

and thus

$$\int_{\mathbb{R}_0^+} \psi_T \; d|\mu_t| \;\leq\; e^{\|c\|_\infty T} \int_{\mathbb{R}_0^+} \Psi_0 \; d|\mu_0| \;<\; \infty.$$

Therefore, the set of all values $\{\mu(t) \,|\, \mu \in \mathscr{N}, t \in [0,T]\} \subset \mathscr{M}(\mathbb{R}_0^+)$ is tight due to Remark 41 (3.) (on page 134).
Furthermore, all total variations $|\mu_t|(\mathbb{R}_0^+)$ are uniformly bounded, i.e.

$$\sup_{\substack{\mu \in \mathscr{N} \\ t \in [0,T]}} |\mu_t|(\mathbb{R}_0^+) < \infty$$

as a consequence of Proposition 65 (iii), Corollary 66 and the piecewise construction of each $\mu(\cdot) \in \mathscr{N}$. Finally the assertion about compactness follows from Proposition 43 (4.) (on page 134). □

Proof (of Proposition 70 on page 165).
Peano's Theorem 18 (on page 114) guarantees the existence of a $\rho_{\mathscr{M}}$-continuous solution $\mu : [0,T] \longrightarrow \mathscr{M}(\mathbb{R}_0^+),\; t \longmapsto \mu_t$ to the mutational equation

$$\overset{\circ}{\mu}_t \ni \vartheta_{\mathbf{F}(\mu_t,t)}$$

with $\mu_0 = \nu_0$. Its proof by means of Euler method reveals that the set of all its values $\{\mu_t \mid t \in [0,T]\} \subset \mathscr{M}(\mathbb{R}_0^+)$ is tight – as a consequence of Lemma 69.
Suppose in addition that $F_1(\nu,t) \in W^{1,\infty}(\mathbb{R}_0^+)$ is nonnegative for any $\nu \in \mathscr{M}^+(\mathbb{R}_0^+)$, $t \in [0,T]$. Then the piecewise Euler approximations used in Peano's Theorem 18 have nonnegative values due to Corollary 64 (on page 155). As $\mathscr{M}^+(\mathbb{R}_0^+)$ is closed in $(\mathscr{M}(\mathbb{R}_0^+), \rho_{\mathscr{M}})$, all values of the resulting solution μ are also in $\mathscr{M}^+(\mathbb{R}_0^+)$.

For the last step, let $\mu : [0,T] \longrightarrow \mathscr{M}(\mathbb{R}_0^+),\; t \longmapsto \mu_t$ denote any solution to the mutational equation

$$\overset{\circ}{\mu}_t \ni \vartheta_{\mathbf{F}(\mu_t,t)}$$

with tight image in $\mathscr{M}(\mathbb{R}_0^+)$. Then $\mu : [0,T] \longrightarrow \mathscr{M}(\mathbb{R}_0^+)$ is narrowly continuous due to Proposition 43 (2.) (on page 134).
We have to verify that μ is a distributional solution to the nonlinear model (2.10). Similarly to the proof for the linear model in § 2.6.2 (Proposition 63 (iii) on page 154), we first choose an arbitrary test function $\psi \in C_c^\infty(\mathbb{R}_0^+)$. Then,

$$\Psi : \; [0,T] \longrightarrow \mathbb{R}, \quad t \longmapsto \int_{\mathbb{R}_0^+} \psi(x) \; d\mu_t(x)$$

is continuous because Proposition 43 (1.) (on page 134) implies

$$\left| \int_{\mathbb{R}_0^+} \psi \, d\mu_t - \int_{\mathbb{R}_0^+} \psi \, d\mu_s \right| \le \max\left\{1, \|\psi\|_\infty, \|\partial_x \psi\|_\infty\right\} \cdot \rho_{\mathscr{M}}(\mu_t, \mu_s).$$

The solution $\mu(\cdot)$ is even Lipschitz continuous with respect to the $W^{1,\infty}$ dual metric $\rho_{\mathscr{M}}$ due to Lemma 10 (on page 107) and thus, Ψ is Lipschitz continuous.
At $\mathscr{L}^1$-almost every time $t \in [0,T[$, the derivative of Ψ is

$$\Psi'(t) = \lim_{h \downarrow 0} \tfrac{1}{h} \cdot \int_{\mathbb{R}_0^+} \psi \; d\left(\vartheta_{F(\mu_t,t)}(h,\mu_t) - \mu_t\right)$$

because Proposition 63 (iii) (on page 154) ensures

$$\begin{aligned}
&\left| \int_{\mathbb{R}_0^+} \psi \; d\mu_{t+h} - \int_{\mathbb{R}_0^+} \psi \; d\mu_t - \int_{\mathbb{R}_0^+} \psi \; d\left(\vartheta_{F(\mu_t,t)}(h,\mu_t) - \mu_t\right) \right| \\
&= \left| \int_{\mathbb{R}_0^+} \psi \; d\left(\mu_{t+h} - \vartheta_{F(\mu_t,t)}(h,\mu_t)\right) \right| \\
&\le \max\{1, \|\psi\|_\infty, \|\partial_x \psi\|_\infty\} \cdot \rho_{\mathscr{M}}\left(\mu_{t+h}, \; \vartheta_{F(\mu_t,t)}(h,\mu_t)\right) \\
&= o(h) \qquad \text{for } h \downarrow 0.
\end{aligned}$$

The special form of $\vartheta_{F(\mu_t,t)}(h,\mu_t)$ has the consequence

$$\begin{aligned}
\Psi'(t) = \lim_{h \downarrow 0} \tfrac{1}{h} \cdot \int_0^h \int_{\mathbb{R}_0^+} \Big(&\psi(0) \cdot F_1(\mu_t,t)(x) + \partial_x \psi(x) \cdot F_2(\mu_t,t)(x) \\
&+ \psi(x) \cdot F_3(\mu_t,t)(x) \Big) \; d\vartheta_{F(\mu_t,t)}(s,\mu_t)(x) \; ds
\end{aligned}$$

for $\mathscr{L}^1$-almost every $t \in [0,T[$.
Finally, this derivative proves to be an integral just with the Radon measure μ_t :

$$\Psi'(t) = \int_{\mathbb{R}_0^+} \Big(\psi(0) \cdot F_1(\mu_t,t)(x) + \partial_x \psi(x) \cdot F_2(\mu_t,t)(x) + \psi(x) \cdot F_3(\mu_t,t)(x) \Big) \, d\mu_t(x).$$

Indeed, using the abbreviation $M := \sup_{t \in [0,T]} \; \sup_{\nu \in \mathscr{M}(\mathbb{R}_0^+)} \|F(\nu,t)\|_{W^{1,\infty}} < \infty$, Proposition 43 (1.) (on page 134) and Proposition 65 (iv) (on page 155) yield for any $s \in]0,1]$

$$\begin{aligned}
&\left| \int_{\mathbb{R}_0^+} \Big(\psi(0) \cdot F_1(\mu_t,t) + \partial_x \psi \cdot F_2(\mu_t,t) + \psi \cdot F_3(\mu_t,t) \Big) \; d\left(\vartheta_{F(\mu_t,t)}(s,\mu_t) - \mu_t\right) \right| \\
&\le \text{const}(M, \|\psi\|_{W^{1,\infty}}) \cdot \rho_{\mathscr{M}}\left(\vartheta_{F(\mu_t,t)}(s,\mu_t), \; \mu_t\right) \\
&\le \text{const}(M, \|\psi\|_{W^{1,\infty}}) \cdot \text{const}(M, \sup_\tau |\mu_\tau|(\mathbb{R}_0^+)) \cdot s.
\end{aligned}$$

The last representation of $\Psi'(t)$ at $\mathscr{L}^1$-almost every time $t \in [0,T]$ leads to

$$\begin{aligned}
&\int_{\mathbb{R}_0^+} \psi \, d\mu_t - \int_{\mathbb{R}_0^+} \psi \, d\nu_0 \\
&= \int_0^t \int_{\mathbb{R}_0^+} \Big(\psi(0) \cdot F_1(\mu_t,t) + \partial_x \psi \cdot F_2(\mu_t,t) + \psi \cdot F_3(\mu_t,t) \Big) \; d\mu_s \; ds
\end{aligned}$$

for every $t \in [0,T]$ and $\psi \in C_c^\infty(\mathbb{R}_0^+)$. The more general interpretation of nonlinear equation (2.10) using *nonautonomous* test functions $\varphi \in C^1(\mathbb{R}_0^+ \times [0,T]) \cap W^{1,\infty}(\mathbb{R}_0^+ \times [0,T])$ results from the chain rule and the continuity with respect to the $W^{1,\infty}$ norm in exactly the same way as for Proposition 63 (iii) (on page 159 f.). □

Proof (of Corollary 71 on page 166). Set $M := \sup_{t\in[0,T]} \sup_{\nu\in\mathscr{M}(\mathbb{R}_0^+)} \|\mathbf{F}(\nu,t)\|_{W^{1,\infty}} < \infty$ as an abbreviation and, consider the subset $\mathscr{N}(\nu_0,T,M)$ of all Euler approximations $[0,T] \longrightarrow \mathscr{M}(\mathbb{R}_0^+)$ as specified in Lemma 69 (on page 165). In fact, the proof of Lemma 69 (on page 167 f.) reveals that the subset

$$\mathscr{N}_{[0,T]} := \{\mu_t \mid t \in [0,T],\ \mu(\cdot) \in \mathscr{N}(\nu_0,T,M)\} \subset \mathscr{M}(\mathbb{R}_0^+)$$

is tight and has uniformly bounded total variations. Hence, narrow convergence and the $W^{1,\infty}$ dual metric $\rho_{\mathscr{M}}$ induce the same topology on $\mathscr{N}_{[0,T]} \subset \mathscr{M}(\mathbb{R}_0^+)$ and, $\mathscr{N}_{[0,T]}$ is relatively compact according to Proposition 43 (on page 134).
Let $\overline{\mathscr{N}_{[0,T]}} \subset \mathscr{M}(\mathbb{R}_0^+)$ denote the closure of $\mathscr{N}_{[0,T]}$ with respect to $\rho_{\mathscr{M}}$. In particular, $\overline{\mathscr{N}_{[0,T]}}$ supplied with the narrow topology is a compact topological space metrized by $\rho_{\mathscr{M}}$. Due to assumption (ii) of this Corollary 71, the restriction

$$\mathbf{F} : \left(\overline{\mathscr{N}_{[0,T]}},\ \rho_{\mathscr{M}}\right) \times [0,T] \longrightarrow \left(W^{1,\infty}(\mathbb{R}_0^+)^3,\ \|\cdot\|_\infty\right)$$

is continuous and, all corresponding transitions on $\left(\overline{\mathscr{N}_{[0,T]}},\ \rho_{\mathscr{M}},\ |\cdot|(\mathbb{R}_0^+)\right)$ have their values in $\overline{\mathscr{N}_{[0,T]}}$. This lays the basis for continuing with the same conclusions as in Proposition 70 (on page 165). □

Proof (of Proposition 72 on page 166). Suppose that

$$\mathbf{F} : \left(\mathscr{M}(\mathbb{R}_0^+), \rho_{\mathscr{M}}\right) \times [0,T] \longrightarrow \left(W^{1,\infty}(\mathbb{R}_0^+)^3,\ \|\cdot\|_\infty\right)$$

is Lipschitz continuous and bounded. Let $\mu : [0,T] \longrightarrow \mathscr{M}(\mathbb{R}_0^+),\ t \longmapsto \mu_t$ denote a narrowly continuous weak solution to the nonlinear population model (2.10) with tight values and $\sup_{t\in[0,T]} |\mu_t|(\mathbb{R}_0^+) < \infty$.
As a consequence of Proposition 43 (2.) (on page 134), $\mu(\cdot)$ is continuous with respect to $\rho_{\mathscr{M}}$. Now we still have to verify for $\mathscr{L}^1$-almost every $t \in [0,T]$

$$\lim_{h\downarrow 0} \tfrac{1}{h} \cdot \rho_{\mathscr{M}}\left(\vartheta_{\mathbf{F}(\mu_t,t)}(h,\mu_t),\ \mu_{t+h}\right) = 0.$$

Choosing any $\psi \in C^1(\mathbb{R}_0^+) \cap W^{1,\infty}(\mathbb{R}_0^+)$ with $\|\psi\|_\infty \le 1,\ \|\partial_x\psi\|_\infty \le 1$, we conclude from the definition of weak solution and Proposition 63 (on page 154 f.) respectively

$$\begin{aligned}
&\left| \int_{\mathbb{R}_0^+} \psi\ d\left(\vartheta_{\mathbf{F}(\mu_t,t)}(h,\mu_t) - \mu_{t+h}\right) \right| \\
&= \Big| \int_t^{t+h} \Big(\int_{\mathbb{R}_0^+} \left(\psi(0)\cdot F_1(\mu_t,t) + \partial_x\psi \cdot F_2(\mu_t,t) + \psi\cdot F_3(\mu_t,t)\right)\ d\mu_t \\
&\qquad - \int_{\mathbb{R}_0^+} \left(\psi(0)\cdot F_1(\mu_s,s) + \partial_x\psi\cdot F_2(\mu_s,s) + \psi\cdot F_3(\mu_s,s)\right)\ d\mu_s \Big)\ ds \Big| \\
&\le \Big| \int_t^{t+h} \int_{\mathbb{R}_0^+} \left(\psi(0)\cdot F_1(\mu_s,s) + \partial_x\psi\cdot F_2(\mu_s,s) + \psi\cdot F_3(\mu_s,s)\right)\ d(\mu_t - \mu_s)\,ds \Big| \\
&\quad + h \cdot \mathrm{const}\left(\|\psi\|_{W^{1,\infty}},\ \mathrm{Lip}\,\mathbf{F}\right) \cdot \left(h + \sup_{t\le s\le t+h} \rho_{\mathscr{M}}(\mu_s,\mu_t)\right) \cdot |\mu_t|(\mathbb{R}_0^+) \\
&\le h \cdot \mathrm{const}\left(\|\psi\|_{W^{1,\infty}},\ \sup \|\mathbf{F}(\cdot,\cdot)\|_\infty\right) \cdot \sup_{t\le s\le t+h} \rho_{\mathscr{M}}(\mu_s,\mu_t) \\
&\quad + h \cdot \mathrm{const}\left(\|\psi\|_{W^{1,\infty}},\ \mathrm{Lip}\,\mathbf{F}\right) \cdot \left(h + \sup_{t\le s\le t+h} \rho_{\mathscr{M}}(\mu_s,\mu_t)\right) \cdot |\mu_t|(\mathbb{R}_0^+) \\
&= o(h) \quad \text{for } h\downarrow 0 \text{ uniformly with respect to } \psi \text{ with } \|\psi\|_\infty \le 1,\ \|\nabla_x\psi\|_\infty \le 1.
\end{aligned}$$

□

2.7 Example: Modified Morphological Equations for Compact Sets via One-Sided Lipschitz Continuous Maps of Linear Growth

In comparison to Aubin's original suggestion in Chapter 1, the extensions of Chapter 2 lay the basis for a more general type of morphological equations.

Indeed, in § 1.9, we have applied the (original) mutational framework to nonempty compact subsets of the Euclidean space $\mathbb{R}^N$ supplied with the Pompeiu-Hausdorff distance $d\!l$ and, we have used reachable sets of differential inclusions as so-called morphological transitions. The set-valued maps in $\mathrm{LIP}(\mathbb{R}^N, \mathbb{R}^N)$ have served as appropriate right-hand side of these differential inclusions as specified in Proposition 1.53. According to Definition 1.49 (on page 60), a map $F \in \mathrm{LIP}(\mathbb{R}^N, \mathbb{R}^N)$ is characterized by the following two conditions:

1. F has nonempty compact values that are uniformly bounded in $\mathbb{R}^N$,
2. F is Lipschitz continuous with respect to the Pompeiu-Hausdorff distance $d\!l$.

Then, in § 1.11, the Lipschitz continuity has been weakened to one-sided Lipschitz continuity in combination with upper semicontinuity. Indeed, the set-valued maps $F : \mathbb{R}^N \rightsquigarrow \mathbb{R}^N$ in $\mathrm{OSLIP}(\mathbb{R}^N, \mathbb{R}^N)$ lead to differential inclusions whose reachable sets are transitions on $(\mathcal{K}(\mathbb{R}^N), d\!l)$ as specified in Proposition 1.97 (on page 98). According to Definition 1.94 (on page 96), every map $F \in \mathrm{OSLIP}(\mathbb{R}^N, \mathbb{R}^N)$ has to satisfy the following three conditions:

1. F has nonempty compact convex values that are uniformly bounded in $\mathbb{R}^N$,
2. F is upper semicontinuous,
3. F is one-sided Lipschitz continuous, i.e. there exists $L \in \mathbb{R}$ such that for every $x, y \in \mathbb{R}^N$, $v \in F(x)$, there is some $w \in F(y)$ with $\langle x - y,\, v - w \rangle \;\leq\; L\, |x - y|^2$.

The condition of uniformly bounded values is still a severe restriction though. In particular, the concept of Chapter 1 does not admit simple *linear* differential inclusions in $\mathbb{R}^N$ for transitions on $\mathcal{K}(\mathbb{R}^N)$. This obstacle is now overcome by means of a linear growth condition (instead of a uniform bound):

Definition 75. $\mathrm{LOSLIP}(\mathbb{R}^N, \mathbb{R}^N)$ consists of all set-valued maps $F : \mathbb{R}^N \rightsquigarrow \mathbb{R}^N$ satisfying the following four conditions:

1. F has nonempty compact convex values,
2. F is upper semicontinuous,
3. F is *locally one-sided Lipschitz continuous*, i.e. for each radius $r > 0$, there is a constant $L_r \in \mathbb{R}$ such that for every $x, y \in \mathbb{B}_r(0) \subset \mathbb{R}^N$ and $v \in F(x)$, there exists some $w \in F(y)$ satisfying
$$\langle x - y,\, v - w \rangle \;\leq\; L_r\, |x - y|^2.$$
The smallest constant $L_r \in \mathbb{R}$ with this property is abbreviated as $\mathrm{Lip}\, F|_{\mathbb{B}_r}$.
4. F has *linear growth*, i.e. there is a constant $c \geq 0$ satisfying for all $x \in \mathbb{R}^N$,
$$\sup_{v \in F(x)} |v| \;\leq\; c \cdot (1 + |x|).$$
The smallest constant $c \geq 0$ with this property is denoted by $\|F\|_{\mathrm{lg}}$.

Basic set	$E := \mathscr{K}(\mathbb{R}^N)$ the set of nonempty compact subsets of the Euclidean space $\mathbb{R}^N$
Distance Absolute value	Pompeiu-Hausdorff metric $d\!l : \mathscr{K}(\mathbb{R}^N) \times \mathscr{K}(\mathbb{R}^N) \longrightarrow \mathbb{R}$, $d\!l(K_1, K_2) := \max\left\{ \sup_{x \in K_1} \operatorname{dist}(x, K_2), \quad \sup_{y \in K_2} \operatorname{dist}(y, K_1) \right\}$ $\lfloor \cdot \rfloor := \lvert \cdot \rvert_\infty$ supremum of Euclidean norms of all elements
Transition	For each $F \in \mathrm{LOSLIP}(\mathbb{R}^N, \mathbb{R}^N)$, i.e. upper semicontinuous and locally one-sided Lipschitz continuous map $F : \mathbb{R}^N \rightsquigarrow \mathbb{R}^N$ with compact convex values and linear growth (Definition 75), define $\vartheta_F : [0,1] \times \mathscr{K}(\mathbb{R}^N) \longrightarrow \mathscr{K}(\mathbb{R}^N)$ by means of reachable sets of the autonomous differential inclusion $x'(\cdot) \in F(x(\cdot))$ a.e.: $\vartheta_F(t,\ K_0) := \big\{ x(t) \bigm\vert \text{there exists } x(\cdot) \in W^{1,1}([0,t], \mathbb{R}^N) : x'(\cdot) \in F(x(\cdot)) \ \mathscr{L}^1\text{-a.e. in } [0,t],\ x(0) \in K_0 \big\}.$
Compactness	Closed bounded balls in $(\mathscr{K}(\mathbb{R}^N), d\!l)$ are always compact: Proposition 1.47 (page 57)
Mutational solutions	Reachable sets of a nonautonomous differential inclusion whose set-valued right-hand side is determined via feedback – if the transitions are induced by additionally continuous maps, i.e. each $F \in \mathrm{CLOSLIP}(\mathbb{R}^N, \mathbb{R}^N)$ (Definition 86): Proposition 87, Corollary 88 (page 177 f.)
List of main results formulated in § 2.7	Existence due to compactness (Peano): Proposition 81 Cauchy-Lipschitz Theorem: Proposition 82 (page 175) Continuity w.r.t. data: Proposition 83 (page 175) Existence under state constraints (Nagumo): Proposition 84 Existence for equations with delay: Proposition 85 (page 176)
Key tools	Filippov-like Theorem A.63 of Donchev and Farkhi [69] about differential inclusions with one-sided Lipschitz continuous right-hand side (page 480) Integral funnel equation for reachable sets of nonautonomous differential inclusions: Proposition A.13 (page 447)

Table 2.4 Brief summary of the example in § 2.7 in mutational terms:
Modified morphological equations for compact sets via one-sided Lipschitz continuous maps of linear growth

Remark 76. Obviously, the following inclusions hold and are even strict:

$$\begin{aligned}\{F \in \mathrm{LIP}(\mathbb{R}^N,\mathbb{R}^N) \mid F \text{ has convex values}\} &\subset \mathrm{OSLIP}(\mathbb{R}^N,\mathbb{R}^N)\\ &\subset \mathrm{LOSLIP}(\mathbb{R}^N,\mathbb{R}^N).\end{aligned}$$

The key advantage of the linear growth condition here is concluded from Gronwall's inequality in the next lemma:

Definition 77. For any nonempty bounded subset $K \subset \mathbb{R}^N$, define

$$|K|_\infty := \sup_{y\in K} |y| \in [0,\infty[$$

Lemma 78. *For every set-valued map $F \in \mathrm{LOSLIP}(\mathbb{R}^N,\mathbb{R}^N)$ and any initial set $K_0 \in \mathscr{K}(\mathbb{R}^N)$, the reachable set at each time $t \geq 0$ fulfills*

$$\big|\vartheta_F(t,K_0)\big|_\infty \leq (|K_0|_\infty + \|F\|_{\mathrm{lg}}\, t)\cdot e^{\|F\|_{\mathrm{lg}}\cdot t}.$$

In particular, $\sup_{t\in[0,1]} \big|\vartheta_F(t,K_0)\big|_\infty \leq (|K_0|_\infty + \|F\|_{\mathrm{lg}}) \cdot e^{\|F\|_{\mathrm{lg}}}.$

Proof. For every point $x_t \in \vartheta_F(t,K_0)$, there exists a solution $x(\cdot) \in W^{1,1}([0,t],\mathbb{R}^N)$ to the differential inclusion $x'(\cdot) \in F(x(\cdot))$ a.e. satisfying $x(0) \in K_0,\ x(t) = x_t$. Then, for every $\tau \in [0,t]$,

$$\begin{aligned}\big|x(\tau)-x(0)\big| &\leq \int_0^\tau |F(x(s))|_\infty\, ds \leq \int_0^\tau \|F\|_{\mathrm{lg}}\,(1+|x(s)|)\, ds\\ &\leq \|F\|_{\mathrm{lg}}\, \tau\,(1+|K_0|_\infty) + \int_0^\tau \|F\|_{\mathrm{lg}}\, \big|x(s)-x(0)\big|\, ds\end{aligned}$$

and, Gronwall's Lemma (Proposition A.1 on page 439) implies

$$\begin{aligned}\big|x(t)-x(0)\big| &\leq \|F\|_{\mathrm{lg}}\, t\,(1+|K_0|_\infty) + \int_0^t e^{\|F\|_{\mathrm{lg}}\cdot(t-s)}\, \|F\|_{\mathrm{lg}}^2\, s\,(1+|K_0|_\infty)\, ds\\ &= (1+|K_0|_\infty)\quad \big(e^{\|F\|_{\mathrm{lg}}\cdot t}-1\big),\\ |x_t| &\leq |K_0|_\infty + (1+|K_0|_\infty)\quad \big(e^{\|F\|_{\mathrm{lg}}\cdot t}-1\big)\\ &\leq |K_0|_\infty\, e^{\|F\|_{\mathrm{lg}}\cdot t} + \|F\|_{\mathrm{lg}}\, t\; e^{\|F\|_{\mathrm{lg}}\cdot t}.\end{aligned}$$

□

Proposition 79. *Choosing arbitrary $r, L > 0$ and $T > 0$, set $R := (r+LT)\, e^{LT}$. For any sets $K_1, K_2 \in \mathscr{K}(\mathbb{R}^N)$ and set-valued maps $F, G \in \mathrm{LOSLIP}(\mathbb{R}^N,\mathbb{R}^N)$ with*

$$\begin{cases} K_1, K_2 \subset \mathbb{B}_r(0),\\ \|F\|_{\mathrm{lg}},\ \|G\|_{\mathrm{lg}} \leq L,\\ \Lambda := \max\{\mathrm{Lip}\, F|_{\mathbb{B}_{R+1}(0)},\ \mathrm{Lip}\, G|_{\mathbb{B}_{R+1}(0)}\} \in \mathbb{R}\end{cases}$$

the reachable sets $\vartheta_F(t,K_1),\ \vartheta_G(t,K_2) \subset \mathbb{R}^N$ are compact subsets of $\mathbb{R}^N$ and, the Pompeiu-Hausdorff distance between the reachable sets at time $t \in [0,T]$ satisfies

$$d\!l\big(\vartheta_F(t,K_1),\ \vartheta_G(t,K_2)\big) \leq \big(d\!l(K_1,K_2) + t\cdot d\!l_\infty(F|_{\mathbb{B}_{R+1}(0)},\ G|_{\mathbb{B}_{R+1}(0)})\big)\cdot e^{\Lambda t}.$$

Proof. Whenever compact initial sets K_0, K_1 are chosen within a ball $\mathbb{B}_r(0) \subset \mathbb{R}^N$ of arbitrarily fixed radius $r > 0$, Lemma 78 provides a joint a priori estimate for any $s, t \in [0,T]$, i.e.

$$\big|\vartheta_F(s,K_0)\big|_\infty,\ \big|\vartheta_F(t,K_1)\big|_\infty \ \le\ (r + \|F\|_{\mathrm{lg}}\, T)\ e^{\|F\|_{\mathrm{lg}}\, T} \stackrel{\text{Def.}}{=} R.$$

Restricting now our considerations to $\mathbb{B}_{R+1}(0) \subset \mathbb{R}^N$, we can draw exactly the same conclusions from Theorem A.63 (on page 480) as we have already done for

- Proposition 1.96 (on page 96) about transitions in $\mathrm{OSLIP}(\mathbb{R}^N, \mathbb{R}^N)$ and for
- Proposition 1.50 (on page 60) about morphological transitions in $\mathrm{LIP}(\mathbb{R}^N, \mathbb{R}^N)$ by means of generalized Filippov's Theorem A.6 respectively. □

In particular, each set-valued map in $\mathrm{LOSLIP}(\mathbb{R}^N, \mathbb{R}^N)$ induces a transition on $(\mathscr{K}(\mathbb{R}^N), d\!l, |\cdot|_\infty)$ and, we identify the relevant parameters of continuity easily:

Proposition 80. *For every set-valued map* $F \in \mathrm{LOSLIP}(\mathbb{R}^N, \mathbb{R}^N)$,

$$\begin{aligned} \vartheta_F : [0,1] \times \mathscr{K}(\mathbb{R}^N) &\longrightarrow \mathscr{K}(\mathbb{R}^N) \\ (t,\ K) &\longmapsto \vartheta_F(t,K) \end{aligned}$$

with $\vartheta_F(t,K) \subset \mathbb{R}^N$ *denoting the reachable set of the initial set* $K \in \mathscr{K}(\mathbb{R}^N)$ *and the differential inclusion* $x' \in F(x)$ *a.e. at time t is a transition on* $(\mathscr{K}(\mathbb{R}^N), d\!l, |\cdot|_\infty)$ *in the sense of Definition 2 (on page 104) with*

$$\begin{aligned} \alpha(\vartheta_F;\, r) &:= \max\big\{0,\ \mathrm{Lip}\, F|_{\mathbb{B}_{r+1}(0)}\big\}, \\ \beta(\vartheta_F;\, r) &:= \|F\|_{\mathrm{lg}} \left(1 + (r + \|F\|_{\mathrm{lg}})\, e^{\|F\|_{\mathrm{lg}}}\right), \\ \gamma(\vartheta_F) &:= \|F\|_{\mathrm{lg}}, \\ D(\vartheta_F,\, \vartheta_G;\, r) &\le d\!l_\infty\big(F|_{\mathbb{B}_{r+1}(0)},\ G|_{\mathbb{B}_{r+1}(0)}\big). \end{aligned}$$

□

As an abbreviation, we again identify each set-valued map $F \in \mathrm{LOSLIP}(\mathbb{R}^N, \mathbb{R}^N)$ with the corresponding transition $\vartheta_F : [0,1] \times \mathscr{K}(\mathbb{R}^N) \longrightarrow \mathscr{K}(\mathbb{R}^N)$.
Now evolving compact subsets of the Euclidean space $\mathbb{R}^N$ are regarded in the recent mutational framework for the tuple $(\mathscr{K}(\mathbb{R}^N), d\!l, |\cdot|_\infty)$ and, the results of § 2.3 provide directly the counterparts of the propositions about existence and continuous dependence in § 1.11 (on page 98 ff.).

Proposition 81 (Peano's Theorem for modified morphological equations). *For* $\mathscr{F} : \mathscr{K}(\mathbb{R}^N) \times [0,T] \longrightarrow \mathrm{LOSLIP}(\mathbb{R}^N, \mathbb{R}^N)$ *and each radius* $r > 0$ *suppose*

(1.) $\displaystyle \sup_{\substack{M \in \mathscr{K}(\mathbb{R}^N) \\ t \in [0,T]}} \big(\|\mathscr{F}(M,t)\|_{\mathrm{lg}} + \max\{0,\ \mathrm{Lip}\, \mathscr{F}(M,t)\big|_{\mathbb{B}_r(0)}\}\big) < \infty,$

(2.) for $\mathscr{L}^1$*-almost every* $t \in [0,T]$ *and every set* $K \in \mathscr{K}(\mathbb{R}^N)$, *the function*

$$\begin{aligned} (\mathscr{K}(\mathbb{R}^N), d\!l) \times [0,T] &\longrightarrow \big(\mathrm{LOSLIP}(\mathbb{R}^N, \mathbb{R}^N),\ d\!l_\infty(\,\cdot\,|_{\mathbb{B}_{r+1}(0)},\ \cdot|_{\mathbb{B}_{r+1}(0)})\big), \\ (M,s) &\longmapsto \mathscr{F}(M,s) \end{aligned}$$

is continuous in (K,t).

Then for every initial set $K_0 \in \mathscr{K}(\mathbb{R}^N)$*, there exists a solution* $K : [0,T] \leadsto \mathbb{R}^N$ *to the modified morphological equation*

$$\overset{\circ}{K}(\cdot) \ni \mathscr{F}\big(K(\cdot), \cdot\big)$$

with $K(0) = K_0$, *i.e.* $K(\cdot)$ *is bounded, continuous with respect to* $d\!l$ *and satisfies for* $\mathscr{L}^1$*-almost every* $t \in [0,T]$

$$\lim_{h \downarrow 0} \ \tfrac{1}{h} \cdot d\!l\big(\vartheta_{\mathscr{F}(K(t),t)}(h, K(t)),\ \ K(t+h)\big) \ = \ 0$$

Proof results from Peano's Theorem 18 for nonautonomous mutational equations (on page 114) in combination with preceding Proposition 80. □

Proposition 82 (Cauchy-Lipschitz for modified morphological equations). *Suppose* $\mathscr{F} : \mathscr{K}(\mathbb{R}^N) \times [0,T] \longrightarrow \mathrm{LOSLIP}(\mathbb{R}^N, \mathbb{R}^N)$ *to satisfy for each radius* $r > 0$

(1.) $\widehat{\alpha}_r := \sup_{\substack{M \in \mathscr{K}(\mathbb{R}^N) \\ t \in [0,T]}} \big(\|\mathscr{F}(M,t)\|_{\mathrm{lg}} + \max\{0,\ \mathrm{Lip}\, \mathscr{F}(M,t)\big|_{\mathbb{B}_r(0)}\}\big) < \infty,$

(2.) *for* $\mathscr{L}^1$*-almost every* $t \in [0,T]$ *and every set* $K \in \mathscr{K}(\mathbb{R}^N)$*, the function*

$$\begin{array}{rcl} \big(\mathscr{K}(\mathbb{R}^N), d\!l\big) \times [0,T] & \longrightarrow & \big(\mathrm{LOSLIP}(\mathbb{R}^N, \mathbb{R}^N),\ d\!l_\infty(\,\cdot\,|_{\mathbb{B}_{r+1}(0)},\ \cdot\,|_{\mathbb{B}_{r+1}(0)})\big), \\ (M,s) & \longmapsto & \mathscr{F}(M,s) \end{array}$$

is continuous in (K,t),

(3.) *there exists* $\lambda_r > 0$ *such that for* $\mathscr{L}^1$*-almost every* $t \in [0,T]$,

$$\begin{array}{rcl} \big(\mathscr{K}(\mathbb{R}^N), d\!l\big) & \longrightarrow & \big(\mathrm{LOSLIP}(\mathbb{R}^N, \mathbb{R}^N),\ d\!l_\infty(\,\cdot\,|_{\mathbb{B}_{r+1}(0)},\ \cdot\,|_{\mathbb{B}_{r+1}(0)})\big), \\ M & \longmapsto & \mathscr{F}(M,t) \end{array}$$

is λ_r*-Lipschitz continuous.*

Then for every initial set $K_0 \in \mathscr{K}(\mathbb{R}^N)$*, the solution* $K : [0,T] \leadsto \mathbb{R}^N$ *to the modified morphological equation* $\overset{\circ}{K}(\cdot) \ni \mathscr{F}\big(K(\cdot), \cdot\big)$ *with* $K(0) = K_0$ *exists and is unique.*

Proof. Existence due to continuity has just been specified in Proposition 81. Uniqueness of solutions results from Corollary 12 (on page 108). □

Proposition 83 (Continuity w.r.t. initial data and the right-hand side). *In addition to the assumptions of Proposition 82 about*

$$\mathscr{F} : \mathscr{K}(\mathbb{R}^N) \times [0,T] \longrightarrow \mathrm{LOSLIP}(\mathbb{R}^N, \mathbb{R}^N),$$

suppose for $\mathscr{G} : \mathscr{K}(\mathbb{R}^N) \times [0,T] \longrightarrow \mathrm{LOSLIP}(\mathbb{R}^N, \mathbb{R}^N)$ *and each* $r > 0$

$$\sup_{M,t}\ d\!l_\infty\big(\mathscr{F}(M,t)|_{\mathbb{B}_r(0)},\ \mathscr{G}(M,t)|_{\mathbb{B}_r(0)}\big) \ < \ \infty.$$

Consider any solutions $K_1(\cdot), K_2(\cdot) : [0,T] \leadsto \mathbb{R}^N$ *to the modified morphological equations*

$$\begin{cases} \overset{\circ}{K}_1(\cdot) \ni \mathscr{F}\big(K_1(\cdot), \cdot\big) \\ \overset{\circ}{K}_2(\cdot) \ni \mathscr{G}\big(K_2(\cdot), \cdot\big) \end{cases}$$

with $\sup\big\{|K_1(t)|_\infty,\ |K_2(t)|_\infty \ \big|\ t \in [0,T]\big\} \le R.$

Then the Pompeiu-Hausdorff distance of $K_1(t)$, $K_2(t)$ satisfies for every $t \in [0,T]$

$$d\!l\big(K_1(t),\, K_2(t)\big) \\ \le \Big(d\!l\big(K_1(0), K_2(0)\big) + t \cdot \sup_{M,s} d\!l_\infty\big(\mathscr{F}(M,s)|_{\mathbb{B}_{R+1}(0)},\, \mathscr{G}(M,s)|_{\mathbb{B}_{R+1}(0)}\big)\Big)\; e^{(\lambda_R + \widehat{\alpha}_R)\, t}.$$

Proof is an immediate consequence of Corollary 12 (on page 108). □

Proposition 84 (Existence of solutions under state constraints).
For $\mathscr{F} : \mathscr{K}(\mathbb{R}^N) \longrightarrow \mathrm{LOSLIP}(\mathbb{R}^N, \mathbb{R}^N)$ and each radius $r > 0$ suppose

(*1.*) $\sup_{M \in \mathscr{K}(\mathbb{R}^N)} \big(\|\mathscr{F}(M)\|_{\mathrm{lg}} + \max\{0,\ \mathrm{Lip}\, \mathscr{F}(M)|_{\mathbb{B}_r(0)}\}\big) < \infty,$

(2.) *the function*

$$\begin{array}{rcl}\big(\mathscr{K}(\mathbb{R}^N),\, d\!l\big) & \longrightarrow & \big(\mathrm{LOSLIP}(\mathbb{R}^N,\mathbb{R}^N),\, d\!l_\infty(\,\cdot\,|_{\mathbb{B}_{r+1}(0)},\, \cdot\,|_{\mathbb{B}_{r+1}(0)})\big), \\ M & \longmapsto & \mathscr{F}(M)\end{array}$$

is continuous.

For the nonempty closed subset $\mathscr{V} \subset \big(\mathscr{K}(\mathbb{R}^N),\, d\!l\big)$ assume the viability condition:

$$\liminf_{h \downarrow 0}\ \tfrac{1}{h} \cdot \inf_{N \in \mathscr{V}}\ d\!l\big(\vartheta_{\mathscr{F}(M)}(h,M),\, N\big) = 0 \qquad \text{for every } M \in \mathscr{V}.$$

Then every compact set $K_0 \in \mathscr{V}$ is the initial compact set of at least one solution $K(\cdot) : [0,1] \longrightarrow \mathscr{K}(\mathbb{R}^N)$ to the modified morphological equation

$$\overset{\circ}{K}(\cdot) \ni \mathscr{F}\big(K(\cdot)\big)$$

with $K(t) \in \mathscr{V}$ for all $t \in [0,1]$.

Proof. It is a corollary of Proposition 24 (on page 123). □

As a new result in comparison with § 1.11, we now obtain the existence of solutions to modified morphological equations *with delay* additionally. Indeed, Proposition 23 (on page 120) implies the following statement:

Proposition 85 (Existence for modified morphological equations with delay).
Assume for some fixed $\tau > 0$, the function

$$\mathscr{F} : \mathrm{BLip}\big([-\tau,0],\, \mathscr{K}(\mathbb{R}^N);\, d\!l,\, |\cdot|_\infty\big) \times [0,T] \longrightarrow \mathrm{LOSLIP}(\mathbb{R}^N, \mathbb{R}^N)$$

and each radius $r > 0$:

(*1.*) $\sup_{M(\cdot),t} \big(\|\mathscr{F}(M(\cdot),t)\|_{\mathrm{lg}} + \max\{0,\ \mathrm{Lip}\, \mathscr{F}(M(\cdot),t)|_{\mathbb{B}_r(0)}\}\big) < \infty,$

(2.) $\lim_{n \to \infty} d\!l_\infty\big(\mathscr{F}(M_n(\cdot),\, t_n)|_{\mathbb{B}_{r+1}(0)},\ \mathscr{F}(M(\cdot),\, t)|_{\mathbb{B}_{r+1}(0)}\big) = 0$

for $\mathscr{L}^1$-almost every $t \in [0,T]$ and any sequences $(M_n(\cdot))_{n \in \mathbb{N}}$, $(t_n)_{n \in \mathbb{N}}$ in $\mathrm{BLip}\big([-\tau,0],\, \mathscr{K}(\mathbb{R}^N);\, d\!l,\, |\cdot|_\infty\big)$ *and $[0,T]$ respectively satisfying*

$$\lim_{n \to \infty} t_n = t, \qquad \lim_{n \to \infty}\ \sup_{s \in [-\tau,0]}\ d\!l\big(M_n(s),\, M(s)\big) = 0.$$

For every function $K_0(\cdot) \in \text{BLip}\big([-\tau,0],\, \mathscr{K}(\mathbb{R}^N);\, d\!l,\, |\cdot|_\infty\big)$*, there exists a curve* $K(\cdot) : [-\tau,T] \longrightarrow \mathscr{K}(\mathbb{R}^N)$ *with the following properties:*

(i) $K(\cdot) \in \text{BLip}\big([-\tau,T],\, \mathscr{K}(\mathbb{R}^N);\, d\!l,\, |\cdot|_\infty\big)$,

(ii) *for* $\mathscr{L}^1$*-almost every* $t \in [0,T]$, $\mathscr{F}\big(K(t+\cdot)\big|_{[-\tau,0]},\, t\big)$ *belongs to* $\overset{\circ}{K}(t)$,

(iii) $K(\cdot)\big|_{[-\tau,0]} = K_0(\cdot)$.

In particular, the restriction $K(\cdot)\big|_{[0,T]}$ *is a solution to the modified morphological equation*

$$\overset{\circ}{K}(t) \ni \mathscr{F}\big(K(t+\cdot)\big|_{[-\tau,0]},\, t\big).$$

In § 1.9.3 and § 1.9.6 (on pages 64, 74 ff. respectively), we have discussed the equivalence between solutions to morphological equations and reachable sets of nonautonomous differential inclusions (whose set-valued right-hand side depends on the wanted tube).
Then in § 1.11, this relationship is extended to modified morphological equations by assuming continuity of set-valued maps additionally. It motivated the definition of $\text{COSLIP}(\mathbb{R}^N,\mathbb{R}^N)$ as abbreviation used in Corollary 1.106 (on page 101).
The same additional hypothesis of continuity for all set-valued maps inducing transitions lays now the foundations for generalizing this equivalence once more – by means of Proposition A.13 (on page 447).
First we introduce the following abbreviation:

Definition 86. $\text{CLOSLIP}(\mathbb{R}^N,\mathbb{R}^N)$ consists of all maps in $\text{LOSLIP}(\mathbb{R}^N,\mathbb{R}^N)$ that are continuous in addition, i.e. every set-valued map $F : \mathbb{R}^N \rightsquigarrow \mathbb{R}^N$ satisfying

1. F has nonempty compact convex values,
2. F is *continuous*,
3. F is locally one-sided Lipschitz continuous, i.e. for each radius $r > 0$, there is a constant $L_r \in \mathbb{R}$ such that for every $x, y \in \mathbb{B}_r(0) \subset \mathbb{R}^N$ and $v \in F(x)$, there exists some $w \in F(y)$ satisfying
$$\langle x-y,\, v-w\rangle \;\le\; L_r\, |x-y|^2.$$
4. F has linear growth, i.e. there is a constant $c \ge 0$ satisfying for all $x \in \mathbb{R}^N$,
$$\sup_{v \in F(x)} |v| \;\le\; c \cdot (1+|x|).$$

Proposition 87 (Modified morphological primitives as reachable sets).
For $\mathscr{G} : [0,T] \longrightarrow \text{CLOSLIP}(\mathbb{R}^N,\mathbb{R}^N)$ *and each radius* $r > 0$ *suppose that*

(1.) $\sup\limits_{t \in [0,T]} \big(\|\mathscr{G}(t)\|_{\text{lg}} + \max\{0,\ \text{Lip}\,\mathscr{G}(t)|_{\mathbb{B}_r(0)}\}\big) < \infty$,

(2.) $[0,T] \longrightarrow \big(\text{CLOSLIP}(\mathbb{R}^N,\mathbb{R}^N),\, d\!l_\infty(\,\cdot\,|_{\mathbb{B}_{r+1}(0)},\, \cdot\,|_{\mathbb{B}_{r+1}(0)})\big), \quad t \longmapsto \mathscr{G}(t)$
is Lebesgue measurable.

Moreover define the set-valued map $\widehat{G} :\ [0,T] \times \mathbb{R}^N \rightsquigarrow \mathbb{R}^N, \quad (t,x) \mapsto \mathscr{G}(t)(x)$.

A tube $K : [0,T] \rightsquigarrow \mathbb{R}^N$ solves the modified morphological equation

$$\overset{\circ}{K}(\cdot) \ni \mathscr{G}(\,\cdot\,)$$

if and only at every time $t \in [0,T]$, its compact value $K(t) \subset \mathbb{R}^N$ coincides with the reachable set of the nonautonomous differential inclusion $x' \in \widehat{G}(\cdot,x)$ a.e.

$$K(t) \;=\; \vartheta_{\widehat{G}}\big(t,\, K(0)\big).$$

Corollary 88 (Solutions to modified morphological equations as reachable sets). *Suppose $\mathscr{F} : \mathscr{K}(\mathbb{R}^N) \times [0,T] \longrightarrow \mathrm{CLOSLIP}(\mathbb{R}^N,\mathbb{R}^N)$ to satisfy for each $r > 0$*

(*1.*) $\displaystyle\sup_{\substack{M\in\mathscr{K}(\mathbb{R}^N)\\ t\in[0,T]}} \Big(\|\mathscr{F}(M,t)\|_{\mathrm{lg}} + \max\{0,\ \mathrm{Lip}\,\mathscr{F}(M,t)\big|_{\mathbb{B}_r(0)}\}\Big) < \infty,$

(2.) $\mathscr{F} : \big(\mathscr{K}(\mathbb{R}^N),\, d\!l\big) \times [0,T] \longrightarrow \big(\mathrm{CLOSLIP}(\mathbb{R}^N,\mathbb{R}^N),\, d\!l_\infty(\,\cdot\,|_{\mathbb{B}_{r+1}(0)}, \cdot\,|_{\mathbb{B}_{r+1}(0)})\big)$
is a Carathéodory function (i.e. here continuous with respect to the first argument and measurable with respect to time).

Then a continuous tube $K : [0,T] \rightsquigarrow \mathbb{R}^N$ is a solution to the modified morphological equation

$$\overset{\circ}{K}(\cdot) \ni \mathscr{F}\big(K(\cdot),\,\cdot\big)$$

if and only if at every time $t \in [0,T]$, the set $K(t) \subset \mathbb{R}^N$ coincides with the reachable set of the initial set $K(0) \subset \mathbb{R}^N$ and the nonautonomous differential inclusion

$$x'(\cdot) \;\in\; \mathscr{F}\big(K(\cdot),\cdot\big)\,(x(\cdot)).$$

Both the recent proposition and its corollary result from Proposition 82 about uniqueness and the following morphological features of reachable sets:

Lemma 89. *In addition to the assumptions of Proposition 87 about $\mathscr{G} : [0,T] \longrightarrow \mathrm{CLOSLIP}(\mathbb{R}^N,\mathbb{R}^N)$, define again $\widehat{G} :\ [0,T] \times \mathbb{R}^N \rightsquigarrow \mathbb{R}^N,\quad (t,x) \mapsto \mathscr{G}(t)(x)$.*

Then for every initial set $K_0 \in \mathscr{K}(\mathbb{R}^N)$, the reachable set

$$K(\cdot) := \vartheta_{\widehat{G}}(\cdot, K_0) :\ [0,T] \longrightarrow \mathscr{K}(\mathbb{R}^N)$$

of the nonautonomous differential inclusion $x' \in \widehat{G}(\cdot,x)$ a.e. is a solution to the modified morphological equation

$$\overset{\circ}{K}(\cdot) \ni \mathscr{G}(\,\cdot\,).$$

Proof. It follows from Proposition A.13 (on page 447) in exactly the same way as Lemma 1.58 (on page 65).
Indeed, $K(\cdot) := \vartheta_{\widehat{G}}(\cdot, K_0) :\ [0,T] \rightsquigarrow \mathbb{R}^N$ has compact values and is Lipschitz continuous with respect to $d\!l$ for the same reasons as in Proposition 80. In particular, $\sup_t\, |K(t)|_\infty < R$ for some $R > 0$ sufficiently large. Thus without loss of generality, we can assume for $\widehat{G}$ additionally that $\|\widehat{G}\|_\infty \le \sup_t\, \|\mathscr{G}(t)\|_{\mathrm{lg}} \cdot (1+R) < \infty$.

Now Proposition A.13 guarantees a set $J \subset [0,T]$ of full Lebesgue measure (i.e. $\mathscr{L}^1([0,T] \setminus J) = 0$) such that at every time $t \in J$ and for any set $M \in \mathscr{K}(\mathbb{R}^N)$,

$$\tfrac{1}{h} \cdot d\!l\Big(\vartheta_{\widehat{G}(t+\cdot,\cdot)}(h,M), \quad \bigcup_{x\in M} \big(x + h \cdot \widehat{G}(t,x)\big)\Big) \longrightarrow 0 \qquad \text{for } h \downarrow 0.$$

Applying the same Proposition A.13 to the autonomous differential inclusion with $\widehat{G}(t,\cdot) : \mathbb{R}^N \leadsto \mathbb{R}^N$ and arbitrary $t \in [0,T]$, we obtain

$$\tfrac{1}{h} \cdot d\!l\Big(\vartheta_{\widehat{G}(t,\cdot)}(h,M), \quad \bigcup_{x\in M} \big(x + h \cdot \widehat{G}(t,x)\big)\Big) \longrightarrow 0 \qquad \text{for } h \downarrow 0.$$

Hence, the triangle inequality of $d\!l$ implies for every $t \in J$ and $M \in \mathscr{K}(\mathbb{R}^N)$

$$\tfrac{1}{h} \cdot d\!l\Big(\vartheta_{\widehat{G}(t+\cdot,\cdot)}(h,M), \quad \vartheta_{\widehat{G}(t,\cdot)}(h,M)\Big) \longrightarrow 0 \qquad \text{for } h \downarrow 0,$$

i.e. for $M := \vartheta_{\widehat{G}}(t,K_0) \in \mathscr{K}(\mathbb{R}^N)$ and each $t \in J$,

$$\tfrac{1}{h} \cdot d\!l\Big(\vartheta_{\widehat{G}}(t+h,K_0), \quad \vartheta_{\mathscr{G}(t)}\big(h,\ \vartheta_{\widehat{G}}(t,K_0)\big)\Big) \longrightarrow 0 \qquad \text{for } h \downarrow 0.$$

□

Remark 90. Corollary 88 differs from Peano's Existence Theorem 81 in the regularity assumptions about $\mathscr{F}$. Indeed, the regularity of a Carathéodory function is weaker than the continuity of $\mathscr{F}$ in every point of $\mathscr{K}(\mathbb{R}^N) \times J$ with a subset $J \subset [0,1]$ of full Lebesgue measure (i.e. $\mathscr{L}^1([0,T] \setminus J) = 0$).
Scorza-Dragoni Theorem A.9 (on page 446), however, provides a very useful link in separable metric spaces for drawing conclusions about existence from this weaker assumption approximatively. Similarly to the proof of Existence Theorem 5.4 (about mutational inclusions on page 396 ff.) below, the following statement results from Theorem 81 together with Corollary 88:

Corollary 91. *Suppose $\mathscr{F} : \mathscr{K}(\mathbb{R}^N) \times [0,T] \longrightarrow \mathrm{CLOSLIP}(\mathbb{R}^N,\mathbb{R}^N)$ to satisfy for each $r > 0$*

(1.) $\sup_{\substack{M\in\mathscr{K}(\mathbb{R}^N)\\ t\in[0,T]}} \big(\|\mathscr{F}(M,t)\|_{\mathrm{lg}} + \max\{0,\ \mathrm{Lip}\, \mathscr{F}(M,t)\big|_{\mathbb{B}_r(0)}\}\big) < \infty,$

(2.) $\mathscr{F} : \big(\mathscr{K}(\mathbb{R}^N), d\!l\big) \times [0,T] \longrightarrow \big(\mathrm{CLOSLIP}(\mathbb{R}^N,\mathbb{R}^N),\ d\!l_\infty(\,\cdot\,|_{\mathbb{B}_{r+1}(0)}, \cdot\,|_{\mathbb{B}_{r+1}(0)})\big)$
is a Carathéodory function (i.e. here continuous with respect to the first argument and measurable with respect to time).

Then for every initial set $K_0 \in \mathscr{K}(\mathbb{R}^N)$, there exists a compact-valued Lipschitz continuous tube $K : [0,T] \leadsto \mathbb{R}^N$ such that at every time $t \in [0,T]$, the set $K(t) \subset \mathbb{R}^N$ coincides with the reachable set of the initial set K_0 and the nonautonomous differential inclusion

$$x'(\cdot) \in \mathscr{F}\big(K(\cdot),\cdot\big)\,(x(\cdot)).$$

Chapter 3
Less Restrictive Conditions on Distance Functions: Continuity Instead of Triangle Inequality

This chapter extends the mutational framework in four essential respects: First, distances do not have to satisfy the triangle inequality any longer and thus, we can also use powers of pseudo-metrics, for example.
Second, we use possibly different families of distances $(d_j)_{j\in\mathcal{J}}$, $(e_j)_{j\in\mathcal{J}}$ for the continuity with respect to state and time. They are to provide the same concept of sequential convergence, but may differ in quantitative features. This extension makes the mutational framework applicable to semilinear evolution equations with a strongly continuous semigroup of bounded linear operators.
Third, the notions of weak convergence and weak compactness are introduced beyond vector spaces – just on the basis of distances with an appropriate structure. The latter serves as a further pillar for proving the existence of solutions.
Fourth, some additional assumptions about Euler curves in any compact time interval make the condition of ω-contractivity on transitions dispensable because we can construct a family of equivalent distances such that each of the modified distances between any two states can grow (at most) exponentially along one and the same transition. This fourth aspect does not change the foundations of the mutational framework, but it extends the class of examples significantly since the additionally assumed inequality about Euler curves holds for many nonautonomous problems.
These four respects have already been sketched as Steps (C) – (F) in § 0.3.4.

In a word, the triangle inequality serves essentially the purpose to estimate the distance between two points by means of a third state. It might be regarded as one of the simplest ways of providing such a relation.
Mutational analysis, however, requires several parameters (for its transitions) so that we can verify the key estimate along transitions in Proposition 2.6, for example:

$$d_j\big(\vartheta(h,x),\ \tau(h,y)\big) \;\le\; \big(d_j(x,y) \,+\, h\cdot D_j(\vartheta,\tau\,;R_j)\big)\cdot e^{\alpha_j(\vartheta;R_j)\,h}$$

with $x,y\in E$ and $R_j := \big(\max\{\lfloor x\rfloor_j,\, \lfloor y\rfloor_j\} + \max\{\gamma_j(\vartheta),\ \gamma_j(\tau)\}\big)\cdot e^{\max\{\gamma_j(\vartheta),\,\gamma_j(\tau)\}}$.
Indeed, the right-hand side of this inequality reflects very well the basic notion of distinguishing between the "initial error" and "first-order terms".
For choosing d_j and D_j suitably in some applications like stochastic analysis, it is recommendable to dispense with the triangle inequality of d_j in its classical form. Instead we *modify* the definitions of D_j and of solutions to mutational equations in such way that the basic structural influence of "initial error" and "transitional error" on comparing estimates is preserved. This "conceptual shift" opens the door to replacing the triangle inequality of d_j and $D_j(\cdot,\cdot;r)$ by appropriate assumptions of continuity. In particular, the results of preceding chapters prove to be special cases.

T. Lorenz, *Mutational Analysis: A Joint Framework for Cauchy Problems In and Beyond Vector Spaces*, Lecture Notes in Mathematics 1996,
DOI 10.1007/978-3-642-12471-6_4,

3.1 General Assumptions of This Chapter

E is always a nonempty set and, $\mathscr{I} \neq \emptyset$ denotes an index set. For each index $j \in \mathscr{I}$,

$$\begin{aligned} d_j, e_j : E \times E &\longrightarrow [0,\infty[, \\ \lfloor \cdot \rfloor_j : \quad E &\longrightarrow [0,\infty[\end{aligned}$$

are supposed to satisfy the following conditions:

(H1) d_j and e_j are reflexive, i.e. for all $x \in E$: $\quad d_j(x,x) = 0 = e_j(x,x)$,

(H2) d_j and e_j are symmetric, i.e. for all $x,y \in E$: $\quad d_j(x,y) = d_j(y,x)$, $\quad e_j(x,y) = e_j(y,x)$,

(H3) $(d_j)_{j\in\mathscr{I}}$ and $(e_j)_{j\in\mathscr{I}}$ induce the same concept of convergence in E and are sequentially (semi-) continuous in the following sense:

(o)
$$\begin{aligned} &\Big(\forall\, j \in \mathscr{I}: \lim_{n\to\infty} d_j(x,x_n) = 0\Big) \\ \Longleftrightarrow\ &\Big(\forall\, j \in \mathscr{I}: \lim_{n\to\infty} e_j(x,x_n) = 0\Big) \end{aligned}$$
for any $x \in E$ and $(x_n)_{n\in\mathbb{N}}$ in E with $\sup_{n\in\mathbb{N}} \lfloor x_n \rfloor_i < \infty$ for each $i \in \mathscr{I}$.

(i)
$$\begin{aligned} d_j(x,y) &= \lim_{n\to\infty} d_j(x_n,y_n), \\ e_j(x,y) &\leq \limsup_{n\to\infty} e_j(x_n,y_n) \end{aligned}$$
for any $x,y \in E$ and $(x_n)_{n\in\mathbb{N}}$, $(y_n)_{n\in\mathbb{N}}$ in E fulfilling for each $i \in \mathscr{I}$,
$$\lim_{n\to\infty} d_i(x,x_n) = 0 = \lim_{n\to\infty} d_i(y_n,y), \qquad \sup_{n\in\mathbb{N}} \{\lfloor x_n \rfloor_i, \lfloor y_n \rfloor_i\} < \infty.$$

(ii)
$$0 = \lim_{n\to\infty} d_j(x,\, x_n)$$
for any $x \in E$ and $(x_n)_{n\in\mathbb{N}}$, $(y_n)_{n\in\mathbb{N}}$ in E fulfilling for each $i \in \mathscr{I}$
$$\lim_{n\to\infty} d_i(x,y_n) = 0 = \lim_{n\to\infty} e_i(y_n,x_n), \qquad \sup_{n\in\mathbb{N}} \{\lfloor x_n \rfloor_i, \lfloor y_n \rfloor_i\} < \infty.$$

(iii)
$$0 = \lim_{n\to\infty} d_j(x,\, x_n)$$
for any $x \in E$ and $(x_n)_{n\in\mathbb{N}}$, $(y_k)_{k\in\mathbb{N}}$, $(z_{k,n})_{k,n\in\mathbb{N}}$ in E fulfilling

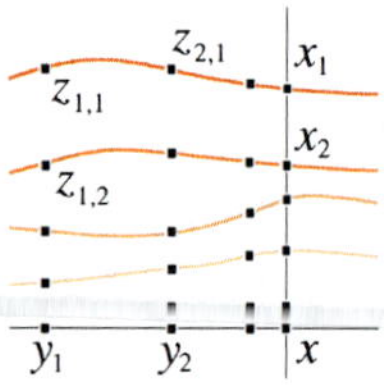

$$\begin{cases} \lim_{k\to\infty} e_i(x,\, y_k) = 0 & \text{for each } i \in \mathscr{I}, \\ \lim_{n\to\infty} d_i(y_k,\, z_{k,n}) = 0 & \text{for each } i \in \mathscr{I}, k \in \mathbb{N}, \\ \lim_{k\to\infty} \sup_{n>k} e_i(z_{k,n},\, x_n) = 0 & \text{for each } i \in \mathscr{I}, \\ \sup_{k,n\in\mathbb{N}} \{\lfloor x_n \rfloor_i, \lfloor y_k \rfloor_i, \lfloor z_{k,n} \rfloor_i\} < \infty & \text{for each } i \in \mathscr{I}. \end{cases}$$

(H4) $\lfloor \cdot \rfloor_j$ is lower semicontinuous with respect to $(d_i)_{i\in\mathscr{I}}$, i.e.,
$$\lfloor x \rfloor_j \leq \liminf_{n\to\infty} \lfloor x_n \rfloor_j$$
for any element $x \in E$ and sequence $(x_n)_{n\in\mathbb{N}}$ in E fulfilling for each $i \in \mathscr{I}$,
$$\lim_{n\to\infty} d_i(x_n,x) = 0, \qquad \sup_{n\in\mathbb{N}} \lfloor x_n \rfloor_i < \infty.$$

Remark 1. In comparison to Chapter 2, these assumptions do not imply the triangle inequality of d_j since d_j does not have to be a pseudo-metric in the sense of Definition 2.1 (on page 104).
But obviously property (H3) is satisfied whenever $d_j \equiv e_j$ is a pseudo-metric for each index $j \in \mathscr{I}$. Hence the topological environment of Chapter 2 is a special case.

A transition $\vartheta : [0,1] \times E \longrightarrow E$ is expected to satisfy essentially the same conditions as in Definition 2.2 (on page 104).
In fact, we can even dispense with the generalized form of semigroup property since estimates will be done "uniformly" along transitions $\vartheta(\cdot,x) : [0,1] \longrightarrow E$ as hypothesis (H7) will clarify in a moment. Indeed, up to now, we have drawn all quantitative conclusions from the "local" features of transitions close to the initial element, i.e., for time tending to 0. (See, for example, Definition 2.5 and Proposition 2.6 on page 105 f.)
As key new aspect about single transitions, we are now free to use different distance functions (namely d_j resp. e_j) for the continuity estimates with respect to initial elements and time. These families of distance functions $(d_j)_{j\in\mathscr{I}}$, $(e_j)_{j\in\mathscr{I}}$ are linked according to hypothesis (H3). In particular, they induce the same concept of convergence, but they might differ in quantitative features.
For extending Definition 2.2, we specify the conditions on a transition — now on the tuple $\big(E,\, (d_j)_{j\in\mathscr{I}},\, (e_j)_{j\in\mathscr{I}},\, (\lfloor\cdot\rfloor_j)_{j\in\mathscr{I}}\big)$:

Definition 2. A function $\vartheta : [0,1] \times E \longrightarrow E$ is called *transition* on the tuple $\big(E,\, (d_j)_{j\in\mathscr{I}},\, (e_j)_{j\in\mathscr{I}},\, (\lfloor\cdot\rfloor_j)_{j\in\mathscr{I}}\big)$ if it has the following properties for each $j \in \mathscr{I}$:

1.) for every $x \in E$: $\quad \vartheta(0,x) = x$

3.) there exists $\alpha_j(\vartheta;\cdot) : [0,\infty[\longrightarrow [0,\infty[$ such that for any $x,y \in E$ with
$$\lfloor x\rfloor_j \le r,\ \lfloor y\rfloor_j \le r: \qquad \limsup_{h\downarrow 0} \frac{d_j(\vartheta(h,x),\, \vartheta(h,y)) - d_j(x,y)}{h} \;\le\; \alpha_j(\vartheta;r)\cdot d_j(x,y)$$

4.′) there exists $\beta_j(\vartheta;\cdot) : [0,\infty[\longrightarrow [0,\infty[$ such that for any $s,t \in [0,1]$ and $x \in E$ with $\lfloor x\rfloor_j \le r$: $\quad e_j\big(\vartheta(s,x),\, \vartheta(t,x)\big) \;\le\; \beta_j(\vartheta;r)\cdot |t-s|$

5.) there exists $\gamma_j(\vartheta) \in [0,\infty[$ such that for any $t \in [0,1]$ and $x \in E$:
$$\lfloor \vartheta(t,x)\rfloor_j \;\le\; \big(\lfloor x\rfloor_j + \gamma_j(\vartheta)\; t\big)\cdot e^{\gamma_j(\vartheta)\,t}$$

The essential new aspect about comparing two transitions comes now into play as counterpart of Definition 2.5 (on page 105): $\widehat{\Theta}\big(E,(d_j)_{j\in\mathscr{I}},(e_j)_{j\in\mathscr{I}},(\lfloor\cdot\rfloor_j)_{j\in\mathscr{I}}\big)$ denotes a nonempty set of transitions on $\big(E,(d_j)_{j\in\mathscr{I}},(e_j)_{j\in\mathscr{I}},(\lfloor\cdot\rfloor_j)_{j\in\mathscr{I}}\big)$ and, for each $j \in \mathscr{I}$, the function
$$\widehat{D}_j:\ \widehat{\Theta}\big(E,(d_j)_j,(e_j)_j,(\lfloor\cdot\rfloor_j)_j\big) \times \widehat{\Theta}\big(E,(d_j)_j,(e_j)_j,(\lfloor\cdot\rfloor_j)_j\big) \times [0,\infty[\;\longrightarrow\; [0,\infty[$$
is *assumed* to satisfy the following conditions:

(H5) for each $r \geq 0$, $\widehat{D}_j(\cdot\,,\,\cdot\,;r)$ is reflexive and symmetric,
for every ϑ, τ, the function $\widehat{D}_j(\vartheta,\tau;\cdot) : [0,\infty[\longrightarrow [0,\infty[$ is nondecreasing,

(H6) for any $r \geq 0$,
$\widehat{D}_j(\cdot,\cdot;r) : \widehat{\Theta}\big(E,(d_j),(e_j),(\lfloor\cdot\rfloor_j)\big) \times \widehat{\Theta}\big(E,(d_j),(e_j),(\lfloor\cdot\rfloor_j)\big) \longrightarrow [0,\infty[$
is sequentially continuous with respect to $(\widehat{D}_i)_{i\in\mathscr{I}}$ in the following sense:

(i) $\widehat{D}_j(\vartheta,\,\tau;\,r) \;=\; \lim_{n\to\infty} \widehat{D}_j(\vartheta_n,\,\tau_n;\,r)$
for any transitions ϑ, τ and sequences $(\vartheta_n)_{n\in\mathbb{N}}$, $(\tau_n)_{n\in\mathbb{N}}$ satisfying for every $i \in \mathscr{I}$ and $R \geq 0$

$$\lim_{n\to\infty} \widehat{D}_i(\vartheta,\,\vartheta_n;\,R) \;=\; 0 \;=\; \lim_{n\to\infty} \widehat{D}_i(\tau,\,\tau_n;\,R)\,.$$

(ii) $\lim_{n\to\infty} \widehat{D}_j(\vartheta,\,\tau_n;\,r) \;=\; 0$
for any transition ϑ and sequences $(\vartheta_n)_{n\in\mathbb{N}}$, $(\tau_n)_{n\in\mathbb{N}}$ satisfying for every $i \in \mathscr{I}$ and $R \geq 0$

$$\lim_{n\to\infty} \widehat{D}_i(\vartheta,\,\vartheta_n;\,R) \;=\; 0 \;=\; \lim_{n\to\infty} \widehat{D}_i(\vartheta_n,\,\tau_n;\,R)\,.$$

(H7)
$$\limsup_{h\downarrow 0} \frac{d_j\big(\vartheta(t_1+h,x),\ \tau(t_2+h,y)\big) \;-\; d_j\big(\vartheta(t_1,x),\ \tau(t_2,y)\big)\cdot e^{\alpha_j(\tau;R_j)\cdot h}}{h}$$
$$\leq\; \widehat{D}_j(\vartheta,\tau;\,R_j) \;<\; \infty$$

for any $\vartheta,\tau \in \widehat{\Theta}\big(E,(d_i)_i,(e_i)_i,(\lfloor\cdot\rfloor_i)_i\big)$, $x,y \in E$, $t_1,t_2 \in [0,1[$, $r \geq 0$, $j \in \mathscr{I}$ with $\lfloor x\rfloor_j, \lfloor y\rfloor_j \leq r$ and $R_j := \big(r+\max\{\gamma_j(\vartheta),\gamma_j(\tau)\}\big)\cdot e^{\max\{\gamma_j(\vartheta),\gamma_j(\tau)\}}$.

Not even $\widehat{D}_j(\cdot,\cdot;r)$ has to satisfy the triangle inequality. Instead we restrict our assumption (H6) to the aspect of continuity. More generally speaking, the triangle inequality can be regarded as the classical tool for simplifying the verification of continuity in metric spaces.

Hypothesis (H7) specifies $\widehat{D}_j(\cdot,\cdot;r)$ in a rather global way whereas Definition 2.5 of $D_j(\cdot,\cdot;r)$ (on page 105) was comparing the evolution of one and the same initial point along two transitions. The criterion here in (H7) is motivated by a question focusing on vanishing times: Which "first-order terms" of the time-dependent distance cannot be estimated just by the initial distance growing exponentially in time ?

Remark 3. If $d_j \equiv e_j$ satisfies the triangle inequality in addition, then the properties (H5) – (H7) can be concluded from Definition 2.5 and from Proposition 2.6 (on page 106). Thus, the results of Chapter 2 prove to be a special case based merely on the additional assumption of the triangle inequality for $d_j \equiv e_j$.

Remark 4 (about separate real time components). In some examples, time is recommendable to be taken into consideration explicitly. One of the easiest ways is to consider tuples in $\widetilde{E} := \mathbb{R} \times E$ with the first real component representing the respective time. In subsequent § 3.5 (on page 221 ff.), we formulate modified hypotheses allowing the same conclusions as in §§ 3.2 – 3.4.

3.2 The Essential Features of Transitions Do Not Change

Appropriate continuity assumptions (instead of the triangle inequality) and two families of distance functions do not have any significant consequences for the features of transitions. We now verify the essential aspects:

Lemma 5. *Let $\vartheta_1 \dots \vartheta_K$ be finitely many transitions on $\big(E, (d_j), (e_j), (\lfloor\cdot\rfloor_j)\big)$ with* $\widehat{\gamma}_j := \sup_{k \in \{1 \dots K\}} \gamma_j(\vartheta_k) < \infty$ *for some $j \in \mathscr{I}$.*

For any $x_0 \in E$ and $0 = t_0 < t_1 < \dots < t_K$ with $\sup_k t_k - t_{k-1} \leq 1$ define the curve $x(\cdot) : [0, t_K] \longrightarrow E$ piecewise as $x(0) := x_0$ and

$$x(t) := \vartheta_k\big(t - t_{k-1},\ x(t_{k-1})\big) \qquad \text{for } t \in \big]t_{k-1}, t_k\big],\ k \in \{1 \dots K\}.$$

Then, $\lfloor x(t) \rfloor_j \leq \big(\lfloor x_0 \rfloor_j + \widehat{\gamma}_j \cdot t\big) \cdot e^{\widehat{\gamma}_j \cdot t}$ *at every time $t \in [0, t_K]$.*

Proof results from exactly the same arguments as Lemma 2.4 (on page 105). □

The following lemma provides the first tool for applying Gronwall's estimate (in Proposition A.2 on page 440). Indeed, it is an immediate consequence of hypotheses (H3) (o), (i) and guarantees that the distance between two continuous curves in E is always continuous with respect to time.
Our version of Gronwall's inequality (in the appendix A.1) has the essential advantage that even lower semicontinuity is sufficient for concluding a global estimate from local properties. (This will be relevant for proving subsequent Proposition 11 on page 189.)

Lemma 6. *Let $x(\cdot), y(\cdot) : [0, T] \longrightarrow E$ be continuous with respect to $(d_i)_{i \in \mathscr{I}}$ (or equivalently with respect to $(e_j)_{j \in \mathscr{I}}$) and bounded with respect to each $\lfloor\cdot\rfloor_j$ ($j \in \mathscr{I}$). Then for each index $j \in \mathscr{I}$, the distance function*

$$[0, T] \longrightarrow [0, \infty[, \qquad t \longmapsto d_j\big(x(t),\ y(t)\big)$$

is continuous. □

Proposition 7. *Let $\vartheta, \tau \in \widehat{\Theta}\big(E, (d_j)_{j \in \mathscr{I}}, (e_j)_{j \in \mathscr{I}}, (\lfloor\cdot\rfloor_j)_{j \in \mathscr{I}}\big)$, $r \geq 0, j \in \mathscr{I}$ and $t_1, t_2 \in [0, 1[$ be arbitrary. For any elements $x, y \in E$ suppose $\lfloor x \rfloor_j \leq r$, $\lfloor y \rfloor_j \leq r$. Then the following estimate holds at each time $h \in [0, 1[$ with $\max\{t_1 + h,\ t_2 + h\} \leq 1$*

$$d_j\big(\vartheta(t_1 + h, x),\ \tau(t_2 + h, y)\big) \leq \Big(d_j\big(\vartheta(t_1, x),\ \tau(t_2, y)\big) + h \cdot \widehat{D}_j(\vartheta, \tau; R_j)\Big)\ e^{\alpha_j(\tau; R_j)\, h}$$

with the constant $R_j := \big(r + \max\{\gamma_j(\vartheta),\ \gamma_j(\tau)\}\big) \cdot e^{\max\{\gamma_j(\vartheta),\, \gamma_j(\tau)\}} < \infty$.

Proof results from Gronwall's inequality (in Proposition A.2 on page 440) applied to the auxiliary function

$$\phi_j : \ h \longmapsto d_j\big(\vartheta(t_1 + h, x),\ \tau(t_2 + h, y)\big)$$

similarly to the proofs of Proposition 1.7 (on page 36) and Proposition 2.6 (on page 106). Indeed, ϕ_j is continuous according to Lemma 6 and the time continuity of transitions (in condition (4.')) of Definition 2). Moreover condition (5.) of Definition 2 ensures $\lfloor \vartheta(h,x) \rfloor_j \le R_j$, $\lfloor \tau(h,y) \rfloor_j \le R_j$ for each $h \in [0,1]$.

Dispensing with the triangle inequality of d_j in this chapter, however, we conclude directly from hypothesis (H7) about $\widehat{D}_j(\cdot,\cdot; R_j)$ for any t and small $h > 0$

$$\begin{aligned}
&\phi_j(t+h) - \phi_j(t) \\
&= \ d_j\big(\vartheta(t_1{+}t{+}h, x),\ \tau(t_2{+}t{+}h, y)\big) \quad - \ d_j\big(\vartheta(t_1{+}t, x),\ \tau(t_2{+}t, y)\big) \\
&\le \ d_j\big(\vartheta(t_1{+}t{+}h, x),\ \tau(t_2{+}t{+}h, y)\big) \quad - \ d_j\big(\vartheta(t_1{+}t, x),\ \tau(t_2{+}t, y)\big)\, e^{\alpha_j(\tau;R_j)\,h} \\
&\quad + d_j\big(\vartheta(t_1{+}t, x),\ \tau(t_2{+}t, y)\big) \cdot e^{\alpha_j(\tau;R_j)\,h} - \ d_j\big(\vartheta(t_1{+}t, x),\ \tau(t_2{+}t, y)\big)
\end{aligned}$$

and thus, $\displaystyle \limsup_{h \downarrow 0} \ \frac{\phi_j(t+h) - \phi_j(t)}{h} \ \le \ \widehat{D}_j(\vartheta, \tau; R_j) \ + \ \alpha_j(\tau; R_j) \cdot \phi_j(t) \ < \ \infty.$

Finally, Gronwall's inequality (in form of Proposition A.2) provides the link to the claimed estimate. □

3.3 Solutions to Mutational Equations

For any single-valued function $f : E \times [0,T] \longrightarrow \widehat{\Theta}\big(E, (d_j)_{j\in\mathcal{I}}, (e_j)_{j\in\mathcal{I}}, (\lfloor\cdot\rfloor_j)_{j\in\mathcal{I}}\big)$, a *solution* $x(\cdot) : [0,T] \longrightarrow E$ to the mutational equation

$$\overset{\circ}{x}(\cdot) \ \ni \ f\big(x(\cdot), \cdot\big)$$

is expected to fulfill the same conditions as in Definition 2.9 (on page 107), i.e., it should satisfy for each index $j \in \mathcal{I}$:

1.) $x(\cdot)$ is continuous with respect to d_j

2.) for $\mathcal{L}^1$-almost every $t \in [0,T[$: $\displaystyle \lim_{h \downarrow 0} \ \tfrac{1}{h} \cdot d_j\big(f(x(t),t)\,(h, x(t)),\ x(t+h)\big) \ = \ 0$

3.) $\displaystyle \sup_{t\in[0,T]} \ \lfloor x(t) \rfloor_j \ < \ \infty.$

Due to the lack of triangle inequality for d_j, however, it is much more difficult to compare such a solution $x(t+\cdot)$ with a transition starting in another "initial point". Indeed, there is no obvious way to draw conclusions about distances d_j vanishing in first order for $h \downarrow 0$.

For the same (rather technical) reason, we have already introduced hypothesis (H7) (on page 184), which is motivated by the earlier estimate in Proposition 2.6 (on page 106) and which has just been used in the proof of Proposition 7 here.

Thus, we specify the term "solution" by a slightly stronger condition (2.′). It is also motivated by the notion that the first-order properties of $x(t+h)$ cannot be distinguished from the features of $f(x(t),t)(h,x(t))$ for $h \downarrow 0$.
As the essential new aspect, however, the direct comparison via d_j, i.e.

$$h \longmapsto d_j\big(f(x(t),t)\,(h,\, x(t)),\;\, x(t+h)\big),$$

is now replaced by the comparisons with $h \mapsto \vartheta(s+h,z) \in E$ for any transition $\vartheta \in \widehat{\Theta}\big(E,(d_j)_{j\in\mathcal{I}},(e_j)_{j\in\mathcal{I}},(\lfloor\cdot\rfloor_j)_{j\in\mathcal{I}}\big)$ and arbitrary initial point $\vartheta(s,z) \in E$.

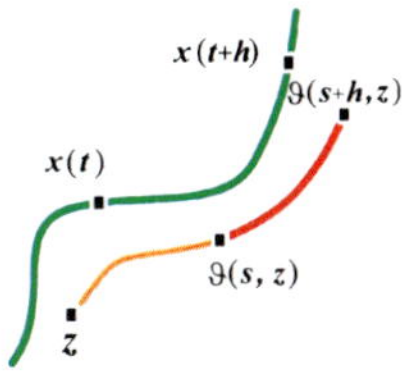

So far the estimate in Proposition 7 and its counterparts in preceding chapters are the main tool for comparing the evolutions along transitions. Now we employ it for specifying the notion of "being indistinguishable up to first order":

Definition 8.
A single-valued function $f : E \times [0,T] \longrightarrow \widehat{\Theta}\big(E,(d_j)_{j\in\mathcal{I}},(e_j)_{j\in\mathcal{I}},(\lfloor\cdot\rfloor_j)_{j\in\mathcal{I}}\big)$ is given. $x(\cdot) : [0,T] \longrightarrow E$ is called a *solution* to the mutational equation

$$\overset{\circ}{x}(\cdot) \ni f\big(x(\cdot),\cdot\big)$$

in $\big(E,(d_j)_{j\in\mathcal{I}},(e_j)_{j\in\mathcal{I}},(\lfloor\cdot\rfloor_j)_{j\in\mathcal{I}},(\widehat{D}_j)_{j\in\mathcal{I}}\big)$ if it satisfies for each $j \in \mathcal{I}$:

1.) $x(\cdot)$ is continuous with respect to e_j, i.e.,
$\lim_{s\to t} e_j\big(x(s),\, x(t)\big) = 0$ for every $t \in [0,T]$,

2.′) there exists $\alpha_j(x;\cdot) : [0,\infty[\, \longrightarrow [0,\infty[$ such that for $\mathcal{L}^1$-a.e. $t \in [0,T[$:
$$\limsup_{h\downarrow 0} \frac{d_j(\vartheta(s+h,z),\, x(t+h)) \;-\; d_j(\vartheta(s,z),\, x(t))\cdot e^{\alpha_j(x;R_j)\,h}}{h} \;\le\; \widehat{D}_j\big(\vartheta,\, f(x(t),t);\, R_j\big)$$
is fulfilled for any $\vartheta \in \widehat{\Theta}\big(E,(d_j),(e_j),(\lfloor\cdot\rfloor_j)\big)$, $s \in [0,1[$, $z \in E$ satisfying $\lfloor\vartheta(\cdot,z)\rfloor_j, \lfloor x(\cdot)\rfloor_j \le R_j$,

3.) $\sup_{t\in[0,T]} \lfloor x(t)\rfloor_j < \infty$.

The continuity with respect to $(e_j)_{j\in\mathcal{I}}$ is equivalent to the continuity with respect to $(d_j)_{j\in\mathcal{I}}$ due to hypothesis (H3) (o) (on page 182).

Furthermore condition (2.′) always implies the preceding property (2.) because d_j and $\widehat{D}_j(\cdot,\cdot,r)$ are assumed to be reflexive. The inverse conclusion "(2.) $\Longrightarrow$ (2.′)" holds if d_j is a pseudo-metric (as in Chapter 2). Indeed, Proposition 2.6 (on page 106) then ensures the equivalence of Definition 2.9 (on page 107) and Definition 8 here.

By means of Gronwall's inequality for lower semicontinuous functions again, essentially the same arguments as for Proposition 7 guarantee that the *local* criterion (2.′) implies a *global* estimate of the same type for comparing solutions and transitions:

Lemma 9. *Let $x(\cdot) : [0,T] \longrightarrow E$ be a solution to the mutational equation*

$$\overset{\circ}{x}(\cdot) \ni f\big(x(\cdot), \cdot\big)$$

in $\big(E, (d_j)_{j\in\mathscr{I}}, (e_j)_{j\in\mathscr{I}}, (\lfloor\cdot\rfloor_j)_{j\in\mathscr{I}}, (\widehat{D}_j)_{j\in\mathscr{I}}\big)$ according to Definition 8. Suppose $\vartheta \in \widehat{\Theta}\big(E, (d_j)_{j\in\mathscr{I}}, (e_j)_{j\in\mathscr{I}}, (\lfloor\cdot\rfloor_j)_{j\in\mathscr{I}}\big)$, $z \in E$, $r \geq 0$, $s \in [0,1[$, $t \in [0,T[$, $j \in \mathscr{I}$ to be arbitrary with $\lfloor z\rfloor_j \leq r$ and the abbreviation

$$R_j := \max\big\{ \sup \lfloor x(\cdot)\rfloor_j,\ \big(r+\gamma_j(\vartheta)\big)\cdot e^{\gamma_j(\vartheta)}\big\} < \infty.$$

Then, $d_j\big(\vartheta(s+h,z),\ x(t+h)\big)$

$$\leq \Big(d_j\big(\vartheta(s,z), x(t)\big) + h\cdot \sup_{[t,t+h]} \widehat{D}_j\big(\vartheta,\ f(x(\cdot),\cdot);\ R_j\big)\Big)\cdot e^{\alpha_j(x;R_j)\,h}$$

for every $h \in [0,1]$ with $s+h \leq 1$ and $t+h \leq T$. □

In particular, the analogy of Lemma 9 and preceding Proposition 7 reflects how we interpret the generalized conceptual goal that a solution $x(t+\cdot)$ cannot be "distinguished" from the curve $f(x(t),t)\,(\cdot\,, x(t)) : [0,1] \longrightarrow E$ along the transition $f(x(t),t)$ "up to first order".

Finally, we focus on the Lipschitz continuity of solutions. For every transition ϑ and initial point $z \in E$, the curve $[0,1] \longrightarrow E,\ t \longmapsto \vartheta(t,z)$ is assumed to be Lipschitz continuous with respect to each e_j. For solutions to mutational equations, the same regularity with respect to d_j $(j \in \mathscr{I})$ can be concluded from Lemma 9 by means of the *identity transition* $\mathbb{I}\mathrm{d}_{\widehat{\Theta}}$ on E:

Corollary 10 (Sufficient conditions for Lipschitz continuity of solutions). *Assume that $\widehat{\Theta}\big(E, (d_i)_{i\in\mathscr{I}}, (e_j)_{j\in\mathscr{I}}, (\lfloor\cdot\rfloor_i)_{i\in\mathscr{I}}\big)$ contains the identity transition*

$$\mathbb{I}\mathrm{d}_{\widehat{\Theta}} : [0,1]\times E \longrightarrow E, \qquad (h,x) \longmapsto x.$$

For $f : E\times[0,T] \longrightarrow \widehat{\Theta}\big(E, (d_j)_j, (e_j)_j, (\lfloor\cdot\rfloor_j)_j\big)$ let $x(\cdot) : [0,T] \longrightarrow E$ be a solution to the mutational equation $\overset{\circ}{x}(\cdot) \ni f(x(\cdot),\cdot)$ in $\big(E, (d_i)_{i\in\mathscr{I}}, (e_i)_{i\in\mathscr{I}}, (\lfloor\cdot\rfloor_i)_{i\in\mathscr{I}}, (\widehat{D}_i)_{i\in\mathscr{I}}\big)$ such that some $j \in \mathscr{I}$ and $L_j, R_j \in \mathbb{R}$ satisfy for all $t \in [0,T]$

$$\lfloor x(t)\rfloor_j \leq R_j, \qquad \widehat{D}_j\big(\mathbb{I}\mathrm{d}_{\widehat{\Theta}},\ f(x(t),t);\ R_j\big) \leq L_j.$$

Then $x(\cdot)$ is Lipschitz continuous with respect to d_j.

Proof. We use arguments very similar to the proof of Lemma 2.10 (on page 107): Fix $s \in [0,T[$ arbitrarily. Then, $\psi_j : [s,T] \longrightarrow \mathbb{R},\quad t \longmapsto d_j\big(x(s),\ x(t)\big)$ is continuous due to hypotheses (H3) (o), (i) and, it satisfies for $\mathscr{L}^1$-a.e. $t \in [s,T]$

$$\begin{aligned}
\limsup_{h\downarrow 0} \frac{\psi_j(t+h)-\psi_j(t)}{h} &= \limsup_{h\downarrow 0} \frac{1}{h}\Big(d_j\big(\mathbb{I}\mathrm{d}_{\widehat{\Theta}}(h,x(s)),\ x(t+h)\big) - d_j\big(x(s),\ x(t)\big)\Big)\\
&\leq \psi_j(t)\cdot \limsup_{h\downarrow 0} \frac{e^{\alpha_j(x;R_j)\ h}-1}{h} + L_j\\
&= \psi_j(t)\cdot \alpha_j(x;R_j) + L_j.
\end{aligned}$$

Finally $\psi_j(t) \leq L_j\, e^{\alpha_j(x;R_j)\,T}\cdot(t-s)$ for all $t \in [s,T]$ results from Gronwall's inequality (Proposition A.2 on page 440) and $\psi_j(s) = 0$. □

3.3.1 Continuity with Respect to Initial States and Right-Hand Side

Dispensing with the triangle inequality of distance functions, we have already faced several difficulties for dealing with further distances vanishing "in first order" for time $h \downarrow 0$. So far the conclusions proved in preceding chapters have usually served as motivation for adapting definitions so that we can bridge the gap due to lacking metric structure.

Now the list of definitions is (almost) completed and, we have to find alternative ways for investigating the continuity of solutions with respect to initial states and right-hand side, for example.
The idea is very similar to our way from property (2.) of solutions to condition (2.′) (in Definition 8): We do not compare two solutions directly by means of d_j as in Proposition 2.11 (on page 108), but we use the respective distances from one and same (arbitrary) state $z \in E$, i.e. we are interested in an upper estimate of the auxiliary distance function $[0,T] \longrightarrow [0,\infty[, t \longmapsto \inf_{z \in E:\, \lfloor z \rfloor_j < \rho} \big(d_j(z,x(t)) + d_j(z,y(t))\big)$.

Proposition 11. *Assume for $f, g : E \times [0,T] \longrightarrow \widehat{\Theta}\big(E,(d_j)_j,(e_j)_j,(\lfloor\cdot\rfloor_j)_j\big)$ and $x, y : [0,T] \longrightarrow E$ that $x(\cdot)$ is a solution to the mutational equation $\overset{\circ}{x}(\cdot) \ni f(x(\cdot),\cdot)$ and $y(\cdot)$ is a solution to the mutational equation $\overset{\circ}{y}(\cdot) \ni g(y(\cdot),\cdot)$ in the tuple $\big(E,(d_j)_{j\in\mathcal{I}},(e_j)_{j\in\mathcal{I}},(\lfloor\cdot\rfloor_j)_{j\in\mathcal{I}},(\widehat{D}_j)_{j\in\mathcal{I}}\big)$.*
For some $j \in \mathcal{I}$, let $\widehat{\alpha}_j, R_j > 0$ and $\varphi_j \in C^0([0,T])$ satisfy for $\mathcal{L}^1$-a.e. $t \in [0,T]$

$$\begin{cases} \lfloor x(t)\rfloor_j,\ \lfloor y(t)\rfloor_j \ < \ R_j \\ \alpha_j(x;R_j),\ \alpha_j(y;R_j) \ \le \ \widehat{\alpha}_j \\ \widehat{D}_j\big(f(x(t),t),\ g(y(t),t);\ R_j\big) \ \le \ \varphi_j(t) \\ \lim_{h\downarrow 0}\ \widehat{D}_j\big(f(x(t),t),\ f(x(t+h),t+h);\ R_j\big) \ = \ 0. \end{cases}$$

Then, $\delta_j : [0,T] \longrightarrow [0,\infty[,\ \ t \longmapsto \inf_{z\in E:\, \lfloor z\rfloor_j < R_j} \big(d_j(z,x(t)) \,+\, d_j(z,y(t))\big)$

fulfills $$\delta_j(t) \ \le \ \Big(\delta_j(0) + \int_0^t \varphi_j(s)\ e^{-\widehat{\alpha}_j\cdot s}\, ds\Big)\cdot e^{\widehat{\alpha}_j\cdot t} \qquad \textit{for every } t\in[0,T].$$

Proof. Due to hypotheses (H3) (o), (i), the auxiliary function $[0,T] \longrightarrow [0,\infty[$, $t \longmapsto d_j(z,x(t)) + d_j(z,y(t))$ is continuous for each element $z \in E$. Hence the infimum $\delta_j(\cdot)$ with respect to all $z \in E$ with $\lfloor z \rfloor < R_j$ is lower semicontinuous.

At $\mathcal{L}^1$-almost every time $t \in [0,T[$, Lemma 9 and the reflexivity of d_j, $\widehat{D}_j(\cdot,\cdot;R_j)$ imply for every $z \in E$ with $\lfloor z \rfloor < R_j$ and any sufficiently small $h \ge 0$

$$\begin{aligned} \delta_j(t+h) \le\, & d_j\big(f(x(t),t)\,(h,z),\ x(t+h)\big) + d_j\big(f(x(t),t)\,(h,z),\ y(t+h)\big) \\ \le & \Big(d_j(z,\,x(t)) \qquad + h\cdot \sup_{[t,t+h]} \widehat{D}_j\big(f(x(t),t),\ f(x(t+\cdot),t+\cdot);R_j\big)\Big)\cdot e^{\widehat{\alpha}_j\cdot h} \\ & + \Big(d_j(z,\,y(t)) \,+\, h\cdot \sup_{[t,t+h]} \varphi_j\Big)\cdot e^{\widehat{\alpha}_j\cdot h}. \end{aligned}$$

The infimum with respect to $z \in E$ satisfying $\lfloor z \rfloor < R_j$ additionally leads to

$$\delta_j(t+h) \le \delta_j(t) \cdot e^{\widehat{\alpha}_j \cdot h} + \sup_{[t,t+h]} \varphi_j \cdot h \cdot e^{\widehat{\alpha}_j \cdot h}$$

$$\limsup_{h \downarrow 0} \frac{\delta_j(t+h) - \delta_j(t)}{h} \le \delta_j(t) \cdot \limsup_{h \downarrow 0} \frac{e^{\widehat{\alpha}_j \cdot h} - 1}{h} + \varphi_j(t) \cdot \limsup_{h \downarrow 0} e^{\widehat{\alpha}_j \cdot h}$$
$$= \delta_j(t) \cdot \widehat{\alpha}_j + \varphi_j(t).$$

Finally the claim results directly from Gronwall's inequality (in Proposition A.2). □

Remark 12. $\delta(t) \le d_j(x(t), y(t))$ results directly from the reflexivity of d_j (due to hypothesis (H1)). If d_j satisfies the triangle inequality in addition, then this infimum $\delta(t)$ is always equal to $d_j(x(t), y(t))$.

3.3.2 Limits of Graphically Converging Solutions: Convergence Theorem

On our way to the existence of solutions, the next step focuses on the question which kind of convergence preserves the solution property.

In Theorem 2.13 (on page 110), pointwise convergence has already proved to be appropriate under the assumptions that all solutions $x_n(\cdot) : [0, T] \longrightarrow E$ are uniformly Lipschitz continuous and that d_j is a pseudo-metric. Now we weaken the conditions on convergence and admit perturbations with respect to time as specified in subsequent assumption (4.) — although d_j does not have to fulfill the triangle inequality any longer.

Here the two families of distance functions $(d_j)_{j \in \mathcal{I}}$, $(e_j)_{j \in \mathcal{I}}$ come into play explicitly for the first time.
In the next theorem, we consider an appropriately converging sequence $(x_n(\cdot))_{n \in \mathbb{N}}$ of solutions, each of which is continuous with respect to every e_j by definition. Concluding the continuity of their limit function usually requires some form of "equi-continuity". For this purpose, the family $(e_j)_{j \in \mathcal{I}}$ is used instead of $(d_j)_{j \in \mathcal{I}}$ and, we suppose uniform Lipschitz continuity with respect to each e_j $(j \in \mathcal{I})$.
Strictly speaking, this Lipschitz continuity is a "quantitative" feature and, we now separate its distance functions from the other quantitative properties of solutions (such as condition (2.′) in Definition 8). "Qualitative" aspects like the topological concepts of convergence and continuity, however, are not concerned — due to hypothesis (H3) (o).
These separate families of distance functions and the continuity assumptions replacing the triangle inequality are two new aspects of the mutational framework in this chapter.

Theorem 13 (Convergence of solutions to mutational equations).
Suppose the following properties of

$$\begin{array}{lll} f_n,\, f : E \times [0,T] & \longrightarrow \widehat{\Theta}\big(E, (d_i)_{i\in\mathcal{I}}, (e_j)_{j\in\mathcal{I}}, (\lfloor\cdot\rfloor_i)_{i\in\mathcal{I}}\big) & (n \in \mathbb{N}) \\ x_n,\, x : \quad [0,T] & \longrightarrow E : & \end{array}$$

1.) $R_j \quad := \sup_{n,t} \ \lfloor x_n(t)\rfloor_j < \infty,$

$\widehat{\alpha}_j(\rho) := \sup_n \ \alpha_j\big(x_n;\rho\big) < \infty$ *for* $\rho \geq 0,$

$\widehat{\beta}_j \quad := \sup_n \ \mathrm{Lip}\,\big(x_n(\cdot) : [0,T] \longrightarrow (E,e_j)\big) < \infty$ *for every* $j \in \mathcal{I}$,

2.) $\overset{\circ}{x}_n(\cdot) \ni f_n(x_n(\cdot),\cdot)$ *(in the sense of Definition 8 on page 187) for every* $n \in \mathbb{N}$,

3.) *Equi-continuity of* $(f_n)_n$ *at* $(x(t),t)$ *at almost every time in the following sense: for* $\mathcal{L}^1$*-almost every* $t \in [0,T]$: $\lim_{n\to\infty} \widehat{D}_j\,(f_n(x(t),t),\ f_n(y_n,t_n);\ r) = 0$ *for each* $j \in \mathcal{I}$, $r \geq 0$ *and any* $(t_n)_{n\in\mathbb{N}}$, $(y_n)_{n\in\mathbb{N}}$ *in* $[t,T]$ *and* E *respectively satisfying* $\lim_{n\to\infty} t_n = t$ *and* $\lim_{n\to\infty} d_i\big(x(t),y_n\big) = 0$, $\sup_{n\in\mathbb{N}} \lfloor y_n\rfloor_i \leq R_i$ *for each* i,

4.) *For* $\mathcal{L}^1$*-almost every* $t \in [0,T[$ $(t = 0$ *inclusive) and any* $\widetilde{t} \in]t,T[$, *there is a sequence* $n_m \nearrow \infty$ *of indices (depending on* $t < \widetilde{t}$*) that satisfies for* $m \longrightarrow \infty$

$$\begin{cases} (i)\ \ \widehat{D}_j\big(f(x(t),t),\ f_{n_m}(x(t),t);\ r\big) \ \longrightarrow\ 0 \qquad \text{for all } r \geq 0,\ j \in \mathcal{I}, \\ (ii)\ \ \text{there is a sequence } \delta_m \searrow 0:\ d_j\big(x(t),\ x_{n_m}(t+\delta_m)\big) \longrightarrow 0 \ \text{for all } j, \\ (iii)\ \text{there is a sequence } \widetilde{\delta}_m \searrow 0:\ d_j\big(x(\widetilde{t}),\ x_{n_m}(\widetilde{t}-\widetilde{\delta}_m)\big) \longrightarrow 0 \ \text{for all } j. \end{cases}$$

Then, $x(\cdot) : [0,T] \longrightarrow E$ *is a solution to the mutational equation* $\overset{\circ}{x}(\cdot) \ni f(x(\cdot),\cdot)$ *in the tuple* $\big(E,\ (d_j)_{j\in\mathcal{I}},\ (e_j)_{j\in\mathcal{I}},\ (\lfloor\cdot\rfloor_j)_{j\in\mathcal{I}},\ (\widehat{D}_j)_{j\in\mathcal{I}}\big)$.

Remark 14. Assumptions (4.ii) and (4.iii) admit small perturbations with respect to time. This is much weaker than pointwise convergence (as in Theorem 2.13 on page 110) and, it can be regarded as a generalized form of converging graphs.
In regard to the influence of index $j \in \mathcal{I}$, however, assumptions (3.) and (4) are slightly stronger than in Theorem 2.13 because we have replaced the triangle inequality of distance functions by hypotheses (H3), (H6), which draw conclusions only from convergence of sequences with respect to all $i \in \mathcal{I}$ simultaneously.

Proof (of Theorem 13). Choose the index $j \in \mathcal{I}$ arbitrarily.
Then $x(\cdot) : [0,T] \longrightarrow (E,e_j)$ is $\widehat{\beta}_j$-Lipschitz continuous. Indeed, for Lebesgue-almost every $t \in [0,T[$ and any $\widetilde{t} \in]t,T]$, assumption (4.) provides a subsequence $\big(x_{n_m}(\cdot)\big)_{m\in\mathbb{N}}$ and sequences $\delta_m \searrow 0, \widetilde{\delta}_m \searrow 0$ satisfying for each $i \in \mathcal{I}$

$$\begin{cases} d_i\big(x(t),\ x_{n_m}(t+\delta_m)\big) \longrightarrow 0 \\ d_i\big(x(\widetilde{t}),\ x_{n_m}(\widetilde{t}-\widetilde{\delta}_m)\big) \longrightarrow 0 \end{cases} \qquad \text{for } m \longrightarrow \infty.$$

The uniform $\widehat{\beta}_j$-Lipschitz continuity of $x_n(\cdot), n \in \mathbb{N}$, with respect to e_j and hypothesis (H3) (i) (on page 182) imply

$$\begin{aligned} e_j\big(x(t),\, x(\widetilde{t})\big) &\le \limsup_{m\to\infty} \;\; e_j\big(x_{n_m}(t+\delta_m),\; x_{n_m}(\widetilde{t}-\widetilde{\delta}_m)\big) \\ &\le \limsup_{m\to\infty} \;\; \widehat{\beta}_j\, |\widetilde{t}-\widetilde{\delta}_m \;-\; t-\delta_m| \\ &\le \widehat{\beta}_j\, |\widetilde{t}-t|\,. \end{aligned}$$

This Lipschitz inequality can be extended to *any* $t, \widetilde{t} \in [0,T]$ due to the lower semicontinuity of e_j (according to hypotheses (H3) (o), (i)). Moreover, hypothesis (H4) about the lower semicontinuity of $\lfloor\cdot\rfloor_j$ ensures

$$\lfloor x(\widetilde{t})\rfloor_j \;\le\; \liminf_{m\to\infty} \lfloor x_{n_m}(\widetilde{t})\rfloor_j \;\le\; R_j.$$

Finally we verify the solution property

$$\limsup_{h\downarrow 0} \frac{d_j(\vartheta(s+h,z),\, x(t+h)) \;-\; d_j(\vartheta(s,z),\, x(t))\cdot e^{\alpha_j(x;\rho)\, h}}{h} \;\le\; \widehat{D}_j\big(\vartheta,\; f(x(t),t);\; \rho\big)$$

for $\mathscr{L}^1$-almost every $t \in [0,T[$ and for any $\vartheta \in \widehat{\Theta}\big(E, (d_i)_{i\in\mathscr{I}}, (e_i)_{i\in\mathscr{I}}, (\lfloor\cdot\rfloor_i)_{i\in\mathscr{I}}\big)$, $s \in [0,1[,\;\; z\in E,\;\; \rho \ge R_j$ with $\lfloor\vartheta(\cdot,z)\rfloor_j \le \rho$,
Indeed, for Lebesgue-almost every $t\in[0,T[$ and any $h \in\,]0,\, T-t[$, assumption (4.) guarantees a subsequence $\big(x_{n_m}(\cdot)\big)_{m\in\mathbb{N}}$ and sequences $\delta_m \searrow 0$, $\widetilde{\delta}_m \searrow 0$ satisfying for each $i\in\mathscr{I}$, $r\ge 0$ and $m \longrightarrow \infty$

$$\begin{cases} \widehat{D}_i\big(f(x(t),t),\; f_{n_m}(x(t),t);\; r\big) \longrightarrow 0 \\ d_i\big(x(t), \qquad x_{n_m}(t+\delta_m)\big) \qquad \longrightarrow 0 \\ d_i\big(x(t+h),\, x_{n_m}(t+h-\widetilde{\delta}_m)\big) \longrightarrow 0. \end{cases}$$

Now we conclude from Lemma 9 (on page 188) and the continuity of d_j (due to hypothesis (H3) (i) on page 182) respectively

$$\begin{aligned} & d_j\big(\vartheta(s+h,z),\;\, x(t+h)\big) \\ &= \lim_{m\to\infty} \;\; d_j\big(\vartheta(s+h-\widetilde{\delta}_m, z),\;\, x_{n_m}(t+h-\widetilde{\delta}_m)\big) \\ &\le \limsup_{m\to\infty} \Big(d_j\big(\vartheta(s+\delta_m,\;\, z),\qquad x_{n_m}(t+\delta_m)\big) \\ &\qquad\qquad + h\cdot \sup_{[t+\delta_m,\; t+h-\widetilde{\delta}_m]} \widehat{D}_j\big(\vartheta,\;\, f_{n_m}(x_{n_m}(\cdot),\cdot);\;\, \rho\big)\Big)\cdot e^{\widehat{\alpha}_j(\rho)\cdot(h-\delta_m-\widetilde{\delta}_m)} \\ &\le \Big(d_j\big(\vartheta(s,z),\, x(t)\big) + h\cdot \limsup_{m\to\infty}\;\; \sup_{[t+\delta_m,\, t+h]} \widehat{D}_j\big(\vartheta,\, f_{n_m}(x_{n_m}(\cdot),\cdot);\, \rho\big)\Big)\cdot e^{\widehat{\alpha}_j(\rho)\, h}. \end{aligned}$$

(In fact, the last inequality justifies why (H3) (i) provides the continuity of d_j and not just its lower semicontinuity as for e_j.) For completing the proof, we verify

$$\limsup_{h\downarrow 0}\; \limsup_{m\to\infty} \;\; \sup_{[t+\delta_m,\; t+h]} \widehat{D}_j\big(\vartheta,\, f_{n_m}(x_{n_m}(\cdot),\cdot);\, \rho\big) \;\le\; \widehat{D}_j\big(\vartheta,\; f(x(t),t);\; \rho\big)$$

for Lebesgue-almost every $t\in[0,T[$ and *any* subsequence $\big(x_{n_m}(\cdot)\big)_{m\in\mathbb{N}}$ satisfying

$$\begin{cases} d_i\big(x(t), \qquad x_{n_m}(t+\delta_m)\big) \qquad \longrightarrow 0 \\ \widehat{D}_i\big(f(x(t),t),\, f_{n_m}(x(t),t);\; r\big) \longrightarrow 0 \end{cases}$$

for $m \longrightarrow \infty$ and each $i \in \mathcal{I}$, $r \geq 0$. Indeed, if this inequality was not correct then we could select some $\varepsilon > 0$ and sequences $(h_l)_{l\in\mathbb{N}}$, $(m_l)_{l\in\mathbb{N}}$, $(s_l)_{l\in\mathbb{N}}$ such that

$$\begin{cases} \widehat{D}_j\big(\vartheta,\ f_{n_{m_l}}(x_{n_{m_l}}(t+s_l), t+s_l);\ \rho\big) \ \geq\ \widehat{D}_j\big(\vartheta,\ f(x(t),t);\ \rho\big) \ +\ \varepsilon \\ \delta_{m_l} \leq s_l \leq h_l \leq \frac{1}{l}, \qquad m_l \geq l \end{cases} \quad \text{for all } l \in \mathbb{N}.$$

Due to property (H3) (ii), the uniform Lipschitz continuity of $(x_{n_m}(\cdot))_{m\in\mathbb{N}}$ implies

$$\lim_{l\to\infty} d_i\big(x(t),\ x_{n_{m_l}}(t+s_l)\big) \ =\ 0$$

for each $i \in \mathcal{I}$. Thus at $\mathcal{L}^1$-almost every time $t \in [0,T[$, assumptions (3.), (4.) (i) and hypothesis (H6) about the continuity of $\widehat{D}_j(\cdot,\cdot;r)$ (on page 184) lead to a contradiction because for any $r \geq 0$,

$$\lim_{l\to\infty} \ \widehat{D}_j\big(\vartheta,\ f_{n_{m_l}}(x_{n_{m_l}}(t+s_l), t+s_l);\ r\big) \ =\ \widehat{D}_j\big(\vartheta,\ f(x(t),t);\ r\big). \qquad \square$$

3.3.3 Existence for Mutational Equations with Delay and without State Constraints

Although the modified topological assumptions (H1)–(H7) have replaced the triangle inequality, Euler method in combination with Euler compactness (almost) leads to the existence of solutions to mutational equations without state constraints. We can even draw our conclusions for mutational equations *with delay* in essentially the same way as in § 2.3.5 (on page 120 ff.). The proofs are again postponed to the end of this section.

Remark 15. (1.) The set $\mathrm{BLip}\big(I, E; (d_j)_{j\in\mathcal{I}}, (\lfloor\cdot\rfloor_j)_{j\in\mathcal{I}}\big)$ consists of all "bounded" and Lipschitz continuous functions $I \longrightarrow E$ as in Definition 2.22 (on page 120).

(2.) The term "Euler compact" was introduced in Definition 2.15 (on page 112) and does not have to be adapted significantly to the modified topological environment in this chapter.
Indeed, $\big(E,\ (d_j)_{j\in\mathcal{I}},\ (e_j)_{j\in\mathcal{I}},\ (\lfloor\cdot\rfloor_j)_{j\in\mathcal{I}},\ \widehat{\Theta}\big(E,(d_i)_{i\in\mathcal{I}},(e_i)_{i\in\mathcal{I}},(\lfloor\cdot\rfloor_i)_{i\in\mathcal{I}}\big)\big)$ is called *Euler compact* if it satisfies the following condition for any initial element $x_0 \in E$, time $T \in]0,\infty[$ and bounds $\widehat{\alpha}_j, \widehat{\beta}_j, \widehat{\gamma}_j > 0$ ($j \in \mathcal{I}$):
Let $\mathcal{N} = \mathcal{N}(x_0, T, (\widehat{\alpha}_j, \widehat{\beta}_j, \widehat{\gamma}_j)_{j\in\mathcal{I}})$ denote the (possibly empty) subset of all curves $y(\cdot) : [0,T] \longrightarrow E$ constructed in the following piecewise way: Choosing an arbitrary equidistant partition $0 = t_0 < t_1 < \ldots < t_n = T$ of $[0,T]$ (with $n > T$) and transitions $\vartheta_1 \ldots \vartheta_n \in \widehat{\Theta}\big(E,(d_i)_{i\in\mathcal{I}},(e_i)_{i\in\mathcal{I}},(\lfloor\cdot\rfloor_i)_{i\in\mathcal{I}}\big)$ with

$$\begin{cases} \sup_k \ \ \gamma_j(\vartheta_k) & \leq\ \widehat{\gamma}_j \\ \sup_k \ \ \alpha_j\big(\vartheta_k;\ (\lfloor x_0\rfloor_j + \widehat{\gamma}_j\, T)\, e^{\widehat{\gamma}_j\, T}\big) & \leq\ \widehat{\alpha}_j \\ \sup_k \ \ \beta_j\big(\vartheta_k;\ (\lfloor x_0\rfloor_j + \widehat{\gamma}_j\, T)\, e^{\widehat{\gamma}_j\, T}\big) & \leq\ \widehat{\beta}_j \end{cases}$$

for each index $j \in \mathcal{I}$, define $y(\cdot) : [0,T] \longrightarrow E$ as

$$y(0) := x_0, \qquad y(t) := \vartheta_k\,(t - t_{k-1},\ y(t_{k-1})) \quad \text{for } t \in]t_{k-1}, t_k],\ k = 1,2\ldots n.$$

Then for each $t \in [0,T]$, every sequence $(z_n)_{n\in\mathbb{N}}$ in $\{y(t) \mid y(\cdot) \in \mathcal{N}\} \subset E$ has a subsequence $(z_{n_m})_{m\in\mathbb{N}}$ and some $z \in E$ with $\lim_{m\to\infty} d_j(z_{n_m}, z) = 0$ for each $j \in \mathcal{I}$.

Since d_j and e_j are now lacking the triangle inequality, we have to cope with a further difficulty: Are curves defined by transitions in a piecewise way like

$$[0,2] \longrightarrow E, \quad t \longmapsto \begin{cases} \vartheta_1(t,x_0) & \text{for } t \in [0,1] \\ \vartheta_2\big(t-1,\ \vartheta_1(1,x_0)\big) & \text{for } t \in]1,2] \end{cases}$$

still always Lipschitz continuous with respect to each e_j ? In particular, Lemma 2.10 (on page 107) might fail if $d_j \equiv e_j$ was not a pseudo-metric.
Corollary 10 (on page 188) has already provided a sufficient condition on the transition set for verifying Lipschitz continuity with respect to d_j, namely via identity transition. In regard to subsequent results about the existence of solutions, however, we prefer introducing a separate assumption focusing on Euler approximations and the distance function e_j $(j \in \mathcal{I})$:

Definition 16.
The tuple $\big(E, (d_j)_{j\in\mathcal{I}}, (e_j)_{j\in\mathcal{I}}, (\lfloor\cdot\rfloor_j)_{j\in\mathcal{I}}, \widehat{\Theta}\big(E,(d_i)_{i\in\mathcal{I}},(e_i)_{i\in\mathcal{I}},(\lfloor\cdot\rfloor_i)_{i\in\mathcal{I}}\big)\big)$ is called *Euler equi-continuous* if it satisfies the following condition for any initial element $x_0 \in E$, time $T \in]0,\infty[$ and bounds $\widehat{\alpha}_j, \widehat{\beta}_j, \widehat{\gamma}_j > 0$ $(j \in \mathcal{I})$:
Let $\mathcal{N} = \mathcal{N}(x_0, T, (\widehat{\alpha}_j, \widehat{\beta}_j, \widehat{\gamma}_j)_{j\in\mathcal{I}})$ denote the (possibly empty) subset of all curves $y(\cdot): [0,T] \longrightarrow E$ constructed in the following piecewise way (as in Definition 2.15 on page 112): Choosing an arbitrary equidistant partition $0 = t_0 < t_1 < \ldots < t_n = T$ of $[0,T]$ (with $n > T$) and transitions $\vartheta_1 \ldots \vartheta_n \in \widehat{\Theta}\big(E,(d_i)_{i\in\mathcal{I}},(e_i)_{i\in\mathcal{I}},(\lfloor\cdot\rfloor_i)_{i\in\mathcal{I}}\big)$ with

$$\begin{cases} \sup_k \ \gamma_j(\vartheta_k) & \le \widehat{\gamma}_j \\ \sup_k \ \alpha_j\big(\vartheta_k;\ (\lfloor x_0\rfloor_j + \widehat{\gamma}_j\, T)\, e^{\widehat{\gamma}_j\, T}\big) & \le \widehat{\alpha}_j \\ \sup_k \ \beta_j\big(\vartheta_k;\ (\lfloor x_0\rfloor_j + \widehat{\gamma}_j\, T)\, e^{\widehat{\gamma}_j\, T}\big) & \le \widehat{\beta}_j \end{cases}$$

for each index $j \in \mathcal{I}$, define $y(\cdot): [0,T] \longrightarrow E$ as

$$y(0) := x_0, \qquad y(t) := \vartheta_k\,(t - t_{k-1},\ y(t_{k-1})) \quad \text{for } t \in]t_{k-1}, t_k],\ k = 1,2\ldots n.$$

Then for each index $j \in \mathcal{I}$, there is a constant $L_j \in [0,\infty[$ such that every curve $y(\cdot) \in \mathcal{N}$ is L_j-Lipschitz continuous with respect to e_j.

Remark 17. If $d_j \equiv e_j$ is a pseudo-metric then Euler equi-continuity (with $L_j := \widehat{\beta}_j$) results directly from the triangle inequality and Lemma 2.10 (on page 107) in a piecewise way.

This additional hypothesis opens the door to selecting "pointwise converging" subsequences of Euler approximations and, we obtain the counterpart of Lemma 2.17 (on page 113) — but with a weaker type of convergence. The subsequent main result about existence is based on this pointwise convergence and specifies continuity assumption (4.) in a stricter way than its counterpart in Proposition 2.23 (on page 120):

Lemma 18 (Euler compact $\wedge$ Euler equi-continuous $\Longrightarrow$ pointwise compact). *Assume* $\left(E, (d_j)_{j\in\mathscr{I}}, (e_j)_{j\in\mathscr{I}}, (\lfloor\cdot\rfloor_j)_{j\in\mathscr{I}}, \widehat{\Theta}\big(E,(d_i)_{i\in\mathscr{I}},(e_i)_{i\in\mathscr{I}},(\lfloor\cdot\rfloor_i)_{i\in\mathscr{I}}\big)\right)$ *to be Euler compact and Euler equi-continuous. Using the notation of Definition 16, choose any initial element* $x_0 \in E$, *time* $T \in]0,\infty[$ *and bounds* $\widehat{\alpha}_j, \widehat{\beta}_j, \widehat{\gamma}_j > 0$ $(j \in \mathscr{I})$.

For every sequence $(y_n(\cdot))_{n\in\mathbb{N}}$ *of curves* $[0,T] \longrightarrow E$ *in* $\mathscr{N}\big(x_0, T, (\widehat{\alpha}_j, \widehat{\beta}_j, \widehat{\gamma}_j)_{j\in\mathscr{I}}\big)$, *there exists a subsequence* $(y_{n_m}(\cdot))_{m\in\mathbb{N}}$ *and a function* $y(\cdot) : [0,T] \longrightarrow E$ *such that* $y(\cdot)$ *is Lipschitz continuous with respect to each* e_j *and for every* $j \in \mathscr{I}$, $t \in [0,T]$,

$$d_j\big(y_{n_m}(t),\, y(t)\big) \longrightarrow 0 \qquad \text{for } m \longrightarrow \infty.$$

Furthermore if $(y_n(t_0))_{n\in\mathbb{N}}$ *is constant for some* $t_0 \in [0,T]$ *then* $y(\cdot)$ *can be chosen with the additional property* $y(t_0) = y_n(t_0)$.

Theorem 19 (Existence of solutions to mutational equations with delay). *Suppose* $\left(E, (d_j)_{j\in\mathscr{I}}, (e_j)_{j\in\mathscr{I}}, (\lfloor\cdot\rfloor_j)_{j\in\mathscr{I}}, \widehat{\Theta}\big(E,(d_i)_{i\in\mathscr{I}},(e_i)_{i\in\mathscr{I}},(\lfloor\cdot\rfloor_i)_{i\in\mathscr{I}}\big)\right)$ *to be Euler compact and Euler equi-continuous. Moreover assume for some fixed* $\tau \geq 0$, *the function*

$$f : \mathrm{BLip}\big([-\tau,0], E; (e_i)_i, (\lfloor\cdot\rfloor_i)_i\big) \times [0,T] \longrightarrow \widehat{\Theta}\big(E,(d_i)_i,(e_i)_i,(\lfloor\cdot\rfloor_i)_i\big)$$

and each $j \in \mathscr{I}$, $R > 0$:

1.) $\sup\limits_{z(\cdot),t} \alpha_j(f(z(\cdot),t);R) < \infty$,
2.) $\sup\limits_{z(\cdot),t} \beta_j(f(z(\cdot),t);R) < \infty$,
3.) $\sup\limits_{z(\cdot),t} \gamma_j(f(z(\cdot),t)) < \infty$,
4.) *for* $\mathscr{L}^1$*-almost every* $t \in [0,T]$: $\lim\limits_{n\to\infty} \widehat{D}_j\big(f(z_n^1(\cdot),t_n^1),\, f(z_n^2(\cdot),t_n^2); R\big) = 0$ *for each* $j \in \mathscr{I}$, $R \geq 0$ *and any sequences* $(t_n^1)_{n\in\mathbb{N}}$, $(t_n^2)_{n\in\mathbb{N}}$ *in* $[0,T]$ *and* $(z_n^1(\cdot))_{n\in\mathbb{N}}$, $(z_n^2(\cdot))_{n\in\mathbb{N}}$ *in* $\mathrm{BLip}\big([-\tau,0], E; (e_j)_{j\in\mathscr{I}}, (\lfloor\cdot\rfloor_j)_{j\in\mathscr{I}}\big)$ *satisfying for every* $i \in \mathscr{I}$ *and* $s \in [-\tau,0]$

$$\lim_{n\to\infty} t_n^1 = t = \lim_{n\to\infty} t_n^2, \quad \lim_{n\to\infty} d_i\big(z_n^1(s), z(s)\big) = 0 = \lim_{n\to\infty} d_i\big(z_n^2(s), z(s)\big)$$
$$\sup_{n\in\mathbb{N}} \sup_{[-\tau,0]} \lfloor z_n^{1,2}(\cdot)\rfloor_i < \infty.$$

For every function $x_0(\cdot) \in \mathrm{BLip}\big([-\tau,0], E; (e_j)_{j\in\mathscr{I}}, (\lfloor\cdot\rfloor_j)_{j\in\mathscr{I}}\big)$, *there exists a curve* $x(\cdot) : [-\tau,T] \longrightarrow E$ *with the following properties:*

(i) $x(\cdot) \in \mathrm{BLip}\big([-\tau,T], E; (e_j)_{j\in\mathscr{I}}, (\lfloor\cdot\rfloor_j)_{j\in\mathscr{I}}\big)$,
(ii) $x(\cdot)\big|_{[-\tau,0]} = x_0(\cdot)$,
(iii) *the restriction* $x(\cdot)\big|_{[0,T]}$ *is a solution to the mutational equation*

$$\overset{\circ}{x}(t) \ni f\big(x(t+\cdot)\big|_{[-\tau,0]}, t\big)$$

in the sense of Definition 8 (on page 187).

Proof (of Lemma 18). Fix $x_0 \in E$, time $T \in]0,\infty[$ and bounds $\widehat{\alpha}_j, \widehat{\beta}_j, \widehat{\gamma}_j > 0$ $(j \in \mathcal{I})$ arbitrarily. Moreover without loss of generality, we assume the set of curves $\mathcal{N} = \mathcal{N}(x_0, T, (\widehat{\alpha}_j, \widehat{\beta}_j, \widehat{\gamma}_j)_{j\in\mathcal{I}})$ to be nonempty. Supposing Euler equi-continuity provides a constant $L_j \in [0,\infty[$ for each index $j \in \mathcal{I}$ such that every curve $y(\cdot) \in \mathcal{N}$ is L_j-Lipschitz constant with respect to e_j. Let $(y_n(\cdot))_{n\in\mathbb{N}}$ be any sequence in $\mathcal{N}$.

We focus on a pointwise converging subsequence and adapt the proof of Lemma 2.17 (on page 114):
For each $t \in [0,T]$, the assumption of Euler compactness ensures a subsequence of $\big(y_n(t)\big)_{n\in\mathbb{N}}$ converging with respect to each d_j. Cantor's diagonal construction provides a subsequence $\big(y_{n_m}(\cdot)\big)_{m\in\mathbb{N}}$ of functions $[0,T] \longrightarrow E$ with the additional property that at every *rational* time $t \in [0,T]$, an element $y(t) \in E$ satisfies

$$d_j\big(y_{n_m}(t),\, y(t)\big) \longrightarrow 0 \qquad \text{for } m \longrightarrow \infty$$

and each $j \in \mathcal{I}$ since the subset $\mathbb{Q} \cap [0,T]$ of rational numbers in $[0,T]$ is countable.

Now we consider any $t \in [0,T] \setminus \mathbb{Q}$. Due to Euler compactness, there exists a subsequence $\big(y_{n_{m_l}}(t)\big)_{l\in\mathbb{N}}$ maybe depending on t, but converging to an element $y(t) \in E$ with respect to each d_j. Lacking the triangle inequality of d_j, however, we conclude from hypothesis (H3) (on page 182)

$$\lim_{m\longrightarrow\infty} d_j\big(y_{n_m}(t),\, y(t)\big) = 0 \qquad \text{for each } j \in \mathcal{I}.$$

Indeed, assumption (H3) (i) implies for every $s \in [0,T] \cap \mathbb{Q}$ and $j \in \mathcal{I}$

$$e_j\big(y(s),\, y(t)\big) \le \limsup_{l\longrightarrow\infty} e_j\big(y_{n_{m_l}}(s),\, y_{n_{m_l}}(t)\big) \le L_j\, |s-t|.$$

Now choose any sequence $(s_k)_{k\in\mathbb{N}}$ in $[0,T] \cap \mathbb{Q}$ with $s_k \longrightarrow t$ $(k \to \infty)$. This implies

$$\sup_{n\in\mathbb{N}} e_j\big(y_n(s_k),\, y_n(t)\big) \le L_j\, |t-s_k| \longrightarrow 0 \qquad \text{for } k \to \infty$$

and each index $j \in \mathcal{I}$. Together with

$$\lim_{m\longrightarrow\infty} d_j\big(y_{n_m}(s_k),\, y(s_k)\big) = 0 \qquad \text{for every } k \in \mathbb{N}, j \in \mathcal{I},$$

we conclude from hypothesis (H3) (iii) directly

$$\lim_{m\longrightarrow\infty} d_j\big(y_{n_m}(t),\, y(t)\big) = 0 \qquad \text{for each } j \in \mathcal{I}.$$

Finally, hypothesis (H3) (i) ensures the L_j-Lipschitz continuity of $y(\cdot)$ w.r.t. e_j. □

Remark 20. In this proof of Lemma 18, we have applied hypothesis (H3) (iii) for the first time. Indeed, all other conclusions are based on hypotheses (H3) (i) or (H3) (ii) in combination with assumption (H3) (o).
For examples with a separate real time component, we are free to draw the same conclusions under the additional assumption that either $s_k \ge t$ for all $k \in \mathbb{N}$ or $s_k \le t$ for every $k \in \mathbb{N}$. This opens the door to taking a form of "time orientation" into consideration as mentioned in Remark 4 (on page 184) and explained in subsequent § 3.5 (on page 221 ff.).

Proof (of Theorem 19). As in the proof of Proposition 2.23 (on page 121 f.), we use a sequence of Euler approximations on equidistant partitions of $[0,T]$. For every $n \in \mathbb{N}$ with $2^n > T$, set

$$\begin{aligned} h_n &:= \tfrac{T}{2^n}, \qquad t_n^k := k\,h_n \qquad \text{for } k = 0\ldots 2^n, \\ x_n(\cdot)\big|_{[-\tau,0]} &:= x_0, \\ x_n(t) &:= f(x_n(t_n^k+\cdot)\big|_{[-\tau,0]}, t_n^k)\big(t-t_n^k,\ x_n(t_n^k)\big) \quad \text{for } t \in\,]t_n^k, t_n^{k+1}],\ k<2^n. \end{aligned}$$

Due to Euler equi-continuity, there is a constant $L_j \in [0,\infty[$ for each index $j \in \mathscr{I}$ such that every curve $x_n(\cdot)$ is L_j-Lipschitz continuous with respect to e_j. Setting $\widehat{\gamma}_j := \sup\ \gamma_j(f(\cdot,\cdot)) < \infty$ as further abbreviation, Lemma 5 (on page 185) provides for every $t \in [0,T],\ n \in \mathbb{N}$ (with $2^n > T$) and each $j \in \mathscr{I}$

$$\lfloor x_n(t)\rfloor_j \ \le\ \big(\lfloor x_0(0)\rfloor_j + \widehat{\gamma}_j\,T\big)\cdot e^{\widehat{\gamma}_j\,T} =: R_j.$$

Assumptions (1.)–(3.) are combined with Euler compactness and Euler equi-continuity. Thus, Lemma 18 guarantees that a subsequence $\big(x_{n_m}(\cdot)\big)_{m\in\mathbb{N}}$ converges to a function $x(\cdot) : [-\tau,\,T] \longrightarrow E$ in the sense that for every $j \in \mathscr{I}$ and $t \in [-\tau,\,T]$,

$$d_j\big(x_{n_m}(t),\, x(t)\big) \longrightarrow 0 \qquad \text{for } m \longrightarrow \infty.$$

In particular, $x(\cdot) = x_0(\cdot)$ in $[-\tau,\,0]$.

For every $t \in [0,T]$, the estimate $\lfloor x(t)\rfloor_j \le R_j$ results from hypothesis (H4) about the lower semicontinuity of $\lfloor\cdot\rfloor_j$ (on page 182) and, $x(\cdot) : [-\tau,\,T] \longrightarrow (E,e_j)$ is also L_j-Lipschitz continuous due to the lower semicontinuity of e_j (in hypothesis (H3) (i)). Hence we obtain

$$x(\cdot) \in \mathrm{BLip}\big([-\tau,T],E;\,(e_j)_{j\in\mathscr{I}},\,(\lfloor\cdot\rfloor_j)_{j\in\mathscr{I}}\big).$$

Finally it is a consequence of Convergence Theorem 13 (on page 191) that

$$\limsup_{h\downarrow 0}\ \frac{d_j\big(\vartheta(s+h,z),x(t+h)\big) - d_j\big(\vartheta(s,z),x(t)\big)\,e^{\widehat{\alpha}_j(\rho)\,h}}{h} \ \le\ \widehat{D}_j\big(\vartheta,\ f\big(x(t+\cdot)\big|_{[-\tau,0]},\,t\big);\,\rho\big)$$

holds for $\mathscr{L}^1$-almost every $t \in [0,T]$ and arbitrary $j \in \mathscr{I},\ \rho \ge R_j,\ s \in [0,1[$, $z \in E,\ \vartheta \in \widehat{\Theta}\big(E,(d_i)_{i\in\mathscr{I}},(e_i)_{i\in\mathscr{I}},(\lfloor\cdot\rfloor_i)_{i\in\mathscr{I}}\big)$ with $\lfloor\vartheta(\cdot,z)\rfloor_j \le \rho$. Indeed, each Euler approximation $x_n(\cdot) : [0,T] \longrightarrow E$, $n \in \mathbb{N}$, can be regarded as a solution of

$$\overset{\circ}{x}_n(\cdot) \ni \widehat{f}_n(\cdot)$$

with the auxiliary function

$$\begin{aligned} \widehat{f}_n : [0,T] &\longrightarrow \widehat{\Theta}\big(E,(d_j)_{j\in\mathscr{I}},(e_j)_{j\in\mathscr{I}},(\lfloor\cdot\rfloor_j)_{j\in\mathscr{I}}\big), \\ \widehat{f}_n(t) &:= f\big(x_n(\cdot)\big|_{[t_n^k-\tau,\,t_n^k]},\,t_n^k\big) \qquad \text{for any } t \in [t_n^k, t_n^{k+1}[,\ k<2^n. \end{aligned}$$

Similarly set
$$\begin{aligned} \widehat{f} : [0,T] &\longrightarrow \widehat{\Theta}\big(E,(d_j)_{j\in\mathscr{I}},(e_j)_{j\in\mathscr{I}},(\lfloor\cdot\rfloor_j)_{j\in\mathscr{I}}\big), \\ t &\longmapsto f\big(x(t+\cdot)\big|_{[-\tau,0]},\,t\big). \end{aligned}$$

At $\mathscr{L}^1$-almost every time $t \in [0,T]$, assumption (4.) has two key consequences. First, with the abbreviation $t_{nm}^k := [\frac{t}{h_{nm}}]\, h_{nm} \in \mathbb{N}\, h_{nm}$,

$$\widehat{D}_j\big(\widehat{f}(t),\ \widehat{f}_{nm}(t);\ \rho\big) \;=\; \widehat{D}_j\big(f(x(t+\cdot)|_{[-\tau,0]},\, t),\ f(x_{nm}(t_{nm}^k+\cdot)|_{[-\tau,0]},\, t_{nm}^k);\ \rho\big) \;\overset{m\to\infty}{\longrightarrow}\; 0,$$

for every $j \in \mathscr{I}$ and $\rho \ge R_j$ because for any index $i \in \mathscr{I}$ and $t \in [0,T]$, $s \in [-\tau,0]$, the pointwise convergence of $(x_{n_m}(\cdot))_{m\in\mathbb{N}}$ and continuity property (H3) (ii) imply

$$d_i\big(x(t+s),\ x_{n_m}(t_{nm}^k+s)\big) \;\overset{m\to\infty}{\longrightarrow}\; 0\,.$$

Second, we obtain for any sequence $t_m \longrightarrow t$ in $[0,T]$ and for every $j \in \mathscr{I}$, $\rho \ge R_j$

$$\begin{aligned}\widehat{D}_j\big(\widehat{f}_{n_m}(t),\ \widehat{f}_{n_m}(t_m);\ \rho\big) \;=\; &\widehat{D}_j\big(f(x_{n_m}(t_{nm}^k+\cdot)|_{[-\tau,0]},\, t_{nm}^k),\\ &f(x_{n_m}(t_{nm}^{l_m}+\cdot)|_{[-\tau,0]},\, t_{nm}^{l_m});\ \rho\big) \;\overset{m\to\infty}{\longrightarrow}\; 0\end{aligned}$$

with the abbreviations $t_{nm}^k := [\frac{t}{h_{nm}}]\, h_{nm}$, $t_{nm}^{l_m} := [\frac{t_m}{h_{nm}}]\, h_{nm}$ because due to continuity property (H3) (ii) again, the following convergence holds for any $i \in \mathscr{I}$, $s \in [-\tau,0]$

$$\begin{cases} d_i\big(x(t+s),\ x_{n_m}(t_{nm}^k+s)\big) \;\overset{m\to\infty}{\longrightarrow}\; 0\\ d_i\big(x(t+s),\ x_{n_m}(t_{nm}^{l_m}+s)\big) \;\overset{m\to\infty}{\longrightarrow}\; 0. \end{cases}$$

Hence the assumptions of Convergence Theorem 13 are satisfied by $\overset{\circ}{x}_n(\cdot) \ni \widehat{f}_n(\cdot)$ and thus, $x(\cdot)|_{[0,T]}$ solves the mutational equation $\overset{\circ}{x}(\cdot) \ni \widehat{f}(\cdot)$ in the tuple $\big(E, (d_j)_{j\in\mathscr{I}}, (e_j)_{j\in\mathscr{I}}, (\lfloor\cdot\rfloor_j)_{j\in\mathscr{I}}, (\widehat{D}_j)_{j\in\mathscr{I}}\big)$, i.e., $x(\cdot)|_{[0,T]}$ is a solution to the mutational equation

$$\overset{\circ}{x}(t) \;\ni\; f\big(x(t+\cdot)\big|_{[-\tau,0]},\, t\big)\,.$$

□

3.3.4 Existence for Systems of Mutational Equations with Delay

Considering mutational equations with delay and without state constraints, the preceding results about existence and convergence of solutions can be extended easily to systems. This feature is regarded as an important advantage in regard to applications as we have already pointed out.

Indeed, starting with the same assumptions as in § 3.3.3 (i.e. Euler compactness and Euler equi-continuity) for each component, Euler method provides a sequences of approximative solutions. Then Lemma 18 (on page 195) is applied to each component successively so that we can extract a subsequence of approximative solutions whose components converge pointwise respectively.
Finally it is to verify that each component of the limit solves the corresponding mutational equation in the sense of Definition 8 (on page 187). For this purpose, we regard the other components as an additional, but known dependence on time — as we have already done successfully in the proof of Theorem 2.20 (on page 118).

Now we formulate the results about two mutational equations in detail and then restrict our considerations of proofs to the aspect of convergence again.

Theorem 21 (Convergence of solutions to systems of mutational equations).
Let the tuples $\left(E_1,\ (d_j^1)_{j\in\mathscr{I}_1},\ (e_j^1)_{j\in\mathscr{I}_1},\ (\lfloor\cdot\rfloor_j^1)_{j\in\mathscr{I}_1},\ (\widehat{D}_j^1)_{j\in\mathscr{I}_1}\right)$
and $\left(E_2,\ (d_j^2)_{j\in\mathscr{I}_2},\ (e_j^2)_{j\in\mathscr{I}_2},\ (\lfloor\cdot\rfloor_j^2)_{j\in\mathscr{I}_2},\ (\widehat{D}_j^2)_{j\in\mathscr{I}_2}\right)$
satisfy the assumptions of § 3.1 (on page 182 ff.) respectively with nonempty sets $\widehat{\Theta}\left(E_1,(d_j^1)_{j\in\mathscr{I}_1},(e_j^1)_{j\in\mathscr{I}_1},(\lfloor\cdot\rfloor_j^1)_{j\in\mathscr{I}_1}\right)$ *and* $\widehat{\Theta}\left(E_2,(d_j^2)_{j\in\mathscr{I}_2},(e_j^2)_{j\in\mathscr{I}_2},(\lfloor\cdot\rfloor_j^2)_{j\in\mathscr{I}_2}\right)$.
Suppose the following properties of

$$\begin{array}{lrl} f_n^1,\ f^1 : E_1\times E_2\times[0,T] & \longrightarrow & \widehat{\Theta}\left(E_1,(d_i^1)_{i\in\mathscr{I}_1},(e_i^1)_{i\in\mathscr{I}_1},(\lfloor\cdot\rfloor_i^1)_{i\in\mathscr{I}_1}\right) \qquad (n\in\mathbb{N})\\ f_n^2,\ f^2 : E_1\times E_2\times[0,T] & \longrightarrow & \widehat{\Theta}\left(E_2,(d_i^2)_{i\in\mathscr{I}_2},(e_i^2)_{i\in\mathscr{I}_2},(\lfloor\cdot\rfloor_i^2)_{i\in\mathscr{I}_2}\right)\\ x_n^1,\ x^1 : \qquad [0,T] & \longrightarrow & E_1\\ x_n^2,\ x^2 : \qquad [0,T] & \longrightarrow & E_2\ : \end{array}$$

1.) for each $j_1\in\mathscr{I}_1, j_2\in\mathscr{I}_2$ *and every* $\rho\ge 0$,

$$\begin{aligned} R_{j_1}^1 &:= \sup_{n,t}\ \lfloor x_n^1(t)\rfloor_{j_1}^1 < \infty, & \widehat{\alpha}_{j_1}^1(\rho) &:= \sup_{n,t,y^1,y^2}\ \alpha_{j_1}^1\left(f_n^1(y^1,y^2,t);\rho\right) < \infty,\\ R_{j_2}^2 &:= \sup_{n,t}\ \lfloor x_n^2(t)\rfloor_{j_2}^2 < \infty, & \widehat{\alpha}_{j_2}^2(\rho) &:= \sup_{n,t,y^1,y^2}\ \alpha_{j_2}^2\left(f_n^2(y^1,y^2,t);\rho\right) < \infty,\\ \widehat{\beta}_{j_1}^1 &:= \sup_{n}\ \mathrm{Lip}\left(x_n^1(\cdot):[0,T]\longrightarrow(E_1,e_{j_1}^1)\right) < \infty,\\ \widehat{\beta}_{j_2}^2 &:= \sup_{n}\ \mathrm{Lip}\left(x_n^2(\cdot):[0,T]\longrightarrow(E_2,e_{j_2}^2)\right) < \infty, \end{aligned}$$

2.) $\overset{\circ}{x}{}_n^1(\cdot)\ \ni\ f_n^1(x_n^1(\cdot),x_n^2(\cdot),\cdot)$
$\overset{\circ}{x}{}_n^2(\cdot)\ \ni\ f_n^2(x_n^1(\cdot),x_n^2(\cdot),\cdot)$ *(in the sense of Definition 8 on p.187) for any* n,

3.) for $\mathscr{L}^1$*-almost every* $t\in[0,T]$:

$$\begin{aligned} \lim_{n\to\infty}\ \widehat{D}_{j_1}^1\left(f_n^1(x^1(t),x^2(t),t),\ \ f_n^1(y_n^1,y_n^2,t_n);\ \rho\right) &= 0\\ \lim_{n\to\infty}\ \widehat{D}_{j_2}^2\left(f_n^2(x^1(t),x^2(t),t),\ \ f_n^2(y_n^1,y_n^2,t_n);\ \rho\right) &= 0 \end{aligned}$$

for each $j_1\in\mathscr{I}_1,\ j_2\in\mathscr{I}_2,\ \rho\ge 0$ *and any sequences* $(t_n)_{n\in\mathbb{N}},\ (y_n^1)_{n\in\mathbb{N}},\ (y_n^2)_{n\in\mathbb{N}}$ *in* $[t,T]$, E_1 *and* E_2 *respectively satisfying*

$$\begin{aligned} \lim_{n\to\infty} t_n = t \quad \text{and} \quad &\lim_{n\to\infty} d_i^1\left(x^1(t),y_n^1\right)=0,\ \sup_{n\in\mathbb{N}}\lfloor y_n^1\rfloor_i^1\le R_i^1 \quad \textit{for each } i\in\mathscr{I}_1,\\ &\lim_{n\to\infty} d_i^2\left(x^2(t),y_n^2\right)=0,\ \sup_{n\in\mathbb{N}}\lfloor y_n^2\rfloor_i^2\le R_i^2 \quad \textit{for each } i\in\mathscr{I}_2, \end{aligned}$$

4.) for Lebesgue-almost every $t\in[0,T]$ ($t=0$ *inclusive) and any* $\widetilde{t}\in\,]t,T[$, *there exist a sequence* $n_m\nearrow\infty$ *of indices and sequences* $\delta_m\searrow 0,\ \widetilde{\delta}_m\searrow 0$ *(depending on* $t,\widetilde{t}$*) satisfying for* $m\longrightarrow\infty$ *and each* $j_1\in\mathscr{I}_1,\ j_2\in\mathscr{I}_2,\ \rho\ge 0$

$$\begin{cases} (i) & \widehat{D}_{j_1}^1\left(f^1(x^1(t),x^2(t),t),\ f_{n_m}^1(x^1(t),x^2(t),t);\ \rho\right) \longrightarrow 0\\ & \widehat{D}_{j_2}^2\left(f^2(x^1(t),x^2(t),t),\ f_{n_m}^2(x^1(t),x^2(t),t);\ \rho\right) \longrightarrow 0\\ (ii) & d_{j_1}^1\left(x^1(t),\ x_{n_m}^1(t+\delta_m)\right)\longrightarrow 0, \qquad d_{j_2}^2\left(x^2(t),\ x_{n_m}^2(t+\delta_m)\right)\longrightarrow 0\\ (iii) & d_{j_1}^1\left(x^1(\widetilde{t}),\ x_{n_m}^1(\widetilde{t}-\widetilde{\delta}_m)\right)\longrightarrow 0, \qquad d_{j_2}^2\left(x^2(\widetilde{t}),\ x_{n_m}^2(\widetilde{t}-\widetilde{\delta}_m)\right)\longrightarrow 0 \end{cases}$$

Then, $x^1(\cdot)$ and $x^2(\cdot)$ are solutions to the mutational equations

$$\overset{\circ}{x}{}^1(\cdot) \ni f^1\big(x^1(\cdot), x^2(\cdot), \cdot\big), \qquad \overset{\circ}{x}{}^2(\cdot) \ni f^2\big(x^1(\cdot), x^2(\cdot), \cdot\big)$$

in $\big(E_1,\ (d_j^1)_{j\in\mathscr{I}_1},\ (e_j^1)_{j\in\mathscr{I}_1},\ (\lfloor\cdot\rfloor_j^1)_{j\in\mathscr{I}_1},\ (\widehat{D}_j^1)_{j\in\mathscr{I}_1}\big)$
and $\big(E_2,\ (d_j^2)_{j\in\mathscr{I}_2},\ (e_j^2)_{j\in\mathscr{I}_2},\ (\lfloor\cdot\rfloor_j^2)_{j\in\mathscr{I}_2},\ (\widehat{D}_j^2)_{j\in\mathscr{I}_2}\big)$
respectively.

Theorem 22 (Existence of solutions to systems with delay).
Suppose each of the tuples

$$\big(E_1,\ (d_j^1)_{j\in\mathscr{I}_1},\ (e_j^1)_{j\in\mathscr{I}_1},\ (\lfloor\cdot\rfloor_j^1)_{j\in\mathscr{I}_1},\ \widehat{\Theta}\big(E_1,(d_i^1)_{i\in\mathscr{I}_1},(e_i^1)_{i\in\mathscr{I}_1},(\lfloor\cdot\rfloor_i^1)_{i\in\mathscr{I}_1}\big)\big)$$
$$\big(E_2,\ (d_j^2)_{j\in\mathscr{I}_2},\ (d_j^2)_{j\in\mathscr{I}_2},\ (\lfloor\cdot\rfloor_j^2)_{j\in\mathscr{I}_2},\ \widehat{\Theta}\big(E_2,(d_i^2)_{i\in\mathscr{I}_2},(e_i^2)_{i\in\mathscr{I}_2},(\lfloor\cdot\rfloor_i^2)_{i\in\mathscr{I}_2}\big)\big)$$

to be Euler compact and Euler equi-continuous. For some fixed $\tau \geq 0$, set

$$\mathscr{BL}^k := \mathrm{BLip}\big([-\tau,0],\ E_k;\ (e_j^k)_{j\in\mathscr{I}_k},\ (\lfloor\cdot\rfloor_j^k)_{j\in\mathscr{I}_k}\big) \qquad (k=1,2).$$

Assume for the functions

$$f^1 : \mathscr{BL}^1 \times \mathscr{BL}^2 \times [0,T] \longrightarrow \widehat{\Theta}\big(E_1,(d_i^1)_{i\in\mathscr{I}_1},(e_i^1)_{i\in\mathscr{I}_1},(\lfloor\cdot\rfloor_i^1)_{i\in\mathscr{I}_1}\big)$$
$$f^2 : \mathscr{BL}^1 \times \mathscr{BL}^2 \times [0,T] \longrightarrow \widehat{\Theta}\big(E_2,(d_i^2)_{i\in\mathscr{I}_2},(e_i^2)_{i\in\mathscr{I}_2},(\lfloor\cdot\rfloor_i^2)_{i\in\mathscr{I}_2}\big)$$

and each $j_1 \in \mathscr{I}_1$, $j_2 \in \mathscr{I}_2$, $R > 0$:

1.) $\sup_{z^1,z^2,t} \alpha_{j_1}^1(f^1(z^1,z^2,t);R) < \infty, \qquad \sup_{z^1,z^2,t} \alpha_{j_2}^2(f^2(z^1,z^2,t);R) < \infty,$
2.) $\sup_{z^1,z^2,t} \beta_{j_1}^1(f^1(z^1,z^2,t);R) < \infty, \qquad \sup_{z^1,z^2,t} \beta_{j_2}^2(f^2(z^1,z^2,t);R) < \infty,$
3.) $\sup_{z^1,z^2,t} \gamma_{j_1}^1(f^1(z^1,z^2,t)) < \infty, \qquad \sup_{z^1,z^2,t} \gamma_{j_2}^2(f^2(z^1,z^2,t)) < \infty,$

4.) *for $\mathscr{L}^1$-almost every $t \in [0,T]$:*

$$\lim_{n\to\infty} D_{j_1}^1\big(f^1(y_n^1,y_n^2,s_n),\ f^1(z_n^1,z_n^2,t_n);\ R\big) = 0$$
$$\lim_{n\to\infty} D_{j_2}^2\big(f^2(y_n^1,y_n^2,s_n),\ f^2(z_n^1,z_n^2,t_n);\ R\big) = 0$$

for every $j_1 \in \mathscr{I}_1$, $j_2 \in \mathscr{I}_2$, $R > 0$ and any sequences $(s_n,t_n)_{n\in\mathbb{N}}$, $(y_n^1,z_n^1)_{n\in\mathbb{N}}$ $(y_n^2,z_n^2)_{n\in\mathbb{N}}$ in $[0,T]$, $\mathscr{BL}^1$, $\mathscr{BL}^2$ respectively satisfying for each $k \in \{1,2\}$, $i \in \mathscr{I}_k$, $s \in [-\tau,0]$,

$$\lim_{n\to\infty} s_n = t = \lim_{n\to\infty} t_n,\quad \lim_{n\to\infty} d_i^k\big(y_n^k(s),\ z^k(s)\big) = 0 = \lim_{n\to\infty} d_i^k\big(z_n^k(s),\ z^k(s)\big)$$
$$\sup_{n\in\mathbb{N}} \sup_{[-\tau,0]} \big\{\lfloor y_n^k(\cdot)\rfloor_i^k,\ \lfloor z_n^k(\cdot)\rfloor_i^k\big\} < \infty.$$

Then for any initial functions $x_0^1 \in \mathscr{BL}^1, x_0^2 \in \mathscr{BL}^2$ given, there exist curves

$$x^1(\cdot) \in \mathrm{BLip}\big([-\tau,T],\ E_1;\ (e_j^1)_{j\in\mathscr{I}_1},\ (\lfloor\cdot\rfloor_j^1)_{j\in\mathscr{I}_1}\big)$$
$$x^2(\cdot) \in \mathrm{BLip}\big([-\tau,T],\ E_2;\ (e_j^2)_{j\in\mathscr{I}_2},\ (\lfloor\cdot\rfloor_j^2)_{j\in\mathscr{I}_2}\big)$$

with $x^1(\cdot)|_{[-\tau,0]} = x_0^1$, $x^2(\cdot)|_{[-\tau,0]} = x_0^2$ whose respective restrictions to $[0,T]$ solve *the two mutational equations with delay*

$$\begin{cases} \overset{\circ}{x}^1(t) \ni f^1\big(x^1(t+\cdot)|_{[-\tau,0]},\ x^2(t+\cdot)|_{[-\tau,0]},\ t\big) \\ \overset{\circ}{x}^2(t) \ni f^2\big(x^1(t+\cdot)|_{[-\tau,0]},\ x^2(t+\cdot)|_{[-\tau,0]},\ t\big) \end{cases}$$

in $\big(E_1,\ (d_j^1)_{j\in\mathscr{I}_1},\ (e_j^1)_{j\in\mathscr{I}_1},\ (\lfloor\cdot\rfloor_j^1)_{j\in\mathscr{I}_1},\ (\widehat{D}_j^1)_{j\in\mathscr{I}_1}\big)$

and $\big(E_2,\ (d_j^2)_{j\in\mathscr{I}_2},\ (e_j^2)_{j\in\mathscr{I}_2},\ (\lfloor\cdot\rfloor_j^2)_{j\in\mathscr{I}_2},\ (\widehat{D}_j^2)_{j\in\mathscr{I}_2}\big).$

Proof (of Theorem 21). We focus on $x^1(\cdot)$ and choose the index $j \in \mathscr{I}_1$ arbitrarily. Then, $x^1(\cdot) : [0,T] \longrightarrow (E_1, e_j^1)$ is $\widehat{\beta}_j^1$-Lipschitz continuous as a consequence of assumption (4.) and the lower semicontinuity of e_j^1 (hypothesis (H3) (i) on page 182). Hypothesis (H4) about the lower semicontinuity of $\lfloor\cdot\rfloor_j^1$ ensures $\sup \lfloor x^1(\cdot)\rfloor_j^1 \le R_j^1$.

Finally we verify the solution property

$$\limsup_{h\downarrow 0}\ \tfrac{1}{h}\cdot\Big(d_j^1\big(\vartheta^1(s+h,z^1),\ x^1(t+h)\big)\ -\ d_j^1\big(\vartheta^1(s,z^1),\ x^1(t)\big)\cdot e^{\widehat{\alpha}_{j_1}^1(\rho)\,h}\Big)$$
$$\le \widehat{D}_j^1\big(\vartheta^1,\ f^1(x^1(t),x^2(t),t);\ \rho\big)$$

for Lebesgue-almost every $t \in [0,T[$ and for any $\vartheta^1 \in \widehat{\Theta}\big(E_1,(d_j^1),(e_j^1),(\lfloor\cdot\rfloor_j^1)\big)$, $s\in[0,1[$, $z^1 \in E_1$, $\rho \ge R_j^1$ with $\lfloor\vartheta^1(\cdot,z^1)\rfloor_j \le \rho$,
Indeed, for Lebesgue-almost every $t\in[0,T[$ and any $h\in]0,T-t[$, assumption (4.) guarantees a subsequence $\big(x_{n_m}(\cdot)\big)_{m\in\mathbb{N}}$ and sequences $\delta_m \searrow 0$, $\widetilde{\delta}_m \searrow 0$ satisfying for each $i_1\in\mathscr{I}_1$, $i_2\in\mathscr{I}_2$, $r\ge 0$ and $m\longrightarrow\infty$

$$\begin{cases} \widehat{D}_{i_1}^1\big(f^1(x^1(t),x^2(t),t),\ f_{n_m}^1(x^1(t),x^2(t),t);\ r\big) \longrightarrow 0 \\ d_{i_1}^1\big(x^1(t),\quad x_{n_m}^1(t+\delta_m)\big) \longrightarrow 0 \\ d_{i_2}^2\big(x^2(t),\quad x_{n_m}^2(t+\delta_m)\big) \longrightarrow 0 \\ d_{i_1}^1\big(x^1(t+h),\ x_{n_m}^1(t+h-\widetilde{\delta}_m)\big) \longrightarrow 0 \end{cases}$$

Now we conclude from Lemma 9 (on page 188) and the continuity of d_j^1 (due to hypothesis (H3) (i)) respectively for each index $j\in\mathscr{I}_1$

$$\begin{aligned} & d_j^1\big(\vartheta^1(s+h,z^1),\ x^1(t+h)\big) \\ = & \lim_{m\to\infty}\ d_j^1\big(\vartheta^1(s+h-\widetilde{\delta}_m,\ z^1),\ x_{n_m}^1(t+h-\widetilde{\delta}_m)\big) \\ \le & \limsup_{m\to\infty}\ \Big(d_j^1\big(\vartheta^1(s+\delta_m,\ z^1),\quad x_{n_m}^1(t+\delta_m)\big) \\ & \qquad + h\cdot \sup_{[t+\delta_m,\ t+h-\widetilde{\delta}_m]} \widehat{D}_j^1\big(\vartheta^1,\ f_{n_m}^1(x_{n_m}^1, x_{n_m}^2,\cdot);\ \rho\big)\Big)\cdot e^{\widehat{\alpha}_j^1(\rho)\cdot h} \\ \le & \Big(d_j^1\big(\vartheta^1(s,z),x^1(t)\big) + h\cdot\limsup_{m\to\infty}\ \sup_{[t+\delta_m,t+h]}\ \widehat{D}_j^1\big(\vartheta^1,f_{n_m}^1(x_{n_m}^1,x_{n_m}^2,\cdot);\rho\big)\Big)\, e^{\widehat{\alpha}_{j_1}^1(\rho)\,h}. \end{aligned}$$

For completing the proof, it is sufficient to verify

$$\limsup_{h\downarrow 0}\ \limsup_{m\to\infty}\ \sup_{[t+\delta_m,\,t+h]} \widehat{D}^1_j\big(\vartheta^1,\ f^1_{n_m}(x^1_{n_m},x^2_{n_m},\cdot);\,\rho\big) \le \widehat{D}^1_j\big(\vartheta^1, f^1(x^1(t),x^2(t),t);\,\rho\big)$$

for Lebesgue-almost every $t\in[0,T[$ and *any* subsequence $n_m \nearrow \infty$ satisfying

$$\begin{cases} d^1_{i_1}\big(x^1(t),\ x^1_{n_m}(t+\delta_m)\big) \longrightarrow 0\\ d^2_{i_2}\big(x^2(t),\ x^2_{n_m}(t+\delta_m)\big) \longrightarrow 0\\ \widehat{D}^1_{i_1}\big(f^1(x^1(t),x^2(t),t),\ f^1_{n_m}(x^1(t),x^2(t),t);\ r\big) \longrightarrow 0\end{cases}$$

for $m\longrightarrow\infty$ and each $i_1\in\mathscr{I}_1,\ i_2\in\mathscr{I}_2,\ r\ge 0$.
Indeed, if this inequality was not correct then we could select some $\varepsilon>0$ and sequences $(h_l)_{l\in\mathbb{N}},\ (m_l)_{l\in\mathbb{N}},\ (s_l)_{l\in\mathbb{N}}$ fulfilling for all $l\in\mathbb{N}$

$$\widehat{D}^1_j\big(\vartheta^1,\ f^1_{n_{m_l}}(x^1_{n_{m_l}}(t+s_l),\, x^2_{n_{m_l}}(t+s_l),\, t+s_l);\,\rho\big) \ge \widehat{D}^1_j\big(\vartheta^1, f^1(x^1(t),\, x^2(t),\, t);\,\rho\big)+\varepsilon,$$
$$\delta_{m_l} \le s_l \le h_l \le \tfrac{1}{l}, \qquad m_l \ge l.$$

Due to property (H3) (ii), the uniform Lipschitz continuity of $(x^1_{n_m}(\cdot))_m,\ (x^2_{n_m}(\cdot))_m$ implies

$$\begin{cases} d^1_{i_1}\big(x^1(t),\ x^1_{n_{m_l}}(t+s_l)\big) \longrightarrow 0\\ d^2_{i_2}\big(x^2(t),\ x^2_{n_{m_l}}(t+s_l)\big) \longrightarrow 0\end{cases}$$

for $l\longrightarrow\infty$ and each $i_1\in\mathscr{I}_1,\ i_2\in\mathscr{I}_2$. Thus at $\mathscr{L}^1$-almost every time $t\in[0,T[$, assumptions (3.), (4.) (i) and hypothesis (H6) about the continuity of $\widehat{D}^1_j(\cdot,\cdot;r)$ would lead to a contradiction because for any $r\ge 0$,

$$\lim_{l\to\infty} \widehat{D}^1_j\big(\vartheta^1,\ f^1_{n_{m_l}}(x^1_{n_{m_l}}(t+s_l),\, x^2_{n_{m_l}}(t+s_l),\, t+s_l);\ r\big) = \widehat{D}^1_j\big(\vartheta^1, f^1(x^1(t),\, x^2(t),\, t);\ r\big).$$

□

3.3.5 Existence Under State Constraints for a Single Index

Similarly to § 2.3.6 (on page 123 f.), we restrict our considerations to the special case that the index set $\mathscr{I}\neq\emptyset$ consists of a single element: $\mathscr{I}=\{0\}$.
Now the goal is to specify sufficient conditions for the existence of solutions to mutational equations with state constraints. Aubin's adaption of Nagumo's Theorem (about ordinary differential equations) formulated in Theorem 1.19 (on page 40) serves as a starting point and provides the viability condition.

In contrast to the counterparts in preceding chapters, we now dispense with assuming sequential compactness of *all* "closed balls" in (E,d_0). Instead we focus on the compactness properties of curves which are constructed via transitions in a piecewise way. But this piecewise construction does not have to be restricted to an equidistant partition of $[0,T]$ as in Definitions 2.15 and 16 about Euler compactness and Euler equi-continuity respectively (on pages 112 and 194).

Definition 23. $\big(E,\ (d_j)_{j\in\mathcal{I}},\ (e_j)_{j\in\mathcal{I}},\ (\lfloor\cdot\rfloor_j)_{j\in\mathcal{I}},\ \widehat{\Theta}(E,(d_i)_i,(e_i)_i,(\lfloor\cdot\rfloor_i)_i)\big)$ is called *nonequidistant Euler compact* if it satisfies the following condition for any initial element $x_0 \in E$, time $T \in]0,\infty[$ and bounds $\widehat{\alpha}_j, \widehat{\beta}_j, \widehat{\gamma}_j, L_j > 0$ $(j \in \mathcal{I})$:
Let $\mathcal{PN} = \mathcal{PN}(x_0, T, (\widehat{\alpha}_j, \widehat{\beta}_j, \widehat{\gamma}_j, L_j)_{j\in\mathcal{I}})$ denote the (possibly empty) subset of all curves $y(\cdot) : [0,T[\longrightarrow E$ with the four following properties

(1.) $y(0) = x_0$,
(2.) for each $j \in \mathcal{I}$, $y : [0,T[\longrightarrow (E, e_j)$ is L_j-Lipschitz continuous,
(3.) for each $j \in \mathcal{I}$, $\sup \lfloor y(\cdot)\rfloor_j \le (\lfloor x_0\rfloor_j + \widehat{\gamma}_j\, T)\cdot e^{\widehat{\gamma}_j T} =: R_j$.
(4.) for any $t \in [0,T[$, there are $s \in]t-1,t]$ and $\vartheta \in \widehat{\Theta}(E,(d_i)_i,(e_i)_i,(\lfloor\cdot\rfloor_i)_i)$ with $y(s+\cdot) = \vartheta(\cdot, y(s))$ in an open neighborhood $I \subset [0,1]$ of $[0,t-s]$ and $\alpha_j(\vartheta; R_j) \le \widehat{\alpha}_j$, $\beta_j(\vartheta; R_j) \le \widehat{\beta}_j$, $\gamma_j(\vartheta) \le \widehat{\gamma}_j$,

Then for each $t \in [0,T[$, every sequence $(z_n)_{n\in\mathbb{N}}$ in $\{y(t)\,|\,y(\cdot)\in\mathcal{PN}\} \subset E$ has a subsequence $(z_{n_m})_{m\in\mathbb{N}}$ and an element $z \in E$ with $d_j(z_{n_m}, z) \longrightarrow 0$ for each $j \in \mathcal{I}$.

The tuple $\big(E,\ (d_j)_{j\in\mathcal{I}},\ (e_j)_{j\in\mathcal{I}},\ (\lfloor\cdot\rfloor_j)_{j\in\mathcal{I}},\ \widehat{\Theta}(E,(d_i)_{i\in\mathcal{I}},(e_i)_{i\in\mathcal{I}},(\lfloor\cdot\rfloor_i)_{i\in\mathcal{I}})\big)$ is called *nonequidistant Euler equi-continuous* if for any initial element $x_0 \in E$, time $T \in]0,\infty[$ and bounds $\widehat{\alpha}_j, \widehat{\beta}_j, \widehat{\gamma}_j > 0$ $(j \in \mathcal{I})$, there exists $\lambda_j > 0$ for each $j \in \mathcal{I}$ such that

$$\mathcal{PN}(x_0,\, T,\, (\widehat{\alpha}_j,\, \widehat{\beta}_j,\, \widehat{\gamma}_j,\, \infty)_{j\in\mathcal{I}}) \;=\; \mathcal{PN}(x_0,\, T,\, (\widehat{\alpha}_j,\, \widehat{\beta}_j,\, \widehat{\gamma}_j,\, \lambda_j)_{j\in\mathcal{I}}),$$

i.e., every curve $y(\cdot) : [0,T[\longrightarrow E$ satisfying preceding conditions (1.), (3.), (4.) is λ_j-Lipschitz continuous with respect to e_j for each $j \in \mathcal{I}$.

Remark 24. We provide two simple implications for the special case $\mathcal{I} = \{0\}$:

(1.) If for every $r_1, r_2 > 0$ and $x_0 \in E$, the set $\{x \in E \mid e_0(x_0,x) \le r_1,\ \lfloor x\rfloor_0 \le r_2\}$ is sequentially compact, then the tuple $\big(E,\, d_0,\, e_0,\, \lfloor\cdot\rfloor_0,\, \widehat{\Theta}\big)$ is always nonequidistant Euler compact.

(2.) If $d_0 \equiv e_0$ is a pseudo-metric, then all curves piecewise constructed by transitions are Lipschitz continuous due to Lemma 2.10 (on page 107). Finally nonequidistant Euler equi-continuity (with $\lambda_0 = \widehat{\beta}_0$) results from the triangle inequality.

Proposition 25 (Existence of solutions under state constraints for $\mathcal{I} = \{0\}$). *In addition to $\mathcal{I} = \{0\}$, let $E \neq \emptyset$ and*

$$\begin{aligned} d_0,\, e_0 : E \times E &\longrightarrow [0,\infty[,\\ \lfloor\cdot\rfloor_0 : E &\longrightarrow [0,\infty[,\\ D_0 : E \times E \times [0,\infty[&\longrightarrow [0,\infty[\end{aligned}$$

satisfy hypotheses (H1) – (H7). Assume $\big(E,\, d_0,\, e_0,\, \lfloor\cdot\rfloor_0,\, \widehat{\Theta}(E,d_0,e_0,\lfloor\cdot\rfloor_0)\big)$ to be nonequidistant Euler compact and nonequidistant Euler equi-continuous.

For each $r > 0$, suppose

$$f : (E, d_0) \longrightarrow \big(\widehat{\Theta}(E, d_0, e_0, \lfloor\cdot\rfloor_0),\ D_0(\cdot,\cdot;r)\big)$$

to be continuous with

$$\begin{aligned}
\widehat{\alpha}(r) &:= \sup_{z\in E}\ \alpha_0(f(z);r) < \infty,\\
\widehat{\beta}(r) &:= \sup_{z\in E}\ \beta_0(f(z);r) < \infty,\\
\widehat{\gamma} &:= \sup_{z\in E}\ \gamma_0(f(z)) \ \ < \infty.
\end{aligned}$$

Let $\mathscr{V} \subset (E, d_0)$ be a closed subset whose projection $E \rightsquigarrow \mathscr{V}$ has always nonempty values and whose distance function $\text{dist}(\cdot, \mathscr{V}) : (E, d_0) \longrightarrow [0, \infty[,\ z \longmapsto \inf_{y\in\mathscr{V}} d_0(y,z)$ *is 1-Lipschitz continuous. Assume the following viability condition*

$$f(z) \in \mathscr{T}_{\mathscr{V}}(z) \qquad \textit{for every } z \in \mathscr{V},$$

i.e.

$$\liminf_{h\downarrow 0}\ \tfrac{1}{h}\cdot \text{dist}\big(f(z)(h,z),\ \mathscr{V}\big) = 0 \qquad \textit{for every } z \in \mathscr{V}.$$

Then every state $x_0 \in \mathscr{V}$ is the initial point of at least one solution $x : [0,1] \longrightarrow E$ to the mutational equation

$$\overset{\circ}{x}(\cdot) \ni f(x(\cdot))$$

in $\big(E, d_0, e_0, \lfloor\cdot\rfloor_0, \widehat{D}_0\big)$ with the state constraint $x(t) \in \mathscr{V}$ for all $t \in [0,1]$.

For proving this proposition, we first construct approximative solutions satisfying weakened forms of the mutational equation and state constraints. Lemma 1.29 (on page 48) and Lemma 2.25 (on page 124) have the following counterpart with $\lambda_0 > 0$ denoting the appropriate Lipschitz constant resulting from nonequidistant Euler equi-continuity and depending on $\widehat{\gamma}$, x_0 essentially.

Lemma 26 (Constructing approximative solutions).
Choose any $\varepsilon > 0$. Under the assumptions of Proposition 25, there always exists a λ_0-Lipschitz continuous function $x_\varepsilon(\cdot) : [0,1] \longrightarrow (E, e_0)$ satisfying

(a) $x_\varepsilon(0) = x_0$,

(b) *for all* $t \in [0,1]$, $\text{dist}\big(x_\varepsilon(t), \mathscr{V}\big) \le \varepsilon\ e^{\widehat{\alpha}}$

(c) *for all* $t \in [0,1[$, *there exist* $\vartheta \in \big\{f(z) \,\big|\, z \in E : d_0(z, x_\varepsilon(t)) \le \varepsilon\ e^{\widehat{\alpha}}\big\} \subset \widehat{\Theta}(E, d_0, e_0, \lfloor\cdot\rfloor_0)$ *and* $s \in [0,t]$ *with* $x_\varepsilon(s+\cdot) = \vartheta(\cdot, x_\varepsilon(s))$ *in an open neighborhood* $I \subset [0,1]$ *of* $[0, t-s]$,

(d) *for all* $t \in [0,1]$, $\lfloor x_\varepsilon(t)\rfloor_0 \le \big(\lfloor x_0\rfloor_0 + \widehat{\gamma}\, t\big)\ e^{\gamma t}$.

Proof (of Lemma 26). For $\varepsilon > 0$ fixed, let $\mathscr{A}_\varepsilon(x_0)$ denote the set of all tuples $(T_x,\ x(\cdot))$ consisting of some $T_x \in [0,1]$ and a λ_0-Lipschitz continuous function $x(\cdot) : [0, T_x] \longrightarrow (E, e_0)$ such that

(a) $x(0) = x_0$,

(b') 1.) $\operatorname{dist}\big(x(T_x),\ \mathscr{V}\big) \le r_\varepsilon(T_x)$ with $r_\varepsilon(t) := \varepsilon\, e^{\widehat{\alpha} t}\, t$,
2.) $\operatorname{dist}\big(x(t),\ \mathscr{V}\big) \le r_\varepsilon(1)$ for all $t \in [0, T_x]$,

(c) for all $t \in [0, T_x[$, there exist $\vartheta \in \big\{ f(z) \,\big|\, z \in E : d_0(z, x_\varepsilon(t)) \le r_\varepsilon(1) \big\} \subset \widehat{\Theta}(E, d_0, e_0, \lfloor\cdot\rfloor_0)$ and $s \in [0,t]$ with $x_\varepsilon(s+\cdot) = \vartheta(\cdot, x_\varepsilon(s))$ in an open neighborhood $I \subset [0, T_x[$ of $[0, t-s]$.

(d) for all $t \in [0, T_x[$, $\lfloor x_\varepsilon(t) \rfloor_0 \le \big(\lfloor x_0 \rfloor_0 + \widehat{\gamma}\, t\big)\ e^{\widehat{\gamma} t}$.

Obviously, $\mathscr{A}_\varepsilon(x_0)$ is not empty since it contains $(0,\ x(\cdot) \equiv x_0)$. Moreover, an order relation $\preceq$ on $\mathscr{A}_\varepsilon(x_0)$ is specified by

$$(T_x, x(\cdot)) \preceq (T_y, y(\cdot)) \quad :\Longleftrightarrow \quad T_x \le T_y \text{ and } x = y\big|_{[0,T_x]}.$$

Hence, Zorn's Lemma provides a maximal element $\big(T,\ x_\varepsilon(\cdot)\big) \in \mathscr{A}_\varepsilon(x_0)$.
As all considered functions with values in E have been supposed to be λ_0-Lipschitz continuous, $x_\varepsilon(\cdot) : [0, T[\longrightarrow (E, e_0)$ is also λ_0-Lipschitz continuous. In particular, $x_\varepsilon(\cdot)$ can always be extended to the closed interval $[0,T] \subset [0,1]$ in a Lipschitz continuous way because the tuple $\big(E,\ d_0,\ e_0,\ \lfloor\cdot\rfloor_0,\ \widehat{\Theta}(E, d_0, e_0, \lfloor\cdot\rfloor_0)\big)$ is assumed to be nonequidistant Euler compact (and for each $k \in \mathbb{N}$, we are free to extend $x(\cdot)\big|_{[0,\,T-\frac{1}{k}]}$ to $[0,T]$ by means of an arbitrarily fixed transition ϑ).

Assuming $T < 1$ for a moment, we obtain a contradiction if $x_\varepsilon(\cdot)$ can be extended to a larger interval $[0, T+\delta] \subset [0,1]$ $(\delta > 0)$ preserving conditions (b'), (c), (d). Due to the assumption about the set-valued projection on $\mathscr{V} \subset E$, the closed set $\mathscr{V}$ contains an element $z \in E$ with $d_0(x_\varepsilon(T), z) = \operatorname{dist}(x_\varepsilon(T),\ \mathscr{V}) \le r_\varepsilon(T)$.
As a consequence of the viability condition, there is a sequence $h_m \downarrow 0$ in $]0, 1-T[$ such that
$$\operatorname{dist}\big(f(z)\,(h_m, z),\ \mathscr{V}\big) \le \varepsilon\, h_m \qquad \text{for all } m \in \mathbb{N}.$$
Now set for each $t \in [T,\ T+h_1]$
$$x_\varepsilon(t) := f(z)\,\big(t-T,\ x_\varepsilon(T)\big).$$
Obviously, this extension of $x_\varepsilon(\cdot)$ satisfies the two conditions (c), (d) in $[0,\ T+h_1]$. Furthermore, the estimate $d_0\big(z,\ x_\varepsilon(T)\big) \le r_\varepsilon(T) < r_\varepsilon(1)$ and the continuity of $x_\varepsilon(\cdot)$ provide some sufficiently small $\delta \in]0, h_1]$ with
$$\operatorname{dist}\big(x_\varepsilon(t),\ \mathscr{V}\big) \le d_0\big(x_\varepsilon(t),\ z\big) \le r_\varepsilon(1) \qquad \text{for every } t \in [T,\ T+\delta]$$
and thus, the extension $x(\cdot)$ fulfills condition (b')(2.) in the interval $[0,\ T+\delta]$.
For any index $m \in \mathbb{N}$ with $h_m < \delta$, we conclude from the 1-Lipschitz continuity of $\operatorname{dist}(\cdot\,, \mathscr{V})$ with respect to d_0 and Proposition 7 (on page 185)
$$\begin{aligned}
\operatorname{dist}\big(x_\varepsilon(T+h_m),\ \mathscr{V}\big) &\le d_0\big(f(z)\,(h_m, x_\varepsilon(T)),\ f(z)\,(h_m, z)\big) + \operatorname{dist}\big(f(z)\,(h_m, z),\ \mathscr{V}\big)\\
&\le d_0\big(x_\varepsilon(T),\ z\big) \cdot e^{\widehat{\alpha}\, h_m} + \varepsilon \cdot h_m\\
&\le \varepsilon\ e^{\widehat{\alpha} T}\ T \quad \cdot e^{\widehat{\alpha}\, h_m} + \varepsilon \cdot h_m\\
&\le r_\varepsilon(T + h_m),
\end{aligned}$$
i.e. condition (b')(1.) is also satisfied at time $t = T + h_m$ with any large $m \in \mathbb{N}$.
Finally, $x_\varepsilon(\cdot)\big|_{[0,\,T+h_m]}$ provides the wanted contradiction and thus, $T = 1$. □

Proof (of Proposition 25). Considering a sequence of approximative solutions $(x_{1/n}(\cdot))_{n\in\mathbb{N}}$ in the sense of Lemma 26, we can select a subsequence $(x_{1/n_j}(\cdot))_{j\in\mathbb{N}}$ that is converging pointwise to a λ_0-Lipschitz continuous curve $x(\cdot) : [0,T] \longrightarrow E$. Indeed, this selection is based on the same arguments as Lemma 18 (on page 195 f.). Moreover, $x(\cdot)$ has all its values in the closed set of constraints $\mathscr{V} \subset E$.
Finally we have to verify that $x(\cdot)$ solves the mutational equation $\overset{\circ}{x}(\cdot) \ni f(x(\cdot))$. It results from Convergence Theorem 13 (on page 191) and the continuity of f. □

3.3.6 Exploiting a Generalized Form of "Weak" Compactness: Convergence and Existence without State Constraints

In § 3.3.3 (on page 193 ff.), the combination of Euler compactness and Euler equi-continuity has laid the foundations for the existence of solutions to the initial value problem without state constraints (in Theorem 19).

This form of compactness with respect to $(d_j)_{j\in\mathscr{I}}$, however, might be very difficult to verify in many applications. In the simple example of a Banach space with affine linear transitions (extending Example 1.2 on page 2), we would have to assume that all transitions have their values (after any positive time) in a finite dimensional subspace. Undoubtedly, it is a very severe restriction.

Similar obstacles have already led to the concepts of weak convergence and weak compactness in functional analysis. They are closely related with linear forms in the considered topological vector space, but such linear functions do not prove to be appropriate for drawing any conclusions in the general tuple $\big(E, (d_j)_{j\in\mathscr{I}}, (e_j)_{j\in\mathscr{I}}\big)$.

In regard to extending the notion of weak convergence to such a tuple, we suggest another well-known relation of linear functional analysis as starting point for bridging the gap between strong and weak topology: In every Banach space $(X, \|\cdot\|_X)$ (with $\mathbb{B}_X$ denoting its closed unit ball), the norm of any element $z \in X$ satisfies

$$\|z\|_X = \sup \big\{ y^*(z) \;\big|\; y^* : X \longrightarrow \mathbb{R} \text{ linear, continuous, } \sup_{x\in\mathbb{B}_X} \|y^*(x)\|_X \le 1 \big\}.$$

Skipping now any aspects of linearity, we realize that the metric on X is represented as supremum of further pseudo-metrics. In particular, weak convergence focuses on the convergence with respect to all these pseudo-metrics instead of their supremum. Such a connection via supremum can be extended easily to $\big(E, (d_j)_{j\in\mathscr{I}}, (e_j)_{j\in\mathscr{I}}\big)$.

Additional assumptions for § 3.3.6.

In addition to the general hypotheses (H1)–(H7) about $d_j, e_j : E \times E \longrightarrow [0,\infty[$ specified in § 3.1 (on page 182 ff.), let $\mathscr{J} \neq \emptyset$ be a further index set. Assume $d_{j,\kappa}, e_{j,\kappa} : E \times E \longrightarrow [0,\infty[$ $(j \in \mathscr{I}, \kappa \in \mathscr{J})$ to satisfy (H1)–(H3) (with index set $\mathscr{I} \times \mathscr{J}$ instead of $\mathscr{I}$ for distance functions) and additionally

(H8)
$$d_j(x,y) = \sup_{\kappa \in \mathscr{J}} d_{j,\kappa}(x,y),$$
$$e_j(x,y) = \sup_{\kappa \in \mathscr{J}} e_{j,\kappa}(x,y) \qquad \text{for all } x,y \in E,\ j \in \mathscr{I}.$$

Moreover, we tighten up hypothesis (H4) in the following form:

(H4') $\lfloor \cdot \rfloor_j$ is lower semicontinuous with respect to $(d_{i,\kappa})_{i \in \mathscr{I}, \kappa \in \mathscr{J}}$, i.e.,
$$\lfloor x \rfloor_j \le \liminf_{n \to \infty} \lfloor x_n \rfloor_j$$
for any $x \in E$ and $(x_n)_{n \in \mathbb{N}}$ in E fulfilling for each $i \in \mathscr{I}, \kappa \in \mathscr{J}$
$$\lim_{n \to \infty} d_{i,\kappa}(x_n, x) = 0, \qquad \sup_{n \in \mathbb{N}} \lfloor x_n \rfloor_i < \infty.$$

Definition 27 (weakly Euler compact).

The tuple $\big(E,\ (d_j)_{j \in \mathscr{I}},\ (d_{j,\kappa})_{j \in \mathscr{I}, \kappa \in \mathscr{J}},\ (e_j)_{j \in \mathscr{I}},\ (e_{j,\kappa})_{j \in \mathscr{I}, \kappa \in \mathscr{J}},\ (\lfloor \cdot \rfloor_j)_{j \in \mathscr{I}},$ $\widehat{\Theta}\big(E, (d_i), (e_i), (\lfloor \cdot \rfloor_i)\big)\big)$ is called *weakly Euler compact* if it satisfies the following condition for any element $x_0 \in E$, time $T \in]0,\infty[$ and bounds $\widehat{\alpha}_j, \widehat{\beta}_j, \widehat{\gamma}_j > 0$ $(j \in \mathscr{I})$:
Let $\mathscr{N} = \mathscr{N}(x_0, T, (\widehat{\alpha}_j, \widehat{\beta}_j, \widehat{\gamma}_j)_{j \in \mathscr{I}})$ denote the (possibly empty) subset of all curves $y(\cdot) : [0,T] \longrightarrow E$ specified in a piecewise way in Definition 2.15 (on page 112) and equivalently in Remark 15 (2.) (on page 193).
Then for each $t \in [0,T]$, every sequence $(z_n)_{n \in \mathbb{N}}$ in $\{y(t) \mid y(\cdot) \in \mathscr{N}\} \subset E$ has a subsequence $(z_{n_m})_{m \in \mathbb{N}}$ and an element $z \in E$ with
$$\lim_{m \to \infty} d_{j,\kappa}(z_{n_m}, z) = 0 \qquad \text{for each } j \in \mathscr{I}, \kappa \in \mathscr{J}.$$

Now the existence of solutions is proved in this modified environment, i.e. on the basis of *weak* Euler compactness, but still by means of Euler approximations. The next theorems about existence and convergence, however, require a form of "weak continuity" (with respect to state) for the transitions which occur on the right-hand of the mutational equation. This is the novelty in assumption (5.) below:

Theorem 28 (Existence due to weak Euler compactness).
Suppose $\big(E,\ (d_j)_{j \in \mathscr{I}},\ (e_j)_{j \in \mathscr{I}},\ (\lfloor \cdot \rfloor_j)_{j \in \mathscr{I}},\ \widehat{\Theta}\big(E, (d_i)_{i \in \mathscr{I}}, (e_i)_{i \in \mathscr{I}}, (\lfloor \cdot \rfloor_i)_{i \in \mathscr{I}}\big)\big)$ *to be Euler equi-continuous (in the sense of Definition 16 on page 194) and the tuple* $\big(E,\ (d_j)_j,\ (d_{j,\kappa})_{j,\kappa},\ (e_j)_j,\ (e_{j,\kappa})_{j,\kappa},\ (\lfloor \cdot \rfloor_j)_j,\ \widehat{\Theta}\big(E, (d_i)_i, (e_i)_i, (\lfloor \cdot \rfloor_i)_i\big)\big)$ *to be weakly Euler compact.*

Moreover assume for some fixed $\tau \ge 0$, *the function*
$$f : \mathrm{BLip}\big([-\tau, 0], E; (e_i)_i, (\lfloor \cdot \rfloor_i)_i\big) \times [0,T] \longrightarrow \widehat{\Theta}\big(E, (d_i)_i, (e_i)_i, (\lfloor \cdot \rfloor_i)_i\big)$$
and each $j \in \mathscr{I}$, $R > 0$:

1.) $\sup_{z(\cdot), t} \alpha_j(f(z(\cdot), t); R) < \infty,$

2.) $\sup_{z(\cdot), t} \beta_j(f(z(\cdot), t); R) < \infty,$

3.) $\sup_{z(\cdot), t} \gamma_j(f(z(\cdot), t)) < \infty,$

4.) *for* $\mathscr{L}^1$*-almost every* $t \in [0,T]$: $\lim_{n\to\infty} \widehat{D}_j\big(f(z_n^1(\cdot), t_n^1),\; f(z_n^2(\cdot), t_n^2); R\big) = 0$ *for each* $j \in \mathscr{I}$, $R \geq 0$ *and any sequences* $(t_n^1)_{n\in\mathbb{N}}$, $(t_n^2)_{n\in\mathbb{N}}$ *in* $[0,T]$ *and* $(z_n^1(\cdot))_{n\in\mathbb{N}}$, $(z_n^2(\cdot))_{n\in\mathbb{N}}$ *in* $\mathrm{BLip}\big([-\tau,0], E; (e_j)_{j\in\mathscr{I}}, (\lfloor\cdot\rfloor_j)_{j\in\mathscr{I}}\big)$ *satisfying for every* $i \in \mathscr{I}$, $\kappa \in \mathscr{J}$ *and* $s \in [-\tau,0]$

$$\lim_{n\to\infty} t_n^1 = t = \lim_{n\to\infty} t_n^2, \quad \lim_{n\to\infty} d_{i,\kappa}\big(z_n^1(s), z(s)\big) = 0 = \lim_{n\to\infty} d_{i,\kappa}\big(z_n^2(s), z(s)\big)$$
$$\sup_{n\in\mathbb{N}} \sup_{[-\tau,0]} \lfloor z_n^{1,2}(\cdot)\rfloor_i < \infty.$$

5.) *for every* $z(\cdot)$ *and* $\mathscr{L}^1$*-a.e.* $t \in [0,T]$, *the function* $f(z(\cdot),t)\,(h,\cdot) : E \longrightarrow E$ *is "weakly" continuous in the following sense:*

$$\lim_{n\to\infty} d_{j,\kappa}\big(f(z(\cdot),t)\,(h,y),\;\; f(z(\cdot),t)\,(h,y_n)\big) = 0$$

for each $\kappa \in \mathscr{J}$, $h \in\,]0,1]$, $y \in E$ *and any sequence* $(y_n)_{n\in\mathbb{N}}$ *in* E *satisfying* $d_{i,\kappa'}(y,y_n) \longrightarrow 0$, $\sup_n \lfloor y_n\rfloor_i < \infty$ *for any* $i \in \mathscr{I}$, $\kappa' \in \mathscr{J}$.

For every function $x_0(\cdot) \in \mathrm{BLip}\big([-\tau,0], E; (e_j)_{j\in\mathscr{I}}, (\lfloor\cdot\rfloor_j)_{j\in\mathscr{I}}\big)$, *there exists a curve* $x(\cdot) : [-\tau,T] \longrightarrow E$ *with the following properties:*

(i) $x(\cdot) \in \mathrm{BLip}\big([-\tau,T], E; (e_j)_{j\in\mathscr{I}}, (\lfloor\cdot\rfloor_j)_{j\in\mathscr{I}}\big)$,
(ii) $x(\cdot)\big|_{[-\tau,0]} = x_0(\cdot)$,
(iii) *For* $\mathscr{L}^1$*-a.e.* $t \in [0,T[$, $\lim_{h\downarrow 0} \frac{1}{h}\cdot d_j\big(f\big(x(t+\cdot)\big|_{[-\tau,0]}, t\big)\,(h, x(t)),\; x(t+h)\big) = 0$.
If each d_j $(j \in J)$ *satisfies the triangle inequality in addition, the restriction* $x(\cdot)\big|_{[0,T]}$ *is a solution to the mutational equation* $\overset{\circ}{x}(t) \ni f\big(x(t+\cdot)\big|_{[-\tau,0]}, t\big)$ *in the sense of Definition 8 (on page 187).*

For constructing a candidate $x(\cdot) : [-\tau,T] \longrightarrow E$, we can follow exactly the same track as for Euler compactness in § 3.3.3 (on page 193 ff.). In particular, the arguments for preceding Lemma 18 (presented on page 196) provide a subsequence of Euler approximations whose restrictions to $[0,T]$ converge to a function $x(\cdot) : [0,T] \longrightarrow E$ pointwise with respect to each $d_{j,\kappa}$ $(j \in \mathscr{I}, \kappa \in \mathscr{J})$.
Now we still have to focus on the solution property of $x(\cdot)\big|_{[0,T]}$:

Proposition 29 (about "weak" pointwise convergence of solutions).
Suppose the following properties of

$$f_n, f : E \times [0,T] \longrightarrow \widehat{\Theta}\big(E, (d_i)_{i\in\mathscr{I}}, (e_j)_{j\in\mathscr{I}}, (\lfloor\cdot\rfloor_i)_{i\in\mathscr{I}}\big) \qquad (n \in \mathbb{N})$$
$$x_n, x : \quad [0,T] \longrightarrow E :$$

1.) $R_j := \sup_{n,t} \lfloor x_n(t)\rfloor_j < \infty$,
$\widehat{\alpha}_j(\rho) := \sup_n \alpha_j\big(x_n; \rho\big) < \infty$ *for* $\rho \geq 0$,
$\widehat{\beta}_j := \sup_n \mathrm{Lip}\,\big(x_n(\cdot) : [0,T] \longrightarrow (E, e_j)\big) < \infty$ *for every* $j \in \mathscr{I}$,

2.) $\overset{\circ}{x}_n(\cdot) \ni f_n(x_n(\cdot), \cdot)$ *(in the sense of Definition 8 on page 187) for every* $n \in \mathbb{N}$,

3.) *Equi-continuity of* $(f_n)_n$ *at* $(x(t),t)$ *at almost every time in the following sense:*

for $\mathscr{L}^1$*-almost every* $t \in [0,T]$: $\quad \lim_{n\to\infty} \widehat{D}_j(f_n(x(t),t),\ f_n(y_n,t_n);\ r) = 0$
for each $j \in \mathscr{I}$, $r \geq 0$ *and any* $(t_n)_{n\in\mathbb{N}}$, $(y_n)_{n\in\mathbb{N}}$ *in* $[t,T]$ *and* E *respectively satisfying* $\lim_{n\to\infty} t_n = t$, $\lim_{n\to\infty} d_{i,\kappa}(x(t), y_n) = 0$, $\sup_{n\in\mathbb{N}} \lfloor y_n \rfloor_i \leq R_i$ *for any* i, κ,

4.) *For* $\mathscr{L}^1$*-almost every* $t \in [0,T[$ $(t = 0$ *inclusive) and any* $\widetilde{t} \in]t,T[$, *there is a sequence* $n_m \nearrow \infty$ *of indices (depending on* $t < \widetilde{t}$*) that satisfies for* $m \longrightarrow \infty$

$$\begin{cases} (i) & \widehat{D}_j\big(f(x(t),t),\ f_{n_m}(x(t),t);\ r\big) \longrightarrow 0 \qquad \text{for all } r \geq 0,\ j \in \mathscr{I}, \\ (ii) & \text{for all } j \in \mathscr{I}, \kappa \in \mathscr{J} :\ d_{j,\kappa}\big(x(t),\ x_{n_m}(t)\big) \longrightarrow 0, \\ (iii) & \text{for all } j \in \mathscr{I}, \kappa \in \mathscr{J} :\ d_{j,\kappa}\big(x(\widetilde{t}),\ x_{n_m}(\widetilde{t})\big) \longrightarrow 0. \end{cases}$$

5.) *Weak continuity of each function* $f(x(t),t)(h,\cdot) : E \longrightarrow E$ *at* $\mathscr{L}^1$*-almost every time* $t \in [0,T]$ *in the following sense:*

$$\lim_{n\to\infty} d_{j,\kappa}\big(f(x(t),t)\,(h,y),\ f(x(t),t)\,(h,y_n)\big) = 0$$

for each $\kappa \in \mathscr{J}$, $h \in]0,1]$, $y \in E$ *and any sequence* $(y_n)_{n\in\mathbb{N}}$ *in* E *satisfying* $d_{i,\kappa'}(y,y_n) \longrightarrow 0$, $\sup_n \lfloor y_n \rfloor_i < \infty$ *for any* $i \in \mathscr{I}, \kappa' \in \mathscr{J}$.

Then, $x(\cdot)$ *is* $\widehat{\beta}_j$*-Lipschitz continuous with respect to* e_j *for each index* $j \in \mathscr{I}$ *and, at* $\mathscr{L}^1$*-almost every time* $t \in [0,T]$,

$$\lim_{h\downarrow 0} \tfrac{1}{h} \cdot d_j\big(f(x(t),t)\,(h,\ x(t)),\ x(t+h)\big) = 0$$

holds for every index $j \in \mathscr{I}$.

If each d_j $(j \in J)$ *satisfies the triangle inequality in addition, then the curve* $x(\cdot) : [0,T] \longrightarrow E$ *is a solution to the mutational equation* $\overset{\circ}{x}(\cdot) \ni f(x(\cdot),\cdot)$ *in the tuple* $\big(E,\ (d_j)_{j\in\mathscr{I}},\ (e_j)_{j\in\mathscr{I}},\ (\lfloor\cdot\rfloor_j)_{j\in\mathscr{I}},\ (\widehat{D}_j)_{j\in\mathscr{I}}\big)$.

Proof (of Proposition 29).
Similarly to the proof of Theorem 13 (on page 191 ff.), choose the index $j \in \mathscr{I}$ arbitrarily.

Then $x(\cdot) : [0,T] \longrightarrow (E,e_j)$ is $\widehat{\beta}_j$-Lipschitz continuous. Indeed, for Lebesgue-almost every $t \in [0,T[$ and any $\widetilde{t} \in]t,T]$, assumption (4.) provides a subsequence $\big(x_{n_m}(\cdot)\big)_{m\in\mathbb{N}}$ satisfying for each $i \in \mathscr{I}$, $\kappa \in \mathscr{J}$

$$\begin{cases} d_{i,\kappa}\big(x(t),\ x_{n_m}(t)\big) \longrightarrow 0 \\ d_{i,\kappa}\big(x(\widetilde{t}),\ x_{n_m}(\widetilde{t})\big) \longrightarrow 0 \end{cases} \qquad \text{for } m \longrightarrow \infty.$$

The uniform $\widehat{\beta}_j$-Lipschitz continuity of $x_n(\cdot)$, $n \in \mathbb{N}$, with respect to e_j and hypothesis (H3) (i) about $(e_{i,\kappa})_{i\in\mathscr{I},\kappa\in\mathscr{J}}$ imply for every $\kappa \in \mathscr{J}$

$$\begin{aligned} e_{j,\kappa}\big(x(t),\, x(\widetilde{t})\big) \;&\le\; \limsup_{m\longrightarrow\infty}\; e_{j,\kappa}\big(x_{n_m}(t),\, x_{n_m}(\widetilde{t})\big) \;\le\; \widehat{\beta}_j\, |\widetilde{t}-t|,\\ e_{j}\big(x(t),\, x(\widetilde{t})\big) \;&=\; \sup_{\kappa\in\mathcal{J}}\; e_{j,\kappa}\big(x(t),\, x(\widetilde{t})\big) \qquad\;\; \le\; \widehat{\beta}_j\, |\widetilde{t}-t|. \end{aligned}$$

This Lipschitz estimate even holds at *any* points of time $t,\widetilde{t}\in[0,T]$ due to the lower semicontinuity of $e_{j,\kappa}$ (hypotheses (H3) (o), (i)). Furthermore, hypothesis (H4') about the lower semicontinuity of $\lfloor\cdot\rfloor_j$ guarantees the bound

$$\lfloor x(\widetilde{t})\rfloor_j \;\le\; \liminf_{m\longrightarrow\infty}\; \lfloor x_{n_m}(\widetilde{t})\rfloor_j \;\le\; R_j.$$

Finally we verify at $\mathscr{L}^1$-almost every time $t\in[0,T[$

$$\limsup_{h\downarrow 0}\; \tfrac{1}{h}\cdot d_j\big(f(x(t),t)\,(h,\, x(t)),\; x(t+h)\big) \;=\; 0.$$

Indeed, for $\mathscr{L}^1$-almost every $t\in[0,T[$ and any $h\in]0,T-t[$, assumption (4.) ensures a subsequence $\big(x_{n_m}(\cdot)\big)_{m\in\mathbb{N}}$ satisfying for each $i\in\mathcal{I}$, $\kappa\in\mathcal{J}$, $r\ge 0$ and $m\longrightarrow\infty$

$$\begin{cases} \widehat{D}_i\big(f(x(t),t),\; f_{n_m}(x(t),t);\; r\big) & \longrightarrow\; 0\\ d_{i,\kappa}\big(x(t),\qquad\quad x_{n_m}(t)\big) & \longrightarrow\; 0\\ d_{i,\kappa}\big(x(t+h),\;\; x_{n_m}(t+h)\big) & \longrightarrow\; 0. \end{cases}$$

For any indices $i\in\mathcal{I}$ and $\kappa\in\mathcal{J}$, we conclude from assumption (5.)

$$\lim_{m\longrightarrow\infty}\; d_{i,\kappa}\big(f(x(t),\,t)\,(h,\,x(t)),\;\; f(x(t),\,t)\,(h,\,x_{n_m}(t))\big) \;=\; 0.$$

Now hypothesis (H3) (i) about $(d_{i,\kappa})_{i\in\mathcal{I},\kappa\in\mathcal{J}}$ implies for every $\kappa\in\mathcal{J}$

$$\begin{aligned} & d_{j,\kappa}\big(f(x(t),t)\,(h,\,x(t)),\;\; x(t+h)\big)\\ &\quad=\; \lim_{m\longrightarrow\infty}\; d_{j,\kappa}\big(f(x(t),t)\,(h,\,x_{n_m}(t)),\;\; x_{n_m}(t+h)\big)\\ &\quad\le\; \limsup_{m\longrightarrow\infty}\; d_{j}\big(f(x(t),t)\,(h,\,x_{n_m}(t)),\;\; x_{n_m}(t+h)\big). \end{aligned}$$

Lemma 9 (on page 188) provides an estimate with $\rho\ge 0$ sufficiently large

$$\begin{aligned} & d_{j,\kappa}\big(f(x(t),t)\,(h,\,x(t)),\;\; x(t+h)\big)\\ &\quad\le h\cdot \limsup_{m\longrightarrow\infty}\; \sup_{[t,\,t+h]}\; \widehat{D}_j\big(f(x(t),t),\; f_{n_m}(x_{n_m}(\cdot),\,\cdot);\; \rho\big)\cdot e^{\widehat{\alpha}_j(\rho)\cdot h}. \end{aligned}$$

For completing the proof, we verify

$$\limsup_{h\downarrow 0}\; \limsup_{m\rightarrow\infty}\; \sup_{[t,\,t+h]}\; \widehat{D}_j\big(f(x(t),t),\; f_{n_m}(x_{n_m}(\cdot),\cdot);\; \rho\big) \;=\; 0$$

for $\mathscr{L}^1$-almost every $t\in[0,T[$ and *any* subsequence $\big(x_{n_m}(\cdot)\big)_{m\in\mathbb{N}}$ satisfying

$$\begin{cases} d_{i,\kappa}\big(x(t),\qquad\quad x_{n_m}(t)\big) & \longrightarrow\; 0\\ \widehat{D}_i\big(f(x(t),t),\, f_{n_m}(x(t),t);\; r\big) & \longrightarrow\; 0 \end{cases}$$

for $m\longrightarrow\infty$ and each $i\in\mathcal{I}$, $\kappa\in\mathcal{J}$, $r\ge 0$. Indeed, if this equation was not correct then we could select some $\varepsilon>0$ and sequences $(h_l)_{l\in\mathbb{N}}$, $(m_l)_{l\in\mathbb{N}}$, $(s_l)_{l\in\mathbb{N}}$ such that

$$\begin{cases} \widehat{D}_j\big(f(x(t),t),\;\; f_{n_{m_l}}(x_{n_{m_l}}(t+s_l),t+s_l);\;\; \rho\big) \;\geq\; \varepsilon \\ 0 \leq s_l \leq h_l \leq \frac{1}{l}, \qquad m_l \geq l \end{cases} \qquad \text{for all } l \in \mathbb{N}.$$

For each $i \in \mathscr{I}$, every curve $x_{n_m} : [0,T] \longrightarrow (E, e_i)$ $(m \in \mathbb{N})$ is $\widehat{\beta}_i$-Lipschitz continuous. Hypothesis (H3) (ii) about $(d_{i,\kappa})_{i,\kappa}$, $(e_{i,\kappa})_{i,\kappa}$ implies for any $i \in \mathscr{I}, \kappa \in \mathscr{J}$

$$\lim_{l \to \infty} \; d_{i,\kappa}\big(x(t),\; x_{n_{m_l}}(t+s_l)\big) \;=\; 0.$$

Thus at $\mathscr{L}^1$-almost every time $t \in [0,T[$, assumptions (3.), (4.) (i) and hypothesis (H6) about the continuity of $\widehat{D}_j(\,\cdot\,,\,\cdot\,;r)$ (on page 184) lead to a contradiction because for any $r \geq 0$,

$$\lim_{l \to \infty} \;\; \widehat{D}_j\big(f(x(t),t),\;\; f_{n_{m_l}}(x_{n_{m_l}}(t+s_l),t+s_l);\; r\big) \;=\; 0.$$

□

3.3.7 Existence of Solutions due to Completeness: Extending the Cauchy-Lipschitz Theorem

In general, many theorems about existence of solutions are based either on a form of *compactness* or on a version of *completeness*. Now we prefer the latter analytical basis and extend the Existence Theorem of Cauchy-Lipschitz to the current mutational framework.
Aubin's adaptation to mutational equations in metric spaces has already been presented in Theorem 1.15 (on page 38). It starts with a compactness assumption about all closed bounded balls (in the metric space) though.
Now the main goal is to formulate its extension assuming merely an appropriate form of completeness. In return for this weaker structural hypothesis, however, the right-hand side of the mutational equation is supposed to be Lipschitz continuous – in an appropriate sense.

Definition 30. The tuple $\big(E,\, (d_j)_{j\in\mathscr{I}},\, (e_j)_{j\in\mathscr{I}},\, (\lfloor\cdot\rfloor_j)_{j\in\mathscr{I}}\big)$ is called *complete* if for every sequence $(x_n)_{n\in\mathbb{N}}$ in E with

$$\begin{cases} \lim\limits_{k\to\infty}\ \sup\limits_{m,n\geq k}\ d_j(x_m,\, x_n) = 0 \\ \qquad\ \ \sup\limits_{n\in\mathbb{N}}\ \lfloor x_n\rfloor_j < \infty \end{cases} \qquad \text{for each } j\in\mathscr{I},$$

there exists an element $x\in E$ fulfilling $\lim\limits_{n\to\infty} d_j(x_n,x) = 0$ for every $j\in\mathscr{I}$.

Theorem 31 (Extended Cauchy-Lipschitz Theorem for mutational equations). *Suppose the tuple $\big(E,\, (d_j)_{j\in\mathscr{I}},\, (e_j)_{j\in\mathscr{I}},\, (\lfloor\cdot\rfloor_j)_{j\in\mathscr{I}}\big)$ to be complete and the tuple $\big(E,\, (d_j)_{j\in\mathscr{I}},\, (e_j)_{j\in\mathscr{I}},\, (\lfloor\cdot\rfloor_j)_{j\in\mathscr{I}},\, \widehat{\Theta}\big(E,(d_i)_{i\in\mathscr{I}},(e_i)_{i\in\mathscr{I}},(\lfloor\cdot\rfloor_i)_{i\in\mathscr{I}}\big)\big)$ to be Euler equi-continuous For $f: E\times[0,T]\longrightarrow\widehat{\Theta}\big(E,\, (d_j)_{j\in\mathscr{I}},\, (e_j)_{j\in\mathscr{I}},\, (\lfloor\cdot\rfloor_j)_{j\in\mathscr{I}}\big)$ assume*

(1.) *For each $j\in\mathscr{I}$ and $R>0$,*
$\widehat{\alpha}_j(R) := \sup_{x,t}\ \alpha_j(f(x,t);\, R) < \infty,$
$\widehat{\beta}_j(R) := \sup_{x,t}\ \beta_j(f(x,t);\, R) < \infty,$
$\widehat{\gamma}_j := \sup_{x,t}\ \gamma_j(f(x,t)) < \infty,$

(2.) *f is Lipschitz continuous w.r.t. state and continuous in the following sense: for each tuple $(r_j)_{j\in\mathscr{I}}$ in $[0,\infty[^{\mathscr{I}}$, there exist constants $\Lambda_j,\mu_j\geq 0$ $(j\in\mathscr{I})$ and moduli of continuity $(\omega_j(\cdot))_{j\in\mathscr{I}}$ such that $\delta_j: E\times E\longrightarrow[0,\infty[$,*

$$\delta_j(x,y) := \inf\big\{d_j(x,z)+\mu_j\cdot e_j(z,y)\ \big|\ z\in E,\ \forall\, i\in\mathscr{I}:\ \lfloor z\rfloor_i\leq r_i\big\}$$

satisfies for every $j\in\mathscr{I}$

$$\widehat{D}_j\big(f(x,s),\ f(y,t);\ r_j\big)\ \leq\ \Lambda_j\cdot\delta_j(x,y)+\omega_j(|t-s|)$$

whenever $(x,s),\ (y,t)\in E\times[0,T]$ fulfill $\max\big\{\lfloor x\rfloor_i,\ \lfloor y\rfloor_i\big\}\leq r_i$ for each i.

Then for every initial element $x_0\in E$, there exists a solution $x(\cdot):[0,T]\longrightarrow E$ to the mutational equation $\overset{\circ}{x}(\cdot)\ni f\big(x(\cdot),\,\cdot\big)$ in the sense of Definition 8 (on page 187) with $x(0)=x_0$.

Proof. We use Euler approximations on equidistant partitions of $[0,T]$ again, but now we conclude their convergence to a candidate $x(\cdot) : [0,T] \longrightarrow E$ (with respect to each distance $d_j, j \in \mathcal{I}$) from completeness. Finally, Convergence Theorem 13 (on page 191) implies that $x(\cdot)$ is a solution to the mutational equation of interest.

For every $n \in \mathbb{N}$ with $2^n > T$, set

$$\begin{aligned} h_n &:= \tfrac{T}{2^n}, \qquad t_n^k := k\, h_n \qquad\qquad \text{for } k = 0 \ldots 2^n, \\ x_n(0) &:= x_0, \\ x_n(t) &:= f(x_n(t_n^k), t_n^k)\big(t - t_n^k,\ x_n(t_n^k)\big) \quad \text{for } t \in]t_n^k, t_n^{k+1}],\ k < 2^n. \end{aligned}$$

Assuming Euler equi-continuity, we obtain a constant $L_j \in [0,\infty[$ for each index j such that every curve $x_n(\cdot)$ is L_j-Lipschitz continuous with respect to e_j. Moreover, Lemma 5 (on page 185) guarantees for every $t \in [0,T],\ n \in \mathbb{N}$ (with $2^n > T$), $j \in \mathcal{I}$

$$\lfloor x_n(t) \rfloor_j \;\le\; \big(\lfloor x_0 \rfloor_j + \widehat{\gamma}_j\, T\big) \cdot e^{\widehat{\gamma}_j\, T} \;=:\; R_j\,.$$

Assumption (2.) provides constants $\Lambda_j, \mu_j \ge 0$ $(j \in \mathcal{I})$ related to the tuple $(R_j)_{j \in \mathcal{I}}$ such that Lipschitz continuity with respect to the corresponding auxiliary function

$$\begin{aligned} \delta_j :\ E \times E &\longrightarrow [0,\infty[, \\ (x,y) &\longmapsto \inf\big\{d_j(x,z) + \mu_j \cdot e_j(z,y) \;\big|\; z \in E,\ \forall\, i \in \mathcal{I} :\ \lfloor z \rfloor_i \le R_i\big\} \end{aligned}$$

holds for every index $j \in \mathcal{I}$. In particular, we conclude from Proposition 7 about estimating evolutions along any two transitions (on page 185) in a piecewise way: For each $j \in \mathcal{I}$ and every $n > m,\ t \in]t_m^k, t_m^{k+1}] \cap]t_n^l, t_n^{l+1}]$,

$$\begin{aligned} & d_j\big(x_m(t),\, x_n(t)\big) \cdot e^{-\widehat{\alpha}_j(R_j)\cdot(t - t_n^l)} \\ &\le d_j\big(x_m(t_n^l),\, x_n(t_n^l)\big) + (t - t_n^l)\cdot\ \widehat{D}_j\big(f(x_m(t_m^k), t_m^k),\ f(x_n(t_n^l), t_n^l);\ R_j\big) \\ &\le d_j\big(x_m(t_n^l),\, x_n(t_n^l)\big) + (t - t_n^l)\cdot\big(\Lambda_j\ \delta_j\big(x_m(t_m^k),\, x_n(t_n^l)\big) + \omega_j(|t_n^l - t_m^k|)\big) \\ &\le d_j\big(x_m(t_n^l),\, x_n(t_n^l)\big) + (t - t_n^l)\cdot\big(\Lambda_j\ \big(d_j(x_m(t_m^k),\, x_n(t_m^k)) + \mu_j \cdot e_j(x_n(t_m^k), x_n(t_n^l))\big) \\ &\qquad\qquad + \omega_j(h_m)\big) \\ &\le d_j\big(x_m(t_n^l),\, x_n(t_n^l)\big) + (t - t_n^l)\cdot\big(\Lambda_j\ \ d_j(x_m(t_m^k), x_n(t_m^k)) + \Lambda_j \mu_j \cdot L_j h_m + \omega_j(h_m)\big) \end{aligned}$$

and thus, $\sup_{s \in [0,t]}\ d_j\big(x_m(s),\, x_n(s)\big) \;\le\; \text{const}(\mu_j, L_j, \Lambda_j) \cdot (h_m + \omega_j(h_m))\ e^{\Lambda_j \cdot t}$

for every $t \in [0,T]$. The sequence of Euler approximation $\big(x_n(\cdot)\big)_{n \in \mathbb{N}}$ is (even) a *uniform* Cauchy sequence with respect to each $d_j,\ j \in \mathcal{I}$.

Due to completeness, there exists an element $x(t) \in E$ at every time $t \in]0,T]$ such that $\lim_{n \to \infty} d_j\big(x_n(t),\, x(t)\big) = 0$ holds for every index $j \in \mathcal{I}$. Setting $x(0) := x_0$ is a rather obvious choice.

As a consequence of Convergence Theorem 13, $x(\cdot) : [0,T] \longrightarrow E$ is a solution to the mutational equation $\overset{\circ}{x}(\cdot) \ni f\big(x(\cdot), \cdot\big)$ in the sense of Definition 8. This results from essentially the same arguments as the proof of Theorem 19 (on page 197 f.).

□

3.4 Local ω-Contractivity of Transitions Can Become Dispensable

Definition 2 of transitions (on page 183) implies the restriction that the initial distance between two points may grow (at most) exponentially while evolving along the same transition ϑ, i.e. for any $x,y \in E$ and $h \in [0,1]$, $j \in \mathcal{I}$,

$$d_j\big(\vartheta(h,x),\ \vartheta(h,y)\big) \le d_j(x,y) \cdot e^{\alpha_j h}$$

with a constant $\alpha_j \in [0,\infty[$ depending on ϑ and $\max\{\lfloor x \rfloor, \lfloor y \rfloor\} < \infty$. The key goal of this subsection is some way out if the candidates for transitions only satisfy

$$d_j\big(\vartheta(h,x),\ \vartheta(h,y)\big) \le C \cdot d_j(x,y) \cdot e^{\alpha_j h}$$

with a constant $C > 1$.
In a very broad sense, we apply the same notion as for the step from Hille-Yosida Theorem (about contractive C^0 semigroups) to the Theorem of Feller, Miyadera and Phillips (about arbitrary C^0 semigroups) (see e.g. [76, Theorem II.3.8]). Indeed, we introduce a suitable auxiliary distance $\check{d}_j$ being "equivalent" to d_j, but in the mutational framework there is no linear resolvent operator available as in the standard proof of the Theorem of Feller, Miyadera and Phillips. Hence we start from a more general inequality of error propagation in finite time intervals instead.

General assumptions and notations for § 3.4.

(A1) $\check{\Theta}\big(E,(d_j)_j,(e_j)_j,(\lfloor\cdot\rfloor)_j\big)$ is a nonempty set of functions $\vartheta : [0,1] \times E \longrightarrow E$ satisfying for each $j \in \mathcal{I}$

(1.) for every $x \in E$: $\quad \vartheta(0,x) = x$

(3.) there is $\beta_j(\vartheta;\cdot) : [0,\infty[\longrightarrow [0,\infty[$ such that for any $s,t \in [0,1]$, $x \in E$ with $\lfloor x \rfloor_j \le r$: $\quad e_j\big(\vartheta(s,x),\ \vartheta(t,x)\big) \le \beta(\vartheta;r) \cdot |t-s|$

(4.) there is $\widehat{\gamma}_j \in [0,\infty[$ (not depending on ϑ) such that for any $t \in [0,1]$ and $x \in E$: $\quad \lfloor \vartheta(t,x) \rfloor_j \le \big(\lfloor x \rfloor_j + \widehat{\gamma}_j\, t\big) \cdot e^{\widehat{\gamma}_j t}$

Moreover, a parameter function $\alpha_j : \check{\Theta}\big(E,(d_j)_j,(e_j)_j,(\lfloor\cdot\rfloor)_j\big) \times [0,\infty[\longrightarrow [0,\infty[$ is nondecreasing with respect to its second argument. (Its purpose is clarified in (A4) below.)

(A2) For any initial element $x_0 \in E$, time $T \in]0,\infty[$ and bounds $\widehat{\alpha}_j, \widehat{\beta}_j > 0$ $(j \in \mathcal{I})$, the set $\mathcal{N} = \mathcal{N}(x_0, T, (\widehat{\alpha}_j, \widehat{\beta}_j, \widehat{\gamma}_j)_{j \in \mathcal{I}})$ consists of all "Euler curves" $[0,T] \longrightarrow E$ related to piecewise constant curves in $\check{\Theta}\big(E,(d_j)_j,(e_j)_j,(\lfloor\cdot\rfloor)_j\big)$ as in Remark 15 (2.) (on page 193) – but with the global bound $\widehat{\gamma}_j < \infty$ (mentioned in (A1) (4.)) instead of $\gamma_j(\vartheta_k)$.

(A3) $\check{D}_j : \check{\Theta}(E,(d_j)_j,(e_j)_j,(\lfloor\cdot\rfloor)_j) \times \check{\Theta}(E,(d_j)_j,(e_j)_j,(\lfloor\cdot\rfloor)_j) \times [0,\infty[\longrightarrow [0,\infty[$ satisfies hypotheses (H5) and (H6) (on page 184), but not necessarily (H7).

(A4) There is a nondecreasing function $\check{C}_j : [0,\infty[\longrightarrow]0,\infty[$ satisfying:
Choose the bounds $\widehat{\alpha}_j, \widehat{\beta}_j, R_j, T > 0$ $(j \in \mathscr{I})$ and initial points $x_0, y_0 \in E$ arbitrarily with $\max\{\lfloor x_0 \rfloor_j, \lfloor y_0 \rfloor_j\} < R_j$ and set $\rho_j(t) := (R_j + \widehat{\gamma}_j t)\, e^{\widehat{\gamma}_j t}$ for each $j \in \mathscr{I}$. Then any curves $x(\cdot) \in \mathscr{N}(x_0, T, (\widehat{\alpha}_j, \widehat{\beta}_j, \widehat{\gamma}_j)_{j \in \mathscr{I}})$, $y(\cdot) \in \mathscr{N}(y_0, T, (\widehat{\alpha}_j, \widehat{\beta}_j, \widehat{\gamma}_j)_{j \in \mathscr{I}})$ and the corresponding piecewise constant $\vartheta, \tau : [0,T] \longrightarrow \check{\Theta}(E, (d_j)_j, (e_j)_j, (\lfloor\cdot\rfloor)_j)$ (as in Remark 15 on page 193) fulfil

$$d_j\big(x(T), y(T)\big) \le \Big(\check{C}_j(0) \quad \cdot d_j(x_0, y_0) + \check{C}_j(T) \cdot \int_0^T \check{D}_j\big(\vartheta(s),\, \tau(s);\, \rho_j(s)\big) \cdot e^{-\check{\alpha}_j^{\rho}(s)}\, ds\Big) \cdot e^{\check{\alpha}_j^{\rho}(T)}$$

with the abbreviation $\check{\alpha}_j^{\rho}(t) := \displaystyle\int_0^t \alpha_j(\tau(s);\, \rho_j(s))\, ds.$
(As T is chosen arbitrarily, the restrictions of $x(\cdot), y(\cdot)$ to $[0,t]$ provide this estimate at even *every* time $t \in [0,T]$.)

In comparison with the preceding general assumptions of this chapter in § 3.1 (on page 182 ff.), the essential new aspect is specified in assumption (A4). Indeed, the details about $\alpha_j(\vartheta;\cdot)$ and $\check{D}_j(\cdot,\cdot;r)$ are now reduced and, we *assume* the structural inequality (of Proposition 7 on page 185) with three modifications:

(i) the initial error is now multiplied by a constant $\check{C}_j(0)$ (possibly > 1),

(ii) we suppose this modified inequality for all "Euler curves" related to *piecewise constant* curves in $\check{\Theta}(E, (d_j)_j, (e_j)_j, (\lfloor\cdot\rfloor)_j)$ in a finite time interval $[0,T]$,

(iii) there is an additional factor $e^{-\check{\alpha}_j^{\rho}(s)}$ in the integral – for technical reasons, but this is no severe restriction because we can usually adapt the choice of $\check{C}_j(T)$.

Constructing Auxiliary Distances with Equivalent Concept of Convergence

Now we bridge the gap between functions in $\check{\Theta}(E, (d_j)_j, (e_j)_j, (\lfloor\cdot\rfloor)_j)$ and transitions (in the strict sense of Definition 2 on page 183) by means of an auxiliary distance function $\check{d}_j$.
Additionally, further real components are introduced for technical reasons. They are essentially to record properly to which "ball" $\{\lfloor\cdot\rfloor_j \le r\} \subset E$ we have to refer for the choice of $\alpha_j, \check{D}_j$ and each $j \in \mathscr{I}$. (Indeed, the tuple $(x, (\rho_j)_{j \in \mathscr{I}}) \in E \times [0,\infty[^{\mathscr{I}}$ is related to $\bigcap_{j \in \mathscr{I}} \{\lfloor\cdot\rfloor_j \le \rho_j \cdot e^{\rho_j}\} \subset E$. This separate exponential factor is just to facilitate updating the radius along transitions.)

Proposition 32. *Consider*
$\check{E} := \{(x, (\rho_j)_{j \in \mathscr{I}}) \in E \times \mathbb{R}^{\mathscr{I}} \mid$ *for each* $j \in \mathscr{I},\ \lfloor x \rfloor_j \le \rho_j \cdot e^{\rho_j}\} \subset E \times [0,\infty[^{\mathscr{I}}$
with the inclusion $E \longrightarrow \check{E}$, $\quad x \longmapsto (x, (\lfloor x \rfloor_j)_{j \in \mathscr{I}})$ *and*
the projection $\pi_i : \check{E} \longrightarrow [0,\infty[$, $\quad (x, (\rho_j)_{j \in \mathscr{I}}) \longmapsto \rho_i$.

Define the extensions of $d_j(\cdot,\cdot)$, $e_j(\cdot,\cdot)$, $\lfloor\cdot\rfloor_j$ *and* $\vartheta \in \check{\Theta}(E, (d_j)_j, (e_j)_j, (\lfloor\cdot\rfloor)_j)$ *as*

$$\begin{array}{rrcll}
d_j: & \check{E}\times\check{E} & \longrightarrow [0,\infty[, & \big((x_1,(\rho_i^1)),\,(x_2,(\rho_i^2))\big) & \longmapsto d_j(x_1,x_2),\\
e_j: & \check{E}\times\check{E} & \longrightarrow [0,\infty[, & \big((x_1,(\rho_i^1)),\,(x_2,(\rho_i^2))\big) & \longmapsto e_j(x_1,x_2),\\
\lfloor\cdot\rfloor_j: & \check{E} & \longrightarrow [0,\infty[, & (x,(\rho_i)) & \longmapsto \lfloor x\rfloor_j,\\
\vartheta: & [0,1]\times\check{E} & \longrightarrow \check{E}, & \big(h,\,(x,(\rho_i))\big) & \longmapsto \big(\vartheta(h,x),\,(\rho_i+\widehat{\gamma}_i\,h)_{i\in\mathcal{I}}\big).
\end{array}$$

For each index $j\in\mathcal{I}$, there exist some $T_j>1$ and a function $\check{d}_j:\check{E}\times\check{E}\longrightarrow[0,\infty[$ satisfying for any $\vartheta,\tau\in\check{\Theta}(E,(d_i)_i,(e_i)_i,(\lfloor\cdot\rfloor_i)_i)$, $\check{x},\check{y}\in\check{E}$, $t_1,t_2,h\geq 0$ with t_1+h, $t_2+h\leq 1$ and the abbreviation $R_j := \big(\max\{\pi_j\check{x},\,\pi_j\check{y}\}+\widehat{\gamma}_j\big)\cdot e^{\max\{\pi_j\check{x},\,\pi_j\check{y}\}+\widehat{\gamma}_j}$

(1.) $d_j(\cdot,\cdot)\leq\check{d}_j(\cdot,\cdot)\leq\check{C}_j(0)\cdot d_j(\cdot,\cdot)$

(2.) $\check{d}_j\big(\vartheta(t_1+h,\check{x}),\,\vartheta(t_2+h,\check{y})\big)\leq\check{d}_j\big(\vartheta(t_1,\check{x}),\,\vartheta(t_2,\check{y})\big)\cdot e^{h\,(1+\alpha_j(\vartheta;R_j))}$,

(3.) $$\check{d}_j\big(\vartheta(t_1+h,\check{x}),\,\tau(t_2+h,\check{y})\big) \leq \Big(\check{d}_j\big(\vartheta(t_1,\check{x}),\,\tau(t_2,\check{y})\big)+h\cdot\check{C}_j(T_j)\,\check{D}_j(\vartheta,\tau;R_j)\Big)\cdot e^{h\,(1+\alpha_j(\tau;R_j))}.$$

In particular, each function $\vartheta\in\check{\Theta}(E,(d_j)_j,(e_j)_j,(\lfloor\cdot\rfloor)_j)$ induces a unique transition on the tuple $\big(\check{E},(\check{d}_j)_j,(e_j)_j,(\lfloor\cdot\rfloor_j)_j\big)$ in the sense of Definition 2 (on page 183).

In a word, the auxiliary distance functions $\check{d}_j$, $j\in\mathcal{I}$, on $\check{E}$ (instead of E) guarantee the form of ω-contractivity that we need for transitions (according to statement (2.)) and, they lead to the same concept of sequential convergence as the original distance functions $(d_j)_{j\in\mathcal{I}}$ (according to statement (1.)).
In particular, statement (3.) indicates explicitly how to adapt the distance function $\check{D}_j$ between transitions and the parameter of error propagation.

If we ensure the general hypotheses (H1) – (H3) for $\check{d}_j:\check{E}\times\check{E}\longrightarrow[0,\infty[$ $(j\in\mathcal{I})$ additionally, then all preceding results about well-posed Cauchy problems in this chapter can be applied to the tuple $\big(\check{E},(\check{d}_j)_j,(e_j)_j,(\lfloor\cdot\rfloor_j)_j\big)$ and its (simply extended) transitions immediately.
As the construction of $\check{d}_j$ in the proof below shows, reflexivity and symmetry of each $\check{d}_j$ $(j\in\mathcal{I})$ result from the corresponding properties of d_j. The "equivalence" between $\check{d}_j$ and d_j lays the basis for verifying hypotheses (H3) (o), (ii), (iii) about converging sequences (on page 182) easily.
The sequential continuity in the sense of hypothesis (H3) (i) is nontrivial though. In the next lemma we prove it under the additional assumption that d_j is uniformly continuous on each "ball". Naturally this form of continuity results from the triangle inequality if d_j is a pseudo-metric, for example. It is worth noticing, however, that the triangle inequality for any power d_j^p $(p>0)$ also proves to be sufficient here. This case will be useful in regard to nonlocal stochastic differential equations (§ 3.6).

Lemma 33 (Sequential continuity of $\check{d}_j$: (H3) (i)).
Suppose for some $j\in\mathcal{I}$ that $d_j:E\times E\longrightarrow[0,\infty[$ is locally uniformly continuous in the following sense: For each $r>0$, there exists a modulus of continuity $\omega_{j,r}(\cdot)$ such that for all $x,x',y,y'\in E$ with $\max\{\lfloor x\rfloor_j,\,\lfloor x'\rfloor_j,\,\lfloor y\rfloor_j,\,\lfloor y'\rfloor_j\}<r$

$$\left|d_j(x,y) - d_j(x',y')\right| \le \omega_{j,r}\left(d_j(x,x') + d_j(y,y')\right).$$

Furthermore assume $\alpha_j(\cdot\,;r)$ to be bounded for each $r \ge 0$.
Then the function $\check{d}_j : \check{E} \times \check{E} \longrightarrow [0,\infty[$ is sequentially continuous in the sense that

$$\lim_{n\to\infty} \check{d}_j(\check{x}_n, \check{y}_n) = \check{d}_j(\check{x}, \check{y})$$

for any $\check{x}, \check{y} \in \check{E}$ and sequences $(\check{x}_n)_{n\in\mathbb{N}}, (\check{y}_n)_{n\in\mathbb{N}}$ in $\check{E}$ fulfilling for each $i \in \mathscr{I}$,

$$\lim_{n\to\infty} d_i(\check{x},\check{x}_n) = 0 = \lim_{n\to\infty} d_i(\check{y}_n,\check{y}), \qquad \sup_{n\in\mathbb{N}} \{\lfloor \check{x}_n \rfloor_i, \lfloor \check{y}_n \rfloor_i\} < \infty.$$

Proof (of Proposition 32 on page 215 f.). Fix $T_j > 1$ with $\check{C}_j(0) \cdot e^{-T_j+1} \le \frac{1}{2}$ and define $\check{d}_j : \check{E} \times \check{E} \longrightarrow [0,\infty[$ as

$$\begin{aligned}
&\check{d}_j(\check{x}_0,\check{y}_0)\\
&= \sup\Big\{ e^{-t}\Big(d_j\big(\check{x}(t),\check{y}(t)\big)\cdot e^{-\check{\alpha}_j^\rho(t)} - \check{C}_j(T_j)\cdot \int_0^t \check{D}_j(\vartheta(s),\tau(s);\rho(s))\cdot e^{-\check{\alpha}_j^\rho(s)}ds\Big) \;\Big|\\
&\qquad t\in[0,T_j], \quad \widehat{\alpha}_i,\widehat{\beta}_i \ge 0 \ (i\in\mathscr{I}),\\
&\qquad \check{x}(\cdot) \in \mathscr{N}(\check{x}_0,\, t,\, (\widehat{\alpha}_i,\widehat{\beta}_i,\widehat{\gamma}_i)_i) \text{ related to piecewise const. } \vartheta(\cdot) : [0,t] \longrightarrow \check{\Theta},\\
&\qquad \check{y}(\cdot) \in \mathscr{N}(\check{y}_0,\, t,\, (\widehat{\alpha}_i,\widehat{\beta}_i,\widehat{\gamma}_i)_i) \text{ related to piecewise const. } \tau(\cdot) : [0,t] \longrightarrow \check{\Theta},\\
&\qquad \rho_j(t') := \big(\max\{\pi_j\,\check{x}_0,\, \pi_j\,\check{y}_0\} + \widehat{\gamma}_j\, t'\big)\cdot e^{\max\{\pi_j\,\check{x}_0,\,\pi_j\,\check{y}_0\}+\widehat{\gamma}_j\,t'},\\
&\qquad \check{\alpha}_j^\rho(t') := \int_0^{t'} \alpha_j(\tau(s);\, \rho_j(s))\; ds \quad \text{for each } t' \in [0,t] \Big\}.
\end{aligned}$$

(1.) $\check{d}_j(\check{x}_0,\check{y}_0) \ge d_j(\check{x}_0,\check{y}_0)$ is obvious for all $\check{x}_0,\check{y}_0 \in \check{E}$ (due to the option $t = 0$). $\check{d}_j(\cdot,\cdot) \le \check{C}_j(0)\cdot d_j(\cdot,\cdot) < \infty$ results directly from assumption (A4).

(2.) This claim is a special case of statement (3.) because $\check{D}_j(\cdot,\cdot;\rho)$ is assumed to be reflexive in hypothesis (A3).

(3.) Choose any $\vartheta_0, \tau_0 \in \check{\Theta}\big(E,(d_j)_j,(e_j)_j,(\lfloor\cdot\rfloor)_j\big)$, $\check{x}_0,\check{y}_0 \in \check{E}$, $t_1,t_2,h \ge 0$ with $t_1+h \le 1$, $t_2+h \le 1$ and for $s \ge -h$, define the abbreviation

$$\begin{aligned}
\rho_j(s) &:= \big(\max\{\pi_j\,\vartheta_0(t_1,\check{x}_0),\ \pi_j\,\tau_0(t_2,\check{y}_0)\} + \widehat{\gamma}_j\cdot(s+h)\big)\cdot\\
&\quad\cdot e^{\max\{\pi_j\,\vartheta_0(t_1,\check{x}_0),\ \pi_j\,\tau_0(t_2,\check{y}_0)\} + \widehat{\gamma}_j\cdot(s+h)} \qquad \le R_j.
\end{aligned}$$

In regard to an upper bound of $\check{d}_j\big(\vartheta_0(t_1+h,\check{x}_0),\ \tau_0(t_2+h,\check{y}_0)\big)$, choose $t \in [0,T_j]$, $\widehat{\alpha}_i,\widehat{\beta}_i \ge 0$ $(i\in\mathscr{I})$ arbitrarily with $\alpha_j(\tau_0;R_j) \le \widehat{\alpha}_j$ (without loss of generality) and select any two "Euler curves"

$$\begin{aligned}
\check{x}(\cdot) &\in \mathscr{N}\big(\vartheta_0(t_1+h,\check{x}),t,\,(\widehat{\alpha}_i,\widehat{\beta}_i,\widehat{\gamma}_i)_{i\in\mathscr{I}}\big),\\
\check{y}(\cdot) &\in \mathscr{N}\big(\tau_0(t_2+h,\check{y}),\,t,\,(\widehat{\alpha}_i,\widehat{\beta}_i,\widehat{\gamma}_i)_{i\in\mathscr{I}}\big)
\end{aligned}$$

related to piecewise constant functions $\vartheta(\cdot),\tau(\cdot) : [0,t] \longrightarrow \check{\Theta}$ respectively.
Extend $\check{x}(\cdot),\check{y}(\cdot)$ and $\vartheta(\cdot),\tau(\cdot)$ to $[-h,t]$ according to

$$\left\{\begin{aligned}
\check{x}(\cdot) &:= \vartheta_0(t_1+h+\cdot\,,\ \check{x}_0), & \vartheta(\cdot) &:= \vartheta_0,\\
\check{y}(\cdot) &:= \tau_0(t_2+h+\cdot\,,\ \check{y}_0), & \tau(\cdot) &:= \tau_0
\end{aligned}\right. \qquad \text{in } [-h,0[.$$

Then,

$$\begin{aligned} & d_j\big(\check{x}(t), \check{y}(t)\big) \cdot e^{-\check{\alpha}_j^{\rho}(t)} \qquad - \check{C}_j(T_j) \int_0^t \check{D}_j(\vartheta, \tau; \rho_j)\, e^{-\check{\alpha}_j^{\rho}}\Big|_s\, ds \\ = {} & e^{h\,\alpha_j(\tau_0;R_j)} \Big(d_j\big(\check{x}(t), \check{y}(t)\big) \cdot e^{-\int_{-h}^{t} \alpha_j(\tau;\rho_j)\, ds} - \check{C}_j(T_j) \int_{-h}^{t} \check{D}_j|_{(\vartheta,\tau;\rho_j)}\, e^{-\int_{-h}^{s} \alpha_j(\tau;\rho_j)} ds \\ & \qquad + \check{C}_j(T_j) \int_{-h}^{0} \check{D}_j|_{(\vartheta,\tau;\rho_j)}\, e^{-\int_{-h}^{s} \alpha_j(\tau;\rho_j)} ds\Big)^{+} \end{aligned}$$

and if we now assume $t+h \le T_j$ in addition,

$$\begin{aligned} & \le e^{h\,\alpha_j(\tau_0;R_j)} \Big(\check{d}_j\big(\check{x}(-h),\, \check{y}(-h)\big) \qquad e^{t+h} + \check{C}_j(T_j) \int_{-h}^{0} \check{D}_j(\vartheta, \tau; \rho_j)\big|_s \cdot 1\, ds\Big) \\ & \le e^{h\,\alpha_j(\tau_0;R_j)} \Big(\check{d}_j\big(\vartheta_0(t_1, \check{x}_0),\ \tau_0(t_2, \check{y}_0)\big) e^{t+h} + \check{C}_j(T_j)\ \ h \cdot \check{D}_j(\vartheta_0, \tau_0; R_j)\Big). \end{aligned}$$

If $t+h > T_j$ (i.e. $T_j - 1 \le T_j - h < t \le T_j$), we conclude from assumption (A4)

$$\begin{aligned} & e^{-t} \ \Big(d_j\big(\check{x}(t), \check{y}(t)\big) \cdot e^{-\check{\alpha}_j^{\rho}(t)} \ - \ \check{C}_j(T_j) \int_0^t \check{D}_j(\vartheta, \tau; \rho_j)\, e^{-\check{\alpha}_j^{\rho}}\Big|_s\, ds\Big) \\ \le {} & e^{-t} \quad \check{C}_j(0) \cdot d_j\big(\check{x}(0),\, \check{y}(0)\big) \\ \le {} & \qquad \tfrac{1}{2} \qquad \cdot \check{d}_j\big(\vartheta_0(t_1{+}h, \check{x}_0),\ \tau_0(t_2{+}h, \check{y}_0)\big) \end{aligned}$$

and so, this second case is not relevant for estimating $\check{d}_j\big(\vartheta_0(t_1{+}h, \check{x}_0),\ \tau_0(t_2{+}h, \check{y}_0)\big)$ as a supremum. Finally, the upper bound for $t+h \le T_j$ leads to the claim. □

Proof (of Lemma 33 on page 216). Assume that $d_j : E \times E \longrightarrow [0, \infty[$ is locally uniformly continuous in the following sense: For each $r > 0$, there exists a non-decreasing modulus of continuity $\omega_{j,r}(\cdot)$ such that

$$\big|d_j(x,y) - d_j(x',y')\big| \ \le \ \omega_{j,r}\big(d_j(x,x') + d_j(y,y')\big)$$

holds for all elements $x, x', y, y' \in E$ with $\max\big\{\lfloor x \rfloor_j,\, \lfloor x' \rfloor_j,\, \lfloor y \rfloor_j,\, \lfloor y' \rfloor_j\big\} < r$.
Now choose $t \in [0, T_j]$ and any piecewise constant functions $\vartheta(\cdot), \tau(\cdot) : [0,t] \longrightarrow \check{\Theta}$ as in the definition of $\check{d}_j(\check{x}_0, \check{y}_0)$ (at the beginning of the previous proof on page 217). For any initial states $\check{x}_0, \check{x}_1, \check{y}_0, \check{y}_1 \in \check{E}$, we can easily construct curves $\check{x}_0(\cdot), \check{x}_1(\cdot) : [0,t] \longrightarrow \check{E}$ related to $\vartheta(\cdot)$ and $\check{y}_0(\cdot), \check{y}_1(\cdot) : [0,t] \longrightarrow \check{E}$ related to $\tau(\cdot)$ in a piecewise way. Then assumption (A4) and the reflexivity of $\check{D}_j$ guarantee

$$\begin{cases} d_j\big(\check{x}_0(t),\ \check{x}_1(t)\big) \ \le \ \check{C}_j(0) \cdot d_j\big(\check{x}_0,\ \check{x}_1\big) \cdot e^{\check{\alpha}_j^{r}(t)} \\ d_j\big(\check{y}_0(t),\ \check{y}_1(t)\big) \ \le \ \check{C}_j(0) \cdot d_j\big(\check{y}_0,\ \check{y}_1\big) \cdot e^{\check{\alpha}_j^{r}(t)} \end{cases}$$

and thus,

$$\big|d_j\big(\check{x}_0(t), \check{y}_0(t)\big) - d_j\big(\check{x}_1(t), \check{y}_1(t)\big)\big| \le \omega_{j,r}\Big(\check{C}_j(0)\, e^{\check{\alpha}_j^{r}(t)} \cdot \big(d_j(\check{x}_0, \check{x}_1) + d_j(\check{y}_0, \check{y}_1)\big)\Big)$$

with sufficiently large $r > 0$ (depending only on $\lfloor \check{x}_0 \rfloor_j$, $\lfloor \check{x}_1 \rfloor_j$, $\lfloor \check{y}_0 \rfloor_j$, $\lfloor \check{y}_1 \rfloor_j$ and $\widehat{\gamma}_j$). In particular, this estimate is uniform with respect to $\vartheta(\cdot), \tau(\cdot)$ and thus, we obtain

$$\big|\check{d}_j(\check{x}_0,\, \check{y}_0) \ - \ \check{d}_j(\check{x}_1,\, \check{y}_1)\big| \ \le \ \omega_{j,r}\Big(\check{C}_j(0)\, e^{\check{\alpha}_j^{r}(T_j)} \cdot \big(d_j(\check{x}_0,\, \check{x}_1) + d_j(\check{y}_0,\, \check{y}_1)\big)\Big). \ \square$$

How to Verify Solutions by Means of the Original Distance Functions $(d_j)_{j\in\mathscr{I}}$

A solution $\check{x} : [0,T] \longrightarrow \check{E}$ to the mutational equation

$$\overset{\circ}{\check{x}}(\cdot) \ni \check{f}(\check{x}(\cdot),\cdot)$$

in the tuple $\big(\check{E}, (\check{d}_j)_j, (e_j)_j, (\lfloor\cdot\rfloor_j)_j, \check{\Theta}\big)$ is obliged to satisfy the condition (2.′) of Definition 8 (on page 187), i.e. there exists $\alpha_j(\check{x};\cdot) : [0,\infty[\longrightarrow [0,\infty[$ such that for $\mathscr{L}^1$-almost every $t \in [0,T[$:

$$\limsup_{h\downarrow 0} \frac{\check{d}_j\big(\vartheta(s+h,z),\ \check{x}(t+h)\big) - \check{d}_j\big(\vartheta(s,z),\ \check{x}(t)\big)\cdot e^{\alpha_j(\check{x};R_j)\,h}}{h} \le \check{C}_j(T_j)\cdot \check{D}_j\big(\vartheta,\ \check{f}(\check{x}(t),t);\ R_j\big)$$

is fulfilled for any $\vartheta \in \check{\Theta}$, $s \in [0,1[$, $\check{z} \in \check{E}$ satisfying $\lfloor\vartheta(\cdot,\check{z})\rfloor_j, \lfloor\check{x}(\cdot)\rfloor_j \le R_j$. This condition, however, can be very difficult to verify in examples because it uses the auxiliary distance functions $\check{d}_j$, $j\in\mathscr{I}$, which might not be known explicitly. Hence we are interested in a sufficient condition on a curve $x : [0,T] \longrightarrow E$ such that its embedded counterpart in $\check{E}$ solves the corresponding mutational equation in $\big(\check{E}, (\check{d}_j)_{j\in\mathscr{I}}, (e_j)_{j\in\mathscr{I}}, (\lfloor\cdot\rfloor_j)_{j\in\mathscr{I}}\big)$.

Proposition 34. *Assume for $x : [0,T] \longrightarrow E$, $\vartheta : [0,T] \longrightarrow \check{\Theta}$ and each $j \in \mathscr{I}$*

(*i*) *$x(\cdot)$ is continuous with respect to d_j and satisfies $R_j := \sup_{[0,T]} \lfloor x(\cdot)\rfloor_j < \infty$,*

(*ii*) *for each $r > 0$, $\vartheta(\cdot)$ is continuous with respect to $\check{D}_j(\cdot,\cdot;r)$ and satisfies $\sup_{[0,T]} \big\{\alpha_j(\vartheta(\cdot);r),\ \beta_j(\vartheta(\cdot);r)\big\} < \infty$,*

(*iii*) *the modified inequality in assumption (A4) (on page 215) holds for these curves $x(\cdot), \vartheta(\cdot)$ and* any *$y(\cdot) \in \mathscr{N}(x(0),\ T, (\widehat{\alpha}_j, \widehat{\beta}_j, \widehat{\gamma}_j)_{j\in\mathscr{I}})$ with its related piecewise constant $\tau : [0,T] \longrightarrow \check{\Theta}$ in the following sense: For each $t \le T$,*

$$d_j\big(x(t), y(t)\big) \le \check{C}_j(T)\cdot \int_0^t \check{D}_j\big(\vartheta(s),\ \tau(s);\ \rho_j(s)\big)\ ds\ \cdot e^{\check{\alpha}_j^{\rho}(t)}$$

with the abbreviations $\rho_j(t) := (R_j + \widehat{\gamma}_j\,t)\,e^{\widehat{\gamma}_j t}$, $\check{\alpha}_j^{\rho}(t) := \int_0^t \alpha_j(\tau;\ \rho_j)\,ds$.

Then $\check{x} : [0,T] \longrightarrow \check{E}$, $t \longmapsto \big(x(t),\ (\lfloor x(t)\rfloor_j)_{j\in\mathscr{I}}\big)$ is a solution to the mutational equation

$$\overset{\circ}{\check{x}}(\cdot) \ni \vartheta(\cdot)$$

in the tuple $\big(\check{E}, (\check{d}_j)_{j\in\mathscr{I}}, (e_j)_{j\in\mathscr{I}}, (\lfloor\cdot\rfloor_j)_{j\in\mathscr{I}}, (\check{C}_j(T_j)\cdot\check{D}_j)_{j\in\mathscr{I}}\big)$ in the sense of Definition 8 (on page 187).

Proof (of Proposition 34). Its basic notion is to approximate $x(\cdot)$ by means of Euler curves, whose embedded counterparts in $\check{E}$ are obviously solutions to some "perturbed" mutational equations. Then the limit process is to preserve the solution property for the original equation — as a consequence of assumption (iii) and Convergence Theorem 13.

Choose the bounds $\widehat{\alpha}_i, \widehat{\beta}_i < \infty$ $(i \in \mathscr{I})$ sufficiently large such that all values of $\vartheta(\cdot)$ are admissible. For each $n \in \mathbb{N}$, the curve $\vartheta_n(\cdot) : [0,T] \longrightarrow \check{\Theta}$ is constructed as the piecewise constant extension of $\vartheta(\cdot)$ restricted to the points of any finite partition of $[0,T]$ with mesh $\leq \frac{1}{n}$. As an indirect consequence of assumption (ii), we obtain for every radius $r > 0$, sequence $\delta_n \downarrow 0$ and index $j \in \mathscr{I}$

$$\lim_{n \longrightarrow \infty} \sup_{\substack{0 \leq s,t \leq T \\ |t-s| \leq \delta_n}} \check{D}_j\big(\vartheta_n(s),\, \vartheta(t);\, r\big) \;=\; 0.$$

Now in a piecewise way, each $\vartheta_n(\cdot)$ $(n \in \mathbb{N})$ induces a continuous curve $x_n(\cdot)$: $[0,T] \longrightarrow E$ starting at the same point $x(0) \in E$ as $x(\cdot)$. Due to assumption (iii), the resulting sequence $\big(x_n(\cdot)\big)_{n \in \mathbb{N}}$ converges uniformly to $x(\cdot)$ with respect to each d_j and thus, the embedded curves $\big(\check{x}_n(\cdot)\big)_{n \in \mathbb{N}}$ converge uniformly to $\check{x}(\cdot)$ with respect to each $\check{d}_j$ $(j \in \mathscr{I})$. For every $n \in \mathbb{N}$, the curve $\check{x}_n(\cdot)$ solves the mutational equation

$$\overset{\circ}{\check{x}}_n(\cdot) \ni \vartheta_n(\cdot)$$

in the tuple $\big(\check{E}, (\check{d}_j)_{j \in \mathscr{I}}, (e_j)_{j \in \mathscr{I}}, (\lfloor \cdot \rfloor_j)_{j \in \mathscr{I}}, (\check{C}_j(T_j) \cdot \check{D}_j)_{j \in \mathscr{I}}\big)$. Finally Convergence Theorem 13 (on page 191) ensures that $\check{x}(\cdot)$ solves the mutational equation

$$\overset{\circ}{\check{x}}(\cdot) \ni \vartheta(\cdot).$$

□

3.5 Considering Tuples with a Separate Real Time Component

In some examples, it is useful to take time (or rather chronological differences) into consideration explicitly. Then the product $\widetilde{E} := \mathbb{R} \times E$ is to play the role of the basic set and, the first real component represents the respective time. The tilde usually reflects that we consider such tuples in $\widetilde{E}$. Now we sketch how this time component can be implemented easily — without changing any essential aspect of the preceding conclusions.

Adapting the Hypotheses About the Distance Functions $\widetilde{d}_j, \widetilde{e}_j \ (j \in \mathscr{I})$

Reflexivity and symmetry of each distance function $\widetilde{d}_j, \widetilde{e}_j : \widetilde{E} \times \widetilde{E} \longrightarrow [0,\infty[\ (j \in \mathscr{I})$ are still obligatory. Thus, hypotheses (H1) and (H2) are not changed.
Continuity hypothesis (H3), however, might be difficult to verify in examples — particularly if $\widetilde{d}_j(\widetilde{x},\widetilde{y})$ or $\widetilde{e}_j(\widetilde{x},\widetilde{y})$ depend on the time components of $\widetilde{x},\widetilde{y} \in \widetilde{E}$. Thus we formulate the following modifications with $\pi_1 : \widetilde{E} \longrightarrow \mathbb{R}$, $\widetilde{x} = (t,x) \longmapsto t$ always denoting the canonical projection on the real time component:

(H3) $(\widetilde{\text{i}})$
$$\widetilde{d}_j(\widetilde{x},\widetilde{y}) = \lim_{n\to\infty} \widetilde{d}_j(\widetilde{x}_n,\widetilde{y}_n), \qquad \widetilde{e}_j(\widetilde{x},\widetilde{y}) \le \limsup_{n\to\infty} \widetilde{e}_j(\widetilde{x}_n,\widetilde{y}_n)$$
for any $\widetilde{x},\widetilde{y} \in \widetilde{E}$ and $(\widetilde{x}_n)_{n\in\mathbb{N}}$, $(\widetilde{y}_n)_{n\in\mathbb{N}}$ in $\widetilde{E}$ fulfilling for each $i \in \mathscr{I}$
$$\lim_{n\to\infty} \widetilde{d}_i(\widetilde{x},\widetilde{x}_n) = 0 = \lim_{n\to\infty} \widetilde{d}_i(\widetilde{y}_n,\widetilde{y}), \qquad \sup_{n\in\mathbb{N}} \{\lfloor \widetilde{x}_n \rfloor_i, \lfloor \widetilde{y}_n \rfloor_i\} < \infty$$
and for all $n \in \mathbb{N}$: $\pi_1\, \widetilde{x}_n \le \pi_1\, \widetilde{y}_n$.

(H3) $(\widetilde{\text{ii}})$
$$0 = \lim_{n\to\infty} \widetilde{d}_j(\widetilde{x},\widetilde{x}_n)$$
for any $\widetilde{x} \in \widetilde{E}$ and $(\widetilde{x}_n)_{n\in\mathbb{N}}$, $(\widetilde{y}_n)_{n\in\mathbb{N}}$ in E fulfilling for each $i \in \mathscr{I}$
$$\lim_{n\to\infty} \widetilde{d}_i(\widetilde{x},\widetilde{y}_n) = 0 = \lim_{n\to\infty} \widetilde{e}_i(\widetilde{y}_n,\widetilde{x}_n), \qquad \sup_{n\in\mathbb{N}} \{\lfloor \widetilde{x}_n \rfloor_i, \lfloor \widetilde{y}_n \rfloor_i\} < \infty,$$
$$\pi_1\, \widetilde{x} \le \pi_1\, \widetilde{y}_n \le \pi_1\, \widetilde{x}_n \ \ \forall\, n \in \mathbb{N} \quad \text{or} \quad \pi_1\, \widetilde{x} \ge \pi_1\, \widetilde{y}_n \ge \pi_1\, \widetilde{x}_n \ \ \forall\, n \in \mathbb{N}.$$

(H3) $(\widetilde{\text{iii}})$
$$0 = \lim_{n\to\infty} \widetilde{d}_j(\widetilde{x},\widetilde{x}_n)$$
for every index $j \in \mathscr{I}$, any element $\widetilde{x} \in \widetilde{E}$ and sequences $(\widetilde{x}_n)_{n\in\mathbb{N}}$, $(\widetilde{y}_k)_{k\in\mathbb{N}}$, $(\widetilde{z}_{k,n})_{k,n\in\mathbb{N}}$ in $\widetilde{E}$ fulfilling

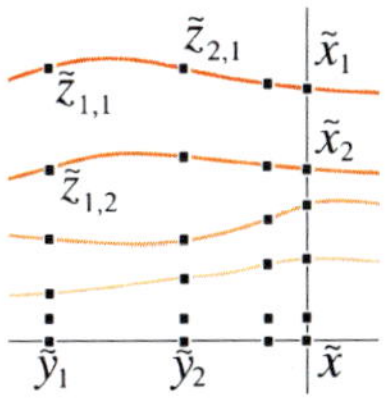

$$\begin{cases} \pi_1\, \widetilde{z}_{k,n} = \pi_1\, \widetilde{y}_k \le \pi_1\, \widetilde{x}_n = \pi_1\, \widetilde{x} & \text{for each } k,n \in \mathbb{N}, \\ \lim\limits_{k\to\infty} \widetilde{e}_i(\widetilde{x},\widetilde{y}_k) = 0 & \text{for each } i \in \mathscr{I}, \\ \lim\limits_{n\to\infty} \widetilde{d}_i(\widetilde{y}_k,\widetilde{z}_{k,n}) = 0 & \text{for each } i \in \mathscr{I}, k \in \mathbb{N}, \\ \lim\limits_{k\to\infty} \sup\limits_{n>k} \widetilde{e}_i(\widetilde{z}_{k,n},\widetilde{x}_n) = 0 & \text{for each } i \in \mathscr{I}, \\ \sup\limits_{k,n\in\mathbb{N}} \{\lfloor \widetilde{x}_n \rfloor_i, \lfloor \widetilde{y}_k \rfloor_i, \lfloor \widetilde{z}_{k,n} \rfloor_i\} < \infty & \text{for each } i \in \mathscr{I}. \end{cases}$$

These assumptions differ from their counterparts in § 3.1 (on page 182) in regard to additional constraints about the time components. They are even "weaker" than original hypotheses (H3) (i)–(iii). Hypothesis (H3) (o) about the equivalence of convergence with respect to $(\widetilde{d}_j)_{j\in\mathscr{I}}$ and $(\widetilde{e}_j)_{j\in\mathscr{I}}$ is not changed.

The Time Components of Transitions and Solutions

Whenever we consider curves $\widetilde{x}(\cdot) : [0,T] \longrightarrow \widetilde{E}$, the time component is expected to reflect the evolution of time properly. Hence we usually demand additivity in the sense of

$$\pi_1\, \widetilde{x}(t) \;=\; \pi_1\, \widetilde{x}(0) \;+\; t$$

for every $t \in [0,T]$. In particular, transitions and solutions are expected to fulfill this condition, i.e., we always assume

$$\pi_1\, \widetilde{\vartheta}(h,\widetilde{x}) \;=\; \pi_1\, \widetilde{x} + h$$

for every transition $\widetilde{\vartheta}$ on $\big(\widetilde{E}, (\widetilde{d}_j)_{j\in\mathscr{I}}, (\widetilde{e}_j)_{j\in\mathscr{I}}, (\lfloor\cdot\rfloor_j)_{j\in\mathscr{I}}\big)$, time $h \in [0,1]$ and $\widetilde{x} \in \widetilde{E}$. Moreover, Definition 8 of solutions (on page 187) is enriched by a further condition:

Definition 35. Let $\widetilde{f} : \widetilde{E} \times [0,T] \longrightarrow \widehat{\Theta}\big(\widetilde{E}, (\widetilde{d}_j)_{j\in\mathscr{I}}, (\widetilde{e}_j)_{j\in\mathscr{I}}, (\lfloor\cdot\rfloor_j)_{j\in\mathscr{I}}\big)$ be given. A curve $\widetilde{x}(\cdot) : [0,T] \longrightarrow \widetilde{E}$ is called a *timed solution* to the mutational equation

$$\overset{\circ}{\widetilde{x}}(\cdot) \;\ni\; \widetilde{f}\big(\widetilde{x}(\cdot), \cdot\big)$$

in $\big(\widetilde{E}, (\widetilde{d}_j)_{j\in\mathscr{I}}, (\widetilde{e}_j)_{j\in\mathscr{I}}, (\lfloor\cdot\rfloor_j)_{j\in\mathscr{I}}, (\widehat{D}_j)_{j\in\mathscr{I}}\big)$ if it satisfies for each $j \in \mathscr{I}$:

1.) $\widetilde{x}(\cdot)$ is continuous with respect to $\widetilde{e}_j$,

2.′) there exists $\alpha_j(\widetilde{x};\cdot) : [0,\infty[\longrightarrow [0,\infty[$ such that for $\mathscr{L}^1$-a.e. $t \in [0,T[$:

$$\limsup_{h\downarrow 0} \frac{\widetilde{d}_j\big(\widetilde{\vartheta}(s+h,\widetilde{z}),\, \widetilde{x}(t+h)\big) \;-\; \widetilde{d}_j(\widetilde{\vartheta}(s,\widetilde{z}),\, \widetilde{x}(t)) \cdot e^{\alpha_j(\widetilde{x};R_j)\, h}}{h} \;\le\; \widehat{D}_j\big(\widetilde{\vartheta},\, \widetilde{f}(\widetilde{x}(t),t);\, R_j\big)$$

for any $\widetilde{\vartheta} \in \widehat{\Theta}\big(\widetilde{E}, (\widetilde{d}_j), (\widetilde{e}_j), (\lfloor\cdot\rfloor_j)\big)$, $s<1$, $\widetilde{z}\in\widetilde{E}$ with $\lfloor\widetilde{\vartheta}(\cdot,\widetilde{z})\rfloor_j, \lfloor\widetilde{x}(\cdot)\rfloor_j \le R_j$,

3.) $\sup_{t\in[0,T]} \lfloor\widetilde{x}(t)\rfloor_j \;<\; \infty$,

4.) for every $t \in [0,T]$, $\quad \pi_1\, \widetilde{x}(t) \;=\; \pi_1\, \widetilde{x}(0) \;+\; t$.

In our subsequent conclusions about existence and stability of solutions, however, we are free to restrict all comparisons to states with identical time components. This leads to a further definition of solution which is slightly weaker than the preceding one and does not have to be equivalent to it:

Definition 36. Let $\widetilde{f} : \widetilde{E} \times [0,T] \longrightarrow \widehat{\Theta}\big(\widetilde{E}, (\widetilde{d}_j)_{j\in\mathscr{I}}, (\widetilde{e}_j)_{j\in\mathscr{I}}, (\lfloor\cdot\rfloor_j)_{j\in\mathscr{I}}\big)$ be given. $\widetilde{x}(\cdot) : [0,T] \longrightarrow \widetilde{E}$ is called a *simultaneously timed solution* of $\overset{\circ}{\widetilde{x}}(\cdot) \ni \widetilde{f}\big(\widetilde{x}(\cdot), \cdot\big)$ in $\big(\widetilde{E}, (\widetilde{d}_j)_{j\in\mathscr{I}}, (\widetilde{e}_j)_{j\in\mathscr{I}}, (\lfloor\cdot\rfloor_j)_{j\in\mathscr{I}}, (\widehat{D}_j)_{j\in\mathscr{I}}\big)$ if for each $j \in \mathscr{I}$, it satisfies conditions (1.), (3.), (4.) of Definition 35 and

2.″) there exists $\alpha_j(\widetilde{x};\cdot) : [0,\infty[\longrightarrow [0,\infty[$ such that for $\mathscr{L}^1$-a.e. $t \in [0,T[$:

$$\limsup_{h \downarrow 0} \frac{\widetilde{d}_j\big(\widetilde{\vartheta}(s+h,\widetilde{z}),\ \widetilde{x}(t+h)\big) - \widetilde{d}_j(\widetilde{\vartheta}(s,\widetilde{z}),\ \widetilde{x}(t)) \cdot e^{\alpha_j(\widetilde{x};R_j)\,h}}{h} \le \widehat{D}_j\big(\widetilde{\vartheta},\ \widetilde{f}(\widetilde{x}(t),t);\ R_j\big)$$

for any $\widetilde{\vartheta} \in \widehat{\Theta}\big(\widetilde{E},(\widetilde{d}_j),(\widetilde{e}_j),(\lfloor\cdot\rfloor_j)\big)$, $s \in [0,1[$ and $\widetilde{z} \in \widetilde{E}$ with $s+\pi_1\,\widetilde{z} = \pi_1\,\widetilde{x}(t)$ and $\lfloor\widetilde{\vartheta}(\cdot,\widetilde{z})\rfloor_j, \lfloor\widetilde{x}(\cdot)\rfloor_j \le R_j$,

Reformulating Some of the Preceding Results for Timed Solutions in $\widetilde{E}$

Now we have laid the foundations for drawing exactly the same conclusions as in the preceding sections 3.2 and 3.3. Some of the results are formulated here explicitly for taking the time component into consideration properly, but we dispense with the detailed proofs.
Furthermore, the step from timed solutions to *simultaneously timed* solutions just requires restricting distance comparisons to states in $\widetilde{E}$ with identical time components, but it does not have any significant influence on the proofs.

Hypothesis (H3)$(\widetilde{\mathrm{i}})$ implies directly the counterpart of Lemma 6 (on page 185):

Lemma 37. *Let $\widetilde{x}(\cdot), \widetilde{y}(\cdot) : [0,T] \longrightarrow \widetilde{E}$ be continuous with respect to $(\widetilde{d}_i)_{i\in\mathscr{I}}$ (or equivalently with respect to $(\widetilde{e}_i)_{i\in\mathscr{I}}$) and bounded with respect to each $\lfloor\cdot\rfloor_j$ ($j \in \mathscr{I}$). Assume $\pi_1\,\widetilde{x}(\cdot) \le \pi_1\,\widetilde{y}(\cdot)$ in $[0,T]$.*

Then for each index $j \in \mathscr{I}$, the distance function

$$[0,T] \longrightarrow [0,\infty[, \qquad t \longmapsto \widetilde{d}_j(\widetilde{x}(t),\ \widetilde{y}(t))$$

is continuous. □

Proposition 38. *Let $\widetilde{\vartheta},\widetilde{\tau} \in \widehat{\Theta}\big(\widetilde{E},(\widetilde{d}_j)_{j\in\mathscr{I}},(\widetilde{e}_j)_{j\in\mathscr{I}},(\lfloor\cdot\rfloor_j)_{j\in\mathscr{I}}\big)$, $r \ge 0$, $j \in \mathscr{I}$ and $t_1,t_2 \in [0,1[$ be arbitrary. For any elements $\widetilde{x},\widetilde{y} \in \widetilde{E}$ suppose $\lfloor\widetilde{x}\rfloor_j \le r$, $\lfloor\widetilde{y}\rfloor_j \le r$. Then the following estimate holds at each time $h \in [0,1[$ with $\max\{t_1+h,\ t_2+h\} \le 1$*

$$\widetilde{d}_j\big(\widetilde{\vartheta}(t_1+h,\widetilde{x}),\ \widetilde{\tau}(t_2+h,\widetilde{y})\big) \le \Big(\widetilde{d}_j\big(\widetilde{\vartheta}(t_1,\widetilde{x}),\ \widetilde{\tau}(t_2,\widetilde{y})\big) + h\cdot\widehat{D}_j(\widetilde{\vartheta},\widetilde{\tau};R_j)\Big)\ e^{\alpha_j(\widetilde{\tau};R_j)\,h}$$

with the constant $R_j := \big(r+\max\{\gamma_j(\widetilde{\vartheta}),\ \gamma_j(\widetilde{\tau})\}\big)\cdot e^{\max\{\gamma_j(\widetilde{\vartheta}),\ \gamma_j(\widetilde{\tau})\}} < \infty.$

Proof is the same as for Proposition 7 (on page 185). □

Essentially the same inequality still holds for the comparison of timed solutions and transitions on $\widetilde{E}$ — correspondingly to Lemma 9 (on page 188):

Corollary 39 (comparing a timed solution and a curve along transition). *Let $\widetilde{x}(\cdot) : [0,T] \longrightarrow \widetilde{E}$ be a timed solution to the mutational equation*

$$\overset{\circ}{\widetilde{x}}(\cdot) \ni \widetilde{f}(\widetilde{x}(\cdot), \cdot)$$

in $\big(\widetilde{E}, (\widetilde{d}_j)_{j\in\mathscr{I}}, (\widetilde{e}_j)_{j\in\mathscr{I}}, (\lfloor\cdot\rfloor_j)_{j\in\mathscr{I}}, (\widehat{D}_j)_{j\in\mathscr{I}}\big)$ according to Definition 35.
Suppose $\widetilde{\vartheta} \in \widehat{\Theta}\big(\widetilde{E}, (\widetilde{d}_j)_{j\in\mathscr{I}}, (\widetilde{e}_j)_{j\in\mathscr{I}}, (\lfloor\cdot\rfloor_j)_{j\in\mathscr{I}}\big)$, $\widetilde{z} \in \widetilde{E}$, $r \geq 0$, $s \in [0,1[$, $t \in [0,T[$, $j \in \mathscr{I}$ to be arbitrary with $\lfloor\widetilde{z}\rfloor_j \leq r$ and the abbreviation

$$R_j := \max\big\{\sup \lfloor\widetilde{x}(\cdot)\rfloor_j,\ \big(r + \gamma_j(\widetilde{\vartheta})\big) \cdot e^{\gamma_j(\widetilde{\vartheta})}\big\} < \infty.$$

Then,

$$\widetilde{d}_j\big(\widetilde{\vartheta}(s+h, \widetilde{z}),\ \widetilde{x}(t+h)\big) \leq \Big(\widetilde{d}_j\big(\widetilde{\vartheta}(s,\widetilde{z}),\ \widetilde{x}(t)\big) + h \cdot \sup_{[t,t+h]} \widehat{D}_j\big(\widetilde{\vartheta},\ \widetilde{f}(\widetilde{x}(\cdot),\cdot);\ R_j\big)\Big) \cdot e^{\alpha_j(\widetilde{x};R_j)\,h}$$

for every $h \in [0,1]$ with $s+h \leq 1$ and $t+h \leq T$. □

For comparing two timed solutions, we formulate the counterpart of Proposition 11 (on page 189):

Proposition 40 (Continuity w.r.t. initial states and right-hand sides). *Assume for $\widetilde{f}, \widetilde{g} : \widetilde{E} \times [0,T] \longrightarrow \widehat{\Theta}\big(\widetilde{E}, (\widetilde{d}_j)_j, (\widetilde{e}_j)_j, (\lfloor\cdot\rfloor_j)_j\big)$ and $\widetilde{x}, \widetilde{y} : [0,T] \longrightarrow \widetilde{E}$ that $\widetilde{x}(\cdot)$ is a timed solution to the mutational equation $\overset{\circ}{\widetilde{x}}(\cdot) \ni \widetilde{f}(\widetilde{x}(\cdot),\cdot)$ and $\widetilde{y}(\cdot)$ is a timed solution to the mutational equation $\overset{\circ}{\widetilde{y}}(\cdot) \ni \widetilde{g}(\widetilde{y}(\cdot),\cdot)$ in the tuple $\big(\widetilde{E}, (\widetilde{d}_j)_{j\in\mathscr{I}}, (\widetilde{e}_j)_{j\in\mathscr{I}}, (\lfloor\cdot\rfloor_j)_{j\in\mathscr{I}}, (\widehat{D}_j)_{j\in\mathscr{I}}\big)$.*
For some $j \in \mathscr{I}$, let $\widehat{\alpha}_j, R_j > 0$ and $\varphi_j \in C^0([0,T])$ satisfy for $\mathscr{L}^1$-a.e. $t \in [0,T]$

$$\begin{cases} \lfloor\widetilde{x}(t)\rfloor_j,\ \lfloor\widetilde{y}(t)\rfloor_j < R_j \\ \alpha_j(\widetilde{x}; R_j),\ \alpha_j(\widetilde{y}; R_j) \leq \widehat{\alpha}_j \\ \widehat{D}_j\Big(\widetilde{f}(\widetilde{x}(t),t),\ \widetilde{g}(\widetilde{y}(t),t);\ R_j\Big) \leq \varphi_j(t) \\ \lim\limits_{h\downarrow 0} \widehat{D}_j\Big(\widetilde{f}(\widetilde{x}(t),t),\ \widetilde{f}(\widetilde{x}(t+h),t+h);\ R_j\Big) = 0. \end{cases}$$

Then, the distance function

$$\delta_j : [0,T] \longrightarrow [0,\infty[, \quad t \longmapsto \inf\big\{\widetilde{d}_j(\widetilde{z},\widetilde{x}(t)) + \widetilde{d}_j(\widetilde{z},\widetilde{y}(t)) \ \big|\ \widetilde{z} \in \widetilde{E} : \lfloor\widetilde{z}\rfloor_j < R_j\big\}$$

fulfills

$$\delta_j(t) \leq \Big(\delta_j(0) + \int_0^t \varphi_j(s)\, e^{-\widehat{\alpha}_j\cdot s}\, ds\Big)\, e^{\widehat{\alpha}_j\cdot t} \qquad \text{for every } t \in [0,T].$$

□

Remark 41. All the preceding inequalities in Proposition 38, Corollary 39 and Proposition 40 do not require identical time components (as long as we do not consider simultaneously timed solutions instead). Thus we can even estimate perturbations with respect to time – rather than state in E.

A similar influence of time has already occurred in Convergence Theorem 13 (on page 191) which we now adapt to timed solutions. In fact, the proof consists of almost the same steps as before and, assumptions (4.ii), (4.iii) provide additional properties which ensure $\pi_1\,\widetilde{x}(t) \;=\; \pi_1\,\widetilde{x}(0) + t$ for every $t \in [0,T]$.

Theorem 42 (Convergence of timed solutions to mutational equations). *Suppose the following properties of*

$$\begin{array}{llll} \widetilde{f}_n,\, \widetilde{f} : \widetilde{E} \times [0,T] & \longrightarrow & \widehat{\Theta}\big(\widetilde{E}, (\widetilde{d}_i)_{i\in\mathscr{I}}, (\widetilde{e}_j)_{j\in\mathscr{I}}, (\lfloor\cdot\rfloor_i)_{i\in\mathscr{I}}\big) & \qquad (n\in\mathbb{N}) \\ \widetilde{x}_n,\, \widetilde{x}: \qquad [0,T] & \longrightarrow & \widetilde{E}\ : & \end{array}$$

1.) $R_j \;:= \sup_{n,t}\ \lfloor \widetilde{x}_n(t)\rfloor_j < \infty,$

$\widehat{\alpha}_j(\rho) := \sup_n\ \alpha_j\big(\widetilde{x}_n;\rho\big) < \infty$ *for* $\rho \geq 0,$

$\widehat{\beta}_j \;:= \sup_n\ \mathrm{Lip}\,\big(\widetilde{x}_n(\cdot) : [0,T] \longrightarrow (\widetilde{E},\widetilde{e}_j)\big) < \infty$ *for every* $j\in\mathscr{I}$,

2.) $\overset{\circ}{\widetilde{x}}_n(\cdot) \ni \widetilde{f}_n(\widetilde{x}_n(\cdot),\cdot)$ *(in the sense of Definition 35 on page 222) for every n,*

3.) Equi-continuity of $(\widetilde{f}_n)_n$ *at* $(\widetilde{x}(t),t)$ *at almost every time in the following sense: for* $\mathscr{L}^1$*-almost every* $t\in[0,T]$: $\lim_{n\to\infty}\ \widehat{D}_j\big(\widetilde{f}_n(\widetilde{x}(t),t),\ \widetilde{f}_n(\widetilde{y}_n,t_n);\ r\big) \;=\; 0$ *for each* $j\in\mathscr{I}$, $r\geq 0$ *and any* $(t_n)_{n\in\mathbb{N}}$, $(\widetilde{y}_n)_{n\in\mathbb{N}}$ *in* $[t,T]$ *and* $\widetilde{E}$ *respectively satisfying* $\lim_{n\to\infty} t_n = t$ *and* $\lim_{n\to\infty} \widetilde{d}_i\big(\widetilde{x}(t),\widetilde{y}_n\big) = 0$, $\sup_{n\in\mathbb{N}} \lfloor\widetilde{y}_n\rfloor_i \leq R_i$ *for each i,*

$$\pi_1\,\widetilde{y}_n \searrow \pi_1\,\widetilde{x}(t) \qquad \text{for } n\longrightarrow\infty,$$

4.) For $\mathscr{L}^1$*-almost every* $t\in[0,T[$ $(t=0$ *inclusive) and any* $\widetilde{t}\in]t,T[$, *there is a sequence* $n_m \nearrow \infty$ *of indices (depending on* $t<\widetilde{t}$*) that satisfies for* $m\longrightarrow\infty$

(i) $\widehat{D}_j\big(\widetilde{f}(\widetilde{x}(t),t),\ \widetilde{f}_{n_m}(\widetilde{x}(t),t);\ r\big) \longrightarrow 0$ *for all* $r\geq 0,\ j\in\mathscr{I}$,

(ii) $\exists\ \delta_m\searrow 0:\ \forall j: \widetilde{d}_j\big(\widetilde{x}(t),\ \widetilde{x}_{n_m}(t+\delta_m)\big) \longrightarrow 0,\ \pi_1\,\widetilde{x}_{n_m}(t+\delta_m)\searrow\pi_1\,\widetilde{x}(t)$

(iii) $\exists\ \widetilde{\delta}_m\searrow 0:\ \forall j: \widetilde{d}_j\big(\widetilde{x}(\widetilde{t}),\ \widetilde{x}_{n_m}(\widetilde{t}-\widetilde{\delta}_m)\big) \longrightarrow 0,\ \pi_1\,\widetilde{x}_{n_m}(\widetilde{t}-\widetilde{\delta}_m)\nearrow\pi_1\,\widetilde{x}(\widetilde{t})$

Then, $\widetilde{x}(\cdot)$ *is always a timed solution to the mutational equation* $\overset{\circ}{\widetilde{x}}(\cdot) \ni \widetilde{f}(\widetilde{x}(\cdot),\cdot)$ *in the tuple* $\big(\widetilde{E},\, (\widetilde{d}_j)_{j\in\mathscr{I}},\, (\widetilde{e}_j)_{j\in\mathscr{I}},\, (\lfloor\cdot\rfloor_j)_{j\in\mathscr{I}},\, (\widehat{D}_j)_{j\in\mathscr{I}}\big)$.

Finally we formulate the counterpart of Existence Theorem 19 (on page 195). As the time component of each timed solution grows at a constant speed of 1, we introduce a further abbreviation:

$\widetilde{\mathrm{B}}\mathrm{Lip}\big(I,\,\widetilde{E};\,(\widetilde{e}_i)_i,\,(\lfloor\cdot\rfloor_i)_i\big)$ consists of all functions $\widetilde{x}(\cdot)\in\mathrm{BLip}\big(I,\,\widetilde{E};\,(\widetilde{e}_i)_i,\,(\lfloor\cdot\rfloor_i)_i\big)$ satisfying $\pi_1\,\widetilde{x}(b) \;=\; \pi_1\,\widetilde{x}(a) + b-a$ for all $a,b\in I$ in addition.

Theorem 43 (Existence of timed solutions to mutational equations with delay). *Suppose* $\big(\widetilde{E},\, (\widetilde{d}_j)_{j\in\mathscr{I}},\, (\widetilde{e}_j)_{j\in\mathscr{I}},\, (\lfloor\cdot\rfloor_j)_{j\in\mathscr{I}},\, \widehat{\Theta}\big(\widetilde{E},(\widetilde{d}_i)_{i\in\mathscr{I}},(\widetilde{e}_i)_{i\in\mathscr{I}},(\lfloor\cdot\rfloor_i)_{i\in\mathscr{I}}\big)\big)$ *to be Euler compact and Euler equi-continuous. Moreover assume for some fixed* $\tau\geq 0$, *the function*

$$\widetilde{f}:\ \widetilde{\mathrm{B}}\mathrm{Lip}\big([-\tau,0],\widetilde{E};\,(\widetilde{e}_i)_i,\,(\lfloor\cdot\rfloor_i)_i\big)\times[0,T]\ \longrightarrow\ \widehat{\Theta}\big(\widetilde{E},(\widetilde{d}_i)_i,(\widetilde{e}_i)_i,(\lfloor\cdot\rfloor_i)_i\big)$$

and each $j\in\mathscr{I},\ R>0$:

1.) $\sup_{\widetilde{z}(\cdot),t}\ \alpha_j(\widetilde{f}(\widetilde{z}(\cdot),t);R)\ <\ \infty$,
2.) $\sup_{\widetilde{z}(\cdot),t}\ \beta_j(\widetilde{f}(\widetilde{z}(\cdot),t);R)\ <\ \infty$,
3.) $\sup_{\widetilde{z}(\cdot),t}\ \gamma_j(\widetilde{f}(\widetilde{z}(\cdot),t))\ <\ \infty$,
4.) *for* $\mathscr{L}^1$*-almost every* $t\in[0,T]$: $\lim_{n\to\infty}\ \widehat{D}_j\big(\widetilde{f}(\widetilde{z}_n^1(\cdot),t_n^1),\ \widetilde{f}(\widetilde{z}_n^2(\cdot),t_n^2);R\big)\ =\ 0$ *for each* $j\in\mathscr{I},\ R\geq 0$ *and any sequences* $(t_n^1)_{n\in\mathbb{N}},\ (t_n^2)_{n\in\mathbb{N}}$ *in* $[0,T]$ *and* $(\widetilde{z}_n^1(\cdot))_{n\in\mathbb{N}},\ (\widetilde{z}_n^2(\cdot))_{n\in\mathbb{N}}$ *in* $\widetilde{\mathrm{B}}\mathrm{Lip}\big([-\tau,0],\widetilde{E};\,(\widetilde{e}_j)_{j\in\mathscr{I}},\,(\lfloor\cdot\rfloor_j)_{j\in\mathscr{I}}\big)$ *satisfying for every* $i\in\mathscr{I}$ *and* $s\in[-\tau,0]$

$$\lim_{n\to\infty} t_n^1 = t = \lim_{n\to\infty} t_n^2,\quad \lim_{n\to\infty}\ \widetilde{d}_i\big(\widetilde{z}_n^1(s),\,\widetilde{z}(s)\big)\ =\ 0\ =\ \lim_{n\to\infty}\ \widetilde{d}_i\big(\widetilde{z}_n^2(s),\,\widetilde{z}(s)\big)$$
$$\sup_{n\in\mathbb{N}}\ \sup_{[-\tau,0]}\ \lfloor\widetilde{z}_n^{1,2}(\cdot)\rfloor_i\ <\ \infty.$$

For every function $\widetilde{x}_0(\cdot)\in\widetilde{\mathrm{B}}\mathrm{Lip}\big([-\tau,0],\widetilde{E};\,(\widetilde{e}_j)_{j\in\mathscr{I}},\,(\lfloor\cdot\rfloor_j)_{j\in\mathscr{I}}\big)$, *there exists a curve* $\widetilde{x}(\cdot):[-\tau,T]\longrightarrow\widetilde{E}$ *with the following properties:*

(i) $\widetilde{x}(\cdot)\in\widetilde{\mathrm{B}}\mathrm{Lip}\big([-\tau,T],\widetilde{E};\,(\widetilde{e}_j)_{j\in\mathscr{I}},\,(\lfloor\cdot\rfloor_j)_{j\in\mathscr{I}}\big)$,
(ii) $\widetilde{x}(\cdot)\big|_{[-\tau,0]}=\widetilde{x}_0(\cdot)$,
(iii) *the restriction* $\widetilde{x}(\cdot)\big|_{[0,T]}$ *is a timed solution to the mutational equation*

$$\overset{\circ}{\widetilde{x}}(t)\ \ni\ \widetilde{f}\big(\widetilde{x}(t+\cdot)\big|_{[-\tau,0]},\,t\big)\,.$$

For verifying the existence of solutions to this mutational equation (via Euler approximations), all the transitions $\widetilde{f}\big(\widetilde{z}(\cdot),t\big)\in\widehat{\Theta}\big(\widetilde{E},(\widetilde{d}_i),(\widetilde{e}_i),(\lfloor\cdot\rfloor_i)\big)$ are required as functions merely on the subset $[0,1]\times\big\{\widetilde{y}\in\widetilde{E}\ \big|\ \pi_1\,\widetilde{y}\geq t\big\}\ \subset\ [0,1]\times\widetilde{E}$.

Implementing the Aspects of "Weak" Convergence in $\widetilde{E}$

Finally, we adapt the concept of *weak* Euler compactness and its consequences in regard to existence of solutions. Correspondingly to § 3.3.6 (on page 206 ff.), let $\mathscr{J}\neq\emptyset$ denote a further index set. For each index $(j,\kappa)\in\mathscr{I}\times\mathscr{J}$, the functions

$$\widetilde{d}_{j,\kappa},\ \widetilde{e}_{j,\kappa}:\ \widetilde{E}\times\widetilde{E}\ \longrightarrow\ [0,\infty[$$

are assumed to fulfill in addition to hypotheses (H1), (H2) and (H3)

(H4') $\lfloor\cdot\rfloor_j$ is lower semicontinuous with respect to $(\widetilde{d}_{i,\kappa})_{i\in\mathscr{I},\kappa\in\mathscr{J}}$, i.e.,

$$\lfloor\widetilde{x}\rfloor_j \;\le\; \liminf_{n\to\infty} \lfloor\widetilde{x}_n\rfloor_j$$

for any $\widetilde{x}\in\widetilde{E}$ and $(\widetilde{x}_n)_{n\in\mathbb{N}}$ in $\widetilde{E}$ fulfilling for each $i\in\mathscr{I},\kappa\in\mathscr{J}$

$$\lim_{n\to\infty} \widetilde{d}_{i,\kappa}(\widetilde{x}_n,\widetilde{x}) \;=\; 0, \qquad \sup_{n\in\mathbb{N}} \lfloor\widetilde{x}_n\rfloor_i < \infty.$$

(H8)

$$\begin{aligned} \widetilde{d}_j(\widetilde{x},\widetilde{y}) &= \sup_{\kappa\in\mathscr{J}} \widetilde{d}_{j,\kappa}(\widetilde{x},\widetilde{y}), \\ \widetilde{e}_j(\widetilde{x},\widetilde{y}) &= \sup_{\kappa\in\mathscr{J}} \widetilde{e}_{j,\kappa}(\widetilde{x},\widetilde{y}) \qquad \text{for all } \widetilde{x},\widetilde{y}\in\widetilde{E},\; j\in\mathscr{I}. \end{aligned}$$

In a word, the separate time component does not have any significant influence on the proofs of the main results in § 3.3.6, i.e., Existence Theorem 28 (on page 207) and Proposition 29 about weakly converging sequences of solutions (on page 208). Just for the sake of subsequent references, we give the formulation in detail:

Theorem 44 (Existence due to weak Euler compactness).
Suppose $\big(\widetilde{E},\, (\widetilde{d}_j)_{j\in\mathscr{I}},\, (\widetilde{e}_j)_{j\in\mathscr{I}},\, (\lfloor\cdot\rfloor_j)_{j\in\mathscr{I}},\, \widehat{\Theta}\big(\widetilde{E},(\widetilde{d}_i)_{i\in\mathscr{I}},(\widetilde{e}_i)_{i\in\mathscr{I}},(\lfloor\cdot\rfloor_i)_{i\in\mathscr{I}}\big)\big)$ *to be Euler equi-continuous (in the sense of Definition 16 on page 194) and the tuple* $\big(\widetilde{E},\, (\widetilde{d}_j)_j,\, (\widetilde{d}_{j,\kappa})_{j,\kappa},\, (\widetilde{e}_j)_j,\, (\widetilde{e}_{j,\kappa})_{j,\kappa},\, (\lfloor\cdot\rfloor_j)_j,\, \widehat{\Theta}\big(\widetilde{E},(\widetilde{d}_i)_i,(\widetilde{e}_i)_i,(\lfloor\cdot\rfloor_i)_i\big)\big)$ *to be weakly Euler compact (in the sense of Definition 27 on page 207).*

Moreover assume for some fixed $\tau\ge 0$, *the function*

$$\widetilde{f}: \widetilde{\mathrm{B}}\mathrm{Lip}\big([-\tau,0],\widetilde{E};(\widetilde{e}_i)_i,(\lfloor\cdot\rfloor_i)_i\big)\times[0,T] \longrightarrow \widehat{\Theta}\big(\widetilde{E},(\widetilde{d}_i)_i,(\widetilde{e}_i)_i,(\lfloor\cdot\rfloor_i)_i\big)$$

and each $j\in\mathscr{I}$, $R>0$:

1.) $\sup_{\widetilde{z}(\cdot),t} \alpha_j(\widetilde{f}(\widetilde{z}(\cdot),t);R) < \infty,$

2.) $\sup_{\widetilde{z}(\cdot),t} \beta_j(\widetilde{f}(\widetilde{z}(\cdot),t);R) < \infty,$

3.) $\sup_{\widetilde{z}(\cdot),t} \gamma_j(\widetilde{f}(\widetilde{z}(\cdot),t)) < \infty,$

4.) for $\mathscr{L}^1$*-almost every* $t\in[0,T]$: $\lim_{n\to\infty} \widehat{D}_j\big(\widetilde{f}(\widetilde{z}_n^1(\cdot),t_n^1),\, \widetilde{f}(\widetilde{z}_n^2(\cdot),t_n^2);R\big) = 0$
for each $j\in\mathscr{I}$, $R\ge 0$ *and any sequences* $(t_n^1)_{n\in\mathbb{N}}$, $(t_n^2)_{n\in\mathbb{N}}$ *in* $[0,T]$ *and* $(\widetilde{z}_n^1(\cdot))_{n\in\mathbb{N}}$, $(\widetilde{z}_n^2(\cdot))_{n\in\mathbb{N}}$ *in* $\widetilde{\mathrm{B}}\mathrm{Lip}\big([-\tau,0],\widetilde{E};(\widetilde{e}_j)_{j\in\mathscr{I}},(\lfloor\cdot\rfloor_j)_{j\in\mathscr{I}}\big)$ *satisfying for every* $i\in\mathscr{I}$, $\kappa\in\mathscr{J}$ *and* $s\in[-\tau,0]$

$$\lim_{n\to\infty} t_n^1 = t = \lim_{n\to\infty} t_n^2,\; \lim_{n\to\infty} \widetilde{d}_{i,\kappa}\big(\widetilde{z}_n^1(s),\widetilde{z}(s)\big) = 0 = \lim_{n\to\infty} \widetilde{d}_{i,\kappa}\big(\widetilde{z}_n^2(s),\widetilde{z}(s)\big)$$
$$\sup_{n\in\mathbb{N}}\sup_{[-\tau,0]} \lfloor\widetilde{z}_n^{1,2}(\cdot)\rfloor_i < \infty.$$

5.) *for every* $\tilde{z}(\cdot)$ *and* $\mathcal{L}^1$*-a.e.* $t \in [0,T]$, *the function* $\widetilde{f}(\tilde{z}(\cdot),t)\,(h,\cdot) : \widetilde{E} \longrightarrow \widetilde{E}$ *is "weakly" continuous in the following sense:*

$$\lim_{n\to\infty} \widetilde{d}_{j,\kappa}\big(\widetilde{f}(\tilde{z}(\cdot),t)\,(h,\tilde{y}),\ \ \widetilde{f}(\tilde{z}(\cdot),t)\,(h,\tilde{y}_n)\big) \ = \ 0$$

for each $\kappa \in \mathcal{J}$, $h \in]0,1]$, $\tilde{y} \in \widetilde{E}$ *and any sequence* $(\tilde{y}_n)_{n\in\mathbb{N}}$ *in* $\widetilde{E}$ *satisfying* $\widetilde{d}_{i,\kappa'}(\tilde{y},\tilde{y}_n) \longrightarrow 0$, $\ \sup_n \lfloor \tilde{y}_n \rfloor_i < \infty$ *for any* $i \in \mathcal{I}, \kappa' \in \mathcal{J}$, $\ \pi_1\, \tilde{y} \le \pi_1\, \tilde{y}_n$.

For every function $\tilde{x}_0(\cdot) \in \widetilde{\mathrm{B}}\mathrm{Lip}\big([-\tau,0],\, \widetilde{E};\, (\tilde{e}_j)_{j\in\mathcal{I}},\, (\lfloor\cdot\rfloor_j)_{j\in\mathcal{I}}\big)$, *there exists a curve* $\tilde{x}(\cdot) : [-\tau,T] \longrightarrow \widetilde{E}$ *with the following properties:*

(*i*) $\tilde{x}(\cdot) \in \widetilde{\mathrm{B}}\mathrm{Lip}\big([-\tau,T],\, \widetilde{E};\, (\tilde{e}_j)_{j\in\mathcal{I}},\, (\lfloor\cdot\rfloor_j)_{j\in\mathcal{I}}\big)$,

(*ii*) $\tilde{x}(\cdot)\big|_{[-\tau,0]} = \tilde{x}_0(\cdot)$,

(*iii*) *For* $\mathcal{L}^1$*-a.e.* $t \in [0,T[$, $\displaystyle\lim_{h\downarrow 0} \tfrac{1}{h}\cdot \widetilde{d}_j\big(\widetilde{f}\big(\tilde{x}(t+\cdot)\big|_{[-\tau,0]},\, t\big)\,(h,\tilde{x}(t)),\ \tilde{x}(t+h)\big) = 0.$

If each $\widetilde{d}_j$ $(j \in J)$ *satisfies the triangle inequality in addition,* $\tilde{x}(\cdot)\big|_{[0,T]}$ *is a timed solution to the mutational equation* $\overset{\circ}{\tilde{x}}(t) \ni \widetilde{f}\big(\tilde{x}(t+\cdot)\big|_{[-\tau,0]},\, t\big)$ *in the sense of Definition 35 (on page 222).*

Proposition 45 (about "weak" pointwise convergence of timed solutions). *Suppose the following properties of*

$$\begin{array}{rrcl} \widetilde{f}_n,\, \widetilde{f} : \widetilde{E} \times [0,T] & \longrightarrow & \widehat{\Theta}\big(\widetilde{E}, (\widetilde{d}_i)_{i\in\mathcal{I}}, (\tilde{e}_j)_{j\in\mathcal{I}}, (\lfloor\cdot\rfloor_i)_{i\in\mathcal{I}}\big) & \quad (n \in \mathbb{N}) \\ \tilde{x}_n,\, \tilde{x} : \qquad [0,T] & \longrightarrow & \widetilde{E} \ : & \end{array}$$

1.) $R_j \quad := \sup_{n,t} \lfloor \tilde{x}_n(t) \rfloor_j < \infty$,

$\widehat{\alpha}_j(\rho) := \sup_n \alpha_j\big(\tilde{x}_n;\, \rho\big) < \infty \quad$ *for* $\rho \ge 0$,

$\widehat{\beta}_j \quad := \sup_n \mathrm{Lip}\,\big(\tilde{x}_n(\cdot) : [0,T] \longrightarrow (\widetilde{E}, \tilde{e}_j)\big) < \infty \quad$ *for every* $j \in \mathcal{I}$,

2.) $\overset{\circ}{\tilde{x}}_n(\cdot) \ni \widetilde{f}_n(\tilde{x}_n(\cdot),\, \cdot)$ *(in the sense of Definition 35 on page 222) for every* $n \in \mathbb{N}$,

3.) Equi-continuity of $(\widetilde{f}_n)_n$ *at* $(\tilde{x}(t),t)$ *at almost every time in the following sense:*

for $\mathcal{L}^1$*-almost every* $t \in [0,T]$: $\qquad \displaystyle\lim_{n\to\infty} \widehat{D}_j\Big(\widetilde{f}_n(\tilde{x}(t),\, t),\ \widetilde{f}_n(\tilde{y}_n, t_n);\ r\Big) = 0$

for each $j \in \mathcal{I}$, $r \ge 0$ *and any* $(t_n)_{n\in\mathbb{N}}$, $(\tilde{y}_n)_{n\in\mathbb{N}}$ *in* $[t,T]$ *and* E *respectively satisfying* $\displaystyle\lim_{n\to\infty} t_n = t$, $\displaystyle\lim_{n\to\infty} \widetilde{d}_{i,\kappa}\big(\tilde{x}(t), \tilde{y}_n\big) = 0$, $\displaystyle\sup_{n\in\mathbb{N}} \lfloor \tilde{y}_n \rfloor_i \le R_i$ *for any* i, κ,

4.) *For $\mathscr{L}^1$-almost every $t \in [0,T[$ ($t = 0$ inclusive) and any $\widetilde{t} \in]t,T[$, there is a sequence $n_m \nearrow \infty$ of indices (depending on $t < \widetilde{t}$) that satisfies for $m \longrightarrow \infty$*

$$\begin{cases} (i) & \widehat{D}_j\big(\widetilde{f}(\widetilde{x}(t),t),\ \widetilde{f}_{n_m}(\widetilde{x}(t),t);\ r\big) \ \longrightarrow\ 0 \qquad \text{for all } r \geq 0,\ j \in \mathscr{I}, \\ (ii) & \forall\, j \in \mathscr{I}, \kappa \in \mathscr{J}:\ \ \widetilde{d}_{j,\kappa}\big(\widetilde{x}(t),\ \widetilde{x}_{n_m}(t)\big) \longrightarrow 0, \quad \pi_1\, \widetilde{x}_{n_m}(t) \searrow \pi_1\, \widetilde{x}(t), \\ (iii) & \forall\, j \in \mathscr{I}, \kappa \in \mathscr{J}:\ \ \widetilde{d}_{j,\kappa}\big(\widetilde{x}(\widetilde{t}),\ \widetilde{x}_{n_m}(\widetilde{t})\big) \longrightarrow 0, \quad \pi_1\, \widetilde{x}_{n_m}(\widetilde{t}) \nearrow \pi_1\, \widetilde{x}(\widetilde{t}), \end{cases}$$

5.) *Weak continuity of each function $\widetilde{f}(\widetilde{x}(t),t)(h,\cdot) : \widetilde{E} \longrightarrow \widetilde{E}$ at $\mathscr{L}^1$-almost every time $t \in [0,T]$ in the following sense:*

$$\lim_{n\to\infty}\ \widetilde{d}_{j,\kappa}\big(\widetilde{f}(\widetilde{x}(t),t)\,(h,\widetilde{y}),\ \ \widetilde{f}(\widetilde{x}(t),t)\,(h,\widetilde{y}_n)\big) \ =\ 0$$

for each $\kappa \in \mathscr{J}$, $h \in]0,1]$, $\widetilde{y} \in \widetilde{E}$ and any sequence $(\widetilde{y}_n)_{n\in\mathbb{N}}$ in $\widetilde{E}$ satisfying $\widetilde{d}_{i,\kappa'}(\widetilde{y},\widetilde{y}_n) \longrightarrow 0$, $\sup_n \lfloor \widetilde{y}_n \rfloor_i < \infty$ for any $i \in \mathscr{I}, \kappa' \in \mathscr{J}$, $\pi_1\, \widetilde{y} \leq \pi_1\, \widetilde{y}_n$.

Then, $\widetilde{x}(\cdot)$ is $\widehat{\beta}_j$-Lipschitz continuous with respect to $\widetilde{e}_j$ for each index $j \in \mathscr{I}$ and, at $\mathscr{L}^1$-almost every time $t \in [0,T]$,

$$\lim_{h\downarrow 0}\ \tfrac{1}{h}\cdot \widetilde{d}_j\big(f(\widetilde{x}(t),t)\,(h,\,\widetilde{x}(t)),\ \ \widetilde{x}(t+h)\big) \ =\ 0$$

holds for every index $j \in \mathscr{I}$.

If each $\widetilde{d}_j$ ($j \in J$) satisfies the triangle inequality in addition, then the curve $\widetilde{x}(\cdot) : [0,T] \longrightarrow \widetilde{E}$ is a timed solution to the mutational equation $\overset{\circ}{\widetilde{x}}(\cdot) \ni \widetilde{f}(\widetilde{x}(\cdot),\cdot)$ in the tuple $\big(\widetilde{E},\ (\widetilde{d}_j)_{j\in\mathscr{I}},\ (\widetilde{e}_j)_{j\in\mathscr{I}},\ (\lfloor\cdot\rfloor_j)_{j\in\mathscr{I}},\ (\widehat{D}_j)_{j\in\mathscr{I}}\big)$.

Extending the Cauchy-Lipschitz Theorem to Timed Solutions

Similarly the results of § 3.3.7 (on page 212 f.) are rather easy to extend to timed solutions in $\widetilde{E}$. The counterpart of Cauchy-Lipschitz Theorem concludes the existence of a timed solution to a given mutational equation from an appropriate form of completeness. In particular, using this property for Euler approximations at a fixed time respectively, we are free to restrict the completeness assumption to sequences in $\widetilde{E}$ with constant time component.

Definition 46. The tuple $\big(\widetilde{E},\ (\widetilde{d}_j)_{j\in\mathscr{I}},\ (\widetilde{e}_j)_{j\in\mathscr{I}},\ (\lfloor\cdot\rfloor_j)_{j\in\mathscr{I}}\big)$ is called *timed complete* if for every sequence $(\widetilde{x}_n)_{n\in\mathbb{N}}$ in $\widetilde{E}$ with

$$\begin{cases} \lim\limits_{k\to\infty}\ \sup\limits_{m,n\geq k}\ \widetilde{d}_j(\widetilde{x}_m,\ \widetilde{x}_n) & = 0 \\ \qquad\ \ \sup\limits_{m,n\in\mathbb{N}}\ \big|\pi_1\,\widetilde{x}_m\ -\ \pi_1\widetilde{x}_n\big| & = 0 \qquad\qquad \text{for each } j \in \mathscr{I}, \\ \qquad\ \ \sup\limits_{n\in\mathbb{N}}\ \lfloor\widetilde{x}_n\rfloor_j & < \infty \end{cases}$$

there exists $\widetilde{x} \in E$ fulfilling $\lim\limits_{n\to\infty} \widetilde{d}_j(\widetilde{x}_n,\widetilde{x}) = 0$ for every $j \in \mathscr{I}$ and $\pi_1\, x = \pi_1\, x_n$.

Theorem 47 (Extended Cauchy-Lipschitz Theorem for timed solutions).
Suppose the tuple $\big(\widetilde{E}, (\widetilde{d}_j)_{j\in\mathscr{I}}, (\widetilde{e}_j)_{j\in\mathscr{I}}, (\lfloor\cdot\rfloor_j)_{j\in\mathscr{I}}\big)$ *to be timed complete and* $\big(\widetilde{E}, (\widetilde{d}_j)_{j\in\mathscr{I}}, (\widetilde{e}_j)_{j\in\mathscr{I}}, (\lfloor\cdot\rfloor_j)_{j\in\mathscr{I}}, \widehat{\Theta}\big(\widetilde{E},(\widetilde{d}_i)_{i\in\mathscr{I}},(\widetilde{e}_i)_{i\in\mathscr{I}},(\lfloor\cdot\rfloor_i)_{i\in\mathscr{I}}\big)\big)$ *to be Euler equi-continuous For* $\widetilde{f} : \widetilde{E}\times[0,T] \longrightarrow \widehat{\Theta}\big(\widetilde{E}, (\widetilde{d}_j)_{j\in\mathscr{I}}, (\widetilde{e}_j)_{j\in\mathscr{I}}, (\lfloor\cdot\rfloor_j)_{j\in\mathscr{I}}\big)$ *assume*

(1.) *For each* $j\in\mathscr{I}$ *and* $R>0$,

$$\begin{array}{lll}\widehat{\alpha}_j(R) &:= \sup_{\widetilde{x},t}\ \alpha_j(\widetilde{f}(\widetilde{x},t);R) &< \infty,\\ \widehat{\beta}_j(R) &:= \sup_{\widetilde{x},t}\ \beta_j(\widetilde{f}(\widetilde{x},t);R) &< \infty,\\ \widehat{\gamma}_j &:= \sup_{\widetilde{x},t}\ \gamma_j(\widetilde{f}(\widetilde{x},t)) &< \infty,\end{array}$$

(2.) $\widetilde{f}$ *is Lipschitz continuous w.r.t. state and continuous in the following sense: for each tuple* $(r_j)_{j\in\mathscr{I}}$ *in* $[0,\infty[^{\mathscr{I}}$, *there exist constants* $\Lambda_j,\mu_j\geq 0$ $(j\in\mathscr{I})$ *and moduli of continuity* $(\omega_j(\cdot))_{j\in\mathscr{I}}$ *such that* $\delta_j:\widetilde{E}\times\widetilde{E}\longrightarrow[0,\infty[$,

$$\delta_j(\widetilde{x},\widetilde{y}) := \inf\big\{\widetilde{d}_j(\widetilde{x},\widetilde{z})+\mu_j\cdot\widetilde{e}_j(\widetilde{z},\widetilde{y})\ \big|\ \widetilde{z}\in\widetilde{E},\ \pi_1\,\widetilde{z}\leq\min\{\pi_1\,\widetilde{x},\ \pi_1\,\widetilde{y}\},\ \forall i\in\mathscr{I}:\ \lfloor\widetilde{z}\rfloor_i\leq r_i\big\}$$

satisfies for every $j\in\mathscr{I}$

$$\widehat{D}_j\big(\widetilde{f}(\widetilde{x},s),\ \widetilde{f}(\widetilde{y},t);\ r_j\big)\ \leq\ \Lambda_j\cdot\delta_j(\widetilde{x},\widetilde{y})\ +\ \omega_j(|t-s|)$$

whenever the tuples $(\widetilde{x},s),\ (\widetilde{y},t)\in\widetilde{E}\times[0,T]$ *fulfill* $\pi_1\,\widetilde{x}\leq\pi_1\,\widetilde{y},\ s\leq t$ *and* $\max\big\{\lfloor\widetilde{x}\rfloor_i,\ \lfloor\widetilde{y}\rfloor_i\big\}\ \leq\ r_i$ *for each index* $i\in\mathscr{I}$.

Then for every initial element $\widetilde{x}_0\in\widetilde{E}$, *there exists a timed solution* $\widetilde{x}(\cdot):[0,T]\longrightarrow\widetilde{E}$ *to the mutational equation* $\overset{\circ}{\widetilde{x}}(\cdot)\ni\widetilde{f}\big(\widetilde{x}(\cdot),\cdot\big)$ *in the sense of Definition 35.*

Remark 48. This existence result can also be extended to systems easily. Now completeness has joined compactness for providing (timed or simultaneously timed) solutions to mutational equations.
With regard to systems of mutational equations, however, combining the preceding Cauchy-Lipschitz Theorem with Peano-like Existence Theorem 43 should be treated with some caution. Indeed, each component based on Euler compactness leads to a *pointwise* converging subsequence of Euler approximations.
When inserting it in the mutational equations of the remaining components (to which Cauchy-Lipschitz Theorem is then to be applied), we usually need to assume some form of uniformity (about the continuity of the right-hand side, for example) for adapting the constructive proof of Theorem 31 (on page 213), i.e. for verifying that these components of Euler approximations induce a Cauchy sequence.

Remark 49. The results of § 3.4 can be adapted easily to tuples with separate real time component, i.e. under suitable additional assumptions the candidates for transitions do not have to be ω-contractive in the sense that

$$\widetilde{d}_j\big(\widetilde{\vartheta}(t_1+h,\widetilde{x}),\ \widetilde{\vartheta}(t_2+h,\widetilde{y})\big)\ \leq\ \widetilde{d}_j\big(\widetilde{\vartheta}(t_1,\widetilde{x}),\ \widetilde{\vartheta}(t_2,\widetilde{y})\big)\cdot e^{\alpha_j(\widetilde{\vartheta};R_j)\,h}$$

for all $\widetilde{x},\widetilde{y}\in\widetilde{E}$, $t_1,t_2,h\in[0,1[$ and sufficiently large $R_j>0$. We dispense with further details here.

3.6 Example: Strong Solutions to Nonlocal Stochastic Differential Equations

Stochastic differential equations are very popular for modelling processes with uncertainties. This is not necessarily restricted to applications in finance, biology or engineering, of course. Whenever competing with opponents, we might now exactly the rules of each participant, but we are possibly lacking precise information about the current state of the competitors and so, some form of estimator (like the expected value modified by means of the second moment) has to come into play. Such a situation leads to stochastic differential equations of the form

$$dX_t = h_1\big(t, \mathbb{E}(\varphi_1(X_t)), \mathbb{E}(|X_t|^2), X_t\big)\, dt + h_2\big(t, \mathbb{E}(\varphi_2(X_t)), \mathbb{E}(|X_t|^2), X_t\big)\, dW_t\,.$$

They differ from what is usually called a "stochastic functional differential equation" (as e.g. in [135, 187]) because its right-hand side can depend on nonlocal features of the current random variable $X_t : \Omega \longrightarrow \mathbb{R}$ (instead of the more popular pathwise dependence).
Now the mutational framework is to provide the (probably) first existence results for so-called *nonlocal stochastic differential equations*.

Even a short glance at the standard literature reveals that stochastic differential equations (in $\mathbb{R}$) are usually considered in combination with the L^2 norm on the corresponding vector space of adapted stochastic processes (with bounded second moments).
Applying the mutational framework, however, our attempts are likely to fail because the Itô integral implies asymptotic properties of $\sqrt{h}$ for short periods $h > 0$. This obstacle has now motivated us to choose the square deviation $\mathbb{E}(|\cdot - \cdot|^2)$ as distance function (instead of its square root). Admittedly, this alternative does not satisfy the triangle inequality, but we can still handle the Cauchy problem by means of the generalizations in this chapter.

First, in § 3.6.3, we sketch rather briefly how to conclude existence from Cauchy-Lipschitz Theorem 31 (on page 212) via the weakening modifications in § 3.4. In regard to mathematical transparency, however, the disadvantage is that distance functions and transitions (with all their parameters) are not specified explicitly in this approach and so, we discuss a (very) special case with fixed additive noise in more detail in § 3.6.4. It deals with stochastic differential equations like

$$dX_t = h\big(t, \mathbb{E}(|X_t|), \mathbb{E}(|X_t|^2)\big)\, dt + b(t)\, dW_t$$

with a bounded Lipschitz continuous function $h(\cdot)$. The main existence result of this case is formulated in subsequent Theorem 56 (on page 240).

Basic set	$\widetilde{E}_{\mathscr{A}} := \{(t,X) \mid t \geq 0,\ X: \Omega \longrightarrow \mathbb{R} \text{ is } \mathscr{A}_t\text{-measurable}, \mathbb{E}(\lvert X\rvert^2) < \infty\}$ with a complete probability space $(\Omega, \mathscr{A}, P)$, a Wiener process $(W_t)_{t\geq 0}$ and a filtration $\mathscr{A} = (\mathscr{A}_t)_{t\geq 0}$ given
Distance	$\widetilde{d}_{\mathscr{A},P} : \widetilde{E}_{\mathscr{A}} \times \widetilde{E}_{\mathscr{A}} \longrightarrow \mathbb{R},\ ((s,X),(t,Y)) \longmapsto \lvert t-s\rvert + \mathbb{E}(\lvert X-Y\rvert^2)$ (mean square deviation does not satisfy the triangle inequality, but $\widetilde{d}_{\mathscr{A},P}$ is still equivalent to the metric of $\lvert\cdot\rvert + \lVert\cdot\rVert_{L^2(\Omega,\mathscr{A},P)}$)
Absolute value	$\lfloor\cdot\rfloor_{\mathscr{A},P} : \widetilde{E}_{\mathscr{A}} \longrightarrow [0,\infty[,\ (t,X) \longmapsto \lvert t\rvert + \mathbb{E}(\lvert X\rvert^2)$
Transitions (i) non-ω-contract. candidates	For $a, b \in W^{1,\infty}(\mathbb{R})$, consider the strong solution $(X_t)_{0\leq t\leq 1}$ to the linear autonomous stochastic ordinary differential equation $dX_t = a(X_t)\,dt + b(X_t)\,dW_t$ with bounded second moment.
(ii) special case in § 3.6.4	Fix $b \in L^\infty(\mathbb{R})$ for additive noise. Each constant $a \in \mathbb{R}$ induces a transition on $\widetilde{E}_{\mathscr{A}}$ via the Itô process $X_t = X_{t_0} + \int_{t_0}^t a\ ds + \int_{t_0}^t b(s)\ dW_s$ with finite second moment.
Compactness	Not available in an obvious way here. Completeness of $L^2(\Omega, \mathscr{A}, P)$ is used instead.
Equi-continuity	Euler equi-continuity w.r.t. $\widetilde{d}_{\mathscr{A},P}$ results from a priori estimates of strong solutions to nonautonomous stochastic differential equations in an arbitrary time interval $[0,T]$.
Mutational solutions	Strong solutions to a stochastic ordinary differential equation: Lemma 54 (page 236)
List of main results formulated in § 3.6	Existence due to completeness (Cauchy-Lipschitz): Theorem 53 Cauchy-Lipschitz Theorem for fixed additive noise in § 3.6.4: Theorem 56 (page 240)
Key tools	Existence and several a priori estimates of pathwise unique strong solutions to stochastic ordinary differential equations: Proposition 51 (page 234)

Table 3.1 Brief summary of the example in § 3.6 in mutational terms:
Strong solutions to nonlocal stochastic differential equations

3.6.1 The General Assumptions for This Example

$(\Omega, \mathscr{A}, P)$ is assumed to be a complete probability space. $(\mathscr{A}_t)_{t\geq 0}$ denotes an increasing family of sub-σ-algebras of $\mathscr{A}$ and, $W = (W_t)_{t\geq 0}$ is a standard scalar Wiener process such that $(\mathscr{A}_t)_{t\geq 0}$ is right continuous, $\mathscr{A}_0$ contains all P-null sets in $\mathscr{A}$ and for all $0 \leq s \leq t$, W_t is $\mathscr{A}_t$-measurable with

$$\mathbb{E}(W_t \,|\, \mathscr{A}_0) = 0, \qquad \mathbb{E}(W_t - W_s \,|\, \mathscr{A}_s) = 0 \qquad \text{with probability 1.}$$

Following the remarks in [106, § 3.2], the σ-algebra $\mathscr{A}_t$ may be thought of as a collection of events that are detectable prior to or at time $t \geq 0$, so that the $\mathscr{A}_t$-measurability of Z_t for a stochastic process $(Z_t)_{t\geq 0}$ indicates its nonanticipativity with respect to the Wiener process W.
For $T \in]0,\infty[$, we define a class $\mathscr{L}^2_{\mathscr{A}}([0,T])$ of functions $f : [0,T] \times \Omega \longrightarrow \mathbb{R}$ with

(1.) f is jointly $\mathscr{L}^1 \times \mathscr{A}$-measurable,
(2.) $\int_{[0,T]} \mathbb{E}\big(|f(t,\cdot)|^2\big)\, dt < \infty$,
(3.) for every $t \in [0,T]$, $\mathbb{E}\big(|f(t,\cdot)|^2\big) < \infty$ and
(4.) for every $t \in [0,T]$, $f(t,\cdot) : \Omega \longrightarrow \mathbb{R}$ is $\mathscr{A}_t$-measurable.

In addition, we consider two functions in $\mathscr{L}^2_{\mathscr{A}}([0,T])$ to be identical if they are equal for all $(t,\omega) \in [0,T] \times \Omega$ except possibly on a subset of $\mathscr{L}^1 \times P$-measure 0. Then with the norm

$$\|f\|_{\mathscr{L}^2_{\mathscr{A}}([0,T])} := \Big(\int_{[0,T]} \mathbb{E}\big(|f(t,\cdot)|^2\big)\, dt\Big)^{\frac{1}{2}},$$

$\mathscr{L}^2_{\mathscr{A}}([0,T])$ (together with the identification mentioned before) is a Banach space. As Kloeden and Platen have already pointed out [106], the characterizing conditions on $f \in \mathscr{L}^2_{\mathscr{A}}([0,T])$ are stronger than $f \in L^2([0,T] \times \Omega, \mathscr{L}^1 \times \mathscr{A}, \mathscr{L} \times P)$. Indeed, Fubini's Theorem guarantees $\mathbb{E}\big(|f(t,\cdot)|^2\big) < \infty$ only for Lebesgue-almost every t.

3.6.2 Some Standard Results About Itô Integrals and Strong Solutions to Stochastic Ordinary Differential Equations

In this subsection, we summarize some well-known properties of the Itô integral and strong solutions. All these results are just quoted and serve as tools for specifying transitions in the mutational framework later on. The proofs can be found in standard references such as the monographs of Friedman [85], Øksendal [147], Karatzas and Shreve [97] or Kloeden and Platen [106].

Proposition 50 ([85, § 4], [106, Theorem 3.2.3], [147, § 3.2]).
The Itô stochastic integral $I(f) : \Omega \longrightarrow \mathbb{R}$, $\omega \longmapsto \int_0^T f(s,\omega)\, dW_s(\omega)$ *has the following properties for every* $f, g \in \mathscr{L}^2_{\mathscr{A}}([0,T])$ *and* $\lambda_1, \lambda_2 \in \mathbb{R}$:

(a) $I(f)$ *is* $\mathscr{A}_T$*-measurable,*

(b) $\mathbb{E}\big(I(f)\big) = 0,$

(c) $I(\lambda_1 f + \lambda_2 g) = \lambda_1 I(f) + \lambda_2 I(g)$ *with propability* 1.

(d) *Itô isometry:* $\mathbb{E}\big(|I(f)|^2\big) = \int_0^T \mathbb{E}\big(|f(t,\cdot)|^2\big)\, dt,$

(e) $\mathbb{E}\big(I(f)\, I(g)\big) = \int_0^T \mathbb{E}\big(f(t,\cdot)\, g(t,\cdot)\big)\, dt,$

(f) *Martingale property:* $\mathbb{E}\big(I(f) \mid \mathscr{A}_t\big) = \int_0^t f(s,\cdot)\, dW_s$ *for any* $t \in [0,T]$.

Proposition 51 (Existence, uniqueness of strong solutions a priori estimates [106, Theorems 4.5.3, 4.5.4]**).**

Suppose

(i) $a,b : [0,T] \times \mathbb{R} \longrightarrow \mathbb{R}$ *are jointly* $\mathscr{L}^2$*-measurable,*

(ii) *there exists a constant* $\Lambda > 0$ *such that for all* $t \in [0,T],\ x,y \in \mathbb{R}$,

$$\begin{cases} |a(t,x) - a(t,y)| \le \Lambda\, |x-y| \\ |b(t,x) - b(t,y)| \le \Lambda\, |x-y| \end{cases}$$

(iii) *there exists a constant* $\widehat{\gamma} < \infty$ *such that for all* $t \in [0,T],\ x \in \mathbb{R}$,

$$|a(t,x)| + |b(t,x)| \le \widehat{\gamma}\,(1+|x|),$$

(iv) $X_0 : \Omega \longrightarrow \mathbb{R}$ *is* $\mathscr{A}_0$*-measurable with* $\mathbb{E}\big(|X_0|^2\big) < \infty$.

Then the stochastic differential equation

$$dX_t = a(t,X_t)\, dt + b(t,X_t)\, dW_t$$

has a pathwise unique strong solution $(X_t)_{0\le t\le T}$ *on* $[0,T]$ *with initial value* X_0 *and*

$$\sup_{0\le t\le T} \mathbb{E}\big(|X_t|^2\big) < \infty,$$

i.e., there exists a function $[0,T]\times\Omega \longrightarrow \mathbb{R},\ (t,\omega) \longmapsto X_t(\omega)$ *in* $\mathscr{L}^2_{\mathscr{A}}([0,T])$ *with*

(1.) *for every* $t \in [0,T],\ X_t = X_0 + \int_0^t a(s,X_s)\, ds + \int_0^t b(s,X_s)\, dW_s,$

(2.) *for every solution* Y_t *of this preceding integral equation with* $Y_0 = X_0$,

$$P\Big(\sup_{0\le t\le T} |X_t - Y_t| > 0\Big) = 0.$$

Moreover, for every $t \in [0,T]$*, it fulfills following estimates with constants* C_1, C_2, C_3 *depending only on* $\widehat{\gamma}, \Lambda, T$

$$\mathbb{E}\big(|X_t|^2\big) \le \big(\mathbb{E}\big(|X_0|^2\big) + C_2\, t\big)\, e^{C_1 t}$$

$$\mathbb{E}\big(|X_t - X_0|^2\big) \le C_3\, \big(\mathbb{E}\big(|X_0|^2\big) + 1\big)\, e^{C_1 t} \cdot t.$$

3.6.3 A Short Cut to Existence of Strong Solutions

Applying the mutational framework, the first essential steps are always to specify the basic set, its distance function(s) and the candidates for transitions.

For taking the filtration $(\mathscr{A}_t)_{t\geq 0}$ into consideration properly, we use a separate real component indicating time and hence, we choose as basic set

$$\widetilde{E}_{\mathscr{A}} := \big\{(t,X) \;\big|\; t\geq 0,\; X:\Omega \longrightarrow \mathbb{R} \text{ is } \mathscr{A}_t\text{-measurable, } \mathbb{E}(|X|^2)<\infty\big\}.$$

Furthermore the last estimate in Proposition 51 indicates that Lipschitz continuity with respect to time is ensured merely for the square deviation (and not for the standard L^2 norm). This observation motivates the following choice:

$$\begin{aligned}
\widetilde{d}_{\mathscr{A},P} : \widetilde{E}_{\mathscr{A}} \times \widetilde{E}_{\mathscr{A}} &\longrightarrow [0,\infty[, & \big((s,X),\,(t,Y)\big) &\longmapsto |t-s| + \mathbb{E}\big(|X-Y|^2\big)\\
\lfloor\cdot\rfloor_{\mathscr{A},P} : \qquad \widetilde{E}_{\mathscr{A}} &\longrightarrow [0,\infty[, & (t,X) &\longmapsto |t| \quad + \mathbb{E}\big(|X|^2\big).
\end{aligned}$$

On the basis of Proposition 51, solutions to *autonomous* stochastic ordinary differential equations provide the candidates for transitions on $\widetilde{E}_{\mathscr{A}}$. There is a significant obstacle though: ω-contractivity is usually lacking – as required in Definition 2 (3.) of transitions (on page 183).
Indeed, the following lemma compares strong solutions to nonautonomous stochastic differential equations by means of their initial values and coefficients. Although the estimate is a quite simple consequence of Gronwall's inequality, there is no obvious way for eliminating the constant coefficient >1 of the initial square deviation on the right-hand side.

Lemma 52. *For $k=1,2$, let $a_k, b_k : [0,T]\times\mathbb{R} \longrightarrow \mathbb{R}$ fulfill the assumptions of Proposition 51 with $\int_0^T \big(\|a_k(s,\cdot)\|_\infty^2 + \|b_k(s,\cdot)\|_\infty^2\big)\,ds < \infty$ and the joint Lipschitz parameter $\Lambda>0$. $(X_t^k)_{0\leq t\leq T}$ denotes a strong solution of*

$$d X_t^k \;=\; a_k(t, X_t^k)\,dt \;+\; b_k(t, X_t^k)\,dW_t$$

with

$$\sup_{0\leq t\leq T} \;\mathbb{E}\big(|X_t^k|^2\big) \;<\; \infty.$$

Then, there exists a constant $C=C(\Lambda)$ such that for every $t\in[0,T]$

$$\begin{aligned}
\mathbb{E}(|X_t^1 - X_t^2|^2) \leq \Big(&3\cdot\mathbb{E}(|X_0^1 - X_0^2|^2)\\
&+C\cdot\int_0^t \big(\|a_1(s,\cdot)-a_2(s,\cdot)\|_\infty^2 + \|b_1(s,\cdot)-b_2(s,\cdot)\|_\infty^2\big)ds\Big)\,e^{C\,t\cdot e^t}.
\end{aligned}$$

Although lacking an error estimate sufficient for ω-contractivity, we can still benefit from the results in § 3.4 (on page 214 ff.) bridging this gap via auxiliary distances. Indeed, Lemma 52 ensures assumption (A4) there (on page 215) if we choose *autonomous* stochastic differential equations with Λ-Lipschitz continuous coefficients as candidates for transitions. In particular, the Lipschitz constant $\Lambda>0$ is fixed. Moreover this estimate is useful only if both $\|a_1-a_2\|_\infty<\infty$ and $\|b_1-b_2\|_\infty<\infty$ and so, we restrict our considerations to *bounded* and Λ-Lipschitz coefficients (in this subsection 3.6.3).

Proposition 51 indicates explicitly how to choose further parameters of the mutational framework for guaranteeing the remaining assumptions of § 3.4. It leads to the main result of this example:

Theorem 53 (Existence of solutions to nonlocal stochastic diff. equations). *Assume for* $\widetilde{\mathbf{f}} = (\widetilde{f}_1, \widetilde{f}_2) : \widetilde{E}_{\mathscr{A}} \longrightarrow W^{1,\infty}(\mathbb{R},\mathbb{R}) \times W^{1,\infty}(\mathbb{R},\mathbb{R})$

(*1.*) $\sup_{\widetilde{Y} \in \widetilde{E}_{\mathscr{A}}} \|\widetilde{\mathbf{f}}(\widetilde{Y})\|_{W^{1,\infty}} < \infty$,

(*2.*) $\widetilde{\mathbf{f}}$ *is locally Lipschitz continuous in the following sense:* *For every* $R > 0$, *there exists a constant* $\lambda_R > 0$ *such that for all* $\widetilde{Y}, \widetilde{Z} \in \widetilde{E}_{\mathscr{A}}$ *with* $\max\big\{\lfloor \widetilde{Y} \rfloor_{\mathscr{A},P}, \lfloor \widetilde{Z} \rfloor_{\mathscr{A},P}\big\} < R$,

$$\Big\| \widetilde{\mathbf{f}}(\widetilde{Y})(\cdot) - \widetilde{\mathbf{f}}(\widetilde{Z})(\cdot) \Big\|^2_{L^\infty} \le \lambda_R \cdot \widetilde{d}_{\mathscr{A},P}(\widetilde{Y}, \widetilde{Z}).$$

Then for every initial tuple $\widehat{X}_0 = (t_0, X_0) \in \widetilde{E}_{\mathscr{A}}$ *and period* $T > 0$, *there exists a continuous curve* $[t_0, t_0+T] \longrightarrow \widetilde{E}_{\mathscr{A}}, \; t \longmapsto \widetilde{X}_t = (t, X_t)$ *such that the stochastic process* $(X_t)_{t_0 \le t \le t_0+T}$ *is a strong solution to the nonlocal stochastic differential equation*

$$\begin{cases} dX_t(\omega) = \widetilde{f}_1(t, X_t)\,(X_t(\omega))\,dt + \widetilde{f}_2(t, X_t)\,(X_t(\omega))\,dW_t(\omega) & \textit{in } [t_0, t_0+T] \\ X_{t_0} = X_0 \end{cases}$$

and, it belongs to $\mathscr{L}^2_{\mathscr{A}}([t_0, t_0+T])$.

Indeed, both Proposition 32 and Lemma 33 (on page 215 ff.) provide a superset $\check{E}_{\mathscr{A}} \subset \widetilde{E}_{\mathscr{A}} \times \mathbb{R}^+_0$ and a distance $\check{d}_{\mathscr{A},P}$ such that the "simultaneously timed" counterpart of extended Cauchy-Lipschitz Theorem 47 (on page 230) can be applied to the tuple $\big(\check{E}_{\mathscr{A}}, \check{d}_{\mathscr{A},P}, \widetilde{d}_{\mathscr{A},P}, \lfloor\cdot\rfloor_{\mathscr{A},P}, \|\cdot\|^2_{L^\infty}\big)$ with all those trivial extensions to $\check{E}_{\mathscr{A}}$.

Hence, the mutational equation $\overset{\circ}{\check{X}} \ni \widetilde{f}\big(\pi_{\widetilde{E}_{\mathscr{A}}} \check{X}\big)$ has a simultaneously timed solution $\check{X} : [t_0, t_0+T] \longrightarrow \check{E}_{\mathscr{A}}, \; t \longmapsto (t, X_t, \rho_t)$ starting in $\big(t_0, X_0, \lfloor (t_0, X_0) \rfloor_{\mathscr{A},P}\big) \in \check{E}_{\mathscr{A}}$, Lipschitz continuous w.r.t. $\widetilde{d}_{\mathscr{A},P}$ and bounded w.r.t. $\lfloor\cdot\rfloor_{\mathscr{A},P}$.
We (just) have to verify that the corresponding stochastic process $(X_t)_{t_0 \le t \le t_0+T}$ solves the original stochastic differential equation:

Lemma 54 (Link from mutational to nonlocal stochastic diff. equations). *In addition to the hypotheses of Theorem 53 about* $\widetilde{\mathbf{f}} = (\widetilde{f}_1, \widetilde{f}_2) : \widetilde{E}_{\mathscr{A}} \longrightarrow W^{1,\infty}(\mathbb{R})^2$, *suppose* $[t_0, t_0+T] \longrightarrow \check{E}_{\mathscr{A}}, \; t \longmapsto \check{X}_t = (t, X_t, \rho_t)$ *to satisfy the four conditions on simultaneously timed solutions of*

$$\overset{\circ}{\check{X}} \ni \widetilde{f}\big(\pi_{\widetilde{E}_{\mathscr{A}}} \check{X}\big)$$

in $\big(\check{E}_{\mathscr{A}}, \check{d}_{\mathscr{A},P}, \widetilde{d}_{\mathscr{A},P}, \lfloor\cdot\rfloor_{\mathscr{A},P}, \|\cdot\|^2_{L^\infty}\big)$ *stated in Definition 36 (on page 222).*

Then, $(X_t)_{t_0 \le t \le t_0+T}$ *is a strong solution to the stochastic differential equation*

$$dX_t(\omega) = \widetilde{f}_1(t, X_t)\,(X_t(\omega))\,dt + \widetilde{f}_2(t, X_t)\,(X_t(\omega))\,dW_t(\omega) \quad \textit{in } [t_0, t_0+T].$$

Proof (of Lemma 54 on page 236). The compositions

$$a : [0,T] \times \mathbb{R} \longrightarrow \mathbb{R}, \quad (t,z) \longmapsto \widetilde{f_1}(\widetilde{X}_t)(z)$$
$$b : [0,T] \times \mathbb{R} \longrightarrow \mathbb{R}, \quad (t,z) \longmapsto \widetilde{f_2}(\widetilde{X}_t)(z)$$

satisfy the hypotheses of Proposition 51 (on page 234). Hence, there exists a pathwise unique strong solution $(Y_t)_{t_0 \le t \le t_0+T}$ to the stochastic differential equation

$$dY_t \;=\; a(t,Y_t)\,dt \;+\; b(t,Y_t)\,dW_t$$

with the same initial value X_0 as $(X_t)_{t_0 \le t \le t_0+T}$ and $\sup_t \lfloor (t,Y_t) \rfloor_{\mathscr{A},P} \le \widehat{R} < \infty$. Then, $[t_0,\, t_0+T] \longrightarrow \check{\widetilde{E}}_{\mathscr{A}},\;\; t \longmapsto \check{Y}_t := \big(t, Y_t,\, \lfloor (t,Y_t) \rfloor_{\mathscr{A}}\big)$ is a simultaneously timed solution to the mutational equation $\overset{\circ}{\check{Y}} \ni \widetilde{\mathbf{f}}(\pi_{\widetilde{E}_{\mathscr{A}}} \check{X})$. This results directly from Proposition 34 (on page 219) in combination with Lemma 52.
Finally we conclude $\pi_2 \check{X} \equiv \pi_2 \check{Y}$ from the "simultaneously timed" counterpart of Proposition 40 (on page 224) since the following auxiliary distance is identical to 0:

$$\begin{aligned}
&[t_0,\, t_0+T] \longrightarrow [0,\infty[,\\
&\qquad t \longmapsto \inf\big\{ \check{d}_{\mathscr{A},P}(\check{Z},\check{X}_t) + \check{d}_{\mathscr{A},P}(\check{Z},\check{Y}_t) \;\big|\; \check{Z} \in \check{\widetilde{E}}_{\mathscr{A}} :\; \pi_1 \check{Z} = t,\\
&\qquad\qquad\qquad\qquad\qquad\qquad \lfloor \check{Z} \rfloor_{\mathscr{A},P} \;<\; 1 + \max\{R, \widehat{R}\} \big\}. \qquad \square
\end{aligned}$$

Proof (of Lemma 52 on page 235). By definition, $(X_t^k)_{0 \le t \le T}$ solves the integral equation $X_t^k \;=\; X_0^k + \int_0^t a_k(s, X_s^k)\,ds + \int_0^t b_k(s, X_s^k)\,dW_s$ at each time $t \in [0,T]$. Due to the simple inequality $(r_1 + r_2)^2 \;=\; r_1^2 + 2\,r_1\,r_2 + r_2^2 \;\le\; 3\,(r_1^2 + r_2^2)$ for all $r_1, r_2 \in \mathbb{R}$, we obtain for each $t \in [0,T]$

$$\begin{aligned}
&\mathbb{E}\big(|X_t^1 - X_t^2|^2\big) \;-\; 3 \cdot \mathbb{E}\big(|X_0^1 - X_0^2|^2\big)\\
&\le 3 \cdot \mathbb{E}\Big(\Big| \int_0^t \; a_1|_{(s,X_s^1)} - a_2|_{(s,X_s^2)} \;\; ds \;+\; \int_0^t \;\; b_1|_{(s,X_s^1)} - b_2|_{(s,X_s^2)} \;\; dW_s \Big|^2\Big)\\
&\le 9 \cdot \mathbb{E}\Big(\Big(\int_0^t \; \big|a_1|_{(s,X_s^1)} - a_2|_{(s,X_s^2)}\big| \;\; ds\Big)^2 + \Big(\int_0^t \;\; \big|b_1|_{(s,X_s^1)} - b_2|_{(s,X_s^2)}\big| \;\; dW_s\Big)^2\Big)\\
&\le 9 \cdot \Big(t \cdot \int_0^t \mathbb{E}\big(\big|a_1|_{(s,X_s^1)} - a_2|_{(s,X_s^2)}\big|^2\big)\,ds \;+\; \int_0^t \mathbb{E}\big(\big|b_1|_{(s,X_s^1)} - b_2|_{(s,X_s^2)}\big|^2\big)\,ds\Big)\\
&\le 9 \cdot e^t \;\Big(\int_0^t \mathbb{E}\big(\big|a_1|_{(s,X_s^1)} - a_2|_{(s,X_s^2)}\big|^2\big)\,ds \;+\; \int_0^t \mathbb{E}\big(\big|b_1|_{(s,X_s^1)} - b_2|_{(s,X_s^2)}\big|^2\big)\,ds\Big)
\end{aligned}$$

as a consequence of Hölder inequality and Itô isometry (in Proposition 50). In fact,

$$\begin{aligned}
&\int_0^t \mathbb{E}\big(\big|a_1(s, X_s^1) - a_2(s, X_s^2)\big|^2\big)\,ds\\
&\qquad \le\; 3 \cdot \int_0^t \Big(\mathbb{E}\big(\big|a_1(s, X_s^1) - a_2(s, X_s^1)\big|^2\big) + \mathbb{E}\big(\big|a_2(s, X_s^1) - a_2(s, X_s^2)\big|^2\big)\Big)\,ds\\
&\qquad \le\; 3 \cdot \int_0^t \Big(\big\|a_1(s,\cdot) - a_2(s,\cdot)\big\|_\infty^2 \qquad + \Lambda^2 \cdot \mathbb{E}\big(|X_s^1 - X_s^2|^2\big)\Big)\,ds
\end{aligned}$$

and, the corresponding estimate holds for $\int_0^t \mathbb{E}\big(\big|b_1(s, X_s^1) - b_2(s, X_s^2)\big|^2\big)\,ds$.

Finally the claimed upper bound of $\mathbb{E}\big(|X_t^1 - X_t^2|^2\big)$ results from Gronwall's inequality (in Proposition A.1) applied to $t \mapsto \mathbb{E}\big(|X_t^1 - X_t^2|^2\big) \cdot e^{-t}$. $\square$

3.6.4 A Special Case with Fixed Additive Noise in More Detail

Now we approach nonlocal stochastic differential equations *without* the auxiliary construction of § 3.4. In particular, distances and transitions on $\widetilde{E}_{\mathscr{A}}$ in use are specified explicitly and verified. These full details imply a disadvantage though. We have to restrict our considerations to the special case of an initially fixed diffusion coefficient $b(\cdot)$ and real-valued drift coefficients (as a function of state in $\widetilde{E}_{\mathscr{A}}$). It leads to nonlocal stochastic differential equations of the form

$$d X_t(\omega) = \widetilde{f}(t, X_t(\cdot))\, dt + b(t)(\omega)\, dW_t(\omega)$$

with a bounded Lipschitz function $\widetilde{f} : (\widetilde{E}_{\mathscr{A}}, \widetilde{d}_{\mathscr{A},P}) \longrightarrow \mathbb{R}$.

Stochastic Ordinary Differential Equations with Fixed Additive Noise Induce Transitions

In contrast to § 3.6.3, we now start from stochastic differential equations only with *constant real-valued drift* and *fixed diffusion coefficient*. Strictly speaking, their strong solutions are "just" Itô processes, but they have all the features we need for timed transitions on $\widetilde{E}_{\mathscr{A}}$.

The only relevant obstacle to Definition 2 (of transitions on page 183) and its timed counterpart is related to the comparison estimate for evolving random variables. We restrict it to *simultaneously* timed states in $\widetilde{E}_{\mathscr{A}}$ so that the Itô integrals do not occur explicitly in the inequalities.

Lemma 55. *Let $a, \widehat{a} \in \mathbb{R}$ satisfy $\max\{|a|, |\widehat{a}|\} \leq \widehat{\gamma}$ and, suppose $b : [0,\infty[\longrightarrow \mathbb{R}$ to be $\mathscr{L}^1$-measurable with $\|b\|_{L^\infty} \leq \widehat{\gamma} < \infty$.*

Then the Itô processes

$$X_t = \widehat{X}_0 + \int_{t_0}^{t} a\ ds + \int_{t_0}^{t} b(s)\ dW_s$$

induce a unique map $\widetilde{\vartheta}_{\mathscr{A},a,b} : [0,1] \times \widetilde{E}_{\mathscr{A}} \longrightarrow \widetilde{E}_{\mathscr{A}},\ \big(h, (t_0, \widehat{X}_0)\big) \longmapsto (t_0 + h, X_{t_0+h})$ with the following properties for all $\widetilde{X}, \widetilde{Y} \in \widetilde{E}_{\mathscr{A}}$, $R \geq 0$, $t, h_1, h_2 \in [0,1]$ $(h_1 + h_2 \leq 1)$

(1.) $\widetilde{\vartheta}_{\mathscr{A},a,b}(0, \cdot) = \mathbb{I}\mathrm{d}_{\widetilde{E}_{\mathscr{A}}}$

(2.) $\widetilde{\vartheta}_{\mathscr{A},a,b}(h_1 + h_2, \cdot) = \widetilde{\vartheta}_{\mathscr{A},a,b}\big(h_2, \widetilde{\vartheta}_{\mathscr{A},a,b}(h_1, \cdot)\big)$

(3.) $\widetilde{d}_{\mathscr{A},P}\big(\widetilde{X}, \widetilde{\vartheta}_{\mathscr{A},a,b}(t, \widetilde{X})\big) \leq \mathrm{const}(\widehat{\gamma}) \cdot (\lfloor \widetilde{X} \rfloor_{\mathscr{A},P} + 1) \cdot t$

(4.) $\lfloor \widetilde{\vartheta}_{\mathscr{A},a,b}(t, \widetilde{X}) \rfloor_{\mathscr{A},P} \leq e^{\,\mathrm{const}(\widehat{\gamma}) \cdot t} \cdot (\lfloor \widetilde{X} \rfloor_{\mathscr{A},P} + \mathrm{const}(\widehat{\gamma}) \cdot t)$

(5.) $\exists\, C = C(R)$: *if* $\pi_1 \widetilde{X} = \pi_1 \widetilde{Y}$ *and* $\max\big\{\lfloor \widetilde{X} \rfloor_{\mathscr{A},P}, \lfloor \widetilde{Y} \rfloor_{\mathscr{A},P}\big\} \leq R$,

$$\lim_{h \downarrow 0} \frac{\widetilde{d}_{\mathscr{A},P}\big(\widetilde{\vartheta}_{\mathscr{A},a,b}(h, \widetilde{X}),\ \widetilde{\vartheta}_{\mathscr{A},\widehat{a},b}(h, \widetilde{Y})\big) - \widetilde{d}_{\mathscr{A},P}(\widetilde{X}, \widetilde{Y})}{h} \leq C \cdot |a - \widehat{a}|.$$

Proof. Statements (1.) and (2.) are obvious because the Itô integral is additive with respect to the interval of integration. Furthermore, statements (3.), (4.) result from the upper bounds of $\mathbb{E}\big(|X_t - X_0|^2\big)$ and $\mathbb{E}\big(|X_t|^2\big)$ in preceding Proposition 51.

Finally, we focus on property (5.) for $\widetilde{X} = (t_0, X)$, $\widetilde{Y} = (t_0, Y) \in \widetilde{E}_{\mathscr{A}}$ with X_t and Y_t denoting the Itô processes

$$\begin{cases} X_t = X + \int_{t_0}^{t} a\ ds + \int_{t_0}^{t} b(s)\ dW_s \\ Y_t = Y + \int_{t_0}^{t} \widehat{a}\ ds + \int_{t_0}^{t} b(s)\ dW_s \end{cases}$$

respectively. Then we obtain at every time $t \in [t_0,\ t_0+1]$

$$\begin{aligned} \mathbb{E}\big(|X_t - Y_t|^2\big) &= \mathbb{E}\big(\big|X - Y + (a-\widehat{a})\cdot(t-t_0)\big|^2\big) \\ &= \mathbb{E}\big(|X-Y|^2\big) + \mathbb{E}\big(2\,(X-Y)\cdot(a-\widehat{a})\,(t-t_0)\big) + |a-\widehat{a}|^2\,|t-t_0|^2 \\ &\leq \mathbb{E}\big(|X-Y|^2\big) + 2\cdot\mathbb{E}(|X|+|Y|)\cdot|a-\widehat{a}|\,|t-t_0| + |a-\widehat{a}|^2\,|t-t_0|^2 \\ &\leq \mathbb{E}\big(|X-Y|^2\big) + \mathbb{E}(|X|^2+|Y|^2)\ \cdot|a-\widehat{a}|\,|t-t_0| + |a-\widehat{a}|^2\,|t-t_0|^2 \end{aligned}$$

and thus,

$$\lim_{h\downarrow 0} \frac{\widetilde{d}_{\mathscr{A},P}\big(\widetilde{\vartheta}_{\mathscr{A},a,b}(h,\widetilde{X}),\ \widetilde{\vartheta}_{\mathscr{A},\widehat{a},b}(h,\widetilde{Y})\big) - \widetilde{d}_{\mathscr{A},P}(\widetilde{X},\widetilde{Y})}{h} \leq \big(\lfloor\widetilde{X}\rfloor_{\mathscr{A},P} + \lfloor\widetilde{Y}\rfloor_{\mathscr{A},P}\big)\cdot|a-\widehat{a}|$$

because in each of these distances, two simultaneous states in $\widetilde{E}_{\mathscr{A}}$ are compared. □

The Step to Nonlocal Stochastic Equations: Existence of Strong Solutions

For every $t \geq 0$, the vector space of $\mathscr{A}_t$-measurable functions $X : \Omega \longrightarrow \mathbb{R}$ with $\mathbb{E}(|X|^2) < \infty$ is known to be complete with respect to its L^2 norm $\sqrt{\mathbb{E}(|\cdot - \cdot|^2)}$. As an obvious consequence, the tuple $\big(\widetilde{E},\ \widetilde{d}_{\mathscr{A},P},\ \widetilde{d}_{\mathscr{A},P},\ \lfloor\cdot\rfloor_{\mathscr{A},P}\big)$ is timed complete in the sense of Definition 46 (on page 229). Moreover, Proposition 51 implies Euler equi-continuity. Hence, these two features are good starting points for concluding the existence of solutions from Cauchy-Lipschitz Theorem.

First, however, we should clarify what kind of stochastic differential equations is considered within the mutational framework and what type of solution is obtained. Indeed, after fixing a bounded $\mathscr{L}^1$-measurable diffusion coefficient $b : [0,\infty[\longrightarrow \mathbb{R}$, we use the transitions $\widetilde{\vartheta}_{\mathscr{A},,a,b} : [0,1]\times\widetilde{E}_{\mathscr{A}} \longrightarrow \widetilde{E}_{\mathscr{A}}$ induced by any constant $a\in\mathbb{R}$ and specified in Lemma 55 (on page 238), i.e., for any initial state $(t_0,\widehat{X}_0) \in \widetilde{E}_{\mathscr{A}}$ given, the second component of $\widetilde{\vartheta}_{\mathscr{A},,a,b}\big(h,\ (t_0,\widehat{X}_0)\big) \in \widetilde{E}_{\mathscr{A}}$ results from the Itô process

$$X_t = \widehat{X}_0 + \int_{t_0}^{t} a\ ds + \int_{t_0}^{t} b(s)\ dW_s$$

for every $t \in [t_0,\ t_0+h]$.

In regard to a mutational equation, we prescribe the real-valued drift $a \in \mathbb{R}$ as a function of time t and $\mathscr{A}_t$-measurable random variable $\Omega \longrightarrow \mathbb{R}$ with bounded second moment in an appropriately continuous way:

$$\widetilde{f} : \widetilde{E}_{\mathscr{A}} \longrightarrow \mathbb{R}.$$

In particular, for any $\widetilde{X} = (t,X) \in \widetilde{E}_{\mathscr{A}}$ given, $\widetilde{f}(\widetilde{X}) \in \mathbb{R}$ might depend on the first or second moment of $X : \Omega \longrightarrow \mathbb{R}$, for example. We interpret such a dependence as a functional relationship and thus, our subsequent initial value problems deal with stochastic *functional* differential equations. But it differs essentially from the other examples in the literature such as [135, 187] because it is not a pathwise dependence. Thus, we regard it as more appropriate to call this problem *nonlocal* stochastic differential equation.

Furthermore, the comparative estimate in Lemma 55 (5.) is restricted to states $\widetilde{X}, \widetilde{Y} \in \widetilde{E}_{\mathscr{A}}$ with identical time components $\pi_1 \widetilde{X} = \pi_1 \widetilde{Y}$ — essentially for preserving the characteristic dependence on "initial error" and "transitional error".
As a consequence, any bounds of distances between Euler approximations are available only at identical points of time and, this constraint leads to *simultaneously timed solutions* to mutational equations in the sense of Definition 36 (on page 222). The aspect of required simultaneity concerns only the distances between states in $\widetilde{E}_{\mathscr{A}}$, but not the distances between transitions when assuming Lipschitz continuity, for example, as the detailed proof of Cauchy-Lipschitz Theorem 31 (on page 213) clarifies.

Theorem 56. *Assume for* $\widetilde{f} : \widetilde{E}_{\mathscr{A}} \longrightarrow \mathbb{R}$

(*1.*) $\sup_{\widetilde{Y} \in \widetilde{E}_{\mathscr{A}}} |\widetilde{f}(\widetilde{Y})| < \infty$,

(2.) $\widetilde{f}$ *is locally Lipschitz continuous in the following sense:*
For every $R > 0$, *there exists a constant* $\lambda_R > 0$ *such that for all* $\widetilde{Y}, \widetilde{Z} \in \widetilde{E}_{\mathscr{A}}$ *with* $\max\{\lfloor \widetilde{Y} \rfloor_{\mathscr{A},P}, \lfloor \widetilde{Z} \rfloor_{\mathscr{A},P}\} < R$,

$$\left|\widetilde{f}(\widetilde{Y}) - \widetilde{f}(\widetilde{Z})\right| \le \lambda_R \cdot \widetilde{d}_{\mathscr{A},P}(\widetilde{Y},\widetilde{Z}).$$

Then for every initial tuple $\widehat{X}_0 = (t_0, X_0) \in \widetilde{E}_{\mathscr{A}}$ *and period* $T > 0$, *there exists a simultaneously timed solution* $[t_0, t_0 + T] \longrightarrow \widetilde{E}_{\mathscr{A}}$, $t \longmapsto \widetilde{X}_t = (t, X_t)$ *to the mutational equation*

$$\overset{\circ}{\widetilde{X}} \ni \widetilde{f}(\widetilde{X})$$

in the sense of Definition 36 (on page 222) with $\widetilde{X}_{t_0} = \widehat{X}_0 = (t_0, X_0)$.

In particular, the stochastic process $(X_t)_{t_0 \le t \le t_0+T}$ *is a strong solution to the non-local stochastic differential equation*

$$\begin{cases} dX_t(\omega) = \widetilde{f}(t, X_t)\, dt + b(t)(\omega)\, dW_t(\omega) & \text{in } [t_0, t_0+T] \\ X_{t_0} = X_0 \end{cases}$$

and, it belongs to $\mathscr{L}^2_{\mathscr{A}}([t_0, t_0 + T])$.

Proof. As mentioned briefly in § 3.5, the existence of simultaneously timed solutions results from exactly the same arguments as Cauchy-Lipschitz Theorem 31 — after restricting the structural estimate (for distances between states evolving along two transitions) in Proposition 7 to simultaneous states in $\widetilde{E}_{\mathscr{A}}$.

Due to the transition properties in Lemma 55, there exists a simultaneously timed solution $[t_0, t_0+T] \longrightarrow \widetilde{E}_{\mathscr{A}}, \; t \longmapsto \widetilde{X}_t = (t, X_t)$ to the mutational equation

$$\overset{\circ}{\widetilde{X}} \ni \widetilde{f}(\widetilde{X})$$

in the sense of Definition 36 (on page 222) with $\widetilde{X}_{t_0} = \widehat{X}_0$ and $\sup_t \lfloor \widetilde{X}_t \rfloor_{\mathscr{A},P} \le R < \infty$. In particular, assumption (1.) provides a constant $L > 0$ with

$$\widetilde{d}_{\mathscr{A},P}\big(\widetilde{X}_s, \widetilde{X}_t\big) \;\le\; L\,|t-s| \qquad \text{for all } s,t \in [t_0, t_0+T].$$

Now the composition

$$a : [0,T] \longrightarrow \mathbb{R}, \quad t \longmapsto \widetilde{f}(\widetilde{X}_t)$$

is Lipschitz continuous and together with $b \in L^\infty(\mathbb{R}_0^+)$, it induces the Itô process

$$Y_t \;=\; X_0 \;+\; \int_{t_0}^{t} a(s)\; ds \;+\; \int_{t_0}^{t} b(s)\; dW_s$$

with the same initial value X_0 as $(X_t)_{t_0 \le t \le t_0+T}$ and $\sup_t \lfloor (t, Y_t) \rfloor_{\mathscr{A},P} \le \widehat{R} < \infty$. Then, $[t_0, t_0+T] \longrightarrow \widetilde{E}_{\mathscr{A}}, \; t \longmapsto \widetilde{Y}_t \overset{\text{Def.}}{=} (t, Y_t)$ is a simultaneously timed solution to the mutational equation

$$\overset{\circ}{\widetilde{Y}} \ni \widetilde{f}(\widetilde{X}).$$

Indeed, choosing any $t \in [t_0, t_0+T[$, $\widehat{a} \in \mathbb{R}$ and $\mathscr{A}_t$-measurable $Z_t : \Omega \longrightarrow \mathbb{R}$ with bounded second moment, let $(Z_s)_{t \le s \le t_0+T}$ denote the auxiliary Itô process

$$Z_s \;=\; Z_t \;+\; \int_t^s \widehat{a}\; ds' \;+\; \int_t^s b(s')\; dW_{s'}\,.$$

Exactly the same arguments as in the proof of Lemma 55 (5.) (on page 239) provide a constant $C > 0$ depending explicitly just on $\lfloor (t_0, \widehat{X}_0) \rfloor_{\mathscr{A},P}$, $\lfloor (t, Z_t) \rfloor_{\mathscr{A},P}$, T, L and the supremum in assumption (1.) such that

$$\begin{aligned} \limsup_{h \downarrow 0} \; \tfrac{1}{h} \cdot \Big(\widetilde{d}_{\mathscr{A},P}\Big(\widetilde{\vartheta}_{\mathscr{A},\widehat{a},b}(h, (t, Z_t)),\; (t+h,\, Y_{t+h})\Big) \;-\; \widetilde{d}_{\mathscr{A},P}\big(Z_t, Y_t\big)\Big) \\ \le C \cdot \limsup_{H \downarrow 0} \quad \big|\widehat{a} - \widetilde{f}(\widetilde{X}_{t+H})\big| \\ = C \cdot \big|\widehat{a} - \widetilde{f}(\widetilde{X}_t)\big| \end{aligned}$$

due to the continuity of $\widetilde{f}$. The "simultaneously timed" counterpart of Proposition 40 (on page 224) implies that the auxiliary distance

$$\begin{aligned} [t_0, t_0+T] \;&\longrightarrow\; [0, \infty[, \\ t \;&\longmapsto\; \inf\Big\{ \widetilde{d}_{\mathscr{A},P}\big(\widetilde{Z}, \widetilde{X}_t\big) \;+\; \widetilde{d}_{\mathscr{A},P}\big(\widetilde{Z}, \widetilde{Y}_t\big) \;\Big|\; \widetilde{Z} \in \widetilde{E}_{\mathscr{A}} :\; \pi_1 \widetilde{Z} = t, \\ &\qquad\qquad \lfloor \widetilde{Z} \rfloor_{\mathscr{A},P} \;<\; 1 + \max\{R, \widehat{R}\} \Big\} \end{aligned}$$

is identical to 0 and thus, $X_t \equiv Y_t$ satisfies the claimed nonlocal stochastic differential equation in the strong sense. □

3.7 Example: Stochastic Morphological Equations for Square Integrable Random Closed Sets in $\mathbb{R}^N$

Many geometric processes in nature share the basic property that aspects of growth and birth interfere simultaneously and, there is often no obvious way to describe the birth of additional shape components in a deterministic way. One of the most popular examples is crystal growth, but such forms of "uncertain shape evolution" also occur in mineralization, solidification and DNA replication.

This observation motivates us to implement morphological set evolutions in a stochastic environment. In other words, we now suggest a stochastic counterpart of morphological equations for describing the evolution of shapes in dependence on both feedback and random effects.

By comparison with the original morphological equations of Aubin (in § 1.9), stochastics comes into play in two respects: First, it is not compact subsets of $\mathbb{R}^N$ that evolve according to a given feedback rule, but we consider *random closed sets* in $\mathbb{R}^N$, i.e. measurable set-valued maps on a complete probability space with nonempty closed values in $\mathbb{R}^N$. Second, the shape evolution is not based on solutions to (deterministic) differential inclusions in $\mathbb{R}^N$, but now we use strong solutions to stochastic differential inclusions. In particular, Theorem A.67 of Da Prato and Frankowska assumes the central role of Filippov's Theorem A.6. These two respects are the starting points of this new example: *Stochastic morphological equations* in $\mathbb{R}^N$.

This approach to stochastic shape evolution has many advantages in common with morphological equations: We are not obliged to guarantee any regularity properties of boundaries (for specifying deformations in normal direction, for example). Moreover, set evolutions may be determined by nonlocal features of the respective sets and do not have to obey the inclusion principle.

Last, but not least, *this proposal is not restricted to convex sets.* In the deterministic case, integrating a set-valued map of time (only) is usually interpreted in the sense of Aumann and thus, it always leads to convex subsets. This "curse of convexity" has been overcome quite easily in § 1.9 by "integrating" along differential inclusions $x' \in F(x,t)$, whose right-hand side may depend on space explicitly: Their reachable sets are not necessarily convex any longer. We benefit from the same advantage of additional feedback now if random closed sets in $\mathbb{R}^N$ evolve along stochastic differential inclusions (instead of just some integration in time).

Random closed reachable sets are always related to their initial set and thus, they can describe only growth processes (§ 3.7.3). Nucleation means here that an additional random closed set initiates such a growth process at a possibly later point of time. This aspect is described as a given set-valued map of time in § 3.7.4.

3.7.1 The General Assumptions for This Example

Morphological equations for nonempty compact sets in $\mathbb{R}^N$ were discussed in § 1.9 (on page 57 ff.) and, their transitions are induced by reachable sets of differential inclusions in the Euclidean space. Filippov's Theorem A.6 (on page 443 f.) proved to be the key tool for verifying the required features in the mutational framework. Now we are going to draw the corresponding conclusions for random closed sets in $\mathbb{R}^N$ from Da Prato-Frankowska Theorem A.67 about stochastic differential inclusions and so, we start with essentially the assumptions of § A.8.1 (on page 482 ff.). For the sake of transparency, however, we neglect the aspect of predictability, i.e. the underlying filtration $(\mathscr{A}_t)_{t\geq 0}$ is constant: $\mathscr{A}_t := \mathscr{A}$ for all $t \geq 0$. (Indeed, this simplification avoids an additional real component, which is just monitoring the respective points of time as in § 3.6.)

General assumptions in § 3.7

(i) $(\Omega, \mathscr{A}, P)$ is a complete probability space.

(ii) $W = (W_t)_{t\geq 0}$ is an m-dimensional Wiener process.

(iii) For finite $T > 0$ fixed, define the class $\mathscr{L}^2_{\mathscr{A}}([0,T], \mathbb{R}^N)$ of functions $f : [0,T] \times \Omega \longrightarrow \mathbb{R}^N$ with

   (1.) f is jointly $\mathscr{L}^1 \times \mathscr{A}$-measurable,

   (2.) $\displaystyle\int_{[0,T]} \mathbb{E}\big(|f(t,\cdot)|^2\big)\, dt < \infty$,

   (3.) for every $t \in [0,T]$, $\mathbb{E}\big(|f(t,\cdot)|^2\big) < \infty$ and

   (4.) for every $t \in [0,T]$, $f(t,\cdot) : \Omega \longrightarrow \mathbb{R}^N$ is $\mathscr{A}$-measurable.

(iv) Let $\mathrm{Lin}(\mathbb{R}^m, \mathbb{R}^N)$ consist of all linear functions $\mathbb{R}^m \longrightarrow \mathbb{R}^N$.
Set $|M|_\infty := \sup_{y\in M} |y| \in [0,\infty]$ for any subset $M \subset \mathbb{R}^N$ (as in Definition 2.77).

(v) $\mathfrak{I}_0(X_0, \gamma, \sigma)$ denotes the Itô process associated with
the initial state $X_0 \in L^2(\Omega, \mathscr{A}, P; \mathbb{R}^N)$,
the drift $\gamma \in \mathscr{L}^2_{\mathscr{A}}([0,T], \mathbb{R}^N)$ and
the diffusion $\sigma \in \mathscr{L}^2_{\mathscr{A}}([0,T], \mathrm{Lin}(\mathbb{R}^m, \mathbb{R}^N))$, i.e. for $t \in [0,T]$,

$$\mathfrak{I}_0(X_0,\gamma,\sigma)\,(t) := X_0 + \int_0^t \gamma(s)\, ds + \int_0^t \sigma(s)\, dW_s,$$

$$\big\|\mathfrak{I}_0(X_0,\gamma,\sigma)\big\|_{\mathfrak{I},[0,t]} := \sqrt{\mathbb{E}(|X_0|^2) + \mathbb{E}\Big(\int_0^t |\gamma|^2\, ds\Big) + \mathbb{E}\Big(\int_0^t |\sigma|^2\, ds\Big)}.$$

In the deterministic example of morphological equations, we consider the metric space $(\mathscr{K}(\mathbb{R}^N), dl)$, i.e. the set of all nonempty compact subset of $\mathbb{R}^N$ supplied with the Pompeiu-Hausdorff metric dl. Stochastic differential equations with a Wiener process, however, imply the obstacle that the pathwise growth of their solutions is difficult to bound explicitly. Thus we modify the type of subsets under consideration and focus on *closed* sets. Furthermore the class of subsets is to be related with the probability space $(\Omega, \mathscr{A}, P)$ and thus, we use the concept of measurable set-valued maps specified in § A.10.1 (on page 489 f.).

Definition 57 (Random closed set). Let $\mathscr{B}$ denote the Borel σ-algebra on $\mathbb{R}^N$. A *random closed set* in $\mathbb{R}^N$ is a measurable set-valued map $(\Omega, \mathscr{A}) \leadsto (\mathbb{R}^N, \mathscr{B})$ with nonempty closed values. $\mathscr{RC}(\Omega, \mathbb{R}^N)$ consists of all these random closed sets.

A random closed set $M \in \mathscr{RC}(\Omega, \mathbb{R}^N)$ is called *square integrable* if there exists a square integrable selection $f : \Omega \longrightarrow \mathbb{R}^N$ of M, i.e. $f(\cdot) \in L^2(\Omega, \mathscr{A}, P; \mathbb{R}^N)$ with

$$f(\omega) \in M(\omega) \qquad \text{for all } \omega \in \Omega.$$

$\mathscr{S}^2_{\mathscr{RC}}(M) \subset L^2(\Omega, \mathscr{A}, P;\, \mathbb{R}^N)$ abbreviates the set of all square integrable selections $\Omega \longrightarrow \mathbb{R}^N$ of M. The set of square integrable random closed sets $\Omega \leadsto \mathbb{R}^N$ is denoted by $\mathscr{RC}^2(\Omega, \mathbb{R}^N)$.

Remark 58. Due to Castaing's Characterization Theorem A.75 (on page 489), a set-valued map $M : (\Omega, \mathscr{A}) \leadsto (\mathbb{R}^N, \mathscr{B})$ with nonempty closed values is measurable if and only if it is the pointwise closure of the union of (at most) countably many measurable functions $f_n : \Omega \longrightarrow \mathbb{R}^N$ $(n \in \mathbb{N})$:

$$M(\omega) = \overline{\bigcup_{n\in\mathbb{N}} f_n(\omega)} \qquad \text{for every } \omega \in \Omega.$$

Remark 59. (1.) As a consequence of the preceding remark, every *square integrable* random closed set $M : \Omega \leadsto \mathbb{R}^N$ can be represented by a sequence $(g_n)_{n\in\mathbb{N}}$ in $L^2(\Omega, \mathscr{A}, P;\, \mathbb{R}^N)$ as

$$M(\omega) = \overline{\bigcup_{n\in\mathbb{N}} g_n(\omega)} \qquad \text{for every } \omega \in \Omega.$$

Indeed, there exists at least one selection $g \in L^2(\Omega, \mathscr{A}, P;\, \mathbb{R}^N)$ of M by definition and for each $m \in \mathbb{N}$, the auxiliary function $g_{n,m} : \Omega \longrightarrow \mathbb{R}^N$ is square integrable with

$$g_{n,m} := \begin{cases} f_n & \text{if } m-1 \le |f_n| < m \\ g & \text{otherwise.} \end{cases}$$

(2.) The essential purpose of subsequent Lemma 60 is that many conclusions about any set $M \in \mathscr{RC}^2(\Omega, \mathbb{R}^N)$ result from just a single pointwise covering family $(g_n)_{n\in\mathbb{N}}$ in $L^2(\Omega, \mathscr{A}, P;\, \mathbb{R}^N)$ (in the sense of statement (1.) above).

(3.) For every $M \in \mathscr{RC}^2(\Omega, \mathbb{R}^N)$, $\mathscr{S}^2_{\mathscr{RC}}(M)$ is closed in $L^2(\Omega, \mathscr{A}, P;\, \mathbb{R}^N)$ and so, each separable closed subset of $L^2(\Omega, \mathscr{A}, P;\, \mathbb{R}^N)$ induces a square integrable random closed set uniquely (see also Proposition 65 below). $M \in \mathscr{RC}^2(\Omega, \mathbb{R}^N)$, however, does *not* have to satisfy $\mathbb{E}\big(|M|^2_\infty\big) \stackrel{\text{Def.}}{=} \int_\Omega \sup_{z\in M(\omega)} |z|^2 \; dP(\omega) < \infty$.

Lemma 60 (Approximation by step-functions [141, Lemma 2.1.3]**).** *Let the sequence $(g_n)_{n\in\mathbb{N}}$ in $L^2(\Omega, \mathscr{A}, P; \mathbb{R}^N)$ satisfy $M(\omega) = \overline{\bigcup_{n\in\mathbb{N}} g_n(\omega)}$ for any ω. Then for every $f \in \mathscr{S}^2_{\mathscr{RC}}(M)$ and $\varepsilon > 0$, there exists a finite measurable partition $A_1, A_2 \ldots A_m$ of Ω such that $\Big\| f - \sum_{j=1}^{m} g_j \cdot \chi_{A_j} \Big\|_{L^2(\Omega)} < \varepsilon$.*

Basic set	$E := \mathscr{RC}^2(\Omega, \mathbb{R}^N)$ set of square integrable random closed sets $M : \Omega \rightsquigarrow \mathbb{R}^N$ $\cong$ decomposable closed subsets of $L^2(\Omega, \mathscr{A}, P; \mathbb{R}^N)$ (as selections) with a complete probability space $(\Omega, \mathscr{A}, P)$ given
Distance	$d\!\!l_{\mathscr{RC}}(M_1, M_2)^2 \stackrel{\text{Def.}}{=} \max\Big\{ \sup_{f \in \mathscr{S}^2_{\mathscr{RC}}(M_1)} \mathbb{E}\big(\text{dist}(f, M_2)^2\big), \sup_{g \in \mathscr{S}^2_{\mathscr{RC}}(M_2)} \mathbb{E}\big(\text{dist}(g, M_1)^2\big) \Big\} \in [0, \infty]$ square of the so-called mean square Pompeiu-Hausdorff distance (Definition 61), which does not satisfy the triangle inequality
Absolute value	$\lfloor \cdot \rfloor := 0$
Transitions: Non-ω-contractive candidates	Fix an m-dimensional Wiener process $(W_t)_{t \geq 0}$ and $\Lambda > 0$. Each measurable/Λ-Lipschitz $(F_1, F_2) : \Omega \times \mathbb{R}^N \rightsquigarrow \mathbb{R}^N \times \text{Lin}(\mathbb{R}^m, \mathbb{R}^N)$ with nonempty compact and uniformly bounded values (Def. 69) induces a possibly non-ω-contractive candidate $\overline{\vartheta_F} : [0,1] \times \mathscr{RC}^2(\Omega, \mathbb{R}^N) \longrightarrow \mathscr{RC}^2(\Omega, \mathbb{R}^N)$ via the random closed reachable set of the autonomous stochastic differential inclusion $dX_t \in F_1(\cdot, X_t)\, dt + F_2(\cdot, X_t)\, dW_t$ (Def. 64)
Compactness	Not available in an obvious way here. Completeness w.r.t. $d\!\!l_{\mathscr{RC}}$ is used instead: Lemma 63 (page 246).
Equi-continuity	Nonequidistant Euler equi-continuity w.r.t. $d\!\!l^2_{\mathscr{RC}}$ results from a priori estimates for random closed reachable sets of nonautonomous stochastic differential inclusions in any interval $[0,T]$: Proposition 68 (1.) (page 249).
Mutational solutions	Random closed reachable sets of nonautonomous stochastic differential inclusions whose set-valued right-hand side is determined via feedback: $K(t) = \overline{\vartheta_{\mathscr{F}(K,\cdot)}}(t, K_0)$ P-almost surely in Ω for every t.
List of main results formulated in § 3.7	Existence for growth process (Cauchy-Lipschitz): Theorem 72 Existence for systems of growth process and nonlocal stochastic differential equation: Corollary 74 (page 253) Existence for birth-and-growth problems: Theorem 76 (p. 255) Existence for continuous or expanding nucleation: Corollary 79
Key tools	The closure of every nonempty decomposable subset of $L^2(\Omega, \mathscr{A}, P; \mathbb{R}^N)$ is the selection set of a square integrable random closed set $\Omega \rightsquigarrow \mathbb{R}^N$: Corollary 66 (page 248) Filippov-like Theorem A.67 of Da Prato and Frankowska about nonautonomous stochastic differential inclusions (page 483)

Table 3.2 Brief summary of the example in § 3.7 in mutational terms:
Stochastic morphological equations for square integrable random closed sets in $\mathbb{R}^N$

Definition 61. The *mean square Pompeiu-Hausdorff excesses* and *distance*

$$e^{\subset}_{\mathscr{RC}},\ e^{\supset}_{\mathscr{RC}},\ d\!l_{\mathscr{RC}} : \mathscr{RC}^2(\Omega,\mathbb{R}^N)\times\mathscr{RC}^2(\Omega,\mathbb{R}^N) \longrightarrow \mathbb{R}\cup\{\infty\}$$

are defined as

$$\begin{aligned} e^{\subset}_{\mathscr{RC}}(M_1,\,M_2) &:= \sup_{f\,\in\,\mathscr{S}^2_{\mathscr{RC}}(M_1)} \sqrt{\mathbb{E}\big(\mathrm{dist}(f,\,M_2)^2\big)},\\ e^{\supset}_{\mathscr{RC}}(M_1,\,M_2) &:= \sup_{g\,\in\,\mathscr{S}^2_{\mathscr{RC}}(M_2)} \sqrt{\mathbb{E}\big(\mathrm{dist}(g,\,M_1)^2\big)},\\ d\!l_{\mathscr{RC}}(M_1,\,M_2) &:= \max\big\{\,e^{\subset}_{\mathscr{RC}}(M_1,\,M_2),\ e^{\supset}_{\mathscr{RC}}(M_1,\,M_2)\big\}\,. \end{aligned}$$

Remark 62. (1.) In the following, we do not exclude explicitly that the mean square Pompeiu-Hausdorff distance between two sets $M_1, M_2 \in \mathscr{RC}^2(\Omega,\mathbb{R}^N)$ is ∞. Indeed, there are some standard transformations for metrics such as $\frac{d_{\mathscr{RC}}(M_1,M_2)}{1+d_{\mathscr{RC}}(M_1,M_2)}$, but they do not provide any additional insight here.

(2.) In combination with Selection Theorem A.74 of Kuratowski and Ryll-Nardzewski (on page 489), Lemma 60 ensures for any sets $M_1, M_2 \in \mathscr{RC}^2(\Omega,\mathbb{R}^N)$ and a sequence $(f_n)_{n\in\mathbb{N}}$ in $L^2(\Omega,\mathscr{A},P;\,\mathbb{R}^N)$ with

$$M_1(\omega) = \overline{\bigcup_{n\in\mathbb{N}} f_n(\omega)}, \qquad \text{for every } \omega\in\Omega:$$

$$\begin{aligned} \sup_{f\,\in\,\mathscr{S}^2_{\mathscr{RC}}(M_1)} \mathbb{E}\big(\mathrm{dist}(f,\,M_2)^2\big) &= \sup_{f\,\in\,\mathscr{S}^2_{\mathscr{RC}}(M_1)}\ \inf_{g\,\in\,\mathscr{S}^2_{\mathscr{RC}}(M_2)} \mathbb{E}\big(|f-g|^2\big)\\ &= \sup_{m\,\in\,\mathbb{N}}\ \inf_{g\,\in\,\mathscr{S}^2_{\mathscr{RC}}(M_2)} \big\|f_m - g\big\|^2_{L^2(\Omega;\mathbb{R}^N)}\,. \end{aligned}$$

As a first consequence, we obtain for any three sets $M_1, M_1', M_2 \in \mathscr{RC}^2(\Omega,\mathbb{R}^N)$

$$e^{\subset}_{\mathscr{RC}}\big(M_1\cup M_1',\,M_2\big) = \max\big\{e^{\subset}_{\mathscr{RC}}(M_1,\,M_2),\ e^{\subset}_{\mathscr{RC}}(M_1',\,M_2)\big\}$$

which will be of technical use for birth processes in § 3.7.4. More significantly here,

$$\begin{aligned} d\!l_{\mathscr{RC}}(M_1,\,M_2) &= \max\Big\{ \sup_{f\,\in\,\mathscr{S}^2_{\mathscr{RC}}(M_1)}\ \inf_{g\,\in\,\mathscr{S}^2_{\mathscr{RC}}(M_2)} \big\|f-g\big\|_{L^2(\Omega;\mathbb{R}^N)},\\ &\qquad\qquad \sup_{g\,\in\,\mathscr{S}^2_{\mathscr{RC}}(M_2)}\ \inf_{f\,\in\,\mathscr{S}^2_{\mathscr{RC}}(M_1)} \big\|f-g\big\|_{L^2(\Omega;\mathbb{R}^N)} \Big\}\\ &= \sup\Big\{ \Big|\inf_{f\,\in\,\mathscr{S}^2_{\mathscr{RC}}(M_1)} \|h-f\|_{L^2} - \inf_{g\,\in\,\mathscr{S}^2_{\mathscr{RC}}(M_2)} \|h-g\|_{L^2}\Big| \ \Big|\ h\in L^2(\Omega;\mathbb{R}^N)\Big\} \end{aligned}$$

is a metric on $\mathscr{RC}^2(\Omega,\mathbb{R}^N)$ — with values in $\mathbb{R}^+_0\cup\{\infty\}$ though.

According to a general result about the Hausdorff metric topology of nonempty closed subsets in any metric space [27, Theorem 3.2.4 (1.)] [40, Theorem II.3], the completeness of $L^2(\Omega,\mathscr{A},P;\,\mathbb{R}^N)$ implies directly:

Lemma 63. *$\mathscr{RC}^2(\Omega,\mathbb{R}^N)$ is complete with respect to $d\!l_{\mathscr{RC}}$.*

3.7.2 Reachable Sets of Stochastic Differential Inclusions are to Induce Transitions on $\mathscr{RC}^2(\Omega, \mathbb{R}^N)$

In the deterministic case, morphological transitions on $(\mathscr{K}(\mathbb{R}^N), dl)$ are based on reachable sets of autonomous differential inclusions, i.e.

$$\vartheta_F(t,\ K_0) := \big\{ x(t) \in \mathbb{R}^N \ \big|\ \text{there exists } x(\cdot) \in W^{1,1}([0,t], \mathbb{R}^N) : \\ x'(\cdot) \in F(x(\cdot))\ \ \mathscr{L}^1\text{-a.e. in } [0,t],\ \ x(0) \in K_0 \big\}$$

for a given set-valued map $F : \mathbb{R}^N \rightsquigarrow \mathbb{R}^N$ according to Definition 1.48 (on page 60). Filippov's Theorem A.6 about differential inclusions implies the properties required for transitions if $F \in \text{LIP}(\mathbb{R}^N, \mathbb{R}^N)$, that means, if $F : \mathbb{R}^N \rightsquigarrow \mathbb{R}^N$ is a Lipschitz continuous set-valued map with nonempty compact and uniformly bounded values.

Now we extend this approach to stochastic differential inclusions with a given m-dimensional Wiener process $W = (W_t)_{t \geq 0}$.
The definition of "reachable set" uses square integrable strong solutions and then, Theorem A.67 of Da Prato-Frankowska (on page 483) is the key tool in regard to transition properties. Hence, as in § A.8, $\mathfrak{I}_0(X_0, \gamma, \sigma) : [0,T] \times \Omega \longrightarrow \mathbb{R}^N$ denotes the Itô process associated with

- the initial state $X_0 \in L^2(\Omega, \mathscr{A}, P;\ \mathbb{R}^N)$,
- the drift $\gamma \in \mathscr{L}^2_{\mathscr{A}}([0,T],\ \mathbb{R}^N)$ and
- the diffusion $\sigma \in \mathscr{L}^2_{\mathscr{A}}([0,T],\ \text{Lin}(\mathbb{R}^m, \mathbb{R}^N))$,

i.e. for $t \in [0,T], \omega \in \Omega$,

$$\mathfrak{I}_0(X_0, \gamma, \sigma)(t, \omega) \ := \ X_0(\omega) + \int_0^t \gamma(s, \omega)\ ds + \int_0^t \sigma(s, \omega)\ dW_s(\omega).$$

Definition 64 (Random reachable set).
Consider an m-dim. Wiener process $W = (W_t)_{t \geq 0}$ on the complete probability space $(\Omega, \mathscr{A}, P)$ and a set-valued map $\widetilde{F} = (\widetilde{F}_1, \widetilde{F}_2) : [0,T] \times \Omega \times \mathbb{R}^N \rightsquigarrow \mathbb{R}^N \times \text{Lin}(\mathbb{R}^m, \mathbb{R}^N)$. The *random reachable set* $\vartheta_{\widetilde{F}}(t, M_0) : \Omega \rightsquigarrow \mathbb{R}^N$ of the stochastic differential inclusion

$$dX_t \ \in\ \widetilde{F}_1(t, \cdot, X_t)\ dt \ +\ \widetilde{F}_2(t, \cdot, X_t)\ dW_t$$

and initial set $M_0 \in \mathscr{RC}^2(\Omega, \mathbb{R}^N)$ at time $t \in [0,T]$ is defined via strong solutions as

$$\vartheta_{\widetilde{F}}(t, M_0) := \big\{ X_t \ \big|\ \exists\ X_0 \in L^2(\Omega, \mathscr{A}, P;\ \mathbb{R}^N),\ \ \gamma \in \mathscr{L}^2_{\mathscr{A}}([0,T],\ \mathbb{R}^N), \\ \sigma \in \mathscr{L}^2_{\mathscr{A}}([0,T],\ \text{Lin}(\mathbb{R}^m, \mathbb{R}^N)) : \\ X = \mathfrak{I}_0(X_0, \gamma, \sigma), \quad X_0 \in M_0\ P\text{-almost surely} \\ \text{and for } (\mathscr{L}^1 \times P)\text{-almost all } (s, \widetilde{\omega}) \in [0,t] \times \Omega, \\ \gamma(s, \widetilde{\omega}) \in \widetilde{F}_1\big(s, \widetilde{\omega}, X_s(\widetilde{\omega})\big), \\ \sigma(s, \widetilde{\omega}) \in \widetilde{F}_2\big(s, \widetilde{\omega}, X_s(\widetilde{\omega})\big) \ \big\}.$$

The *random closed reachable set* $\overline{\vartheta_{\widetilde{F}}}(t, M_0) : \Omega \rightsquigarrow \mathbb{R}^N$ is the random closed set whose set of selections in $L^2(\Omega, \mathscr{A}, P;\ \mathbb{R}^N)$ coincides with the closure of all these solutions $X_t \in L^2(\Omega, \mathscr{A}, P;\ \mathbb{R}^N)$ starting in M_0 P-almost surely.

Obviously, $\overline{\vartheta_{\widetilde{F}}}(0,M_0) = M_0$ holds for every $M_0 \in \mathscr{RC}^2(\Omega,\mathbb{R}^N)$ and set-valued map $\widetilde{F} : [0,T] \times \mathbb{R}^N \rightsquigarrow \mathbb{R}^N \times \mathrm{Lin}(\mathbb{R}^m,\mathbb{R}^N)$ because M_0 is characterized by $\mathscr{S}^2_{\mathscr{RC}}(M_0) \neq \emptyset$.

The next step is to verify that $\overline{\vartheta_{\widetilde{F}}}(t,M_0) : \Omega \rightsquigarrow \mathbb{R}^N$ is well-defined in $\mathscr{RC}^2(\Omega,\mathbb{R}^N)$ – essentially due to pathwise structure of the stochastic differential inclusion. For this purpose, we quote a general criterion for subsets of $L^2(\Omega,\mathscr{A},P;\,\mathbb{R}^N)$ inducing random closed sets:

Proposition 65 (Decomposable sets and selections [141, Theorem 2.1.6]**).**
Let Ξ be a nonempty closed subset of $L^2(\Omega,\mathscr{A},P;\,\mathbb{R}^N)$.
Then there exists a random closed set $M : \Omega \rightsquigarrow \mathbb{R}^N$ with $\mathscr{S}^2_{\mathscr{RC}}(M) = \Xi$ if and only if Ξ is decomposable *in the following sense: For any $\xi_1,\xi_2 \in \Xi$ and $A \in \mathscr{A}$, the function $\chi_A \cdot \xi_1 + (1-\chi_A)\cdot \xi_2 : \Omega \longrightarrow \mathbb{R}^N$ also belongs to Ξ (with $\chi_A : \Omega \longrightarrow \{0,1\}$ denoting the characteristic function of $A \subset \Omega$).*

The proof of this equivalence as it is presented in the monograph of Molchanov [141], for example, provides the following implication for the case that the nonempty set $\Xi \subset L^2(\Omega,\mathscr{A},P;\,\mathbb{R}^N)$ is *not* assumed to be closed:

Corollary 66. *Let Ξ be a nonempty subset of $L^2(\Omega,\mathscr{A},P;\,\mathbb{R}^N)$.*
If Ξ is decomposable then there exists a square integrable random closed set $M : \Omega \rightsquigarrow \mathbb{R}^N$ whose set $\mathscr{S}^2_{\mathscr{RC}}(M)$ of L^2 selections is equal to the closure of Ξ.

Corollary 67. *For every $\widetilde{F} : [0,T] \times \mathbb{R}^N \rightsquigarrow \mathbb{R}^N \times \mathrm{Lin}(\mathbb{R}^m,\mathbb{R}^N)$, $M_0 \in \mathscr{RC}^2(\Omega,\mathbb{R}^N)$ and $t \in [0,T]$, the set-valued map $\overline{\vartheta_{\widetilde{F}}}(t,M_0) : \Omega \rightsquigarrow \mathbb{R}^N$ defined via the L^2 closure of solutions $X_t \in L^2(\Omega,\mathscr{A},P;\,\mathbb{R}^N)$ starting in M_0 is a square integrable random closed set.*

Proof. At time $t \in [0,T]$ consider the set Ξ of all $X_t \in L^2(\Omega,\mathscr{A},P;\,\mathbb{R}^N)$ induced by an Itô process $X = \mathfrak{I}_0(X_0,\gamma,\sigma)$ which solves the stochastic differential inclusion $dX_s \in \widetilde{F}_1(s,\cdot,X_s)\,ds + \widetilde{F}_2(s,\cdot,X_s)\,dW_s$ and starts in M_0, i.e. $X_0 \in \mathscr{S}^2_{\mathscr{RC}}(M_0)$ and for $(\mathscr{L}^1 \times P)$-almost all $(s,\omega) \in [0,T]\times\Omega$,

$$\begin{cases} \gamma(s,\omega) \in \widetilde{F}_1\big(s,\,\omega,\,X_s(\omega)\big), \\ \sigma(s,\omega) \in \widetilde{F}_2\big(s,\,\omega,\,X_s(\omega)\big). \end{cases}$$

This subset Ξ is decomposable because each integral related to the Itô process $\mathfrak{I}_0(X_0,\gamma,\sigma)$ is evaluated pathwise. Hence there exists a set $M_t \in \mathscr{RC}^2(\Omega,\mathbb{R}^N)$ with $\mathscr{S}^2_{\mathscr{RC}}(M_t) = \overline{\Xi} \subset L^2(\Omega,\mathscr{A},P;\,\mathbb{R}^N)$ and by definition, $\overline{\vartheta_{\widetilde{F}}}(t,M_0) := M_t$. □

Now the continuity properties of $\overline{\vartheta_{\widetilde{F}}} : [0,T] \times \mathscr{RC}^2(\Omega,\mathbb{R}^N) \longrightarrow \mathscr{RC}^2(\Omega,\mathbb{R}^N)$ decide whether random closed reachable sets induce transitions on $\mathscr{RC}^2(\Omega,\mathbb{R}^N)$ with respect to the mean square Pompeiu-Hausdorff distance $dl_{\mathscr{RC}}$.
The Filippov-like Theorem A.67 of Da Prato-Frankowska (on page 483) is the analytical basis for comparing the evolution of two initial sets in $\mathscr{RC}^2(\Omega,\mathbb{R}^N)$ — in essentially the same way as we used Filippov's Theorem A.6 for morphological transitions in Proposition 1.50 (on page 60).

Proposition 68. *Let $\widetilde{F},\widetilde{G} : [0,T] \times \Omega \times \mathbb{R}^N \leadsto \mathbb{R}^N \times \mathrm{Lin}(\mathbb{R}^m,\mathbb{R}^N)$ satisfy*
(i) $\widetilde{F},\widetilde{G}$ have nonempty compact values,
(ii) for every $x \in \mathbb{R}^N$, $\widetilde{F}(\cdot,\cdot,x)$ and $\widetilde{G}(\cdot,\cdot,x)$ are measurable,
(iii) $\exists\, \Lambda > 0$: for any $t \in [0,T], \omega \in \Omega$: $\widetilde{F}(t,\omega,\cdot)$, $\widetilde{G}(t,\omega,\cdot)$ are Λ-Lipschitz,
(iv) $\exists\, \widehat{\gamma} > 0$: $\max\{|\widetilde{F}(t,\omega,x)|_\infty, |\widetilde{G}(t,\omega,x)|_\infty\} \le \widehat{\gamma}$ holds for all t,ω,x.

Then the following statements hold for any $0 \le s \le t$ and $M, M_1, M_2 \in \mathscr{RC}^2(\Omega,\mathbb{R}^N)$

(1.) $dl_{\mathscr{RC}}\big(\overline{\vartheta_{\widetilde{F}}}(s,M),\ \overline{\vartheta_{\widetilde{F}}}(t,M)\big)^2 \ \le\ 18\ \widehat{\gamma}^2\ e^t \cdot (t-s)$

(2.) $$dl_{\mathscr{RC}}\big(\overline{\vartheta_{\widetilde{F}}}(t,M_1),\ \overline{\vartheta_{\widetilde{G}}}(t,M_2)\big)^2 \le C \cdot \Big(dl_{\mathscr{RC}}(M_1,M_2)^2 + \int_0^t \sup_{Y \in L^2} dl_{\mathscr{RC}}\big(\widetilde{F}(s,\cdot,Y),\widetilde{G}(s,\cdot,Y)\big)^2 ds\Big) \cdot e^{C\cdot(2+t)t}$$

with a constant $C \ge 1$ depending only on Λ.

The proof will be given in a moment. We now focus on the consequences:
Firstly, the estimates reveal that Lipschitz continuity with respect to time is verified for the *square* distance $dl_{\mathscr{RC}}(\cdot,\cdot)^2$ (rather than the metric $dl_{\mathscr{RC}}$). Secondly, random closed reachable sets of autonomous stochastic differential inclusions might not be ω-contractive with respect to $dl^2_{\mathscr{RC}}$ because the constant C in the second estimate could be larger than 1.
Statement (2.), however, concerns *nonautonomous* stochastic differential inclusions and, it enables us to bridge this gap by means of the distance construction in § 3.4 (on page 214 ff.). Indeed, the arguments for Proposition 32 (in the simplified case $\lfloor\cdot\rfloor \equiv 0$) lead directly to the following result about autonomous stochastic differential inclusions with "measurable/Lipschitz" right-hand side:

Definition 69. For $\Lambda > 0$ fixed, $\mathrm{MLIP}_\Lambda(\Omega,\mathbb{R}^N;\mathbb{R}^{m_1},\mathbb{R}^{m_2\times m_3})$ consists of all set-valued maps $F = (F_1,F_2) : \Omega \times \mathbb{R}^N \leadsto \mathbb{R}^{m_1} \times \mathbb{R}^{m_2\times m_3}$ with the following properties:

(1.) F has nonempty compact values,
(2.) for every $x \in \mathbb{R}^N$, $F(\cdot,x)$ is measurable,
(3.) for every $\omega \in \Omega$, $F_1(\omega,\cdot)$ and $F_2(\omega,\cdot)$ are Λ-Lipschitz continuous,
(4.) $\|F\|_\infty \stackrel{\text{Def.}}{=} \sup\{|F(\omega,x)|_\infty \mid \omega \in \Omega,\, x \in \mathbb{R}^N\} < \infty.$

Corollary 70. *Fix the parameter $\Lambda \geq 0$ arbitrarily. Then there exist*

$$\check{d}^2_{\mathscr{RC}} : \mathscr{RC}^2(\Omega, \mathbb{R}^N) \times \mathscr{RC}^2(\Omega, \mathbb{R}^N) \longrightarrow [0, \infty[$$

and a constant $\check{c} \geq 1$ such that for any $F, G \in \mathrm{MLIP}_\Lambda\big(\Omega, \mathbb{R}^N;\, \mathbb{R}^N, \mathrm{Lin}(\mathbb{R}^m, \mathbb{R}^N)\big)$ and all sets $M_1, M_2 \in \mathscr{RC}^2(\Omega, \mathbb{R}^N)$, $t_1, t_2, h \geq 0$ with $t_1 + h$, $t_2 + h \leq 1$

(1.) $d\!l_{\mathscr{RC}}(\cdot,\cdot)^2 \leq \check{d}^2_{\mathscr{RC}}(\cdot,\cdot) \leq \check{c} \cdot d\!l_{\mathscr{RC}}(\cdot,\cdot)^2$

(2.) $\check{d}^2_{\mathscr{RC}}\big(\overline{\vartheta_F}(t_1{+}h, M_1),\ \overline{\vartheta_F}(t_2{+}h, M_2)\big) \leq \check{d}^2_{\mathscr{RC}}\big(\overline{\vartheta_F}(t_1, M_1),\ \overline{\vartheta_F}(t_2, M_2)\big) \cdot e^{\check{c}\,h}$,

(3.) $\check{d}^2_{\mathscr{RC}}\big(\overline{\vartheta_F}(t_1{+}h, M_1),\ \overline{\vartheta_G}(t_2{+}h, M_2)\big)$
$$\leq \Big(\check{d}^2_{\mathscr{RC}}\big(\overline{\vartheta_F}(t_1, M_1),\ \overline{\vartheta_G}(t_2, M_2)\big) + h \cdot \check{c} \cdot d\!l_\infty\big(F, G\big)^2\Big) \cdot e^{\check{c}\,h}.$$

Random closed reachable sets of each map $F \in \mathrm{MLIP}_\Lambda\big(\Omega, \mathbb{R}^N;\, \mathbb{R}^N, \mathrm{Lin}(\mathbb{R}^m, \mathbb{R}^N)\big)$ induce transitions on the tuple $\big(\mathscr{RC}^2(\Omega, \mathbb{R}^N),\, \check{d}^2_{\mathscr{RC}},\, d\!l^2_{\mathscr{RC}},\, 0\big)$.

Proof (of Proposition 68).
(1.) Remark A.66 (on page 483) mentions the general estimate

$$\mathbb{E}\big(\big|\mathfrak{I}_0(X_0, \gamma, \sigma)\,(t)\big|^2\big) \;\leq\; 9\ e^t\ \big\|\mathfrak{I}_0(X_0, \gamma, \sigma)\big\|^2_{\mathfrak{I},[0,t]}$$
$$\stackrel{\text{Def.}}{=} 9\ e^t\ \Big(\mathbb{E}(|X_0|^2) + \mathbb{E}\Big(\int_0^t |\gamma|^2\,ds\Big) + \mathbb{E}\Big(\int_0^t |\sigma|^2\,ds\Big)\Big)$$

for any initial $X_0 \in L^2(\Omega, \mathscr{A}, P; \mathbb{R}^N)$, drift $\gamma \in \mathscr{L}^2_{\mathscr{A}}([0,T], \mathbb{R}^N)$ and diffusion $\sigma \in \mathscr{L}^2_{\mathscr{A}}([0,T], \mathrm{Lin}(\mathbb{R}^m, \mathbb{R}^N))$. This implies for a solution

$$X = \mathfrak{I}_0(X_s, \gamma, \sigma) : [s,t] \longrightarrow L^2(\Omega, \mathscr{A}, P; \mathbb{R}^N)$$

to the stochastic differential inclusion $dX_t \in \widetilde{F}_1(t, X_t)\,dt + \widetilde{F}_2(t, X_t)\,dW_t$ starting in X_s at time s

$$\mathbb{E}\big(\big|X_t - X_s\big|^2\big) \;\leq\; 9\ e^t \cdot \big\|\mathfrak{I}_0(0, \gamma, \sigma)\big\|^2_{\mathfrak{I},[s,t]}$$
$$\leq 9\ e^t \cdot \Big(\mathbb{E}\Big(\int_s^t |\gamma|^2\,ds\Big) + \mathbb{E}\Big(\int_s^t |\sigma|^2\,ds\Big)\Big)$$
$$\leq 9\ e^t \cdot 2\,\widehat{\gamma}^2\,(t-s).$$

(2.) Due to the symmetry of $d\!l_{\mathscr{RC}}$ and the definition of $\overline{\vartheta_{\widetilde{F}}}(t, M_1)$ via L^2 closure, it is sufficient to prove for every solution $X = \mathfrak{I}_0(X_0, \gamma, \sigma) : [0,t] \longrightarrow L^2(\Omega, \mathscr{A}, P; \mathbb{R}^N)$ to the stochastic differential inclusion $dX_s \in \widetilde{F}_1(s, \cdot\, X_s)\,ds + \widetilde{F}_2(s, \cdot\, X_s)\,dW_s$ starting in a selection $X_0 \in L^2(\Omega, \mathscr{A}, P;\, \mathbb{R}^N)$ of $M_1 \in \mathscr{RC}^2(\Omega, \mathbb{R}^N)$:

$$\mathbb{E}\big(\mathrm{dist}(X_t,\ \vartheta_{\widetilde{G}}(t, M_2))^2\big)$$
$$\leq 9\,C\Big(\mathbb{E}\big(\mathrm{dist}(X_0, M_2)^2\big) + \int_0^t d\!l_{\mathscr{RC}}\big(\widetilde{F}(s, \cdot, X_s),\ \widetilde{G}(s, \cdot, X_s)\big)^2\,ds\Big) \cdot e^{C\,(2+t)\,t}$$

with a constant $C \geq 1$ depending only on Λ.

Choose a selection $Y_0 \in \mathscr{S}^2_{\mathscr{RC}}(M_2)$ with $\|X_0 - Y_0\|^2_{L^2} = \mathbb{E}\big(\mathrm{dist}(X_0, M_2)^2\big)$ by means of Proposition A.80 about marginal maps and Selection Theorem A.74 (on page 489 f.).

According to Theorem A.67 of Da Prato-Frankowska (on page 483), there exist a constant $C = C(\Lambda) \geq 1$ and a strong solution $Y : [0,t] \longrightarrow L^2(\Omega, \mathscr{A}, P; \mathbb{R}^N)$ of

$$dY_s \in \widetilde{G}_1(s, \cdot, Y_s)\, ds + \widetilde{G}_2(s, \cdot, Y_s)\, dW_s$$

starting in Y_0 such that

$$\|Y - X\|^2_{\mathfrak{I},[0,t]} \leq C \cdot \Big(\mathbb{E}\big(|Y_0 - X_0|^2\big) + \int_0^t \mathbb{E}\Big(\mathrm{dist}\big((\gamma(s,\cdot),\sigma(s,\cdot)),\ \widetilde{G}(s,\cdot,X_s)\big)^2\Big)\, ds\Big) \cdot e^{C\cdot(1+t)t}$$

Remark A.66 (on page 483) and the definition of $d\!l_{\mathscr{RC}}(\cdot,\cdot)^2$ via L^2 selections imply

$$\begin{aligned}\mathbb{E}\big(\mathrm{dist}(X_t,\ \vartheta_{\widetilde{G}}(t,M_2))^2\big) &\leq \mathbb{E}\big(|Y_t - X_t|^2\big)\\ &\leq 9\, e^t\ C \cdot \Big(\mathbb{E}\big(\mathrm{dist}(X_0,M_2)^2\big) + \int_0^t d\!l_{\mathscr{RC}}\Big(\widetilde{F}(s,\cdot,X_s),\ \widetilde{G}(s,\cdot,X_s)\Big)^2\, ds\Big) \cdot e^{C\cdot(1+t)\,t}.\end{aligned}$$

□

3.7.3 The Main Conclusions About Stochastic Growth Processes

We have just laid the foundations for applying the mutational framework to the tuple $\big(\mathscr{RC}^2(\Omega,\mathbb{R}^N), \check{d}^2_{\mathscr{RC}}, d\!l^2_{\mathscr{RC}}, 0\big)$ and the random closed reachable sets $\overline{\vartheta_F}(\cdot,\cdot)$ induced by the set-valued maps in $\mathrm{MLIP}_\Lambda\big(\Omega, \mathbb{R}^N;\ \mathbb{R}^N,\ \mathrm{Lin}(\mathbb{R}^m,\mathbb{R}^N)\big)$ with an initially fixed Lipschitz bound $\Lambda \geq 0$.
Proposition 68 and Corollary 70 indicate two essential features: Firstly, the transition parameters can be chosen as a function of the parameter Λ and secondly, the corresponding transition distance is defined as $\check{c}(\Lambda) \cdot d\!l_\infty(\cdot,\cdot)^2 < \infty$.

Now we can handle the stochastic counterparts of morphological equations (originally introduced by Aubin for $(\mathscr{K}(\mathbb{R}^N), d\!l)$ and discussed in § 1.9 on page 57 ff.), i.e. the evolution of square integrable random closed sets is prescribed as a function of their current shape.
The geometric interpretation of a solution plays an important role again. In regard to (deterministic) morphological equations, Proposition 1.57 (on page 64) provides the equivalent characterization as reachable sets of *nonautonomous* differential inclusions in $\mathbb{R}^N$. Now a similar link with nonautonomous *stochastic* differential inclusions results directly from Proposition 34 (on page 219) due to Proposition 68:

Lemma 71. *Suppose* $\mathscr{F} : [0,T] \longrightarrow \big(\mathrm{MLIP}_\Lambda\big(\Omega,\mathbb{R}^N;\ \mathbb{R}^N, \mathrm{Lin}(\mathbb{R}^m,\mathbb{R}^N)\big),\ d\!l_\infty\big)$ *to be continuous with* $\sup_{[0,T]} \|\mathscr{F}(t)\|_\infty < \infty$ *and define the set-valued map*

$$\widehat{F}:\ [0,T] \times \Omega \times \mathbb{R}^N \leadsto \mathbb{R}^N \times \mathrm{Lin}(\mathbb{R}^m,\mathbb{R}^N), \qquad (t,\omega,x) \mapsto \mathscr{F}(t)(\omega,x).$$

Then for any $K_0 \in \mathscr{RC}^2(\Omega,\mathbb{R}^N)$, *the random closed reachable set* $K(\cdot) := \overline{\vartheta_{\widehat{F}}}(\cdot,K_0)$: $[0,T] \longrightarrow \mathscr{RC}^2(\Omega,\mathbb{R}^N)$ *is a solution to the mutational equation* $\overset{\circ}{K}(t) \ni \mathscr{F}(t)$ *in the tuple* $\big(\mathscr{RC}^2(\Omega,\mathbb{R}^N),\ \check{d}^2_{\mathscr{RC}},\ d\!l^2_{\mathscr{RC}},\ 0,\ \check{c}(\Lambda)\cdot d\!l_\infty(\cdot,\cdot)^2\big)$.

The extended Cauchy-Lipschitz Theorem 31 (on page 212) guarantees existence of solutions due to completeness:

Theorem 72 (Cauchy-Lipschitz Theorem for stochastic morpholog. equations). *Assume for $\Lambda > 0$ and*

$$\mathscr{F} = (\mathscr{F}_1, \mathscr{F}_2) : \mathscr{RC}^2(\Omega, \mathbb{R}^N) \times [0,T] \longrightarrow \mathrm{MLIP}_\Lambda\big(\Omega, \mathbb{R}^N; \mathbb{R}^N, \mathrm{Lin}(\mathbb{R}^m, \mathbb{R}^N)\big) :$$

(1.) *there exists $\widehat{\gamma} > 0$ such that* $\sup \|\mathscr{F}(\cdot)\|_\infty \le \widehat{\gamma}$,

(2.) *$\mathscr{F}$ is Lipschitz continuous w.r.t. state and continuous in the following sense: There exist a constant $\lambda > 0$ and a modulus of continuity $\omega(\cdot)$ such that for any $M_1, M_2 \in \mathscr{RC}^2(\Omega, \mathbb{R}^N)$ and $t_1, t_2 \in [0,T]$,*

$$d\!l_\infty\big(\mathscr{F}(M_1,t_2),\ \mathscr{F}(M_2,t_2)\big)^2 \ \le\ \lambda \cdot d\!l_{\mathscr{RC}}\big(M_1, M_2\big)^2 \ +\ \omega(|t_1 - t_2|).$$

Then, for any initial set $K_0 \in \mathscr{RC}^2(\Omega, \mathbb{R}^N)$, there exists a unique curve $K(\cdot) : [0,T] \longrightarrow \mathscr{RC}^2(\Omega, \mathbb{R}^N)$ which is Lipschitz continuous with respect to $d\!l^2_{\mathscr{RC}}$ and has the two following (equivalent) properties:

(*i*) *for Lebesgue-almost every $t \in [0,T]$:* $\overset{\circ}{K}(t) \ni \mathscr{F}(K(t), t)$,

i.e. for any random closed set $M \in \mathscr{RC}^2(\Omega, \mathbb{R}^N)$ and bounded set-valued map $G \in \mathrm{MLIP}_\Lambda\big(\Omega, \mathbb{R}^N; \mathbb{R}^N, \mathrm{Lin}(\mathbb{R}^m, \mathbb{R}^N)\big)$, it satisfies

$$\begin{aligned}\limsup_{h \downarrow 0}\ \tfrac{1}{h} \cdot \Big(\check{d}^2_{\mathscr{RC}}\big(\overline{\vartheta_G}(h,M),\ K(t+h)\big) \ -\ \check{d}^2_{\mathscr{RC}}\big(M,\, K(t)\big) \cdot e^{\check{c}\,h}\Big)\\ \le\ \check{c} \cdot d\!l_\infty\big(G,\ \mathscr{F}(K(t),t)\big)^2\end{aligned}$$

where $\check{d}^2_{\mathscr{RC}} : \mathscr{RC}^2(\Omega, \mathbb{R}^N) \times \mathscr{RC}^2(\Omega, \mathbb{R}^N) \longrightarrow [0,\infty[$ denotes the distance function and $\check{c} = \check{c}(\Lambda)$ the constant mentioned in Corollary 70 (on page 250),

(*ii*) *for every $t \in [0,T]$:* $K(t) \ =\ \overline{\vartheta_{\mathscr{F}(K,\cdot)}(t, K_0)}$,

i.e. $K(t)$ always coincides with the random closed reachable set of the non-autonomous stochastic differential inclusion

$$dX_t(\omega)\ \in\ \mathscr{F}_1(K(t),t)\,(\omega, X_t(\omega))\ dt\ +\ \mathscr{F}_2(K(t),t)\,(\omega, X_t(\omega))\ dW_t(\omega)$$

and initial set $K_0 \in \mathscr{RC}^2(\Omega, \mathbb{R}^N)$ (in the sense of Definition 64 on page 247) P-almost everywhere in Ω.

Remark 73. Property (i) generalizes the criterion for the time derivative of a curve in $\mathscr{RC}^2(\Omega, \mathbb{R}^N)$ as we have already discussed in the beginning of § 3.3.

Property (ii) can be regarded as "integral" characterization. In a figurative sense, it is the set-valued counterpart of the well-known integral criterion for absolutely continuous functions $[0,T] \longrightarrow \mathbb{R}^N$.

An advantage of this existence result by means of the mutational framework is that we are immediately free to combine these stochastic growth processes in $\mathscr{RC}^2(\Omega, \mathbb{R}^N)$ with any other example.
Here we select exemplarily a system with a nonlocal stochastic differential equation. This kind of dynamic problem occurs in models where a stochastic differential equation depends on an uncertain control parameter whose control set in $\mathscr{RC}^2(\Omega, \mathbb{R}^N)$ is prescribed in a closed-loop way.
For the sake of simplicity (merely), we restrict the next corollary to an autonomous system with $\mathscr{A}_t = \mathscr{A}$ for all $t \in [0,T]$.

Alternatively, we can apply the same existence arguments to systems with nonlocal reaction-diffusion equations in fixed cylindrical domains (discussed in § 3.9 below), for example.

Corollary 74 (System with a nonlocal stochastic differential equation).
Assume for $\Lambda > 0$ *and*

$$\begin{aligned}
\mathbf{f} = (f_1,\ f_2) &: L^2(\Omega;\mathbb{R}) \times \mathscr{RC}^2(\Omega,\mathbb{R}^N) \longrightarrow W^{1,\infty}(\mathbb{R},\mathbb{R}) \times W^{1,\infty}(\mathbb{R},\mathbb{R}),\\
\mathscr{F} = (\mathscr{F}_1, \mathscr{F}_2) &: L^2(\Omega;\mathbb{R}) \times \mathscr{RC}^2(\Omega,\mathbb{R}^N) \longrightarrow \mathrm{MLIP}_\Lambda\big(\Omega, \mathbb{R}^N; \mathbb{R}^N, \mathrm{Lin}(\mathbb{R}^m, \mathbb{R}^N)\big):
\end{aligned}$$

(*1.*) $\sup\limits_{(Y,M)} \big(\|\mathbf{f}(Y,M)\|_{W^{1,\infty}} + \|\mathscr{F}(Y,M)\|_\infty\big) < \infty,$

(*2.*) $\mathbf{f}$ *and* $\mathscr{F}$ *are Lipschitz continuous in the following sense:*
There exists a constant $\lambda > 0$ *such that for all* $Y_1, Y_2 \in L^2(\Omega, \mathscr{A}, P; \mathbb{R})$ *and* $M_1, M_2 \in \mathscr{RC}^2(\Omega, \mathbb{R}^N)$,

$$\begin{aligned}
&\big\|\mathbf{f}(Y_1,M_1)(\cdot) - \mathbf{f}(Y_2,M_2)(\cdot)\big\|_{L^\infty}^2 + d\!l_\infty\big(\mathscr{F}(Y_1,M_1),\ \mathscr{F}(Y_2,M_2)\big)^2\\
&\quad \le \lambda \cdot \Big(\mathbb{E}\big(|Y_1 - Y_2|^2\big) + d\!l_{\mathscr{RC}}(M_1,M_2)^2\Big).
\end{aligned}$$

Then for any initial states $X_0 \in L^2(\Omega, \mathscr{A}, P; \mathbb{R})$ *and* $K_0 \in \mathscr{RC}^2(\Omega, \mathbb{R}^N)$, *there exist unique curves* $X : [0,T] \longrightarrow L^2(\Omega, \mathscr{A}, P; \mathbb{R})$ *and* $K : [0,T] \longrightarrow \mathscr{RC}^2(\Omega, \mathbb{R}^N)$ *with the following properties:*

(*i*) $X : [0,T] \longrightarrow L^2(\Omega, \mathscr{A}, P;\ \mathbb{R})$ *is Lipschitz continuous w.r.t.* $\mathbb{E}\big(|\cdot - \cdot|^2\big)$,
$K : [0,T] \longrightarrow \mathscr{RC}^2(\Omega, \mathbb{R}^N)$ *is Lipschitz continuous w.r.t.* $d\!l_{\mathscr{RC}}(\cdot,\cdot)^2$.

(*ii*) *X is a strong solution to the stochastic differential equation*

$$dX_s(\omega) = f_1(X_s, K(s))\,(X_s(\omega))\,ds + f_2(X_s, K(s))\,(X_s(\omega))\,dW_s(\omega)$$

(*iii*) *for every* $t \in [0,T]$, $K(t) \in \mathscr{RC}^2(\Omega, \mathbb{R}^N)$ *coincides with the random closed reachable set of the stochastic differential inclusion*

$$dY_s(\omega) \in \mathscr{F}_1(X_s, K(s))\,\big(\omega, Y_s(\omega)\big)\,ds + \mathscr{F}_2(X_s, K(s))\,\big(\omega, Y_s(\omega)\big)\,dW_s(\omega)$$

and the initial set K_0 *(P-almost everywhere in* Ω*).*

3.7.4 Extensions to Stochastic Birth-and-Growth Processes

The evolution of random closed sets along stochastic differential inclusions covers a quite broad class of growth processes. In particular, it is not restricted to convex-valued or expanding set evolutions in $\mathscr{RC}^2(\Omega,\mathbb{R}^N)$.
Theorem 72 and Corollary 74 do not consider any form of *nucleation* though. Roughly speaking, the growth process is initiated completely by the initial random closed set $K_0 \in \mathscr{RC}^2(\Omega,\mathbb{R}^N)$ at time $t=0$ — and not by some additional random process starting maybe elsewhere and later.

In this section, we suggest how this restriction can be overcome in some cases by means of approximation. Obviously, there is no significant difficulty in solving the following problem in a piecewise way:

$$\begin{cases} \overset{\circ}{K}(\cdot) \ni \mathscr{F}\big(K(\cdot),\cdot\big) & \mathscr{L}^1\text{-a.e. in } [0,T] \\ K(0) = K_0 \\ K(t_1) = \lim\limits_{t\uparrow t_1} K(t) \;\cup\; N_1 \end{cases}$$

with given random closed sets $K_0, N_1 \in \mathscr{RC}^2(\Omega,\mathbb{R}^N)$, the time of nucleation $t_1 \in]0,T[$ and the function

$$\mathscr{F} = (\mathscr{F}_1,\mathscr{F}_2) : \mathscr{RC}^2(\Omega,\mathbb{R}^N) \times [0,T] \longrightarrow \mathrm{MLIP}_\Lambda\big(\Omega,\mathbb{R}^N;\mathbb{R}^N,\mathrm{Lin}(\mathbb{R}^m,\mathbb{R}^N)\big)$$

satisfying the assumptions of Theorem 72. Here the limit on the right-hand side is understood in $\mathscr{RC}^2(\Omega,\mathbb{R}^N)$ with respect to $d\!l_{\mathscr{RC}}$ and, its existence results from the Lipschitz continuity of solutions with respect to $d\!l^2_{\mathscr{RC}}$ (due to Proposition 68 (1.)). Lemma 71 applied first to $[0,t_1]$ and then to $[t_1,T]$ leads to the geometric characterization of the unique solution $K(\cdot) : [0,T] \longrightarrow \mathscr{RC}^2(\Omega,\mathbb{R}^N)$:

$$K(t) = \begin{cases} \overline{\vartheta_{\mathscr{F}(K,\cdot)}}\big(t,\, K_0\big) & \text{if } 0 \le t < t_1 \\ \overline{\vartheta_{\mathscr{F}(K,\cdot)}}\big(t,\, K_0\big) \;\cup\; \overline{\vartheta_{\mathscr{F}(K(t_1+\cdot),t_1+\cdot)}}\big(t-t_1,\, N_1\big) & \text{if } t_1 \le t \le T. \end{cases}$$

This simple example is based on a single additional nucleation at time $t_1 \in]0,T[$. Now the central question is which type of convergence of the time-dependent nucleation is appropriate for extending this piecewise construction approximatively.

Proposition 75 (A priori estimate for countably many nucleation processes). *Fix $\Lambda > 0$ and assume for $k=1,2,$*

$$\mathscr{F}^k = (\mathscr{F}^k_1,\mathscr{F}^k_2) : \mathscr{RC}^2(\Omega,\mathbb{R}^N) \times [0,T] \longrightarrow \mathrm{MLIP}_\Lambda\big(\Omega,\mathbb{R}^N;\mathbb{R}^N,\mathrm{Lin}(\mathbb{R}^m,\mathbb{R}^N)\big) :$$

(1.) *there exists $\widehat{\gamma} > 0$ such that* $\sup \|\mathscr{F}^k(\cdot,\cdot)\|_\infty \le \widehat{\gamma}$,

(2.) *$\mathscr{F}^k$ is Lipschitz continuous w.r.t. state in the following sense: There exists a constant $\lambda > 0$ such that for any $M_1, M_2 \in \mathscr{RC}^2(\Omega,\mathbb{R}^N)$ and $t \in [0,T]$,*

$$d\!l_\infty\big(\mathscr{F}^k(M_1,t),\, \mathscr{F}^k(M_2,t)\big)^2 \;\le\; \lambda \cdot d\!l_{\mathscr{RC}}\big(M_1,M_2\big)^2.$$

Let $I_1, I_2 \subset [0,T]$ be (at most) countable and contain 0. Assume for the functions $N_1 : I_1 \longrightarrow \mathscr{RC}^2(\Omega,\mathbb{R}^N)$ and $N_2 : I_2 \longrightarrow \mathscr{RC}^2(\Omega,\mathbb{R}^N)$ that the Hausdorff excesses between their graphs are bounded with respect to time in the following sense:

$$I_1 \longmapsto \mathbb{R}_0^+, \quad t \longmapsto \Delta_1(t) := \inf_{s \in I_2 \cap [0,t]} \Big(|t-s| + e^{\subset}_{\mathscr{RC}}\big(N_1(t), N_2(s)\big)^2 \Big)$$
$$I_2 \longmapsto \mathbb{R}_0^+, \quad t \longmapsto \Delta_2(t) := \inf_{s \in I_1 \cap [0,t]} \Big(|t-s| + e^{\subset}_{\mathscr{RC}}\big(N_2(t), N_1(s)\big)^2 \Big)$$

are bounded.

Suppose $K_1, K_2 : [0,T] \longrightarrow \mathscr{RC}^2(\Omega,\mathbb{R}^N)$ to fulfill for every $t \in [0,T]$ and $k = 1,2$

$$K_k(t) = \bigcup_{s \in I_k \cap [0,t]} \overline{\vartheta_{\mathscr{F}^k(K_k(s+\cdot),\, s+\cdot)}}\big(t-s,\, N_k(s)\big).$$

Then, there is a constant $\widehat{C} = \widehat{C}(\Lambda, \widehat{\gamma}, T) > 0$ such that the following estimate holds for every $t \in [0,T]$

$$d\!\!l_{\mathscr{RC}}\big(K_1(t), K_2(t)\big)^2 \le \widehat{C}\, e^{\widehat{C}\,t} \cdot \Big(\sup_{k=1,2} \Delta_k|_{I_k \cap [0,t]} + t \cdot \sup\, d\!\!l_\infty\big(\mathscr{F}^1, \mathscr{F}^2\big)^2 \Big).$$

This proposition clarifies which "distance" between nucleation rules is relevant for the corresponding solutions. So far we prescribe the nucleation by means of a function $I \longrightarrow \mathscr{RC}^2(\Omega,\mathbb{R}^N)$ with an (at most) countable domain $I \subset [0,T]$.
Now this results is the main tool for solving a birth-and-growth problem approximatively if the domain I of the nucleation function is possibly not countable. We formulate the characterizing comparison via $d\!\!l_{\mathscr{RC}}$ because it is not immediately clear then whether the union for all $s \in I \cap [0,t]$ is a random *closed* set (Remark 77).

Theorem 76 (Existence of solutions to some birth-and-growth problems).
As in Cauchy-Lipschitz Theorem 72 (on page 252), fix $\Lambda > 0$ and assume for

$$\mathscr{F} = (\mathscr{F}_1, \mathscr{F}_2) : \mathscr{RC}^2(\Omega,\mathbb{R}^N) \times [0,T] \longrightarrow \mathrm{MLIP}_\Lambda\big(\Omega,\mathbb{R}^N; \mathbb{R}^N, \mathrm{Lin}(\mathbb{R}^m,\mathbb{R}^N)\big):$$

(1.) *there exists $\widehat{\gamma} > 0$ such that $\sup \|\mathscr{F}(\cdot,\cdot)\|_\infty \le \widehat{\gamma}$,*
(2.) *$\mathscr{F}$ is Lipschitz continuous w.r.t. state and continuous in the following sense: There exist a constant $\lambda > 0$ and a modulus of continuity $\omega(\cdot)$ such that for any $M_1, M_2 \in \mathscr{RC}^2(\Omega,\mathbb{R}^N)$ and $t_1, t_2 \in [0,T]$,*

$$d\!\!l_\infty\big(\mathscr{F}(M_1,t_1),\, \mathscr{F}(M_2,t_2)\big)^2 \le \lambda \cdot d\!\!l_{\mathscr{RC}}\big(M_1, M_2\big)^2 + \omega(|t_1 - t_2|).$$

Let $I \subset [0,T]$ contain 0 and, assume for $N : I \longrightarrow \mathscr{RC}^2(\Omega,\mathbb{R}^N)$ that a sequence $(s_n)_{n\in\mathbb{N}}$ in I satisfies $s_1 = 0$ and

$$\sup_{t \in I \cap [0,T]} \ \inf_{\substack{m \in \{1\ldots n\}: \\ s_m \le t}} \Big(|t - s_m| + e^{\subset}_{\mathscr{RC}}\big(N(t), N(s_m)\big)^2 \Big) \longrightarrow 0 \quad \text{for } n \longrightarrow \infty.$$

Then there exists a function $K(\cdot) : [0,T] \longrightarrow \mathscr{RC}^2(\Omega,\mathbb{R}^N)$ with

$$\sup_{t \in [0,T]} d\!\!l_{\mathscr{RC}}\Big(K(t),\ \bigcup_{s \in I \cap [0,t]} \overline{\vartheta_{\mathscr{F}(K(s+\cdot),\, s+\cdot)}}\big(t-s,\, N(s)\big)\Big) = 0.$$

Remark 77. The union

$$\bigcup_{s \in I \cap [0,t]} \overline{\vartheta_{\mathscr{F}(K(s+\cdot),\, s+\cdot)}}\big(t-s,\, N(s)\big) \;:\; \Omega \leadsto \mathbb{R}^N$$

consists of possibly uncountably many sets in $\mathscr{RC}^2(\Omega,\mathbb{R}^N)$ and, it is not obvious if it also belongs to $\mathscr{RC}^2(\Omega,\mathbb{R}^N)$. Each of these random closed reachable sets, however, is constructed by means of solutions $X_t \in L^2(\Omega,\mathscr{A},P;\, \mathbb{R}^N)$ to a stochastic differential inclusion in a time interval $[s,t]$ and, each set of these L^2 functions is decomposable as mentioned in Corollary 67 (on page 248).
Corollary 66 states the existence of $M_t \in \mathscr{RC}^2(\Omega,\mathbb{R}^N)$ such that $\mathscr{S}^2_{\mathscr{RC}}(M_t)$ coincides with the L^2 closure of the union of all solutions $X_t \in L^2(\Omega,\mathscr{A},P;\, \mathbb{R}^N)$ inducing $\vartheta_{\mathscr{F}(K(s+\cdot),\, s+\cdot)}\big(t-s,\, N(s)\big)$ for any initial time $s \in I \cap [0,t]$. In this selection-wise sense, M_t can be regarded as the closure of the union above:

$$\overline{\bigcup_{s \in I \cap [0,t]} \overline{\vartheta_{\mathscr{F}(K(s+\cdot),\, s+\cdot)}}\big(t-s,\, N(s)\big)} \;=\; M_t \,\in \mathscr{RC}^2(\Omega,\mathbb{R}^N).$$

Finally, the condition on $K(\cdot) : [0,T] \longrightarrow \mathscr{RC}^2(\Omega,\mathbb{R}^N)$ in Theorem 76 is equivalent to $K(t) = M_t$ for every $t \in [0,T]$.

Remark 78 (extension to possibly empty sets of nucleation). An assumption of Theorem 76, namely $N(t) \in \mathscr{RC}^2(\Omega,\mathbb{R}^N)$ for each $t \in I$, implies that the sets $N(t)(\omega) \subset \mathbb{R}^N$ are nonempty for all $t \in I$ and $\omega \in \Omega$. It serves essentially the rather technical purpose that the mean square Pompeiu-Hausdorff excess $e^{\subset}_{\mathscr{RC}}\big(N(t), N(s_m)\big)$ is well-defined in $[0,\infty]$ via L^2 selections of $N(t)$.
For many applications in modelling, however, this hypothesis is not a significant obstacle. If all considerations are restricted to random closed sets in a fixed closed subset $K \subsetneq \mathbb{R}^N$, for example, then we should assume

$$\mathscr{F}(M,t)(\cdot,\cdot) \;=\; \{0\} \,\subset\, \mathbb{R}^N \times \mathrm{Lin}(\mathbb{R}^m,\mathbb{R}^N) \qquad \text{in } \Omega \times (\mathbb{R}^N \setminus K)$$

for all $M \in \mathscr{RC}^2(\Omega,\mathbb{R}^N)$ and $t \in [0,T]$ anyway so that the admissible solutions to stochastic differential inclusions cannot move outside K. Now we choose some arbitrary point $x_0 \in \mathbb{R}^N \setminus K$ and always consider $N(t) \in \mathscr{RC}^2(\Omega,\mathbb{R}^N)$ with the additional feature $x_0 \in N(t,\omega)$ for all $\omega \in \Omega$. This modification does not have any explicit influence on the evolution of the random closed sets $K(t) \in \mathscr{RC}^2(\Omega,\mathbb{R}^N)$.

Corollary 79 (Two special cases of birth-and-growth processes).
In addition to the hypotheses about Λ and $\mathscr{F}$ in Theorem 76, suppose $N : [0,T] \longrightarrow \mathscr{RC}^2(\Omega,\mathbb{R}^N)$ to fulfill one of the following conditions:

(a) *$N(\cdot)$ is continuous with respect to $d\!l_{\mathscr{RC}}$ or,*
(b) *$N(\cdot)$ is measurable w.r.t. $d\!l_{\mathscr{RC}}$ and expanding in the sense that $N(t_1) \subset N(t_2)$ P-almost surely in Ω whenever $0 \le t_1 \le t_2 \le T$.*

Then there exists a function $K(\cdot) : [0,T] \longrightarrow \mathscr{RC}^2(\Omega,\mathbb{R}^N)$ with

$$\sup_{t \in [0,T]} \; d\!l_{\mathscr{RC}}\Big(K(t),\; \bigcup_{s \in I \cap [0,t]} \overline{\vartheta_{\mathscr{F}(K(s+\cdot),\, s+\cdot)}}\big(t-s,\, N(s)\big)\Big) \;=\; 0.$$

Now we complete this subsection with the missing proofs:

Proof (of Proposition 75 on page 254). The main tools are the explicit continuity estimates of random closed reachable sets with respect to time and initial set as specified in Proposition 68 (on page 249).
Fix $\varepsilon > 0$ arbitrarily. For each $s \in I_1$, there exists a point of time $r \in I_2 \cap [0,s]$ with

$$|s-r| + e^{\subset}_{\mathscr{RC}}\big(N_1(s),\, N_2(r)\big)^2 \;\le\; \Delta_1(s) + \varepsilon$$

by definition of $\Delta_1(s)$. Moreover, Proposition 68 (1.) states

$$dl_{\mathscr{RC}}\Big(K_2(r),\ \overline{\vartheta_{\mathscr{F}^2(K_2(r+\cdot),\,r+\cdot)}}\big(s-r,\ K_2(r)\big)\Big)^2 \;\le\; 18\ \widehat{\gamma}^2\ e^s \cdot (s-r)$$

and, the inclusions

$$\begin{aligned} N_2(r) \;&\subset\; K_2(r)\\ \overline{\vartheta_{\mathscr{F}^2(K_2(r+\cdot),\,r+\cdot)}}\big(s-r,\ K_2(r)\big) \;&\subset\; K_2(s) \end{aligned}$$

hold P-almost surely in Ω due to the characterizing assumption about $K_2(\cdot)$.
Thus, we obtain

$$\begin{aligned} e^{\subset}_{\mathscr{RC}}\big(N_1(s),\, K_2(s)\big)^2 &\le 3\cdot\Big(e^{\subset}_{\mathscr{RC}}\big(N_1(s),\, N_2(r)\big)^2 + e^{\subset}_{\mathscr{RC}}\big(N_2(r),\, K_2(s)\big)^2\Big)\\ &\le 3\cdot \max\big\{1,\ 18\ \widehat{\gamma}^2\ e^T\big\}\ \big(\Delta_1(s)+\varepsilon\big). \end{aligned}$$

The detailed proof of Proposition 68 (2.) (on page 250 f.) and assumption (2.) about the Lipschitz continuity of $\mathscr{F}$ ensure for every $t \in [s,T]$

$$\begin{aligned} &e^{\subset}_{\mathscr{RC}}\Big(\overline{\vartheta_{\mathscr{F}^1(K_1(s+\cdot),\,s+\cdot)}}\big(t-s,\ N_1(s)\big),\ \ \overline{\vartheta_{\mathscr{F}^2(K_2(s+\cdot),\,s+\cdot)}}\big(t-s,\ K_2(s)\big)\Big)^2\\ &\le C\cdot\Big(e^{\subset}_{\mathscr{RC}}\big(N_1(s),K_2(s)\big)^2 + \int_s^t dl_\infty\big(\mathscr{F}^1|_{(K_1(s'),s')},\mathscr{F}^2|_{(K_2(s'),s')}\big)^2 ds'\Big)\cdot e^{C(2+T)T}\\ &\le C\cdot\Big(e^{\subset}_{\mathscr{RC}}\big(N_1(s),K_2(s)\big)^2 + \int_s^t 3\cdot\Big(\sup dl_\infty\big(\mathscr{F}^1,\mathscr{F}^2\big)^2\\ &\qquad\qquad +\lambda\cdot dl_{\mathscr{RC}}\big(K_1(s'),K_2(s')\big)^2\Big)ds'\Big)\cdot e^{C(2+T)T} \end{aligned}$$

with a constant $C \ge 1$ depending only on Λ. As $\varepsilon > 0$ is arbitrarily small, we can specify a larger constant $\widehat{C} = \widehat{C}(\Lambda,\widehat{\gamma},T) \ge 1$ such that for every $s \in I_1$ and $t \in [s,T]$, the following estimate holds:

$$\begin{aligned} &e^{\subset}_{\mathscr{RC}}\Big(\overline{\vartheta_{\mathscr{F}^1(K_1(s+\cdot),\,s+\cdot)}}\big(t-s,\ N_1(s)\big),\ \ K_2(t)\Big)^2\\ &\le\ e^{\subset}_{\mathscr{RC}}\Big(\overline{\vartheta_{\mathscr{F}^1(K_1(s+\cdot),\,s+\cdot)}}\big(t-s,\ N_1(s)\big),\ \ \overline{\vartheta_{\mathscr{F}^2(K_2(s+\cdot),\,s+\cdot)}}\big(t-s,\ K_2(s)\big)\Big)^2\\ &\le\ \widehat{C}\cdot\Big(\Delta_1(s) + (t-s)\cdot \sup\ dl_\infty\big(\mathscr{F}^1,\mathscr{F}^2\big)^2 + \lambda\cdot\int_s^t dl_{\mathscr{RC}}\big(K_1(s'),\ K_2(s')\big)^2 ds'\Big). \end{aligned}$$

Remark 62 (2.) (on page 246) has already mentioned that the excess of a (finite) union from a fixed set in $\mathscr{RC}^2(\Omega,\mathbb{R}^N)$ is the supremum of the excesses. For the same reasons, countable unions of random closed sets share this property and hence,

$$\begin{aligned}
e^{\subset}_{\mathscr{RC}}\big(K_1(t),\,K_2(t)\big)^2 &= e^{\subset}_{\mathscr{RC}}\Big(\bigcup_{s\in I_1\cap[0,t]} \overline{\vartheta_{\mathscr{F}(K_1(s+\cdot),s+\cdot)}\big(t-s,\,N_1(s)\big)},\quad K_2(t)\Big)^2\\
&\le \sup_{s\in I_1\cap[0,t]} e^{\subset}_{\mathscr{RC}}\Big(\overline{\vartheta_{\mathscr{F}(K_1(s+\cdot),s+\cdot)}\big(t-s,\,N_1(s)\big)},\quad K_2(t)\Big)^2\\
&\le \widehat{C}\cdot \sup_{s\in I_1\cap[0,t]} \Big(\Delta_1(s)+(t-s)\cdot \sup\, d\!\!l_\infty(\mathscr{F}^1,\mathscr{F}^2)^2+\lambda\cdot\int_s^t d\!\!l_{\mathscr{RC}}\big(K_1,\,K_2\big)^2\,ds'\Big)\\
&\le \widehat{C}\cdot\Big(\sup_{s\in I_1\cap[0,t]} \Delta_1(s)+t\cdot \sup\, d\!\!l_\infty(\mathscr{F}^1,\mathscr{F}^2)^2+\lambda\cdot\int_0^t d\!\!l_{\mathscr{RC}}\big(K_1,\,K_2\big)^2\,ds'\Big)
\end{aligned}$$

is satisfied for every $t\in[0,T]$. The same arguments lead to

$$\begin{aligned}
&e^{\supset}_{\mathscr{RC}}\big(K_1(t),\,K_2(t)\big)^2\\
&\quad\le \widehat{C}\cdot\Big(\sup_{s\in I_2\cap[0,t]} \Delta_2(s)+t\cdot \sup\, d\!\!l_\infty(\mathscr{F}^1,\mathscr{F}^2)^2+\lambda\cdot\int_0^t d\!\!l_{\mathscr{RC}}\big(K_1,\,K_2\big)^2\,ds'\Big).
\end{aligned}$$

Finally, the claim about

$$d\!\!l_{\mathscr{RC}}\big(K_1(t),\,K_2(t)\big)^2 \stackrel{\text{Def.}}{=} \max\Big\{e^{\subset}_{\mathscr{RC}}\big(K_1(t),\,K_2(t)\big)^2,\;\; e^{\supset}_{\mathscr{RC}}\big(K_1(t),\,K_2(t)\big)^2\Big\}$$

results from Gronwall's inequality in Proposition A.1 (on page 439).

□

Proof (of Existence Theorem 76 on page 255).
For each $n\in\mathbb{N}$, set $I_n := \{s_1\ldots s_n\}\subset I$, and Cauchy-Lipschitz Theorem 72 (on page 252) provides a solution $K_n : [0,T]\longrightarrow \mathscr{RC}^2(\Omega,\mathbb{R}^N)$ to the problem

$$K_n(t) \;=\; \bigcup_{s\in I_n\cap[0,t]} \overline{\vartheta_{\mathscr{F}(K_n(s+\cdot),\,s+\cdot)}\big(t-s,\,N(s)\big)}$$

for every $t\in[0,T]$ — in a piecewise way as described at the beginning of this subsection. Proposition 75 and the asymptotic assumption about $(s_n)_{n\in\mathbb{N}}$ in I imply the Cauchy property of $\big(K_n(\cdot)\big)_{n\in\mathbb{N}}$ in the following sense:

$$\sup_{n_1,n_2\ge m}\;\sup_{t\in[0,T]}\; d\!\!l_{\mathscr{RC}}\big(K_{n_1}(t),\;K_{n_2}(t)\big)^2 \longrightarrow 0 \qquad \text{for } m\longrightarrow\infty.$$

Hence, the completeness of the metric space $\big(\mathscr{RC}^2(\Omega,\mathbb{R}^N),\,d\!\!l_{\mathscr{RC}}\big)$ guarantees a limit function $K(\cdot):[0,T]\longrightarrow\mathscr{RC}^2(\Omega,\mathbb{R}^N)$ with

$$\sup_{t\in[0,T]}\; d\!\!l_{\mathscr{RC}}\big(K_n(t),\quad K(t)\big)^2 \longrightarrow 0 \qquad \text{for } n\longrightarrow\infty.$$

The asymptotic hypothesis about $(s_n)_{n\in\mathbb{N}}$ is uniform w.r.t. $t\in[0,T]$ and so, we conclude from the detailed proof of Proposition 75 that

$$e^{\subset}_{\mathscr{RC}}\Big(\bigcup_{s\in I\cap[0,t]} \overline{\vartheta_{\mathscr{F}(K(s+\cdot),\,s+\cdot)}\big(t-s,\,N(s)\big)},\;\;\bigcup_{s\in I_n\cap[0,t]} \overline{\vartheta_{\mathscr{F}(K(s+\cdot),\,s+\cdot)}}\big(t-s,\,N(s)\big)\Big)$$

converges to 0 for $n\longrightarrow\infty$ and each $t\in[0,T]$. Obviously the second argument is contained in the first union and thus, the corresponding excess $e^{\supset}_{\mathscr{RC}}$ is 0.

Furthermore Proposition 68 (on page 249) and the Lipschitz continuity of $\mathscr{F}(\cdot,s)$ in assumption (2.) lead to the convergence of

$$dl_{\mathscr{RC}}\Big(\bigcup_{s\in I_n\cap[0,t]} \overline{\vartheta_{\mathscr{F}(K(s+\cdot),\,s+\cdot)}}\big(t-s,\, N(s)\big),\; \bigcup_{s\in I_n\cap[0,t]} \overline{\vartheta_{\mathscr{F}(K_n(s+\cdot),\,s+\cdot)}}\big(t-s,\, N(s)\big)\Big)$$

to 0 for $n \longrightarrow \infty$ and every $t \in [0,T]$. Finally the claimed identity results from the triangle inequality of $dl_{\mathscr{RC}}$.

□

Proof (of Corollary 79 on page 256).
(1.) If $N(\cdot)$ is continuous with respect to $dl_{\mathscr{RC}}$ then it is even uniformly continuous in the compact interval $[0,T]$. This implies directly the assumptions of Theorem 76 concerning $N(\cdot)$ and a sequence $(s_n)_{n\in\mathbb{N}}$.

(2.) Let $N(\cdot) : [0,T] \longrightarrow \mathscr{RC}^2(\Omega,\mathbb{R}^N)$ be measurable and expanding.

Now we use Lusin's Theorem, that relates measurability to continuity of metric space-valued curves (Lebesgue-almost everywhere) and that is a special case of Scorza-Dragoni Theorem A.9 (on page 446) for functions with just one argument (see e.g. [5, Theorem 1.45], [30, Theorems 7.1.13, 7.14.25], [84, Theorem 2B]): For every $\varepsilon > 0$, there is a compact subset S_ε of $[0,T]$ such that $\mathscr{L}^1([0,T]\setminus S_\varepsilon) < \varepsilon$ and the restriction of $N(\cdot)$ to S_ε is continuous with respect to $dl_{\mathscr{RC}}$.

The uniform continuity of $N|_{S_\varepsilon}$ provides finitely many points of time $s_1 \ldots s_n$ in S_ε with $\quad \min_{i=1\ldots n} \big(|t-s_i| + dl_{\mathscr{RC}}(N(t),N(s_i))^2\big) < \varepsilon \quad$ for every $t \in S_\varepsilon \subset [0,T]$.
Set $s_{n+1} := T$ additionally. At each time t in the complement (i.e. $t \in [0,T]\setminus S_\varepsilon$), we can find some $t' \in S_\varepsilon \cup \{T\}$ with $t < t' \le t+\varepsilon$ due to $\mathscr{L}^1([0,T]\setminus S_\varepsilon) < \varepsilon$. Now the expanding property of $N(\cdot)$ implies $N(t) \subset N(t')$ P-almost surely in Ω and thus,

$$\begin{aligned} &\min_{i=1\ldots n+1} \big(|t-s_i| \qquad\;\; + e^{\subset}_{\mathscr{RC}}\big(N(t),\, N(s_i)\big)^2\big) \\ \le\; &\min_{i=1\ldots n+1} \big(\varepsilon + |t'-s_i| + e^{\subset}_{\mathscr{RC}}\big(N(t'),\, N(s_i)\big)^2\big) \;<\; 2\,\varepsilon. \end{aligned}$$

Finally, we replace the fixed parameter $\varepsilon > 0$ by a sequence $\varepsilon_m \downarrow 0$ and continue this supplementary selection of finitely many points of time in $[0,T]$ inductively. This procedure provides a (not necessarily monotone) sequence $(s_n)_{n\in\mathbb{N}}$ in $[0,T]$ with

$$\sup_{t\in I\cap[0,T]} \; \inf_{\substack{m\in\{1\ldots n\}:\\ s_m\le t}} \Big(|t-s_m| + e^{\subset}_{\mathscr{RC}}\big(N(t),\, N(s_m)\big)^2\Big) \longrightarrow 0 \qquad \text{for } n \longrightarrow \infty.$$

□

3.8 Example: Nonlinear Continuity Equations with Coefficients of Bounded Variation for $\mathscr{L}^N$-Absolutely Continuous Measures

The continuity equation

$$\tfrac{d}{dt}\,\mu \;+\; \mathrm{div}_x\,(\widetilde{\mathbf{b}}\,\mu) \;=\; 0 \qquad\qquad (\text{in } \mathbb{R}^N \times\,]0,T[)$$

is the classical analytical tool for describing the conservation of some real-valued quantity $\mu = \mu(t,x)$ while "flowing" (or, rather, evolving) along a given vector field $\widetilde{\mathbf{b}} : \mathbb{R}^N \times [0,T] \longrightarrow \mathbb{R}^N$. Thus, it is playing a key role in many applications of modelling like fluid dynamics and, it has been investigated under completely different types of assumptions about $\widetilde{\mathbf{b}}(\cdot,\cdot)$.

In § 2.5 (on page 132 ff.), we have already focused on the nonlinear transport equation for Radon measures on $\mathbb{R}^N$. Its coefficients were bounded and Lipschitz continuous vector fields on $\mathbb{R}^N$ prescribed as a function of time and the current Radon measure.
The main goal now is to weaken the regularity conditions on the vector fields considered as coefficients in the continuity equation. In particular, spatial vector fields $\mathbf{b}(\cdot)$ of bounded variation have aroused interest for weakening the assumption of (local) Lipschitz continuity.

Recent results of Ambrosio [3, 4] make a suggestion how to specify a flow $\mathbf{X} : [0,T] \times \mathbb{R}^N \longrightarrow \mathbb{R}^N$ along certain vector fields of bounded (spatial) variation in a unique way. This uniqueness is based on an additional condition of regularity, i.e. the absolute continuity with respect to Lebesgue measure $\mathscr{L}^N$ is preserved uniformly: For any nonnegative function $\rho \in L^1(\mathbb{R}^N) \cap L^\infty(\mathbb{R}^N)$, the measure $\mu_0 := \rho\,\mathscr{L}^N$ satisfies $\mathbf{X}(t,\cdot)_\sharp\,\mu_0 \leq C\,\mathscr{L}^N$ for all $t \in [0,T]$ with a constant C independent of t.

This result of Ambrosio about the so-called Lagrangian flow serves as starting point of this example and thus, it motivates to replace the set $\mathscr{M}(\mathbb{R}^N)$ of finite Radon measures by

$$\mathbb{L}^{\infty\cap 1}(\mathbb{R}^N) \;:=\; \big\{\rho\,\mathscr{L}^N \;\big|\; \rho \in L^1(\mathbb{R}^N) \cap L^\infty(\mathbb{R}^N),\ \rho \geq 0\big\}.$$

After summarizing some features of the Lagrangian flow, we exploit the corresponding vector fields of (locally) bounded spatial variation for inducing transitions on these measures. It allows us to deal with nonlinear continuity equations in the mutational framework.
The main conclusions presented in subsequent § 3.8.4 consist in sufficient conditions for existence, uniqueness and stability of distributional solutions $\mu(\cdot) : [0,T] \longrightarrow \mathbb{L}^{\infty\cap 1}(\mathbb{R}^N)$ to the Cauchy problem

$$\begin{cases} \tfrac{d}{dt}\,\mu \;+\; \mathrm{div}_x\,(\mathbf{f}(\mu,\cdot)\,\mu) = 0 & \text{in } \mathbb{R}^N \times\,]0,T[\\ \qquad\qquad\qquad \mu(0) = \rho_0\,\mathscr{L}^N \in \mathbb{L}^{\infty\cap 1}(\mathbb{R}^N) \end{cases}$$

for a given functional relationship in the form of

$$\mathbf{f} :\ \mathbb{L}^{\infty\cap 1}(\mathbb{R}^N) \times [0,T] \;\longrightarrow\; \mathrm{BV}_{\mathrm{loc}}(\mathbb{R}^N,\mathbb{R}^N) \cap L^\infty(\mathbb{R}^N,\mathbb{R}^N).$$

Basic set	$\mathbb{L}^{\infty\cap 1}(\mathbb{R}^N) := \{\rho\,\mathscr{L}^N \mid \rho \in L^1(\mathbb{R}^N)\cap L^\infty(\mathbb{R}^N),\ \rho \geq 0\}$
Distances	$d_{j,\mathbb{L}^{\infty\cap 1}}(\mu,\nu) := \|\varphi_j\cdot(\mu-\nu)\|(\mathbb{R}^N)$ with a suitably dense family $(\varphi_j)_{j\in\mathscr{J}}$ of smooth positive Schwartz functions satisfying $\|\nabla\varphi_j(\cdot)\| \leq \lambda_j\cdot\varphi_j(\cdot)$ (Lemma 86)
Absolute values	$\lfloor\mu\rfloor := \|\sigma\|_{L^1(\mathbb{R}^N)} + \|\sigma\|_{L^\infty(\mathbb{R}^N)}$ for $\mu = \sigma\,\mathscr{L}^N \in \mathbb{L}^{\infty\cap 1}(\mathbb{R}^N)$
Transition	Each $\mathbf{b}\in \mathrm{BV}_{\mathrm{loc}}(\mathbb{R}^N,\mathbb{R}^N)\cap L^\infty$ with $D\cdot\mathbf{b} = \mathrm{div}\,\mathbf{b}\ \mathscr{L}^N \ll \mathscr{L}^N$ and $\mathrm{div}\,\mathbf{b}\in L^\infty(\mathbb{R}^N)$ induces the unique Lagrangian flow $\mathbf{X_b}(\cdot,\cdot)$. Set $\vartheta_{\mathbb{L}^{\infty\cap 1},\mathbf{b}}: [0,1]\times\mathbb{L}^{\infty\cap 1}(\mathbb{R}^N) \longrightarrow \mathbb{L}^{\infty\cap 1}(\mathbb{R}^N),\ (h,\mu_0) \longmapsto \mathbf{X_b}(h,\cdot)_\sharp\,\mu_0$
Compactness	weak Euler compactness with respect to $d_{j,\kappa,\kappa',\mathbb{L}^{\infty\cap 1}}(\mu,\nu) := \left\|\int_{\mathbb{R}^N}\varphi_j\,(\varphi_\kappa-\varphi_{\kappa'})\ d(\mu-\nu)\right\|$ (Def. 87) essentially due to Prokhorov's Theorem: Lemma 95 (page 272)
Equi-continuity	Euler equi-continuity results from uniform Lipschitz continuity of transitions and the triangle inequality of $d_{j,\mathbb{L}^{\infty\cap 1}}$.
Mutational solutions	Narrowly continuous distributional solution to nonlin. continuity equation $\partial_t\,\mu_t + \mathrm{div}_x\,(\mathbf{f}(\mu_t,t)\ \mu_t) = 0$ in $\mathbb{R}^N\times\,]0,T[$
List of main results formulated in § 3.8	Existence due to weak* compactness: Theorem 96 (page 272) Uniqueness due to Lipschitz continuity: Theorem 99 (page 274) Continuous dependence of solutions on data: Theorem 100
Key tools	The Lagrangian flow in the sense of Ambrosio is specified by the linear continuity equations with coefficients of bounded spatial variation: § 3.8.1 Explicit solutions to the linear continuity equations with coefficients in $W^{1,\infty}_{\mathrm{loc}}(\mathbb{R}^N,\mathbb{R}^N)\cap L^\infty$ (for approximating the Lagrangian flow): Lemma 92 (page 267)

Table 3.3 Brief summary of the example in § 3.8 in mutational terms:
Nonlinear continuity equations with coefficients of bounded variation for $\mathscr{L}^N$-absolutely continuous measures

3.8.1 The Lagrangian Flow in the Sense of Ambrosio

Considering the linear continuity equation

$$\tfrac{d}{dt}\,\mu + \operatorname{div}_x(\widetilde{\mathbf{b}}\,\mu) = 0 \qquad (\text{in } \mathbb{R}^N \times\,]0,T[),$$

the regularity of the coefficient $\widetilde{\mathbf{b}} : \mathbb{R}^N \times [0,T] \longrightarrow \mathbb{R}^N$ plays the decisive role in the question if the method of characteristics provides an explicit solution directly. Proposition 2.47 (on page 138), for example, guarantees such a solution if $\widetilde{\mathbf{b}}$ is bounded, Lipschitz continuous with respect to space and Lebesgue integrable with respect to time.
Motivated by the results of DiPerna and Lions [64], Ambrosio has suggested how to specify characteristics under weaker assumptions about spatial regularity [3, 4]. Now we summarize the properties relevant for our subsequent conclusions in the following proposition:

Proposition 80 (Ambrosio [3, 4]).
Assume $\widetilde{\mathbf{b}} : [0,T] \times \mathbb{R}^N \longrightarrow \mathbb{R}^N$ *to be in* $L^1\big([0,T],\, \mathrm{BV}_{\mathrm{loc}}(\mathbb{R}^N,\mathbb{R}^N)\big)$ *satisfying*
(1.) $\frac{|\widetilde{\mathbf{b}}|}{1+|x|} \in L^1\big([0,T], L^1(\mathbb{R}^N)\big) + L^1\big([0,T], L^\infty(\mathbb{R}^N)\big)$,
(2.) $\operatorname{div}_x \widetilde{\mathbf{b}}(t,\cdot)\ \mathscr{L}^N \ll \mathscr{L}^N$ *for* $\mathscr{L}^1$*-almost every* $t \in [0,T]$,
(3.) $[\operatorname{div}_x \widetilde{\mathbf{b}}]^- \in L^1\big([0,T],\, L^\infty(\mathbb{R}^N)\big)$.

Then there exists a so-called Lagrangian flow $\mathbf{X} : [0,T] \times \mathbb{R}^N \longrightarrow \mathbb{R}^N$ *such that*

(a) $\mathbf{X}(\cdot,x) : [0,T] \longrightarrow \mathbb{R}^N$ *is absolutely continuous for* $\mathscr{L}^N$*-almost every* $x \in \mathbb{R}^N$,

$$\mathbf{X}(t,x) = x + \int_0^t \widetilde{\mathbf{b}}\big(s, \mathbf{X}(s,x)\big)\, ds \qquad \text{for all } t \in [0,T],$$

(b) *there is a constant* $C > 0$ *satisfying* $\mathbf{X}(t,\cdot)_\sharp\,(\sigma\,\mathscr{L}^N) \le C\,\|\sigma\|_\infty\,\mathscr{L}^N$ *for all* $\sigma \in L^1(\mathbb{R}^N) \cap L^\infty(\mathbb{R}^N)$, $\sigma \ge 0$, *and* $t \in [0,T]$.

$\mathbf{X}(t,\cdot) : \mathbb{R}^N \longrightarrow \mathbb{R}^N$ *is unique up to* $\mathscr{L}^N$*-negligible sets for every* $t \in [0,T]$ *and,* $\mu(t) := \mathbf{X}(t,\cdot)_\sharp\,\mu_0$ *is the unique distributional solution to the continuity equation*

$$\tfrac{d}{dt}\,\mu + \operatorname{div}_x(\widetilde{\mathbf{b}}\,\mu) = 0 \qquad \text{in } \mathbb{R}^N \times\,]0,T[$$

for every initial measure $\mu_0 := \sigma\,\mathscr{L}^N$ *with* $\sigma \in L^1(\mathbb{R}^N) \cap L^\infty(\mathbb{R}^N)$, $\sigma \ge 0$.

Mollifying each $\mu(t)$ *with a joint Gaussian kernel* $\rho \in C^1(\mathbb{R}^N,]0,\infty[)$, *the measures* $\mu_\delta(t) := \mu(t) * \rho_\delta$ *solve the continuity equation*

$$\tfrac{d}{dt}\,\mu_\delta + \operatorname{div}_x(\widetilde{\mathbf{b}}_\delta\,\mu_\delta) = 0 \qquad (\text{in the distributional sense})$$

with $\widetilde{\mathbf{b}}_\delta(t,\cdot) := \frac{(\widetilde{\mathbf{b}}(t,\cdot)\,\mu(t)) * \rho_\delta}{\mu_\delta(t)}$ *being in* $L^1\big([0,T],\, W^{1,\infty}_{\mathrm{loc}}(\mathbb{R}^N,\mathbb{R}^N)\big)$.
In particular, at every time $t \in [0,T]$, $\mu_\delta(t) \longrightarrow \mu(t)$ *narrowly (i.e. with respect to the duality of bounded continuous functions) for* $\delta \downarrow 0$.

Remark 81 (about the proof of Proposition 80). This proposition collects several results of Ambrosio in [3, 4], but it is not formulated in this summarizing form there. The arguments of its proof are rather widespread in the lecture notes [3].

Indeed, extending [3, Theorem 4.3] to vector fields of locally bounded spatial variation (as stated at the end of [3, § 5]), there exists a Lagrangian flow $\mathbf{X} : [0,T] \times \mathbb{R}^N \longrightarrow \mathbb{R}^N$ with properties (a),(b) and, it is unique (up to $\mathscr{L}^N$-negligible sets).

The proof of [3, Theorem 3.5] bridges the gap between the Lagrangian flow and the measure-valued solution to the continuity equation (by means of push-forward). The uniqueness of $\mu(\cdot)$ results from the comparison principle of the continuity equation (due to the assumptions about $\widetilde{\mathbf{b}}$) according to [3, Theorem 4.1].

Finally the proof of [3, Theorem 3.2] implies the narrow sequential compactness of $\eta_\delta := \big(x, \mathbf{X}_{\widetilde{\mathbf{b}}_\delta}(\cdot,x)\big)_\sharp\, \mu_\delta(0)$ (using Prokhorov compactness theorem). In particular, its equation (3.3) implies the narrow convergence of $\mu_\delta(t)$ to its unique limit $\mu(t)$.

Similarly, [3, Theorem 4.4] and the remarks at the end of [3, § 5] guarantee:

Proposition 82 (Stability of Lagrangian flows, Ambrosio [3]). *Assume* $\widetilde{\mathbf{b}}, \widetilde{\mathbf{b}}_n : [0,T] \times \mathbb{R}^N \longrightarrow \mathbb{R}^N$ $(n \in \mathbb{N})$ *to be in* $L^1\big([0,T],\ \mathrm{BV}_{\mathrm{loc}}(\mathbb{R}^N,\mathbb{R}^N)\big)$ *satisfying conditions (1.)–(3.) of Proposition 80. Furthermore suppose*

(i) $\widetilde{\mathbf{b}}_n \longrightarrow \widetilde{\mathbf{b}}$ *in* $L^1_{\mathrm{loc}}(]0,T[\, \times \mathbb{R}^N)$ *for* $n \longrightarrow \infty$,
(ii) *there exists a constant* $C > 0$ *such that for all* $n \in \mathbb{N}$, $|\widetilde{\mathbf{b}}_n| \le C$,
(iii) $\big\{ [\mathrm{div}_x\, \widetilde{\mathbf{b}}_n]^- \;\big|\; n \in \mathbb{N} \big\}$ *is bounded in* $L^1\big([0,T],\ L^\infty(\mathbb{R}^N)\big)$.

Let $\mathbf{X}_{\widetilde{\mathbf{b}}}, \mathbf{X}_{\widetilde{\mathbf{b}}_n}$ $(n \in \mathbb{N})$ *denote the Lagrangian flows relative to* $\widetilde{\mathbf{b}}, \widetilde{\mathbf{b}}_n$ *respectively and, choose* $\mu = \rho\, \mathscr{L}^N$ *with* $\rho \in L^1(\mathbb{R}^N), \rho \ge 0$ *arbitrarily.*

Then, $$\lim_{n \to \infty} \int_{\mathbb{R}^N} \max_{[0,T]} \min\big\{ \big|\mathbf{X}_{\widetilde{\mathbf{b}}_n}(\cdot,x) - \mathbf{X}_{\widetilde{\mathbf{b}}}(\cdot,x)\big|,\ \rho(x) \big\}\ d\mathscr{L}^N x \;=\; 0.$$

Remark 83. In comparison with the nonlinear transport equation investigated in § 2.5 (on page 132 ff.), it is remarkable that the linear problem here is stable with respect to L^1 perturbations of the coefficient field whereas all estimates in § 2.5 are taking the L^∞ norm into consideration (see e.g. Lemma 2.49 (f) on page 139 and consequently Theorem 2.53 on page 142).

Corollary 84. *In addition to the hypotheses of Proposition 82, let* $t \in [0,T]$ *and* $\mu_0 = \sigma_0\, \mathscr{L}^N$ *be arbitrary with* $\sigma_0 \in L^1(\mathbb{R}^N)$. *Then,*

$$\mathbf{X}_{\widetilde{\mathbf{b}}_n}(t,\cdot)_\sharp\, \mu_0 \;\longrightarrow\; \mathbf{X}_{\widetilde{\mathbf{b}}}(t,\cdot)_\sharp\, \mu_0 \qquad \textit{narrowly for } n \longrightarrow \infty,$$

i.e., for any bounded and continuous $\psi : \mathbb{R}^N \longrightarrow \mathbb{R}$,

$$\int_{\mathbb{R}^N} \psi\big(\mathbf{X}_{\widetilde{\mathbf{b}}_n}(t,x)\big)\ \sigma_0(x)\ d\mathscr{L}^N x \;\longrightarrow\; \int_{\mathbb{R}^N} \psi\big(\mathbf{X}_{\widetilde{\mathbf{b}}}(t,x)\big)\ \sigma_0(x)\ d\mathscr{L}^N x. \qquad \square$$

3.8.2 The Subset $\mathbb{L}^{\infty\cap 1}(\mathbb{R}^N)$ of Measures and its Pseudo-Metrics

In this example, Proposition 80 of Ambrosio is to provide the measure-valued solutions to the linear continuity equation. It motivates our choice of both coefficient functions and measures on $\mathbb{R}^N$.

Definition 85. Set $\mathbb{L}^{\infty\cap 1}(\mathbb{R}^N) := \{\rho\, \mathscr{L}^N \mid \rho \in L^1(\mathbb{R}^N) \cap L^\infty(\mathbb{R}^N),\ \rho \geq 0\}$.

In regard to distance functions on $\mathbb{L}^{\infty\cap 1}(\mathbb{R}^N)$, we suggest the weighted total variation – with a countable family $(\varphi_j)_{j\in\mathscr{I}}$ of smooth positive weight functions whose gradient can be estimated by the function itself. In comparison with the $W^{1,\infty}$ dual metric used in § 2.5, this last property proves to be particularly useful for estimating the effects of distributional derivatives via initial data.

Lemma 86. *There exists a countable family $(\varphi_j)_{j\in\mathscr{I}}$ of smooth Schwartz functions $\mathbb{R}^N \longrightarrow [0,\infty[$ with the following properties*

(1.) $(\varphi_j)_{j\in\mathscr{I}}$ is dense in $\big(C_0^0(\mathbb{R}^N, [0,\infty[),\ \|\cdot\|_\infty\big)$,
(2.) $C_c^\infty(\mathbb{R}^N, [0,\infty[)$ is contained in the closure of $(\varphi_j)_{j\in\mathscr{I}}$ w.r.t. the C^1 norm
(3.) for each $j \in \mathscr{I}$, there exists $\lambda_j > 0$ with $|\nabla \varphi_j(\cdot)| \leq \lambda_j \cdot \varphi_j(\cdot)$ in $\mathbb{R}^N$,

Definition 87. Let $(\varphi_j)_{j\in\mathscr{I}}$ be a family of Schwartz functions as described in Lemma 86 and, $\mathscr{J} \subset \mathscr{I}$ denotes the subset of all indices $\kappa \in \mathscr{I}$ with $0 < \varphi_\kappa \leq 1$. For each indices $j \in \mathscr{I}$ and $\kappa, \kappa' \in \mathscr{J}$, define

$$d_{j,\mathbb{L}^{\infty\cap 1}},\ d_{j,\kappa,\kappa',\mathbb{L}^{\infty\cap 1}} :\ \mathbb{L}^{\infty\cap 1}(\mathbb{R}^N) \times \mathbb{L}^{\infty\cap 1}(\mathbb{R}^N) \longrightarrow [0,\infty[$$

as

$$\begin{aligned} d_{j,\mathbb{L}^{\infty\cap 1}}(\mu,\nu) &:= \big|\varphi_j \cdot (\mu - \nu)\big|(\mathbb{R}^N) \\ &\overset{\text{Def.}}{=} \sup\Big\{ \sum_{k=0}^{\infty} \Big| \int_{E_k} \varphi_j\, d(\mu-\nu)\Big| \ \Big|\ (E_k)_{k\in\mathbb{N}} \text{ pairwise disjoint} \\ &\qquad\qquad \text{Borel sets, } \mathbb{R}^N = \bigcup_{k\in\mathbb{N}} E_k \Big\}, \\ d_{j,\kappa,\kappa',\mathbb{L}^{\infty\cap 1}}(\mu,\nu) &:= \Big| \int_{\mathbb{R}^N} \varphi_j\, (\varphi_\kappa - \varphi_{\kappa'})\ d(\mu-\nu)\Big|. \end{aligned}$$

Remark 88. Obviously, Gronwall's Lemma implies $\varphi_j > 0$ in $\mathbb{R}^N$ unless $\varphi_j \equiv 0$. Assuming $\varphi_j \not\equiv 0$ for all $j \in \mathscr{I}$ from now on, each $d_{j,\mathbb{L}^{\infty\cap 1}}$ takes all points of $\mathbb{R}^N$ into consideration – in a weighted form.
Moreover, all functions $d_{j,\mathbb{L}^{\infty\cap 1}}, d_{j,\kappa,\kappa',\mathbb{L}^{\infty\cap 1}}$ $(j \in \mathscr{I},\ \kappa, \kappa' \in \mathscr{J})$ are pseudo-metrics on $\mathbb{L}^{\infty\cap 1}(\mathbb{R}^N)$, i.e. in particular, they satisfy the triangle inequality.

Before presenting lacking proofs, we specify the relation between the functions $d_{j,\mathbb{L}^{\infty\cap 1}}$, $d_{j,\kappa,\kappa',\mathbb{L}^{\infty\cap 1}}$ $(j \in \mathscr{I},\ \kappa, \kappa' \in \mathscr{J})$ and more popular topologies of Radon measures mentioned in § 2.5.1 (on page 132 ff.). The next lemma enables us to apply the existence results of § 3.3.6 (concluded from a generalized form of "weak" compactness on page 206 ff.) later on.

Lemma 89. *For every finite Radon measure $\mu \in \mathscr{M}(\mathbb{R}^N)$ and open set $A \subset \mathbb{R}^N$, the total variation satisfies*

$$|\mu|(A) = \sup\left\{ \int_{\mathbb{R}^N} \psi \, d\mu \;\middle|\; \psi \in C_c^0(A),\ \|\psi\|_\infty \leq 1 \right\}$$

and thus, for all $\mu, \nu \in \mathbb{L}^{\infty\cap 1}(\mathbb{R}^N)$,

$$d_{j,\mathbb{L}^{\infty\cap 1}}(\mu,\nu) = \sup_{\kappa,\kappa' \in \mathscr{J}} d_{j,\kappa,\kappa',\mathbb{L}^{\infty\cap 1}}(\mu,\nu).$$

Lemma 90. *(i) Let $(\mu_n)_{n\in\mathbb{N}}$ be in $\mathbb{L}^{\infty\cap 1}(\mathbb{R}^N)$ with bounded total variation. $(\mu_n)_{n\in\mathbb{N}}$ converges weakly* to $\mu \in \mathbb{L}^{\infty\cap 1}(\mathbb{R}^N)$ with respect to $\big(C_0^0(\mathbb{R}^N),\ \|\cdot\|_{\sup}\big)$ if and only if for every indices $j \in \mathscr{I},\ \kappa, \kappa' \in \mathscr{J}$,*

$$\lim_{n\to\infty} d_{j,\kappa,\kappa',\mathbb{L}^{\infty\cap 1}}\big(\mu_n,\ \mu\big) = 0.$$

Assuming in addition that $\{\mu_n \mid n \in \mathbb{N}\}$ is tight (in the sense of Definition 2.40), this equivalence can be extended to narrow convergence of $(\mu_n)_{n\in\mathbb{N}}$ (in the sense of Definition 2.39 on page 132).

(ii) Let $(\mu_n = \sigma_n \mathscr{L}^N)_{n\in\mathbb{N}}$ be a tight sequence in $\mathbb{L}^{\infty\cap 1}(\mathbb{R}^N)$ with bounded total variation and consider $\mu = \sigma\,\mathscr{L}^N \in \mathbb{L}^{\infty\cap 1}(\mathbb{R}^N)$.
Then, $\sigma_n \longrightarrow \sigma$ in $L^1_{\text{loc}}(\mathbb{R}^N)$ for $n \longrightarrow \infty$ if and only if for every index $j \in \mathscr{I}$,

$$\lim_{n\to\infty} d_{j,\mathbb{L}^{\infty\cap 1}}\big(\mu_n,\ \mu\big) = 0.$$

Proof (of Lemma 86). Such a family of functions $\varphi_j \in C^\infty(\mathbb{R}^N, [0,\infty[)$ can be generated by means of convolution.
Indeed, $C_0^\infty(\mathbb{R}^N, [0,\infty[)$ is known to be separable with respect to $\|\cdot\|_\infty$. Now consider a countable dense subset $(f_k)_{k\in\mathbb{N}}$ of $C_c^\infty(\mathbb{R}^N, [0,\infty[)$ together with

$$\psi_\delta : \mathbb{R}^N \longrightarrow\ ]0,\infty[,\ x \longmapsto c_{\delta,N} \cdot \exp(-\,\delta\,\tfrac{|x|^2}{1+|x|})$$

for arbitrarily large $\delta > 0$ and the constant $c_{\delta,N} > 0$ such that $\|\psi_\delta\|_{L^1(\mathbb{R}^N)} = 1$.
Then, each convolution $f_k * \psi_\delta : \mathbb{R}^N \longrightarrow \mathbb{R}$ is smooth, nonnegative and satisfies

$$|\nabla(f_k * \psi_\delta)| = |f_k * (\nabla\psi_\delta)| \leq \delta\ f_k * \psi_\delta$$

since the auxiliary function $\widehat{\psi}_\delta : [0,\infty[\ \longrightarrow\ ]0,1],\quad r \longmapsto c_{\delta,N} \cdot \exp(-\,\delta\,\frac{r^2}{1+r})$ is smooth with

$$\tfrac{d}{dr}\,\widehat{\psi}_\delta(r) = -\,\delta\,\tfrac{r\,(r+2)}{(r+1)^2}\ \widehat{\psi}_\delta(r) \in [-\delta,0]\cdot\widehat{\psi}_\delta(r)$$

and thus, $\quad \frac{d}{dr}\,\widehat{\psi}_\delta(r) = O(r) \quad$ for $r \longrightarrow 0^+$.

Furthermore, $f_k * \psi_\delta$ is a Schwartz function because so is ψ_δ and f_k is assumed to have compact support. $(f_k * \psi_\delta)_{k,\delta\in\mathbb{N}}$ is dense in $\big(C_0^0(\mathbb{R}^N,\, [0,\infty[),\, \|\cdot\|_\infty\big)$ since so is $(f_k)_{k\in\mathbb{N}}$ and $(\psi_\delta)_{\delta\in\mathbb{N}}$ is a Dirac sequence.
Finally it satisfies the second required property because for any $g \in C_c^\infty(\mathbb{R}^N,\, [0,\infty[)$ and subsequence $(f_{k_j})_{j\in\mathbb{N}}$ with $\|g - f_{k_j}\|_\infty \longrightarrow 0\ (j \longrightarrow \infty)$, we obtain for $j \longrightarrow \infty$

$$\nabla(f_{k_j} * \psi_\delta) \;=\; f_{k_j} * (\nabla\psi_\delta) \;\longrightarrow\; g * (\nabla\psi_\delta) = (\nabla g) * \psi_\delta \qquad \text{uniformly}$$

and, the last convolution converges uniformly to ∇g for $\delta \longrightarrow \infty$. □

Proof (of Lemma 89). The representation of total variation as supremum is proven in [5, Proposition 1.47], for example.
As a consequence of Lemma 86, the set $\{\varphi_\kappa \mid \kappa \in \mathscr{J}\}$ is dense in $C_0^0(\mathbb{R}^N,\, [0,1])$ with respect to the supremum norm. Thus, $\{\varphi_\kappa - \varphi_{\kappa'} \mid \kappa, \kappa' \in \mathscr{J}\}$ is dense in $C_0^0(\mathbb{R}^N,\, [-1,1])$ with respect to the supremum norm. Finally the first equality in this Lemma 89 implies for every finite Radon measure $\mu \in \mathscr{M}(\mathbb{R}^N)$

$$\int_{\mathbb{R}^N} \varphi_j \; d\,|\mu| \;=\; \sup_{\kappa,\kappa'\in\mathscr{J}} \int_{\mathbb{R}^N} \varphi_j\; (\varphi_\kappa - \varphi_{\kappa'})\; d\mu. \qquad \square$$

Proof (of Lemma 90). (i) Due to Lemma 86, $\{\varphi_\kappa - \varphi_{\kappa'} \mid \kappa, \kappa' \in \mathscr{J}\}$ is dense in $C_0^0(\mathbb{R}^N,\, [-1,1])$ with respect to the supremum norm and thus, $\{\varphi_j\; (\varphi_\kappa - \varphi_{\kappa'}) \mid j \in \mathscr{I},\; \kappa, \kappa' \in \mathscr{J}\}$ is dense in $\big(C_0^0(\mathbb{R}^N),\, \|\cdot\|_{\sup}\big)$.
Hence the first claimed equivalence is just a special case of a standard characterization of weak* convergence by means of strongly dense subsets (see e.g. [188, Theorem V.1.10]). The equivalence of narrow and weak* convergence for tight sequences has already been mentioned in Remark 2.41 (1.) (on page 134).

(ii) It is a direct consequence of tightness and Lemma 86. □

3.8.3 Autonomous Linear Continuity Problems Induce Transitions on $\mathbb{L}^{\infty\cap 1}(\mathbb{R}^N)$ via Lagrangian Flows

Motivated by Proposition 80 of Ambrosio (on page 262) again, we introduce an abbreviation for suitable autonomous vector fields on $\mathbb{R}^N$ and specify candidates for their associated transitions on $\mathbb{L}^{\infty\cap 1}(\mathbb{R}^N)$:

Definition 91.
$\mathrm{BV}^{\infty,\mathrm{div}}_{\mathrm{loc}}(\mathbb{R}^N)$ denotes the set of all functions $\mathbf{b} \in \mathrm{BV}_{\mathrm{loc}}(\mathbb{R}^N, \mathbb{R}^N) \cap L^\infty(\mathbb{R}^N, \mathbb{R}^N)$ satisfying $D\cdot\mathbf{b} = \operatorname{div}\mathbf{b}\; \mathscr{L}^N \ll \mathscr{L}^N$ and $\operatorname{div}\mathbf{b} \in L^\infty(\mathbb{R}^N)$.

For each vector field $\mathbf{b} \in \mathrm{BV}^{\infty,\mathrm{div}}_{\mathrm{loc}}(\mathbb{R}^N)$, define

$$\vartheta_{\mathbb{L}^{\infty\cap 1},\mathbf{b}} :\; [0,1] \times \mathbb{L}^{\infty\cap 1}(\mathbb{R}^N) \longrightarrow \mathbb{L}^{\infty\cap 1}(\mathbb{R}^N), \qquad (h,\mu_0) \longmapsto \mathbf{X}_\mathbf{b}(h,\cdot)_\sharp\, \mu_0$$

with $\mathbf{X}_\mathbf{b}(\cdot,\cdot)$ denoting its Lagrangian flow according to Proposition 80.

Now we first investigate the regularity features of $\vartheta_{\mathbb{L}^{\infty\cap 1},\mathbf{b}}(\cdot,\cdot)$ for more regular vector fields $\mathbf{b} \in W^{1,\infty}_{\text{loc}}(\mathbb{R}^N,\mathbb{R}^N) \cap L^\infty$ with respect to each pseudo-metric $d_{j,\mathbb{L}^{\infty\cap 1}}$ ($j \in \mathscr{I}$). Afterwards the approximation via convolution and Ambrosio's stability result in Proposition 82 lead to the estimates for $\mathbf{b} \in \mathrm{BV}^{\infty,\mathrm{div}}_{\text{loc}}(\mathbb{R}^N)$ in Proposition 94 below.

Lemma 92. *Suppose* $\mathbf{b}, \mathbf{b}_1, \mathbf{b}_2 \in W^{1,\infty}_{\text{loc}}(\mathbb{R}^N,\mathbb{R}^N) \cap L^\infty$.
Then, for any $\mu_0 = \rho\, \mathscr{L}^N$, $\nu_0 \in \mathbb{L}^{\infty\cap 1}(\mathbb{R}^N)$ *and* $j \in \mathscr{I}$, $s,t,h \in [0,1]$ *with* $t+h \leq 1$,

$$
\begin{aligned}
&(1.)\ \vartheta_{\mathbb{L}^{\infty\cap 1},\mathbf{b}}(0,\cdot) \;=\; \mathrm{Id}_{\mathbb{L}^{\infty\cap 1}(\mathbb{R}^N)},\\
&(2.)\ \vartheta_{\mathbb{L}^{\infty\cap 1},\mathbf{b}}\big(h,\ \vartheta_{\mathbb{L}^{\infty\cap 1},\mathbf{b}}(t,\mu_0)\big) \;=\; \vartheta_{\mathbb{L}^{\infty\cap 1},\mathbf{b}}(t+h,\ \mu_0),\\
&(3.)\ \limsup_{h\downarrow 0} \frac{d_{j,\mathbb{L}^{\infty\cap 1}}\big(\vartheta_{\mathbb{L}^{\infty\cap 1},\mathbf{b}}(h,\mu_0),\ \vartheta_{\mathbb{L}^{\infty\cap 1},\mathbf{b}}(h,\nu_0)\big) - d_{j,\mathbb{L}^{\infty\cap 1}}(\mu_0,\nu_0)}{h \quad d_{j,\mathbb{L}^{\infty\cap 1}}(\mu_0,\nu_0)} \;\leq\; \lambda_j\, \|\mathbf{b}\|_\infty,\\
&(4.)\ \big|\varphi_j\ \vartheta_{\mathbb{L}^{\infty\cap 1},\mathbf{b}}(t,\mu_0)\big|\,(\mathbb{R}^N) \;\leq\; \big|\varphi_j\,\mu_0\big|\,(\mathbb{R}^N)\cdot e^{\lambda_j\,\|\mathbf{b}\|_\infty\cdot t},\\
&(5.)\ d_{j,\mathbb{L}^{\infty\cap 1}}\big(\vartheta_{\mathbb{L}^{\infty\cap 1},\mathbf{b}}(s,\mu_0),\ \vartheta_{\mathbb{L}^{\infty\cap 1},\mathbf{b}}(t,\mu_0)\big) \leq |t-s|\cdot\lambda_j\,\|\mathbf{b}\|_\infty\, e^{\lambda_j\,\|\mathbf{b}\|_\infty}\,\big|\varphi_j\,\mu_0\big|(\mathbb{R}^N),\\
&(6.)\ \limsup_{h\downarrow 0} \frac{d_{j,\mathbb{L}^{\infty\cap 1}}\big(\vartheta_{\mathbf{b}_1}(h,\mu_0),\ \vartheta_{\mathbf{b}_2}(h,\mu_0)\big)}{h} \;\leq\; \lambda_j\,\big|\varphi_j\,|\mathbf{b}_1-\mathbf{b}_2|\,\mu_0\big|\,(\mathbb{R}^N)\\
&\qquad\qquad \leq\; \lambda_j\,\|\rho\|_\infty\cdot\big\|\varphi_j\,|\mathbf{b}_1-\mathbf{b}_2|\big\|_{L^1(\mathbb{R}^N)}.
\end{aligned}
$$

In regard to the choice of $\lfloor\cdot\rfloor_j$ ($j \in \mathscr{I}$), there are even two candidates now. The first one is the weighted total variation (as mentioned here in Lemma 92 (4.)). Dispensing with the weight function φ_j, however, we find the total variation as an alternative whose growth also proves to be bounded in the required way. Statement (6.) in Lemma 92 motivates us to take the L^∞ norm into consideration (if possible) and thus, we introduce for $\mu = \sigma\,\mathscr{L}^N \in \mathbb{L}^{\infty\cap 1}(\mathbb{R}^N)$

$$\lfloor\mu\rfloor \;:=\; |\mu|\,(\mathbb{R}^N) + \big\|\tfrac{\mu}{\mathscr{L}^N}\big\|_\infty \;=\; \|\sigma\|_{L^1(\mathbb{R}^N)} + \|\sigma\|_{L^\infty(\mathbb{R}^N)}\,.$$

Supplying $\mathbb{L}^{\infty\cap 1}(\mathbb{R}^N)$ with the weak* topology (w.r.t. $C^0_0(\mathbb{R}^N)$), this functional $\lfloor\cdot\rfloor$ is lower semicontinuous and thus, hypothesis (H4') (on page 207) is fulfilled.

Lemma 93. *For every vector field* $\mathbf{b} \in \mathrm{BV}^{\infty,\mathrm{div}}_{\text{loc}}(\mathbb{R}^N)$ *and initial measure* $\mu = \sigma\,\mathscr{L}^N \in \mathbb{L}^{\infty\cap 1}(\mathbb{R}^N)$, *the Radon-Nikodym derivative* σ_t *of* $\vartheta_{\mathbb{L}^{\infty\cap 1},\mathbf{b}}(t,\mu)$ *with respect to Lebesgue measure* $\mathscr{L}^N$ *satisfies*

$$
\begin{aligned}
\|\sigma_t\|_\infty &\leq \|\sigma\|_\infty \quad e^{\|\mathrm{div}\,\mathbf{b}\|_\infty\, t},\\
\big|\vartheta_b(t,\mu)\big|(\mathbb{R}^N) = \|\sigma_t\|_{L^1} &\leq \|\sigma\|_{L^1} \quad e^{2\,\|\mathrm{div}\,\mathbf{b}\|_\infty\, t}.
\end{aligned}
$$

The gap between vector fields in $W^{1,\infty}_{\text{loc}}(\mathbb{R}^N,\mathbb{R}^N) \cap L^\infty$ (as assumed in Lemma 92) and $\mathrm{BV}^{\infty,\mathrm{div}}_{\text{loc}}(\mathbb{R}^N)$ can be bridged by means of mollifying as indicated in Proposition 80. The stability result presented in Corollary 84 implies about the limit for $\delta \downarrow 0$:

Proposition 94. *For every vector field* $\mathbf{b} \in \mathrm{BV}^{\infty,\mathrm{div}}_{\mathrm{loc}}(\mathbb{R}^N)$, *the function*

$$\vartheta_{\mathbb{L}^{\infty\cap 1},\mathbf{b}}: [0,1] \times \mathbb{L}^{\infty\cap 1}(\mathbb{R}^N) \longrightarrow \mathbb{L}^{\infty\cap 1}(\mathbb{R}^N)$$

is a transition on the tuple $\big(\mathbb{L}^{\infty\cap 1}(\mathbb{R}^N), (d_{j,\mathbb{L}^{\infty\cap 1}})_{j\in\mathscr{I}}, (d_{j,\mathbb{L}^{\infty\cap 1}})_{j\in\mathscr{I}}, \lfloor\cdot\rfloor\big)$ *with*

$$\begin{aligned}
\alpha(\vartheta_{\mathbb{L}^{\infty\cap 1},\mathbf{b}}; r) &:= \lambda_j \ \|\mathbf{b}\|_\infty \\
\beta(\vartheta_{\mathbb{L}^{\infty\cap 1},\mathbf{b}}; r) &:= \lambda_j \ \|\mathbf{b}\|_\infty \ \|\varphi_j\|_\infty \ e^{\lambda_j \|\mathbf{b}\|_\infty} \cdot r \\
\gamma(\vartheta_{\mathbb{L}^{\infty\cap 1},\mathbf{b}}) &:= 2\, \|\mathrm{div}\, \mathbf{b}\|_\infty \\
\widehat{D}_j(\vartheta_{\mathbb{L}^{\infty\cap 1},\mathbf{b}}, \vartheta_{\mathbb{L}^{\infty\cap 1},\widehat{\mathbf{b}}}; r) &:= \lambda_j \cdot r\, e^{3\,\|\mathrm{div}\,\mathbf{b}\|_\infty} \cdot \big\|\varphi_j\, |\mathbf{b} - \widehat{\mathbf{b}}|\big\|_{L^1(\mathbb{R}^N)}.
\end{aligned}$$

Moreover, for every $h \in [0,1]$ *and indices* $j \in \mathscr{I}$, $\kappa, \kappa \in \mathscr{J}$, *the function*

$$\vartheta_{\mathbb{L}^{\infty\cap 1},\mathbf{b}}(h,\cdot): \big(\mathbb{L}^{\infty\cap 1}(\mathbb{R}^N), \text{weakly* w.r.t.} C^0_0\big) \longrightarrow \big(\mathbb{L}^{\infty\cap 1}(\mathbb{R}^N), d_{j,\kappa,\kappa',\mathbb{L}^{\infty\cap 1}}\big)$$

is continuous. From now on, the set of these transitions is abbreviated as

$$\widehat{\Theta}\big(\mathbb{L}^{\infty\cap 1}(\mathbb{R}^N), (d_{j,\mathbb{L}^{\infty\cap 1}})_{j\in\mathscr{I}}, (d_{j,\mathbb{L}^{\infty\cap 1}})_{j\in\mathscr{I}}, \lfloor\cdot\rfloor\big).$$

The lacking proofs in detail are to complete this section:

Proof (of Lemma 92).
The measure-valued flow $\vartheta_{\mathbb{L}^{\infty\cap 1},\mathbf{b}}: [0,1] \times \mathbb{L}^{\infty\cap 1}(\mathbb{R}^N) \longrightarrow \mathbb{L}^{\infty\cap 1}(\mathbb{R}^N)$ still satisfies the semigroup property and thus statements (1.), (2.).
For any $\mu_0 = \rho\,\mathscr{L}^N$, $\nu_0 = \sigma\,\mathscr{L}^N \in \mathbb{L}^{\infty\cap 1}(\mathbb{R}^N)$, the definitions of total variation and push-forward imply

$$\begin{aligned}
& d_{j,\mathbb{L}^{\infty\cap 1}}\big(\vartheta_{\mathbb{L}^{\infty\cap 1},\mathbf{b}}(h,\mu_0),\ \vartheta_{\mathbb{L}^{\infty\cap 1},\mathbf{b}}(h,\nu_0)\big) \\
&= \big| \ \varphi_j \cdot \big(\mathbf{X_b}(h,\cdot)_\sharp\, \mu_0 - \mathbf{X_b}(h,\cdot)_\sharp\, \nu_0\big)\big|\,(\mathbb{R}^N) \\
&\le \int_{\mathbb{R}^N} \varphi_j(\mathbf{X_b}(h,\cdot)) \qquad |\rho - \sigma|\ d\mathscr{L}^N \\
&\le \int_{\mathbb{R}^N} \big|\,\varphi_j(\mathbf{X_b}(h,\cdot)) - \varphi_j\big|\ |\rho - \sigma|\ d\mathscr{L}^N + \big|\varphi_j \cdot (\mu_0 - \nu_0)\big|\,(\mathbb{R}^N).
\end{aligned}$$

The choice of φ_j (in Lemma 86) has the consequence

$$\begin{aligned}
& \limsup_{h\downarrow 0}\ \tfrac{1}{h} \cdot \Big(d_{j,\mathbb{L}^{\infty\cap 1}}\big(\vartheta_{\mathbb{L}^{\infty\cap 1},\mathbf{b}}(h,\mu_0),\ \vartheta_{\mathbb{L}^{\infty\cap 1},\mathbf{b}}(h,\nu_0)\big) - d_{j,\mathbb{L}^{\infty\cap 1}}(\mu_0,\nu_0)\Big) \\
&\le \limsup_{h\downarrow 0}\ \tfrac{1}{h} \cdot \int_{\mathbb{R}^N} \big|\varphi_j(\mathbf{X_b}(h,\cdot)) - \varphi_j\big|\ |\rho - \sigma|\ d\mathscr{L}^N \\
&\le \int_{\mathbb{R}^N} |\nabla\varphi_j(x)\cdot \mathbf{b}(x)| \qquad |\rho - \sigma|\ d\mathscr{L}^N \\
&\le \|\mathbf{b}\|_\infty \int_{\mathbb{R}^N} \lambda_j\ \varphi_j \qquad |\rho - \sigma|\ d\mathscr{L}^N \\
&\le \|\mathbf{b}\|_\infty \quad \lambda_j \cdot d_{j,\mathbb{L}^{\infty\cap 1}}(\mu_0,\nu_0).
\end{aligned}$$

Applying this estimate to $\nu_0 \equiv 0$ and $\vartheta_{\mathbb{L}^{\infty\cap 1},\mathbf{b}}(t,\mu_0)$ (instead of μ_0), we conclude property (4.) from Gronwall's inequality (in Proposition A.2 on page 440) because the lower semicontinuous auxiliary function

$$\delta_\varepsilon: [0,1] \longrightarrow \mathbb{R},\ t \longmapsto \big|\varphi_j\ \vartheta_{\mathbb{L}^{\infty\cap 1},\mathbf{b}}(t,\mu_0)\big|\,(\mathbb{R}^N) = \big|\varphi_j(\mathbf{X}_b(t,\cdot))\ \mu_0\big|\,(\mathbb{R}^N)$$

is one-sided differentiable and satisfies $\frac{d^+}{dt^+}\ \delta_\varepsilon(\cdot) \le \lambda_j\ \|\mathbf{b}\|_\infty \cdot \delta_\varepsilon(\cdot)$.

Correspondingly we obtain statement (5.) by estimating the auxiliary function

$$\widehat{\delta}_\varepsilon : [s,1] \longrightarrow \mathbb{R}, \ t \longmapsto \left|\varphi_j \left(\vartheta_{\mathbb{L}^{\infty\cap 1},\mathbf{b}}(t,\mu_0) - \vartheta_{\mathbb{L}^{\infty\cap 1},\mathbf{b}}(s,\mu_0)\right)\right|(\mathbb{R}^N) = \left|(\varphi_j(\mathbf{X}_b(t-s,\cdot)) - \varphi_j)\ \vartheta_{\mathbb{L}^{\infty\cap 1},\mathbf{b}}(s,\mu_0)\right|(\mathbb{R}^N)$$

with $s \in [0,1[$ fixed and

$$\tfrac{d^+}{dt^+}\ \widehat{\delta}_\varepsilon(t) \le \lambda_j \,\|\mathbf{b}\|_\infty \left|\varphi_j\ \vartheta_{\mathbb{L}^{\infty\cap 1},\mathbf{b}}(t,\mu_0)\right|(\mathbb{R}^N) \le \lambda_j \,\|\mathbf{b}\|_\infty\, e^{\lambda_j \|\mathbf{b}\|_\infty} \left|\varphi_j\, \mu_0\right|(\mathbb{R}^N).$$

In regard to property (6.), choose any $\mathbf{b}_1, \mathbf{b}_2 \in W^{1,\infty}_{\mathrm{loc}}(\mathbb{R}^N,\mathbb{R}^N) \cap L^\infty$ and initial measure $\mu_0 = \rho\,\mathscr{L}^N \in \mathbb{L}^{\infty\cap 1}(\mathbb{R}^N)$. Then, for every $h \in [0,1]$,

$$\begin{aligned}
\limsup_{h\downarrow 0}\ & \tfrac{1}{h}\cdot d_{j,\mathbb{L}^{\infty\cap 1}}\left(\vartheta_{\mathbb{L}^{\infty\cap 1},\mathbf{b}_1}(h,\mu_0),\ \vartheta_{\mathbb{L}^{\infty\cap 1},\mathbf{b}_2}(h,\mu_0)\right) \\
&\le \int_{\mathbb{R}^N} \limsup_{h\downarrow 0} \frac{|\varphi_j(\mathbf{X}_{\mathbf{b}_1}(h,\cdot)) - \varphi_j(\mathbf{X}_{\mathbf{b}_2}(h,\cdot))|}{h}\ |\rho|\ d\mathscr{L}^N \\
&\le \int_{\mathbb{R}^N} \lambda_j\ \varphi_j\ |\mathbf{b}_1 - \mathbf{b}_2|\ |\rho|\ d\mathscr{L}^N \\
&\le \lambda_j\,\|\rho\|_\infty \cdot \big\|\varphi_j\ |\mathbf{b}_1-\mathbf{b}_2|\big\|_{L^1(\mathbb{R}^N)}.
\end{aligned}$$

□

Proof (of Lemma 93). As mentioned in Proposition 80, mollifying with a Gaussian kernel leads to approximating vector fields $\mathbf{b}_\delta \in W^{1,\infty}_{\mathrm{loc}}(\mathbb{R}^N,\mathbb{R}^N)$, $\delta > 0$, with $\operatorname{div}\mathbf{b}_\delta \in L^\infty$. [4, Remark 6.3] implies for all $t \ge 0$ and $\mathscr{L}^N$-a.e. $x \in \mathbb{R}^N$

$$\exp\left(-t\ \big\|\,[\operatorname{div}_x \mathbf{b}_\delta]^-\,\big\|_\infty\right) \le \det D_x \mathbf{X}_{\mathbf{b}_\delta}(t,x) \le \exp\left(t\ \big\|\,[\operatorname{div}_x \mathbf{b}_\delta]^+\,\big\|_\infty\right).$$

Now we conclude from the area formula and the transformation of Lebesgue integrals that for any $\mu = \sigma\,\mathscr{L}^N$ with $\sigma \in L^1(\mathbb{R}^N)\cap L^\infty(\mathbb{R}^N)$,

$$\begin{aligned}
\left|\vartheta_{\mathbb{L}^{\infty\cap 1},\mathbf{b}_\delta}(t,\mu)\right|(\mathbb{R}^N) &= \left|\mathbf{X}_{\mathbf{b}_\delta}(t,\cdot)_\sharp\,\mu\right|(\mathbb{R}^N) \\
&= \int_{\mathbb{R}^N} \left|\frac{\sigma}{|\det D_x \mathbf{X}_{\mathbf{b}_\delta}(t,\cdot)|} \circ \mathbf{X}_{\mathbf{b}_\delta}(t,\cdot)^{-1}\right| d\mathscr{L}^N \\
&\le \int_{\mathbb{R}^N} \left|\sigma \circ \left(\mathbf{X}_{\mathbf{b}_\delta}(t,\cdot)^{-1}\right)\right|\ d\mathscr{L}^N \cdot \exp\left(t\ \big\|\,[\operatorname{div}_x \mathbf{b}_\delta]^-\,\big\|_\infty\right) \\
&\le \int_{\mathbb{R}^N} |\sigma|\ d\mathscr{L}^N \cdot \big\|\det D_x \mathbf{X}_{\mathbf{b}_\delta}(t,\cdot)\big\|_\infty \cdot \exp\left(t\ \big\|\,[\operatorname{div}_x \mathbf{b}_\delta]^-\,\big\|_\infty\right).
\end{aligned}$$

According to Corollary 84, $\vartheta_{\mathbb{L}^{\infty\cap 1},\mathbf{b}_\delta}(t,\mu)$ converges narrowly to $\vartheta_{\mathbb{L}^{\infty\cap 1},\mathbf{b}}(t,\mu)$ for $\delta \downarrow 0$. In particular, the total variation is lower semicontinuous with respect to weak* convergence (see e.g. [5, Theorem 1.59]) and thus,

$$\left|\vartheta_{\mathbb{L}^{\infty\cap 1},\mathbf{b}}(t,\mu)\right|(\mathbb{R}^N) \le \liminf_{\delta\downarrow 0} \left|\vartheta_{\mathbb{L}^{\infty\cap 1},\mathbf{b}_\delta}(t,\mu)\right|(\mathbb{R}^N) \le \|\sigma\|_{L^1}\ e^{2\,\|\operatorname{div}\mathbf{b}\|_\infty\, t}.$$

For proving the first statement, we start with the duality relation between L^1 and L^∞ and then use the area formula. Indeed, the L^∞ norm of σ_t is equal to

$$
\begin{aligned}
&\sup\Big\{ \int \psi\sigma_t \;\, d\mathscr{L}^N \Big|\; \psi \in C_0^\infty(\mathbb{R}^N),\;\, \|\psi\|_{L^1} \leq 1 \Big\} \\
&\quad = \sup\Big\{ \limsup_{\delta\downarrow 0} \int \psi \; d\,\vartheta_{\mathbb{L}^{\infty\cap 1},\mathbf{b}_\delta}(t,\mu) \Big|\; \psi \in C_0^\infty(\mathbb{R}^N),\;\, \|\psi\|_{L^1} \leq 1 \Big\} \\
&\quad = \sup\Big\{ \limsup_{\delta\downarrow 0} \int \psi\Big(\frac{\sigma}{\det D_x \mathbf{X}_{\mathbf{b}_\delta}(t,\cdot)}\Big)\Big|_{\mathbf{X}_{\mathbf{b}_\delta}(t,\cdot)^{-1}} \; d\mathscr{L}^N \Big|\; \psi \in C_0^\infty(\mathbb{R}^N), \|\psi\|_{L^1} \leq 1 \Big\} \\
&\quad \leq \sup\Big\{ \limsup_{\delta\downarrow 0} \int \;\psi \|\sigma\|_\infty \;\, e^{\|\mathrm{div}\, \mathbf{b}_\delta\|_\infty\, t} d\mathscr{L}^N \Big|\; \psi \in C_0^\infty(\mathbb{R}^N),\;\, \|\psi\|_{L^1} \leq 1 \Big\} \\
&\quad \leq \|\sigma\|_\infty \;\, e^{\|\mathrm{div}\, \mathbf{b}\|_\infty\, t}\,.
\end{aligned}
$$
□

Proof (of Proposition 94). Choose a Gaussian kernel $\rho \in C^1(\mathbb{R}^N,]0,\infty[)$ and set $\rho_\delta(x) := \delta^{-N}\rho(\frac{x}{\delta})$ for $\delta > 0$. Each vector field $\mathbf{b}_\delta := \mathbf{b} * \rho_\delta$ belongs to $W^{1,\infty}_{\mathrm{loc}}(\mathbb{R}^N,\mathbb{R}^N)$ and satisfies $\|\mathbf{b}_\delta\|_\infty \leq \|\mathbf{b}\|_\infty < \infty$, $\|\mathrm{div}_x\, \mathbf{b}_\delta\|_\infty \leq \|\mathrm{div}_x\, \mathbf{b}\|_\infty < \infty$.

Hence, for each $\mathbf{b} \in \mathrm{BV}^{\infty,\mathrm{div}}_{\mathrm{loc}}(\mathbb{R}^N)$ and $\delta > 0$, Lemmas 92 and 93 imply the transition properties of $\vartheta_{\mathbb{L}^{\infty\cap 1},\mathbf{b}_\delta}(\cdot,\cdot) : [0,1] \times \mathbb{L}^{\infty\cap 1}(\mathbb{R}^N) \longrightarrow \mathbb{L}^{\infty\cap 1}(\mathbb{R}^N)$ with the parameters

$$
\begin{array}{llll}
\alpha(\vartheta_{\mathbb{L}^{\infty\cap 1},\mathbf{b}_\delta};\, r) & := \;\lambda_j \;\|\mathbf{b}_\delta\|_\infty & & \leq\; \lambda_j \;\|\mathbf{b}\|_\infty, \\
\beta(\vartheta_{\mathbb{L}^{\infty\cap 1},\mathbf{b}_\delta};\, r) & := \;\lambda_j \;\|\mathbf{b}_\delta\|_\infty \;\|\varphi_j\|_\infty \; e^{\lambda_j\, \|\mathbf{b}_\delta\|_\infty}\, r, & & \leq\; \lambda_j \;\|\mathbf{b}\|_\infty \;\|\varphi_j\|_\infty \; e^{\lambda_j\, \|\mathbf{b}\|_\infty}\, r \\
\gamma(\vartheta_{\mathbb{L}^{\infty\cap 1},\mathbf{b}_\delta}) & := \;2 \;\;\|\mathrm{div}\, \mathbf{b}_\delta\|_\infty & & \leq\; 2 \;\;\|\mathrm{div}\, \mathbf{b}\|_\infty.
\end{array}
$$

Moreover for arbitrary $\mathbf{b}, \widehat{\mathbf{b}} \in \mathrm{BV}^{\infty,\mathrm{div}}_{\mathrm{loc}}(\mathbb{R}^N)$, $\mu_1, \mu_2 \in \mathbb{L}^{\infty\cap 1}(\mathbb{R}^N)$ and $\delta, \widehat{\delta} > 0$, $h \in [0,1]$, we conclude

$$
\begin{aligned}
&d_{j,\mathbb{L}^{\infty\cap 1}}\big(\vartheta_{\mathbb{L}^{\infty\cap 1},\mathbf{b}_\delta}(h,\, \mu_1),\;\; \vartheta_{\mathbb{L}^{\infty\cap 1},\widehat{\mathbf{b}}_{\widehat{\delta}}}(h,\, \mu_2)\big) \\
&\leq \Big(d_{j,\mathbb{L}^{\infty\cap 1}}(\mu_1,\mu_2) \;+\; \lambda_j \cdot \sup_{[0,1]} \Big\| \frac{\vartheta_{\mathbb{L}^{\infty\cap 1},\mathbf{b}_\delta}(\cdot,\mu_1)}{\mathscr{L}^N} \Big\|_\infty \cdot \big\|\varphi_j\, |\mathbf{b}_\delta - \widehat{\mathbf{b}}_{\widehat{\delta}}|\big\|_{L^1(\mathbb{R}^N)}\Big)\; e^{\lambda_j\, \|\mathbf{b}_\delta\|_\infty\, h} \\
&\leq \Big(d_{j,\mathbb{L}^{\infty\cap 1}}(\mu_1,\mu_2) \;+\; \lambda_j \cdot\; \lfloor \mu_1 \rfloor \; e^{\|\mathrm{div}\, \mathbf{b}\|_\infty} \qquad\qquad \cdot \big\|\varphi_j\, |\mathbf{b}_\delta - \widehat{\mathbf{b}}_{\widehat{\delta}}|\big\|_{L^1(\mathbb{R}^N)}\Big)\; e^{\lambda_j\, \|\mathbf{b}_\delta\|_\infty\, h}
\end{aligned}
$$

from Lemma 92 (6.), Lemma 93 and Gronwall's inequality in exactly the same way as for Proposition 2.6 (on page 106). In particular, this estimate motivates

$$
\begin{aligned}
\widehat{D}_j(\vartheta_{\mathbb{L}^{\infty\cap 1},\mathbf{b}_\delta},\; \vartheta_{\mathbb{L}^{\infty\cap 1},\widehat{\mathbf{b}}_{\widehat{\delta}}};\, r) \;&:=\; \lambda_j \;\cdot\, r\, e^{3\, \|\mathrm{div}\, \mathbf{b}_\delta\|_\infty} \cdot \big\|\varphi_j\, |\mathbf{b}_\delta - \widehat{\mathbf{b}}_{\widehat{\delta}}|\big\|_{L^1(\mathbb{R}^N)} \\
&\leq\; \lambda_j \;\cdot\, r\, e^{3\, \|\mathrm{div}\, \mathbf{b}\|_\infty} \;\cdot \big\|\varphi_j\, |\mathbf{b}_\delta - \widehat{\mathbf{b}}_{\widehat{\delta}}|\big\|_{L^1(\mathbb{R}^N)}\,.
\end{aligned}
$$

For arbitrary vector fields $\mathbf{b}, \widehat{\mathbf{b}} \in \mathrm{BV}^{\infty,\mathrm{div}}_{\mathrm{loc}}(\mathbb{R}^N)$ and measures $\mu_1, \mu_2 \in \mathbb{L}^{\infty\cap 1}(\mathbb{R}^N)$, we now consider the limit for $\delta \downarrow 0$ and conclude from the narrow convergence mentioned in Corollary 84

$$
\begin{aligned}
&d_{j,\mathbb{L}^{\infty\cap 1}}\big(\vartheta_{\mathbb{L}^{\infty\cap 1},\mathbf{b}}(h,\, \mu_1), \vartheta_{\mathbb{L}^{\infty\cap 1},\widehat{\mathbf{b}}}(h,\, \mu_2)\big) \\
&\quad = \sup_{\kappa,\kappa' \in \mathscr{J}} \; d_{j,\kappa,\kappa',\mathbb{L}^{\infty\cap 1}}\big(\vartheta_{\mathbb{L}^{\infty\cap 1},\mathbf{b}}(h,\, \mu_1), \vartheta_{\mathbb{L}^{\infty\cap 1},\widehat{\mathbf{b}}}(h,\, \mu_2)\big) \\
&\quad = \sup_{\kappa,\kappa' \in \mathscr{J}} \; \lim_{\delta\downarrow 0} d_{j,\kappa,\kappa',\mathbb{L}^{\infty\cap 1}}\big(\vartheta_{\mathbb{L}^{\infty\cap 1},\mathbf{b}_\delta}(h,\, \mu_1), \vartheta_{\mathbb{L}^{\infty\cap 1},\widehat{\mathbf{b}}_{\widehat{\delta}}}(h,\, \mu_2)\big) \\
&\quad \leq \limsup_{\delta\downarrow 0} d_{j,\mathbb{L}^{\infty\cap 1}}\big(\vartheta_{\mathbb{L}^{\infty\cap 1},\mathbf{b}_\delta}(h,\, \mu_1),\;\; \vartheta_{\mathbb{L}^{\infty\cap 1},\widehat{\mathbf{b}}_{\widehat{\delta}}}(h,\, \mu_2)\big) \\
&\quad \leq \Big(d_{j,\mathbb{L}^{\infty\cap 1}}(\mu_1,\mu_2) \;+\; \lambda_j \cdot\; \lfloor \mu_1 \rfloor \; e^{\|\mathrm{div}\, \mathbf{b}\|_\infty} \cdot \big\|\varphi_j\, |\mathbf{b} - \widehat{\mathbf{b}}|\big\|_{L^1(\mathbb{R}^N)}\Big)\; e^{\lambda_j\, \|\mathbf{b}\|_\infty\, h}\,.
\end{aligned}
$$

As a consequence of Lemma 93, $\vartheta_{\mathbb{L}^{\infty\cap 1},\mathbf{b}}(\cdot,\cdot) : [0,1] \times \mathbb{L}^{\infty\cap 1}(\mathbb{R}^N) \longrightarrow \mathbb{L}^{\infty\cap 1}(\mathbb{R}^N)$ fulfills all conditions on a transition with

$$\begin{aligned}
\alpha(\vartheta_{\mathbb{L}^{\infty\cap 1},\mathbf{b}};\, r) &:= \lambda_j \ \|\mathbf{b}\|_\infty \\
\beta(\vartheta_{\mathbb{L}^{\infty\cap 1},\mathbf{b}};\, r) &:= \lambda_j \ \|\mathbf{b}\|_\infty \ \|\varphi_j\|_\infty \ e^{\lambda_j \|\mathbf{b}\|_\infty} \cdot r \\
\gamma(\vartheta_{\mathbb{L}^{\infty\cap 1},\mathbf{b}}) &:= 2 \, \|\mathrm{div}\, \mathbf{b}\|_\infty \\
\widehat{D}_j(\vartheta_{\mathbb{L}^{\infty\cap 1},\mathbf{b}},\, \vartheta_{\mathbb{L}^{\infty\cap 1},\widehat{\mathbf{b}}};\, r) &:= \lambda_j \cdot r\, e^{3\,\|\mathrm{div}\,\mathbf{b}\|_\infty} \cdot \big\| \varphi_j \, |\mathbf{b} - \widehat{\mathbf{b}}| \big\|_{L^1(\mathbb{R}^N)}.
\end{aligned}$$

Finally, we have to verify that for every $h \in [0,1]$ and indices $j \in \mathscr{I},\ \kappa, \kappa \in \mathscr{J}$, the function

$$\vartheta_{\mathbb{L}^{\infty\cap 1},\mathbf{b}}(h,\cdot) : \ \big(\mathbb{L}^{\infty\cap 1}(\mathbb{R}^N), \text{ weakly* w.r.t. } C_0^0\big) \longrightarrow \big(\mathbb{L}^{\infty\cap 1}(\mathbb{R}^N),\, d_{j,\kappa,\kappa',\mathbb{L}^{\infty\cap 1}}\big)$$

is continuous.
Let $\big(\mu_n = \sigma_n \mathscr{L}^N\big)_{n\in\mathbb{N}}$ be any sequence in $\mathbb{L}^{\infty\cap 1}(\mathbb{R}^N)$ converging weakly* to $\mu = \sigma \mathscr{L}^N \in \mathbb{L}^{\infty\cap 1}(\mathbb{R}^N)$. Choose $h \in]0,1]$, $\delta > 0$ and $\varphi \in C_0^0(\mathbb{R}^N)$ arbitrarily.

Using a smooth Gaussian kernel ρ as described in Proposition 80 (on page 262), the mollified measure $\mu_\delta(t) := \vartheta_{\mathbb{L}^{\infty\cap 1},\mathbf{b}}(t,\mu) * \rho_\delta$ solves the nonautonomous continuity equation

$$\tfrac{d}{dt}\, \mu_\delta \; + \; \mathrm{div}_x\,(\widetilde{\mathbf{b}}_\delta\, \mu_\delta) \;=\; 0 \qquad \text{(in the distributional sense)}$$

with the time-dependent vector field $\widetilde{\mathbf{b}}_\delta(t,\cdot) := \frac{(\mathbf{b}\,\mu(t)) * \rho_\delta}{\mu_\delta(t)}$ belonging to the function space $L^1\big([0,T],\, W^{1,\infty}_{\mathrm{loc}}(\mathbb{R}^N,\mathbb{R}^N)\big)$. In comparison to the Lagrangian flow of $\mathbf{b} \in \mathrm{BV}^{\infty,\mathrm{div}}_{\mathrm{loc}}(\mathbb{R}^N)$, the flow $\mathbf{X}_{\widetilde{\mathbf{b}}_\delta} : [0,T] \times \mathbb{R}^N \longrightarrow \mathbb{R}^N$ along $\widetilde{\mathbf{b}}_\delta$ has the supplementary advantage of being continuous and, the solution can be represented as push-forward

$$\mu_\delta(t) \;=\; \mathbf{X}_{\mathbf{b}_\delta}(t,\cdot)_\sharp\,(\mu(0) * \rho_\delta).$$

Now we conclude from the well-known features of convolution

$$\begin{aligned}
\int_{\mathbb{R}^N} \varphi * \rho_\delta \ \ d\,\vartheta_{\mathbb{L}^{\infty\cap 1},\mathbf{b}}(h,\mu) &= \int_{\mathbb{R}^N} \varphi \ d\,\big(\vartheta_{\mathbb{L}^{\infty\cap 1},\mathbf{b}}(h,\mu) * \rho_\delta\big) \\
&= \int_{\mathbb{R}^N} \varphi \ \ d\,\mu_\delta(h) \\
&= \int_{\mathbb{R}^N} \varphi\big(\mathbf{X}_{\mathbf{b}_\delta}(h,\cdot)\big) * \rho_\delta \ \ \sigma \ \ d\mathscr{L}^N \\
&= \lim_{n\to\infty} \int_{\mathbb{R}^N} \varphi\big(\mathbf{X}_{\mathbf{b}_\delta}(h,\cdot)\big) * \rho_\delta \ \ \sigma_n \ \ d\mathscr{L}^N = \ldots \\
&= \lim_{n\to\infty} \int_{\mathbb{R}^N} (\varphi * \rho_\delta) \ d\,\vartheta_{\mathbb{L}^{\infty\cap 1},\mathbf{b}}(h,\mu_n),
\end{aligned}$$

i.e., $\displaystyle\int_{\mathbb{R}^N} \psi \quad d\,\vartheta_{\mathbb{L}^{\infty\cap 1},\mathbf{b}}(h,\mu) \;=\; \lim_{n\to\infty} \int_{\mathbb{R}^N} \psi \quad d\,\vartheta_{\mathbb{L}^{\infty\cap 1},\mathbf{b}}(h,\mu_n)$

for all functions ψ in a dense subset of $\big(C_0^0(\mathbb{R}^N),\, \|\cdot\|_{\sup}\big)$. Due to the uniform bound of total variation, i.e. $\sup_n\, |\vartheta_{\mathbb{L}^{\infty\cap 1},\mathbf{b}}(h,\mu_n)|(\mathbb{R}^N) \;\le\; \sup_n\, |\mu_n|(\mathbb{R}^N) \cdot e^{2\,\|\mathrm{div}\,\mathbf{b}\|_\infty} < \infty$, we obtain $\vartheta_{\mathbb{L}^{\infty\cap 1},\mathbf{b}}(h,\mu_n) \longrightarrow \vartheta_{\mathbb{L}^{\infty\cap 1},\mathbf{b}}(h,\mu)$ weakly* with respect to $C_0^0(\mathbb{R}^N)$ and, thus the claimed continuity of $\vartheta_{\mathbb{L}^{\infty\cap 1},\mathbf{b}}(h,\cdot)$ w.r.t. every $d_{j,\kappa,\kappa',\mathbb{L}^{\infty\cap 1}}$. □

3.8.4 Conclusions About Nonlinear Continuity Equations

Now we specify sufficient conditions on the functional coefficient

$$\mathbf{f}: \mathbb{L}^{\infty \cap 1}(\mathbb{R}^N) \times [0,T] \longrightarrow \mathrm{BV}^{\infty,\mathrm{div}}_{\mathrm{loc}}(\mathbb{R}^N)$$

for the nonlinear Cauchy problem

$$\begin{cases} \frac{d}{dt}\,\mu + \mathrm{div}_x\,(\mathbf{f}(\mu,\cdot)\,\mu) \;=\; 0 & \text{in } \mathbb{R}^N \times\,]0,T[\\ \mu(0) \;=\; \rho_0\, \mathscr{L}^N \in \mathbb{L}^{\infty\cap 1}(\mathbb{R}^N) \end{cases}$$

being well-posed in the distributional sense. The transitions introduced in Definition 91 (on page 266) and the general results of § 3.3.6 (about solving mutational equations via a generalized form of "weak" compactness) are to provide the required tools for existence. In particular, the additional hypothesis (H4') (on page 207) results from the lower semicontinuity of total variation.
After formulating the main results of this example, we collect all proofs at the end.

Lemma 95. *(1.) The tuple* $\big(\mathbb{L}^{\infty\cap 1}(\mathbb{R}^N),\ (d_{j,\mathbb{L}^{\infty\cap 1}})_{j\in\mathscr{I}},\ (d_{j,\kappa,\kappa',\mathbb{L}^{\infty\cap 1}})_{j,\kappa,\kappa'},$ $(d_{j,\mathbb{L}^{\infty\cap 1}})_{j\in\mathscr{I}},\ (d_{j,\kappa,\kappa',\mathbb{L}^{\infty\cap 1}})_{j,\kappa,\kappa'},\ \lfloor\cdot\rfloor,\ \widehat{\Theta}\big(\mathbb{L}^{\infty\cap 1}(\mathbb{R}^N), (d_{j,\mathbb{L}^{\infty\cap 1}}), (d_{j,\mathbb{L}^{\infty\cap 1}}), \lfloor\cdot\rfloor\big)\big)$ *with the pseudo-metrics specified in Definition 87 (on page 264) and the transitions of Proposition 94 (on page 268) is weakly Euler compact (in the sense of Definition 27 on page 207).*

(2.) The tuple $\big(\mathbb{L}^{\infty\cap 1}(\mathbb{R}^N),\ (d_{j,\mathbb{L}^{\infty\cap 1}})_{j\in\mathscr{I}},\ (d_{j,\mathbb{L}^{\infty\cap 1}})_{j\in\mathscr{I}},\ \lfloor\cdot\rfloor\big)$ *in combination with the transitions in* $\widehat{\Theta}\big(\mathbb{L}^{\infty\cap 1}(\mathbb{R}^N), (d_{j,\mathbb{L}^{\infty\cap 1}}), (d_{j,\mathbb{L}^{\infty\cap 1}}), \lfloor\cdot\rfloor\big)$ *is Euler equi-continuous (in the sense of Definition 16 on page 194).*

Theorem 96 (Existence of $\mathbb{L}^{\infty\cap 1}(\mathbb{R}^N)$-valued solutions).
For $\mathbf{f}: \mathbb{L}^{\infty\cap 1}(\mathbb{R}^N) \times [0,T] \longrightarrow \mathrm{BV}^{\infty,\mathrm{div}}_{\mathrm{loc}}(\mathbb{R}^N)$ *suppose*

(i) $\sup_{\mu,t}\ \big(\|\mathbf{f}(\mu,t)\|_{L^\infty} + \|\mathrm{div}_x\,\mathbf{f}(\mu,t)\|_{L^\infty}\big) < \infty,$

(ii) $\mathbf{f}$ *is continuous in the following sense: For* $\mathscr{L}^1$*-almost every* $t \in [0,T]$ *and any sequences* $(t_m)_{m\in\mathbb{N}},\ (\mu_m = \sigma_m\,\mathscr{L}^N)_{m\in\mathbb{N}}$ *in* $[0,T],\ \mathbb{L}^{\infty\cap 1}(\mathbb{R}^N)$ *respectively with*

$$\begin{cases} t_m \longrightarrow t & \text{for } m \longrightarrow \infty, \\ \mu_m \longrightarrow \mu \ \text{ weakly* with respect to } C^0_0(\mathbb{R}^N) & \text{for } m \longrightarrow \infty, \\ \sup_{m\in\mathbb{N}}\ \big(\|\sigma_m\|_{L^1} + \|\sigma_m\|_{L^\infty}\big) < \infty, \end{cases}$$

it fulfills $\mathbf{f}(\mu_m,t_m) \longrightarrow \mathbf{f}(\mu,t)$ *in* $L^1_{\mathrm{loc}}(\mathbb{R}^N,\mathbb{R}^N)$ *for* $m \longrightarrow \infty$.

Then for every initial measure $\mu_0 = \sigma_0\,\mathscr{L}^N \in \mathbb{L}^{\infty\cap 1}(\mathbb{R}^N)$, *there exists a solution* $\mu(\cdot): [0,T] \longrightarrow \mathbb{L}^{\infty\cap 1}(\mathbb{R}^N)$ *to the mutational equation*

$$\overset{\circ}{\mu}(\cdot) \ni \vartheta_{\mathbb{L}^{\infty\cap 1},\,\mathbf{f}(\mu(\cdot),\cdot)}$$

on the tuple $\big(\mathbb{L}^{\infty\cap 1}(\mathbb{R}^N),\ (d_{j,\mathbb{L}^{\infty\cap 1}})_{j\in\mathscr{I}},\ (d_{j,\mathbb{L}^{\infty\cap 1}})_{j\in\mathscr{I}},\ \lfloor\cdot\rfloor,\ (\widehat{D}_j)_{j\in\mathscr{I}}\big)$ *satisfying* $\mu(0) = \mu_0$ *and, all its values in* $\mathbb{L}^{\infty\cap 1}(\mathbb{R}^N)$ *are tight.*

Moreover every solution $\mu(\cdot) : [0,T] \longrightarrow \mathbb{L}^{\infty\cap 1}(\mathbb{R}^N)$ *(to this mutational equation) with tight values in* $\mathbb{L}^{\infty\cap 1}(\mathbb{R}^N)$ *is a narrowly continuous distributional solution to the nonlinear continuity equation*

$$\partial_t \, \mu_t \; + \; \mathrm{div}_x \, (\mathbf{f}(\mu_t,t) \; \mu_t) \; = \; 0 \qquad \text{in } \mathbb{R}^N \times]0,T[$$

in the sense that for every $t \in [0,T]$ *and any test function* $\varphi \in C_c^\infty(\mathbb{R}^N, \mathbb{R})$,

$$\int_{\mathbb{R}^N} \varphi \; d\mu_t - \int_{\mathbb{R}^N} \varphi \; d\mu_0 = \int_0^t \int_{\mathbb{R}^N} \nabla\varphi(x) \cdot \mathbf{f}_1(\mu_s,s)(x) \; d\mu_s(x) \; ds.$$

Remark 97. In § 3.3.6, Theorem 28 (on page 207) states the existence of solutions to mutational equations *with delay*. Strictly speaking, we can even handle $\mathbb{L}^{\infty\cap 1}(\mathbb{R}^N)$-valued solutions to nonlinear continuity equations with delay.

The uniqueness of $\mathbb{L}^{\infty\cap 1}(\mathbb{R}^N)$-valued solutions to the linear, but nonautonomous continuity equation is guaranteed by Proposition 80 of Ambrosio and, it is the starting point for the opposite implication:

Proposition 98 (Distributional solutions satisfy mutational equation).
For $\mathbf{f} : \mathbb{L}^{\infty\cap 1}(\mathbb{R}^N) \times [0,T] \longrightarrow \mathrm{BV}_{\mathrm{loc}}^{\infty,\mathrm{div}}(\mathbb{R}^N)$ *suppose*

(*i*) $\sup_{\mu,t} \; \left(\big\|\mathbf{f}(\mu,t)\big\|_{L^\infty} + \big\|\mathrm{div}_x \, \mathbf{f}(\mu,t)\big\|_{L^\infty}\right) < \infty$,

(*ii'*) $\mathbf{f}$ *is continuous in the following sense: For* $\mathscr{L}^1$*-almost every* $t \in [0,T]$ *and any sequences* $(t_m)_{m\in\mathbb{N}}$, $(\mu_m = \sigma_m \mathscr{L}^N)_{m\in\mathbb{N}}$ *in* $[0,T]$, $\mathbb{L}^{\infty\cap 1}(\mathbb{R}^N)$ *respectively,* $\mu = \sigma \mathscr{L}^N \in \mathbb{L}^{\infty\cap 1}(\mathbb{R}^N)$ *with*

$$\begin{cases} t_m \longrightarrow t & \text{for } m \longrightarrow \infty, \\ \sigma_m \longrightarrow \sigma \quad \text{in } L^1_{\mathrm{loc}}(\mathbb{R}^N) & \text{for } m \longrightarrow \infty, \\ \sup\limits_{m\in\mathbb{N}} \; \left(\|\sigma_m\|_{L^1} + \|\sigma_m\|_{L^\infty}\right) < \infty, \end{cases}$$

it fulfills $\mathbf{f}(\mu_m,t_m) \longrightarrow \mathbf{f}(\mu,t)$ *in* $L^1_{\mathrm{loc}}(\mathbb{R}^N,\mathbb{R}^N)$ *for* $m \longrightarrow \infty$.

Let $\mu(\cdot) = \sigma(\cdot)\,\mathscr{L}^N : [0,T] \longrightarrow \mathbb{L}^{\infty\cap 1}(\mathbb{R}^N)$ *be a distributional solution of*

$$\partial_t \, \mu_t \; + \; \mathrm{div}_x \, (\mathbf{f}(\mu_t,t) \; \mu_t) \; = \; 0$$

with the properties

(*a*) $\{\mu(t) \mid 0 \le t \le T\} \subset \mathbb{L}^{\infty\cap 1}(\mathbb{R}^N)$ *is tight,*
(*b*) $\sigma(\cdot) : [0,T] \longrightarrow L^1_{\mathrm{loc}}(\mathbb{R}^N)$ *is continuous,*
(*c*) $\|\sigma(\cdot)\|_{L^1(\mathbb{R}^N)} + \|\sigma(\cdot)\|_{L^\infty(\mathbb{R}^N)}$ *is bounded in* $[0,T]$.

Then, $\mu(\cdot)$ *solves the mutational equation*

$$\overset{\circ}{\mu}(\cdot) \ni \vartheta_{\mathbb{L}^{\infty\cap 1},\, \mathbf{f}(\mu(\cdot),\cdot)}$$

on the tuple $\left(\mathbb{L}^{\infty\cap 1}(\mathbb{R}^N),\, (d_{j,\mathbb{L}^{\infty\cap 1}})_{j\in\mathscr{I}},\, (d_{j,\mathbb{L}^{\infty\cap 1}})_{j\in\mathscr{I}},\, \lfloor\cdot\rfloor,\, (\widehat{D}_j)_{j\in\mathscr{I}}\right)$.

Uniqueness and continuous dependence on data result directly from the general statements about mutational equations (in § 3.3.1 on page 189 f.) and the local specification of transitions in Proposition 94 (on page 268). Thus we even dispense with their proofs in detail.

Theorem 99 (Uniqueness of solution to nonlinear continuity equation). *For* $\mathbf{f} : \mathbb{L}^{\infty\cap 1}(\mathbb{R}^N) \times [0,T] \longrightarrow \mathrm{BV}_{\mathrm{loc}}^{\infty,\mathrm{div}}(\mathbb{R}^N)$ *suppose*

(*i*) $\sup_{\mu,t} \left(\|\mathbf{f}(\mu,t)\|_{L^\infty} + \|\mathrm{div}_x \mathbf{f}(\mu,t)\|_{L^\infty}\right) < \infty$,

(*ii'*) $\mathbf{f}$ *is continuous in the sense specified in assumption* (*ii'*) *of Proposition 98.*

(*iii*) $\mathbf{f}$ *is Lipschitz continuous with respect to state in the following sense: For each* $j \in \mathscr{I}$, *there exists a constant* $\Lambda_j > 0$ *such that for* $\mathscr{L}^1$*-almost every* $t \in [0,T]$ *and every* $\nu_1, \nu_2 \in \mathbb{L}^{\infty\cap 1}(\mathbb{R}^N)$,

$$\left\|\varphi_j \left|\mathbf{f}(\nu_1,t) - \mathbf{f}(\nu_2,t)\right|\right\|_{L^1(\mathbb{R}^N)} \le \Lambda_j \cdot d_{j,\mathbb{L}^{\infty\cap 1}}(\nu_1, \nu_2).$$

Then for every $\mu_0 \in \mathbb{L}^{\infty\cap 1}(\mathbb{R}^N)$, *the distributional solution* $[0,T] \longrightarrow \mathbb{L}^{\infty\cap 1}(\mathbb{R}^N)$, $t \longmapsto \mu_t = \sigma(t)\,\mathscr{L}^N$ *to the nonlinear continuity equation*

$$\partial_t\, \mu_t \;+\; \mathrm{div}_x \left(\mathbf{f}(\mu_t,t)\; \mu_t\right) \;=\; 0 \qquad \text{in } \mathbb{R}^N \times\,]0,T[$$

being continuous w.r.t. $L^1_{\mathrm{loc}}(\mathbb{R}^N)$, *bounded w.r.t.* $\|\cdot\|_{L^1(\mathbb{R}^N)} + \|\cdot\|_{L^\infty(\mathbb{R}^N)}$, *having initial measure* μ_0 *at time* $t = 0$ *and tight values in* $\mathbb{L}^{\infty\cap 1}(\mathbb{R}^N)$ *is unique.*

Theorem 100 (Continuous dependence on initial data and coefficients). *For* $\mathbf{f}, \mathbf{g} : \mathbb{L}^{\infty\cap 1}(\mathbb{R}^N) \times [0,T] \longrightarrow \mathrm{BV}_{\mathrm{loc}}^{\infty,\mathrm{div}}(\mathbb{R}^N)$ *suppose*

(*i*) $\sup_{\mu,t} \left(\|\mathbf{f}(\mu,t)\|_{L^\infty} + \|\mathrm{div}_x \mathbf{f}(\mu,t)\|_{L^\infty}\right) < \infty$,
$\sup_{\mu,t} \left(\|\mathbf{g}(\mu,t)\|_{L^\infty} + \|\mathrm{div}_x \mathbf{g}(\mu,t)\|_{L^\infty}\right) < \infty$,

(*ii*) $\mathbf{f}$ *and* $\mathbf{g}$ *are continuous in the sense specified in assumption* (*ii*) *of preceding Existence Theorem 96.*

(*iii*) $\mathbf{f}$ *is Lipschitz continuous with respect to state as in Uniqueness Theorem 99.*

Let $\mu(\cdot) : [0,T] \longrightarrow \mathbb{L}^{\infty\cap 1}(\mathbb{R}^N)$, $t \longmapsto \rho(t)\,\mathscr{L}^N$ *be a distributional solution of*

$$\partial_t\, \mu_t \;+\; \mathrm{div}_x \left(\mathbf{f}(\mu_t,t)\; \mu_t\right) \;=\; 0 \qquad \text{in } \mathbb{R}^N \times\,]0,T[$$

being continuous w.r.t. $L^1_{\mathrm{loc}}(\mathbb{R}^N)$, *bounded w.r.t.* $\|\cdot\|_{L^1(\mathbb{R}^N)} + \|\cdot\|_{L^\infty(\mathbb{R}^N)}$ *and having tight values in* $\mathbb{L}^{\infty\cap 1}(\mathbb{R}^N)$.

For any parameter $R > 0$, *there exist constants* $C_j > 0$ ($j \in \mathscr{I}$) *depending only on* $\mathbf{f}$, $\mathbf{g}$, $\lfloor \mu_0 \rfloor$, R *with the following property:*

For every measure $\nu_0 = \sigma_0\,\mathscr{L}^N \in \mathbb{L}^{\infty\cap 1}(\mathbb{R}^N)$ *with* $\|\sigma_0\|_{L^1(\mathbb{R}^N)} + \|\sigma_0\|_{L^\infty(\mathbb{R}^N)} \le R$, *there is a narrowly continuous distributional solution* $\nu(\cdot) : [0,T] \longrightarrow \mathbb{L}^{\infty\cap 1}(\mathbb{R}^N)$, $t \longmapsto \sigma(t)\,\mathscr{L}^N$ *to the continuity equation*

$$\partial_t\, \nu_t \;+\; \mathrm{div}_x \left(\mathbf{g}(\nu_t,t)\; \nu_t\right) \;=\; 0 \qquad \text{in } \mathbb{R}^N \times\,]0,T[$$

being bounded w.r.t. $\|\cdot\|_{L^1(\mathbb{R}^N)} + \|\cdot\|_{L^\infty(\mathbb{R}^N)}$, *having initial measure* ν_0 *at time* $t = 0$ *and satisfying for every* $t \in [0,T]$ *and* $j \in \mathscr{I}$ *additionally*

$$\left\|\varphi_j\,(\rho(t) - \sigma(t))\right\|_{L^1} \le \left(\left\|\varphi_j\,(\rho_0 - \sigma_0)\right\|_{L^1(\mathbb{R}^N)} + C_j \cdot \sup \left\|\varphi_j\,(\mathbf{f} - \mathbf{g})\right\|_{L^1(\mathbb{R}^N)}\right) e^{C_j t}.$$

Proof (of Lemma 95). (1.) In regard to Definition 27 (on page 207) and Lemma 90 (on page 265), choose $\mu_0 \in \mathbb{L}^{\infty\cap 1}(\mathbb{R}^N)$, $T > 0$ and $R > 0$ arbitrarily and let $\mathscr{N} = \mathscr{N}(\mu_0, T, R)$ denote the subset of all curves $\mu(\cdot) : [0,T] \longrightarrow \mathbb{L}^{\infty\cap 1}(\mathbb{R}^N)$ constructed in the following piecewise way: Choosing an arbitrary equidistant partition $0 = t_0 < t_1 < \ldots < t_n = T$ of $[0,T]$ $(n > T)$ and $\mathbf{b}_1 \ldots \mathbf{b}_n \in \mathrm{BV}^{\infty,\mathrm{div}}_{\mathrm{loc}}(\mathbb{R}^N)$ with

$$\max \left\{ \|\mathbf{b}_k\|_{L^\infty},\ \|\mathrm{div}_x\, \mathbf{b}_k\|_{L^\infty} \;\middle|\; 1 \le k \le n \right\} \;\le\; R,$$

define $\mu(\cdot) : [0,T] \longrightarrow \mathbb{L}^{\infty\cap 1}(\mathbb{R}^N)$, $t \longmapsto \mu_t$ as

$$\mu_t := \vartheta_{\mathbb{L}^{\infty\cap 1},\mathbf{b}_k}\left(t - t_{k-1},\ \mu_{t_{k-1}}\right) \qquad \text{for } t \in]t_{k-1}, t_k],\ k = 1,2\ldots n.$$

Then we have to verify at each time $t \in [0,T]$: The set $\{\mu_t \,|\, \mu(\cdot) \in \mathscr{N}\} \subset \mathbb{L}^{\infty\cap 1}(\mathbb{R}^N) \subset \mathscr{M}(\mathbb{R}^N)$ is relatively sequentially compact with respect to the weak* topology (w.r.t. $(C^0_0(\mathbb{R}^N), \|\cdot\|_{\sup})$).

Due to Lemma 93 (on page 267), the total variation $|\nu|(\mathbb{R}^N)$ is uniformly bounded for all measures $\nu \in \{\mu_t \,|\, t \in [0,T],\ \mu(\cdot) \in \mathscr{N}\} \subset \mathscr{M}(\mathbb{R}^N)$:

$$|\nu|(\mathbb{R}^N) \;\le\; e^{2RT}\, |\mu_0|(\mathbb{R}^N).$$

Finally, all these measures are tight as a consequence of the inequality

$$\left|\mathbf{X}_{\mathbf{b}_k}(t,x) \;-\; x\right| \;\le\; R\, t$$

(for a.e. $x \in \mathbb{R}^N$ and all $t \in [0,T]$) and essentially the same arguments as the proof of Lemma 2.52 (on page 144) although the Lagrangian flow $\mathbf{X}_{\mathbf{b}_k}(t,\cdot) : \mathbb{R}^N \longrightarrow \mathbb{R}^N$ does not have to be continuous.

(2.) Euler equi-continuity with respect to the pseudo-metrics $(d_{j,\mathbb{L}^{\infty\cap 1}})_{j\in\mathscr{I}}$ is a direct consequence of Proposition 94 (on page 268) and the triangle inequality of each $d_{j,\mathbb{L}^{\infty\cap 1}}$. This implication has already been pointed out in Remark 17 (on page 194). □

Proof (of Existence Theorem 96).
The existence of a solution to the mutational equation results from Theorem 28 (on page 207) due to the preparations in Lemma 90 (on page 265), Proposition 94 (on page 268) and Lemma 95 (on page 272).

In addition, with $R > 0$ denoting the bound in assumption (i), the proof of Lemma 95 (1.) implies that the values of all Euler approximations in $\mathscr{N}(\mu_0, T, R)$,

$$\left\{\nu_t \,\middle|\, t \in [0,T],\ \nu(\cdot) \in \mathscr{N}(\mu_0, T, R)\right\} \subset \mathbb{L}^{\infty\cap 1}(\mathbb{R}^N),$$

are tight. Thus for every $\varepsilon > 0$, there exists a compact set $K_\varepsilon \subset \mathbb{R}^N$ satisfying

$$|\nu_t|(\mathbb{R}^N \setminus K_\varepsilon) \;<\; \varepsilon \qquad \text{for all } t \in [0,T] \text{ and } \nu(\cdot) \in \mathscr{N}(\mu_0, T, R).$$

Since the solution $\mu(\cdot) : t \longmapsto \mu_t$ is constructed by means of Euler approximations, each measure μ_t is the weak* limit of a sequence in $\left\{\nu_t \;\middle|\; \nu(\cdot) \in \mathscr{N}(\mu_0, T, R)\right\}$ due to Lemma 90. The lower semicontinuity of total variation implies $|\mu_t|(\mathbb{R}^N \setminus K_\varepsilon) < \varepsilon$. Therefore, $\{\mu_t \,|\, t \in [0,T]\} \subset \mathbb{L}^{\infty\cap 1}(\mathbb{R}^N) \subset \mathscr{M}(\mathbb{R}^N)$ is tight.

Now we verify the claimed distributional property of any solution $t \mapsto \mu_t = \sigma(t,\cdot)\,\mathscr{L}^N$ to the mutational equation

$$\overset{\circ}{\mu}(\cdot) \ni \vartheta_{\mathbb{L}^{\infty\cap 1},\, \mathbf{f}(\mu(\cdot),\cdot)}$$

on the tuple $\big(\mathbb{L}^{\infty\cap 1}(\mathbb{R}^N),\, (d_{j,\mathbb{L}^{\infty\cap 1}})_{j\in\mathscr{I}},\, (d_{j,\mathbb{L}^{\infty\cap 1}})_{j\in\mathscr{I}},\, \lfloor\cdot\rfloor,\, (\widehat{D}_j)_{j\in\mathscr{I}}\big)$.
Indeed, due to Definition 8 (on page 187), $\mu(\cdot)$ is continuous with respect to each pseudo-metric $d_{j,\mathbb{L}^{\infty\cap 1}}$ $(j\in\mathscr{I})$ and satisfies for each index $j\in\mathscr{I}$

$$\lim_{h\downarrow 0}\ \tfrac{1}{h}\cdot\big|\varphi_j\cdot\big(\mathbf{X}_{\mathbf{f}(\mu_t,t)}(h,\cdot)_\sharp\,\mu_t\ -\ \mu_{t+h}\big)\big|(\mathbb{R}^N)\ =\ 0.$$

at $\mathscr{L}^1$-almost every time $t\in[0,T[$.
Assuming tight values in addition implies the continuity of $\mu(\cdot)$ with respect to narrow convergence as a consequence of Lemma 90.
Furthermore, the Lagrangian flow $\mathbf{X}_{\mathbf{f}(\mu_t,t)}: [0,1]\times\mathbb{R}^N \longrightarrow \mathbb{R}^N$ of the vector field $\mathbf{f}(\mu_t,t)\in \mathrm{BV}^{\infty,\mathrm{div}}_{\mathrm{loc}}(\mathbb{R}^N)$ satisfies for $\mathscr{L}^N$-almost every $x\in\mathbb{R}^N$

$$\mathbf{X}_{\mathbf{f}(\mu_t,t)}(h,x)\ =\ x+\int_0^h \mathbf{f}(\mu_t,t)\big(\mathbf{X}_{\mathbf{f}(\mu_t,t)}(s,x)\big)\ ds \qquad \text{for all } h\in[0,1]$$

according to Proposition 80 (a) (on page 262). Hence there exists a set $I\subset[0,T]$ of full Lebesgue measure such that for every $t\in I$, the following right Dini derivative exists and is uniformly bounded in I

$$\begin{aligned}\tfrac{d^+}{dt^+}\int_{\mathbb{R}^N}\varphi_j\ d\mu_t\ &=\ \lim_{h\downarrow 0}\ \tfrac{1}{h}\cdot\int_{\mathbb{R}^N}\big(\varphi_j(\mathbf{X}_{\mathbf{f}(\mu_t,t)}(h,x))\ -\ \varphi_j(x)\big)\ \ \sigma(t,x)\ \ d\mathscr{L}^N x\\ &=\ \int_{\mathbb{R}^N}\nabla\varphi_j(x)\cdot\mathbf{f}(\mu_t,t)(x)\qquad \sigma(t,x)\ \ d\mathscr{L}^N x.\end{aligned}$$

The continuous function $[0,T[\,\longrightarrow\mathbb{R}_0^+,\ t\longmapsto\int_{\mathbb{R}^N}\varphi_j\ d\mu_t$ is even Lipschitz continuous as a consequence of Gronwall's estimate (in Proposition A.2 on page 440) and, its weak derivative is

$$\tfrac{d}{dt}\int_{\mathbb{R}^N}\varphi_j\ d\mu_t\ =\ \int_{\mathbb{R}^N}\nabla\varphi_j(x)\cdot\mathbf{f}(\mu_t,t)(x)\ \ d\mu_t(x).$$

Now every nonnegative test function $\varphi\in C_c^\infty(\mathbb{R}^N)$, $\varphi\geq 0$, can be approximated by $(\varphi_j)_{j\in\mathscr{I}}$ with respect to the C^1 norm due to Lemma 86 (on page 264). Thus,

$$[0,T[\ \longrightarrow\ \mathbb{R}_0^+,\qquad t\longmapsto\int_{\mathbb{R}^N}\varphi\ d\mu_t$$

is also absolutely continuous and satisfies at $\mathscr{L}^1$-almost every time $t\in[0,T[$

$$\tfrac{d}{dt}\int_{\mathbb{R}^N}\varphi\ d\mu_t\ =\ \int_{\mathbb{R}^N}\nabla\varphi(x)\cdot\mathbf{f}(\mu_t,t)(x)\ \ d\mu_t(x).$$

Moreover the condition $\varphi\geq 0$ is not required, i.e., the same features are guaranteed for any $\varphi\in C_c^\infty(\mathbb{R}^N)$. Indeed, choosing any nonnegative auxiliary function $\xi\in C_c^\infty(\mathbb{R}^N)$ with $\xi\equiv\|\varphi\|_\infty+1$ in $\mathbb{B}_1(\operatorname{supp}\varphi)\subset\mathbb{R}^N$, we apply the previous results (about absolute continuity and its derivative) to both $\varphi(\cdot)+\xi(\cdot)\geq 0$ and $\xi(\cdot)\geq 0$. □

Proof (of Proposition 98). Let $\mu(\cdot) = \sigma(\cdot)\,\mathscr{L}^N : [0,T] \longrightarrow \mathbb{L}^{\infty\cap 1}(\mathbb{R}^N)$ be any distributional solution to the nonlinear continuity equation

$$\partial_t\, \mu_t \;+\; \mathrm{div}_x\, (\mathbf{f}(\mu_t,t)\;\mu_t) \;=\; 0$$

with the additional properties

$$\begin{array}{ll} \text{(a)} & \{\mu(t) \mid 0 \le t \le T\} \;\subset\; \mathbb{L}^{\infty\cap 1}(\mathbb{R}^N) \text{ is tight,} \\ \text{(b)} & \sigma(\cdot) : [0,T] \longrightarrow L^1_{\mathrm{loc}}(\mathbb{R}^N) \quad \text{is continuous,} \\ \text{(c)} & \|\sigma(\cdot)\|_{L^1(\mathbb{R}^N)} + \|\sigma(\cdot)\|_{L^\infty(\mathbb{R}^N)} \text{ is bounded in } [0,T]. \end{array}$$

Hence $\mu(\cdot)$ is continuous with respect to each of the weighted L^1 distances $d_{j,\mathbb{L}^{\infty\cap 1}}$ $(j \in \mathscr{I})$ due to Lemma 90 (on page 265).
Continuity assumption (ii') and the transitional distances $\widehat{D}_j(\cdot,\cdot;r)$ $(j \in \mathscr{I})$ specified in Proposition 94 (on page 268) imply that the function of time

$$\begin{array}{rcl} \tau :\; [0,T] & \longrightarrow & \Big(\widehat{\Theta}\big(\mathbb{L}^{\infty\cap 1}(\mathbb{R}^N),\, (d_{i,\mathbb{L}^{\infty\cap 1}})_{i\in\mathscr{I}},\, (d_{i,\mathbb{L}^{\infty\cap 1}})_{i\in\mathscr{I}},\, \lfloor\cdot\rfloor\big),\;\; \widehat{D}_j(\cdot,\cdot;r)\Big) \\ t & \longmapsto & \vartheta_{\mathbb{L}^{\infty\cap 1},\, \mathbf{f}(\mu_t,t)}(\cdot,\cdot) \end{array}$$

is continuous for each radius $r > 0$ and index $j \in \mathscr{I}$. Theorem 96 (on page 272) thus provides a solution $\nu(\cdot) : [0,T] \longrightarrow \mathbb{L}^{\infty\cap 1}(\mathbb{R}^N)$ to the mutational equation

$$\overset{\circ}{\nu}(\cdot) \;\ni\; \tau(\cdot)$$

on the tuple $\big(\mathbb{L}^{\infty\cap 1}(\mathbb{R}^N),\, (d_{j,\mathbb{L}^{\infty\cap 1}})_{j\in\mathscr{I}},\, (d_{j,\mathbb{L}^{\infty\cap 1}})_{j\in\mathscr{I}},\, \lfloor\cdot\rfloor,\, (\widehat{D}_j)_{j\in\mathscr{I}}\big)$ with initial measure $\nu_0 = \mu_0$ and tight values in $\mathbb{L}^{\infty\cap 1}(\mathbb{R}^N)$. Furthermore, it is a narrowly continuous distributional solution to the nonautonomous, but linear equation

$$\partial_t\, \nu_t \;+\; \mathrm{div}_x\, (\mathbf{f}(\mu_t,t)\;\nu_t) \;=\; 0 \qquad\qquad \text{in } \mathbb{R}^N \times\,]0,T[.$$

Proposition 80 of Ambrosio (on page 262) guarantees that the Cauchy problem of such a nonautonomous linear continuity equation always has unique solutions with values in $\mathbb{L}^{\infty\cap 1}(\mathbb{R}^N)$ and thus, $\nu(\cdot) \equiv \mu(\cdot)$, i.e. $\mu(\cdot)$ solves the mutational equation

$$\overset{\circ}{\mu}(\cdot) \;\ni\; \vartheta_{\mathbb{L}^{\infty\cap 1},\, \mathbf{f}(\mu(\cdot),\cdot)}$$

on the tuple $\big(\mathbb{L}^{\infty\cap 1}(\mathbb{R}^N),\, (d_{j,\mathbb{L}^{\infty\cap 1}})_{j\in\mathscr{I}},\, (d_{j,\mathbb{L}^{\infty\cap 1}})_{j\in\mathscr{I}},\, \lfloor\cdot\rfloor,\, (\widehat{D}_j)_{j\in\mathscr{I}}\big)$.

□

3.9 Example: Nonlocal Parabolic Equations in Cylindrical Domains

3.9.1 *Motivation: Smoothing an Image, but Preserving its Edges*

In the field of mathematical image processing, the challenge of smoothing images and detecting edges simultaneously was discussed thoroughly in the popular article [41] of Catté et al. in 1992.
Indeed, Perona and Malik had proposed to let the given image grey values $u_0(\cdot)$ evolve according to a nonlinear diffusion equation whose smoothing diffusion term depends on the absolute value of the (spatial) gradient [155, 156]. Indeed, $|\nabla_x u|$ large indicates a potential edge and thus, diffusion is to be reduced, maybe even "switched off" (if possible). $|\nabla_x u|$ being small, however, indicates where to smooth the image locally. Their experimental results had been remarkable in comparison with other algorithms. From the mathematical point of view, however, the underlying initial value problem of

$$\tfrac{\partial}{\partial t} u \;-\; \operatorname{div}_x\bigl(g(|\nabla u|)\;\nabla u\bigr) \;=\; 0$$

with a smooth weight function $g(\cdot) \geq 0$ is not well-posed in general. This drawback was among the motivations of Catté et al. for suggesting the smoothing of ∇u by means of a convolution:

$$\begin{cases} \tfrac{\partial}{\partial t} u \;-\; \operatorname{div}_x\bigl(g(|u * \nabla G_\sigma|)\;\nabla u\bigr) \;=\; 0 & \text{in }]0,T[\,\times\Omega \\ \nabla u \cdot \nu_\Omega \;=\; 0 & \text{on }]0,T[\,\times\partial\Omega \\ u(0,\cdot) \;=\; u_0 & \text{in } \Omega \subset \mathbb{R}^N \end{cases}$$

with the Gaussian kernel $G_\sigma(x) := \frac{1}{(2\pi\sigma)^{N/2}} \cdot \exp\bigl(-\frac{|x|^2}{4\sigma}\bigr)$. This nonlinear problem is rather a *nonlocal* diffusion equation because the convolution implies the immediate dependence on all other simultaneous grey-values (in a weighted form though). Existence of strong solutions, their uniqueness and continuous dependence on data belong to the main contributions of article [41].
Nonlinear diffusion equations have become the pillar of several image smoothing algorithms which are not to loose relevant edges. In monograph [183], for example, Weickert discusses continuous and discrete methods whose weight functions $g(\cdot)$ take also the direction of the approximated gradient into consideration (and not just its absolute value). The final diffusion equations are *anisotropic*.

Now we want to cover a class of nonlinear parabolic equations in a fixed cylindrical domain $]0,T[\,\times\Omega$ which generalize the functional dependence on the wanted function:

$$\begin{cases} -\frac{\partial u}{\partial t} + \operatorname{div}_x\bigl(\mathbf{A}|_{(t,u(t,\cdot))}\cdot\nabla u\bigr) + \operatorname{div}_x\bigl(\mathbf{b}|_{(t,u(t,\cdot))}\cdot u\bigr) \\ \quad + \;\; \mathbf{c}|_{(t,u(t,\cdot))}\cdot\nabla u \;+\; c_0|_{(t,u(t,\cdot))}\,u \;=\; \operatorname{div}_x \mathbf{f}|_{(t,u(t,\cdot))} + g|_{(t,u(t,\cdot))} \\ \bigl(\mathbf{A}|_{(t,u(t,\cdot))}\,\nabla u \;+\; \mathbf{b}|_{(t,u(t,\cdot))}\,u \;-\; \mathbf{f}|_{(t,u(t,\cdot))}\bigr)\cdot\nu_\Omega \\ \qquad\qquad + \; b_0|_{(t,u(t,\cdot))}\,u \;=\; \psi|_{(t,u(t,\cdot))} \qquad \text{on } \partial\Omega \\ \qquad\qquad u(0,\cdot) \;=\; u_0 \qquad \text{in } \Omega \end{cases}$$

The solutions are always understood in the distributional sense and, ν_Ω denotes the inner normal of a bounded domain $\Omega \subset \mathbb{R}^N$ with Lipschitz boundary. The coefficients are given as suitably bounded continuous functions of time t and the current spatial function $u(t,\cdot) \in W^{1,2}(\Omega)$, i.e.

$$\begin{array}{rlcl}
\mathbf{A}: & [0,T] \times \big(W^{1,2}(\Omega), \|\cdot\|_{L^2(\Omega)}\big) & \longrightarrow & W^{1,\infty}(\Omega, \mathrm{Sym}(\mathbb{R}^{N\times N}))\\
\mathbf{b}: & [0,T] \times \big(W^{1,2}(\Omega), \|\cdot\|_{L^2(\Omega)}\big) & \longrightarrow & W^{1,\infty}(\Omega, \mathbb{R}^N)\\
\mathbf{c}: & [0,T] \times \big(W^{1,2}(\Omega), \|\cdot\|_{L^2(\Omega)}\big) & \longrightarrow & L^{\infty}(\Omega, \mathbb{R}^N)\\
b_0: & [0,T] \times \big(W^{1,2}(\Omega), \|\cdot\|_{L^2(\Omega)}\big) & \longrightarrow & L^{\infty}(\partial\Omega)\\
c_0: & [0,T] \times \big(W^{1,2}(\Omega), \|\cdot\|_{L^2(\Omega)}\big) & \longrightarrow & L^{\infty}(\Omega, \mathbb{R})\\
\mathbf{f}: & [0,T] \times \big(W^{1,2}(\Omega), \|\cdot\|_{L^2(\Omega)}\big) & \longrightarrow & W^{1,2}(\Omega, \mathbb{R}^N)\\
g: & [0,T] \times \big(W^{1,2}(\Omega), \|\cdot\|_{L^2(\Omega)}\big) & \longrightarrow & L^{2}(\Omega, \mathbb{R})\\
\psi: & [0,T] \times \big(W^{1,2}(\Omega), \|\cdot\|_{L^2(\Omega)}\big) & \longrightarrow & L^{2}(\partial\Omega)
\end{array}$$

Starting from the corresponding linear problems, mutational analysis leads directly to conditions sufficient for the existence of strong solutions, their uniqueness and continuous dependence on data.

3.9.2 The Main Result

First we specify the parabolic problem of interest and the concept of weak solution used in this example. Such a linear parabolic initial-boundary value problem is called a *conormal problem* and, it is known to have a weak solution [114, § VI.10].

Definition 101 ([114, § VI.10]).
Let $\Omega \subset \mathbb{R}^N$ be a bounded domain with Lipschitz boundary and $T > 0$, $u_0 \in L^2(\Omega)$,

$$\begin{array}{rlrl}
\check{\mathbf{A}} \in & L^\infty([0,T]\times\Omega, \mathrm{Sym}(\mathbb{R}^{N\times N})), & \check{\mathbf{f}} \in & L^2([0,T]\times\Omega, \mathbb{R}^N),\\
\check{\mathbf{b}}, \check{\mathbf{c}} \in & L^\infty([0,T]\times\Omega, \mathbb{R}^N), & \check{g} \in & L^2([0,T]\times\Omega, \mathbb{R}),\\
\check{b}_0 \in & L^\infty([0,T]\times\partial\Omega), & \check{\psi}_0 \in & L^2([0,T]\times\partial\Omega).\\
\check{c}_0 \in & L^\infty([0,T]\times\Omega, \mathbb{R}), & &
\end{array}$$

A function $u : [0,T] \longrightarrow L^2(\Omega)$ is called *weak solution* to the conormal problem

$$\begin{cases}
-\frac{\partial u}{\partial t} + \mathrm{div}_x\big(\check{\mathbf{A}}\cdot\nabla u\big) + \mathrm{div}_x\big(\check{\mathbf{b}}\cdot u\big) + \check{\mathbf{c}}\cdot\nabla u + \check{c}_0\, u = \mathrm{div}_x\,\check{\mathbf{f}} + \check{g}\\
\big(\check{\mathbf{A}}\,\nabla u + \check{\mathbf{b}}\,u - \check{\mathbf{f}}\big)\cdot\nu_\Omega + \check{b}_0\,u = \check{\psi}_0 \text{ on } [0,T]\times\partial\Omega\\
u(0,\cdot) = u_0 \text{ in } \Omega
\end{cases}$$

if it satisfies

(i) $u \in L^2([0,T]\times\Omega)$ and $\nabla u \in L^2([0,T]\times\Omega)$,

(ii) for each $t \in [0,T]$, $u(t,\cdot) \in L^2(\Omega)$ and $\sup_{0\le s\le T} \|u(s,\cdot)\|_{L^2(\Omega)} < \infty$,

(iii) for every test function $\varphi \in C^1([0,T]\times\Omega)$ vanishing at $\{T\}\times\Omega$,

$$\int_{[0,T]}\int_\Omega \Big(-u\cdot\partial_t\varphi + \big(\check{\mathbf{A}}\,\nabla u + \check{\mathbf{b}}\;u - \check{\mathbf{f}}\big)\cdot\nabla\varphi - \big(\check{\mathbf{c}}\cdot\nabla u + \check{c}_0\,u - \check{g}\big)\cdot\;\varphi\Big)\,dx\,dt$$

$$= \int_{[0,T]}\int_{\partial\Omega} \big(\check{b}_0\,u + \check{\psi}_0\big)\cdot\varphi\;d\omega_x\,dt + \int_\Omega u_0\cdot\varphi(0,\cdot)\,dx$$

Basic set	$E_\Omega := W^{1,2}(\Omega)$ for a bounded domain $\Omega \subset \mathbb{R}^N$ with boundary $\partial\Omega$ of class C^2
Distances Absolute values	$d_\Omega := \|\cdot - \cdot\|^2_{L^2(\Omega)}$ $e_\Omega := \|\cdot - \cdot\|^2_{L^1(\Omega)}$ $\lfloor \cdot \rfloor_\Omega := \|\cdot\|_{W^{1,2}(\Omega)}$
Transitions: Non-ω-contract. candidates	For the autonomous coefficients $\check{\mathbf{A}} \in W^{1,\infty}(\Omega, \mathrm{Sym}(\mathbb{R}^{N\times N})), \quad \check{\mathbf{f}} \in W^{1,2}(\Omega, \mathbb{R}^N),$ $\check{\mathbf{b}} \in W^{1,\infty}(\Omega, \mathbb{R}^N), \quad \check{g} \in L^2(\Omega, \mathbb{R}),$ $\check{\mathbf{c}} \in L^\infty(\Omega, \mathbb{R}^N), \quad \check{\psi}_0 \in L^2(\partial\Omega),$ $\check{c}_0 \in L^\infty(\Omega, \mathbb{R}), \quad \check{b}_0 \in L^\infty(\partial\Omega)$ with the condition of uniform parabolicity, consider the weak solution $u(\cdot) : [0,1] \longrightarrow W^{1,2}(\Omega)$ to the conormal problem $\begin{cases} -\frac{\partial u}{\partial t} + \mathrm{div}_x\big(\check{\mathbf{A}} \cdot \nabla u\big) + \mathrm{div}_x\big(\check{\mathbf{b}} \cdot u\big) + \check{\mathbf{c}} \cdot \nabla u + \check{c}_0\, u = \mathrm{div}_x\, \check{\mathbf{f}} + \check{g} \\ \big(\check{\mathbf{A}}\, \nabla u + \check{\mathbf{b}}\, u - \check{\mathbf{f}}\big) \cdot \nu_\Omega + \check{b}_0\, u = \check{\psi}_0 \\ u(0) = u_0 \end{cases}$
Compactness Equi-continuity	Every closed norm-bounded ball in $W^{1,2}(\Omega)$ is compact in $L^2(\Omega)$ due to Sobolev's Embedding Theorem. It implies Euler compactness: Corollary 112 (page 286) Euler equi-continuity results from standard a priori estimates for nonautonomous conormal problems in finite time intervals.
Mutational solutions	Weak solutions to nonautonomous conormal problems with spatially nonlocal dependence of coefficients (Link to strong solutions: Remark 113 on page 288)
List of main results formulated in § 3.9	Existence due to compactness, uniqueness due to Lipschitz continuity and continuous dependence of solutions on data: Theorem 103 (on page 281 f.)
Key tools	Existence of weak solutions to nonautonomous linear conormal problems with some a priori estimates: Proposition 102 (page 281) Higher regularity of weak solutions to nonautonomous linear conormal problems and corresponding a priori estimates: Proposition 105 (page 283) Comparative estimate for two solutions to nonautonomous linear conormal problems: Corollary 108 (page 284)

Table 3.4 Brief summary of the example in § 3.9 in mutational terms:
Nonlocal parabolic equations in cylindrical domains

Proposition 102 ([114, Theorems 6.38, 6.39]).
Let $\Omega \subset \mathbb{R}^N$ be a domain with Lipschitz boundary, $T > 0$ and $\check{\mathbf{A}}$, $\check{\mathbf{b}}$, $\check{b}_0$, $\check{\mathbf{c}}$, $\check{c}_0$, $\check{\mathbf{f}}$, $\check{g}$, $\check{\psi}_0$ be measurable functions as in Definition 101. Furthermore assume $\check{\mathbf{A}}(\cdot)$ to be uniformly parabolic in the sense that a constant $\lambda > 0$ satisfies

$$\xi^T \cdot \check{\mathbf{A}}(t,x)\,\xi \;\geq\; \lambda\,|\xi|^2$$

for all $(t,x) \in [0,T] \times \Omega$ and $\xi \in \mathbb{R}^N$.

Then for every $u_0 \in L^2(\Omega)$, there exists a weak solution $u : [0,T] \longrightarrow L^2(\Omega)$ of

$$\begin{cases} -\frac{\partial u}{\partial t} + \operatorname{div}_x\big(\check{\mathbf{A}} \cdot \nabla u\big) + \operatorname{div}_x\big(\check{\mathbf{b}} \cdot u\big) + \check{\mathbf{c}} \cdot \nabla u + \check{c}_0\, u = \operatorname{div}_x \check{\mathbf{f}} + \check{g} \\ \big(\check{\mathbf{A}}\, \nabla u + \check{\mathbf{b}}\, u - \check{\mathbf{f}}\big) \cdot \nu_\Omega + \check{b}_0\, u = \check{\psi}_0 \text{ on } [0,T] \times \partial\Omega \\ u(0,\cdot) = u_0 \text{ in } \Omega \end{cases}$$

Every weak solution $u(\cdot)$ satisfies

$$\big\|\nabla u\big\|_{L^2([0,T]\times\Omega)} + \sup_{[0,T]} \|u(t,\cdot)\|_{L^2(\Omega)} \leq C \cdot e^{C\,T}\,\big(\|\check{\mathbf{f}}\|_{L^2} + \|\check{g}\|_{L^2} + \|\check{\psi}_0\|_{L^2} + \|u_0\|_{L^2}\big)$$

with a constant $C > 0$ depending only on λ, N and the L^∞ norm of $\check{\mathbf{A}}$, $\check{\mathbf{b}}$, $\check{b}_0$, $\check{\mathbf{c}}$, $\check{c}_0$.

Now these results about the *linear* problem are extended to the more general case that the coefficients are prescribed in dependence on the current spatial functions. This relationship may have nonlocal character in space, of course.

Theorem 103. *Let $\Omega \subset \mathbb{R}^N$ be a bounded domain with C^2 boundary and $T, \lambda > 0$. Assume the coefficient functions*

$$\begin{aligned} \mathbf{A} :&\ [0,T] \times \big(W^{1,2}(\Omega), \|\cdot\|_{L^2(\Omega)}\big) \longrightarrow W^{1,\infty}(\Omega, \operatorname{Sym}(\mathbb{R}^{N\times N})) \\ \mathbf{b} :&\ [0,T] \times \big(W^{1,2}(\Omega), \|\cdot\|_{L^2(\Omega)}\big) \longrightarrow W^{1,\infty}(\Omega, \mathbb{R}^N) \\ \mathbf{c} :&\ [0,T] \times \big(W^{1,2}(\Omega), \|\cdot\|_{L^2(\Omega)}\big) \longrightarrow L^\infty(\Omega, \mathbb{R}^N) \\ b_0 :&\ [0,T] \times \big(W^{1,2}(\Omega), \|\cdot\|_{L^2(\Omega)}\big) \longrightarrow L^\infty(\partial\Omega) \\ c_0 :&\ [0,T] \times \big(W^{1,2}(\Omega), \|\cdot\|_{L^2(\Omega)}\big) \longrightarrow L^\infty(\Omega, \mathbb{R}) \\ \mathbf{f} :&\ [0,T] \times \big(W^{1,2}(\Omega), \|\cdot\|_{L^2(\Omega)}\big) \longrightarrow W^{1,2}(\Omega, \mathbb{R}^N) \\ g :&\ [0,T] \times \big(W^{1,2}(\Omega), \|\cdot\|_{L^2(\Omega)}\big) \longrightarrow L^2(\Omega, \mathbb{R}) \\ \psi :&\ [0,T] \times \big(W^{1,2}(\Omega), \|\cdot\|_{L^2(\Omega)}\big) \longrightarrow L^2(\partial\Omega) \end{aligned}$$

to be uniformly bounded w.r.t. the respective Sobolev norms of their values,
continuous w.r.t. the L^∞ or L^2 norm of their values respectively and
satisfy the condition of uniform parabolicity $\xi^T \cdot \mathbf{A}(t,v)(x)\,\xi \geq \lambda\,|\xi|^2$
for all $t \in [0,T]$, $v \in W^{1,2}(\Omega)$, $x \in \Omega$, $\xi \in \mathbb{R}^N$.

Then for every initial $u_0 \in W^{1,2}(\Omega)$, there exists a weak solution $u(\cdot) : [0,T] \longrightarrow W^{1,2}(\Omega)$ to the nonlinear conormal problem with the composed coefficients

$$\begin{aligned} \check{\mathbf{A}} &:= \mathbf{A}(\cdot, u(\cdot)) : [0,T] \times \Omega \longrightarrow \mathbb{R}^{N\times N}, \quad (t,x) \longmapsto \mathbf{A}(t, u(t,\cdot))(x) \\ \check{\mathbf{b}} &:= \mathbf{b}(\cdot, u(\cdot)) : [0,T] \times \Omega \longrightarrow \mathbb{R}^N, \quad (t,x) \longmapsto \mathbf{b}(t, u(t,\cdot))(x) \qquad \text{etc.} \end{aligned}$$

In fact, $u \in C^0\big([0,T], L^2(\Omega)\big) \cap L^2\big([0,T], W^{1,2}(\Omega)\big)$, $\partial_t u \in L^2\big([0,T], W^{1,2}(\Omega)^*\big)$.

Lipschitz continuity of the coefficients $\mathbf{A}, \mathbf{b}, b_0, \mathbf{c}, c_0, \mathbf{f}, g$ *(w.r.t. the* L^∞ *or* L^2 *norms of their values respectively) implies uniqueness of this weak solution and its continuous dependence on the given data. As a consequence, the existence result can then be extended to initial functions* $u_0 \in L^2(\Omega)$ *approximatively.*

3.9.3 The Underlying Details in Terms of Mutational Analysis

Now we give the detailed proof of Theorem 103 via the mutational framework. The use of these analytical tools lay the basis for coupling nonlocal conormal problems and any other example of mutational equations in systems. In particular, nonlocal diffusion equations in image processing (mentioned in § 3.9.1 on page 278 and discussed in [41, 183]) can be combined with morphological equations for compact sets in $\mathbb{R}^N$, which are discussed in § 1.9 and applied to image segmentation in § 1.10.

The first step is to specify the basic set and the transitions.
Due to Proposition 102, the linear conormal problem serves as a starting point and so, $L^2(\Omega)$ is a suitable choice for the basic set E – at first glance. It causes difficulties in regard to time regularity though. Indeed, according to the general theory of parabolic equations, rather moderate regularity assumptions about the coefficients guarantee distributional time derivatives of each weak solution in L^2, but this does not lead to a form of Lipschitz continuity in time as it is needed for transitions.
Hence, we prefer another space of measurable functions $\Omega \longrightarrow \mathbb{R}$ with two essential advantages:

(1.) In each of its "bounded" balls, the L^2 norm proves to be equivalent to any other L^p norm with $1 \leq p < \infty$ so that we will be able to benefit from the Hölder inequality in regard to time regularity.

(2.) General a priori estimates about parabolic conormal problems provide some finite maximal radius for all weak solutions (and potential Euler approximations) related to a given initial function $u_0 : \Omega \longrightarrow \mathbb{R}$.

Furthermore, the regularity of weak solutions required for transitions will be concluded from the general theory of parabolic differential equations [112, 114]. These additional aspects motivate us to consider $E_\Omega := W^{1,2}(\Omega)$.

Lemma 104. *For a nonempty bounded domain* $\Omega \subset \mathbb{R}^N$ *with Lipschitz boundary consider*

$$\begin{aligned} E_\Omega &:= W^{1,2}(\Omega), \\ d_\Omega &:= \|\cdot - \cdot\|^2_{L^2(\Omega)} \\ e_\Omega &:= \|\cdot - \cdot\|^2_{L^1(\Omega)} \\ \lfloor\cdot\rfloor_\Omega &:= \|\cdot\|_{W^{1,2}(\Omega)} \end{aligned}$$

Then the tuple $(E_\Omega, d_\Omega, e_\Omega, \lfloor\cdot\rfloor_\Omega)$ *satisfies the general assumptions (H1) – (H4) in § 3.1 (on page 182 ff.).*

Proposition 102 (on page 281) quotes some results about weak solutions to the linear conormal problem from Lieberman's monograph [114, § VI]. Now we complete the general statements which we use for discussing this nonlocal example in the mutational framework afterwards:

Proposition 105 (Weak derivatives of solutions, [112, Remark III.6.3]**).**
In addition to the hypotheses of Proposition 102 (on page 281), let $\Omega \subset \mathbb{R}^N$ be a bounded domain with C^2 boundary. Suppose the weak spatial derivatives $\nabla \check{\mathbf{A}}$, $\nabla \check{\mathbf{b}}$ to be in L^∞ and $\operatorname{div}_x \check{\mathbf{f}} \in L^2([0,T] \times \Omega)$ respectively.

Then every weak solution $u(\cdot) \in L^2([0,T] \times \Omega)$ to the linear conormal problem has square integrable weak partial derivatives w.r.t. time up to order 1 and w.r.t. space up to order 2, i.e.

$$\begin{aligned} \partial_t u &\in L^2([0,T] \times \Omega, \mathbb{R}), \\ \nabla u &\in L^2([0,T] \times \Omega, \mathbb{R}^N), \\ \nabla^2 u &\in L^2([0,T] \times \Omega, \mathbb{R}^{N \times N}). \end{aligned}$$

u satisfies the parabolic differential equation almost everywhere in $[0,T] \times \Omega$.

Furthermore there exists a constant $C > 0$ depending on λ, N, Ω and the L^∞ norms of the coefficients $\check{\mathbf{A}}$, $\nabla\check{\mathbf{A}}$, $\check{\mathbf{b}}$, $\nabla\check{\mathbf{b}}$, $\check{b}_0$, $\check{\mathbf{c}}$, $\check{c}_0$ such that for every $t \in [0,T]$,

$$\begin{aligned} &\|\partial_t u\|^2_{L^2([0,t]\times\Omega)} + \|\nabla u(t,\cdot)\|^2_{L^2(\Omega)} + \|\nabla^2 u\|^2_{L^2([0,t]\times\Omega)} \\ &\leq C \cdot e^{C\,t} \cdot \Big(\|\nabla u(0,\cdot)\|^2_{L^2(\Omega)} + \| -\partial_t u + \operatorname{div}_x (\check{\mathbf{A}} \cdot \nabla u)\|^2_{L^2([0,t]\times\Omega)} \Big). \end{aligned}$$

Remark 106. (1.) The weak partial derivative $\partial_t u$ will be used for verifying the required Lipschitz continuity of (potential) transitions with respect to time. The partial derivatives in space are the tools for proving Euler compactness.

(2.) Due to Cauchy's inequality, the preceding estimate and Proposition 102 ensure a constant $\widehat{C} > 0$ depending on the same parameters as C such that each weak solution $u(\cdot) \in L^2([0,T] \times \Omega)$ satisfies at every time $t \in [0,T]$

$$\begin{aligned} &\|\partial_t u\|^2_{L^2([0,t]\times\Omega)} + \|\nabla u(t,\cdot)\|^2_{L^2(\Omega)} + \|\nabla^2 u\|^2_{L^2([0,t]\times\Omega)} \\ &\leq \widehat{C}\, e^{\widehat{C} t} \cdot \Big(\|\nabla u(0,\cdot)\|^2_{L^2(\Omega)} + \|\operatorname{div}_x \check{\mathbf{f}}\|^2_{L^2([0,t]\times\Omega)} + \|\check{\mathbf{f}}\|^2_{L^2([0,t]\times\Omega)} + \|\check{g}\|^2_{L^2([0,t]\times\Omega)} \\ &\qquad\qquad + \|u(0,\cdot)\|^2_{L^2(\Omega)} \quad + \|\check{\psi}_0\|^2_{L^2([0,t]\times\partial\Omega)} \Big). \end{aligned}$$

(3.) The regularity assumptions about the coefficients concern only derivatives in space and are still rather moderate (see also details in [111, § III.4], [112, § III.6]). This will be useful for the step from autonomous linear conormal problems (inducing candidates for transitions) to corresponding Euler approximations.

Corollary 107 (Lipschitz continuity w.r.t. time).
Under the assumptions of Proposition 105, every weak solution $u(\cdot) \in L^2([0,T] \times \Omega)$ to the linear conormal problem satisfies for any $s,t \in [0,T]$

$$\begin{aligned} &e_\Omega\big(u(s,\cdot),\, u(t,\cdot)\big) \\ &\le\ C\,|t-s| \cdot \Big(\|u_0\|^2_{W^{1,2}(\Omega)} + \|\check{\mathbf{f}}\|^2_{W^{1,2}([0,T]\times\Omega)} + \|\check{g}\|^2_{L^2([0,T]\times\Omega)} + \|\check{\psi}_0\|^2_{L^2([0,T]\times\partial\Omega)} \Big) \end{aligned}$$

with a constant C depending only on λ, N, Ω and the L^∞ norms of the coefficients $\check{\mathbf{A}}$, $\nabla\check{\mathbf{A}}$, $\check{\mathbf{b}}$, $\nabla\check{\mathbf{b}}$, $\check{b}_0$, $\check{\mathbf{c}}$, $\check{c}_0$.

Corollary 108 (Comparing solutions of two linear conormal problems).
In addition to the hypotheses of Proposition 105, let $v \in L^2([0,T] \times \Omega)$ denote the weak solution to the conormal problem associated with further coefficients $\hat{\mathbf{A}}$, $\hat{\mathbf{b}}$, $\hat{b}_0$, $\hat{\mathbf{c}}$, $\hat{c}_0$, $\hat{\mathbf{f}}$, $\hat{g}$, $\hat{\psi}_0$ of the corresponding regularity and the initial function $v_0 \in E_\Omega$.

Then there exists a constant $\widehat{C}$ depending only on λ, N, Ω, $\|\check{\mathbf{A}}\|_{L^\infty}$, $\|\nabla\check{\mathbf{A}}\|_{L^\infty}$, $\|\check{\mathbf{b}}\|_{L^\infty}$, $\|\nabla\check{\mathbf{b}}\|_{L^\infty}$, $\|\check{b}_0\|_{L^\infty}$, $\|\check{\mathbf{c}}\|_{L^\infty}$, $\|\check{c}_0\|_{L^\infty}$ and $\|\hat{\mathbf{A}}\|_{L^\infty}$, $\|\nabla\hat{\mathbf{A}}\|_{L^\infty}$, $\|\hat{\mathbf{b}}\|_{L^\infty}$, $\|\nabla\hat{\mathbf{b}}\|_{L^\infty}$, $\|\hat{b}_0\|_{L^\infty}$, $\|\hat{\mathbf{c}}\|_{L^\infty}$, $\|\hat{c}_0\|_{L^\infty}$ such that for every $t \in [0,T]$,

$$\begin{aligned} &d_\Omega\big(u(t),\, v(t))\big) \\ &\le\ \widehat{C}\,e^{\widehat{C}t} \cdot \Big(d_\Omega(u_0,v_0) + \big(\|(\check{\mathbf{A}}, \check{\mathbf{b}}, \check{\mathbf{c}}, \check{c}_0) - (\hat{\mathbf{A}}, \hat{\mathbf{b}}, \hat{\mathbf{c}}, \hat{c}_0)\|^2_{L^\infty([0,t]\times\Omega)} \cdot t \\ &\qquad + \|(\check{\mathbf{f}}, \check{g}) - (\hat{\mathbf{f}}, \hat{g})\|^2_{L^2([0,t]\times\Omega)} \\ &\qquad + \|\check{b}_0 - \hat{b}_0\|^2_{L^\infty([0,t]\times\partial\Omega)} \cdot t + \|\check{\psi}_0 - \hat{\psi}_0\|^2_{L^2([0,t]\times\partial\Omega)} \big) \\ &\qquad \cdot \big(1 + \|v_0\|_{W^{1,2}}\big)^2 \cdot \widehat{C}\,e^{\widehat{C}t} \Big). \end{aligned}$$

Now we specify the autonomous linear conormal problem inducing the candidates for transitions on the tuple $\big(E_\Omega, d_\Omega, e_\Omega, \lfloor\cdot\rfloor_\Omega\big)$. All proofs are again postponed to the end of this section.

Proposition 109. *Let $\Omega \subset \mathbb{R}^N$ be a nonempty bounded domain with C^2 boundary and define E_Ω, d_Ω, e_Ω, $\lfloor\cdot\rfloor_\Omega$ as in Lemma 104. For the autonomous coefficients*

$$\begin{aligned} \check{\mathbf{A}} &\in W^{1,\infty}(\Omega, \mathrm{Sym}(\mathbb{R}^{N\times N})), & \check{\mathbf{f}} &\in W^{1,2}(\Omega, \mathbb{R}^N), \\ \check{\mathbf{b}} &\in W^{1,\infty}(\Omega, \mathbb{R}^N), & \check{g} &\in L^2(\Omega, \mathbb{R}), \\ \check{\mathbf{c}} &\in L^\infty(\Omega, \mathbb{R}^N), & \check{\psi}_0 &\in L^2(\partial\Omega), \\ \check{c}_0 &\in L^\infty(\Omega, \mathbb{R}), & \check{b}_0 &\in L^\infty(\partial\Omega) \end{aligned}$$

with the condition of uniform parabolicity $\xi^T \cdot \check{\mathbf{A}}(x)\,\xi \ge \lambda\,|\xi|^2$ for all x, ξ and any initial function $u_0 \in E_\Omega$, the unique weak solution $\vartheta(\cdot, u_0) = u(\cdot) : [0,1] \longrightarrow L^2(\Omega)$ to the linear conormal problem

$$\begin{cases} -\frac{\partial u}{\partial t} + \mathrm{div}_x\big(\check{\mathbf{A}} \cdot \nabla u\big) + \mathrm{div}_x\big(\check{\mathbf{b}} \cdot u\big) + \check{\mathbf{c}} \cdot \nabla u + \check{c}_0\, u = \mathrm{div}_x\, \check{\mathbf{f}} + \check{g} \\ \big(\check{\mathbf{A}}\, \nabla u + \check{\mathbf{b}}\, u - \check{\mathbf{f}}\big) \cdot \nu_\Omega + \check{b}_0\, u = \check{\psi}_0 \quad \text{on } [0,1]\times\partial\Omega \\ u(0) = u_0 \end{cases}$$

has the following properties for every $s,t \in [0,1]$, $v_0 \in E_\Omega$:

$$\begin{aligned}
&(1.)\ \vartheta(t,u_0) && \in W^{1,2}(\Omega) \stackrel{\text{Def.}}{=} E_\Omega,\\
&\|\vartheta(t,u_0)\|_{L^2(\Omega)} && \le C\cdot e^{Ct}\cdot\Big(\|u_0\|_{L^2(\Omega)} + \|\check{\mathbf{f}}\|_{L^2} + \|\check{g}\|_{L^2} + \|\check{\psi}_0\|_{L^2(\partial\Omega)}\Big)\\
&\|\nabla\vartheta(t,u_0)\|^2_{L^2(\Omega)} && \le C\cdot e^{Ct}\cdot\Big(\|u_0\|^2_{W^{1,2}(\Omega)} + \|\check{\mathbf{f}}\|^2_{W^{1,2}} + \|\check{g}\|^2_{L^2} + \|\check{\psi}_0\|^2_{L^2(\partial\Omega)}\Big)
\end{aligned}$$

(2.) $\vartheta(\cdot,u_0)$ *has the semigroup property,*

(3.) $d_\Omega\big(\vartheta(t,u_0),\ \vartheta(t,v_0)\big) \le C\cdot d_\Omega(u_0,\ v_0)$

(4.) $e_\Omega\big(\vartheta(s,u_0),\ \vartheta(t,u_0)\big) \le |s-t|\cdot C\cdot\Big(\|u_0\|^2_{W^{1,2}} + \|\check{\mathbf{f}}\|^2_{W^{1,2}} + \|\check{g}\|^2_{L^2} + \|\check{\psi}_0\|^2_{L^2}\Big)$

with a constant C depending on λ, N, Ω, $\|\check{\mathbf{A}}\|_{W^{1,\infty}}$, $\|\check{\mathbf{b}}\|_{W^{1,\infty}}$, $\|\check{b}_0\|_{L^\infty}$, $\|\check{\mathbf{c}}\|_{L^\infty}$, $\|\check{c}_0\|_{L^\infty}$.

Furthermore let $\vartheta_2(\cdot,v_0) = v(\cdot) : [0,1] \longrightarrow L^2(\Omega)$ denote the solution associated with further autonomous coefficients $\hat{\mathbf{A}}$, $\hat{\mathbf{b}}$, $\hat{b}_0$, $\hat{\mathbf{c}}$, $\hat{c}_0$, $\hat{\mathbf{f}}$, $\hat{g}$, $\hat{\psi}_0$ of this kind and the initial function $v_0 \in E_\Omega$.
Then there exists a constant $\widehat{C}$ depending only on λ, N, Ω and the respective Sobolev norm of all the coefficients such that for every $t \in [0,1]$,

$$\begin{aligned}
d_\Omega\big(\vartheta(t,u_0),\vartheta_2(t,v_0)\big) \le \widehat{C}\, e^{\widehat{C}t}\cdot\Big(& d_\Omega(u_0,v_0)\\
&+t\cdot\Big(\big\|(\check{\mathbf{A}},\check{\mathbf{b}},\check{b}_0,\check{\mathbf{c}},\check{c}_0) - (\hat{\mathbf{A}},\hat{\mathbf{b}},\hat{b}_0,\hat{\mathbf{c}},\hat{c}_0)\big\|^2_{L^\infty}\\
&\qquad + \big\|(\check{\mathbf{f}},\check{g}) - (\hat{\mathbf{f}},\hat{g})\big\|^2_{L^2} + \|\check{\psi}_0 - \hat{\psi}_0\|^2_{L^2}\Big)\\
&\qquad \times\big(1+\|v_0\|_{W^{1,2}}\big)^2\ \widehat{C}e^{\widehat{C}t}\Big).
\end{aligned}$$

In the following, $\check{\Theta}_\Omega$ denotes the set of all weak solutions $\vartheta : [0,1]\times E_\Omega \longrightarrow E_\Omega$ to linear conormal problems related to coefficients $\check{\mathbf{A}}$, $\check{\mathbf{b}}$, $\check{b}_0$, $\check{\mathbf{c}}$, $\check{c}_0$, $\check{\mathbf{f}}$, $\check{g}$, $\check{\psi}_0$ as in Proposition 109.

Remark 110. (1.) This proposition summarizes essential properties in regard to transitions on the tuple $\big(E_\Omega, d_\Omega, e_\Omega, \lfloor\cdot\rfloor_\Omega\big)$ (in the sense of Definition 2 on page 183). There are two obstacles though. First, $\vartheta : [0,1]\times E_\Omega \longrightarrow E_\Omega$ is not ω-contractive, but satisfies

$$d_\Omega\big(\vartheta(t,u_0),\ \vartheta(t,v_0)\big)\ \le\ C\cdot d_\Omega(u_0,\ v_0)$$

for all $u_0, v_0 \in E_\Omega$ and $t \in [0,1]$ with a constant C possibly larger than 1. The results of § 3.4 (on page 214 ff.) based on an auxiliary distance equivalent to d_Ω will prove to be useful for this gap.
Second, statement (1.) does not provide an obvious choice for the parameter $\gamma(\vartheta)$ such that $\lfloor\vartheta(t,u_0)\rfloor_\Omega \stackrel{\text{Def.}}{=} \|\vartheta(t,u_0)\|_{W^{1,2}(\Omega)} \le \big(\|u_0\|_{W^{1,2}(\Omega)} + \gamma_j(\vartheta)\ t\big)\cdot e^{\gamma_j(\vartheta)\,t}$ holds for every $u_0 \in E_\Omega$ and $t \in [0,1]$. Possibly the standard inequalities quoted from [112, 114] can be "improved" slightly, but now we just point out that these explicit a priori estimates for *any* time interval $[0,T]$ are sufficient for excluding "explosions" of any Euler curves in finite time. Hence all the essential arguments in the mutational framework can still be applied to this example although we cannot use Lemma 5 (about general a priori bounds of Euler curves on page 185).

(2.) Statement (4.) formulates that $\vartheta(\cdot, u_0) : [0,1] \longrightarrow E_\Omega$ is Lipschitz continuous with respect to e_Ω, but it is just a special case of Corollary 107 above. In mutational terms, this corollary even guarantees that the tuple $\big(E_\Omega, d_\Omega, e_\Omega, \lfloor \cdot \rfloor_\Omega, \check{\Theta}_\Omega\big)$ is *nonequidistant Euler equi-continuous* in the sense of Definition 23 (on page 203).

(3.) Similarly, Corollary 108 is stronger than just the last estimate in Proposition 109. It ensures that hypothesis (A4) in § 3.4 (on page 215) is satisfied – with the slight modification that we have explicit a priori estimates for any Euler curves in each finite time interval (instead of the conclusions via the exponential growth parameter $\widehat{\gamma}$).
This lays the foundation for applying the approach of § 3.4 and considering transitions on $\check{E}_\Omega := E_\Omega \times \mathbb{R}$ with an auxiliary distance $\check{d}_\Omega : \check{E}_\Omega \times \check{E}_\Omega \longrightarrow [0,\infty[$ equivalent to d_Ω in the sense of Proposition 32 (on page 215 f.)

In regard to compactness, we conclude from Sobolev's Embedding Theorem and the fact that each norm-closed ball in the Hilbert space $W^{1,2}(\Omega)$ is weakly compact:

Lemma 111.
Every closed ball $\big\{w \in E_\Omega \overset{\text{Def.}}{=} W^{1,2}(\Omega) \;\big|\; \|w\|_{W^{1,2}(\Omega)} \leq r\big\}$ *with finite radius* $r \geq 0$ *is sequentially compact with respect to the* $L^2(\Omega)$ *norm.*

Then the a priori estimates for the nonautonomous linear conormal problem in Propositions 102 and 105 guarantee immediately:

Corollary 112 (Nonequidistant Euler compactness).
The tuple $\big(E_\Omega, d_\Omega, e_\Omega, \lfloor \cdot \rfloor_\Omega\big)$ *in combination with the candidates for transitions in* $\check{\Theta}_\Omega$ *is nonequidistant Euler compact in the sense of Definition 23 (on page 203). In fact, the following (slightly more general) statement holds:*

Fix any bounds $\lambda, M > 0$ *and choose sequences of coefficients*

$$
\begin{aligned}
&\check{\mathbf{A}}^n \in L^\infty([0,T],\, W^{1,\infty}(\Omega, \mathrm{Sym}(\mathbb{R}^{N\times N}))), && \check{\mathbf{f}}^n \in L^\infty([0,T],\, W^{1,2}(\Omega, \mathbb{R}^N)),\\
&\check{\mathbf{b}}^n \in L^\infty([0,T],\, W^{1,\infty}(\Omega, \mathbb{R}^N)), && \check{g}^n \in L^\infty([0,T],\quad L^2(\Omega, \mathbb{R})),\\
&\check{\mathbf{c}}^n \in L^\infty([0,T],\quad L^\infty(\Omega, \mathbb{R}^N)), && \check{\psi}_0^n \in L^\infty([0,T],\quad L^2(\partial\Omega)),\\
&\check{c}_0^n \in L^\infty([0,T],\quad L^\infty(\Omega, \mathbb{R})) && \check{b}_0^n \in L^\infty([0,T],\quad L^\infty(\partial\Omega))
\end{aligned}
$$

with the condition of uniform parabolicity $\xi^T \cdot \check{\mathbf{A}}^n(t,x)\, \xi \;\geq\; \lambda\, |\xi|^2$ *for all* n,t,x,ξ,

$$
\begin{aligned}
\max \big\{ &\|\check{\mathbf{A}}^n\|_{L^\infty(W^{1,\infty})},\ \|\check{\mathbf{b}}^n\|_{L^\infty(W^{1,\infty})},\ \|\check{b}_0^n\|_{L^\infty(L^\infty)},\ \|\check{\mathbf{c}}^n\|_{L^\infty(L^\infty)},\ \|\check{c}_0^n\|_{L^\infty(L^\infty)},\\
&\|\check{\mathbf{f}}^n\|_{L^\infty(W^{1,2})},\ \|\check{g}^n\|_{L^\infty(L^2)},\ \|\check{\psi}_0^n\|_{L^\infty(L^2)} \big\} \leq M.
\end{aligned}
$$

For each $n \in \mathbb{N}$, *let* $u^n(\cdot) \in L^2([0,T] \times \Omega)$ *denote the unique weak solution to the respective nonautonomous linear conormal problem with a joint initial* $u_0 \in E_\Omega$.

Then for every $t \in [0,T]$, *a subsequence of* $\big(u^n(t,\cdot)\big)_{n\in\mathbb{N}}$ *converges to a function in* $E_\Omega \overset{\text{Def.}}{=} W^{1,2}(\Omega)$ *with respect to* d_Ω.

In regard to the proof of Theorem 103 about the full nonlinear conormal problem (on page 281), we can now draw conclusions from the mutational framework:

Assume the coefficient functions

$$\begin{array}{rlcl}
\mathbf{A}: & [0,T]\times\big(W^{1,2}(\Omega),\,\|\cdot\|_{L^2(\Omega)}\big) & \longrightarrow & W^{1,\infty}(\Omega,\,\mathrm{Sym}(\mathbb{R}^{N\times N}))\\
\mathbf{b}: & [0,T]\times\big(W^{1,2}(\Omega),\,\|\cdot\|_{L^2(\Omega)}\big) & \longrightarrow & W^{1,\infty}(\Omega,\,\mathbb{R}^N)\\
\mathbf{c}: & [0,T]\times\big(W^{1,2}(\Omega),\,\|\cdot\|_{L^2(\Omega)}\big) & \longrightarrow & L^{\infty}(\Omega,\,\mathbb{R}^N)\\
b_0: & [0,T]\times\big(W^{1,2}(\Omega),\,\|\cdot\|_{L^2(\Omega)}\big) & \longrightarrow & L^{\infty}(\partial\Omega)\\
c_0: & [0,T]\times\big(W^{1,2}(\Omega),\,\|\cdot\|_{L^2(\Omega)}\big) & \longrightarrow & L^{\infty}(\Omega,\,\mathbb{R})\\
\mathbf{f}: & [0,T]\times\big(W^{1,2}(\Omega),\,\|\cdot\|_{L^2(\Omega)}\big) & \longrightarrow & W^{1,2}(\Omega,\,\mathbb{R}^N)\\
g: & [0,T]\times\big(W^{1,2}(\Omega),\,\|\cdot\|_{L^2(\Omega)}\big) & \longrightarrow & L^2(\Omega,\,\mathbb{R})\\
\psi: & [0,T]\times\big(W^{1,2}(\Omega),\,\|\cdot\|_{L^2(\Omega)}\big) & \longrightarrow & L^2(\partial\Omega)
\end{array}$$

to be uniformly bounded w.r.t. the respective Sobolev norms of their values,
continuous w.r.t. the L^∞ or L^2 norm of their values respectively and
satisfy the condition of uniform parabolicity $\xi^T\cdot\mathbf{A}(t,v)(x)\,\xi\;\geq\;\lambda\,|\xi|^2$
for all $t\in[0,T]$, $v\in W^{1,2}(\Omega)$, $x\in\Omega$, $\xi\in\mathbb{R}^N$.

Let $\mathscr{F}_\Omega(\cdot,\cdot)$ abbreviate the tuple of these coefficient functions on $[0,T]\times E_\Omega$.

Now the construction in § 3.4 provides both the basic set $\check{E}_\Omega \stackrel{\text{Def.}}{=} E_\Omega\times\mathbb{R}$ related to $E_\Omega \stackrel{\text{Def.}}{=} W^{1,2}(\Omega)$ and the auxiliary distance $\check{d}_\Omega$ on $\check{E}_\Omega$ equivalent to d_Ω such that the solutions in $\check{\Theta}_\Omega$ (to autonomous linear conormal problems) induce transitions on $\big(\check{E}_\Omega,\check{d}_\Omega,e_\Omega,\lfloor\cdot\rfloor_\Omega\big)$. Furthermore this tuple is both nonequidistant Euler compact and nonequidistant Euler equi-continuous due to Corollary 112 and Remark 110 (3.) respectively — if we use the explicit a priori estimates for each Euler curve (instead of the exponential growth parameter $\widehat{\gamma}<\infty$).

Hence Existence Theorem 19 (on page 195) states that the mutational equation

$$\overset{\circ}{\check{u}}(t)\;\ni\;\mathscr{F}_\Omega\big(t,\,u(t,\cdot)\big)$$

has a solution $\check{u}:\,[0,T]\longrightarrow\check{E}_\Omega,\;t\longmapsto(u(t,\cdot),\,\rho(t))$ in the tuple $\big(\check{E}_\Omega,\,\check{d}_\Omega,\,e_\Omega,\,\lfloor\cdot\rfloor_\Omega,\,\mathrm{const}\cdot(\|\cdot\|_{L^\infty}+\|\cdot\|_{L^2})\big)$ for each given initial state $(u_0,\lfloor u_0\rfloor_\Omega)\in\check{E}_\Omega$.

Finally we have to verify that $u(\cdot,\cdot)$ is a weak solution to the nonlinear conormal problem with the composed coefficients $(t,x)\longmapsto\mathscr{F}(t,u(t,\cdot))\,(x)$. Indeed, the nonautonomous, but linear conormal problem in w with these coefficients, i.e.

$$\begin{cases}
-\frac{\partial w}{\partial t}+\mathrm{div}_x\big(\mathbf{A}|_{(t,u(t,\cdot))}\cdot\nabla w\big)+\mathrm{div}_x\big(\mathbf{b}|_{(t,u(t,\cdot))}\cdot w\big) & \\
\quad+\;\;\mathbf{c}|_{(t,u(t,\cdot))}\cdot\nabla w\;+\;c_0|_{(t,u(t,\cdot))}\,w\;=\;\mathrm{div}_x\,\mathbf{f}|_{(t,u(t,\cdot))}+g|_{(t,u(t,\cdot))} & \\
\big(\mathbf{A}|_{(t,u(t,\cdot))}\,\nabla w\;+\;\mathbf{b}|_{(t,u(t,\cdot))}\,w\;-\;\mathbf{f}|_{(t,u(t,\cdot))}\big)\cdot\nu_\Omega & \\
\qquad\qquad+\;b_0|_{(t,u(t,\cdot))}\,w\;=\;\psi|_{(t,u(t,\cdot))} & \text{on }\partial\Omega\\
\qquad\qquad\qquad w(0,\cdot)\;=\;u_0 & \text{in }\;\Omega
\end{cases}$$

has a unique weak solution $w\in L^2([0,T]\times\Omega)$ according to Proposition 102. Due to Proposition 105, $w:[0,T]\longrightarrow W^{1,2}(\Omega)$ is continuous with respect to the $L^1(\Omega)$ norm and bounded with respect to the $W^{1,2}(\Omega)$ norm. The comparative estimate in Corollary 108 assures the condition sufficient for mutational solutions in

Proposition 34 (on page 219) and thus,

$$\check{w}: [0,T] \longrightarrow \check{E}_\Omega, \qquad t \longmapsto \big(w(t,\cdot),\ \|w(t,\cdot)\|_{W^{1,2}(\Omega)}\big)$$

is a solution to the mutational equation

$$\overset{\circ}{\check{w}}(t) \ni \mathscr{F}_\Omega\big(t,\ u(t,\cdot)\big)$$

in the tuple $\big(\check{E}_\Omega, \check{d}_\Omega, e_\Omega, \lfloor\cdot\rfloor_\Omega, \text{const}\cdot(\|\cdot\|_{L^\infty}+\|\cdot\|_{L^2})\big)$. Due to the joint initial value $u(0) = u_0 = w(0,\cdot)$, Proposition 11 (about the continuous dependence of mutational solutions on given data on page 189) implies $d_\Omega\big(u(t), w(t)\big) = 0$ for all $t \in [0,T]$. This completes the proof of Theorem 103.

Remark 113 (about strong solutions). In a word, the proof of Theorem 103 reflects a basic strategy of the mutational framework: The general existence theorem provides a continuous curve $\check{u} = (u,\rho) : [0,T] \longrightarrow E_\Omega$ as a solution to the mutational equation. Its first component $u(\cdot)$ is inserted in the coefficient functions and, the resulting nonautonomous linear problem has a weak solution $w(\cdot)$ which solves a (simpler) mutational equation. Thus, $u(\cdot) \equiv w(\cdot)$.
The general regularity properties stated in Proposition 105 imply that $w(\cdot)$ is even a *strong* solution to the nonautonomous linear conormal problem and so, strictly speaking, the proof of Theorem 103 guarantees the existence of strong solutions to the functional conormal problem (on page 281).

The Remaining Proofs About the Basic Set E_Ω and the Transition Properties

Proof (of Lemma 104 on page 282). Hypothesis (H1) about reflexivity and (H2) about symmetry are obvious.

(H3)(o) The claimed equivalence between sequential convergence w.r.t. d_Ω and e_Ω results essentially from Sobolev's Embedding Theorem because all considered sequences are assumed to be bounded in $W^{1,2}(\Omega)$.
Indeed, let $u \in W^{1,2}(\Omega)$ and $(u_n)_{n\in\mathbb{N}}$ be a bounded sequence in $W^{1,2}(\Omega)$ with $\|u_n - u\|_{L^1(\Omega)} \longrightarrow 0$ for $n \longrightarrow \infty$. Then a subsequence $(u_{n_k})_{k\in\mathbb{N}}$ converges to u almost everywhere in Ω. Due to Sobolev's Embedding Theorem, a further subsequence (again denoted by) $(u_{n_k})_{k\in\mathbb{N}}$ converges to some $v \in L^2(\Omega)$ with respect to the $L^2(\Omega)$ norm. Then $v \equiv u$ almost everywhere in Ω and, we obtain indirectly that the whole sequence $(u_n)_{n\in\mathbb{N}}$ converges to u in $L^2(\Omega)$. Hence, the convergence of $(u_n)_{n\in\mathbb{N}}$ w.r.t. e_Ω implies the convergence w.r.t. d_Ω.
The opposite implication results directly from Hölder inequality since $\mathscr{L}^N(\Omega) < \infty$.

(H3) (i) – (iii) can now be concluded from the triangle inequality of $\sqrt{d_\Omega}$ and $\sqrt{e_\Omega}$ respectively.

(H4) The lower semicontinuity of $\lfloor\cdot\rfloor_\Omega \overset{\text{Def.}}{=} \|\cdot\|_{W^{1,2}(\Omega)}$ w.r.t. the $L^2(\Omega)$ norm results indirectly from the general facts that each closed ball in the Hilbert space $W^{1,2}(\Omega)$ is weakly compact and the norm is always lower semicontinuous with respect to weak convergence. □

Proof (of Corollary 107 on page 284). It is based on the estimate in Remark 106 (2.) and the Hölder inequality. Indeed, for any $0 \leq s \leq t \leq T$,

$$\begin{aligned}
&\sqrt{e_\Omega\big(u(s,\cdot),\ u(t,\cdot)\big)} \\
&\overset{\text{Def.}}{=} \big\|u(s,\cdot) - u(t,\cdot)\big\|_{L^1(\Omega)} \\
&\leq \big\|\partial_t u\big\|_{L^1([s,t]\times\Omega)} \\
&\leq \sqrt{\mathscr{L}^N(\Omega)\cdot|t-s|}\ \big\|\partial_t u\big\|_{L^2([0,T]\times\Omega)} \\
&\leq \sqrt{\mathscr{L}^N(\Omega)\cdot|t-s|}\ \ \sqrt{\widehat{C}\cdot\Big(\|u_0\|^2_{W^{1,2}} + \|\check{\mathbf{f}}\|^2_{W^{1,2}} + \|\check{g}\|^2_{L^2} + \|\check{\psi}_0\|^2_{L^2}\Big)}.
\end{aligned}$$

□

Proof (of Corollary 108 on page 284). For an initial function $v_0 \in E_\Omega$ and further coefficients $\hat{\mathbf{A}}$, $\hat{\mathbf{b}}$, $\hat{b}_0$, $\hat{\mathbf{c}}$, $\hat{c}_0$, $\hat{\mathbf{f}}$, $\hat{g}$, $\hat{\psi}_0$, let $v \in L^2([0,T]\times\Omega)$ be the weak solution of

$$\begin{cases}
-\frac{\partial v}{\partial t} + \operatorname{div}_x\big(\hat{\mathbf{A}}\cdot\nabla v\big) + \operatorname{div}_x\big(\hat{\mathbf{b}}\cdot v\big) + \hat{\mathbf{c}}\cdot\nabla v + \hat{c}_0\, v = \operatorname{div}_x\hat{\mathbf{f}} + \hat{g} \\
\qquad\big(\hat{\mathbf{A}}\ \nabla v + \hat{\mathbf{b}}\ v - \hat{\mathbf{f}}\big)\cdot\nu_\Omega + \hat{b}_0\, v = \hat{\psi}_0 \quad \text{on } [0,T]\times\partial\Omega \\
\qquad v(0,\cdot) = v_0 \quad \text{in } \Omega.
\end{cases}$$

The difference $w := u - v \in L^2([0,T]\times\Omega)$ satisfies the linear conormal problem

$$\begin{cases}
-\frac{\partial w}{\partial t} + \operatorname{div}_x\big(\check{\mathbf{A}}\cdot\nabla w\big) + \operatorname{div}_x\big(\check{\mathbf{b}}\cdot w\big) + \check{\mathbf{c}}\cdot\nabla w + \check{c}_0\, w = \operatorname{div}_x\bar{\mathbf{f}} + \bar{g} \\
\qquad\big(\check{\mathbf{A}}\ \nabla w + \check{\mathbf{b}}\ w - \bar{\mathbf{f}}\big)\cdot\nu_\Omega + \check{b}_0\, w = \bar{\psi}_0 \quad \text{on } [0,T]\times\partial\Omega \\
\qquad w(0,\cdot) = u_0 - v_0
\end{cases}$$

with the coefficients

$$\begin{aligned}
\bar{\mathbf{f}} &:= \check{\mathbf{f}} - \hat{\mathbf{f}} - (\check{\mathbf{A}} - \hat{\mathbf{A}})\cdot\nabla v - (\check{\mathbf{b}} - \hat{\mathbf{b}})\cdot v, \\
\bar{g} &:= \check{g} - \hat{g} - (\check{\mathbf{c}} - \hat{\mathbf{c}})\cdot\nabla v - (\check{c}_0 - \hat{c}_0)\cdot v, \\
\bar{\psi}_0 &:= \check{\psi}_0 - \hat{\psi}_0 - (\check{b}_0 - \hat{b}_0)\cdot v.
\end{aligned}$$

Due to the estimates in Proposition 102 and 105 (on pages 281, 283),

$$\|v(t,\cdot)\|_{W^{1,2}(\Omega)} \leq C\, e^{Ct}\cdot\big(1 + \|v_0\|_{W^{1,2}(\Omega)}\big)$$

holds for every $t \in [0,T]$ with a constant C depending only on λ, N, Ω, $\|\hat{\mathbf{A}}\|_{L^\infty}$, $\|\nabla\hat{\mathbf{A}}\|_{L^\infty}$, $\|\hat{\mathbf{b}}\|_{L^\infty}$, $\|\nabla\hat{\mathbf{b}}\|_{L^\infty}$, $\|\hat{b}_0\|_{L^\infty}$, $\|\hat{\mathbf{c}}\|_{L^\infty}$, $\|\hat{c}_0\|_{L^\infty}$. Hölder's inequality and the continuous trace operator on $\partial\Omega \in C^2$ imply for each $t \in [0,T]$

$$\begin{aligned}
\|\bar{\mathbf{f}}\|_{L^2([0,t]\times\Omega)} &\leq \|\check{\mathbf{f}} - \hat{\mathbf{f}}\|_{L^2([0,t]\times\Omega)} + \Big(\|\check{\mathbf{A}} - \hat{\mathbf{A}}\|_{L^\infty([0,t]\times\Omega)} + \|\check{\mathbf{b}} - \hat{\mathbf{b}}\|_{L^\infty([0,t]\times\Omega)}\Big)\,\rho_t \\
\|\bar{g}\|_{L^2([0,t]\times\Omega)} &\leq \|\check{g} - \hat{g}\|_{L^2([0,t]\times\Omega)} + \Big(\|\check{\mathbf{c}} - \hat{\mathbf{c}}\|_{L^\infty([0,t]\times\Omega)} + \|\check{c}_0 - \hat{c}_0\|_{L^\infty([0,t]\times\Omega)}\Big)\,\rho_t \\
\|\bar{\psi}_0\|_{L^2([0,t]\times\partial\Omega)} &\leq \|\check{\psi}_0 - \hat{\psi}_0\|_{L^2([0,t]\times\partial\Omega)} + C\ \|\check{b}_0 - \hat{b}_0\|_{L^\infty([0,t]\times\partial\Omega)}\,\rho_t
\end{aligned}$$

with the abbreviation

$$\rho_t := C\, e^{Ct}\cdot\big(1 + \|v_0\|_{W^{1,2}(\Omega)}\big)\cdot\sqrt{\mathscr{L}^N(\Omega)\, t}.$$

The estimate in Proposition 102 ensures a modified constant $\widehat{C} > 0$ (determined merely by λ, N, Ω, $\|\check{\mathbf{A}}\|_{L^\infty}$, $\|\check{\mathbf{b}}\|_{L^\infty}$, $\|\check{b}_0\|_{L^\infty}$,$\|\check{\mathbf{c}}\|_{L^\infty}$ and $\|\check{c}_0\|_{L^\infty}$) such that for any t,

$$\|w(t,\cdot)\|^2_{L^2(\Omega)} \le \widehat{C}\, e^{\widehat{C}t} \Big(\|u_0 - v_0\|^2_{L^2(\Omega)} + \|\bar{\mathbf{f}}\|^2_{L^2([0,t]\times\Omega)} + \|\bar{g}\|^2_{L^2([0,t]\times\Omega)} + \|\bar{\psi}_0\|^2_{L^2([0,t]\times\partial\Omega)}\Big).$$

Together with the preceding inequalities, we obtain the claimed bound of

$$d_\Omega\big(u(t,\cdot), v(t,\cdot)\big) = \|w(t,\cdot)\|^2_{L^2(\Omega)}. \qquad \square$$

Proof (of Proposition 109 on page 284). Existence and uniqueness of the weak solution $u(\cdot)$ are guaranteed explicitly by Proposition 102 (on page 281). This implies the semigroup property claimed in statement (2.).
Statement (1.) results directly from Proposition 105 (on page 283) in combination with the estimate in Proposition 102.
Statement (3.), i.e. $\big\|\vartheta(t,u_0) - \vartheta(t,v_0)\big\|_{L^2(\Omega)} \le C \cdot \|u_0 - v_0\|_{L^2(\Omega)}$ for all $t \in [0,1]$, is a consequence of Proposition 102: Due to the linearity of the conormal problem, $\vartheta(\cdot,u_0) - \vartheta(\cdot,v_0)$ solves the homogeneous conormal problem and starts at $u_0 - v_0$.
Statement (4.) and the final estimate comparing solutions to two problems result immediately from Corollary 107 and 108 (on page 284) respectively. $\square$

Remark 114 (about extending the results to $u_0 \in L^2(\Omega)$).
Essentially the same proof as for Corollary 108 provides the following estimate

$$\begin{aligned} d_\Omega\big(u(t), v(t))\big) \le \widehat{C}\, e^{\widehat{C}t} \cdot \Big(& d_\Omega(u_0,v_0) + \big(\big\|(\check{\mathbf{A}}, \check{\mathbf{b}}, \check{\mathbf{c}}, \check{c}_0) - (\hat{\mathbf{A}}, \hat{\mathbf{b}}, \hat{\mathbf{c}}, \hat{c}_0)\big\|^2_{L^\infty([0,t]\times\Omega)} \\ & + \big\|(\check{\mathbf{f}}, \check{g}) - (\hat{\mathbf{f}}, \hat{g})\big\|^2_{L^2([0,t]\times\Omega)} + \big\|\check{b}_0 - \hat{b}_0\big\|^2_{L^\infty([0,t]\times\partial\Omega)} \\ & + \|\check{\psi}_0 - \hat{\psi}_0\|^2_{L^2([0,t]\times\partial\Omega)}\big) \\ & \times \big(1 + \|v_0\|_{L^2} + \|\hat{\mathbf{f}}\|_{L^2} + \|\hat{g}\|_{L^2} + \|\hat{\psi}_0\|_{L^2}\big)^2 \cdot \widehat{C}\, e^{\widehat{C}t}\Big) \end{aligned}$$

due to $\|v\|_{L^2([0,t]\times\Omega)} + \|\nabla v\|_{L^2([0,t]\times\Omega)} \le C e^{Ct} \cdot \big(\|v_0\|_{L^2} + \|\hat{\mathbf{f}}\|_{L^2} + \|\hat{g}\|_{L^2} + \|\hat{\psi}_0\|_{L^2}\big)$.
In comparison with Corollary 108, the right-hand side depends just on $\|v_0\|_{L^2(\Omega)}$ instead of $\|v_0\|_{W^{1,2}(\Omega)}$, but it is lacking the factor t, which indicates first-order terms in the mutational framework.

This modified inequality implies the additional statement in Theorem 103: If the coefficients $\mathbf{A}, \mathbf{b}, b_0, \mathbf{c}, c_0, \mathbf{f}, g$ are Lipschitz continuous (w.r.t. the L^∞ or L^2 norms of their values respectively), then the existence result can then be extended to initial functions $u_0 \in L^2(\Omega)$ approximatively. Indeed, any sequence $(u_0^n)_{n\in\mathbb{N}}$ in $W^{1,2}(\Omega)$ converging to u_0 in $L^2(\Omega)$ induces a sequence of weak solutions

$$u^n(\cdot) : [0,T] \longrightarrow W^{1,2}(\Omega) \qquad (n \in \mathbb{N}),$$

which proves to be a Cauchy sequence in $C^0([0,T], L^2(\Omega))$. Hence, there exists a limit $u \in C^0([0,T], L^2(\Omega))$. The coefficients being composed with $u(\cdot)$ specify a nonautonomous, but linear conormal problem with a unique weak solution $w \in L^2([0,T]\times\Omega)$ due to Prop. 102. Finally the estimate applied to w, u^n ensures $u \equiv w$.

3.10 Example: Semilinear Evolution Equations in Arbitrary Banach Spaces

Now we consider semilinear evolution equations again

$$\tfrac{d}{dt}\, u(t) \;=\; A\, u(t) \,+\, f\big(u(t),t\big)$$

with a fixed generator A of a strongly continuous semigroup $(S(t))_{t\geq 0}$ on a Banach space X. The goal is to specify sufficient conditions on the semigroup and the function $f : X \times [0,T] \longrightarrow X$ so that initial value problems can be solved in the mutational framework.
In contrast to the example in § 2.4 (on page 125 ff.), however, we dispense with any hypotheses about the Banach space X (such as reflexivity and separability) and, we prefer topological assumptions about the semigroup or the image of f instead.

In particular, a single distance function on X is to cover the strong continuity of the semigroup appropriately. This challenge has been the main motivation for introducing two distance functions d,e in the mutational framework recently. The first distance refers to comparing (mostly simultaneous) states whereas the second one is rather related to changes in time (as explained in § 45, Step (C) on page 22 f. and § 3.1 on page 182 ff.).
The required regularity of transitions with respect to time makes now tuples with a separate real time component (as in § 3.5 on page 221 ff.) very useful indeed.

Assumptions for § 3.10.

(1.) $(X, \|\cdot\|_X)$ is a $\mathbb{R}$-Banach space, $\widetilde{X} := \mathbb{R} \times X$ and $\pi_1 : \widetilde{X} \longrightarrow \mathbb{R},\ (t,x) \longmapsto t$.

(2.) The linear operator A generates a C^0 semigroup $(S(t))_{t\geq 0}$ of bounded linear operators on X.

(3.) $(S(t))_{t\geq 0}$ is ω-contractive, i.e., there exists a constant $\omega > 0$ such that $\|S(t)\,x\|_X \leq e^{\omega t}\ \|x\|_X$ for all $x \in X,\ t \geq 0$.

Remark 115. All the essential results about semilinear evolution equations in this section 3.10 can be extended easily to *non-ω-contractive* C^0 semigroups $(S(t))_{t\geq 0}$. For the sake of transparency only, we dispense with the detailed statements here.
There are two arguments why the ω-contractivity of $(S(t))_{t\geq 0}$ is not really needed. First, we have discussed in § 3.4 (on page 214 ff.) that non-ω-contractive candidates for transitions can fulfill all required conditions in the general mutational framework if the distance function d on E is replaced appropriately. This conclusion holds in this example because every C^0 semigroup $(S(t))_{t\geq 0}$ satisfies an estimate of the form $\|S(t)x\|_X \leq M\ e^{\omega t}\ \|x\|_X$ for all $x \in X,\ t \geq 0$ with fixed constants $M \geq 1, \omega > 0$.
Second, the theory of one-parameter semigroups of bounded linear operators on Banach spaces even provides an equivalent norm on X with respect to which $(S(t))_{t\geq 0}$ is contractive (i.e. $\omega = 0, M = 1$). This explicit construction (via powers of the resolvent operator) is the key ingredient in the standard proof of the Generation Theorem of Feller, Miyadera and Phillips [76, Theorem II.3.8], which provides the link to Hille-Yosida Theorem about contractive semigroups.

Basic set	$\widetilde{X} = \mathbb{R} \times X$ with any real Banach space X (The additional real component indicates the respective point of time for simplifying the aspects of regularity in time.) $(S(t))_{t\geq 0}$ is an ω-contractive C^0 semigroup of bounded linear operators on X with the generator A. (By means of § 3.4, the main results can be extended easily to strongly continuous semigroups which are *not* ω-contractive.)
Distances	$\widetilde{d}_0 : \widetilde{X} \times \widetilde{X} \longrightarrow \mathbb{R}_0^+,\ \big((s,x), (t,y)\big) \longmapsto \|t-s\| + \big\|x - y\big\|_X$ $\widetilde{e}_0 : \widetilde{X} \times \widetilde{X} \longrightarrow \mathbb{R}_0^+,\ \big((s,x), (t,y)\big) \longmapsto \|t-s\| + \big\|S(t-s)\,x - y\big\|_X$ (if $s \leq t$)
Absolute value	$\\|\cdot\\|_{\widetilde{X}} : \widetilde{X} \longrightarrow \mathbb{R}_0^+,\ (t,x) \longmapsto \|t\| + \\|x\\|_X$
Transitions:	For each vector $v \in X$, the variation of constants formula induces $\widetilde{\tau}_v : [0,1] \times \widetilde{X} \longrightarrow \widetilde{X},$ $\big(h,\ (t,x)\big) \longmapsto \Big(t+h,\ S(h)\,x + \int_0^h S(h-s)\,v\,ds\Big)$
Compactness	(i) Assume $(S(t))_{t\geq 0}$ to be immediately compact in addition: § 3.10.3 (page 300 ff.) — the "standard" situation [76, 154] (ii) Suppose f to have relatively (weakly) compact image set: § 3.10.4 (page 306 ff.)
Equi-continuity	Nonequidistant Euler equi-continuity results from representing any Euler curve by means of the variation of constants formula.
Mutational solutions	Mild solutions to the semilinear evolution equation $\frac{d}{dt}\,x(\cdot) = A\,x(\cdot) + f\big(x(\cdot), \cdot\big)$
List of main results formulated in § 3.10	(i) in § 3.10.3: Existence due to compactness: Theorem 129 (page 301) Existence for equations with delay: Corollary 130 Existence for systems with modified morph. equations: Corollary 132 (ii) in § 3.10.4: Existence for equations with delay and norm topology: Theorem 134 (page 307) Existence for equations with delay and weak topology: Theorem 136 (page 308)
Key tools	Variation of constants formula Integral representation of the linear resolvent operator

Table 3.5 Brief summary of the example in § 3.10 in mutational terms:
Semilinear evolution equations in arbitrary Banach spaces

3.10.1 The Distance Functions $(\widetilde{d}_j)_{j\in\mathbb{R}_0^+}, (\widetilde{e}_j)_{j\in\mathbb{R}_0^+}$ on $\widetilde{X} = \mathbb{R} \times X$

In this example, the essential aspect is to take the strong continuity of $(S(t))_{t\geq 0}$ into consideration properly. This regularity has influence on the chronological features and thus on the family $(\widetilde{e}_j)_j$ of distance functions (rather than $(\widetilde{d}_j)_j$). In particular, it is the main motivation for considering tuples with separate time component, i.e., $\widetilde{X}$ instead of X. As abbreviations, set $\mathbb{R}_0^+ := [0,\infty[$ and $\mathbb{R}^+ :=]0,\infty[$.

Definition 116.
Under the general assumptions of § 3.10, we define for each index $j \in \mathbb{R}_0^+$

$$\begin{array}{llll} \widetilde{d}_j : \widetilde{X}\times\widetilde{X} \longrightarrow [0,\infty[, & ((s,x),\,(t,y)) & \longmapsto & |t-s| + \|S(j)\,x - S(j)\,y\|_X \\ \|\cdot\|_{\widetilde{X}} : \quad \widetilde{X} \longrightarrow [0,\infty[, & (t,x) & \longmapsto & |t| + \|x\|_X \,. \end{array}$$

$$\widetilde{e}_j : \quad \widetilde{X}\times\widetilde{X} \longrightarrow [0,\infty[,$$
$$((s,x),\,(t,y)) \longmapsto |t-s| + \begin{cases} \|S(j+t-s)\,x - S(j)\,y\|_X & \text{if } s<t \\ \|S(j)\,x - S(j+s-t)\,y\|_X & \text{if } s\geq t \end{cases}$$

Obviously, $\widetilde{d}_0(\cdot,\cdot) \equiv \|\cdot - \cdot\|_{\widetilde{X}}$ holds in $\widetilde{X}\times\widetilde{X}$. In fact, the convergence of norm bounded sequences with respect to $(\widetilde{d}_j)_{j\in\mathbb{R}^+}$ is equivalent to norm convergence in $\widetilde{X}$ as proved in following Proposition 117. The detour via $j \in \mathbb{R}^+$ (instead of $j = 0$) serves merely the purpose of concluding the convergence with respect to $\widetilde{d}_0$ from $\widetilde{e}_0$.

Proposition 117. *For every element $\widetilde{x} \in \widetilde{X}$ and any bounded sequence $(\widetilde{x}_n)_{n\in\mathbb{N}}$ in $(\widetilde{X}, \|\cdot\|_{\widetilde{X}})$, the following properties are equivalent:*

$$\begin{array}{lll} (i) & \lim_{n\to\infty} \|\widetilde{x} - \widetilde{x}_n\|_{\widetilde{X}} & = 0 \\ (ii)\ \forall\, j\in\mathbb{R}^+ : & \lim_{n\to\infty} \widetilde{d}_j(\widetilde{x},\widetilde{x}_n) & = 0 \\ (iii)\ \forall\, j\in\mathbb{R}^+ : & \lim_{n\to\infty} \widetilde{e}_j(\widetilde{x},\widetilde{x}_n) & = 0 \\ (iv) & \lim_{n\to\infty} \widetilde{e}_0(\widetilde{x},\widetilde{x}_n) & = 0. \end{array}$$

This equivalence and subsequent Lemmas 119 – 121 imply directly

Corollary 118. *The tuple $(\widetilde{X},\widetilde{d}_0,\widetilde{e}_0)$ satisfies hypotheses (H1), (H2), (H3) (o), (H4) (on page 182) and hypotheses (H3) $(\widetilde{i})$–$(\widetilde{iii})$ (on page 221).* □

Proof (of Proposition 117). "$(i) \Longrightarrow (ii)$" and "$(iv) \Longrightarrow (iii)$" are obvious consequences of Definition 116 since each linear operator $S(j) : X \longrightarrow X$ $(j \in \mathbb{R}_0^+)$ of the C^0 semigroup is continuous.

"$(ii) \Longrightarrow (i)$" Assume for $\widetilde{x} = (t,x)$ and the bounded sequence $(\widetilde{x}_n = (t_n,x_n))_{n\in\mathbb{N}}$ in $\widetilde{X}$ that $\quad \widetilde{d}_j(\widetilde{x},\widetilde{x}_n) \stackrel{\text{Def.}}{=} |t-t_n| + \|S(j)\,x - S(j)\,x_n\|_X \longrightarrow 0 \qquad (n\longrightarrow\infty)$

holds for every $j \in \mathbb{R}^+$. The resolvent $R(\lambda, A)$ of the generator A of $(S(t))_{t\geq 0}$ is known to have the representation as limit of Bochner integrals

$$R(\lambda, A)\, y \;=\; \lim_{\tau \to \infty} \int_0^\tau e^{-\lambda t}\; S(t)\, y\; dt$$

for every $y \in X$ and $\lambda \in \mathbb{C}$ with $\mathrm{Re}\,\lambda > \omega$ (see [76, Theorem II.1.10], for example). As a consequence, Lebesgue's Theorem about Dominated Convergence leads to

$$\big\| R(\omega+2, A)\, \big(x - x_n\big) \big\|_X \longrightarrow 0 \qquad \text{for } n \longrightarrow \infty.$$

It implies $\|x - x_n\|_X \longrightarrow 0$ since $R(\omega+2, A): X \longrightarrow X$ is a bijective contraction with $\|R(\omega+2, A)\| \leq \frac{1}{2}$.

"$(iii) \Longrightarrow (iv)$" It also results from the integral representation of the resolvent $R(\omega+2, A)$. Indeed, assuming for a norm bounded sequence $\big(\widetilde{x}_n = (t_n, x_n)\big)_{n \in \mathbb{N}}$

$$\widetilde{e}_j(\widetilde{x}, \widetilde{x}_n) \overset{\text{Def.}}{=} |t - t_n| + \big\| S(j + (t_n - t)^+)\, x - S(j + (t - t_n)^+)\, x_n \big\|_X \overset{n \to \infty}{\longrightarrow} 0$$

for every $j \in \mathbb{R}^+$ (with the abbreviation $r^+ := \max\{r, 0\}$ for $r \in \mathbb{R}$) implies

$$\big\| R(\omega+2, A)\; \big(S((t_n - t)^+)\, x - S((t - t_n)^+)\, x_n\big) \big\|_X \overset{n \to \infty}{\longrightarrow} 0$$

and thus, $\widetilde{e}_0(\widetilde{x}, \widetilde{x}_n) \overset{\text{Def.}}{=} |t - t_n| + \big\| S((t_n - t)^+)\, x - S((t - t_n)^+)\, x_n \big\|_X \overset{n \to \infty}{\longrightarrow} 0.$

"$(ii) \Longrightarrow (iii)$" Let the sequence $\big(\widetilde{x}_n = (t_n, x_n)\big)_{n \in \mathbb{N}}$ and $\widetilde{x} = (t, x) \in \widetilde{X}$ be arbitrary with $\widetilde{d}_j(\widetilde{x}, \widetilde{x}_n) \longrightarrow 0$ for each $j \in \mathbb{R}^+$.
First we assume $t_n \geq t$ for all $n \in \mathbb{N}$ in addition. Then,

$$\begin{aligned}
\widetilde{e}_j(\widetilde{x}, \widetilde{x}_n) &= |t - t_n| + \big\| S(j + t_n - t)\, x - S(j)\, x_n \big\|_X \\
&\leq |t - t_n| + \big\| S(j)\, x - S(j)\, x_n \big\|_X + \big\| S(j)\, x - S(j + t_n - t)\, x \big\|_X \\
&= \widetilde{d}_j(\widetilde{x}, \widetilde{x}_n) + e^{\omega j}\, \big\| x - S(t_n - t)\, x \big\|_X \\
&\longrightarrow 0 \qquad \text{for } n \longrightarrow \infty \text{ and each } j \in \mathbb{R}^+.
\end{aligned}$$

Similarly we obtain under the additional assumption $t_n \leq t$ for all $n \in \mathbb{N}$

$$\begin{aligned}
\widetilde{e}_j(\widetilde{x}, \widetilde{x}_n) &= |t - t_n| + \big\| S(j)\, x - S(j + t - t_n)\, x_n \big\|_X \\
&\leq |t - t_n| + \big\| S(j + t - t_n)\, x - S(j + t - t_n)\, x_n \big\|_X + \big\| S(j)\, x - S(j + t - t_n)\, x \big\|_X \\
&\leq |t - t_n| + e^{\omega\, (t - t_n)}\, \big\| S(j)\, x - S(j)\, x_n \big\|_X + e^{\omega j}\, \big\| x - S(t - t_n)\, x \big\|_X \\
&\leq e^{\omega\, |t - t_n|}\; \widetilde{d}_j(\widetilde{x}, \widetilde{x}_n) + e^{\omega j}\, \big\| x - S(t - t_n)\, x \big\|_X \\
&\longrightarrow 0 \qquad \text{for } n \longrightarrow \infty \text{ and each } j \in \mathbb{R}^+.
\end{aligned}$$

Applying these cases to subsequences, we conclude without additional assumptions

$$\widetilde{e}_j\big(\widetilde{x}, \widetilde{x}_n\big) \longrightarrow 0 \qquad \text{for } n \longrightarrow \infty \text{ and each } j \in \mathbb{R}^+.$$

"$(iii) \Longrightarrow (ii)$" Let the sequence $\big(\widetilde{x}_n = (t_n, x_n)\big)_{n \in \mathbb{N}}$ and $\widetilde{x} = (t, x) \in \widetilde{X}$ be arbitrary with $\widetilde{e}_j(\widetilde{x}, \widetilde{x}_n) \longrightarrow 0$ for each $j \in \mathbb{R}^+$.
First we suppose $t_n \geq t$ for all $n \in \mathbb{N}$ in addition. Then,

$$\begin{aligned}
\widetilde{d}_j(\widetilde{x}, \widetilde{x}_n) &= |t - t_n| + \big\|S(j)\,x - S(j)\,x_n\big\|_X \\
&\leq |t - t_n| + \big\|S(j+t_n-t)\,x - S(j)\,x_n\big\|_X + \big\|S(j)\,x - S(j+t_n-t)\,x\big\|_X \\
&= \widetilde{e}_j(\widetilde{x}, \widetilde{x}_n) + e^{\omega j}\,\big\|x - S(t_n-t)\,x\big\|_X \\
&\longrightarrow 0 \text{ for } n \longrightarrow \infty \text{ and each } j \in \mathbb{R}^+.
\end{aligned}$$

Complementarily we conclude under the additional assumption $t_n \leq t$ for all $n \in \mathbb{N}$

$$\begin{aligned}
\widetilde{d}_j(\widetilde{x}, \widetilde{x}_n) &= |t - t_n| + \big\|S(j)\,x_n - S(j)\,x\big\|_X \\
&\leq |t - t_n| + \big\|S(\tfrac{j}{2} - t + t_n)\big\|\ \big\|S(\tfrac{j}{2}+t-t_n)\,x_n - S(\tfrac{j}{2}+t-t_n)\,x\big\|_X \\
&\leq |t - t_n| + e^{\omega\,(\frac{j}{2}-t+t_n)} \Big(\big\|S(\tfrac{j}{2}+t-t_n)\,x_n - S(\tfrac{j}{2})\,x\big\|_X \\
&\qquad + \big\|S(\tfrac{j}{2}+t-t_n)\,x - S(\tfrac{j}{2})\,x\big\|_X\Big) \\
&\leq e^{\omega\,(\frac{j}{2}+|t-t_n|)} \Big(\widetilde{e}_{\frac{j}{2}}(\widetilde{x}, \widetilde{x}_n) + \big\|S(\tfrac{j}{2}+t-t_n)\,x - S(\tfrac{j}{2})\,x\big\|_X\Big) \\
&\longrightarrow 0 \text{ for } n \longrightarrow \infty \text{ and each } j \in \mathbb{R}^+.
\end{aligned}$$

Hence, $\widetilde{d}_j\big(\widetilde{x}, \widetilde{x}_n\big) \longrightarrow 0$ holds for $n \longrightarrow \infty$ and every index $j \in \mathbb{R}^+$ in general. □

Lemma 119. *The tuple $(\widetilde{X}, \widetilde{d}_0, \widetilde{e}_0)$ fulfills hypothesis (H3) $\widetilde{\text{(i)}}$ (on page 221).*

Proof. Choose any $\widetilde{x} = (s,x),\ \widetilde{y} = (t,y) \in \widetilde{X}$ and sequences $\big(\widetilde{x}_n = (s_n, x_n)\big)_{n\in\mathbb{N}}$, $\big(\widetilde{y}_n = (t_n, y_n)\big)_{n\in\mathbb{N}}$ with

$$\lim_{n\to\infty} \widetilde{d}_0(\widetilde{x}, \widetilde{x}_n) = 0 = \lim_{n\to\infty} \widetilde{d}_0(\widetilde{y}, \widetilde{y}_n).$$

Obviously, $\widetilde{d}_0$ satisfies the triangle inequality and thus,

$$\widetilde{d}_0(\widetilde{x}, \widetilde{y}) = \lim_{n\to\infty} \widetilde{d}_0\big(\widetilde{x}_n, \widetilde{y}_n\big).$$

For verifying the same continuity property of $\widetilde{e}_0$, we assume $s_n \leq t_n$ for all $n \in \mathbb{N}$ sufficiently large. Then, $s \leq t$ and, we conclude from the semigroup property and ω-contractivity of $(S(\cdot))$

$$\begin{aligned}
\big|\widetilde{e}_0(\widetilde{x},\widetilde{y}) - \widetilde{e}_0(\widetilde{x}_n,\widetilde{y}_n)\big| &\leq \Big|\big||s-t| - |s_n - t_n|\big| + \big|\|S(t-s)x - y\|_X \\
&\qquad - \|S(t_n - s_n)x_n - y_n\|_X\big| \\
&\leq \big|s - t - (s_n - t_n)\big| + \big\|S(t_n - s_n)x_n - S(t-s)x\big\|_X \\
&\qquad + \big\|y_n - y\big\|_X \\
&\leq |s - s_n| + |t - t_n| + \big\|S(t_n - s_n)x_n - S(t_n - s_n)x\big\|_X \\
&\qquad + \big\|S(t_n - s_n)x - S(t-s)x\big\|_X \\
&\qquad + \big\|y_n - y\big\|_X \\
&\leq e^{\omega\,|t_n - s_n|}\,\widetilde{d}_0(\widetilde{x}, \widetilde{x}_n) + \big\|S(t_n - s_n)x - S(t-s)x\big\|_X + \widetilde{d}_0(\widetilde{y}, \widetilde{y}_n) \\
&\longrightarrow 0 \text{ for } n \longrightarrow \infty.
\end{aligned}$$

Finally, property (H3) $\widetilde{\text{(i)}}$ is fulfilled. □

Lemma 120. *The distance functions $\widetilde{d}_j, \widetilde{e}_j : \widetilde{X} \times \widetilde{X} \longrightarrow [0,\infty[$ $(j \in \mathbb{R}^+)$ fulfill hypothesis (H3) $(\widetilde{\text{ii}})$ (on page 221).*

Proof. Let $\widetilde{x} = (s,x) \in \widetilde{X}$ and the sequences $\big(\widetilde{x}_n = (s_n,x_n)\big)_{n\in\mathbb{N}}$, $\big(\widetilde{y}_n = (t_n,y_n)\big)_{n\in\mathbb{N}}$ in $\widetilde{X}$ be arbitrary with

$$\lim_{n\to\infty} \widetilde{d}_j(\widetilde{x},\widetilde{y}_n) = 0 = \lim_{n\to\infty} \widetilde{e}_j(\widetilde{y}_n,\widetilde{x}_n) \quad \text{for every } j \in \mathbb{R}^+.$$

In particular, $t_n \longrightarrow s$ and thus, $s_n \longrightarrow s$ for $n \longrightarrow \infty$.
Under the additional assumption $s \le t_n \le s_n$ for all $n \in \mathbb{N}$, we obtain for every $j \in \mathbb{R}^+$

$$\begin{aligned}
\widetilde{d}_j(\widetilde{x}_n, \widetilde{x}) &= s_n - s + \big\|S(j)\,x_n - S(j)\,x\big\|_X \\
&\le s_n - t_n + \big\|S(j)\,x_n - S(j+s_n-t_n)\,y_n\big\|_X \\
&\quad + \big\|S(j+s_n-t_n)\,y_n - S(j+s_n-t_n)\,x\big\|_X \\
&\quad + t_n - s + \big\|S(j+s_n-t_n)\,x - S(j)\,x\big\|_X \\
&\le \widetilde{e}_j(\widetilde{x}_n, \widetilde{y}_n) + e^{\omega\,|s_n-t_n|} \cdot \widetilde{d}_j(\widetilde{y}_n,\widetilde{x}) \\
&\quad + t_n - s + \big\|S(j+s_n-t_n)\,x - S(j)\,x\big\|_X \\
&\longrightarrow 0 \text{ for } n \longrightarrow \infty.
\end{aligned}$$

Correspondingly, the supplementary hypothesis $s \ge t_n \ge s_n$ for all $n \in \mathbb{N}$ leads to

$$\begin{aligned}
\widetilde{d}_j(\widetilde{x}_n, \widetilde{x}) &= s - s_n + \big\|S(j)\,x_n - S(j)\,x\big\|_X \\
&\le s - s_n + \|S(\tfrac{j}{2}+s_n-t_n)\|_{\mathscr{L}(X,X)} \cdot \big\|S(\tfrac{j}{2}+t_n-s_n)\,x_n - S(\tfrac{j}{2})\,y_n\big\|_X \\
&\quad + \|S(\tfrac{j}{2}+s_n-t_n)\|_{\mathscr{L}(X,X)} \cdot \big\|S(\tfrac{j}{2})\,y_n - S(\tfrac{j}{2})\,x\big\|_X \\
&\quad + \big\|S(j+s_n-t_n)\,x - S(j)\,x\big\|_X \\
&\le s - s_n + e^{\omega\,j}\,\big(\widetilde{e}_{j/2}(\widetilde{x}_n, \widetilde{y}_n) + \widetilde{d}_{j/2}(\widetilde{y}_n, \widetilde{x})\big) \\
&\quad + \big\|S(j+s_n-t_n)\,x - S(j)\,x\big\|_X \\
&\longrightarrow 0 \text{ for } n \longrightarrow \infty.
\end{aligned}$$

Finally, property (H3) $(\widetilde{\text{ii}})$ also holds. □

Lemma 121. *The tuple $(\widetilde{X}, \widetilde{d}_0, \widetilde{e}_0)$ fulfills hypothesis (H3) $(\widetilde{\text{iii}})$ (on page 221).*

Proof. Choose any element $\widetilde{x} \in \widetilde{X}$ and sequences $(\widetilde{x}_n)_{n\in\mathbb{N}}$, $(\widetilde{y}_k)_{k\in\mathbb{N}}$, $(\widetilde{z}_{k,n})_{k,n\in\mathbb{N}}$ in $\widetilde{X}$ fulfilling

$$\begin{cases}
\pi_1\,\widetilde{z}_{k,n} = \pi_1\,\widetilde{y}_k \le \pi_1\,\widetilde{x}_n = \pi_1\,\widetilde{x} & \text{for each } k,n \in \mathbb{N}, \\
\lim_{k\to\infty} \widetilde{d}_0(\widetilde{x},\widetilde{y}_k) = 0, & \\
\lim_{n\to\infty} \widetilde{d}_0(\widetilde{y}_k,\widetilde{z}_{k,n}) = 0 & \text{for each } k \in \mathbb{N}, \\
\lim_{k\to\infty} \sup_{n>k} \widetilde{e}_0(\widetilde{z}_{k,n},\widetilde{x}_n) = 0, & \\
\sup_{k,n\in\mathbb{N}} \big\{ \lfloor\widetilde{x}_n\rfloor_i, \lfloor\widetilde{y}_k\rfloor_i, \lfloor\widetilde{z}_{k,n}\rfloor_i \big\} < \infty. &
\end{cases}$$

As abbreviations, set $\widetilde{x} = (t,x)$, $\widetilde{x}_n = (t,x_n)$, $\widetilde{y}_k = (t_k,y_k)$, $\widetilde{z}_{k,n} = (t_k,z_{k,n}) \in \widetilde{X}$. Then, $\lim_{k\to\infty} t_k = t$ results directly from $\lim_{k\to\infty} \widetilde{d}_0(\widetilde{x},\widetilde{y}_k) = 0$. The auxiliary elements $\widetilde{\xi}_n = (t_n,x_n) \in \widetilde{X}$ $(n \in \mathbb{N})$ fulfill

$$\begin{aligned}
\widetilde{e}_0(\widetilde{\xi}_n,\, \widetilde{x}) &= |t_n - t| + \big\| S(t-t_n)\, x_n - x \big\|_X \\
&\le t - t_n + \big\| S(t-t_n)\, x_n - S(2\,(t-t_n))\, z_{k,n} \big\|_X \\
&\quad + \big\| S(2\,(t-t_n))\, z_{k,n} - S(2\,(t-t_n))\, y_k \big\|_X \\
&\quad + \big\| S(2\,(t-t_n))\, y_k - S(2\,(t-t_n))\, x \big\|_X + \big\| S(2\,(t-t_n))\, x - x \big\|_X \\
&\le e^{\omega\,|t-t_n|}\; \widetilde{e}_0(\widetilde{x}_n,\, \widetilde{z}_{k,n}) + e^{\omega\,2\,|t-t_n|}\; \big(\widetilde{d}_0(\widetilde{z}_{k,n},\, \widetilde{y}_k) \;+\; \widetilde{d}_0(\widetilde{y}_k,\, \widetilde{x})\big) \\
&\quad + \big\| S(2\,(t-t_n))\, x - x \big\|_X .
\end{aligned}$$

Choosing first $k \in \mathbb{N}$ and then $n \in \mathbb{N}$ sufficiently large leads to

$$\lim_{n\to\infty} \widetilde{e}_0(\widetilde{\xi}_n,\, \widetilde{x}) = 0$$

and due to Proposition 117, $\limsup_{n\to\infty} \widetilde{d}_0(\widetilde{x}_n,\, \widetilde{x}) \le \lim_{n\to\infty} \widetilde{d}_0(\widetilde{\xi}_n,\, \widetilde{x}) = 0.$ □

3.10.2 The Variation of Constants Induces Transitions on $\widetilde{X}$

Similarly to the preceding example in § 2.4 (on page 125 ff.), a simple affine linear initial value problem motivates the choice of candidates for transitions. Definition 2.28 on the basis of the variation of constants formula is now extended to tuples in $\widetilde{X} = \mathbb{R} \times X$:

Definition 122. For each $v \in X$, the function $\tau_v : [0,1] \times X \longrightarrow X$ is defined as mild solution to the initial value problem $\frac{d}{dt}\, u(t) = A\, u(t) + v,\ u(0) = x \in X$, i.e.

$$\tau_v(h,x) \;:=\; S(h)\,x + \int_0^h S(h-s)\; v\; ds.$$

Furthermore, set $\widetilde{\tau}_v : [0,1] \times \widetilde{X} \longrightarrow \widetilde{X}, \quad \big(h,\, (t,x)\big) \longmapsto \big(t+h,\ \tau_v(h,x)\big).$

Lemma 123. *For every vector $v, w \in X$, the functions $\widetilde{\tau}_v, \widetilde{\tau}_w : [0,1] \times \widetilde{X} \longrightarrow \widetilde{X}$ have the following properties for every $j \in \mathbb{R}_0^+$, $\widetilde{x}, \widetilde{y} \in \widetilde{X}$ and $s,h \in [0,1]$ with $s+h \le 1$*

(1.) $\widetilde{\tau}_v(0,\widetilde{x}) = \widetilde{x}$

(2.) $\widetilde{\tau}_v(s+h,\widetilde{x}) = \widetilde{\tau}_v\big(h, \widetilde{\tau}_v(s,\widetilde{x})\big)$

(3.) $\widetilde{e}_j\big(\widetilde{x}, \widetilde{\tau}_v(h,\widetilde{x})\big) \le h \cdot \big(1 + e^{\omega\,(j+1)}\, \|v\|_X\big)$

(4.) $\|\widetilde{\tau}_v(h,\widetilde{x})\|_{\widetilde{X}} \le \big(\|\widetilde{x}\|_{\widetilde{X}} + h \cdot (1 + \|v\|_X)\big)\ e^{\omega\, h}$

(5.) $\widetilde{d}_j\big(\widetilde{\tau}_v(h,\widetilde{x}),\ \widetilde{\tau}_w(h,\widetilde{y})\big) \le \widetilde{d}_j(\widetilde{x},\widetilde{y}) \cdot e^{\omega\, h} + h \cdot e^{\omega\,(j+h)}\, \|v-w\|_X.$

Postponing its proof for a moment, we conclude directly from these estimates in combination with the semigroup property of $\widetilde{\tau}_v$:

Proposition 124. *For each vector $v \in X$, the function $\widetilde{\tau}_v : [0,1] \times \widetilde{X} \longrightarrow \widetilde{X}$ specified in Definition 122 is a transition on $\big(\widetilde{X}, (\widetilde{d}_j)_{j\in\mathbb{R}^+}, (\widetilde{e}_j)_{j\in\mathbb{R}^+}, (\|\cdot\|_{\widetilde{X}})_{j\in\mathbb{R}^+}\big)$ in the sense of Definition 2 (on page 183) with*

$$\begin{aligned} \alpha_j(\widetilde{\tau}_v;\, r) &:= \omega \\ \beta_j(\widetilde{\tau}_v;\, r) &:= 1 + \|v\|_X \cdot e^{\omega\,(j+1)} \\ \gamma_j(\widetilde{\tau}_v) &:= \max\big\{1 + \|v\|_X,\ \omega\big\} \end{aligned}$$

and the additional property $\pi_1\, \widetilde{\tau}_v(h,\widetilde{x}) = \pi_1\, \widetilde{x} + h$ *for all* $\widetilde{x} \in X,\ h \in [0,1]$. □

Inequality (5.) in Lemma 123, applied to $j = 0$, however, reveals an alternative to the family $(\widetilde{d}_j)_{j\in\mathbb{R}^+}$, which is even more popular: the norm of $\widetilde{X}$.
In fact, we even have transitions on the simpler tuple $\big(\widetilde{X}, \widetilde{d}_0, \widetilde{e}_0, \|\cdot\|_{\widetilde{X}}\big)$ and, the norm instead of the family $(\widetilde{d}_j)_{j\in\mathbb{R}^+}$ will provide a direct link between timed solutions (to mutational equations) and mild solutions (to semilinear evolution equations) in

subsequent § 3.10.3. In regard to the preceding topological results of § 3.10.1, the hypotheses (H1) – (H4) are also fulfilled by the latter tuple — due to the equivalence of convergence in Proposition 117 (on page 293).

Corollary 125. *For each vector $v \in X$, the function $\widetilde{\tau}_v : [0,1] \times \widetilde{X} \longrightarrow \widetilde{X}$ specified in Definition 122 is a transition on the tuple $\big(\widetilde{X}, \widetilde{d}_0, \widetilde{e}_0, \|\cdot\|_{\widetilde{X}}\big)$ with*

$$\begin{aligned} \alpha_0(\widetilde{\tau}_v; r) &:= \omega \\ \beta_0(\widetilde{\tau}_v; r) &:= 1 + \|v\|_X \cdot e^{\omega} \\ \gamma_0(\widetilde{\tau}_v) &:= \max\big\{1 + \|v\|_X,\ \omega\big\} \end{aligned}$$

and the additional property $\pi_1\, \widetilde{\tau}_v(h, \widetilde{x}) = \pi_1\, \widetilde{x} + h$ for all $\widetilde{x} \in \widetilde{X}$, $h \in [0,1]$.

Furthermore setting

$$\widehat{D}_0\big(\widetilde{\tau}_v, \widetilde{\tau}_w, r\big) := \|v - w\|_X$$

for any vectors $v, w \in X$ and radius $r \ge 0$, the function $\widehat{D}_0(\cdot, \cdot; r)$ is a metric of these transitions on $\widetilde{X}$ and, hypotheses (H5) – (H7) (on page 184) are fulfilled.

□

Proof (of Lemma 123). Statements (1.) and (2.) result from the semigroup property of $(S(t))_{t \ge 0}$ in a quite obvious way.

(3.) For every $\widetilde{x} = (t,x) \in \widetilde{X}$, $h \in [0,1]$ and $j \in \mathbb{R}_0^+$,

$$\begin{aligned} \widetilde{e}_j\big((t,x), \widetilde{\tau}_v(h,(t,x))\big) &= t + h - t + \Big\| S(j)\Big(S(h)x + \int_0^h S(h-r) v\, dr\Big) \\ &\qquad - S(j + t + h - t)x \Big\|_X \\ &= h + \Big\| \int_0^h S(j + h - r) v\, dr \Big\|_X \\ &\le h + h \quad e^{\omega (j+h)} \quad \|v\|_X . \end{aligned}$$

(4.) In regard to the norm $\|\cdot\|_{\widetilde{X}}$, we obtain for every $\widetilde{x} = (t,x) \in \widetilde{X}$, $h \in [0,1]$

$$\begin{aligned} \|\, \widetilde{\tau}_v(h, \widetilde{x})\, \|_{\widetilde{X}} &= |t + h| + \Big\| S(h)\, x + \int_0^h S(h - r)\ v\ dr \Big\|_X \\ &\le |t| + h + e^{\omega h} \|x\|_X + h \quad e^{\omega h} \quad \|v\|_X \\ &\le e^{\omega h} \quad \big(\|\widetilde{x}\|_{\widetilde{X}} + h \cdot (1 + \|v\|_X)\big). \end{aligned}$$

(5.) Finally, the definitions imply for any $\widetilde{x} = (s,x)$, $\widetilde{y} = (t,y) \in \widetilde{X}$ and $h \in [0,1]$

$$\begin{aligned} \widetilde{d}_j\big(\widetilde{\tau}_v(h,(s,x)), \widetilde{\tau}_w(h,(t,y))\big) &= |t - s| + \Big\| S(j)\Big(S(h)x + \int_0^h S(h-r) v\, dr\Big) \\ &\qquad - S(j)\Big(S(h)y + \int_0^h S(h-r) w\, dr\Big) \Big\|_X \\ &\le |t - s| + e^{\omega h} \big\| S(j)(x - y) \big\|_X + h \quad e^{\omega(j+h)} \|v - w\|_X \\ &\le \widetilde{d}_j\big(\widetilde{x}, \widetilde{y}\big) \cdot e^{\omega h} + h \quad e^{\omega(j+h)} \|v - w\|_X . \end{aligned}$$

□

3.10.3 Mild Solutions to Semilinear Evolution Equations in X — Using an Immediately Compact Semigroup

The recently proposed transitions on $(\widetilde{X}, \widetilde{d}_0, \widetilde{e}_0, \|\cdot\|_{\widetilde{X}})$ are based on autonomous linear evolution equations. Now the mutational framework provides the tools for the step to nonautonomous semilinear evolution equations and their mild solutions. For this purpose, we first prove the existence of timed solutions to the corresponding mutational equations by means of Theorem 43 (on page 226). Then we focus on the connection between these timed solutions and the more popular concept of mild solutions (to the underlying semilinear evolution equation in X).

Existence Theorem 43 is based on assuming Euler compactness and Euler equi-continuity. For the tuple $(\widetilde{X}, \widetilde{d}_0, \widetilde{e}_0, \|\cdot\|_{\widetilde{X}})$, however, even the nonequidistant counterparts of these two properties (specified in Definition 23 on page 203) are not difficult to verify because the variation of constants formula provides a useful integral representation of every (nonequidistant) Euler approximation.
If the ω-contractive C^0 semigroup $(S(t))_{t\geq 0}$ on X is immediately compact in addition, then nonequidistant Euler compactness also holds.

Lemma 126 (Characterization of nonequidistant Euler approximations).
Suppose for $\widetilde{x}_0 = (t_0, x_0) \in \widetilde{X}$, $\widehat{\gamma} \geq 0$ and a $\|\cdot\|_{\widetilde{X}}$-continuous curve $\widetilde{y}(\cdot) : [0,T[\longrightarrow \widetilde{X}$

(1.) $\widetilde{y}(0) = \widetilde{x}_0$,
(2.) *for any* $t \in [0,T[$, *there exist* $s \in]t-1,t]$ *and* $v \in X$ *with* $\|v\|_X \leq \widehat{\gamma}$ *and* $\widetilde{y}(s+\cdot) = \widetilde{\tau}_v(\cdot, \widetilde{y}(s))$ *in an open neighborhood* $I \subset [0,1]$ *of* $[0, t-s]$.

Then there exists $v(\cdot) \in L^\infty([0,T], X)$ with $\|v\|_{L^\infty} \leq \widehat{\gamma}$ and for every $t \in [0,T[$,

$$\widetilde{y}(t) = \Big(t_0 + t, \quad S(t)\, x_0 + \int_0^t S(t-r)\, v(r)\, dr\Big)$$

This representation of an Euler approximation in combination with the proof of Lemma 123 (3.) implies directly its Lipschitz continuity with respect to $\widetilde{e}_0$:

Corollary 127 (nonequidistant Euler equi-continuous).
Every $\|\cdot\|_{\widetilde{X}}$-continuous curve $\widetilde{y} : [0,T[\longrightarrow \widetilde{X}$ satisfying conditions (1.), (2.) in Lemma 126 is Lipschitz continuous with respect to $\widetilde{e}_0$ and, its Lipschitz constant is $\leq 1 + \widehat{\gamma} \cdot e^{\omega T}$.
Thus $(\widetilde{X}, \widetilde{d}_0, \widetilde{e}_0, \|\cdot\|_{\widetilde{X}})$ together with all the transitions of Corollary 125 is nonequidistant Euler equi-continuous in the sense of Definition 23 (on page 203).
□

Lemma 128 (nonequidistant Euler compact).
Assume in addition that $(S(t))_{t\geq 0}$ is immediately compact, i.e., for every $t > 0$, the linear operator $S(t) : X \longrightarrow X$ is compact.
Then the tuple $(\widetilde{X}, \widetilde{d}_0, \widetilde{e}_0, \|\cdot\|_{\widetilde{X}})$ together with all the transitions of Corollary 125 is nonequidistant Euler compact in the sense of Definition 23 (on page 203).

Now preceding Theorem 43 (on page 226) provides the existence of timed solutions to mutational equations in $\big(\widetilde{X}, \widetilde{d}_0, \widetilde{e}_0, \|\cdot\|_{\widetilde{X}}\big)$. They prove to induce mild solutions to the underlying semilinear evolution equation in X:

Theorem 129 (Existence of mild solutions to semilin. evolution equations in X). *Let* $\pi_2 : \widetilde{X} = \mathbb{R} \times X \longrightarrow X,\ (t,x) \longmapsto x$ *abbreviates the canonical projection on the second component and, A denotes the generator of an immediately compact, ω-contractive C^0 semigroup $(S(t))_{t\geq 0}$ of linear operators on X. Assume for* $f : X \times [0,T] \longrightarrow X$

(i) $\sup_{x,t} \|f(x,t)\|_X < \infty$,
(ii) *for $\mathscr{L}^1$-almost every $t \in [0,T]$, the function $f(\cdot,t) : X \longrightarrow X$ is continuous with respect to $\|\cdot\|_X$.*

Then for every $\widetilde{x}_0 = (t_0, x_0) \in \widetilde{X}$, there exists a timed solution $\widetilde{x}(\cdot) : [0,T] \longrightarrow \widetilde{X}$ to the mutational equation $\overset{\circ}{\widetilde{x}}(\cdot) \ni \widetilde{\tau}_{f(\pi_2\,\widetilde{x}(\cdot),\,\cdot)}$ *in* $\big(\widetilde{X}, \widetilde{d}_0, \widetilde{e}_0, \|\cdot\|_{\widetilde{X}}\big)$.

Moreover if $\widetilde{x}(\cdot) : [0,T] \longrightarrow \widetilde{X}$ is a timed solution to this mutational equation, then $x(\cdot) := \pi_2\,\widetilde{x}(\cdot) : [0,T] \longrightarrow X$ is a mild solution to the semilinear evolution equation

$$\tfrac{d}{dt}\, x(\cdot) \;=\; A\, x(\cdot) \;+\; f\big(x(\cdot), \cdot\big).$$

In fact, Theorem 43 takes even delays into consideration. Its full generality and the preceding relation to mild solutions (mentioned in Theorem 129) lead to the following existence result.

Corollary 130 (Existence of mild solutions to semilinear equations with delay). *Let* $\pi_2 : \widetilde{X} = \mathbb{R} \times X \longrightarrow X,\ (t,x) \longmapsto x$ *abbreviates the canonical projection on the second component and, A denotes the generator of an immediately compact, ω-contractive C^0 semigroup $(S(t))_{t\geq 0}$ of linear operators on X. Moreover assume for fixed $\tau \geq 0$ and*

$$f : C^0\big([-\tau,0],\ (X, \|\cdot\|_X)\big) \times [0,T] \longrightarrow X$$

(i) $\sup_{z(\cdot),t} \|f(z(\cdot),t)\|_X < \infty$,
(ii) *for $\mathscr{L}^1$-almost every $t \in [0,T]$,* $\displaystyle\lim_{n\longrightarrow\infty} \big\|f(z_n^1(\cdot),t_n^1) - f(z_n^2(\cdot),t_n^2)\big\|_X = 0$
for any sequences $(t_n^1)_{n\in\mathbb{N}}$, $(t_n^2)_{n\in\mathbb{N}}$ in $[0,T]$ and $(z_n^1(\cdot))_{n\in\mathbb{N}}$, $(z_n^2(\cdot))_{n\in\mathbb{N}}$ in $C^0\big([-\tau,0], (X, \|\cdot\|_X)\big)$ satisfying for every $s \in [-\tau,0]$

$$\lim_{n\to\infty} t_n^1 = t = \lim_{n\to\infty} t_n^2, \quad \lim_{n\to\infty} \big\|z_n^1(s) - z(s)\big\|_X = 0 = \lim_{n\to\infty} \big\|z_n^2(s) - z(s)\big\|_X$$
$$\sup_{n\in\mathbb{N}}\ \sup_{[-\tau,0]} \|z_n^{1,2}(\cdot)\|_X < \infty.$$

For every Lipschitz continuous function $x_0(\cdot) : [-\tau,0] \longrightarrow (X, \|\cdot\|_X)$, there exists a curve $\widetilde{x}(\cdot) : [-\tau,T] \longrightarrow \widetilde{X}$ with the following properties:

(*i*) $\widetilde{x}(\cdot) \in \widetilde{\text{BLip}}\big([-\tau,T],\ \widetilde{X};\ \widetilde{e}_0,\ \|\cdot\|_{\widetilde{X}}\big)$,

(*ii*) $\widetilde{x}(t) = (t,\ x_0(t))$ *for every* $t \in [-\tau,0]$,

(*iii*) *the restriction* $\widetilde{x}(\cdot)\big|_{[0,T]}$ *is a timed solution to the mutational equation*

$$\overset{\circ}{\widetilde{x}}(t) \ni \widetilde{\tau}_{f\left(\pi_2\,\widetilde{x}(t+\cdot)\big|_{[-\tau,0]},\ t\right)}$$

in the sense of Definition 35.

In particular, the projected restriction $\pi_2\,\widetilde{x}(\cdot)\big|_{[0,T]} : [0,T] \longrightarrow X$ *is a mild solution to the semilinear evolution equation with delay*

$$\tfrac{d}{dt}\,x(t) = A\,x(t) + f\big(x(t+\cdot)\big|_{[-\tau,0]},\ t\big) \qquad \text{in } [0,T].$$

□

Remark 131. In comparison with standard literature about evolution equations, neither Theorem 129 nor Corollary 130 are completely new results. The essential point is, however, that these semilinear evolution equations are solved in the mutational framework — just by adding a separate time component temporarily and introducing distance function $\widetilde{e}_0$ suitable for handling the strong continuity of $(S(t))_{t\geq 0}$.

In particular, we are free to combine this type of dynamical problem with any other example fitting in this mutational framework. Correspondingly to Proposition 2.37 (on page 131), we conclude from Existence Theorem 22 about systems of mutational equations and from the example in § 1.11 (on page 96 ff.) immediately:

Corollary 132 (Existence of solutions to a system with semilinear evolution equation and modified morphological equation).
Suppose A to be the generator of an immediately compact, ω-contractive C^0 *semigroup* $(S(t))_{t\geq 0}$ *of linear operators on X and, assume for*

$$f : X \times \mathscr{K}(\mathbb{R}^N) \times [0,T] \longrightarrow X,$$
$$\mathscr{G} : X \times \mathscr{K}(\mathbb{R}^N) \times [0,T] \longrightarrow \text{OSLIP}(\mathbb{R}^N,\mathbb{R}^N)$$

(*i*) $\sup_{x,M,t} \big(\|f(x,M,t)\|_X + \|\mathscr{G}(x,M,t)\|_\infty + \max\{0,\ \text{Lip}\,\mathscr{G}(x,M,t)\}\big) < \infty.$

(*ii*) *f and* $\mathscr{G}$ *are continuous in the following sense:*

$$\begin{cases} \big\|f(y_n,M_n,t_n) - f(y,M,t)\big\|_X \longrightarrow 0 \\ dl_\infty\big(\mathscr{G}(y_n,M_n,t_n),\ \mathscr{G}(y,M,t)\big) \longrightarrow 0 \end{cases} \qquad \text{for } n \longrightarrow \infty$$

holds for $\mathscr{L}^1$*-almost every* $t \in [0,T]$ *and any sequences* $(t_n)_{n\in\mathbb{N}}$, $(M_n)_{n\in\mathbb{N}}$ *and* $(y_n)_{n\in\mathbb{N}}$ *in* $[0,T], \mathscr{K}(\mathbb{R}^N), X$ *respectively satisfying* $t_n \longrightarrow t$, $dl(M_n,M) \longrightarrow 0$ *and* $\|y_n - y\|_X \longrightarrow 0$ *for* $n \longrightarrow \infty$.

Then for every initial vector $x_0 \in X$ *and set* $K_0 \in \mathscr{K}(\mathbb{R}^N)$, *there exist curves* $x(\cdot) : [0,T] \longrightarrow X$ *and* $K(\cdot) : [0,T] \longrightarrow \mathscr{K}(\mathbb{R}^N)$ *with* $x(0) = x_0$, $K(0) = K_0$ *and the following properties:*

(*1.*) *$x(\cdot) : [0,T] \longrightarrow X$ is a mild solution to the evolution equation*
$$\tfrac{d}{dt}\, x(t) \;=\; A\, x(t) \,+\, f(x(t),\, K(t),\, t).$$

(*2.*) *$K(\cdot)$ is Lipschitz continuous w.r.t. $d\!l$ and satisfies for $\mathscr{L}^1$-almost every t*
$$\lim_{h \downarrow 0} \; \tfrac{1}{h} \cdot d\!l\big(\vartheta_{\mathscr{G}(x(t),K(t),t)}(h,\, K(t)),\;\; K(t+h)\big) \;=\; 0.$$

(*3.*) *If, in addition, the set-valued map $\mathscr{G}(x(t),K(t),t) : \mathbb{R}^N \rightsquigarrow \mathbb{R}^N$ is continuous for each $t \in [0,T]$, then the set $K(t) \subset \mathbb{R}^N$ coincides with the reachable set $\vartheta_{\mathscr{G}(x(\cdot),K(\cdot),\cdot)}(t, K_0)$ of the nonautonomous differential inclusion*
$$y'(\cdot) \;\in\; \mathscr{G}\big(x(\cdot),\, K(\cdot),\, \cdot\,\big)\,(y(\cdot))$$
at every time $t \in [0,T]$. □

Finally, we close the gap of lacking proofs.

Proof (of Lemma 126). Due to assumption (2.) and the finite Lebesgue measure of the domain $[0,T[$, there exists an (at most countable) set of pairs (s_l,t_l) $(l \in N \subset \mathbb{N})$ with the following properties:

(i) for every $l \in N$, $0 \le s_l < t_l < T$ and $t_l - s_l \le 1$,
for some $l_0 \in N$, $s_{l_0} = 0$,
(ii) the intervals $]s_l,t_l[$ $(l \in N)$ are pairwise disjoint,
(iii) $\bigcup_{l \in N} [s_l,t_l] \;=\; [0,T[$,
(iv) for every $l \in N$, there exists a vector $v_l \in X$ with $\|v_l\|_X \le \widehat{\gamma}$ and
$\widetilde{y}(\cdot) \;=\; \widetilde{\tau}_{v_l}\big(\,\cdot\, - s_l,\; \widetilde{y}(s_l)\big)$ in $[s_l,t_l[$.

Setting $v(t) := v_l$ for $t \in [s_l,t_l[$ $(l \in N)$, the function $v(\cdot)$ is well-defined Lebesgue-almost everywhere in $[0,T[$ and belongs to $L^\infty([0,T[,X)$. Then the definition of $\widetilde{\tau}_{v_l}(\cdot,\cdot)$ and the continuity of $\widetilde{y}(\cdot)$ (with respect to $\|\cdot\|_X$ by assumption) lead to the claimed integral representation in $[0,T[$. □

Proof (of Lemma 128). We claim that $\big(\widetilde{X},\, \widetilde{d_0},\, \widetilde{e}_0,\, \|\cdot\|_{\widetilde{X}}\big)$ is nonequidistant Euler compact in the sense of Definition 23 (on page 203). Due to the integral representation in Lemma 126 (on page 300), it is sufficient to verify the following statement:

Choose $x_0 \in X$ and $T \in]0,\infty[$ arbitrarily. Let $(v_n(\cdot))_{n\in\mathbb{N}}$ be a bounded sequence in $L^\infty([0,T],X)$ and, set
$$\begin{aligned} y_n :\; [0,T] \longrightarrow X, \quad t \longmapsto\;\; & S(t)\, x_0 \,+\, \int_0^t S(t-r)\, v_n(r)\;\, dr \\ =\;\; & S(t)\, x_0 \,+\, \int_0^t S(s)\;\, v_n(t-s)\;\, ds \end{aligned}$$
for each $n \in \mathbb{N}$. Then for every $\widehat{t} \in]0,T]$, there exists a subsequence of $\big(y_n(\widehat{t})\big)_{n\in\mathbb{N}}$ converging strongly in X.

This proof is based on the supplementary assumption that the semigroup $(S(t))_{t\ge 0}$ is immediately compact, i.e., for every $t > 0$, the operator $S(t) : X \longrightarrow X$ is compact.

For each $k \in \mathbb{N}$ with $\frac{1}{k} < \widehat{t}$, the sequence

$$y_n(\widehat{t}) \;-\; \int_0^{\frac{1}{k}} S(s)\; v_n(t-s)\; ds \;=\; S(\tfrac{1}{k}) \Big(\int_{\frac{1}{k}}^{\widehat{t}} S(s-\tfrac{1}{k})\; v_n(t-s)\; ds\Big) \qquad (n \in \mathbb{N})$$

has a subsequence converging with respect to $\|\cdot\|_X$. Cantor's diagonal construction provides a strictly increasing sequence $(n_l)_{l\in\mathbb{N}}$ of indices and a sequence $(z_k)_{k\in\mathbb{N}}$ in X such that for every $k \in \mathbb{N}$ with $\frac{1}{k} < \widehat{t}$,

$$y_{n_l}(\widehat{t}) \;-\; \int_0^{\frac{1}{k}} S(s)\; v_{n_l}(t-s)\; ds \;\longrightarrow\; z_k \qquad \text{for } l \longrightarrow \infty.$$

In particular,

$$\begin{aligned} \limsup_{l\to\infty} \big\|y_{n_l}(\widehat{t}) \;-\; z_k\big\|_X \;&\leq\; \limsup_{l\to\infty} \Big\|\int_0^{\frac{1}{k}} S(s)\; v_{n_l}(t-s)\; ds\Big\|_X \\ &\leq\; \tfrac{1}{k} \cdot e^{\frac{\omega}{k}} \cdot\; \sup_n \|v_n\|_{L^\infty}\,. \end{aligned}$$

Furthermore, $(z_k)_{k\in\mathbb{N}}$ is a Cauchy sequence in X since for any $k_1, k_2 \in \mathbb{N} \cap \,]\frac{1}{\widehat{t}}, \infty[$,

$$\begin{aligned} &\|z_{k_1} \;-\; z_{k_2}\|_X \\ &= \lim_{l\to\infty} \Big\|y_{n_l}(\widehat{t}) - \int_0^{\frac{1}{k_1}} S(s)\; v_{n_l}(t-s)\; ds - y_{n_l}(\widehat{t}) + \int_0^{\frac{1}{k_2}} S(s) v_{n_l}(t-s) ds\Big\|_X \\ &\leq \sup_{l\in\mathbb{N}} \Big(\frac{1}{k_1} \quad e^{\frac{\omega}{k_1}} \quad \|v_{n_l}\|_{L^\infty} + \frac{1}{k_2} \quad e^{\frac{\omega}{k_2}} \quad \|v_{n_l}\|_{L^\infty}\Big)\,. \end{aligned}$$

Hence, $(z_k)_{k\in\mathbb{N}}$ converges to a limit $z \in X$ and, $\|z_k - z\|_X \leq \frac{e^{\omega} \cdot \sup_n \|v_n\|_{L^\infty}}{k}$ for all large $k \in N$. Finally we obtain $\|y_{n_l}(\widehat{t}) - z\|_X \longrightarrow 0$ for $l \longrightarrow \infty$ simply by means of the triangle inequality. □

Proof (of Theorem 129).
The existence of a timed solution to the mutational equation

$$\overset{\circ}{\widetilde{x}}(\cdot) \;\ni\; \widetilde{\tau}_{f(\pi_2\, \widetilde{x}(\cdot),\, \cdot)}$$

in $\big(\widetilde{X}, \widetilde{d}_0, \widetilde{e}_0, \|\cdot\|_{\widetilde{X}}\big)$ results from Theorem 43 (on page 226) due to Corollary 127 and Lemma 128 (on page 300). Indeed, the projection $\pi_2 : (\widetilde{X}, \|\cdot\|_{\widetilde{X}}) \longrightarrow (X, \|\cdot\|_X)$ is continuous and thus, the composition $\widetilde{X} \times [0,T] \longrightarrow X,\; (\widetilde{z},t) \longmapsto f(\pi_2\, \widetilde{z},\, t)$ fulfills the continuity assumptions of Theorem 43.

Now we focus on the second part of the claim: If $\widetilde{x}(\cdot) : [0,T] \longrightarrow \widetilde{X}$ is a timed solution to this mutational equation, then $x(\cdot) := \pi_2\, \widetilde{x}(\cdot) : [0,T] \longrightarrow X$ is a mild solution to the semilinear evolution equation

$$\tfrac{d}{dt}\, x(\cdot) \;=\; A\, x(\cdot) \;+\; f\big(x(\cdot), \cdot\big)\,.$$

Indeed, the composition $[0,T] \longrightarrow (X, \|\cdot\|_X),\; t \longmapsto f(x(t),\, t)$ is continuous and, $[0,T] \longrightarrow \mathscr{L}(X,X),\; t \longmapsto S(t)$ is bounded with respect to the operator norm. Thus, the auxiliary function

$$y(\cdot):\; [0,T] \longrightarrow (X, \|\cdot\|_X), \quad t \longmapsto S(t)\, x(0) \;+\; \int_0^t S(t-s)\; f\big(x(s),\, s\big)\; ds$$

is continuous, bounded and, it satisfies for every $t \in [0,T[$, $h \in [0,1]$

$$\begin{aligned}
\tau_{f(x(t),t)}\big(h,\, y(t)\big) &\stackrel{\text{Def.}}{=} S(h)\, y(t) + \int_0^h S(h-s)\, f(x(t),t)\, ds \\
&= S(t+h)\, x(0) + \int_0^t S(t+h-s)\, f\big(x(s),\, s\big)\, ds \\
&\quad + \int_0^h S(h-s)\, f(x(t),\, t)\; ds \\
&= S(t+h)\, x(0) + \int_0^{t+h} S(t+h-s) \\
&\quad \times f\big(x(\min\{s,t\}),\, \min\{s,t\}\big)\; ds\,.
\end{aligned}$$

It implies at Lebesgue-almost every time $t \in [0,T]$

$$\begin{aligned}
&\frac{1}{h} \cdot \big\| y(t+h) \;-\; \tau_{f(x(t),t)}\big(h,\, y(t)\big) \big\|_X \\
&\quad = \frac{1}{h} \, \Big\| \int_t^{t+h} S(t+h-s)\, \big(f\big(x(s),\, s\big) \;-\; f\big(x(t),\, t\big)\big) \; ds \Big\|_X \\
&\quad \le e^{\omega\,(T+1)} \;\cdot\; \sup_{[t,\, t+h]} \big\| f\big(x(\cdot),\, \cdot\big) \;-\; f\big(x(t),\, t\big) \big\|_X \longrightarrow 0 \quad \text{for } h \downarrow 0.
\end{aligned}$$

As a consequence, this auxiliary function supplied with a real time component, i.e.,

$$\widetilde{y}(\cdot):\; [0,T] \longrightarrow \widetilde{X},\;\; t \longmapsto \Big(\pi_1\, \widetilde{x}(0) \,+\, t,\;\; S(t)\, x(0) \,+\, \int_0^t S(t-s)\, f\big(x(s),\, s\big)\; ds \Big)$$

is a timed solution to the mutational equation

$$\overset{\circ}{\widetilde{y}}(\cdot) \;\ni\; \widetilde{\tau}_{f(\pi_2\, \widetilde{x}(\cdot),\, \cdot)}$$

in $\big(\widetilde{X}, \widetilde{d}_0, \widetilde{e}_0, \|\cdot\|_{\widetilde{X}}\big)$. Finally Proposition 40 (on page 224) ensures

$$\begin{aligned}
0 &= \inf\Big\{ \big\|\widetilde{z} - \widetilde{x}(t)\big\|_{\widetilde{X}} + \big\|\widetilde{z} - \widetilde{y}(t)\big\|_{\widetilde{X}} \;\big|\; \widetilde{z} \in \widetilde{X} : \|\widetilde{z}\|_{\widetilde{X}} < 1 + \sup\big\{ \|\widetilde{x}(\cdot)\|_{\widetilde{X}},\, \|\widetilde{y}(\cdot)\|_{\widetilde{X}} \big\} \Big\} \\
&= \big\| \widetilde{x}(t) \;-\; \widetilde{y}(t) \big\|_{\widetilde{X}}
\end{aligned}$$

for every $t \in [0,T]$, i.e., $x(\cdot) \equiv y(\cdot)$. □

3.10.4 Exploiting Relatively Compact Terms of Inhomogeneity

Considering an immediately compact semigroups $(S(t))_{t\geq 0}$ on X in the preceding section 3.10.3 has served essentially one single purpose, namely to guarantee Euler compactness (as formulated in Lemma 128 on 300 and proved on page 303).
In particular, all other conclusions like the connection between mutational equations and mild solutions to semilinear evolution equations do not require this supplementary assumption explicitly.
Now we suggest an alternative aspect for compactness to come into play, i.e., the image of the function f in the semilinear evolution equation

$$\tfrac{d}{dt}\, x(\cdot) \;=\; A\, x(\cdot) \;+\; f\big(x(\cdot), \cdot\big).$$

Lemma 133 (nonequidistant Euler compact).
Let $W \neq \emptyset$ be a compact subset of the Banach space $(X, \|\cdot\|_X)$.

Then the tuple $\big(\widetilde{X}, \widetilde{d}_0, \widetilde{e}_0, \|\cdot\|_{\widetilde{X}}\big)$ together with the transitions $\widetilde{\tau}_v : [0,1] \times \widetilde{X} \longrightarrow \widetilde{X}$ induced by any vector $v \in W$ as in Definition 122 is nonequidistant Euler compact in the sense of Definition 23 (on page 203).

Proof. According to Lemma 126 (on page 300), every nonequidistant Euler approximation $\widetilde{y}(\cdot) : [0,T[\longrightarrow \widetilde{X}$ is characterized by a function $w(\cdot) \in L^\infty([0,T], X)$ satisfying $\|w\|_{L^\infty} \leq \widehat{\gamma}$ and for every $t \in [0,T[$,

$$\widetilde{y}(t) \;=\; \Big(t_0 + t, \quad S(t)\, x_0 \;+\; \int_0^t S(t-r)\, w(r)\; dr\Big).$$

$W_0 := \overline{\mathrm{co}}\,(W \cup \{0\})$ is convex and compact in X due to [168, II.4.3 Corollary].
At each time $t \in \,]0,T]$, the state $\pi_2\, \widetilde{y}(t)$ is contained in the subset

$$V_t \;:=\; S(t)\, x_0 \;+\; \int_0^t S(s)\; W_0\; ds \;\subset\; X$$

whose set-valued integral is understood in the sense of Aumann.
V_t is totally bounded in X. Indeed, the function $[0,t] \times W_0 \longrightarrow X,\; (s,x) \longmapsto S(s)\, x$ is uniformly continuous due to [76, Lemma I.5.2]. Hence for each $\varepsilon > 0$, there exists a sufficiently large integer $n_\varepsilon \in \mathbb{N}$ such that

$$\big\| S([n_\varepsilon\, s]\, \tfrac{1}{n_\varepsilon})\, x \;-\; S(s)\, x \big\|_X \;\leq\; \tfrac{\varepsilon}{t}$$

for every $x \in W_0$ and $s \in [0,t]$ (with $[r]$ always denoting the largest integer $\leq r$).
This piecewise constant approximation of $S(\cdot)|_{W_0}$ and $0 \in W_0$ lead to

$$\begin{aligned} V_t \;&\subset\; S(t)\, x_0 + \int_0^t \big(S([n_\varepsilon\, r]\, \tfrac{1}{n_\varepsilon})\, W_0 + \mathbb{B}_{\frac{\varepsilon}{t}}(0)\big)\; ds \\ &\subset\; S(t)\, x_0 + \tfrac{1}{n_\varepsilon} \cdot \sum_{j=0}^{[\frac{t}{n_\varepsilon}]+1} S(\tfrac{j}{n_\varepsilon})\, W_0 + \mathbb{B}_\varepsilon(0) \end{aligned}$$

and, the last superset can be covered by finitely many balls of radius 2ε due to the compactness of W_0. Finally, the set V_t is relatively compact in X for each $t \in [0,T]$. □

Theorem 134 (Existence of mild solutions to semilinear equations with delay). *Let $\pi_2 : \widetilde{X} = \mathbb{R} \times X \longrightarrow X$, $(t,x) \longmapsto x$ abbreviates the canonical projection on the second component and, A denotes the generator of an ω-contractive C^0 semigroup $(S(t))_{t \geq 0}$ of linear operators on X. Moreover assume for some fixed $\tau \geq 0$ and*

$$f : C^0\big([-\tau,0],\ (X, \|\cdot\|_X)\big) \times [0,T] \longrightarrow X$$

(i) the image of f is relatively compact in X and (thus, in particular) it satisfies $\sup_{z(\cdot),t} \|f(z(\cdot),t)\|_X < \infty$,

(ii) for $\mathscr{L}^1$-almost every $t \in [0,T]$, $\displaystyle\lim_{n \to \infty} \big\|f(z_n^1(\cdot),t_n^1) - f(z_n^2(\cdot),t_n^2)\big\|_X = 0$
for any sequences $(t_n^1)_{n\in\mathbb{N}}$, $(t_n^2)_{n\in\mathbb{N}}$ in $[0,T]$ and $(z_n^1(\cdot))_{n\in\mathbb{N}}$, $(z_n^2(\cdot))_{n\in\mathbb{N}}$ in $C^0\big([-\tau,0], (X, \|\cdot\|_X)\big)$ satisfying for every $s \in [-\tau,0]$

$$\lim_{n\to\infty} t_n^1 = t = \lim_{n\to\infty} t_n^2, \quad \lim_{n\to\infty} \big\|z_n^1(s) - z(s)\big\|_X = 0 = \lim_{n\to\infty} \big\|z_n^2(s) - z(s)\big\|_X$$
$$\sup_{n\in\mathbb{N}} \sup_{[-\tau,0]} \|z_n^{1,2}(\cdot)\|_X < \infty.$$

For every Lipschitz continuous function $x_0(\cdot) : [-\tau,0] \longrightarrow (X, \|\cdot\|_X)$, there exists a curve $\widetilde{x}(\cdot) : [-\tau,T] \longrightarrow \widetilde{X}$ with the following properties:

(i) $\widetilde{x}(\cdot) \in \widetilde{\mathrm{B}}\mathrm{Lip}\big([-\tau,T], \widetilde{X};\ \widetilde{e}_0,\ \|\cdot\|_{\widetilde{X}}\big)$,

(ii) $\widetilde{x}(t) = (t,\ x_0(t))$ for every $t \in [-\tau,0]$,

(iii) the restriction $\widetilde{x}(\cdot)\big|_{[0,T]}$ is a timed solution to the mutational equation

$$\overset{\circ}{\widetilde{x}}(t) \ni \widetilde{\tau}_{f(\pi_2 \widetilde{x}(t+\cdot)|_{[-\tau,0]},\, t)}$$

in the sense of Definition 35.

In particular, the projected restriction $\pi_2\, \widetilde{x}(\cdot)\big|_{[0,T]} : [0,T] \longrightarrow X$ is a mild solution to the semilinear evolution equation with delay

$$\tfrac{d}{dt}\, x(t) = A\, x(t) + f\big(x(t+\cdot)\big|_{[-\tau,0]},\ t\big) \qquad \text{in } [0,T].$$

□

In regard to an alternative topology on X, Ülger formulated a criterion sufficient for the relative weak compactness of Bochner-integrable functions in the 1990s and, we quote it in Proposition A.85 here. It is used for verifying the following lemma about *weak* Euler compactness:

Lemma 135 (weakly Euler compact).
Let $W \neq \emptyset$ be a weakly compact subset of the Banach space X.

Then the tuple $\big(\widetilde{X}, \widetilde{d}_0, \widetilde{e}_0, \|\cdot\|_{\widetilde{X}}\big)$ together with the transitions $\widetilde{\tau}_v : [0,1] \times \widetilde{X} \longrightarrow \widetilde{X}$ induced by any vector $v \in W$ as in Definition 122 is weakly Euler compact in the sense of Definition 27 (on page 207).

Proof. According to Lemma 126 (on page 300), every nonequidistant Euler approximation $\widetilde{y}(\cdot) : [0,T[\longrightarrow \widetilde{X}$ is characterized by a function $w(\cdot) \in L^\infty([0,T], X)$ satisfying $\|w\|_{L^\infty} \le \widehat{\gamma}$ and for every $t \in [0,T[$,

$$\widetilde{y}(t) = \Big(t_0 + t, \quad S(t)\, x_0 + \int_0^t S(t-r)\, w(r)\, dr\Big).$$

Now we benefit from the additional property that the values of $w(\cdot)$ belong to the weakly compact set $W \subset X$. Due to Proposition A.85 of Ülger (on page 492),

$$\big\{ w(\cdot) \in L^1([0,T], X) \;\big|\; \text{for all } t \in [0,T]: \; w(t) \in W \big\}$$

is relatively weakly compact in the space $L^1([0,T],X)$ of Bochner-integrable functions with values in Banach space X.
Hence, for any sequence $\big(\widetilde{y}_n(\cdot)\big)_{n\in\mathbb{N}}$ of nonequidistant Euler approximations in $\mathscr{PN} = \mathscr{PN}(\widetilde{x}_0, T, \widehat{\alpha}, \widehat{\beta}, \widehat{\gamma}, L) \neq \emptyset$, there always exists a sequence $n_k \nearrow \infty$ of indices such that the corresponding characterizing functions $w_{n_k}(\cdot)$, $k \in \mathbb{N}$, in $L^\infty([0,T],W)$ converge weakly in $L^1([0,T],X)$. Their weak limit is denoted by $w(\cdot) \in L^1([0,T],X)$. In particular, Proposition A.86 implies $\|w\|_{L^\infty([0,T],X)} \le \widehat{\gamma}$.
As the linear operators $S(t) : X \longrightarrow X$, $0 \le t \le T$, are uniformly bounded, the weak convergence of $\big(w_{n_k}(\cdot)\big)_{k\in\mathbb{N}}$ to $w(\cdot)$ has the consequence for each $t \in [0,T]$

$$\widetilde{y}_{n_k}(t) - \Big(t_0 + t, \; S(t)\, x_0 + \int_0^t S(t-r)\, w(r)\, dr\Big) \stackrel{k\to\infty}{\longrightarrow} 0 \qquad \text{weakly in } \mathbb{R} \times X.$$

□

In regard to mild solutions, we now apply Existence Theorem 44 (on page 227 f.) and use the link between mutational equations and semilinear evolution equations presented in § 3.10.3.
Indeed, the transitions (based on the variation of constants formula in Definition 122) are always weakly continuous with respect to state because they are linear with respect to the initial state and each operator $S(t) : X \longrightarrow X$, $t \ge 0$, is assumed to be bounded.
The final result is very similar to preceding Theorem 134, but takes the weak topology on X into consideration.

Theorem 136 (Existence of mild solutions to semilinear equations with delay). *Let* $\pi_2 : \widetilde{X} = \mathbb{R} \times X \longrightarrow X$, $(t,x) \longmapsto x$ *abbreviates the canonical projection on the second component and, A denotes the generator of an* ω*-contractive* C^0 *semigroup* $(S(t))_{t\ge 0}$ *of linear operators on X. Moreover assume for some fixed* $\tau \ge 0$ *and*

$$f : C^0([-\tau, 0], \; (X, \|\cdot\|_X)) \times [0,T] \longrightarrow X$$

(i) *the image of f is relatively* weakly *compact in X and (thus, in particular)* $\sup_{z(\cdot),t} \|f(z(\cdot),t)\|_X < \infty$, (*ii*) *for* $\mathscr{L}^1$*-almost every* $t \in [0,T]$: $f(z_n^1(\cdot), t_n^1) - f(z_n^2(\cdot), t_n^2) \overset{n\to\infty}{\longrightarrow} 0$ *weakly in* X

for any sequences $(t_n^1)_{n\in\mathbb{N}}$, $(t_n^2)_{n\in\mathbb{N}}$ *in* $[0,T]$ *and* $(z_n^1(\cdot))_{n\in\mathbb{N}}$, $(z_n^2(\cdot))_{n\in\mathbb{N}}$ *in* $C^0\big([-\tau,0], (X, \|\cdot\|_X)\big)$ *satisfying for some* $z(\cdot) \in C^0\big([-\tau,0], X\big)$ *and every* $s \in [-\tau,0]$

$$\lim_{n\to\infty} t_n^1 = t = \lim_{n\to\infty} t_n^2, \qquad \sup_{n\in\mathbb{N}} \sup_{[-\tau,0]} \|z_n^{1,2}(\cdot)\|_X < \infty,$$
$$z_n^1(s) \longrightarrow z(s) \text{ weakly } (n \to \infty)$$
$$z_n^2(s) \longrightarrow z(s) \text{ weakly } (n \to \infty).$$

For every Lipschitz continuous function $x_0(\cdot) : [-\tau,0] \longrightarrow (X, \|\cdot\|_X)$, *there exists a curve* $\widetilde{x}(\cdot) : [-\tau,T] \longrightarrow \widetilde{X}$ *with the following properties:*

(*i*) $\widetilde{x}(\cdot) \in \widetilde{\mathrm{B}}\mathrm{Lip}\big([-\tau,T], \widetilde{X}; \widetilde{e}_0, \|\cdot\|_{\widetilde{X}}\big)$,

(*ii*) $\widetilde{x}(t) = (t, x_0(t))$ *for every* $t \in [-\tau,0]$,

(*iii*) *the restriction* $\widetilde{x}(\cdot)\big|_{[0,T]}$ *is a timed solution to the mutational equation*

$$\overset{\circ}{\widetilde{x}}(t) \ni \widetilde{\tau}_{f(\pi_2 \widetilde{x}(t+\cdot)|_{[-\tau,0]},\, t)}$$

in the sense of Definition 35.

In particular, the projected restriction $\pi_2 \widetilde{x}(\cdot)\big|_{[0,T]} : [0,T] \longrightarrow X$ *is a mild solution to the semilinear evolution equation with delay*

$$\tfrac{d}{dt}\, x(t) = A\, x(t) + f\big(x(t+\cdot)\big|_{[-\tau,0]},\, t\big) \qquad \text{in } [0,T].$$

□

3.11 Example: Strong Solutions to Parabolic Differential Equations with Zero Dirichlet Boundary Conditions in Noncylindrical Domains

Applying the previous examples of the mutational framework to partial differential equations, we can usually handle problems in *fixed* domains in the Euclidean space. In particular, the coupling with set evolutions has been restricted to the coefficients of the differential equations so far – but not via their domains. Proposition 2.37 (on page 131) and Corollary 132 (on page 302), for example, focus on the system

$$\begin{cases} \frac{d}{dt}\, x(t) \;=\; A\, x(t) \;+\; f(x(t), K(t), t) \\ \overset{\circ}{K}(t) \;\ni\; \mathscr{G}\big(x(t), K(t), t\big) \end{cases}$$

with mild solutions $x(\cdot) : [0,T] \longrightarrow X$ to a semilinear evolution equation, but fixed generator A of a C^0 semigroup.

The next example is a step in the direction of coupling via time-dependent domain. Indeed, we want to draw conclusions about strong solutions to the semilinear initial-boundary value problem of parabolic type

$$\begin{cases} \Big(\sum\limits_{k,l=1}^{N} a_{kl}(t,\cdot)\, \frac{\partial^2}{\partial x_k\, \partial x_l} + \sum\limits_{k=1}^{N} b_k(t,\cdot)\, \frac{\partial}{\partial x_k} + c(t,\cdot) - \frac{\partial}{\partial t} \Big)\, u \;=\; \mathscr{F}(t,u) & \text{in} \quad \Omega(t) \\ u \;=\; 0 & \text{on } \partial\Omega(t) \\ u(0,\cdot) \;=\; u_0 & \text{in} \quad \overline{\Omega(0)} \end{cases}$$

with a set-valued map $\Omega(\cdot) : [0,T] \leadsto \mathbb{R}^N$ that might be determined by a morphological equation. In particular, the set $\Omega(t) \subset \mathbb{R}^N$ will be free to change some of its topological properties while time t is increasing. The typical approach based on time-dependent transformations to a fixed reference domain (as in [32, 107, 116], for example) is to fail here.

3.11.1 The General Assumptions for This Example

The coefficients

$$\begin{aligned} a_{kl} : [S,T] \times \mathbb{R}^N &\longrightarrow \mathbb{R} & (k,l = 1 \ldots N) \\ b_k : [S,T] \times \mathbb{R}^N &\longrightarrow \mathbb{R} & (k \;\;= 1 \ldots N) \\ c : [S,T] \times \mathbb{R}^N &\longrightarrow \,]-\infty, 0] \end{aligned}$$

are assumed to be bounded, continuous and uniformly elliptic, i.e., there is some $\mu > 0$ such that for any $x, y \in \mathbb{R}^N$ and $t \in [S,T]$,

$$\sum_{k,l=1}^{N} a_{kl}(t,x)\; y_k\; y_l \;\geq\; \mu\; |y|^2 .$$

Basic set	$\widetilde{E}_{\widetilde{\Omega}} := \mathbb{R} \times C_0^0(\widetilde{\Omega}_S)$ This example applies essentially the results about semilinear evolution equations in § 3.10 to parabolic differential equations in a fixed noncylindrical domain $\widetilde{\Omega}_S$ with zero Dirichlet boundary condition. Due to Lumer and Schnaubelt [133], the linear parabolic problem with fixed $L, \widetilde{\Omega}$ induces a contractive C^0 semigroup $(\mathscr{S}(t))_{t\geq 0}$ on $C_0^0(\widetilde{\Omega}_S)$ by means of suitable extensions.
Distances	$\widetilde{d}_{\widetilde{\Omega}} : \widetilde{E}_{\widetilde{\Omega}}^2 \longrightarrow \mathbb{R}_0^+,\ ((s,u),(t,v)) \longmapsto \lvert t-s\rvert + \lVert u - v \rVert_{\sup}$ $\widetilde{e}_{\widetilde{\Omega}} : \widetilde{E}_{\widetilde{\Omega}}^2 \longrightarrow \mathbb{R}_0^+,\ ((s,u),(t,v)) \longmapsto \lvert t-s\rvert + \lVert \mathscr{S}(t-s)\,u - v \rVert_{\sup}$ (if $s \leq t$)
Absolute value	$\lvert\cdot\rvert_{\widetilde{\Omega}} : \widetilde{E}_{\widetilde{\Omega}} \longrightarrow \mathbb{R}_0^+,\ (t,u) \longmapsto \lvert t\rvert + \lVert u \rVert_{\sup}$
Transitions:	For each function $F \in C_0^0(\widetilde{\Omega}_S)$, consider $\widetilde{\vartheta}_{\widetilde{\Omega},F} : [0,1] \times \widetilde{E}_{\widetilde{\Omega}} \longrightarrow \widetilde{E}_{\widetilde{\Omega}},\quad (h,(t,u)) \longmapsto \Big(t+h,\ \mathscr{S}(h)\,u + \int_0^h \mathscr{S}(h-s)\,F\,ds\Big)$ which is related to the linear parabolic problem $\begin{cases} Lv = F & \text{in } \widetilde{\Omega}_s \\ v(s,\cdot) = u(s,\cdot) & \text{in } \widetilde{\Omega}(s) \subset \mathbb{R}^N \\ v = 0 & \text{on } \partial\widetilde{\Omega}_s \setminus (\{s\} \times \widetilde{\Omega}(s)) \end{cases}$ (Theorem 139, Remark 140 and explanations on page 318)
Compactness	We apply the existence results in § 3.10.4 (page 306 ff.) based on relatively compact terms of inhomogeneity.
Equi-continuity	Nonequidistant Euler equi-continuity results from representing any Euler curve by means of the variation of constants formula.
Mutational solutions	Strong solution $u \in C^0(\overline{\widetilde{\Omega}_0}) \cap W^{1;2}_{p,\mathrm{loc}}(\widetilde{\Omega}_0)$ (with any $p > N+2$) to an initial-boundary value problem of parabolic type $\begin{cases} Lu(t,\cdot) = \mathscr{F}(t,u)(\cdot) & \text{in } \widetilde{\Omega}(t) \text{ for a.e. } t \in\ ]0,\widehat{T}[, \\ u(0,\cdot) = u_0 & \text{in } \widetilde{\Omega}(0) \subset \mathbb{R}^N, \\ u = 0 & \text{on } \partial\widetilde{\Omega}_0 \setminus (\{0\} \times \widetilde{\Omega}(0)). \end{cases}$
List of main results formulated in § 3.11	Existence Theorem 146 (page 317)
Key tools	Semigroup approach of Lumer and Schnaubelt [133] to nonautonomous linear parabolic differential equations in noncylindrical domains: § 3.11.2 (page 313 ff.) The uniform tusk condition on $\widetilde{\Omega}_S$ is a sufficient condition for approximative Cauchy barriers: Proposition 158 (page 326)

Table 3.6 Brief summary of the example in § 3.11 in mutational terms:
Strong solutions to parabolic differential equations with zero Dirichlet boundary conditions in noncylindrical domains

As an abbreviation set $L := \sum_{k,l=1}^{N} a_{kl} \frac{\partial^2}{\partial x_k \, \partial x_l} + \sum_{k=1}^{N} b_k \frac{\partial}{\partial x_k} + c - \frac{\partial}{\partial t}$.

Fixing $p > N+2$ arbitrarily, we define for any nonempty open set $\widetilde{\Omega} \subset [S,T] \times \mathbb{R}^N$

$$\widetilde{\Omega}_s := \widetilde{\Omega} \cap \big(]s,T] \times \mathbb{R}^N\big),$$

$$\widetilde{\Omega}(s) := \big\{ y \in \mathbb{R}^N \;\big|\; (s,y) \in \widetilde{\Omega} \big\} \qquad \text{for } s \in [S,T],$$

$$W^{1;2}_{p,\mathrm{loc}}(\widetilde{\Omega}_S) := \big\{ u \in L^p_{\mathrm{loc}}(\widetilde{\Omega}_S) \;\big|\; \forall\ \widetilde{V} \subset \widetilde{\Omega} \cap (]S,T[\times \mathbb{R}^N) \text{ with compact closure}: \ \tfrac{\partial u}{\partial t}, \tfrac{\partial u}{\partial x_k}, \tfrac{\partial^2 u}{\partial x_k \, \partial x_l} \in L^p(\widetilde{V}) \ \text{ for } k,l = 1 \dots N \big\}$$

$$D(L, \widetilde{\Omega}_S) := \big\{ u \in C^0(\widetilde{\Omega}_S) \;\big|\; u \in W^{1;2}_{p,\mathrm{loc}}(\widetilde{\Omega}_S) \text{ and } \exists\, g \in C^0(\widetilde{\Omega}_S): \ Lu = g \ \ \mathscr{L}^N\text{-a.e. in } \widetilde{\Omega} \cap (]S,T[\times \mathbb{R}^N) \big\}$$

3.11.2 Some Results of Lumer and Schnaubelt About Parabolic Problems in Noncylindrical Domains

In [133], Lumer and Schnaubelt present a very sophisticated approach for time-dependent parabolic problems in noncylindrical domains. It is based on Lumer's earlier results about so-called *local operators* and provides a successive construction of a so-called *variable space propagator* which can be regarded as a generalization of *strongly continuous evolution families* (in the sense of [76, § VI.9]).

In this section, we summarize some of their results in regard to parabolic differential equations on noncylindrical domains. They serve as tools for specifying transitions in the mutational framework later on.

Definition 137 ([133, Definition 4.8]). Let $I \subset \mathbb{R}$ be an interval and for each $t \in I$, $Y(t)$ denotes a real Banach space which is isomorphic to a subspace $Y(t)^\sharp$ of a fixed Banach space $Y^\sharp$.
A family of linear operators $U(t,s): \ Y(s) \longrightarrow Y(t), \ (s,t) \in I^2, \ s \le t,$ is called *variable space propagator* if it satisfies the following conditions:

(*i*) $U(s,s) = \mathbb{I}\mathrm{d}_{Y(s)}$ for every $s \in I$,

(*ii*) $U(t,s) = U(t,r) \circ U(r,s)$ for every $r,s,t \in I$ with $s < r \le t$,

(*iii*) $\{(s,t) \in I^2) \,|\, s \le t\} \longrightarrow Y^\sharp, \ (s,t) \longmapsto \big(U(t,s)\, f(s)\big)^\sharp$ is continuous for any function $t \mapsto f(t) \in Y(t)$ whose transformed counterpart $I \longrightarrow Y^\sharp, \ t \longmapsto f(t)^\sharp$ is continuous.

The propagator is called *bounded* if $\sup_{s \le t} \|U(t,s)\|_{\mathrm{Lin}(Y(s),Y(t))} < \infty$.

Definition 138 ([133, special case of Definition 3.1]). A nonempty open set $\widetilde{\Omega} \subset]S,T] \times \mathbb{R}^N$ possesses a so-called *Cauchy barrier* with respect to L if there exist a compact set $\widetilde{K} \subset \widetilde{\Omega}$ and a function $h \in D(L, \widetilde{\Omega} \setminus \widetilde{K})$ satisfying

(i) $h > 0$ and $(L - \lambda)\, h \le 0$ in $\widetilde{\Omega} \setminus \widetilde{K}$ for some $\lambda \ge 0$,

(ii) for every $\varepsilon > 0$, there exists a compact set $\widetilde{K}_\varepsilon$ with $\widetilde{K} \subset \widetilde{K}_\varepsilon \subset \widetilde{\Omega}$ and $0 \le h \le \varepsilon$ in $\widetilde{\Omega} \setminus \widetilde{K}_\varepsilon$.

Now we formulate a special case of [133, Theorem 6.1] restricted to bounded subsets of $[S,T] \times \mathbb{R}^N$ and Dirichlet boundary conditions:

Theorem 139 ([133]). *Let $\widetilde{\Omega}$ be a bounded open subset of $[S,T] \times \mathbb{R}^N$, $s \in [S,T[$, $f \in C_0^0(\widetilde{\Omega}(s))$ and the function F satisfy*

(i) $\widetilde{\Omega} \cap (\{t\} \times \mathbb{R}^N) \neq \emptyset$ *for every* $t \in [S,T]$,

(ii) $\widetilde{\Omega}_S$ *is the intersection of finitely many open subsets of* $]S,T] \times \mathbb{R}^N$ *each of which admits a Cauchy barrier with respect to L,*

(iii) $F \in C^0(\overline{\widetilde{\Omega}_s})$, $F = 0$ *on* $\partial \widetilde{\Omega}_s \setminus (\{s\} \times \widetilde{\Omega}(s))$ *if* $S < s < T$,
$F \in C_0^0(\widetilde{\Omega}_S)$ *if* $s = S$.

Then there exists a unique function $u \in C^0(\overline{\widetilde{\Omega}_s}) \cap W^{1;2}_{p,\mathrm{loc}}(\widetilde{\Omega}_s)$ solving

$$\begin{cases} Lu = F & \text{in } \widetilde{\Omega}_s \\ u(s,\cdot) = f & \text{in } \widetilde{\Omega}(s) \subset \mathbb{R}^N \\ u = 0 & \text{on } \partial\widetilde{\Omega}_s \setminus (\{s\} \times \widetilde{\Omega}(s)) \end{cases}$$

If $F = 0$ in addition, then $\|u\|_{\sup} \le \|f\|_{\sup}$.
If f and $-F$ are nonnegative in addition, then u is also nonnegative.

Furthermore, there exists a bounded variable space propagator $(U_{\widetilde{\Omega}}(t,s))_{S \le s \le t \le T}$ depending only on $\widetilde{\Omega}$ and L such that assuming an extension $F_0 \in C_0^0(\widetilde{\Omega}_S)$ of F to $\widetilde{\Omega}_S$ provides the representation

$$u(t,\cdot) \;=\; U_{\widetilde{\Omega}}(t,s)\, f \;-\; \int_s^t U_{\widetilde{\Omega}}(t,\tau)\, F_0(\tau,\cdot)\, d\tau \quad \text{in } \widetilde{\Omega}(t) \subset \mathbb{R}^N.$$

More generally, considering trivial extensions to $\mathbb{R}^N$ by 0 respectively (and indicating it via $\sharp$), there is a bounded variable space propagator $(U^\sharp_{\widetilde{\Omega}}(t,s))_{S \le s \le t \le T}$ depending just on L and $\widetilde{\Omega}$ such that the solution $u \in C^0(\overline{\widetilde{\Omega}_s}) \cap W^{1;2}_{p,\mathrm{loc}}(\widetilde{\Omega}_s)$ is the restriction of the continuous function

$$v : [s,T] \times \mathbb{R}^N \longrightarrow \mathbb{R}$$

with

$$v(t,\cdot) \;=\; U^\sharp_{\widetilde{\Omega}}(t,s)\, f^\sharp \;-\; \int_s^t U^\sharp_{\widetilde{\Omega}}(t,\tau)\, F^\sharp(\tau,\cdot)\, d\tau \qquad \text{in } \mathbb{R}^N.$$

□

Remark 140. According to [133, Proposition 4.18], this bounded variable space propagator $(U^{\sharp}_{\widetilde{\Omega}}(t,s))_{S \le s \le t \le T}$ is related to a contractive C^0 semigroup $(\mathscr{S}(\tau))_{\tau \ge 0}$ on the Banach space $\big(C^0_0(\widetilde{\Omega}_S),\ \|\cdot\|_{\sup}\big)$ in the sense of

$$\mathscr{S}(\tau)\,F:\ \widetilde{\Omega}_S \longrightarrow \mathbb{R}, \quad (t,x) \longmapsto U^{\sharp}_{\widetilde{\Omega}}(t,t-\tau)\ F^{\sharp}(t-\tau,\cdot)$$

for every function $F \in C^0_0(\widetilde{\Omega}_S)$ and its trivial extension $F^{\sharp}: \mathbb{R} \times \mathbb{R}^N \longrightarrow \mathbb{R}$ (by 0). This close relation provides the link with the results in § 3.10.

For applying this existence theorem, a key question is how to guarantee Cauchy barriers as required in hypothesis (ii). Lumer and Schnaubelt prove the following sufficient geometric condition:

Proposition 141 ([133, Proposition 6.4]). *In addition to the assumptions about coefficients in § 3.11.1, let $\widetilde{\Omega}$ be a bounded open subset of $[S,T] \times \mathbb{R}^N$ satisfying*

(*i*) $\widetilde{\Omega} \cap \big(\{t\} \times \mathbb{R}^N\big) \neq \emptyset$ *for every* $t \in [S,T]$,

(*ii*) *the boundary $\partial\widetilde{\Omega}$ is given by $x_i = \phi_k(t, x_1 \ldots x_{i-1}, x_{i+1} \ldots x_n)$ for some $i \in \{1 \ldots n\}$ and finitely many functions ϕ_k that are defined on open subsets of $[S,T] \times \mathbb{R}^{N-1}$, continuously differentiable with respect to t and twice continuously differentiable with respect to x,*

(*iii*) *$\widetilde{\Omega}$ is locally on one side of its boundary.*

Then, $\widetilde{\Omega}_S \stackrel{\text{Def.}}{=} \widetilde{\Omega} \cap \big(]S,T] \times \mathbb{R}^N\big)$ possesses a Cauchy barrier with respect to L. □

Their characterization of well-posed Cauchy problems by means of so-called *excessive barriers* is the basis for concluding from [133, Corollary 3.26] directly:

Lemma 142 ([133]). *If the nonempty open set $\widetilde{\Omega}$ is the intersection of finitely many open sets each of which admits a Cauchy barrier with respect to L, then $\widetilde{\Omega}$ possesses a Cauchy barrier with respect to L.* □

In their joint publications [132, 133], however, Lumer and Schnaubelt do not specify any method for extending such results to *countably many* intersections or to *merely local* geometric criteria similar to the exterior cone condition, for example, which has proved to be very useful for strong solutions to elliptic partial differential equations of second order (see e.g. [87, Theorem 9.30]).
Roughly speaking, the essential challenge is to construct a global function satisfying both the zero boundary condition and the differential inequality. For this reason, we replace the assumption of Cauchy barriers by a weaker condition which serves exactly the same purposes in the proofs of Lumer and Schnaubelt. The basic idea is to guarantee the auxiliary "barrier" function not *globally* (as in Definition 138), but depending on the special approximative features needed for the respective conclusions *close* to the boundary.

Definition 143. A nonempty open set $\widetilde{\Omega} \subset]S,T] \times \mathbb{R}^N$ is said to possess *a family of approximative Cauchy barriers* with respect to L if there exists a compact set $\widetilde{K} \subset \widetilde{\Omega}$ with the following property: For every compact set $\widetilde{K}'$ with $\widetilde{K} \subset \widetilde{K}' \subset \widetilde{\Omega}$ and any scalar $0 < \varepsilon_1 \le \varepsilon_2$, there exists a function $h \in D(L, \widetilde{\Omega} \setminus \widetilde{K})$ satisfying

(i) $h > 0$ and $(L - \lambda)\, h \le 0$ in $\widetilde{\Omega} \setminus \widetilde{K}$ for some $\lambda \ge 0$,

(ii) $h \ge \varepsilon_2$ in $\widetilde{K}'$,

(iii) there exists a compact set $\widetilde{K}''$ with $\widetilde{K}' \subset \widetilde{K}'' \subset \widetilde{\Omega}$ and $h \le \varepsilon_1$ in $\widetilde{\Omega} \setminus \widetilde{K}''$.

Studying the general proof of [133, Theorem 3.25] reveals that assuming a family of approximative Cauchy barriers (instead of a single Cauchy barrier) also implies the well-posedness of the linear homogeneous Cauchy problems considered in [133, § 3]. Finally we conclude from the same arguments as for preceding Theorem 139 quoting a special case of [133, Theorem 6.1]:

Corollary 144. *Theorem 139 holds if its assumption (ii) is replaced by*

(ii') $\widetilde{\Omega}_S$ *possesses a family of approximative Cauchy barriers with respect to L.*

□

Remark 145 (about the proof of Corollary 144). Strictly speaking, we have to verify that a family of approximative Cauchy barriers enables us to draw essentially the same conclusions as Lumer and Schnaubelt did in regard to well-posedness and its consequences. Most of their steps are based on local approximation and comparison and thus, it is to check whether their "global" Cauchy barrier can be adapted to the required "accuracy" locally.

In particular, [133, Theorem 3.25] applied to our parabolic problem in a nonempty bounded open set $\widetilde{\mathscr{O}} \subset \widetilde{\Omega}_S$ states that the Cauchy problem induced by L is well-posed in $C_0^0(\widetilde{\mathscr{O}})$ if and only if $\widetilde{\mathscr{O}}$ has a Cauchy barrier with respect to L. We focus on the sufficient aspect of Cauchy barriers (providing existence of solutions). Although all sets under consideration here are bounded, we avoid applying [133, Lemma 3.24] immediately, but instead we start with selecting an expanding sequence $\widetilde{\mathscr{W}_n} \uparrow \widetilde{\mathscr{O}}$ of open sets and functions $\widetilde{h}_n$ $(n \in \mathbb{N})$ in the family of approximative Cauchy barriers in an alternating way such that $\widetilde{h}_n > n$ in $\overline{\widetilde{\mathscr{W}_n}}$ and $0 \le \widetilde{h}_n < \frac{1}{n}$ in $\widetilde{\Omega} \setminus \widetilde{\mathscr{W}}_{n+1}$.

In a word, $\widetilde{h}_n$ is to take the role of the "global" Cauchy barrier h whenever we consider restrictions to $\mathscr{W}_{n+2} \subset \widetilde{\mathscr{O}}$. Then we can follow essentially the conclusions of Lumer and Schnaubelt for constructing so-called *locally excessive barriers* as in [133, Lemma 3.24]. For initial functions with compact support in $\widetilde{\mathscr{O}}$, the approximative solutions in [133, Corollary 3.9] form a Cauchy sequence due to the parabolic maximum principle in [133, Theorem 2.29] and, its limit solves the parabolic Cauchy problem of interest in [133, Theorem 3.25].

This existence of solutions due to approximative Cauchy barriers provides the tools for verifying further statements in [133, Proposition 3.17 and Theorems 4.11 – 4.14].

3.11.3 Semilinear Parabolic Differential Equations in a Fixed Noncylindrical Domain

In this subsection, we consider $S < 0 < \widehat{T} < T$ and assume $\widetilde{\Omega} \subset [S,T] \times \mathbb{R}^N$ to be a *fixed* open subset of $[S,T] \times \mathbb{R}^N$ satisfying the assumptions (i), (ii') of Theorem 139 and Corollary 144, i.e.,

(i) $\widetilde{\Omega} \cap (\{t\} \times \mathbb{R}^N) \neq \emptyset$ for every $t \in [S,T]$,
(ii') $\widetilde{\Omega}_S$ possesses a family of approximative Cauchy barriers with respect to L.

The results of Lumer and Schnaubelt focus on existence and uniqueness of solutions $u \in C^0(\overline{\widetilde{\Omega}_s}) \cap W^{1;2}_{p,\mathrm{loc}}(\widetilde{\Omega}_s)$ to the inhomogeneous linear parabolic problem

$$\begin{cases} Lu = F & \text{in } \widetilde{\Omega}_s \\ u(s,\cdot) = f & \text{in } \widetilde{\Omega}(s) \subset \mathbb{R}^N \\ u = 0 & \text{on } \partial\widetilde{\Omega}_s \setminus (\{s\} \times \widetilde{\Omega}(s)) \end{cases}$$

for given $s \in [0,T[$, $f \in C^0_0(\widetilde{\Omega}(s))$, $F \in C^0(\overline{\widetilde{\Omega}_s})$ with $F = 0$ on $\partial\widetilde{\Omega}_s \setminus (\{s\} \times \widetilde{\Omega}(s))$.

Our goal is to obtain similar existence results for the *semilinear* parabolic differential equations in the smaller time interval $[0,\widehat{T}]$, i.e., the function F on the right-hand side is prescribed as a function of time t and the current solution $u(t,\cdot) : \widetilde{\Omega}(t) \longrightarrow \mathbb{R}$. The results are essentially direct conclusions of § 3.10 about evolution equations. Nevertheless we discuss the steps of proof in detail afterwards.

Theorem 146 (Existence of solutions to semilinear parabolic problem in $\widetilde{\Omega}$). *In addition to the hypotheses of § 3.11.1 (on page 311 f.) and $S < 0 < \widehat{T} < T$, assume $\widetilde{\Omega} \subset [S,T] \times \mathbb{R}^N$ to be a nonempty bounded open subset of $[S,T] \times \mathbb{R}^N$ satisfying*

(i) $\widetilde{\Omega} \cap (\{t\} \times \mathbb{R}^N) \neq \emptyset$ *for every* $t \in [S,T]$,
(ii') $\widetilde{\Omega}_S$ *possesses a family of approximative Cauchy barriers with respect to L.*

Furthermore, let $\mathscr{F} : \bigcup_{t \in [0,\widehat{T}]} \big(\{t\} \times C^0_0(\widetilde{\Omega}(t))\big) \longrightarrow C^0_c(\mathbb{R}^N)$ *fulfill*

(iii) for all $t \in [0,\widehat{T}]$ *and* $v \in C^0_0(\widetilde{\Omega}(t))$: $\operatorname{supp} \mathscr{F}(t,v) \subset \overline{\widetilde{\Omega}(t)} \subset \mathbb{R}^N$,

(iv) the image $\{\mathscr{F}(t,v) \,|\, t \in [0,\widehat{T}],\ v \in C^0_0(\widetilde{\Omega}(t))\} \subset C^0_c(\mathbb{R}^N)$ *is bounded, equicontinuous and, there exist constants* $\alpha \in]0,1]$, $C_{\mathscr{F}} \in [0,\infty[$ *such that for all* (t,v) *of the domain,*

$$|\mathscr{F}(t,v)| \;\le\; C_{\mathscr{F}} \cdot \operatorname{dist}\big((t,\cdot),\, \mathbb{R}^{1+N} \setminus \widetilde{\Omega}_S\big)^{\alpha},$$

(v) $\mathscr{F}$ *is continuous in the following sense:* $\big\|\mathscr{F}(t,v)^\sharp - \mathscr{F}(t_n,v_n)^\sharp\big\|_{\sup} \longrightarrow 0$ *for any* $t \in [0,\widehat{T}]$, $v \in C^0_0(\widetilde{\Omega}(t))$ *and sequences* $(t_n)_{n\in\mathbb{N}}$, $(v_n)_{n\in\mathbb{N}}$ *satisfying* $v_n \in C^0_0(\widetilde{\Omega}(t_n))$ *for all* $n \in \mathbb{N}$ *and* $t_n \longrightarrow t$, $\|v_n^\sharp - v^\sharp\|_{\sup} \longrightarrow 0$ *for* $n \longrightarrow \infty$.

Then, for every initial function $u_0 \in C_0^0(\widetilde{\Omega}(0))$*, there exists a strong solution* $u \in C^0\big(\overline{\widetilde{\Omega}_0}\big) \cap W^{1;2}_{p,\mathrm{loc}}(\widetilde{\Omega}_0)$ *to the initial-boundary value problem of parabolic type*

$$\begin{cases} Lu(t,\cdot) = \mathscr{F}(t,u)(\cdot) & \textit{in } \; \widetilde{\Omega}(t) \; \textit{ for a.e. } t \in \,]0,\widehat{T}[, \\ u(0,\cdot) = u_0 & \textit{in } \; \widetilde{\Omega}(0) \subset \mathbb{R}^N, \\ u = 0 & \textit{on } \; \partial\widetilde{\Omega}_0 \setminus (\{0\} \times \widetilde{\Omega}(0)). \end{cases}$$

Specifying the Set $\widetilde{E}_{\widetilde{\Omega}}$ and Its Distances via the Related Semigroup $(\mathscr{S}(\tau))_{\tau \geq 0}$

The vector space $C_0^0(\widetilde{\Omega}(t))$ $(t \in [0,T])$ supplied with the supremum norm is a very obvious choice indeed.
Due to the obstacles of strong continuity and time-dependent domains $\widetilde{\Omega}(t)$, however, we would prefer a fixed Banach space supplied with a separate real time component and use the results of § 3.10 (on page 291 ff.). This motivates the choice of $C_0^0(\widetilde{\Omega}_S)$ and the supremum norm, but it might lead to difficulties in regard to defining transitions for all periods $h \in [0,1]$ because $t+h$ might be larger than T.

Hence, we return to Remark 140 (on page 315) and use the contractive C^0 semigroup $(\mathscr{S}(\tau))_{\tau \geq 0}$ on the Banach space $\big(C_0^0(\widetilde{\Omega}_S),\ \|\cdot\|_{\sup}\big)$ specified by

$$\mathscr{S}(\tau)\,v : \; \widetilde{\Omega}_S \longrightarrow \mathbb{R}, \quad (t,x) \longmapsto \begin{cases} U^{\sharp}_{\widetilde{\Omega}}(t,t-\tau)\; v^{\sharp}(t-\tau,\cdot) & \text{if } \; t-\tau \geq S \\ 0 & \text{if } \; t-\tau < S \end{cases}$$

for every function $v \in C_0^0(\widetilde{\Omega}_S)$ and its trivial extension $v^{\sharp} : \mathbb{R} \times \mathbb{R}^N \longrightarrow \mathbb{R}$ (by 0).

In other words, after defining

$$\widetilde{\Omega}(s') := \widetilde{\Omega}(S) \subset \mathbb{R}^N \qquad\qquad \text{for every } s' < S$$

additionally and extending the coefficients of L to $]-\infty,S] \times \widetilde{\Omega}(S) \subset \mathbb{R} \times \mathbb{R}^N$ constantly (with respect to time), the respective function $\big(\mathscr{S}(\tau)\,v\big)(t,\cdot) : \widetilde{\Omega}(t) \longrightarrow \mathbb{R}$ at time $t \leq T$ is induced by the unique solution $u \in C^0\big(\overline{\widetilde{\Omega}_s}\big) \cap W^{1;2}_{p,\mathrm{loc}}(\widetilde{\Omega}_s)$ to the *homogeneous* linear parabolic problem starting at time $s := t - \tau \in \,]-\infty, T]$

$$\begin{cases} Lu = 0 & \text{in } \; \widetilde{\Omega}_s \\ u(s,\cdot) = v^{\sharp}(s,\cdot) & \text{in } \; \widetilde{\Omega}(s) \subset \mathbb{R}^N \\ u = 0 & \text{on } \; \partial\widetilde{\Omega}_s \setminus (\{s\} \times \widetilde{\Omega}(s)) \end{cases}$$

In the case of $s \stackrel{\text{Def.}}{=} t - \tau \geq S$, existence and uniqueness of this solution result directly from Theorem 139 of Lumer and Schnaubelt and, otherwise (i.e. if $t - \tau < S$), the parabolic maximum principle excludes any alternative to the trivial solution.

Strictly speaking, we consider the set

$$\widetilde{E}_{\widetilde{\Omega}} := \mathbb{R} \times C_0^0(\widetilde{\Omega}_S)$$

supplied with the functions

$$\begin{array}{llcl}
|\cdot|_{\widetilde{\Omega}}: & \widetilde{E}_{\widetilde{\Omega}} & \longrightarrow & [0,\infty[, \\
& \widetilde{u} = (t,u) & \longmapsto & |t| + \|u\|_{\sup}, \\
\widetilde{d}_{\widetilde{\Omega}}: & \widetilde{E}_{\widetilde{\Omega}} \times \widetilde{E}_{\widetilde{\Omega}} & \longrightarrow & [0,\infty[, \\
& \big((s,u),\ (t,v)\big) & \longmapsto & |s-t| \ + \ \big\|u - v\big\|_{\sup}, \\
\widetilde{e}_{\widetilde{\Omega}}: & \widetilde{E}_{\widetilde{\Omega}} \times \widetilde{E}_{\widetilde{\Omega}} & \longrightarrow & [0,\infty[, \\
& \big((s,u),\ (t,v)\big) & \longmapsto & |s-t| \ + \ \big\|\mathscr{S}\big((t-s)^+\big)\, u \ - \ \mathscr{S}\big((s-t)^+\big)\, v\big\|_{\sup}.
\end{array}$$

using the general abbreviation $r^+ := \max\{r, 0\}$ for every $r \in \mathbb{R}$.
Obviously, $\widetilde{d}_{\widetilde{\Omega}}$ satisfies the triangle inequality. Furthermore, $\widetilde{e}_{\widetilde{\Omega}}$ fulfills the so-called *timed* triangle inequality, i.e. whenever $\widetilde{u}, \widetilde{v}, \widetilde{w} \in \widetilde{E}_{\widetilde{\Omega}}$ satisfy $\pi_1\, \widetilde{u} \le \pi_1\, \widetilde{v} \le \pi_1\, \widetilde{w}$, then

$$\widetilde{e}_{\widetilde{\Omega}}(\widetilde{u},\widetilde{w}) \ \le \ \widetilde{e}_{\widetilde{\Omega}}(\widetilde{u},\widetilde{v}) \ + \ \widetilde{e}_{\widetilde{\Omega}}(\widetilde{v},\widetilde{w}).$$

The analytical "detour" via the contractive C^0 semigroup $(\mathscr{S}(\tau))_{\tau\ge 0}$ on the fixed Banach space $\big(C_0^0(\widetilde{\Omega}_S),\ \|\cdot\|_{\sup}\big)$ has the essential advantage that we can apply the results of § 3.10 (on page 291 ff.). In particular, the arguments for Corollary 118 ensure that the tuple $\big(\widetilde{E}_{\widetilde{\Omega}}, \widetilde{d}_{\widetilde{\Omega}}, \widetilde{e}_{\widetilde{\Omega}}, |\cdot|_{\widetilde{\Omega}}\big)$ fulfills hypotheses (H1), (H2), (H3), (H4) required for the mutational framework in § 3.5 (on page 221 ff.).

Specifying Transitions on $\widetilde{E}_{\widetilde{\Omega}}$

Due to a glance at mild solutions to semilinear evolution equations (in § 3.10), the variation of constants formula serves as starting point for specifying transitions on $\widetilde{E}_{\widetilde{\Omega}}$. The results of § 3.10.2 (on page 298 ff.) lead to:

Definition 147. For any function $F \in C_0^0(\widetilde{\Omega}_S)$, define

$$\widetilde{\vartheta}_{\widetilde{\Omega},F}:\ [0,1] \times \widetilde{E}_{\widetilde{\Omega}} \longrightarrow \widetilde{E}_{\widetilde{\Omega}}, \quad \big(h,\ (t,u)\big) \longmapsto \big(t+h,\ \vartheta_{\widetilde{\Omega},F}(h,\ (t,u))\big)$$

with the function $\vartheta_{\widetilde{\Omega},F}(h,\ (t,u)) : \widetilde{\Omega}_S \longrightarrow \mathbb{R}$,

$$\vartheta_{\widetilde{\Omega},F}(h,\ (t,u)) \ := \ \mathscr{S}(h)\, u \ + \ \int_0^h \mathscr{S}(h-s)\, F\, ds.$$

Lemma 148. *For every $F \in C_0^0(\widetilde{\Omega}_S)$, the function $\widetilde{\vartheta}_{\widetilde{\Omega},F} : [0,1] \times \widetilde{E}_{\widetilde{\Omega}} \longrightarrow \widetilde{E}_{\widetilde{\Omega}}$ is well-defined and, the continuous function $\vartheta_{\widetilde{\Omega},F}(h,(t,u)) : \widetilde{\Omega}_S \longrightarrow \mathbb{R}$ maps*

$$(s,x) \longmapsto \Big(U^\sharp_{\widetilde{\Omega}}(s,\ s-h)\ u^\sharp(s-h,\cdot) \ - \ \int_{s-h}^{s} U^\sharp_{\widetilde{\Omega}}(s,\tau)\ \ F^\sharp(\tau,\cdot)\ d\tau\Big)\,(x).$$

with the variable space propagator $(U^\sharp_{\widetilde{\Omega}}(t,s))_{S\le s\le t\le T}$ mentioned in Theorem 139.

It has the properties for all $\widetilde{u}, \widetilde{v} \in \widetilde{E}_{\widetilde{\Omega}}$, $G \in C_0^0(\widetilde{\Omega}_S)$, $h, h_1, h_2 \in [0,1]$ *with* $h_1 + h_2 \le 1$:

$$\begin{array}{lll}
(1.) & \widetilde{\vartheta}_{\widetilde{\Omega},F}(0,\cdot) = \mathbb{I}\mathrm{d}_{\widetilde{E}_{\widetilde{\Omega}}}, & \\
(2.) & \widetilde{\vartheta}_{\widetilde{\Omega},F}(h_1,\ \widetilde{\vartheta}_{\widetilde{\Omega},F}(h_2,\cdot)) \;=\; \widetilde{\vartheta}_{\widetilde{\Omega},F}(h_1+h_2,\cdot), & \\
(3.) & \widetilde{d}_{\widetilde{\Omega}}\big(\widetilde{\vartheta}_{\widetilde{\Omega},F}(h,\widetilde{u}),\ \widetilde{\vartheta}_{\widetilde{\Omega},G}(h,\widetilde{v})\big) & \le\ \widetilde{d}_{\widetilde{\Omega}}(\widetilde{u},\widetilde{v}) + \|F-G\|_{\sup}\, h, \\
(4.) & \widetilde{e}_{\widetilde{\Omega}}\big(\widetilde{u},\ \widetilde{\vartheta}_{\widetilde{\Omega},F}(h,\widetilde{u})\big) & \le\ (1+\|F\|_{\sup})\ h \\
(5.) & \big|\,\widetilde{\vartheta}_{\widetilde{\Omega},F}(h,\widetilde{u})\,\big|_{\widetilde{\Omega}} & \le\ |\widetilde{u}|_{\widetilde{\Omega}} + (1+\|F\|_{\sup})\ h.
\end{array}$$

□

Corollary 149. *For every* $F \in C_0^0(\widetilde{\Omega}_S)$, *the function* $\widetilde{\vartheta}_{\widetilde{\Omega},F} : [0,1] \times \widetilde{E}_{\widetilde{\Omega}} \longrightarrow \widetilde{E}_{\widetilde{\Omega}}$ *is a transition on* $\big(\widetilde{E}_{\widetilde{\Omega}}, \widetilde{d}_{\widetilde{\Omega}}, \widetilde{e}_{\widetilde{\Omega}}, |\cdot|_{\widetilde{\Omega}}\big)$ *in the sense of Definition 2 (on page 183) with*

$$\begin{aligned}
\alpha(\widetilde{\vartheta}_{\widetilde{\Omega},F}; r) &:= 0 \\
\beta(\widetilde{\vartheta}_{\widetilde{\Omega},F}; r) &:= 1 + \|F\|_{\sup} \\
\gamma(\widetilde{\vartheta}_{\widetilde{\Omega},F}) &:= 1 + \|F\|_{\sup} \\
\widehat{D}(\widetilde{\vartheta}_{\widetilde{\Omega},F}, \widetilde{\vartheta}_{\widetilde{\Omega},G}; r) &:= \|F-G\|_{\sup}
\end{aligned}$$

and the property $\pi_1\, \widetilde{\vartheta}_{\widetilde{\Omega},F}(h,\widetilde{u}) = \pi_1\, \widetilde{u} + h$ *for all* $\widetilde{u} \in \widetilde{E}_{\widetilde{\Omega}}$, $h \in [0,1]$. □

Remark 150. The timed triangle inequality of distance function $\widetilde{e}_{\widetilde{\Omega}}$ and semi-group property (2.) in Lemma 148 imply directly: The tuple $\big(\widetilde{E}_{\widetilde{\Omega}}, \widetilde{d}_{\widetilde{\Omega}}, \widetilde{e}_{\widetilde{\Omega}}, |\cdot|_{\widetilde{\Omega}}\big)$ together with the transitions in Definition 147 is Euler equi-continuous in the sense of Definition 16 (on page 194).

Existence of a Timed Solution to the Mutational Equation

Up to now, we are lacking suitable global a priori estimates (for $\widetilde{\Omega}$ and L) implying that the C^0 semigroup $(\mathscr{S}(\tau))_{\tau \ge 0}$ is immediately compact. This gap prevents us from applying the existence results of § 3.10.3 (on page 300 ff.) and thus, we prefer the conclusions of § 3.10.4 (on page 306 ff.).

The closure of $\widetilde{\Omega}_S$ is compact as $\widetilde{\Omega}$ is assumed to be bounded. As a consequence, Arzelà-Ascoli Theorem A.82 (on page 491) provides a characterization of relatively compact subsets in $\big(C_0^0(\widetilde{\Omega}_S), \|\cdot\|_{\sup}\big)$ in terms of boundedness and equi-continuity. Now we conclude from Existence Theorem 134 (on page 307) about timed solutions to the mutational equation and their corresponding mild solutions to the semilinear evolution equation:

Proposition 151 (Existence of timed solutions to the mutational equation). *In addition to the hypotheses of § 3.11.1 (on page 311 f.) and* $S < 0 < \widehat{T} < T$, *assume* $\widetilde{\Omega} \subset [S,T] \times \mathbb{R}^N$ *to be a nonempty bounded open subset of* $[S,T] \times \mathbb{R}^N$ *satisfying*

(*i*) $\widetilde{\Omega} \cap (\{t\} \times \mathbb{R}^N) \neq \emptyset$ *for every* $t \in [S,T]$,
(*ii'*) $\widetilde{\Omega}_S$ *possesses a family of approximative Cauchy barriers with respect to L.*

$(\mathscr{S}(\tau))_{\tau \geq 0}$ *denotes the contractive* C^0 *semigroup on* $C_0^0(\widetilde{\Omega}_S)$ *related to differential operator L as specified on page 318. Furthermore, let* $\widetilde{f} : \widetilde{E}_{\widetilde{\Omega}} \longrightarrow C_0^0(\widetilde{\Omega}_S)$ *fulfill*

(*iii*) *the image of* $\widetilde{f}$ *is bounded in* $\big(C_0^0(\widetilde{\Omega}_S), \|\cdot\|_{\sup}\big)$ *and equi-continuous,*
(*iv*) $\widetilde{f} : \big(\widetilde{E}_{\widetilde{\Omega}}, \widetilde{d}_{\widetilde{\Omega}}\big) \longrightarrow \big(C_0^0(\widetilde{\Omega}_S), \|\cdot\|_{\sup}\big)$ *is continuous.*

Then for every initial element $\widetilde{u}_0 = (t_0, u_0) \in \widetilde{E}_{\widetilde{\Omega}}$, *there exists a timed solution* $\widetilde{u} : [0,\widehat{T}] \longrightarrow \widetilde{E}_{\widetilde{\Omega}}$ *to the mutational equation* $\overset{\circ}{\widetilde{u}}(\cdot) \ni \widetilde{\vartheta}_{\widetilde{\Omega},\widetilde{f}(\widetilde{u}(\cdot))}$ *in* $\big(\widetilde{E}_{\widetilde{\Omega}}, \widetilde{d}_{\widetilde{\Omega}}, \widetilde{e}_{\widetilde{\Omega}}, |\cdot|_{\widetilde{\Omega}}, \widehat{D}\big)$ *with* $\widetilde{u}(0) = \widetilde{u}_0$. *Its second component is a mild solution to the corresponding semi-linear evolution equation in* $\big(C_0^0(\widetilde{\Omega}_S), \|\cdot\|_{\sup}\big)$. □

The Step from Mutational Equations to Parabolic Differential Equations

Strictly speaking, we are taking more information into consideration than we need for the semilinear initial-boundary value problem

$$\begin{cases} Lu(t,\cdot) = \mathscr{F}(t,u)(\cdot) & \text{in } \widetilde{\Omega}(t) \text{ for } \mathscr{L}^1\text{-a.e. } t \in]0,\widehat{T}[, \\ u(0,\cdot) = u_0 & \text{in } \widetilde{\Omega}(0) \subset \mathbb{R}^N \\ u = 0 & \text{on } \partial\widetilde{\Omega}_0 \setminus (\{0\} \times \widetilde{\Omega}(0)) \end{cases}$$

Indeed, the wanted functions $u(t,\cdot) \in C_0^0(\widetilde{\Omega}(t))$, $t \in [0,T]$, have been replaced by the states in $\widetilde{E}_{\widetilde{\Omega}} \overset{\text{Def.}}{=} \mathbb{R} \times C_0^0(\widetilde{\Omega}_S)$ providing information about the whole domain $\widetilde{\Omega}$ in space-time (and not just about the spatial set $\widetilde{\Omega}(t) \subset \mathbb{R}^N$ at time $t \in [0,T]$). Now the suitable "section" in the cylinder $[0,T] \times \widetilde{\Omega} \subset \mathbb{R}^{2+N}$ is to lay the basis for the step "back" to the original parabolic problems in the noncylindrical domain $\widetilde{\Omega} \cap ([0,T] \times \mathbb{R}^N)$.

For identifying such an appropriate section, we focus on the approximative construction leading to the timed solution in preceding Proposition 151. Indeed, the proof of Theorem 134 starts with equidistant Euler approximations.

Similarly to Lemma 126 preparing mild solutions to semilinear evolution equations (on page 300), the variation of constants formula provides an integral characterization of all Euler approximations. The proof uses exactly the same (piecewise) conclusions as for Lemma 126 (on page 303 f.) and thus, it is skipped here.

Lemma 152 (Characterization of nonequidistant Euler approximations). *Assume for* $\widetilde{u}_0 = (t_0, u_0) \in \widetilde{E}_{\widetilde{\Omega}}$, $M \geq 0$ *and a continuous curve* $\widetilde{u} : [0, \widehat{T}] \longrightarrow \widetilde{E}_{\widetilde{\Omega}}$

(1.) $\widetilde{u}(0) = \widetilde{u}_0$,

(2.) *for any* $t \in [0, \widehat{T}]$, *there exist* $s \in]t-1, t]$ *and* $F \in C_0^0(\widetilde{\Omega}_S)$ *with* $\|F\|_{\sup} \leq M$, $\widetilde{u}(s + \cdot) = \widetilde{\vartheta}_{\widetilde{\Omega},F}(\cdot, \widetilde{u}(s))$ *in an open neighborhood* $I \subset [0,1]$ *of* $[0, t-s]$.

Then there exists a piecewise constant function $G(\cdot) \in L^\infty([0, \widehat{T}], C_0^0(\widetilde{\Omega}_S))$ *with (at most) countably many points of discontinuity in* $[0, \widehat{T}]$, $\|G\|_{L^\infty} \leq M$ *and*

$$\widetilde{u}(t) = \big(t_0 + t,\ u(t)\big) \in \widetilde{E}_{\widetilde{\Omega}}$$

$$u(t)\,(s,x) = \Big(U^\sharp_{\widetilde{\Omega}}(s, s-t)\ u_0^\sharp(s-t, \cdot) - \int_{s-t}^{s} U^\sharp_{\widetilde{\Omega}}(s, \tau)\ \ G(\tau - (s-t))^\sharp(\tau, \cdot)\ d\tau\Big)\,(x)$$

for every $t \in [0, \widehat{T}]$ *and* $(s,x) \in \widetilde{\Omega}$.

If, in addition, assumption (2.) holds with a finite *partition of* $[0, \widehat{T}]$, *then* $G(\cdot)$ *is piecewise constant with respect to the same finite partition of* $[0, \widehat{T}]$, *i.e.,* $G(\cdot)$ *has at most finitely many points of discontinuity in* $[0, \widehat{T}]$.

□

Lumer and Schnaubelt's characterization of unique solutions to the linear problem (in Theorem 139 on page 314) can be applied to finitely many time intervals successively. Thus, it provides a link between Euler approximations with finite partition of $[0, \widehat{T}]$ on the one hand and parabolic initial-boundary value problems on the other hand (by focusing on $s - t =$ const, in short).

Corollary 153 (Euler approximations solve parabolic initial value problems). *For any initial state* $\widetilde{u}_0 \in \widetilde{E}_{\widetilde{\Omega}}$ *and bounds* $\widehat{\alpha}, \widehat{\beta}, \widehat{\gamma} > 0$ *let* $\mathscr{N} = \mathscr{N}\big(\widetilde{u}_0, \widehat{T}, (\widehat{\alpha}, \widehat{\beta}, \widehat{\gamma})\big)$ *denote the (possibly empty) subset of all curves* $\widetilde{u}(\cdot) : [0, \widehat{T}] \longrightarrow \widetilde{E}_{\widetilde{\Omega}}$ *constructed via transitions in the piecewise way as specified in Remark 15 (2.) (on page 193).*

Then for each curve $\widetilde{u}(\cdot) \in \mathscr{N}\big(\widetilde{u}_0, \widehat{T}, (\widehat{\alpha}, \widehat{\beta}, \widehat{\gamma})\big)$ *and time parameter* $t_0 \in]-\infty, \widehat{T}[$, *the function*

$$\widetilde{\Omega} \cap \big([t_0,\ t_0 + \widehat{T}] \times \mathbb{R}^N\big) \longrightarrow \mathbb{R}, \qquad (t,x) \longmapsto \widetilde{u}(t - t_0)(t,x)$$

is a strong solution $u(\cdot,\cdot)$ *to the linear parabolic initial-boundary value problem*

$$\begin{cases} Lu(t,x) = G(t - t_0)^\sharp(t,x) & \text{for almost every } (t,x) \in \widetilde{\Omega} \cap \big(]t_0,\ t_0 + \widehat{T}] \times \mathbb{R}^N\big) \\ u(t_0, \cdot) = u_0^\sharp(t_0, \cdot) & \text{in } \ \widetilde{\Omega}(t_0) \subset \mathbb{R}^N \\ u = 0 & \text{on } \ \partial\widetilde{\Omega}_{t_0} \setminus \big(\{t_0\} \times \widetilde{\Omega}(t_0)\big) \end{cases}$$

with a piecewise constant function $G : [0, \widehat{T}] \longrightarrow C_0^0(\widetilde{\Omega}_S)$, $\|G\|_{L^\infty([0,\widehat{T}], L^\infty)} \leq \widehat{\gamma}$.

□

Finally, we have to check whether such a relationship also holds for the limit as the step size of Euler approximations is tending to 0. The main analytical tool is the following local a priori estimate. In fact, the initial assumption $p > N+2$ comes into play here (again).

Proposition 154 (Interior a priori estimate [112, § IV.10], [114, Th. VII.7.22]**).** *In addition to the general assumptions of § 3.11.1 (on page 311 f.), let $\widetilde{\Omega}' \subset \mathbb{R}^N$ be any bounded subdomain of $\widetilde{\Omega}$ with $\overline{\widetilde{\Omega}'} \subset \widetilde{\Omega}$.*
Then there exists a constant $C_{\widetilde{\Omega}'}$ such that every function $v \in W^{1;2}_{p,\mathrm{loc}}(\widetilde{\Omega}_S) \cap L^p(\widetilde{\Omega})$ satisfies

$$\|\partial_t v\|_{L^p(\widetilde{\Omega}')} + \|\partial_x v\|_{L^p(\widetilde{\Omega}')} + \|\partial_x^2 v\|_{L^p(\widetilde{\Omega}')} \;\le\; C_{\widetilde{\Omega}'} \cdot \left(\|v\|_{L^p(\widetilde{\Omega})} + \|Lv\|_{L^p(\widetilde{\Omega})}\right).$$

Proposition 155. *Suppose the assumptions of Proposition 151 (on page 321) for $\widetilde{\Omega}$, L and $\widetilde{f} : \widetilde{E}_{\widetilde{\Omega}} \longrightarrow C^0_0(\widetilde{\Omega}_S)$.*

Then for every initial element $\widetilde{u}_0 = (0, u_0) \in \widetilde{E}_{\widetilde{\Omega}}$, there exist a continuous curve $\widetilde{u} = (\cdot, u) : [0, \widehat{T}] \longrightarrow (\widetilde{E}_{\widetilde{\Omega}}, \widetilde{d}_{\widetilde{\Omega}})$ and a strong solution $\check{u} \in C^0(\overline{\widetilde{\Omega}_0}) \cap W^{1;2}_{p,\mathrm{loc}}(\widetilde{\Omega}_0)$ to the initial-boundary value problem of parabolic type

$$\begin{cases} L\check{u}(t,\cdot) = \widetilde{f}(\widetilde{u}(t))(t,\cdot) & \text{in } \widetilde{\Omega}(t) \text{ for } \mathscr{L}^1\text{-a.e. } t \in \,]0,\widehat{T}[, \\ \check{u}(0,\cdot) = u_0(0,\cdot) & \text{in } \widetilde{\Omega}(0) \subset \mathbb{R}^N \\ \check{u} = 0 & \text{on } \partial\widetilde{\Omega}_0 \setminus (\{0\} \times \widetilde{\Omega}(0)) \end{cases}$$

with $$\check{u}(t,x) = u(t)(t,x) \quad \text{for all } t \in [0,\widehat{T}],\ x \in \mathbb{R}^N \text{ with } (t,x) \in \widetilde{\Omega}.$$

Proof (of Proposition 155). Let $\widetilde{u}_n(\cdot) = (\cdot, u_n(\cdot)) : [0,\widehat{T}] \longrightarrow \widetilde{E}_{\widetilde{\Omega}}$, $n \in \mathbb{N}$, denote the sequence of equidistant Euler approximations starting in $\widetilde{u}_0 = (0,u_0) \in \widetilde{E}_{\widetilde{\Omega}}$ and related with step size $h_n := \frac{\widehat{T}}{2^n}$ (as e.g. in the proof of Existence Theorem 19).
Then for each index n, Corollary 153 always provides a piecewise constant function $G_n \in L^\infty\left([0,\widehat{T}], C^0_0(\widetilde{\Omega}_S)\right)$ with $G_n(t) \in \left\{\widetilde{f}(\widetilde{u}_n(s)) \,\middle|\, s \in [0,t] \cap \mathbb{B}_{h_n}(t)\right\}$ for each t.

There is a sequence $n_k \nearrow \infty$ of indices such that there exist $w(\cdot) \in C^0\left([0,\widehat{T}], C^0_0(\widetilde{\Omega}_S)\right)$ and $G(\cdot) \in L^1\left([0,\widehat{T}], C^0_0(\widetilde{\Omega}_S)\right)$ with

(*i*) $u_{n_k}(\cdot) \longrightarrow u(\cdot)$ strongly in $\left(C^0\left([0,\widehat{T}], C^0_0(\widetilde{\Omega}_S)\right), \|\cdot\|_{\sup}\right)$,
(*ii*) $G_{n_k}(\cdot) \longrightarrow G(\cdot)$ weakly in $L^1\left([0,\widehat{T}], C^0_0(\widetilde{\Omega}_S)\right)$ for $k \longrightarrow \infty$.

Indeed, assumption (iii) and Arzelà-Ascoli Theorem A.82 (on page 491) ensure that the set of trivial extensions $\left\{\widetilde{f}(\widetilde{v})^\sharp|_{\overline{\widetilde{\Omega}_S}} \,\middle|\, \widetilde{v} \in \widetilde{E}_{\widetilde{\Omega}}\right\}$ is compact in $\left(C^0(\overline{\widetilde{\Omega}_S}), \|\cdot\|_{\sup}\right)$. As a first consequence, the functions $u_n(\cdot)$, $n \in \mathbb{N}$, are equi-continuous. Moreover, they are pointwise relatively compact due to Lemma 133 (on page 306) and so, we conclude relative compactness from Arzelà-Ascoli Theorem A.82 (on page 491).

Second, Ülger's Proposition A.85 guarantees that $\{G_n \,|\, n \in \mathbb{N}\}$ is relatively weakly compact in $L^1\big([0,\widehat{T}], C_0^0(\widetilde{\Omega}_S)\big)$. At every time $t \in [0,\widehat{T}]$, we obtain

$$u(t) \;=\; \mathscr{S}(t)\, u_0 \;+\; \int_0^t \mathscr{S}(t-s)\, G(s)\; ds \;\in\; C_0^0(\widetilde{\Omega}_S).$$

In particular, $[0,\widehat{T}] \longmapsto \widetilde{E}_{\widetilde{\Omega}},\; t \longmapsto (t, u(t))$ is exactly the timed solution to the corresponding mutational equation mentioned in Proposition 151.

Further results about convergence, however, can be concluded from Mazur's Lemma about strong approximations of weak limits (e.g. [188, Theorem V.1.2]) and the interior a priori estimate in Proposition 154.

According to the well-known Lemma of Mazur, there exists a sequence $(H_k)_{k\in\mathbb{N}}$ in $L^1\big([0,\widehat{T}], C_0^0(\widetilde{\Omega}_S)\big)$ converging strongly to $G(\cdot)$ and satisfying

$$H_k(\cdot) \in \operatorname{co}\big\{G_{n_k}(\cdot),\, G_{n_{k+1}}(\cdot)\,\ldots\big\} \subset L^1\big([0,\widehat{T}], C_0^0(\widetilde{\Omega}_S)\big).$$

An appropriate subsequence (again denoted by) $(H_k)_{k\in\mathbb{N}}$ instead ensures in addition that for Lebesgue-almost every $t \in [0,\widehat{T}]$, $\big\|H_k(t) - G(t)\big\|_{\sup} \longrightarrow 0$ for $k \longrightarrow \infty$.

Due to $H_k(t) \in \overline{\operatorname{co}}\big\{\widetilde{f}(\widetilde{u}_{n_k}(s)),\, \widetilde{f}(\widetilde{u}_{n_{k+1}}(s))\,\ldots \,\big|\, (t-h_{n_k})^+ \le s \le t\big\}$, assumption (iv) about the continuity of $\widetilde{f}$ guarantees $G(t) = \widetilde{f}\big(\widetilde{u}(t)\big)$ for $\mathscr{L}^1$-a.e. $t \in [0,\widehat{T}]$.

As a further consequence, each function $H_k(\cdot)$, $k \in \mathbb{N}$, is also piecewise constant,

$$v_k : [0,\widehat{T}] \longrightarrow C_0^0(\widetilde{\Omega}_S), \qquad t \longmapsto \mathscr{S}(t)\, u_0 \;+\; \int_0^t \mathscr{S}(t-s)\, H_{n_k}(s)\; ds$$

belongs to the convex hull of Euler approximations $u_{n_k}(\cdot),\, u_{n_{k+1}}(\cdot)\,\ldots$ for each $k\in\mathbb{N}$ and thus, at every time $t \in [0,\widehat{T}]$, $\big\|v_k(t) - u(t)\big\|_{\sup} \longrightarrow 0$ for $k \longrightarrow \infty$.

For the same reasons as in Corollary 153, the function

$$\check{v}_k : \overline{\widetilde{\Omega}} \cap ([0,\widehat{T}] \times \mathbb{R}^N) \longrightarrow \mathbb{R}, \quad (t,x) \longmapsto v_k(t)(t,x) \qquad (k \in \mathbb{N}),$$

is a strong solution to the linear parabolic initial-boundary value problem

$$\begin{cases} L\,\check{v}_k\,(t,x) \;=\; H_k(t)\,(t,x) & \text{for almost every } (t,x) \in \widetilde{\Omega} \cap ([0,\widehat{T}] \times \mathbb{R}^N) \\ \check{v}_k(0,\cdot) \;=\; u_0(0,\cdot) & \text{in } \;\widetilde{\Omega}(0) \subset \mathbb{R}^N \\ \check{v}_k \;=\; 0 & \text{on } \;\partial\widetilde{\Omega}_0 \setminus \big(\{0\} \times \widetilde{\Omega}(0)\big). \end{cases}$$

For $k \longrightarrow \infty$, the sequence $\big(\check{v}_k(\cdot,\cdot)\big)_{k\in\mathbb{N}}$ converges pointwise to

$$\check{u} : \overline{\widetilde{\Omega}} \cap ([0,\widehat{T}] \times \mathbb{R}^N) \longrightarrow \mathbb{R}, \quad (t,x) \longmapsto u(t)(t,x).$$

Finally the interior a priori estimate in Proposition 154 and Lebesgue's Theorem of Dominated Convergence guarantee for any bounded subdomain $\widetilde{\Omega}'$ of $\widetilde{\Omega}$ with $\overline{\widetilde{\Omega}'} \subset \widetilde{\Omega} \cap ([0,\widehat{T}] \times \mathbb{R}^N)$ that the following Cauchy property holds

$$\sup_{k,l \ge K} \Big(\big\|\partial_t\,(\check{v}_k - \check{v}_l)\big\|_{L^p(\widetilde{\Omega}')} + \big\|\partial_x\,(\check{v}_k - \check{v}_l)\big\|_{L^p(\widetilde{\Omega}')} + \big\|\partial_x^2\,(\check{v}_k - \check{v}_l)\big\|_{L^p(\widetilde{\Omega}')} \Big) \overset{K\to\infty}{\longrightarrow} 0.$$

Thus, $\check{u} \in C^0\big(\overline{\widetilde{\Omega}_0}\big) \cap W^{1;2}_{p,\mathrm{loc}}(\widetilde{\Omega}_0)$ and for almost every $(t,x) \in \widetilde{\Omega} \cap ([0,\widehat{T}] \times \mathbb{R}^N)$,

$$L\,\check{u}\,(t,x) \;=\; G(t)\,(t,x) \;=\; \widetilde{f}(t, u(t))\,(t,x).$$

□

Extending the Functions Prescribed by $\mathscr{F}$ from a Subset of $C_0^0(\widetilde{\Omega}(t))$ to $C_0^0(\widetilde{\Omega}_S)$

The last essential gap between Existence Theorem 146 (on page 317) and Proposition 155 is due to the type of prescribed data.

Existence Theorem 146 focuses on strong solutions $u \in C^0\big(\overline{\widetilde{\Omega}_0}\big) \cap W^{1;2}_{p,\mathrm{loc}}(\widetilde{\Omega}_0)$ to the semilinear initial-boundary value problem of parabolic type

$$\begin{cases} Lu(t,\cdot) = \mathscr{F}(t,u)(\cdot) & \text{in } \widetilde{\Omega}(t) \ \text{ for a.e. } t \in]0,\widehat{T}[, \\ u(0,\cdot) = u_0 & \text{in } \widetilde{\Omega}(0) \subset \mathbb{R}^N, \\ u = 0 & \text{on } \partial\widetilde{\Omega}_0 \setminus (\{0\} \times \widetilde{\Omega}(0)). \end{cases}$$

Here for every $t \in [0,\widehat{T}]$ and $v \in C_0^0(\widetilde{\Omega}(t))$, we have to specify the function $\mathscr{F}(t,v) \in C_0^0(\widetilde{\Omega}(t))$ for the right-hand side of the partial differential equation. Strictly speaking, it is again a functional relationship because it does not have to be based on pointwise composition.

In contrast, Proposition 155 assumes a function $\widetilde{f} : \widetilde{E}_{\widetilde{\Omega}} \longrightarrow C_0^0(\widetilde{\Omega}_S)$ for the right-hand side of the corresponding mutational equation. The comparison of the values reveals that more information (namely on whole $\widetilde{\Omega} \subset \mathbb{R} \times \mathbb{R}^N$ instead of $\widetilde{\Omega}(t) \subset \mathbb{R}^N$) is required here.

The following lemma suggests a very easy way to bridge this gap by extending. The price to pay for its analytical simplicity, however, consists in stronger assumptions about the decay close to the topological boundary of $\widetilde{\Omega}_S$. Indeed, by assumption, there exist constants $\alpha \in]0,1]$ and $C_{\mathscr{F}} \in [0,\infty[$ such that

$$|\mathscr{F}(t,v)(\cdot)| \ \leq \ C_{\mathscr{F}} \cdot \operatorname{dist}\big((t,\cdot),\, \mathbb{R}^{1+N} \setminus \widetilde{\Omega}_S\big)^{\alpha}$$

holds for all $t \in [0,\widehat{T}]$ and $v \in C_0^0(\widetilde{\Omega}(t))$. This very restrictive condition can surely be weakened whenever an extension operator preserves boundedness and equi-continuity in an appropriate way. We complete the proof of Existence Theorem 146.

Lemma 156. *Let $d_{\complement\widetilde{\Omega}_S}(\cdot)$ denote the Euclidean distance from the complement of $\widetilde{\Omega}_S \stackrel{\text{Def.}}{=} \widetilde{\Omega} \cap (]S,T] \times \mathbb{R}^N)$, i.e.*

$$d_{\complement\widetilde{\Omega}_S}(\cdot) : \ \mathbb{R} \times \mathbb{R}^N \longrightarrow \mathbb{R}, \quad (t,x) \longmapsto \inf\big\{|(s,y)-(t,x)| \ \big| \ (s,y) \in \mathbb{R}^{1+N} \setminus \widetilde{\Omega}_S\big\}.$$

For each $\alpha \in]0,1]$ and $C \geq 0$, the operator $\displaystyle\bigcup_{t \in [0,\widehat{T}]} \big(\{t\} \times C_0^0(\widetilde{\Omega}(t))\big) \longrightarrow C_0^0(\widetilde{\Omega}_S)$

mapping any $(t,\, v) \in \{t\} \times C_0^0(\widetilde{\Omega}(t))$ to the continuous function

$$\widetilde{\Omega}_S \longrightarrow \mathbb{R}, \quad (s,y) \longmapsto \max\Big\{\min\big\{v(y),\ C \cdot d_{\complement\widetilde{\Omega}_S}(s,y)^{\alpha}\big\},\ -C \cdot d_{\complement\widetilde{\Omega}_S}(s,y)^{\alpha}\Big\}$$

is continuous with respect to the supremum norm.

Whenever the trivial extensions of some functions (to $\mathbb{R}^N$) are uniformly bounded or equi-continuous, the set of their images shares the respective property. □

3.11.4 The Tusk Condition for Approximative Cauchy Barriers

Effros and Kazdan investigated sufficient conditions for the continuity of solutions to the heat equation at the boundary in [75] and, they formulated a counterpart of the classical cone condition known for elliptic differential equations of second order. Later Lieberman took up their boundary condition geometrically similar to a tusk and extended it to more general parabolic differential equations in 1989 [115]. His essential contribution was to construct a function that serves as *local* barrier *from earlier time* and vanishes (merely) at the peak of the tusk.

In this subsection, we use Lieberman's local barrier function for concluding a family of approximative Cauchy barriers (with respect to L) merely from the uniform exterior tusk condition.
Now we specify the tusk condition as in Definition A.48 (on page 473) and then formulate the main result of this subsection:

Definition 157 (Exterior tusk condition [114, § 3], [115]**).**
A nonempty subset $M \subset \mathbb{R} \times \mathbb{R}^N$ is called *tusk* in $(t_0, x_0) \in \mathbb{R} \times \mathbb{R}^N$ if there exist constants $R, \tau > 0$ and a point $x_1 \in \mathbb{R}^N$ with

$$M = \Big\{(t,x) \in \mathbb{R} \times \mathbb{R}^N \;\Big|\; t_0 - \tau < t < t_0, \; \big|(x - x_0) - \sqrt{t_0 - t} \cdot x_1\big| < R\,\sqrt{t_0 - t}\Big\}.$$

A nonempty subset $\widetilde{\Omega} \subset \mathbb{R} \times \mathbb{R}^N$ satisfies the so-called *exterior tusk condition* if for every point $(t,x) \subset \partial\widetilde{\Omega}$ belonging to the parabolic boundary of $\widetilde{\Omega}$ (i.e.

$$\Big\{(s,y) \in \mathbb{R} \times \mathbb{R}^N \;\Big|\; |x - y| \leq \varepsilon,\; t - \varepsilon < s < t\Big\} \setminus \widetilde{\Omega} \neq \emptyset \qquad \text{for any } \varepsilon > 0),$$

there exists a tusk $M \subset \mathbb{R} \times \mathbb{R}^N$ in (t,x) with $\overline{M} \cap \overline{\widetilde{\Omega}} = \{(t,x)\}$.

A nonempty subset $\widetilde{\Omega} \subset \mathbb{R} \times \mathbb{R}^N$ is said to fulfill the *uniform exterior tusk condition* if it satisfies the exterior tusk conditions and if the scalar geometric parameters $R, \tau > 0$ of the tusks can be chosen independently of the respective points (t,x) of the parabolic boundary of $\widetilde{\Omega}$.

Proposition 158. *Let $\widetilde{\Omega}$ be a nonempty open subset of $[S,T] \times \mathbb{R}^N$ satisfying*

(i) *$\widetilde{\Omega}$ is bounded,*
(ii) *$\widetilde{\Omega} \cap \big(\{t\} \times \mathbb{R}^N\big) \neq \emptyset$ for every $t \in [S,T]$,*
(iii) *$\widetilde{\Omega}_S \stackrel{\text{Def.}}{=} \widetilde{\Omega} \cap \big(]S,T] \times \mathbb{R}^N\big)$ fulfills the uniform exterior tusk condition.*

Then $\widetilde{\Omega}_S$ possesses a family of approximative Cauchy barriers with respect to L (in the sense of Definition 143 on page 316).

The proof of this proposition is based on subsequent Lemma 159.
In fact, [115, Lemma 12.2] implies the following existence of a local barrier function for a single boundary point — even under weaker assumptions about the coefficients than the general hypotheses in § 3.11.1 (on page 311 f.):

Lemma 159 (Tusk condition provides local barrier from earlier time [115]**).**
Let $\widetilde{\Omega} \subset]-\infty, 0[\times \mathbb{R}^N$ *be a nonempty bounded open set such that the complement of* $\widetilde{\Omega}$ *contains a tusk in its boundary point* $(0,0)$.
Then for every $\sigma > 0$ *sufficiently small, there exist positive constants* η, γ_1, γ_2 *and a continuous function* $w : \overline{\widetilde{\Omega}} \setminus \{(0,0)\} \longrightarrow \mathbb{R}$ *which is continuously differentiable with respect to time and twice continuously differentiable with respect to space such that for every* $(t,x) \in \widetilde{\Omega}$,

$$\begin{cases} & Lw(t,x) \le -\eta \cdot \max\{|x|, |t|^{\frac{1}{2}}\}^{\sigma-2} \\ \eta \cdot \max\{|x|, |t|^{\frac{1}{2}}\}^{\sigma} \le & w(t,x) \le \max\{|x|, |t|^{\frac{1}{2}}\}^{\sigma} \\ & |Dw(t,x)| \le \max\{|x|, |t|^{\frac{1}{2}}\}^{\sigma-1} \\ & w(0,y) = \gamma_1 \cdot \left(1 - e^{-\gamma_2 |y|^{\sigma}}\right) \quad \text{if } (0,y) \in \overline{\widetilde{\Omega}} \setminus \{(0,0)\}. \end{cases}$$

The successive choice of admissible $\sigma > 0$ *and then of* $\eta, \gamma_1, \gamma_2 > 0$ *depends only on the supremum norms of the coefficients of L, its constant of uniform ellipticity, the diameter of* $\widetilde{\Omega}$ *and the geometric parameters* $R, \tau > 0$ *of the tusk in* $(0,0)$.

□

In [115], Lieberman then applies this local barrier from earlier time to parabolic problems with locally Hölder continuous coefficients for proving the existence of classical solutions to the first initial-boundary value problem by means of Perron method.
Now we leave this track of Lieberman and, we focus on merely continuous coefficients and strong solutions in $C^0 \cap W^{1;2}_{p,\mathrm{loc}}$ (with $p > N+2$) instead.

For each $T' > 0$ and any smooth cut-off function $\psi \in C_c^\infty(\mathbb{R}, [0,1])$, the problem

$$\begin{cases} L\widetilde{w} = -1 & \text{in }]0,T'] \times \mathbb{R}^N \\ \widetilde{w}(0,y) = \gamma_1 \cdot \left(1 - e^{-\gamma_2 |y|^{\sigma}}\right) \cdot \psi(|y|^2) & \text{for } y \in \mathbb{R}^N \end{cases}$$

is known to have a solution $\widetilde{w} \in C^0([0,T'] \times \mathbb{R}^N) \cap W^{1;2}_{p,\mathrm{loc}}(]0,T'[\times \mathbb{R}^N)$ vanishing at infinity [133]. Due to the parabolic maximum principle quoted in Proposition 161 below, the auxiliary function

$$(t,x) \longmapsto \widetilde{w}(t,x) - \varepsilon_1 \, t - \varepsilon_2 \, |x|^2 \, \psi(|x|^2)$$

(with $\varepsilon_1, \varepsilon_2 > 0$ sufficiently small) is nonnegative in any compact neighborhood of $(0,0)$ in $[0,T'] \times \mathbb{B}_{|\mathrm{supp}\,\psi|_\infty}(0) \subset [0,T'] \times \mathbb{R}^N$. In combination with the local barrier function from earlier time in Lemma 159, we conclude:

Corollary 160 (Tusk condition implies local barrier not just from earlier time). *Let $\widetilde{\Omega} \subset \mathbb{R} \times \mathbb{R}^N$ be a nonempty bounded open set such that the complement of $\widetilde{\Omega}$ contains a tusk in its boundary point* $(0,0)$.

Then there exist constants $\gamma, \delta, \eta, \sigma > 0$ and a function $w \in C^0(\overline{\widetilde{\Omega}}) \cap W^{1;2}_{p,\mathrm{loc}}(\widetilde{\Omega})$ such that for Lebesgue-almost every $(t,x) \in \widetilde{\Omega}$,

$$\begin{cases} \qquad\qquad Lw(t,x) < 0 & \\ \eta \cdot \max\{|x|, |t|^{\frac{1}{2}}\}^\sigma \le w(t,x) \le \max\{|x|, |t|^{\frac{1}{2}}\}^\sigma & \text{if } t \le 0 \\ \gamma \cdot (|x|^2 + t) \le w(t,x) & \text{if } t > 0. \end{cases}$$

The suitable choice of $\gamma, \delta, \eta, \sigma > 0$ depends only on the supremum norms of the continuous coefficients of L, its constant of uniform ellipticity, the diameter of $\widetilde{\Omega}$ and the geometric parameters $R, \tau > 0$ of the tusk in $(0,0)$.

For the sake of completeness, the following parabolic maximum principle on cylindrical domains has served as a tool:

Proposition 161 (Bony maximum principle for parabolic PDEs [70, Th.VII.28]**).** *Let O be a bounded domain in $\mathbb{R}^N$ and $Q :=]0,T] \times O$. Suppose $u \in W^{1;2}_{N+1,\mathrm{loc}}(Q)$,*

$$\widehat{L}u := \Big(\sum_{k,l=1}^{N} \widehat{a}_{kl}(t,\cdot)\, \tfrac{\partial^2}{\partial x_k\, \partial x_l} + \sum_{k=1}^{N} \widehat{b}_k(t,\cdot)\, \tfrac{\partial}{\partial x_k} + \widehat{c}(t,\cdot) - \tfrac{\partial}{\partial t} \Big)\, u$$

where $\widehat{a}_{kl}, \widehat{b}_k, \widehat{c} : Q \longrightarrow \mathbb{R}$ are bounded measurable, $\big(\widehat{a}_{kl}\big)_{k,l=1\ldots N} \ge 0$ and $\widehat{c} \le 0$. If u attains a nonpositive minimum at $(t_0, x_0) \in Q$, then

$$\lim \operatorname{ess\,inf}_{(s,y) \to (t_0,x_0)} \widehat{L}u(s,y) \ge 0.$$

□

Proof (of Proposition 158). Due to the assumptions of Proposition 158, $\widetilde{\Omega}_S$ fulfills the uniform exterior tusk condition. Hence, there exist strictly increasing moduli of continuity $\omega_1(\cdot), \omega_2(\cdot) :]0,\infty[\longrightarrow]0,\infty[$ (i.e. $\omega_1(r) + \omega_2(r) \longrightarrow 0$ for $r \downarrow 0$) such that for each boundary point $\widetilde{x} = (t,x) \in \partial\widetilde{\Omega}$ with $t > S$, Corollary 160 provides a function $w_{\widetilde{x}} \in C^0(\overline{\widetilde{\Omega}}) \cap W^{1;2}_{p,\mathrm{loc}}(\widetilde{\Omega})$ satisfying for Lebesgue-almost every $(s,y) \in \widetilde{\Omega}$,

$$\begin{cases} \qquad\qquad Lw_{\widetilde{x}}(s,y) < 0 \\ \omega_1\big(|y-x| + |s-t|^{\frac{1}{2}}\big) \le w_{\widetilde{x}}(s,y) \le \omega_2\big(|y-x| + |s-t|^{\frac{1}{2}}\big) \end{cases}$$

In regard to a family of approximative Cauchy barriers with respect to L, choose $0 < \varepsilon_1 \le \varepsilon_2$ and a compact subset $\widetilde{K}' \subset [S,T] \times \mathbb{R}^N$ with $\widetilde{K}' \subset \widetilde{\Omega}_S$ arbitrarily.
The boundary of the bounded set $\widetilde{\Omega}$ is compact. As a consequence, firstly,

$$\rho := \inf \Big\{ |y-x| + |s-t|^{\frac{1}{2}} \;\Big|\; (s,y) \in \widetilde{K}',\ (t,x) \in \partial\widetilde{\Omega} \Big\} > 0.$$

Secondly we can select finitely many points $\widetilde{x}_1 = (t_1, x_1) \ldots \widetilde{x}_k = (t_k, x_k) \in \partial\widetilde{\Omega}$ with

$$\partial\widetilde{\Omega} \subset \bigcup_{j=1}^{k} \left\{ (s,y) \in \mathbb{R} \times \mathbb{R}^N \,\middle|\, \omega_2\big(|y-x_j| + |s-t_j|^{\frac{1}{2}}\big) \leq \varepsilon_1 \, \frac{\omega_1(\rho)}{\varepsilon_2} \right\} =: N_{\partial\widetilde{\Omega}}.$$

Then, $w := \frac{\varepsilon_2}{\omega_1(\rho)} \cdot \min_{j=1\ldots k} w_{\widetilde{x}_j} : \overline{\widetilde{\Omega}} \longrightarrow [0,\infty[$ also belongs to $C^0(\overline{\widetilde{\Omega}}) \cap W^{1;2}_{p,\mathrm{loc}}(\widetilde{\Omega})$ and, it satisfies for $\mathscr{L}^1$-almost every $(s,y) \in \widetilde{\Omega}$ and its related index $j \in \{1 \ldots k\}$

$$\begin{cases} Lw(s,y) < 0 \\ \frac{\varepsilon_2}{\omega_1(\rho)} \cdot \omega_1\big(|y-x_j| + |s-t_j|^{\frac{1}{2}}\big) \leq w(s,y) \leq \frac{\varepsilon_2}{\omega_1(\rho)} \cdot \omega_2\big(|y-x_j| + |s-t_j|^{\frac{1}{2}}\big). \end{cases}$$

In fact, $w(s,y) \geq \frac{\varepsilon_2}{\omega_1(\rho)} \cdot \inf_l \omega_1\big(|y-x_l| + |s-t_l|^{\frac{1}{2}}\big) \geq \varepsilon_2$ for $(s,y) \in \widetilde{K}'$

and $w(s,y) \leq \frac{\varepsilon_2}{\omega_1(\rho)} \cdot \omega_2\big(|y-x_j| + |s-t_j|^{\frac{1}{2}}\big) \leq \varepsilon_1$ for $(s,y) \in \overline{\widetilde{\Omega}_S} \cap N_{\partial\widetilde{\Omega}}$.

□

3.11.5 *Successive Coupling of Nonlinear Parabolic Problem and Morphological Equation*

We restrict our consideration to a rather simple way of coupling an initial-boundary value problem of parabolic type with a morphological equation.
If the morphological equation does not depend on the wanted solution to the parabolic problem, we are free to solve it by means of § 1.9.6 first. This leads to a time-dependent reachable set of a nonautonomous differential inclusion and, then its graph provides a noncylindrical domain for the parabolic problem.

In regard to appropriate assumptions, however, we should prefer considerations in the opposite direction. Indeed, Theorem 146 (on page 317) always guarantees a strong solution to the parabolic problem if the noncylindrical domain $\widetilde{\Omega}_S \subset]S,T] \times \mathbb{R}^N$ has a family of approximative Cauchy barriers with respect to L. Proposition 158 (on page 326) provides a geometric condition sufficient for such a family, namely the uniform exterior tusk condition.

Finally we need an appropriate link between this tusk condition and reachable sets of differential inclusions in $\mathbb{R}^N$ because every solution to a morphological equation is a reachable set of a nonautonomous differential inclusion (according to Proposition 1.70 on page 74).
In fact, Corollary A.50 (on page 474) provides conditions on the differential inclusion sufficient for such a connection, but we obtain the exterior tusk condition for the *complements of graphs* of reachable sets.
Moreover, their exterior tusks are guaranteed to be uniform *only after* the reachable sets have evolved for an arbitrarily small period. For "imitating" such an evolution in the past (i.e., before the initial time $t_0 = 0$), we suppose the uniform exterior ball condition on the open initial set Ω_0 (whose complement starts deforming along a differential inclusion at time $t_0 = 0$).

For the sake of transparency, we prefer summarizing this notion in terms of reachable sets of nonautonomous differential inclusions (rather than noncompact-valued solutions to morphological equations). As in § 3.11.3, we suppose $S < 0 < \widehat{T} < T$.

Proposition 162. *Let $\Omega_0 \subset \mathbb{R}^N$ be a nonempty bounded open subset satisfying the uniform exterior ball condition at its boundary.*
In regard to Corollary A.50 (on page 474), suppose for $\widetilde{G} : [0,T] \times \mathbb{R}^N \rightsquigarrow \mathbb{R}^N$

(a) *every value of $\widetilde{G}$ is nonempty, compact, convex and has positive erosion of uniform radius $\rho > 0$ (see Definition A.25 on page 455),*

(b) *the Hamiltonian of $\widetilde{G}(t,\cdot)$ at each time $t \in [0,T]$*

$$\mathscr{H}_{\widetilde{G}}(t,\cdot,\cdot) : \mathbb{R}^N \times \mathbb{R}^N \longrightarrow \mathbb{R}, \quad (x,p) \longmapsto \sup_{z \in \widetilde{G}(t,x)} p \cdot z$$

is twice continuously differentiable in $\mathbb{R}^N \times (\mathbb{R}^N \setminus \{0\})$

(c) *there exists $\lambda_{\widetilde{G}} > 0$ such that for $\mathscr{L}^1$-almost every $t \in [0,T]$,*

$$\|\mathscr{H}_{\widetilde{G}}(t,\cdot,\cdot)\|_{C^{1,1}(\mathbb{R}^N \times \partial\mathbb{B}_1)} < \lambda_{\widetilde{G}}$$

(iv) *for every $t \in [0,T]$, the reachable set $\vartheta_{\widetilde{G}}(t, \mathbb{R}^N \setminus \Omega_0)$ is not identical to $\mathbb{R}^N$.*

Then the complement of the graph $t \longmapsto \vartheta_{\widetilde{G}}(t, \mathbb{R}^N \setminus \Omega_0)$ induces the set

$$\widetilde{\Omega} := \big([S,0] \times \Omega_0\big) \;\cup\; \bigcup_{t \in [0,T]} \big(\{t\} \times (\mathbb{R}^N \setminus \vartheta_{\widetilde{G}}(t, \mathbb{R}^N \setminus \Omega_0))\big) \;\subset\; [S,T] \times \mathbb{R}^N$$

fulfilling the uniform exterior tusk condition with respect to L and thus, $\widetilde{\Omega}$ satisfies the assumptions (i), (ii') of Existence Theorem 146 (on page 317).

In addition, let $\mathscr{F} : \bigcup_{t \in [0,\widehat{T}]} \big(\{t\} \times C^0_0(\widetilde{\Omega}(t))\big) \longrightarrow C^0_c(\mathbb{R}^N)$ satisfy the hypotheses (iii) – (v) of Theorem 146.

Then, for every initial function $u_0 \in C^0_0(\widetilde{\Omega}(0))$, there exists a strong solution $u \in C^0(\overline{\widetilde{\Omega}_0}) \cap W^{1;2}_{p,\mathrm{loc}}(\widetilde{\Omega}_0)$ to the initial-boundary value problem of parabolic type

$$\begin{cases} Lu(t,\cdot) = \mathscr{F}(t,u)(\cdot) & \text{in } \widetilde{\Omega}(t) \text{ for a.e. } t \in]0,\widehat{T}[, \\ u(0,\cdot) = u_0 & \text{in } \widetilde{\Omega}(0) \subset \mathbb{R}^N, \\ u = 0 & \text{on } \partial\widetilde{\Omega}_0 \setminus (\{0\} \times \widetilde{\Omega}(0)). \end{cases}$$

□

Chapter 4
Introducing Distribution-Like Solutions to Mutational Equations

In this chapter, we focus on examples of evolving compact sets in the Euclidean space and draw them on new useful aspects for generalizing the mutational framework.

Now the normal cones of the compact sets are to have an explicit influence on the geometric evolution. Reachable sets of differential inclusions still induce the transitions on $\mathscr{K}(\mathbb{R}^N)$, but we leave the typical metric space of $\mathscr{K}(\mathbb{R}^N)$ supplied with the Pompeiu-Hausdorff metric $d\!l$ (as in the preceding sections 1.9, 1.11, 2.7). Additionally we take the graphs of limiting normal cones into consideration.

This type of problems reveals two obstacles which motivate the main aspects of generalizing in comparison with Chapter 3. Analytically speaking, these extensions have a weakening effect on how "uniform" the continuity parameters $\alpha_j(\vartheta;r)$, $\beta_j(\vartheta;r)$ of transitions have to be.

For the Regularity in Time: Distance Functions Do Not Have to Be Symmetric

Let us consider first the consequences of the boundary for the continuity of $\vartheta_F : [0,1] \times \mathscr{K}(\mathbb{R}^N) \longrightarrow \mathscr{K}(\mathbb{R}^N)$ with respect to time.

The key aspect is illustrated easily by an annulus $K_{\circledcirc}$ expanding isotropically at a constant speed. After a positive finite time t_3, the "hole" in the center has disappeared of course.

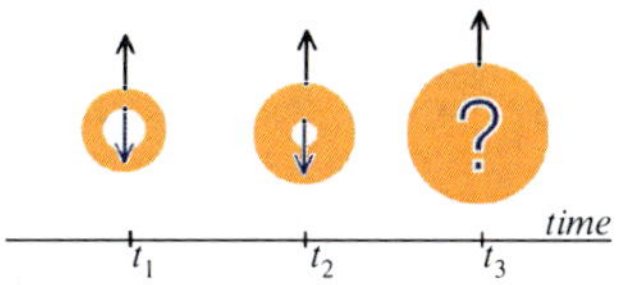

In general, the topological boundary of a time-dependent reachable set $\vartheta_F(\cdot,K) : [0,\infty[\rightsquigarrow \mathbb{R}^N$ (with $K \in \mathscr{K}(\mathbb{R}^N)$) is not continuous with respect to $d\!l$. Furthermore, the normals of *later* sets find close counterparts among the normals of *earlier* sets, but usually not vice versa.

For this reason, we dispense with the symmetry condition (H2) on distance functions. Whenever we consider distances in this chapter, their first arguments refer to the earlier state and their second arguments to the later state. For the sake of transparency, all general results about mutational equations are formulated for tuples with separate real time component.

T. Lorenz, *Mutational Analysis: A Joint Framework for Cauchy Problems In and Beyond Vector Spaces*, Lecture Notes in Mathematics 1996,
DOI 10.1007/978-3-642-12471-6_5, © Springer-Verlag Berlin Heidelberg 2010

For the Regularity with Respect to Initial States: The Distributional Notion

Applying now the typical steps of mutational analysis, we encounter analytical obstacles soon. In particular, $[0,1] \longrightarrow [0,\infty[, \quad t \longmapsto d_j\big(\vartheta(t,x_1),\ \vartheta(t,x_2)\big)$ does not have to be continuous for arbitrary initial elements x_1, x_2.

Consider e.g. reachable sets $\vartheta_F(t,K_1),\ \vartheta_F(t,K_2)$ of a differential inclusion $x'(\cdot) \in F(x(\cdot))$ with initial sets $K_1, K_2 \in \mathscr{K}(\mathbb{R}^N)$ and a given map $F \in \mathrm{LIP}(\mathbb{R}^N, \mathbb{R}^N)$. The figure on the right-hand side sketches a situation in which the distance between topological boundaries

$$[0,1] \longrightarrow \mathbb{R}_0^+, \quad t \longmapsto \mathrm{dist}(\partial\vartheta_F(t,K_2),\ \partial\vartheta_F(t,K_1))$$

cannot be continuous.

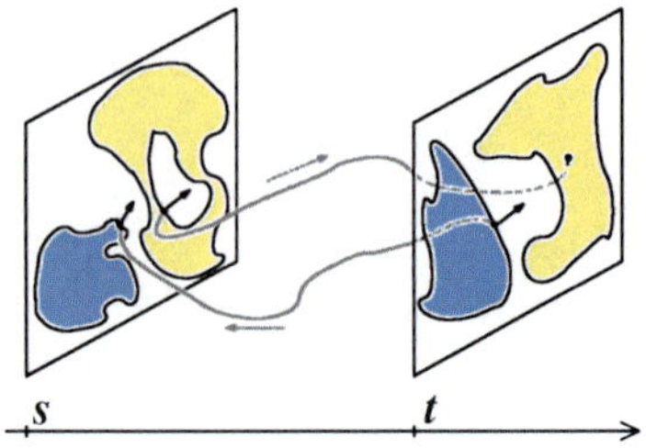

Even if we do not take normal cones into account explicitly, it is difficult to find a (possibly nonsymmetric) distance function on $\mathscr{K}(\mathbb{R}^N)$ depending on the boundary, but without such a lack of continuity.

As a first important consequence, we require a form of Gronwall's inequality which starts from weaker assumptions than its continuous counterpart in standard textbooks like [10, 92, 181]. The essential advantage of Proposition A.2 (on page 440) is that only lower semicontinuity of the real-valued function is supposed. For estimating the distance d_j between transitions and $(e_j)_{j\in\mathscr{J}}$-continuous curves, we will use an additional semicontinuity condition on transitions rather than a general hypothesis about distances.

Nevertheless, we have to exclude such a discontinuity of evolving boundaries – for short times at least. In the first subsequent geometric example (in § 4.4 on page 359 ff.), additional assumptions about K_1 are needed. Suitable conditions on $F \in \mathrm{LIP}(\mathbb{R}^N, \mathbb{R}^N)$ can guarantee that compact sets with $C^{1,1}$ boundary preserve this regularity for short times (see Appendix A.5.3 on page 458 ff.) and, their topological properties do not change essentially.

Assuming restrictive conditions on one of the sets $K_1, K_2 \in \mathscr{K}(\mathbb{R}^N)$ prevents us from applying the recent mutational framework, though. Thus we want to introduce a form of distributional solution.

For a set with families of distance functions, however, there are no obvious generalizations of linear forms or partial integration and hence, distributions in their widespread sense cannot be introduced. This gap makes a more general interpretation of distributional solutions indispensable. In fact, *their basic idea is to select an important property and preserve it (only) for all elements of a given fixed "test set" – instead of the whole "basic set".*
Usually this important feature is the rule of partial integration and, it is preserved for smooth test functions with compact support (or Schwartz functions).

In the mutational framework, one of the most important properties so far has been the estimate comparing two states while evolving along two transitions, i.e., according to Proposition 3.7 (on page 185)

$$d_j\big(\vartheta(t_1+h,x),\ \tau(t_2+h,y)\big) \le \Big(d_j\big(\vartheta(t_1,x),\ \tau(t_2,y)\big) + h\cdot \widehat{D}_j(\vartheta,\tau;R_j)\Big)\ e^{\alpha_j(\tau;R_j)\,h}$$

with radius $R_j := \big(\max\{\lfloor x \rfloor_j,\ \lfloor y \rfloor_j\} + \max\{\gamma_j(\vartheta),\ \gamma_j(\tau)\}\big)\cdot e^{\max\{\gamma_j(\vartheta),\ \gamma_j(\tau)\}} < \infty$. As explained in the beginning of § 3.3, it has even laid the foundations for adapting the definition of solution to a mutational equation in Definition 3.8 (on page 187) — in form of the condition:

2.′) there exists $\alpha_j(x;\cdot) : [0,\infty[\longrightarrow [0,\infty[$ such that for $\mathscr{L}^1$-a.e. $t \in [0,T[$:

$$\limsup_{h\downarrow 0} \frac{d_j(\vartheta(s+h,z),\ x(t+h)) - d_j(\vartheta(s,z),\ x(t))\cdot e^{\alpha_j(x;R_j)\,h}}{h} \le \widehat{D}_j\big(\vartheta,\ f(x(t),t);\ R_j\big)$$

is fulfilled for any $\vartheta \in \widehat{\Theta}\big(E,(d_j),(e_j),(\lfloor\cdot\rfloor_j)\big)$, $s \in [0,1[$, $z \in E$ satisfying $\lfloor \vartheta(\cdot,z)\rfloor_j, \lfloor x(\cdot)\rfloor_j \le R_j$,

These key estimates should be preserved while comparing with all elements z of a given fixed "test set" $\mathscr{D} \neq \emptyset$ (instead of all $z \in E$ as in Chapter 3). It is plausible to demand that such an element $z \in \mathscr{D}$ stays in the test set $\mathscr{D}$ for a short time while evolving along a transition so that the comparison is feasible for this short period (at least). This notion leads to a form of distributional solution in the mutational framework and, it still dispenses with any linear structure.

In addition, it opens the door to making the continuity parameter α_j and the transitional distance $\widehat{D}_j$ "less uniform" — in the sense that they are free to depend on the respective test element of $\mathscr{D}$. In other words, admissible transitions can now be "less regular" than in Chapter 3.

Motivated by the finite element methods of Petrov-Galerkin in numerics (e.g. [21]), we do not assume that the fixed test set $\mathscr{D}$ has to be a subset of the basic set E. This additional aspect of freedom will be very useful in the second subsequent geometric example in § 4.5 (on page 372 ff.).

4.1 General Assumptions of This Chapter

$\mathscr{D}$ and E are always nonempty sets and, $\widetilde{\mathscr{D}} := \mathbb{R} \times \mathscr{D}$, $\widetilde{E} := \mathbb{R} \times E$. ($\mathscr{D} \subset E$ is not required in general.) $\pi_1 : \widetilde{\mathscr{D}} \cup \widetilde{E} \longrightarrow \mathbb{R}$, $\widetilde{x} = (t,x) \longmapsto t$ abbreviates the canonical projection on the real component. $\mathscr{I} \neq \emptyset$ denotes an index set. For each $j \in \mathscr{I}$,

$$\begin{array}{rcl} \widetilde{d}_j, \widetilde{e}_j : (\widetilde{\mathscr{D}} \cup \widetilde{E}) \times (\widetilde{\mathscr{D}} \cup \widetilde{E}) & \longrightarrow & [0,\infty[, \\ \lfloor \cdot \rfloor_j : \qquad \widetilde{\mathscr{D}} \cup \widetilde{E} & \longrightarrow & [0,\infty[\end{array}$$

are supposed to satisfy the following conditions:

(H1) $\widetilde{d}_j$ and $\widetilde{e}_j$ are reflexive, i.e. for all $\widetilde{x} \in \widetilde{\mathscr{D}} \cup \widetilde{E}$: $\widetilde{d}_j(\widetilde{x},\widetilde{x}) = 0 = \widetilde{e}_j(\widetilde{x},\widetilde{x})$.

(H3') $(\widetilde{d}_j)_{j\in\mathscr{I}}$ and $(\widetilde{e}_j)_{j\in\mathscr{I}}$ induce the same concept of convergence in E and are sequentially (semi-) continuous in the following sense:

($\widetilde{\text{o}}_l$)
$$\begin{array}{rl} & \big(\forall\, j \in \mathscr{I}: \ \lim\limits_{n\to\infty} \ \widetilde{d}_j(\widetilde{x}, \widetilde{x}_n) = 0\big) \\ \Longleftrightarrow & \big(\forall\, j \in \mathscr{I}: \ \lim\limits_{n\to\infty} \ \widetilde{e}_j(\widetilde{x}, \widetilde{x}_n) = 0\big) \end{array}$$
for any $\widetilde{x} \in \widetilde{\mathscr{D}} \cup \widetilde{E}$ and $(\widetilde{x}_n)_{n\in\mathbb{N}}$ in $\widetilde{\mathscr{D}} \cup \widetilde{E}$ with $\pi_1\, \widetilde{x} \leq \pi_1\, \widetilde{x}_n$ for all n and $\sup\limits_{n\in\mathbb{N}} \lfloor \widetilde{x}_n \rfloor_i < \infty$ for each $i \in \mathscr{I}$.

($\widetilde{\text{o}}_r$)
$$\begin{array}{rl} & \big(\forall\, j \in \mathscr{I}: \ \lim\limits_{n\to\infty} \ \widetilde{d}_j(\widetilde{x}_n, \widetilde{x}) = 0\big) \\ \Longleftrightarrow & \big(\forall\, j \in \mathscr{I}: \ \lim\limits_{n\to\infty} \ \widetilde{e}_j(\widetilde{x}_n, \widetilde{x}) = 0\big) \end{array}$$
for any $\widetilde{x} \in \widetilde{\mathscr{D}} \cup \widetilde{E}$ and $(\widetilde{x}_n)_{n\in\mathbb{N}}$ in $\widetilde{\mathscr{D}} \cup \widetilde{E}$ with $\pi_1\, \widetilde{x}_n \leq \pi_1\, \widetilde{x}$ for all n and $\sup\limits_{n\in\mathbb{N}} \lfloor \widetilde{x}_n \rfloor_i < \infty$ for each $i \in \mathscr{I}$.

($\widetilde{\text{i}}$')
$$\begin{array}{rcl} \widetilde{d}_j(\widetilde{x}, \widetilde{y}) & \leq & \limsup\limits_{n\to\infty} \ \widetilde{d}_j(\widetilde{x}_n, \widetilde{y}_n), \\ \widetilde{e}_j(\widetilde{x}, \widetilde{y}) & \leq & \limsup\limits_{n\to\infty} \ \widetilde{e}_j(\widetilde{x}_n, \widetilde{y}_n) \end{array}$$
for any $\widetilde{x},\widetilde{y} \in \widetilde{\mathscr{D}} \cup \widetilde{E}$ and $(\widetilde{x}_n)_{n\in\mathbb{N}}$, $(\widetilde{y}_n)_{n\in\mathbb{N}}$ in $\widetilde{\mathscr{D}} \cup \widetilde{E}$ s.t. for each $i \in \mathscr{I}$
$$\lim_{n\to\infty} \widetilde{d}_i(\widetilde{x}, \widetilde{x}_n) = 0 = \lim_{n\to\infty} \widetilde{d}_i(\widetilde{y}_n, \widetilde{y}), \quad \sup_{n\in\mathbb{N}} \{\lfloor \widetilde{x}_n \rfloor_i, \lfloor \widetilde{y}_n \rfloor_i\} < \infty$$
and for all $n \in \mathbb{N}$: $\pi_1\, \widetilde{x} \leq \pi_1\, \widetilde{x}_n \leq \pi_1\, \widetilde{y}_n \leq \pi_1\, \widetilde{y}$.

($\widetilde{\text{i}}$")
$$\widetilde{d}_j(\widetilde{z}, \widetilde{y}) \geq \limsup_{n\to\infty} \ \widetilde{d}_j(\widetilde{z}, \widetilde{y}_n),$$
for any $\widetilde{z} \in \widetilde{\mathscr{D}}$, $\widetilde{y} \in \widetilde{E}$ and $(\widetilde{y}_n)_{n\in\mathbb{N}}$ in $\widetilde{E}$ fulfilling for each $i \in \mathscr{I}$
$$\lim_{n\to\infty} \widetilde{d}_i(\widetilde{y}, \widetilde{y}_n) = 0, \qquad \sup_{n\in\mathbb{N}} \lfloor \widetilde{y}_n \rfloor_i < \infty$$
and for all $n \in \mathbb{N}$: $\pi_1\, z \leq \pi_1\, \widetilde{y} \leq \pi_1\, \widetilde{y}_n$.

($\widetilde{\text{ii}}_l$)
$$0 = \lim_{n\to\infty} \ \widetilde{d}_j(\widetilde{x},\widetilde{x}_n)$$
for any $\widetilde{x} \in \widetilde{E}$ and $(\widetilde{x}_n)_{n\in\mathbb{N}}$, $(\widetilde{y}_n)_{n\in\mathbb{N}}$ in $\widetilde{E}$ fulfilling for each $i \in \mathscr{I}$
$$\lim_{n\to\infty} \ \widetilde{d}_i(\widetilde{x}, \widetilde{y}_n) = 0 = \lim_{n\to\infty} \ \widetilde{e}_i(\widetilde{y}_n, \widetilde{x}_n), \qquad \sup_{n\in\mathbb{N}} \{\lfloor \widetilde{x}_n \rfloor_i, \lfloor \widetilde{y}_n \rfloor_i\} < \infty,$$
$$\pi_1\, \widetilde{x} \leq \pi_1\, \widetilde{y}_n \leq \pi_1\, \widetilde{x}_n \quad \text{for all } n \in \mathbb{N}.$$

($\widetilde{\text{iii}}_l$) $0 = \lim_{n\to\infty} \widetilde{d}_j(\widetilde{x}, \widetilde{x}_n)$

for every index $j \in \mathcal{I}$, any element $\widetilde{x} \in \widetilde{E}$ and sequences $(\widetilde{x}_n)_{n\in\mathbb{N}}$, $(\widetilde{y}_k)_{k\in\mathbb{N}}$, $(\widetilde{z}_{k,n})_{k,n\in\mathbb{N}}$ in $\widetilde{E}$ fulfilling

$$\begin{cases} \pi_1\,\widetilde{x} \le \pi_1\,\widetilde{z}_{k,n} = \pi_1\,\widetilde{y}_k \le \pi_1\,\widetilde{x}_n & \text{for each } k,n\in\mathbb{N},\\ \lim_{k\to\infty} \widetilde{d}_i(\widetilde{x},\widetilde{y}_k) = 0 & \text{for each } i\in\mathcal{I},\\ \lim_{n\to\infty} \widetilde{d}_i(\widetilde{y}_k,\widetilde{z}_{k,n}) = 0 & \text{for each } i\in\mathcal{I}, k\in\mathbb{N},\\ \lim_{k\to\infty}\sup_{n>k} \widetilde{e}_i(\widetilde{z}_{k,n},\widetilde{x}_n) = 0 & \text{for each } i\in\mathcal{I},\\ \sup_{k,n\in\mathbb{N}} \{\lfloor\widetilde{x}_n\rfloor_i, \lfloor\widetilde{y}_k\rfloor_i, \lfloor\widetilde{z}_{k,n}\rfloor_i\} < \infty & \text{for each } i\in\mathcal{I}. \end{cases}$$

($\widetilde{\text{iii}}_r$) $0 = \lim_{n\to\infty} \widetilde{d}_j(\widetilde{x}_n, \widetilde{x})$

for every index $j \in \mathcal{I}$, any element $\widetilde{x} \in \widetilde{E}$ and sequences $(\widetilde{x}_n)_{n\in\mathbb{N}}$, $(\widetilde{y}_k)_{k\in\mathbb{N}}$, $(\widetilde{z}_{k,n})_{k,n\in\mathbb{N}}$ in $\widetilde{E}$ fulfilling

$$\begin{cases} \pi_1\,\widetilde{x}_n \le \pi_1\,\widetilde{z}_{k,n} = \pi_1\,\widetilde{y}_k \le \pi_1\,\widetilde{x} & \text{for each } k,n\in\mathbb{N},\\ \lim_{k\to\infty} \widetilde{d}_i(\widetilde{y}_k,\widetilde{x}) = 0 & \text{for each } i\in\mathcal{I},\\ \lim_{n\to\infty} \widetilde{d}_i(\widetilde{z}_{k,n},\widetilde{y}_k) = 0 & \text{for each } i\in\mathcal{I}, k\in\mathbb{N},\\ \lim_{k\to\infty}\sup_{n>k} \widetilde{e}_i(\widetilde{x}_n,\widetilde{z}_{k,n}) = 0 & \text{for each } i\in\mathcal{I},\\ \sup_{k,n\in\mathbb{N}} \{\lfloor\widetilde{x}_n\rfloor_i, \lfloor\widetilde{y}_k\rfloor_i, \lfloor\widetilde{z}_{k,n}\rfloor_i\} < \infty & \text{for each } i\in\mathcal{I}. \end{cases}$$

(H4) $\lfloor\cdot\rfloor_j$ is lower semicontinuous with respect to $(\widetilde{d}_i)_{i\in\mathcal{I}}$, i.e.,

$$\lfloor\widetilde{x}\rfloor_j \le \liminf_{n\to\infty} \lfloor\widetilde{x}_n\rfloor_j$$

for any element $\widetilde{x} \in \widetilde{E}$ and sequence $(\widetilde{x}_n)_{n\in\mathbb{N}}$ in $\widetilde{E}$ fulfilling for each $i \in \mathcal{I}$,

$$\lim_{n\to\infty} \widetilde{d}_i(\widetilde{x}_n,\widetilde{x}) = 0, \qquad \pi_1\,\widetilde{x}_n \nearrow \pi_1\,\widetilde{x} \text{ for } n\to\infty \qquad \text{and} \qquad \sup_{n\in\mathbb{N}} \lfloor\widetilde{x}_n\rfloor_i < \infty.$$

Now we adapt the definition of transition and admit different properties of the time component for elements of basic set $\widetilde{E}$ and the test set $\widetilde{\mathcal{D}}$:

Definition 1. A function $\widetilde{\vartheta} : [0,1] \times (\widetilde{\mathscr{D}} \cup \widetilde{E}) \longrightarrow (\widetilde{\mathscr{D}} \cup \widetilde{E})$ is called *timed transition* on the tuple $\big(\widetilde{E},\ \widetilde{\mathscr{D}},\ (\widetilde{d}_j)_{j\in\mathscr{I}},\ (\widetilde{e}_j)_{j\in\mathscr{I}},\ (\lfloor\cdot\rfloor_j)_{j\in\mathscr{I}}\big)$ if it satisfies for each $j \in \mathscr{I}$:

1.) for every $\widetilde{x} \in \widetilde{E}$: $\quad \widetilde{\vartheta}(0,\widetilde{x}) = \widetilde{x}$

3.') for every $\widetilde{z} \in \widetilde{\mathscr{D}}$, there are $\mathbb{T}_j = \mathbb{T}_j(\widetilde{\vartheta},\widetilde{z}) \in]0,1]$, $\alpha_j(\widetilde{\vartheta};\widetilde{z},\cdot) : [0,\infty[\longrightarrow [0,\infty[$ such that for any $\widetilde{y} \in \widetilde{E}$, $t \in [0,\mathbb{T}_j[$ with $\lfloor\widetilde{y}\rfloor_j \le r$ and $t + \pi_1\,\widetilde{z} \le \pi_1\,\widetilde{y}$:

$$\limsup_{h\downarrow 0} \frac{\widetilde{d}_j(\widetilde{\vartheta}(t+h,\widetilde{z}),\ \widetilde{\vartheta}(h,\widetilde{y})) - \widetilde{d}_j(\widetilde{\vartheta}(t,\widetilde{z}),\widetilde{y})}{h} \le \alpha_j(\widetilde{\vartheta};\widetilde{z},r)\cdot \widetilde{d}_j\big(\widetilde{\vartheta}(t,\widetilde{z}),\ \widetilde{y}\big)$$

4.') there exists $\beta_j(\widetilde{\vartheta};\cdot) : [0,\infty[\longrightarrow [0,\infty[$ such that for any $r \ge 0$, $s,t \in [0,1]$ and $\widetilde{x} \in \widetilde{E}$ with $\lfloor\widetilde{x}\rfloor_j \le r$: $\quad \widetilde{e}_j\big(\widetilde{\vartheta}(s,\widetilde{x}),\ \widetilde{\vartheta}(t,\widetilde{x})\big) \le \beta_j(\widetilde{\vartheta};r)\cdot|t-s|$

5.) there exists $\gamma_j(\widetilde{\vartheta}) \in [0,\infty[$ such that for any $t \in [0,1]$ and $\widetilde{x} \in \widetilde{E}$:

$$\lfloor\widetilde{\vartheta}(t,\widetilde{x})\rfloor_j \le \big(\lfloor\widetilde{x}\rfloor_j + \gamma_j(\widetilde{\vartheta})\ t\big)\cdot e^{\gamma_j(\widetilde{\vartheta})\,t},$$

$$\limsup_{h\downarrow 0}\ \sup_{\widetilde{z}\in\widetilde{\mathscr{D}}}\ \big(\lfloor\widetilde{\vartheta}(h,\widetilde{z})\rfloor_j - \lfloor\widetilde{z}\rfloor_j\ e^{\gamma_j(\widetilde{\vartheta})\,h}\big) \le 0,$$

6.) for every $\widetilde{z} \in \widetilde{\mathscr{D}}$: $\quad \widetilde{\vartheta}(h,\widetilde{z}) \in \widetilde{\mathscr{D}}$ for all $h \in [0,\ \mathbb{T}_j(\widetilde{\vartheta},\widetilde{z})[$, $\ \sup_{[0,\mathbb{T}_j[} \lfloor\widetilde{\vartheta}(\cdot,\widetilde{z})\rfloor_j < \infty$

7.) for every $\widetilde{y} \in \widetilde{E}$: $\quad \widetilde{\vartheta}(h,\ \widetilde{y}) \in \{h + \pi_1\,\widetilde{y}\} \times E \subset \widetilde{E} \quad$ for all $h \in [0,1]$,
for every $\widetilde{z} \in \widetilde{\mathscr{D}}$: $\quad \pi_1\,\widetilde{\vartheta}(h',\widetilde{z}) \le \pi_1\,\widetilde{\vartheta}(h,\ \widetilde{z}) \le h + \pi_1\,\widetilde{z}\ $ for all $h' \le h \le 1$

8.) for every $\widetilde{z} \in \widetilde{\mathscr{D}}$, $t < \mathbb{T}_j(\widetilde{\vartheta},\widetilde{z})$: $\ \widetilde{d}_j\big(\widetilde{\vartheta}(t,\widetilde{z}),\widetilde{y}\big) \le \limsup_{n\to\infty} \widetilde{d}_j\big(\widetilde{\vartheta}(t-h_n,\widetilde{z}),\widetilde{y}_n\big)$
for any $(h_n)_{n\in\mathbb{N}}$, $(\widetilde{y}_n)_{n\in\mathbb{N}}$ in $\mathbb{R}_0^+,\widetilde{E}$ and $\widetilde{y} \in \widetilde{E}$ with $h_n \longrightarrow 0$, $\widetilde{e}_i(\widetilde{y}_n,\widetilde{y}) \longrightarrow 0$ for each $i \in \mathscr{I}$ and $\pi_1\,\widetilde{\vartheta}(t-h_n,\widetilde{z}) \le \pi_1\,\widetilde{y}_n \nearrow \pi_1\,\widetilde{y}$.

Remark 2. (i) Four additional assumptions lead to almost the same environment as in Chapter 3 (see § 3.5 on page 221 ff. in particular):

(i) $\widetilde{\mathscr{D}} = \widetilde{E}$,
(ii) $\mathbb{T}_j(\cdot,\cdot) \equiv 1$,
(iii) each function $\widetilde{d}_j,\widetilde{e}_j$ ($j \in \mathscr{I}$) is symmetric,
(iv) continuity parameter $\alpha_j(\widetilde{\vartheta};\widetilde{z},r) \ge 0$ does not depend on $\widetilde{z} \in \widetilde{\mathscr{D}}$.

Indeed, the only relevant difference is that condition (3.') here is restricted to comparisons with merely *earlier* test elements. This is indicated by the constraint $t + \pi_1\,\widetilde{z} \le \pi_1\,\widetilde{y}$ and, it is consistent with our general intention to sort the arguments of distances by time.
There is no corresponding condition on time components in Definition 3.35 of timed solutions (on page 222), for example. Hence, all variants of the mutational framework presented in preceding chapters prove to be special cases.

(ii) Hypothesis (H3') is to make the timed triangle inequality (p. 319) dispensable. Condition (8.), however, does not result directly from the timed triangle inequality. We will need it essentially for applying a semicontinuous version of Gronwall's inequality later on (see e.g. Lemma 5 and Proposition 6).

$\widehat{\Theta}\big(\widetilde{E}, \widetilde{\mathscr{D}}, (\widetilde{d}_j)_{j\in\mathscr{I}}, (\widetilde{e}_j)_{j\in\mathscr{I}}, (\lfloor\cdot\rfloor_j)_{j\in\mathscr{I}}\big)$ denotes a nonempty set of timed transitions on $\big(\widetilde{E}, \widetilde{\mathscr{D}}, (\widetilde{d}_j)_{j\in\mathscr{I}}, (\widetilde{e}_j)_{j\in\mathscr{I}}, (\lfloor\cdot\rfloor_j)_{j\in\mathscr{I}}\big)$ and, for each $j \in \mathscr{I}$, the function

$$\widehat{D}_j : \ \widehat{\Theta}\big(\widetilde{E}, \widetilde{\mathscr{D}}, (\widetilde{d}_j)_j, (\widetilde{e}_j)_j, (\lfloor\cdot\rfloor_j)_j\big)^2 \times \widetilde{\mathscr{D}} \times [0,\infty[\ \longrightarrow \ [0,\infty[$$

is assumed to satisfy the following conditions:

(H5') for each $\widetilde{z} \in \widetilde{\mathscr{D}}, r \geq 0$, $\widehat{D}_j(\cdot\,,\cdot\,;\widetilde{z}, r)$ is reflexive (but possibly nonsymmetric), for any $\widetilde{z} \in \widetilde{\mathscr{D}}$ and timed transitions $\widetilde{\vartheta}, \widetilde{\tau}$, the function $\widehat{D}_j(\widetilde{\vartheta}, \widetilde{\tau}; \widetilde{z}, \cdot) : [0,\infty[\longrightarrow [0,\infty[$ is nondecreasing,

(H6') for each $\widetilde{z} \in \widetilde{\mathscr{D}}$ and any $r \geq 0$,
$\widehat{D}_j(\cdot,\cdot;\widetilde{z}, r) : \widehat{\Theta}\big(\widetilde{E}, (\widetilde{d}_j), (\widetilde{e}_j), (\lfloor\cdot\rfloor_j)\big) \times \widehat{\Theta}\big(\widetilde{E}, (\widetilde{d}_j), (\widetilde{e}_j), (\lfloor\cdot\rfloor_j)\big) \longrightarrow [0,\infty[$ is sequentially continuous with respect to $(\widehat{D}_i)_{i\in\mathscr{I}}$ in the following sense:

(i) $\widehat{D}_j(\widetilde{\vartheta}, \widetilde{\tau}; \widetilde{z}, r) \ = \ \lim\limits_{n\to\infty} \widehat{D}_j(\widetilde{\vartheta}_n, \widetilde{\tau}_n; \widetilde{z}, r)$
for any timed transitions $\widetilde{\vartheta}, \widetilde{\tau}$ and sequences $(\widetilde{\vartheta}_n)_{n\in\mathbb{N}}, (\widetilde{\tau}_n)_{n\in\mathbb{N}}$ satisfying for every $i \in \mathscr{I}$, $\widetilde{z}' \in \widetilde{\mathscr{D}}$ and $R \geq 0$

$$\lim_{n\to\infty} \widehat{D}_i(\widetilde{\vartheta}, \widetilde{\vartheta}_n; \widetilde{z}', R) \ = \ 0 \ = \ \lim_{n\to\infty} \widehat{D}_i(\widetilde{\tau}_n, \widetilde{\tau}; \widetilde{z}', R).$$

(ii) $\lim\limits_{n\to\infty} \widehat{D}_j(\widetilde{\vartheta}, \widetilde{\tau}_n; \widetilde{z}, r) \ = \ 0$
for any timed transition $\widetilde{\vartheta}$ and sequences $(\widetilde{\vartheta}_n)_{n\in\mathbb{N}}, (\widetilde{\tau}_n)_{n\in\mathbb{N}}$ satisfying for every $i \in \mathscr{I}$, $\widetilde{z}' \in \widetilde{\mathscr{D}}$ and $R \geq 0$

$$\lim_{n\to\infty} \widehat{D}_i(\widetilde{\vartheta}, \widetilde{\vartheta}_n; \widetilde{z}', R) \ = \ 0 \ = \ \lim_{n\to\infty} \widehat{D}_i(\widetilde{\vartheta}_n, \widetilde{\tau}_n; \widetilde{z}', R).$$

(H7') $\limsup\limits_{h\downarrow 0} \dfrac{\widetilde{d}_j\big(\widetilde{\vartheta}(t_1+h,\widetilde{z}),\ \widetilde{\tau}(t_2+h,\widetilde{y})\big) - \widetilde{d}_j\big(\widetilde{\vartheta}(t_1,\widetilde{z}),\ \widetilde{\tau}(t_2,\widetilde{y})\big) \cdot e^{\alpha_j(\widetilde{\tau};\widetilde{z},R_j)\cdot h}}{h} \ \leq \ \widehat{D}_j(\widetilde{\vartheta}, \widetilde{\tau}; \widetilde{z}, R_j)$
for any $\widetilde{\vartheta}, \widetilde{\tau} \in \widehat{\Theta}\big(\widetilde{E}, \widetilde{\mathscr{D}}, (\widetilde{d}_i)_i, (\widetilde{e}_i)_i, (\lfloor\cdot\rfloor_i)_i\big)$, $\widetilde{z} \in \widetilde{\mathscr{D}}$, $\widetilde{y} \in \widetilde{E}$, $t_1, t_2 \in [0,1[$, $r \geq 0$, $j \in \mathscr{I}$ with $t_1 < \mathbb{T}_j(\widetilde{\vartheta}, \widetilde{z})$, $t_1 + \pi_1\,\widetilde{z} \leq t_2 + \pi_1\,\widetilde{y}$, $\lfloor\widetilde{y}\rfloor_j \leq r$ and $R_j := \big(r + \gamma_j(\widetilde{\tau})\big) \cdot e^{\gamma_j(\widetilde{\tau})}$.

Remark 3. In this chapter, all general results about mutational equations are formulated for elements in $\widetilde{E}$ and $\widetilde{\mathscr{D}}$ respectively, i.e. for states with a separate real time component.
If this time component is not relevant to distances or transitions, however, we are free to skip it. Indeed, the step from transitions on $(E, \mathscr{D})$ to $(\widetilde{E}, \widetilde{\mathscr{D}})$ by means of

$$\widetilde{\vartheta}\big(h, (t,x)\big) \ = \ \big(t+h, \ \vartheta(h,x)\big)$$

has already been indicated in § 3.5 (on page 221 ff.). For the sake of consistency, we then skip the adjective "timed" as well. In particular, we will benefit from this simplification in the geometric example of § 4.4 (on page 359 ff.), but not in the second example in § 4.5 (on page 372 ff.).

4.2 Comparing with "Test Elements" of $\widetilde{\mathscr{D}}$ along Timed Transitions

Following the typical "mutational track" similarly to § 3.2 (on page 185 f.), we first mention briefly that the "absolute value" of states in $\widetilde{E}$ evolving along finitely many transitions is bounded in exactly the same way because the generalizations do not have any effect on the simple arguments having proved Lemma 2.4 (on page 105).

Lemma 4. *Let $\widetilde{\vartheta}_1 \dots \widetilde{\vartheta}_K$ be finitely many timed transitions on $\big(\widetilde{E}, \widetilde{\mathscr{D}}, (\widetilde{d}_j)_{j\in\mathscr{I}}, (\widetilde{e}_j)_{j\in\mathscr{I}}, (\lfloor\cdot\rfloor_j)_{j\in\mathscr{I}}\big)$ with $\widehat{\gamma}_j := \sup_{k\in\{1\dots K\}} \gamma_j(\widetilde{\vartheta}_k) < \infty$ for some $j \in \mathscr{I}$.*
For any $\widetilde{x}_0 \in \widetilde{E}$ and $0 = t_0 < t_1 < \dots < t_K$ with $\sup_k t_k - t_{k-1} \le 1$ define the curve $\widetilde{x}(\cdot) : [0, t_K] \longrightarrow \widetilde{E}$ piecewise as $\widetilde{x}(0) := \widetilde{x}_0$ and

$$\widetilde{x}(t) := \widetilde{\vartheta}_k\big(t - t_{k-1},\ \widetilde{x}(t_{k-1})\big) \qquad \text{for } t \in \big]t_{k-1}, t_k\big],\ k \in \{1 \dots K\}.$$

Then, $\lfloor\widetilde{x}(t)\rfloor_j \le \big(\lfloor\widetilde{x}_0\rfloor_j + \widehat{\gamma}_j \cdot t\big) \cdot e^{\widehat{\gamma}_j \cdot t}$ *at every time $t \in [0, t_K]$.* □

Due to the possible lack of symmetry of $\widetilde{d}_j$ ($j \in \mathscr{I}$), we now conclude from condition (8.) on timed transitions (in Definition 1) – instead of the global hypothesis (H3') about continuity of distance functions:

Lemma 5. *Let $\widetilde{x}(\cdot) : [0, T] \longrightarrow \widetilde{E}$ be any curve satisfying $\pi_1 \widetilde{x}(t) = t + \pi_1 x(0)$,*

$$\lim_{h\downarrow 0} \widetilde{e}_j(\widetilde{x}(t-h),\ \widetilde{x}(t)) = 0 \qquad \text{for every } t \in]0, T],\ j \in \mathscr{I}.$$

Choose any timed transition $\widetilde{\vartheta}$ on $\big(\widetilde{E}, \widetilde{\mathscr{D}}, (\widetilde{d}_j)_{j\in\mathscr{I}}, (\widetilde{e}_j)_{j\in\mathscr{I}}, (\lfloor\cdot\rfloor_j)_{j\in\mathscr{I}}\big)$, element $\widetilde{z} \in \widetilde{\mathscr{D}}$ and points of time $t_1 \in [0, \mathbb{T}_j(\widetilde{\vartheta}, \widetilde{z})[$, $t_2 \in [0, T[$ with $t_1 + \pi_1 \widetilde{z} \le \pi_1 \widetilde{x}(t_2)$.

Then each distance function

$$\begin{aligned} \big[0, \min\{\mathbb{T}_j(\widetilde{\vartheta}, \widetilde{z}) - t_1,\ T - t_2\}\big[&\longrightarrow [0, \infty[, \\ s &\longmapsto \widetilde{d}_j\big(\widetilde{\vartheta}(t_1 + s, \widetilde{z}),\ \widetilde{x}(t_2 + s)\big) \end{aligned}$$

($j \in \mathscr{I}$) fulfills the following condition of lower semicontinuity at every time s

$$\widetilde{d}_j\big(\widetilde{\vartheta}(t_1 + s, \widetilde{z}),\ \widetilde{x}(t_2 + s)\big) \le \liminf_{h\downarrow 0} \widetilde{d}_j\big(\widetilde{\vartheta}(t_1 + s - h, \widetilde{z}),\ \widetilde{x}(t_2 + s - h)\big). \qquad \square$$

Proposition 6. *Let $\widetilde{\vartheta}, \widetilde{\tau} \in \widetilde{\Theta}\big(E, \mathscr{D}, (\widetilde{d}_j)_{j\in\mathscr{I}}, (\widetilde{e}_j)_{j\in\mathscr{I}}, (\lfloor\cdot\rfloor_j)_{j\in\mathscr{I}}\big)$, $r \ge 0$, $j \in \mathscr{I}$ and $t_1, t_2 \in [0, 1[$ be arbitrary. For any elements $\widetilde{y} \in \widetilde{E}$ and $\widetilde{z} \in \widetilde{\mathscr{D}}$ suppose $\lfloor\widetilde{y}\rfloor_j \le r$, $t_1 \le \mathbb{T}_j(\widetilde{\vartheta}, \widetilde{z})$ and $t_1 + \pi_1 \widetilde{z} \le t_2 + \pi_1 \widetilde{y}$. Set $R_j := \big(r + \gamma_j(\tau)\big) \cdot e^{\gamma_j(\tau)} < \infty$.*

Then at each time $h \ge 0$ with $t_1 + h \le \mathbb{T}_j(\widetilde{\vartheta}, \widetilde{z})$ and $t_2 + h \le 1$,

$$\widetilde{d}_j\big(\widetilde{\vartheta}(t_1{+}h, \widetilde{z}),\ \widetilde{\tau}(t_2{+}h, \widetilde{y})\big) \le \Big(\widetilde{d}_j\big(\widetilde{\vartheta}(t_1, \widetilde{z}),\ \widetilde{\tau}(t_2, \widetilde{y})\big) + h \cdot \widehat{D}_j(\widetilde{\vartheta}, \widetilde{\tau}; \widetilde{z}, R_j)\Big)\ e^{\alpha_j(\widetilde{\tau}; \widetilde{z}, R_j)\, h}.$$

Proof. It is based on essentially the same arguments as corresponding Proposition 3.7 (on page 185), but now the rather weak regularity assumptions of Gronwall's inequality in Proposition A.2 (on page 440) are exploited to their full extent.
Consider the auxiliary function

$$\phi_j : \big[0, \min\{\mathbb{T}_j(\widetilde{\vartheta},\widetilde{z}) - t_1, 1 - t_2\}\big] \longrightarrow \mathbb{R}, \quad h \longmapsto \widetilde{d}_j\big(\widetilde{\vartheta}(t_1 + h, \widetilde{x}), \; \widetilde{\tau}(t_2 + h, \widetilde{y})\big).$$

Indeed, ϕ_j satisfies $\phi_j(t) \le \limsup_{h\downarrow 0} \phi_j(t-h)$ according to preceding Lemma 5. Furthermore condition (5.) of Definition 1 ensures $\lfloor \widetilde{\tau}(h,\widetilde{y}) \rfloor_j \le R_j$ for each $h \in [0,1]$ and due to condition (7.) on timed transitions,

$$\pi_1\, \widetilde{\vartheta}(t_1 + h, \widetilde{z}) \;\le\; t_1 + h + \pi_1 \widetilde{z} \;\le\; t_2 + h + \pi_1 \widetilde{y} \;=\; \pi_1\, \widetilde{\tau}(t_2 + h, \widetilde{y}).$$

Hypothesis (H7') about $\widehat{D}_j(\cdot,\cdot; R_j)$ (on page 337) implies for every t in the interior of the domain of ϕ_j

$$\begin{aligned}
&\phi_j(t+h) - \phi_j(t)\\
&= \widetilde{d}_j\big(\widetilde{\vartheta}(t_1 + t + h, \widetilde{z}), \; \widetilde{\tau}(t_2 + t + h, \widetilde{y})\big) \qquad - \widetilde{d}_j\big(\widetilde{\vartheta}(t_1 + t, \widetilde{z}), \; \widetilde{\tau}(t_2 + t, \widetilde{y})\big)\\
&\le \widetilde{d}_j\big(\widetilde{\vartheta}(t_1 + t + h, \widetilde{z}), \; \widetilde{\tau}(t_2 + t + h, \widetilde{y})\big) \qquad - \widetilde{d}_j\big(\widetilde{\vartheta}(t_1 + t, \widetilde{z}), \; \widetilde{\tau}(t_2 + t, \widetilde{y})\big)\, e^{\alpha_j(\widetilde{\tau};\widetilde{z},R_j)\,h}\\
&\quad + \widetilde{d}_j\big(\widetilde{\vartheta}(t_1 + t, \widetilde{z})), \; \widetilde{\tau}(t_2 + t, \widetilde{y}))\big) \cdot e^{\alpha_j(\widetilde{\tau};\widetilde{z},R_j)\,h} - \widetilde{d}_j\big(\widetilde{\vartheta}(t_1 + t, \widetilde{z}), \; \widetilde{\tau}(t_2 + t, \widetilde{y})\big)
\end{aligned}$$

and thus, $\displaystyle \limsup_{h\downarrow 0} \frac{\phi_j(t+h) - \phi_j(t)}{h} \;\le\; \widehat{D}_j(\widetilde{\vartheta}, \widetilde{\tau}; \widetilde{z}, R_j) \;+\; \alpha_j(\widetilde{\tau}; \widetilde{z}, R_j) \cdot \phi_j(t) \;<\; \infty.$

Finally, the claimed inequality results directly from Gronwall's inequality (in form of Proposition A.2). □

4.3 Timed Solutions to Mutational Equations

In comparison with Definition 3.35 of timed solutions (on page 222) in the mutational framework of Chapter 3, the essential differences are based on two aspects: First, the arguments of distances are sorted by time and second, only "test elements" of $\widetilde{\mathscr{D}}$ evolving along transitions are admissible for comparing distances.
This leads to the following definition:

Definition 7. Let $\widetilde{f} : \widetilde{E} \times [0,T] \longrightarrow \widehat{\Theta}\big(\widetilde{E}, (\widetilde{d}_j)_{j\in\mathscr{I}}, (\widetilde{e}_j)_{j\in\mathscr{I}}, (\lfloor\cdot\rfloor_j)_{j\in\mathscr{I}}\big)$ be given. A curve $\widetilde{x}(\cdot) : [0,T] \longrightarrow \widetilde{E}$ is called a *timed solution* to the mutational equation

$$\overset{\circ}{\widetilde{x}}(\cdot) \;\ni\; \widetilde{f}\big(\widetilde{x}(\cdot), \cdot\big)$$

in $\big(\widetilde{E}, \; \widetilde{\mathscr{D}}, \; (\widetilde{d}_j)_{j\in\mathscr{I}}, \; (\widetilde{e}_j)_{j\in\mathscr{I}}, \; (\lfloor\cdot\rfloor_j)_{j\in\mathscr{I}}, \; (\widehat{D}_j)_{j\in\mathscr{I}}\big)$ if it satisfies for each $j \in \mathscr{I}$:

1.′) $\widetilde{x}(\cdot)$ is continuous with respect to $\widetilde{e}_j$ in the sense that there exists a modulus of continuity $\omega_j(\widetilde{x};\cdot) : [0,\infty[\longrightarrow [0,\infty[$ with $\lim_{\rho\downarrow 0} \omega_j(\widetilde{x};\rho) = 0$ and

$$\widetilde{e}_j\big(\widetilde{x}(s),\, \widetilde{x}(t)\big) \;\le\; \omega_j(\widetilde{x},\, t-s) \qquad \text{for every } 0 \le s \le t \le T,$$

2.″) for each element $\widetilde{z} \in \widetilde{\mathscr{D}}$, there exists $\alpha_j(\widetilde{x};\widetilde{z},\cdot) : [0,\infty[\longrightarrow [0,\infty[$ such that for $\mathscr{L}^1$-a.e. $t \in [0,T[$:

$$\limsup_{h\downarrow 0} \frac{\widetilde{d}_j\big(\widetilde{\vartheta}(s+h,\widetilde{z}),\, \widetilde{x}(t+h)\big) - \widetilde{d}_j\big(\widetilde{\vartheta}(s,\widetilde{z}),\, \widetilde{x}(t)\big)\cdot e^{\alpha_j(\widetilde{x};\widetilde{z},R_j)\,h}}{h} \le \widehat{D}_j\big(\widetilde{\vartheta},\, \widetilde{f}(\widetilde{x}(t),t);\, \widetilde{z}, R_j\big)$$

for any $\widetilde{\vartheta} \in \widehat{\Theta}\big(\widetilde{E}, \widetilde{\mathscr{D}}, (\widetilde{d}_j), (\widetilde{e}_j), (\lfloor\cdot\rfloor_j)\big)$, $s \in \big[0,\, \mathbb{T}_j(\widetilde{\vartheta},\widetilde{z})\big[$ with $\lfloor\widetilde{x}(\cdot)\rfloor_j < R_j$ and $s + \pi_1\,\widetilde{z} \;\le\; \pi_1\,\widetilde{x}(t)$,

3.) $\sup_{t\in[0,T]} \;\lfloor\widetilde{x}(t)\rfloor_j \;<\; \infty$,

4.) for every $t \in [0,T]$, $\quad \pi_1\,\widetilde{x}(t) \;=\; \pi_1\,\widetilde{x}(0) \,+\, t$.

In combination with Lemma 5, the same arguments at $\mathscr{L}^1$-almost every time as for Proposition 6 (on page 339) lead to the following estimate:

Lemma 8 (comparing timed solution and curve in $\widetilde{\mathscr{D}}$ along transition). *Let $\widetilde{x}(\cdot) : [0,T] \longrightarrow \widetilde{E}$ be a timed solution to the mutational equation*

$$\overset{\circ}{\widetilde{x}}(\cdot) \;\ni\; \widetilde{f}\big(\widetilde{x}(\cdot),\, \cdot\big)$$

in the tuple $\big(\widetilde{E},\, \widetilde{\mathscr{D}},\, (\widetilde{d}_j)_{j\in\mathscr{I}},\, (\widetilde{e}_j)_{j\in\mathscr{I}},\, (\lfloor\cdot\rfloor_j)_{j\in\mathscr{I}},\, (\widehat{D}_j)_{j\in\mathscr{I}}\big)$.

Suppose $\widetilde{\vartheta} \in \widehat{\Theta}\big(\widetilde{E},\, \widetilde{\mathscr{D}},\, (\widetilde{d}_i)_{i\in\mathscr{I}}, (\widetilde{e}_i)_{i\in\mathscr{I}}, (\lfloor\cdot\rfloor_i)_i\big)$, $j \in \mathscr{I}$, $\widetilde{z} \in \widetilde{\mathscr{D}}$, $s \in \big[0,\, \mathbb{T}_j(\widetilde{\vartheta},\widetilde{z})\big[$, $t \in [0,T[$ to be arbitrary with $s + \pi_1\,\widetilde{z} \le \pi_1\,\widetilde{x}(t)$ and set $R_j := 1 + \sup \lfloor\widetilde{x}(\cdot)\rfloor_j < \infty$ as an abbreviation.

Then,

$$\widetilde{d}_j\big(\widetilde{\vartheta}(s+h,\widetilde{z}),\;\widetilde{x}(t+h)\big) \le \Big(\widetilde{d}_j\big(\widetilde{\vartheta}(s,\widetilde{z}),\, \widetilde{x}(t)\big) + h\cdot \sup_{[t,t+h]} \widehat{D}_j\big(\widetilde{\vartheta},\, \widetilde{f}(\widetilde{x}(\cdot),\cdot)\widetilde{z},\, R_j\big)\Big)\cdot e^{\alpha_j(\widetilde{x};\,\widetilde{z},\,R_j)\,h}$$

for every $h \in [0,1]$ with $s+h \le \mathbb{T}_j(\widetilde{\vartheta},\widetilde{z})$ and $t+h \le T$. □

4.3.1 Continuity with Respect to Initial States and Right-Hand Side

In § 3.3.1 (on page 189 f.), we suggested the auxiliary distance function

$$[0,T] \longrightarrow [0,\infty[, \quad t \longmapsto \inf_{z \in E:\ \lfloor z \rfloor_j < R_j} \big(d_j(z,x(t)) + d_j(z,y(t))\big)$$

for comparing two solutions $x(\cdot),\ y(\cdot) : [0,T] \longrightarrow E$ to mutational equations. For taking the separate time component into consideration, this proposal was modified in Proposition 3.40 (on page 224):

$$[0,T] \longrightarrow [0,\infty[, \quad t \longmapsto \inf \big\{\widetilde{d}_j(\widetilde{z},\widetilde{x}(t)) + \widetilde{d}_j(\widetilde{z},\widetilde{y}(t)) \;\big|\; \widetilde{z} \in \widetilde{E}:\ \lfloor \widetilde{z} \rfloor_j < R_j\big\}.$$

Now we have to obey in addition that arguments of distances are sorted by time and that timed solutions are characterized by comparing with evolving test elements of $\widetilde{\mathscr{D}}$ shortly. Thus, it is plausible to consider the auxiliary distance function

$$t \longmapsto \inf \big\{\widetilde{d}_j(\widetilde{z},\widetilde{x}(t)) + \widetilde{d}_j(\widetilde{z},\widetilde{y}(t)) \;\big|\; \widetilde{z} \in \widetilde{\mathscr{D}}: \lfloor \widetilde{z} \rfloor_j < R_j, \quad \pi_1\,\widetilde{z} < \min\{\pi_1\,\widetilde{x}(t),\ \pi_1\,\widetilde{y}(t)\}\big\}.$$

This infimum at time $t \in [0,T[$ is approximated by a minimal sequence $(\widetilde{z}_n)_{n\in\mathbb{N}}$ in $\widetilde{\mathscr{D}}$ whose elements evolve along the transition $\widetilde{f}\big(\widetilde{x}(t),t\big)$ characterizing $\widetilde{x}(t+\cdot)$.

An additional assumption about its time parameters $\mathbb{T}_j\big(\widetilde{f}\big(\widetilde{x}(t),t\big),\ \widetilde{z}_n\big)$, $n \in \mathbb{N}$, however, is required so that we can compare the evolutions for a sufficiently long time. Indeed, without such a lower bound providing a form of uniformity, the typical approach to a global estimate by means of Gronwall's inequality might fail because two limit processes are exchanged.
The detailed analysis leads to the following versions:

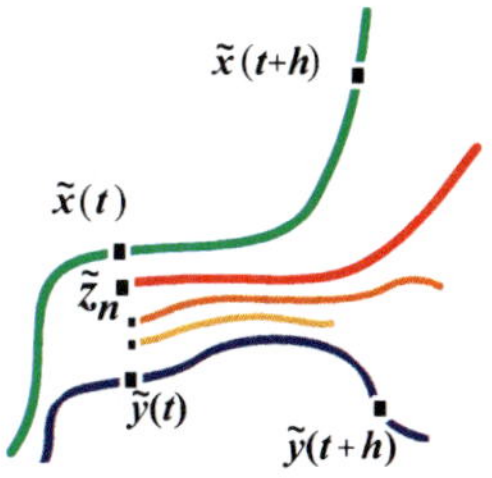

Proposition 9.
Assume for $\widetilde{f},\widetilde{g} : \widetilde{E} \times [0,T] \longrightarrow \widehat{\Theta}\big(\widetilde{E},\widetilde{\mathscr{D}},(\widetilde{d}_j)_j,(\widetilde{e}_j)_j,(\lfloor\cdot\rfloor_j)_j\big)$ *and* $\widetilde{x},\widetilde{y} : [0,T] \longrightarrow \widetilde{E}$ *that* $\widetilde{x}(\cdot)$ *is a timed solution to the mutational equation* $\overset{\circ}{\widetilde{x}}(\cdot) \ni \widetilde{f}(\widetilde{x}(\cdot),\cdot)$ *and*

$\widetilde{y}(\cdot)$ *is a timed solution to the mutational equation* $\overset{\circ}{\widetilde{y}}(\cdot) \ni \widetilde{g}(\widetilde{y}(\cdot),\cdot)$

in the tuple $\big(\widetilde{E},\ \widetilde{\mathscr{D}},\ (\widetilde{d}_j)_{j\in\mathscr{I}},\ (\widetilde{e}_j)_{j\in\mathscr{I}},\ (\lfloor\cdot\rfloor_j)_{j\in\mathscr{I}},\ (\widehat{D}_j)_{j\in\mathscr{I}}\big)$.

For some $j \in \mathscr{I}$, *let* $\widehat{\alpha}_j, \widehat{\gamma}_j, R_j > 0$ *and* $\varphi_j \in C^0([0,T])$ *satisfy for every* $t \in [0,T]$

$$\begin{cases} & & \lfloor \widetilde{x}(t) \rfloor_j,\ \lfloor \widetilde{y}(t) \rfloor_j & < & R_j \\ & \sup\limits_{\widetilde{z}\in\widetilde{\mathscr{D}}:\ \lfloor \widetilde{z} \rfloor_j < R_j} & \big\{\alpha_j(\widetilde{x};\ \widetilde{z},\ R_j),\ \ \alpha_j(\widetilde{y};\ \widetilde{z},\ R_j)\big\} & \le & \widehat{\alpha}_j \\ & & \gamma_j\big(\widetilde{f}(\widetilde{x}(t),t)\big) & \le & \widehat{\gamma}_j \\ \limsup\limits_{h\downarrow 0} & \sup\limits_{\widetilde{z}\in\widetilde{\mathscr{D}}:\ \lfloor \widetilde{z} \rfloor_j < R_j} & \widehat{D}_j\Big(\widetilde{f}(\widetilde{x}(t),t),\ \ \widetilde{g}(\widetilde{y}(t+h),t+h);\ \widetilde{z},\ R_j\Big) & \le & \varphi_j(t) \\ \limsup\limits_{h\downarrow 0} & \sup\limits_{\widetilde{z}\in\widetilde{\mathscr{D}}:\ \lfloor \widetilde{z} \rfloor_j < R_j} & \widehat{D}_j\Big(\widetilde{f}(\widetilde{x}(t),t),\ \ \widetilde{f}(\widetilde{x}(t+h),t+h);\ \widetilde{z},\ R_j\Big) & = & 0 \end{cases}$$

For some $\widetilde{\vartheta} \in \widehat{\Theta}\big(\widetilde{E}, \widetilde{\mathscr{D}}, (\widetilde{d}_i)_i, (\widetilde{e}_i)_i, (\lfloor\cdot\rfloor_i)_i\big)$ *assume* $\inf_{\widetilde{z} \in \widetilde{\mathscr{D}}:\, \lfloor\widetilde{z}\rfloor_j < R_j} \mathbb{T}_j(\widetilde{\vartheta}, \widetilde{z}) > 0$ *and*

$$\sup_{t \in [0,T]} \sup_{\widetilde{z} \in \widetilde{\mathscr{D}}:\, \lfloor\widetilde{z}\rfloor_j < R_j} \widehat{D}_j\left(\widetilde{\vartheta},\ \widetilde{f}(\widetilde{x}(t),t);\ \widetilde{z},\ R_j\right) < \infty.$$

Considering the distance function

$$\begin{aligned} \delta_j : [0,T] &\longrightarrow [0,\infty[, \\ t &\longmapsto \inf\big\{\widetilde{d}_j\big(\widetilde{z},\widetilde{x}(t)\big) + \widetilde{d}_j\big(\widetilde{z},\widetilde{y}(t)\big) \,\big|\, \widetilde{z} \in \widetilde{\mathscr{D}}:\ \lfloor\widetilde{z}\rfloor_j < R_j, \\ &\qquad \pi_1\,\widetilde{z} < \min\{\pi_1\,\widetilde{x}(t),\ \pi_1\,\widetilde{y}(t)\}\big\}, \end{aligned}$$

suppose at Lebesgue-almost every time $t \in [0,T]$ *that the infimum of* $\delta_j(t)$ *can be approximated by a minimal sequence* $(\widetilde{z}_n)_{n\in\mathbb{N}}$ *in* $\widetilde{\mathscr{D}}$ *satisfying*

$$\begin{aligned} \sup\nolimits_n \lfloor\widetilde{z}_n\rfloor_j &< R_j, \\ \pi_1\,\widetilde{z}_n \le \pi_1\,\widetilde{z}_{n+1} &< \min\big\{\pi_1\,\widetilde{x}(t),\ \pi_1\,\widetilde{y}(t)\big\} \quad \text{for every } n \in \mathbb{N}, \\ \inf_{n\in\mathbb{N}} \mathbb{T}_j(\widetilde{f}(\widetilde{x}(t),t),\ \widetilde{z}_n) &> 0. \end{aligned}$$

Then, $$\delta_j(t) \le \Big(\delta_j(0) + \int_0^t \varphi_j(s)\, e^{-\widehat{\alpha}_j\cdot s}\, ds\Big)\, e^{\widehat{\alpha}_j\cdot t} \qquad \text{for every } t \in [0,T].$$

Proposition 10.

Let $\widetilde{f},\widetilde{g} : \widetilde{E} \times [0,T] \longrightarrow \widehat{\Theta}\big(\widetilde{E}, \widetilde{\mathscr{D}}, (\widetilde{d}_j)_j, (\widetilde{e}_j)_j, (\lfloor\cdot\rfloor_j)_j\big)$, $\widetilde{x},\widetilde{y} : [0,T] \longrightarrow \widetilde{E}$, $j \in \mathscr{I}$, $\widehat{\alpha}_j, \widehat{\gamma}_j, R_j > 0$ *and* $\varphi_j \in C^0([0,T])$ *fulfill the same assumptions as in Proposition 9.*

Considering the same distance function

$$\begin{aligned} \delta_j : [0,T] &\longrightarrow [0,\infty[, \\ t &\longmapsto \inf\big\{\widetilde{d}_j\big(\widetilde{z},\widetilde{x}(t)\big) + \widetilde{d}_j\big(\widetilde{z},\widetilde{y}(t)\big) \,\big|\, \widetilde{z} \in \widetilde{\mathscr{D}}:\ \lfloor\widetilde{z}\rfloor_j < R_j, \\ &\qquad \pi_1\,\widetilde{z} < \min\{\pi_1\,\widetilde{x}(t),\ \pi_1\,\widetilde{y}(t)\}\big\}, \end{aligned}$$

suppose at every *time* $t \in [0,T]$ *that the infimum of* $\delta_j(t)$ *can be approximated by a minimal sequence* $(\widetilde{z}_n)_{n\in\mathbb{N}}$ *in* $\widetilde{\mathscr{D}}$ *satisfying*

$$\begin{aligned} \sup\nolimits_n \lfloor\widetilde{z}_n\rfloor_j &< R_j, \\ \pi_1\,\widetilde{z}_n \le \pi_1\,\widetilde{z}_{n+1} &< \min\big\{\pi_1\,\widetilde{x}(t),\ \pi_1\,\widetilde{y}(t)\big\} \quad \text{for every } n \in \mathbb{N}, \\ \frac{\widetilde{d}_j(\widetilde{z}_n, \widetilde{x}(t)) + \widetilde{d}_j(\widetilde{z}_n, \widetilde{y}(t)) - \delta_j(t)}{\mathbb{T}_j(\widetilde{f}(\widetilde{x}(t),t),\ \widetilde{z}_n)} &\longrightarrow 0 \quad \text{for } n \longrightarrow \infty. \end{aligned}$$

Furthermore assume the local equi-continuity of the distance family

$$\widetilde{d}_j(\widetilde{z},\cdot) :\ ]\pi_1\,\widetilde{z},\ \infty[\ \times E \longrightarrow \mathbb{R} \qquad (\widetilde{z} \in \widetilde{\mathscr{D}},\ \lfloor\widetilde{z}\rfloor_j < R_j)$$

in the following sense: Every sequence $(\widetilde{\xi}_n)_{n\in\mathbb{N}}$ *in* $\widetilde{E}$ *and element* $\widetilde{\xi} \in \widetilde{E}$ *with* $\lim_{n\to\infty} \widetilde{e}_i(\widetilde{\xi}_n, \widetilde{\xi}) = 0$ *for each* $i \in \mathscr{I}$ *and* $\pi_1\,\widetilde{\xi}_n \le \pi_1\,\widetilde{\xi}_{n+1} \nearrow \pi_1\,\widetilde{\xi}$ *for* $n \longrightarrow \infty$ *have the asymptotic property*

$$\lim_{n\to\infty} \sup\Big\{\widetilde{d}_j(\widetilde{z},\ \widetilde{\xi}) - \widetilde{d}_j(\widetilde{z},\ \widetilde{\xi}_n) \,\Big|\, \widetilde{z} \in \widetilde{\mathscr{D}}:\ \pi_1\,\widetilde{z} < \pi_1\,\widetilde{\xi}_n,\ \lfloor\widetilde{z}\rfloor_j < R_j\Big\} = 0.$$

Then, $$\delta_j(t) \le \Big(\delta_j(0) + \int_0^t \varphi_j(s)\, e^{-\widehat{\alpha}_j\cdot s}\, ds\Big)\, e^{\widehat{\alpha}_j\cdot t} \qquad \text{for every } t \in [0,T].$$

Remark 11. On the basis of Remark 2 (i) (on page 336), Proposition 9 implies the estimates of Propositions 3.11 and 3.40 (on pages 189, 224) as special cases.

Advantageously, Proposition 10 dispenses with supposing a positive bound of the time parameters like $\mathbb{T}_j(\widetilde{f}(\widetilde{x}(t),t), \widetilde{z}_n)$, but it makes assumptions about the relative features of $\mathbb{T}_j(\widetilde{f}(\widetilde{x}(t),t), \widetilde{z}_n)$ and $\widetilde{d}_j(\widetilde{z}_n, \widetilde{x}(t)) + \widetilde{d}_j(\widetilde{z}_n, \widetilde{y}(t)) - \delta_j(t)$ for $n \to \infty$. This conclusion, however, results from another semicontinuous version of Gronwall's inequality specified in Proposition A.4 (on page 442) and thus, it requires further assumptions about the equi-continuity of $\widetilde{d}_j(\widetilde{z},\cdot) : \widetilde{E} \longrightarrow \mathbb{R}$ $(\widetilde{z} \in \widetilde{\mathscr{D}},\ \lfloor \widetilde{z} \rfloor_j < R_j)$. Note that the *timed triangle inequality* of $\widetilde{d}_j(\cdot,\cdot)$, i.e.

$$\widetilde{d}_j(\widetilde{u},\widetilde{w}) \ \leq\ \widetilde{d}_j(\widetilde{u},\widetilde{v}) \ +\ \widetilde{d}_j(\widetilde{v},\widetilde{w})$$

whenever $\widetilde{u},\widetilde{v},\widetilde{w} \in \widetilde{E}$ satisfy $\pi_1\, \widetilde{u} \ \leq\ \pi_1\, \widetilde{v} \ \leq\ \pi_1\, \widetilde{w}$, is always sufficient for this supplementary hypothesis.

Proof (of Proposition 9). It is based on the same notion as Proposition 3.11. Choosing a timed transition $\widetilde{\vartheta}$ with $\tau_{\widetilde{\vartheta}} := \inf_{\widetilde{z} \in \widetilde{\mathscr{D}}: \lfloor \widetilde{z} \rfloor_j < R_j} \mathbb{T}_j(\widetilde{\vartheta},\widetilde{z}) > 0$ and

$$\sup_{t\in[0,T]}\ \sup_{\widetilde{z}\in\widetilde{\mathscr{D}}:\,\lfloor\widetilde{z}\rfloor_j<R_j}\ \widehat{D}_j\Big(\widetilde{\vartheta},\ \widetilde{f}(\widetilde{x}(t),t);\ \widetilde{z},\ R_j\Big) \ <\ \infty,$$

Lemma 8 (on page 340) provides a constant $C = C(t, j, \widetilde{f}, \widehat{\alpha}_j) < \infty$ for each $t \in]0,T[$ such that for every $h \in]0,\ \tau_{\widetilde{\vartheta}}[$ and $\widetilde{z} \in \widetilde{\mathscr{D}}$ with $h + \pi_1\, \widetilde{z} \ <\ \min\{\pi_1\, \widetilde{x}(t),\ \pi_1\, \widetilde{y}(t)\}$, $\lfloor \widetilde{z} \rfloor_i < R_i$, the following estimates hold

$$\begin{cases} \widetilde{d}_j\big(\widetilde{\vartheta}(h,\widetilde{z}),\ \widetilde{x}(t)\big) \ \leq\ \big(\widetilde{d}_j\big(\widetilde{z},\ \widetilde{x}(t-h)\big) \ +\ C\,h\big) \cdot e^{C\,h} \\ \widetilde{d}_j\big(\widetilde{\vartheta}(h,\widetilde{z}),\ \widetilde{y}(t)\big) \ \leq\ \big(\widetilde{d}_j\big(\widetilde{z},\ \widetilde{y}(t-h)\big) \ +\ C\,h\big) \cdot e^{C\,h}. \end{cases}$$

Due to property (5.) of timed transitions, it implies $\delta(t) \ \leq\ \limsup_{h\downarrow 0}\ \delta_j(t-h)$.

At $\mathscr{L}^1$-a.e. time $t \in [0,T[$, we can choose a sequence $(\widetilde{z}_n)_{n\in\mathbb{N}}$ in $\widetilde{\mathscr{D}}$ and $\tau > 0$ with

$$\begin{cases} \sup_n\ \lfloor \widetilde{z}_n \rfloor_j \ <\ R_j, \\ \pi_1\, \widetilde{z}_n \ \leq\ \pi_1\, \widetilde{z}_{n+1} \ <\ \min\big\{\pi_1\, \widetilde{x}(t),\ \ \pi_1\, \widetilde{y}(t)\big\}, \\ \mathbb{T}_j(\widetilde{f}(\widetilde{x}(t),t), \widetilde{z}_n) \ \geq\ \tau. \end{cases}$$

Lemma 8 (on page 340) implies for each $n \in \mathbb{N}$ and $h \in \big[0,\ \mathbb{T}_j(\widetilde{f}(\widetilde{x}(t),t), \widetilde{z}_n)\big[$

$$\begin{aligned} &\widetilde{d}_j\big(\widetilde{f}(\widetilde{x}(t),t)\,(h,\widetilde{z}_n),\ \ \widetilde{x}(t+h)\big) \\ &\quad \leq\ \Big(\widetilde{d}_j\big(\widetilde{z}_n,\ \widetilde{x}(t)\big) \ +\ h \cdot \sup_{[t,t+h]}\ \widehat{D}_j\big(\widetilde{f}(\widetilde{x}(t),t),\ \widetilde{f}(\widetilde{x}(\cdot),\cdot);\ \widetilde{z}_n, R_j\big)\Big) \cdot e^{\widehat{\alpha}_j\,h} \end{aligned}$$

and

$$\begin{aligned} &\widetilde{d}_j\big(\widetilde{f}(\widetilde{x}(t),t)\,(h,\widetilde{z}_n),\ \ \widetilde{y}(t+h)\big) \\ &\quad \leq\ \Big(\widetilde{d}_j\big(\widetilde{z}_n,\ \widetilde{y}(t)\big) \ +\ h \cdot \sup_{[t,t+h]}\ \widehat{D}_j\big(\widetilde{f}(\widetilde{x}(t),t),\ \widetilde{g}(\widetilde{y}(\cdot),\cdot);\ \widetilde{z}_n, R_j\big)\Big) \cdot e^{\widehat{\alpha}_j\,h}. \end{aligned}$$

Hence, we obtain an upper bound of

$$\delta_j(t+h) \ \leq\ \widetilde{d}_j\big(\widetilde{f}(\widetilde{x}(t),t)\,(h,\widetilde{z}_n),\ \widetilde{x}(t+h)\big) \ +\ \widetilde{d}_j\big(\widetilde{f}(\widetilde{x}(t),t)\,(h,\widetilde{z}_n),\ \widetilde{y}(t+h)\big)$$

for every $h \in [0, \tau[\subset \big[0, \inf_m \mathbb{T}_j(\widetilde{f}(\widetilde{x}(t),t), \widetilde{z}_m)\big[$ and, $n \longrightarrow \infty$ leads to

$$\delta_j(t+h) \le \Big(\delta_j(t) + h \cdot \sup_{[t,t+h]} \sup_{\widetilde{z}\in\widetilde{\mathscr{D}}} \widehat{D}_j\big(\widetilde{f}(\widetilde{x}(t),t),\ \widetilde{f}(\widetilde{x}(\cdot),\cdot);\ \widetilde{z}, R_j\big)$$
$$+ h \cdot \sup_{[t,t+h]} \sup_{\widetilde{z}\in\widetilde{\mathscr{D}}} \widehat{D}_j\big(\widetilde{f}(\widetilde{x}(t),t),\ \widetilde{g}(\widetilde{y}(\cdot),\cdot);\ \widetilde{z}, R_j\big)\Big)\ e^{\widehat{\alpha}_j h}.$$

Thus,

$$\limsup_{h\downarrow 0} \frac{\delta_j(t+h) - \delta_j(t)}{h} \le \widehat{\alpha}_j \cdot \delta_j(t) + 0 + \varphi_j(t) < \infty.$$

Finally Gronwall's inequality in Proposition A.2 (on page 440) implies the claim. □

Proof (of Proposition 10). It draws conclusions very similarly to the preceding proof of Proposition 9, but cannot rely on uniform positive bounds of the transition parameter $\mathbb{T}_j(\cdot,\cdot)$. For this reason, it uses the modified Gronwall's inequality in Proposition A.4 (on page 442) for the first time so far.

Choosing any sequence $h_n \downarrow 0$, the assumption about local equi-continuity of $\widetilde{d}_j(\widetilde{z},\cdot)$ ensures for every $t \in]0,T[$

$$\begin{cases} \lim\limits_{n\to\infty} \sup\limits_{\widetilde{z}\in\widetilde{\mathscr{D}}} \Big\{\widetilde{d}_j(\widetilde{z},\ \widetilde{x}(t)) - \widetilde{d}_j(\widetilde{z},\ \widetilde{x}(t-h_n))\ \Big|\ \pi_1\,\widetilde{z} < \pi_1\,\widetilde{x}(t) - h_n,\ \lfloor \widetilde{z}\rfloor_j < R_j\Big\} = 0 \\ \lim\limits_{n\to\infty} \sup\limits_{\widetilde{z}\in\widetilde{\mathscr{D}}} \Big\{\widetilde{d}_j(\widetilde{z},\ \widetilde{y}(t)) - \widetilde{d}_j(\widetilde{z},\ \widetilde{y}(t-h_n))\ \Big|\ \pi_1\,\widetilde{z} < \pi_1\,\widetilde{y}(t) - h_n,\ \lfloor \widetilde{z}\rfloor_j < R_j\Big\} = 0 \end{cases}$$

and, it implies $\delta_j(t) \le \liminf\limits_{h\downarrow 0} \delta_j(t-h)$ for every $t \in]0,T[$.

At *every* time $t \in [0,T[$, we can choose a sequence $(\widetilde{z}_n)_{n\in\mathbb{N}}$ in $\widetilde{\mathscr{D}}$ with

$$\begin{cases} \sup_n \lfloor \widetilde{z}_n\rfloor_j < R_j, \\ \pi_1\,\widetilde{z}_n \le \pi_1\,\widetilde{z}_{n+1} < \min\big\{\pi_1\,\widetilde{x}(t),\ \pi_1\,\widetilde{y}(t)\big\}, \\ \widetilde{d}_j(\widetilde{z}_n, \widetilde{x}(t)) + \widetilde{d}_j(\widetilde{z}_n, \widetilde{y}(t)) - \delta_j(t) \le \frac{1}{n^2} \cdot \mathbb{T}_j(\widetilde{f}(\widetilde{x}(t),t), \widetilde{z}_n). \end{cases}$$

In exactly the same way as for Proposition 9, Lemma 8 (on page 340) provides an upper bound of

$$\delta_j(t+h) \le \widetilde{d}_j\big(\widetilde{f}(\widetilde{x}(t),t)\,(h,\widetilde{z}_n),\ \widetilde{x}(t+h)\big) + \widetilde{d}_j\big(\widetilde{f}(\widetilde{x}(t),t)\,(h,\widetilde{z}_n),\ \widetilde{y}(t+h)\big)$$

for every $h \in \big[0,\ \mathbb{T}_j(\widetilde{f}(\widetilde{x}(t),t), \widetilde{z}_n)\big[$ now still depending on $n \in \mathbb{N}$ though:

$$\delta_j(t+h) \le \Big(\delta_j(t) + \frac{\mathbb{T}_j(\widetilde{f}(\widetilde{x}(t),t), \widetilde{z}_n)}{n^2} + h \cdot \sup_{[t,t+h]} \widehat{D}_j\big(\widetilde{f}(\widetilde{x}(t),t),\ \widetilde{f}(\widetilde{x}(\cdot),\cdot);\ \widetilde{z}_n, R_j\big)$$
$$+ h \cdot \sup_{[t,t+h]} \widehat{D}_j\big(\widetilde{f}(\widetilde{x}(t),t),\ \widetilde{g}(\widetilde{y}(\cdot),\cdot);\ \widetilde{z}_n, R_j\big)\Big)\ e^{\widehat{\alpha}_j h}.$$

Setting $h := \frac{\mathbb{T}_j(\widetilde{f}(\widetilde{x}(t),t), \widetilde{z}_n)}{n} \le \frac{1}{n}$ for each $n \in \mathbb{N}$ respectively, the assumptions about $(\widetilde{z}_n)_{n\in\mathbb{N}}$ ensure for $n \longrightarrow \infty$

$$\liminf_{h\downarrow 0} \frac{\delta_j(t+h) - \delta_j(t)}{h} \le \widehat{\alpha}_j \cdot \delta_j(t) + 0 + \varphi_j(t) < \infty.$$

Gronwall's inequality in Proposition A.4 (on page 442) bridges the gap to the claimed bound for every $t \in [0,T]$. □

4.3.2 Convergence of Timed Solutions

In spite of all the conceptual generalizations presented in Chapter 4 so far, the characterization of timed solutions is stable with respect to the same type of graphical convergence as in § 3.3.2 (on page 190 ff.) and § 3.5 (on page 221 ff.).
The following theorem lays the foundations for constructing timed solutions to initial value problems by means of Euler approximations in the next section.

Theorem 12 (Convergence of timed solutions to mutational equations).
Suppose the following properties of

$$\begin{array}{ll} \widetilde{f}_n, \widetilde{f} : \widetilde{E} \times [0,T] \longrightarrow \widehat{\Theta}\big(\widetilde{E}, \widetilde{\mathscr{D}}, (\widetilde{d}_i)_{i\in\mathscr{I}}, (\widetilde{e}_j)_{j\in\mathscr{I}}, (\lfloor\cdot\rfloor_i)_{i\in\mathscr{I}}\big) & (n \in \mathbb{N}) \\ \widetilde{x}_n, \widetilde{x} : \quad [0,T] \longrightarrow \widetilde{E} : & \end{array}$$

1.) $R_j := \sup_{n,t} \lfloor \widetilde{x}_n(t) \rfloor_j + 1 < \infty,$

$\widehat{\alpha}_j(\widetilde{z},\rho) := \sup_n \alpha_j\big(\widetilde{x}_n; \widetilde{z}, \rho\big) < \infty$ *for each* $\widetilde{z} \in \widetilde{\mathscr{D}}, \rho \geq 0,$

$\widehat{\beta}_j := \sup_n \operatorname{Lip}\big(\widetilde{x}_n(\cdot) : [0,T] \longrightarrow (\widetilde{E}, \widetilde{e}_j)\big) < \infty$ *for every* $j \in \mathscr{I}$.

2.) $\overset{\circ}{\widetilde{x}}_n(\cdot) \ni \widetilde{f}_n(\widetilde{x}_n(\cdot), \cdot)$ *(in the sense of Definition 7 on page 339) for every n.*

3.) *Equi-continuity of* $(\widetilde{f}_n)_n$ *at* $(\widetilde{x}(t),t)$ *at almost every time in the following sense: for any* $\widetilde{z} \in \widetilde{\mathscr{D}}$ *and* $\mathscr{L}^1$*-a.e.* $t \in [0,T]$: $\lim_{n\to\infty} \widehat{D}_j\big(\widetilde{f}_n(\widetilde{x}(t),t),\ \widetilde{f}_n(\widetilde{y}_n,t_n);\ \widetilde{z},\ r\big) = 0$ *for each* $j \in \mathscr{I}$, $r \geq 0$ *and any* $(t_n)_{n\in\mathbb{N}}$, $(\widetilde{y}_n)_{n\in\mathbb{N}}$ *in* $[t,T]$ *and* $\widetilde{E}$ *respectively satisfying* $\lim_{n\to\infty} t_n = t$ *and* $\lim_{n\to\infty} \widetilde{d}_i\big(\widetilde{x}(t), \widetilde{y}_n\big) = 0$, $\sup_{n\in\mathbb{N}} \lfloor \widetilde{y}_n \rfloor_i \leq R_i$ *for each i,*

$$\pi_1\, \widetilde{y}_n \searrow \pi_1\, \widetilde{x}(t) \qquad \textit{for } n \longrightarrow \infty.$$

4.) *For* $\mathscr{L}^1$*-almost every* $s \in [0,T[$ *and any* $t < t'$ *in* $[0,T]$, *there is a sequence* $n_m \nearrow \infty$ *of indices (depending on* s,t,t'*) that satisfies for* $m \longrightarrow \infty$

(i) $\widehat{D}_j\big(\widetilde{f}(\widetilde{x}(s),s),\ \widetilde{f}_{n_m}(\widetilde{x}(s),s);\ \widetilde{z},\ r\big) \longrightarrow 0$ *for all* $\widetilde{z} \in \widetilde{\mathscr{D}}$, $r \geq 0$, $j \in \mathscr{I}$,

(ii) $\exists\ \delta_m \searrow 0 :\ \forall\, j : \widetilde{d}_j\big(\widetilde{x}(t),\ \widetilde{x}_{n_m}(t+\delta_m)\big) \longrightarrow 0,\ \pi_1\, \widetilde{x}_{n_m}(t+\delta_m) \searrow \pi_1\, \widetilde{x}(t)$

(iii) $\exists\ \widetilde{\delta}_m \searrow 0 :\ \forall\, j : \widetilde{d}_j\big(\widetilde{x}_{n_m}(t'-\widetilde{\delta}_m), \widetilde{x}(t')\big) \longrightarrow 0,\ \pi_1\, \widetilde{x}_{n_m}(t'-\widetilde{\delta}_m) \nearrow \pi_1\, \widetilde{x}(t')$

Then, $\widetilde{x}(\cdot)$ *is always a timed solution to the mutational equation* $\overset{\circ}{\widetilde{x}}(\cdot) \ni \widetilde{f}(\widetilde{x}(\cdot), \cdot)$ *in the tuple* $\big(\widetilde{E}, \widetilde{\mathscr{D}}, (\widetilde{d}_j)_{j\in\mathscr{I}}, (\widetilde{e}_j)_{j\in\mathscr{I}}, (\lfloor\cdot\rfloor_j)_{j\in\mathscr{I}}, (\widehat{D}_j)_{j\in\mathscr{I}}\big)$.

Remark 13. In comparison with Convergence Theorem 3.13 and its timed counterpart (i.e. Theorem 3.42 on page 225), assumption (4.) is slightly stronger because convergence property (ii) is now supposed for *every* $t \in [0,T[$. This modification is required for proving the Lipschitz continuity of $\widetilde{x}(\cdot)$ w.r.t. each $\widetilde{e}_j$.
It is caused by two differences in general assumptions between Chapter 4 and § 3.5: First, $\widetilde{d}_j, \widetilde{e}_j$ do not have to be symmetric. Second, hypothesis (H3') (i') (on page 334) considers only sequences with a stronger condition on their sorted time components than hypothesis $(\widetilde{\text{H3}})$ (i') (on page 221).

Proof. In comparison with the proof of Theorem 3.13 (on page 191 ff.), we just have to take two key aspects into consideration properly: Arguments of distances are sorted by time and, timed solutions are characterized by means of comparisons with evolving earlier test elements of $\widetilde{\mathscr{D}}$.
For the sake of transparency, the analogous formulation is to underline the parallels.

Choose the index $j \in \mathscr{I}$ arbitrarily.
Then $\widetilde{x}(\cdot) : [0,T] \longrightarrow (\widetilde{E}, \widetilde{e}_j)$ is $\widehat{\beta}_j$-Lipschitz continuous. Indeed, for any $t < t'$ in $[0,T]$, assumption (4.) provides a subsequence $\big(\widetilde{x}_{n_m}(\cdot)\big)_{m\in\mathbb{N}}$ and sequences $\delta_m \searrow 0$, $\widetilde{\delta}_m \searrow 0$ satisfying for each index $i \in \mathscr{I}$

$$\left\{\begin{array}{lll} \widetilde{d}_i\big(\widetilde{x}(t), \quad \widetilde{x}_{n_m}(t+\delta_m)\big) \longrightarrow 0, & \pi_1\, \widetilde{x}_{n_m}(t+\delta_m) \searrow \pi_1\, \widetilde{x}(t) & \\ \widetilde{d}_i\big(\widetilde{x}_{n_m}(t'-\widetilde{\delta}_m), \widetilde{x}(t')\big) \longrightarrow 0, & \pi_1\, \widetilde{x}_{n_m}(t'-\widetilde{\delta}_m) \nearrow \pi_1\, \widetilde{x}(t') & \end{array}\right. \quad \text{for } m \longrightarrow \infty.$$

Firstly, we conclude $\pi_1\, \widetilde{x}(t') = t' - t + \pi_1\, \widetilde{x}(t) = \pi_1\, \widetilde{x}_{n_m}(t')$ for each $m \in \mathbb{N}$. Secondly, the uniform $\widehat{\beta}_j$-Lipschitz continuity of $\widetilde{x}_n(\cdot), n \in \mathbb{N}$, with respect to $\widetilde{e}_j$ and hypothesis (H3') $\widetilde{\text{(i')}}$ (on page 334) imply

$$\begin{aligned} \widetilde{e}_j\big(\widetilde{x}(t),\, \widetilde{x}(t')\big) &\le \limsup_{m\to\infty}\ \widetilde{e}_j\big(\widetilde{x}_{n_m}(t+\delta_m),\ \widetilde{x}_{n_m}(t'-\widetilde{\delta}_m)\big) \\ &\le \limsup_{m\to\infty}\ \widehat{\beta}_j\, |t'-\widetilde{\delta}_m - t - \delta_m| \\ &\le \widehat{\beta}_j\, |t'-t|\,. \end{aligned}$$

Moreover, hypothesis (H4) about the lower semicontinuity of $\lfloor\cdot\rfloor_j$ ensures

$$\lfloor \widetilde{x}(t')\rfloor_j \le \liminf_{m\to\infty} \lfloor \widetilde{x}_{n_m}(t'-\widetilde{\delta}_m)\rfloor_j \le R_j - 1.$$

Finally we verify the solution property

$$\limsup_{h\downarrow 0} \frac{\widetilde{d}_j\big(\widetilde{\vartheta}(s+h,\widetilde{z}),\, \widetilde{x}(t+h)\big) - \widetilde{d}_j\big(\widetilde{\vartheta}(s,\widetilde{z}),\, \widetilde{x}(t)\big)\cdot e^{\alpha_j(\widetilde{x};\rho)\,h}}{h} \le \widehat{D}_j\big(\widetilde{\vartheta},\ \widetilde{f}(\widetilde{x}(t),t);\ \widetilde{z},\, R_j\big)$$

for $\mathscr{L}^1$-almost every $t \in [0,T[$ and any $\widetilde{\vartheta} \in \widehat{\Theta}\big(\widetilde{E}, \widetilde{\mathscr{D}}, (\widetilde{d}_i)_{i\in\mathscr{I}}, (\widetilde{e}_i)_{i\in\mathscr{I}}, (\lfloor\cdot\rfloor_i)_{i\in\mathscr{I}}\big)$, $\widetilde{z} \in \widetilde{\mathscr{D}}$, $s \in [0, \mathbb{T}_j(\widetilde{\vartheta},\widetilde{z})[$ with $s + \pi_1\,\widetilde{z} \le \pi_1\,\widetilde{x}(t)$.
Indeed, for Lebesgue-almost every $t \in [0,T[$ and any $h \in]0, T-t[$, assumption (4.) guarantees a subsequence $\big(\widetilde{x}_{n_m}(\cdot)\big)_{m\in\mathbb{N}}$ and sequences $\delta_m \searrow 0$, $\widetilde{\delta}_m \searrow 0$ satisfying for each $\widetilde{z} \in \widetilde{\mathscr{D}}$, $i \in \mathscr{I}$, $r \ge 0$ and $m \longrightarrow \infty$

$$\left\{\begin{array}{lll} \widehat{D}_i\big(\widetilde{f}(\widetilde{x}(t),t),\ \widetilde{f}_{n_m}(\widetilde{x}(t),t);\ \widetilde{z},\, r\big) \longrightarrow 0, & & \\ \widetilde{d}_i\big(\widetilde{x}(t), \quad \widetilde{x}_{n_m}(t+\delta_m)\big) \longrightarrow 0, & \pi_1\, \widetilde{x}_{n_m}(t+\delta_m) & \searrow \pi_1\, \widetilde{x}(t), \\ \widetilde{d}_i\big(\widetilde{x}_{n_m}(t+h-\widetilde{\delta}_m),\, \widetilde{x}(t+h)\big) \longrightarrow 0, & \pi_1\, \widetilde{x}_{n_m}(t+h-\widetilde{\delta}_m) & \nearrow \pi_1\, \widetilde{x}(t+h)\,. \end{array}\right.$$

For every test element $\widetilde{z} \in \widetilde{\mathscr{D}}$ and each time $s \ge 0$ with $s + \pi_1\,\widetilde{z} \le \pi_1\,\widetilde{x}(t)$ and $s + h < \mathbb{T}_j(\widetilde{\vartheta},\widetilde{z})$, we conclude from condition (8.) on timed transitions that for all $k \in]0,h[$ sufficiently small (depending on $h,s,t,\widetilde{z}$)

$$\widetilde{d}_j\big(\widetilde{\vartheta}(s+h,\widetilde{z}),\ \widetilde{x}(t+h)\big) \le \widetilde{d}_j\big(\widetilde{\vartheta}(s+h-k,\widetilde{z}),\ \widetilde{x}(t+h)\big) + h^2.$$

Lemma 8 (on page 340) and the semicontinuity of $\widetilde{d_j}$ (in the sense of hypothesis (H3') $\widetilde{\text{(i')}}$ on page 334) imply

$$\begin{aligned}
&\widetilde{d_j}\big(\widetilde{\vartheta}(s+h,\widetilde{z}),\ \widetilde{x}(t+h)\big) \ - \ h^2 \\
\leq\ & \qquad \widetilde{d_j}\big(\widetilde{\vartheta}(s+h-k,\widetilde{z}),\ \widetilde{x}(t+h)\big) \\
\leq\ & \limsup_{m\to\infty} \Big(\widetilde{d_j}\big(\widetilde{\vartheta}(s,\ \widetilde{z}), \qquad \widetilde{x}_{n_m}(t+k-\widetilde{\delta}_m)\big) + \\
& \qquad (h-k)\cdot \sup_{[t+k-\widetilde{\delta}_m,\ t+h-\widetilde{\delta}_m]} \widehat{D}_j\big(\widetilde{\vartheta},\ \widetilde{f}_{n_m}(\widetilde{x}_{n_m},\cdot);\ \widetilde{z},R_j\big)\Big)\cdot e^{\widehat{\alpha}_j(\widetilde{z},R_j)\cdot(h-k)}.
\end{aligned}$$

Choosing now suitable subsequences $(\delta_{m_l})_{l\in\mathbb{N}}$, $(\widetilde{\delta}_{m_l})_{l\in\mathbb{N}}$ and a sequence $(k_l)_{l\in\mathbb{N}}$ such that the preceding limit superior for $m\to\infty$ coincides with the limit for $l\to\infty$ and $\delta_{m_l} < k_l - \widetilde{\delta}_{m_l} < \frac{1}{l}$ for each $l\in\mathbb{N}$, we obtain successively

$$\begin{aligned}
\lim_{l\to\infty}\ & \widetilde{d_j}\big(\widetilde{x}(t), \qquad \widetilde{x}_{n_{m_l}}(t+k_l-\widetilde{\delta}_{m_l})\big) \ = \ 0, \\
\limsup_{l\to\infty}\ & \widetilde{d_j}\big(\widetilde{\vartheta}(s,\widetilde{z}),\ \widetilde{x}_{n_{m_l}}(t+k_l-\widetilde{\delta}_{m_l})\big) \ \leq\ \widetilde{d_j}\big(\widetilde{\vartheta}(s,\widetilde{z}),\ \widetilde{x}(t)\big)
\end{aligned}$$

as consequences of hypotheses (H3') $(\widetilde{\text{ii}_l})$, $\widetilde{\text{(i'')}}$ (on page 334). Now $l \longrightarrow \infty$ leads to

$$\begin{aligned}
&\widetilde{d_j}\big(\widetilde{\vartheta}(s+h,\widetilde{z}),\ \widetilde{x}(t+h)\big) \ - \ 2h^2 \ - \ \widetilde{d_j}\big(\widetilde{\vartheta}(s,\widetilde{z}),\ \widetilde{x}(t)\big)\cdot e^{\widehat{\alpha}_j(\widetilde{z},R_j)\,h} \\
&\quad \leq h\cdot \limsup_{m\to\infty}\ \sup_{[t+\delta_m,\ t+h]} \widehat{D}_j\big(\widetilde{\vartheta},\ \widetilde{f}_{n_m}(\widetilde{x}_{n_m}(\cdot),\cdot);\ \widetilde{z},R_j\big)\cdot e^{\widehat{\alpha}_j(\widetilde{z},R_j)\,h}.
\end{aligned}$$

For completing the proof, we verify

$$\limsup_{h\downarrow 0}\ \limsup_{m\to\infty}\ \sup_{[t+\delta_m,\ t+h]} \widehat{D}_j\big(\widetilde{\vartheta},\ \widetilde{f}_{n_m}(\widetilde{x}_{n_m}(\cdot),\cdot);\ \widetilde{z},R_j\big) \ \leq\ \widehat{D}_j\big(\widetilde{\vartheta},\ \widetilde{f}(\widetilde{x}(t),t);\ \widetilde{z},R_j\big)$$

for $\mathscr{L}^1$-almost every $t\in[0,T[$ and *any* subsequence $\big(\widetilde{x}_{n_m}(\cdot)\big)_{m\in\mathbb{N}}$ satisfying

$$\begin{cases} \widetilde{d_i}\big(\widetilde{x}(t), \qquad \widetilde{x}_{n_m}(t+\delta_m)\big) & \longrightarrow\ 0 \\ \widehat{D}_i\big(\widetilde{f}(\widetilde{x}(t),t),\ \widetilde{f}_{n_m}(\widetilde{x}(t),t);\ \widetilde{z},\ r\big) & \longrightarrow\ 0 \end{cases}$$

for $m\longrightarrow\infty$ and each $i\in\mathscr{I}$, $r\geq 0$. Indeed, if this inequality was not correct then we could select $\varepsilon>0$ and sequences $(h_l)_{l\in\mathbb{N}}$, $(m_l)_{l\in\mathbb{N}}$, $(s_l)_{l\in\mathbb{N}}$ s.t. for all $l\in\mathbb{N}$,

$$\begin{cases} \widehat{D}_j\big(\widetilde{\vartheta},\ \widetilde{f}_{n_{m_l}}(\widetilde{x}_{n_{m_l}}(t+s_l),\,t+s_l);\ \widetilde{z},\ R_j\big) \ \geq\ \widehat{D}_j\big(\widetilde{\vartheta},\ \widetilde{f}(\widetilde{x}(t),\,t);\ \widetilde{z},\ R_j\big)\ +\ \varepsilon, \\ \delta_{m_l}\ \leq\ s_l\ \leq\ h_l\ \leq\ \frac{1}{l}, \qquad m_l\ \geq\ l\,. \end{cases}$$

Due to property (H3') $(\widetilde{\text{ii}_l})$, the uniform Lipschitz continuity of $(\widetilde{x}_{n_m}(\cdot))_{m\in\mathbb{N}}$ implies

$$\lim_{l\to\infty}\ \widetilde{d_i}\big(\widetilde{x}(t),\ \widetilde{x}_{n_{m_l}}(t+s_l)\big) \ = \ 0$$

for each $i\in\mathscr{I}$. Hence, at $\mathscr{L}^1$-a.e. time t, assumptions (3.), (4.) $\widetilde{\text{(i)}}$ and hypothesis (H6') (on page 337) lead to a contradiction with regard to $\widehat{D}_j\big(\widetilde{\vartheta},\ \widetilde{f}(\widetilde{x}(t),\,t);\ \widetilde{z},\ r\big)$ for any $r\geq 0$.

□

4.3.3 Existence for Mutational Equations with Delay and without State Constraints

Euler approximations in combination with a suitable form of sequential compactness have proved to be very useful for verifying the existence of solutions to mutational equations.

The concept of Euler compactness as specified in Definition 2.15 (on page 112) and Remark 3.15 (2.) (on page 193) focuses on pointwise sequential compactness, i.e., the convergence of Euler approximations is considered at an arbitrary, but fixed point of time $t \in [0,T]$.

Preceding Convergence Theorem 12, however, admits vanishing perturbations with respect to time. In general, this notion of convergence is weaker than pointwise convergence if we dispense with the symmetry of distances and, it may be rather associated with "graphical" convergence of curves in $\widetilde{E}$.

Assuming compactness of Euler approximations with respect to this modified convergence can be of particular interest whenever the transitions have "smoothening" effects on the elements of $\widetilde{E}$ instantaneously. Indeed, in subsequent § 4.4 (on page 359 ff.), we consider geometric evolutions along reachable sets of differential inclusion which exploit such an effect (see Proposition 36 on page 368).

Definition 14 (transitionally Euler compact).
$\big(\widetilde{E}, \widetilde{\mathscr{D}}, (\widetilde{d}_j)_{j\in\mathscr{I}}, (\widetilde{e}_j)_{j\in\mathscr{I}}, (\lfloor\cdot\rfloor_j)_{j\in\mathscr{I}}, \widehat{\Theta}\big(\widetilde{E}, \widetilde{\mathscr{D}}, (\widetilde{d}_i)_{i\in\mathscr{I}}, (\widetilde{e}_i)_{i\in\mathscr{I}}, (\lfloor\cdot\rfloor_i)_{i\in\mathscr{I}}\big)\big)$ is called *transitionally Euler compact* if it satisfies the following condition for any element $\widetilde{x}_0 \in \widetilde{E}$, time $T \in]0,\infty[$ and bounds $\widehat{\alpha}_j : \widetilde{\mathscr{D}} \longrightarrow [0,\infty[$, $\widehat{\beta}_j, \widehat{\gamma}_j > 0$ $(j \in \mathscr{I})$:

Let $\mathscr{N} = \mathscr{N}(\widetilde{x}_0, T, (\widehat{\alpha}_j, \widehat{\beta}_j, \widehat{\gamma}_j)_{j\in\mathscr{I}})$ denote the (possibly empty) subset of all curves $\widetilde{y}(\cdot) : [0,T] \longrightarrow \widetilde{E}$ constructed in the following piecewise way: Choosing an arbitrary equidistant partition $0 = t_0 < t_1 < \ldots < t_n = T$ of $[0,T]$ (with $n > T$) and timed transitions $\widetilde{\vartheta}_1 \ldots \widetilde{\vartheta}_n \in \widehat{\Theta}\big(\widetilde{E}, \widetilde{\mathscr{D}}, (\widetilde{d}_i)_{i\in\mathscr{I}}, (\widetilde{e}_i)_{i\in\mathscr{I}}, (\lfloor\cdot\rfloor_i)_{i\in\mathscr{I}}\big)$ with

$$\begin{cases} \sup_k & \gamma_j(\widetilde{\vartheta}_k) & \le \widehat{\gamma}_j \\ \sup_k & \alpha_j\big(\widetilde{\vartheta}_k;\, \widetilde{z}, (\lfloor\widetilde{x}_0\rfloor_j + \widehat{\gamma}_j\, T)\, e^{\widehat{\gamma}_j\, T}\big) & \le \widehat{\alpha}_j(\widetilde{z}) \\ \sup_k & \beta_j\big(\widetilde{\vartheta}_k;\quad (\lfloor\widetilde{x}_0\rfloor_j + \widehat{\gamma}_j\, T)\, e^{\widehat{\gamma}_j\, T}\big) & \le \widehat{\beta}_j \end{cases}$$

for each index $j \in \mathscr{I}$ and test element $\widetilde{z} \in \widetilde{\mathscr{D}}$, define $\widetilde{y}(\cdot) : [0,T] \longrightarrow \widetilde{E}$ as

$$\widetilde{y}(0) := \widetilde{x}_0, \qquad \widetilde{y}(t) := \widetilde{\vartheta}_k\,(t - t_{k-1},\, \widetilde{y}(t_{k-1})) \quad \text{for } t \in]t_{k-1}, t_k],\ k = 1, 2 \ldots n.$$

Then for each time $t \in [0,T[$ and sequence $h_m \downarrow 0$, every sequence $(\widetilde{y}_n(\cdot))_{n\in\mathbb{N}}$ in $\mathscr{N}$ has a subsequence $(\widetilde{y}_{n_m}(\cdot))_{m\in\mathbb{N}}$ and some element $\widetilde{x} \in \widetilde{E}$ satisfying for each $j \in \mathscr{I}$,

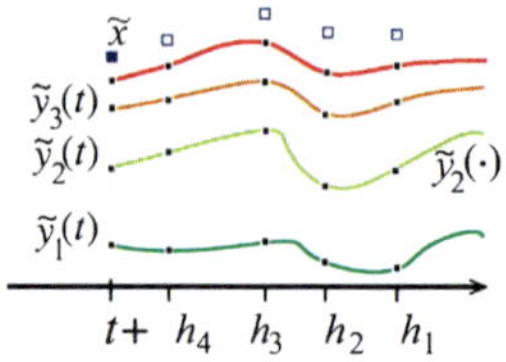

$$\begin{cases} \pi_1\, \widetilde{y}_{n_m}(t) \;=\; t + \pi_1\, \widetilde{x}_0 \;=\; \pi_1\, \widetilde{x} \\ \lim\limits_{m\to\infty} \ \widetilde{d}_j\big(\widetilde{y}_{n_m}(t),\ \widetilde{x}\big) = 0 \\ \lim\limits_{k\to\infty}\ \sup\limits_{m\ge k}\ \widetilde{d}_j\big(\widetilde{x},\quad \widetilde{y}_{n_m}(t + h_k)\big) = 0 \end{cases}$$

Remark 15. If each distance function $\widetilde{d}_j$ $(j \in \mathscr{I})$ is symmetric in addition, then Euler compactness (in the form of Remark 3.15 (2.)) always implies transitional Euler compactness — due to hypothesis (H3') $(\widetilde{\text{ii}}_l)$ (on page 334).

Just for avoiding misunderstandings, we reformulate the definition of "Euler equi-continuous" for the current case of possibly nonsymmetric distance functions. The main idea coincides with Definition 3.16 (on page 194), but now the arguments of $\widetilde{e}_j$ are always sorted by time.

Definition 16.
$\big(\widetilde{E}, \widetilde{\mathscr{D}}, (\widetilde{d}_j)_{j\in\mathscr{I}}, (\widetilde{e}_j)_{j\in\mathscr{I}}, (\lfloor\cdot\rfloor_j)_{j\in\mathscr{I}}, \widehat{\Theta}\big(\widetilde{E}, \widetilde{\mathscr{D}}, (\widetilde{d}_i)_{i\in\mathscr{I}}, (\widetilde{e}_i)_{i\in\mathscr{I}}, (\lfloor\cdot\rfloor_i)_{i\in\mathscr{I}}\big)\big)$ is called *Euler equi-continuous* if it satisfies the following condition for any element $\widetilde{x}_0 \in \widetilde{E}$, time $T \in]0,\infty[$ and bounds $\widehat{\alpha}_j : \widetilde{\mathscr{D}} \longrightarrow [0,\infty[,\ \widehat{\beta}_j, \widehat{\gamma}_j > 0$ $(j \in \mathscr{I})$:

Let $\mathscr{N} = \mathscr{N}(\widetilde{x}_0, T, (\widehat{\alpha}_j, \widehat{\beta}_j, \widehat{\gamma}_j)_{j\in\mathscr{I}})$ denote the (possibly empty) subset specified in Definition 14. Then, for each index $j \in \mathscr{I}$, there exists a constant $L_j \in [0,\infty[$ such that every curve $\widetilde{y}(\cdot) \in \mathscr{N}$ satisfies for all $s,t \in [0,T]$ with $s \le t$

$$\widetilde{e}_j\big(\widetilde{y}(s),\ \widetilde{y}(t)\big) \ \le\ L_j \cdot (t-s).$$

In this particular sense of Lipschitz continuity (i.e. always with the arguments of $\widetilde{e}_j$ sorted by time), we also consider $\widetilde{\text{BLip}}\big(I, \widetilde{E};\ (\widetilde{e}_i)_i,\ (\lfloor\cdot\rfloor_i)_i\big)$ from now on.
Finally the counterpart of Existence Theorem 3.43 (on page 226) states:

Theorem 17 (Existence of timed solutions to mutational equations with delay). *Suppose* $\big(\widetilde{E}, \widetilde{\mathscr{D}}, (\widetilde{d}_j)_{j\in\mathscr{I}}, (\widetilde{e}_j)_{j\in\mathscr{I}}, (\lfloor\cdot\rfloor_j)_{j\in\mathscr{I}}, \widehat{\Theta}\big(\widetilde{E}, \widetilde{\mathscr{D}}, (\widetilde{d}_i)_{i\in\mathscr{I}}, (\widetilde{e}_i)_{i\in\mathscr{I}}, (\lfloor\cdot\rfloor_i)_{i\in\mathscr{I}}\big)\big)$ *to be transitionally Euler compact and Euler equi-continuous. Moreover assume for a fixed period* $\tau \ge 0$, *the function*

$$\widetilde{f} :\ \widetilde{\text{BLip}}\big([-\tau,0], \widetilde{E};\ (\widetilde{e}_i)_i,\ (\lfloor\cdot\rfloor_i)_i\big) \times [0,T] \ \longrightarrow\ \widehat{\Theta}\big(\widetilde{E},\ \widetilde{\mathscr{D}},\ (\widetilde{d}_i)_i,\ (\widetilde{e}_i)_i,\ (\lfloor\cdot\rfloor_i)_i\big)$$

and each $\widetilde{z} \in \widetilde{\mathscr{D}},\ j \in \mathscr{I},\ R > 0$:

1.) $\sup_{\widetilde{y}(\cdot),t}\ \alpha_j(\widetilde{f}(\widetilde{y}(\cdot),t);\widetilde{z},R) < \infty$,
2.) $\sup_{\widetilde{y}(\cdot),t}\ \beta_j(\widetilde{f}(\widetilde{y}(\cdot),t);\quad R) < \infty$,
3.) $\sup_{\widetilde{y}(\cdot),t}\ \gamma_j(\widetilde{f}(\widetilde{y}(\cdot),t)) < \infty$,
4.) *for* $\mathscr{L}^1$*-almost every* $t \in [0,T]$: $\lim_{n\to\infty} \widehat{D}_j\big(\widetilde{f}(\widetilde{y}_n^1(\cdot),t_n^1),\ \widetilde{f}(\widetilde{y}_n^2(\cdot),t_n^2);\widetilde{z},R\big) = 0$ *holds for each* $j \in \mathscr{I}$, $R \ge 0$ *and any sequences* $(t_n^1)_{n\in\mathbb{N}}$, $(t_n^2)_{n\in\mathbb{N}}$ *in* $[0,T]$ *and* $(\widetilde{y}_n^1(\cdot))_{n\in\mathbb{N}}$, $(\widetilde{y}_n^2(\cdot))_{n\in\mathbb{N}}$ *in* $\widetilde{\text{BLip}}\big([-\tau,0], \widetilde{E};\ (\widetilde{e}_j)_{j\in\mathscr{I}},\ (\lfloor\cdot\rfloor_j)_{j\in\mathscr{I}}\big)$ *satisfying for every* $i \in \mathscr{I}$ *and* $s \in [-\tau,0]$

$$\lim_{n\to\infty} t_n^1 = t = \lim_{n\to\infty} t_n^2,\quad \lim_{n\to\infty} \widetilde{d}_i\big(\widetilde{y}(s), \widetilde{y}_n^1(s)\big) \ =\ 0\ =\ \lim_{n\to\infty} \widetilde{d}_i\big(\widetilde{y}(s), \widetilde{y}_n^2(s)\big)$$
$$\sup_{n\in\mathbb{N}}\ \sup_{[-\tau,0]}\ \lfloor \widetilde{y}_n^{1,2}(\cdot)\rfloor_i \ <\ \infty.$$

For every function $\widetilde{x}_0(\cdot) \in \widetilde{\mathrm{BLip}}\big([-\tau,0],\, \widetilde{E};\, (\widetilde{e}_j)_{j\in\mathscr{I}},\, (\lfloor\cdot\rfloor_j)_{j\in\mathscr{I}}\big)$*, there exists a curve* $\widetilde{x}(\cdot) : [-\tau,T] \longrightarrow \widetilde{E}$ *with the following properties:*

(i) $\widetilde{x}(\cdot) \in \widetilde{\mathrm{BLip}}\big([-\tau,T],\, \widetilde{E};\, (\widetilde{e}_j)_{j\in\mathscr{I}},\, (\lfloor\cdot\rfloor_j)_{j\in\mathscr{I}}\big)$,

(ii) $\widetilde{x}(\cdot)\big|_{[-\tau,0]} = \widetilde{x}_0(\cdot)$,

(iii) the restriction $\widetilde{x}(\cdot)\big|_{[0,T]}$ *is a timed solution to the mutational equation*

$$\overset{\circ}{\widetilde{x}}(t) \;\ni\; \widetilde{f}\big(\widetilde{x}(t+\cdot)\big|_{[-\tau,0]},\, t\big)$$

in the sense of Definition 7 (on page 339 f.).

Proof. Similarly to the proof of Theorem 3.19 (on page 197 f.), we use a subsequence of Euler approximations for constructing a limit curve $\widetilde{x} : [-\tau,T] \longrightarrow \widetilde{E}$ and, Convergence Theorem 12 (on page 345) ensures that the restriction $\widetilde{x}(\cdot)|_{[0,T]}$ is a timed solution to the given mutational equation.

For every $n \in \mathbb{N}$ with $2^n > T$, set

$$\begin{array}{rcll} h_n &:=& \frac{T}{2^n}, \qquad t_n^k \quad := k\, h_n & \text{for } k = 0 \ldots 2^n, \\ \widetilde{x}_n(\cdot)\big|_{[-\tau,0]} &:=& \widetilde{x}_0, & \\ \widetilde{x}_n(t) &:=& \widetilde{f}\big(\widetilde{x}_n(t_n^k+\cdot)\big|_{[-\tau,0]},\, t_n^k\big)\big(t-t_n^k,\, \widetilde{x}_n(t_n^k)\big) & \text{for } t \in\,]t_n^k,\, t_n^{k+1}],\ k < 2^n, \\ \widetilde{x}_n(t) &:=& \widetilde{f}\big(\widetilde{x}_n(T+\cdot)\big|_{[-\tau,0]},\, T\big)\big(t-T,\, \widetilde{x}_n(T)\big) & \text{for } t \in\,]T,\, T+1]. \end{array}$$

Due to Euler equi-continuity, there is a constant $L_j \in [0,\infty[$ for each index $j \in \mathscr{I}$ such that every curve $\widetilde{x}_n(\cdot)$ is L_j-Lipschitz continuous with respect to $\widetilde{e}_j$. Setting $\widehat{\gamma}_j := \sup\ \gamma_j(\widetilde{f}(\cdot,\cdot)) < \infty$, Lemma 4 (on page 338) guarantees for every $t \in [0,\, T+1]$ and each $j \in \mathscr{I}$

$$\lfloor\widetilde{x}_n(t)\rfloor_j \;\le\; \big(\lfloor\widetilde{x}_0(0)\rfloor_j + \widehat{\gamma}_j\cdot(T+1)\big)\cdot e^{\widehat{\gamma}_j\,(T+1)} \;=:\; R_j.$$

The next step focuses on selecting subsequences $(\widetilde{x}_{n_m}(\cdot))_{m\in\mathbb{N}}$, $(h_{n'_m})_{m\in\mathbb{N}}$ such that some $\widetilde{x}(\cdot) : [-\tau,T] \longrightarrow \widetilde{E}$ satisfies $\widetilde{x}(\cdot)|_{[-\tau,0]} = \widetilde{x}_0$ and for every $t \in [0,T]$, $j \in \mathscr{I}$

$$\left\{\begin{array}{rcl} \lim\limits_{m\to\infty}\ \widetilde{d}_j\big(\widetilde{x}_{n_m}(t-h_{n'_m}),\ \widetilde{x}(t)\big) &=& 0 \\ \lim\limits_{m\to\infty}\ \widetilde{d}_j\big(\widetilde{x}(t),\ \widetilde{x}_{n_m}(t+h_{n'_m})\big) &=& 0 \\ \pi_1\,\widetilde{x}(t) &=& t\,+\,\pi_1\,\widetilde{x}_0(0). \end{array}\right.$$

Indeed, at every time $t \in [0,T+1[$, transitional Euler compactness provides a sequence $n_k \nearrow \infty$ of indices and an element $\widetilde{x}(t) \in \widetilde{E}$ satisfying for every index $j \in \mathscr{I}$

$$\left\{\begin{array}{rcl} \lim\limits_{k\to\infty}\ \widetilde{d}_j\big(x_{n_k}(t),\ \widetilde{x}(t)\big) &=& 0 \\ \lim\limits_{k\to\infty}\ \sup\limits_{l\ge k}\ \widetilde{d}_j\big(\widetilde{x}(t),\ \widetilde{x}_{n_l}(t+h_k)\big) &=& 0. \end{array}\right.$$

Now Cantor's diagonal construction lays the foundations for extending this selection to countably many points of time simultaneously. In particular, there exists a joint sequence $n_k \nearrow \infty$ and a function $\widetilde{x}(\cdot) : [0,T]\cap\mathbb{Q} \longrightarrow \widetilde{E}$ such that for every rational $t \in [0,T]$ and each index $j \in \mathscr{I}$,

$$\begin{cases} \lim\limits_{k\to\infty} \widetilde{d}_j\big(\widetilde{x}_{n_k}(t),\ \widetilde{x}(t)\big) & = 0 \\ \lim\limits_{k\to\infty} \sup\limits_{l\geq k} \widetilde{d}_j\big(\widetilde{x}(t),\ \widetilde{x}_{n_l}(t+h_k)\big) & = 0 \\ \pi_1\,\widetilde{x}(t) & = t + \pi_1\,\widetilde{x}_0(0). \end{cases}$$

Choose $t \in [0,T]\setminus\mathbb{Q}$ arbitrarily. As a consequence of transitional Euler compactness again, there exists a subsequence $n_{k_l} \nearrow \infty$ possibly depending on t such that an element $\widetilde{x}(t) \in \widetilde{E}$ fulfills for every index $j \in \mathscr{I}$

$$\begin{cases} \lim\limits_{l\to\infty} \widetilde{d}_j\big(\widetilde{x}_{n_{k_l}}(t),\ \widetilde{x}(t)\big) & = 0 \\ \lim\limits_{l\to\infty} \sup\limits_{l'\geq l} \widetilde{d}_j\big(\widetilde{x}(t),\ \widetilde{x}_{n_{k_{l'}}}(t+h_l)\big) & = 0. \\ \pi_1\,\widetilde{x}(t) & = t + \pi_1\,\widetilde{x}_0(0). \end{cases}$$

Hypothesis (H3') (on page 334 f.) even ensures the convergence of $\big(\widetilde{x}_{n_k}(\cdot)\big)_{k\in\mathbb{N}}$ at this time $t \in [0,T]\setminus\mathbb{Q}$ in the following sense for each index $j \in \mathscr{I}$

$$\begin{cases} \lim\limits_{k\to\infty} \widetilde{d}_j\big(\widetilde{x}_{n_k}(t-h_k),\ \widetilde{x}(t)\big) & = 0 \\ \lim\limits_{k\to\infty} \widetilde{d}_j\big(\widetilde{x}(t),\ \widetilde{x}_{n_k}(t+2h_k)\big) & = 0. \end{cases} \qquad (*)$$

Indeed, assumption (H3') $\widetilde{\text{(i')}}$ implies for every $s \in [0,t[\,\cap\mathbb{Q}$ and $j \in \mathscr{I}$

$$\widetilde{e}_j\big(\widetilde{x}(s),\ \widetilde{x}(t)\big) \;\leq\; \limsup_{l\to\infty}\ \widetilde{e}_j\big(\widetilde{x}_{n_{k_l}}(s+h_{k_l}),\ \widetilde{x}_{n_{k_l}}(t)\big) \;\leq\; L_j\,|s-t|\,.$$

Choosing any sequence $(s_l)_{l\in\mathbb{N}}$ in $[0,t[\,\cap\mathbb{Q}$ with $t-h_l < s_l < t$ for all $l \in \mathbb{N}$, we obtain for every index $j \in \mathscr{I}$

$$\begin{aligned} \lim_{l\to\infty} \widetilde{d}_j\big(\widetilde{x}(s_l),\ \widetilde{x}(t)\big) &= 0, \\ \lim_{k\to\infty} \widetilde{d}_j\big(\widetilde{x}_{n_k}(s_l),\ \widetilde{x}(s_l)\big) &= 0 \qquad \text{for each } l \in \mathbb{N}, \\ \lim_{l\to\infty}\ \sup_{k\in\mathbb{N}}\ \widetilde{e}_j\big(\widetilde{x}_{n_k}(t-h_l),\ \widetilde{x}_{n_k}(s_l)\big) \;\leq\; \lim_{l\to\infty} L_j\,h_l &= 0. \end{aligned}$$

and thus, hypothesis (H3') $(\widetilde{\text{iii}}_r)$ (on page 335) guarantees

$$\lim_{l\to\infty} \widetilde{d}_j\big(\widetilde{x}_{n_l}(t-h_l),\ \widetilde{x}(t)\big) \;=\; 0 \qquad \text{for each } j \in \mathscr{I}.$$

Similarly any sequence $(s'_l)_{l\in\mathbb{N}}$ in $]t, T+1]\cap\mathbb{Q}$ with $t < s'_l < t+h_l$ for all $l\in\mathbb{N}$ leads to

$$\begin{aligned} \lim_{l\to\infty} \widetilde{d}_j\big(\widetilde{x}(t),\ \widetilde{x}(s'_l)\big) &= 0, \\ \lim_{k\to\infty} \widetilde{d}_j\big(\widetilde{x}(s'_l),\ \widetilde{x}_{n_k}(s'_l+h_l)\big) &= 0 \qquad \text{for each } l \in \mathbb{N}, \\ \lim_{l\to\infty}\ \sup_{k\in\mathbb{N}}\ \widetilde{e}_j\big(\widetilde{x}_{n_k}(s'_l+h_l),\ \widetilde{x}_{n_k}(t+2h_l)\big) \;\leq\; \lim_{l\to\infty} L_j\,h_l &= 0 \end{aligned}$$

for every index $j \in \mathscr{I}$ and thus, hypothesis (H3') $(\widetilde{\text{iii}}_l)$ (on page 335) implies

$$\lim_{l\to\infty} \widetilde{d}_j\big(\widetilde{x}(t),\ \widetilde{x}_{n_l}(t+2h_l)\big) \;=\; 0 \qquad \text{for each } j \in \mathscr{I}.$$

In a word, preceding statement $(*)$ about the convergence of $\big(\widetilde{x}_{n_k}(\cdot)\big)_{k\in\mathbb{N}}$ holds at every time $t \in [0,T]$.

For every $t \in [0,T]$, the estimate $\lfloor \widetilde{x}(t) \rfloor_j \le R_j$ results from hypothesis (H4') about the lower semicontinuity of $\lfloor \cdot \rfloor_j$ (on page 335) and, $\widetilde{x}(\cdot) : [0,T] \longrightarrow (E, e_j)$ is also L_j-Lipschitz continuous (in time direction) due to the lower semicontinuity of e_j (in hypothesis (H3') $\widetilde{\text{(i')}}$). Defining $\widetilde{x}(\cdot)\big|_{[-\tau,0]} := \widetilde{x}_0(\cdot)$, we obtain

$$\widetilde{x}(\cdot) \in \widetilde{\mathrm{B}}\mathrm{Lip}\big([-\tau,T], \widetilde{E}; (\widetilde{e}_j)_{j\in\mathscr{I}}, (\lfloor\cdot\rfloor_j)_{j\in\mathscr{I}}\big).$$

Finally, Convergence Theorem 12 (on page 345) is to guarantee that $\widetilde{x}(\cdot)|_{[0,T]}$ is a timed solution to the mutational equation

$$\overset{\circ}{\widetilde{x}}(t) \ni \widetilde{f}\big(\widetilde{x}(t+\cdot)\big|_{[-\tau,0]}, t\big)$$

in the tuple $\big(\widetilde{E}, \widetilde{\mathscr{D}}, (\widetilde{d}_j)_{j\in\mathscr{I}}, (\widetilde{e}_j)_{j\in\mathscr{I}}, (\lfloor\cdot\rfloor_j)_{j\in\mathscr{I}}, (\widehat{D}_j)_{j\in\mathscr{I}}\big)$.
Indeed, each shifted Euler approximation $\widetilde{x}_n(\cdot + 3h_n) : [0, T-3h_n] \longrightarrow \widetilde{E}$, $n \in \mathbb{N}$, can be regarded as a timed solution of $\overset{\circ}{\widetilde{y}}(\cdot) \ni \widehat{f}_n(\cdot)$ with the auxiliary function

$$\widehat{f}_n : [0,T] \longrightarrow \widehat{\Theta}\big(\widetilde{E}, \widetilde{\mathscr{D}}, (\widetilde{d}_j)_{j\in\mathscr{I}}, (\widetilde{e}_j)_{j\in\mathscr{I}}, (\lfloor\cdot\rfloor_j)_{j\in\mathscr{I}}\big),$$
$$\widehat{f}_n(t) := \widetilde{f}\big(\widetilde{x}_n(\cdot)\big|_{[t_n^{k+3}-\tau,\, t_n^{k+3}]}, t_n^{k+3}\big) \quad \text{for any } t \in [t_n^k, t_n^{k+1}[,\ k < 2^n.$$

(The time shift here is caused by convergence statement $(*)$ and ensures that all arguments below are sorted by time properly.)
Similarly set

$$\widehat{f} : [0,T] \longrightarrow \widehat{\Theta}\big(\widetilde{E}, \widetilde{\mathscr{D}}, (\widetilde{d}_j)_{j\in\mathscr{I}}, (\widetilde{e}_j)_{j\in\mathscr{I}}, (\lfloor\cdot\rfloor_j)_{j\in\mathscr{I}}\big),$$
$$t \longmapsto \widetilde{f}\big(\widetilde{x}(t+\cdot)\big|_{[-\tau,0]}, t\big).$$

At $\mathscr{L}^1$-almost every time $t \in [0,T]$, assumption (4.) has two essential consequences. First, with the abbreviation $t_{n_k}^l := \big([\frac{t}{h_{n_k}}] + 3\big)\, h_{n_k} \in\,]t + 2h_{n_k},\ t + 3h_{n_k}]$,

$$\widehat{D}_j\big(\widehat{f}(t), \widehat{f}_{n_k}(t); \widetilde{z}, \rho\big) = \widehat{D}_j\big(\widetilde{f}(\widetilde{x}(t+\cdot)|_{[-\tau,0]}, t),\ \widetilde{f}(\widetilde{x}_{n_k}(t_{n_k}^l + \cdot)|_{[-\tau,0]}, t_{n_k}^l);\ \widetilde{z}, \rho\big) \overset{k\to\infty}{\longrightarrow} 0,$$

for every $j \in \mathscr{I}$, $\widetilde{z} \in \widetilde{\mathscr{D}}$ and $\rho > 0$ because for any $i \in \mathscr{I}$ and $t \in [0,T]$, $s \in [-\tau, 0]$, statement $(*)$ about the convergence of $(\widetilde{x}_{n_k}(\cdot))_{m\in\mathbb{N}}$ and hypothesis (H3') $\widetilde{(\text{ii}_l)}$ imply

$$\widetilde{d}_i\big(\widetilde{x}(t+s),\ \widetilde{x}_{n_k}(t_{n_k}^l + s)\big) \overset{k\to\infty}{\longrightarrow} 0.$$

Second, we obtain for any sequence $t_k \longrightarrow t$ in $[t,T]$ and $\widetilde{z} \in \widetilde{\mathscr{D}}$, $j \in \mathscr{I}$, $\rho \ge 0$

$$\widehat{D}_j\big(\widehat{f}_{n_k}(t), \widehat{f}_{n_k}(t_k); \widetilde{z}, \rho\big) = \widehat{D}_j\big(\widetilde{f}(\widetilde{x}_{n_k}(t_{n_k}^l + \cdot)|_{[-\tau,0]}, t_{n_k}^l),$$
$$\widetilde{f}(\widetilde{x}_{n_k}(t_{n_k}^{l_k} + \cdot)|_{[-\tau,0]}, t_{n_k}^{l_k});\ \widetilde{z}, \rho\big) \overset{k\to\infty}{\longrightarrow} 0$$

with the abbreviations $t_{n_k}^l := \big([\frac{t}{h_{n_k}}] + 3\big)\, h_{n_k} \le t_{n_k}^{l_k} := \big([\frac{t_k}{h_{n_k}}] + 3\big)\, h_{n_k}$ because due to hypothesis (H3') $\widetilde{(\text{ii}_l)}$ and statement $(*)$ again, the following convergences hold for any $i \in \mathscr{I}$, $s \in [-\tau, 0]$

$$\widetilde{d}_i\big(\widetilde{x}(t+s),\ \widetilde{x}_{n_k}(t_{n_k}^l + s)\big) \overset{k\to\infty}{\longrightarrow} 0, \qquad \widetilde{d}_i\big(\widetilde{x}(t+s),\ \widetilde{x}_{n_k}(t_{n_k}^{l_k} + s)\big) \overset{k\to\infty}{\longrightarrow} 0.$$

Hence, the assumptions of Convergence Theorem 12 are satisfied and, $\widetilde{x}(\cdot)|_{[0,T]}$ solves the mutational equation $\overset{\circ}{\widetilde{x}}(\cdot) \ni \widehat{f}(\cdot)$. □

4.3.4 Existence of Timed Solutions without State Constraints due to Another Form of "Weak" Euler Compactness

Now we formulate the counterparts of the results in § 3.3.6 (on page 206 ff.). The main idea is again that firstly, each distance function $\widetilde{d}_j$, $\widetilde{e}_j$ $(j \in \mathscr{I})$ can be represented as supremum of further distance functions $\widetilde{d}_{j,\kappa}$, $\widetilde{e}_{j,\kappa}$ $(\kappa \in \mathscr{J})$ and secondly, the assumptions about sequential compactness focus on the *right* convergence with respect to $\widetilde{d}_{j,\kappa}$ $(j \in \mathscr{I}, \kappa \in \mathscr{J})$.

In contrast to § 3.3.6, however, we consider the *left* convergence with respect to each $\widetilde{d}_j$ $(j \in \mathscr{I})$. This difference in regard to topology is particularly useful for proving the adapted Convergence Theorem (in Proposition 21 on page 355 below) and, it motivates the term "strong-weak" for the current form of transitional Euler compactness in Definition 18.

Additional assumptions for § 4.3.4.

In addition to the general hypotheses (H1), (H3'), (H5')–(H7') about the distance functions $\widetilde{d}_j, \widetilde{e}_j : (\widetilde{\mathscr{D}} \cup E) \times (\widetilde{\mathscr{D}} \cup E) \longrightarrow [0,\infty[$ specified in § 4.1 (on page 334 ff.), let $\mathscr{J} \neq \emptyset$ denote a further index set. For each index $(j,\kappa) \in \mathscr{I} \times \mathscr{J}$, the functions $\widetilde{d}_{j,\kappa}$, $\widetilde{e}_{j,\kappa} : \widetilde{E} \times \widetilde{E} \longrightarrow \mathbb{R}_0^+$ are assumed to fulfill in addition to hypotheses (H1),(H3')

(H4') $\lfloor\cdot\rfloor_j$ is lower semicontinuous with respect to $(\widetilde{d}_{i,\kappa})_{i \in \mathscr{I}, \kappa \in \mathscr{J}}$, i.e.,

$$\lfloor\widetilde{x}\rfloor_j \le \liminf_{n\to\infty} \lfloor\widetilde{x}_n\rfloor_j$$

for any $\widetilde{x} \in \widetilde{E}$ and $(\widetilde{x}_n)_{n\in\mathbb{N}}$ in $\widetilde{E}$ fulfilling for each $i \in \mathscr{I}, \kappa \in \mathscr{J}$

$$\lim_{n\to\infty} \widetilde{d}_{i,\kappa}(\widetilde{x}_n,\widetilde{x}) = 0, \qquad \pi_1\,\widetilde{x}_n \nearrow \pi_1\,\widetilde{x} \text{ for } n \to \infty, \qquad \sup_{n\in\mathbb{N}} \lfloor\widetilde{x}_n\rfloor_i < \infty.$$

(H8') $$\widetilde{d}_j(\cdot,\cdot) = \sup_{\kappa\in\mathscr{J}} \widetilde{d}_{j,\kappa}(\cdot,\cdot), \quad \widetilde{e}_j(\cdot,\cdot) = \sup_{\kappa\in\mathscr{J}} \widetilde{e}_{j,\kappa}(\cdot,\cdot) \quad \text{for each } j \in \mathscr{I}.$$

Definition 18 (strongly-weakly transitionally Euler compact).
The tuple $\big(\widetilde{E},\, \widetilde{\mathscr{D}},\, (\widetilde{d}_j)_{j\in\mathscr{I}},\, (\widetilde{d}_{j,\kappa})_{j\in\mathscr{I},\kappa\in\mathscr{J}},\, (\widetilde{e}_j)_{j\in\mathscr{I}},\, (\widetilde{e}_{j,\kappa})_{j\in\mathscr{I},\kappa\in\mathscr{J}},\, (\lfloor\cdot\rfloor_j)_{j\in\mathscr{I}},$ $\widehat{\Theta}\big(\widetilde{E},\widetilde{\mathscr{D}},(\widetilde{d}_i)_{i\in\mathscr{I}},(\widetilde{e}_i)_{i\in\mathscr{I}},(\lfloor\cdot\rfloor_i)_{i\in\mathscr{I}}\big)\big)$ is called *strongly-weakly transitionally Euler compact* if it satisfies the following condition for any $\widetilde{x}_0 \in \widetilde{E}$, time $T \in]0,\infty[$ and bounds $\widehat{\alpha}_j : \widetilde{\mathscr{D}} \longrightarrow [0,\infty[$, $\widehat{\beta}_j, \widehat{\gamma}_j > 0$ $(j \in \mathscr{I})$:

Let $\mathscr{N} = \mathscr{N}(\widetilde{x}_0, T, (\widehat{\alpha}_j, \widehat{\beta}_j, \widehat{\gamma}_j)_{j\in\mathscr{I}})$ denote the (possibly empty) subset specified in Definition 14 (on page 348). Then for each time $t \in [0,T[$ and sequence $h_m \downarrow 0$, every sequence $(\widetilde{y}_n(\cdot))_{n\in\mathbb{N}}$ in $\mathscr{N}$ has a subsequence $(\widetilde{y}_{n_m}(\cdot))_{m\in\mathbb{N}}$ and some element $\widetilde{x} \in \widetilde{E}$ satisfying for each $j \in \mathscr{I}$ and $\kappa \in \mathscr{J}$,

$$\begin{cases} \pi_1\,\widetilde{y}_{n_m}(t) = t + \pi_1\,\widetilde{x}_0 = \pi_1\,\widetilde{x} \\ \lim_{m\to\infty} \widetilde{d}_{j,\kappa}\big(\widetilde{y}_{n_m}(t),\ \widetilde{x}\big) = 0 \\ \lim_{k\to\infty} \sup_{m\ge k} \widetilde{d}_j\big(\widetilde{x},\ \widetilde{y}_{n_m}(t+h_k)\big) = 0. \end{cases}$$

Remark 19. The essential difference between Definition 18 and its counterpart in Definition 3.27 (on page 207) used in Theorem 3.44 and Proposition 3.45 (on page 227 ff.) is that $\widetilde{d}_{j,\kappa}$ is considered only for the right convergence, i.e. for all j, κ,

$$\lim_{m \longrightarrow \infty} \widetilde{d}_{j,\kappa}\big(\widetilde{y}_{n_m}(t),\, \widetilde{x}\big) \;=\; 0,$$

whereas the left convergence is formulated with respect to $\widetilde{d}_j$, i.e. for all $j \in \mathscr{I}$,

$$\lim_{k \longrightarrow \infty} \sup_{m \geq k} \widetilde{d}_j\big(\widetilde{x},\, \widetilde{y}_{n_m}(t+h_k)\big) \;=\; 0.$$

The main advantage of this stronger type of convergence is that we obtain existence and convergence results about timed solutions to the mutational equations — without assuming the triangle inequality for each $\widetilde{d}_j$ $(j \in \mathscr{I})$ in addition (as in Theorem 3.44). In the geometric example of subsequent § 4.5 (on page 372 ff.), this special form of compactness proves to be appropriate indeed.

Theorem 20 (Existence due to strong-weak transitional Euler compactness). *Suppose the tuple* $\big(\widetilde{E},\, \widetilde{\mathscr{D}},\, (\widetilde{d}_j)_{j\in\mathscr{I}},\, (\widetilde{d}_{j,\kappa})_{j\in\mathscr{I},\kappa\in\mathscr{J}},\, (\widetilde{e}_j)_{j\in\mathscr{I}},\, (\widetilde{e}_{j,\kappa})_{j\in\mathscr{I},\kappa\in\mathscr{J}},\, (\lfloor\cdot\rfloor_j)_j,$ $\widehat{\Theta}\big(\widetilde{E},\widetilde{\mathscr{D}},(\widetilde{d}_i)_i,(\widetilde{e}_i)_i,(\lfloor\cdot\rfloor_i)_i\big)\big)$ *to be strongly-weakly transitionally Euler compact and* $\big(\widetilde{E},\, \widetilde{\mathscr{D}},\, (\widetilde{d}_j)_{j\in\mathscr{I}},\, (\widetilde{e}_j)_{j\in\mathscr{I}},\, (\lfloor\cdot\rfloor_j)_{j\in\mathscr{I}},\, \widehat{\Theta}\big(\widetilde{E},\widetilde{\mathscr{D}},(\widetilde{d}_i)_i,(\widetilde{e}_i)_i,(\lfloor\cdot\rfloor_i)_i\big)\big)$ *to be Euler equi-continuous (in the sense of Definition 16 on page 349).*

Moreover assume for a fixed period $\tau \geq 0$, *the function*

$$\widetilde{f}:\ \widetilde{\mathrm{BLip}}\big([-\tau,0],\widetilde{E};\,(\widetilde{e}_i)_i,\,(\lfloor\cdot\rfloor_i)_i\big)\times[0,T] \;\longrightarrow\; \widehat{\Theta}\big(\widetilde{E},\,\widetilde{\mathscr{D}},\,(\widetilde{d}_i)_i,\,(\widetilde{e}_i)_i,\,(\lfloor\cdot\rfloor_i)_i\big)$$

and each $\widetilde{z} \in \widetilde{\mathscr{D}},\ j \in \mathscr{I},\ R > 0$:

1.) $\sup_{\widetilde{y}(\cdot),t}\ \alpha_j\big(\widetilde{f}(\widetilde{y}(\cdot),t);\,\widetilde{z},\,R\big) < \infty,$

2.) $\sup_{\widetilde{y}(\cdot),t}\ \beta_j\big(\widetilde{f}(\widetilde{y}(\cdot),t);\quad R\big) < \infty,$

3.) $\sup_{\widetilde{y}(\cdot),t}\ \gamma_j\big(\widetilde{f}(\widetilde{y}(\cdot),t)\big) < \infty,$

4.) *for* $\mathscr{L}^1$*-almost every* $t \in [0,T]$: $\lim_{n\longrightarrow\infty} \widehat{D}_j\big(\widetilde{f}(\widetilde{y}_n^1(\cdot),t_n^1),\ \widetilde{f}(\widetilde{y}_n^2(\cdot),t_n^2);\,\widetilde{z},\,R\big) = 0$
for each $j \in \mathscr{I},\ R \geq 0$ *and any sequences* $(t_n^1)_{n\in\mathbb{N}},\ (t_n^2)_{n\in\mathbb{N}}$ *in* $[0,T]$ *and* $(\widetilde{y}_n^1(\cdot))_{n\in\mathbb{N}},\ (\widetilde{y}_n^2(\cdot))_{n\in\mathbb{N}}$ *in* $\widetilde{\mathrm{BLip}}\big([-\tau,0],\widetilde{E};\,(\widetilde{e}_j)_{j\in\mathscr{I}},\,(\lfloor\cdot\rfloor_j)_{j\in\mathscr{I}}\big)$ *satisfying for every* $i \in \mathscr{I}$ *and* $s \in [-\tau,0]$

$$\lim_{n\longrightarrow\infty} t_n^1 = t = \lim_{n\longrightarrow\infty} t_n^2,\quad \lim_{n\longrightarrow\infty} \widetilde{d}_i\big(\widetilde{y}(s),\,\widetilde{y}_n^1(s)\big) \;=\; 0 \;=\; \lim_{n\longrightarrow\infty} \widetilde{d}_i\big(\widetilde{y}(s),\,\widetilde{y}_n^2(s)\big)$$
$$\sup_{n\in\mathbb{N}}\ \sup_{[-\tau,0]}\ \lfloor\widetilde{y}_n^{1,2}(\cdot)\rfloor_i \;<\; \infty.$$

For every function $\widetilde{x}_0(\cdot) \in \widetilde{\mathrm{BLip}}\big([-\tau,0],\widetilde{E};\,(\widetilde{e}_j)_{j\in\mathscr{I}},\,(\lfloor\cdot\rfloor_j)_{j\in\mathscr{I}}\big)$, *there exists a curve* $\widetilde{x}(\cdot): [-\tau,T] \longrightarrow \widetilde{E}$ *with the following properties:*

(i) $\widetilde{x}(\cdot) \in \widetilde{\mathrm{BLip}}\big([-\tau,T],\widetilde{E};\,(\widetilde{e}_j)_{j\in\mathscr{I}},\,(\lfloor\cdot\rfloor_j)_{j\in\mathscr{I}}\big),$

(ii) $\widetilde{x}(\cdot)\big|_{[-\tau,0]} = \widetilde{x}_0(\cdot),$

(iii) *the restriction* $\widetilde{x}(\cdot)\big|_{[0,T]}$ *is a timed solution of* $\overset{\circ}{\widetilde{x}}(t) \ni \widetilde{f}\big(\widetilde{x}(t+\cdot)\big|_{[-\tau,0]},\,t\big)$.

The *proof* of this Existence Theorem is based on exactly the same conclusions as the one of preceding Theorem 17 (on page 350 ff.). Indeed, the first key difference is due to considering $\widetilde{d}_{j,\kappa}$ $(j \in \mathscr{I}, \kappa \in \mathscr{J})$ for all statements about right convergence. Second, we need an adapted form of Convergence Theorem:

Proposition 21 (about "strong-weak" convergence of timed solutions).
Suppose the following properties of

$$\begin{array}{lll} \widetilde{f}_n, \widetilde{f} : \widetilde{E} \times [0,T] & \longrightarrow \widehat{\Theta}\big(\widetilde{E}, \widetilde{\mathscr{D}}, (\widetilde{d}_i)_{i \in \mathscr{I}}, (\widetilde{e}_j)_{j \in \mathscr{I}}, (\lfloor \cdot \rfloor_i)_{i \in \mathscr{I}}\big) & (n \in \mathbb{N}) \\ \widetilde{x}_n, \widetilde{x} : \qquad [0,T] & \longrightarrow \widetilde{E} \; : & \end{array}$$

1.)
$$\begin{array}{lll} R_j & := \sup\limits_{n,t} \; \lfloor \widetilde{x}_n(t) \rfloor_j + 1 < \infty, & \\ \widehat{\alpha}_j(\widetilde{z}, \rho) & := \sup\limits_{n,t} \; \alpha_j\big(\widetilde{x}_n; \widetilde{z}, \rho\big) < \infty & \textit{for each } \widetilde{z} \in \widetilde{\mathscr{D}}, \rho \geq 0, \\ \widehat{\beta}_j & := \sup\limits_{n} \; \mathrm{Lip}\,\big(\widetilde{x}_n(\cdot) : [0,T] \longrightarrow (\widetilde{E}, \widetilde{e}_j)\big) < \infty & \textit{for every } j \in \mathscr{I}. \end{array}$$

2.) $\overset{\circ}{\widetilde{x}}_n(\cdot) \ni \widetilde{f}_n(\widetilde{x}_n(\cdot), \cdot)$ *(in the sense of Definition 7 on page 339) for every n.*

3.) Equi-continuity of $(\widetilde{f}_n)_n$ at $(\widetilde{x}(t), t)$ at almost every time in the following sense:
for any $\widetilde{z} \in \widetilde{\mathscr{D}}$ and $\mathscr{L}^1$-a.e. $t \in [0,T]$: $\lim\limits_{n \to \infty} \widehat{D}_j\big(\widetilde{f}_n(\widetilde{x}(t), t),\, \widetilde{f}_n(\widetilde{y}_n, t_n);\, \widetilde{z}, r\big) = 0$
for each $j \in \mathscr{I}$, $r \geq 0$ and any $(t_n)_{n \in \mathbb{N}}$, $(\widetilde{y}_n)_{n \in \mathbb{N}}$ in $[t,T]$ and $\widetilde{E}$ respectively with $\lim\limits_{n \to \infty} t_n = t$ *and* $\lim\limits_{n \to \infty} \widetilde{d}_i\big(\widetilde{x}(t), \widetilde{y}_n\big) = 0$, $\sup\limits_{n \in \mathbb{N}} \lfloor \widetilde{y}_n \rfloor_i \leq R_i$ *for each i,*
$\pi_1\, \widetilde{y}_n \searrow \pi_1\, \widetilde{x}(t)$ *for $n \longrightarrow \infty$.*

4'.) For $\mathscr{L}^1$-almost every $s \in [0,T[$ and any $t < t'$ in $[0,T]$, there is a sequence $n_m \nearrow \infty$ of indices (depending on s, t, t') that satisfies for $m \longrightarrow \infty$
(i) $\widehat{D}_j\big(\widetilde{f}(\widetilde{x}(s), s),\, \widetilde{f}_{n_m}(\widetilde{x}(s), s);\, \widetilde{z}, r\big) \longrightarrow 0$ *for all $\widetilde{z} \in \widetilde{\mathscr{D}}$, $r \geq 0$, $j \in \mathscr{I}$,*
(ii) $\exists\, \delta_m \searrow 0 : \forall\, j :$ $\widetilde{d}_j\big(\widetilde{x}(t),\; \widetilde{x}_{n_m}(t+\delta_m)\big) \longrightarrow 0,\ \pi_1\, \widetilde{x}_{n_m}(t+\delta_m) \searrow \pi_1\, \widetilde{x}(t)$
(iii) $\exists\, \widetilde{\delta}_m \searrow 0 : \forall\, j, \kappa : \widetilde{d}_{j,\kappa}\big(\widetilde{x}_{n_m}(t'-\widetilde{\delta}_m), \widetilde{x}(t')\big) \longrightarrow 0,\ \pi_1\, \widetilde{x}_{n_m}(t'-\widetilde{\delta}_m) \nearrow \pi_1\, \widetilde{x}(t').$

Then, $\widetilde{x}(\cdot)$ is always a timed solution to the mutational equation $\overset{\circ}{\widetilde{x}}(\cdot) \ni \widetilde{f}(\widetilde{x}(\cdot), \cdot)$ in the tuple $\big(\widetilde{E},\, \widetilde{\mathscr{D}},\, (\widetilde{d}_j)_{j \in \mathscr{I}},\, (\widetilde{e}_j)_{j \in \mathscr{I}},\, (\lfloor \cdot \rfloor_j)_{j \in \mathscr{I}},\, (\widehat{D}_j)_{j \in \mathscr{I}}\big)$.

Proof (of Proposition 21). It imitates the proof of Convergence Theorem 12 (on page 346 f.), but takes the right convergence with respect to $\widetilde{d}_{j,\kappa}$ $(\kappa \in \mathscr{J})$ into consideration appropriately.

Choose the index $j \in \mathscr{I}$ arbitrarily.
Then $\widetilde{x}(\cdot) : [0,T] \longrightarrow (\widetilde{E}, \widetilde{e}_j)$ is $\widehat{\beta}_j$-Lipschitz continuous. Indeed, for any $t < t'$ in $[0,T]$, assumption (4'.) provides a subsequence $\big(\widetilde{x}_{n_m}(\cdot)\big)_{m \in \mathbb{N}}$ and sequences $\delta_m \searrow 0$,

$\widetilde{\delta}_m \searrow 0$ satisfying for any indices $i \in \mathscr{I}, \kappa \in \mathscr{J}$

$$\begin{cases} \widetilde{d}_i\big(\widetilde{x}(t), \quad \widetilde{x}_{n_m}(t+\delta_m)\big) \longrightarrow 0, & \pi_1\, \widetilde{x}_{n_m}(t+\delta_m) \searrow \pi_1\, \widetilde{x}(t) \\ \widetilde{d}_{i,\kappa}\big(\widetilde{x}_{n_m}(t'-\widetilde{\delta}_m), \widetilde{x}(t')\big) \longrightarrow 0, & \pi_1\, \widetilde{x}_{n_m}(t'-\widetilde{\delta}_m) \nearrow \pi_1\, \widetilde{x}(t') \end{cases} \quad \text{for } m \to \infty.$$

First, we conclude $\pi_1\, \widetilde{x}(t') = t' - t + \pi_1\, \widetilde{x}(t) = \pi_1\, \widetilde{x}_{n_m}(t')$ for each $m \in \mathbb{N}$. Second, the uniform $\widehat{\beta}_j$-Lipschitz continuity of $\widetilde{x}_n(\cdot), n \in \mathbb{N}$, with respect to $\widetilde{e}_j$ and hypothesis (H3') $\widetilde{(\text{i'})}$ about $(\widetilde{e}_{j,\kappa})_{j\in\mathscr{I},\kappa\in\mathscr{J}}$ (on page 334) imply for each $\kappa \in \mathscr{J}$

$$\begin{aligned} \widetilde{e}_{j,\kappa}\big(\widetilde{x}(t), \widetilde{x}(t')\big) &\le \limsup_{m\to\infty} \; \widetilde{e}_{j,\kappa}\big(\widetilde{x}_{n_m}(t+\delta_m),\, \widetilde{x}_{n_m}(t'-\widetilde{\delta}_m)\big) \\ &\le \limsup_{m\to\infty} \; \widehat{\beta}_j\, |t' - \widetilde{\delta}_m - t - \delta_m| \\ &\le \widehat{\beta}_j\, |t'-t|, \\ \widetilde{e}_j\big(\widetilde{x}(t), \widetilde{x}(t')\big) &\le \widehat{\beta}_j\, |t'-t|. \end{aligned}$$

Moreover, hypothesis (H4') about the lower semicontinuity of $\lfloor\cdot\rfloor_j$ ensures

$$\lfloor \widetilde{x}(t')\rfloor_j \le \liminf_{m\to\infty} \lfloor \widetilde{x}_{n_m}(t'-\widetilde{\delta}_m)\rfloor_j \le R_j - 1.$$

Finally we verify the solution property

$$\limsup_{h\downarrow 0} \frac{\widetilde{d}_j\big(\widetilde{\vartheta}(s+h,\widetilde{z}),\, \widetilde{x}(t+h)\big) - \widetilde{d}_j\big(\widetilde{\vartheta}(s,\widetilde{z}),\, \widetilde{x}(t)\big)\cdot e^{\alpha_j(\widetilde{x};\rho)\, h}}{h} \le \widehat{D}_j\big(\widetilde{\vartheta},\, \widetilde{f}(\widetilde{x}(t),t);\, \widetilde{z},\, R_j\big)$$

for $\mathscr{L}^1$-almost every $t \in [0,T[$ and any $\widetilde{\vartheta} \in \widehat{\Theta}\big(\widetilde{E}, \widetilde{\mathscr{D}}, (\widetilde{d}_i)_{i\in\mathscr{I}}, (\widetilde{e}_i)_{i\in\mathscr{I}}, (\lfloor\cdot\rfloor_i)_{i\in\mathscr{I}}\big)$, $\widetilde{z} \in \widetilde{\mathscr{D}}$, $s \in [0, \mathbb{T}_j(\widetilde{\vartheta},\widetilde{z})[$ with $s + \pi_1\,\widetilde{z} \le \pi_1\,\widetilde{x}(t)$.
Indeed, for Lebesgue-almost every $t \in [0,T[$ and any $h \in]0, T-t[$, assumption (4.) guarantees a subsequence $\big(\widetilde{x}_{n_m}(\cdot)\big)_{m\in\mathbb{N}}$ and sequences $\delta_m \searrow 0$, $\widetilde{\delta}_m \searrow 0$ satisfying for each $\widetilde{z} \in \widetilde{\mathscr{D}}$, $i \in \mathscr{I}$, $\kappa \in \mathscr{J}$, $r \ge 0$ and $m \longrightarrow \infty$

$$\begin{cases} \widehat{D}_i\big(\widetilde{f}(\widetilde{x}(t),t),\; \widetilde{f}_{n_m}(\widetilde{x}(t),t);\, \widetilde{z},\, r\big) \longrightarrow 0, & \\ \widetilde{d}_i\big(\widetilde{x}(t), \quad \widetilde{x}_{n_m}(t+\delta_m)\big) \longrightarrow 0, & \pi_1\, \widetilde{x}_{n_m}(t+\delta_m) \searrow \pi_1\, \widetilde{x}(t), \\ \widetilde{d}_{i,\kappa}\big(\widetilde{x}_{n_m}(t+h-\widetilde{\delta}_m),\, \widetilde{x}(t+h)\big) \longrightarrow 0, & \pi_1\, \widetilde{x}_{n_m}(t+h-\widetilde{\delta}_m) \nearrow \pi_1\, \widetilde{x}(t+h). \end{cases}$$

For every test element $\widetilde{z} \in \widetilde{\mathscr{D}}$ and each time $s \ge 0$ with $s + \pi_1\,\widetilde{z} \le \pi_1\,\widetilde{x}(t)$ and $s+h < \mathbb{T}_j(\widetilde{\vartheta},\widetilde{z})$, we conclude from condition (8.) on timed transitions that for all $k \in]0,h[$ sufficiently small (depending on $h,s,t,\widetilde{z}$)

$$\widetilde{d}_j\big(\widetilde{\vartheta}(s+h,\widetilde{z}),\; \widetilde{x}(t+h)\big) \le \widetilde{d}_j\big(\widetilde{\vartheta}(s+h-k,\widetilde{z}),\; \widetilde{x}(t+h)\big) + \tfrac{h^2}{2}.$$

Due to Lemma 8 (on page 340) and the semicontinuity of $\widetilde{d}_{j,\kappa}$ (in the sense of hypothesis (H3') $\widetilde{(\text{i'})}$ on page 334), the index $\kappa \in \mathscr{J}$ depending on $h,k,s,t,\widetilde{z}$ can be selected such that

$$
\begin{aligned}
\widetilde{d}_j\big(\widetilde{\vartheta}(s+h,\widetilde{z}),\ \widetilde{x}(t+h)\big) - h^2 \le\ & \widetilde{d}_j\big(\widetilde{\vartheta}(s+h-k,\widetilde{z}),\ \widetilde{x}(t+h)\big) \ - \ \frac{h^2}{2} \\
\le\ & \widetilde{d}_{j,\kappa}\big(\widetilde{\vartheta}(s+h-k,\widetilde{z}),\ \widetilde{x}(t+h)\big) \\
\le & \limsup_{m\to\infty} \ \Big(\widetilde{d}_{j,\kappa}\big(\widetilde{\vartheta}(s+h-k,\widetilde{z}), \widetilde{x}_{n_m}(t+h-\widetilde{\delta}_m)\big)\Big) \\
\le & \limsup_{m\to\infty} \ \Big(\widetilde{d}_{j}\big(\widetilde{\vartheta}(s,\ \widetilde{z}), \widetilde{x}_{n_m}(t+k-\widetilde{\delta}_m)\big) \\
& + (h-k)\cdot \sup_{[t+k-\widetilde{\delta}_m,\ t+h-\widetilde{\delta}_m]} \widehat{D}_j\big(\widetilde{\vartheta},\ \widetilde{f}_{n_m}(\widetilde{x}_{n_m},\cdot);\ \widetilde{z},R_j\big)\Big) \\
& \qquad \times e^{\widehat{\alpha}_j(\widetilde{z},R_j)\cdot(h-k)}.
\end{aligned}
$$

From now on, the influence of the index $\kappa \in \mathcal{J}$ is of no further relevance and, we continue exactly as in the proof of Convergence Theorem 12:

Indeed, choosing suitable subsequences $(\delta_{m_l})_{l\in\mathbb{N}}$, $(\widetilde{\delta}_{m_l})_{l\in\mathbb{N}}$ and a sequence $(k_l)_{l\in\mathbb{N}}$ such that the preceding limit superior for $m \to \infty$ coincides with the limit for $l \to \infty$ and $\delta_{m_l} < k_l - \widetilde{\delta}_{m_l} < \frac{1}{l}$ for each $l \in \mathbb{N}$, we obtain successively

$$
\begin{aligned}
\lim_{l\to\infty} \ \ & \widetilde{d}_j\big(\widetilde{x}(t),\quad \widetilde{x}_{n_{m_l}}(t+k_l-\widetilde{\delta}_{m_l})\big) \ = \ 0, \\
\limsup_{l\to\infty} \ \ & \widetilde{d}_j\big(\widetilde{\vartheta}(s,\widetilde{z}),\ \widetilde{x}_{n_{m_l}}(t+k_l-\widetilde{\delta}_{m_l})\big) \ \le \ \widetilde{d}_j\big(\widetilde{\vartheta}(s,\widetilde{z}),\ \widetilde{x}(t)\big)
\end{aligned}
$$

as consequences of hypotheses (H3') ($\widetilde{\text{ii}}_l$), ($\widetilde{\text{i}}$") (on page 334). Now $l \longrightarrow \infty$ leads to

$$
\begin{aligned}
& \widetilde{d}_j\big(\widetilde{\vartheta}(s+h,\widetilde{z}),\ \widetilde{x}(t+h)\big) \ - \ 2\,h^2 \ - \ \widetilde{d}_j\big(\widetilde{\vartheta}(s,\widetilde{z}),\ \widetilde{x}(t)\big) \cdot e^{\widehat{\alpha}_j(\widetilde{z},R_j)\,h} \\
& \quad \le h \cdot \limsup_{m\to\infty} \ \sup_{[t+\delta_m,\ t+h]} \ \widehat{D}_j\big(\widetilde{\vartheta},\ \widetilde{f}_{n_m}(\widetilde{x}_{n_m}(\cdot),\cdot);\ \widetilde{z},R_j\big) \ \cdot e^{\widehat{\alpha}_j(\widetilde{z},R_j)\,h}.
\end{aligned}
$$

For completing the proof, we verify

$$
\limsup_{h\downarrow 0}\ \limsup_{m\to\infty} \ \sup_{[t+\delta_m,\ t+h]} \ \widehat{D}_j\big(\widetilde{\vartheta},\ \widetilde{f}_{n_m}(\widetilde{x}_{n_m}(\cdot),\cdot);\ \widetilde{z},R_j\big) \ \le \ \widehat{D}_j\big(\widetilde{\vartheta},\ \widetilde{f}(\widetilde{x}(t),t);\ \widetilde{z},R_j\big)
$$

for $\mathcal{L}^1$-almost every $t \in [0,T[$ and *any* subsequence $\big(\widetilde{x}_{n_m}(\cdot)\big)_{m\in\mathbb{N}}$ satisfying

$$
\begin{cases}
\widetilde{d}_i\big(\widetilde{x}(t),\qquad \widetilde{x}_{n_m}(t+\delta_m)\big) & \longrightarrow \ 0 \\
\widehat{D}_i\big(\widetilde{f}(\widetilde{x}(t),t),\ \widetilde{f}_{n_m}(\widetilde{x}(t),t);\ \widetilde{z},\ r\big) & \longrightarrow \ 0
\end{cases}
$$

for $m \longrightarrow \infty$ and each $i \in \mathcal{I}$, $r \ge 0$. Indeed, if this inequality was not correct then we could select $\varepsilon > 0$ and sequences $(h_l)_{l\in\mathbb{N}}$, $(m_l)_{l\in\mathbb{N}}$, $(s_l)_{l\in\mathbb{N}}$ s.t. for all $l \in \mathbb{N}$,

$$
\begin{cases}
\widehat{D}_j\big(\widetilde{\vartheta},\ \widetilde{f}_{n_{m_l}}(\widetilde{x}_{n_{m_l}}(t+s_l),t+s_l);\ \widetilde{z},\ R_j\big) \ \ge \ \widehat{D}_j\big(\widetilde{\vartheta},\ \widetilde{f}(\widetilde{x}(t),\ t);\ \widetilde{z},\ R_j\big) \ + \ \varepsilon, \\
\delta_{m_l} \ \le \ s_l \ \le \ h_l \ \le \ \frac{1}{l}, \qquad m_l \ge l.
\end{cases}
$$

Due to property (H3') ($\widetilde{\text{ii}}_l$), the uniform Lipschitz continuity of $(\widetilde{x}_{n_m}(\cdot))_{m\in\mathbb{N}}$ implies

$$
\lim_{l\to\infty} \ \widetilde{d}_i\big(\widetilde{x}(t),\ \widetilde{x}_{n_{m_l}}(t+s_l)\big) \ = \ 0
$$

for each $i \in \mathcal{I}$. At $\mathcal{L}^1$-a.e. time $t \in [0,T[$, assumptions (3.), (4.') (i) and hypothesis (H6') (on page 337) lead to a contradiction with regard to $\widehat{D}_j\big(\widetilde{\vartheta},\ \widetilde{f}(\widetilde{x}(t),\ t);\ \widetilde{z},\ r\big)$ for any $r \ge 0$. □

4.4 Example: Mutational Equations for Compact Sets in $\mathbb{R}^N$ Depending on the Normal Cones

$\mathscr{K}(\mathbb{R}^N)$ consists of all nonempty compact subsets of $\mathbb{R}^N$. One of the main goals in this chapter is to take the normal cones at the topological boundary of the respective compact set into consideration explicitly. The introduction has already revealed that there are some obstacles which we want to overcome by means of nonsymmetric distance functions and the notion of distribution-like (timed) solutions.
In this section, we present a geometric example in detail. It also uses reachable sets of autonomous differential inclusions for inducing transitions. A separate time component, however, is of no additional use here and thus, we simply skip it.

4.4.1 Limiting Normal Cones Induce Distance $d_{\mathscr{K},N}$ on $\mathscr{K}(\mathbb{R}^N)$

The so-called *Pompeiu-Hausdorff excess* is an example of a nonsymmetric distance function on $\mathscr{K}(\mathbb{R}^N)$ that is very similar to Pompeiu-Hausdorff distance dl:

$$e^{\subset}(K_1, K_2) := \sup_{x \in K_1} \operatorname{dist}(x, K_2)$$
$$e^{\supset}(K_1, K_2) := \sup_{y \in K_2} \operatorname{dist}(y, K_1).$$

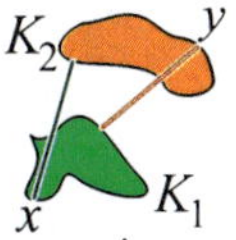

for $K_1, K_2 \in \mathscr{K}(\mathbb{R}^N)$. Obviously, the link to the Pompeiu-Hausdorff distance is

$$\mathit{dl}(K_1, K_2) = \max\{e^{\subset}(K_1, K_2),\ e^{\supset}(K_1, K_2)\}$$

(see also [10, § 3.2] and [162, § 4.C], for example).

In the following, we prefer taking the boundaries into consideration explicitly. The Pompeiu-Hausdorff excess $e^{\supset}(K_1, K_2)$, however, does not distinguish between boundary points and interior points of the compact sets K_1, K_2. Thus, a new distance function $d_{\mathscr{K},N}$ on $\mathscr{K}(\mathbb{R}^N)$ is defined in a moment. Strictly speaking, we even use the first-order approximation of the boundary represented by the limiting normal cones of a set. Following the standard definitions as in [162, 180], the proximal normal cone $N_C^P(x)$ and the limiting normal cone $N_C(x)$ of any nonempty closed subset $C \subset \mathbb{R}^N$ are introduced in Definition A.23 (on page 454).
As a further abbreviation, we set ${}^\flat N_C(x) := N_C(x) \cap \mathbb{B} = \{v \in N_C(x) : |v| \leq 1\}$.

Definition 22. Set $d_{\mathscr{K},N} : \mathscr{K}(\mathbb{R}^N) \times \mathscr{K}(\mathbb{R}^N) \longrightarrow [0, \infty[$,

$$d_{\mathscr{K},N}(K_1, K_2) := \mathit{dl}(K_1, K_2) + e^{\supset}(\operatorname{Graph} {}^\flat N_{K_1},\ \operatorname{Graph} {}^\flat N_{K_2}).$$

Obviously, the function $d_{\mathscr{K},N}$ is an example of a so-called *quasi-metric* on the set $\mathscr{K}(\mathbb{R}^N)$, i.e., it is positive definite and satisfies the triangle inequality, but in general, it is not symmetric.

The properties of $d_{\mathscr{K},N}$ with respect to convergence depend on the relation between the normal cones of compact sets K_n $(n \in \mathbb{N})$ and their limit $K = \mathrm{Lim}_{n\to\infty} K_n$ in the sense of Painlevé-Kuratowski (if it exists).
In general, they do not coincide of course, but each limiting normal vector of K can be approximated by limiting normal vectors of a subsequence $(K_{n_j})_{j\in\mathbb{N}}$. This asymptotic inclusion is formulated in the next proposition and, its proofs results from Proposition A.72 (on page 487), [14, Theorem 8.4.6], [50, Lemma 4.1] or [162, Example 6.18], for example. But the inclusion might be strict.

Proposition 23. *Let $(M_k)_{k\in\mathbb{N}}$ be a sequence of closed subsets of $\mathbb{R}^N$ and set $M := \mathrm{Limsup}_{k\to\infty} M_k$ in the sense of Painlevé-Kuratowski. Then,*

(1.) $\mathrm{Graph}\, N_M^P \subset \mathrm{Limsup}_{k\to\infty}\, \mathrm{Graph}\, N_{M_k}^P$,
(2.) $\mathrm{Graph}\, N_M \subset \mathrm{Limsup}_{k\to\infty}\, \mathrm{Graph}\, N_{M_k}$.

Corollary 24. *Let $(M_k)_{k\in\mathbb{N}}$ be a sequence of closed subsets of $\mathbb{R}^N$ whose limit $M := \mathrm{Lim}_{k\to\infty} M_k$ exists in the sense of Painlevé-Kuratowski. Then*

$$\mathrm{Graph}\, N_M \subset \mathrm{Liminf}_{k\to\infty}\, \mathrm{Graph}\, N_{M_k}.$$

In particular, $$\partial M \subset \mathrm{Liminf}_{k\to\infty}\, \partial M_k.$$

Proof is an indirect consequence of Proposition 23 due to $M = \mathrm{Lim}_{k\to\infty} M_k$. □

4.4.2 Reachable Sets of Differential Inclusions Provide Transitions

Now we focus on reachable sets of a differential inclusion $x'(\cdot) \in F(x(\cdot))$ and the evolution of limiting normal cones at the topological boundary. In particular, we use the *Hamilton condition* as a key tool. It implies that roughly speaking, every boundary point x_0 of $\vartheta_F(t_0, K)$ and normal vector $\nu \in N_{\vartheta_F(t_0,K)}(x_0)$ have a solution of $x'(\cdot) \in F(x(\cdot))$ and an adjoint arc linking x_0 to some $z \in \partial K$ and ν to $N_K(z)$, respectively.

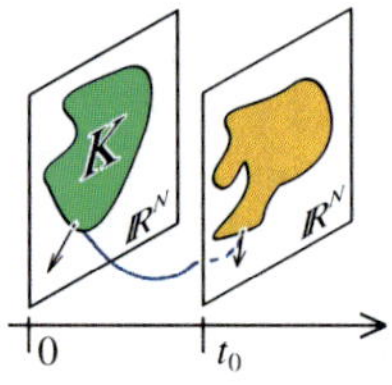

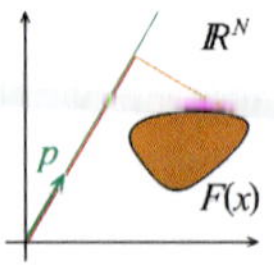

Furthermore the solution and its adjoint arc fulfill a system of partial differential equations with the so-called *(upper) Hamiltonian* of the set-valued map $F : \mathbb{R}^N \rightsquigarrow \mathbb{R}^N$,

$$\mathscr{H}_F : \mathbb{R}^N \times \mathbb{R}^N \longrightarrow \mathbb{R}^N, \quad (x,p) \longmapsto \sup_{y\in F(x)} p\cdot y.$$

Although the Hamilton condition is known in much more general forms (consider e.g. [180, Theorem 7.7.1] applied to proximal balls), we use only the following "smooth" version — due to later regularity conditions on F.

Basic set Test set	$E := \mathscr{K}(\mathbb{R}^N)$ the set of nonempty compact subsets of the Euclidean space $\mathbb{R}^N$ $\mathscr{D} := \mathscr{K}_{C^{1,1}}(\mathbb{R}^N)$ the set of all nonempty compact N-dimensional $C^{1,1}$ submanifolds of $\mathbb{R}^N$
Distance Absolute value	$d_{\mathscr{K},N}(K_1,K_2) := dl(K_1,K_2) + e^{\supset}(\text{Graph}\ {}^\flat N_{K_1},\ \text{Graph}\ {}^\flat N_{K_2})$ with the limiting normal cone $N_{K_1}(\cdot)$ (Definition A.23, page 454), ${}^\flat N_{K_1}(x) := N_{K_1}(x) \cap \mathbb{B} = \{v \in N_{K_1}(x) : \|v\| \le 1\}$ and Pompeiu-Hausdorff excess $e^{\supset}(M_1,M_2) := \sup_{y \in M_2} \text{dist}(y,M_1)$ (It is not symmetric, but satisfies the triangle inequality.) $\lfloor \cdot \rfloor := 0$
Transition	For each $F \in \text{LIP}_\lambda^{(\mathscr{H})}(\mathbb{R}^N,\mathbb{R}^N)$, i.e. set-valued map $F : \mathbb{R}^N \rightsquigarrow \mathbb{R}^N$ with compact convex values and $C^{1,1}$ Hamiltonian (Def.26), set $\vartheta_F : [0,1] \times \mathscr{K}(\mathbb{R}^N) \longrightarrow \mathscr{K}(\mathbb{R}^N)$ by means of reachable sets of the autonomous differential inclusion $x'(\cdot) \in F(x(\cdot))$ a.e.: $\vartheta_F(t,\ K_0) := \{x(t) \mid \text{there exists } x(\cdot) \in W^{1,1}([0,t],\mathbb{R}^N) : x'(\cdot) \in F(x(\cdot))\ \mathscr{L}^1\text{-a.e. in } [0,t],\ x(0) \in K_0\}.$
Compactness	transitionally Euler compact — if the transitions are restricted to maps $F \in \text{LIP}_\lambda^{(\mathscr{H}_\circ^\rho)}(\mathbb{R}^N,\mathbb{R}^N) \subset \text{LIP}_\lambda^{(\mathscr{H})}(\mathbb{R}^N,\mathbb{R}^N)$ (Definition 34) due to smoothing effects of C^2 Hamiltonians on interior spheres in reachable sets (§ A.5.5): Proposition 36 (page 368)
Mutational solutions	Reachable sets of a nonautonomous differential inclusion whose set-valued right-hand side is determined via feedback — if all set values are $C^{1,1}$ submanifolds of $\mathbb{R}^N$: Remark 39
List of main results formulated in § 4.4	Existence due to transitional Euler compactness: Corollary 37 (page 369) Existence for equations with delay: Corollary 38 (No appropriate results about uniqueness or continuity w.r.t. data, however, because lower bounds of $\mathbb{T}(\vartheta_F,M) > 0$ are lacking for $M \in \mathscr{D}$ so far, see § 4.3.1. This gap motivates the next example in § 4.5 below.)
Key tools	The adjoint arc is a necessary condition on boundary points and their limiting normals: Proposition 25 (page 362) Boundary regularity of reachable sets of differential inclusions (by means of adjoint arcs) in Appendix A.5 (page 454 ff.)

Table 4.1 Brief summary of the example in § 4.4 in mutational terms:
Mutational equations for compact sets in $\mathbb{R}^N$ depending on the normal cones

Proposition 25. *Suppose for the set-valued map* $F : \mathbb{R}^N \rightsquigarrow \mathbb{R}^N$

1. $F(\cdot)$ *has nonempty convex compact values,*
2. $\mathscr{H}_F(\cdot,\cdot)$ *is continuously differentiable in* $\mathbb{R}^N \times (\mathbb{R}^N \setminus \{0\})$,
3. *the derivative of* $\mathscr{H}_F$ *has linear growth in* $\mathbb{R}^N \times (\mathbb{R}^N \setminus \mathbb{B}_1)$, *i.e.* $\|D\mathscr{H}_F(x,p)\| \leq const \cdot (1+|x|+|p|)$ *for all* $x,p \in \mathbb{R}^N$, $|p|>1$.

Let $K \in \mathscr{K}(\mathbb{R}^N)$ *be any initial set and* $t_0 > 0$.

For every boundary point $x_0 \in \partial\,\vartheta_F(t_0,K)$ *and normal* $\nu \in N_{\vartheta_F(t_0,K)}(x_0) \setminus \{0\}$, *there exist a solution* $x(\cdot) \in C^1([0,t_0],\mathbb{R}^N)$ *and its adjoint arc* $p(\cdot) \in C^1([0,t_0],\mathbb{R}^N)$ *with*

$$\begin{cases} x'(t) = \frac{\partial}{\partial p}\, \mathscr{H}_F(x(t),\, p(t)) \in F(x(t)), & x(t_0) = x_0, \quad x(0) \in \partial K, \\ p'(t) = -\frac{\partial}{\partial x}\, \mathscr{H}_F(x(t),\, p(t)), & p(t_0) = \nu, \quad p(0) \in N_K(x(0)). \end{cases}$$

These assumptions give a first hint about adequate conditions on $F : \mathbb{R}^N \rightsquigarrow \mathbb{R}^N$ for transitions with respect to $d_{\mathscr{K},N}$. Supposing $D\mathscr{H}_F$ to be Lipschitz continuous (in addition) provides some technical advantages such as global existence of unique solutions to the Hamiltonian system (see also Remark 30 (a) below).

Definition 26. For any parameter $\lambda > 0$, the set $\mathrm{LIP}_\lambda^{(\mathscr{H})}(\mathbb{R}^N,\mathbb{R}^N)$ contains all set-valued maps $F : \mathbb{R}^N \rightsquigarrow \mathbb{R}^N$ with

(1.) $F : \mathbb{R}^N \rightsquigarrow \mathbb{R}^N$ has nonempty compact convex values,

(2.) $\mathscr{H}_F(\cdot,\cdot) \in C^{1,1}(\mathbb{R}^N \times (\mathbb{R}^N \setminus \{0\}))$,

(3.) $\|\mathscr{H}_F\|_{C^{1,1}(\mathbb{R}^N \times\, \partial\mathbb{B}_1)} \stackrel{\text{Def.}}{=} \|\mathscr{H}_F\|_{C^1(\mathbb{R}^N \times\, \partial\mathbb{B}_1)} + \mathrm{Lip}\; D\mathscr{H}_F|_{\mathbb{R}^N \times\, \partial\mathbb{B}_1} < \lambda$.

The Lipschitz continuity with respect to time is a first (and still rather simple) example how the Hamiltonian system in combination with the bounds on the Hamiltonian can be used:

Lemma 27. *For every* $F \in \mathrm{LIP}_\lambda^{(\mathscr{H})}(\mathbb{R}^N,\mathbb{R}^N)$ *and* $K \in \mathscr{K}(\mathbb{R}^N)$, $0 \leq s \leq t \leq T$,

$$d_{\mathscr{K},N}\Big(\vartheta_F(s,K),\, \vartheta_F(t,K)\Big) \leq \lambda\,(e^{\lambda T}+2) \cdot (t-s).$$

Proof. Obviously, the Pompeiu-Hausdorff distance satisfies for every $s,t \geq 0$

$$d\!l\,(\vartheta_F(s,K),\, \vartheta_F(t,K)) \leq \sup_{\mathbb{R}^N} \|F(\cdot)\|_\infty \cdot |t-s| \leq \lambda\,|t-s|.$$

Proposition 25 guarantees that for every $0 \leq s < t$, $x \in \partial\,\vartheta_F(t,K)$ and $p \in {}^\flat N_{\vartheta_F(t,K)}(x) \setminus \{0\}$, there exist a solution $x(\cdot) \in C^1([s,t],\mathbb{R}^N)$ and its adjoint arc $p(\cdot) \in C^1([s,t],\mathbb{R}^N)$ satisfying

$$\begin{cases} x'(\tau) = \frac{\partial}{\partial p}\, \mathscr{H}_F(x(\tau),\, p(\tau)) \in F(x(\tau)), & x(t) = x, \quad x(s) \in \partial\vartheta_F(s,K), \\ p'(\tau) = -\frac{\partial}{\partial x}\, \mathscr{H}_F(x(\tau),\, p(\tau)), & p(t) = p, \quad p(s) \in N_{\vartheta_F(s,K)}(x(s)). \end{cases}$$

Obviously, $\mathscr{H}_F$ is positively homogeneous with respect to its second argument and thus, $|p'(\tau)| \le \lambda\ |p(\tau)|$ for all τ. Moreover $|p| \le 1$ implies that the projection of p on any cone is also contained in $\mathbb{B}_1$. Finally we obtain

$$\begin{aligned} \operatorname{dist}\Big((x,p),\ \operatorname{Graph}{}^\flat N_{\vartheta_F(s,K)}\Big) &\le |x - x(s)| \ +\ |p - p(s)| \\ &\le \sup_{s\le\tau\le t} \Big(|\frac{\partial}{\partial p}\mathscr{H}_F| + |\frac{\partial}{\partial x}\mathscr{H}_F|\Big)\Big|_{(x(\tau),p(\tau))} \cdot (t-s) \\ &\le \Big(\lambda + \lambda\, e^{\lambda t}\Big)\cdot (t-s). \end{aligned}$$

□

Now the next question considers the choice of suitable "test sets".
The difficulties in regard to regularity usually occur when the topological boundary of the reachable set is not continuous. This rather qualitative observation motivates the question for which type of compact subsets and differential inclusions we can exclude such discontinuities — within short periods at least.
In subsequent Appendix A.5 (on page 454 ff.), the regularity of reachable sets is investigated. Let us summarize some results which are of special interest here:

Definition 28. $\mathscr{K}_{C^{1,1}}(\mathbb{R}^N)$ abbreviates the set of all nonempty compact N-dimensional $C^{1,1}$ submanifolds of $\mathbb{R}^N$ with boundary.
A closed subset $C \subset \mathbb{R}^N$ is said to have *positive erosion of radius* $\rho > 0$ if for every $r \in]0,\rho[$, there is a closed set $M \subset \mathbb{R}^N$ with

$$\begin{cases} C = \{x \in \mathbb{R}^N \mid \operatorname{dist}(x,M) \le r\}, \\ M = \{x \in C \mid \operatorname{dist}(x,\partial C) \ge r\}. \end{cases}$$

$\mathscr{K}_\circ^\rho(\mathbb{R}^N)$ consists of all sets with positive erosion of radius $\rho > 0$ and, set $\mathscr{K}_\circ(\mathbb{R}^N) := \bigcup_{\rho>0} \mathscr{K}_\circ^\rho(\mathbb{R}^N)$.

Proposition 29. *Let $F : \mathbb{R}^N \rightsquigarrow \mathbb{R}^N$ be a map of $\mathrm{LIP}_\lambda^{(\mathscr{H})}(\mathbb{R}^N,\mathbb{R}^N)$. For every compact N-dimensional $C^{1,1}$ submanifold K of $\mathbb{R}^N$ with boundary, there exist a time $\mathbb{T} = \mathbb{T}(\vartheta_F, K) > 0$ and a radius $\rho > 0$ such that for all $t \in [0,\mathbb{T}[$,*

(1.) $\vartheta_F(t,K) \in \mathscr{K}_{C^{1,1}}(\mathbb{R}^N)$ *with radius of curvature* $\ge \rho$,

(2.) $K = \mathbb{R}^N \Big\backslash\ \vartheta_{-F}(t,\ \mathbb{R}^N \setminus \vartheta_F(t,K))$.

Remark 30. (a) A complete proof is presented in Propositions A.34 and A.36. For statement (1.), we use the evolution of Graph $(N_K(\cdot) \cap \partial\mathbb{B}) \subset \mathbb{R}^N \times \mathbb{R}^N$ along the Hamiltonian system with $\mathscr{H}_F$.
Indeed, Lemma A.35 (on page 458) specifies sufficient conditions on the system so that graphs of Lipschitz continuous functions preserve this regularity for short times. Applying this lemma to unit normals to reachable sets of $K \in \mathscr{K}_{C^{1,1}}(\mathbb{R}^N)$ requires the Hamiltonian $\mathscr{H}_F$ to be in $C^{1,1}(\mathbb{R}^N \times (\mathbb{R}^N \setminus \{0\}))$ instead of C^1.
In fact, this Lemma A.35 is an analytical reason for choosing $\mathscr{K}_{C^{1,1}}(\mathbb{R}^N)$ as "test subset" of $\mathscr{K}(\mathbb{R}^N)$ — instead of compact sets with C^1 boundary, for example.

(b) Together with Proposition 25, statement (2.) provides a connection between the boundaries ∂K and $\partial\, \vartheta_F(t,K)$ — now in both forward and backward time direction.

(c) Sets of positive erosion are closely related to "sets of positive reach" (in the sense of Federer). Further details are presented in § A.5.1 (on page 454 ff.).

Lemma 31. *Assume for $F, G \in \mathrm{LIP}^{(\mathscr{H})}_\lambda(\mathbb{R}^N,\mathbb{R}^N)$, $K_1, K_2 \in \mathscr{K}(\mathbb{R}^N)$ and $\rho, T > 0$ that all the sets $\vartheta_F(t,K_1) \in \mathscr{K}_{C^{1,1}}(\mathbb{R}^N)$ $(0 \le t \le T)$ have positive reach $\ge \rho$ (in the sense of Definition A.26 on page 455).*

Then, for every $t \in [0,T[$,

$$\begin{aligned} &d_{\mathscr{K},N}\big(\vartheta_F(t,K_1),\ \vartheta_G(t,K_2)\big) \\ &\quad\le\ e^{(\Lambda_F+\lambda)\,t} \cdot \Big(d_{\mathscr{K},N}(K_1,\,K_2) + 6\,N\,t\ \|\mathscr{H}_F - \mathscr{H}_G\|_{C^1(\mathbb{R}^N\times\partial\mathbb{B}_1)}\Big) \end{aligned}$$

with $\qquad \Lambda_F \ :=\ 9\,e^{2\lambda T}\ \|\mathscr{H}_F\|_{C^{1,1}(\mathbb{R}^N\times\partial\mathbb{B}_1)} \ \le\ 9\,e^{2\lambda T}\,\lambda\ <\ \infty.$

Postponing the proof for a moment, we now obtain all the parameters needed for a transition on $\mathscr{K}(\mathbb{R}^N)$:

Proposition 32. *For every $\lambda \ge 0$, the reachable sets of the set-valued maps in $\mathrm{LIP}^{(\mathscr{H})}_\lambda(\mathbb{R}^N,\mathbb{R}^N)$ induce transitions on $(\mathscr{K}(\mathbb{R}^N),\ \mathscr{K}_{C^{1,1}}(\mathbb{R}^N),\ d_{\mathscr{K},N},\ d_{\mathscr{K},N},\ 0)$ in the sense of Definition 1 and Remark 3 (on page 336 f.) with*

$$\begin{aligned} \alpha(\vartheta_F;\,\cdot,\cdot) &\overset{\text{Def.}}{=}\ 10\ \lambda,\\ \beta(\vartheta_F;\,\cdot) &\overset{\text{Def.}}{=}\ \quad\lambda\ (e^\lambda + 2),\\ \gamma(\vartheta_F) &\overset{\text{Def.}}{=}\ 0,\\ \widehat{D}(\vartheta_F,\vartheta_G;\,\cdot,\cdot) &\overset{\text{Def.}}{=}\ 6\ N\ \|\mathscr{H}_F - \mathscr{H}_G\|_{C^1(\mathbb{R}^N\times\partial\mathbb{B}_1)}\,. \end{aligned}$$

Proof (of Lemma 31). Proposition 1.50 (on page 60) concludes the following estimate of the Pompeiu-Hausdorff distance from Filippov's Theorem A.6 about differential inclusions (with Lipschitz continuous right-hand side)

$$\begin{aligned} d\!l\big(\vartheta_F(t,K_1),\ \vartheta_G(t,K_2)\big) &\le\ d\!l(K_1,K_2)\cdot e^{\lambda t} + \sup_{\mathbb{R}^N} d\!l\big(F(\cdot),G(\cdot)\big) \quad \cdot \tfrac{e^{\lambda t}-1}{\lambda}\\ &\le\ d\!l(K_1,K_2)\cdot e^{\lambda t} + \sup_{\mathbb{R}^N\times\partial\mathbb{B}_1} |\mathscr{H}_F - \mathscr{H}_G| \ \cdot t\ e^{\lambda t}. \end{aligned}$$

Now we still need an upper bound of $e^{\supset}\big(\mathrm{Graph}\,{}^\flat N_{\vartheta_F(t,K_1)},\ \mathrm{Graph}\,{}^\flat N_{\vartheta_G(t,K_2)}\big)$.

Choose $x \in \partial\,\vartheta_G(t,K_2)$, $p \in N_{\vartheta_G(t,K_2)}(x) \cap \partial\mathbb{B}_1$ and $\delta > 0$ arbitrarily. According to Proposition 25 (on page 362), there exist a solution $x(\cdot) \in C^1([0,t],\mathbb{R}^N)$ relative to G and its adjoint arc $p(\cdot) \in C^1([0,t],\mathbb{R}^N)$ with

$$\begin{cases} x'(\cdot) = \frac{\partial}{\partial p}\mathscr{H}_G(x(\cdot),\,p(\cdot)) \in G(x(\cdot)), & p'(\cdot) = -\frac{\partial}{\partial x}\mathscr{H}_G(x(\cdot),\,p(\cdot)) \in \lambda\,|p(\cdot)|\cdot\mathbb{B}\\ x(0) \in \partial K_2, & p(0) \in N_{K_2}(x(0)),\\ x(t) = x, & p(t) = p. \end{cases}$$

Gronwall's inequality guarantees

$$0 < e^{-\lambda t} \le |p(\cdot)| \le e^{\lambda t}$$

and hence,
$$p(0)\, e^{-\lambda t} \in {}^\flat N_{K_2}(x(0)) \setminus \{0\}.$$

Now let $(y_0, \widehat{q}_0)$ denote an element of Graph ${}^\flat N_{K_1}$ with $\widehat{q}_0 \neq 0$ and

$$\big|(y_0, \widehat{q}_0) - \big(x(0),\, p(0)\, e^{-\lambda t}\big)\big| \le$$
$$\le e^{\supset}\Big(\text{Graph}\, {}^\flat N_{K_1},\ \text{Graph}\, {}^\flat N_{K_2}\Big) + \delta.$$

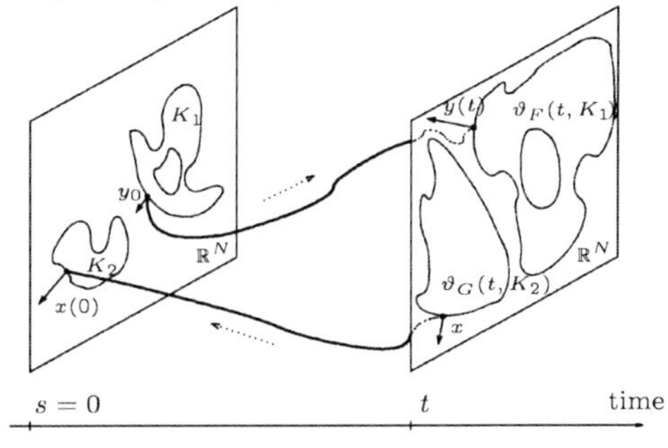

Assuming that all sets $\vartheta_F(s, K_1) \in \mathscr{K}(\mathbb{R}^N)$ $(s \in [0,t])$ have uniform positive reach implies the reversibility in time due to Proposition A.36 (on page 461):
$\mathbb{R}^N \setminus K_1 = \vartheta_{-F}(t,\, \mathbb{R}^N \setminus \vartheta_F(t, K_1)).$

In particular, y_0 is a boundary point of the (not bounded) N-dimensional $C^{1,1}$ submanifold $\mathbb{R}^N \setminus \overset{\circ}{K_1} = \vartheta_{-F}(t, \overline{\mathbb{R}^N \setminus \vartheta_F(t, K_1)})$ with boundary and, $-\widehat{q}_0$ belongs to its limiting normal cone at y_0. As a consequence of Proposition 25 again and due to $\mathscr{H}_{-F}(z,v) = \mathscr{H}_F(z,-v)$ for all z, v, we obtain a solution $y(\cdot) \in C^1([0,t], \mathbb{R}^N)$ and its adjoint arc $q(\cdot)$ satisfying

$$\begin{cases} y'(\cdot) = \frac{\partial}{\partial p} \mathscr{H}_F(y(\cdot), q(\cdot)), & q'(\cdot) = -\frac{\partial}{\partial y} \mathscr{H}_F(y(\cdot), q(\cdot)), \\ y(0) = y_0, & q(0) = \widehat{q}_0\, e^{\lambda t} \neq 0, \\ y(t) \in \partial\, \vartheta_F(t, K_1), & q(t) \in N_{\vartheta_F(t,K_1)}(y(t)). \end{cases}$$

According to Lemma 33 below, the derivative of $\mathscr{H}_F$ is Λ_F-Lipschitz continuous on $\mathbb{R}^N \times (\mathbb{B}_{e^{\lambda T}} \setminus \overset{\circ}{\mathbb{B}}_{e^{-\lambda T}})$. Thus, the Theorem of Cauchy-Lipschitz leads to

$$\text{dist}\big((x,p),\ \text{Graph}\ {}^\flat N_{\vartheta_F(t,K_1)}\big) \le \big|(x,\, p) - (y(t), q(t))\big|$$
$$\le e^{\Lambda_F \cdot t} \cdot \big|(x(0),\, p(0)) - (y_0,\, \widehat{q}_0\, e^{\lambda t})\big| + \frac{e^{\Lambda_F \cdot t} - 1}{\Lambda_F} \cdot \sup_{0 \le s \le t} \big|D\mathscr{H}_F - D\mathscr{H}_G\big|\Big|_{(x(s),\, p(s))}.$$

$\mathscr{H}_F$ and $\mathscr{H}_G$ are positively homogeneous w.r.t. the second argument and thus,

$$\Big|\tfrac{\partial}{\partial x_j}\, (\mathscr{H}_F - \mathscr{H}_G)|_{(x(s),\, p(s))}\Big| \le e^{\lambda t}\, \|D\mathscr{H}_F - D\mathscr{H}_G\|_{C^0(\mathbb{R}^N \times \partial \mathbb{B}_1)},$$
$$\Big|\tfrac{\partial}{\partial p_j}\, (\mathscr{H}_F - \mathscr{H}_G)|_{(x(s),\, p(s))}\Big| \le 3 \cdot \|\mathscr{H}_F - \mathscr{H}_G\|_{C^1(\mathbb{R}^N \times \partial \mathbb{B}_1)}.$$

as the partial derivatives in the proof of Lemma 33 below reveal. Now we obtain

$$\text{dist}\big((x,p),\ \text{Graph}\ {}^\flat N_{\vartheta_F(t,K_1)}\big)$$
$$\le e^{(\Lambda_F + \lambda)\, t}\, \big|(x(0),\, p(0)\, e^{-\lambda t}) - (y_0, \widehat{q}_0)\big| + e^{\Lambda_F t} t \cdot 6 N e^{\lambda t}\, \|\mathscr{H}_F - \mathscr{H}_G\|_{C^1(\mathbb{R}^N \times \partial \mathbb{B}_1)}$$

and, since $\delta > 0$ is arbitrarily small and $|p| = 1$,

$$e^{\supset}\big(\text{Graph}\ {}^\flat N_{\vartheta_F(t,K_1)},\ \text{Graph}\ {}^\flat N_{\vartheta_G(t,K_2)}\big)$$
$$\le e^{(\Lambda_F + \lambda)\, t} \cdot \Big\{ e^{\supset}\big(\text{Graph}\, {}^\flat N_{K_1},\ \text{Graph}\, {}^\flat N_{K_2}\big) + 6 N t \cdot \|\mathscr{H}_F - \mathscr{H}_G\|_{C^1(\mathbb{R}^N \times \partial \mathbb{B}_1)} \Big\}. \qquad \square$$

Lemma 33. *For every* $F \in \mathrm{LIP}_\lambda^{(\mathscr{H})}(\mathbb{R}^N,\mathbb{R}^N)$ *and radius* $R > 1$, *the product* $9R^2\lambda$ *is a Lipschitz constant of the derivative* $D\mathscr{H}_F$ *restricted to* $\mathbb{R}^N \times (\mathbb{B}_R \setminus \overset{\circ}{\mathbb{B}}_{\frac{1}{R}})$.

Proof (of Lemma 33). It results essentially from the fact that $\mathscr{H}_F(x,p)$ is positively homogeneous with respect to p:
For every $(x,p) \in \mathbb{R}^N \times (\mathbb{B}_R \setminus \overset{\circ}{\mathbb{B}}_{\frac{1}{R}})$, we conclude from $\mathscr{H}_F(x,p) = |p|\ \mathscr{H}_F(x, \frac{p}{|p|})$

$$\begin{aligned}
\frac{\partial\,\mathscr{H}_F(x,p)}{\partial p_j} &= \frac{\partial}{\partial p_j}|p| \cdot \mathscr{H}_F(x,\tfrac{p}{|p|}) + |p| \cdot \sum_{k=1}^N \frac{\partial}{\partial p_k}\mathscr{H}_F|_{(x,\frac{p}{|p|})} \cdot \frac{\partial}{\partial p_j}\frac{p_k}{|p|} \\
&= \frac{p_j}{|p|} \cdot \mathscr{H}_F(x,\tfrac{p}{|p|}) + |p| \cdot \sum_{k=1}^N \frac{\partial}{\partial p_k}\mathscr{H}_F|_{(x,\frac{p}{|p|})} \cdot \Big(-\frac{p_j\,p_k}{|p|^3} + \frac{\delta_{jk}}{|p|}\Big) \\
&= \frac{p_j}{|p|} \cdot \Big(\mathscr{H}_F(x,\tfrac{p}{|p|}) - \frac{p}{|p|} \cdot \frac{\partial}{\partial p}\mathscr{H}_F|_{(x,\frac{p}{|p|})}\Big) + \frac{\partial}{\partial p_j}\mathscr{H}_F|_{(x,\frac{p}{|p|})}.
\end{aligned}$$

Thus, the Lipschitz constant of $p \longmapsto \frac{\partial}{\partial p_j}\mathscr{H}_F(x,p)$ has the upper bound

$$\begin{aligned}
&\mathrm{Lip}\,(p \mapsto \frac{p_j}{|p|}) \cdot \Big(\|\mathscr{H}_F\|_{C^0(\mathbb{R}^N\times\partial\mathbb{B}_1)} + 1 \cdot \|\frac{\partial}{\partial p}\mathscr{H}_F\|_{C^0(\mathbb{R}^N\times\partial\mathbb{B}_1)}\Big) \\
&\quad +1 \cdot \mathrm{Lip}\,(p \mapsto \frac{p}{|p|})\Big(\mathrm{Lip}\,\mathscr{H}_F|_{\mathbb{R}^N\times\partial\mathbb{B}_1} + \|\frac{\partial}{\partial p}\mathscr{H}_F\|_{C^0(\mathbb{R}^N\times\partial\mathbb{B}_1)} + 1 \cdot \mathrm{Lip} \\
&\qquad\qquad \times \frac{\partial}{\partial p}\mathscr{H}_F|_{\mathbb{R}^N\times\partial\mathbb{B}_1}\Big) + \mathrm{Lip}\,(p \mapsto \frac{p}{|p|}) \cdot \mathrm{Lip}\,\frac{\partial}{\partial p}\mathscr{H}_F|_{\mathbb{R}^N\times\partial\mathbb{B}_1} \\
&\le 2R \cdot \|\mathscr{H}_F\|_{C^1(\mathbb{R}^N\times\partial\mathbb{B}_1)} + 2R \cdot 2\,\|D\mathscr{H}_F\|_{C^0(\mathbb{R}^N\times\partial\mathbb{B}_1)} + 2R \cdot 2\ \mathrm{Lip}\,\frac{\partial}{\partial p}\mathscr{H}_F|_{\mathbb{R}^N\times\partial\mathbb{B}_1} \\
&\le 6R \cdot \|\mathscr{H}_F\|_{C^{1,1}(\mathbb{R}^N\times\partial\mathbb{B}_1)}.
\end{aligned}$$

Correspondingly the Lipschitz constant of $x \longmapsto \frac{\partial}{\partial p_j}\mathscr{H}_F(x,p)$ is bounded from above by $2\,\|D\mathscr{H}_F\|_{C^{0,1}(\mathbb{R}^N\times\partial\mathbb{B}_1)} \le 2\lambda$ and thus,

$$\begin{aligned}
\mathrm{Lip}\,\frac{\partial\,\mathscr{H}_F}{\partial p_j} &\le \max\Big\{\mathrm{Lip}\,\big(x \mapsto \frac{\partial}{\partial p_j}\mathscr{H}_F(x,p)\big),\ \mathrm{Lip}\,\big(p \mapsto \frac{\partial}{\partial p_j}\mathscr{H}_F(x,p)\big)\Big\} \\
&\overset{R>1}{\le} 6R \cdot \|\mathscr{H}_F\|_{C^{1,1}(\mathbb{R}^N\times\partial\mathbb{B}_1)}.
\end{aligned}$$

Furthermore, $\frac{\partial}{\partial x_j}\mathscr{H}_F(x,p) = |p| \cdot \frac{\partial}{\partial x_j}\mathscr{H}_F|_{(x,\frac{p}{|p|})}$ has the consequence

$$\begin{aligned}
\mathrm{Lip}\Big(x \mapsto \frac{\partial\mathscr{H}_F}{\partial x_j}\Big) &\le R \cdot \mathrm{Lip}\,\frac{\partial}{\partial x}\mathscr{H}_F|_{\mathbb{R}^N\times\partial\mathbb{B}_1} \le R \cdot \lambda, \\
\mathrm{Lip}\Big(p \mapsto \frac{\partial\mathscr{H}_F}{\partial x_j}\Big) &\le 1 \cdot \|\frac{\partial\,\mathscr{H}_F}{\partial x}\|_{C^0(\mathbb{R}^N\times\partial\mathbb{B}_1)} + R \cdot \mathrm{Lip}\,\frac{\partial}{\partial x}\mathscr{H}_F|_{\mathbb{R}^N\times\partial\mathbb{B}_1} \cdot \mathrm{Lip}\,(p \mapsto \tfrac{p}{|p|}) \\
&\le \lambda + R \cdot \lambda \cdot 2R \overset{R>1}{\le} 3R^2\lambda.
\end{aligned}$$

□

Proof (of Proposition 32 on page 364).
The semigroup property of reachable sets implies again

$$\begin{aligned}
d_{\mathscr{K},N}\big(\vartheta_F(h,\ \vartheta_F(t,K)),\quad \vartheta_F(t+h,\ K)\big) &= 0, \\
d_{\mathscr{K},N}\big(\vartheta_F(t+h,\ K),\quad \vartheta_F(h,\ \vartheta_F(t,K))\big) &= 0
\end{aligned}$$

for all $F \in \mathrm{LIP}_\lambda^{(\mathscr{H})}(\mathbb{R}^N,\mathbb{R}^N)$, $K \in \mathscr{K}(\mathbb{R}^N)$, $h,t \ge 0$ since $d_{\mathscr{K},N}$ is a quasi-metric.

According to Proposition 29 (on page 363), every map $F \in \mathrm{LIP}_\lambda^{(\mathscr{H})}(\mathbb{R}^N, \mathbb{R}^N)$ and initial set $K_1 \in \mathscr{K}_{C^{1,1}}(\mathbb{R}^N)$ lead to a time $\mathbb{T}(\vartheta_F, K_1) > 0$ and a radius $\rho > 0$ such that $\vartheta_F(t, K_1) \in \mathscr{K}_{C^{1,1}}(\mathbb{R}^N)$ has positive reach of radius $\geq \rho$ for any $t < \mathbb{T}(\vartheta_F, K_1)$. Lemma 31 guarantees for all $K_1 \in \mathscr{K}_{C^{1,1}}(\mathbb{R}^N)$ and $K_2 \in \mathscr{K}(\mathbb{R}^N)$ with $K_1 \neq K_2$

$$\begin{aligned}
&\limsup_{h \downarrow 0} \frac{d_{\mathscr{K},N}(\vartheta_F(h,K_1),\, \vartheta_F(h,K_2)) - d_{\mathscr{K},N}(K_1,K_2)}{h \quad d_{\mathscr{K},N}(K_1,K_2)} \\
\leq\ &\limsup_{h \downarrow 0} \frac{1}{h} \left(e^{(9\,e^{2\lambda h}\,\lambda + \lambda)\cdot h} - 1 \right) \quad = \quad 10\,\lambda \quad \stackrel{\text{Def.}}{=} \quad \alpha(\vartheta_F; \cdot, \cdot)
\end{aligned}$$

and for every $F, G \in \mathrm{LIP}_\lambda^{(\mathscr{H})}(\mathbb{R}^N, \mathbb{R}^N)$

$$\begin{aligned}
&\limsup_{h \downarrow 0} \ \frac{1}{h} \left(d_{\mathscr{K},N}(\vartheta_F(h, K_1),\ \vartheta_G(h, K_2)) \ - \ d_{\mathscr{K},N}(K_1,\, K_2) \cdot e^{10\,\lambda\, h} \right) \\
\leq\ &\limsup_{h \downarrow 0} \ \left(d_{\mathscr{K},N}(K_1,\, K_2) \cdot \frac{1}{h} \left(e^{(9\,e^{2\lambda h}\,\lambda + \lambda)\cdot h} - e^{10\,\lambda\, h} \right) \right. \\
&\qquad\qquad \left. + \ 6\,N \cdot \|\mathscr{H}_F - \mathscr{H}_G\|_{C^1(\mathbb{R}^N \times \partial \mathbb{B}_1)} \cdot e^{(9\,e^{2\lambda h}\,\lambda + \lambda)\cdot h} \right) \\
=\ &\qquad\qquad 6\,N \cdot \|\mathscr{H}_F - \mathscr{H}_G\|_{C^1(\mathbb{R}^N \times \partial \mathbb{B}_1)}.
\end{aligned}$$

This estimate justifies the definition

$$\widehat{D}(\vartheta_F,\, \vartheta_G; \cdot, \cdot) \ \stackrel{\text{Def.}}{=} \ 6\,N \cdot \|\mathscr{H}_F - \mathscr{H}_G\|_{C^1(\mathbb{R}^N \times \partial \mathbb{B}_1)}.$$

Moreover Lemma 27 (on page 362) states the uniform Lipschitz continuity with respect to time

$$d_{\mathscr{K},N}\Big(\vartheta_F(s, K),\ \vartheta_F(t, K) \Big) \ \leq \ \lambda\, (e^\lambda + 2) \cdot (t - s)$$

for any $0 \leq s \leq t \leq 1$ and $K \in \mathscr{K}(\mathbb{R}^N)$.

Finally we verify

$$\limsup_{h \downarrow 0} \ d_{\mathscr{K},N}\big(\vartheta_F(t-h,\, K_1),\, K_2\big) \ \geq \ d_{\mathscr{K},N}\big(\vartheta_F(t, K_1),\, K_2\big)$$

for all $F \in \mathrm{LIP}_\lambda^{(\mathscr{H})}(\mathbb{R}^N, \mathbb{R}^N)$, $K_1 \in \mathscr{K}_{C^{1,1}}(\mathbb{R}^N)$, $K_2 \in \mathscr{K}(\mathbb{R}^N)$ and $0 < t < \mathbb{T}(\vartheta_F, K_1)$ because in combination with the triangle inequality of $d_{\mathscr{K},N}$, it implies condition (8.) on (timed) transitions in Definition 1 (on page 336).
Proposition A.36 (on page 461) ensures the reversibility in time in $[0, \mathbb{T}(\vartheta_F, K_1)[$, i.e. for every $0 < h < t < \mathbb{T}(\vartheta_F, K_1)$,

$$\mathbb{R}^N \setminus \vartheta_F(t - h, K_1) \ = \ \vartheta_{-F}\big(h,\ \mathbb{R}^N \setminus \vartheta_F(t, K_1)\big).$$

Assuming $F \in \mathrm{LIP}_\lambda^{(\mathscr{H})}(\mathbb{R}^N, \mathbb{R}^N)$ (in the sense of Definition 26 on page 362), the flow of the Hamiltonian system even induces a Lipschitz homeomorphism between Graph $N_{\vartheta_F(t-h, K_1)}$ and Graph $N_{\vartheta_F(t, K_1)}$ since each limiting normal cone contains exactly one direction and $N_{\vartheta_F(t,K_1)}(\cdot) = -N_{\overline{\mathbb{R}^N \setminus \vartheta_F(t,K_1)}}(\cdot)$.
Thus, Corollary 24 (on page 360) implies

$$\text{Graph } N_{\vartheta_F(t,K_1)} \ = \ \mathrm{Lim}_{h \downarrow 0} \text{ Graph } N_{\vartheta_F(t-h,K_1)}$$

and finally, $\quad d_{\mathscr{K},N}\big(\vartheta_F(t, K_1),\ \vartheta_F(t-h,\, K_1)\big) \ \longrightarrow \ 0 \quad$ for $h \downarrow 0$.

The last claim results from the triangle inequality of $d_{\mathscr{K},N}$. □

4.4.3 Existence of Solutions due to Transitional Euler Compactness

For applying the existence results of § 4.3.3 (on page 348 ff.), we now have to focus on an essential question: What are sufficient conditions on set-valued maps $F \in \mathrm{LIP}_\lambda^{(\mathscr{H})}(\mathbb{R}^N, \mathbb{R}^N)$ for the transitional Euler compactness with respect to $d_{\mathscr{K},N}$?

Definition 34. For any $\lambda > 0$ and $\rho > 0$, the set $\mathrm{LIP}_\lambda^{(\mathscr{H}_\circ^\rho)}(\mathbb{R}^N, \mathbb{R}^N)$ consists of all set-valued maps $F : \mathbb{R}^N \rightsquigarrow \mathbb{R}^N$ satisfying

(1.) $F : \mathbb{R}^N \rightsquigarrow \mathbb{R}^N$ has compact convex values in $\mathscr{K}_\circ^\rho(\mathbb{R}^N)$.
(2.) $\mathscr{H}_F(\cdot,\cdot) \in C^2(\mathbb{R}^N \times (\mathbb{R}^N \setminus \{0\}))$,
(3.) $\|\mathscr{H}_F\|_{C^{1,1}(\mathbb{R}^N \times \partial\mathbb{B}_1)} \stackrel{\text{Def.}}{=} \|\mathscr{H}_F\|_{C^1(\mathbb{R}^N \times \partial\mathbb{B}_1)} + \mathrm{Lip}\, D\mathscr{H}_F|_{\mathbb{R}^N \times \partial\mathbb{B}_1} < \lambda$.

Remark 35. $\mathrm{LIP}_\lambda^{(\mathscr{H}_\circ^\rho)}(\mathbb{R}^N, \mathbb{R}^N)$ is a subset of $\mathrm{LIP}_\lambda^{(\mathscr{H})}(\mathbb{R}^N, \mathbb{R}^N)$ introduced in Definition 26 (on page 362).
Its set-valued maps, however, even fulfill standard hypothesis $(\widetilde{\mathscr{H}_\circ^\rho})$ (specified in Definition A.39 on page 464). In particular, they make points evolve into convex reachable sets of positive erosion for short times according to Proposition A.41. This is the "geometrically smoothening" effect on reachable sets which we are now using for verifying transitional Euler compactness.

Proposition 36.
For any $\lambda, \rho > 0$, consider the maps $F \in \mathrm{LIP}_\lambda^{(\mathscr{H}_\circ^\rho)}(\mathbb{R}^N, \mathbb{R}^N)$ (i.e. their reachable sets, strictly speaking) as transitions on $(\mathscr{K}(\mathbb{R}^N),\ \mathscr{K}_{C^{1,1}}(\mathbb{R}^N),\ d_{\mathscr{K},N},\ d_{\mathscr{K},N},\ 0)$ in the sense of Definition 1 and Remark 3 (on page 336 f.).

Then, $\big(\mathscr{K}(\mathbb{R}^N),\ \mathscr{K}_{C^{1,1}}(\mathbb{R}^N),\ d_{\mathscr{K},N},\ d_{\mathscr{K},N},\ 0,\ \mathrm{LIP}_\lambda^{(\mathscr{H}_\circ^\rho)}(\mathbb{R}^N, \mathbb{R}^N)\big)$ is transitionally Euler compact in the following sense (see Definition 14 on page 348):
Suppose each $G_n : [0,1] \longrightarrow \mathrm{LIP}_\lambda^{(\mathscr{H}_\circ^\rho)}(\mathbb{R}^N, \mathbb{R}^N)$ to be piecewise constant $(n \in \mathbb{N})$ and set with arbitrarily fixed $K_0 \in \mathscr{K}(\mathbb{R}^N)$

$$\widetilde{G}_n : [0,1] \times \mathbb{R}^N \rightsquigarrow \mathbb{R}^N,\ (t,x) \longmapsto G_n(t)(x),$$
$$K_n(h) := \vartheta_{\widetilde{G}_n}(h, K_0) \qquad \text{for } h \geq 0.$$

Furthermore let $(h_j)_{j\in\mathbb{N}}$ be a sequence in $]0,1[$ with $h_j \downarrow 0$ and choose $t \in [0,1[$. Then there exist a sequence $n_k \nearrow \infty$ of indices and a set $K(t) \in \mathscr{K}(\mathbb{R}^N)$ satisfying

$$\limsup_{k\to\infty}\ d_{\mathscr{K},N}\big(K_{n_k}(t),\quad K(t)\big) = 0,$$
$$\limsup_{j\to\infty}\ \sup_{k\geq j}\ d_{\mathscr{K},N}\big(K(t),\ K_{n_k}(t+h_j)\big) = 0.$$

In fact, we obtain as an immediate consequence of Theorem 17 (on page 349 f.):

Corollary 37 (Existence of compact-valued solutions w.r.t. $d_{\mathscr{K},N}$).
Let $f : \mathscr{K}(\mathbb{R}^N) \times [0,T] \longrightarrow \mathrm{LIP}_\lambda^{(\mathscr{H}_\circ^\rho)}(\mathbb{R}^N,\mathbb{R}^N)$ *satisfy*

$$\left\|\mathscr{H}_{f(K_1,t_1)} - \mathscr{H}_{f(K_2,t_2)}\right\|_{C^1(\mathbb{R}^N\times\partial\mathbb{B}_1)} \le \omega(d_{\mathscr{K},N}(K_1,K_2) + t_2 - t_1)$$

for all $K_1, K_2 \in \mathscr{K}(\mathbb{R}^N)$ *and* $0 \le t_1 \le t_2 \le T$ *with a modulus* $\omega(\cdot)$ *of continuity and consider the reachable sets of maps in* $\mathrm{LIP}_\lambda^{(\mathscr{H}_\circ^\rho)}(\mathbb{R}^N,\mathbb{R}^N)$ *as transitions on* $(\mathscr{K}(\mathbb{R}^N),\ \mathscr{K}_{C^{1,1}}(\mathbb{R}^N),\ d_{\mathscr{K},N},\ d_{\mathscr{K},N},\ 0)$ *according to Proposition 32 (on page 364).*

Then for every initial compact set $K_0 \in \mathscr{K}(\mathbb{R}^N)$, *there always exists a solution* $K : [0,T] \longrightarrow \mathscr{K}(\mathbb{R}^N)$ *to the mutational equation* $\overset{\circ}{K}(\cdot) \ni f(K(\cdot),\cdot)$ *(in the sense of Definition 7 on page 339 and Remark 3 on page 337) with* $K(0) = K_0$, *i.e. here,*

(a) $\limsup_{h\downarrow 0} \frac{1}{h}\cdot\left(d_{\mathscr{K},N}\big(\vartheta_{f(K(t),\,t)}(h,M),\ K(t+h)\big) - d_{\mathscr{K},N}(M,\ K(t))\cdot e^{10\,\lambda\,h}\right) \le 0$
for every compact N-dimensional submanifold $M \subset \mathbb{R}^N$ *with* $C^{1,1}$ *boundary and* $\mathscr{L}^1$*-almost every* $t \in [0,T[$.

(b) $d_{\mathscr{K},N}(K(s),\,K(t)) \le \mathit{const}(\lambda,T)\cdot(t-s)$ *for all* $0 \le s < t < T$.

□

Corollary 38 (Existence of compact-valued solutions to equations with delay).
Let $\tau > 0$ *be a fixed period,* $\lambda > 0$ *and assume for*

$$f:\ \mathrm{BLip}\big([-\tau,0],\ \mathscr{K}(\mathbb{R}^N);\ d_{\mathscr{K},N},\ 0\big)\times[0,T] \longrightarrow \mathrm{LIP}_\lambda^{(\mathscr{H}_\circ^\rho)}(\mathbb{R}^N,\mathbb{R}^N)$$

and $\mathscr{L}^1$*-almost every* $t \in [0,T[$:

$$\lim_{n\to\infty} \left\|\mathscr{H}_{f(M_n(\cdot),t_n)} - \mathscr{H}_{f(M(\cdot),t)}\right\|_{C^1(\mathbb{R}^N\times\partial\mathbb{B}_1)} = 0$$

holds for any curve $M(\cdot) \in \mathrm{BLip}\big([-\tau,0],\ \mathscr{K}(\mathbb{R}^N);\ d_{\mathscr{K},N},\ 0\big)$ *and sequences* $(t_n)_{n\in\mathbb{N}}$, $(M_n(\cdot))_{n\in\mathbb{N}}$ *in* $[0,T]$ *and* $\mathrm{BLip}\big([-\tau,0],\ \mathscr{K}(\mathbb{R}^N);\ d_{\mathscr{K},N},\ 0\big)$ *respectively satisfying*

$$\lim_{n\to\infty} t_n = t, \qquad \lim_{n\to\infty} d_{\mathscr{K},N}\big(M(s),\,M_n(s)\big) = 0 \quad \text{for every } s \in [-\tau,0].$$

For every function $K_0(\cdot) \in \mathrm{BLip}\big([-\tau,0],\ \mathscr{K}(\mathbb{R}^N);\ d_{\mathscr{K},N},\ 0\big)$, *there exists a curve* $K(\cdot) \in \mathrm{BLip}\big([-\tau,T],\ \mathscr{K}(\mathbb{R}^N);\ d_{\mathscr{K},N},\ 0\big)$ *with* $K(\cdot)\big|_{[-\tau,0]} = K_0(\cdot)$ *and*

$$\limsup_{h\downarrow 0} \frac{1}{h}\cdot\left(d_{\mathscr{K},N}\big(\vartheta_{f(K(t+\cdot)|_{[-\tau,0]},\,t)}(h,M),\ K(t+h)\big) - d_{\mathscr{K},N}(M,\ K(t))\cdot e^{10\,\lambda\,h}\right) \le 0$$

for $\mathscr{L}^1$*-almost every* $t \in [0,T[$ *and any compact N-dimensional submanifold M of* $\mathbb{R}^N$ *with* $C^{1,1}$ *boundary.*

□

Remark 39. We hesitate using the term "morphological equations" here because we have usually reserved it for mutational equations in the metric space $(\mathscr{K}(\mathbb{R}^N), d\!l)$ with transitions induced by LIP$(\mathbb{R}^N, \mathbb{R}^N)$ — as introduced by Aubin (see § 1.9 on page 57 ff.). In this section, however, $\mathscr{K}(\mathbb{R}^N)$ is supplied with the other distance function $d_{\mathscr{K},N}$ and we apply the mutational framework with "test elements".
The characterization reveals that every solution to a mutational equation in this recent generalized sense solves the morphological equation in the sense of Aubin (see § 1.9.6 on page 74 ff.) whenever all its values are in $\mathscr{K}_{C^{1,1}}(\mathbb{R}^N)$.

Proof (of Proposition 36).
Every closed bounded ball in $(\mathscr{K}(\mathbb{R}^N), d\!l)$ is compact according to Proposition 1.47 (on page 57). Hence, there exist a sequence of indices $n_k \nearrow \infty$ and a set $K(t) \in \mathscr{K}(\mathbb{R}^N)$ with

$$d\!l(K_{n_k}(t), K(t)) \longrightarrow 0 \qquad (k \longrightarrow \infty).$$

Thus, $d\!l(K(t), K_{n_k}(t+h)) \le d\!l(K(t), K_{n_k}(t)) + \lambda\, h \longrightarrow \lambda\, h$ for $k \to \infty$. Furthermore Corollary 24 (on page 360) implies

$$d_{\mathscr{K},N}(K_{n_k}(t), K(t)) \longrightarrow 0.$$

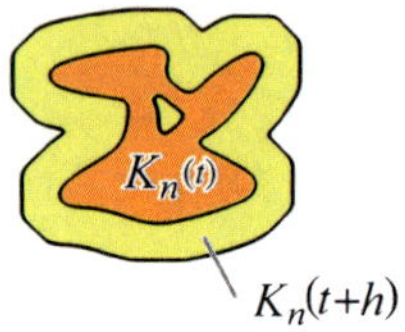

Now we want to prove that $K(t)$ satisfies the claim by selecting subsequences of $(n_k)_{k\in\mathbb{N}}$ for countably many times and finally applying Cantor's diagonal construction.

An important tool is Proposition A.41 (on page 464). After choosing radius $\widehat{r} > 0$ sufficiently large with $\bigcup_{t\in[0,T],\, n\in\mathbb{N}} K_n(t) \subset \mathbb{B}_{\widehat{r}-1}(0) \subset \mathbb{R}^N$, it ensures the existence of $\sigma = \sigma(\lambda,\rho,\widehat{r}) > 0$ and $\widehat{h} = \widehat{h}(\lambda,\rho,\widehat{r}) > 0$ such that the reachable set $\vartheta_{-\widetilde{G}_n(t+h-\cdot,\cdot)}(h, z)$ is convex and has positive erosion of radius $\sigma\, h$ for every $h \in\,]0,\widehat{h}]$ and $z \in \mathbb{B}_{\widehat{r}}(0)$.
In the following, we assume $0 < h_j < \widehat{h}$ for all $j \in \mathbb{N}$ without loss of generality. Moreover, each set $K_n(t)$ at time $t > 0$ is the closed r-neighborhood of a compact set with a sufficiently small radius $r = r(n,t) > 0$.

Now the asymptotic properties of

$$e^{\supset}\Big(\text{Graph}\ {}^{\flat}N_{K(t)},\ \text{Graph}\ {}^{\flat}N_{K_{n_k}(t+h)}\Big) \qquad (k \longrightarrow \infty)$$

have to be investigated for each $h \in\,]0,\widehat{h}]$.
According to Definition A.23 (on page 454), every limiting normal cone results from the neighboring proximal normal cones, i.e.

$$N_C(x) \stackrel{\text{Def.}}{=} \text{Limsup}_{\substack{y\to x\\ y\in C}} N_C^P(y)$$

for every nonempty set $C \subset \mathbb{R}^N$ and point $x \in \partial C$. Thus, Graph N_C = $\overline{\text{Graph } N_C^P}$ and from now on, we confine our considerations to the excess

$$e^{\supset}\Big(\text{Graph}\ {}^{\flat}N_{K(t)},\ \text{Graph}\ {}^{\flat}N^P_{K_{n_k}(t+h)}\Big)$$

for any $h \in\,]0,\widehat{h}]$.

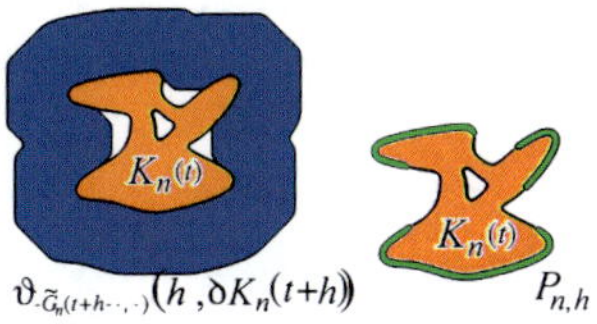

$P_{n,h} := K_n(t) \cap \vartheta_{-\widetilde{G}_n(t+h-\cdot,\cdot)}(h, \partial K_n(t+h))$ is a subset of $\partial K_n(t)$. More precisely, it consists of all points $x \in K_n(t)$ such that a solution of $\widetilde{G}_n$ starts in x at time t and reaches $\partial K_n(t+h)$ at time $t+h$. In addition, every boundary point y of $K_n(t+h)$ is attained by such a solution.

By means of boundary solutions and their adjoint arcs, the Hamiltonian system in Proposition 25 (on page 362) leads to the following estimate for every $n \in \mathbb{N}$ (similarly to Lemma 27)

$$e^{\supset}\Big(\text{Graph } {}^\flat N_{K_n(t)}\Big|_{P_{n,h}}, \ \text{Graph } {}^\flat N^P_{K_n(t+h)}\Big) \ \le \ \text{const}(\lambda) \cdot h.$$

In fact, whenever such an adjoint arc traces a proximal normal vector of $K_n(t+h)$ back to the boundary of $K_n(t)$, it ends up in a *proximal* normal vector to $K_n(t)$ (and not just a limiting normal vector) because each point of the corresponding boundary solution has evolved into convex sets of positive erosion shortly while time is going back. Hence, we even obtain the estimate

$$e^{\supset}\Big(\text{Graph } {}^\flat N^P_{K_n(t)}\Big|_{P_{n,h}}, \ \text{Graph } {}^\flat N^P_{K_n(t+h)}\Big) \ \le \ \text{const}(\lambda) \cdot h.$$

The proximal normal cones $N^P_{\overline{\mathbb{R}^N \setminus K_n(t)}}(x) = -N^P_{K_n(t)}(x)$ contain exactly one direction for every point $x \in P_{n,h}$ as a consequence of Lemma A.24 (on page 454): Indeed, first, $N^P_{\overline{\mathbb{R}^N \setminus K_n(t)}}(x) \neq \emptyset$ for all $x \in \partial K_n(t)$ since $K_n(t)$ is a r-neighborhood.
Second, $\quad N^P_{K_n(t)}(x) \quad \neq \emptyset$ for all $x \in P_{n,h}$
because $\vartheta_{-\widetilde{G}_n(t+h-\cdot,\cdot)}(h, \partial K_n(t+h))$ is a closed σh-neighborhood of a compact set (Proposition A.41) and $K_n(t) \cap \big(\vartheta_{-\widetilde{G}_n(t+h-\cdot,\cdot)}(h, \partial K_n(t+h))\big)^\circ = \emptyset$.
For the same reason, the proximal radius of $K_n(t)$ at each $x \in P_{n,h}$ (in its unique proximal direction) is $\ge \sigma h$. As this lower bound of proximal radius does not depend on $n \in \mathbb{N}$ (but merely on $h, \lambda, \rho, \widehat{r}$), Proposition A.72 (1.) (on page 487) ensures

$$e^{\supset}\Big(\text{Graph } {}^\flat N_{K(t)}, \ \text{Graph } {}^\flat N^P_{K_{n_k}(t)}\Big|_{P_{n,h}}\Big) \ \longrightarrow \ 0 \qquad (k \longrightarrow \infty)$$

for every $h \in \,]0, \widehat{h}]$. The triangle inequality of $e^{\supset}$ leads to the estimate for every h,

$$\limsup_{k\to\infty} \ e^{\supset}\Big(\text{Graph } {}^\flat N_{K(t)}, \ \text{Graph } {}^\flat N^P_{K_{n_k}(t+h)}\Big) \ \le \ \text{const}(\lambda) \cdot h.$$

For completing the proof of transitional Euler compactness, a sequence $(h_j)_{j \in \mathbb{N}}$ in $]0, \widehat{h}]$ with $h_j \longrightarrow 0$ is given. By means of Cantor's diagonal construction, we obtain a subsequence (again denoted by) $(n_k)_{k \in \mathbb{N}}$ satisfying for every $j \in \mathbb{N}$, $k \ge j$

$$e^{\supset}\big(\text{Graph } {}^\flat N_{K(t)}, \ \text{Graph } {}^\flat N^P_{K_{n_k}(t+h_j)}\big) \ \le \ \text{const}(\lambda) \cdot h_j + \tfrac{1}{k},$$

and thus, $\quad \limsup\limits_{j\to\infty} \ \sup\limits_{k \ge j} \ d_{\mathcal{K},N}(K(t), \ K_{n_k}(t+h_j)) \ = \ 0.$

□

4.5 Further Example: Mutational Equations for Compact Sets Depending on the Normal Cones

In the preceding section 4.4, we consider a geometric example with the evolution of compact subsets of $\mathbb{R}^N$ depending on their respective normal cones. Indeed, the set $\mathscr{K}(\mathbb{R}^N)$ of all nonempty compact subsets of $\mathbb{R}^N$ is supplied with the quasi-metric

$$d_{\mathscr{K},N}(K_1,K_2) \stackrel{\text{Def.}}{=} dl(K_1,K_2) + e^{\supset}(\text{Graph } {}^\flat N_{K_1}, \text{ Graph } {}^\flat N_{K_2}).$$

$\mathscr{K}_{C^{1,1}}(\mathbb{R}^N)$ consisting of all nonempty compact subsets with $C^{1,1}$ boundary is used for "test elements". Then for any parameter $\lambda > 0$ fixed, the set-valued maps $F : \mathbb{R}^N \rightsquigarrow \mathbb{R}^N$ satisfying

(1.) $F : \mathbb{R}^N \rightsquigarrow \mathbb{R}^N$ has nonempty compact convex values,

(2.) $\mathscr{H}_F(x,p) \stackrel{\text{Def.}}{=} \sup_{v \in F(x)} p \cdot v$ belongs to $C^{1,1}(\mathbb{R}^N \times (\mathbb{R}^N \setminus \{0\}))$,

(3.) $\|\mathscr{H}_F\|_{C^{1,1}(\mathbb{R}^N \times \partial\mathbb{B}_1)} \stackrel{\text{Def.}}{=} \|\mathscr{H}_F\|_{C^1(\mathbb{R}^N \times \partial\mathbb{B}_1)} + \text{Lip}\, D\mathscr{H}_F|_{\mathbb{R}^N \times \partial\mathbb{B}_1} < \lambda$

induce transitions on $\big(\mathscr{K}(\mathbb{R}^N), \mathscr{K}_{C^{1,1}}(\mathbb{R}^N), d_{\mathscr{K},N}, d_{\mathscr{K},N}, 0\big)$ by means of their reachable sets of differential inclusions.
Under stronger assumptions about the Hamiltonian $\mathscr{H}_F$, the required properties of transitional Euler compactness are also verified in Proposition 36 (on page 368) and thus, we obtain the existence of solutions to the corresponding mutational equations (in the sense of Definition 1 and Remark 3 on page 336 f.)

The estimates between solutions (presented in § 4.3.1 on page 341 ff.) do not provide uniqueness though. Indeed, the smooth sets in $\mathscr{K}_{C^{1,1}}(\mathbb{R}^N)$ stay smooth for short times while evolving along such a differential inclusion, but there is no obvious lower bound of this period satisfying the approximating hypotheses of Proposition 9 or 10 (on page 341 f.).
Lacking results about uniqueness are the key obstacle motivating a further example.

In this section, we introduce another distance function for describing evolutions of compact subsets of $\mathbb{R}^N$ in Definition 41 below. In contrast to the example of § 4.4, the substantial idea is now to

1. use *all* nonempty compact subsets as "test elements" (instead of $\mathscr{K}_{C^{1,1}}(\mathbb{R}^N)$), but

2. take only the proximal normals with an exterior ball of radius $\geq j$ into consideration simultaneously. Choosing the parameter j here as positive real number induces a family of distance functions specified in subsequent Definition 41.

The essential geometric advantage is that Proposition A.46 (on page 469) provides an upper estimate how fast these exterior balls can shrink (at most) and thus, the corresponding time parameter $\mathbb{T}_j(\cdot,\cdot)$ may depend on j, but not on the "test element".

3. "record" the period $h > 0$ how long the compact set $K(s+h) \subset \mathbb{R}^N$ and the "test set" $\vartheta_F(h, K(s))$ have been evolving while being compared. This period determines the radii of exterior balls that are related with each other for calculating the "distance" between these two sets.

   The separate time component is to provide information about period h : The compact set $K(s+h)$ is supplied with a linearly increasing time component whereas all "test sets" preserve their initial time components. Then the wanted period results from their difference.

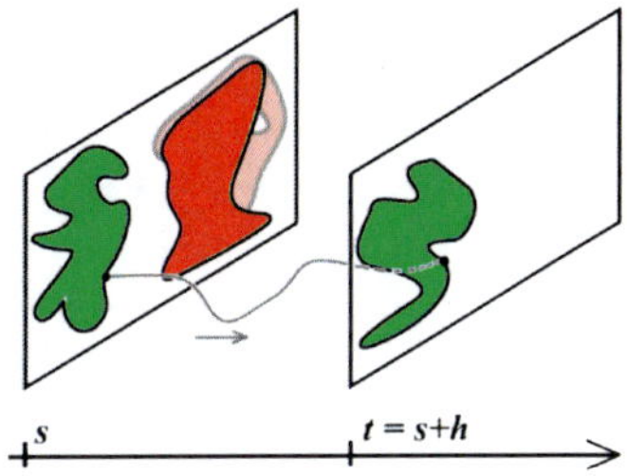

   For implementing this notion in the mutational framework, we introduce an additional component being either 0 (for "test elements") or 1 (otherwise) and indicating the growth of the time component while evolving (see Definition 44 on page 376 below).

4.5.1 Specifying Sets and Distance Functions

Now we consider

$$E := \{1\} \times \mathscr{K}(\mathbb{R}^N), \qquad \qquad \widetilde{E} := \mathbb{R} \times \{1\} \times \mathscr{K}(\mathbb{R}^N),$$
$$\mathscr{D} := \{0\} \times \mathscr{K}(\mathbb{R}^N) \qquad \text{and thus,} \qquad \widetilde{\mathscr{D}} := \mathbb{R} \times \{0\} \times \mathscr{K}(\mathbb{R}^N).$$

In comparison with the earlier geometric example in § 4.4, the main advantage of this second approach is the uniqueness stated in subsequent Proposition 51 (on page 383).

From now on, fix the parameter $\Lambda > 0$ arbitrarily. It is used for both the distance function $\widetilde{d}_{\mathscr{K},j}$ in Definition 41 and the set-valued maps (whose reachable sets induce candidates for timed transitions) in Definition 43.

Definition 40. Let $C \subset \mathbb{R}^N$ be a nonempty closed set.

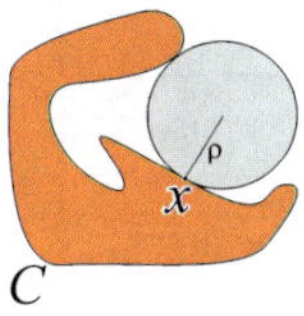

For any $\rho > 0$, the set $N^P_{C,\rho}(x) \subset \mathbb{R}^N$ consists of all proximal normal vectors $\eta \in N^P_C(x) \setminus \{0\}$ with the proximal radius $\geq \rho$ (and thus might be empty). Furthermore ${}^\flat N^P_{C,\rho}(x) := N^P_{C,\rho}(x) \cap \mathbb{B}$.

Definition 41. Set

$$\widetilde{\mathscr{K}}^{\rightarrow}(\mathbb{R}^N) := \mathbb{R} \times \{1\} \times \mathscr{K}(\mathbb{R}^N),$$
$$\widetilde{\mathscr{K}}^{\curlyvee}(\mathbb{R}^N) := \mathbb{R} \times \{0\} \times \mathscr{K}(\mathbb{R}^N).$$

Basic set	$\widetilde{E} := \widetilde{\mathscr{K}}^{\rightarrow}(\mathbb{R}^N) \stackrel{\text{Def.}}{=} \mathbb{R} \times \{1\} \times \mathscr{K}(\mathbb{R}^N)$
Test set	$\widetilde{\mathscr{D}} := \widetilde{\mathscr{K}}^{\curlyvee}(\mathbb{R}^N) \stackrel{\text{Def.}}{=} \mathbb{R} \times \{0\} \times \mathscr{K}(\mathbb{R}^N)$ (The second component in $\{0,1\}$ is just to indicate if the real time component is increasing linearly along transitions or not.)
Distances	$\widetilde{d}_{\mathscr{K},j}\big((s,\mu,C),\,(t,\nu,D)\big) :=$ $\displaystyle d\!\!l(C,D) + \limsup_{\kappa\downarrow 0} \int_j^\infty \psi(\rho+\kappa+200\,\Lambda\,\|t-s\|) \cdot e^{\subset}\Big(\text{Graph } {}^\flat N^P_{D,\,(\rho+\kappa+200\,\Lambda\,\|t-s\|)},\ \text{Graph } {}^\flat N^P_{C,\rho}\Big)\, d\rho$ with a fixed nonincreasing weight function $\psi \in C_0^\infty([0,2[,\,[0,\infty[)$ and each $j \in [0,1]$. (It satisfies the timed triangle inequality.)
Absolute value	$\lfloor\cdot\rfloor := 0$
Timed transition	For each $F \in \mathrm{LIP}_\Lambda^{(C^2)}(\mathbb{R}^N,\mathbb{R}^N)$, i.e. set-valued map $F: \mathbb{R}^N \rightsquigarrow \mathbb{R}^N$ with compact convex values and C^2 Hamiltonian (Def.43), define $\widetilde{\vartheta}_F : [0,1] \times \big(\widetilde{\mathscr{K}}^{\curlyvee}(\mathbb{R}^N) \cup \widetilde{\mathscr{K}}^{\rightarrow}(\mathbb{R}^N)\big) \longrightarrow \widetilde{\mathscr{K}}^{\curlyvee}(\mathbb{R}^N) \cup \widetilde{\mathscr{K}}^{\rightarrow}(\mathbb{R}^N)$ as $\widetilde{\vartheta}_F\big(h,\,(t,\mu,K)\big) := \begin{cases} \big(t+h,\ 1,\ \vartheta_F(h,K)\big) & \text{if } \mu = 1 \\ \big(t,\ \ 0,\ \vartheta_F(h,K)\big) & \text{if } \mu = 0 \end{cases}$ by means of reachable sets $\vartheta_F(h,\,K_0)$ of the autonomous differential inclusion $x'(\cdot) \in F(x(\cdot))$ a.e.
Compactness	strongly-weakly transitionally Euler compact (due to asymptotic properties of proximal normals with positive proximal radii bounded from below, § A.9): Lemma 49 (p. 381)
Mutational solutions	Reachable sets of a nonautonomous differential inclusion whose set-valued right-hand side is determined via feedback: a consequence of Proposition 1.57 (page 64) (This geometric relation holds *without* the restriction of preceding § 4.4 that all set values have to be $C^{1,1}$ submanifolds of $\mathbb{R}^N$ as in Remark 39, page 370.)
List of main results formulated in § 4.5	Existence due to strong-weak transitional Euler compactness: Proposition 50 (page 381) Uniqueness due to Lipschitz continuity: Proposition 51
Key tools	The adjoint arc is a necessary condition on boundary points and their limiting normals: Proposition 25 (page 362) Proximal balls at the boundary of reachable sets of differential inclusions (by means of adjoint arcs and matrix Riccati equations) in Appendix A.5.6 (page 469 ff.)

Table 4.2 Brief summary of the example in § 4.5 in mutational terms:
Mutational equations for compact sets in $\mathbb{R}^N$ depending on the normal cones

For each index $j, \kappa \in [0,1]$, define

$$\widetilde{d}_{\mathscr{K},j,\kappa} : (\widetilde{\mathscr{K}^{\Upsilon}}(\mathbb{R}^N) \cup \widetilde{\mathscr{K}^{\rightarrow}}(\mathbb{R}^N)) \times (\widetilde{\mathscr{K}^{\Upsilon}}(\mathbb{R}^N) \cup \widetilde{\mathscr{K}^{\rightarrow}}(\mathbb{R}^N)) \longrightarrow [0,\infty[,$$

by

$$\begin{aligned} \widetilde{d}_{\mathscr{K},j,\kappa}\big((s,\mu,C),\,(t,\nu,D)\big) \;:=&\\ dl(C,D) \,+\, \int_j^\infty \psi(\rho+\kappa+200\,\Lambda\,|t-s|)\,\cdot\, e^{\subset}\Big(&\text{Graph } {}^\flat N^P_{D,\,(\rho+\kappa+200\,\Lambda\,|t-s|)},\\ &\text{Graph } {}^\flat N^P_{C,\rho}\Big)\; d\rho \end{aligned}$$

with a fixed nonincreasing weight function $\psi \in C_0^\infty([0,2[),\ \psi \geq 0$. Furthermore set

$$\begin{aligned} \widetilde{d}_{\mathscr{K},j}\big((s,\mu,C),\,(t,\nu,D)\big) \;:=&\; \sup_{\kappa\in\,]0,1]} \; \widetilde{d}_{\mathscr{K},j,\kappa}\big((s,\mu,C),\,(t,\nu,D)\big)\\ =&\; \limsup_{\kappa\downarrow 0} \; \widetilde{d}_{\mathscr{K},j,\kappa}\big((s,\mu,C),\,(t,\nu,D)\big). \end{aligned}$$

In fact, the second component (being either 0 or 1) does not have any influence on $\widetilde{d}_{\mathscr{K},j}$ and $\widetilde{d}_{\mathscr{K},j,\kappa}$. Its purpose will only be to determine the evolution of time components for "test elements" and "normal" elements in a different way (as specified in subsequent Definition 44).

Lemma 42. *For each $j \in [0,1]$, the function $\widetilde{d}_{\mathscr{K},j}$ is reflexive and satisfies the timed triangle inequality on $\widetilde{\mathscr{K}^{\Upsilon}}(\mathbb{R}^N) \cup \widetilde{\mathscr{K}^{\rightarrow}}(\mathbb{R}^N)$. Moreover, $(\widetilde{d}_{\mathscr{K},j,\kappa})_{\kappa\in\,]0,1]}$ satisfies the following generalization of the timed triangle inequality:*

$$\widetilde{d}_{\mathscr{K},j,\kappa+\kappa'}(\widetilde{K}_1,\,\widetilde{K}_3) \;\leq\; \widetilde{d}_{\mathscr{K},j,\kappa'}(\widetilde{K}_1,\,\widetilde{K}_2) \,+\, \widetilde{d}_{\mathscr{K},j,\kappa}(\widetilde{K}_2,\,\widetilde{K}_3)$$

for any $\kappa, \kappa' \in]0,1]$, $\widetilde{K}_1, \widetilde{K}_2, \widetilde{K}_3 \in \widetilde{\mathscr{K}^{\Upsilon}}(\mathbb{R}^N) \cup \widetilde{\mathscr{K}^{\rightarrow}}(\mathbb{R}^N)$ with $\pi_1\,\widetilde{K}_1 \leq \pi_1\,\widetilde{K}_2 \leq \pi_1\,\widetilde{K}_3$.
Thus, $(\widetilde{d}_{\mathscr{K},j})_{j\in\,]0,1]}$ and $(\widetilde{d}_{\mathscr{K},j,\kappa})_{j,\kappa\in\,]0,1]}$ fulfill the hypotheses (H1), (H3') of § 4.1.

Proof. Reflexivity is obvious. For verifying the timed triangle inequality, choose any $(t_1,\mu_1,K_1),\ (t_2,\mu_2,K_2),\ (t_3,\mu_3,K_3) \in \mathbb{R}\times\{0,1\}\times\mathscr{K}(\mathbb{R}^N)$ with $t_1 \leq t_2 \leq t_3$. Then, we obtain for every $\kappa, \kappa' > 0$

$$\begin{aligned} & e^{\subset}\Big(\text{Graph } {}^\flat N^P_{K_3,\,(\rho+\kappa+\kappa'+200\,\Lambda\,(t_3-t_1))},\ \text{Graph } {}^\flat N^P_{K_1,\rho}\Big)\\ \leq\;& e^{\subset}\Big(\text{Graph } {}^\flat N^P_{K_3,\,(\rho+\kappa+\kappa'+200\,\Lambda\,(t_3-t_1))},\ \text{Graph } {}^\flat N^P_{K_2,\,(\rho+\kappa+200\,\Lambda\,(t_2-t_1))}\Big)\\ &+ e^{\subset}\Big(\text{Graph } {}^\flat N^P_{K_2,\,(\rho+\kappa+200\,\Lambda\,(t_2-t_1))},\ \ \text{Graph } {}^\flat N^P_{K_1,\rho}\Big). \end{aligned}$$

With regard to the weighted integral in $\widetilde{d}_{\mathscr{K},j,\kappa+\kappa'}\big((t_1,\mu_1,K_1),\,(t_3,\mu_3,K_3)\big)$, a simple translation of coordinates (for the first distance term) and the monotonicity of ψ (related with the second distance term) imply

$$\begin{aligned} &\widetilde{d}_{\mathscr{K},j,\kappa+\kappa'}\big((t_1,\mu_1,K_1),\,(t_3,\mu_3,K_3)\big)\\ &\quad\leq \widetilde{d}_{\mathscr{K},j,\kappa'}\big((t_1,\mu_1,K_1),\,(t_2,\mu_2,K_2)\big) + \widetilde{d}_{\mathscr{K},j,\kappa}\big((t_2,\mu_2,K_2),\,(t_3,\mu_3,K_3)\big)\\ &\quad\leq \widetilde{d}_{\mathscr{K},j}\,\big((t_1,\mu_1,K_1),\,(t_2,\mu_2,K_2)\big) + \widetilde{d}_{\mathscr{K},j}\big((t_2,\mu_2,K_2),\,(t_3,\mu_3,K_3)\big). \end{aligned}$$

□

4.5.2 Reachable Sets Induce Timed Transitions on $(\widetilde{\mathscr{K}}^{\rightarrow}(\mathbb{R}^N), \widetilde{\mathscr{K}}^{\curlyvee}(\mathbb{R}^N))$

The Hamilton condition in Proposition 25 (on page 362) is to bridge the gap between the geometric evolution of proximal normal cones and its analytical description. In particular, Corollary A.47 (on page 470) gives a bound how fast the exterior ball in a proximal direction can change its radius at most. For applying this result as a tool in a moment, we choose the following class of set-valued maps:

Definition 43. For $\Lambda > 0$ fixed, the set $\mathrm{LIP}_{\Lambda}^{(C^2)}(\mathbb{R}^N, \mathbb{R}^N)$ consists of all set-valued maps $F : \mathbb{R}^N \rightsquigarrow \mathbb{R}^N$ satisfying

1.) $F : \mathbb{R}^N \rightsquigarrow \mathbb{R}^N$ has nonempty compact convex values,
2.) $\mathscr{H}_F(x,p) := \sup_{v \in F(x)} p \cdot v$ is twice continuously differentiable in $\mathbb{R}^N \times (\mathbb{R}^N \setminus \{0\})$,
3.) $\|\mathscr{H}_F\|_{C^2(\mathbb{R}^N \times \partial\mathbb{B}_1)} < \Lambda$.

These set-valued maps of $\mathrm{LIP}_{\Lambda}^{(C^2)}(\mathbb{R}^N, \mathbb{R}^N)$ induce the candidates for timed transitions on $(\widetilde{\mathscr{K}}^{\rightarrow}(\mathbb{R}^N), \widetilde{\mathscr{K}}^{\curlyvee}(\mathbb{R}^N), (\widetilde{d}_{\mathscr{K},j})_{j \in]0,1]}, (\widetilde{d}_{\mathscr{K},j})_{j \in]0,1]}, 0)$ in the following sense:

Definition 44. For any set-valued map $F \in \mathrm{LIP}_{\Lambda}^{(C^2)}(\mathbb{R}^N, \mathbb{R}^N)$, element $(t, \mu, K) \in \mathbb{R} \times \{0,1\} \times \mathscr{K}(\mathbb{R}^N) = \widetilde{\mathscr{K}}^{\curlyvee}(\mathbb{R}^N) \cup \widetilde{\mathscr{K}}^{\rightarrow}(\mathbb{R}^N)$ and time $h > 0$, set

$$\widetilde{\vartheta}_F(h, (t, \mu, K)) := (t + \mu\, h,\ \mu,\ \vartheta_F(h, K))$$

with the reachable set $\vartheta_F(h, K) \subset \mathbb{R}^N$ of the differential inclusion $x(\cdot) \in F(x(\cdot))$ a.e.

Proposition 45. *The maps*

$$\widetilde{\vartheta}_F : [0,1] \times (\widetilde{\mathscr{K}}^{\curlyvee}(\mathbb{R}^N) \cup \widetilde{\mathscr{K}}^{\rightarrow}(\mathbb{R}^N)) \longrightarrow \widetilde{\mathscr{K}}^{\curlyvee}(\mathbb{R}^N) \cup \widetilde{\mathscr{K}}^{\rightarrow}(\mathbb{R}^N)$$

of all $F \in \mathrm{LIP}_{\Lambda}^{(C^2)}(\mathbb{R}^N, \mathbb{R}^N)$ introduced in Definition 44 induce timed transitions on the tuple $(\widetilde{\mathscr{K}}^{\rightarrow}(\mathbb{R}^N), \widetilde{\mathscr{K}}^{\curlyvee}(\mathbb{R}^N), (\widetilde{d}_{\mathscr{K},j})_{j \in]0,1]}, (\widetilde{d}_{\mathscr{K},j})_{j \in]0,1]}, 0)$ with

$$\begin{aligned}
\alpha_j(\widetilde{\vartheta}_F; \cdot, \cdot) &\overset{\text{Def.}}{=} 10\ \Lambda\ e^{2\Lambda \cdot \tau(j,\Lambda)},\\
\beta_j(\widetilde{\vartheta}_F; \cdot) &\overset{\text{Def.}}{=} \Lambda\ (1 + \|\psi\|_{L^1}\ (e^{\Lambda} + 1)),\\
\gamma_j(\widetilde{\vartheta}_F) &\overset{\text{Def.}}{=} 0,\\
\mathbb{T}_j(\widetilde{\vartheta}_F, \cdot) &\overset{\text{Def.}}{=} \min\{\tau(j,\Lambda), 1\} \quad \textit{(mentioned in Corollary A.47)},\\
\widehat{D}_j(\widetilde{\vartheta}_F, \widetilde{\vartheta}_G; \cdot, \cdot) &\overset{\text{Def.}}{=} (1 + 6N\, \|\psi\|_{L^1}) \cdot \|\mathscr{H}_F - \mathscr{H}_G\|_{C^1(\mathbb{R}^N \times \partial\mathbb{B}_1)}.
\end{aligned}$$

The proof consists of several steps which we first summarize and then verify in detail. They are very similar to the proofs in § 4.4.2 indeed, but take the proximal radii into consideration additionally.

Lemma 46. *For every set-valued map* $F \in \mathrm{LIP}^{(C^2)}_{\Lambda}(\mathbb{R}^N, \mathbb{R}^N)$, *initial element* $\widetilde{K} = (b, 1, K) \in \widetilde{\mathscr{K}}^{\to}(\mathbb{R}^N)$ *and any times* $0 \le s < t \le 1$,

$$\widetilde{d}_{\mathscr{K},j}\Big(\widetilde{\vartheta}_F(s, \widetilde{K}),\ \widetilde{\vartheta}_F(t, \widetilde{K})\Big) \ \le\ \Lambda\,(1 + \|\psi\|_{L^1}\,(e^{\Lambda} + 1)) \cdot |t - s|.$$

Lemma 47. *For any* $j \in]0,1]$, *let* $\tau(j,\Lambda) > 0$ *denote the time period mentioned in Corollary A.47 (on page 470). Choose any maps* $F, G \in \mathrm{LIP}^{(C^2)}_{\Lambda}(\mathbb{R}^N, \mathbb{R}^N)$, *initial elements* $\widetilde{K}_1 = (t_1, 0, K_1) \in \widetilde{\mathscr{K}}^{\curlyvee}(\mathbb{R}^N)$, $\widetilde{K}_2 = (t_2, 1, K_2) \in \widetilde{\mathscr{K}}^{\to}(\mathbb{R}^N)$ *with* $t_1 \le t_2$.

Then for all $h \in [0, \tau(j,\Lambda)[$, *the following inequality holds*

$$\begin{aligned}&\widetilde{d}_{\mathscr{K},j}\Big(\widetilde{\vartheta}_F(h, \widetilde{K}_1), \widetilde{\vartheta}_G(h, \widetilde{K}_2)\Big)\\ &\quad\le e^{(\lambda_{\mathscr{H}}+\Lambda)h} \cdot \Big(\widetilde{d}_{\mathscr{K},j}(\widetilde{K}_1, \widetilde{K}_2) + (1 + 6N\,\|\psi\|_{L^1}) \cdot h \cdot \|\mathscr{H}_F - \mathscr{H}_G\|_{C^1(\mathbb{R}^N \times \partial\mathbb{B}_1)}\Big)\end{aligned}$$

with the abbreviation $\lambda_{\mathscr{H}} := 9\,\Lambda\, e^{2\Lambda\cdot\tau(j,\Lambda)}$.

Corollary 48. *Under the assumptions of Lemma 47,*

$$\begin{aligned}&\widetilde{d}_{\mathscr{K},j}\Big(\widetilde{\vartheta}_F(t{+}h, \widetilde{K}_1),\ \widetilde{\vartheta}_G(h, \widetilde{K}_2)\Big)\\ &\quad\le\ e^{(\lambda_{\mathscr{H}}+\Lambda)h} \cdot \Big(\widetilde{d}_{\mathscr{K},j}(\widetilde{\vartheta}_F(t, \widetilde{K}_1), \widetilde{K}_2) + (1{+}6N\,\|\psi\|_{L^1})h\|\mathscr{H}_F - \mathscr{H}_G\|_{C^1(\mathbb{R}^N \times \partial\mathbb{B}_1)}\Big)\end{aligned}$$

for all $h, t \ge 0$ *with* $t + h < \tau(j,\Lambda)$ *and*
$\widetilde{K}_1 = (t_1, 0, K_1) \in \widetilde{\mathscr{K}^{\curlyvee}}(\mathbb{R}^N)$, $\widetilde{K}_2 = (t_2, 1, K_2) \in \widetilde{\mathscr{K}^{\to}}(\mathbb{R}^N)$ *with* $t_1 \le t_2$.

Proof (of Lemma 46). Obviously, the Pompeiu-Hausdorff distance satisfies for every $s, t \ge 0$

$$d\!l\big(\vartheta_F(s,K),\ \vartheta_F(t,K)\big) \ \le\ \sup_{\mathbb{R}^N} \|F(\cdot)\|_\infty \cdot |t-s| \ \le\ \Lambda\ |t-s|.$$

Let $\tau(j,\Lambda) > 0$ denote the time period mentioned in Corollary A.47 (on page 470). Without loss of generality, we can now assume $0 < t - s < \frac{1}{200\,\Lambda}\ \tau(j,\Lambda)$ as a consequence of the timed triangle inequality.

For any $(x,p) \in \mathrm{Graph}\ {}^{\flat}N^P_{\vartheta_F(t,K),\,(\rho+200\,\Lambda\,(t-s))}$ and $\rho \ge j$ with $\rho + 200\,\Lambda\,(t-s) \le 2$, Corollary A.47 and Proposition 25 (on page 362) provide both a solution $x(\cdot) \in C^1([s,t], \mathbb{R}^N)$ and its adjoint arc $p(\cdot) \in C^1([s,t], \mathbb{R}^N)$ satisfying

$$\begin{cases} x'(\sigma) = \ \ \frac{\partial}{\partial p}\,\mathscr{H}_F(x(\sigma), p(\sigma)) \in F(x(\sigma)), & x(t) = x, \quad x(s) \in \partial\vartheta_F(s,K),\\ p'(\sigma) = -\frac{\partial}{\partial x}\,\mathscr{H}_F(x(\sigma), p(\sigma)), & p(t) = p, \quad p(s) \in N^P_{\vartheta_F(s,K)}(x(s)) \end{cases}$$

and, $p(s)$ has proximal radius $\ge \rho + 200\,\Lambda\,(t-s) - 81\,\Lambda\,(t-s) > \rho$.

Obviously, $\mathscr{H}_F$ is positively homogeneous with respect to its second argument and thus, its definition implies $|p'(\sigma)| \leq \Lambda\ |p(\sigma)|$ for all σ. Moreover $|p| \leq 1$ implies that the projection of p on any cone is also contained in $\mathbb{B}_1$ and so finally, we obtain similarly to Lemma 27 (on page 362)

$$\begin{aligned} e^{\subset}\Big((x,p),\ \mathrm{Graph}\ {}^{\flat}N^P_{\vartheta_F(s,K),\rho}\Big) &\leq |x - x(s)| + |p - p(s)| \\ &\leq \sup_{s\leq\sigma\leq t} \Big(|\frac{\partial}{\partial p}\mathscr{H}_F| + |\frac{\partial}{\partial x}\mathscr{H}_F|\Big)\Big|_{(x(\sigma),p(\sigma))} \cdot (t-s) \\ &\leq \Big(\Lambda + \Lambda\, e^{\Lambda t}\Big)\cdot(t-s). \end{aligned}$$

□

Proof (of Lemma 47). Proposition 1.50 (on page 60) concludes the following estimate of the Pompeiu-Hausdorff distance from Filippov's Theorem A.6 about differential inclusions (with Lipschitz continuous right-hand side)

$$\begin{aligned} d\!l\Big(\vartheta_F(h,K_1),\ \vartheta_G(h,K_2)\Big) &\leq d\!l(K_1,K_2)\cdot e^{\Lambda h} + \sup_{\mathbb{R}^N} d\!l\Big(F(\cdot),G(\cdot)\Big)\ \cdot \frac{e^{\Lambda h}-1}{\Lambda} \\ &\leq d\!l(K_1,K_2)\cdot e^{\Lambda h} + \sup_{\mathbb{R}^N\times\partial\mathbb{B}_1} |\mathscr{H}_F - \mathscr{H}_G|\ \cdot h\ e^{\Lambda h}. \end{aligned}$$

According to Definition 44,

$$\begin{aligned} \widetilde{\vartheta}_F(h,\widetilde{K}_1) &\in \quad \{t_1\}\times\{0\}\times\mathscr{K}(\mathbb{R}^N) \subset \widetilde{\mathscr{K}}^{\curlyvee}(\mathbb{R}^N), \\ \widetilde{\vartheta}_G(h,\widetilde{K}_2) &\in \{t_2+h\}\times\{1\}\times\mathscr{K}(\mathbb{R}^N) \subset \widetilde{\mathscr{K}}^{\rightharpoonup}(\mathbb{R}^N). \end{aligned}$$

Now for any $\kappa\in\,]0,1]$ and $\rho \geq j$ with $\rho+\kappa+200\,\Lambda\,(t_2-t_1+h) \leq 2$, we need an upper bound of $e^{\subset}\big(\mathrm{Graph}\,{}^{\flat}N^P_{\vartheta_G(h,K_2),\,(\rho+\kappa+200\,\Lambda\,(t_2-t_1+h))},\ \mathrm{Graph}\,{}^{\flat}N^P_{\vartheta_F(h,K_1),\,\rho}\big)$:

Choose any $\delta>0$, $x\in\partial\,\vartheta_G(h,K_2)$ and $p\in N^P_{\vartheta_G(h,K_2)}(x)\cap\partial\mathbb{B}_1$ with proximal radius $\geq \rho+\kappa+200\,\Lambda\,(t_2-t_1+h)$ arbitrarily. According to Corollary A.47 and Proposition 25, there exist a solution $x(\cdot)\in C^1([0,h],\mathbb{R}^N)$ and its adjoint arc $p(\cdot)\in C^1([0,h],\mathbb{R}^N)$ fulfilling

$$\begin{cases} x'(\cdot) = \frac{\partial}{\partial p}\mathscr{H}_G(x(\cdot),\,p(\cdot)) \in G(x(\cdot)), & p'(\cdot) = -\frac{\partial}{\partial x}\mathscr{H}_G(x(\cdot),\,p(\cdot)) \in \Lambda\,|p(\cdot)|\cdot\mathbb{B} \\ x(0)\in\partial K_2, & p(0)\in N^P_{K_2}(x(0)), \\ x(h) = x, & p(h) = p, \end{cases}$$

and, the proximal radius at $x(0)$ in direction $p(0)$ is

$$\geq\ \rho+\kappa+200\,\Lambda\,(t_2-t_1+h) - 81\,\Lambda\,h\ >\ \rho+\kappa+100\,\Lambda\,h+200\,\Lambda\,(t_2-t_1).$$

Gronwall's inequality ensures $e^{-\Lambda h}\leq|p(\cdot)|\leq e^{\Lambda h}$ in $[0,h]$ and hence,

$$p(0)\,e^{-\Lambda h}\in {}^{\flat}N^P_{K_2}(x(0))\setminus\{0\}.$$

Now let $(y_0,\widehat{q}_0)$ denote an element of $\mathrm{Graph}\,{}^{\flat}N^P_{K_1,\,(\rho+100\,\Lambda\,h)}$ with $\widehat{q}_0\neq 0$ and

$$\begin{aligned} &\Big|(y_0,\widehat{q}_0) - \big(x(0),\ p(0)\,e^{-\Lambda h}\big)\Big| \\ &\quad\leq e^{\subset}\Big(\mathrm{Graph}\,{}^{\flat}N^P_{K_2,\,(\rho+\kappa+100\,\Lambda\,h+200\,\Lambda\,(t_2-t_1))},\mathrm{Graph}\,{}^{\flat}N^P_{K_1,\,(\rho+100\,\Lambda\,h)}\Big)+\delta. \end{aligned}$$

As another consequence of Corollary A.47, we get a solution $y(\cdot) \in C^1([0,h],\mathbb{R}^N)$ and its adjoint arc $q(\cdot)$ satisfying

$$\begin{cases} y'(\cdot) = \frac{\partial}{\partial p}\mathscr{H}_F(y(\cdot),q(\cdot)), & q'(\cdot) = -\frac{\partial}{\partial y}\mathscr{H}_F(y(\cdot),q(\cdot)) \in \Lambda\,|q(\cdot)|\cdot\mathbb{B} \\ y(0) = y_0, & q(0) = \widehat{q}_0\, e^{\Lambda h} \neq 0, \\ y(h) \in \partial\,\vartheta_F(h,K_1), & q(h) \in N^P_{\vartheta_F(h,K_1)}(y(h)) \end{cases}$$

and the proximal radius at $y(h)$ in direction $q(h)$ is $\geq \rho + 100\,\Lambda\,h - 81\,\Lambda\,h > \rho$. $\mathscr{H}_F$ is assumed to be *twice* continuously differentiable with $\|\mathscr{H}_F\|_{C^2(\mathbb{R}^N\times\,\partial\mathbb{B}_1)} < \Lambda$. Moreover, $\mathscr{H}_F(x,p)$ is positively homogeneous with respect to p and thus, the derivative of $\mathscr{H}_F$ is $\lambda_{\mathscr{H}}$-Lipschitz continuous in $\mathbb{R}^N \times (\mathbb{B}_{e^{\Lambda\cdot\tau(j,\Lambda)}} \setminus \overset{\circ}{\mathbb{B}}_{e^{-\Lambda\cdot\tau(j,\Lambda)}})$ with the abbreviation $\lambda_{\mathscr{H}} := 9\,\Lambda\,e^{2\Lambda\cdot\tau(j,\Lambda)}$ (due to Lemma 33 on page 366). Correspondingly to the proof of Lemma 31 (on page 364), the Theorem of Cauchy-Lipschitz applied to the Hamiltonian system leads to

$$\begin{aligned} &\boldsymbol{e}^{\subset}\Big((x,p),\ \mathrm{Graph}\ {}^{\flat}N^P_{\vartheta_F(h,K_1),\rho}\Big) \\ &\leq \Big|(x,\,p)-(y(h),q(h))\Big| \\ &\leq e^{\lambda_{\mathscr{H}}\cdot h}\cdot\Big|(x(0),p(0))-(y_0,\widehat{q}_0 e^{\Lambda h})\Big| + \tfrac{e^{\lambda_{\mathscr{H}}\cdot h}-1}{\lambda_{\mathscr{H}}}\cdot\sup_{[0,h]}\big|D\mathscr{H}_F - D\mathscr{H}_G\big|\Big|_{(x(\cdot),p(\cdot))}. \end{aligned}$$

$\mathscr{H}_F$ and $\mathscr{H}_G$ are positively homogeneous with respect to the second argument and thus,

$$\begin{aligned} \Big|\tfrac{\partial}{\partial x_j}(\mathscr{H}_F-\mathscr{H}_G)|_{(x(s),\,p(s))}\Big| &\leq e^{\Lambda h}\,\|D\mathscr{H}_F - D\mathscr{H}_G\|_{C^0(\mathbb{R}^N\times\partial\mathbb{B}_1)}, \\ \Big|\tfrac{\partial}{\partial p_j}(\mathscr{H}_F-\mathscr{H}_G)|_{(x(s),\,p(s))}\Big| &\leq 3\cdot\|\mathscr{H}_F-\mathscr{H}_G\|_{C^1(\mathbb{R}^N\times\partial\mathbb{B}_1)}. \end{aligned}$$

We obtain

$$\begin{aligned} \boldsymbol{e}^{\subset}\Big((x,p),\ \mathrm{Graph}\ {}^{\flat}N^P_{\vartheta_F(h,K_1),\rho}\Big) \leq e^{(\lambda_{\mathscr{H}}+\Lambda)\,h}\Big(\Big|(x(0),\,p(0)\,e^{-\Lambda h})-(y_0,\widehat{q}_0)\Big| \\ +h\cdot 6N\,\|\mathscr{H}_F-\mathscr{H}_G\|_{C^1(\mathbb{R}^N\times\partial\mathbb{B}_1)}\Big) \end{aligned}$$

and, since $\delta > 0$ is arbitrarily small and $|p| = 1$,

$$\begin{aligned} &\boldsymbol{e}^{\subset}\Big(\mathrm{Graph}\,{}^{\flat}N^P_{\vartheta_G(h,K_2),\,(\rho+\kappa+200\Lambda\,(t_2-t_1+h))},\ \mathrm{Graph}\,{}^{\flat}N^P_{\vartheta_F(h,K_1),\,\rho}\Big) \\ &\leq e^{(\lambda_{\mathscr{H}}+\Lambda)\,h}\cdot\Big\{\boldsymbol{e}^{\subset}\Big(\mathrm{Graph}\,{}^{\flat}N^P_{K_2,\,(\rho+\kappa+100\Lambda\,h+200\Lambda\,(t_2-t_1))},\ \mathrm{Graph}\,{}^{\flat}N^P_{K_1,\,(\rho+100\Lambda\,h)}\Big) \\ &\qquad + 6Nh\cdot\|\mathscr{H}_F-\mathscr{H}_G\|_{C^1(\mathbb{R}^N\times\partial\mathbb{B}_1)}\Big\}. \end{aligned}$$

With regard to $\widetilde{d}_{\mathscr{K},j,\kappa}\Big(\widetilde{\vartheta}_F(h,\widetilde{K}_1),\ \widetilde{\vartheta}_G(h,\widetilde{K}_2)\Big)$, integrating over ρ and the monotonicity of the weight function ψ (supposed in Definition 40) leads to the claimed estimate for all $h \in [0,\tau(j,\Lambda)[$. □

Proof (of Corollary 48). It results directly from Lemma 47 since

$$\begin{aligned} \widetilde{\vartheta}_F(t+h,\widetilde{K}_1) &= \{t_1\}\times\{0\}\times\vartheta_F(t+h,K_1) = \widetilde{\vartheta}_F\big(h,\ \widetilde{\vartheta}_F(t,\widetilde{K}_1)\big), \\ \widetilde{\vartheta}_F(t,\widetilde{K}_1) &= \{t_1\}\times\{0\}\times\vartheta_F(t,K_1) \quad \in \widetilde{\mathscr{K}^{\Upsilon}}(\mathbb{R}^N). \end{aligned}$$

□

Proof (of Proposition 45). The semigroup property $\widetilde{\vartheta}_F\big(h,\, \widetilde{\vartheta}_F(t,\widetilde{K})\big) = \widetilde{\vartheta}_F(t+h,\widetilde{K})$ holds for all $F \in \mathrm{LIP}_\lambda^{(C^2)}(\mathbb{R}^N,\mathbb{R}^N)$, $\widetilde{K} \in \widetilde{\mathscr{K}^{\curlyvee}}(\mathbb{R}^N) \cup \widetilde{\mathscr{K}^{\rightarrow}}(\mathbb{R}^N)$, $h,t \ge 0$.
Moreover, Definition 44 has the immediate consequences for every $\widetilde{K} \in \widetilde{\mathscr{K}^{\rightarrow}}(\mathbb{R}^N)$, $\widetilde{Z} \in \widetilde{\mathscr{K}^{\curlyvee}}(\mathbb{R}^N)$ and $h \in [0,1]$

$$\begin{array}{rcl} \widetilde{\vartheta}_F(0,\widetilde{K}) &=& \widetilde{K} \\ \widetilde{\vartheta}_F(h,\widetilde{Z}) &\in& \{\pi_1\,\widetilde{Z}\}\times\{0\}\times\mathscr{K}(\mathbb{R}^N) \subset \widetilde{\mathscr{K}^{\curlyvee}}(\mathbb{R}^N) \\ \widetilde{\vartheta}_F(h,\widetilde{K}) &\in& \{h+\pi_1\,\widetilde{K}\}\times\{1\}\times\mathscr{K}(\mathbb{R}^N) \subset \widetilde{\mathscr{K}^{\rightarrow}}(\mathbb{R}^N) \end{array}$$

i.e., conditions (1.), (6.), (7.) of Definition 1 (on page 336) are also satisfied.

Set $\mathbb{T}_j(\widetilde{\vartheta}_F,\cdot) \stackrel{\text{Def.}}{=} \min\{\tau(j,\Lambda),1\}$ with the time parameter $\tau(j,\Lambda) > 0$ mentioned in Corollary A.47 (on page 470). Then, Corollary 48 guarantees for all $\widetilde{Z} \in \widetilde{\mathscr{K}^{\curlyvee}}(\mathbb{R}^N)$, $\widetilde{K} \in \widetilde{\mathscr{K}^{\rightarrow}}(\mathbb{R}^N)$, $t \in [0,\mathbb{T}_j(\widetilde{\vartheta}_F,\widetilde{Z})[$ with $t+\pi_1\,\widetilde{Z} \le \pi_1\,\widetilde{K}$

$$\limsup_{h\downarrow 0} \frac{\widetilde{d}_{\mathscr{K},j}\big(\widetilde{\vartheta}_F(t+h,\widetilde{Z}),\, \widetilde{\vartheta}_F(h,\widetilde{K})\big) - \widetilde{d}_{\mathscr{K},j}(\widetilde{\vartheta}_F(t,\widetilde{Z}),\widetilde{K})}{h \quad \widetilde{d}_{\mathscr{K},j}(\widetilde{\vartheta}_F(t,\widetilde{Z}),\widetilde{K})} \le \lambda_{\mathscr{H}} + \Lambda \le 10\,\Lambda\, e^{2\Lambda\cdot\tau(j,\Lambda)}.$$

Lemma 46 implies condition (4.') of Definition 1 with the Lipschitz constant

$$\beta_j(\widetilde{\vartheta}_F;\cdot) \stackrel{\text{Def.}}{=} \Lambda\,\big(1+\|\psi\|_{L^1}\,(e^\Lambda+1)\big).$$

Setting for all $\widetilde{Z} \in \widetilde{\mathscr{K}^{\curlyvee}}(\mathbb{R}^N)$ and $F,G \in \mathrm{LIP}_\Lambda^{(C^2)}(\mathbb{R}^N,\mathbb{R}^N)$,

$$\widehat{D}_j(\widetilde{\vartheta}_F,\,\widetilde{\vartheta}_G;\,\widetilde{Z},\cdot) \stackrel{\text{Def.}}{=} (1+6N\,\|\psi\|_{L^1})\,\|\mathscr{H}_F - \mathscr{H}_G\|_{C^1(\mathbb{R}^N\times\partial\mathbb{B}_1)}\cdot$$

hypotheses (H5') – (H7') (on page 337) are fulfilled due to Corollary 48.

Finally condition (8.) of Definition 1 has to be verified, i.e.,

$$\limsup_{n\to\infty}\ \widetilde{d}_{\mathscr{K},j}\big(\widetilde{\vartheta}_F(t-h_n,\widetilde{Z}),\,\widetilde{K}_n\big) \ \ge\ \widetilde{d}_{\mathscr{K},j}\big(\widetilde{\vartheta}_F(t,\widetilde{Z}),\,\widetilde{K}\big)$$

for all $\widetilde{Z} \in \widetilde{\mathscr{K}^{\curlyvee}}(\mathbb{R}^N)$, $\widetilde{K},\widetilde{K}_n \in \widetilde{\mathscr{K}^{\rightarrow}}(\mathbb{R}^N)$, $t \in [0,\mathbb{T}_j(\widetilde{\vartheta}_F,\widetilde{Z})]$ and $h_n \downarrow 0$ satisfying $\pi_1\,\widetilde{Z} \le \pi_1\,\widetilde{K}_n \nearrow \pi_1\,\widetilde{K}$ and $\widetilde{d}_{\mathscr{K},j}(\widetilde{K}_n,\widetilde{K}) \longrightarrow 0$ for each $j \in]0,1]$.
Indeed, $d\!l\big(\vartheta_F(t-h,Z),\ \vartheta_F(t,Z)\big) \longrightarrow 0$ holds for $h\downarrow 0$ and any set $Z \in \mathscr{K}(\mathbb{R}^N)$. Proposition A.72 (on page 487) states for any $0<r<\rho$ and $(K_n)_{n\in\mathbb{N}}$ tending to $K \in \mathscr{K}(\mathbb{R}^N)$

$$\begin{array}{lll} \mathrm{Limsup}_{n\to\infty}\ \mathrm{Graph}\ {}^\flat N^P_{\vartheta_F(t-h_n,Z),\,\rho} & \subset & \mathrm{Graph}\ {}^\flat N^P_{\vartheta_F(t,Z),\,\rho} \\ \mathrm{Liminf}_{n\to\infty}\ \mathrm{Graph}\ {}^\flat N^P_{K_n,\,r} & \supset & \mathrm{Graph}\ {}^\flat N^P_{K,\,\rho} \end{array}$$

Thus, we obtain for every $\widetilde{Z} = (a,0,Z) \in \widetilde{\mathscr{K}^{\curlyvee}}(\mathbb{R}^N)$, $\widetilde{K} = (b,1,K)$, $\widetilde{K}_n = (b_n,1,K_n) \in \widetilde{\mathscr{K}^{\rightarrow}}(\mathbb{R}^N)$, $\rho>0$, $\kappa \in]0,1]$ and $t \in [0,\mathbb{T}_j(\widetilde{\vartheta}_F,\widetilde{Z})]$ with $a \le b_n \nearrow b$

$$\begin{array}{l} \liminf\limits_{n\to\infty}\ e^{\subset}\Big(\mathrm{Graph}\ {}^\flat N^P_{K_n,\,(\rho+\frac{\kappa}{2}+200\Lambda\,|b_n-a|)},\ \mathrm{Graph}\ {}^\flat N^P_{\vartheta_F(t-h_n,Z),\,\rho}\Big) \\ \qquad \ge e^{\subset}\Big(\mathrm{Graph}\ {}^\flat N^P_{K,\ (\rho+\kappa+200\Lambda\,|b-a|)},\ \mathrm{Graph}\ {}^\flat N^P_{\vartheta_F(t,Z),\,\rho}\Big). \end{array}$$

Due to $\pi_1\,\widetilde{\vartheta}_F(t-h,\widetilde{Z}) = a = \pi_1\,\widetilde{\vartheta}_F(t,\widetilde{Z})$, this inequality, the monotonicity of $\psi(\cdot)$ and Fatou's Lemma imply the wanted relation with respect to $\widetilde{d}_{\mathscr{K},j}$. □

4.5.3 Existence due to Strong-Weak Transitional Euler Compactness

In §§ 4.3.3, 4.3.4, the results about existence of timed solutions to mutational equations are based on two appropriate forms of transitional Euler compactness (see Definitions 14, 18). Considering a converging sequence of compact sets, some features of their proximal cones are summarized in Appendix A.9 (on page 487 f.). In particular, the inclusion

$$\text{Graph}\, N^P_{K,\rho} \;\subset\; \text{Limsup}_{n\to\infty}\, \text{Graph}\, N^P_{K_n,\rho}$$

does *not* hold for every radius $\rho > 0$ in general. This rather technical aspect is the obstacle why we now prefer the second approach of § 4.3.4 using "strongly-weakly transitionally Euler compact" and Existence Theorem 20 (on page 354).

In fact, each timed solution $\widetilde{K}(\cdot) = (\cdot, 1, K(\cdot)) : [0,T] \longrightarrow \widetilde{\mathscr{K}^{\rightarrow}}(\mathbb{R}^N)$ induces a solution to the underlying morphological equation in the sense of Aubin (due to $\widetilde{\mathscr{K}^{\Upsilon}}(\mathbb{R}^N) \cong \widetilde{\mathscr{K}^{\rightarrow}}(\mathbb{R}^N) \cong \mathbb{R} \times \mathscr{K}(\mathbb{R}^N)$).

Lemma 49. *The tuple* $(\widetilde{\mathscr{K}^{\rightarrow}}(\mathbb{R}^N),\ \widetilde{\mathscr{K}^{\Upsilon}}(\mathbb{R}^N),\ (\widetilde{d}_{\mathscr{K},j})_{j\in]0,1]},\ (\widetilde{d}_{\mathscr{K},j,\kappa})_{j,\kappa\in]0,1]},$ $(\widetilde{d}_{\mathscr{K},j})_{j\in]0,1]},\ (\widetilde{d}_{\mathscr{K},j,\kappa})_{j,\kappa\in]0,1]},\ 0,\ \text{LIP}^{(C^2)}_{\Lambda}(\mathbb{R}^N,\mathbb{R}^N))$ *is strongly-weakly transitionally Euler compact (in the sense of Definition 18 on page 353), i.e. here:*

Suppose each function $G_n : [0,1] \longrightarrow \text{LIP}^{(C^2)}_{\Lambda}(\mathbb{R}^N,\mathbb{R}^N)$ $(n \in \mathbb{N})$ *to be piecewise constant and set with some arbitrarily fixed* $\widetilde{K}_0 = (t_0, 1, K_0) \in \widetilde{\mathscr{K}^{\rightarrow}}(\mathbb{R}^N)$

$$\widetilde{G}_n : [0,1]\times\mathbb{R}^N \leadsto \mathbb{R}^N, \quad (t,x) \longmapsto G_n(t)(x),$$
$$\widetilde{K}_n(h) := \{t_0+h\}\times\{1\}\times\vartheta_{\widetilde{G}_n}(h,K_0) \in \widetilde{\mathscr{K}^{\rightarrow}}(\mathbb{R}^N) \qquad \text{for } h\in[0,1].$$

For any $t\in[0,1[$ *and sequence* $h_m \searrow 0$, *there exist a sequence* $n_k \nearrow \infty$ *of indices and an element* $\widetilde{K} = (t,1,K) \in \widetilde{\mathscr{K}^{\rightarrow}}(\mathbb{R}^N)$ *satisfying for every* $j,\kappa\in]0,1]$

$$\lim_{k\to\infty}\ \widetilde{d}_{\mathscr{K},j,\kappa}(\widetilde{K}_{n_k}(t),\ \ \widetilde{K}) = 0,$$
$$\lim_{m\to\infty}\ \sup_{k\ge m}\ \widetilde{d}_{\mathscr{K},j}(\widetilde{K},\ \widetilde{K}_{n_k}(t+h_m)) = 0.$$

Proposition 50.

Regard the maps $\widetilde{\vartheta}_F$ *of all set-valued maps* $F\in \text{LIP}^{(C^2)}_{\Lambda}(\mathbb{R}^N,\mathbb{R}^N)$ *(as in Definitions 43, 44) as timed transitions on* $(\widetilde{\mathscr{K}^{\rightarrow}}(\mathbb{R}^N),\ \widetilde{\mathscr{K}^{\Upsilon}}(\mathbb{R}^N),\ (\widetilde{d}_{\mathscr{K},j})_{j\in]0,1]},\ (\widetilde{d}_{\mathscr{K},j})_{j\in]0,1]},\ 0)$ *according to Proposition 45.*

For $\widetilde{f} : \widetilde{\mathscr{K}^{\rightarrow}}(\mathbb{R}^N)\times[0,T] \longrightarrow \text{LIP}^{(C^2)}_{\Lambda}(\mathbb{R}^N,\mathbb{R}^N)$, *suppose continuity in the sense that*

$$\|\mathscr{H}_{\widetilde{f}(\widetilde{K},t)} - \mathscr{H}_{\widetilde{f}(\widetilde{K}_m,t_m)}\|_{C^1(\mathbb{R}^N\times\partial\mathbb{B}_1)} \stackrel{m\to\infty}{\longrightarrow} 0$$

whenever $t_m \searrow t$ *and* $\widetilde{d}_{\mathscr{K},0}(\widetilde{K},\widetilde{K}_m) \longrightarrow 0$ $(\widetilde{K},\widetilde{K}_m \in \widetilde{\mathscr{K}^{\rightarrow}}(\mathbb{R}^N),\ \pi_1\,\widetilde{K}\le \pi_1\,\widetilde{K}_m)$.

Then for every initial element $\widetilde{K}_0 \in \widetilde{\mathscr{K}^{\rightarrow}}(\mathbb{R}^N)$, *there exists a timed solution* $\widetilde{K} = (\tau,1,K) : [0,T] \longrightarrow \widetilde{\mathscr{K}^{\rightarrow}}(\mathbb{R}^N)$ *to the mutational equation* $\overset{\circ}{\widetilde{K}}(\cdot) \ni \widetilde{f}(\widetilde{K}(\cdot),\cdot)$.

In particular, $\limsup_{h\downarrow 0}\ \frac{1}{h}\cdot dl\Big(\vartheta_{\widetilde{f}(\widetilde{K}(t),\,t)}(h,\,K(t)),\ K(t+h)\Big) = 0$ *for* $\mathscr{L}^1$*-a.e.* t

and, $K(t)\subset\mathbb{R}^N$ *coincides with the reachable set* $\vartheta_{\widetilde{f}(\widetilde{K}(\cdot),\,\cdot)}\big(t,\,K(0)\big)$ *for every* t.

Proof (of Lemma 49). It is very similar to the proof of Proposition 36 (on page 370 ff.), but takes the proximal radii into consideration additionally.

Each closed bounded ball in $(\mathscr{K}(\mathbb{R}^N), dl)$ is compact due to Proposition 1.47 (on page 57). Hence, there exist a sequence $n_k \nearrow \infty$ of indices and $\widetilde{K} = (t, 1, K) \in \widetilde{\mathscr{K}}^{\rightarrow}(\mathbb{R}^N)$ with $dl(K_{n_k}(t), K) \longrightarrow 0 \;\; (k \longrightarrow \infty)$. Proposition A.72 (3.) (on page 487) ensures for all $\rho, \kappa > 0$

$$e^{\subset}\big(\text{Graph}\ {}^{\flat}N^P_{K,\rho+\kappa},\ \text{Graph}\ {}^{\flat}N^P_{K_{n_k}(t),\rho}\big) \longrightarrow 0 \qquad (k \longrightarrow \infty)$$

and thus, $$\widetilde{d}_{\mathscr{K},j,\kappa}\big(\widetilde{K}_{n_k}(t),\ \widetilde{K}\big) \longrightarrow 0 \qquad \text{for every } j, \kappa \in]0,1].$$

Now we prove $$\sup_{k \geq m}\ \widetilde{d}_{\mathscr{K},j}\big(\widetilde{K},\ \widetilde{K}_{n_k}(t+h_m)\big) \longrightarrow 0 \qquad \text{for } m \longrightarrow \infty,$$

i.e. in particular, the convergence is uniform with respect to $\kappa \in]0,1]$.

Indeed, $$e^{\subset}\Big(\text{Graph}\ {}^{\flat}N^P_{K_{n_k}(t),\rho},\ \text{Graph}\ {}^{\flat}N^P_{K,\rho}\Big) \longrightarrow 0 \qquad (k \longrightarrow \infty)$$

results from Proposition A.72 (1.) (on page 487) for every $\rho > 0$ and hence, Lebesgue's Theorem of Dominated Convergence guarantees

$$\int_0^2 e^{\subset}\Big(\text{Graph}\ {}^{\flat}N^P_{K_{n_k}(t),\rho},\ \text{Graph}\ {}^{\flat}N^P_{K,\rho}\Big)\, d\rho \longrightarrow 0 \qquad (k \longrightarrow \infty).$$

Thus,

$$\begin{aligned} &\widetilde{d}_{\mathscr{K},j}\big(\widetilde{K},\ \widetilde{K}_{n_k}(t)\big) \\ &\quad \leq dl\big(K, K_{n_k}(t)\big) + \|\psi\|_{L^\infty} \cdot \int_0^2 e^{\subset}\Big(\text{Graph}\ {}^{\flat}N^P_{K_{n_k}(t),\rho},\ \text{Graph}\ {}^{\flat}N^P_{K,\rho}\Big)\, d\rho \\ &\longrightarrow \quad 0 \qquad\qquad (k \longrightarrow \infty). \end{aligned}$$

Finally the timed triangle inequality of $\widetilde{d}_{\mathscr{K},j}$ (according to Lemma 42 on page 375) and the uniform Lipschitz continuity in time (according to Lemma 46 on page 377) imply for any sequence $h_m \searrow 0$

$$\sup_{k \geq m}\ \widetilde{d}_{\mathscr{K},j}\big(\widetilde{K},\ \widetilde{K}_{n_k}(t+h_m)\big) \longrightarrow 0 \qquad (m \longrightarrow \infty).$$ □

Proof (of Proposition 50). It results from Existence Theorem 20 (on page 354). Indeed, $\widetilde{d}_{\mathscr{K},0}$ and $\widetilde{d}_{\mathscr{K},j}$ $(j \in]0,1])$ satisfy

$$\begin{aligned} dl(K_1, K_2) \leq \widetilde{d}_{\mathscr{K},j}(\widetilde{K}_1, \widetilde{K}_2) &\leq \widetilde{d}_{\mathscr{K},0}(\widetilde{K}_1, \widetilde{K}_2) \\ &\leq \widetilde{d}_{\mathscr{K},j}(\widetilde{K}_1, \widetilde{K}_2) + \|\psi\|_{L^\infty}\,(\|K_1\|_\infty + \|K_2\|_\infty + 2)\, j \end{aligned}$$

for all $\widetilde{K}_1 = (t_1, \mu_1, K_1),\ \widetilde{K}_2 = (t_2, \mu_2, K_2) \in \widetilde{\mathscr{K}}^{\rightarrow}(\mathbb{R}^N) \cup \widetilde{\mathscr{K}}^{\curlyvee}(\mathbb{R}^N)$.
For any sequence $\big(\widetilde{K}_m = (t_m, 1, K_m)\big)_{m \in \mathbb{N}}$ in $\widetilde{\mathscr{K}}^{\rightarrow}(\mathbb{R}^N)$ and $\widetilde{K} = (t, 1, K) \in \widetilde{\mathscr{K}}^{\rightarrow}(\mathbb{R}^N)$ suppose $t_m \searrow t$ and $\widetilde{d}_{\mathscr{K},j}(\widetilde{K}, \widetilde{K}_m) \longrightarrow 0 \;\; (m \to \infty)$ for each $j \in]0,1]$. Then,

$$\widetilde{d}_{\mathscr{K},0}(\widetilde{K}, \widetilde{K}_m) = \limsup_{j \downarrow 0}\ \widetilde{d}_{\mathscr{K},j}(\widetilde{K}, \widetilde{K}_m) \overset{m \to \infty}{\longrightarrow} 0$$

and finally $\|\mathscr{H}_{\widetilde{f}(\widetilde{K},t)} - \mathscr{H}_{\widetilde{f}(\widetilde{K}_m,t_m)}\|_{C^1(\mathbb{R}^N \times \partial\mathbb{B}_1)} \overset{m \to \infty}{\longrightarrow} 0$ – as needed for Theorem 20.
The claimed link to reachable sets of the nonautonomous differential inclusion $y' \in \widetilde{f}(\widetilde{K}(\cdot), \cdot)(y)$ results from Proposition 1.57 (on page 64). □

4.5.4 Uniqueness of Timed Solutions

In comparison with the preceding geometric example in § 4.4, an essential advantage of the current tuple

$$(\widetilde{\mathscr{K}^{\rightharpoondown}}(\mathbb{R}^N),\ \widetilde{\mathscr{K}^{\curlyvee}}(\mathbb{R}^N),\ (\widetilde{d}_{\mathscr{K},j})_{j\in]0,1]},\ (\widetilde{d}_{\mathscr{K},j})_{j\in]0,1]},\ 0)$$

is that Proposition 9 (on page 341) leads to sufficient conditions (on the right-hand side $\widetilde{f}$) for the uniqueness of timed solutions to the mutational initial value problem.

Proposition 51. *For* $\widetilde{f}:(\widetilde{\mathscr{K}^{\rightharpoondown}}(\mathbb{R}^N)\cup\widetilde{\mathscr{K}^{\curlyvee}}(\mathbb{R}^N))\times[0,T]\longrightarrow \mathrm{LIP}_{\Lambda}^{(C^2)}(\mathbb{R}^N,\mathbb{R}^N)$, *suppose that there exist a modulus* $\widehat{\omega}(\cdot)$ *of continuity and a constant* $L\geq 0$ *with*

$$\|\mathscr{H}_{\widetilde{f}(\widetilde{Z},s)}-\mathscr{H}_{\widetilde{f}(\widetilde{K},t)}\|_{C^1(\mathbb{R}^N\times\partial\mathbb{B}_1)}\ \leq\ L\cdot\widetilde{d}_{\mathscr{K},0}(\widetilde{Z},\widetilde{K})\ +\ \widehat{\omega}(t-s)$$

for all $0\leq s\leq t\leq T$ *and* $\widetilde{Z}\in\widetilde{\mathscr{K}^{\curlyvee}}(\mathbb{R}^N),\ \ \widetilde{K}\in\widetilde{\mathscr{K}^{\rightharpoondown}}(\mathbb{R}^N)\ \ (\pi_1\widetilde{Z}\leq\pi_1\widetilde{K})$.

Then for every initial $\widetilde{K}_0\in\widetilde{\mathscr{K}^{\rightharpoondown}}(\mathbb{R}^N)$, *the timed solution* $\widetilde{K}:[0,T]\longrightarrow\widetilde{\mathscr{K}^{\rightharpoondown}}(\mathbb{R}^N)$ *to the mutational equation* $\overset{\circ}{\widetilde{K}}(\cdot)\ni\widetilde{f}(\widetilde{K}(\cdot),\cdot)$ *with* $\widetilde{K}(0)=\widetilde{K}_0$ *is unique.*

Proof. It results from the arguments for Proposition 9 (on page 341 f.) in combination with the Lipschitz continuity of $\widetilde{f}$:

For any element $\widetilde{K}_0=(t_0,1,K_0)\in\widetilde{\mathscr{K}^{\rightharpoondown}}(\mathbb{R}^N)$ fixed, let $\widetilde{K}_1(\cdot)=(t_0+\cdot,1,K_1(\cdot))$ and $\widetilde{K}_2(\cdot)=(t_0+\cdot,1,K_2(\cdot))$ denote two timed solutions $[0,T]\longrightarrow\widetilde{\mathscr{K}^{\rightharpoondown}}(\mathbb{R}^N)$ to the mutational equation $\overset{\circ}{\widetilde{K}_n}(\cdot)\ni\widetilde{f}(\widetilde{K}_n(\cdot),\cdot)$ with $\widetilde{K}_1(0)=\widetilde{K}_0=\widetilde{K}_2(0)$.

Then the continuity of $\widetilde{K}_1(\cdot),\widetilde{K}_2(\cdot)$ with respect to each $\widetilde{d}_{\mathscr{K},j}$ (in forward time direction) implies the continuity of the tubes $K_1(\cdot),K_2(\cdot):[0,T]\longrightarrow\mathscr{K}(\mathbb{R}^N)$ w.r.t. $d\!\!l$. Hence, $R>1$ can be chosen sufficiently large with

$$K_1(t)\cup K_2(t)\ \subset\ \mathbb{B}_{R-1}(0)\ \subset\ \mathbb{R}^N \qquad \text{for all } t\in[0,T[.$$

Set $\widehat{R}:=4\,(R+1)\,(\|\psi\|_{L^1}+1)>R$ as an additional abbreviation.

Without loss of generality, we can restrict our considerations to compact subsets M_1,M_2 of the closed ball $\mathbb{B}_{\widehat{R}}(0)\subset\mathbb{R}^N$. In particular, for all $j\in]0,1]$, we obtain

$$\widetilde{d}_{\mathscr{K},0}((t_1,0,M_1),\ (t_2,1,M_2))\ \leq\ \widetilde{d}_{\mathscr{K},j}((t_1,0,M_1),\ (t_2,1,M_2))+\|\psi\|_{L^\infty}2\,(\widehat{R}+1)\,j$$

implying

$$\|\mathscr{H}_{\widetilde{f}(\widetilde{Z},s)}-\mathscr{H}_{\widetilde{f}(\widetilde{K},t)}\|_{C^1(\mathbb{R}^N\times\partial\mathbb{B}_1)}\leq\ L\cdot\widetilde{d}_{\mathscr{K},j}(\widetilde{Z},\widetilde{K})\ +\ L\,\|\psi\|_{L^\infty}2\,(\widehat{R}+1)\cdot j\ +\ \widehat{\omega}(t-s)$$

for all $s\leq t\leq T,\ \widetilde{Z}\in\widetilde{\mathscr{K}^{\curlyvee}}(\mathbb{R}^N),\ \widetilde{K}\in\widetilde{\mathscr{K}^{\rightharpoondown}}(\mathbb{R}^N)$ with $\pi_1\widetilde{Z}\leq\pi_1\widetilde{K},\ Z,K\subset\mathbb{B}_{\widehat{R}}(0)$.

In regard to Proposition 9, the auxiliary function $\delta_j : [0,T] \longrightarrow [0,\infty[$

$$\delta_j(t) := \inf_{\substack{\widetilde{Z} \in \widetilde{\mathscr{K}^\Upsilon}(\mathbb{R}^N), \\ \pi_1 \widetilde{Z} < t_0+t}} \left(\widetilde{d}_{\mathscr{K},j}(\widetilde{Z}, \widetilde{K}_1(t)) + \widetilde{d}_{\mathscr{K},j}(\widetilde{Z}, \widetilde{K}_2(t)) \right)$$

has the obvious upper bound

$$\delta_j(t) \leq d\!l(K_1(t), K_2(t)) + \|\psi\|_{L^1} (2R+2) < \tfrac{1}{2}\widehat{R}$$

as the choice of "test element" $\widetilde{Z} := (t_0+t-\delta, 0, K_1(t))$ with any small $\delta > 0$ shows. Thus, $\delta_j(t)$ can be described as infimum over all $\widetilde{Z} = (s, 0, Z) \in \widetilde{\mathscr{K}^\Upsilon}(\mathbb{R}^N)$ satisfying $Z \subset \mathbb{B}_{\widehat{R}}(0) \subset \mathbb{R}^N$ additionally:

$$\delta_j(t) = \inf_{\substack{\widetilde{Z} \in \widetilde{\mathscr{K}^\Upsilon}(\mathbb{R}^N): \\ \pi_1 \widetilde{Z} < t_0+t,\ \|Z\|_\infty \leq \widehat{R}}} \left(\widetilde{d}_{\mathscr{K},j}(\widetilde{Z}, \widetilde{K}_1(t)) + \widetilde{d}_{\mathscr{K},j}(\widetilde{Z}, \widetilde{K}_2(t)) \right).$$

Furthermore, the time parameter $\mathbb{T}_j(\cdot,\cdot)$ (specified in Proposition 45 on page 376 and characterized in Corollary A.47 on page 470) depends only on $j \in]0,1]$ and Λ. Due to $\widetilde{K}_1(0) = \widetilde{K}_2(0)$, Proposition 9 and the Lipschitz continuity of $\mathscr{H}_{\widetilde{f}(\cdot,s)}$ mentioned before guarantee for each $t \in [0,T]$ and $j \in]0,1]$

$$\delta_j(t) \leq \text{const}(L, \Lambda, \|\psi\|_{L^\infty}, \widehat{R}, T) \cdot j \xrightarrow{j \downarrow 0} 0$$

in the same way as we have already proved Proposition 1.24 (on page 44). Finally, the triangle inequality of the Pompeiu-Hausdorff distance $d\!l$ implies

$$d\!l(K_1(t), K_2(t)) \leq \inf_{j>0} \delta_j(t) = 0.$$

□

Chapter 5
Mutational Inclusions in Metric Spaces

After specifying sufficient conditions for the existence of solutions to mutational *equations* (in the successively generalized framework of the preceding chapters), the next step of interest is based on the notion of admitting more than just one transition for the mutation of the wanted curve at (almost) every state of the basic set $\widetilde{E}$. This goal corresponds to the step from ordinary differential *equations* to differential *inclusions* in the Euclidean space, for example.

In this chapter, we are going to discuss two situations.
First we investigate mutational inclusions with continuous right-hand side in § 5.1. This direction is motivated by the classical results of Antosiewicz and Cellina [8], but has to pass the traditional border of vector spaces.
To be more precise, we extend the conclusions of Kisielewicz from separable Banach spaces in [103] to metric spaces here. In particular, the existence of measurable selections of set-valued maps is a key tool and thus, we restrict these considerations to the mutational framework with transitions in a metric space.

Second we provide existence results for solutions to inclusions with state constraints in § 5.2. Following the classical approximation of Haddad for differential inclusions in $\mathbb{R}^N$, we need more "structure" of "transition curves". Indeed, this concept uses weak sequential compactness of curves whose values are transitions. For this rather technical reason, we focus on morphological inclusions in $(\mathcal{K}(\mathbb{R}^N), dl)$ and find a counterpart for the well-known viability theorem about differential inclusions in $\mathbb{R}^N$ [14].

Whenever sufficient conditions for the existence of solutions with state constraints are available, it is not really difficult to formulate and solve control problems whose states are not in vector spaces. Subsequent § 5.3 gives more details about the special case of morphological control problems in $(\mathcal{K}(\mathbb{R}^N), dl)$.

T. Lorenz, *Mutational Analysis: A Joint Framework for Cauchy Problems In and Beyond Vector Spaces*, Lecture Notes in Mathematics 1996,
DOI 10.1007/978-3-642-12471-6_6,

5.1 Mutational Inclusions without State Constraints

In a word, we return to the topological environment of metric spaces and in contrast to Chapter 2, we take only one metric on E into consideration:

General assumptions for § 5.1

Let (E,d) be a nonempty separable metric space. $\lfloor\cdot\rfloor : E \longrightarrow [0,\infty[$ is supposed to be lower semicontinuous with respect to d.
$\Theta(E,d,\lfloor\cdot\rfloor)$ denotes a set of transitions in the sense of Definition 2.2 (on page 104). Supply the transition set $\Theta(E,d,\lfloor\cdot\rfloor)$ with the topology induced by $\big(D(\cdot,\cdot;r)\big)_{r\geq 0}$, i.e., $\vartheta_n \longrightarrow \vartheta \ (n \longrightarrow \infty)$ is equivalent to $\lim\limits_{n\to\infty} D(\vartheta_n,\vartheta;r) = 0$ for each $r \geq 0$.
In addition, $\Theta(E,d,\lfloor\cdot\rfloor)$ is supposed to be Hausdorff, separable and complete.

Due to Definition 2.5 (on page 105), each function $D(\vartheta_1,\vartheta_2;\cdot) : [0,\infty[\longrightarrow [0,\infty[$ $\big(\vartheta_1,\vartheta_2 \in \Theta(E,d,\lfloor\cdot\rfloor)\big)$ is nondecreasing and thus, the topology of $\Theta(E,d,\lfloor\cdot\rfloor)$ is induced by a pseudo-metric like, for example,

$$\check{D}(\vartheta_1,\vartheta_2) := \sum_{n=1}^{\infty} 2^{-n} \frac{D(\vartheta_1,\vartheta_2;n)}{1 + D(\vartheta_1,\vartheta_2;n)}.$$

The supplementary hypothesis about the Hausdorff separation property implies that $\check{D}(\cdot,\cdot)$ is positive definite in addition and thus, $\check{D}(\cdot,\cdot)$ is a metric on $\Theta(E,d,\lfloor\cdot\rfloor)$. Finally, $\Theta(E,d,\lfloor\cdot\rfloor)$ is a complete separable metric space.

5.1.1 Solutions to Mutational Inclusions: Definition and Existence

Solutions to mutational inclusions extend Definition 2.9 (on page 107) about solutions to mutational equations. In particular, they are to satisfy the same conditions with respect to continuity and boundedness.

Definition 1. Let the set-valued map $\mathscr{F} : E \times [0,T] \rightsquigarrow \Theta(E,d,\lfloor\cdot\rfloor)$ be given. A curve $x : [0,T] \longrightarrow E$ is called a *solution* to the mutational inclusion

$$\overset{\circ}{x}(\cdot) \cap \mathscr{F}\big(x(\cdot),\cdot\big) \neq \emptyset$$

in $\big(E,d,\lfloor\cdot\rfloor\big)$ if it satisfies the following conditions:

(1.) $x(\cdot)$ is continuous with respect to d,

(2.) for $\mathscr{L}^1$-almost every $t \in [0,T[$, there exists a transition $\vartheta \in \mathscr{F}(x(t),t) \subset \Theta(E,d,\lfloor\cdot\rfloor)$ with
$$\lim_{h\downarrow 0} \tfrac{1}{h} \cdot d\big(\vartheta(h,x(t)),\ x(t+h)\big) = 0,$$

(3.) $\sup\limits_{t\in[0,T]} \lfloor x(t)\rfloor < \infty.$

At first glance, the term "inclusion" and the symbol $\cap$ might make a contradictory impression, but the mutation $\overset{\circ}{x}(t)$ is defined as *set* of all transitions providing a first-order approximation (in Definition 2.7 on page 106). The curve of interest, $x(\cdot) : [0,T] \longrightarrow E$, is characterized by the existence of a joint transition in both $\overset{\circ}{x}(t)$ and the prescribed transition set $\mathscr{F}(x(t),t) \subset \Theta(E,d,\lfloor\cdot\rfloor)$ at $\mathscr{L}^1$-almost every time t — denoted correctly as an intersection condition.
Every solution $x(\cdot) : [0,T] \longrightarrow E$ to a mutational inclusion can be characterized by an appropriate measurable selection of $\mathscr{F}(x(\cdot),\cdot) : [0,T] \leadsto \Theta(E,d,\lfloor\cdot\rfloor)$.

Proposition 2. *Suppose the set-valued map $\mathscr{F} : E \times [0,T] \leadsto \Theta(E,d,\lfloor\cdot\rfloor)$ to have the image set $\mathscr{F}(E \times [0,T])$ contained in a compact subset $\mathscr{C} \subset \Theta(E,d,\lfloor\cdot\rfloor)$ with $\sup_{\vartheta\in\mathscr{C}} \alpha(\vartheta;R) < \infty$ for each $R > 0$ and $\sup_{\vartheta\in\mathscr{C}} \gamma(\vartheta) < \infty$.*
$x : [0,T] \longrightarrow E$ is a solution to the mutational inclusion $\overset{\circ}{x}(\cdot) \cap \mathscr{F}\big(x(\cdot),\cdot\big) \neq \emptyset$ in $\big(E,d,\lfloor\cdot\rfloor\big)$ if and only if it has the following properties:

(*i*) *$x(\cdot)$ is continuous with respect to d,*

(*ii*) *there exists a measurable function $\vartheta(\cdot) : [0,T] \longrightarrow \Theta(E,d,\lfloor\cdot\rfloor)$ with*
$$\begin{cases} \vartheta(t) \in \overset{\circ}{x}(t) & \text{for Lebesgue-almost every } t \in [0,T] \\ \vartheta(t) \in \mathscr{F}\big(x(t),\, t\big) & \text{for every } t \in [0,T] \end{cases}$$

(*iii*) $\sup_{t\in[0,T]} \lfloor x(t) \rfloor < \infty.$

The equivalence results from Selection Theorem A.74 of Kuratowski and Ryll-Nardzewski (on page 489) if the intersection
$$[0,T] \leadsto \Theta(E,d,\lfloor\cdot\rfloor), \quad t \mapsto \overset{\circ}{x}(t) \cap \mathscr{F}\big(x(t),t\big)$$
proves to be measurable. The aspect of measurability does not change for this set-valued map by modifying it on a subset of Lebesgue measure 0. Hence, this feature can be concluded from the next lemma and Proposition A.77 (on page 490):

Lemma 3. *Assume for $x(\cdot) : [0,T] \longrightarrow E$ and $\mathscr{C} : [0,T] \leadsto \Theta(E,d,\lfloor\cdot\rfloor)$*

(*1.*) *$x(\cdot)$ is continuous with respect to d,*

(*2.*) $R := 1 + \sup \lfloor x(\cdot) \rfloor < \infty$,

(*3.*) *$\mathscr{C}$ is upper semicontinuous with nonempty compact values and satisfies* $\sup \big\{\alpha(\vartheta;R),\ \gamma(\vartheta) \bigm| \vartheta \in \mathscr{C}(t),\ t \in [0,T]\big\} < \infty$,

(*4.*) *$\overset{\circ}{x}(t) \cap \mathscr{C}(t)$ is nonempty for Lebesgue-almost every $t \in [0,T]$.*

Then the map $[0,T] \leadsto \Theta(E,d,\lfloor\cdot\rfloor),\ t \mapsto \overset{\circ}{x}(t) \cap \mathscr{C}(t)$ is Lebesgue-measurable in the sense of Definition A.73 (on page 489).

Its detailed proof is postponed to § 5.1.3 (on page 395 ff.).
The main result of this section 5.1 is the following existence theorem for mutational inclusions without state constraints:

Theorem 4. *Assume $\big(E, d, \lfloor\cdot\rfloor, \Theta(E,d,\lfloor\cdot\rfloor)\big)$ to be Euler compact in the sense of Definition 2.15 (on page 112). Let $\mathscr{F} : E \times [0,T] \rightsquigarrow \Theta(E,d,\lfloor\cdot\rfloor)$ be an integrably bounded compact-valued Carathéodory map in the following sense:*

(*i*) *all values of $\mathscr{F}$ are nonempty, compact and satisfy for each $r \geq 0$*
$$\sup\big\{\alpha(\vartheta;r),\ \beta(\vartheta;r),\ \gamma(\vartheta)\ \big|\ \vartheta \in \mathscr{F}(x,t),\ x \in E,\ t \in [0,T]\big\} < \infty,$$
(*ii*) *for every $x \in E$, $\mathscr{F}(x,\cdot) : [0,T] \rightsquigarrow \Theta(E,d,\lfloor\cdot\rfloor)$ is measurable,*

(*iii*) *for almost every $t \in [0,T]$, $\mathscr{F}(\cdot,t) : \big(E,d\big) \rightsquigarrow \big(\Theta(E,d,\lfloor\cdot\rfloor), \check{D}\big)$ is continuous,*

(*iv*) *for each $R > 0$, there exist $\widehat{m}_R(\cdot) \in L^1([0,T])$ and $\vartheta_R \in \Theta(E,d,\lfloor\cdot\rfloor)$ such that for $\mathscr{L}^1$-almost every t,*
$$\sup\big\{D(\vartheta_R,\vartheta;R)\ \big|\ \vartheta \in \mathscr{F}(x,t),\ x \in E,\ \lfloor x\rfloor \leq R\big\} \leq \widehat{m}_R(t).$$

Then for every initial state $x_0 \in E$, there exists a solution $x(\cdot) : [0,T] \longrightarrow E$ to the mutational inclusion
$$\overset{\circ}{x}(\cdot) \cap \mathscr{F}\big(x(\cdot),\cdot\big) \neq \emptyset$$
in the tuple $\big(E,d,\lfloor\cdot\rfloor\big)$ with $x(0) = x_0$.

5.1.2 A Selection Principle Generalizing the Theorem of Antosiewicz-Cellina

In their classical paper [8] in 1975, Antosiewicz and Cellina showed for differential inclusions $x' \in G(x,\cdot)$ in finite space dimensions that the Carathéodory regularity of the set-valued map $G(\cdot,\cdot)$ is sufficient for the existence of useful selections on the way of proving existence of solutions. Indeed, their new essential aspect was to focus on continuous functions $g : \mathbb{R}^N \longrightarrow L^1([0,T],\mathbb{R}^N)$ with $g(x)(t) \in G(x,t)$ for $\mathscr{L}^1$-almost every t and every x.
Kisielewicz extended their results to separable Banach spaces in [103] in 1982. Now we generalize it to the separable metric spaces (E,d), $\big(\Theta(E,d,\lfloor\cdot\rfloor),\check{D}\big)$ and adapt essentially the arguments of Kisielewicz, but avoid measures of noncompactness. Strictly speaking, the statements in this subsection do not use that $\Theta(E,d,\lfloor\cdot\rfloor)$ consists of transitions on E and so, they hold for any pair of separable metric spaces whose second component is complete.

Proposition 5. *Let the set-valued map $\mathscr{F} : E \times [0,T] \rightsquigarrow \Theta(E,d,\lfloor\cdot\rfloor)$ fulfill the following conditions:*

(*i*) *all values of $\mathscr{F}$ are nonempty, compact and satisfy for each $r \geq 0$*
$$\sup\big\{\alpha(\vartheta;r),\ \beta(\vartheta;r),\ \gamma(\vartheta)\ \big|\ \vartheta \in \mathscr{F}(x,t),\ x \in E,\ t \in [0,T]\big\} < \infty,$$
(*ii*) *for every $x \in E$, $\mathscr{F}(x,\cdot) : [0,T] \rightsquigarrow \Theta(E,d,\lfloor\cdot\rfloor)$ is measurable,*

(*iii*) *for $\mathscr{L}^1$-almost every $t \in [0,T]$, $\mathscr{F}(\cdot,t) : \big(E,d\big) \rightsquigarrow \Theta(E,d,\lfloor\cdot\rfloor)$ is continuous,*

(*iii'*) *the family $(\mathscr{F}(\cdot,t))_{t\in[0,T]}$ of maps $E \rightsquigarrow \Theta(E,d,\lfloor\cdot\rfloor)$ is equi-continuous.*

Then there exists a single-valued function $f : E \times [0,T] \longrightarrow \Theta(E,d,\lfloor\cdot\rfloor)$ *satisfying*

(a) *at Lebesgue-almost every time* $t \in [0,T]$: $f(x,t) \in \mathscr{F}(x,t)$ *for every* $x \in E$,

(b) *for every* $x \in E$, $f(x,\cdot) : [0,T] \longrightarrow \Theta(E,d,\lfloor\cdot\rfloor)$ *is measurable,*

(c) $\lim_{l\to\infty} \int_{[0,T]} \check{D}\big(f(x,t),\ f(x_l,t)\big)\, dt \;=\; 0$
whenever a sequence $(x_l)_{l\in\mathbb{N}}$ *in* E *converges to* $x \in E$ *with respect to* d.

Corollary 6. *If the set-valued map* $\mathscr{F} : E \times [0,T] \rightsquigarrow \Theta(E,d,\lfloor\cdot\rfloor)$ *satisfies the Carathéodory conditions (i)–(iii) of Theorem 4, then there exists a single-valued function* $f : E \times [0,T] \longrightarrow \Theta(E,d,\lfloor\cdot\rfloor)$ *with*

(a) *at Lebesgue-almost every time* $t \in [0,T]$: $f(x,t) \in \mathscr{F}(x,t)$ *for every* $x \in E$,

(b) *for every* $x \in E$, $f(x,\cdot) : [0,T] \longrightarrow \Theta(E,d,\lfloor\cdot\rfloor)$ *is measurable,*

(c) $\lim_{l\to\infty} \int_{[0,T]} \check{D}\big(f(x,t),\ f(x_l,t)\big)\, dt \;=\; 0$
whenever a sequence $(x_l)_{l\in\mathbb{N}}$ *in* E *converges to* $x \in E$ *with respect to* d.

The proof follows the approximative arguments initiated by Antosiewicz-Cellina and continued by Kisielewicz. All these subsequent conclusions do not require the linear structure of a Banach space and thus, we can apply them in the metric spaces (E,d), $\big(\Theta(E,d,\lfloor\cdot\rfloor),\check{D}\big)$:

Lemma 7. *Suppose the assumptions of Proposition 5 about* $\mathscr{F}(\cdot,\cdot)$.
For each $\varepsilon > 0$, *there exists a function* $f_\varepsilon : E \times [0,T] \longrightarrow \Theta(E,d,\lfloor\cdot\rfloor)$ *satisfying*

(a) $\operatorname{dist}\big(f_\varepsilon(x,t),\ \mathscr{F}(x,t)\big) \stackrel{\text{Def.}}{=} \inf_{\vartheta \in \mathscr{F}(x,t)} \check{D}\big(f_\varepsilon(x,t),\ \vartheta\big) \le \varepsilon$ *for any* $x \in E$, $t < T$,

(b) *for every* $x \in E$, $f_\varepsilon(x,\cdot) : [0,T] \longrightarrow \Theta(E,d,\lfloor\cdot\rfloor)$ *is measurable,*

(c) $\lim_{l\to\infty} \int_{[0,T]} \check{D}\big(f_\varepsilon(x,t),\ f_\varepsilon(x_l,t)\big)\, dt \;=\; 0$
whenever a sequence $(x_l)_{l\in\mathbb{N}}$ *in* E *converges to* $x \in E$ *with respect to* d.

Proof (of Lemma 7). Fix $\varepsilon > 0$ and choose $x \in E$ arbitrarily. As in the proof of [103, Lemma 3.2], the equi-continuity of the set-valued maps $\mathscr{F}(\cdot,t) : E \rightsquigarrow \Theta(E,d,\lfloor\cdot\rfloor)$, $t \in [0,T]$, provides some $\delta(x,\varepsilon) > 0$ with

$$dl_{\check{D}}\big(\mathscr{F}(y_1,t),\ \mathscr{F}(y_2,t)\big) \;<\; \varepsilon$$

for all $t \in [0,T]$, $y_1, y_2 \in \mathbb{B}_{\delta(x,\varepsilon)}(x) \subset E$. Here $dl_{\check{D}}$ denotes the Pompeiu-Hausdorff distance between nonempty subsets of $\Theta(E,d,\lfloor\cdot\rfloor)$ with respect to the metric $\check{D}$ specified in § 5.1 (on page 386), i.e.,

$$d\!l_{\check{D}}\big(\mathscr{M}_1, \mathscr{M}_2\big) \overset{\text{Def.}}{=} \max\Big\{ \sup_{\vartheta_1 \in \mathscr{M}_1} \inf_{\vartheta_2 \in \mathscr{M}_2} \check{D}(\vartheta_1, \vartheta_2), \\ \sup_{\vartheta_2 \in \mathscr{M}_2} \inf_{\vartheta_1 \in \mathscr{M}_1} \check{D}(\vartheta_1, \vartheta_2) \Big\}$$

for any nonempty sets $\mathscr{M}_1, \mathscr{M}_2 \subset \Theta(E,d,\lfloor\cdot\rfloor)$.

The open balls $\mathbb{B}_{\delta(x,\varepsilon)}(x)^\circ \subset E$ (with respect to d), $x \in E$, cover E. By assumption (on page 386), (E,d) is separable and thus, we can select a countable cover of E from these balls. As a consequence of Stone's Theorem, a further selection even provides a countable cover $\big(\mathbb{B}_{\delta(x_m,\varepsilon)}(x_m)^\circ\big)_{m \in \mathbb{N}}$ that is locally finite in addition. There exists a subordinated continuous partition of unity, $\zeta_m : E \longrightarrow [0,1]$ $(m \in \mathbb{N})$.

This lays the basis for a countable partition of the interval $[0,T[$ depending on the element $x \in E$ in a continuous way. Indeed, we set for $x \in E$ and $m = 1, 2 \ldots,$

$$\begin{aligned} t_0(x) &:= 0, \\ t_m(x) &:= t_{m-1}(x) + \zeta_m(x) \cdot T \\ J_m(x) &:= \big[t_{m-1}(x),\ t_m(x)\big[. \end{aligned}$$

For each index $m \in \mathbb{N}$, Selection Theorem A.74 of Kuratowski and Ryll-Nardzewski (on page 489) provides a measurable function

$$\vartheta_m : [0,T] \longrightarrow \Theta(E,d,\lfloor\cdot\rfloor)$$

satisfying the condition $\qquad \vartheta_m(t) \in \mathscr{F}(x_m,t) \qquad$ for every $t \in [0,T]$ due to assumption (ii) about the measurability of each $\mathscr{F}(x,\cdot) : [0,T] \rightsquigarrow \Theta(E,d,\lfloor\cdot\rfloor)$ and the general hypothesis that the metric space $(\Theta(E,d,\lfloor\cdot\rfloor),\ \check{D})$ is complete and separable.

Now define $f_\varepsilon : E \times [0,T] \longrightarrow \Theta(E,d,\lfloor\cdot\rfloor)$ in a piecewise way with respect to time:

$$\begin{aligned} f_\varepsilon(x,\, t\,) &:= \vartheta_m(t) && \text{if} \quad t \in J_m(x), \\ f_\varepsilon(x,T) &:= \vartheta_M(T) && \text{with } M := \inf\big\{m \in \mathbb{N} \,\big|\, t_m(x) = T\big\} < \infty. \end{aligned}$$

Obviously, $f_\varepsilon(x,\cdot) : [0,T] \longrightarrow \Theta(E,d,\lfloor\cdot\rfloor)$ is measurable for every $x \in E$.

Furthermore, $\ \operatorname{dist}\big(f_\varepsilon(x,t),\ \mathscr{F}(x,t)\big) \le \varepsilon \ $ holds for every $x \in E$ and $t \in [0,T[$. Indeed, we can choose the unique index $m \in \mathbb{N}$ with $t \in J_m(x) = [t_{m-1}(x), t_m(x)[$. This implies $x \in \mathbb{B}_{\delta(x_m,\varepsilon)}(x_m)$ and $f_\varepsilon(x,t) = \vartheta_m(t) \in \mathscr{F}(x_m,t)$. Now we conclude from the triangle inequality of $\check{D}$

$$\begin{aligned} \operatorname{dist}\big(f_\varepsilon(x,t),\ \mathscr{F}(x,t)\big) &\le \operatorname{dist}\big(f_\varepsilon(x,t),\ \mathscr{F}(x_m,t)\big) + d\!l_{\check{D}}\big(\mathscr{F}(x_m,t),\ \mathscr{F}(x,t)\big) \\ &\le 0 + \varepsilon, \end{aligned}$$

i.e., $f_\varepsilon(\cdot,\cdot)$ satisfies the claimed property (a).

We still have to verify property (c), i.e.,

$$\lim_{l\to\infty} \int_{[0,T]} \check{D}\big(f_\varepsilon(x,t),\ f_\varepsilon(x_l,t)\big)\,dt \;=\; 0$$

whenever a sequence $(x_l)_{l\in\mathbb{N}}$ in E converges to $x \in E$ with respect to d.
Indeed, as the partition of unity $(\zeta_m)_{m\in\mathbb{N}}$ is locally finite, there exist a neighborhood U_x of x and finitely many indices $\{m_1 \ldots m_{\eta_x}\} \subset \mathbb{N}$ with

$$\sum_{k=1}^{\eta_x} \zeta_{m_k}(\cdot) \;=\; 1 \quad \text{ in } U_x.$$

Due to the continuity of each auxiliary function $t_m : E \longrightarrow [0,T]$ $(m \in \mathbb{N})$, we obtain for every sequence $(x_l)_{l\in\mathbb{N}}$ converging to x

$$\sup\big\{|t_{m_k}(x) - t_{m_k}(x_l)| \;\big|\; k \in \{1 \ldots \eta_x\}\big\} \;\longrightarrow\; 0 \qquad \text{for } l \longrightarrow \infty.$$

As $\check{D}\big(f_\varepsilon(x,t),\ f_\varepsilon(x_l,t)\big) = 0$ holds for any $t \in J_{m_k}(x) \cap J_{m_k}(x_l)$, we conclude from $\check{D}(\cdot,\cdot) \le 1$ that for all large $l \in \mathbb{N}$,

$$\begin{aligned}
&\int_{[0,T]} \check{D}\big(f_\varepsilon(x,t),\ f_\varepsilon(x_l,t)\big)\,dt\\
&\quad \le \int_{[0,T]} \sum_{k=1}^{\eta_x} \big(\chi_{J_{m_k}(x)\setminus J_{m_k}(x_l)}(t) + \chi_{J_{m_k}(x_l)\setminus J_{m_k}(x)}(t)\big)\ \check{D}\big(f_\varepsilon(x,t),\ f_\varepsilon(x_l,t)\big)\,dt\\
&\quad \le \sum_{k=1}^{\eta_x} \Big(\int_{J_{m_k}(x)\setminus J_{m_k}(x_l)} 1\ dt \;+\; \int_{J_{m_k}(x_l)\setminus J_{m_k}(x)} 1\ dt\Big)\\
&\longrightarrow\; 0 \qquad\qquad \text{for } l \longrightarrow \infty.
\end{aligned}$$

□

Proof (of Proposition 5 on page 388). For every $\varepsilon > 0$, Lemma 7 guarantees a function $f_\varepsilon : E \times [0,T] \longrightarrow \Theta(E,d,\lfloor\cdot\rfloor)$ satisfying both

$$\operatorname{dist}\big(f_\varepsilon(x,t),\ \mathscr{F}(x,t)\big) \overset{\text{Def.}}{=} \inf_{\vartheta\in\mathscr{F}(x,t)} \check{D}\big(f_\varepsilon(x,t),\ \vartheta\big) \;\le\; \varepsilon$$

for all $x \in E$, $t \in [0,T[$, the measurability of each $f_\varepsilon(x,\cdot) : [0,T] \longrightarrow \Theta(E,d,\lfloor\cdot\rfloor)$ and the continuity condition that for every $x \in E$ and $\delta > 0$, there exists a positive radius $\rho(x,\delta) > 0$ such that all $y \in \mathbb{B}_{\rho(x,\delta)}(x) \subset E$ fulfill

$$\mathscr{L}^1\big(\{t \in [0,T] \mid \check{D}(f_\varepsilon(x,t),\ f_\varepsilon(y,t)) > \delta\}\big) \;<\; \delta\,.$$

In particular, the preceding proof of Lemma 7 motivates the following inductive construction of approximative selections $(f_k)_{k\in\mathbb{N}}$:

There exists such a function $f_1 : E \times [0,T] \longrightarrow \Theta(E,d,\lfloor\cdot\rfloor)$ with

$$\operatorname{dist}\big(f_1(x,t),\ \mathscr{F}(x,t)\big) \;\le\; \tfrac{1}{2^2}$$

for every $x \in E$ and $t \in [0,T[$. In combination with assumption (iii') about the equi-continuity of $\mathscr{F}(\cdot,t)$, $t \in [0,T]$, we can even find a radius $\delta_1(x) > 0$ for each $x \in E$ with

$$\begin{cases} d\!\!l_{\check{D}}\big(\mathscr{F}(x,t),\ \mathscr{F}(y,t)\big) < \tfrac{1}{2^3} & \text{for all } y\in\mathbb{B}_{\delta_1(x)}(x),\ t\in[0,T],\\ \mathscr{L}^1\big(\{t \in [0,T] \mid \check{D}(f_1(x,t),\ f_1(y,t)) > \tfrac{1}{2^2}\}\big) < \tfrac{1}{2^2} & \text{for all } y\in\mathbb{B}_{\delta_1(x)}(x). \end{cases}$$

The same arguments as in the proof of Lemma 7 lead now to a locally finite partition of unity $(\zeta_m^1)_{m\in\mathbb{N}}$ and a sequence $(x_m^1)_{m\in\mathbb{N}}$ such that the support of $\zeta_m^1(\cdot) \in C^0(E)$ is contained in $\mathbb{B}_{\delta_1(x_m^1)}(x_m^1) \subset E$ for each index $m \in \mathbb{N}$.

Due to Proposition A.80 about measurability of marginal maps (on page 490), there exists a measurable selection $\vartheta_m^1(\cdot) : [0,T] \longrightarrow \Theta(E,d,\lfloor\cdot\rfloor)$ for each $m \in \mathbb{N}$ satisfying at every time $t \in [0,T]$,

$$\begin{cases} \vartheta_m^1(t) \in \mathscr{F}(x_m^1,t) \\ \check{D}\big(\vartheta_m^1(t),\, f_1(x_m^1,t)\big) = \operatorname{dist}\big(f_1(x_m^1,t),\, \mathscr{F}(x_m^1,t)\big) \end{cases}$$

because each set-valued map $\mathscr{F}(x_m^1,\cdot) : [0,T] \leadsto \Theta(E,d,\lfloor\cdot\rfloor)$, $m \in \mathbb{N}$, is measurable with nonempty compact values by assumption.

Now we set for $x \in E$ and $m = 1, 2\ldots$ successively

$$\begin{aligned} t_0^1(x) &:= 0, \\ t_m^1(x) &:= t_{m-1}^1(x) + \zeta_m^1(x)\cdot T \\ J_m^1(x) &:= \big[t_{m-1}^1(x),\; t_m^1(x)\big[\end{aligned}$$

and define $f_2 : E \times [0,T] \longrightarrow \Theta(E,d,\lfloor\cdot\rfloor)$ in a piecewise way again

$$\begin{aligned} f_2(x,\,t) &:= \vartheta_m^1(t) && \text{if} \quad t \in J_m^1(x), \\ f_2(x,T) &:= \vartheta_M^1(T) && \text{with } M := \inf\big\{m \in \mathbb{N} \,\big|\, t_m^1(x) = T\big\} < \infty. \end{aligned}$$

Obviously, $f_2(x,\cdot) : [0,T] \longrightarrow \Theta(E,d,\lfloor\cdot\rfloor)$ is measurable for every $x \in E$.
The arguments of the preceding proof even imply continuity property (c) for this auxiliary function $f_2(\cdot,\cdot)$, i.e.,

$$\lim_{l\to\infty} \int_{[0,T]} \check{D}\big(f_2(x,t),\, f_2(x_l,t)\big)\, dt = 0$$

whenever a sequence $(x_l)_{l\in\mathbb{N}}$ in E converges to $x \in E$ with respect to d.

Moreover, $\operatorname{dist}\big(f_2(x,t),\, \mathscr{F}(x,t)\big) \le \frac{1}{2^3}$ holds for every $x \in E$ and $t \in [0,T[$. Indeed, there always exists a unique index $m \in \mathbb{N}$ with $t \in J_m^1(x) = [t_{m-1}^1(x), t_m^1(x)[$. Thus, $x \in \mathbb{B}_{\delta_1(x_m^1)}(x_m^1)$, $f_2(x,t) = \vartheta_m^1(t) \in \mathscr{F}(x_m^1,t)$ and last, but not least,

$$\begin{aligned} \operatorname{dist}\big(f_2(x,t),\, \mathscr{F}(x,t)\big) &\le \operatorname{dist}\big(f_2(x,t),\, \mathscr{F}(x_m^1,t)\big) + d\!l_{\check{D}}\big(\mathscr{F}(x_m^1,t),\, \mathscr{F}(x,t)\big) \\ &\le 0 + \frac{1}{2^3}. \end{aligned}$$

Finally,

$$\begin{aligned} \check{D}\big(f_2(x,t),\, f_1(x,t)\big) &\le \check{D}\big(\vartheta_m^1(t),\, f_1(x_m^1,t)\big) + \check{D}\big(f_1(x_m^1,t), f_1(x,t)\big) \\ &\le \operatorname{dist}\big(f_1(x_m^1,t),\, \mathscr{F}(x_m^1,t)\big) + \check{D}\big(f_1(x_m^1,t),\, f_1(x,t)\big) \\ &\le \frac{1}{2^2} + \check{D}\big(f_1(x_m^1,t),\, f_1(x,t)\big) \end{aligned}$$

for every $t \in [0,T[$ has the consequence

$$\mathscr{L}^1\big(\{t \in [0,T] \mid \check{D}(f_2(x,t),\, f_1(x,t)) > \tfrac{1}{2}\}\big) < \tfrac{1}{2^2}.$$

By means of induction, we now construct a sequence $(f_n)_{n\in\mathbb{N}}$ of functions $E \times [0,T] \longrightarrow \Theta(E,d,\lfloor\cdot\rfloor)$ with properties (b), (c) and

$$\begin{cases} \operatorname{dist}\big(f_n(x,t),\ \mathscr{F}(x,t)\big) \ \le\ \frac{1}{2^{n+1}} & \text{for all } x \in E,\ t \in [0,T[,\\ \mathscr{L}^1\big(\{t \in [0,T] \,|\, \check{D}(f_n(x,t),\ f_{n-1}(x,t)) > \frac{1}{2^{n-1}}\}\big) \ <\ \frac{1}{2^n} & \text{for all } x \in E\,. \end{cases}$$

In particular, due to $\sum\limits_{k=n}^{N} 2^{-k} = 2^{1-n} - 2^{-N}$ for all $n < N$, the inequality

$$\begin{aligned} &\mathscr{L}^1\big(\big\{t \in [0,T] \ \big|\ \check{D}(f_N(x,t),\ f_n(x,t)) > \tfrac{1}{2^{n-2}}\big\}\big)\\ &\le \mathscr{L}^1\Big(\bigcup_{k=n+1}^{N} \big\{t \in [0,T] \ \big|\ \check{D}(f_k(x,t),\ f_{k-1}(x,t)) > \tfrac{1}{2^{k-1}}\big\}\Big)\\ &\le \sum_{k=n+1}^{N} \tfrac{1}{2^k} \ \le\ \tfrac{1}{2^n} \end{aligned}$$

holds for every element $x \in E$ and all indices $n < N$. Due to the completeness of L^1 spaces for metric space valued functions [6, § 5.4], there exists a function

$$f : E \times [0,T] \ \longrightarrow\ \Theta(E,d,\lfloor\cdot\rfloor)$$

such that for every element $x \in E$, $f(x,\cdot) : [0,T] \longrightarrow \Theta(E,d,\lfloor\cdot\rfloor)$ is measurable and

$$\int_0^T \check{D}(f(x,t),\ f_n(x,t))\ dt \ \longrightarrow\ 0 \qquad\qquad \text{for } n \longrightarrow \infty.$$

All values of $\mathscr{F}$ are assumed to be closed and so, a modification of $f(x,\cdot)$ on a set of Lebesgue measure 0 leads to the additional property $f(x,t) \in \mathscr{F}(x,t)$ for all x,t.

Finally we have to verify continuity property (c) of $f(\cdot,\cdot)$, i.e.,

$$\lim_{l \to \infty} \int_{[0,T]} \check{D}\big(f(x,t),\ f(x_l,t)\big)\ dt \ =\ 0$$

whenever a sequence $(x_l)_{l\in\mathbb{N}}$ in E converges to $x \in E$ with respect to d.
Indeed, each function $f_n(\cdot,\cdot)$ $(n \in \mathbb{N})$ has this feature by construction. Considering the last inequality for $N \longrightarrow \infty$ leads to the estimate

$$\mathscr{L}^1\big(\big\{t \in [0,T] \ \big|\ \check{D}(f(x,t),\ f_n(x,t)) > \tfrac{1}{2^{n-2}}\big\}\big) \ \le\ \tfrac{1}{2^n}$$

being uniform with respect to $x \in E$. It implies the current claim about continuity of $f(\cdot,t) : E \longrightarrow \Theta(E,d,\lfloor\cdot\rfloor)$ due to $\check{D}(\cdot,\cdot) \le 1$. □

Proof (of Corollary 6 on page 389).
The assumptions of this corollary differ from their counterparts of Proposition 5 (on page 388) in just one relevant respect: We dispense with hypothesis (iii'), i.e., the family $\big(\mathscr{F}(\cdot,t)\big)_{t\in[0,T]}$ of set-valued maps $E \rightsquigarrow \Theta(E,d,\lfloor\cdot\rfloor)$ is not supposed to be equi-continuous.
Now Scorza-Dragoni Theorem quoted in Proposition A.9 (on page 446) provides the tool for bridging this gap approximatively.

Indeed, for every $\varepsilon > 0$, Proposition A.9 guarantees a closed subset $I_\varepsilon \subset [0,T]$ with $\mathscr{L}^1\big([0,T] \setminus I_\varepsilon\big) < \varepsilon$ such that the restriction

$$\mathscr{F}(\cdot,\cdot)|_{E\times I_\varepsilon} : \ (E,d) \times I_\varepsilon \longrightarrow \big(\mathscr{K}(\Theta(E,d,\lfloor\cdot\rfloor)),\ d\!\!l_{\check{D}}\big)$$

is continuous. As I_ε is compact, we conclude easily that the family $\big(\mathscr{F}(\cdot,t)\big)_{t\in I_\varepsilon}$ of set-valued maps $E \leadsto \Theta(E,d,\lfloor\cdot\rfloor)$ is equi-continuous.

This construction leads to a sequence $(I_n)_{n\in\mathbb{N}}$ of closed subsets of $[0,T]$ with $\mathscr{L}^1\big([0,T] \setminus I_n\big) < 2^{-n}$ such that each family $\big(\mathscr{F}(\cdot,t)\big)_{t\in I_n}$ is equi-continuous. Setting $S_1 := I_1,\ S_{n+1} := I_{n+1} \setminus \bigcup_{k=1}^n I_k$ for each $n \in \mathbb{N}$ and choosing an arbitrary transition $\vartheta_0 \in \Theta(E,d,\lfloor\cdot\rfloor)$, the auxiliary maps $\mathscr{F}_n : E \times [0,T] \leadsto \Theta(E,d,\lfloor\cdot\rfloor)$, $n \in \mathbb{N}$, with

$$\mathscr{F}_n(x,t) := \begin{cases} \mathscr{F}(x,t) & \text{if } t \in I_n,\\ \{\vartheta_0\} & \text{if } t \in [0,T] \setminus I_n \end{cases}$$

fulfill the assumptions of Proposition 5. For each $n \in \mathbb{N}$, there exists a selection $f_n : E \times [0,T] \longrightarrow \Theta(E,d,\lfloor\cdot\rfloor)$ of $\mathscr{F}_n(\cdot,\cdot)$ satisfying measurability condition (b) and

$$\lim_{l\to\infty} \int_{[0,T]} \check{D}\big(f_n(x,t),\ f_n(x_l,t)\big)\ dt \ = \ 0$$

whenever a sequence $(x_l)_{l\in\mathbb{N}}$ in E converges to some element $x \in E$ w.r.t. d. Now the function $f : E \times [0,T] \longrightarrow \Theta(E,d,\lfloor\cdot\rfloor)$ defined by

$$f(\cdot,t) := \begin{cases} f_n(\cdot,t) & \text{if } t \in S_n \text{ for some (and then unique) } n \in \mathbb{N}\\ \vartheta_0 & \text{if } t \in [0,T] \setminus \bigcup_{n\in\mathbb{N}} S_n \ = \ [0,T] \setminus \bigcup_{n\in\mathbb{N}} I_n \end{cases}$$

shares property (b) of measurability with each f_n and fulfills condition (c) of continuity as well. Indeed, the construction of $(I_n)_{n\in\mathbb{N}}$ and $(S_n)_{n\in\mathbb{N}}$ ensures

$$\mathscr{L}^1\Big([0,T] \setminus \bigcup_{n\in\mathbb{N}} S_n\Big) \ \le\ \limsup_{n\to\infty} \mathscr{L}^1([0,T] \setminus I_n) \ = \ 0$$

and thus, for any $\varepsilon > 0$, we can select an index $N_\varepsilon \in \mathbb{N}$ such that

$$\mathscr{L}^1\Big([0,T] \setminus \bigcup_{n=1}^{N_\varepsilon} S_n\Big) \ \le\ \varepsilon\,.$$

Finally, we obtain for every converging sequence $(x_l)_{l\in\mathbb{N}}$ in E and its limit $x \in E$

$$\begin{aligned}
&\limsup_{l\to\infty} \int_{[0,T]} \check{D}\big(f(x,t),\ f(x_l,t)\big)\ dt\\
&\le \limsup_{l\to\infty} \Big(\sum_{n=1}^{N_\varepsilon} \int_{S_n} \check{D}\big(f(x,t),\ f(x_l,t)\big)\ dt + \int_{[0,T]\setminus \bigcup_{n=1}^{N_\varepsilon} S_n} 1\ dt\Big)\\
&\le \limsup_{l\to\infty} \sum_{n=1}^{N_\varepsilon} \int_{S_n} \check{D}\big(f_n(x,t), f_n(x_l,t)\big)\ dt + \mathscr{L}^1\Big([0,T] \setminus \bigcup_{n=1}^{N_\varepsilon} S_n\Big)\\
&\le \sum_{n=1}^{N_\varepsilon} 0 + \varepsilon\,.
\end{aligned}$$

□

5.1.3 Proofs on the Way to Existence Theorem 5.4

Now we give two proofs missing in this section 5.1. In particular, we focus on Lemma 3 (on page 387) stating that the intersection of the mutation and a compact-valued upper semicontinuous map is always a measurable map $[0,T] \rightsquigarrow \Theta(E,d,\lfloor\cdot\rfloor)$ and Existence Theorem 4 (on page 388).

Proof (of Lemma 3 on page 387).

Without loss of generality, we can assume in addition that there exists a transition $\vartheta_0 \in \Theta(E,d,\lfloor\cdot\rfloor) \setminus \overline{\mathscr{C}([0,T])}$. (It can be constructed by means of a supplementary auxiliary component, for example – similarly to the disjoint sum of topological spaces.)
From now on, we mostly consider the union of transition sets with $\{\vartheta_0\}$ so that all closed sets in $\Theta(E,d,\lfloor\cdot\rfloor)$ are nonempty and thus, the general results about measurability in Appendix A.10 (on page 489 f.) can be applied directly.

Now for each $m,n \in \mathbb{N}$, define the set-valued map $\mathscr{M}_{m,n} : [0,T] \rightsquigarrow \Theta(E,d,\lfloor\cdot\rfloor)$ in the following way: $\mathscr{M}_{m,n}(t)$ consists of ϑ_0 and all transitions $\vartheta \in \mathscr{C}(t) \subset \Theta(E,d,\lfloor\cdot\rfloor)$ such that

$$d\big(\vartheta(h,\, x(t)),\, x(t+h)\big) \;\leq\; \tfrac{1}{m}\, h \qquad \text{for all } h \in [0,\tfrac{1}{n}].$$

The graph of $\mathscr{M}_{m,n}$ is closed. Indeed, let $((t_k,\vartheta_k))_{k\in\mathbb{N}}$ be any convergent sequence in Graph $\mathscr{M}_{m,n} \subset [0,T] \times \Theta(E,d,\lfloor\cdot\rfloor)$ with the limit (t,ϑ). If $\vartheta = \vartheta_0$, then we conclude $\vartheta_k = \vartheta_0$ for all large $k \in \mathbb{N}$ due to $\vartheta_0 \notin \overline{\mathscr{C}([0,T])}$. Hence, we can restrict our considerations to $\{\vartheta_k, \vartheta \mid k \in \mathbb{N}\} \subset \mathscr{C}([0,T])$ and in particular, for each $k \in \mathbb{N}$,

$$d\left(\vartheta_k(h,\, x(t_k)),\, x(t_k+h)\right) \;\leq\; \tfrac{1}{m}\, h \qquad \text{for all } h \in [0,\tfrac{1}{n}].$$

The standard estimate about two solutions in Proposition 2.11 (on page 108) implies

$$d\left(\vartheta(h,\, x(t)), x(t+h)\right) \;=\; \lim_{k\to\infty} d\left(\vartheta_k(h,\, x(t_k)),\, x(t_k+h)\right) \;\leq\; \tfrac{1}{m}\, h \quad \text{for all } h \in [0,\tfrac{1}{n}],$$

i.e. $\vartheta \in \mathscr{M}_{m,n}(t)$. Thus, Graph $\mathscr{M}_{m,n}$ is closed in $[0,T] \times \Theta(E,d,\lfloor\cdot\rfloor)$.

Furthermore, all values of $\mathscr{M}_{m,n}$ are nonempty and closed. As $\mathscr{C}$ is supposed to be upper semicontinuous with compact values, [19, Proposition 1.4.9] ensures that

$$\mathscr{M}_{m,n} \;=\; \mathscr{M}_{m,n} \cap (\mathscr{C} \cup \{\vartheta_0\}) : \; [0,T] \;\rightsquigarrow\; \Theta(E,d,\lfloor\cdot\rfloor)$$

is upper semicontinuous (in the sense of Bouligand and Kuratowski). Finally, it implies the measurability of $\mathscr{M}_{m,n}$ for each $m,n \in \mathbb{N}$ due to Corollary A.76 (on page 489).

Now we bridge the gap between the countable family $(\mathscr{M}_{m,n})_{m,n\in\mathbb{N}}$ of measurable set-valued maps and $[0,T[\rightsquigarrow \Theta(E,d,\lfloor\cdot\rfloor),\ t \mapsto \overset{\circ}{x}(t) \cap \mathscr{C}(t)$ considered in the claim:

Due to the definition of $\mathscr{M}_{m,n}$,

$$\bigcup_{n\in\mathbb{N}} \mathscr{M}_{m,n}(t) \subset \Big\{\vartheta \in \mathscr{C}(t) \;\Big|\; \limsup_{h\downarrow 0} \tfrac{1}{h}\cdot d(\vartheta(h,x(t)),\, x(t+h)) \le \tfrac{1}{m}\Big\} \cup \{\vartheta_0\}$$

$$\bigcup_{n\in\mathbb{N}} \mathscr{M}_{m,n}(t) \supset \Big\{\vartheta \in \mathscr{C}(t) \;\Big|\; \limsup_{h\downarrow 0} \tfrac{1}{h}\cdot d(\vartheta(h,x(t)),\, x(t+h)) < \tfrac{1}{m}\Big\} \cup \{\vartheta_0\}.$$

Now the standard estimate about evolutions along transitions in Proposition 2.6 (on page 106) implies for every $t \in [0,T[$

$$\begin{aligned}
\overline{\bigcup_{n\in\mathbb{N}} \mathscr{M}_{m,n}(t)} &\subset \overline{\Big\{\vartheta \in \mathscr{C}(t) \;\Big|\; \limsup_{h\downarrow 0} \tfrac{1}{h}\cdot d(\vartheta(h,x(t)),\, x(t+h)) \le \tfrac{1}{m}\Big\} \cup \{\vartheta_0\}}\\
&\subset \Big\{\vartheta \in \mathscr{C}(t) \;\Big|\; \limsup_{h\downarrow 0} \tfrac{1}{h}\cdot d(\vartheta(h,x(t)),\, x(t+h)) \le \tfrac{2}{m}\Big\} \cup \{\vartheta_0\},\\
\overline{\bigcup_{n\in\mathbb{N}} \mathscr{M}_{m,n}(t)} &\supset \overline{\Big\{\vartheta \in \mathscr{C}(t) \;\Big|\; \limsup_{h\downarrow 0} \tfrac{1}{h}\cdot d(\vartheta(h,x(t)),\, x(t+h)) \le 0\Big\}} \cup \{\vartheta_0\}\\
&= \Big\{\vartheta \in \mathscr{C}(t) \;\Big|\; \limsup_{h\downarrow 0} \tfrac{1}{h}\cdot d(\vartheta(h,x(t)),\, x(t+h)) \le 0\Big\} \cup \{\vartheta_0\}.
\end{aligned}$$

Finally,

$$\begin{aligned}
\bigcap_{m\in\mathbb{N}} \overline{\bigcup_{n\in\mathbb{N}} \mathscr{M}_{m,n}(t)} &= \Big\{\vartheta \in \mathscr{C}(t) \;\Big|\; \lim_{h\downarrow 0} \tfrac{1}{h}\cdot d(\vartheta(h,x(t)),\, x(t+h)) = 0\Big\} \cup \{\vartheta_0\}\\
&= \Big(\overset{\circ}{x}(t) \cap \mathscr{C}(t)\Big) \qquad\qquad \cup \{\vartheta_0\}.
\end{aligned}$$

Proposition A.77 (on page 490) ensures that the closure of a countable union and the countable intersection preserve measurability of set-valued maps. It completes the proof of Lemma 3. □

Proof (of Existence Theorem 4 on page 388).

In a word, we use the selection principle in Corollary 6 (on page 389) for a connection between the mutational inclusion here and the mutational equation discussed in § 2.3 (on page 104 ff.).
The tuple $\big(E,\, d,\, \lfloor\cdot\rfloor,\, \Theta(E,d,\lfloor\cdot\rfloor)\big)$ is Euler compact by assumption.

Let $f : E \times [0,T] \longrightarrow \Theta(E,d,\lfloor\cdot\rfloor)$ denote the selection of the set-valued map $\mathscr{F} : E \times [0,T] \rightsquigarrow \Theta(E,d,\lfloor\cdot\rfloor)$ whose existence is stated in Corollary 6.

Strictly speaking, just one obstacle is preventing us from applying Peano's Existence Theorem 2.18 for nonautonomous mutational equations (on page 114) immediately, namely its assumption (4.) about continuity:
For each $R > 0$, there is a set $I \subset [0,T]$ of $\mathscr{L}^1$ measure 0 such that for any $t \in [0,T]\setminus I$,

$$\lim_{n\to\infty} D\big(f(x_n,t_n),\, f(x,t);\, R\big) = 0$$

holds for any sequences $(t_n)_{n\in\mathbb{N}}$ in $[0,T]$ and $(x_n)_{n\in\mathbb{N}}$ in E satisfying $\lim_{n\to\infty} t_n = t$ and $\lim_{n\to\infty} d(x_n,x) = 0$, $\sup_{n\in\mathbb{N}} \lfloor x_n \rfloor < \infty$. (In particular, I should not depend on $x \in E$.)

Similarly to the proof of Corollary 6, Scorza-Dragoni Theorem (in Proposition A.9 on page 446) ensures for each $\varepsilon > 0$ that there exists a closed subset $I_\varepsilon \subset [0,T]$ with $\mathscr{L}^1\big([0,T] \setminus I_\varepsilon\big) < \varepsilon$ such that the restriction $f(\cdot,\cdot)|_{E\times I_\varepsilon} : E \times I_\varepsilon \longrightarrow \Theta(E,d,\lfloor\cdot\rfloor)$ is continuous (with respect to the metric $\check{D}$ on $\Theta(E,d,\lfloor\cdot\rfloor)$).

Now $f_\varepsilon : E \times [0,T] \longrightarrow \Theta(E,d,\lfloor\cdot\rfloor)$ is defined as the extension of $f(\cdot,\cdot)|_{E\times I_\varepsilon}$ with

$$f_\varepsilon(x,t) := f(x,s_t)$$

for $x \in E,\ t \in [0,T] \setminus I_\varepsilon$ and $s_t := \sup\{s \in I_\varepsilon \mid s \le t\} \in I_\varepsilon$.

Obviously, this extension f_ε is continuous in the open subset $E \times ([0,T] \setminus \partial I_\varepsilon)$ (with respect to the metric $\check{D}$ on $\Theta(E,d,\lfloor\cdot\rfloor)$ again). In particular, ∂I_ε is (at most) countable because $[0,T] \setminus \partial I_\varepsilon$ is open with finite Lebesgue measure. As a consequence, f_ε satisfies continuity assumption (4.) of Peano's Theorem 2.18.

Fixing the initial state $x_0 \in E$ arbitrarily, there exists a solution $x_\varepsilon : [0,T] \longrightarrow E$ to the mutational equation

$$\overset{\circ}{x}_\varepsilon(\cdot) \ni f_\varepsilon\big(x_\varepsilon(\cdot), \cdot\big)$$

in the tuple $\big(E,d,\lfloor\cdot\rfloor\big)$ with $x_\varepsilon(0) = x_0$ and

$$\sup_{[0,T]} \lfloor x_\varepsilon(\cdot)\rfloor < (\lfloor x_0\rfloor + \widehat{\gamma}\,T)\, e^{\widehat{\gamma}\,T} + 1 =: R$$

using the abbreviation $\widehat{\gamma} := \sup\big\{\gamma(\vartheta) \bigm| \vartheta \in \mathscr{F}(x,t),\ x \in E,\ t \in [0,T]\big\} < \infty$.

Finally we choose any sequence $(\varepsilon_n)_{n\in\mathbb{N}}$ in $]0,1[$ with $\sum_{n=1}^\infty \varepsilon_n < \infty$. Then Proposition 2.11 about the continuity of solutions with respect to data (on page 108) implies that $\big(x_{\varepsilon_n}(\cdot)\big)_{n\in\mathbb{N}}$ is a Cauchy sequence in $C^0([0,T],E)$ with respect to the uniform topology. According to Remark 2.19 (on page 116), there exists a pointwise limit curve $x(\cdot) : [0,T] \longrightarrow E$.
Furthermore $(x_{\varepsilon_n}(\cdot))_{n\in\mathbb{N}}$ is uniformly Lipschitz continuous with respect to d due to assumption (i) and thus, $x(\cdot) : [0,T] \longrightarrow (E,d)$ is also Lipschitz continuous. The lower semicontinuity of $\lfloor\cdot\rfloor : (E,d) \longrightarrow [0,\infty[$ (by assumption) ensures

$$\sup_{[0,T]} \lfloor x(\cdot)\rfloor \le \sup_{\varepsilon\in]0,1[}\ \sup_{[0,T]} \lfloor x_\varepsilon(\cdot)\rfloor < R < \infty.$$

$x(\cdot)$ is a solution to the mutational equation $\overset{\circ}{x}(\cdot) \ni f\big(x(\cdot), \cdot\big)$ in the tuple $\big(E,d,\lfloor\cdot\rfloor\big)$. Indeed, Proposition 2.11 (extended to Lebesgue-integrable distances between transitions approximatively) implies for every $m \in \mathbb{N}$, $t \in [0,T[$ and $h \in [0,1]$ with $t+h \le T$ and $J_m := \bigcap_{n\ge m} I_{\varepsilon_n} \subset [0,T]$

$$\begin{aligned} d\big(f(x(t),t)\,(h,x(t)),\ x(t+h)\big) &= \lim_{n\to\infty}\ d\big(f(x(t),t)\,(h,x(t)),\ x_{\varepsilon_n}(t+h)\big) \\ &\le \liminf_{n\to\infty}\ \Big(d\big(x(t),\ x_{\varepsilon_n}(t)\big) + \int_t^{t+h} \\ &\qquad \times D\big(f(x(t),t),\ f_{\varepsilon_n}(x_{\varepsilon_n}(s),s);\ R\big)\ ds\Big)\ e^{\widehat{\alpha}\,h} \\ &\le \liminf_{n\to\infty} e^{\widehat{\alpha}\,h}\cdot\Big(\int_{[t,t+h]\setminus J_m} 2\ \widehat{m}_R(s)\,ds\ + \\ &\quad \int_{[t,t+h]\cap J_m} D\big(f(x(t),t),\ f(x_{\varepsilon_n}(s),s);R\big)\ ds\Big). \end{aligned}$$

with the bound $\widehat{m}_R(\cdot)$ mentioned in assumption (iv). The continuity of the restriction $f(\cdot,\cdot)|_{E\times J_m}$ guarantees for all $m\in\mathbb{N}$, $t\in[0,T[$, $h\in[0,1]$ with $t+h\le T$

$$\begin{aligned} d\big(f(x(t),t)\,(h,x(t)),x(t+h)\big) \le \Big(&\int_{[t,t+h]\setminus J_m} 2\ \widehat{m}_R(s)\,ds \\ &+\int_{[t,t+h]\cap J_m} D\big(f(x(t),t),\ f(x(s),s);R\big)\ ds\Big)e^{\widehat{\alpha}\,h}. \end{aligned}$$

Moreover, the set $\widehat{J_m}$ of all Lebesgue points of the integrable product $\chi_{[0,T[\setminus J_m}\ \widehat{m}_R$: $[0,T]\longrightarrow\mathbb{R}$ has full Lebesgue measure due to [189, Theorem 1.3.8] and, these two properties imply for every $m\in\mathbb{N}$ and $t\in J_m\cap\widehat{J_m}$,

$$\begin{cases} \lim\limits_{h\downarrow 0}\ \frac{1}{h}\displaystyle\int_{[t,t+h]\setminus J_m} \widehat{m}_R(s)\ ds & = \big(\chi_{[0,T[\setminus J_m}\,\widehat{m}_R\big)(t)\ =\ 0,\\ \lim\limits_{h\downarrow 0}\ \frac{1}{h}\displaystyle\int_{[t,t+h]\cap J_m} D\big(f(x(t),t),\ f(x(s),s);\ R\big)\ ds & = 0. \end{cases}$$

In combination with

$$\begin{aligned} J_m\ &\subset J_{m+1} && \text{for each } m\in\mathbb{N},\\ \mathscr{L}^1\big([0,T]\setminus J_m\big) &\le \sum_{n=m}^{\infty}\mathscr{L}^1\big([0,T]\setminus I_{\varepsilon_n}\big) &&\le \sum_{n=m}^{\infty}\varepsilon_n\ \stackrel{m\to\infty}{\longrightarrow}\ 0, \end{aligned}$$

we obtain for $\mathscr{L}^1$-almost every $t\in[0,T]$

$$\limsup_{h\downarrow 0}\ \tfrac{1}{h}\cdot d\big(f(x(t),t)\,(h,x(t)),\ x(t+h)\big)\ \le\ 0.$$

□

Remark 8. This proof of Existence Theorem 4 has essentially two pillars: The selection principle in Corollary 6 connects the existence problem of mutational inclusions with mutational equations and, then we apply Peano's Existence Theorem 2.18 (based on Euler compactness).
This selection principle is formulated only for separable metric spaces here (as we have already pointed out at the beginning of § 5.1.2), but we are still free to combine it with the existence results in § 3.3. This lays the foundations for extending Existence Theorem 4 to several examples that require the generalizations in Chapter 3. In particular, a curve is (sequentially) continuous with respect to a metric if and only if it is sequentially continuous with respect to some power of an equivalent metric (no matter whether the latter satisfies the triangle inequality).

5.2 Morphological Inclusions with State Constraints: A Viability Theorem

In this section, we focus on the geometric example of the metric space $(\mathscr{K}(\mathbb{R}^N), dl)$ and consider transitions induced by reachable sets of differential inclusions whose set-valued right-hand sides belong to LIP$(\mathbb{R}^N, \mathbb{R}^N)$. The corresponding mutational equations are usually called *morphological equations* and, they are discussed in § 1.9 (on page 57 ff.).

Now *morphological inclusions* are based on the goal to admit more than just one transition for each compact subset of $\mathbb{R}^N$. In contrast to the preceding § 5.1, however, additional state constraints $K(t) \in \mathscr{V}$ on the wanted tube $K(\cdot) : [0,T] \longrightarrow \mathscr{K}(\mathbb{R}^N)$ are to come into play. This difficulty is handled just by means of the supplementary "structure" of the morphological transition set LIP$(\mathbb{R}^N, \mathbb{R}^N)$.

The problems of invariance and viability have already been investigated for transitions induced by bounded Lipschitz vector fields (instead of the set-valued maps in LIP$(\mathbb{R}^N, \mathbb{R}^N)$).

Indeed, Doyen [74] has given sufficient and some necessary conditions on $\mathscr{F}(\cdot)$ and $\mathscr{V} \subset \mathscr{K}(\mathbb{R}^N)$ for the *invariance* of $\mathscr{V}$ (i.e. *all continuous* solutions starting in $\mathscr{V}$ stay in $\mathscr{V}$). His key notion is first to extend Filippov's existence theorem from differential inclusions (in $\mathbb{R}^N$) to morphological inclusions in $\mathscr{K}(\mathbb{R}^N)$ [74, Theorem 7.1] and then to verify dist$(K(\cdot), \mathscr{V}) \le 0$ (under the assumption that the values of $\mathscr{F}(\cdot)$ are contained in the respective contingent transition set to $\mathscr{V}$) [74, Theorem 8.2].

The corresponding question about viability of $\mathscr{V}$ (i.e. *at least one* continuous solution has to stay in $\mathscr{V}$) was pointed out as open by Aubin in [10, § 2.3.3]. A first answer was given in [122] – but only for transitions induced by bounded Lipschitz vector fields.

Now we consider the viability problem for morphological inclusions with transitions in LIP$_{\overline{\text{co}}}(\mathbb{R}^N, \mathbb{R}^N)$ in their full generality (as in [121]).

Definition 9. LIP$_{\overline{\text{co}}}(\mathbb{R}^N, \mathbb{R}^N)$ consists of all set-valued maps $F \in$ LIP$(\mathbb{R}^N, \mathbb{R}^N)$ whose values are convex in addition, i.e., every map $F : \mathbb{R}^N \rightsquigarrow \mathbb{R}^N$ in LIP$_{\overline{\text{co}}}(\mathbb{R}^N, \mathbb{R}^N)$ satisfies the following conditions:

1.) F has nonempty compact convex values that are uniformly bounded in $\mathbb{R}^N$,
2.) F is Lipschitz continuous with respect to the Pompeiu-Hausdorff distance dl.

In fact, the main result of this section, i.e. Theorem 12 (on page 401) below, is very similar to the viability theorem for differential inclusions in $\mathbb{R}^N$ (discussed in [14] and quoted here in Theorem 11).

5.2.1 (Well-Known) Viability Theorem for Differential Inclusions

The situation has already been investigated intensively for differential inclusions in $\mathbb{R}^N$ (see e.g. [14, 15]). For clarifying the new aspects of morphological inclusions, we now quote the corresponding result from [14, Theorems 3.3.2, 3.3.5].

Definition 10 ([14, Definition 2.2.4]). Let X and Y be normed vector spaces. A set-valued map $F : X \leadsto Y$ is called *Marchaud map* if it has the following properties:

1. F is nontrivial, i.e. Graph $F \neq \emptyset$,
2. F is upper semicontinuous, i.e. for any $x \in X$ and neighborhood $V \supset F(x)$, there is a neighborhood $U \subset X$ of x: $F(U) \subset V$,
3. F has compact convex values,
4. F has linear growth, i.e. $\sup_{v \in F(x)} |v| \leq C\,(1+|x|)$ for all $x \in X$.

Theorem 11 (Viability theorem for diff. inclusions [14, Theorems 3.3.2, 3.3.5]**).** *Consider a Marchaud map $F : \mathbb{R}^N \leadsto \mathbb{R}^N$ and a nonempty closed subset $V \subset \mathbb{R}^N$ with $F(x) \neq \emptyset$ for all $x \in V$. Then for any finite time $T \in \,]0,\infty[$, the following two statements are equivalent:*

1. For every point $x_0 \in V$, there is at least one solution $x(\cdot) \in W^{1,1}([0,T],\, \mathbb{R}^N)$ of $x'(\cdot) \in F(x(\cdot))$ (almost everywhere) with $x(0) = x_0$ and $x(t) \in V$ for all t.

2. $F(x) \cap T_V(x) \neq \emptyset$ for all $x \in V$.

The implication (1.) $\Longrightarrow$ (2.) is rather obvious. For proving (2.) $\Longrightarrow$ (1.), a standard approach uses an "approximating" sequence $\big(x_n(\cdot)\big)_{n\in\mathbb{N}}$ in $W^{1,\infty}([0,1],\mathbb{R}^N)$ such that $\sup_t \operatorname{dist}(x_n(t),\, V) \longrightarrow 0$ $(n \to \infty)$ and $\big(x_n(t),\, \frac{d}{dt}x_n(t)\big)$ is close to Graph $F \subset \mathbb{R}^N \times \mathbb{R}^N$ for almost every t. Then the theorems of Arzelà-Ascoli and Alaoglu provide a subsequence $\big(x_{n_j}(\cdot)\big)_{j\in\mathbb{N}}$ and limits $x \in C^0([0,1],\mathbb{R}^N)$, $w \in L^\infty([0,1],\mathbb{R}^N)$ with

$$x_{n_j}(\cdot) \longrightarrow x(\cdot) \text{ uniformly}, \qquad \tfrac{d}{dt}x_{n_j}(\cdot) \longrightarrow w(\cdot) \text{ weakly* in } L^\infty([0,1],\mathbb{R}^N).$$

Due to the continuous embedding $L^\infty([0,1],\mathbb{R}^N) \subset L^1([0,1],\mathbb{R}^N)$, we even obtain the convergence $\frac{d}{dt}x_{n_j}(\cdot) \longrightarrow w(\cdot)$ weakly in $L^1([0,1],\mathbb{R}^N)$. Thus, $w(\cdot)$ is the weak derivative of $x(\cdot)$ in $[0,1]$ and, $x(\cdot)$ is Lipschitz continuous. Finally Mazur's Lemma implies

$$w(t) \in \bigcap_{\varepsilon>0} \overline{\mathrm{co}}\Big(\bigcup_{z\in\mathbb{B}_\varepsilon(x(t))} F(z)\Big) = F(x(t)) \qquad \text{for almost every } t.$$

Considering now morphological inclusions on $(\mathscr{K}(\mathbb{R}^N), d\!l)$ (instead of differential inclusions), an essential aspect changes: The derivative of a curve is not represented as a function in $L^1([0,1],\mathbb{R}^N)$ any longer, but we are dealing with $\mathrm{LIP}(\mathbb{R}^N,\mathbb{R}^N)$-valued curves here. Now the classical theorems of Arzelà-Ascoli, Alaoglu and Mazur might have to be replaced by their counterparts concerning functions with their values in a Banach space (instead of $\mathbb{R}^N$).

5.2.2 Adapting This Concept to Morphological Inclusions: The Main Theorem

Now $\mathscr{F} : \mathscr{K}(\mathbb{R}^N) \rightsquigarrow \text{LIP}(\mathbb{R}^N, \mathbb{R}^N)$ and a set of constraints $\mathscr{V} \subset \mathscr{K}(\mathbb{R}^N)$ are given. Correspondingly to Theorem 11 about differential inclusions, we focus on the so-called *viability condition* demanding from each compact set $K \in \mathscr{V}$ that the value $\mathscr{F}(K)$ and the contingent transition set $\mathscr{T}_{\mathscr{V}}(K) \subset \text{LIP}(\mathbb{R}^N, \mathbb{R}^N)$ have at least one morphological transition in common. Lacking a concrete counterpart of Aumann integral in the metric space $(\mathscr{K}(\mathbb{R}^N), dl)$, the question of its necessity (for the existence of "in $\mathscr{V}$ viable" solutions) is more complicated than for differential inclusions in $\mathbb{R}^N$ and thus, we skip it here deliberately.
The main result of this section 5.2 is that in combination with appropriate assumptions about $\mathscr{F}(\cdot)$ and $\mathscr{V}$, the viability condition is *sufficient*.

Convexity comes into play again, but we have to distinguish between (at least) two respects:
First, assuming $\mathscr{F}$ to have convex values in $\text{LIP}(\mathbb{R}^N, \mathbb{R}^N)$ and second, supposing each set-valued map $G \in \mathscr{F}(K) \subset \text{LIP}(\mathbb{R}^N, \mathbb{R}^N)$ (with $K \in \mathscr{K}(\mathbb{R}^N)$) to have convex values in $\mathbb{R}^N$. The latter, however, does not really provide a geometric restriction on morphological transitions. Indeed, Relaxation Theorem A.19 of Filippov-Ważewski (on page 453) implies $\vartheta_G(t, K) = \vartheta_{\overline{co}\, G}(t, K)$ for every map $G \in \text{LIP}(\mathbb{R}^N, \mathbb{R}^N)$, initial set $K \in \mathscr{K}(\mathbb{R}^N)$ and time $t \geq 0$.
Thus, we suppose the values of $\mathscr{F}$ to be in $\text{LIP}_{\overline{co}}(\mathbb{R}^N, \mathbb{R}^N)$:

Theorem 12 (Viability theorem for morphological inclusions).
Let $\mathscr{F} : \mathscr{K}(\mathbb{R}^N) \rightsquigarrow \text{LIP}_{\overline{co}}(\mathbb{R}^N, \mathbb{R}^N)$ be a set-valued map and $\mathscr{V} \subset \mathscr{K}(\mathbb{R}^N)$ a nonempty closed subset satisfying the following conditions:

1.) *all values of $\mathscr{F}$ are nonempty and convex (i.e. for any $G_1, G_2 \in \mathscr{F}(K) \subset \text{LIP}_{\overline{co}}(\mathbb{R}^N, \mathbb{R}^N)$ and $\lambda \in [0,1]$, the set-valued map*
$$\mathbb{R}^N \rightsquigarrow \mathbb{R}^N, \quad x \mapsto \lambda \cdot G_1(x) + (1-\lambda) \cdot G_2(x)$$
also belongs to $\mathscr{F}(K)$),

2.) $$A := \sup_{M \in \mathscr{K}(\mathbb{R}^N)} \sup_{G \in \mathscr{F}(M)} \text{Lip}\, G < \infty,$$
$$B := \sup_{M \in \mathscr{K}(\mathbb{R}^N)} \sup_{G \in \mathscr{F}(M)} \|G\|_\infty < \infty,$$

3.) *the graph of $\mathscr{F}$ is closed (w.r.t. locally uniform convergence in* $\text{LIP}(\mathbb{R}^N, \mathbb{R}^N)$*),*

4.) $\mathscr{T}_{\mathscr{V}}(K) \cap \mathscr{F}(K) \neq \emptyset \qquad$ *for all $K \in \mathscr{V}$.*

Then for every initial set $K_0 \in \mathscr{V}$, there exists a compact-valued Lipschitz continuous solution $K(\cdot) : [0,1] \rightsquigarrow \mathbb{R}^N$ to the morphological inclusion

$$\overset{\circ}{K}(\cdot) \cap \mathscr{F}(K(\cdot)) \neq \emptyset$$

with $K(0) = K_0$ and $K(t) \in \mathscr{V}$ for all $t \in [0,1]$.

Remark 13. In assumption (3.), the topology on $\text{LIP}(\mathbb{R}^N, \mathbb{R}^N)$ is specified. A sequence $(G_n)_{n\in\mathbb{N}}$ in $\text{LIP}(\mathbb{R}^N, \mathbb{R}^N)$ is said to converge "locally uniformly" to $G \in \text{LIP}(\mathbb{R}^N, \mathbb{R}^N)$ if for every nonempty compact set $M \subset \mathbb{R}^N$,

$$d\!l_\infty(G_n(\cdot)|_M, G(\cdot)|_M) \overset{\text{Def.}}{=} \sup_{x\in M} d\!l(G_n(x), G(x)) \longrightarrow 0 \qquad \text{for } n \longrightarrow \infty$$

using here the Pompeiu-Hausdorff distance $d\!l$ on $\mathcal{K}(\mathbb{R}^N)$. This topology can be regarded as an example induced by the metric $\check{D}$ in § 5.1 (on page 386).

Due to the uniform bounds in assumption (2.), the image set $\mathcal{F}(\mathcal{K}(\mathbb{R}^N))$ is sequentially compact in $\text{LIP}_{\overline{\text{co}}}(\mathbb{R}^N, \mathbb{R}^N)$ with respect to this topology (as we prove in subsequent Lemma 19). Hence, $\mathcal{F}$ is upper semicontinuous (in the sense of Bouligand and Kuratowski) according to [19, Proposition 1.4.8].

Now Viability Theorem 12 is applied to two very special forms of constraints:

$$\mathcal{V}_1 := \{K \in \mathcal{K}(\mathbb{R}^N) \mid K \cap M \neq \emptyset\}$$
$$\mathcal{V}_2 := \{K \in \mathcal{K}(\mathbb{R}^N) \mid K \subset M\}$$

with some (arbitrarily fixed) nonempty closed subset $M \subset \mathbb{R}^N$. Indeed, Gorre has already characterized the corresponding contingent transition sets — as discussed in Example 1.64 (on page 68) and quoted in Proposition 1.66. Thus, we conclude directly:

Corollary 14 (Solutions having nonempty intersection with fixed $M \subset \mathbb{R}^N$).
Let $\mathcal{F} : \mathcal{K}(\mathbb{R}^N) \rightsquigarrow \text{LIP}_{\overline{\text{co}}}(\mathbb{R}^N, \mathbb{R}^N)$ be a set-valued map and $M \subset \mathbb{R}^N$ a closed subset satisfying:

1.) *all values of $\mathcal{F}$ are nonempty, convex with global bounds (as in Theorem 12),*
2.) *the graph of $\mathcal{F}$ is closed (w.r.t. locally uniform convergence in $\text{LIP}(\mathbb{R}^N, \mathbb{R}^N)$),*
3.) *for any $K \in \mathcal{K}(\mathbb{R}^N)$ with $K \cap M \neq \emptyset$, there exist $G \in \mathcal{F}(K)$, $x \in K \cap M$ with $G(x) \cap P_M^K(x) \neq \emptyset$.*

Then for every compact set $K_0 \subset \mathbb{R}^N$ with $K_0 \cap M \neq \emptyset$, there exists a compact-valued Lipschitz continuous solution $K(\cdot) : [0,1] \rightsquigarrow \mathbb{R}^N$ to the morphological inclusion $\overset{\circ}{K}(\cdot) \cap \mathcal{F}(K(\cdot)) \neq \emptyset$ with $K(0) = K_0$ and $K(t) \cap M \neq \emptyset$ for all t.

Corollary 15 (Solutions being contained in fixed $M \subset \mathbb{R}^N$).
Let $\mathcal{F} : \mathcal{K}(\mathbb{R}^N) \rightsquigarrow \text{LIP}_{\overline{\text{co}}}(\mathbb{R}^N, \mathbb{R}^N)$ be a set-valued map and $M \subset \mathbb{R}^N$ a closed subset satisfying:

1.) *all values of $\mathcal{F}$ are nonempty, convex with global bounds (as in Theorem 12),*
2.) *the graph of $\mathcal{F}$ is closed (w.r.t. locally uniform convergence in $\text{LIP}(\mathbb{R}^N, \mathbb{R}^N)$),*
3.) *for any compact set $K \subset M$, there exist $G \in \mathcal{F}(K)$ with $G(x) \subset T_M(x)$ for every $x \in K$.*

Then for every nonempty compact set $K_0 \subset M$, there exists a compact-valued Lipschitz continuous solution $K(\cdot) : [0,1] \rightsquigarrow \mathbb{R}^N$ to the morphological inclusion

$$\overset{\circ}{K}(\cdot) \cap \mathcal{F}(K(\cdot)) \neq \emptyset \qquad \text{with } K(0) = K_0 \text{ and } K(t) \subset M \text{ for all } t \in [0,1].$$

5.2.3 The Steps for Proving the Morphological Viability Theorem

The proof of Viability Theorem 12 uses a concept of approximation developed by Haddad and others for differential inclusions in $\mathbb{R}^N$ (and sketched in § 5.2.1).

For any given "threshold" $\varepsilon > 0$, we verify the existence of an approximative solution $K_\varepsilon(\cdot) : [0,1] \longrightarrow \mathscr{K}(\mathbb{R}^N)$ such that its values have distance $\leq \varepsilon$ from the set of constraints $\mathscr{V} \subset \mathscr{K}(\mathbb{R}^N)$.
In addition, each $K_\varepsilon(\cdot)$ is induced by a piecewise constant function

$$f_\varepsilon(\cdot) : [0,1[\longrightarrow \mathrm{LIP}_{\overline{\mathrm{co}}}(\mathbb{R}^N, \mathbb{R}^N)$$

of morphological transitions such that $(K_\varepsilon(t),\, f_\varepsilon(t))$ is close to Graph $\mathscr{F}$ at every time $t \in [0,T[$ (Lemma 16). Proposition A.81 about parameterization (on page 491) bridges the gap between $f_\varepsilon(\cdot) : [0,1[\longrightarrow \mathrm{LIP}_{\overline{\mathrm{co}}}(\mathbb{R}^N, \mathbb{R}^N)$ and the auxiliary function $\widehat{f_\varepsilon}(\cdot) : [0,1[\longrightarrow \mathrm{BLip}(\mathbb{R}^N \times \mathbb{B}_1,\, \mathbb{R}^N)$ whose single values are in the Banach space $\big(C^0(\mathbb{R}^N \times \mathbb{B}_1,\, \mathbb{R}^N),\, \|\cdot\|_\infty\big)$ additionally.

Then, letting $\varepsilon > 0$ tend to 0, we obtain subsequences $(K_n(\cdot))_{n\in\mathbb{N}}$, $\big(\widehat{f_n}(\cdot)\big)_{n\in\mathbb{N}}$ that are converging to some $K(\cdot) : [0,1] \longrightarrow \mathscr{K}(\mathbb{R}^N)$ and $\widehat{f} : [0,1[\longrightarrow \mathrm{BLip}(\mathbb{R}^N \times \mathbb{B}_1, \mathbb{R}^N)$, respectively, in an appropriate sense – due to compactness (see Lemma 18).

Last, but not least, we prove that these limits satisfy for $\mathscr{L}^1$-almost every $t \in [0,T[$

$$\widehat{f}(t)(\cdot,\mathbb{B}_1) \in \overset{\circ}{K}(t) \cap \mathscr{F}(K(t)) \neq \emptyset.$$

Indeed, Lemma 20 concludes $\widehat{f}(t)(\cdot,\mathbb{B}_1) \in \mathscr{F}(K(t))$ for $\mathscr{L}^1$-almost every $t \in [0,T[$ from Lemma 19 stating that the graph of $\mathscr{F}$ is sequentially compact. Furthermore, $K(\cdot)$ can be characterized as reachable set, i.e. $\vartheta_{\widehat{f}(\cdot)(\cdot,\mathbb{B}_1)}(t, K_0) = K(t)$ for every t (Lemma 21). Finally, preceding Proposition 1.57 (on page 64) implies

$$\widehat{f}(t)(\cdot,\mathbb{B}_1) \in \overset{\circ}{K}(t) \qquad \text{for } \mathscr{L}^1\text{-}\textit{almost} \text{ every } t \in]0,1[.$$

Let us now formulate these steps in detail and then prove them.

Lemma 16 (Constructing approximative solutions). *Choose any* $\varepsilon > 0$.
Under the assumptions of Viability Theorem 12, there are a B-Lipschitz continuous function $K_\varepsilon(\cdot) : [0,1] \longrightarrow \mathscr{K}(\mathbb{R}^N)$ *and a function* $f_\varepsilon(\cdot) : [0,1[\longrightarrow \mathrm{LIP}_{\overline{\mathrm{co}}}(\mathbb{R}^N, \mathbb{R}^N)$ *satisfying with* $R_\varepsilon := \varepsilon\, e^A$

a) $K_\varepsilon(0) = K_0$,

b) $\mathrm{dist}\big(K_\varepsilon(t),\, \mathscr{V}\big) \leq R_\varepsilon$ *for all* $t \in [0,1]$,

c) $f_\varepsilon(t) \in \overset{\circ}{K_\varepsilon}(t) \cap \mathscr{F}\big(\mathbb{B}_{R_\varepsilon}(K_\varepsilon(t))\big) \neq \emptyset$ *for all* $t \in [0,1[$,

d) $f_\varepsilon(\cdot)$ *is piecewise constant in the following sense: for each* $t \in [0,1[$, *there exists some* $\delta > 0$ *such that* $f_\varepsilon(\cdot)|_{[t,t+\delta[}$ *is constant.*

Remark 17. As a direct consequence of property (d), the function $f_\varepsilon : [0,1[\longrightarrow \mathrm{LIP}_{\overline{\mathrm{co}}}(\mathbb{R}^N, \mathbb{R}^N)$ can have at most countably many points of discontinuity. This enables us to apply earlier results about autonomous morphological equations (§ 1.9 on page 57 ff.) to the approximations $K_\varepsilon(\cdot), f_\varepsilon(\cdot)$ in a "piecewise" way.

Now the "threshold of accuracy" $\varepsilon > 0$ is tending to 0. The "detour" of parameterization (Proposition A.81) and the subsequent statements about sequential compactness lay the basis for extracting subsequences with additional features of convergence:

Lemma 18 (Selecting an approximative subsequence).
Under the assumptions of Viability Theorem 12, there are a constant $c = c(N,A,B)$, sequences $K_n(\cdot) : [0,1] \longrightarrow \mathscr{K}(\mathbb{R}^N)$, $\widehat{f}_n(\cdot) : [0,1[\longrightarrow \mathrm{BLip}(\mathbb{R}^N \times \mathbb{B}_1, \mathbb{R}^N)$ $(n \in \mathbb{N})$ and $K(\cdot) : [0,1] \longrightarrow \mathscr{K}(\mathbb{R}^N)$, $\widehat{f}(\cdot) : [0,1[\longrightarrow \mathrm{BLip}(\mathbb{R}^N \times \mathbb{B}_1, \mathbb{R}^N)$ such that for every $j,n \in \mathbb{N}$, $t \in [0,1[$, $x \in \mathbb{R}^N$, $u \in \mathbb{B}_1 \subset \mathbb{R}^N$

a) $K_0 = K_n(0) = K(0)$,
b) $K(\cdot)$ *and* $K_n(\cdot)$ *are B-Lipschitz continuous w.r.t.* $d\!l$,
c) $\widehat{f}_n(\cdot)(x,u)$ *is piecewise constant (in the sense of Lemma 16 (d)),*
$\|\widehat{f}_n(t)(\cdot,\cdot)\|_\infty + \mathrm{Lip}\,\widehat{f}_n(t)(\cdot,\cdot) \le c < \infty$,
d) $\mathrm{dist}\big(K_n(t), \mathscr{V}\big) \le \frac{1}{n}$
e) $\widehat{f}_n(t)(\cdot,\mathbb{B}_1) \in \overset{\circ}{K}_n(t) \cap \mathscr{F}\big(\mathbb{B}_{1/n}(K_n(t))\big) \neq \emptyset$
f) $d\!l\big(K_m(\cdot), K(\cdot)\big) \longrightarrow 0$ *uniformly in* $[0,1]$ *for* $m \to \infty$,
g) $\widehat{f}_m(\cdot)|_{\widetilde{K}_j \times \mathbb{B}_1} \longrightarrow \widehat{f}(\cdot)|_{\widetilde{K}_j \times \mathbb{B}_1}$ *weakly in* $L^1\big([0,1], C^0(\widetilde{K}_j \times \mathbb{B}_1, \mathbb{R}^N)\big)$ *for* $m \to \infty$,
h) $\|\widehat{f}(t)(\cdot,\cdot)\|_\infty + \mathrm{Lip}\,\widehat{f}(t)(\cdot,\cdot) \le c < \infty$,
i) $K(t) \in \mathscr{V}$

with the abbreviation $\widetilde{K}_j := \mathbb{B}_{j+B}(K_0) \overset{\text{Def.}}{=} \big\{x \in \mathbb{R}^N \,\big|\, \mathrm{dist}(x, K_0) \le j+B\big\} \in \mathscr{K}(\mathbb{R}^N)$.

Lemma 19 (Sequential compactness in the image and graph of $\mathscr{F}(\cdot)$).
In addition to the hypotheses of Viability Theorem 12, let $(G_k)_{k \in \mathbb{N}}$ be an arbitrary sequence in the image set $\mathscr{F}(\mathscr{K}(\mathbb{R}^N)) = \bigcup_{M \in \mathscr{K}(\mathbb{R}^N)} \mathscr{F}(M) \subset \mathrm{LIP}_{\overline{\mathrm{co}}}(\mathbb{R}^N, \mathbb{R}^N)$.

Then, there exist a subsequence $(G_{k_j})_{j \in \mathbb{N}}$ and a map $G \in \mathrm{LIP}_{\overline{\mathrm{co}}}(\mathbb{R}^N, \mathbb{R}^N)$ such that for any compact set $M \subset \mathbb{R}^N$, $\sup_{x \in M} d\!l(G_{k_j}(x), G(x)) \longrightarrow 0$ $(j \longrightarrow \infty)$ and

$$\mathrm{Lip}\, G \le A, \qquad \|G\|_\infty \le B.$$

Let now $(K_k)_{k \in \mathbb{N}}$ be an arbitrary sequence in $\mathscr{K}(\mathbb{R}^N)$ such that $\bigcup_{k \in \mathbb{N}} K_k \subset \mathbb{R}^N$ is bounded and $G_k \in \mathscr{F}(K_k)$ for each $k \in \mathbb{N}$. Then there exist subsequences $(K_{k_j})_{j \in \mathbb{N}}$, $(G_{k_j})_{j \in \mathbb{N}}$, a set $K \in \mathscr{K}(\mathbb{R}^N)$ and a map $G \in \mathscr{F}(K) \subset \mathrm{LIP}_{\overline{\mathrm{co}}}(\mathbb{R}^N, \mathbb{R}^N)$ with

$$d\!l(K_{k_j}, K) \overset{j \to \infty}{\longrightarrow} 0 \qquad \sup_{x \in M} d\!l(G_{k_j}(x), G(x)) \overset{j \to \infty}{\longrightarrow} 0 \quad \textit{for each } M \in \mathscr{K}(\mathbb{R}^N).$$

Lemma 20.
Let the sequences $K_n : [0,1] \longrightarrow \mathscr{K}(\mathbb{R}^N)$, $\widehat{f_n} : [0,1[\longrightarrow \mathrm{BLip}(\mathbb{R}^N \times \mathbb{B}_1, \mathbb{R}^N)$ $(n \in \mathbb{N})$ *and the functions* $K(\cdot) : [0,1] \longrightarrow \mathscr{K}(\mathbb{R}^N)$, $\widehat{f}(\cdot) : [0,1[\longrightarrow \mathrm{BLip}(\mathbb{R}^N \times \mathbb{B}_1, \mathbb{R}^N)$ *be as in Lemma 18 above.*

Then, for $\mathscr{L}^1$*-almost every* $t \in [0,1[$,

$$\operatorname{dist}\Big(\widehat{f}(t)(x,\mathbb{B}_1),\ co\,\big\{\widehat{f_n}(t)(x,\mathbb{B}_1),\ \widehat{f_{n+1}}(t)(x,\mathbb{B}_1)\dots\big\}\Big) \overset{n\to\infty}{\longrightarrow} 0$$

locally uniformly in $x \in \mathbb{R}^N$ *and, the coefficients of the approximating convex combinations can be chosen independently of* t, x.
In particular, $\widehat{f}(t)(\cdot,\mathbb{B}_1) \in \mathscr{F}(K(t)) \subset \mathrm{LIP}_{\overline{co}}(\mathbb{R}^N,\mathbb{R}^N)$.

Last, but not least, we have to prove $\widehat{f}(t)(\cdot,\mathbb{B}_1) \in \overset{\circ}{K}(t)$ at $\mathscr{L}^1$-almost every time t. Due to Proposition 1.57 (on page 64), we can restrict our considerations to describing $K(t)$ as reachable set of a nonautonomous differential inclusion, i.e.

$$\vartheta_{\widehat{f}(\cdot)(\cdot,\mathbb{B}_1)}(t,K_0) = K(t) \qquad \text{for every } t \in]0,1].$$

Lemma 21 ($K(t)$ as a reachable set of $\widehat{f}(\cdot)(\cdot,\mathbb{B}_1)$).
Let the sequences $K_n : [0,1] \longrightarrow \mathscr{K}(\mathbb{R}^N)$, $\widehat{f_n} : [0,1[\longrightarrow \mathrm{BLip}(\mathbb{R}^N \times \mathbb{B}_1, \mathbb{R}^N)$ $(n \in \mathbb{N})$ *and the functions* $K(\cdot) : [0,1] \longrightarrow \mathscr{K}(\mathbb{R}^N)$, $\widehat{f}(\cdot) : [0,1[\longrightarrow \mathrm{BLip}(\mathbb{R}^N \times \mathbb{B}_1, \mathbb{R}^N)$ *be as in Lemma 18.*

Then, for any $x(\cdot) \in C^0([0,1],\mathbb{R}^N)$ *and Lebesgue measurable set* $J \subset [0,1]$,

$$d\!\!l\Big(\int_J \widehat{f_n}(s)(x(s),\mathbb{B}_1)\,ds,\ \int_J \widehat{f}(s)(x(s),\mathbb{B}_1)\,ds\Big) \overset{n\to\infty}{\longrightarrow} 0.$$

In particular, $\vartheta_{\widehat{f}(\cdot)(\cdot,\mathbb{B}_1)}(t,K_0) = K(t)$ *for every* $t \in]0,1]$.

The next proposition serves as tool for proving Lemma 21 and focuses on solutions of nonautonomous differential inclusions in $\mathbb{R}^N$. In a word, this earlier theorem of Stassinopoulos and Vinter [176] characterizes perturbations (of the set-valued right-hand side) that have vanishing effect on the sets of continuous solutions.

Proposition 22 (Stassinopoulos and Vinter [176, Theorem 7.1]**).**
Let $D : [0,1] \times \mathbb{R}^N \rightsquigarrow \mathbb{R}^N$ *and each* $D_n : [0,1] \times \mathbb{R}^N \rightsquigarrow \mathbb{R}^N$ $(n \in \mathbb{N})$ *satisfy the following assumptions:*

1. D *and* D_n *have nonempty convex compact values,*
2. $D(\cdot,x), D_n(\cdot,x) : [0,1] \rightsquigarrow \mathbb{R}^N$ *are measurable for every* $x \in \mathbb{R}^N$,
3. *there exists* $k(\cdot) \in L^1([0,1])$ *such that* $D(t,\cdot), D_n(t,\cdot) : \mathbb{R}^N \rightsquigarrow \mathbb{R}^N$ *are* $k(t)$*-Lipschitz for* $\mathscr{L}^1$*-almost every* $t \in [0,1]$,
4. *there exists* $h(\cdot) \in L^1([0,1])$ *such that* $\sup_{v \in D(t,x) \cup D_n(t,x)} |v| \le h(t)$ *for every* $x \in \mathbb{R}^N$ *and* $\mathscr{L}^1$*-almost every* $t \in [0,1]$.

Fixing the initial point $a \in \mathbb{R}^N$ *arbitrarily, the absolutely continuous solutions of*

$$\begin{cases} y'(\cdot) \in D_n(\cdot, y(\cdot)) & \text{a.e. in } [0,1] \\ y(0) = a \end{cases} \quad \text{and} \quad \begin{cases} y'(\cdot) \in D(\cdot, y(\cdot)) & \text{a.e. in } [0,1] \\ y(0) = a \end{cases}$$

respectively form compact subsets of $\big(C^0([0,1],\mathbb{R}^N), \|\cdot\|_\infty\big)$ *denoted by* $\mathscr{D}_n$ $(n \in \mathbb{N})$ *and* $\mathscr{D}$.

Then, $\mathscr{D}_n$ *converges to* $\mathscr{D}$ *(w.r.t. the Pompeiu-Hausdorff metric on compact subsets of* $C^0([0,1],\mathbb{R}^N)$*) if and only if for every solution* $d(\cdot) \in \mathscr{D}$, $D_n(\cdot, d(\cdot)) : [0,1] \rightsquigarrow \mathbb{R}^N$ *converges to* $D(\cdot, d(\cdot)) : [0,1] \rightsquigarrow \mathbb{R}^N$ *weakly in the following sense*

$$dl\Big(\int_J D_n(s, d(s))\; ds, \quad \int_J D(s, d(s))\; ds\Big) \stackrel{n\to\infty}{\longrightarrow} 0$$

for every measurable subset $J \subset [0,1]$. □

Now let the proofs begin:

Proof (of Lemma 16 on page 403). It imitates the proof of Lemma 1.29 (on page 48 f.) and uses Zorn's Lemma: For $\varepsilon > 0$ fixed, let $\mathscr{A}_\varepsilon(K_0)$ denote the set of all tuples $(\tau_K,\, K(\cdot),\, f(\cdot))$ consisting of some $\tau_K \in [0,1]$, a B-Lipschitz continuous function $K(\cdot) : [0, \tau_K] \longrightarrow (\mathscr{K}(\mathbb{R}^N), dl)$ and a piecewise constant function $f(\cdot) : [0,1[\longrightarrow \mathrm{LIP}_{\overline{\mathrm{co}}}(\mathbb{R}^N, \mathbb{R}^N)$ such that

a) $K(0) = K_0$,

b') 1.) $\operatorname{dist}\big(K(\tau_K),\, \mathscr{V}\big) \le r_\varepsilon(\tau_K)$ with $r_\varepsilon(t) := \varepsilon\, e^{At}\, t$,
2.) $\operatorname{dist}\big(K(t),\, \mathscr{V}\big) \le R_\varepsilon$ for all $t \in [0, \tau_K]$,

c) $f(t) \in \overset{\circ}{K}(t) \cap \mathscr{F}\big(\mathbb{B}_{R_\varepsilon}(K(t))\big) \neq \emptyset$ for all $t \in [0, \tau_K[$.

Obviously, $\mathscr{A}_\varepsilon(K_0) \neq \emptyset$ since it contains $(0,\, K(\cdot) \equiv K_0,\, f(\cdot) \equiv f_0)$ with arbitrary $f_0 \in \mathrm{LIP}_{\overline{\mathrm{co}}}(\mathbb{R}^N, \mathbb{R}^N)$. Moreover, an order relation $\preceq$ on $\mathscr{A}_\varepsilon(K_0)$ is specified by

$$(\tau_K, K(\cdot), f(\cdot)) \preceq (\tau_M, M(\cdot), g(\cdot)) \quad :\Longleftrightarrow \quad \tau_K \le \tau_M,\ M\big|_{[0,\tau_K]} = K,\ g\big|_{[0,\tau_K[} = f.$$

Hence, Zorn's Lemma provides a maximal element $\big(\tau,\, K_\varepsilon(\cdot),\, f_\varepsilon(\cdot)\big) \in \mathscr{A}_\varepsilon(K_0)$. As all considered functions with values in $\mathscr{K}(\mathbb{R}^N)$ have been supposed to be B-Lipschitz continuous, $K_\varepsilon(\cdot)$ is well-defined on the closed interval $[0, \tau] \subset [0,1]$.

Assuming $\tau < 1$ for a moment, we obtain a contradiction if $K_\varepsilon(\cdot), f_\varepsilon(\cdot)$ can be extended to a larger interval $[0, \tau + \delta] \subset [0,1]$ $(\delta > 0)$ preserving conditions (b'), (c). Since closed bounded balls of $(\mathscr{K}(\mathbb{R}^N), dl)$ are compact, the closed set $\mathscr{V}$ contains an element $Z \in \mathscr{K}(\mathbb{R}^N)$ with $dl(K_\varepsilon(\tau), Z) = \operatorname{dist}(K_\varepsilon(\tau), \mathscr{V}) \le r_\varepsilon(\tau)$ and, assumption (4.) of Viability Theorem 12 provides a set-valued map

$$G \in \mathscr{T}_{\mathscr{V}}(Z) \cap \mathscr{F}(Z) \subset \mathrm{LIP}_{\overline{\mathrm{co}}}(\mathbb{R}^N, \mathbb{R}^N).$$

Due to Definition 1.16 of the contingent transition set $\mathscr{T}_{\mathscr{V}}(Z)$, there is a sequence $h_m \downarrow 0$ in $]0, 1-\tau[$ such that $\operatorname{dist}(\vartheta_G(h_m, Z),\, \mathscr{V}) \le \varepsilon\, h_m$ for all $m \in \mathbb{N}$. Now set

$$K_\varepsilon(t) := \vartheta_G\big(t - \tau,\, K_\varepsilon(\tau)\big), \qquad f_\varepsilon(t) := G \quad \text{for each } t \in [\tau,\, \tau + h_1[.$$

Obviously, $G \in \overset{\circ}{K_\varepsilon}(t)$ holds for all $t \in [\tau,\ \tau + h_1[$. Moreover, it leads to

$$\begin{array}{llll} d\!l\big(K_\varepsilon(t),\, Z\big) & \leq\ d\!l\big(\vartheta_G(t-\tau,\, K_\varepsilon(\tau)),\ K_\varepsilon(\tau)\big) & +\ d\!l\big(K_\varepsilon(\tau),\ Z\big) & \\ & \leq\ B \cdot (t-\tau) & +\ \varepsilon\ e^{A\,\tau}\ \tau & \leq\ R_\varepsilon \end{array}$$

for every $t \in [\tau,\ \tau+\delta[$ with $\delta := \min\big\{h_1,\ \varepsilon\, e^A\, \frac{1-\tau}{1+B}\big\}$, i.e. conditions (b')(2.) and (c) hold in the interval $[\tau, \tau+\delta]$. For any index $m \in \mathbb{N}$ with $h_m < \delta$,

$$\begin{array}{lll} \mathrm{dist}\big(K_\varepsilon(\tau+h_m),\ \mathscr{V}\big) & \leq\ d\!l\big(\vartheta_G(h_m, K_\varepsilon(\tau)),\ \vartheta_G(h_m, Z)\big) & +\ \mathrm{dist}\big(\vartheta_G(h_m, Z),\ \mathscr{V}\big) \\ & \leq\ d\!l\big(K_\varepsilon(\tau), Z\big)\ \cdot\ e^{A\,h_m} & +\ \varepsilon \cdot h_m \\ & \leq\ \varepsilon\ e^{A\,\tau}\ \tau \qquad \cdot\ e^{A\,h_m} & +\ \varepsilon \cdot h_m\ \leq\ r_\varepsilon(\tau + h_m), \end{array}$$

i.e. condition (b')(1.) is also satisfied at time $t = \tau + h_m$ with any large $m \in \mathbb{N}$. Finally, $K_\varepsilon(\cdot)\big|_{[0,\ \tau+h_m]}$ and $f_\varepsilon(\cdot)\big|_{[0,\ \tau+h_m[}$ provide the wanted contradiction, i.e. $\tau = 1$. □

Proof (of Lemma 18 on page 404).

For each $n \in \mathbb{N}$, Lemma 16 provides

$$\begin{array}{l} K_n(\cdot) : [0,1] \longrightarrow \mathscr{K}(\mathbb{R}^N), \\ f_n(\cdot) : [0,1[\longrightarrow \mathrm{LIP}_{\overline{\mathrm{co}}}(\mathbb{R}^N, \mathbb{R}^N) \end{array}$$

corresponding to $\varepsilon := \frac{1}{n}\, e^{-A}$. Now according to Proposition A.81 (on page 491), the set-valued map $[0,1[\times\mathbb{R}^N \rightsquigarrow \mathbb{R}^N,\ (t,x) \mapsto f_n(t)(x)$ has a parameterization $[0,1[\times\mathbb{R}^N \times \mathbb{B}_1 \longrightarrow \mathbb{R}^N$ that we interpret as $\widehat{f_n} : [0,1[\longrightarrow \mathrm{BLip}(\mathbb{R}^N \times \mathbb{B}_1, \mathbb{R}^N)$. Obviously, they satisfy the claimed properties (a) – (e).
In particular, these features stay correct whenever we consider subsequences instead and again abbreviate them as $(K_n(\cdot))_{n\in\mathbb{N}}$, $(\widehat{f_n}(\cdot))_{n\in\mathbb{N}}$ respectively.

For property (f) about uniform convergence of $(K_n(\cdot))$ with respect to $d\!l$:

The B-Lipschitz continuity of each $K_n(\cdot)$ has two important consequences, i.e.
1. all curves $K_n(\cdot) : [0,1] \longrightarrow \big(\mathscr{K}(\mathbb{R}^N), d\!l\big)$ $(n \in \mathbb{N})$ are equi-continuous and
2. $\bigcup_{n\in\mathbb{N},\ t\in[0,1]} \big\{K_n(t)\big\}$ is contained in the compact subset $\mathbb{B}_B(K_0)$ of $\big(\mathscr{K}(\mathbb{R}^N),\ d\!l\big)$.

Theorem A.82 of Arzelà-Ascoli (on page 491) provides a subsequence (again denoted by) $(K_n(\cdot))_n$ converging uniformly to a function $K(\cdot) : [0,1] \longrightarrow (\mathscr{K}(\mathbb{R}^N), d\!l)$. In particular, $K(\cdot)$ is also B-Lipschitz continuous with $K(0) = K_0$, i.e. properties (a) – (f) are fulfilled completely.

For property (g) about weak convergence of $f_n(\cdot)|_{\widetilde{K}}$ with a fixed compact $\widetilde{K} \subset \mathbb{R}^N$:

We cannot follow the same steps as for differential inclusions in $\mathbb{R}^N$ any longer. Indeed, the functions $\widehat{f_n}(\cdot)$ of morphological transitions have their values in $\mathrm{BLip}(\mathbb{R}^N \times \mathbb{B}_1,\ \mathbb{R}^N)$, which cannot be regarded as a dual space in an obvious way. Thus, Alaoglu's Theorem (stating that closed balls of dual Banach spaces are weakly* compact) cannot be applied similarly to differential inclusions (§ 5.2.1).

Alternatively, we restrict our considerations to a compact neighborhood $\widetilde{K}$ of $\bigcup_{\substack{n\in\mathbb{N}\\ t\in[0,1]}} K_n(t) \subset \mathbb{R}^N$ and use a sufficient condition on relatively weakly compact sets in $L^1\big([0,1],\, C^0(\widetilde{K}\times\mathbb{B}_1,\,\mathbb{R}^N)\big)$. Here $C^0(\widetilde{K}\times\mathbb{B}_1,\,\mathbb{R}^N)$ (supplied with the supremum norm $\|\cdot\|_\infty$) denotes the Banach space of all continuous functions $\widetilde{K}\times\mathbb{B}_1 \longrightarrow \mathbb{R}^N$. According to Proposition A.85 of Ülger (on page 492), if $W \subset C^0(\widetilde{K}\times\mathbb{B}_1,\,\mathbb{R}^N)$ is weakly compact then the subset

$$\Big\{ h \in L^1\big([0,1],\, C^0(\widetilde{K}\times\mathbb{B}_1,\,\mathbb{R}^N)\big) \;\Big|\; h(t)\in W \text{ for } \mathscr{L}^1\text{-almost every } t\in[0,1] \Big\}$$

is relatively weakly compact in $L^1\big([0,1],\, C^0(\widetilde{K}\times\mathbb{B}_1,\,\mathbb{R}^N)\big)$.
In fact, the set $\big\{\widehat{f}_n(t) \,\big|\, n\in\mathbb{N},\, t\in[0,1]\big\} \subset C^0(\mathbb{R}^N\times\mathbb{B}_1,\,\mathbb{R}^N)$ is uniformly bounded and equi-continuous (due to property (c)). Due to Theorem A.82 of Arzelà-Ascoli, the set of their restrictions to the compact set $\widetilde{K}\times\mathbb{B}_1 \subset \mathbb{R}^N\times\mathbb{R}^N$

$$W := \Big\{ \widehat{f}_n(t)\big|_{\widetilde{K}\times\mathbb{B}_1} \;\Big|\; n\in\mathbb{N},\, t\in[0,1] \Big\} \subset C^0(\widetilde{K}\times\mathbb{B}_1,\,\mathbb{R}^N)$$

is relatively compact with respect to $\|\cdot\|_\infty$. Thus, $\big\{\widehat{f}_n(\cdot)|_{\widetilde{K}\times\mathbb{B}_1} \,\big|\, n\in\mathbb{N}\big\}$ is relatively weakly compact in $L^1\big([0,1],\, C^0(\widetilde{K}\times\mathbb{B}_1,\,\mathbb{R}^N)\big)$ and, we obtain a subsequence (again denoted by) $(\widehat{f}_n(\cdot))_{n\in\mathbb{N}}$ and some $g(\cdot)\in L^1\big([0,1],\, C^0(\widetilde{K}\times\mathbb{B}_1,\,\mathbb{R}^N)\big)$ with

$$\widehat{f}_n(\cdot)|_{\widetilde{K}\times\mathbb{B}_1} \stackrel{n\to\infty}{\longrightarrow} g(\cdot) \qquad \text{weakly in } L^1\big([0,1],\, C^0(\widetilde{K}\times\mathbb{B}_1,\,\mathbb{R}^N)\big).$$

For property (g) about $f_n(\cdot)|_{\widetilde{K}_j}$ with every compact $\widetilde{K}_j \stackrel{\text{Def.}}{=} \mathbb{B}_{j+B}(K_0) \subset \mathbb{R}^N$ $(j\in\mathbb{N})$:

Now this construction of subsequences is applied to

$$\widetilde{K}_j \stackrel{\text{Def.}}{=} \mathbb{B}_{j+B}(K_0) = \big\{x\in\mathbb{R}^N \,\big|\, \operatorname{dist}(x,K_0) \le j+B\big\}$$

for $j = 1,2,3\ldots$ successively.
By means of Cantor's diagonal construction, we obtain a subsequence (again denoted by) $(\widehat{f}_n(\cdot))_{n\in\mathbb{N}}$ and some $g_j(\cdot)\in L^1\big([0,1],\, C^0(\widetilde{K}_j\times\mathbb{B}_1,\,\mathbb{R}^N)\big)$ (for each $j\in\mathbb{N}$) such that for every index $j\in\mathbb{N}$,

$$\widehat{f}_n(\cdot)|_{\widetilde{K}_j\times\mathbb{B}_1} \stackrel{n\to\infty}{\longrightarrow} g_j(\cdot) \qquad \text{weakly in } L^1\big([0,1],\, C^0(\widetilde{K}_j\times\mathbb{B}_1,\,\mathbb{R}^N)\big).$$

As restrictions to $\widetilde{K}_j\times\mathbb{B}_1$ of one and the same subsequence $(\widehat{f}_n(\cdot))_{n\in\mathbb{N}}$ converge weakly for each $j\in\mathbb{N}$, the inclusion $\widetilde{K}_j\subset\widetilde{K}_{j+1}$ implies for any indices $j<k$

$$g_j(t)(\cdot) = g_k(t)(\cdot)|_{\widetilde{K}_j\times\mathbb{B}_1} \in C^0(\widetilde{K}_j\times\mathbb{B}_1,\,\mathbb{R}^N) \quad \text{for } \mathscr{L}^1 \text{ a.e. } t\in[0,1].$$

Hence, $(g_j(\cdot))_{j\in\mathbb{N}}$ induces a single function $\widehat{f}: [0,1[\longrightarrow C^0(\mathbb{R}^N\times\mathbb{B}_1,\mathbb{R}^N)$ defined as

$$\widehat{f}(t)(x,u) := g_j(t)(x,u) \text{ for } x\in\widetilde{K}_j,\ u\in\mathbb{B}_1 \text{ and } \mathscr{L}^1\text{-a.e. } t\in[0,1[.$$

For property (h) about Lipschitz continuity and bounds of limit function $f(\cdot)$:

Finally, we verify $\widehat{f}(t) \in \mathrm{BLip}(\mathbb{R}^N \times \mathbb{B}_1, \mathbb{R}^N)$, $\|\widehat{f}(t,\cdot,\cdot)\|_\infty + \mathrm{Lip}\, \widehat{f}(t,\cdot,\cdot) \le c$ for almost every $t \in [0,1[$. Indeed, as in the case of differential inclusions (§ 5.2.1), Mazur's Lemma (e.g. [188, Theorem V.1.2]) ensures for each fixed index $j \in \mathbb{N}$

$$\widehat{f}(\cdot)|_{\widetilde{K}_j \times \mathbb{B}_1} \in \bigcap_{n \in \mathbb{N}} \overline{co} \left\{ \widehat{f}_n(\cdot)|_{\widetilde{K}_j \times \mathbb{B}_1}, f_{n+1}(\cdot)|_{\widetilde{K}_j \times \mathbb{B}_1} \dots \right\} \text{ in } L^1\big([0,1], C^0(\widetilde{K}_j \times \mathbb{B}_1, \mathbb{R}^N)\big).$$

Thus, $\widehat{f}(\cdot)|_{\widetilde{K}_j \times \mathbb{B}_1}$ can be approximated by convex combinations of $\{\widehat{f}_1(\cdot)|_{\widetilde{K}_j \times \mathbb{B}_1}, \widehat{f}_2(\cdot)|_{\widetilde{K}_j \times \mathbb{B}_1} \dots\}$ with respect to the L^1 norm. A further subsequence (of these convex combinations) converges to $\widehat{f}(\cdot)|_{\widetilde{K}_j \times \mathbb{B}_1}$ $\mathscr{L}^1$-almost everywhere in $[0,1]$. For $\mathscr{L}^1$-almost every $t \in [0,1]$, $\widehat{f}(t)|_{\widetilde{K}_j \times \mathbb{B}_1}$ belongs to the same compact convex subset of $\big(C^0(\widetilde{K}_j \times \mathbb{B}_1, \mathbb{R}^N), \|\cdot\|_\infty\big)$ as $\widehat{f}_1(t)|_{\widetilde{K}_j \times \mathbb{B}_1}$, $\widehat{f}_2(t)|_{\widetilde{K}_j \times \mathbb{B}_1} \dots$, namely $\big\{w \in \mathrm{BLip}(\widetilde{K}_j \times \mathbb{B}_1, \mathbb{R}^N) \,\big|\, \|w\|_\infty + \mathrm{Lip}\, w \le c\big\}$. As the index $j \in \mathbb{N}$ is fixed arbitrarily, we obtain property (h).

Property (i), i.e. $K(t) \in \mathscr{V}$ for every $t \in [0,1]$, results directly from statements (d), (f) and the assumption that $\mathscr{V}$ is closed in $\big(\mathscr{K}(\mathbb{R}^N), d\!l\big)$. This completes the proof of Lemma 18. □

The last step is to verify at Lebesgue-almost every time $t \in [0,1[$ that $\widehat{f}(t)(\cdot,\mathbb{B}_1) : \mathbb{R}^N \rightsquigarrow \mathbb{R}^N$ belongs to both $\mathscr{F}(K(t))$ and the morphological mutation $\overset{\circ}{K}(t)$.

First we interpret the weak convergence of $\widehat{f}_n(\cdot)|_{\widetilde{K}_j \times \mathbb{B}_1} \longrightarrow \widehat{f}(\cdot)|_{\widetilde{K}_j \times \mathbb{B}_1}$ (in L^1) with respect to the corresponding set-valued maps $[0,1[\times \widetilde{K}_j \rightsquigarrow \mathbb{R}^N$ and meet the topology of locally uniform convergence in $\mathrm{LIP}(\mathbb{R}^N, \mathbb{R}^N)$.
As a rather technical tool, Lemma 19 (on page 404) clarifies how the uniform Lipschitz bounds of $\mathscr{F}(\mathscr{K}(\mathbb{R}^N)) \subset \mathrm{LIP}_{\overline{co}}(\mathbb{R}^N, \mathbb{R}^N)$ (due to assumption (2.)) imply useful compactness features which ensure that the limit map $\widehat{f}(t)(\cdot,\mathbb{B}_1) : \mathbb{R}^N \rightsquigarrow \mathbb{R}^N$ is related to $\mathscr{F}(K(t))$ at $\mathscr{L}^1$-almost every time t.

Proof (of Lemma 19 on page 404).

Applying Parameterization Theorem A.81 (on page 491) to the autonomous maps $G_k : \mathbb{R}^N \rightsquigarrow \mathbb{R}^N$ provides a sequence $(g_k)_{k \in \mathbb{N}}$ of Lipschitz functions $\mathbb{R}^N \times \mathbb{B}_1 \longrightarrow \mathbb{R}^N$ with $g_k(\cdot,\mathbb{B}_1) = G_k$ for each $k \in \mathbb{N}$ and $\sup_k (\|g_k\|_\infty + \mathrm{Lip}\, g_k) \le \mathrm{const}(A,B) < \infty$.

For any nonempty compact set $K \subset \mathbb{R}^N$, Theorem A.82 of Arzelà-Ascoli guarantees a subsequence $(g_{k_j})_{j \in \mathbb{N}}$ converging uniformly in $K \times \mathbb{B}_1$. In combination with Cantor's diagonal construction, we obtain even a subsequence (again denoted by) $(g_{k_j})_{j \in \mathbb{N}}$ converging uniformly in each of the countably many compact sets $\mathbb{B}_m(0) \times \mathbb{B}_1 \subset \mathbb{R}^N \times \mathbb{R}^N$ $(m \in \mathbb{N})$.

Let $h_m : \mathbb{R}^N \times \mathbb{B}_1 \longrightarrow \mathbb{R}^N$ denote an arbitrary Lipschitz function with

$$\sup_{\mathbb{B}_m(0)\times\mathbb{B}_1} |g_{k_j}(\cdot) - h_m(\cdot)| \stackrel{j\to\infty}{\longrightarrow} 0.$$

Then we obtain the unique function $h : \mathbb{R}^N \times \mathbb{B}_1 \longrightarrow \mathbb{R}^N$ by setting $h(x,\cdot) := h_m(x,\cdot)$ for all $x \in \mathbb{B}_m(0)$, $m \in \mathbb{N}$ and, $g_{k_j} \longrightarrow h \ (j \to \infty)$ locally uniformly in $\mathbb{R}^N \times \mathbb{B}_1$.

In particular, $h(\cdot)$ is also Lipschitz continuous and has the same global Lipschitz bounds as $(g_k)_{k\in\mathbb{N}}$. Hence, $G := h(\cdot, \mathbb{B}_1) : \mathbb{R}^N \rightsquigarrow \mathbb{R}^N$ provides a set-valued map that is Lipschitz continuous and satisfies

$$\sup_{x\in M} dl(G_{k_j}(x), G(x)) \le \sup_{x\in M} \sup_{u\in\mathbb{B}_1} |g_{k_j}(x,u) - h(x,u)| \longrightarrow 0 \qquad (j \to \infty)$$

for any $M \in \mathscr{K}(\mathbb{R}^N)$. This convergence of $(G_{k_j})_{j\in\mathbb{N}}$ implies directly $\operatorname{Lip} G \le A$, $\|G\|_\infty \le B$ and the convexity of all values of G. Now the first claim is proved.

For verifying the second claim, we extract a convergent subsequence $(K_{k_l})_{l\in\mathbb{N}}$ as all sets $K_k, k \in \mathbb{N}$, are contained in one and the same compact subset of $\mathbb{R}^N$. Hence, there is $K \in \mathscr{K}(\mathbb{R}^N)$ with $dl(K_{k_l}, K) \stackrel{l\to\infty}{\longrightarrow} 0$. The same arguments as in the first part lead to subsequences (again denoted by) $(K_{k_j})_{j\in\mathbb{N}}$, $(G_{k_j})_{j\in\mathbb{N}}$ such that in addition, the latter converges to a map $G \in \mathrm{LIP}_{\overline{\mathrm{co}}}(\mathbb{R}^N, \mathbb{R}^N)$ locally uniformly. According to assumption (3.) of Viability Theorem 12, Graph $\mathscr{F} \subset \mathscr{K}(\mathbb{R}^N) \times \mathrm{LIP}_{\overline{\mathrm{co}}}(\mathbb{R}^N, \mathbb{R}^N)$ is closed with respect to these topologies and thus, it contains (K, G).

□

Proof (of Lemma 20 on page 405).

Lemma 18 (g) specifies the convergence resulting directly from construction

$$\widehat{f}_n(\cdot)|_{\widetilde{K}_j\times\mathbb{B}_1} \stackrel{n\to\infty}{\longrightarrow} \widehat{f}(\cdot)|_{\widetilde{K}_j\times\mathbb{B}_1} \qquad \text{weakly in } L^1\big([0,1], C^0(\widetilde{K}_j \times \mathbb{B}_1, \mathbb{R}^N)\big)$$

for each $j\in\mathbb{N}$ with the abbreviation $\widetilde{K}_j := \mathbb{B}_{j+B}(K_0) \stackrel{\text{Def.}}{=} \big\{x\in\mathbb{R}^N \,\big|\, \operatorname{dist}(x, K_0) \le j+B\big\}$.

Fixing the index $j \in \mathbb{N}$ of compact sets arbitrarily, Mazur's Lemma provides a sequence $\big(h_{j,n}(\cdot)\big)_{n\in\mathbb{N}}$ with

$$\begin{aligned} h_{j,n}(\cdot) &\in co\big\{\widehat{f}_n(\cdot)|_{\widetilde{K}_j\times\mathbb{B}_1}, \widehat{f}_{n+1}(\cdot)|_{\widetilde{K}_j\times\mathbb{B}_1} \cdots\big\} \subset L^1\big([0,1], C^0(\widetilde{K}_j\times\mathbb{B}_1, \mathbb{R}^N)\big), \\ h_{j,n}(\cdot) &\longrightarrow \widehat{f}(\cdot)|_{\widetilde{K}_j\times\mathbb{B}_1} \quad (n\to\infty) \quad \text{strongly in } L^1\big([0,1], C^0(\widetilde{K}_j \times \mathbb{B}_1, \mathbb{R}^N)\big). \end{aligned}$$

For a subsequence $\big(h_{j,n_k}(\cdot)\big)_{k\in\mathbb{N}}$, we even obtain convergence for $\mathscr{L}^1$-a.e. $t \in [0,1]$,

$$h_{j,n_k}(t) \longrightarrow \widehat{f}(t)|_{\widetilde{K}_j\times\mathbb{B}_1} \quad (k\to\infty) \qquad \text{in } \big(C^0(\widetilde{K}_j\times\mathbb{B}_1, \mathbb{R}^N), \|\cdot\|_\infty\big),$$

i.e. uniformly in $\widetilde{K}_j \times \mathbb{B}_1 \subset \mathbb{R}^N \times \mathbb{R}^N$. Now the first claim is proved.
In particular, all values of $\widehat{f}(t)(\cdot, \mathbb{B}_1) : \mathbb{R}^N \rightsquigarrow \mathbb{R}^N$ are convex since each map $\widehat{f}_n(t)(\cdot, \mathbb{B}_1) \in \operatorname{im} \mathscr{F} \subset \mathrm{LIP}_{\overline{\mathrm{co}}}(\mathbb{R}^N, \mathbb{R}^N)$ has convex values.

Furthermore, we obtain the following inclusions for $\mathscr{L}^1$-almost every $t \in [0,1]$ (and each index $j \in \mathbb{N}$) in a pointwise way

$$\begin{aligned}
\widehat{f}(t)(\cdot,\mathbb{B}_1)\big|_{\widetilde{K}_j} &\in \bigcap_{n\in\mathbb{N}} \overline{h_{j,n}(t)(\cdot,\mathbb{B}_1)\big|_{\widetilde{K}_j} \cup h_{j,n+1}(t)(\cdot,\mathbb{B}_1)\big|_{\widetilde{K}_j} \cup \ldots} \\
&\subset \bigcap_{n\in\mathbb{N}} \overline{co} \bigcup_{m\geq n} \widehat{f}_m(t)(\cdot,\mathbb{B}_1)\big|_{\widetilde{K}_j} \\
&\subset \bigcap_{n\in\mathbb{N}} \overline{co} \bigcup_{m\geq n} \mathscr{F}\big(\mathbb{B}_{1/m}(K_m(t))\big)\big|_{\widetilde{K}_j} \\
&\subset \bigcap_{\varepsilon>0} \overline{co}\ \mathscr{F}\big(\mathbb{B}_\varepsilon(K(t))\big)\big|_{\widetilde{K}_j}
\end{aligned}$$

due to Lemma 18 (e) and $dl(K_m(t),K(t)) \longrightarrow 0$ for $m \to \infty$ respectively. Here, to be more precise, the closed convex hull (in the last line) denotes the following set-valued map

$$\widetilde{K}_j \rightsquigarrow \mathbb{R}^N, \qquad x \mapsto \overline{co} \bigcup_{\substack{M\in\mathscr{K}(\mathbb{R}^N)\\ dl(K(t),M)\leq\varepsilon}} \ \bigcup_{G\in\mathscr{F}(M)} G(x).$$

Fixing now $j \in \mathbb{N}$ and $\delta > 0$ arbitrarily, we introduce the abbreviation

$$\begin{aligned}
\mathscr{B}_\delta\Big(\mathscr{F}(K(t));\,\widetilde{K}_j\Big) := \Big\{G \in \mathrm{LIP}_{\overline{co}}(\mathbb{R}^N,\mathbb{R}^N)\ \Big|\ & \\
\delta \ \geq\ \mathrm{dist}\Big(G(\cdot)\big|_{\widetilde{K}_j},\ \mathscr{F}(K(t))\big|_{\widetilde{K}_j}\Big) & \\
\stackrel{\text{Def.}}{=} \inf_{Z\in\mathscr{F}(K(t))}\ \sup_{x\in\widetilde{K}_j}\ dl(G(x),\,Z(x))\Big\} &
\end{aligned}$$

for the "ball" around the set $\mathscr{F}(K(t))$ containing all maps $G \in \mathrm{LIP}_{\overline{co}}(\mathbb{R}^N,\mathbb{R}^N)$ whose restriction to $\widetilde{K}_j$ has the "uniform distance" $\leq \delta$ from $\mathscr{F}(K(t))$.

For any $\delta > 0$ and each $j \in \mathbb{N}$, there exists a radius $\rho > 0$ with

$$\mathscr{F}\big(\mathbb{B}_\rho(K(t))\big) \subset \mathscr{B}_\delta\big(\mathscr{F}(K(t));\,\widetilde{K}_j\big)$$

because otherwise there would be two sequences $(M_k)_{k\in\mathbb{N}}$, $(G_k)_{k\in\mathbb{N}}$ in $\mathscr{K}(\mathbb{R}^N)$ and $\mathrm{LIP}_{\overline{co}}(\mathbb{R}^N,\mathbb{R}^N)$ with $dl(M_k,\,K(t)) \leq \frac{1}{k}$, $G_k \in \mathscr{F}(M_k) \setminus \mathscr{B}_\delta\big(\mathscr{F}(K(t));\,\widetilde{K}_j\big)$ for each $k \in \mathbb{N}$ and, Lemma 19 would lead to a contradiction (similarly to [19, Proposition 1.4.8] about closed graph and upper semicontinuity of set-valued maps between metric spaces).

Obviously, $\mathscr{B}_\delta\big(\mathscr{F}(K(t));\,\widetilde{K}_j\big) \subset \mathrm{LIP}_{\overline{co}}(\mathbb{R}^N,\mathbb{R}^N)$ is closed with respect to locally uniform convergence. Moreover, it is convex with regard to pointwise convex combinations because $\mathscr{F}(K(t))$ is supposed to be convex.

Thus, we even obtain the inclusion $\overline{co}\,\mathscr{F}(\mathbb{B}_\rho(K(t))) \subset \mathscr{B}_\delta\big(\mathscr{F}(K(t));\,\widetilde{K}_j\big)$, i.e.

$$\widehat{f}(t)(\cdot,\mathbb{B}_1)\big|_{\widetilde{K}_j} \in \bigcap_{\delta>0} \mathscr{B}_\delta\big(\mathscr{F}(K(t));\,\widetilde{K}_j\big) \qquad \text{for } \mathscr{L}^1\text{-a.e. } t \text{ and each } j\in\mathbb{N}.$$

In particular, there exists some $Z_j \in \mathscr{F}(K(t))$ satisfying

$$\sup_{x\in\widetilde{K}_j} dl\,(f(t)(x,\mathbb{B}_1),\,Z_j(x)) \leq \tfrac{1}{j}$$

and, the compactness property of Lemma 19 implies for $\mathscr{L}^1$-almost every time t

$$\widehat{f}(t)(\cdot,\mathbb{B}_1) \in \mathscr{F}(K(t)). \qquad \square$$

Proof (of Lemma 21 on page 405).

According to the definition of Aumann integral (e.g. [19, § 8.6]),

$$\int_J \widehat{f}(s)(x(s),\mathbb{B}_1)\ ds \stackrel{\text{Def.}}{=} \Big\{ \int_J \widehat{f}(s)(x(s),u(s))\ ds \ \Big|\ u(\cdot) \in L^1(J,\mathbb{B}_1) \Big\}.$$

Fixing $u(\cdot) \in L^1(J,\mathbb{B}_1)$ and $x(\cdot) \in C^0([0,1],\mathbb{R}^N)$ arbitrarily, we conclude from Lemma 18 (g)

$$\int_J \widehat{f_n}(s)(x(s),u(s))\ ds \longrightarrow \int_J \widehat{f}(s)(x(s),u(s))\ ds \qquad \text{for } n \to \infty$$

since $L^1\big([0,1], C^0(\widetilde{K}_j \times \mathbb{B}_1, \mathbb{R}^N)\big) \longrightarrow \mathbb{R},\ h \longmapsto \int_J h(s)(x(s),u(s))\ ds$ is continuous and linear whenever $x([0,1]) \subset \widetilde{K}_j$. This implies

both $$\mathrm{dist}\Big(\int_J \widehat{f_n}(s)(x(s),\mathbb{B}_1)\ ds,\ \int_J \widehat{f}(s)(x(s),\mathbb{B}_1)\ ds\Big) \longrightarrow 0$$

and $$\mathrm{dist}\Big(\int_J \widehat{f}(s)(x(s),\mathbb{B}_1)\ ds,\ \int_J \widehat{f_n}(s)(x(s),\mathbb{B}_1)\ ds\Big) \longrightarrow 0.$$

Hence, the first claim holds.

Due to Lemma 18 (c), each $\widehat{f_n}(\cdot)(x,\mathbb{B}_1) : [0,1[\rightsquigarrow \mathbb{R}^N$ $(n \in \mathbb{N}, x \in \mathbb{R}^N)$ is piecewise constant and thus, it has at most countably many points of discontinuity. We conclude from Lemma 18 (e) and Proposition 1.57 (about the equivalence between morphological primitives and reachable sets on page 64)

$$\vartheta_{\widehat{f_n}(\cdot)(\cdot,\mathbb{B}_1)}(t,K_0) = K_n(t) \qquad \text{for every } t \in]0,1] \text{ and } n \in \mathbb{N}.$$

$d\!l(K_n(t),\ K(t)) \longrightarrow 0$ has already been mentioned in Lemma 18 (f). Now we still have to verify

$$d\!l\Big(\vartheta_{\widehat{f_n}(\cdot)(\cdot,\mathbb{B}_1)}(t,K_0),\ \ \vartheta_{\widehat{f}(\cdot)(\cdot,\mathbb{B}_1)}(t,K_0)\Big) \longrightarrow 0 \qquad \text{for every } t \in]0,1] \text{ and } n \to \infty.$$

If $K_0 \subset \mathbb{R}^N$ consists of only one point, then this convergence results directly from Proposition 22 of Stassinopoulos and Vinter (on page 405).

For extending it to arbitrary initial sets $K_0 \in \mathcal{K}(\mathbb{R}^N)$, we exploit two features: first, the reachable set of a union is always the union of the corresponding reachable sets and second, the Lipschitz dependence (of reachable sets) on the initial sets in the sense of Proposition 1.50 (on page 60), i.e., for any $M_1, M_2 \in \mathcal{K}(\mathbb{R}^N)$ and $t \in [0,1]$

$$\begin{cases} d\!l\big(\vartheta_{\widehat{f_n}(\cdot)(\cdot,\mathbb{B}_1)}(t,M_1),\ \vartheta_{\widehat{f_n}(\cdot)(\cdot,\mathbb{B}_1)}(t,M_2)\big) \le e^A\ d\!l(M_1,M_2) \\ d\!l\big(\vartheta_{\widehat{f}(\cdot)(\cdot,\mathbb{B}_1)}(t,M_1),\ \vartheta_{\widehat{f}(\cdot)(\cdot,\mathbb{B}_1)}(t,M_2)\big) \le e^A\ d\!l(M_1,M_2). \end{cases}$$

This second general property for *nonautonomous* differential inclusions is covered by Filippov's Theorem A.6 (on page 443 f.) correspondingly to Proposition 1.50.

□

5.3 Morphological Control Problems for Compact Sets in $\mathbb{R}^N$ with State Constraints

Similarly to classical control theory in $\mathbb{R}^N$, a metric space (U, d_U) of control parameter and a single-valued function $f : \mathscr{K}(\mathbb{R}^N) \times U \longrightarrow \mathrm{LIP}(\mathbb{R}^N, \mathbb{R}^N)$ of state and control are given. For each initial set $K(0) \in \mathscr{K}(\mathbb{R}^N)$, we are looking for a Lipschitz continuous curve $K(\cdot) : [0,T] \longrightarrow \big(\mathscr{K}(\mathbb{R}^N), d\!l\big)$ solving the following nonautonomous morphological equation

$$\overset{\circ}{K}(t) \ni f(K(t), u(t)) \qquad \text{in } [0,T[$$

with a measurable control function $u(\cdot) : [0,T] \longrightarrow U$, i.e. by definition

$$\lim_{h \downarrow 0} \ \tfrac{1}{h} \cdot d\!l\left(\vartheta_{f(K(t),u(t))}(h,\ K(t)),\ K(t+h)\right) \ = \ 0 \qquad \text{for } \mathscr{L}^1\text{-a.e. } t \in [0,T].$$

This is an open-loop control problem in the metric space $(\mathscr{K}(\mathbb{R}^N), d\!l)$.

The existence of solutions is closely related to the corresponding morphological inclusion for which we take all admitted controls into consideration simultaneously. We introduce the set-valued map

$$\mathscr{F}_U : \mathscr{K}(\mathbb{R}^N) \rightsquigarrow \mathrm{LIP}(\mathbb{R}^N, \mathbb{R}^N), \quad K \mapsto \{f(K,u) \,|\, u \in U\} \subset \mathrm{LIP}(\mathbb{R}^N, \mathbb{R}^N)$$

and consider the morphological inclusion $\overset{\circ}{K}(\cdot) \cap \mathscr{F}_U(K(\cdot)) \neq \emptyset$ in $[0,T[$.
In § 5.3.2, Proposition 25 (on page 416) specifies sufficient conditions on U and f such that solutions to this morphological inclusion solve the morphological control problem and vice versa.
The step from inclusion to control problem requires the existence of a measurable control function and, it is concluded here from a well-known selection principle of Filippov whose Euclidean special case is usually applied to differential inclusions in $\mathbb{R}^N$ and classical control theory.

All available results about morphological inclusions can then be used for morphological control problems. In the following, Viability Theorem 12 (on page 401) plays a key role. It concerns a morphological inclusion $\overset{\circ}{K}(\cdot) \cap \mathscr{F}(K(\cdot)) \neq \emptyset$ with state constraints $K(t) \in \mathscr{V} \subset \mathscr{K}(\mathbb{R}^N)$ at every time t.
This viability theorem specifies sufficient conditions on $\mathscr{F}$ and the nonempty set of constraints $\mathscr{V} \subset \mathscr{K}(\mathbb{R}^N)$ such that at least one solution $K(\cdot) : [0,1] \longrightarrow \mathscr{V} \subset \mathscr{K}(\mathbb{R}^N)$ starts at each initial set $K(0) \in \mathscr{V}$. In § 5.3.3 (on page 418 ff.), the close relationship between morphological inclusions and control problems provides directly sufficient conditions on a morphological control system with state constraints for the existence of solutions (Proposition 28).
In § 5.3.4, essentially the same approach is then used for solving *relaxed* control problems in the morphological framework. They are based on replacing the metric space U of control parameters by the set of Borel probability measures on U (supplied with the linear Wasserstein metric). As immediate analytical benefit, we can weaken some conditions of convexity in Proposition 35 (on page 422).

The Step to Closed-Loop Control Problems for Compact Sets in $\mathbb{R}^N$

Consider morphological control problems with state constraints

$$\begin{cases} \overset{\circ}{K}(\cdot) \ni f(K(\cdot),u), \quad u \in U & \text{a.e. in } [0,T[\\ K(t) \in \mathscr{V} & \text{for every } t \in [0,T[. \end{cases}$$

The metric space (U, d_U) of control, the function $f : \mathscr{K} \times U \longrightarrow \mathrm{LIP}(\mathbb{R}^N, \mathbb{R}^N)$ and the closed set of constraints $\mathscr{V} \subset \mathscr{K}(\mathbb{R}^N)$ are given. The morphological viability condition mentioned before indicates where candidates for a closed-loop control $u : \mathscr{V} \longrightarrow U$ can be found, namely among those controls $u \in U$ whose reachable sets $\vartheta_{f(K,u)}(\cdot, K)$ are "contingent" to $\mathscr{V}$. This reflects the notion of *regulation maps* defined by Aubin for control problems in finite-dimensional vector spaces [14, § 6].

In § 5.3.7 (on page 436 ff.), we specify sufficient conditions on $U, f, \mathscr{V}$ such that Michael's famous selection theorem implies the existence of a continuous closed-loop control (Proposition 52 on page 436). Michael's selection theorem (quoted here in Proposition 53), however, focuses on lower semicontinuous set-valued maps. Now we need information about the semicontinuity properties of these regulation maps.

In this regard, the classical results about finite-dimensional vector spaces serve as motivation again. The Clarke tangent cone $T_V^C(x) \subset \mathbb{R}^N$, $x \in V$, to a nonempty closed set $V \subset \mathbb{R}^N$ (alias circatangent set, see Definition 37) is known to have closed graph whereas the Bouligand contingent cone to the same set does not have such a semicontinuity feature in general [19, 162]. Furthermore, Rockafellar characterized the interior of the convex Clarke tangent cone $T_V^C(x) \subset \mathbb{R}^N$ by a topological criterion leading to the so-called hypertangent cone ([161, Theorem 2], [46, § 2,4] and quoted here in § 5.3.6). The set-valued map of hypertangent cones to a fixed set $V \subset \mathbb{R}^N$ is lower semicontinuous whenever all these cones are nonempty.

These two concepts, i.e. Clarke tangent cone and hypertangent cone to a given closed set in $\mathbb{R}^N$, are extended to the morphological framework where the metric space $(\mathscr{K}(\mathbb{R}^N), dl)$ has replaced the Euclidean space.
In § 5.3.5, we apply Aubin's definition of "circatangent transition set" [10, Definition 1.5.4] to $(\mathscr{K}(\mathbb{R}^N), dl)$ together with reachable sets of differential inclusions. The result proves to be a nonempty closed cone in $\mathrm{LIP}(\mathbb{R}^N, \mathbb{R}^N)$ and, its intersection with $\mathrm{BLip}(\mathbb{R}^N, \mathbb{R}^N)$ is convex.
In § 5.3.6, the so-called hypertangent transition set is introduced for a nonempty closed subset $\mathscr{V} \subset \mathscr{K}(\mathbb{R}^N)$. Its graph is identical to the interior of the graph of circatangent transition sets in $\mathscr{V} \times \mathrm{LIP}(\mathbb{R}^N, \mathbb{R}^N)$.
In particular, this topological characterization proves to be helpful for constructing closed-loop controls on the basis of Michael's selection principle in subsequent Proposition 52 (on page 436).

5.3.1 Formulation

Now a control parameter is to come into play. Indeed, the so-called control problems

$$\begin{cases} \frac{d}{dt}\, x(t) &= f(x(t),u) \\ u &\in U \end{cases} \tag{5.1}$$

have been studied thoroughly in both finite-dimensional and infinite-dimensional vector spaces. Our contribution now is to formulate the corresponding problem in the metric space $(\mathcal{K}(\mathbb{R}^N), d\!\!l)$ using the morphological framework for derivatives.

Definition 23.
Let (U, d_U) denote a metric space and $f : \mathcal{K}(\mathbb{R}^N) \times U \longrightarrow \mathrm{LIP}(\mathbb{R}^N, \mathbb{R}^N)$ be given. A tube $K : [0,T] \leadsto \mathbb{R}^N$ is called a *solution* to the *morphological control problem*

$$\begin{cases} \overset{\circ}{K}(\cdot) &\ni f(K(\cdot),u) \qquad \text{a.e. in } [0,T] \\ u &\in U \end{cases} \tag{5.2}$$

if there exists a measurable function $u(\cdot) : [0,T[\longrightarrow U$ such that $K(\cdot)$ solves the nonautonomous morphological equation $\overset{\circ}{K}(\cdot) \ni f(K(\cdot), u(\cdot))$, i.e. satisfying

1. $K(\cdot) : [0,T] \leadsto \mathbb{R}^N$ is continuous with respect to $d\!\!l$ and
2. for $\mathcal{L}^1$-almost every $t \in [0,T[$, $f(K(t),u(t)) \in \mathrm{LIP}(\mathbb{R}^N, \mathbb{R}^N)$ belongs to $\overset{\circ}{K}(t)$ or, equivalently, $\quad \lim_{h \downarrow 0} \frac{1}{h} \cdot d\!\!l\left(\vartheta_{f(K(t),u(t))}(h, K(t)),\; K(t+h)\right) = 0.$

Proposition 24 (Solutions as reachable sets).
Assume the metric space (U, d_U) to be complete and separable and, consider $\mathrm{LIP}_{\overline{\mathrm{co}}}(\mathbb{R}^N, \mathbb{R}^N)$ with the topology of locally uniform convergence. Suppose $f : \mathcal{K}(\mathbb{R}^N) \times U \longrightarrow \mathrm{LIP}_{\overline{\mathrm{co}}}(\mathbb{R}^N, \mathbb{R}^N)$ to be continuous with

$$\sup_{\substack{M \in \mathcal{K}(\mathbb{R}^N) \\ u \in U}} \left(\|f(M,u)\|_\infty + \mathrm{Lip}\, f(M,u) \right) < \infty.$$

Let $K : [0,T] \leadsto \mathbb{R}^N$ be any compact-valued solution to the morphological control problem (5.2).

Then there is a measurable function $u(\cdot) : [0,T] \longrightarrow U$ such that at every time $t \in [0,T]$, the compact set $K(t) \subset \mathbb{R}^N$ coincides with the reachable set $\vartheta_{f(K(\cdot),u(\cdot))}(t, K(0)) \subset \mathbb{R}^N$ of the nonautonomous differential inclusion

$$\tfrac{d}{d\tau}\, x(\tau) \in f(K(\tau), u(\tau))\,\big(x(\tau)\big) \subset \mathbb{R}^N \quad \mathcal{L}^1\text{-a.e.}$$

Proof. It results from Proposition 1.57 (on page 64) stating the equivalence between morphological primitives and reachable sets because the composition

$$f(K(\cdot), u(\cdot)) : [0,T] \longrightarrow \mathrm{LIP}_{\overline{\mathrm{co}}}(\mathbb{R}^N, \mathbb{R}^N)$$

is Lebesgue measurable. □

5.3.2 The Link to Morphological Inclusions

In vector spaces, the close relationship between control problem (5.1) and the corresponding differential inclusion

$$\tfrac{d}{dt}\, x(t) \;\in\; \bigcup_{u \in U} f(x(t), u) \qquad \mathscr{L}^1\text{-a.e.}$$

had been realized soon. A measurable selection provides the same link now for morphological inclusions. In a word, the classical techniques using appropriate measurable selections (which had been developed for differential inclusions in the Euclidean space) can also be used in the morphological framework because the transitions are in a complete separable metric space, namely $\mathrm{LIP}(\mathbb{R}^N, \mathbb{R}^N)$.
A main result of this section is the following equivalence:

Proposition 25. *Assume the metric space (U, d_U) to be complete and separable. Consider the set $\mathrm{LIP}(\mathbb{R}^N, \mathbb{R}^N)$ with the topology of locally uniform convergence. Let $f : \mathscr{K}(\mathbb{R}^N) \times U \longrightarrow \mathrm{LIP}(\mathbb{R}^N, \mathbb{R}^N)$ be continuous with*

$$\sup_{\substack{M \in \mathscr{K}(\mathbb{R}^N)\\ u \in U}} \big(\|f(M,u)\|_\infty + \mathrm{Lip}\, f(M,u)\big) \;<\; \infty.$$

Set $\mathscr{F}_U : \mathscr{K}(\mathbb{R}^N) \leadsto \mathrm{LIP}(\mathbb{R}^N, \mathbb{R}^N),\;\; K \mapsto \{f(K,u) \,|\, u \in U\} \subset \mathrm{LIP}(\mathbb{R}^N, \mathbb{R}^N)$.

A tube $K(\cdot) : [0,T] \leadsto \mathbb{R}^N$ is a solution to the morphological control problem

$$\begin{cases} \overset{\circ}{K}(\cdot) \;\ni\; f(K(\cdot), u) & \text{a.e. in } [0,T] \\ \quad u \;\in\; U \end{cases}$$

if and only if $K(\cdot)$ is a solution to the morphological inclusion $\overset{\circ}{K}(\cdot) \cap \mathscr{F}_U(K(\cdot)) \neq \emptyset$ (in the sense of Definition 1 on page 386).

Obviously, every morphological control problem leads to a morphological inclusion. For proving Proposition 25, we require the inverse connection (i.e. from inclusion to control problem). In the literature about differential inclusions in vector spaces, it is usually based on a selection result that is said to go back to Filippov.

Lemma 26 (Filippov [19, Theorem 8.2.10]**).**
Consider a complete σ-finite measure space (Ω, A, μ), complete separable metric spaces X, Y and a measurable set-valued map $H : \Omega \leadsto X$ with closed nonempty images. Let $g : X \times \Omega \longrightarrow Y$ be a Carathéodory function (i.e. continuous in the first argument and measurable in the second one).
Then for every measurable function $k : \Omega \longrightarrow Y$ satisfying

$$k(\omega) \,\in\, g(H(\omega), \omega) \qquad \text{for } \mu\text{-almost all } \omega \in \Omega,$$

there exists a measurable selection $h(\cdot) : \Omega \longrightarrow X$ of $H(\cdot)$ such that

$$k(\omega) \,=\, g(h(\omega), \omega) \qquad \text{for } \mu\text{-almost all } \omega \in \Omega.$$

For applying Lemma 26 to morphological inclusions, we focus on two aspects: First, $\mathrm{LIP}(\mathbb{R}^N, \mathbb{R}^N)$ is regarded as a separable metric space. Indeed, we supply $\mathrm{LIP}(\mathbb{R}^N, \mathbb{R}^N)$ with the topology of locally uniform convergence as in § 5.2. Similarly to the beginning of § 5.1 (on page 386), this topology can be metrized by

$$\begin{aligned} d_{\mathrm{LIP}} : \mathrm{LIP}(\mathbb{R}^N, \mathbb{R}^N) \times \mathrm{LIP}(\mathbb{R}^N, \mathbb{R}^N) &\longrightarrow [0,1], \\ (G,H) &\longmapsto \sum_{j=1}^{\infty} 2^{-j} \frac{dl_\infty\big(G(\cdot)|_{\mathbb{B}_j(0)},\, H(\cdot)|_{\mathbb{B}_j(0)}\big)}{1 + dl_\infty\big(G(\cdot)|_{\mathbb{B}_j(0)},\, H(\cdot)|_{\mathbb{B}_j(0)}\big)} \end{aligned}$$

with the abbreviation $dl_\infty\big(G(\cdot)|_{\mathbb{B}_j(0)},\, H(\cdot)|_{\mathbb{B}_j(0)}\big) \stackrel{\text{Def.}}{=} \sup_{\substack{x \in \mathbb{R}^N, \\ |x| \le j}} dl(G(x), H(x)) < \infty.$

Moreover, $\mathrm{LIP}(\mathbb{R}^N, \mathbb{R}^N)$ is separable with respect to d_{LIP} due to the (global) Lipschitz continuity of each of its set-valued maps and because both domains and values belong to the separable Euclidean space $\mathbb{R}^N$.

Second, we study measurability of the "derivatives" for any compact-valued solution $K(\cdot) : [0,T] \rightsquigarrow \mathbb{R}^N$. Indeed for real-valued functions, it is well-known that Lipschitz continuity implies a Lebesgue-integrable weak derivative and, the latter coincides with the differential quotient at Lebesgue-almost every time (as a consequence of Rademacher's Theorem [162, Theorem 9.60]). In the morphological framework, however, the derivative is described as a subset of $\mathrm{LIP}(\mathbb{R}^N, \mathbb{R}^N)$, i.e., the mutation (in the sense of Definition 1.10 on page 37).
In combination with Arzelà-Ascoli Theorem A.82 in metric spaces, we conclude directly from Lemma 3 (on page 387):

Lemma 27 (Measurability of compact mutation subsets).
For every threshold $B \in [0, \infty[$ and continuous tube $K(\cdot) : [0,T] \rightsquigarrow \mathbb{R}^N$ with values in $\mathscr{K}(\mathbb{R}^N)$, the following set-valued map of transitions

$$[0,T] \rightsquigarrow \mathrm{LIP}(\mathbb{R}^N, \mathbb{R}^N), \quad t \longmapsto \overset{\circ}{K}(t) \cap \{G \in \mathrm{LIP}(\mathbb{R}^N, \mathbb{R}^N) \mid \|G\|_\infty + \mathrm{Lip}\, G \le B\}$$

is Lebesgue-measurable. □

Proof (of Proposition 25).

"$\Longleftarrow$" Let the compact-valued tube $K(\cdot) : [0,T] \rightsquigarrow \mathbb{R}^N$ be a solution to the morphological inclusion $\overset{\circ}{K}(\cdot) \cap \mathscr{F}_U(K(\cdot)) \neq \emptyset$ (in the sense of Definition 1), i.e.

1.) $K(\cdot) : [0,T] \rightsquigarrow \mathbb{R}^N$ is continuous with respect to dl and
2.) $\mathscr{F}_U(K(t)) \cap \overset{\circ}{K}(t) \neq \emptyset$ for $\mathscr{L}^1$-almost every t, i.e. there is some $u \in U$ such that the set-valued map $f(K(t), u) \in \mathscr{F}_U(K(t)) \subset \mathrm{LIP}(\mathbb{R}^N, \mathbb{R}^N)$ belongs to the mutation $\overset{\circ}{K}(t)$ or, equivalently,

$$\lim_{h \downarrow 0} \tfrac{1}{h} \cdot dl\left(\vartheta_{f(K(t),u)}(h, K(t)),\; K(t+h)\right) = 0.$$

Setting $B := \sup_{M \in \mathscr{K}(\mathbb{R}^N),\, u \in U} (\|f(M,u)\|_\infty + \operatorname{Lip} f(M,u)) < \infty$, the set-valued map

$$[0,T] \rightsquigarrow \mathrm{LIP}(\mathbb{R}^N,\mathbb{R}^N), \quad t \mapsto \overset{\circ}{K}(t) \cap \{G \in \mathrm{LIP}(\mathbb{R}^N,\mathbb{R}^N) \mid \|G\|_\infty + \operatorname{Lip} G \le B\}$$

is Lebesgue-measurable according to Lemma 27. As a consequence of Proposition A.77 and Selection Theorem A.74 (on page 489 f.), the intersection

$$[0,T] \rightsquigarrow \mathrm{LIP}(\mathbb{R}^N,\mathbb{R}^N), \quad t \mapsto \overset{\circ}{K}(t) \cap \mathscr{F}_U(K(t))$$

is also Lebesgue-measurable (with nonempty values at $\mathscr{L}^1$-almost every time) and thus, it has a measurable selection

$$k(\cdot) : [0,T] \longrightarrow \big(\mathrm{LIP}(\mathbb{R}^N,\mathbb{R}^N),\, d_{\mathrm{LIP}}\big).$$

Finally, Lemma 26 of Filippov provides a measurable selection $u(\cdot) : [0,T] \longrightarrow U$ of the constant map $H(\cdot) \equiv U : [0,T] \rightsquigarrow U$ such that $k(t) = f(K(t),u(t))$ for $\mathscr{L}^1$-almost every $t \in [0,T]$. □

5.3.3 Application to Control Problems with State Constraints

The relationship between morphological control problems and morphological inclusions opens the door to applying Viability Theorem 12 immediately. Now we can specify sufficient conditions on a morphological control problem with state constraints for having at least one viable solution:

Proposition 28 (Viability theorem for morphological control problems). *Assume the metric space (U,d_U) to be compact and separable and, consider the set $\mathrm{LIP}_{\overline{\mathrm{co}}}(\mathbb{R}^N,\mathbb{R}^N)$ with the topology of locally uniform convergence. Suppose for $f : \mathscr{K}(\mathbb{R}^N) \times U \longrightarrow \mathrm{LIP}_{\overline{\mathrm{co}}}(\mathbb{R}^N,\mathbb{R}^N)$ and the nonempty closed subset $\mathscr{V} \subset \mathscr{K}(\mathbb{R}^N)$:*

1.) *for any $K \in \mathscr{K}(\mathbb{R}^N)$, the set $\{f(K,u) \mid u \in U\} \subset \mathrm{LIP}_{\overline{\mathrm{co}}}(\mathbb{R}^N,\mathbb{R}^N)$ is convex, i.e. for any $u_1, u_2 \in U$ and $\lambda \in [0,1]$, there exists some $u \in U$ such that $f(K,u) \in \mathrm{LIP}_{\overline{\mathrm{co}}}(\mathbb{R}^N,\mathbb{R}^N)$ is identical to the set-valued map*
$$\mathbb{R}^N \rightsquigarrow \mathbb{R}^N, \quad x \mapsto \lambda \cdot f(K,u_1)(x) + (1-\lambda)\cdot f(K,u_2)(x),$$

2.) $\sup_{K \in \mathscr{K}(\mathbb{R}^N),\, u \in U} (\|f(K,u)\|_\infty + \operatorname{Lip} f(K,u)) < \infty,$

3.) *f is continuous,*

4.) *for each $K \in \mathscr{V}$, there exists some $u \in U$ with $f(K,u) \in \mathscr{T}_{\mathscr{V}}(K)$.*

Then for every initial set $K_0 \in \mathscr{V}$, there exists a compact-valued Lipschitz continuous solution $K(\cdot) : [0,1] \rightsquigarrow \mathbb{R}^N$ to the morphological control problem

$$\overset{\circ}{K}(\cdot) \ni f(K(\cdot),u), \quad u \in U$$

with $K(0) = K_0$ and $K(t) \in \mathscr{V}$ for all $t \in [0,1]$.

Proof. Define the set-valued map

$$\mathscr{F}_U : \mathscr{K}(\mathbb{R}^N) \rightsquigarrow \mathrm{LIP}_{\overline{\mathrm{co}}}(\mathbb{R}^N, \mathbb{R}^N), \quad K \mapsto \{f(K,u) \mid u \in U\}.$$

Obviously, it has nonempty convex values due to assumption (1.). Moreover, the graph of $\mathscr{F}_U$ is a closed subset of $\mathscr{K}(\mathbb{R}^N) \times \mathrm{LIP}(\mathbb{R}^N, \mathbb{R}^N)$ because f is continuous and U is compact. Hence, $\mathscr{F}_U$ satisfies the assumptions of Viability Theorem 12 and thus, for every initial set $K_0 \in \mathscr{V}$, there exists a compact-valued Lipschitz continuous solution $K(\cdot) : [0,1] \rightsquigarrow \mathbb{R}^N$ to the morphological inclusion

$$\overset{\circ}{K}(\cdot) \cap \mathscr{F}_U(K(\cdot)) \neq \emptyset$$

with $K(0) = K_0$ and $K(t) \in \mathscr{V}$ for all $t \in [0,1]$.
Due to Proposition 25, $K(\cdot)$ is a solution to the morphological control problem

$$\overset{\circ}{K}(\cdot) \ni f(K(\cdot), u), \quad u \in U.$$

□

For a given closed subset $M \subset \mathbb{R}^N$, we conclude from Gorre's characterization in Example 1.64 (on page 68) directly:

Corollary 29.
Assume the metric space (U, d_U) to be compact and separable and, consider the set $\mathrm{LIP}_{\overline{\mathrm{co}}}(\mathbb{R}^N, \mathbb{R}^N)$ with the topology of locally uniform convergence. Suppose for $f : \mathscr{K}(\mathbb{R}^N) \times U \longrightarrow \mathrm{LIP}_{\overline{\mathrm{co}}}(\mathbb{R}^N, \mathbb{R}^N)$ and the nonempty closed subset $M \subset \mathbb{R}^N$:

1.) *for any $K \in \mathscr{K}(\mathbb{R}^N)$, the set $\{f(K,u) \mid u \in U\} \subset \mathrm{LIP}_{\overline{\mathrm{co}}}(\mathbb{R}^N, \mathbb{R}^N)$ is convex (as in Proposition 28),*
2.) $\sup_{K \in \mathscr{K}(\mathbb{R}^N),\, u \in U} \left(\|f(K,u)\|_\infty + \mathrm{Lip}\, f(K,u) \right) < \infty,$
3.) *f is continuous,*
4.) *for each nonempty compact set $K \subset M$, there exists $u \in U$ with*
$$f(K,u)(x) \subset T_M(x) \qquad \text{for all } x \in K.$$

Then for every nonempty compact subset $K_0 \subset M$, there exists a compact-valued Lipschitz continuous solution $K : [0,1] \rightsquigarrow \mathbb{R}^N$ to the morphological control problem

$$\begin{cases} \overset{\circ}{K}(\cdot) \ni f(K(\cdot), u) \\ \quad u \in U \end{cases}$$

with $K(0) = K_0$ and $K(t) \subset M$ for all $t \in [0,1]$. □

5.3.4 Relaxed Control Problems with State Constraints

Considering the morphological control problem

$$\begin{cases} \overset{\circ}{K}(\cdot) \ni f(K(\cdot),u) & \text{in } [0,T[\\ u \in U \end{cases}$$

(and the statements in Proposition 28 or Corollary 29, for example), the convexity of $\{f(K,u) \mid u \in U\} \subset \mathrm{LIP}_{\overline{\mathrm{co}}}(\mathbb{R}^N, \mathbb{R}^N)$ is a hypothesis that can be difficult to verify. For basically the same reason, the concept of "relaxed control" has been established for classical control problems in vector spaces. In a word, it is based on replacing the metric space U of control parameters by the set of Borel probability measures on U, from now on denoted by $\mathscr{P}(U)$.
Now the goal is to adapt "relaxed controls" to the morphological framework.

Definition 30. Let (U, d_U) be a metric space and consider $\mathrm{LIP}(\mathbb{R}^N, \mathbb{R}^N)$ with the topology of locally uniform convergence (metrized by d_{LIP} as in § 5.3.2, page 417). Suppose $g : U \longrightarrow \mathrm{LIP}(\mathbb{R}^N, \mathbb{R}^N)$ to be continuous.
For any probability measure $\mu \in \mathscr{P}(U)$, the integral $\int_U g(u)\ d\mu(u)$ is defined as set-valued map by

$$\int_U g(u)\ d\mu(u) : \quad \mathbb{R}^N \rightsquigarrow \mathbb{R}^N, \quad x \mapsto \int_U g(u)(x)\ d\mu(u).$$

Remark 31. Using the notation of Definition 30, for each point $x \in \mathbb{R}^N$ fixed, the set-valued map $U \rightsquigarrow \mathbb{R}^N,\ u \mapsto g(u)(x)$ is compact-valued and continuous in the sense of Bouligand and Kuratowski. Thus the integral $\int_U g(u)(x)\ d\mu(u) \subset \mathbb{R}^N$ is well-defined in the sense of Aumann.

Definition 32.
Let (U, d_U) denote a metric space and $f : \mathscr{K}(\mathbb{R}^N) \times U \longrightarrow \mathrm{LIP}(\mathbb{R}^N, \mathbb{R}^N)$ be given. A tube $K(\cdot) : [0,T] \longrightarrow \mathscr{K}(\mathbb{R}^N)$ is called a *solution* to the *morphological relaxed control problem*

$$\begin{cases} \overset{\circ}{K}(\cdot) \ni f(K(\cdot),u) & \mathscr{L}^1\text{-a.e. in } [0,T] \\ u \in U \end{cases}$$

if there is a measurable function $\mu : [0,T[\longrightarrow \mathscr{P}(U), t \longmapsto \mu_t$ such that $K(\cdot)$ solves the nonautonomous morphological equation $\overset{\circ}{K}(t) \ni \overline{\int_U f(K(t),u)\ d\mu_t(u)}$ in $[0,T]$, i.e., satisfying

1.) $K(\cdot) : [0,T] \rightsquigarrow \mathbb{R}^N$ is continuous with respect to $d\!l$ and
2.) for $\mathscr{L}^1$-a.e. $t \in [0,T]$, the closure $\overline{\int_U f(K(t),u)\ d\mu_t(u)} \in \mathrm{LIP}(\mathbb{R}^N, \mathbb{R}^N)$ belongs to the mutation $\overset{\circ}{K}(t)$.

The first question is now: Which effects do probability measures (on U) instead of U have on the corresponding set-valued map $\mathscr{F}_U : \mathscr{K}(\mathbb{R}^N) \rightsquigarrow \mathrm{LIP}(\mathbb{R}^N, \mathbb{R}^N)$?

Proposition 33. *Assume the metric space (U, d_U) to be compact and separable. Consider the set $\mathrm{LIP}_{\overline{\mathrm{co}}}(\mathbb{R}^N, \mathbb{R}^N)$ with the topology of locally uniform convergence and the set $\mathscr{P}(U)$ of Borel probability measures on U with the topology of narrow convergence (i.e. the dual setting with continuous and thus bounded functions $U \longrightarrow \mathbb{R}$). Let $f : \mathscr{K}(\mathbb{R}^N) \times U \longrightarrow \mathrm{LIP}_{\overline{\mathrm{co}}}(\mathbb{R}^N, \mathbb{R}^N)$ be continuous with*

$$\sup_{\substack{K \in \mathscr{K}(\mathbb{R}^N) \\ u \in U}} \big(\|f(K,u)\|_\infty + \mathrm{Lip}\, f(K,u) \big) < \infty$$

and, set for each $K \in \mathscr{K}(\mathbb{R}^N)$

$$\mathscr{F}_U(K) := \{ \, f(K,u) \mid u \in U \},$$

$$\widetilde{\mathscr{F}_U}(K) := \Big\{ \overline{\int_U f(K,u) \, d\mu(u)} \;\Big|\; \mu \in \mathscr{P}(U) \Big\}.$$

Then,

1.) *$\widetilde{\mathscr{F}_U}(\cdot)$ is a set-valued map $\mathscr{K}(\mathbb{R}^N) \rightsquigarrow \mathrm{LIP}_{\overline{\mathrm{co}}}(\mathbb{R}^N, \mathbb{R}^N)$ with $\mathscr{F}_U(K) \subset \widetilde{\mathscr{F}_U}(K)$ for every $K \in \mathscr{K}(\mathbb{R}^N)$.*
2.) *$\widetilde{\mathscr{F}_U}(\cdot)$ has closed convex values with $\overline{\mathrm{co}}\; \mathscr{F}_U(K) = \widetilde{\mathscr{F}_U}(K) \subset \mathrm{LIP}_{\overline{\mathrm{co}}}(\mathbb{R}^N, \mathbb{R}^N)$ for every $K \in \mathscr{K}(\mathbb{R}^N)$.*
3.) *The graph of $\widetilde{\mathscr{F}_U}(\cdot)$ is closed.*

The proof of this proposition uses some tools about Borel probability measures and Aumann integrals. It is postponed to the end of this section (on page 423 ff.).
The main notion is now to consider $\mathscr{P}(U)$ as control set instead of U. For applying Proposition 25 about the relationship between control problem and morphological inclusion, however, the parameter space has to be metric. We need the following lemma for obtaining the counterparts to Proposition 28 and Corollary 29.
Proposition 35 and Corollary 36 are the main results of this section.

Lemma 34 ([6, §§ 5.1, 7.1])**.**
Let $U \neq \emptyset$ be a Polish space (i.e. complete and separable metric space) with a bounded metric d_U.

Then the set $\mathscr{P}(U)$ of Borel probability measures on U supplied with the topology of narrow convergence is metrizable and separable. An example for a suitable metric on $\mathscr{P}(U)$ is the linear Wasserstein distance (in its dual representation)

$$d_{\mathscr{P}(U)}(\mu, \nu) := \sup \Big\{ \int_U \psi \; d(\mu - \nu) \;\Big|\; \psi : U \longrightarrow \mathbb{R} \;\; 1\text{-Lipschitz continuous} \Big\}.$$

A subset $\mathscr{M} \subset \mathscr{P}(U)$ is relatively compact in $\mathscr{P}(U)$ if and only if $\mathscr{M}$ is tight, i.e. for every $\varepsilon > 0$, there exists a compact subset $C \subset U$ with $\mu(U \setminus C) \leq \varepsilon$ for all $\mu \in \mathscr{M}$ (known as Prokhorov's Theorem).

Proposition 35 (Viability theorem for morphological relaxed control problems). *Assume the metric space* (U, d_U) *to be compact and separable. Consider the set* $\mathrm{LIP}_{\overline{\mathrm{co}}}(\mathbb{R}^N, \mathbb{R}^N)$ *with the topology of locally uniform convergence and the set* $\mathscr{P}(U)$ *of Borel probability measures on U with the topology of narrow convergence.*

Suppose for $f : \mathscr{K}(\mathbb{R}^N) \times U \longrightarrow \mathrm{LIP}_{\overline{\mathrm{co}}}(\mathbb{R}^N, \mathbb{R}^N)$ *and the nonempty closed subset* $\mathscr{V} \subset \mathscr{K}(\mathbb{R}^N)$*:*

(i) $\sup_{\substack{K \in \mathscr{K}(\mathbb{R}^N) \\ u \in U}} \left(\|f(K,u)\|_\infty + \mathrm{Lip}\, f(K,u) \right) < \infty,$

(ii) *f is continuous,*

(iii) $\mathscr{T}_{\mathscr{V}}(K) \cap \overline{\mathrm{co}}\ \{f(K,u) \,|\, u \in U\} \neq \emptyset$ *for each* $K \in \mathscr{V}$.

Then for every initial set $K_0 \in \mathscr{V}$, *there exists a compact-valued Lipschitz continuous solution* $K(\cdot) : [0,1] \rightsquigarrow \mathbb{R}^N$ *to the morphological* relaxed *control problem*

$$\overset{\circ}{K}(\cdot) \ni f(K(\cdot), u), \quad u \in U$$

(in the sense of Definition 32) with $K(0) = K_0$ *and* $K(t) \in \mathscr{V}$ *for all* $t \in [0,1]$.

Proof. Considering $\left(\mathscr{P}(U),\, d_{\mathscr{P}(U)}\right)$ as metric parameter space instead of (U, d_U), the set-valued map

$$\widetilde{\mathscr{F}_U} : \mathscr{K}(\mathbb{R}^N) \rightsquigarrow \mathrm{LIP}_{\overline{\mathrm{co}}}(\mathbb{R}^N, \mathbb{R}^N), \quad K \mapsto \left\{ \overline{\int_U f(K,u)\ d\mu(u)} \;\middle|\; \mu \in \mathscr{P}(U) \right\}$$

satisfies the assumptions of Viability Theorem 12 according to Proposition 33. For each $K_0 \in \mathscr{V}$, there exists a compact-valued Lipschitz continuous solution $K(\cdot) : [0,1] \rightsquigarrow \mathbb{R}^N$ to the morphological inclusion $\overset{\circ}{K}(\cdot) \cap \widetilde{\mathscr{F}_U}(K(\cdot)) \neq \emptyset$ with $K(0) = K_0$ and $K(t) \in \mathscr{V}$ for all $t \in [0,1]$.
Finally Proposition 25 guarantees that $K(\cdot)$ is a solution to the morphological control problem

$$\overset{\circ}{K}(\cdot) \ni \overline{\int_U f(K(\cdot), u)\ d\mu(u)}, \qquad \mu \in \mathscr{P}(U),$$

i.e., it solves the *relaxed* control problem. □

Corollary 36. *Assume the metric space* (U, d_U) *to be compact and separable. Consider the set* $\mathrm{LIP}_{\overline{\mathrm{co}}}(\mathbb{R}^N, \mathbb{R}^N)$ *with the topology of locally uniform convergence and the set* $\mathscr{P}(U)$ *of Borel probability measures on U with the topology of narrow convergence. Suppose for* $f : \mathscr{K}(\mathbb{R}^N) \times U \longrightarrow \mathrm{LIP}_{\overline{\mathrm{co}}}(\mathbb{R}^N, \mathbb{R}^N)$ *and the nonempty closed subset* $M \subset \mathbb{R}^N$*:*

(i) $\sup_{\substack{K \in \mathscr{K}(\mathbb{R}^N) \\ u \in U}} \left(\|f(K,u)\|_\infty + \mathrm{Lip}\, f(K,u) \right) < \infty,$

(ii) *f is continuous,*

(iii) *for each compact* $K \subset M$, *there is a set-valued map* $G \in \overline{\mathrm{co}}\ \{f(K,u) \,|\, u \in U\} \subset \mathrm{LIP}_{\overline{\mathrm{co}}}(\mathbb{R}^N, \mathbb{R}^N)$ *satisfying* $G(x) \subset T_M(x)$ *for every* $x \in K$.

Then for every nonempty compact subset $K_0 \subset M$, there exists a compact-valued Lipschitz continuous solution $K(\cdot) : [0,1] \rightsquigarrow \mathbb{R}^N$ *to the morphological* relaxed *control problem* $\overset{\circ}{K}(\cdot) \ni f(K(\cdot),u),\ u \in U$ *(in the sense of Definition 32) with* $K(0) = K_0$ *and* $K(t) \subset M$ *for all* $t \in [0,1]$.

□

Now we close this section with the proof of Proposition 33.

Proof (of Proposition 33). (1.) As mentioned in Remark 31, the integral $\int_U f(K,u)\, d\mu(u)$ is a well-defined set-valued map $\mathbb{R}^N \rightsquigarrow \mathbb{R}^N$ for each $K \in \mathscr{K}(\mathbb{R}^N)$, $u \in U$ and $\mu \in \mathscr{P}(U)$.
Moreover, its closure is convex since all set-valued maps $f(K,u) \in \mathrm{LIP}_{\overline{\mathrm{co}}}(\mathbb{R}^N,\mathbb{R}^N)$ have convex values and due to the general properties of Aumann integral (see e.g. [141, Theorem 2.1.17] or for the special case of nonatomic measures, [19, § 8.6]).
Due to the assumption $B := \sup_{K,u} \big(\|f(K,u)\|_\infty + \mathrm{Lip}\, f(K,u)\big) < \infty$, all nonempty compact sets $f(K,u)(x) \subset \mathbb{R}^N$ (with $K \in \mathscr{K}(\mathbb{R}^N), u \in U, x \in \mathbb{R}^N$) are contained in the closed convex ball $\{y \in \mathbb{R}^N \mid |y| \leq B\}$ and so are all values of the closures of $\int_U f(K,u)\, d\mu(u)$.

Finally we prove that $\overline{\int_U f(K,u)\, d\mu(u)} : \mathbb{R}^N \rightsquigarrow \mathbb{R}^N$ is B-Lipschitz continuous for each $K \in \mathscr{K}(\mathbb{R}^N)$. For any $x_1, x_2 \in \mathbb{R}^N$, the inclusion

$$f(K,u)(x_1) \ \subset\ f(K,u)(x_2) + \mathbb{B}_{B\cdot|x_1-x_2|}(0)\ \subset\ \mathbb{R}^N$$

holds for every $u \in U$ and we conclude from [19, Proposition 8.6.2]

$$\begin{aligned}\overline{\int_U f(K,u)(x_1)\, d\mu(u)} \ &\subset\ \overline{\int_U \big(f(K,u)(x_2) + \mathbb{B}_{B\cdot|x_1-x_2|}(0)\big)\, d\mu(u)}\\ &\subset\ \overline{\int_U f(K,u)(x_2)\, d\mu(u)} \ +\ \mathbb{B}_{B\cdot|x_1-x_2|}(0).\end{aligned}$$

(2.) The convexity of $\widetilde{\mathscr{F}}(K) \subset \mathrm{LIP}_{\overline{\mathrm{co}}}(\mathbb{R}^N,\mathbb{R}^N)$ (with respect to pointwise convex combinations as in Theorem 12, assumption (1.) on page 401) results from the convexity of $\mathscr{P}(U)$. Furthermore, $\mathrm{co}\,\mathscr{F}(K) \subset \widetilde{\mathscr{F}}(K) \subset \overline{\mathrm{co}}\,\mathscr{F}(K)$ can be concluded easily from the fact that finite convex combinations of Dirac masses are dense in $\mathscr{P}(U)$ (since U is compact separable and due to [26, Corollary 30.5]).

Now we prove that $\widetilde{\mathscr{F}}(K) \subset \mathrm{LIP}_{\overline{\mathrm{co}}}(\mathbb{R}^N,\mathbb{R}^N)$ is closed (with respect to locally uniform convergence) for every $K \in \mathscr{K}(\mathbb{R}^N)$. Indeed, let $(\mu_n)_{n\in\mathbb{N}}$ be any sequence in $\mathscr{P}(U)$ such that

$$\overline{\int_U f(K,u)\, d\mu_n(u)} \ \overset{n\to\infty}{\longrightarrow}\ G \in \mathrm{LIP}_{\overline{\mathrm{co}}}(\mathbb{R}^N,\mathbb{R}^N) \qquad \text{locally uniformly in } \mathbb{R}^N.$$

As U is assumed to be compact, the sequence $(\mu_n)_{n\in\mathbb{N}}$ is tight and thus relatively compact in $\mathscr{P}(U)$ according to Lemma 34. Hence, a subsequence $(\mu_{n_j})_{j\in\mathbb{N}}$ converges narrowly to a measure $\mu_\infty \in \mathscr{P}(U)$. We want to verify for every $x \in \mathbb{R}^N$

$$\overline{\int_U f(K,u)(x)\ d\mu_\infty(u)} \;=\; G(x) \subset \mathbb{R}^N.$$

Indeed, the set-valued map $f(K,\cdot)(x) : U \leadsto \mathbb{R}^N$ is continuous with nonempty compact convex values. Both the closed integral in the recent claim and $G(x)$ are nonempty, compact and convex. [19, Proposition 8.6.2] states closed Aumann integral and support function commute with each other, i.e. here for any vector $p \in \mathbb{R}^N$ and any measure $\nu \in \mathscr{P}(U)$,

$$\sup\Big(p\cdot \overline{\int_U f(K,u)(x)\ d\nu(u)}\Big) \;=\; \int_U \sup\,(p\cdot f(K,u)(x))\ d\nu(u).$$

Here the single-valued function $\sup\,(p\cdot f(K,\cdot)(x)) : U \longrightarrow \mathbb{R}$ is continuous and bounded. On the one hand, we conclude from the narrow convergence $\mu_{n_j} \longrightarrow \mu_\infty$ for each $p \in \mathbb{R}^N$

$$\sup\Big(p\cdot \overline{\int_U f(K,u)(x)\ d\mu_{n_j}(u)}\Big) \;\stackrel{j\to\infty}{\longrightarrow}\; \sup\Big(p\cdot \overline{\int_U f(K,u)(x)\ d\mu_\infty(u)}\Big).$$

On the other hand, the initial assumption of locally uniform convergence to $G(\cdot)$ implies for each $p \in \mathbb{R}^N$

$$\sup\Big(p\cdot \overline{\int_U f(K,u)(x)\ d\mu_{n_j}(u)}\Big) \;\stackrel{j\to\infty}{\longrightarrow}\; \sup\,(p\cdot G(x)).$$

Hence, the two following convex sets coincide for every $x \in \mathbb{R}^N$

$$\overline{\int_U f(K,u)(x)\ d\mu_\infty(u)} \;=\; G(x) \subset \mathbb{R}^N.$$

Finally we have verified that $\widetilde{\mathscr{F}}(K) \subset \mathrm{LIP}_{\overline{\mathrm{co}}}(\mathbb{R}^N,\mathbb{R}^N)$ is closed.

(3.) For proving that $\mathrm{Graph}\ \widetilde{\mathscr{F}} \subset \mathscr{K}(\mathbb{R}^N) \times \mathrm{LIP}_{\overline{\mathrm{co}}}(\mathbb{R}^N,\mathbb{R}^N)$ is closed, let $(K_n)_{n\in\mathbb{N}}$, $(\mu_n)_{n\in\mathbb{N}}$ be any sequences in $\mathscr{K}(\mathbb{R}^N)$ and $\mathscr{P}(U)$ respectively such that

$$\begin{cases} K_n \stackrel{n\to\infty}{\longrightarrow} K \in \mathscr{K}(\mathbb{R}^N) & \text{with respect to } d\!\!l, \\ \overline{\int_U f(K_n,u)\ d\mu_n(u)} \stackrel{n\to\infty}{\longrightarrow} G \in \mathrm{LIP}(\mathbb{R}^N,\mathbb{R}^N) & \text{locally uniformly in } \mathbb{R}^N. \end{cases}$$

Our goal is to verify $G \in \widetilde{\mathscr{F}}(K)$.

Due to the compactness of U, the set $\{\mu_n \mid n \in \mathbb{N}\} \subset \mathscr{P}(U)$ is tight and, there exists a subsequence (again denoted by) $(\mu_n)_{n\in\mathbb{N}}$ converging narrowly to some $\mu_\infty \in \mathscr{P}(U)$. In the proof of statement (2.), we have already drawn the conclusion that for each $x \in \mathbb{R}^N$,

$$\overline{\int_U f(K,u)(x)\ d\mu_n(u)} \;\stackrel{n\to\infty}{\longrightarrow}\; \overline{\int_U f(K,u)(x)\ d\mu_\infty(u)} \subset \mathbb{R}^N$$

Now it is sufficient to verify for each $x \in \mathbb{R}^N$

$$\overline{\int_U f(K_n,u)(x)\ d\mu(u)} \stackrel{n\to\infty}{\longrightarrow} \overline{\int_U f(K,u)(x)\ d\mu(u)} \qquad \text{uniformly in } \mu \in \mathscr{P}(U)$$

since it ensures the wanted convergence for every $x \in \mathbb{R}^N$

$$\overline{\int_U f(K_n,u)(x)\ d\mu_n(u)} \stackrel{n\to\infty}{\longrightarrow} \overline{\int_U f(K,u)(x)\ d\mu_\infty(u)} \subset \mathbb{R}^N$$

Indeed, the continuous function $f : \mathscr{K}(\mathbb{R}^N) \times U \longrightarrow \mathrm{LIP}_{\overline{\mathrm{co}}}(\mathbb{R}^N, \mathbb{R}^N)$ (between metric spaces) is uniformly continuous on the compact product set $\{K, K_n \,|\, n \in \mathbb{N}\} \times U$. Evaluating the set-valued maps at a fixed point $x \in \mathbb{R}^N$ respectively, we obtain for each $\varepsilon > 0$ that a small radius $\delta = \delta(\varepsilon) > 0$ satisfies

$$d\!\!l(K_n,K) + d_U(u_1,u_2) \le \delta \quad \Longrightarrow \quad d\!\!l\big(f(K_n,u_1)(x),\ f(K,u_2)(x)\big) \le \varepsilon.$$

In particular, there is some $m = m(\varepsilon) \in \mathbb{N}$ with

$$d\!\!l\big(f(K_n,u)(x),\ f(K,u)(x)\big) \le \varepsilon \qquad \text{for all } n \ge m,\ u \in U.$$

Since $f(K_n,u)(x)$ and $f(K,u)(x)$ are compact convex subsets of $\mathbb{R}^N$, it implies for the closure of the Aumann integral with respect to any probability measure $\mu \in \mathscr{P}(U)$ [141, Theorem 2.1.17 (i)]

$$d\!\!l\Big(\overline{\int_U f(K_n,u)\ d\mu(u)},\ \overline{\int_U f(K,u)\ d\mu(u)}\Big) \le \varepsilon \qquad \text{for all } n \ge m(\varepsilon).$$

□

5.3.5 Clarke Tangent Cone in the Morphological Framework: The Circatangent Transition Set

The invariance condition of Nagumo (in Theorem 1.19 on page 40) has already served Aubin as motivation for extending the contingent cone $T_V(x)$ in a normed vector space to the mutational framework (see Definition 1.16 on page 39).

In this section, we start with the classical definition of Clarke tangent cone introduced by Frank H. Clarke in the seventies (see [46] for details) and extend it to the morphological framework. Following the alternative nomenclature of Aubin and Frankowska in [19, Definition 4.1.5 (2)], its counterpart will be called *circatangent transition set* – just because this term fits to the established "contingent transition set".

Indeed, Aubin introduced circatangent transition sets in the more general framework of metric spaces in [10, Definition 1.5.4] and, Definition 38 below is equivalent to the special case of $(\mathcal{K}(\mathbb{R}^N), d\!\!l)$ and morphological transitions.
Murillo Hernández applied this concept to tuples $(v, K) \in \mathbb{R}^N \times \mathcal{K}(\mathbb{R}^N)$ with $v \in K$ and proved an asymptotic relationship between their contingent and circatangent transition set implying that the latter is closed [144, Theorem 4.6].
In this section we generalize further features from the Euclidean space to the metric space $(\mathcal{K}(\mathbb{R}^N), d\!\!l)$.

Definition 37 ([46, § 2.4], [19, § 4.1.3], [162, § 6.F]). Let K be a nonempty subset of a normed vector space X and $x \in X$ belong to the closure of K.
The *Clarke tangent cone* or *circatangent cone* $T_K^C(x)$ is defined (equivalently) by

$$\begin{aligned} T_K^C(x) &:= \operatorname{Liminf}_{\substack{h \downarrow 0,\\ y \xrightarrow[K]{} x}} \frac{K-y}{h} \\ &= \Big\{ v \in X \;\Big|\; \forall\, h_n \downarrow 0,\ y_n \to x \text{ with } y_n \in K:\ \operatorname{dist}\big(v, \tfrac{K-y_n}{h_n}\big) \overset{n\to\infty}{\longrightarrow} 0 \Big\} \\ &= \Big\{ v \in X \;\Big|\; \forall\, h_n \downarrow 0,\ y_n \to x \text{ with } y_n \in K:\ \frac{\operatorname{dist}(y_n + h_n \cdot v,\ K)}{h_n} \overset{n\to\infty}{\longrightarrow} 0 \Big\}. \end{aligned}$$

Definition 38. For a nonempty subset $\mathscr{V} \subset \mathcal{K}(\mathbb{R}^N)$ and any element $K \in \mathscr{V}$,

$$\begin{aligned} \mathscr{T}_{\mathscr{V}}^C(K) := \Big\{ F \in \mathrm{LIP}(\mathbb{R}^N, \mathbb{R}^N) \;\Big|\; & \forall\, h_n \downarrow 0,\ K_n \to K \text{ with } K_n \in \mathscr{V} \subset \mathcal{K}(\mathbb{R}^N): \\ & \tfrac{1}{h_n} \cdot \operatorname{dist}\big(\vartheta_F(h_n, K_n),\ \mathscr{V}\big) \overset{n\to\infty}{\longrightarrow} 0 \Big\} \end{aligned}$$

is called *circatangent transition set* of $\mathscr{V}$ at K (in the metric space $(\mathcal{K}(\mathbb{R}^N), d\!\!l)$).

In fact, we do not have to restrict our considerations to arbitrary sequences $(K_n)_{n\in\mathbb{N}}$ in $\mathscr{V} \subset \mathcal{K}(\mathbb{R}^N)$. An equivalent characterization of $\mathscr{T}_{\mathscr{V}}^C(K)$ uses all sequences in $\mathcal{K}(\mathbb{R}^N)$ converging to K:

Lemma 39. *For every nonempty closed subset $\mathscr{V} \subset \big(\mathscr{K}(\mathbb{R}^N), dl\big)$ and $K \in \mathscr{V}$,*

$$\mathscr{T}^C_{\mathscr{V}}(K) = \Big\{ F \in \mathrm{LIP}(\mathbb{R}^N, \mathbb{R}^N) \;\Big|\; \forall\ h_n \downarrow 0,\ K_n \to K : \limsup_{n \to \infty} \frac{\mathrm{dist}(\vartheta_F(h_n, K_n), \mathscr{V}) - \mathrm{dist}(K_n, \mathscr{V})}{h_n} \leq 0 \Big\}.$$

So far, the circatangent transition set has been characterized by two sequences providing the arbitrarily fixed relation between "step size" $h_n > 0$ and neighboring sets $K_n \in \mathscr{K}(\mathbb{R}^N)$. The following condition proves to be equivalent and avoids the aspect of countability:

Lemma 40. *Let $K \in \mathscr{K}(\mathbb{R}^N)$ be any element of the closed set $\mathscr{V} \subset \big(\mathscr{K}(\mathbb{R}^N), dl\big)$. Then, a set-valued map $F \in \mathrm{LIP}(\mathbb{R}^N, \mathbb{R}^N)$ belongs to the circatangent transition set $\mathscr{T}^C_{\mathscr{V}}(K)$ if and only if there is a function $\omega : [0, \infty[\longrightarrow [0, \infty[$ with $\lim_{\delta \to 0} \omega(\delta) = 0$,*

$$\tfrac{1}{h} \cdot \Big(\mathrm{dist}\big(\vartheta_F(h, M), \mathscr{V}\big) - \mathrm{dist}(M, \mathscr{V})\Big) \;\leq\; \omega\big(dl(M, K) + h\big)$$

for all $h \in]0, 1]$, $M \in \mathscr{K}(\mathbb{R}^N)$.

The next proposition indicates further properties which the circatangent transition set shares with the Clarke tangent cone in normed vector spaces. Indeed, it is a nonempty closed cone in $\mathrm{LIP}(\mathbb{R}^N, \mathbb{R}^N)$.
Convexity, however, is verified here only for morphological transitions in $\mathscr{T}^C_{\mathscr{V}}(K)$ which are induced by $\mathrm{BLip}(\mathbb{R}^N, \mathbb{R}^N)$, i.e. bounded Lipschitz continuous vector fields $\mathbb{R}^N \longrightarrow \mathbb{R}^N$ and their ordinary differential equations (rather than set-valued maps in $\mathrm{LIP}(\mathbb{R}^N, \mathbb{R}^N)$ and reachable sets of their respective differential inclusions).

Proposition 41. *For every element $K \in \mathscr{K}(\mathbb{R}^N)$ of a closed set $\mathscr{V} \subset \big(\mathscr{K}(\mathbb{R}^N), dl\big)$,*

1. *the circatangent transition set $\mathscr{T}^C_{\mathscr{V}}(K) \subset \mathrm{LIP}(\mathbb{R}^N, \mathbb{R}^N)$ is a nonempty cone, i.e., for any $G \in \mathscr{T}^C_{\mathscr{V}}(K)$ and $\lambda \geq 0$, the set-valued map $\mathbb{R}^N \leadsto \mathbb{R}^N$, $x \mapsto \lambda \cdot G(x)$ (in the Minkowski sense) also belongs to $\mathscr{T}^C_{\mathscr{V}}(K)$.*

2. *for every threshold $B \in [0, \infty[$, the intersection*

$$\mathscr{T}^C_{\mathscr{V}}(K) \cap \{G \in \mathrm{LIP}(\mathbb{R}^N, \mathbb{R}^N) \mid \|G\|_\infty + \mathrm{Lip}\, G \leq B\}$$

is closed in $\mathrm{LIP}(\mathbb{R}^N, \mathbb{R}^N)$ with the topology of locally uniform convergence.

Proposition 42. *Let $K \in \mathscr{K}(\mathbb{R}^N)$ be in the closed set $\mathscr{V} \subset \big(\mathscr{K}(\mathbb{R}^N), dl\big)$.*

Then, $\mathscr{T}^C_{\mathscr{V}}(K) \cap \mathrm{BLip}(\mathbb{R}^N, \mathbb{R}^N)$ is convex,

i.e., for any $g_1, g_2 \in \mathscr{T}^C_{\mathscr{V}}(K) \cap \mathrm{BLip}(\mathbb{R}^N, \mathbb{R}^N)$ and $\lambda \in [0, 1]$, the Lipschitz continuous function $\mathbb{R}^N \longrightarrow \mathbb{R}^N$, $x \longmapsto \lambda \cdot g_1(x) + (1 - \lambda) \cdot g_2(x)$ also belongs to $\mathscr{T}^C_{\mathscr{V}}(K)$.

Now we provide the missing proofs in regard to the circatangent transition set.

Proof (of Lemma 39). "$\supset$" is an obvious consequence of Definition 38.
"$\subset$" For any $F \in \mathscr{T}_{\mathscr{V}}^{C}(K) \subset \mathrm{LIP}(\mathbb{R}^N,\mathbb{R}^N)$ choose the arbitrary sequences $(h_n)_{n\in\mathbb{N}}$, $(K_n)_{n\in\mathbb{N}}$ in $]0,\infty[$ and $\mathscr{K}(\mathbb{R}^N)$ respectively with $h_n \longrightarrow 0$, $d\!l(K_n,K) \longrightarrow 0$ for $n \longrightarrow \infty$. Since closed balls in $(\mathscr{K}(\mathbb{R}^N), d\!l)$ are known to be compact, there exists a set $M_n \in \mathscr{V} \subset \mathscr{K}(\mathbb{R}^N)$ for each $n \in \mathbb{N}$ satisfying

$$d\!l(K_n,M_n) \;=\; \mathrm{dist}\big(K_n,\mathscr{V}\big) \longrightarrow 0\,.$$

$F \in \mathscr{T}_{\mathscr{V}}^{C}(K)$ implies $\quad \frac{1}{h_n}\cdot \mathrm{dist}\big(\vartheta_F(h_n,M_n),\,\mathscr{V}\big) \longrightarrow 0 \quad$ for $n \longrightarrow \infty$
and, Proposition 1.50 ensures $d\!l\big(\vartheta_F(h_n,K_n),\,\vartheta_F(h_n,M_n)\big) \;\le\; d\!l(K_n,M_n)\cdot e^{\mathrm{Lip}\,F\cdot h_n}$ for each $n \in \mathbb{N}$. Finally, we obtain

$$\begin{aligned}
&\tfrac{1}{h_n}\cdot \Big(\mathrm{dist}\big(\vartheta_F(h_n,K_n),\,\mathscr{V}\big) - \mathrm{dist}\big(K_n,\,\mathscr{V}\big)\Big)\\
&\quad\le \tfrac{1}{h_n}\cdot \Big(d\!l\big(\vartheta_F(h_n,K_n),\,\vartheta_F(h_n,M_n)\big) + \mathrm{dist}\big(\vartheta_F(h_n,M_n),\,\mathscr{V}\big) - d\!l(K_n,M_n)\Big)\\
&\quad\le d\!l(K_n,M_n)\,\cdot\, \frac{e^{\mathrm{Lip}\,F\cdot h_n}-1}{h_n} + \frac{\mathrm{dist}(\vartheta_F(h_n,M_n),\,\mathscr{V})}{h_n}
\end{aligned}$$

and thus, its limit superior for $n \longrightarrow \infty$ is nonpositive. □

Proof (of Lemma 40 on page 427).
"$\Longleftarrow$" is an immediate consequence of Lemma 39.
"$\Longrightarrow$" The triangle inequality of $d\!l$ and Lemma 1.51 (on page 61) guarantee

$$\mathrm{dist}\big(\vartheta_F(h,M),\,\mathscr{V}\big) \;-\; \mathrm{dist}(M,\,\mathscr{V}) \;\le\; d\!l\big(M,\,\vartheta_F(h,M)\big) \;\le\; \|F\|_\infty\, h$$

for all $h > 0$ and $M \in \mathscr{K}(\mathbb{R}^N)$. Hence the auxiliary function $\omega : [0,\infty[\longrightarrow [0,\infty[$,

$$\begin{aligned}
\omega(\delta) \;:=\; \sup\Big\{ \tfrac{1}{h}\cdot \big(\mathrm{dist}\big(\vartheta_F(h,M),\,\mathscr{V}\big) \;-\; \mathrm{dist}(M,\,\mathscr{V})\big) \;\Big|\; \\
M \in \mathscr{K}(\mathbb{R}^N),\; h \in]0,1],\; d\!l(M,K)+h \le \delta \Big\}
\end{aligned}$$

is well-defined and bounded for any set-valued map $F \in \mathrm{LIP}(\mathbb{R}^N,\mathbb{R}^N)$.

For $F \in \mathscr{T}_{\mathscr{V}}^{C}(K)$, however, we still have to verify $\omega(\delta) \longrightarrow 0$ for $\delta \longrightarrow 0$. If this asymptotic feature was not correct, there would exist some $\varepsilon > 0$ and sequences $(h_n)_{n\in\mathbb{N}}$, $(M_n)_{n\in\mathbb{N}}$ in $]0,1]$, $\mathscr{K}(\mathbb{R}^N)$ respectively satisfying for all $n \in \mathbb{N}$

$$\begin{cases}
d\!l(M_n,K) + h_n \;\le\; \tfrac{1}{n}\\
\tfrac{1}{h_n}\cdot \big(\mathrm{dist}\big(\vartheta_F(h_n,M_n),\,\mathscr{V}\big) \;-\; \mathrm{dist}(M_n,\,\mathscr{V})\big) \;\ge\; \varepsilon > 0.
\end{cases}$$

Due to $h_n \downarrow 0$ and $M_n \longrightarrow K$, it would contradict $F \in \mathscr{T}_{\mathscr{V}}^{C}(K)$ due to Lemma 39. □

Proof (of Proposition 41 on page 427).

(1.) Obviously, the constant set-valued map $G_0(\cdot) := \{0\} : \mathbb{R}^N \leadsto \mathbb{R}^N$ belongs to both $\mathrm{LIP}(\mathbb{R}^N,\mathbb{R}^N)$ and $\mathscr{T}_{\mathscr{V}}^{C}(K)$ because $\vartheta_{G_0}(h,K) = K$ for every $K \in \mathscr{K}(\mathbb{R}^N)$ and $h \ge 0$. Thus, $\mathscr{T}_{\mathscr{V}}^{C}(K) \neq \emptyset$.

For proving the cone property, choose any $K \in \mathscr{V} \subset \mathscr{K}(\mathbb{R}^N)$, $G \in \mathscr{T}^C_{\mathscr{V}}(K) \subset \mathrm{LIP}(\mathbb{R}^N, \mathbb{R}^N)$ and $\lambda > 0$. Moreover, let $(h_n)_{n\in\mathbb{N}}$ and $(K_n)_{n\in\mathbb{N}}$ be arbitrary sequences in $]0,\infty[$ and $\mathscr{V} \subset \mathscr{K}(\mathbb{R}^N)$ respectively with $h_n \longrightarrow 0$, $dl(K_n, K) \longrightarrow 0$ $(n \to \infty)$. Every solution $x(\cdot) \in W^{1,1}([0,h_n], \mathbb{R}^N)$ of $x'(\cdot) \in \lambda\, G(x(\cdot))$ induces a solution $y(\cdot) \in W^{1,1}([0, \frac{h_n}{\lambda}], \mathbb{R}^N)$ of $y'(\cdot) \in G(y(\cdot))$ (and vice versa) by time scaling, i.e. $x(t) = y(\lambda \cdot t)$. Hence,

$$\vartheta_{\lambda\, G}(h_n, K_n) \;=\; \vartheta_G(\tfrac{h_n}{\lambda}, K_n).$$

The assumption $G \in \mathscr{T}^C_{\mathscr{V}}(K)$ guarantees now

$$\tfrac{1}{h_n} \cdot \mathrm{dist}\big(\vartheta_{\lambda\, G}(h_n, K_n),\, \mathscr{V}\big) \;=\; \tfrac{1}{\lambda}\, \tfrac{\lambda}{h_n} \cdot \mathrm{dist}\big(\vartheta_G(\tfrac{h_n}{\lambda}, K_n),\, \mathscr{V}\big) \;\longrightarrow 0 \qquad \text{for } n \to \infty.$$

(2.) Let $(G^j)_{j\in\mathbb{N}}$ be a sequence in $\mathscr{T}^C_{\mathscr{V}}(K)$ with $\|G^j\|_\infty + \mathrm{Lip}\, G^j \leq B$ for each $j \in \mathbb{N}$ and converging to $G(\cdot) \in \mathrm{LIP}(\mathbb{R}^N, \mathbb{R}^N)$ locally uniformly in $\mathbb{R}^N$. Obviously, $\|G\|_\infty + \mathrm{Lip}\, G \leq B$ holds. Our aim is to verify $G \in \mathscr{T}^C_{\mathscr{V}}(K)$.

Let $(h_n)_{n\in\mathbb{N}}$ and $(K_n)_{n\in\mathbb{N}}$ be any sequences in $]0,1]$ and $\mathscr{V} \subset \mathscr{K}(\mathbb{R}^N)$ respectively with $h_n \longrightarrow 0$ and $dl(K_n, K) \longrightarrow 0$ (for $n \to \infty$). The last convergence implies that all K_n, $n \in \mathbb{N}$, and $K \in \mathscr{K}(\mathbb{R}^N)$ are contained in a ball $\mathbb{B}_R(0) \subset \mathbb{R}^N$ of sufficiently large radius $R < \infty$. Due to $\sup_n h_n \leq 1$,

$$\bigcup_{j,n\in\mathbb{N}} \; \bigcup_{0\leq t\leq h_n} \big(\vartheta_{G^j}(t, K_n) \cup \vartheta_G(t, K_n)\big) \;\subset\; \mathbb{B}_{R+B}(0) \;\subset\; \mathbb{R}^N.$$

On the basis of Proposition 1.50 (on page 60), we obtain the estimate for all $j, n \in \mathbb{N}$

$$\begin{aligned}
&\tfrac{1}{h_n} \cdot \mathrm{dist}\big(\vartheta_G(h_n, K_n),\, \mathscr{V}\big) \\
&\quad\leq \tfrac{1}{h_n} \cdot dl\big(\vartheta_G(h_n, K_n),\, \vartheta_{G^j}(h_n, K_n)\big) \;+\; \tfrac{1}{h_n} \cdot \mathrm{dist}\big(\vartheta_{G^j}(h_n, K_n),\, \mathscr{V}\big) \\
&\quad\leq e^{B\, h_n} \cdot \sup_{|x|\leq R+B} dl(G(x),\, G^j(x)) \;+\; \tfrac{1}{h_n} \cdot \mathrm{dist}\big(\vartheta_{G^j}(h_n, K_n),\, \mathscr{V}\big).
\end{aligned}$$

For any $\varepsilon > 0$ given, we can fix $j \in \mathbb{N}$ sufficiently large with

$$\sup_{|x|\leq R+B} dl(G(x),\, G^j(x)) < \varepsilon$$

and, $G^j \in \mathscr{T}^C_{\mathscr{V}}(K)$ guarantees

$$\limsup_{n\to\infty} \tfrac{1}{h_n} \cdot \mathrm{dist}\big(\vartheta_{G^j}(h_n, K_n),\, \mathscr{V}\big) \;\leq\; \varepsilon$$

with arbitrarily small $\varepsilon > 0$, i.e.,

$$\limsup_{n\to\infty} \tfrac{1}{h_n} \cdot \mathrm{dist}\big(\vartheta_G(h_n, K_n),\, \mathscr{V}\big) \;=\; 0. \qquad \square$$

The subsequent proof of Proposition 42 uses the following auxiliary result about representing a constant λ as integral mean. A similar statement cannot hold for the L^1 deviation because any integrable function $\mu : [0,1] \longrightarrow \{0,1\}$ satisfies for every $t \in]0,1]$ and $\lambda \in [0,1]$

$$\tfrac{1}{t} \cdot \int_0^t |\mu(s) - \lambda| \; ds \;\geq\; \min\{\lambda, 1-\lambda\}.$$

Lemma 43. *For every $\lambda \in]0,1[$, there exists $\mu \in L^1([0,1])$ satisfying*

$$\begin{cases} \frac{1}{t} \cdot \int_0^t (\mu(s) - \lambda)\, ds \longrightarrow 0 & \text{for } t \downarrow 0, \\ \mu(\cdot) \in \{0,1\} & \text{piecewise constant in }]0,1[. \end{cases}$$

Proof (of Lemma 43). $\mu(\cdot)$ is defined piecewise in each interval $\left[\frac{1}{\sqrt{n+1}}, \frac{1}{\sqrt{n}}\right[$.

Set $\mu(t) := \begin{cases} 0 & \text{for } \frac{1}{\sqrt{n+1}} \le t < \frac{\lambda}{\sqrt{n+1}} + \frac{1-\lambda}{\sqrt{n}} \\ 1 & \text{for } \frac{\lambda}{\sqrt{n+1}} + \frac{1-\lambda}{\sqrt{n}} \le t < \frac{1}{\sqrt{n}} \end{cases}$ for each $n \in \mathbb{N}$.

Then, $\int_{\frac{1}{\sqrt{n+1}}}^{\frac{1}{\sqrt{n}}} (\mu(s) - \lambda)\, ds = 0$ and thus, $\int_0^{\frac{1}{\sqrt{n}}} (\mu(s) - \lambda)\, ds = 0$.

Moreover, $\int_{\frac{1}{\sqrt{n+1}}}^{\frac{1}{\sqrt{n}}} |\mu(s) - \lambda|\, ds = 2\,\lambda\,(1-\lambda) \left(\frac{1}{\sqrt{n}} - \frac{1}{\sqrt{n+1}}\right)$ implies

$$\sup_{\frac{1}{\sqrt{n+1}} \le t \le \frac{1}{\sqrt{n}}} \frac{1}{t} \cdot \left| \int_0^t (\mu(s) - \lambda)\, ds \right| \le \sqrt{n+1} \cdot \int_{\frac{1}{\sqrt{n+1}}}^{\frac{1}{\sqrt{n}}} |\mu(s) - \lambda|\, ds \stackrel{n \to \infty}{\longrightarrow} 0.$$

□

Proof (of Proposition 42 on page 427).

For any functions $g_1, g_2 \in \mathscr{T}_{\mathscr{V}}^C(K) \cap \mathrm{BLip}(\mathbb{R}^N, \mathbb{R}^N)$ and $\lambda \in]0,1[$, we verify that

$$g : \mathbb{R}^N \longrightarrow \mathbb{R}^N, \qquad x \longmapsto \lambda \cdot g_1(x) + (1-\lambda) \cdot g_2(x)$$

also belongs to $\mathscr{T}_{\mathscr{V}}^C(K)$.

Obviously, $g(\cdot)$ is bounded, Lipschitz continuous and thus, $g \in \mathrm{BLip}(\mathbb{R}^N, \mathbb{R}^N)$. According to Lemma 43, there exists $\mu \in L^1([0,1])$ satisfying

$$\begin{cases} \frac{1}{t} \cdot \int_0^t (\mu(s) - \lambda)\, ds \longrightarrow 0 & \text{for } t \downarrow 0, \\ \mu(\cdot) \in \{0,1\} & \text{piecewise constant in }]0,1[. \end{cases}$$

First we compare the evolution of an arbitrary set $M \in \mathscr{K}(\mathbb{R}^N)$ along the autonomous differential equation with the right-hand side

$$g : \mathbb{R}^N \longrightarrow \mathbb{R}^N, \quad x \longmapsto \lambda \cdot g_1(x) + (1-\lambda) \cdot g_2(x)$$

and along the nonautonomous differential equation with the right-hand side

$$f : \mathbb{R}^N \times [0,1] \longrightarrow \mathbb{R}^N, \quad (x,t) \longmapsto \mu(t) \cdot g_1(x) + (1-\mu(t)) \cdot g_2(x).$$

In particular, we prove

$$\lim_{t \downarrow 0} \frac{1}{t} \cdot d\!l\big(\vartheta_f(t,M),\ \vartheta_g(t,M)\big) = 0 \quad \text{uniformly in } M \in \mathscr{K}(\mathbb{R}^N).$$

Let $x(\cdot) \in W^{1,1}([0,1], \mathbb{R}^N)$ denote any solution to the nonautonomous differential equation $x'(\cdot) \in f(x(\cdot), \cdot)$. There exists a solution $y(\cdot) \in W^{1,1}([0,1], \mathbb{R}^N)$ to the initial value problem $y'(\cdot) = g(y(\cdot))$, $y(0) = x(0)$ and, we estimate the difference

$$\begin{aligned}
|y(t)-x(t)| &= \Big| \int_0^t \Big(\lambda g_1\big(y(s)\big) - \mu(s) g_1\big(x(s)\big) + \\
&\qquad (1-\lambda) g_2\big(y(s)\big) - (1-\mu(s)) g_2\big(x(s)\big)\Big)\, ds \Big| \\
&\le \Big| \int_0^t \Big((\lambda-\mu(s))\; g_1\big(y(s)\big) \;+\; (\mu(s)-\lambda)\, g_2\big(y(s)\big)\Big)\; ds \Big| \\
&\quad + \int_0^t \mu(s)\cdot \mathrm{Lip}\, g_1 \cdot |x(s)-y(s)|\, ds \\
&\quad + \int_0^t (1-\mu(s))\cdot \mathrm{Lip}\, g_2 \cdot |x(s)-y(s)|\, ds \\
&\le \Big| \int_0^t (\lambda-\mu(s))\cdot \big(g_1(x(0)) - g_2(x(0))\big)\; ds \Big| \\
&\quad + \int_0^t |\lambda-\mu(s)|\; (\mathrm{Lip}\, g_1 + \mathrm{Lip}\, g_2)\; |y(s)-x(0)|\; ds \\
&\quad + \max\{\mathrm{Lip}\, g_1,\, \mathrm{Lip}\, g_2\} \cdot \int_0^t |x(s)-y(s)|\, ds \\
&\le c\cdot \Big(\Big| \int_0^t (\lambda-\mu(s))\; ds \Big| + \int_0^t \|g\|_{\sup}\cdot s\; ds \;+\; \int_0^t |x(s)-y(s)|\, ds \Big)
\end{aligned}$$

with a constant $c > 0$ depending only on $g_1(\cdot)$, $g_2(\cdot)$. Due to Gronwall's inequality, $|x(t)-y(t)| \le o(t)$ for $t \downarrow 0$ uniformly with respect to the initial point $x(0) = y(0)$. (In particular, the estimate of Filippov's Theorem is difficult to be applied here directly as the integral mean of $\mu(\cdot)-\lambda$ tends to 0 for $t \downarrow 0$, but not of $|\mu(\cdot)-\lambda|$.)

Thus, for any initial set $M \in \mathscr{K}(\mathbb{R}^N)$, the reachable sets satisfy

$$\lim_{t\downarrow 0} \; \tfrac{1}{t} \cdot e^{\subset}\big(\vartheta_f(t,M),\; \vartheta_g(t,M)\big) \;=\; 0 \qquad \text{uniformly in } M \in \mathscr{K}(\mathbb{R}^N).$$

The same uniform estimates hold for $e^{\subset}\big(\vartheta_g(t,M),\; \vartheta_f(t,M)\big)$ since the preceding solutions $x(\cdot)$ and $y(\cdot)$ have required only the joint initial point at time 0. Hence,

$$\lim_{t\downarrow 0} \; \tfrac{1}{t} \cdot d\!l\big(\vartheta_f(t,M),\; \vartheta_g(t,M)\big) \;=\; 0 \qquad \text{uniformly in } M \in \mathscr{K}(\mathbb{R}^N).$$

Finally, we focus on the asymptotic features of $\vartheta_f(\cdot,\cdot)$ in regard to the circatangent transition set $\mathscr{T}_{\mathscr{V}}^{C}(K)$, i.e. for any $\varepsilon > 0$, we verify the existence of a radius $r > 0$ in a moment such that all $h \in\,]0,r]$ and sets $M \in \mathscr{K}(\mathbb{R}^N)$ with $d\!l(M,K) \le r$ satisfy

$$\mathrm{dist}\big(\vartheta_f(h,M),\; \mathscr{V}\big) \;-\; \mathrm{dist}(M,\; \mathscr{V}) \;\le\; \varepsilon\, h.$$

Then, for any sequences $h_n \downarrow 0$ and $(K_n)_{n\in\mathbb{N}}$ in $\mathscr{V} \subset \mathscr{K}(\mathbb{R}^N)$ converging to K

$$\tfrac{1}{h_n} \cdot \mathrm{dist}\big(\vartheta_f(h_n,K_n),\; \mathscr{V}\big) \longrightarrow 0 \qquad\qquad \text{for } n \longrightarrow \infty$$

and in combination with the uniform convergence mentioned before, we conclude

$$\tfrac{1}{h_n} \cdot \mathrm{dist}\big(\vartheta_g(h_n,K_n),\; \mathscr{V}\big) \longrightarrow 0 \qquad\qquad \text{for } n \longrightarrow \infty,$$

i.e., $g \in \mathscr{T}_{\mathscr{V}}^{C}(K)$ due to Definition 38.

Indeed, applying Lemma 40 (on page 427) to $g_1, g_2 \in \mathscr{T}^C_{\mathscr{V}}(K) \cap \mathrm{BLip}(\mathbb{R}^N, \mathbb{R}^N)$, we obtain a joint function $\omega : [0,\infty[\longrightarrow [0,\infty[$ satisfying $\lim_{\delta \to 0} \omega(\delta) = 0$ and

$$\tfrac{1}{h} \cdot \big(\mathrm{dist}\big(\vartheta_{g_j}(h,M), \mathscr{V}\big) - \mathrm{dist}(M, \mathscr{V})\big) \;\le\; \omega\big(dl(M,K) + h\big)$$

for all $j \in \{1,2\}$, $h \in]0,1]$ and $M \in \mathscr{K}(\mathbb{R}^N)$.
Fixing $\varepsilon > 0$ arbitrarily small, there exist a radius $R > 0$ with $\sup_{[0,R]} \omega(\cdot) \le \varepsilon$ and additionally, some $r \in]0, \frac{R}{2}]$ such that $r \cdot \big(1 + \|g_1\|_\infty + \|g_2\|_\infty\big) \le \frac{R}{2}$.
Then, each $j \in \{1,2\}$ and every $h \in]0,r]$, $M \in \mathscr{K}(\mathbb{R}^N)$ with $dl(M,K) \le r$ satisfy

$$\begin{cases} dl\big(\vartheta_{g_j}(h,M), K\big) \;\le\; dl(M,K) + \|g_j\|_\infty\, h \;\le\; \frac{R}{2} \\ \mathrm{dist}\big(\vartheta_{g_j}(h,M), \mathscr{V}\big) - \mathrm{dist}(M, \mathscr{V}) \;\le\; \omega\big(dl(M,K) + h\big) \cdot h \;\le\; \varepsilon\, h. \end{cases}$$

For drawing now conclusions about $\vartheta_f(h,M)$, we exploit the piecewise constant structure of auxiliary function $\mu(\cdot) : [0,1] \longrightarrow \{0,1\}$ (introduced in Lemma 43). Indeed, there is a sequence $(t_k)_{k \in \mathbb{N}}$ tending to 0 monotonically such that $\mu(\cdot)$ is constant in every interval $[t_{k+1}, t_k[$, $k \in \mathbb{N}$. The last estimate in each of these subintervals leads to the following inequalities for every $h \in]0,r]$, $M \in \mathscr{K}(\mathbb{R}^N)$ with $dl(M,K) \le r$ and sufficiently large $k \in \mathbb{N}$ with $t_{k+1} < h \le t_k$

$$\begin{aligned} & \mathrm{dist}\big(\vartheta_f(h,M), \mathscr{V}\big) - \mathrm{dist}(M, \mathscr{V}) \\ \le\; & \mathrm{dist}\big(\vartheta_f(h - t_{k+1}, \vartheta_f(t_{k+1}, M)), \mathscr{V}\big) - \mathrm{dist}\big(\vartheta_f(t_{k+1}, M), \mathscr{V}\big) \\ & + \mathrm{dist}\big(\vartheta_f(t_{k+1}, M), \mathscr{V}\big) \qquad\qquad - \mathrm{dist}\big(\vartheta_f(t_{k+2}, M), \mathscr{V}\big) \pm \ldots \\ & - \mathrm{dist}(M, \mathscr{V}) \\ \le\; & \varepsilon \cdot (h - t_{k+1}) \quad + \quad \varepsilon \cdot (t_{k+1} - t_{k+2}) \qquad\qquad + \ldots \\ \le\; & \varepsilon \cdot h. \end{aligned}$$

□

5.3.6 The Hypertangent Transition Set

For any closed subset of the Euclidean space, the interior of the Clarke tangent cone has been characterized by Rockafellar in 1979 [161]:

Proposition 44 (Rockafellar [161, Theorem 2], [162, Theorem 6.36]**).** *Let $K \subset \mathbb{R}^N$ be a closed set and $x \in K$. Then the interior of Clarke tangent cone to K at x satisfies*

$$T_K^C(x)^\circ = \{v \in \mathbb{R}^N \mid \exists\, \varepsilon > 0:\ (K \cap \mathbb{B}_\varepsilon(x)) +]0,\varepsilon[\cdot \mathbb{B}_\varepsilon(v) \subset K\}$$
$$= \{v \in \mathbb{R}^N \mid \exists\, \varepsilon > 0\ \forall\, y \in K \cap \mathbb{B}_\varepsilon(x),\ w \in \mathbb{B}_\varepsilon(v),\ \tau \in]0,\varepsilon[:\ y + \tau w \in K\}$$

with $\mathbb{B}_\varepsilon(v)$ abbreviating the closed ball $\mathbb{B}_\varepsilon(v) := \{w \in \mathbb{R}^N \mid |w - v| \le \varepsilon\}$ and U° denoting always the interior of a set U.

This equivalence is the motivation for introducing "hypertangent cones":

Definition 45 ([46, § 2, 4])**.** A vector v in a Banach space X is said to be *hypertangent* to the set $K \subset X$ at the point $x \in K$ if for some $\varepsilon > 0$, all vectors $y \in \mathbb{B}_\varepsilon(x) \cap K$, $w \in \mathbb{B}_\varepsilon(v) \subset X$ and real $t \in]0,\varepsilon[$ satisfy $y + t \cdot w \in K$.

We now focus on a similar description in the morphological framework. To be more precise, we are going to specify subsets $\mathscr{T}_{\mathscr{V}}^H(K) \subset \mathrm{LIP}(\mathbb{R}^N,\mathbb{R}^N)$ of the circatangent transition sets $\mathscr{T}_{\mathscr{V}}^C(K)$, $K \in \mathscr{V}$, whose graph $\mathscr{V} \rightsquigarrow \mathrm{LIP}(\mathbb{R}^N,\mathbb{R}^N)$, $K \mapsto \mathscr{T}_{\mathscr{V}}^H(K)$ is identical to the interior of the graph of $\mathscr{T}_{\mathscr{V}}^C(\cdot)$ in $\mathscr{V} \times \mathrm{LIP}(\mathbb{R}^N,\mathbb{R}^N)$.

There is an essential difference between the vector space $\mathbb{R}^N$ and the metric space $(\mathscr{K}(\mathbb{R}^N), d\!\!l)$, however, preventing us from applying Definition 45 directly. Indeed, considering the neighborhood of a vector $y + t \cdot v$ (with $y, v \in \mathbb{R}^N$, $t > 0$), *each* of its points can be represented as $y + t\,w$ with a "perturbed" vector w close to v. The corresponding statement does not hold for reachable sets of differential inclusions in general: For given $F \in \mathrm{LIP}(\mathbb{R}^N,\mathbb{R}^N)$, $K \in \mathscr{K}(\mathbb{R}^N)$, $t > 0$, *not every* compact set $M \subset \mathbb{R}^N$ with arbitrarily small Hausdorff distance from $\vartheta_F(t,K)$ can be represented as reachable set $\vartheta_{\widetilde{G}}(t,K)$ with some $\widetilde{G} \in \mathrm{LIP}(\mathbb{R}^N,\mathbb{R}^N)$ "close to" F. As a typical example, we can consider $M := \vartheta_F(t,K) \setminus \mathbb{B}_\varepsilon(x_0)^\circ \in \mathscr{K}(\mathbb{R}^N)$ with an interior point x_0 of $\vartheta_F(t,K)$ and sufficiently small $\varepsilon > 0$.

For this reason, we prefer a different approach to the interior of Graph $\mathscr{T}_{\mathscr{V}}^C(\cdot)$, but use the terminology of hypertangents:

Definition 46. Consider the set $\mathrm{LIP}(\mathbb{R}^N,\mathbb{R}^N)$ with the topology of locally uniform convergence. For a nonempty subset $\mathscr{V} \subset \mathscr{K}(\mathbb{R}^N)$ and any element $K \in \mathscr{V}$,

$$\mathscr{T}_{\mathscr{V}}^H(K) := \Big\{ F \in \mathrm{LIP}(\mathbb{R}^N,\mathbb{R}^N) \;\Big|\; \begin{array}{l} \exists\ \varepsilon > 0, \text{ neighborhood } U \subset \mathrm{LIP}(\mathbb{R}^N,\mathbb{R}^N) \text{ of } F \\ \forall\ G \in U:\ \lim\limits_{h \downarrow 0} \frac{1}{h} \cdot \mathrm{dist}\big(\vartheta_G(h,M),\ \mathscr{V}\big) = 0 \\ \text{uniformly in } M \in \mathscr{V} \cap \mathbb{B}_\varepsilon(K) \end{array} \Big\}$$

is called *hypertangent transition set* of $\mathscr{V}$ at K (in the metric space $(\mathscr{K}(\mathbb{R}^N), d\!\!l)$).

Lemma 47. *Let $K \in \mathscr{K}(\mathbb{R}^N)$ be in the nonempty closed set $\mathscr{V} \subset (\mathscr{K}(\mathbb{R}^N), d\!l)$. Then, a set-valued map $F \in \mathrm{LIP}(\mathbb{R}^N, \mathbb{R}^N)$ belongs to the hypertangent transition set $\mathscr{T}_{\mathscr{V}}^H(K)$ if and only if there exist a radius $\varepsilon > 0$ and a neighborhood $U \subset \mathrm{LIP}(\mathbb{R}^N, \mathbb{R}^N)$ of F such that for each map $G \in U$, a modulus of continuity $\omega : [0,1] \longrightarrow [0,\infty[$ (i.e. $\lim_{\delta \to 0} \omega(\delta) = 0$) satisfies*

$$\tfrac{1}{h} \cdot \big(\mathrm{dist}\big(\vartheta_G(h,M), \mathscr{V}\big) \; - \; \mathrm{dist}(M, \mathscr{V})\big) \;\leq\; \omega(h)$$

for all $h \in]0,1]$ and $M \in \mathbb{B}_\varepsilon(K) \subset \mathscr{K}(\mathbb{R}^N)$.

The proof results from essentially the same arguments as Lemma 40 about the circatangent transition set (on page 427). Furthermore, in combination with Lemma 40, we conclude immediately:

Lemma 48. *For every nonempty closed subset $\mathscr{V} \subset \mathscr{K}(\mathbb{R}^N)$ and element $K \in \mathscr{V}$, the hypertangent transition set $\mathscr{T}_{\mathscr{V}}^H(K)$ is contained in the interior of the circatangent transition set $\mathscr{T}_{\mathscr{V}}^C(K)$.* □

For the same reason, we obtain an even more general result:

Lemma 49. *Consider the set $\mathrm{LIP}(\mathbb{R}^N, \mathbb{R}^N)$ with the topology of locally uniform convergence. For every nonempty closed subset $\mathscr{V} \subset \mathscr{K}(\mathbb{R}^N)$, the graph of hypertangent transition sets*

$$\mathscr{V} \rightsquigarrow \mathrm{LIP}(\mathbb{R}^N, \mathbb{R}^N), \quad K \mapsto \mathscr{T}_{\mathscr{V}}^H(K)$$

is contained in the interior of the graph of $\mathscr{V} \rightsquigarrow \mathrm{LIP}(\mathbb{R}^N, \mathbb{R}^N), \;\; K \mapsto \mathscr{T}_{\mathscr{V}}^C(K)$. □

In fact, also the opposite inclusion holds and thus, we have a complete characterization of the interior of Graph $\mathscr{T}_{\mathscr{V}}^C(\cdot)$ in $\mathscr{V} \times \mathrm{LIP}(\mathbb{R}^N, \mathbb{R}^N)$:

Proposition 50. *Let $\mathscr{V} \subset \mathscr{K}(\mathbb{R}^N)$ be nonempty and closed with respect to $d\!l$. Then,* Graph $\mathscr{T}_{\mathscr{V}}^H(\cdot) \subset \mathscr{V} \times \mathrm{LIP}(\mathbb{R}^N, \mathbb{R}^N)$ *is equal to the interior of* Graph $\mathscr{T}_{\mathscr{V}}^C(\cdot)$ *in $\mathscr{V} \times \mathrm{LIP}(\mathbb{R}^N, \mathbb{R}^N)$.*

Proof. Due to Lemma 49, we just have to show: If (K, F) belongs to the interior of Graph $\mathscr{T}_{\mathscr{V}}^C(\cdot)$ in $\mathscr{V} \times \mathrm{LIP}(\mathbb{R}^N, \mathbb{R}^N)$, then $F \in \mathscr{T}_{\mathscr{V}}^H(K)$.

There exist a radius $\rho > 0$ and a neighborhood $U \subset \mathrm{LIP}(\mathbb{R}^N, \mathbb{R}^N)$ of F (with respect to locally uniform convergence) such that all tuples $(M, G) \in (\mathscr{V} \cap \mathbb{B}_\rho(K)) \times U \subset \mathscr{K}(\mathbb{R}^N) \times \mathrm{LIP}(\mathbb{R}^N, \mathbb{R}^N)$ belong to Graph $\mathscr{T}_{\mathscr{V}}^C(\cdot)$. For an arbitrary set-valued map $G \in U$, we now prove indirectly

$$\limsup_{h \downarrow 0} \; \tfrac{1}{h} \cdot \mathrm{dist}\big(\vartheta_G(h,M), \mathscr{V}\big) \;=\; 0 \qquad \text{uniformly in } M \in \mathscr{V} \cap \mathbb{B}_\rho(K).$$

Otherwise there exist $\delta > 0$ and sequences $(h_n)_{n\in\mathbb{N}}$, $(M_n)_{n\in\mathbb{N}}$ in $]0,1[$ and $\mathscr{V} \subset \mathscr{K}(\mathbb{R}^N)$ respectively satisfying for all $n \in \mathbb{N}$,

$$\begin{cases} \operatorname{dist}\big(\vartheta_G(h_n, M_n),\ \mathscr{V}\big) \ \geq\ \delta \cdot h_n, \\ \qquad\qquad\qquad 0\ <\ h_n\ <\ \frac{1}{n}, \\ \qquad\qquad\quad dl(M_n, K)\ \leq\ \rho. \end{cases}$$

In the metric space $(\mathscr{K}(\mathbb{R}^N), dl)$, all bounded closed balls are compact according to Proposition 1.47 (on page 57). Thus, there is a subsequence $(M_{n_j})_{j\in\mathbb{N}}$ converging to a compact set $M \in \mathscr{V} \cap \mathbb{B}_\rho(K)$. Due to the choice of ρ and U, we obtain $G \in \mathscr{T}^C_{\mathscr{V}}(M)$ in particular. This contradicts, however,

$$\begin{cases} \liminf\limits_{j\to\infty} \ \frac{1}{h_{n_j}} \cdot \operatorname{dist}\big(\vartheta_G(h_{n_j}, M_{n_j}),\ \mathscr{V}\big) \geq \delta\ > 0 \\ \ \lim\limits_{j\to\infty}\ \ dl\big(M_{n_j}, M\big) \qquad\qquad\qquad\qquad = 0 \end{cases}$$

completing the indirect proof.

□

Remark 51. Circatangent transition set $\mathscr{T}^C_{\mathscr{V}}(K)$ and hypertangent transition set $\mathscr{T}^H_{\mathscr{V}}(K)$ differ from each other in an essential feature:

The condition on a map $F \in \mathscr{T}^C_{\mathscr{V}}(K)$ depends on $\mathscr{V} \subset \mathscr{K}(\mathbb{R}^N)$ close to K, of course, but only on reachable sets of the set-valued map F. In particular, it does not have any influence on this condition if we replace such a map $F \in \mathrm{LIP}(\mathbb{R}^N, \mathbb{R}^N)$ by its pointwise convex hull $\mathbb{R}^N \rightsquigarrow \mathbb{R}^N$, $x \mapsto \overline{\mathrm{co}}\ F(x)$ – due to Relaxation Theorem A.19 of Filippov-Ważewski and its Corollary A.21 (on page 453).

The condition on $F \in \mathscr{T}^H_{\mathscr{V}}(K)$, however, takes all set-valued maps $G \in \mathrm{LIP}(\mathbb{R}^N, \mathbb{R}^N)$ in a neighborhood of F into account. Considering the topology of locally uniform convergence in $\mathrm{LIP}(\mathbb{R}^N, \mathbb{R}^N)$, the values of these neighboring set-valued maps G do not have to be convex even if F belongs to $\mathrm{LIP}_{\overline{\mathrm{co}}}(\mathbb{R}^N, \mathbb{R}^N)$.

5.3.7 Closed Control Loops for Problems with State Constraints

In this section, we specify sufficient conditions on the morphological control system and state constraints for the existence of a closed-loop control, i.e., a continuous function $u(\cdot) : \mathscr{V} \longrightarrow U$ is to provide a feedback law such that for any initial set $K_0 \in \mathscr{V} \subset \mathscr{K}(\mathbb{R}^N)$, every solution $K(\cdot) : [0,T] \leadsto \mathbb{R}^N$ to the morphological equation

$$\begin{cases} \overset{\circ}{K}(\cdot) \ni f(K(\cdot),\, u(K(\cdot))) & \mathscr{L}^1\text{-a.e. in } [0,T] \\ K(0) \in K_0 \end{cases}$$

solves the morphological control problem with state constraints

$$\begin{cases} \overset{\circ}{K}(\cdot) \ni f(K(\cdot),u), \quad u \in U & \mathscr{L}^1\text{-a.e. in } [0,T] \\ K(t) \in \mathscr{V} & \text{for each } t \in [0,T]. \end{cases}$$

Corresponding to Aubin's notion of *regulation maps* [14, § 6], Nagumo's Theorem 1.74 (on page 76) motivates us to construct the wanted closed-loop control $u(\cdot) : \mathscr{V} \longrightarrow U$ as a continuous selection of the set-valued map

$$\mathscr{V} \leadsto U, \quad K \mapsto \{u \in U \mid f(K,u) \in \mathscr{T}_{\mathscr{V}}(K)\}$$

indicating "consistent" control parameters for preserving values in $\mathscr{V}$.

Applying Michael's famous Selection Theorem for lower semicontinuous, this approach has been developed for constrained control problems in the Euclidean space [14, § 6.6.1]. Our contribution now is to extend it to the morphological framework.

The key challenge is to specify appropriate subsets of the contingent transition set $\mathscr{T}_{\mathscr{V}}(K) \subset \mathrm{LIP}(\mathbb{R}^N,\mathbb{R}^N)$ so that "convenient" assumptions about them ensure the existence of a closed-loop control. For this purpose, we use circatangent transition set $\mathscr{T}_{\mathscr{V}}^C(K)$ and hypertangent transition set $\mathscr{T}_{\mathscr{V}}^H(K)$ introduced in § 5.3.5 and § 5.3.6. There is a close relation between these two subsets of the contingent transition set: Graph $\mathscr{T}_{\mathscr{V}}^H(\cdot)$ is the interior of the graph of $\mathscr{T}_{\mathscr{V}}^C(\cdot) : \mathscr{V} \leadsto \mathrm{LIP}(\mathbb{R}^N,\mathbb{R}^N)$ due to Proposition 50.

Now we can formulate the main result of this section:

Proposition 52 (Closed-loop control for morphological equations).
Let U be a separable Banach space and, consider the set $\mathrm{LIP}(\mathbb{R}^N,\mathbb{R}^N)$ *with the topology of locally uniform convergence. For a nonempty closed set $\mathscr{V} \subset (\mathscr{K}(\mathbb{R}^N), d\!l)$ and $f : \mathscr{K}(\mathbb{R}^N) \times U \longrightarrow \mathrm{LIP}(\mathbb{R}^N,\mathbb{R}^N)$ suppose:*

(1.) f is continuous and bounded in the sense that

$$\sup \big\{ \|f(M,u)\|_\infty + \mathrm{Lip}\, f(M,u) \;\big|\; M \in \mathscr{K}(\mathbb{R}^N),\ u \in U \big\} < \infty.$$

(2.) $R^H : \mathscr{V} \leadsto U,\ K \mapsto \{u \in U \mid f(K,u) \in \mathscr{T}_{\mathscr{V}}^H(K)\}$ *has nonempty convex values.*

Then, the pointwise closure $\overline{R}^H : \mathscr{V} \rightsquigarrow U,\ K \mapsto \overline{R^H(K)}$ *has a selection* $u \in C^0(\mathscr{V}, U)$. *In particular, every continuous and compact-valued solution* $K(\cdot) : [0,T] \rightsquigarrow \mathbb{R}^N$ *to the morphological equation*

$$\begin{cases} \overset{\circ}{K}(\cdot) \ni f(K(\cdot),\, u(K(\cdot))) & \text{a.e. in } [0,T[\\ K(0) \in K_0 \end{cases}$$

with initial set $K_0 \in \mathscr{V}$ *is viable in* $\mathscr{V}$, *i.e.* $K(t) \in \mathscr{V}$ *for all* $t \in [0,T]$.

In combination with Nagumo's theorem 1.74 (on page 76), Michael's well-known selection theorem lays the analytical basis. In particular, it requires a Banach space for the control set U (instead of a metric space as in the preceding subsections of § 5.3).

Proposition 53 (Michael [138],[15, Theorem 1.11.1], [19, Theorem 9.1.2]**).**
Let $R : X \rightsquigarrow Y$ *be a lower semicontinuous set-valued map with nonempty closed convex values from a compact metric space* X *to a Banach space* Y.
Then R *has a continuous selection, i.e. there exists a continuous single-valued function* $r : X \longrightarrow Y$ *with* $r(x) \in R(x)$ *for every* $x \in X$.

Proof (of Proposition 52).
Similarly to the proof of [14, Proposition 6.3.2], we first verify the lower semicontinuity of

$$R^H : \mathscr{V} \rightsquigarrow U, \quad K \mapsto \{u \in U \mid f(K,u) \in \mathscr{T}^H_{\mathscr{V}}(K)\}$$

(in the sense of Bouligand and Kuratowski).
Indeed, choose any $K \in \mathscr{V}$ and $u \in R^H(K)$. Graph $\mathscr{T}^H_{\mathscr{V}}$ is open in $\mathscr{V} \times \mathrm{LIP}(\mathbb{R}^N, \mathbb{R}^N)$ as a direct consequence of Definition 46. Hence, there is a radius $r > 0$ with

$$\big(\mathbb{B}_r(K) \times \mathbb{B}_r\big(f(K,u)\big)\big) \cap \big(\mathscr{V} \times \mathrm{LIP}(\mathbb{R}^N,\mathbb{R}^N)\big) \subset \text{Graph } \mathscr{T}^H_{\mathscr{V}},$$

i.e.
$$\mathbb{B}_r\big(f(K,u)\big) \subset \mathscr{T}^H_{\mathscr{V}}(M) \quad \text{for all } M \in \mathbb{B}_r(K) \cap \mathscr{V} \subset \mathscr{K}(\mathbb{R}^N).$$

Finally the continuity of f provides a smaller radius $\rho \in]0,r[$ with

$$f(M,v) \in \mathbb{B}_r\big(f(K,u)\big) \subset \mathscr{T}^H_{\mathscr{V}}(M)$$

for all $v \in \mathbb{B}_\rho(u) \subset U$ and $M \in \mathbb{B}_\rho(K) \cap \mathscr{V} \subset \mathscr{K}(\mathbb{R}^N)$. In particular, the intersection of the sets $R^H(M) \overset{\text{Def.}}{=} \{v \in U \mid f(M,v) \in \mathscr{T}^H_{\mathscr{V}}(M)\}$ for all $M \in \mathbb{B}_\rho(K) \cap \mathscr{V}$ contains the ball $\mathbb{B}_\rho(u) \subset U$ and thus, it is a neighborhood of $u \in R^H(K)$.
As a consequence, $R^H(\cdot) : \mathscr{V} \rightsquigarrow U$ is lower semicontinuous.

Now we consider the pointwise closure of R^H, i.e.

$$\overline{R}^H : \mathscr{V} \rightsquigarrow U, \quad K \mapsto \overline{\{u \in U \mid f(K,u) \in \mathscr{T}^H_{\mathscr{V}}(K)\}}.$$

Obviously, $\overline{R}^H(\cdot)$ has nonempty closed convex values in the Banach space U. Additionally, it inherits lower semicontinuity from $R^H(\cdot)$ as the topological criterion of lower semicontinuity (via neighborhoods) reveals easily.

For any nonempty compact ball $B \subset (\mathcal{K}(\mathbb{R}^N), d\!l)$, Michael's Theorem (quoted in Proposition 53) provides a continuous selection $u_B : B \cap \mathcal{V} \longrightarrow U$ of the set-valued restriction $\overline{R}^H\Big|_{B\cap\mathcal{V}} : B \cap \mathcal{V} \rightsquigarrow U$.

Finally we cover the metric space $(\mathcal{K}(\mathbb{R}^N), d\!l)$ with countably many balls and, a locally finite continuous partition of unity leads to a selection $u \in C^0(\mathcal{V}, U)$ of $\overline{R}^H : \mathcal{V} \rightsquigarrow U$ because all values of $\overline{R}^H$ are convex.

□

Appendix A
Tools

A.1 The Lemma of Gronwall and its Generalizations

Gronwall's estimate plays a key role whenever the growth of a function is bounded by linear terms of the function itself. Such a bound of the growth can be described by an integral inequality or a differential inequality.
First we consider the estimate resulting from an integral inequality. It is very popular indeed for continuous functions and thus can be found in many standard textbooks such as [10, 92, 181]. Subsequent Proposition A.1, however, provides a similar estimate (almost everywhere) for any nonnegative function that is merely Lebesgue integrable.

Proposition 1 (Lemma of Gronwall : Integral version**).**
Let $\psi, g \in L^1([a,b],\mathbb{R})$, $f \in C^0([a,b])$ *satisfy* $\psi(\cdot), f(\cdot) \geq 0$ *and*

$$\psi(t) \;\leq\; g(t) + \int_a^t f(s)\,\psi(s)\,ds \qquad \text{for } \mathscr{L}^1\text{-almost every } t \in [a,b].$$

Then, for $\mathscr{L}^1$*-almost every* $t \in [a,b]$,

$$\psi(t) \;\leq\; g(t) \;+\; \int_a^t e^{\mu(t)-\mu(s)}\, f(s)\, g(s)\, ds$$

with $\mu(t) := \int_a^t f(s)\,ds$.
Assuming in addition that $g(\cdot)$ *is upper semicontinuous and that* $\psi(\cdot)$ *is lower semicontinuous or monotone, then this inequality holds for any* $t \in\,]a,b[$.

Proof. The function $\varphi : [a,b] \longrightarrow \mathbb{R},\; t \longmapsto \int_a^t f(s)\,\psi(s)\,ds$ is absolutely continuous and satisfies for almost every $t \in [a,b]$ (since $f(\cdot) \geq 0$)

$$\varphi'(t) \;=\; f(t)\,\psi(t) \;\leq\; f(t)\,g(t) \;+\; f(t)\,\varphi(t).$$

Thus, $t \longmapsto e^{-\mu(t)}\,\varphi(t)$ is also absolutely continuous and has the weak derivative

$$\tfrac{d}{dt}\left(e^{-\mu(t)}\,\varphi(t)\right) = e^{-\mu(t)}\left(\varphi'(t) - f(t)\,\varphi(t)\right) \le e^{-\mu(t)}\,f(t)\,g(t).$$

Now we obtain for any $t \in [a,b]$

$$\begin{aligned} e^{-\mu(t)}\,\varphi(t) &\le e^{-\mu(a)}\,\varphi(a) + \int_a^t e^{-\mu(s)}\, f(s)\,g(s)\,ds \\ \varphi(t) &\le 0 + \int_a^t e^{\mu(t)-\mu(s)}\, f(s)\,g(s)\,ds \end{aligned}$$

and this estimate implies the assertion for Lebesgue-almost every $t \in [a,b]$.

Now suppose that $g(\cdot)$ is upper semicontinuous and that $\psi(\cdot)$ is lower semicontinuous or monotone. Then for every $t \in]a,b[$, there exists a sequence $(t_n)_{n\in\mathbb{N}}$ in $]a,b[$ such that $t_n \longrightarrow t \;\; (n \longrightarrow \infty)$ and

$$\begin{aligned} \psi(t) &\le \limsup_{n\to\infty}\; \psi(t_n), \\ \psi(t_n) &\le g(t_n) + \int_a^{t_n} e^{\mu(t_n)-\mu(s)} f(s)\,g(s)\,ds \end{aligned}$$

for each $n \in \mathbb{N}$. As an easy consequence, we obtain

$$\begin{aligned} \psi(t) &\le \limsup_{n\to\infty}\left(g(t_n) + \int_a^{t_n} e^{\mu(t_n)-\mu(s)}\, f(s)\,g(s)\,ds\right) \\ &\le g(t) + \int_a^{t} e^{\mu(t)-\mu(s)}\, f(s)\,g(s)\,ds. \end{aligned}$$

□

This integral version of Gronwall's Lemma now leads to a subdifferential version which has two new aspects: First, the nonnegative function $\psi(\cdot)$ does not have be continuous, but just lower semicontinuous (as in [130]). Second, the hypothesis about an affine linear bound of the upper Dini derivative is not required in the whole time interval, but just at Lebesgue-almost every time. The proof is based on a connection to Proposition A.1 by means of a nondecreasing auxiliary function (in combination with Fatou's Lemma):

Proposition 2. *Let $\psi : [a,b] \longrightarrow \mathbb{R}$ and $f,g \in C^0([a,b],\mathbb{R})$ satisfy $f(\cdot), g(\cdot) \ge 0$ and*

$$\begin{aligned} 0 \le \psi(t) &\le \limsup_{h\downarrow 0}\; \psi(t-h), && \text{for every } t \in\,]a,\,b], \\ \psi(t) &\ge \limsup_{h\downarrow 0}\; \psi(t+h), && \text{for every } t \in [a,\,b[, \end{aligned}$$

$$\limsup_{h\downarrow 0} \frac{\psi(t+h)-\psi(t)}{h} \le f(t)\; \limsup_{h\downarrow 0}\, \psi(t-h) + g(t) \quad \text{for almost every } t \in\,]a,\,b[.$$

Then, for every $t \in [a,b]$, the function $\psi(\cdot)$ fulfills the upper estimate

$$\psi(t) \le \psi(a)\cdot e^{\mu(t)} + \int_a^t e^{\mu(t)-\mu(s)}\, g(s)\,ds$$

with $\mu(t) := \int_a^t f(s)\,ds.$

Proof. Obviously, the auxiliary function $\xi : [a,b] \longrightarrow \mathbb{R}_0^+,\ t \longmapsto \sup_{[a,t]} \psi(\cdot)$ is nonnegative and nondecreasing. The second assumption about $\psi(\cdot)$ implies the continuity of $\xi(\cdot)$. Furthermore, it satisfies for $\mathscr{L}^1$-almost every $t \in]a,b[$

$$\limsup_{h \downarrow 0} \frac{\xi(t+h) - \xi(t)}{h} \le f(t) \cdot \xi(t) + g(t).$$

Indeed, choose any $t \in]a,b[$ for which the third assumption about ψ is satisfied. Then for any $\delta > 0$, there exists some $h_0 \in]0, b-t[$ such that for all $h \in]0, h_0]$,

$$\begin{aligned} \frac{\psi(t+h) - \psi(t)}{h} &\le f(t) \cdot \xi(t) + g(t) + \delta \\ \text{i.e.} \qquad \psi(t+h) &\le \big(f(t) \cdot \xi(t) + g(t) + \delta\big) \cdot h + \psi(t) \\ &\le \big(f(t) \cdot \xi(t) + g(t) + \delta\big) \cdot h + \xi(t). \end{aligned}$$

Hence, $\xi(t+h) = \max\big\{\xi(t),\ \sup_{[t,t+h]} \psi(\cdot)\big\}$ fulfills this estimate for all $h \in]0, h_0]$:

$$\begin{aligned} \xi(t+h) &\le \big(f(t) \cdot \xi(t) + g(t) + \delta\big) \cdot h + \xi(t) \\ \frac{\xi(t+h) - \xi(t)}{h} &\le f(t) \cdot \xi(t) + g(t) + \delta. \end{aligned}$$

As $\delta > 0$ was chosen arbitrarily, we obtain the claimed estimate for the upper Dini derivative of $\xi(\cdot)$ at t.

In particular, the continuous function $\xi(\cdot)$ is bounded in the compact interval $[a,b]$ and thus, so is $\psi(\cdot)$. The auxiliary function

$$[a,b[\longrightarrow \mathbb{R}_0^+, \qquad t \longmapsto \limsup_{h \downarrow 0} \frac{\xi(t+h) - \xi(t)}{h}$$

is Lebesgue-measurable and bounded Lebesgue-almost everywhere. The well-known Lemma of Fatou implies for every $T \in [a,b[$

$$\limsup_{h \downarrow 0} \int_0^T \frac{\xi(t+h) - \xi(t)}{h}\, dt \le \int_0^T \limsup_{h \downarrow 0} \frac{\xi(t+h) - \xi(t)}{h}\, dt$$

and thus lays the basis for estimating $\xi(T) - \xi(0)$:

$$\begin{aligned} \limsup_{h \downarrow 0} \int_0^T \frac{\xi(t+h) - \xi(t)}{h}\, dt &= \limsup_{h \downarrow 0} \frac{1}{h} \cdot \Big(\int_0^T \xi(t+h)\, dt - \int_0^T \xi(t)\, dt\Big) \\ &= \limsup_{h \downarrow 0} \frac{1}{h} \cdot \Big(\int_T^{T+h} \xi(t)\, dt - \int_0^h \xi(t)\, dt\Big) \\ &= \qquad \xi(T) - \xi(0) \end{aligned}$$

due to the continuity of $\xi(\cdot)$. Now we obtain an estimate for $\xi(T)$ for *every* $T \in [a,b[$

$$\xi(T) - \xi(0) \le \int_0^T \limsup_{h \downarrow 0} \frac{\xi(t+h) - \xi(t)}{h}\, dt \le \int_0^T \big(f(t) \cdot \xi(t) + g(t)\big)\, dt.$$

Finally, the claim results from Proposition A.1. □

Remark 3. 1. This subdifferential version of Gronwall's Lemma also holds if $f, g : [a, b[\longrightarrow \mathbb{R}_0^+$ are only upper semicontinuous (instead of continuous). The proof is based on upper approximations of $f(\cdot)$, $g(\cdot)$ by continuous functions.

2. The condition $\limsup_{h \downarrow 0} \frac{\psi(t+h)-\psi(t)}{h} \le f(t) \cdot \psi(t) + g(t)$ (supposed in the widespread forms of Gronwall's Lemma) is stronger than the third assumption of Proposition A.2 due to the semicontinuity condition $\psi(t) \le \limsup_{h \downarrow 0} \psi(t-h)$.

A similar statement holds with limits inferior replacing the limits superior — under the additional assumption, however, that the growth condition is fulfilled at *every* time (instead of $\mathscr{L}^1$-almost every time). The proof presented by the author in [130] is based on a simple indirect argument and thus, it is completely independent of the integral version in Proposition A.1:

Proposition 4. *Let $\psi : [a, b] \longrightarrow \mathbb{R}$ and $f, g \in C^0([a, b], \mathbb{R})$ satisfy $f(\cdot) \ge 0$ and*

$$
\begin{aligned}
0 \le \psi(t) &\le \liminf_{h \downarrow 0} \psi(t-h), && \text{for every } t \in]a, b], \\
\psi(t) &\ge \liminf_{h \downarrow 0} \psi(t+h), && \text{for every } t \in [a, b[, \\
\liminf_{h \downarrow 0} \frac{\psi(t+h)-\psi(t)}{h} &\le f(t) \cdot \liminf_{h \downarrow 0} \psi(t-h) + g(t) && \text{for every } t \in]a, b[.
\end{aligned}
$$

Then, for every $t \in [a, b]$, the function $\psi(\cdot)$ fulfills the upper estimate

$$
\psi(t) \le \psi(a) \cdot e^{\mu(t)} + \int_a^t e^{\mu(t)-\mu(s)} g(s)\, ds
$$

with $\mu(t) := \int_a^t f(s)\, ds.$

Proof. Let $\delta > 0$ be arbitrarily small. The proof is based on comparing ψ with the auxiliary function $\varphi_\delta : [a, b] \longrightarrow \mathbb{R}$ that uses $\psi(a) + \delta$, $g(\cdot) + \delta$ instead of $\psi(a)$, $g(\cdot)$:

$$
\varphi_\delta(t) := \big(\psi(a) + \delta\big)\, e^{\mu(t)} + \int_a^t e^{\mu(t)-\mu(s)} \big(g(s) + \delta\big)\, ds.
$$

Then,
$$
\begin{aligned}
\varphi_\delta'(t) &= f(t)\, \varphi_\delta(t) + g(t) + \delta && \text{in } [a, b[, \\
\varphi_\delta(s_n) &> \psi(s_n) && \text{for some sequence } s_n \downarrow a.
\end{aligned}
$$

Assume now that there exists some $t_0 \in]a, b]$ such that $\varphi_\delta(t_0) < \psi(t_0)$. Setting

$$
t_1 := \inf \big\{ t \in [a, t_0] \;\big|\; \varphi_\delta(\cdot) < \psi(\cdot) \text{ in } [t, t_0] \big\} \ge s_1 > a,
$$

we conclude $t_1 < t_0$ from the condition $\psi(t_0) \le \liminf_{h \downarrow 0} \psi(t_0 - h)$ and the continuity of $\varphi_\delta(\cdot)$. Moreover, $\varphi_\delta(t_1) = \psi(t_1)$ is a consequence of

$$
\begin{aligned}
\varphi_\delta(t_1) &= \lim_{h \downarrow 0} \varphi_\delta(t_1 - h) \ge \liminf_{h \downarrow 0} \psi(t_1 - h) \ge \psi(t_1), \\
\varphi_\delta(t_1) &= \lim_{h \downarrow 0} \varphi_\delta(t_1 + h) \le \liminf_{h \downarrow 0} \psi(t_1 + h) \le \psi(t_1).
\end{aligned}
$$

Thus, the definition of t_1 implies

$$\begin{aligned}\liminf_{h\downarrow 0} \frac{\varphi_\delta(t_1+h)-\varphi_\delta(t_1)}{h} &\le \liminf_{h\downarrow 0} \frac{\psi(t_1+h)-\psi(t_1)}{h}\\ \varphi_\delta'(t_1) &\le f(t_1)\cdot \liminf_{h\downarrow 0} \;\; \psi(t_1-h) + g(t_1)\\ f(t_1)\,\varphi_\delta(t_1)+g(t_1)+\delta &\le f(t_1)\cdot \limsup_{h\downarrow 0}\; \varphi_\delta(t_1-h) + g(t_1)\\ &\le f(t_1)\cdot \varphi_\delta(t_1) \qquad\qquad\;\; + g(t_1)\end{aligned}$$

— a contradiction. Finally, $\varphi_\delta(\cdot) \ge \psi(\cdot)$ for any $\delta > 0$. □

A.2 Filippov's Theorem for Differential Inclusions

According to the well-known convention, we define the solutions to a differential inclusion in the sense of Carathéodory as it is described e.g. in [15, 19]. The Theorem of Filippov represents the counterpart of the Cauchy-Lipschitz Theorem about ordinary differential equations.

Definition 5. Let $\widetilde{F} : [0,T]\times\mathbb{R}^N \leadsto \mathbb{R}^N$ be a set-valued map. A curve $x : [0,T] \longrightarrow \mathbb{R}^N$ is called a *solution* to the differential inclusion $x'(\cdot) \in \widetilde{F}(\cdot,x(\cdot))$ a.e. if $x(\cdot)$ is absolutely continuous and its (weak) derivative $x'(\cdot)$ satisfies $x'(t) \in \widetilde{F}(t,x(t))$ for Lebesgue-almost every $t \in [0,T]$.

The *reachable set* of $\widetilde{F}$ and a nonempty initial set $M \subset \mathbb{R}^N$ at time $t \in [0,T]$ contains the points $x(t)$ of all solutions $x(\cdot)$ of $x'(\cdot) \in \widetilde{F}(\cdot,x(\cdot))$ a.e. starting in M, i.e.

$$\vartheta_{\widetilde{F}}(t,M) := \Big\{ x(t)\in\mathbb{R}^N \;\Big|\; \begin{array}{l} x(\cdot) \in W^{1,1}([0,t],\mathbb{R}^N), \quad x(0)\in M,\\ x'(\cdot) \in \widetilde{F}(\cdot,x(\cdot)) \;\; \mathscr{L}^1\text{-almost everywhere in } [0,t] \end{array}\Big\}.$$

Theorem 6 (Generalized Theorem of Filippov).
Let $\mathscr{O}$ be a relatively open subset of $[0,T]\times\mathbb{R}^N$. Take a set-valued map $\widetilde{F} : \mathscr{O} \leadsto \mathbb{R}^N$, an arc $y(\cdot)\in W^{1,1}([0,T],\mathbb{R}^N)$, a point $\eta\in\mathbb{R}^N$ and $\delta\in\,]0,\infty]$ such that

$$\mathscr{N}(y,\delta) := \bigcup_{0\le t\le T} \{t\}\times\mathbb{B}_\delta(y(t)) \subset \mathscr{O}.$$

Assume that

(i) $\widetilde{F}(t,z) \neq \emptyset$ *is closed for every* $(t,z) \in \mathcal{N}(y,\delta)$ *and*
Graph $\widetilde{F}$ *is* $\mathcal{L}^1 \times \mathcal{B}^N$ *measurable,*

(ii) *there exists* $k(\cdot) \in L^1([0,T])$ *such that* $\widetilde{F}(t,z_1) \subset \widetilde{F}(t,z_2) + k(t)\,|z_1 - z_2| \cdot \mathbb{B}_1$ *for all* $z_1, z_2 \in \mathbb{B}_\delta(y(t))$ *and Lebesgue-almost every* $t \in [0,T]$.

Suppose further

$$e^{\|k\|_{L^1}} \cdot \Big(|\eta - y(0)| + \int_0^T \operatorname{dist}\Big(y'(t),\ \widetilde{F}(t,y(t))\Big)\, dt \Big) \le \delta.$$

Then there exists a solution $x(\cdot) \in W^{1,1}([0,T],\mathbb{R}^N)$ *of* $x'(\cdot) \in \widetilde{F}(\cdot, x(\cdot))$ *a.e. satisfying* $x(0) = \eta$ *and*

$$\|x - y\|_{L^\infty} \le |\eta - y(0)|\; e^{\|k\|_{L^1}} + \int_0^T e^{\int_t^T k(s)\,ds} \operatorname{dist}\Big(y'(t),\ \widetilde{F}(t,y(t))\Big)\, dt.$$

Now assume that (i) *and* (ii) *are replaced by the stronger hypotheses:*

(i′) $\widetilde{F}(t,z) \neq \emptyset$ *is convex and compact for every* $(t,z) \in \mathcal{N}(y,\delta)$,

(ii′) *there exist* $\omega(\cdot) : [0,\infty[\longrightarrow [0,\infty[$ *and* $k_\infty \in]0,\infty[$ *such that* $\lim_{h \downarrow 0} \omega(h) = 0$,
$$\widetilde{F}(t_1,z_1) \subset \widetilde{F}(t_2,z_2) + \Big(k_\infty\,|z_1 - z_2| + \omega(|t_1 - t_2|)\Big)\, \mathbb{B}_1$$
for all $(t_1,z_1),\ (t_2,z_2) \in \mathcal{N}(y,\delta)$.

If $y(\cdot)$ *is continuously differentiable, then the solution* $x(\cdot)$ *can be chosen as a continuously differentiable function too.*

Proof is given in [180, Theorem 2.4.3], for example.

For applying Filippov's Theorem to *compact* reachable sets in $\mathbb{R}^N$, we combine some global properties of a set-valued map $\widetilde{F} : [0,T] \times \mathbb{R}^N \rightsquigarrow \mathbb{R}^N$ of space and time and coin the new term "Filippov continuous". It reflects the gist of the feature "measurable/Lipschitz" defined in [19, Definition 9.5.1] – but in a more detailed formulation.

Definition 7. A set-valued map $\widetilde{F} : [0,T] \times \mathbb{R}^N \rightsquigarrow \mathbb{R}^N$ is called *Filippov continuous* if it satisfies the following conditions:

1.) all values of $\widetilde{F}$ are nonempty closed subsets of $\mathbb{R}^N$,
2.) Graph $\widetilde{F} \subset [0,T] \times \mathbb{R}^N \times \mathbb{R}^N$ belongs to $\mathcal{L}^1 \otimes \mathcal{L}^N \otimes \mathcal{B}^N$,
3.) $\widetilde{F}$ has at most linear growth, i.e. $\sup_{(t,x) \in [0,T] \times \mathbb{R}^N} \ \sup_{v \in \widetilde{F}(t,x)} \frac{|v|}{|x|+|t|+1} < \infty$.
4.) there is $\lambda(\cdot) \in L^1([0,T],\mathbb{R})$ such that at Lebesgue-almost every time $t \in [0,T]$, the set-valued map $\widetilde{F}(t,\cdot) : \mathbb{R}^N \rightsquigarrow \mathbb{R}^N$ is $\lambda(t)$-Lipschitz w.r.t. dl.

Here $\mathcal{L}^N$ consists of all Lebesgue subsets of $\mathbb{R}^N$ and, $\mathcal{B}^N$ denotes the set of all Borel subsets of $\mathbb{R}^N$. Condition (2.) is equivalent to the measurability of the set-valued map $\widetilde{F}$ according to Characterization Theorem A.75 (on page 489) below. Furthermore, the linear growth condition (3.) implies first that all values of $\widetilde{F}$ are

compact and second that Gronwall's Lemma provides locally uniform bounds for solutions to the corresponding nonautonomous differential inclusion.
These conditions are slightly stronger than the assumptions of Theorem A.6. Indeed, Theorem A.6 does not assume the linear growth condition (3.) and, Lipschitz continuity with respect to space is supposed only locally. These distinctions result from different emphases: Theorem A.6 focuses on spatially local aspects of existence of solutions to a differential inclusion. We, however, aim for conclusions about reachable sets in the whole Euclidean space. The additional linear growth condition (3.), for example, is to ensure that we can restrict our geometric considerations to compact neighborhoods of compact initial sets.

Proposition 8 (Invariance Theorem). *Let $\widetilde{F} : [0,T] \times \mathbb{R}^N \leadsto \mathbb{R}^N$ be Filippov continuous. Assume the nonempty closed set $K \subset \mathbb{R}^N$ to satisfy*

$$F(t,x) \subset T_K(x) \qquad \text{for every } x \in K \text{ and } \mathscr{L}^1\text{-almost every } t \in [0,T].$$

with $T_K(x) \subset \mathbb{R}^N$ denoting the contingent cone of K at x in the sense of Bouligand.

Then every solution $x(\cdot) \in W^{1,1}([t_1,t_2],\mathbb{R}^N)$ to the differential inclusion $x'(\cdot) \in \widetilde{F}(\cdot,x(\cdot))$ a.e. with $[t_1,t_2] \subset [0,T]$ and $x(t_1) \in K$ has all its values in K.

Proof. It adapts the standard proof of [14, Theorem 5.3.4] that deals with autonomous differential inclusions.
Every solution $x(\cdot) \in W^{1,1}([t_1,t_2],\mathbb{R}^N)$ of $x'(\cdot) \in \widetilde{F}(\cdot,x(\cdot))$ a.e. is even Lipschitz continuous due to the linear growth condition on $\widetilde{F}$ (and Gronwall's Lemma). The auxiliary distance function $\delta : [t_1,t_2] \longrightarrow \mathbb{R},\ t \longmapsto \operatorname{dist}(x(t), K)$ is Lipschitz continuous. Whenever $x(\cdot)$ and $\delta(\cdot)$ are differentiable at time $t \in [t_1,t_2]$, it satisfies with a projection point $y_t \in K$ of $x(t)$ (i.e. $|x(t)-y_t| = \operatorname{dist}(x(t),K)$) and any $v \in \mathbb{R}^N$

$$\begin{aligned}
\delta'(t) &\le \liminf_{h\downarrow 0} \tfrac{1}{h}\cdot \big(\operatorname{dist}(x(t+h),K) - |x(t)-y_t|\big)\\
&\le \liminf_{h\downarrow 0} \tfrac{1}{h}\cdot \ \operatorname{dist}\Big(y_t + \int_t^{t+h} x'(s)\,ds,\ K\Big)\\
&\le \liminf_{h\downarrow 0} \tfrac{1}{h}\cdot \Big(\operatorname{dist}\big(y_t + h\,v,\ K\big) + \Big|h\,v - \int_t^{t+h} x'(s)\,ds\Big|\Big)\\
&\le \liminf_{h\downarrow 0} \tfrac{1}{h}\cdot \ \operatorname{dist}\big(y_t + h\,v,\ K\big) + \ \big|v - x'(t)\big|.
\end{aligned}$$

Selecting now $v \in \widetilde{F}(t,y_t)$ with $|x'(t)-v| \le d\!l(\widetilde{F}(t,x(t)), \widetilde{F}(t,y_t))$, we conclude from $\widetilde{F}(t,y_t) \subset T_K(y_t)$ and the $\lambda(t)$-Lipschitz continuity of $\widetilde{F}(t,\cdot)$ the estimate

$$\delta'(t) \ \le\ 0 + d\!l(\widetilde{F}(t,x(t)), \widetilde{F}(t,y_t)) \ \le\ \lambda(t)\,|x(t)-y_t| \ =\ \lambda(t)\,\delta(t)$$

for $\mathscr{L}^1$-almost every $t \in [t_1,t_2]$. According to Gronwall's Lemma (Proposition A.2), $\delta(0) = 0$ implies $\delta(\cdot) \equiv 0$ and thus, every value $x(t)$ belongs to the closed set K. □

A.3 Scorza-Dragoni Theorem and Applications to Reachable Sets

The classical theorem of Scorza-Dragoni [167] can be extended to functions between metric spaces as shown by Ricceri and Villani. A so-called Carathéodory function depends on two arguments, namely "time" (in a topological space like $\mathbb{R}$) and "state" (in a metric space). By definition, it is measurable with respect to time and continuous with respect to state. The key point of Scorza-Dragoni is to guarantee continuity with respect to both arguments on "almost" the whole domain in the following sense:

Proposition 9 ([160, Theorem 1]). *Let S be a compact Hausdorff topological space, μ a Radon measure on S and X, Y metric spaces. Suppose X to be separable.*

Then every Carathéodory function $g : S \times X \longrightarrow Y$ satisfies the so-called Scorza-Dragoni property, i.e. for every $\varepsilon > 0$, there exists a closed subset $S_\varepsilon \subset S$ with $\mu(S \setminus S_\varepsilon) < \varepsilon$ such that the restriction $g|_{S_\varepsilon \times X}$ is continuous.

Now this proposition can be regarded as a counterpart of well-known Lusin's Theorem (relating measurability to continuity almost everywhere) – but now for functions with two arguments.
In 1977 Jarnik and Kurzweil published an extension of the Scorza-Dragoni Theorem to set-valued maps which are measurable in time and upper semicontinuous in space [96]:

Proposition 10 ([83, Corollary 2.2], [96]). *Let X be a separable metric space. Suppose that $\widetilde{F} : [0,T] \times X \rightsquigarrow \mathbb{R}^N$ has convex closed values and for $\mathscr{L}^1$-almost all $t \in [0,T]$, $\widetilde{F}(t,\cdot)$ is upper semicontinuous. Assume that $\widetilde{F}$ is measurably bounded, i.e. there is a measurable function $\beta : [0,T] \longrightarrow \mathbb{R}$ such that for $\mathscr{L}^1$-almost all $t \in [0,T]$ and every $x \in X$, $|\widetilde{F}(t,x)|_\infty \leq \beta(t)$.*

Then there exists a set-valued map $\widehat{F} : [0,T] \times X \rightsquigarrow \mathbb{R}^N$ with closed convex values satisfying the following conditions:

1. *For $\mathscr{L}^1$-almost all $t \in [0,T]$ and for all $x \in X$, $\widehat{F}(t,x) \subset \widetilde{F}(t,x)$.*
2. *For every measurable set $\Lambda \subset [0,T]$ and any measurable maps $u : \Lambda \longrightarrow X$, $v : \Lambda \longrightarrow \mathbb{R}^N$ with $v(\cdot) \in \widetilde{F}(\cdot, u(\cdot))$ $\mathscr{L}^1$-a.e. in Λ, we have $v(\cdot) \in \widehat{F}(\cdot, u(\cdot))$ a.e.*
3. *For any $\varepsilon > 0$, there is a closed set $J_\varepsilon \subset [0,T]$ such that $\mathscr{L}^1([0,T] \setminus J_\varepsilon) < \varepsilon$ and $\widehat{F}|_{J_\varepsilon \times X}$ is upper semicontinuous.*

This proposition provides a useful tool for investigating nonautonomous differential inclusions with set-valued maps being measurable in time and upper semicontinuous in space. Indeed, it bridges the gap to differential inclusions with upper semicontinuous right-hand side. Motivated by the nomenclature of Aubin in [14], we introduce the following abbreviating term for this type of set-valued maps:

Definition 11. A set-valued map $\widetilde{F} : [0,T] \times \mathbb{R}^N \rightsquigarrow \mathbb{R}^N$, $(t,x) \mapsto \widetilde{F}(t,x)$ is called *nonautonomous Marchaud map* if it has the following properties:

1. $\widetilde{F}$ is nontrivial (i.e. Graph $\widetilde{F} \neq \emptyset$),
2. $\widetilde{F}(t,\cdot)$ is upper semicontinuous for Lebesgue-almost every $t \in [0,T]$,
3. $\widetilde{F}(\cdot,x)$ is measurable for every $x \in \mathbb{R}^N$,
4. $\widetilde{F}$ has compact convex values and
5. there exists $\mu(\cdot) \in L^1([0,T])$ such that $\widetilde{F}(t,x) \subset \mu(t)\,(1+|x|)\,\mathbb{B}$ for all $x \in \mathbb{R}^N$ and Lebesgue-almost every $t \in [0,T]$.

Such a Scorza-Dragoni type theorem also holds for set-valued maps being continuous with respect to space at Lebesgue-almost every time. Frankowska, Plaskacz and Rzeżuchowski concluded the following version from their counterpart of Proposition A.10 by means of a single-valued parameterization [83]. Alternatively, it can be regarded as a special case of Proposition A.9 with values in the metric space $Y := (\mathscr{K}(\mathbb{R}^N), d\!l)$.

Proposition 12 ([83, Theorem 2.4]). *Let the set-valued map $\widetilde{F} : [0,T] \times \mathbb{R}^N \rightsquigarrow \mathbb{R}^N$, $(t,x) \mapsto \widetilde{F}(t,x)$ have nonempty compact values, be measurable with respect to t and continuous with respect to x.*
Then for every $\varepsilon > 0$, there exists a closed set $J_\varepsilon \subset [0,T]$ with $\mathscr{L}^1([0,T] \setminus J_\varepsilon) < \varepsilon$ for which the restriction $\widetilde{F}|_{J_\varepsilon \times \mathbb{R}^N}$ is continuous.

Applications to Reachable Sets: Integral Funnel Equation

Considering a nonautonomous differential inclusion, the set-valued map on its right-hand side provides a first-order approximation of the reachable set starting in an arbitrary point. For various nonautonomous differential inclusions with continuous right-hand side, this result is well-known as *integral funnel equation* due to papers of Kurzhanski, Filippova, Panasyuk, Tolstonogov and others (e.g. [110, 152]).

In [83], Frankowska, Plaskacz and Rzeżuchowski extended such approximating results to differential inclusions whose right-hand sides are just measurable in time. Their detailed estimates of the Hausdorff distances, however, are formulated for an arbitrary initial point in space (rather than initial sets). Now we verify that these estimates hold even locally uniformly in space and time:

Proposition 13. *Let the set-valued map $\widetilde{F} : [0,T] \times \mathbb{R}^N \rightsquigarrow \mathbb{R}^N$ satisfy*
1. $\widetilde{F}$ has nonempty closed convex values,
2. for $\mathscr{L}^1$-almost all $t \in [0,T]$, the map $\mathbb{R}^N \rightsquigarrow \mathbb{R}^N$, $x \mapsto \widetilde{F}(t,x)$ is continuous,
3. for every $x \in \mathbb{R}^N$, the map $[0,T] \rightsquigarrow \mathbb{R}^N$, $t \mapsto \widetilde{F}(t,x)$ is measurable,
4. there exists $\mu(\cdot) \in L^1([0,T])$ with $\big|\widetilde{F}(t,x)\big|_\infty \leq \mu(t)$ for all $x \in \mathbb{R}^N$ and a.e. t.

Then, there exists a set $J \subset [0,T]$ of full Lebesgue measure (i.e. $\mathscr{L}^1([0,T] \setminus J) = 0$) such that for every $t \in J$ and $K \in \mathscr{K}(\mathbb{R}^N)$,

$$\tfrac{1}{h} \cdot dl\Big(\vartheta_{\widetilde{F}(t+\cdot,\cdot)}(h, K), \; \bigcup_{x \in K} \big(x + h \cdot \widetilde{F}(t,x)\big)\Big) \longrightarrow 0 \qquad \text{for } h \downarrow 0.$$

Proof consists of subsequent Corollary A.15 and Lemma A.16 focusing on the Pompeiu-Hausdorff excesses

$$\begin{aligned} h &\longmapsto \operatorname{dist}\Big(\vartheta_{\widetilde{F}(t+\cdot,\cdot)}(h, K), \quad \bigcup_{x \in K} \big(x + h \cdot \widetilde{F}(t,x)\big)\Big), \\ h &\longmapsto \operatorname{dist}\Big(\bigcup_{x \in K} \big(x + h \cdot \widetilde{F}(t,x)\big), \; \vartheta_{\widetilde{F}(t+\cdot,\cdot)}(h, K), \Big) \end{aligned}$$

respectively. Indeed, the subsequent inclusions are locally uniform with respect to the initial point $x \in K$ and small time $h > 0$.

Lemma 14. *Let $\widetilde{F} : [0,T] \times \mathbb{R}^N \leadsto \mathbb{R}^N$ be a nonautonomous Marchaud map with nonempty (compact convex) values.*
Then there exists a set $J \subset [0,T]$ of full measure (i.e. $\mathscr{L}^1([0,T] \setminus J) = 0$) with the following property: For every $t_0 \in J$, $x_0 \in \mathbb{R}^N$ and $\varepsilon \in]0,1[$, there are $t_1 > 0$ and $\delta > 0$ satisfying for all $x \in \mathbb{B}_\delta(x_0)$, $h \in]0,t_1[$.

$$\vartheta_{\widetilde{F}(t_0+\cdot,\cdot)}(h,x) \;\subset\; x + h\,\big(\widetilde{F}(t_0,x_0) + \varepsilon\,\mathbb{B}\big).$$

Applying this result to every time $t_0 \in J \subset [0,T]$ at which $\widetilde{F}(t,\cdot) : \mathbb{R}^N \leadsto \mathbb{R}^N$ is continuous in addition, we obtain directly:

Corollary 15. *Under the assumptions of Proposition A.13, there exists a subset $J \subset [0,T]$ of full measure (i.e. $\mathscr{L}^1([0,T] \setminus J) = 0$) with the following property: For every $t_0 \in J$, $x_0 \in \mathbb{R}^N$ and $\varepsilon \in]0,1[$, there are $t_1 > 0$ and $\delta > 0$ satisfying*

$$\vartheta_{\widetilde{F}(t_0+\cdot,\cdot)}(h,x) \;\subset\; x + h\,\big(\widetilde{F}(t_0,x) + 2\,\varepsilon\,\mathbb{B}\big)$$

for all $x \in \mathbb{B}_\delta(x_0)$, $h \in]0,t_1[$. □

Before proving Lemma A.14 in detail, we formulate the opposite inclusion correctly. This completes the proof of Proposition A.13.

Lemma 16. *Under the assumptions of Proposition A.13, there exists a subset $J \subset [0,T]$ of full measure (i.e. $\mathscr{L}^1([0,T] \setminus J) = 0$) with the following property: For every $t_0 \in J$, $x_0 \in \mathbb{R}^N$ and $\varepsilon \in]0,1[$, there are $t_1 > 0$ and $\delta > 0$ satisfying*

$$x + h\,\widetilde{F}(t_0,x) \;\subset\; \vartheta_{\widetilde{F}(t_0+\cdot,\cdot)}(h,x) \;+\; \varepsilon\,h\,\mathbb{B}$$

for all $x \in \mathbb{B}_\delta(x_0)$, $h \in]0,t_1[$.

Finally we now discuss the missing proofs of Lemmas A.14 and A.16:

Proof (of Lemma A.14). It follows the same arguments of [83, Lemma 2.6] and thus uses the basic idea of Rzeżuchowski in [166].
Let $\widehat{F} : [0,T] \times \mathbb{R}^N \rightsquigarrow \mathbb{R}^N$ denote the set-valued map according to Scorza-Dragoni type Proposition A.10. For any $\gamma > 0$, there exists a closed subset $\widetilde{J}_\gamma \subset [0,T]$ with $\mathscr{L}^1([0,T] \setminus \widetilde{J}_\gamma) < \gamma$ such that $\widehat{F}|_{\widetilde{J}_\gamma \times \mathbb{R}^N}$ is upper semicontinuous and

$$\text{Graph}\, \widehat{F}|_{\widetilde{J}_\gamma \times \mathbb{R}^N} \subset \text{Graph}\, \widetilde{F}.$$

Now let $J_\gamma \subset \widetilde{J}_\gamma$ denote the set of density points of $\widetilde{J}_\gamma$ that are also Lebesgue points of $\mu(\cdot) \cdot \chi_{[0,T] \setminus \widetilde{J}_\gamma}(\cdot) : [0,T] \longrightarrow \mathbb{R}$. It satisfies $\mathscr{L}^1(J_\gamma) = \mathscr{L}^1(\widetilde{J}_\gamma)$ because Lebesgue points of each Lebesgue-integrable function always have full Lebesgue measure [189, Theorem 1.3.8] and thus, in particular, density points of any measurable set also have full Lebesgue measure.

For arbitrary $t_0 \in J_\gamma$, $x_0 \in \mathbb{R}^N$ and $\varepsilon \in]0,1]$, the upper semicontinuity of $\widehat{F}|_{J_\gamma \times \mathbb{R}^N}$ and the construction of J_γ provide $r, \delta, t_1 > 0$ satisfying for every $t \in [t_0, t_0 + t_1]$

$$\begin{cases} \widehat{F}\big(J_\gamma \cap [t_0,t],\, \mathbb{B}_r(x_0)\big) \subset \widehat{F}(t_0,x_0) + \frac{\varepsilon}{3}\,\mathbb{B} \subset \widetilde{F}(t_0,x_0) + \frac{\varepsilon}{3}\,\mathbb{B}, \\ \vartheta_{\widetilde{F}(t_0+\cdot,\cdot)}\big(t - t_0,\, \mathbb{B}_\delta(x_0)\big) \subset x_0 + r\,\mathbb{B}, \\ \frac{\mathscr{L}^1([t_0,t] \cap \widetilde{J}_\gamma)}{t - t_0}\, \widetilde{F}(t_0,x_0) \subset \widetilde{F}(t_0,x_0) + \frac{\varepsilon}{3}\,\mathbb{B}, \\ \frac{1}{t - t_0} \int_{[t_0,t] \setminus \widetilde{J}_\gamma} \mu(s)\, ds \le \frac{\varepsilon}{3} \cdot (1 + |x_0| + r)^{-1}. \end{cases}$$

Then for any $x \in \mathbb{B}_\delta(x_0)$ and $h \in [0, t_1]$, we obtain

$$\begin{aligned} &\vartheta_{\widetilde{F}(t_0+\cdot,\cdot)}(h,x) - x \subset \\ &\subset \int_{[t_0,t_0+h] \cap \widetilde{J}_\gamma} \widehat{F}\big(s,\, \mathbb{B}_r(x_0)\big)\, ds + \int_{[t_0,t_0+h] \setminus \widetilde{J}_\gamma} \widehat{F}\big(s,\, \mathbb{B}_r(x_0)\big)\, ds \\ &\subset \mathscr{L}^1([t_0,t_0+h] \cap \widetilde{J}_\gamma) \cdot \Big(\widetilde{F}(t_0,x_0) + \tfrac{\varepsilon}{3}\,\mathbb{B}\Big) + \int_{[t_0,t_0+h] \setminus \widetilde{J}_\gamma} \mu(s)\, (1 + |x_0| + r)\, ds \cdot \mathbb{B} \\ &\subset h \Big(\widetilde{F}(t_0,x_0) + \tfrac{\varepsilon}{3}\,\mathbb{B} + \tfrac{\varepsilon}{3}\,\mathbb{B}\Big) + \tfrac{\varepsilon}{3}\, h\, \mathbb{B} \\ &= h \big(\widetilde{F}(t_0,x_0) + \varepsilon\, \mathbb{B}\big). \end{aligned}$$

□

Proof (of Lemma A.16). Choosing $\gamma > 0$ arbitrarily small, Proposition A.12 (on page 447) provides a closed subset $\widetilde{J}_\gamma \subset [0,T]$ with $\mathscr{L}^1([0,T] \setminus \widetilde{J}_\gamma) < \gamma$ such that the set-valued restriction $\widetilde{F}|_{\widetilde{J}_\gamma \times \mathbb{R}^N}$ is continuous.
As in the proof of Lemma A.14, let $J_\gamma \subset \widetilde{J}_\gamma$ denote the set of density points of $\widetilde{J}_\gamma$ that are Lebesgue points of $\mu(\cdot) \cdot \chi_{[0,T] \setminus \widetilde{J}_\gamma}(\cdot) \in L^1([0,T])$ in addition. It also satisfies

$$\mathscr{L}^1(J_\gamma) = \mathscr{L}^1(\widetilde{J}_\gamma) > T - \gamma.$$

For arbitrary $t_0 \in J_\gamma$, $x_0 \in \mathbb{R}^N$ and $\varepsilon \in]0,1]$, the continuity of $\widetilde{F}|_{J_\gamma \times \mathbb{R}^N}$ and the construction of J_γ guarantee parameters $r, \delta, t_1 \in]0,1]$ successively such that for every $t \in [t_0, t_0 + t_1] \cap J_\gamma$, $x \in \mathbb{B}_\delta(x_0)$, $y \in \mathbb{B}_r(x_0)$

$$\begin{cases} d\!l\big(\widetilde{F}(t,y),\ \widetilde{F}(t_0,x_0)\big) \ \le\ \frac{\varepsilon}{8} \\ x\ +\ (t-t_0)\cdot \widetilde{F}(t_0,x)\ \subset\ x_0 + r\,\mathbb{B}, \\ \frac{\mathscr{L}^1([t_0,t]\setminus \widetilde{J}_\gamma)}{t-t_0}\ \ \widetilde{F}(t_0,x_0)\ \subset\ \frac{\varepsilon}{4}\,\mathbb{B}, \\ \frac{1}{t-t_0}\displaystyle\int_{[t_0,t]\setminus \widetilde{J}_\gamma} \mu(s)\ ds\ \le\ \frac{\varepsilon}{4} \\ \delta\ +\displaystyle\int_{[t_0,t]} \mu(s)\ ds\ \le\ r. \end{cases}$$

Choose now any $x \in \mathbb{B}_\delta(x_0)$ and $v \in \widetilde{F}(t_0,x)$. We want to verify for all $h \in [0,t_1]$

$$x\ +\ h\,v\ \in\ \vartheta_{\widetilde{F}(t_0+\cdot,\cdot)}(h,\,x)\ +\ \varepsilon\,h\,\mathbb{B}.$$

Since all values of $\widetilde{F}$ are assumed to be convex, the projection of v on $\widetilde{F}(\cdot,\cdot)$

$$[0,T]\times\mathbb{R}^N\ \leadsto\ \mathbb{R}^N,\ \ (t,y)\ \mapsto\ \Pi_{\widetilde{F}(t,y)}(v) \stackrel{\text{Def.}}{=} \big\{w\in\widetilde{F}(t,y)\ \big|\ \operatorname{dist}(v,\,\widetilde{F}(t,y)) = |w-v|\big\}$$

is single-valued and thus denoted by $f : [0,T]\times\mathbb{R}^N \longrightarrow \mathbb{R}^N$.
Moreover, $f(\cdot,y) : [0,T] \longrightarrow \mathbb{R}^N$ is measurable for every $y\in\mathbb{R}^N$ due to Proposition A.80 (on page 490). Whenever $\widetilde{F}(t,\cdot) : \mathbb{R}^N \leadsto \mathbb{R}^N$ is continuous, its composition with the projection mapping is upper semicontinuous in the sense of Painlevé-Kuratowski according to [162, Proposition 4.9] and thus, the single-valued function $f(t,\cdot) : \mathbb{R}^N \longrightarrow \mathbb{R}^N$ is continuous. As a consequence, f is a Carathéodory function in $[0,T]\times\mathbb{R}^N$ with the time-dependent absolute bound $\mu(\cdot)\in L^1([0,T])$ and, its restriction $f|_{J_\gamma\times\mathbb{R}^N}$ is continuous because $\widetilde{F}|_{J_\gamma\times\mathbb{R}^N}$ is continuous.
There exists an absolutely continuous solution $y(\cdot) : [t_0, t_0+t_1] \longrightarrow \mathbb{R}^N$ to the ordinary differential equations $y'(\cdot) = f\big(\cdot, y(\cdot)\big)$ a.e. with $y(t_0) = x$. Then, $y(\cdot)$ solves the differential inclusion $y'(\cdot) \in \widetilde{F}(\cdot, y(\cdot))$ a.e. and satisfies for all $h\in[0,t_1]$

$$\begin{aligned} &\big|x+h\,v\ -\ y(t_0+h)\big| \\ &\le \int_{[t_0,t_0+h]\cap J_\gamma} \big|v - f(s,y(s))\big|\ ds\ \ +\int_{[t_0,t_0+h]\setminus J_\gamma} \big(|v|\ +\ \mu(s)\big)\ ds \\ &\le \int_{[t_0,t_0+h]\cap J_\gamma} \operatorname{dist}\big(v,\ \widetilde{F}(s,y(s))\big)\ ds\ +\int_{[t_0,t_0+h]\setminus J_\gamma} \big(|v|\ +\ \mu(s)\big)\ ds \\ &\le\ \ 2\,\tfrac{\varepsilon}{8}\cdot h\ \ +\ \ 2\,\tfrac{\varepsilon}{4}\cdot h\ +\ \tfrac{\varepsilon}{4}\cdot h\ \ = \varepsilon\cdot h. \end{aligned}$$ □

This proof of Lemma A.16 is quite easy to adapt to the following statement whose autonomous counterpart is used for verifying Proposition 1.68 (2.) (on page 70):

Lemma 17. *In addition to the assumptions of Proposition A.13, let $K\subset\mathbb{R}^N$ be a nonempty compact subset and $R>0$.*

Then there exists a subset $J\subset[0,T]$ of full Lebesgue measure such that for every $t_0\in J$, $x_0\in K$ and $\varepsilon\in\,]0,1[$, there are $t_1,\delta>0$ with

$$x\ +\ h\cdot\widetilde{F}(t_0,x)\ +\ h\cdot\big(T_K^C(x_0)\cap\mathbb{B}_R\big)\ \subset\ \vartheta_{\widetilde{F}(t_0+\cdot,\cdot)}(h,K)\ +\ 2\,\varepsilon\,h\,\mathbb{B}$$

for all $x\in\mathbb{B}_\delta(x_0)\cap K,\ \ h\in\,]0,t_1[$.

The contribution of the circatangent cone $T_K^C(x_0)$ to this modification is summarized in the next lemma:

Lemma 18. *Let K, M be nonempty compact subsets of $\mathbb{R}^N$.*
For each point $x_0 \in K$ and every $\varepsilon > 0$, there exists a radius $\rho > 0$ such that

$$\operatorname{dist}(y + h\,w,\ K) \ -\ \operatorname{dist}(y,\ K) \ \leq\ \varepsilon\, h$$

is satisfied for all $w \in T_K^C(x_0) \cap M$, $h \in [0, \rho]$ and $y \in \mathbb{R}^N$ with $|y - x_0| \leq \rho$.

Proof (of Lemma A.18). Equivalently to Definition 1.63 (on page 68), a vector $v \in \mathbb{R}^N$ belongs to the circatangent cone $T_K^C(x)$ in $x_0 \in K$ if and only if for every $\varepsilon > 0$, there exists a radius $\rho(x_0, \varepsilon, v) > 0$ with

$$\operatorname{dist}(y + h\,v,\ K) \ -\ \operatorname{dist}(y,\ K) \ \leq\ \tfrac{\varepsilon}{2}\, h$$

for all $h \in [0, \rho]$ and $y \in \mathbb{R}^N$ with $|y - x_0| \leq \rho$. This is easy to prove indirectly by means of the projection on the compact set $K \subset \mathbb{R}^N$ – similarly to the morphological analogue in Lemma 5.39 (on page 427, see also the Clarke's "original" definition of tangents via "generalized directional derivative" in [46, § 2]).
In particular, all vectors $w \in \mathbb{B}_{\frac{\varepsilon}{2}}(v) \subset \mathbb{R}^N$ have in common:

$$\operatorname{dist}(y + h\,w,\ K) \ -\ \operatorname{dist}(y,\ K) \ \leq\ \varepsilon\, h$$

for every $h \in [0, \rho]$ and $y \in \mathbb{R}^N$ with $|y - x_0| \leq \rho$ due to the triangle inequality. Hence, the radius $\rho > 0$ can be chosen locally uniformly with respect to $v \in T_K^C(x_0)$, i.e. for every compact $M \subset \mathbb{R}^N$ and $\varepsilon > 0$, there is $\rho = \rho(x_0, \varepsilon, M) > 0$ with

$$\operatorname{dist}(y + h\,w,\ K) \ -\ \operatorname{dist}(y,\ K) \ \leq\ \varepsilon\, h$$

for all $w \in M \cap T_K^C(x_0)$, $h \in [0, \rho]$ and $y \in \mathbb{R}^N$ with $|y - x_0| \leq \rho$. □

Proof (of Lemma A.17). Fix any $\gamma > 0$ and construct closed subsets $J_\gamma \subset \widetilde{J}_\gamma \subset [0, T]$ as in the proof of Lemma A.16.
For arbitrary $t_0 \in J_\gamma$, $x_0 \in K$ and $\varepsilon \in]0, 1]$, the continuity of $\widetilde{F}|_{J_\gamma \times \mathbb{R}^N}$, the selection of J_γ and Lemma A.18 (in addition now) provide $r, \delta, t_1 \in]0, 1]$ successively such that for every $t \in [t_0,\ t_0 + t_1] \cap J_\gamma$, $h \in [0, t_1]$, $x \in \mathbb{B}_\delta(x_0)$, $y \in \mathbb{B}_r(x_0)$ and $w \in T_K^C(x_0) \cap \mathbb{B}_R$,

$$\left\{ \begin{array}{rcl} \operatorname{dist}(y + h\,w,\ K) \ -\ \operatorname{dist}(y,\ K) & \leq & \varepsilon\, h \\ d\!\!l\big(\widetilde{F}(t, y),\ \widetilde{F}(t_0, x_0)\big) & \leq & \frac{\varepsilon}{8} \\ x \ +\ h \cdot \widetilde{F}(t_0, x) & \subset & x_0 + r\,\mathbb{B}, \\ \frac{\mathscr{L}^1([t_0,t] \setminus \widetilde{J}_\gamma)}{t - t_0}\ \ \widetilde{F}(t_0, x_0) & \subset & \frac{\varepsilon}{4}\,\mathbb{B}, \\ \frac{1}{t - t_0} \displaystyle\int_{[t_0,t] \setminus \widetilde{J}_\gamma} \mu(s)\ ds & \leq & \frac{\varepsilon}{4} \\ \delta + h\,(R + \varepsilon) + \displaystyle\int_{[t_0,t]} \mu(s)\ ds & \leq & r. \end{array} \right.$$

For all $x \in \mathbb{B}_\delta(x_0)$, $v \in \widetilde{F}(t_0,x)$, $w \in T_K^C(x_0) \cap \mathbb{B}_R$ and $h \in [0,t_1]$, we are now to check

$$x + h\,(v+w) \;\in\; \vartheta_{\widetilde{F}(t_0+\cdot,\cdot)}(h,K) + 2\,\varepsilon\,h\,\mathbb{B}.$$

As in the proof of Lemma A.16, the projection of v on $\widetilde{F}(\cdot,\cdot)$

$$[0,T]\times\mathbb{R}^N \rightsquigarrow \mathbb{R}^N,\;\; (t,y) \mapsto \Pi_{\widetilde{F}(t,y)}(v) \stackrel{\text{Def.}}{=} \big\{w \in \widetilde{F}(t,y) \;\big|\; \operatorname{dist}(v, \widetilde{F}(t,y)) = |w-v|\big\}$$

induces a single-valued Carathéodory function $f: [0,T]\times\mathbb{R}^N \longrightarrow \mathbb{R}^N$.
The new essential aspect is to base the comparison on an absolutely continuous solution $y(\cdot): [t_0, t_0+t_1] \longrightarrow \mathbb{R}^N$ that does not start in x, but in a possibly different point of K:
Choose $z = z(x,h,w) \in K$ with $|x+h\,w-z| = \operatorname{dist}(x+h\,w, K) \le \varepsilon\,h$. In particular, $|z-x_0| \le \delta + h\,(R+\varepsilon) < r$. Then there exists an absolutely continuous solution $y(\cdot): [t_0,t_0+t_1] \longrightarrow \mathbb{R}^N$ to the ordinary differential equations $y'(\cdot) = f\big(\cdot, y(\cdot)\big)$ a.e. with $y(t_0) = z \in K$. $y(\cdot)$ has all values in $\mathbb{B}_r(x_0)$ and solves the differential inclusion $y'(\cdot) \in \widetilde{F}(\cdot, y(\cdot))$ a.e. again (but depends now on x,h,w).

The comparison with $t \mapsto x + h\,w + (t-t_0)\,v$ at time t_0+h leads to the estimate

$$\begin{aligned}
&\operatorname{dist}\big(x+h\,(v+w),\; \vartheta_{\widetilde{F}(t_0+\cdot,\cdot)}(h,K)\big) \;\le\; \big|x+h\,(v+w) - y(t_0+h)\big| \\
&\le |x+h\,w-z| + \int_{[t_0,t_0+h]\cap J_\gamma} \big|v - f(s,y(s))\big|\,ds + \int_{[t_0,t_0+h]\setminus J_\gamma} \big(|v|+\mu(s)\big)\,ds \\
&\le \varepsilon\,h + \int_{[t_0,t_0+h]\cap J_\gamma} \operatorname{dist}\big(v, \widetilde{F}(s,y(s))\big)\,ds + \int_{[t_0,t_0+h]\setminus J_\gamma} \big(|v|+\mu(s)\big)\,ds \\
&\le \varepsilon\,h + 2\,\tfrac{\varepsilon}{8}\cdot h + 2\,\tfrac{\varepsilon}{4}\cdot h + \tfrac{\varepsilon}{4}\cdot h \\
&\le 2\,\varepsilon\,h.
\end{aligned}$$

□

A.4 Relaxation Theorem of Filippov-Ważewski for Differential Inclusions

The so-called Relaxation Theorem bridges the gap between a differential inclusion

$$x'(\cdot) \in \widetilde{F}(\cdot, x(\cdot))$$

and its *relaxed* counterpart with (pointwise) convexified values on the right-hand side, i.e.,

$$y'(\cdot) \in \overline{\text{co}}\,\widetilde{F}(\cdot, y(\cdot)).$$

In particular, it provides sufficient conditions on the set-valued map $\widetilde{F}: [0,T]\times\mathbb{R}^N \rightsquigarrow \mathbb{R}^N$ which make the additional assumption of convex values dispensable in regard to compact reachable sets.

Theorem 19 (Relaxation Theorem of Filippov-Ważewski). *Suppose for the set-valued map* $\widetilde{F} : [0,T] \times \mathbb{R}^N \rightsquigarrow \mathbb{R}^N$ *and the curve* $y(\cdot) \in W^{1,1}([0,T], \mathbb{R}^N)$:

(1.) *the values* $\widetilde{F}$ *are nonempty closed subsets of* $\mathbb{R}^N$,

(2.) *for every* $x \in \mathbb{R}^N$, $F(\cdot, x) : [0,T] \rightsquigarrow \mathbb{R}^N$ *is measurable,*

(3.) *there exist* $\rho > 0$ *and* $\lambda(\cdot) \in L^1([0,T], \mathbb{R}_0^+)$ *such that for* $\mathscr{L}^1$*-almost every* $t \in [0,T]$, *the restriction* $F(t,\cdot)\big|_{\mathbb{B}_\rho(y(t))} : \mathbb{B}_\rho(y(t)) \rightsquigarrow \mathbb{R}^N$ *is* $\lambda(t)$*-Lipschitz continuous w.r.t.* $d\!l$,

(4.) *there is* $\mu(\cdot) \in L^1([0,T])$ *with* $\big|\widetilde{F}(t, y(t))\big|_\infty \le \mu(t)$ *for* $\mathscr{L}^1$*-almost every* t.

(5.) $[0,T] \longrightarrow \mathbb{R}, \; t \longmapsto \operatorname{dist}\big(y'(t), \; \widetilde{F}(t, y(t))\big)$ *is Lebesgue-integrable,*

(6.) $e^{\|k\|_{L^1}} \cdot \displaystyle\int_0^T \operatorname{dist}\Big(y'(t), \; \widetilde{F}(t, y(t))\Big) \, dt \; \le \; \rho,$

(7.) $y'(t) \in \overline{\operatorname{co}}\, \widetilde{F}(t, y(t))$ *for* $\mathscr{L}^1$*-almost every* $t \in [0,T]$.

Then for every $\delta > 0$, *there exists a solution* $x(\cdot) \in W^{1,1}([0,T], \mathbb{R}^N)$ *to the differential inclusion* $x'(\cdot) \in F(\cdot, x(\cdot))$ *a.e. satisfying* $x(0) = y(0)$ *and* $\|x(\cdot) - y(\cdot)\|_{L^\infty} \le \delta$.

Proof is given in [80, Theorem 1.36], for example, as a consequence of Filippov's Theorem A.6 and an appropriate selection principle. The autonomous counterpart and its proof can be found in [15, Theorem 2.4.2].

Aubin and Frankowska have already pointed out a well-known consequence in [19, Theorem 10.4.4]:

Corollary 20. *In addition to the hypotheses of Relaxation Theorem A.19 with* $\rho = \infty$, *assume that* $R(\cdot) \in L^1([0,T])$ *satisfies* $\widetilde{F}(t,x) \subset R(t)\, \mathbb{B}$ *for every* $x \in \mathbb{R}^N$ *and a.e.* t.

Then the solutions to the differential inclusion $x'(\cdot) \in \widetilde{F}(\cdot, x(\cdot))$ *a.e. are dense in the set of solutions to the relaxed inclusion* $y'(\cdot) \in \overline{\operatorname{co}}\, \widetilde{F}(\cdot, y(\cdot))$ *a.e. with respect to the supremum norm.* □

Considering now reachable sets of differential inclusions, we obtain

Corollary 21. *Let* $\widetilde{F} : [0,T] \times \mathbb{R}^N \rightsquigarrow \mathbb{R}^N$ *be Filippov continuous (according to Definition A.7 on page 444).*

Then, $\vartheta_{\widetilde{F}}(t,K) \; = \; \vartheta_{\overline{\operatorname{co}}\,\widetilde{F}}(t,K)$ *for every* $K \in \mathscr{K}(\mathbb{R}^N)$ *and* $t \in [0,T]$.

Proof. Relaxation Theorem A.19 implies

$$\overline{\vartheta_{\widetilde{F}}(t,M)} \; = \; \overline{\vartheta_{\overline{\operatorname{co}}\,\widetilde{F}}(t,M)}$$

for every nonempty (not necessarily closed) subset $M \subset \mathbb{R}^N$ and any $t \in [0,T]$.
In addition, the reachable set $\vartheta_{\widetilde{F}}(t,K) \subset \mathbb{R}^N$ is closed as a consequence of Filippov's Theorem A.6 (on page 443). Finally, $\overline{\operatorname{co}}\, \widetilde{F} : [0,T] \times \mathbb{R}^N \rightsquigarrow \mathbb{R}^N$ has Filippov continuity in common with $\widetilde{F}$ and thus, $\vartheta_{\overline{\operatorname{co}}\,\widetilde{F}}(t,K) \subset \mathbb{R}^N$ is also closed. □

A.5 Regularity of Reachable Sets of Differential Inclusions

In this section, we focus on the boundary of reachable sets of differential inclusions. Adjoint arcs are used for describing the time-dependent limiting normal cones. They serve as tools for sufficient conditions on the differential inclusion for preserving smooth boundaries shortly, for example.
First we prove in Proposition A.34 that $C^{1,1}$ boundaries are preserved for short times. Then according to Proposition A.36, the same hypothesis guarantees that the evolution of smooth sets is reversible in time. Afterwards, the conditions on the Hamiltonian function $\mathscr{H}_F$ are supposed to be stronger for guaranteeing that points evolve into sets of positive erosion (see Proposition A.41). Finally, we estimate the maximal shrinking of exterior or interior balls and focus on exterior tusks.

Definition 22. For any set-valued map $\widetilde{F} : [0,T] \times \mathbb{R}^N \rightsquigarrow \mathbb{R}^N$, the support function

$$\begin{aligned} \mathscr{H}_{\widetilde{F}} : [0,T] \times \mathbb{R}^N \times \mathbb{R}^N &\longmapsto \mathbb{R} \\ (t,x,p) &\longmapsto \sigma\big(p, \widetilde{F}(t,x)\big) \stackrel{\text{Def.}}{=} \sup\big\{\langle p,v\rangle \;\big|\; v \in \widetilde{F}(t,x)\big\} \end{aligned}$$

is called *(upper) Hamiltonian* of $\widetilde{F}$.

A.5.1 Normal Cones and Compact Sets: Definitions and Notation

This section serves mainly the purpose of clarifying the notation in regard to normal cones and summarizing some features of compact subsets of $\mathbb{R}^N$.

Definition 23. Let $C \subset \mathbb{R}^N$ be a nonempty closed set.
A vector $\eta \in \mathbb{R}^N$, $\eta \neq 0$, is said to be a *proximal normal vector* to C at $x \in C$ if there exists $\rho > 0$ with $\mathbb{B}_\rho(x + \rho\,\frac{\eta}{|\eta|}) \cap C = \{x\}$.
The supremum of all ρ with this property is called *proximal radius* of C at x in direction η. The cone of all proximal normal vectors is called the *proximal normal cone* to C at x and is abbreviated as $N_C^P(x)$.

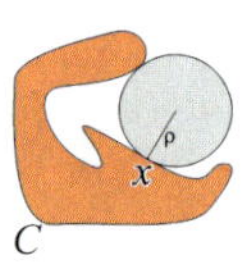

The so-called *limiting normal cone* $N_C(x)$ to C at x consists of all vectors $\eta \in \mathbb{R}^N$ that can be approximated by sequences $(\eta_n)_{n\in\mathbb{N}}$, $(x_n)_{n\in\mathbb{N}}$ satisfying

$$x_n \longrightarrow x, \quad \eta_n \longrightarrow \eta, \qquad x_n \in C, \quad \eta_n \in N_C^P(x_n),$$

i.e. $\quad N_C(x) \stackrel{\text{Def.}}{=} \operatorname{Limsup}_{\substack{y \longrightarrow x \\ y\in C}} N_C^P(y) \quad$ (in the sense of Painlevé-Kuratowski).
As a further abbreviation, we set ${}^\flat N_C(x) := N_C(x) \cap \mathbb{B} = \big\{v \in N_C(x) \;\big|\; |v| \leq 1\big\}$.

Convention. In the following we restrict ourselves to normal directions at boundary points, i.e. strictly speaking, Graph N_C and Graph ${}^\flat N_C$ are the abbreviations of Graph $N_C|_{\partial C}$ and Graph ${}^\flat N_C|_{\partial C}$, respectively.

Lemma 24 ([47, Lemma 6.4]). *For a nonempty closed subset $M \subset \mathbb{R}^N$, assume $\eta \in N^P_{\overline{\mathbb{R}^N\setminus M}}(x) \setminus \{0\}$ and $N_M^P(x) \neq \{0\}$. Then, $N^P_{\overline{\mathbb{R}^N\setminus M}}(x) = -N_M^P(x) = \mathbb{R}_0^+\,\eta$.*

Definition 25. $\mathscr{K}_{C^{1,1}}(\mathbb{R}^N)$ abbreviates the set of all nonempty compact N-dimensional $C^{1,1}$ submanifolds of $\mathbb{R}^N$ with boundary.

A closed subset $C \subset \mathbb{R}^N$ is said to have *positive erosion of radius* $\rho > 0$ if for each $r \in]0, \rho[$, there exists a closed set $M \subset \mathbb{R}^N$ with

$$\begin{cases} C = \{x \in \mathbb{R}^N \mid \operatorname{dist}(x, M) \le r\}, \\ M = \{x \in C \mid \operatorname{dist}(x, \partial C) \ge r\}. \end{cases}$$

$\mathscr{K}_\circ^\rho(\mathbb{R}^N)$ consists of all sets with positive erosion of radius $\rho > 0$ and, set $\mathscr{K}_\circ(\mathbb{R}^N) := \bigcup_{\rho>0} \mathscr{K}_\circ^\rho(\mathbb{R}^N)$.

Definition 26 (Sets of positive reach [79], [55, Definition 4.7.1]**).**
A nonempty set $M \subset \mathbb{R}^N$ is said to have *positive reach* if there exists $h > 0$ such that the projection $\Pi_M(x) \overset{\text{Def.}}{=} \{y \in \overline{M} \mid |x-y| = \operatorname{dist}(x, M)\}$ is single-valued for every $x \in \mathbb{R}^N$ with $\operatorname{dist}(x, M) < h$. The maximum $h > 0$ for which this property holds is called the *reach* of M.

Remark 27. The morphological term "erosion" is motivated by the fact that a set $C = \overline{C^\circ} \subset \mathbb{R}^N$ has positive erosion if and only if the closure $\overline{\mathbb{R}^N \setminus C}$ of its complement has positive reach. This implies a collection of interesting regularity properties presented (for closed subsets of a Hilbert space) in [47, 48, 159]. Here we summarize some of the features for subsets of $\mathbb{R}^N$:

Proposition 28 ([47], [48, Theorem 4.1], [55, Theorem 4.7.1], [159, Theorem 4.1]**).**
Given a nonempty closed subset $M \subset \mathbb{R}^N$, the following conditions are equivalent:

(1.) *M has positive reach $\ge \rho > 0$,*

(2.) $\operatorname{dist}(\cdot, M)$ *belongs to* $C^{1,1}_{\text{loc}}(\{0 < \operatorname{dist}(\cdot, M) < \rho\})$,

(3.) $\operatorname{dist}(\cdot, M)$ *belongs to* $C^{1}_{\text{loc}}(\{0 < \operatorname{dist}(\cdot, M) < \rho\})$,

(4.) *$\Pi_M(x) \subset M$ is single-valued for all points $x \in \mathbb{R}^N$ with* $0 < \operatorname{dist}(x, M) < \rho$,

(5.) *$\Pi_M(x) \subset M$ is single-valued for all $x \in \mathbb{R}^N$ with* $0 < \operatorname{dist}(x, M) < \rho$ *and, Π_M belongs to* $C^{0,1}_{\text{loc}}(\{0 < \operatorname{dist}(\cdot, M) < \rho\})$,

(6.) $\operatorname{dist}(\cdot, M)^2$ *belongs to* $C^{1,1}_{\text{loc}}(\{0 < \operatorname{dist}(\cdot, M) < \rho\})$,

(7.) *for every $r \in]0, \rho[$, all points $x \in \mathbb{R}^N$ with $0 < \operatorname{dist}(x, M) < r$ satisfy*
$$\operatorname{dist}(x, M) + \operatorname{dist}\big(x, \mathbb{R}^N \setminus \mathbb{B}_r(M)^\circ\big) = r,$$

(8.) *for any $r \in]0, \rho[$,* $\{\operatorname{dist}(\cdot, \overline{\mathbb{R}^N \setminus \mathbb{B}_r(M)}) \ge r\} = M$,

(9.) *every proximal normal vector $\neq 0$ at any $x \in \partial M$ has proximal radius $\ge \rho$,*

(10.) *for any $r \in]0, \rho[$, each $x \in \mathbb{R}^N$ with $\operatorname{dist}(x, M) = r$ satisfies* $N^P_{\mathbb{B}_r(M)}(x) \neq \{0\}$,

(11.) *$N_M(\cdot) \cap \mathbb{B}^\circ_\rho$ is hypermonotone, i.e. whenever $x_1, x_2 \in M$ and $v_k \in N_M(x_k)$ with $|v_k| < \rho$ $(k = 1, 2)$, then* $(v_1 - v_2) \cdot (x_1 - x_2) \ge -|x_1 - x_2|^2$,

(12.) $\operatorname{dist}(y - x, T_M(x)) \le \frac{1}{2\rho}\, |y - x|^2$ *for any $y, x \in M$ (global Shapiro property).*

Corollary 29 ([48, Corollary 4.15]).
Every nonempty closed set $M \subset \mathbb{R}^N$ with positive reach $\geq \rho > 0$ fulfills:

(a) *the proximal, limiting and Clarke normal cone coincide at each point $x \in M$,*

(b) *for every $r \in]0, \rho[$ and each point $x \in \mathbb{R}^N$ with $\operatorname{dist}(x, M) = r$,*
$$N^P_{\mathbb{B}_r(M)}(x) \;=\; N_{\mathbb{B}_r(M)}(x) \;=\; \mathbb{R}^+_0 \cdot \big(x - p(x)\big)$$
where $p(x) \in M$ is the unique closest point to x in M,

(c) *for every $r \in]0, \rho[$, the topological boundary of $\{\operatorname{dist}(\cdot, M) \leq r\}$ is a $C^{1,1}$ submanifold of codimension 1 in $\mathbb{R}^N$.*

A.5.2 Adjoint Arcs for Evolving Normal Cones to Reachable Sets

The so-called Hamilton condition is known under very mild assumptions using the tools of nonsmooth functions. First we quote the version of Vinter's monograph [180]. Applying these results to proximal balls leads to a necessary condition on boundary points of reachable sets and their proximal normal vectors. Approximating sequences then lay the basis for extending this result to limiting normal vectors in subsequent Proposition A.32. In particular, it is formulated only for Hamiltonian functions with continuous partial derivatives $\partial_x \mathscr{H}_{\widetilde{F}}, \partial_y \mathscr{H}_{\widetilde{F}}$ because we exploit the regularity of solutions to ordinary differential equations in the next sections.

Proposition 30 (Extended Hamilton Condition).
Let $x(\cdot) \in W^{1,1}([S,T], \mathbb{R}^N)$ be a local minimizer (with respect to perturbations in $W^{1,1}([0,T], \mathbb{R}^N)$) of the problem
$$\begin{aligned}
&g(y(S), y(T)) \;\longrightarrow\; \min \\
&\textit{over} \;\; y(\cdot) \in W^{1,1}([S,T], \mathbb{R}^N) \;\textit{satisfying} \\
&\qquad y'(t) \in \widetilde{F}(t, y(t)) \qquad \textit{for Lebesgue-almost every } t \in [S,T], \\
&\qquad (y(S), y(T)) \in C \subset \mathbb{R}^N \times \mathbb{R}^N.
\end{aligned}$$
Assume also that

(G1) *g is locally Lipschitz continuous;*

(G2)′ *$\widetilde{F}(t,x) \neq \emptyset$ is convex for each (t,x), $\widetilde{F}$ is $\mathscr{L}^{1+N} \times \mathscr{B}^N$ measurable, and* Graph *$\widetilde{F}(t, \cdot)$ is closed for each $t \in [S,T]$.*

Suppose, furthermore, that either of the following hypotheses is satisfied:

(a) *There exist $k \in L^1([S,T])$ and $\varepsilon > 0$ such that for almost every t*
$$\widetilde{F}(t, x_1) \cap \big(x'(t) + \varepsilon\, k(t)\, \mathbb{B}\big) \;\subset\; \widetilde{F}(t, x_2) + k(t)\, |x_1 - x_2|\, \mathbb{B}$$
for all $x_1, x_2 \in \mathbb{B}_\varepsilon(x(t))$.

(b) *There exist $k \in L^1([S,T])$, $\overline{K} > 0$ and $\varepsilon > 0$ such that the following two conditions are satisfied for almost every $t \in [S,T]$ and all $x_1, x_2 \in \mathbb{B}_\varepsilon(x(t))$*
$$\widetilde{F}(t, x_1) \cap \big(x'(t) + \varepsilon\, \mathbb{B}\big) \subset \widetilde{F}(t, x_2) + k(t)\, |x_1 - x_2|\, \mathbb{B},$$
$$\inf\big\{\, |v - x'(t)| \;\big|\; v \in \widetilde{F}(t, x_1) \big\} \leq \overline{K}\, |x_1 - x(t)|\,.$$

Then there exist an arc $p(\cdot) \in W^{1,1}([S,T],\mathbb{R}^N)$ *and a constant* $\lambda \geq 0$ *such that*

$$\begin{array}{ll} (i) & (p(\cdot),\lambda) \neq (0,0), \\ (ii) & p'(t) \in co\,\{\eta \in \mathbb{R}^N \,|\, (\eta, p(t)) \in N_{\mathrm{Graph}\,\widetilde{F}(t,\cdot)}(x(t), x'(t))\} \quad \textit{for } \mathscr{L}^1\textit{-a.e. } t \\ (iii) & \big(p(S), -p(T)\big) \in \lambda\, \partial^L g(x(S),x(T)) + N_C(x(S),x(T)). \end{array}$$

Condition (ii) implies

$$\begin{array}{lll} (iv) & p(t)\cdot x'(t) = \sup\big(p(t)\cdot \widetilde{F}(t,x(t))\big) & \textit{for } \mathscr{L}^1\textit{-a.e. } t \\ (v) & p'(t) \in co\,\big\{-q \in \mathbb{R}^N \,\big|\, (q,x'(t)) \in \partial^L \mathscr{H}_{\widetilde{F}}(t,\cdot,\cdot)|_{(x(t),p(t))}\big\} & \textit{for } \mathscr{L}^1\textit{-a.e. } t. \end{array}$$

Proof is presented in [180, Theorem 7.7.1], for example.

Remark 31. This adjoint $p(\cdot)$ also satisfies $|p'(t)| \leq k(t)\,|p(t)|$ for almost every t as an immediate consequence of statement (ii) and the so-called *Mordukhovich criterion* (see e.g. [162, Theorem 9.40]).

Proposition 32. *Suppose for the set-valued map* $\widetilde{F} : [0,T] \times \mathbb{R}^N \rightsquigarrow \mathbb{R}^N$

1. $\widetilde{F}(\cdot)$ *is measurable with nonempty convex compact values,*
2. *for* $\mathscr{L}^1$*-almost every* $t \in [0,T]$, $\mathscr{H}_{\widetilde{F}}(t,\cdot,\cdot)$ *is continuously differentiable in* $\mathbb{R}^N \times (\mathbb{R}^N \setminus \{0\})$,
3. *there exists* $k(\cdot) \in L^1([0,T]$ *such that for* $\mathscr{L}^1$*-almost every* $t \in [0,T]$, $\|\partial_{(x,p)} \mathscr{H}_{\widetilde{F}}(x,p)\| \leq k(t)\cdot(1+|x|+|p|)$ *for all* $x,p \in \mathbb{R}^N$, $|p| > 1$.

Let $K \in \mathscr{K}(\mathbb{R}^N)$ *be any initial set and* $t_0 > 0$.

For every boundary point $x_0 \in \partial\, \vartheta_{\widetilde{F}}(t_0,K)$ *and normal* $\nu \in N_{\vartheta_{\widetilde{F}}(t_0,K)}(x_0) \setminus \{0\}$, *there exist a solution* $x(\cdot) \in W^{1,1}([0,t_0],\mathbb{R}^N)$ *and its adjoint* $p(\cdot) \in W^{1,1}([0,t_0],\mathbb{R}^N)$ *with*

$$\begin{cases} x'(t) = \frac{\partial}{\partial p}\mathscr{H}_{\widetilde{F}}(t,x(t),p(t)) \in \widetilde{F}(t,x(t)), & x(t_0) = x_0, \quad x(0) \in \partial K, \\ p'(t) = -\frac{\partial}{\partial x}\mathscr{H}_{\widetilde{F}}(t,x(t),p(t)), & p(t_0) = \nu, \quad p(0) \in N_K(x(0)). \end{cases}$$

□

A.5.3 Hamiltonian System Helps Preserving $C^{1,1}$ Boundaries Shortly

Definition 33. For a set-valued map $\widetilde{F} : [0,T] \times \mathbb{R}^N \rightsquigarrow \mathbb{R}^N$, the standard hypothesis $(\widetilde{\mathscr{H}})$ comprises the following conditions on $\mathscr{H}_{\widetilde{F}}(t,x,p) := \sup\ p \cdot \widetilde{F}(t,x)$

1. $\widetilde{F}$ is measurable and has nonempty compact convex values,
2. $\mathscr{H}_{\widetilde{F}}(t,\cdot,\cdot) : \mathbb{R}^N \times (\mathbb{R}^N \setminus \{0\}) \longrightarrow \mathbb{R}$ is continuously differentiable for every t,
3. for every $R > 1$, there exists $\lambda_R(\cdot) \in L^1([0,T])$ such that the derivative of $\mathscr{H}_{\widetilde{F}}(t,\cdot,\cdot)$ restricted to $\mathbb{B}_R \times (\mathbb{B}_R \setminus \overset{\circ}{\mathbb{B}}_{\frac{1}{R}})$ is $\lambda_R(t)$-Lipschitz continuous for Lebesgue-almost every $t \in [0,T]$,
4. there is $k_{\widetilde{F}} \in L^1([0,T])$ such that for a.e. $t \in [0,T]$ and all $x, p \in \mathbb{R}^N$ $(|p| \geq 1)$,
$$\left\|\partial_{(x,p)} \mathscr{H}_{\widetilde{F}}(t,x,p)\right\|_{\mathrm{Lin}(\mathbb{R}^N \times \mathbb{R}^N, \mathbb{R})} \leq k_{\widetilde{F}}(t) \cdot (1 + |x| + |p|).$$

Proposition 34. *Assume standard hypothesis $(\widetilde{\mathscr{H}})$ for $\widetilde{F} : [0,T] \times \mathbb{R}^N \rightsquigarrow \mathbb{R}^N$. For every initial compact set $K \in \mathscr{K}_{C^{1,1}}(\mathbb{R}^N)$, there exist $\tau = \tau(\widetilde{F},K) > 0$ and $\rho = \rho(\widetilde{F},K) > 0$ such that $\vartheta_{\widetilde{F}}(t,K)$ is also a N-dimensional $C^{1,1}$ submanifold of $\mathbb{R}^N$ with boundary for all $t \in [0,\tau]$ and, its radius of curvature is $\geq \rho$ at every boundary point. In particular, $\vartheta_{\widetilde{F}}(t,K)$ has both positive reach and erosion.*

The proof of Proposition A.34 is based on the following lemma:

Lemma 35. *Suppose for $H : [0,T] \times \mathbb{R}^N \times \mathbb{R}^N \longrightarrow \mathbb{R}$, $\psi : \mathbb{R}^N \longrightarrow \mathbb{R}^N$ and the Hamiltonian system*

$$\begin{cases} y'(t) = \frac{\partial}{\partial q} H(t, y(t), q(t)), & y(0) = y_0 \\ q'(t) = -\frac{\partial}{\partial y} H(t, y(t), q(t)), & q(0) = \psi(y_0) \end{cases} \qquad (*)$$

the following properties:

1. *$H(t,\cdot,\cdot)$ is differentiable for every $t \in [0,T]$,*
2. *for every $R > 0$, there exists $k_R \in L^1([0,T])$ such that the derivative of $H(t,\cdot,\cdot)$ is $k_R(t)$-Lipschitz continuous on $\mathbb{B}_R \times \mathbb{B}_R$ for $\mathscr{L}^1$-almost every t,*
3. *ψ is locally Lipschitz continuous,*
4. *every solution $(y(\cdot),q(\cdot))$ to the Hamiltonian system $(*)$ can be extended to $[0,T]$ and depends continuously on the initial data in the following sense: Let each $(y_n(\cdot),q_n(\cdot))$ be a solution satisfying $y_n(t_n) \longrightarrow z_0$, $q_n(t_n) \longrightarrow q_0$ for some $t_n \longrightarrow t_0$, $z_0, q_0 \in \mathbb{R}^N$.*
Then $(y_n(\cdot),q_n(\cdot))_{n\in\mathbb{N}}$ converges uniformly to a solution $(y(\cdot),q(\cdot))$ to the Hamiltonian system with $y(t_0) = z_0$, $q(t_0) = q_0$.

For a compact set $K \subset \mathbb{R}^N$ and $t \in [0,T]$, define

$$M_t^{\mapsto}(K) := \left\{ (y(t), q(t)) \,\middle|\, (y(\cdot), q(\cdot)) \text{ solves system } (*),\ y_0 \in K \right\} \subset \mathbb{R}^N \times \mathbb{R}^N.$$

Then there exist $\delta > 0$ and $\lambda > 0$ such that $M_t^{\mapsto}(K)$ is the graph of a λ-Lipschitz continuous function for every $t \in [0,\delta]$.

Proof (of Lemma A.35). It is based on the indirect proof of [80, Lemma 5.5] about the same Hamiltonian system with $y(T) = y_T, \; q(T) = q_T$ given (without mentioning the uniform Lipschitz constant λ explicitly).

Suppose that the claim is false. Then there exists a sequence $(t_n)_{n\in\mathbb{N}}$ in $]0,T]$ with $t_n \longrightarrow 0$ such that either $M_{t_n}^{\mapsto}(K)$ is not the graph of a Lipschitz function or the corresponding Lipschitz constants converge to ∞. In both cases, we can find distinct solutions $(y_n^1(\cdot), q_n^1(\cdot)), \; (y_n^2(\cdot), q_n^2(\cdot)), \; n\in\mathbb{N}$, to the Hamiltonian system $(*)$ with

$$\varepsilon_n := \frac{|y_n^1(t_n) - y_n^2(t_n)|}{|q_n^1(t_n) - q_n^2(t_n)|} \longrightarrow 0 \qquad \text{for } n \longrightarrow \infty.$$

Assumption (4.) and $K \in \mathscr{K}(\mathbb{R}^N)$ imply $\bigcup_{0\le t\le T} M_t^{\mapsto}(K) \subset \mathbb{B}_R \times \mathbb{B}_R$ for some $R > 0$.

Assumption (2.) provides the estimate

$$\begin{aligned}
&|y_n^1(t) - y_n^2(t)| \\
&\le \; |y_n^1(t_n) - y_n^2(t_n)| + \int_t^{t_n} k_R(s) \Big(|y_n^1(s) - y_n^2(s)| + |q_n^1(s) - q_n^2(s)|\Big)\, ds \\
&\le \varepsilon_n\, |q_n^1(t_n) - q_n^2(t_n)| + \int_t^{t_n} k_R(s) \Big(|y_n^1(s) - y_n^2(s)| + |q_n^1(s) - q_n^2(s)|\Big)\, ds
\end{aligned}$$

for all $t \in [0, t_n]$, and the integral version of Gronwall's inequality (Proposition A.1) leads to a constant $C_1 > 0$ (independent of n) with

$$|y_n^1(t) - y_n^2(t)| \;\le\; C_1 \Big(\varepsilon_n\, |q_n^1(t_n) - q_n^2(t_n)| + \int_t^{t_n} k_R(s)\; |q_n^1(s) - q_n^2(s)|\; ds\Big).$$

Due to $\sup_n \varepsilon_n < \infty$, we obtain a constant $C_2 > 0$ such that for all $n \in \mathbb{N}, \; t \in [0, t_n]$,

$$\begin{aligned}
&|q_n^1(t) - q_n^2(t)| \\
&\le \; |q_n^1(t_n) - q_n^2(t_n)| + \int_t^{t_n} k_R(s) \Big(|y_n^1(s) - y_n^2(s)| + |q_n^1(s) - q_n^2(s)|\Big) ds \\
&\le C_2 \Big(|q_n^1(t_n) - q_n^2(t_n)| + \int_t^{t_n} k_R(s)\; |q_n^1(s) - q_n^2(s)|\; ds\Big).
\end{aligned}$$

As a consequence of Gronwall's Proposition A.1 again, there is a constant $C_3 > 0$ (independent of n) with $|q_n^1(t) - q_n^2(t)| \le C_3\, |q_n^1(t_n) - q_n^2(t_n)|$ for all $n, \; t \in [0, t_n]$. In particular,

$$\varepsilon_n' := \sup_{0 \le t \le t_n} \frac{|y_n^1(t) - y_n^2(t)|}{|q_n^1(t_n) - q_n^2(t_n)|} \;\le\; C_1 \Big(\varepsilon_n + C_3 \int_0^{t_n} k_R(s)\, ds\Big) \overset{n\to\infty}{\longrightarrow} 0.$$

Similarly we get a constant $C_4 = C_4(\|k_R\|_{L^1}) > 0$ fulfilling

$$|q_n^1(t_n) - q_n^2(t_n)| \;\le\; C_4\, |q_n^1(0) - q_n^2(0)| \;=\; C_4\, |\psi(y_n^1(0)) - \psi(y_n^2(0))|$$

for all $n \in \mathbb{N}$ sufficiently large. Indeed, for all $t \in [0, t_n]$, assumption (2.) ensures

$$\begin{aligned}
&|q_n^1(t) - q_n^2(t)| \\
&\le |q_n^1(0) - q_n^2(0)| + \int_0^t k_R(s) \Big(|y_n^1(s) - y_n^2(s)| + |q_n^1(s) - q_n^2(s)|\Big)\, ds \\
&\le |q_n^1(0) - q_n^2(0)| + \int_0^t k_R(s) \Big(\varepsilon_n'\, |q_n^1(t_n) - q_n^2(t_n)| + |q_n^1(s) - q_n^2(s)|\Big)\, ds
\end{aligned}$$

and Gronwall's inequality (Proposition A.1) provides $C_5 = C_5(\|k_R\|_{L^1}) > 0$ such that for every $n \in \mathbb{N}$,

$$|q_n^1(t_n) - q_n^2(t_n)| \le \frac{C_5}{2}\,|q_n^1(0) - q_n^2(0)| + \text{const}(\|k_R\|_{L^1})\ \varepsilon_n'\ |q_n^1(t_n) - q_n^2(t_n)|.$$

Due to $\varepsilon_n' \longrightarrow 0$, we obtain $|q_n^1(t_n) - q_n^2(t_n)| \le C_5\,|q_n^1(0) - q_n^2(0)|$ for all $n \in \mathbb{N}$ large enough. Finally,

$$\begin{aligned}\frac{|\,\psi(y_n^1(0)) - \psi(y_n^2(0))\,|}{|\,y_n^1(0) - y_n^2(0)\,|} &= \frac{|\,q_n^1(0) - q_n^2(0)\,|}{|\,q_n^1(t_n) - q_n^2(t_n)\,|} \cdot \frac{|\,q_n^1(t_n) - q_n^2(t_n)\,|}{|\,y_n^1(0) - y_n^2(0)\,|}\\ &\ge \frac{1}{C_5} \cdot \frac{1}{\varepsilon_n'}\\ &\longrightarrow \infty \qquad \text{for } n \longrightarrow \infty\end{aligned}$$

— contradicting the local Lipschitz continuity of ψ at each joint cluster point of $(y_n^1(0))_{n\in\mathbb{N}}$ and $(y_n^2(0))_{n\in\mathbb{N}}$ in K. □

Proof (of Proposition A.34). Assuming that $K \in \mathscr{K}(\mathbb{R}^N)$ is a N-dimensional $C^{1,1}$ submanifold of $\mathbb{R}^N$ with boundary, the *exterior* unit normal vectors to K (restricted to ∂K) can be extended to a Lipschitz continuous function $\psi : \mathbb{R}^N \longrightarrow \mathbb{R}^N$. Choosing some cut-off function $\varphi \in C^\infty([0,\infty[, [0,1])$ with $\varphi|_{[0,\frac14]} \equiv 0$, $\varphi|_{[\frac12,\infty[} \equiv 1$, $H(t,x,p) := \mathscr{H}_{\widetilde{F}}(t,x,p) \cdot \varphi(|p|)$ satisfies condition (1.), (2.), (4.) of Lemma A.35 due to standard hypothesis $(\widetilde{\mathscr{H}})$.

For arbitrary $x_0 \in \partial K$, consider the differential equations

$$\begin{cases} x'(t) = \frac{\partial}{\partial p} H(t, x(t), p(t)), & x(0) = x_0,\\ p'(t) = -\frac{\partial}{\partial x} H(t, x(t), p(t)), & p(0) = \psi(x_0). \end{cases} \qquad (**)$$

Due to $|\psi(\cdot)| = 1$ on ∂K and $H \in C^{1,1}$, there exists some $\tau_1 > 0$ such that $|p(t)| > \frac12$ for any $t \in [0,\tau_1]$ and all solutions $(x(\cdot), p(\cdot))$ of $(**)$ with $x_0 \in \partial K$. Thus, $H = \mathscr{H}_F$ close to $(x(t), p(t))$. Now Proposition A.32 can be reformulated as

$$\text{Graph } N_{\vartheta_F(t,K)}(\cdot) \subset \big\{ (x(t),\ \lambda\ p(t)) \,\big|\, (x(\cdot), p(\cdot)) \text{ solves system } (**),\ x_0 \in \partial K,\ \lambda \ge 0 \big\},$$

for all $t \in [0,\tau_1]$. Lemma A.35 yields $\tau \in\,]0,\tau_1[$ and $\lambda_M > 0$ such that

$$M_t^{\mapsto}(\partial K) := \big\{ (x(t),\ p(t)) \,\big|\, (x(\cdot),\ p(\cdot)) \text{ solves system } (**),\ x_0 \in \partial K \big\}$$

is the graph of a λ_M-Lipschitz continuous function for each $t \in [0,\tau]$.

Then for every point $z \in \partial\,\vartheta_{\widetilde{F}}(t,K)$, the limiting normal cone $N_{\vartheta_{\widetilde{F}}(t,K)}(z)$ contains exactly one direction and, its unit vector depends on z in a Lipschitz continuous way. (The Lipschitz constant is uniformly bounded by a constant depending on λ_M because the choice of τ_1 ensures $|p(\cdot)| > \frac12$ on $[0,\tau_1]$ for each solution of $(**)$.) Hence, the compact set $\vartheta_{\widetilde{F}}(t,K)$ is N-dimensional $C^{1,1}$ submanifold of $\mathbb{R}^N$ with boundary for all $t \in [0,\tau]$ and, its radius of curvature has a uniform lower bound. □

A.5.4 How to Guarantee Reversibility of Reachable Sets in Time

The Hamilton condition has led to a necessary condition on boundary points $x \in \partial\, \vartheta_{\widetilde{F}}(t,K)$ and their limiting normal cones in Proposition A.32 (on page 457). If each set $\vartheta_{\widetilde{F}}(t,K)$ $(0 \le t \le T)$ has positive reach $\ge \rho$, then standard hypothesis $(\widetilde{\mathscr{H}})$ turns adjoint arcs into sufficient conditions and, we conclude that the evolution of reachable sets is reversible with respect to time — in the following sense:

Proposition 36. *Suppose standard hypothesis $(\widetilde{\mathscr{H}})$ for $\widetilde{F} : [0,T] \times \mathbb{R}^N \rightsquigarrow \mathbb{R}^N$. Assume for $K_0 \in \mathscr{K}(\mathbb{R}^N)$ and $\rho > 0$ that every compact reachable set $K_t := \vartheta_{\widetilde{F}}(t,K_0)$ $(0 \le t \le T)$ has positive reach $\ge \rho$ (in the sense of Definition A.26). Then for every $0 \le s \le t < T$, $K_s = \mathbb{R}^N \setminus \vartheta_{-\widetilde{F}(t-\cdot,\cdot)}(t-s,\, \mathbb{R}^N \setminus K_t)$.*

Remark 37. 1. $\mathscr{K}(\mathbb{R}^N) \rightsquigarrow \mathbb{R}^N,\ K_0 \longmapsto \mathbb{R}^N \setminus \vartheta_{-\widetilde{F}(t-\cdot,\cdot)}(t,\, \mathbb{R}^N \setminus \vartheta_{\widetilde{F}}(t,K_0))$ generalizes the morphological operation of closing (of sets in $\mathscr{K}(\mathbb{R}^N)$) that was introduced by Minkowski and is usually defined as

$$\mathscr{P}(X) \rightsquigarrow X, \qquad K \longmapsto (K - t\,B) \ominus (-t\,B) \stackrel{\text{Def.}}{=} \{y \in X \mid y - t\,B \subset K - t\,B\}$$

for a vector space X and fixed $B \subset X$, $t > 0$ (see e.g. [10, Definition 3.3.1]).

2. In [25], viscosity solutions to the Hamilton-Jacobi-Bellman equation $\partial_t\, u + H(t,x,Du) = 0$ are investigated and in a word, the continuous differentiability of u is concluded from the reversibility in time:
If $u \in C^0([0,T] \times \mathbb{R}^N, \mathbb{R})$ is a viscosity solution of $\partial_t\, u + H(t, \cdot, Du) = 0$ and $v(t,x) := u(T-t,x)$ is a viscosity solution of $\partial_t\, v - H(T-t,\cdot,Dv) = 0$ then adequate assumptions about H ensure $u \in C^1(]0,T[\, \times \mathbb{R}^N)$.
Referring to the relation between reachable sets and level sets of viscosity solutions, we draw an inverse conclusion since we assume smoothness and obtain reversibility in time.

3. The reversibility in time (in the sense of Proposition A.36) can also be regarded as recovering the initial data. Further results about this problem have already been published by Rzeżuchowski in [164, 165], for example, but they usually assume other conditions. Either the initial set consists of only one point or the Hamiltonian function $\mathscr{H}_F$ is of class C^2.

In Proposition 36, we even suppose a uniform radius ρ of positive reach for $K_t \stackrel{\text{Def.}}{=} \vartheta_{\widetilde{F}}(t,K_0)$. The essential advantage for the proof is the relation between the boundaries of $K_t \subset \mathbb{R}^N$ and Graph $(t \longmapsto K_t) \subset \mathbb{R} \times \mathbb{R}^N$ stated in the next lemma:

Lemma 38. *Suppose for* $\widetilde{F} : [0,T] \times \mathbb{R}^N \rightsquigarrow \mathbb{R}^N$, $K \in \mathscr{K}(\mathbb{R}^N)$ *and* $\rho > 0$ *that the map* $[0,T] \rightsquigarrow \mathbb{R}^N$, $t \mapsto \vartheta_{\widetilde{F}}(t,K)$ *is* λ*-Lipschitz continuous (with respect to* $d\!\!l$*) and each set* $\vartheta_{\widetilde{F}}(t,K)$ $(0 \le t \le T)$ *has positive reach of radius* ρ.

Then the topological boundary of Graph $\vartheta_{\widetilde{F}}(\cdot,K)|_{[0,T]}$ *in* $\mathbb{R} \times \mathbb{R}^N$ *is*

$$(\{0\} \times K) \;\cup \bigcup_{0<t<T} (\{t\} \times \partial \vartheta_{\widetilde{F}}(t,K)) \;\cup\; (\{T\} \times \vartheta_{\widetilde{F}}(T,K)).$$

Proof (of Lemma 38). The inclusion

$$(\{0\} \times K) \;\cup \bigcup_{0<t<T} (\{t\} \times \partial \vartheta_{\widetilde{F}}(t,K)) \;\cup\; (\{T\} \times \vartheta_{\widetilde{F}}(T,K)) \;\subset\; \partial\,\text{Graph}\; \vartheta_{\widetilde{F}}(\cdot,K)$$

is obvious. Due to the Lipschitz continuity of $\vartheta_{\widetilde{F}}(\cdot,K)$, we only have to show

$$\partial\,\text{Graph}\; \vartheta_{\widetilde{F}}(\cdot,K) \;\cap\; (]0,T[\times \mathbb{R}^N) \;\subset \bigcup_{0<t<T} (\{t\} \times \partial\,\vartheta_{\widetilde{F}}(t,K)).$$

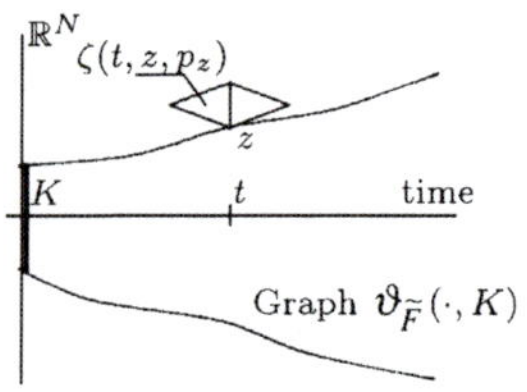

Every $z \in \partial\,\vartheta_{\widetilde{F}}(t,K)$ $(0 \le t \le T)$ and any unit vector $p_z \in N^P_{\vartheta_{\widetilde{F}}(t,K)}(z) = N_{\vartheta_{\widetilde{F}}(t,K)}(z)$ satisfy

$$\overset{\circ}{\mathbb{B}}_\rho\,(z + \rho\, p_z) \cap \vartheta_{\widetilde{F}}(t,K) = \emptyset$$

and thus,

$$(\{t\} \times \overset{\circ}{\mathbb{B}}_\rho\,(z + \rho\, p_z)) \;\cap\; \text{Graph}\; \vartheta_{\widetilde{F}}(\cdot,K) \;=\; \emptyset.$$

The λ-Lipschitz continuity of $\vartheta_{\widetilde{F}}(\cdot,K)$ implies

$$\zeta(t,z,p_z) \cap \text{Graph}\; \vartheta_{\widetilde{F}}(\cdot,K) \;=\; \emptyset$$

for the open set $\zeta(t,z,p_z) := \big\{ (s,y) \in \mathbb{R}^{1+N} \,\big|\, |z + \rho\, p_z - y| < \rho - \lambda\,|s-t| \big\}$.

Now choose $(t,x) \in \partial\,\text{Graph}\; \vartheta_{\widetilde{F}}(\cdot,K)$ with $0 < t < T$ arbitrarily. The continuity of $\vartheta_{\widetilde{F}}(\cdot,K)$ guarantees that Graph $\vartheta_{\widetilde{F}}(\cdot,K)$ is closed and thus, it contains (t,x). Moreover there are sequences $(t_n)_{n\in\mathbb{N}}$, $(x_n)_{n\in\mathbb{N}}$ in $]0,T[$, $\mathbb{R}^N$ respectively with

$$\begin{array}{ll} (t_n,x_n) \notin \text{Graph}\; \vartheta_{\widetilde{F}}(\cdot,K) & \text{for every } n \in \mathbb{N}, \\ (t_n,x_n) \longrightarrow (t,x) & \text{for } n \longrightarrow \infty. \end{array}$$

For each $n \in \mathbb{N}$, let z_n be an element of the projection $\Pi_{\vartheta_{\widetilde{F}}(t_n,K)}(x_n) \subset \partial\vartheta_{\widetilde{F}}(t_n,K)$.

Then, $\quad 0 < |x_n - z_n| \;=\; \text{dist}(x_n,\, \vartheta_{\widetilde{F}}(t_n,K)) \;\le\; |x_n - x| + \text{dist}(x,\, \vartheta_{\widetilde{F}}(t_n,K)) \;\longrightarrow\; 0$

and $\quad p_n := \frac{x_n - z_n}{|x_n - z_n|} \;\in\; N^P_{\vartheta_{\widetilde{F}}(t_n,K)}(z_n) \cap \partial\mathbb{B}_1$.

As mentioned before, we obtain $\zeta(t_n,z_n,p_n) \cap \text{Graph}\; \vartheta_{\widetilde{F}}(\cdot,K) \;=\; \emptyset$ for each $n \in \mathbb{N}$. Adequate subsequences (again denoted by) $(t_n)_{n\in\mathbb{N}}$, $(x_n)_{n\in\mathbb{N}}$, $(p_n)_{n\in\mathbb{N}}$ lead to the additional convergence $p_n \longrightarrow p \in \partial\mathbb{B}_1$ $(n \longrightarrow \infty)$. Finally,

$$\zeta(t,x,p) \cap \text{Graph}\; \vartheta_{\widetilde{F}}(\cdot,K) \;=\; \emptyset.$$

In particular, $\overset{\circ}{\mathbb{B}}_\rho\,(x + \rho\, p) \cap \vartheta_{\widetilde{F}}(t,K) \;=\; \emptyset$ implies $x \in \partial\,\vartheta_{\widetilde{F}}(t,K)$.

□

Proof (of Proposition A.36). $\vartheta_{\widetilde{F}}(s, K_0) \subset \mathbb{R}^N \setminus \vartheta_{-\widetilde{F}(t-\cdot,\cdot)}(t-s,\ \mathbb{R}^N \setminus K_t)$ is an easy indirect consequence of definitions since it is equivalent to

$$\vartheta_{\widetilde{F}}(s, K_0) \ \cap \ \vartheta_{-\widetilde{F}(t-\cdot,\cdot)}(t-s,\ \mathbb{R}^N \setminus K_t) \ = \ \emptyset.$$

For proving the inverse inclusion indirectly at time $s = 0$ (w.l.o.g.), we assume the existence of $t \in [0, T[$ and $y_0 \in \mathbb{R}^N$ with $y_0 \notin K_0 \cup \vartheta_{-\widetilde{F}(t-\cdot,\cdot)}(t,\ \mathbb{R}^N \setminus K_t)$.
As an immediate consequence of $y_0 \notin \vartheta_{-\widetilde{F}(t-\cdot,\cdot)}(t,\ \mathbb{R}^N \setminus K_t)$, the reachable set $\vartheta_{\widetilde{F}}(t, y_0)$ is contained in $K_t \stackrel{\text{Def.}}{=} \vartheta_{\widetilde{F}}(t, K_0)$. Now set

$$\tau := \inf \big\{ s \in [0, t] \ \big| \ \vartheta_{\widetilde{F}}(s, y_0) \subset \vartheta_{\widetilde{F}}(s, K_0) \big\}.$$

In particular, $\tau > 0$ due to $y_0 \notin K_0$.
and $\vartheta_{\widetilde{F}}(\tau, y_0) \subset \vartheta_{\widetilde{F}}(\tau, K_0)$ due to the continuity of the reachable sets.
There are sequences $\tau_n \nearrow \tau$ and $(x_n(\cdot))_{n \in \mathbb{N}}$ in $W^{1,1}([0,T], \mathbb{R}^N)$ satisfying

$$x_n'(\cdot) \in \widetilde{F}(\cdot, x_n(\cdot)) \quad \mathscr{L}^1\text{-a.e.}, \qquad x_n(0) = y_0, \qquad x_n(\tau_n) \notin \vartheta_{\widetilde{F}}(\tau_n, K_0).$$

Standard hypothesis $(\widetilde{\mathscr{H}})$ and the compactness of solutions (as formulated in [180, Theorem 2.5.3]) lead to subsequences (again denoted by) $(\tau_n)_{n \in \mathbb{N}}$, $(x_n(\cdot))_{n \in \mathbb{N}}$ and a solution $x(\cdot) \in W^{1,1}([0,T], \mathbb{R}^N)$ of $x'(\cdot) \in \widetilde{F}(\cdot, x(\cdot))$ ($\mathscr{L}^1$-almost everywhere) with

$$x_n(\cdot) \longrightarrow x(\cdot) \ \text{ uniformly in } [0,T], \qquad x_n'(\cdot) \longrightarrow x'(\cdot) \ \text{ weakly in } L^1([0,T], \mathbb{R}^N).$$

In particular, $(\tau, x(\tau))$ has to be in the boundary of Graph $\vartheta_{\widetilde{F}}(\cdot, K_0)$. Lemma A.38 and $0 < \tau \le t < T$ ensure $x_\tau := x(\tau) \in \partial K_\tau \stackrel{\text{Def.}}{=} \partial\, \vartheta_{\widetilde{F}}(\tau, K_0)$.

Moreover, $K_\tau \stackrel{\text{Def.}}{=} \vartheta_{\widetilde{F}}(\tau, K_0)$ is supposed to have positive reach. Its limiting and proximal normal cone coincide at each boundary point due to Corollary A.29. Thus,

$$\emptyset \ \neq \ N_{\vartheta_{\widetilde{F}}(\tau, K_0)}(x_\tau) \ = \ N^P_{\vartheta_{\widetilde{F}}(\tau, K_0)}(x_\tau) \ \subset \ N^P_{\vartheta_{\widetilde{F}}(\tau, y_0)}(x_\tau).$$

For every unit normal vector $\nu \in N_{\vartheta_{\widetilde{F}}(\tau, K_0)}(x_\tau)$, Proposition A.32 provides a solution $z(\cdot) \in W^{1,1}([0,\tau], \mathbb{R}^N)$ and its adjoint arc $q(\cdot) \in W^{1,1}([0,\tau], \mathbb{R}^N)$ satisfying the corresponding Hamiltonian system and $z(0) \in K_0$, $z(\tau) = x_\tau$, $q(\tau) = \nu$.
The same Cauchy problem is solved by $x(\cdot)$ and its adjoint arc as well. Standard hypothesis $(\widetilde{\mathscr{H}})$ implies the uniqueness of solutions and, its consequence $z(0) = x(0) = y_0 \notin K_0$ leads to a contradiction. □

A.5.5 How to Make Points Evolve into Convex Sets of Positive Erosion

Our aim consists in sufficient assumptions for the interior ball condition on $\vartheta_F(t, K)$ — without any regularity assumptions about the initial set $K \in \mathscr{K}(\mathbb{R}^N)$. In particular, we focus on K consisting just of a single point. For this purpose, we are willing to tolerate stronger assumptions about the set-valued map $\widetilde{F} : [0,T] \times \mathbb{R}^N \rightsquigarrow \mathbb{R}^N$ than standard hypothesis $(\widetilde{\mathscr{H}})$ (specified in Definition A.33 on page 458).

Definition 39. For any $\rho > 0$, a set-valued map $\widetilde{F} : [0,T] \times \mathbb{R}^N \rightsquigarrow \mathbb{R}^N$ satisfies the so-called *standard hypothesis* $(\widetilde{\mathscr{H}_\circ^\rho})$ if it has the following properties:

1. $\widetilde{F}$ is measurable and, all its values are nonempty convex compact subsets of positive erosion of radius ρ,
2. for every $t \in [0,T]$, $\mathscr{H}_{\widetilde{F}}(t,\cdot,\cdot) \in C^2(\mathbb{R}^N \times (\mathbb{R}^N \setminus \{0\}))$,
3. for every $R > 1$, there exists $\lambda_R(\cdot) \in L^1([0,T])$ such that the derivative of $\mathscr{H}_{\widetilde{F}}(t,\cdot,\cdot)$ restricted to $\mathbb{B}_R \times (\mathbb{B}_R \setminus \overset{\circ}{\mathbb{B}}_{\frac{1}{R}})$ is $\lambda_R(t)$-Lipschitz continuous for Lebesgue-almost every $t \in [0,T]$,
4. there is $k_{\widetilde{F}} \in L^1([0,T])$ such that for a.e. $t \in [0,T]$ and all $x, p \in \mathbb{R}^N$ $(|p| \geq 1)$,
$$\left\| \partial_{(x,p)} \mathscr{H}_{\widetilde{F}}(t,x,p) \right\|_{\mathrm{Lin}(\mathbb{R}^N \times \mathbb{R}^N, \mathbb{R})} \leq k_{\widetilde{F}}(t) \cdot (1 + |x| + |p|).$$

Remark 40. Standard hypothesis $(\widetilde{\mathscr{H}_\circ^\rho})$ differs from its counterpart $(\widetilde{\mathscr{H}})$ in two respects: The values of $\widetilde{F}$ have uniform positive erosion (additionally) and, its Hamiltonian $\mathscr{H}_{\widetilde{F}}(t,\cdot,\cdot)$ is even twice continuously differentiable in $\mathbb{R}^N \times (\mathbb{R}^N \setminus \{0\})$. This second restriction has the advantage that we can apply the tools of matrix Riccati equation (mentioned in subsequent Lemmas A.43 and A.44).

Proposition 41. *In addition to standard hypothesis* $(\widetilde{\mathscr{H}_\circ^\rho})$*, assume for the set-valued map* $\widetilde{F} : [0,T] \times \mathbb{R}^N \rightsquigarrow \mathbb{R}^N$ *that some* $\lambda(\cdot) \in L^1([0,T])$ *satisfies*
$$\begin{aligned} \|\mathscr{H}_{\widetilde{F}}(t,\cdot,\cdot)\|_{C^{1,1}(\mathbb{R}^N \times \partial \mathbb{B}_1)} &\overset{\text{Def.}}{=} \|\mathscr{H}_{\widetilde{F}}(t,\cdot,\cdot)\|_{C^1(\mathbb{R}^N \times \partial \mathbb{B}_1)} + \mathrm{Lip}\, \partial \mathscr{H}_{\widetilde{F}}(t,\cdot,\cdot)|_{\mathbb{R}^N \times \partial \mathbb{B}_1} \\ &< \lambda(t) \end{aligned}$$
at $\mathscr{L}^1$*-almost every time* $t \in [0,T]$*. Choose* $K \in \mathscr{K}(\mathbb{R}^N)$ *arbitrarily.*

Then there exist $\sigma > 0$ *and a time* $\widehat{\tau} \in]0,T]$ *(depending only on* $\|\lambda\|_{L^1}, \rho, K$*) such that the reachable set* $\vartheta_{\widetilde{F}}(t,x_0)$ *is convex and has positive erosion of radius* σt *for any* $t \in]0,\widehat{\tau}[$, $x_0 \in K$.

As a direct consequence, the reachable set $\vartheta_{\widetilde{F}}(t,K_1)$ *is the closed* (σt)*-neighborhood of a compact set for all* $t \in]0,\widehat{\tau}[$ *and each nonempty compact subset* $K_1 \subset K$.

The proof of this proposition uses matrix Riccati equations for Hamiltonian systems, but these tools of subsequent Lemma A.43 consider initial values induced by a Lipschitz function ψ. First we specify how to exchange the two components $(x(\cdot), p(\cdot))$ (of a solution and its adjoint arc) for preserving the Hamiltonian structure of their differential equations:

Lemma 42. *Assume the Hamiltonian system for* $x(\cdot),\, p(\cdot) \in W^{1,1}([0,T],\mathbb{R}^N)$
$$\begin{cases} x'(t) = \frac{\partial}{\partial p} H_1(t, x(t), p(t)) \\ p'(t) = -\frac{\partial}{\partial x} H_1(t, x(t), p(t)) \end{cases} \quad \text{a.e. in } [0,T]$$
with sufficiently smooth $H_1 : [0,T] \times \mathbb{R}^N \times \mathbb{R}^N \longrightarrow \mathbb{R}$*. Moreover set*
$$y(t) := -p(t), \qquad q(t) := x(t), \qquad H_2(t,\xi,\zeta) := H_1(t,\zeta,-\xi).$$

Then the absolutely continuous functions $(y(\cdot), q(\cdot))$ satisfy the Hamiltonian system

$$\begin{cases} y'(t) = \frac{\partial}{\partial q} H_2(t, y(t), q(t)) \\ q'(t) = -\frac{\partial}{\partial y} H_2(t, y(t), q(t)) \end{cases} \qquad a.e. \text{ in } [0,T].$$

□

Lemma 43.
In addition to the assumptions (2.)–(4.) of Lemma A.35 (on page 458), suppose for $\psi : \mathbb{R}^N \longrightarrow \mathbb{R}^N$, $H : [0,T] \times \mathbb{R}^N \times \mathbb{R}^N \longrightarrow \mathbb{R}$ and the Hamiltonian system

$$\begin{cases} y'(t) = \frac{\partial}{\partial q} H(t, y(t), q(t)), & y(0) = y_0 \\ q'(t) = -\frac{\partial}{\partial y} H(t, y(t), q(t)), & q(0) = \psi(y_0) \end{cases} \qquad (*)$$

1'. $H(t,\cdot,\cdot)$ is twice *continuously differentiable for every $t \in [0,T]$.*

Then for every initial set $K \in \mathscr{K}(\mathbb{R}^N)$, the following statements are equivalent:

(i) For all $t \in [0,T]$, $M_t^{\mapsto}(K) := \{(y(t), q(t)) \mid (y(\cdot), q(\cdot)) \text{ solves } (*),\ y_0 \in K\}$ is the graph of a locally Lipschitz continuous function,

(ii) For any solution $(y(\cdot), q(\cdot)) : [0,T] \longrightarrow \mathbb{R}^N \times \mathbb{R}^N$ to initial value problem $(*)$ and each cluster point $Q_0 \in \operatorname{Limsup}_{z \to y_0} \{\nabla \psi(z)\} \subset \mathbb{R}^{N \times N}$, the following matrix Riccati equation has a solution $Q(\cdot)$ on $[0,T]$

$$\begin{cases} \partial_t Q + \frac{\partial^2 H}{\partial p\, \partial x}(t, y(t), q(t))\, Q + Q\, \frac{\partial^2 H}{\partial x\, \partial p}(t, y(t), q(t)) \\ \quad + Q\, \frac{\partial^2 H}{\partial p^2}(t, y(t), q(t))\, Q + \frac{\partial^2 H}{\partial x^2}(t, y(t), q(t)) = 0, \\ \hfill Q(0) = Q_0. \end{cases}$$

If one of these equivalent properties is satisfied and if ψ is (continuously) differentiable, then $M_t^{\mapsto}(K)$ is even the graph of a (continuously) differentiable function.

Proof is given in [80, Theorem 5.3], for the same Hamiltonian system but with $y(T) = y_T$, $q(T) = q_T$ given. Hence, this lemma is a direct consequence considering $-H(T - \cdot, \cdot, \cdot)$ and $(y(T - \cdot), q(T - \cdot))$. □

For preventing singularities of $Q(\cdot)$, the following comparison principle provides a bridge to a *scalar* Riccati equation.

Lemma 44 (Comparison theorem for the matrix Riccati equation, [163, Th.2]**).**
Let $A_j, B_j, C_j : [0,T[\longrightarrow \mathbb{R}^{N \times N}$ $(j = 0,1,2)$ be bounded continuous matrix-valued functions such that each $M_j(t) := \begin{pmatrix} A_j(t) & B_j(t) \\ B_j(t)^T & C_j(t) \end{pmatrix}$ is symmetric.
Assume that U_0, $U_2 : [0,T[\longrightarrow \mathbb{R}^{N \times N}$ are solutions to the matrix Riccati equation

$$\tfrac{d}{dt} U_j = A_j + B_j\, U_j + U_j\, B_j^T + U_j\, C_j\, U_j$$

with $M_2(\cdot) \geq M_0(\cdot)$ (i.e. $M_2(t) - M_0(t)$ is positive semi-definite for every t).

For symmetric $U_1(0) \in \mathbb{R}^{N \times N}$ with $U_2(0) \geq U_1(0) \geq U_0(0)$, $M_2(\cdot) \geq M_1(\cdot) \geq M_0(\cdot)$, given, there exists a solution $U_1 : [0,T[\longrightarrow \mathbb{R}^{N \times N}$ to the Riccati equation with matrix $M_1(\cdot)$. Moreover, $U_2(t) \geq U_1(t) \geq U_0(t)$ for all $t \in [0,T[$.

Proof (of Proposition A.41).
The integrable bound of $t \mapsto \|\mathscr{H}_{\widetilde{F}}(t,\cdot,\cdot)\|_{C^{1,1}(\mathbb{R}^N \times \partial \mathbb{B}_1)}$ and Gronwall's Lemma lead to a radius $R = R(\|\lambda\|_{L^1}, K) > 1$ and a time $\widehat{T} = \widehat{T}(\|\lambda\|_{L^1}, K) \in\,]0,T]$ such that

1. $\vartheta_{\widetilde{F}}(t,K) \subset \mathbb{B}_R$ for all $t \in [0,T]$,
2. for every solution $x(\cdot)$ of $x'(\cdot) \in \widetilde{F}(\cdot, x(\cdot))$ starting in K and each adjoint $p(\cdot)$ with $\frac{1}{2} \le |p(0)| \le 2$ fulfills $\frac{1}{R} < |p(\cdot)| < R, \;\; |p(\cdot) - p(0)| < \frac{1}{4R}$ on $[0,\widehat{T}]$.

A smooth cut-off function provides a map $H_1 : [0,\widehat{T}] \times \mathbb{R}^N \times \mathbb{R}^N \longrightarrow \mathbb{R}$ that fulfills the assumptions of Lemma A.43 and

$$H_1 = \mathscr{H}_{\widetilde{F}} \qquad \text{in } [0,\widehat{T}] \times \mathbb{R}^N \times (\mathbb{R}^N \setminus \mathbb{B}_{\frac{1}{2R}}).$$

Using the transformation of the preceding Lemma A.42, the auxiliary function

$$H_2 : [0,T] \times \mathbb{R}^N \times \mathbb{R}^N \longrightarrow \mathbb{R}, \qquad (t,\xi,\zeta) \longmapsto H_1(t,\zeta,-\xi)$$

is still holding the conditions of Lemma A.43. As a consequence, we obtain for any initial point $x_0 \in K$ and time $\tau \in\,]0,\widehat{T}]$ that the following statements are equivalent:

(*i*) For all $t \in [0,\tau]$, the set M_t^1 of all points $(p(t), x(t))$ with solutions $(x(\cdot), p(\cdot)) \in W^{1,1}([0,t], \mathbb{R}^N \times \mathbb{R}^N)$ of

$$\begin{cases} x'(s) = \frac{\partial}{\partial p} H_1(s, x(s), p(s)), & x(0) = x_0 \\ p'(s) = -\frac{\partial}{\partial x} H_1(s, x(s), p(s)), & p(0) \in \mathbb{B}_2 \setminus \overset{\circ}{\mathbb{B}}_{\frac{1}{2}} \end{cases}$$

is the graph of a continuously differentiable function f_t.

(*ii*) For all $t \in [0,\tau]$, the set M_t^2 of all points $(y(t), q(t))$ with solutions $(y(\cdot), q(\cdot)) \in W^{1,1}([0,t], \mathbb{R}^N \times \mathbb{R}^N)$ of

$$\begin{cases} y'(s) = \frac{\partial}{\partial q} H_2(s, y(s), q(s)), & y(0) \in \mathbb{B}_2 \setminus \overset{\circ}{\mathbb{B}}_{\frac{1}{2}} \\ q'(s) = -\frac{\partial}{\partial y} H_2(s, y(s), q(s)), & q(0) = x_0 \end{cases}$$

is the graph of a C^1 function g_t (and $g_t(\xi) = f_t(-\xi)$).

(*iii*) For any solution $(y,q) : [0,t] \longrightarrow \mathbb{R}^N \times \mathbb{R}^N$ to the initial value problem (*ii*) $(t \le \tau)$, there is a solution $Q : [0,t] \longrightarrow \mathbb{R}^{N \times N}$ to the Riccati equation

$$\begin{cases} Q' + \frac{\partial^2 H_2}{\partial q\, \partial y}(s, y(s), q(s))\, Q + Q\, \frac{\partial^2 H_2}{\partial y\, \partial q}(s, y(s), q(s)) \\ \quad + Q\, \frac{\partial^2 H_2}{\partial q^2}(t, y(s), q(s))\, Q + \frac{\partial^2 H_2}{\partial y^2}(s, y(s), q(s)) = 0, \\ \qquad Q(0) = 0. \end{cases}$$

(*iv*) For any solution $(x,p) : [0,t] \longrightarrow \mathbb{R}^N \times \mathbb{R}^N$ to the initial value problem (*i*) $(t \le \tau)$, there is a solution $Q : [0,t] \longrightarrow \mathbb{R}^{N \times N}$ to the Riccati equation

$$\begin{cases} Q' - \frac{\partial^2 H_1}{\partial x\, \partial p}(s, x(s), p(s))\, Q - Q\, \frac{\partial^2 H_1}{\partial p\, \partial x}(s, x(s), p(s)) \\ \quad + Q\, \frac{\partial^2 H_1}{\partial x^2}(s, x(s), p(s))\, Q + \frac{\partial^2 H_1}{\partial p^2}(s, x(s), p(s)) = 0, \\ \qquad Q(0) = 0. \end{cases}$$

Now we give a criterion for the choice of $\widehat{\tau} \in \,]0, \widehat{T}]$. Setting

$$\mu(t) \;:=\; \sup_{\substack{|x| \le R \\ \frac{1}{R} \le |p| \le R}} \left\| \begin{pmatrix} \frac{\partial^2}{\partial p^2}\, \mathscr{H}_{\widetilde{F}}(t,x,p) & -\, \frac{\partial^2}{\partial x\, \partial p}\, \mathscr{H}_{\widetilde{F}}(t,x,p) \\ -\, \frac{\partial^2}{\partial p\, \partial x}\, \mathscr{H}_{\widetilde{F}}(t,x,p) & \frac{\partial^2}{\partial x^2}\, \mathscr{H}_{\widetilde{F}}(t,x,p) \end{pmatrix} \right\|_{\mathrm{Lin}(\mathbb{R}^{2N},\mathbb{R}^{2N})}$$

the comparison theorem for matrix Riccati equations (Lemma A.44 extended to integrable coefficients via Lusin's Theorem and approximation, see also [80, § 5.2]) guarantees existence and uniqueness of such a solution $Q \in W^{1,1}([0,t],\, \mathbb{R}^{N\times N})$ for every $t < \min\{T,\, \frac{\pi}{2\,\|\mu\|_{L^1}}\}$. Indeed, for $a(\cdot) = \pm \mu(\cdot) \in L^1([0,T])$, the scalar Riccati equation

$$\tfrac{d}{dt}\, u(t) \;=\; a(t) + a(t)\, u(t)^2, \qquad u(0) = 0$$

has the solution $u(t) \;=\; \tan\big(\int_0^t a(s)\, ds\big)$ in $[0, \frac{\pi}{2\,\|a\|_{L^1}}[$. Furthermore we obtain the upper bound $\|Q(t)\| \;\le\; \tan\, \|\,\mu|_{[0,t]}\,\|_{L_1}$.

All values of $\widetilde{F}$ are compact convex sets with positive erosion of radius ρ due to standard hypothesis $(\widetilde{\mathscr{H}_\circ^\rho})$. It implies a constant $\widehat{\sigma} = \widehat{\sigma}(\rho,K,R) > 0$ with

$$\xi \cdot \tfrac{\partial^2}{\partial p^2}\, \mathscr{H}_{\widetilde{F}}(t,x,p)\; \xi \;\ge\; 9\, \widehat{\sigma}\, \Big|\, \xi \;-\; \tfrac{\xi \cdot p}{|p|^2}\; p\, \Big|^2$$

for all $t \in [0,T],\; |x| \le R,\; \frac{1}{R} \le |p| \le R,\; \xi$. Using the matrix abbreviation

$$\begin{aligned} D(t,x,p) \;:=\; &-\; \frac{\partial^2 \mathscr{H}_{\widetilde{F}}}{\partial x\, \partial p}(t,x,p)\; Q(t) \;-\; Q(t)\; \frac{\partial^2 \mathscr{H}_{\widetilde{F}}}{\partial p\, \partial x}(t,x,p) \\ &+\; Q(t)\; \frac{\partial^2 \mathscr{H}_{\widetilde{F}}}{\partial x^2}(t,x,p)\; Q(t), \end{aligned}$$

choose $\widehat{\tau} = \widehat{\tau}(\lambda,\rho,K) > 0$ small enough such that

$$\begin{cases} \qquad\qquad \widehat{\tau} \;<\; \min\{\widehat{T},\; \frac{\pi}{2\,\|\mu\|_{L^1}}\}, \\ \displaystyle\int_0^{\widehat{\tau}} \lambda(t)\; dt \;<\; 1, \\ \|D(t,x,p)\| \;\le\; \widehat{\sigma} \qquad \text{for every } t \in [0,\widehat{\tau}],\; |x| \le R,\; \frac{1}{R} \le |p| \le R. \end{cases}$$

As a next step, we conclude that the solution $Q(t)$ of (iv) (restricted to $[0,\widehat{\tau}]$) satisfies $Q(t) \le -\,\widehat{\sigma}\, t \cdot \mathbb{I}\mathrm{d}$ in the $(N-1)$-dimensional subspace of $\mathbb{R}^N$ perpendicular to $p(t)$. Indeed, let $(x(\cdot),\, p(\cdot)) \in W^{1,1}([0,\widehat{\tau}],\, \mathbb{R}^N \times \mathbb{R}^N)$ be a solution to the Hamiltonian system (i) and choose an arbitrary unit vector $\xi \in \mathbb{R}^N$ with $|\xi \cdot p(0)| < \frac{1}{4R}$. Then the auxiliary function

$$\varphi : [0,\widehat{\tau}] \longrightarrow \mathbb{R}^N,\; t \longmapsto \xi \cdot Q(t)\, \xi \;+\; \widehat{\sigma}\, t\, \Big|\, \xi \;-\; \tfrac{\xi \cdot p(t)}{|p(t)|^2}\; p(t)\, \Big|^2$$

satisfies $\varphi(0) = 0$ and is absolutely continuous with $\varphi(\cdot) \le 0$. Indeed,

$$\begin{aligned} \varphi'(t) &= \xi \cdot Q'(t)\, \xi + \widehat{\sigma}\, \Big|\, \xi - \tfrac{\xi \cdot p(t)}{|p(t)|^2}\, p(t) \Big|^2 - 2\, \widehat{\sigma}\, t\, \Big(\xi - \tfrac{\xi \cdot p(t)}{|p(t)|^2}\, p(t)\Big) \cdot \tfrac{d}{dt}\Big(\tfrac{\xi \cdot p(t)}{|p(t)|^2}\, p(t)\Big) \\ &= \xi \cdot Q'(t)\, \xi + \widehat{\sigma}\, \Big|\, \xi - \tfrac{\xi \cdot p(t)}{|p(t)|^2}\, p(t) \Big|^2 - 2\, \widehat{\sigma}\, t\, \Big(\xi - \tfrac{\xi \cdot p(t)}{|p(t)|^2}\, p(t)\Big) \cdot \;\; \tfrac{\xi \cdot p(t)}{|p(t)|^2} \;\; p'(t) \end{aligned}$$

because $\xi - \frac{\xi \cdot p(t)}{|p(t)|^2}\, p(t)$ is perpendicular to $p(t)$.

Now $|p(t) - p(0)| < \frac{1}{4R}$, $\frac{1}{R} \le |p(t)| \le R$ and $|\xi \cdot p(0)| < \frac{1}{4R}$ imply $\left|\frac{\xi \cdot p(t)}{|p(t)|}\right| < \frac{1}{2}$ and $\frac{1}{2}\,|\xi| = 1 - \frac{1}{2} \le \left|\xi - \frac{\xi \cdot p(t)}{|p(t)|^2}\,p(t)\right| \le 1 + \frac{1}{2}$. Thus,

$$\begin{aligned}
\varphi'(t) &\le (-9+4+1)\,\widehat{\sigma}\,\left|\xi - \tfrac{\xi \cdot p(t)}{|p(t)|^2}\,p(t)\right|^2 + 2\,\widehat{\sigma}\,t\,\left|\xi - \tfrac{\xi \cdot p(t)}{|p(t)|^2}\,p(t)\right| \tfrac{|\xi|\,|p(t)|}{|p(t)|^2}\,|p'(t)| \\
&\le -4\,\widehat{\sigma}\,\left|\xi - \tfrac{\xi \cdot p(t)}{|p(t)|^2}\,p(t)\right|^2 + 2\,\widehat{\sigma}\,t\,\left|\xi - \tfrac{\xi \cdot p(t)}{|p(t)|^2}\,p(t)\right| \lambda(t) \\
&\le 2\,\widehat{\sigma}\,\left|\xi - \tfrac{\xi \cdot p(t)}{|p(t)|^2}\,p(t)\right| \cdot \Big(-2\,\left|\xi - \tfrac{\xi \cdot p(t)}{|p(t)|^2}\,p(t)\right| + \lambda(t)\,t\Big) \\
&\le 2\,\widehat{\sigma}\,\left|\xi - \tfrac{\xi \cdot p(t)}{|p(t)|^2}\,p(t)\right| \cdot \Big(-2\,\Big(1 - \left|\tfrac{\xi \cdot p(t)}{|p(t)|}\right|\Big) + \lambda(t)\,t\Big) \\
&\le 2\,\widehat{\sigma}\,\left|\xi - \tfrac{\xi \cdot p(t)}{|p(t)|^2}\,p(t)\right| \cdot \Big(-2\,\Big(1 - \tfrac{1}{2}\Big) + \lambda(t)\,\widehat{\tau}\Big) \\
&\le \widehat{\sigma} \cdot 3 \cdot \big(-1 + \lambda(t)\,\widehat{\tau}\big).
\end{aligned}$$

Now we obtain $\varphi(t) \le 0$ for all $t \in [0, \widehat{\tau}]$ and as a consequence, $Q(t) \le -\widehat{\sigma}\,t \cdot \mathbb{I}\mathrm{d}$ is fulfilled in the subspace of $\mathbb{R}^N$ perpendicular to $p(t)$.

Finally we need the geometric interpretation for concluding convexity and positive erosion of $\vartheta_{\widetilde{F}}(t, x_0)$ (of radius $\widehat{\sigma}\,t$) for each $t \in \,]0, \widehat{\tau}[$ and $x_0 \in K$.
As mentioned before, the existence of the solution $Q(\cdot)$ on $[0, \widehat{\tau}[$ implies for all $t \in [0, \widehat{\tau}[$ that the set M_t^1 is the graph of a C^1 function f_t. Moreover Proposition A.32 (on page 457) guarantees

$$\begin{aligned}
\mathrm{Graph}\,N_{\vartheta_{\widetilde{F}}(t,x_0)} &\subset \Big\{(x(t),\, \lambda\,p(t))\,\Big|\,(x(\cdot), p(\cdot))\ \text{solves}\ (i),\ \lambda \ge 0\Big\} \\
&\overset{\text{Def.}}{=} \bigcup_{\lambda \ge 0} \mathrm{Graph}\ (\lambda\,f_t^{-1}).
\end{aligned}$$

Now we obtain at every time $t \in \,]0, \widehat{\tau}[$ that each $p \in \mathbb{R}^N \setminus \{0\}$ belongs to the limiting normal cone of a unique boundary point $z \in \partial\,\vartheta_{\widetilde{F}}(t, x_0)$ and, $z = z(p)$ is continuously differentiable.
In particular, every supporting hyperplane of the closed convex hull $\overline{\mathrm{co}}\ \vartheta_{\widetilde{F}}(t, x_0)$ may have at most one point in common with the compact reachable set $\vartheta_{\widetilde{F}}(t, x_0)$. Thus, $\overline{\mathrm{co}}\ \vartheta_{\widetilde{F}}(t, x_0) \subset \mathbb{R}^N$ is even *strictly* convex and coincides with $\vartheta_{\widetilde{F}}(t, x_0)$ at each time $t \in \,]0, \widehat{\tau}[$. It is sufficient to consider the limiting normal cones of $\vartheta_{\widetilde{F}}(t, x_0)$ *locally* at every boundary point.

Well-known properties of variational equations (see e.g. [80]) and the uniqueness of solutions to the matrix Riccati equation (iv) imply that $-Q(s)$ is the derivative of the C^1 function f_s for $0 < s \le t < \widehat{\tau}$. Indeed, for each solution $(x(\cdot), p(\cdot))$ to the Hamiltonian system (i), set $(y(\cdot), q(\cdot)) := (-p(\cdot), x(\cdot))$ again and let $(U(\cdot), V(\cdot)) : [0, t] \longrightarrow \mathbb{R}^{N \times N} \times \mathbb{R}^{N \times N}$ denote the solution to the linearized system

$$\begin{cases}
U'(s) = \frac{\partial^2}{\partial y\,\partial q} H_2(s, y(s), q(s))\,U(s) + \frac{\partial^2}{\partial q^2} H_2(s, y(s), q(s))\,V(s), \\
V'(s) = -\frac{\partial^2}{\partial y^2} H_2(s, y(s), q(s))\,U(s) - \frac{\partial^2}{\partial q\,\partial y} H_2(s, y(s), q(s))\,V(s), \\
U(0) = \mathbb{I}\mathrm{d}_{\mathbb{R}^{N \times N}}, \qquad V(0) = 0.
\end{cases}$$

Then for any $s \in]0,t]$ and initial direction $u_0 \in \mathbb{R}^N \setminus \{0\}$, $(U(s)\,u_0,\ V(s)\,u_0)$ belongs to the contingent cone of $M_s^2 \subset \mathbb{R}^N \times \mathbb{R}^N$ at $(y(s),q(s))$ (due to the variational equations, see e.g. [80]).
Since M_s^2 is the graph of a continuously differentiable function g_s, we conclude that firstly, this cone $T_{M_s^2}(y(s),q(s))$ is a N-dimensional subspace of $\mathbb{R}^N \times \mathbb{R}^N$ and secondly, $|V(s)\,u_0| \leq \text{const} \cdot \lambda(s) \cdot |U(s)\,u_0|$ (due to Remark A.31 on page 457).
The latter property and the uniqueness of the linearized system ensure $U(s)\,u_0 \neq 0$ for all $u_0 \neq 0$ and thus, $U(s)$ is invertible. Comparing the dimensions leads to

$$T_{M_s^2}(y(s),q(s)) \;=\; (U(s),\, V(s))\ \mathbb{R}^N$$

and $\quad V(s)\,U(s)^{-1}$ is the derivative of g_s at $y(s)$.
Hence, $-V(s)\,U(s)^{-1}$ is the derivative of $f_s = g_s(-\,\cdot)$ at $p(s) = -y(s)$.
Moreover it is easy to check that $V(s)\,U(s)^{-1}$ satisfies the matrix Riccati equation (iii) and thus, its uniqueness implies $V(s)\,U(s)^{-1} = Q(s)$ for $0 < s \leq t < \widehat{\tau}$.

Thus for every time $t \in]0, \widehat{\tau}[$, the derivative of f_t at $p(t)$ is bounded by $\widehat{\sigma}\,t$ from below in a $(N-1)$-dimensional subspace of $\mathbb{R}^N$.
Since $\vartheta_{\widetilde{F}}(t,x_0)$ is convex, it implies that $\vartheta_{\widetilde{F}}(t,x_0)$ has positive erosion of radius increasing (at least) linearly in time. □

A.5.6 Reachable Sets of Balls and Their Complements

In this section, we investigate the proximal radius of boundary points while sets are evolving along differential inclusions. Compact balls and their complements exemplify the key features for short times (as stated in subsequent Proposition A.46). They lead to the main results about proximal radii in both forward and backward time direction as a corollary.
The proofs are based on the Hamiltonian system and its regularity — in the same way as in § A.5.5.

Definition 45. For $\Lambda > 0$ fixed, the set $\text{LIP}_\Lambda^{(C^2)}(\mathbb{R}^N,\mathbb{R}^N)$ consists of all set-valued maps $F : \mathbb{R}^N \rightsquigarrow \mathbb{R}^N$ satisfying

1. $F : \mathbb{R}^N \rightsquigarrow \mathbb{R}^N$ has nonempty compact convex values,
2. $\mathscr{H}_F(x,p) := \sup_{v \in F(x)} p \cdot v$ is twice continuously differentiable in $\mathbb{R}^N \times (\mathbb{R}^N \setminus \{0\})$,
3. $\|\mathscr{H}_F\|_{C^2(\mathbb{R}^N \times\, \partial\mathbb{B}_1)} < \Lambda$.

Proposition 46. *Let F be any set-valued map of $\text{LIP}_\Lambda^{(C^2)}(\mathbb{R}^N,\mathbb{R}^N)$ and $B := \mathbb{B}_r(x_0) \subset \mathbb{R}^N$ a compact ball of positive radius r.*
Then there exists a time $\tau = \tau(r,\Lambda) > 0$ such that for all times $t \in [0, \tau(r,\Lambda)[$,

1.) $\vartheta_F(t,B)$ is convex and has radius of curvature $\geq r - 9\,\Lambda\,(1+r)^2\,t$,
2.) $\vartheta_F(t,\mathbb{R}^N \setminus B)$ is concave and has radius of curvature $\geq r - 9\,\Lambda\,(1+r)^2\,t$.

Restricting ourselves to $0 < r \le 2$, the time $\tau(r,\Lambda) > 0$ can be chosen as an increasing function of r. The claim of Proposition A.46 does not include, however, that $r - 9\Lambda\,(1+r)^2\,t \ge 0$ for all $t \in [0, \tau(r,\Lambda)[$ (because then it is not immediately clear how to choose $\tau(r,\Lambda) > 0$ as increasing with respect to all $r \in]0,2]$).

As an equivalent formulation of statement (1.), the convex set $\vartheta_F(t,B)$ has *positive erosion* of radius $\rho(t) \ge r - 9\Lambda\,(1+r)^2\,t$, i.e. there is some $K_t \subset \mathbb{R}^N$ with $\vartheta_F(t,B) = \mathbb{B}_{\rho(t)}(K_t)$.
Strictly speaking, statement (2.) is of more interest here: $\vartheta_F(t,\, \mathbb{R}^N \setminus B) \subset \mathbb{R}^N$ has *positive reach* $\ge \rho(t) \ge r - 9\Lambda\,(1+r)^2\,t$ (in the sense of Federer, see Def. A.26). Roughly speaking, the proofs of these two statements just differ in a sign and thus, both of them are mentioned here.

Applying Proposition A.46 to adequate proximal balls, the inclusion principle of reachable sets and Proposition A.32 (on page 457) have the immediate consequence:

Corollary 47. *For every map* $F \in \mathrm{LIP}^{(C^2)}_\Lambda(\mathbb{R}^N,\mathbb{R}^N)$ *and radius* $r_0 \in]0,2]$, *there exists some* $\tau = \tau(r_0,\Lambda) > 0$ *such that for any* $K \in \mathscr{K}(\mathbb{R}^N)$, $r \in [r_0, 2]$ *and* $t \in [0,\tau[$,

1. *each* $x_1 \in \partial\vartheta_F(t,K)$ *and* $v_1 \in N^P_{\vartheta_F(t,K)}(x_1)$ *with proximal radius* r *are linked to some* $x_0 \in \partial K$ *and* $v_0 \in N^P_K(x_0)$ *with proximal radius* $\ge r - 81\,\Lambda\,t$ *by a solution to* $x'(\cdot) \in F(x(\cdot))$ *and its adjoint arc, respectively.*

2. *each* $x_0 \in \partial K$ *and* $v_0 \in N^P_K(x_0)$ *with proximal radius* r *are linked to some* $x_1 \in \partial\vartheta_F(t,K)$ *and* $v_1 \in N^P_{\vartheta_F(t,K)}(x_1)$ *with proximal radius* $\ge r - 81\,\Lambda\,t$ *by a solution to* $x'(\cdot) \in F(x(\cdot))$ *and its adjoint arc, respectively.* □

For describing the time-dependent limiting normals, we use adjoint arcs and benefit from the Hamiltonian system they are satisfying together with the solutions (as formulated in preceding Proposition A.32 on page 457).
In short, the graph of normal cones at time t, Graph $N_{\vartheta_F(t,K)}(\cdot)|_{\partial\,\vartheta_F(t,K)}$, can be traced back to the beginning by means of the Hamiltonian system with $\mathscr{H}_F$.

As in § A.5.5, we take the next order into consideration and, the matrix Riccati equation provides an analytical access to geometric properties like curvature. In particular, Lemma A.43 (on page 465) motivates the assumption that $\mathscr{H}_F$ is twice continuously differentiable in $\mathbb{R}^N \times (\mathbb{R}^N \setminus \{0\})$ for all maps $F \in \mathrm{LIP}^{(C^2)}_\Lambda(\mathbb{R}^N,\mathbb{R}^N)$.

For preventing singularities of the matrix solution $Q(\cdot)$ to the Riccati equation, the comparison principle in Lemma A.44 (on page 465) provides a connection with solutions to a *scalar* Riccati equation again.

Proof (of Proposition A.46). Similarly to Proposition A.41 (on page 464), statement (1.) is based on applying Lemma A.43 (on page 465) to the boundary $K := \partial\,\mathbb{B}_r(0)$ and its exterior unit normals, i.e. $\psi(x) := \frac{x}{r}$, after assuming $B = \mathbb{B}_r(0)$ without loss of generality. Obviously, ψ can be extended to $\psi \in C^1(\mathbb{R}^N, \mathbb{R}^N)$. (Statement (2.) of Proposition 46 is shown in the same way – just with inverse signs, i.e. $\widehat{\psi}(x) := -\frac{x}{r}$ instead. Hence, we do not formulate this part in detail.)

For every point $y_0 \in \partial\,\mathbb{B}_r$, there exist a solution $y(\cdot) \in C^1([0,\infty[, \mathbb{R}^N)$ and its adjoint $q(\cdot) \in C^1([0,\infty[, \mathbb{R}^N)$ satisfying

$$\begin{cases} y'(t) = \frac{\partial}{\partial q}\, \mathscr{H}_F(y(t), q(t)) \in F(y(t)), & y(0) = y_0, \\ q'(t) = -\frac{\partial}{\partial y}\, \mathscr{H}_F(y(t), q(t)), & q(0) = \psi(y_0) \end{cases} \qquad (*)$$

and, $F \in \mathrm{LIP}^{(C^2)}_{\Lambda}(\mathbb{R}^N, \mathbb{R}^N)$ implies the a priori bounds

$$\begin{aligned} |y(t) - y_0| &\le \Lambda\, t, \\ e^{-\Lambda t} \le |q(t)| &\le e^{\Lambda t}. \end{aligned}$$

After restricting to the finite time interval $I_r = [0, t_r[$ (specified explicitly later), a simple cut-off function provides a twice continuously differentiable extension $\mathscr{H} : \mathbb{R}^N \times \mathbb{R}^N \longrightarrow \mathbb{R}$ of $\mathscr{H}_F|_{\mathbb{R}^N \times (\mathbb{R}^N \setminus \mathbb{B}^\circ_{\exp(-\Lambda t_r)}(0))}$ and finally, Lemma A.43 can be applied to $\partial\mathbb{B}_r$, ψ and $\mathscr{H}_F$.
Furthermore $\mathscr{H}_F(x,p) \overset{\text{Def.}}{=} \sup_{v \in F(x)} p \cdot v$ is positively homogeneous with respect to p and thus, the second derivatives of $\mathscr{H}_F$ are bounded by $9\,\Lambda\, R^2$ on $\mathbb{R}^N \times (\mathbb{B}_R \setminus \overset{\circ}{\mathbb{B}}_{\frac{1}{R}})$ (according to Lemma 4.33 on page 366). Together with the preceding a priori bounds, we obtain

$$\left\| D^2\, \mathscr{H}_F(y(t), q(t)) \right\|_{\mathrm{Lin}(\mathbb{R}^{2N}, \mathbb{R}^{2N})} \le 9\,\Lambda\, e^{2\Lambda t}.$$

Let $Q(\cdot)$ denote the solution to the matrix Riccati equation

$$\begin{cases} \partial_t\, Q + \frac{\partial^2 \mathscr{H}_F}{\partial p\, \partial x}(y(t), q(t))\, Q + Q\, \frac{\partial^2 \mathscr{H}_F}{\partial x\, \partial p}(y(t), q(t)) \\ \quad + Q\, \frac{\partial^2 \mathscr{H}_F}{\partial p^2}(y(t), q(t))\, Q + \frac{\partial^2 \mathscr{H}_F}{\partial x^2}(y(t), q(t)) = 0, \\ \qquad Q(0) = \nabla \psi(y_0) = \frac{1}{r} \cdot \mathrm{Id}_{\mathbb{R}^N}. \end{cases}$$

Due to the comparison principle in Lemma A.44 (on page 465), $Q(\cdot)$ exists (at least) as long as the two scalar Riccati equations

$$\partial_t\, u_\pm = \pm 9\,\Lambda\, e^{2\Lambda t} \pm 9\,\Lambda\, e^{2\Lambda t}\, u_\pm^2, \qquad u_\pm(0) = \tfrac{1}{r}$$

have finite solutions and within this period, they fulfill

$$u_-(t) \cdot \mathrm{Id}_{\mathbb{R}^N} \le Q(t) \le u_+(t) \cdot \mathrm{Id}_{\mathbb{R}^N}.$$

In fact, we get the explicit solutions in $I_r := \big[0,\ \frac{1}{2\Lambda} \cdot \log\big(1 + \frac{\pi}{9} - \frac{2}{9} \cdot \arctan \frac{1}{r}\big)\big[$, namely $\qquad u_\pm(t) = \tan\big(\pm \frac{9}{2}(e^{2\Lambda t} - 1) + \arctan \frac{1}{r}\big),$

Hence, $Q(t)$ is positive definite with eigenvalues $\geq u_-(t)$ at every time t of the (maybe smaller) interval $I'_r := I_r \cap \left[0, \frac{1}{2\Lambda} \cdot \log\left(1 + \frac{2}{9} \cdot \arctan \frac{1}{r}\right)\right[$.

Now we focus on the geometric interpretation of $Q(\cdot)$.
Due to Lemma A.43 (on page 465),

$$M_t^{\mapsto}(\partial\,\mathbb{B}_r) := \left\{ (y(t), q(t)) \,\middle|\, (y(\cdot), q(\cdot)) \text{ solves system } (*),\ |y_0| = r \right\}$$

is graph of a continuously differentiable function and, $Q(t)$ is related to its derivative at $y(t)$ as we clarified in the proof of Proposition A.41 (on page 466 ff.). Furthermore the Hamilton condition of Proposition A.32 (on page 457) ensures

$$\text{Graph } N_{\vartheta_F(t,\mathbb{B}_r)}(\cdot) \subset \left\{ (y(t), \lambda\, q(t)) \,\middle|\, (y(\cdot), q(\cdot)) \text{ solves } (*),\ |y_0| = r,\ \lambda \geq 0 \right\}$$

and thus, the graph property of $M_t^{\mapsto}(\partial\,\mathbb{B}_r)$ implies that each $q(t)$ is a normal vector to the smooth reachable set $\vartheta_F(t,\mathbb{B}_r)$ at $y(t)$.
As $q(t) \neq 0$ might not have norm 1, the eigenvalues of $Q(t)$ are not always identical to the principal curvatures $(\kappa_j)_{j=1\ldots N}$ of $\vartheta_F(t,\mathbb{B}_r)$ at $y(t)$, but they provide bounds:

$$e^{-\Lambda t} \cdot u_-(t) \leq \kappa_j \leq e^{\Lambda t} \cdot u_+(t)$$

due to $e^{-\Lambda t} \leq |q(t)| \leq e^{\Lambda t}$. Thus, $\vartheta_F(t,\mathbb{B}_r)$ is convex for all times $t \in I'_r$ and, the *local* properties of principal curvatures have the *nonlocal* consequence that $\vartheta_F(t,\mathbb{B}_r) \subset \mathbb{R}^N$ has positive erosion of radius

$$\rho(t) \geq \frac{1}{e^{\Lambda t} \cdot u_+(t)} \geq r - 9\,\Lambda\,(1+r)^2\, t \qquad \text{for all } t \in I'_r.$$

Indeed, the linear estimate at the end is shown by means of the auxiliary function

$$t \longmapsto \frac{1}{e^{\Lambda t} \cdot u_+(t)} - r + 9\,\Lambda\,(1+r)^2\, t$$

that is 0 at $t = 0$, has positive derivative at $t = 0$ and is convex (due to nonnegative second derivative in I'_r).

The time $\tau(r,\Lambda) > 0$ is chosen as minimum of $\frac{1}{2\Lambda} \cdot \log\left(1 + \frac{\pi}{9} - \frac{2}{9} \cdot \arctan \frac{1}{r}\right)$, $\frac{1}{2\Lambda} \cdot \log\left(1 + \frac{2}{9} \cdot \arctan \frac{1}{r}\right)$. The linear estimate does not have to be positive in $[0, \tau(r,\Lambda)[$ though. □

A.5.7 The (Uniform) Tusk Condition for Graphs of Reachable Sets

The so-called exterior tusk condition is an essential tool for verifying the boundary regularity of solutions to parabolic differential equations of second order. Indeed, its role is comparable to the exterior cone condition for elliptic differential equations of second order. Effros and Kazdan investigated it in connection with the heat equation in [75] and, Lieberman extended it to more general parabolic equations in [115].

Definition 48 ([114, § 3], [115]). A nonempty subset $M \subset \mathbb{R} \times \mathbb{R}^N$ is called *tusk* in $(t_0, x_0) \in \mathbb{R} \times \mathbb{R}^N$ if there exist constants $R, \tau > 0$ and a point $x_1 \in \mathbb{R}^N$ with

$$M = \left\{ (t,x) \in \mathbb{R} \times \mathbb{R}^N \;\middle|\; t_0 - \tau < t < t_0, \; \left|(x - x_0) - \sqrt{t_0 - t} \cdot x_1\right| < R\,\sqrt{t_0 - t} \right\}.$$

A nonempty subset $\Omega \subset \mathbb{R} \times \mathbb{R}^N$ satisfies the so-called *exterior tusk condition* if for every point $(t,x) \subset \partial\Omega$ belonging to the parabolic boundary of Ω (i.e.

$$\Omega \cap \left\{ (s,y) \in \mathbb{R} \times \mathbb{R}^N \;\middle|\; |x - y| \le \varepsilon, \; t - \varepsilon < s < t \right\} \neq \emptyset \qquad \text{for any } \varepsilon > 0),$$

there exists a tusk $M \subset \mathbb{R} \times \mathbb{R}^N$ in (t,x) with $\overline{M} \cap \overline{\Omega} = \{(t,x)\}$.

A nonempty subset $\Omega \subset \mathbb{R} \times \mathbb{R}^N$ is said to fulfill the *uniform exterior tusk condition* if it satisfies the exterior tusk conditions and if the scalar geometric parameters $R, \tau > 0$ of the tusks can be chosen independently of the respective points (t,x) of the parabolic boundary of Ω.

Now we focus on the exterior tusk condition for graphs of reachable sets.
In particular, its uniform version can be verified for parts of the complement if the differential inclusion makes every point evolve into convex sets with positive erosion of increasing radius for short times. Thus, Proposition A.41 (on page 464) provides sufficient conditions on the nonautonomous differential inclusion — independently of the compact initial set.

Proposition 49. *For $\widetilde{F} : [0,T] \times \mathbb{R}^N \rightsquigarrow \mathbb{R}^N$ suppose standard hypothesis $(\widetilde{\mathscr{H}})$ with uniform linear growth of $\partial_{(x,p)} \mathscr{H}_{\widetilde{F}}(t,\cdot,\cdot)$ (i.e. $k_{\widetilde{F}} \in L^\infty([0,T])$ in Definition A.33) and the following property:*
For every set $\widetilde{K} \in \mathscr{K}([0,T] \times \mathbb{R}^N)$, there exist $\widehat{\tau} \in]0,T]$ and some nondecreasing $\sigma : [0, \widehat{\tau}] \longrightarrow [0, \infty[$ such that the reachable set $\vartheta_{\widetilde{F}(t_0 + \cdot, \cdot)}(s, x_0) \subset \mathbb{R}^N$ is convex and has positive erosion of radius $\sigma(s) > 0$ for any $s \in]0, \widehat{\tau}]$, $(t_0, x_0) \in \widetilde{K}$ with $t_0 + s \le T$.

Then for every initial set $K_0 \in \mathscr{K}(\mathbb{R}^N)$ and any time parameter $\tau_{\min} \in]0,T[$, the complement of the graph of $[0,T] \rightsquigarrow \mathbb{R}^N$, $t \mapsto \vartheta_{\widetilde{F}}(t, K_0)$ (as a subset of $\mathbb{R} \times \mathbb{R}^N$) satisfies the uniform exterior tusk condition in all boundary points in $]\tau_{\min}, T[\, \times \mathbb{R}^N$.

Corollary 50. *In addition to standard hypothesis* ($\widetilde{\mathscr{H}_\circ^\rho}$) *(on page 464), assume for the set-valued map* $\widetilde{F} : [0,T] \times \mathbb{R}^N \rightsquigarrow \mathbb{R}^N$ *that some* $\lambda(\cdot) \in L^\infty([0,T])$ *satisfies*

$$\|\mathscr{H}_{\widetilde{F}}(t,\cdot,\cdot)\|_{C^{1,1}(\mathbb{R}^N \times \partial\mathbb{B}_1)} \stackrel{\text{Def.}}{=} \|\mathscr{H}_{\widetilde{F}}(t,\cdot,\cdot)\|_{C^1(\mathbb{R}^N \times \partial\mathbb{B}_1)} + \operatorname{Lip} \partial\mathscr{H}_{\widetilde{F}}(t,\cdot,\cdot)|_{\mathbb{R}^N \times \partial\mathbb{B}_1} < \lambda(t)$$

at $\mathscr{L}^1$*-almost every time* $t \in [0,T]$.

Then for every initial set $K_0 \in \mathscr{K}(\mathbb{R}^N)$ *and any time parameter* $\tau_{\min} \in]0,T[$, *the complement of the graph of* $[0,T] \rightsquigarrow \mathbb{R}^N,\ t \mapsto \vartheta_{\widetilde{F}}(t,K_0)$ *(as a subset of* $\mathbb{R}\times\mathbb{R}^N$*) satisfies the uniform exterior tusk condition in all boundary points in* $]\tau_{\min},T[\,\times\mathbb{R}^N$. □

For proving Proposition A.49, we conclude the exterior tusk condition from a similar property about truncated cones (alias conical frustums). In particular, the possibility of choosing geometric parameters *uniformly* does not depend on the shape of a tusk or a conical frustum. The latter condition, however, is easier to verify for graphs of reachable sets by means of boundary solutions and their adjoints (in the sense of Proposition A.32 on page 457).

Lemma 51 (Conical frustum provides suitable tusk).
Let $\Omega \subset \mathbb{R}\times\mathbb{R}^N$ *be nonempty. Assume* $(t_0,x_0) \in \partial\Omega$ *and* $x_1 \in \mathbb{R}^N$, $h,\lambda > 0$ *to satisfy* $\lambda\, h < |x_0 - x_1|$ *and*

$$\overline{\Omega} \cap \big\{(s,y) \in \mathbb{R}\times\mathbb{R}^N \,\big|\, t_0 - h \le s \le t_0,\ |y - x_1| \le |x_0 - x_1| - \lambda\,(t_0 - s)\big\} = \big\{(t_0,x_0)\big\}.$$

Then there exists a tusk in (t_0,x_0) *whose closure has only* (t_0,x_0) *in common with* $\overline{\Omega}$. *Furthermore the scalar geometric parameters of this tusk depend merely on* h,λ.

Lemma 52 (Graphs of reachable sets have interior conical frustums).
Under the assumptions of Proposition A.49, every accumulation point (t_0,x_0) *of* $\partial\left(\text{Graph }\vartheta_{\widetilde{F}}(\cdot,K_0)\big|_{[0,T]}\right) \cap \left(]0,T[\,\times\mathbb{R}^N\right)$ *with* $t_0 > 0$ *has an open conical frustum*

$$\big\{(s,y) \in \mathbb{R}\times\mathbb{R}^N \,\big|\, t_0 - h < s < t_0,\ |y - x_1| < |x_0 - x_1| - \lambda\,(t_0 - s)\big\}$$

(with suitable parameters $h,\lambda > 0$ *and* $x_1 \in \mathbb{R}^N$*) whose closure has only* (t_0,x_0) *in common with the closed complement of* Graph $\vartheta_{\widetilde{F}}(\cdot,K_0)\big|_{[0,T]} \subset \mathbb{R}\times\mathbb{R}^N$.

If $t_0 > \tau_{\min}$ *with an arbitrarily fixed parameter* $\tau_{\min}$ *in addition, the parameters* $h,\lambda > 0$ *can be chosen independently of* (t_0,x_0), *but just depending on* K_0, $\widetilde{F}, T, \tau_{\min}$.

Proof (of Lemma A.51). Consider the following tusk with $R := \frac{|x_0 - x_1| - \lambda\, h}{\sqrt{h}} > 0$

$$M := \Big\{(s,y) \in \mathbb{R}\times\mathbb{R}^N \,\Big|\, t_0 - h < s < t_0,\ \Big|(y - x_0) - \sqrt{t_0 - s}\cdot\tfrac{x_1 - x_0}{\sqrt{h}}\Big| < R\,\sqrt{t_0 - s}\Big\}.$$

As a simple consequence of the triangle inequality in $\mathbb{R}^N$, M is contained in the given conical frustum and thus, $\overline{\Omega} \cap \overline{M} = \{(t_0,x_0)\}$. □

Proof (of Lemma A.52). As an accumulation point, $(t_0, x_0) \in]0,T] \times \mathbb{R}^N$ can be approximated by a sequence of points in $\partial \big(\text{Graph } \vartheta_{\widetilde{F}}(\cdot, K_0)\big|_{[0,T]}\big) \cap \big(]0,T[\times \mathbb{R}^N\big)$. Applying preceding Proposition A.32 (on page 457) to each of these boundary points, an appropriate subsequence provides a solution $x(\cdot) \in W^{1,1}([0,t_0], \mathbb{R}^N)$ and its adjoint $p(\cdot) \in W^{1,1}([0,t_0], \mathbb{R}^N)$ satisfying

$$\begin{cases} x'(t) = \ \frac{\partial}{\partial p} \mathscr{H}_{\widetilde{F}}(t, x(t), p(t)) \in \widetilde{F}(t, x(t)), & x(t_0) \ = x_0, \\ p'(t) = -\frac{\partial}{\partial x} \mathscr{H}_{\widetilde{F}}(t, x(t), p(t)), & |p(t_0)| = 1 \end{cases}$$

and the additional properties for every $s \in [0, t_0[$

$$\begin{cases} x(s) \ \in \ \partial\, \vartheta_{\widetilde{F}}(s, K_0) \\ p(s) \ \in \ N_{\vartheta_{\widetilde{F}}(s,K_0)}\big(x(s)\big) \setminus \{0\} \end{cases}$$

due to regularity and uniqueness of the Hamiltonian initial value problem.

Choose any compact neighborhood $\widetilde{C}$ of the graph of $\vartheta_{\widetilde{F}}(\cdot, K_0) : [0,T] \leadsto \mathbb{R}^N$ in $[0,T] \times \mathbb{R}^N$. Due to the assumption of Proposition A.49, there exist $\widehat{\tau} \in]0,T]$ and a nondecreasing function $\sigma : [0, \widehat{\tau}] \longrightarrow [0, \infty[$ such that $\vartheta_{\widetilde{F}(t+\cdot,\cdot)}(s,y) \subset \mathbb{R}^N$ is convex and has positive erosion of radius $\sigma(s)$ for any $s \in]0, \widehat{\tau}]$, $(t,y) \in \widetilde{C}$ with $t+s \le T$. (If some $\tau_{\min} > 0$ with $\tau_{\min} \le t_0$ is fixed additionally, replace $\widehat{\tau}$ by $\min\{\widehat{\tau}, \tau_{\min}\} > 0$.) Without loss of generality, we assume $\widehat{\tau} < t_0$, $(t_0 - \widehat{\tau}, x(t_0 - \widehat{\tau})) \in \widetilde{C}$.

Set $t_1 := t_0 - \widehat{\tau} > 0$ and $t_2 := t_0 - \frac{\widehat{\tau}}{2} \in]t_1, t_0[$.
At every time $s \in [t_2, t_0[$, the point $x(s)$ belongs to the topological boundary of the convex set $\vartheta_{\widetilde{F}(t_1+\cdot,\cdot)}\big(s - t_1,\ x(t_1)\big)$ with positive erosion of radius $\ge \sigma(\frac{\widehat{\tau}}{2}) =: \rho_{\widehat{\tau}}$. Furthermore the inclusion $\vartheta_{\widetilde{F}(t_1+\cdot,\cdot)}\big(s - t_1,\ x(t_1)\big) \subset \vartheta_{\widetilde{F}}(s, K_0)$ and the convexity of the reachable set $\vartheta_{\widetilde{F}(t_1+\cdot,\cdot)}\big(s - t_1,\ x(t_1)\big)$ imply

$$p(s) \in N_{\vartheta_{\widetilde{F}}(s,K_0)}\big(x(s)\big) \setminus \{0\} \subset N^P_{\vartheta_{\widetilde{F}(t_1+\cdot,\cdot)}(s-t_1,\, x(t_1))}\big(x(s)\big).$$

Now the aspects of (uniform) positive erosion and continuity ensure

$$\mathbb{B}_{\rho_{\widehat{\tau}}}\big(x(s) - \rho_{\widehat{\tau}} \tfrac{p(s)}{|p(s)|}\big) \subset \vartheta_{\widetilde{F}(t_1+\cdot,\cdot)}\big(s - t_1,\ x(t_1)\big) \subset \vartheta_{\widetilde{F}}(s, K_0)$$

for every $s \in [t_2, t_0]$. Moreover, due to the uniform linear growth of $\partial_{(x,p)} \mathscr{H}_{\widetilde{F}}(t,\cdot,\cdot)$, the set-valued map $[t_2, t_0] \leadsto \mathbb{R}^N$, $s \mapsto \mathbb{B}_{\rho_{\widehat{\tau}}}\big(x(s) - \rho_{\widehat{\tau}} \frac{p(s)}{|p(s)|}\big)$ is Lipschitz continuous with convex values and, its Lipschitz constant Λ depends only on $\widetilde{C}, \widetilde{F}, T, \widehat{\tau}$.

Finally comparing graphs of Lipschitz set-valued maps implies for any $\gamma > \Lambda$ that the truncated cone

$$C_\gamma := \Big\{(s,y) \in \mathbb{R}^{1+N} \ \Big| \ t_0 - \tfrac{\rho_{\widehat{\tau}}}{\gamma} \le s < t_0,\ \big|x_0 - \rho_{\widehat{\tau}} \tfrac{p(t_0)}{|p(t_0)|} - y\big| < \rho_{\widehat{\tau}} - \gamma \cdot (t_0 - s)\Big\}$$

is a subset of $\bigcup_{s \in [t_2, t_0]} \big(\{s\} \times \mathbb{B}_{\rho_{\widehat{\tau}}}\big(x(s) - \rho_{\widehat{\tau}} \frac{p(s)}{|p(s)|}\big)\big) \subset \mathbb{R} \times \mathbb{R}^N$.
Obviously the modified truncated cone $C_{2\gamma}$ is contained in the interior of its counterpart C_γ and thus, $C_{2\gamma}$ belongs to the interior of Graph $\vartheta_{\widetilde{F}}(\cdot, K_0)|_{[0,T]} \subset \mathbb{R} \times \mathbb{R}^N$.
□

A.6 Reynolds Transport Theorem for Differential Inclusions with Carathéodory Maps

Reynolds Transport Theorem concerns the time derivative of a Lebesgue integral whose domain is deformed due to a sufficiently smooth vector field (e.g. [55, § 8.3]):

Theorem 53 (Reynolds Transport Theorem). *Suppose $w \in C^1(\mathbb{R}^N, \mathbb{R}^N)$. For a nonempty compact set $K_0 \subset \mathbb{R}^N$, let $K(t) \subset \mathbb{R}^N$ contain all points $x(t)$ of solutions $x(\cdot) \in C^1([0,t], \mathbb{R}^N)$ of $x' = w(x)$, $x(0) \in K_0$.*
Then for every $\Psi \in C^1(\mathbb{R} \times \mathbb{R}^N)$, the function $\mathbb{I}_w : t \longmapsto \int_{K(t)} \Psi(t,x)\, dx$ fulfills

$$\tfrac{d^+}{dt^+}\, \mathbb{I}_w(0) \overset{\text{Def.}}{=} \lim_{t \downarrow 0} \tfrac{\mathbb{I}_w(t) - \mathbb{I}_w(0)}{t} = \int_{K_0} \Big(\partial_t \Psi(0,x) + \operatorname{div}\,(\Psi(0,x)\, w(x)) \Big)\, dx.$$

If, in addition, K_0 satisfies the assumptions of Gauss' Integral Theorem then

$$\tfrac{d^+}{dt^+}\, \mathbb{I}_w(0) = \int_{K_0} \partial_t \Psi(0,x)\, dx + \int_{\partial K_0} \Psi(0,x)\, w(x) \cdot \nu_{K_0}(x)\, d\sigma_x$$

with the exterior unit normal ν_{K_0} to K_0.

Although the name of Osborne Reynolds (1842 – 1912) is used mainly in continuum mechanics this theorem has broad applications, e.g. in shape optimization and free boundary problems.
Now we focus on the integrals over compact reachable sets of differential inclusions, i.e. for a given function $\psi \in L^1_{\text{loc}}(\mathbb{R}^N)$, we consider

$$\mathbb{I}_{\widetilde{F}} : [0,T] \longrightarrow \mathbb{R}, \qquad t \longmapsto \int_{\vartheta_{\widetilde{F}}(t,K_0)} \psi(x)\, dx.$$

As a key point, a priori assumptions about the regularity of ∂K_0 are avoided completely. However, $\widetilde{F} : [0,T] \times \mathbb{R}^N \rightsquigarrow \mathbb{R}^N$ has to fill the gap concerning sufficient conditions. In particular, any generalization of Theorem A.53 (with a boundary integral) has to exclude the example that a nonrectifiable set $K_0 \subset \mathbb{R}^N$ is simply translated. For this reason, $\widetilde{F}$ is supposed to have a continuous selection of its interior $\widetilde{F}(\cdot,\cdot)^\circ$ and, the main result of this section is

Theorem 54. *Assume $N \geq 2$. Let $\rho_{\widetilde{F}}, \mu_{\widetilde{F}} > 0$, $v_{\widetilde{F}} \in C^0([0,T] \times \mathbb{R}^N, \mathbb{R}^N)$ and $\widetilde{F} : [0,T] \times \mathbb{R}^N \rightsquigarrow \mathbb{R}^N$ be a Carathéodory map with compact convex values and*

$$\mathbb{B}_{\rho_{\widetilde{F}}}(v_{\widetilde{F}}(t,x)) \subset \widetilde{F}(t,x) \subset \mu_{\widetilde{F}}\,(1+|x|) \cdot \mathbb{B}$$

for every $(t,x) \in [0,T] \times \mathbb{R}^N$. Furthermore assume $K_0 \in \mathscr{K}(\mathbb{R}^N)$, $\psi \in C^0(\mathbb{R}^N)$.

Then $\mathbb{I}_{\widetilde{F}} : [0,T] \longrightarrow \mathbb{R}$ is absolutely continuous and has the weak derivative

$$\tfrac{d}{dt}\, \mathbb{I}_{\widetilde{F}}(t) = \int_{\partial\, \vartheta_{\widetilde{F}}(t,K_0)} \psi(x)\, \sup \Big(\widetilde{F}(t,x) \cdot {}^{\flat}N^B_{\vartheta_{\widetilde{F}}(t,K_0)}(x) \Big)\, d\mathscr{H}^{N-1}x.$$

Here ${}^{\flat}N^B_K(x)$ denotes the set of Bouligand normal vectors in the unit ball $\mathbb{B}$, i.e.

$${}^{\flat}N^B_K(x) \overset{\text{Def.}}{=} \big\{ v \in \mathbb{B}_1(0) \,\big|\, v \cdot w \leq 0 \ \text{for all } w \in T_K(x) \big\} \subset \mathbb{R}^N.$$

Corollary 55 (Reynolds Transport Theorem for differential inclusions).
Let $\widetilde{F} : [0,T] \times \mathbb{R}^N \rightsquigarrow \mathbb{R}^N$ satisfy the assumptions of Theorem 54. Moreover suppose $K_0 \in \mathscr{K}(\mathbb{R}^N)$ and $\Psi \in C^1([0,T] \times \mathbb{R}^N)$.
Then the function $[0,T] \longrightarrow \mathbb{R}, \; t \longmapsto \int_{\vartheta_{\widetilde{F}}(t,K_0)} \Psi(t,x)\, dx$ is absolutely continuous and has the weak derivative

$$\int_{\partial\, \vartheta_{\widetilde{F}}(t,K_0)} \Psi(t,x)\ \sup\left(\widetilde{F}(t,x) \cdot {}^{\flat}N^{B}_{\vartheta_{\widetilde{F}}(t,K_0)}(x)\right)\ d\mathscr{H}^{N-1}x \;+\; \int_{\vartheta_{\widetilde{F}}(t,K_0)} \tfrac{\partial}{\partial t}\Psi(t,x)\ dx.$$

Corollary 56 (for autonomous differential inclusions, [127, Corollary 3.4]**).**
The absolute continuity in the preceding Corollary A.55 also holds for an autonomous Lipschitz continuous map $F : \mathbb{R}^N \rightsquigarrow \mathbb{R}^N$ with nonempty compact convex values if for each $x \in \mathbb{R}^N$, either $0 \in F(x)^\circ$ or $F(x) = \{0\}$.

Sketch of the Proof for the Special Case of Strictly Expanding Sets ($v_{\widetilde{F}} \equiv 0$)

In the special case $v_{\widetilde{F}} \equiv 0$, the vector 0 belongs to the interior of each value of $\widetilde{F}$. Then the reachable sets represent strict expansions in the sense that for every $s < t$,

$$\vartheta_{\widetilde{F}}(s,K_0) \subset \left(\vartheta_{\widetilde{F}}(t,K_0)\right)^\circ.$$

Due to this observation, we can describe both the reachable sets and their topological boundaries easily via the so-called *minimal time function* $\tau_{\widetilde{F}} : \mathbb{R}^N \longrightarrow [0,\infty]$,

$$\begin{aligned} \tau_{\widetilde{F}}(x) &:= \inf\{\, t \in [0,T] \mid x \in \vartheta_{\widetilde{F}}(t,K_0)\,\}, \\ &= \inf\{\, t \in [0,T] \mid K_0 \cap \vartheta_{-\widetilde{F}(t-\cdot,\cdot)}(t,x) \neq \emptyset\,\}. \end{aligned}$$

In many papers about minimal time functions (e.g. [23, 34, 35, 82, 186]), the condition on admitted solutions usually concerns their final points, i.e.

$$x \longmapsto \inf\left\{ t \in [0,T] \;\middle|\; \exists \text{ solution } z(\cdot) : \; z(0) = x, \; z(t) \in K_0 \right\}.$$

Here we consider a state constraint for the initial point instead : $z(0) \in K_0, \; z(\underset{\sim}{t}) = x$. These two definitions can be regarded as equivalent *only* if the function $\widetilde{F}$ does not depend on time explicitly. For an autonomous Lipschitz map $G : \mathbb{R}^N \rightsquigarrow \mathbb{R}^N$ with compact convex values, the properties of $\tau_G(\cdot)$ have already been investigated extensively. In particular, $\tau_G(\cdot)$ is the viscosity solution of the Eikonal equation

$$\begin{cases} \sup\left(G(x) \cdot \nabla \tau_G(x)\right) = 1 & \text{in } \vartheta_G([0,T],K_0)^\circ, \\ \tau_G = 0 & \text{in } K_0. \end{cases}$$

In [127], the detailed proof of Theorem A.54 has a rather geometric character and verifies the subsequent properties of $\tau_{\widetilde{F}}(\cdot)$ (only). In particular, no results about viscosity solutions are used there. As we consider just the points of differentiability for a locally Lipschitz continuous function, we do not need stronger regularity assumptions about $\widetilde{F}$. Further characterizations of reachable sets by means of normals can be found in [45, 65, 68].

Lemma 57. *Let $\widetilde{F} : [0,T] \times \mathbb{R}^N \rightsquigarrow \mathbb{R}^N$ satisfy the assumptions of Theorem A.54 with $v_{\widetilde{F}} \equiv 0$, i.e. for some $\mu_{\widetilde{F}}, \rho_{\widetilde{F}} > 0$, $\widetilde{F}$ is a Carathéodory set-valued map with compact convex values and $\rho_{\widetilde{F}}\,\mathbb{B} \subset \widetilde{F}(t,x) \subset \mu_{\widetilde{F}}\,(1+|x|)\,\mathbb{B}$ for every (t,x), and assume $K_0 \in \mathscr{K}(\mathbb{R}^N)$.*
Then the corresponding minimum time function $\tau_{\widetilde{F}} : \vartheta_{\widetilde{F}}(T,K_0) \longrightarrow [0,\infty[$ has the properties for every $t \in [0,T]$:

1. *$\tau_{\widetilde{F}}$ is Lipschitz continuous in $\vartheta_{\widetilde{F}}(T,K_0)$,*
2. *$\vartheta_{\widetilde{F}}(t,K_0) = \tau_{\widetilde{F}}^{-1}([0,t])$,*
3. *the topological boundary $\partial\,\vartheta_{\widetilde{F}}(t,K_0)$ is contained in the level set $\tau_{\widetilde{F}}^{-1}(t)$.*
4. *Let $D_{\widetilde{F}}$ consist of all points in $\vartheta_{\widetilde{F}}(T,K_0)^\circ \setminus K_0$ at which $\tau_{\widetilde{F}}$ is differentiable.*
   Then for every $x \in D_{\widetilde{F}}$, $\quad |\nabla \tau_{\widetilde{F}}(x)| \ge \frac{1}{\mu_{\widetilde{F}}\cdot(1+|x|)}$,
   $$N^B_{\vartheta_{\widetilde{F}}(\tau_{\widetilde{F}}(x),K_0)}(x) = [0,\infty[\,\cdot\nabla\tau_{\widetilde{F}}(x)$$
   and for every $t \in\,]0,T[$, $\quad D_{\widetilde{F}} \cap \tau_{\widetilde{F}}^{-1}(\{t\}) \subset \partial\,\vartheta_{\widetilde{F}}(t,K_0)$.
5. $|\nabla\tau_{\widetilde{F}}(x)| \cdot \sup\Big(\widetilde{F}(\tau_{\widetilde{F}}(x),x)\cdot {}^{\flat}N^B_{\vartheta_{\widetilde{F}}(\tau_{\widetilde{F}}(x),K_0)}(x)\Big) = 1$ *for $\mathscr{L}^N$-a.e. $x \in D_{\widetilde{F}}$.*

The ball assumption about the values of $\widetilde{F}$ is to guarantee the properties (1.), (4.). Then the other statements result from the strict expansion property of $\vartheta_{\widetilde{F}}(\cdot,K_0)$.
Now Theorem A.54 is a consequence of the so-called co-area formula, which is an important tool in (geometric) measure theory. For any Lipschitz continuous map $f : \mathbb{R}^m \longrightarrow \mathbb{R}^n$ $(m > n)$, let $C_n f(x)$ denote its *n-dimensional co-area factor* if f is differentiable at x:

$$C_n f(x) \overset{\text{Def.}}{=} \sqrt{\det(Df(x)\cdot Df(x)^T)}.$$

For the minimum time function $\tau_{\widetilde{F}}(\cdot)$ in particular, the dimension $n = 1$ implies $C_1\tau_{\widetilde{F}}(x) = |\nabla\tau_{\widetilde{F}}(x)|$ for every $x \in D_F$.

Proposition 58 (Co-area formula, [77, § 3.4.3, Theorem 2], [78, Theorem 3.2.12]**).** *If $f : \mathbb{R}^m \longrightarrow \mathbb{R}^n$ is Lipschitz continuous and $m > n$, then*

$$\int_{\mathbb{R}^m} g(x)\; C_n f(x)\; d\mathscr{L}^m x = \int_{\mathbb{R}^n}\int_{f^{-1}(\{y\})} g(x)\; d\mathscr{H}^{m-n} x\; d\mathscr{L}^n y$$

for every (Lebesgue) $\mathscr{L}^m$ integrable function $g : \mathbb{R}^m \longrightarrow [-\infty,\infty]$. Here $\mathscr{H}^{m-n}$ denotes the $(m-n)$-dimensional Hausdorff measure of nonempty subsets in $\mathbb{R}^m$.

Indeed, from a merely formal point of view, this formula applied to

$$g(x) := \psi(x)\cdot \sup\Big(\widetilde{F}(\tau_{\widetilde{F}}(x),x)\cdot {}^{\flat}N^B_{\vartheta_{\widetilde{F}}(\tau_{\widetilde{F}}(x),K_0)}(x)\Big)$$

leads to

$$\begin{aligned}\int_{\vartheta_{\widetilde{F}}(t,K_0)\setminus K_0} \psi\, dx &= \int_0^t \int_{\tau_{\widetilde{F}}^{-1}(s)} \psi(y)\cdot\sup\Big(\widetilde{F}(s,y)\cdot {}^{\flat}N^B_{\vartheta_{\widetilde{F}}(s,K_0)}(y)\Big)\; d\mathscr{H}^{N-1}y\; ds\\ &= \int_0^t \int_{\partial\vartheta_{\widetilde{F}}(s,K_0)} \psi(y)\cdot\sup\Big(\widetilde{F}(s,y)\cdot {}^{\flat}N^B_{\vartheta_{\widetilde{F}}(s,K_0)}(y)\Big)\; d\mathscr{H}^{N-1}y\; ds.\end{aligned}$$

Sketch of the Proof for the General Case Via Coordinate Transformation

Restricting to sufficiently short time intervals in $[0,T]$, the continuous selection $v_{\widetilde{F}}(\cdot,\cdot)$ can be approximated locally by autonomous functions $w_{\widetilde{F}} \in C_0^\infty(\mathbb{R}^N,\mathbb{R}^N)$. The flow $T_t(x) := \vartheta_{-w_{\widetilde{F}}}(t,x)$ induced by the velocity field $-w_{\widetilde{F}}(\cdot)$ is a diffeomorphism and maps the reachable set $\vartheta_F(t,K_0)$ to $T_t(\vartheta_F(t,K_0)) = \vartheta_G(t,K_0)$ with the set-valued map

$$\widetilde{G}(t,y) := -w_{\widetilde{F}}(y) + DT_t\big(T_t^{-1}(y)\big) \cdot F\big(t,\, T_t^{-1}(y)\big) \subset \mathbb{R}^N.$$

Moreover this map $\widetilde{G}(\cdot,\cdot)$ satisfies the assumptions of Theorem A.54 at any point of an (initially fixed) compact neighborhood of $K_0 \subset \mathbb{R}^N$ and for $t \in [0,T]$ sufficiently small. Additionally a fixed ball with center at 0 is contained in each value of $\widetilde{G}$ and so, the preceding special case can be applied to integrals over $\vartheta_{\widetilde{G}}(t,K_0)$.
The remaining challenge is now to verify that the coordinate transformation via T_t preserves the structure of the integral representation. In particular, all Lebesgue *and* Hausdorff integrals have to be well-defined and finite (almost everywhere). This requires changes of variables for Hausdorff integrals: the so-called area formula.

Proposition 59 (Generalized area formula, [5, Theorem 2.91], [78, Cor. 3.2.20]**).** *Let $f : \mathbb{R}^m \longrightarrow \mathbb{R}^n$ be a Lipschitz continuous function and $E \subset \mathbb{R}^m$ a countably $\mathscr{H}^k$-rectifiable set ($k \le n$). Then, the multiplicity function*

$$\mathbb{R}^n \longrightarrow \mathbb{R}, \quad y \longmapsto \mathscr{H}^0(E \cap f^{-1}(y))$$

is $\mathscr{H}^k$-measurable in $\mathbb{R}^n$ and for every Borel function $g : \mathbb{R}^n \longrightarrow \mathbb{R}$,

$$\int_{\mathbb{R}^n} g(y) \cdot \mathscr{H}^0\big(E \cap f^{-1}(y)\big)\, d\mathscr{H}^k y \;=\; \int_E g(f(x)) \cdot J_k d^E f_x \, d\mathscr{H}^k x.$$

Here $d^E f_x$ denotes the approximate tangential differential of f at x and, J_k abbreviates the k-dimensional Jacobian, i.e. here $J_k d^E f_x \overset{\text{Def.}}{=} \sqrt{\det(d^E f_x^T \cdot d^E f_x)}$.

The generalized Gauss-Green Theorem is a further tool used in [127, § 6] for investigating the level sets of $\tau_{\widetilde{F}}$, $\tau_{\widetilde{G}}$ and their boundaries. It involves the so-called *measure theoretic boundary* $\partial_* M$ of a nonempty $\mathscr{L}^n$-measurable set $M \subset \mathbb{R}^n$, i.e.

$$\partial_* M := \Big\{ x \in \mathbb{R}^n \;\Big|\; \limsup_{r \to 0} \tfrac{\mathscr{L}^n(\mathbb{B}_r(x) \cap M)}{r^N} > 0, \quad \limsup_{r \to 0} \tfrac{\mathscr{L}^n(\mathbb{B}_r(x) \setminus M)}{r^N} > 0 \Big\} \subset \partial M.$$

Proposition 60 (Generalized Gauss-Green Theorem, [77, § 5.8, Theorem 1]**).** *Let $M \subset \mathbb{R}^n$ have locally finite perimeter. Then for $\mathscr{H}^{n-1}$-a.e. $x \in \partial_* M$, there is a unique measure theoretic unit outer normal $\nu_M(x)$ such that for all $\varphi \in C_c^1(\mathbb{R}^n,\mathbb{R}^n)$*

$$\int_M \operatorname{div} \varphi \, dx \;=\; \int_{\partial_* M} \varphi \cdot \nu_M \, d\mathscr{H}^{n-1}$$

Indeed the reachable sets $\vartheta_{\widetilde{F}}(t,K_0)$, $\vartheta_{\widetilde{G}}(t,K_0)$ have locally finite perimeter at $\mathscr{L}^1$-almost every time t due to their finite $\mathscr{H}^{N-1}$ measures and [77, § 5.11, Theorem 1]. Moreover, every point $x \in \partial \vartheta_{\widetilde{G}}(t,K_0)$ at which $\tau_{\widetilde{G}}$ is differentiable proves to belong even to the measure theoretic boundary $\partial_* \vartheta_{\widetilde{G}}(t,K_0)$ and thus we can use the Eikonal equation in Lemma A.57 (5.).

A.7 Differential Inclusions with One-Sided Lipschitz Continuous Maps

In [69], Donchev and Farkhi prove the existence of solutions to another type of differential inclusions – with a stability estimate as in Filippov's Theorem A.6 (on page 443) included. Their essential aspect is to replace the classical Lipschitz condition with respect to space by a weakened form (called one-sided Lipschitz condition) in combination with upper semicontinuity and convex values:

Definition 61 ([69, Definition 2.1]). A set-valued map $\widetilde{F} : [0,T] \times \mathbb{R}^N \rightsquigarrow \mathbb{R}^N$, $(t,x) \mapsto F(t,x)$ is called *one-sided Lipschitz continuous* with respect to x if there is a function $L(\cdot) \in L^1([0,T])$ such that for every $x, y \in \mathbb{R}^N$, $t \in [0,T]$ and $v \in \widetilde{F}(t,x)$, there exists an element $w \in \widetilde{F}(t,y)$ satisfying

$$\langle x-y,\ v-w\rangle \ \leq\ L(t)\ |x-y|^2.$$

Remark 62. 1. As Donchev has already pointed out in several of his papers, $\widetilde{F} : [0,T] \times \mathbb{R}^N \rightsquigarrow \mathbb{R}^N$ is one-sided Lipschitz continuous with respect to x if and only if some $L(\cdot) \in L^1([0,T])$ satisfies

$$\mathscr{H}_{\widetilde{F}}\big(x-y,\ \widetilde{F}(t,x)\big) \ -\ \mathscr{H}_{\widetilde{F}}\big(x-y,\ \widetilde{F}(t,y)\big) \ \leq\ L(t)\ |x-y|^2$$

for every $x, y \in \mathbb{R}^N$ and $t \in [0,T]$.

2. Obviously, every Lipschitz continuous map is also one-sided Lipschitz continuous, but not vice versa in general. In particular, one-sided Lipschitz continuous maps do not have to be upper or lower semicontinuous.

3. The function $L(\cdot) \in L^1([0,T])$ is assumed to be real-valued, but we do not restrict our considerations to $L(\cdot) \geq 0$. The special case of strictly negative $L(\cdot)$ admits interesting conclusions about asymptotic features which usually do not have counterparts of the (classically) Lipschitz continuous maps.

Theorem 63 (Filippov-like existence for one-sided Lipschitz maps [69, Th. 3.2]**).** *Let* $\widetilde{F} : [0,T] \times \mathbb{R}^N \rightsquigarrow \mathbb{R}^N$, $(t,x) \mapsto \widetilde{F}(t,x)$ *be a nonautonomous Marchaud map (in the sense of Definition A.11 on page 447) being one-sided Lipschitz continuous with respect to x. For* $y(\cdot) \in W^{1,1}([0,T], \mathbb{R}^N)$ *and* $g(\cdot) \in L^1([0,T])$ *suppose*

$$\operatorname{dist}\big(y'(t),\ \widetilde{F}(t,\ y(t))\big) \ \leq\ g(t)$$

at Lebesgue-almost every time $t \in [0,T]$.

Then for every initial point $x_0 \in \mathbb{R}^N$*, there exists a solution* $x(\cdot) \in W^{1,1}([0,T], \mathbb{R}^N)$ *of* $x'(\cdot) \in \widetilde{F}(\cdot, x(\cdot))$ *a.e. satisfying* $x(0) = x_0$ *and for every* $t \in [0,T]$

$$\big|x(t) - y(t)\big| \ \leq\ |x_0 - y(0)|\ e^{\int_0^t L(r)\,dr} \ +\ \int_0^t e^{\int_s^t L(r)\,dr}\ g(s)\ ds.$$

Remark 64. The existence results of Theorem A.63 and Filippov's Theorem A.6 differ from each other in an essential aspect:
Under the assumptions of Theorem A.63, not every point $x_0 \in \mathbb{R}^N$ and vector $v_0 \in \widetilde{F}(0,x_0)$ has to be related to a solution $x(\cdot) \in W^{1,1}([0,T],\mathbb{R}^N)$ of $x'(\cdot) \in \widetilde{F}(\cdot,x(\cdot))$ satisfying $x(0) = x_0$ and

$$\lim_{h \downarrow 0} \tfrac{1}{h} \cdot \big(x(h) - x(0)\big) \;=\; v_0.$$

An example is given by the following map $\widetilde{F}$ and the initial data $x_0 := 0 \in \mathbb{R}$, $v_0 := \frac{1}{2}$

$$\widetilde{F}: \;[0,1] \times \mathbb{R} \rightsquigarrow \mathbb{R}, \quad (t,x) \mapsto \begin{cases} -1 & \text{for } x > 0 \\ [-1,1] & \text{for } x = 0 \\ 1 & \text{for } x < 0 \end{cases}$$

Proposition 65. *As in Theorem A.63, let* $\widetilde{F} : [0,T] \times \mathbb{R}^N \rightsquigarrow \mathbb{R}^N$, $(t,x) \mapsto \widetilde{F}(t,x)$ *be a nonautonomous Marchaud map (in the sense of Definition A.11 on page 447) being one-sided Lipschitz continuous with respect to x.*
In addition suppose $\widetilde{F}(\cdot,\cdot)$ *to be lower semicontinuous at each* $(t,x) \in \{0\} \times \mathbb{R}^N$.

Then for any $x_0 \in \mathbb{R}^N$ *and* $v_0 \in \widetilde{F}(0,x_0)$, *there is a solution* $x(\cdot) \in W^{1,1}([0,T],\mathbb{R}^N)$ *of* $x'(\cdot) \in \widetilde{F}(\cdot,x(\cdot))$ *a.e. satisfying* $x(0) = x_0$ *and*

$$\lim_{h \downarrow 0} \tfrac{1}{h} \cdot \big(x(t) - x_0\big) \;=\; v_0.$$

Proof. Theorem A.63 applied to $y(t) := x_0 + t\, v_0$ provides a solution $x(\cdot) \in W^{1,1}([0,T],\mathbb{R}^N)$ of $x'(\cdot) \in \widetilde{F}(\cdot,x(\cdot))$ a.e. satisfying $x(0) = x_0$ and

$$\begin{aligned} \big|x(h) - x_0 - h\, v_0\big| \;&\le\; \int_0^h e^{\int_s^h L(r)\,dr} \;\; \mathrm{dist}\big(v_0,\; \widetilde{F}(s,\, x_0 + s\, v_0)\big)\; ds \\ &\le\; e^{\|L\|_{L^1([0,T])}} \int_0^h \mathrm{dist}\big(v_0,\; \widetilde{F}(s,\, x_0 + s\, v_0)\big)\; ds. \end{aligned}$$

In particular, the lower semicontinuity of $\widetilde{F}$ in $(0,x_0)$ implies

$$\mathrm{dist}\big(v_0,\; \widetilde{F}(s,\, x_0 + s\, v_0)\big) \longrightarrow 0 \qquad \text{for } s \searrow 0$$

and thus,

$$\limsup_{h \downarrow 0} \tfrac{1}{h} \cdot \big|x(h) - x_0 - h\, v_0\big| \;\le\; 0. \qquad \square$$

A.8 Stochastic Differential Inclusions in $\mathbb{R}^N$

A.8.1 Filippov-Like Theorem of Da Prato and Frankowska

Now the focus of interest is an existence theorem for *stochastic* differential inclusions in the Euclidean space $\mathbb{R}^N$. In regard to the mutational framework, we need a priori estimates that compare a given curve with a solution to the inclusion and, they should have a form similar to Filippov's Theorem A.6 (on page 443 f.).
In 1994, Da Prato and Frankowska presented such an existence result for stochastic differential inclusions with globally Lipschitz continuous drift and diffusion terms [53]. Their main statements even concern Itô integral inclusions with a strongly continuous semigroup on a separable Hilbert space. Just one year later, Motyl published independent existence and uniqueness results about stochastic differential inclusions in the Euclidean space under some assumptions of dissipative type [143].
Now we consider only the finite-dimensional case and prove such a Filippov-like theorem essentially by means of the arguments of Aubin, Da Prato and Frankowska in [18, Theorem 4.1]. The comparative estimate, however, is slightly modified so that we can use it more easily in the example of § 3.7 (on page 242 ff.) and thus, the proof is presented completely here.

General assumptions in § A.8

(i) $(\Omega, \mathscr{A}, P)$ is a complete probability space.

(ii) $(\mathscr{A}_t)_{t\geq 0}$ denotes a filtration with the usual conditions, i.e. $(\mathscr{A}_t)_{t\geq 0}$ is a right continuous and increasing family of sub-σ-algebras of $\mathscr{A}$ and, $\mathscr{A}_0$ contains all P-null sets.

(iii) $W = (W_t)_{t\geq 0}$ is an m-dimensional Wiener process.

(iv) For finite $T > 0$ fixed, define the class $\mathscr{L}^2_{\mathscr{A}}([0,T], \mathbb{R}^N)$ of functions $f : [0,T] \times \Omega \longrightarrow \mathbb{R}^N$ with

  (1.) f is jointly $\mathscr{L}^1 \times \mathscr{A}$-measurable,
  (2.) $\int_{[0,T]} \mathbb{E}\big(|f(t,\cdot)|^2\big)\, dt < \infty$,
  (3.) for every $t \in [0,T]$, $\mathbb{E}\big(|f(t,\cdot)|^2\big) < \infty$ and
  (4.) for every $t \in [0,T]$, $f(t,\cdot) : \Omega \longrightarrow \mathbb{R}^N$ is $\mathscr{A}_t$-measurable.

(v) Let $\mathrm{Lin}(\mathbb{R}^m, \mathbb{R}^N)$ consist of all linear functions $\mathbb{R}^m \longrightarrow \mathbb{R}^N$.

(vi) $\mathfrak{I}_0(X_0, \gamma, \sigma)$ denotes the Itô process associated with
the initial state $X_0 \in L^2(\Omega, \mathscr{A}_0, P; \mathbb{R}^N)$,
the drift $\gamma \in \mathscr{L}^2_{\mathscr{A}}([0,T], \mathbb{R}^N)$ and
the diffusion $\sigma \in \mathscr{L}^2_{\mathscr{A}}([0,T], \mathrm{Lin}(\mathbb{R}^m, \mathbb{R}^N))$, i.e. for $t \in [0,T]$,

$$\mathfrak{I}_0(X_0, \gamma, \sigma)(t) := X_0 + \int_0^t \gamma(s)\, ds + \int_0^t \sigma(s)\, dW_s,$$

$$\|\mathfrak{I}_0(X_0, \gamma, \sigma)\|_{\mathfrak{I},[0,t]} := \sqrt{\mathbb{E}(|X_0|^2) + \mathbb{E}\Big(\int_0^t |\gamma|^2\, ds\Big) + \mathbb{E}\Big(\int_0^t |\sigma|^2\, ds\Big)}.$$

Remark 66. The general estimate for every Itô process

$$\begin{aligned}\mathbb{E}\big(\big|\mathfrak{I}_0(X_0,\gamma,\sigma)(t)\big|^2\big) &\le 9\,(1+t)\ \big\|\mathfrak{I}_0(X_0,\gamma,\sigma)\big\|^2_{\mathfrak{I},[0,t]}\\ &\le 9\ e^t\ \ \big\|\mathfrak{I}_0(X_0,\gamma,\sigma)\big\|^2_{\mathfrak{I},[0,t]}\end{aligned}$$

results from the Hölder inequality, the Itô isometry (quoted for one dimension in Proposition 3.50 (d) on page 233 f.) and the simple inequality $(r+s)^2 \le 3\,(r^2+s^2)$ for any $r,s \in \mathbb{R}$.

Theorem 67 (Da Prato-Frankowska for stochastic differential inclusions). *Suppose for the set-valued map* $\widetilde{F} = (\widetilde{F}_1, \widetilde{F}_2) : [0,T] \times \Omega \times \mathbb{R}^N \rightsquigarrow \mathbb{R}^N \times \mathrm{Lin}(\mathbb{R}^m, \mathbb{R}^N)$:

(i) $\widetilde{F}$ *has nonempty compact values,*

(ii) *for every* $x \in \mathbb{R}^N$, $\widetilde{F}(\cdot,\cdot,x)$ *is measurable,*

(iii) *there is* $\Lambda > 0$ *such that for each* $t \in [0,T]$, $\omega \in \Omega$, $\widetilde{F}(t,\omega,\cdot)$ *is* Λ*-Lipschitz,*

(iv) *there is* $\gamma > 0$ *such that* $|\widetilde{F}(t,\omega,0)|_\infty \le \gamma$ *holds for all* $t \in [0,T]$, $\omega \in \Omega$.

Furthermore let $Y := \mathfrak{I}_0(Y_0, \gamma_0, \sigma_0)$ *be any Itô process with drift* $\gamma_0 \in \mathscr{L}^2_{\mathscr{A}}([0,T],\ \mathbb{R}^N)$ *and diffusion* $\sigma_0 \in \mathscr{L}^2_{\mathscr{A}}([0,T],\ \mathrm{Lin}(\mathbb{R}^m, \mathbb{R}^N))$.

For every initial random variable $X_0 \in L^2(\Omega, \mathscr{A}_0, P;\ \mathbb{R}^N)$*, there exist a drift* $\gamma \in \mathscr{L}^2_{\mathscr{A}}([0,T],\ \mathbb{R}^N)$ *and a diffusion* $\sigma \in \mathscr{L}^2_{\mathscr{A}}([0,T],\ \mathrm{Lin}(\mathbb{R}^m, \mathbb{R}^N))$ *such that the related Itô process* $X := \mathfrak{I}_0(X_0, \gamma, \sigma)$ *satisfies both*

$$\begin{cases} \gamma(t,\omega) \in \widetilde{F}_1\big(t,\ \omega,\ X_t(\omega)\big)\\ \sigma(t,\omega) \in \widetilde{F}_2\big(t,\ \omega,\ X_t(\omega)\big)\end{cases}$$

for $(\mathscr{L}^1 \times P)$*-almost all* $(t,\omega) \in [0,T] \times \Omega$ *and for each* $t \in [0,T]$

$$\|X - Y\|^2_{\mathfrak{I},[0,t]} \le C \cdot \Big(\mathbb{E}\big(|X_0 - Y_0|^2\big) + \int_0^t \mathbb{E}\Big(\mathrm{dist}\big((\gamma_0(s), \sigma_0(s)),\ \widetilde{F}(s, Y_s)\big)^2\Big)\ ds\Big) \cdot e^{C\cdot(1+t)\ t}$$

with a constant $C > 0$ *depending merely on* Λ.

The proof is based on essentially the same iterative construction of approximate solutions as [18, Theorem 4.1].
We use the following lemma, which is easy to verify by means of partial integration:

Lemma 68 **([10, Lemma 1.4.3], [18, Lemma 4.2]).** *Every Lebesgue-integrable function* $g : [0,T] \longrightarrow \mathbb{R}$ *fulfills for each* $n \in \mathbb{N}$

$$\int_0^T \int_0^t g(s)\ \tfrac{(t-s)^{n-1}}{(n-1)!}\ ds\ dt \;=\; \int_0^T g(s)\ \tfrac{(T-s)^n}{n!}\ ds$$

Proof (of Theorem A.67).
Proposition A.80 about measurable marginal maps (on page 490) and Selection Theorem A.74 (of Kuratowski and Ryll-Nardzewski on page 489) guarantee the existence of $\gamma_1 \in \mathscr{L}^2_{\mathscr{A}}([0,T],\ \mathbb{R}^N)$ and $\sigma_1 \in \mathscr{L}^2_{\mathscr{A}}([0,T],\ \mathrm{Lin}(\mathbb{R}^m,\mathbb{R}^N))$ satisfying for every $(t,\omega) \in [0,T] \times \Omega$

$$\begin{cases} \gamma_1(t,\omega) \in \widetilde{F}_1\big(t,\omega,Y_t(\omega)\big), & \big|\,\gamma_0|_{(t,\omega)} - \gamma_1|_{(t,\omega)}\big| = \mathrm{dist}\big(\,\gamma_0|_{(t,\omega)},\ \widetilde{F}_1|_{(t,\omega,Y_t(\omega))}\big) \\ \sigma_1(t,\omega) \in \widetilde{F}_2\big(t,\omega,Y_t(\omega)\big), & \big|\sigma_0|_{(t,\omega)} - \sigma_1|_{(t,\omega)}\big| = \mathrm{dist}\big(\sigma_0|_{(t,\omega)},\ \widetilde{F}_2|_{(t,\omega,Y_t(\omega))}\big). \end{cases}$$

Then the Itô process $X^1 := \mathfrak{I}_0(X_0,\gamma_1,\sigma_1)$ fulfills

$$\begin{aligned} \big\|X^1 - Y\big\|^2_{\mathfrak{I},[0,t]} &= \mathbb{E}\big(|X_0 - Y_0|^2\big) + \mathbb{E}\Big(\int_0^t |\gamma_1 - \gamma_0|^2\,ds\Big) + \mathbb{E}\Big(\int_0^t |\sigma_1 - \sigma_0|^2\,ds\Big) \\ &= \mathbb{E}\big(|X_0 - Y_0|^2\big) + \mathbb{E}\Big(\int_0^t \mathrm{dist}\big((\gamma_0,\sigma_0),\ \widetilde{F}(s,\cdot,Y_s)\big)^2\,ds\Big) \end{aligned}$$

at every time $t \in [0,T]$. Now we iterate this construction and obtain three sequences $(\gamma_n)_{n\in\mathbb{N}}$, $(\sigma_n)_{n\in\mathbb{N}}$, $(X^n)_{n\in\mathbb{N}}$ in $\mathscr{L}^2_{\mathscr{A}}([0,T],\ \mathbb{R}^N)$, $\mathscr{L}^2_{\mathscr{A}}([0,T],\ \mathrm{Lin}(\mathbb{R}^m,\mathbb{R}^N))$ and $\mathscr{L}^2_{\mathscr{A}}([0,T],\ \mathbb{R}^N)$ respectively with

$$\begin{cases} \gamma_{n+1}|_{(t,\omega)} \in \widetilde{F}_1|_{(t,\omega,X^n_t(\omega))}, & \big|\,\gamma_n|_{(t,\omega)} - \gamma_{n+1}|_{(t,\omega)}\big| = \mathrm{dist}\big(\,\gamma_n|_{(t,\omega)},\ \widetilde{F}_1|_{(t,\omega,X^n_t(\omega))}\big) \\ \sigma_{n+1}|_{(t,\omega)} \in \widetilde{F}_2|_{(t,X^n_t(\omega))}, & \big|\sigma_n|_{(t,\omega)} - \sigma_{n+1}|_{(t,\omega)}\big| = \mathrm{dist}\big(\sigma_n|_{(t,\omega)},\ \widetilde{F}_2|_{(t,\omega,X^n_t(\omega))}\big) \\ X^{n+1} = \mathfrak{I}_0\big(X_0,\ \gamma_{n+1},\ \sigma_{n+1}\big) \end{cases}$$

for all $(t,\omega) \in [0,T] \times \Omega$. Then the uniform Λ-Lipschitz continuity of $F(t,\omega,\cdot)$ and Remark A.66 imply

$$\begin{aligned} \big\|X^{n+1} - X^n\big\|^2_{\mathfrak{I},[0,t]} &= \mathbb{E}\Big(\int_0^t |\gamma_{n+1} - \gamma_n|^2\,ds\Big) + \mathbb{E}\Big(\int_0^t |\sigma_{n+1} - \sigma_n|^2\,ds\Big) \\ &= \mathbb{E}\Big(\int_0^t \mathrm{dist}\Big(\big(\gamma_n(s,\cdot),\ \sigma_n(s,\cdot)\big),\ \ \widetilde{F}(s,\cdot,X^n_s)\Big)^2\,ds\Big) \\ &\le \mathbb{E}\Big(\int_0^t \mathit{d\!l}\big(\widetilde{F}(s,\cdot,X^{n-1}_s),\ \widetilde{F}(s,\cdot,X^n_s)\big)^2\,ds\Big) \\ &\le \Lambda^2 \cdot \mathbb{E}\Big(\int_0^t |X^{n-1}_s - X^n_s|^2\,ds\Big) \\ &\le \Lambda^2 \cdot 9\,(1+t) \cdot \int_0^t \big\|X^n - X^{n-1}\big\|^2_{\mathfrak{I},[0,s]}\,ds. \end{aligned}$$

By means of induction with respect to n, we obtain for every $n \in \mathbb{N}$ and $t \in [0,T]$

$$\begin{aligned} &\big\|X^{n+1} - X^n\big\|^2_{\mathfrak{I},[0,t]} \\ &\le \big(9\,\Lambda^2\,(1+t)\big)^n \cdot \int_0^t ds_n \int_0^{s_n} ds_{n-1} \ldots \int_0^{s_2} \big\|X^1 - Y\big\|^2_{\mathfrak{I},[0,s_1]}\,ds_1 \\ &\le \big(9\,\Lambda^2\,(1+t)\big)^n \cdot \int_0^t \big\|X^1 - Y\big\|^2_{\mathfrak{I},[0,s_n]}\,\frac{(t-s_n)^{n-1}}{(n-1)!}\,ds_n \\ &\le \tfrac{1}{n!}\,\big(9\,\Lambda^2\,(1+t)\,t\big)^n \cdot \big\|X^1 - Y\big\|^2_{\mathfrak{I},[0,t]} \\ &\le \tfrac{1}{n!}\,\big(3\,\Lambda\ \ (1+t)\big)^{2n} \cdot \big\|X^1 - Y\big\|^2_{\mathfrak{I},[0,t]} \end{aligned}$$

due to Lemma A.68.

The series

$$c(t) := \sum_{n=0}^{\infty} \tfrac{1}{\sqrt{n!}} \,(3\,\Lambda\,(1+t))^n$$

is absolutely convergent for every $t \in \mathbb{R}$ as d'Alembert's ratio test reveals. Hence, $(X^n)_{n\in\mathbb{N}}$ is a Cauchy sequence with respect to $\|\cdot\|_{\mathfrak{I},[0,t]}$ for each $t \in [0,T]$.

Then there exist limits $\gamma \in \mathscr{L}^2_{\mathscr{A}}([0,T],\, \mathbb{R}^N)$, $\sigma \in \mathscr{L}^2_{\mathscr{A}}([0,T],\, \mathrm{Lin}(\mathbb{R}^m, \mathbb{R}^N))$ and $X \in \mathscr{L}^2_{\mathscr{A}}([0,T],\, \mathbb{R}^N)$ of $(\gamma_n)_{n\in\mathbb{N}}$, $(\sigma_n)_{n\in\mathbb{N}}$, $(X^n)_{n\in\mathbb{N}}$ respectively with

$$X = \mathfrak{I}_0\big(X_0, \gamma, \sigma\big).$$

Furthermore, we conclude

$$\gamma(t,\omega) \in \widetilde{F}_1\big(t,\,\omega,\,X_t(\omega)\big), \qquad \sigma(t,\omega) \in \widetilde{F}_2\big(t,\,\omega,\,X_t(\omega)\big)$$

for $(\mathscr{L}^1 \times P)$-*almost* all $(t,\omega) \in [0,T] \times \Omega$ from the facts that some subsequences of $(\gamma_n)_{n\in\mathbb{N}}$, $(\sigma_n)_{n\in\mathbb{N}}$, $(X^n)_{n\in\mathbb{N}}$ converge to their respective limits pointwise almost everywhere in $[0,T] \times \Omega$ and that $\widetilde{F}(t,\omega,\cdot)$ is continuous by assumption (iii).

Finally, X satisfies at each time $t \in [0,T]$

$$\begin{aligned}
\big\|X - Y\big\|_{\mathfrak{I},[0,t]} &\le \sum_{n=1}^{\infty} \big\|X^{n+1} - X^n\big\|_{\mathfrak{I},[0,t]} && + \big\|X^1 - Y\big\|_{\mathfrak{I},[0,t]} \\
&\le \sum_{n=1}^{\infty} \tfrac{(3\,\Lambda\,(1+t))^n}{\sqrt{n!}} \,\big\|X^1 - Y\big\|_{\mathfrak{I},[0,t]} && + \big\|X^1 - Y\big\|_{\mathfrak{I},[0,t]} \\
&= \sum_{n=0}^{\infty} \tfrac{(3\,\Lambda\,(1+t))^n}{\sqrt{n!}} \,\big\|X^1 - Y\big\|_{\mathfrak{I},[0,t]},
\end{aligned}$$

$$\begin{aligned}
\text{i.e. } \big\|X - Y\big\|^2_{\mathfrak{I},[0,t]} &\le c(t)^2 \qquad \big\|X^1 - Y\big\|^2_{\mathfrak{I},[0,t]} \\
&= c(t)^2 \,\Big(\mathbb{E}\big(|X_0 - Y_0|^2\big) + \mathbb{E}\Big(\int_0^t \mathrm{dist}\big((\gamma_0,\sigma_0),\, \widetilde{F}(s,\,Y_s)\big)^2\, ds\Big)\Big).
\end{aligned}$$

In regard to the claimed estimate, we have to verify $c(t)^2 \le \mathrm{const} \cdot e^{\,\mathrm{const}\cdot(1+t)\,t}$. Due to absolute convergence, $c(\cdot) > 0$ is analytic in $[0,\infty[$ and,

$$\begin{aligned}
0 \le \tfrac{d}{dt}\, c(t) &= \sum_{n=1}^{\infty} \tfrac{n}{\sqrt{n!}} \,(3\,\Lambda)^n \,(1+t)^{n-1} \\
&= 3\,\Lambda + \sum_{n=2}^{\infty} \sqrt{\tfrac{n}{(n-1)!}} \,(3\,\Lambda)^n \,(1+t)^{n-1} \\
&\le 3\,\Lambda + \sum_{n=2}^{\infty} \sqrt{\tfrac{2}{(n-2)!}} \,(3\,\Lambda)^n \,(1+t)^{n-1} \\
&= 3\,\Lambda + \sqrt{2}\,(3\,\Lambda)^2\,(1+t) \cdot \sum_{m=0}^{\infty} \tfrac{1}{\sqrt{m!}} \,(3\,\Lambda)^m \,(1+t)^m \\
&= 3\,\Lambda + \sqrt{2}\,(3\,\Lambda)^2\,(1+t) \cdot c(t)
\end{aligned}$$

implies

$$c(t) \le \big(c(0) + 3\,\Lambda\,t\big) \cdot e^{18\,\Lambda^2\,(1+t)\,t} \le \big(c(0) + 1\big) \cdot e^{3\,\Lambda\,(1+6\Lambda)\,(1+t)\,t}$$

for all $t \ge 0$ by means of Gronwall's inequality. (This upper bound is quite simple to prove, but obviously not optimal.) □

A.8.2 A Sufficient Condition on Invariant Subsets

In the field of stochastic differential inclusions, the aspects of invariance and viability have been investigated thoroughly by several authors like Aubin, Da Prato and Frankowska [18], Michta [139, 140], Motyl [142], Truong-Van & Truong [178]. A broad survey is presented in [104].

For the sake of completeness, we introduce the required notions of contingent cone briefly and then just quote the sufficient result, which can be regarded as the stochastic counterpart of Proposition A.8 (on page 445).

Definition 69 (Stochastic contingent set [17, Definition 1.1], [18]**).**
Let $K : \Omega \rightsquigarrow \mathbb{R}^N, \omega \mapsto K_\omega$ be a random closed set, i.e. here: K is an $\mathscr{A}_0$-measurable set-valued map with nonempty closed values. For some $t \geq 0$, consider a $\mathscr{A}_t$-measurable selection $x : \Omega \longrightarrow \mathbb{R}^N$ of K.
The *stochastic contingent set* $T_K^S(t,x)$ to K at x with respect to $\mathscr{A}_t$ is defined as the set of pairs (η, v) of $\mathscr{A}_t$-random variables satisfying the following property: There exist sequences $(h_n)_{n\in\mathbb{N}}$ in $]0,\infty[$ and $(a_n)_{n\in\mathbb{N}}$, $(b_n)_{n\in\mathbb{N}}$ of $\mathscr{A}_{t+h_n}$-random variables such that

$$\begin{cases} h_n \longrightarrow 0 & \text{for } n \longrightarrow \infty, \\ \mathbb{E}\big(|a_n|^2\big) \longrightarrow 0 & \text{for } n \longrightarrow \infty, \\ \mathbb{E}\big(|b_n|^2\big) \longrightarrow 0 & \text{for } n \longrightarrow \infty, \\ \mathbb{E}(b_n) = 0, & \\ b_n \text{ is independent of } \mathscr{A}_t & \end{cases}$$

and for each $n \in \mathbb{N}$, the sum $x + v\,\big(W_{t+h_n} - W_t\big) + h_n\,\eta + h_n\,a_n + \sqrt{h_n}\,b_n$ is a square integrable selection of K.

Proposition 70 (Sufficient condition for invariance, [18, Theorem 5.1]**).**
Let $F = (F_1, F_2) : \mathbb{R}^N \rightsquigarrow \mathbb{R}^N \times \mathrm{Lin}(\mathbb{R}^m, \mathbb{R}^N)$ be a Lipschitz continuous set-valued map with nonempty compact values. Suppose $K : \Omega \rightsquigarrow \mathbb{R}^N$ to be an $\mathscr{A}_0$-measurable set-valued map with nonempty closed values satisfying for all $t \geq 0$ and each $\mathscr{A}_t$-measurable selection $x : \Omega \longrightarrow \mathbb{R}^N$ of K

$$F(x) \subset T_K^S(t,x) \qquad \textit{almost everywhere in } \Omega.$$

Then the random closed set K is invariant *under F in the following sense: Every Itô process $X := \mathfrak{I}_0(X_0, \gamma, \sigma)$ starting in K and solving the stochastic diff. inclusion*

$$dX_t \in F_1(X_t)\,dt + F_2(X_t)\,dW_t$$

i.e. $X_0 \in L^2(\Omega, \mathscr{A}_0, P; \mathbb{R}^N)$, $\gamma \in \mathscr{L}^2_{\mathscr{A}}([0,T], \mathbb{R}^N)$ and $\sigma \in \mathscr{L}^2_{\mathscr{A}}([0,T], \mathrm{Lin}(\mathbb{R}^m, \mathbb{R}^N))$ with

$$\begin{cases} X_0 \in K \quad a.e., \\ \gamma(t,\omega) \in F_1\big(X_t(\omega)\big) \\ \sigma(t,\omega) \in F_2\big(X_t(\omega)\big) \end{cases}$$

for $(\mathscr{L}^1 \times P)$-almost all $(t,\omega) \in [0,T] \times \Omega$, satisfies $X_t \in K$ almost everywhere in Ω for each $t \geq 0$.

A.9 Proximal Normals of Set Sequences in $\mathbb{R}^N$

Comparing the proximal normals of a converging sequence $(K_n)_{n\in\mathbb{N}}$ in $(\mathscr{K}(\mathbb{R}^N), dl)$ with the normals of its limit $K \in \mathscr{K}(\mathbb{R}^N)$, the following inclusion is not difficult to prove by means of exterior balls and, it has already been quoted in Proposition 4.23 (on page 360)

$$\text{Graph}\, N_K^P \;\subset\; \text{Limsup}_{n\to\infty}\, \text{Graph}\, N_{K_n}^P$$

(see e.g. [50, Lemma 4.1]). Of course, the equality here is not fulfilled in general. A key advantage of the subset $N_{K,\rho}^P$ $(\rho > 0)$ specified equivalently in Definition 4.40 (on page 373) is that an inverse inclusion is satisfied.
The following proposition provides the inclusions in both directions and their proofs.

Definition 71. Let $C \subset \mathbb{R}^N$ be a nonempty closed set.

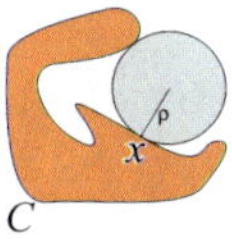

C

For any $\rho > 0$, the set $N_{C,\rho}^P(x) \subset \mathbb{R}^N$ consists of all proximal normal vectors $\eta \in N_C^P(x) \setminus \{0\}$ with the proximal radius $\geq \rho$ (and thus might be empty).
Furthermore define ${}^{\flat}N_{C,\rho}^P(x) := N_{C,\rho}^P(x) \cap \mathbb{B}$.

Proposition 72. *Let $(K_n)_{n\in\mathbb{N}}$ be a converging sequence in $\mathscr{K}(\mathbb{R}^N)$ and K its limit. Π_{K_n}, $\Pi_K : \mathbb{R}^N \rightsquigarrow \mathbb{R}^N$ denote the projections on K_n, K $(n \in \mathbb{N})$ respectively,*

i.e., $\qquad \Pi_K : \mathbb{R}^N \rightsquigarrow \mathbb{R}^N, \quad x \mapsto \{y \in K \mid |y - x| = \operatorname{dist}(x, K)\} \subset \mathbb{R}^N.$

Then,

(1.) $\text{Limsup}_{n\to\infty}\, \text{Graph}\, {}^{\flat}N_{K_n,\rho}^P \subset \text{Graph}\, {}^{\flat}N_{K,\rho}^P$ *for any* $\rho > 0$,

(2.) $\text{Limsup}_{\substack{y\to x\\ n\to\infty}}\, \Pi_{K_n}(y) \subset \Pi_K(x)$ *for any* $x \in \mathbb{R}^N$,

(3.) $\text{Graph}\, {}^{\flat}N_{K,\rho}^P \subset \text{Liminf}_{n\to\infty}\, \text{Graph}\, {}^{\flat}N_{K_n,r}^P$ *for any* $0 < r < \rho$.

Proof.
(1.) Choose any converging sequence $\big((x_{n_j}, p_{n_j})\big)_{j\in\mathbb{N}}$ with $p_{n_j} \in N_{K_{n_j},\rho}^P(x_{n_j}) \cap \partial\mathbb{B}$ and set $x := \lim_{j\to\infty} x_{n_j} \in K$, $p := \lim_{j\to\infty} p_{n_j} \in \partial\mathbb{B}$. According to Definition A.23 (on page 454), each K_{n_j} is contained in the complement of the open ball with center $x_{n_j} + \rho\, p_{n_j}$ and radius ρ,

$$K_{n_j} \;\subset\; \mathbb{R}^N \setminus \mathring{\mathbb{B}}_\rho\big(x_{n_j} + \rho\, p_{n_j}\big).$$

As an indirect consequence, $j \longrightarrow \infty$ leads to

$$K \;\subset\; \mathbb{R}^N \setminus \mathring{\mathbb{B}}_\rho(x + \rho\, p),$$

i.e.

$$p \in N_{K,\rho}^P(x).$$

(2.) Let $r > 0$ and $n \in \mathbb{N}$ be arbitrary. For $y \in \mathbb{B}_r(x)$ given, choose any $z \in \Pi_{K_n}(y)$ and $\xi \in \Pi_K(z)$. Then,

$$|\xi - z| \le d\!l(K_n, K)$$

and

$$\begin{aligned} |x-\xi| &\le |x-y| + |y-z| + |z-\xi| \\ &\le |x-y| + \operatorname{dist}(y,K) + d\!l(K,K_n) + |z-\xi| \\ &\le |x-y| + |y-x| + \operatorname{dist}(x,K) + d\!l(K,K_n) + d\!l(K_n,K) \\ &\le 2r + \operatorname{dist}(x,K) + 2\, d\!l(K_n,K). \end{aligned}$$

Thus, $\Pi_{K_n}(y) \subset \mathbb{B}_{d\!l(K_n,K)}\Big(K \cap \mathbb{B}_{2r+\operatorname{dist}(x,K)+2d\!l(K_n,K)}(x)\Big)$ for any $y \in \mathbb{B}_r(x)$.

The set-valued map $[0,\infty[\rightsquigarrow \mathbb{R}^N,\ r \mapsto K \cap \mathbb{B}_r(x)$ is upper semicontinuous (due to [19, Corollary 1.4.10]) and in the closed interval $[\operatorname{dist}(x,K),\infty[$, it has nonempty compact values. For every $\eta > 0$, there exists $\rho = \rho(x,\eta) \in\,]0,\eta[$ such that

$$K \cap \mathbb{B}_{r'}(x) \subset \mathbb{B}_\eta\big(\Pi_K(x)\big)$$

for all $r' \in \big[\operatorname{dist}(x,K),\ \operatorname{dist}(x,K) + 2\rho\big]$. Due to $d\!l(K_n,K) \longrightarrow 0\ (n \longrightarrow \infty)$, there is an index $m \in \mathbb{N}$ with $d\!l(K_n,K) \le \frac{\rho}{4}$ for all $n \ge m$. Thus we obtain for every point $y \in \mathbb{B}_{\rho/4}(x) \cap \mathbb{B}_r(x)$ and index $n \ge m$

$$\begin{aligned} \Pi_{K_n}(y) &\subset \mathbb{B}_{\frac{\rho}{4}}\big(K \cap \mathbb{B}_{2\frac{\rho}{4}+\operatorname{dist}(x,K)+2\frac{\rho}{4}}(x)\big) &&= \mathbb{B}_{\frac{\rho}{4}}\big(K \cap \mathbb{B}_{\operatorname{dist}(x,K)+\rho}(x)\big) \\ &\subset \mathbb{B}_{\frac{\rho}{4}}\big(\mathbb{B}_\eta(\Pi_K(x))\big) &&\subset \mathbb{B}_{2\eta}\big(\Pi_K(x)\big), \end{aligned}$$

i.e. $\operatorname{Limsup}_{\substack{y\to x\\ n\to\infty}} \Pi_{K_n}(y) \subset \Pi_K(x)$.

(3.) Choose any $x \in \partial K$ and $p \in N^P_{K,\rho}(x) \neq \emptyset$ with $|p| = 1$.

Then x is the unique projection of $x + \delta\, p$ on the set K for every $\delta \in\,]0,\rho[$. Considering now a sequence $(x_n)_{n\in\mathbb{N}}$ with $x_n \in \Pi_{K_n}(x + \delta\, p) \subset K_n$, the preceding statement (2.) implies $x_n \longrightarrow x$ and, the definition of proximal normal ensures

$$p_n := \frac{x + \delta\, p - x_n}{|x + \delta\, p - x_n|} \in {}^{\flat}N^P_{K_n}(x_n)$$

converging to p for $n \longrightarrow \infty$.
Finally the proximal radius of p_n is $\ge |x + \delta\, p - x_n| \ge \delta - |x - x_n|$, and thus,

$$(x,p) \in \operatorname{Liminf}_{n\to\infty} \operatorname{Graph} {}^{\flat}N^P_{K_n,r} \quad \text{for every } 0 < r < \delta < \rho.$$

□

A.10 Tools for Set-Valued Maps

A.10.1 Measurable Set-Valued Maps

In this section we summarize some useful results about set-valued maps in regard to measurability. The monograph of Castaing and Valadier [40] is usually regarded as a standard reference providing many of the well-known results. Here we quote the corresponding theorems from the monograph of Aubin and Frankowska [19].

Definition 73 ([19, Definition 8.1.1]). Consider a measurable space $(\Omega, \mathscr{A})$, a complete separable metric space E and a set-valued map $F : \Omega \leadsto E$ with closed images.
F is called *measurable* if the inverse image of each open set is a measurable set, i.e., for every open set $O \subset E$,

$$F^{-1}(O) \stackrel{\text{Def.}}{=} \{\omega \in \Omega \mid F(\omega) \cap O \neq \emptyset\} \in \mathscr{A}.$$

Theorem 74 (Kuratowski and Ryll-Nardzewski [109], [19, Theorem 8.1.3]**).** *Let E be a complete separable metric space, $(\Omega, \mathscr{A})$ a measurable space, $F : \Omega \leadsto E$ a measurable set-valued map with nonempty closed values.*
Then there exists a measurable selection of F, i.e., a measurable single-valued function $f : \Omega \longrightarrow E$ satisfying $f(\omega) \in F(\omega)$ for every $\omega \in \Omega$.

Theorem 75 (Characterization Theorem [19, Theorem 8.1.4]**).** *Let $(\Omega, \mathscr{A}, \mu)$ be a complete σ-finite measure space, E a complete separable metric space and $F : \Omega \leadsto E$ a set-valued map with nonempty closed values.*
Then the following properties are equivalent:

(i) *F is measurable.*
(ii) *The graph of F belongs to $\mathscr{A} \otimes \mathscr{B}$.*
(iii) *$F^{-1}(C) \in \mathscr{A}$ for every closed set $C \subset E$.*
(iv) *$F^{-1}(B) \in \mathscr{A}$ for every Borel set $B \subset E$.*
(v) *For each element $x \in E$, the function $\operatorname{dist}(x, F(\cdot)) : \Omega \longrightarrow [0, \infty[$ is measurable.*
(vi) *There exists a sequence $(f_n)_{n \in \mathbb{N}}$ of measurable selections of F such that*

$$F(\omega) = \overline{\bigcup_{n \in \mathbb{N}} f_n(\omega)} \qquad \text{for every } \omega \in \Omega.$$

Corollary 76 (Upper and lower semicontinuous maps [19, Proposition 8.2.1]**).** *Consider a metric space Ω and a complete σ-finite measure space $(\Omega, \mathscr{A}, \mu)$ such that $\mathscr{A}$ contains all open subsets of Ω. Let E be a complete separable metric space and $F : \Omega \leadsto E$ a set-valued map with nonempty closed images.*

If F is upper semicontinuous, then F is measurable.
If F is lower semicontinuous, then F is measurable.

Proposition 77 (Closed union and intersection [19, Theorem 8.2.4]**).**
Let $(\Omega, \mathscr{A}, \mu)$ be a complete σ-finite measure space, E a complete separable metric space and $F_n : \Omega \rightsquigarrow E$ $(n \in \mathbb{N})$ set-valued maps with nonempty closed values.

Then the following set-valued maps are measurable:

$$\Omega \rightsquigarrow E, \quad \omega \mapsto \overline{\bigcup_{n \in \mathbb{N}} F_n(\omega)},$$
$$\Omega \rightsquigarrow E, \quad \omega \mapsto \bigcap_{n \in \mathbb{N}} F_n(\omega).$$

Proposition 78 (Direct image [19, Theorem 8.2.8]**).**
Let $(\Omega, \mathscr{A}, \mu)$ be a complete σ-finite measure space, E_1, E_2 complete separable metric spaces and $F : \Omega \rightsquigarrow E_1$ a measurable set-valued map with nonempty closed values. Consider a Carathéodory set-valued map $G : \Omega \times E_1 \rightsquigarrow E_2$, i.e., for every $x \in E_1$, the map $G(\cdot, x) : \Omega \rightsquigarrow E_2$ is measurable and for every $\omega \in \Omega$, the map $G(\omega, \cdot) : E_1 \rightsquigarrow E_2$ is continuous.

Then the set-valued map $\Omega \rightsquigarrow E_2$, $\omega \mapsto \overline{G(\omega, F(\omega))}$ is measurable.

Proposition 79 (Inverse image, Filippov selection [19, Theorems 8.2.9, 8.2.10]**).**
Consider a complete σ-finite measure space $(\Omega, \mathscr{A}, \mu)$, complete separable metric spaces E_1, E_2 and measurable set-valued maps $F : \Omega \rightsquigarrow E_1$, $G : \Omega \rightsquigarrow E_2$ with nonempty closed values. Let $g : \Omega \times E_1 \longrightarrow E_2$ be a Carathéodory function.

Then the set-valued map

$$\Omega \rightsquigarrow E_1, \quad \omega \mapsto \{x \in F(\omega) \mid g(\omega, x) \in G(\omega)\} \subset E_1$$

is measurable.

Consequently, if $g(\omega, F(\omega)) \cap G(\omega)$ is nonempty for every $\omega \in \Omega$, then there exists a measurable selection $f : \Omega \longrightarrow E_1$ of F such that for every $\omega \in \Omega$, the element $g(\omega, f(\omega))$ belongs to $G(\omega)$.

In particular, for every measurable function $h : \Omega \longrightarrow E_2$ with $h(\omega) \in g(\omega, F(\omega))$ for almost all $\omega \in \Omega$, there exists a measurable selection $f : \Omega \longrightarrow E_1$ of F with $h = g(\cdot, f(\cdot))$ almost everywhere in Ω.

Proposition 80 (Marginal map [19, Theorem 8.2.11]**).**
Consider a complete σ-finite measure space $(\Omega, \mathscr{A}, \mu)$, a complete separable metric space E, a measurable set-valued map $F : \Omega \rightsquigarrow E$ with nonempty closed values and a real-valued Carathéodory function $f : \Omega \times E \longrightarrow \mathbb{R}$.

Then the so-called marginal function

$$\Omega \longrightarrow \mathbb{R} \cup \{-\infty\}, \quad \omega \longmapsto \inf_{x \in F(\omega)} f(\omega, x)$$

is measurable. Furthermore the so-called marginal map

$$\Omega \rightsquigarrow E, \quad \omega \mapsto \left\{ x \in F(x) \;\middle|\; f(\omega, x) = \inf_{y \in F(\omega)} f(\omega, y) \right\} \subset E$$

is measurable.

A.10.2 Parameterization of Set-Valued Maps

Proposition 81 ([19, Theorem 9.7.2]).
Consider a metric space X and a set-valued map $G : [a,b] \times X \rightsquigarrow \mathbb{R}^N$ satisfying

1. *G has nonempty compact convex values,*
2. *$G(\cdot,x) : [a,b] \rightsquigarrow \mathbb{R}^N$ is measurable for every $x \in X$,*
3. *there exists $k(\cdot) \in L^1([a,b])$ such that for every $t \in [a,b]$, the set-valued map $G(t,\cdot) : X \rightsquigarrow \mathbb{R}^N$ is $k(t)$-Lipschitz continuous.*

Then there exists a single-valued function $g : [a,b] \times X \times \mathbb{B}_1 \longrightarrow \mathbb{R}^N$ (with the closed unit ball $\mathbb{B}_1 \subset \mathbb{R}^N$) fulfilling for all $t \in [a,b]$, $x \in X$, $u, v \in \mathbb{B}_1$ respectively

1. *$G(t,x) = \bigcup_{w \in \mathbb{B}_1} g(t,x,w)$,*
2. *$g(\cdot,x,u) : [a,b] \longrightarrow \mathbb{R}^N$ is measurable,*
3. *$g(t,\cdot,u) : X \longrightarrow \mathbb{R}^N$ is $c \cdot k(t)$-Lipschitz continuous*
4. *$|g(t,x,u) - g(t,x,v)| \le c\, \|G(t,x)\|_\infty\, |u-v|$*

with a constant $c > 0$ independent of G.

A.11 Compactness of Continuous Functions Between Metric Spaces

The essential compactness result about continuous functions between metric spaces is the Arzelà-Ascoli Theorem. We use it in the following version of Green and Valentine:

Theorem 82 (Arzelà-Ascoli in metric spaces [88]**).**
Let (E_1,d_1), (E_2,d_2) be two precompact metric spaces, i.e. for any $\varepsilon > 0$, each set E_i $(i = 1,2)$ can be covered by finitely many ε-balls with respect to metric d_i. Moreover, suppose the sequence $(f_n)_{n\in\mathbb{N}}$ of functions $E_1 \longrightarrow E_2$ to be uniformly equi-continuous (i.e. with a common modulus of continuity in E_1).
Then there exists a subsequence $(f_{n_j})_{j\in\mathbb{N}}$ being Cauchy sequence with respect to uniform convergence. If (E_2,d_2) is complete in addition, then $(f_{n_j})_{j\in\mathbb{N}}$ converges uniformly to a continuous function $E_1 \longrightarrow E_2$.

Kisielewicz characterized *weakly* compact sets in the space of Banach-valued continuous functions. His result can be interpreted as a "weak counterpart" of the Arzelà-Ascoli Theorem.

Proposition 83 (Kisielewicz [102, Theorem 4]**).**
Let S be a compact Hausdorff space and X a Banach space.
A subset $W \subset C^0(S,X)$ is weakly compact in $\big(C^0(S,X), \|\cdot\|_{\sup}\big)$ if it is bounded, equi-continuous and if for every $s \in S$, the set $\{f(s) \mid s \in S\}$ is relatively weakly compact in X.

A.12 Bochner Integrals and Weak Compactness in L^1

The so-called Bochner integral extends the familiar concept of integration from real-valued functions to Banach-valued functions on the basis of "simple" functions.

Definition 84 ([63]). Let (Ω, Σ, μ) be a finite measure space and X a Banach space. A function $f : \Omega \longrightarrow X$ is called *simple* if there exist $x_1, x_2 \ldots x_n \in X$ and $E_1, E_2 \ldots E_n \in \Sigma$ such that $f = \sum_{j=1}^{n} x_j \, \chi_{E_j}$ with $\chi_{E_j} : \Omega \longrightarrow \{0,1\}$ denoting the characteristic function of $E_j \subset \Omega$.
A function $f : \Omega \longrightarrow X$ is called *μ-measurable* if there exists a sequence $(f_n)_{n \in \mathbb{N}}$ of simple functions $\Omega \longrightarrow X$ with $\|f - f_n\|_X \longrightarrow 0$ μ-almost everywhere for $n \to \infty$.
A μ-measurable function $f : \Omega \longrightarrow X$ is called *Bochner integrable* if there exists a sequence $(f_n)_{n \in \mathbb{N}}$ of simple functions $\Omega \longrightarrow X$ such that

$$\lim_{n \to \infty} \int_\Omega \|f - f_n\|_X \, d\mu \;=\; 0.$$

Then, the *Bochner integral* of f over $E \in \Sigma$ is defined by $\int_E f \, d\mu := \lim_{n \to \infty} \int_E f_n \, d\mu$.

Let $L^1(\mu, X)$ denote the Banach space of Bochner integrable functions $\Omega \longrightarrow X$ equipped with its usual L^1 norm.

In the nineties, Ülger proved that restricting the values of Bochner integrable functions to a weakly compact subset of X implies the relative weak compactness of these functions in $L^1(\mu, X)$. For real-valued Lebesgue integrable functions, this is closely related with Alaoglu's Theorem and a compact embedding.

Proposition 85 ([179, Proposition 7]). *Let (Ω, Σ, μ) be a probabilistic space, X an arbitrary Banach space. For any weakly compact subset $W \subset X$, the set*

$$\left\{ h \in L^1(\mu, X) \;\middle|\; h(\omega) \in W \text{ for } \mu\text{-almost every } \omega \in \Omega \right\}$$

is relatively weakly compact.

An earlier version of this result is presented in [61] and, [62] considers weak compactness of Bochner integrable functions with values in an arbitrary Banach space under weaker assumptions (see also [22]). The next proposition of Ülger provides a "weakly pointwise" characterization of weakly convergent sequences in $L^1(\mu, X)$.

Proposition 86 ([179, Corollary 5]). *Let (Ω, Σ, μ) be a probabilistic space and X an arbitrary Banach space as in preceding Proposition A.85.*
Set $W := \left\{ g \in L^1(\mu, X) \;\middle|\; |g(\omega)| \le 1 \text{ for } \mu\text{-almost every } \omega \in \Omega \right\}$.

A sequence $\left(g_n(\cdot)\right)_{n \in \mathbb{N}}$ in $W \subset L^1(\mu, X)$ converges weakly to $g \in L^1(\mu, X)$ if and only if for any subsequence $\left(g_{n_k}(\cdot)\right)_{k \in \mathbb{N}}$ given, there exists a sequence $\left(h_k(\cdot)\right)_{k \in \mathbb{N}}$ with $h_k \in co\left\{g_{n_k}, \, g_{n_{k+1}} \ldots\right\}$ such that for μ-almost every $\omega \in \Omega$,

$$h_k(\omega) \longrightarrow g(\omega) \qquad (k \longrightarrow \infty) \qquad \text{weakly in } X.$$

Appendix B
Bibliographical Notes

Chapter 1

This chapter reflects the theory of mutational equations as it was introduced by Jean-Pierre Aubin in the 1990s [10, 12, 13]. It extends earlier results about integral funnel equations – for describing set evolutions with feedback. Similar concepts have been introduced by Russian mathematicians in the 1980s and 1990s. Among the more popular examples for metric spaces are the so-called *quasidifferential equations* of Panasyuk (see [150, 153] and references there). Further approaches to generalized differential equations in metric spaces are suggested in [31, 108, 113, 146] later.
Both the structure and the proofs in Chapter 1 are adapted to the generalizations in subsequent chapters so that the new aspects there are easier to identify.

§ 1.9.3 provides new results in comparison with Aubin's monograph [10]: The link between morphological primitives and reachable sets of nonautonomous differential inclusions. The analytical tools are summarized in Appendix A.3.
The examples of morphological primitives in § 1.9.4 are motivated by several questions of Robert Baier during our joint research stay at the Hausdorff Research Institute for Mathematics (HIM) in Bonn in spring 2008.
§ 1.9.5 is mostly based on earlier results of Anne Gorre quoted in Aubin's monograph [13]. Proposition 69 provides a partial answers to an open question that Jean-Pierre Aubin posed the author in November 2007. The closely related conclusions are drawn in Corollary 78.

§ 1.10 presents a set-valued approach to image segmentation that was published by the author in [131] in 2001.

§ 1.11 was developed during the stay at HIM in Bonn after the author had learned more about one-sided Lipschitz maps in the survey lectures of Tzanko Donchev.

Chapter 2

This chapter provides the first extensions of the mutational framework in comparison with Aubin's monograph [10]. They are based on the key notion that the parameters of transitions are just locally uniform.
Continuity parameters *with linear growth* were introduced in the first version of preprint [129] about transport equations for Radon measures in 2005. Later the linear growth condition was weakened to locally uniform bounds as in this chapter. These details were presented in the preprint [126] for the first time and used in [91]. The results about existence with delay and under state constraints in § 2.3.5 and § 2.3.6 respectively have been developed in the initial version of this monograph, i.e. habilitation thesis [117].

The example in § 2.4 dealing with semilinear evolution equations (and the weak topology) in the mutational framework has already been suggested in the author's Ph.D. thesis [130] in 2004.

The Cauchy problem of nonlinear transport equations for Radon measures on $\mathbb{R}^N$ was discussed in the preprint [126] with the same kind of transitions, but another metric and restricted to positive Radon measures with compact support. Hence the results of § 2.5 using the $W^{1,\infty}$ dual metric and solutions in the mutational framework are new in the initial version [117] of this book.

The nonlinear structured population model in § 2.6 provides the main conclusions of [91], which was jointly elaborated with Piotr Gwiazda (Warsaw) and Anna Marciniak-Czochra (Heidelberg).

In § 2.7, morphological equations are modified in a very "natural" way as transitions on $\mathscr{K}(\mathbb{R}^N)$ are now induced by reachable sets of differential inclusions *with linear growth*. In particular, this opens the door to applying the mutational framework to reachable sets of *linear* differential inclusions.

Chapter 3

It provides three substantial contributions of this monograph to mutational analysis:

1. Continuity conditions on distances make the triangle inequality dispensable,
2. sequential continuity of transitions with respect to state and time are handled by separate families of distances,
3. ω-contractivity of transitions (in the sense that the initial distance between states may grow at most exponentially while evolving along one and the same transition) proves to be dispensable under additional assumptions.

Currently the author is not aware of any other approach similar to mutational or quasidifferential equations beyond metric *spaces.*

Nonlocal stochastic differential equations as discussed in § 3.6 were introduced in the initial version [117] of this monograph (to the best of our knowledge). A relevant extension to nonadditive noise is sketched in [119] and generalized in [105].

In comparison with the thesis [117], the statements about time-dependent random closed sets in § 3.7 and nonlocal parabolic equations in cylindrical domains in § 3.9 belong to the new contributions here.

The only further suggestion (for constructing stochastic birth-and-growth processes via random closed sets in continuous time) which the author has found in the literature so far was made by Aletti, Bongiorno and Capasso in the preprint [1] in 2008. Their proposal is based on random closed sets in a reflexive Banach space and takes the aspect of predictability (via filtration) into consideration explicitly. Due to the Aumann integral as an essential ingredient, however, it is restricted to both convex-valued and expanding growth processes [1, Theorem 1.3.2]. Moreover, it assumes the nucleation process to be expanding and so, the final birth-and-growth processes are expanding P-almost surely in Ω [1, Theorem 1.4.9]. The set-valued approach in § 3.7 here has been developed independently (as a part of a DFG project, the author applied for in 2007). From our current point of view, this concept can be adapted to random closed sets in a separable Banach space Y if $\mathscr{RC}^2(\Omega, Y)$ proves to be complete with respect to $dl_{\mathscr{RC}}$ and if a counterpart of the Da Prato-Frankowska Theorem A.67 holds (see e.g. the original article [53]).

With regard to § 3.8, nonlinear continuity equations with coefficients of bounded variation were investigated as examples of mutational equations in the preprint [129] after attending the lectures of Prof. Ambrosio in a C.I.M.E. summer school in 2005.

The conclusions about semilinear evolution equations in § 3.10 and about parabolic differential equations in noncylindrical domains in § 3.11 respectively are also developed originally in the author's thesis [117] and published here now.

During the Czech-German-French Conference on Optimization in Heidelberg in September 2007 and a workshop at HIM Bonn in March 2008, José Alberto Murillo Hernández (Cartagena, Spain) reported about the heat equation in a domain governed by a morphological equation — similarly to § 3.11.5.
His conclusions were based on the results [116] of Límaco, Medeiros and Zuazua and thus, the noncylindrical domain had to obey bi-Lipschitz transformations to a reference domain. As a consequence, the morphological transitions were restricted to bounded Lipschitz continuous vector fields (instead of the set-valued maps in $\mathrm{LIP}(\mathbb{R}^N, \mathbb{R}^N)$).

Chapters 4 and 5

The author suggested the notion of distribution-like solutions in his Ph.D. thesis [130], but still for tuples with non-symmetric distance functions which fulfill the timed triangle inequality. The example in § 4.4 was also presented in [124, 130]. The second geometric example here in § 4.5 was introduced in [120] in 2008.

In regard to mutational inclusions, the existence results of § 5.1 have been developed in connection with the thesis [117] recently and, they are published in [118]. § 5.2 about the viability theorem for morphological inclusions was prepared in [122] and published in its full generality in [121].
The corresponding approach to control problems (in § 5.3) has its origin in preprint [125] and was motivated by conversations with Zvi Artstein at Weizmann Institute of Science in Rehovot (Israel) in summer 2007.

Appendix A

The generalizations of Gronwall's inequality are essentially new. In particular, Proposition A.2 has less restrictive assumptions than all the other versions which the author found in the literature. It lays the foundations for concluding global estimates from local properties (Lebesgue-almost everywhere). Proposition A.4 has already been presented in Ph.D. thesis [130].

Section A.3 provides the tools for the link between morphological primitives and reachable sets: the integral funnel equation in Proposition A.13. Following a strategy close to the one of Frankowska, Plaskacz and Rzeżuchowski in [83], the author has proved this connection in 2006 and reused these arguments in [121, Corollary 3.14] and [122] later. He developed these proofs independently from earlier results of Tolstonogov [177], which the author found while writing his thesis [117] since 2008.

Most of the results in section A.5 were introduced and proved in [120, 124, 130]. In particular, they were developed by the author independently from the article [33] of Cannarsa and Frankowska (about the interior sphere property of reachable sets of control equations). The consequences of the uniform tusk condition in A.5.7 are presented in [117] and published here for the first time.
Originally Reynolds Transport Theorem in § A.6 was extended to differential inclusions for applications in image segmentation [131]. The author published its complete proof in [127].

Sections A.2, A.4, A.7, A.8 and A.10 – A.12 summarize standard results which are mostly quoted and prove to be useful in this monograph.

References

1. Aletti, G., Bongiorno, E.G. & Capasso, V. (2008): *A set-valued framework for birth-and-growth process*, submitted preprint at `arXiv:0805.3912v1`
2. Alt, H.-W. (2006): *Lineare Funktionalanalysis*, Springer, fifth edition
3. Ambrosio, L. (2008): Transport Equation and Cauchy Problem for Non-Smooth Vector Fields, in: Dacorogna, B. & Marcellini, P. (Eds.), *Calculus of Variations and Nonlinear Partial Differential Equations*, Springer, Lecture Notes in Mathematics 1927, pp.1-41
4. Ambrosio, L. (2004): Transport equation and Cauchy problem for BV vector fields, *Invent. Math.* **158**, no.2, pp.227-260
5. Ambrosio, L., Fusco, N. & Pallara, D. (2000): *Functions of bounded variation and free discontinuity problems*, Oxford Mathematical Monographs
6. Ambrosio, L., Gigli, N. & Savaré, G. (2005): *Gradient flows in metric spaces and in the space of probability measures*, Birkhäuser, ETH Lecture Notes in Mathematics
7. Antosiewicz, H.A. & Cellina, A. (1977): Continuous extensions of multifunctions, *Ann. Pol. Math.* **34**, pp.107-111
8. Antosiewicz, H.A. & Cellina, A. (1975): Continuous selections and differential relations, *J. Differ. Equations* **19**, pp.386-398
9. Artstein, Z. (1994): First-order approximations for differential inclusions, *Set-Valued Anal.* **2**, pp.7-18
10. Aubin, J.-P. (1999): *Mutational and Morphological Analysis : Tools for Shape Evolution and Morphogenesis*, Birkhäuser 1999
11. Aubin, J.-P. (1997): *Dynamic Economic Theory. A Viability Approach*, Springer
12. Aubin, J.-P. (1993): Mutational equations in metric spaces, *Set-Valued Anal.* **1**, pp.3-46
13. Aubin, J.-P. (1992): A note on differential calculus in metric spaces and its applications to the evolution of tubes, *Bull. Pol. Acad. Sci., Math.* **40**, no.2, pp.151-162
14. Aubin, J.-P. (1991): *Viability Theory*, Birkhäuser
15. Aubin, J.-P. & Cellina, A. (1984): *Differential Inclusions*, Springer, Grundlehren der mathematischen Wissenschaften 264
16. Aubin, J.-P. & Da Prato, G. (1998): The viability theorem for stochastic differential inclusions, *Stochastic Anal. Appl.* **16**, no.1, pp.1-15
17. Aubin, J.-P. & Da Prato, G. (1990): Stochastic viability and invariance, *Ann. Scuola Norm. Sup. Pisa Cl. Sci. (4)* **17**, no.4, pp.595-613
18. Aubin, J.-P., Da Prato, G. & Frankowska, H. (2000): Stochastic invariance for differential inclusions, *Set-Valued Anal.* **8**, pp.181-201
19. Aubin, J.-P. & Frankowska, H. (1990): *Set-Valued Analysis*, Birkhäuser
20. Ball, J.M. (1977): Strongly continuous semigroups, weak solutions, and the variation of constants formula, *Proc. Am. Math. Soc.* **63**, pp.370-373
21. Bangerth, W. & Rannacher, R. (2003): *Adaptive finite element methods for differential equations*, ETH Zürich, Birkhäuser
22. Bárcenas, D. (1991): Weak compactness criteria for set-valued integrals and Radon–Nikodym theorem for vector-valued multimeasures, *Czech. Math. J.* **51** (126), pp.493-504
23. Bardi, M. & Capuzzo-Dolcetta, I. (1997): *Optimal control and viscosity solutions of Hamilton-Jacobi-Bellman equations*, Birkhäuser
24. Barles, G., Soner, H.M. & Souganidis, P.E. (1993): Front propagation and phase field theory, *SIAM J. Control Optim.* **31**, no.2, pp.439-469
25. Barron, E.N., Cannarsa, P., Jensen, R. & C. Sinestrari (1999): Regularity of Hamilton–Jacobi equations when forward is backward, *Indiana Univ. Math. J.* **48**, no.2, pp.385-409
26. Bauer, H. (1992): *Mass– und Integrationstheorie*, de Gruyter
27. Beer, G. (1993): *Topologies on closed and closed convex sets*, Kluwer
28. Bellettini, G. & Novaga, M. (1998): Comparison results between minimal barriers and viscosity solutions for geometric evolutions, *Ann. Sc. Norm. Super. Pisa, Cl. Sci., IV.Ser.* **26**, no.1, pp.97-131

29. Bellettini, G. & Novaga, M. (1997): Minimal barriers for geometric evolutions, *J. Differ. Equations* **139**, no.1, pp.76-103
30. Bogachev, V.I. (2007): *Measure theory*, Vol. II, Springer
31. Calcaterra, C. & Bleecker, D. (2000): Generating flows on metric spaces, *J. Math. Anal. Appl.* **248**, no.2, pp.645-677
32. Cannarsa, P., Da Prato, G. & Zolésio, J.-P. (1989): Evolution equations in non-cylindrical domains, *Atti Accad. Naz. Lincei, VIII. Ser., Rend., Cl. Sci. Fis. Mat. Nat.* **83**, pp.73-77
33. Cannarsa, P. & Frankowska, H. (2006): Interior sphere property of attainable sets and time optimal control problems, *ESAIM Control Optim. Calc. Var.* **12**, no.2, pp.350-370
34. Cannarsa, P., Frankowska, H. & Sinestrari, C. (2000): Optimality conditions and synthesis for the minimum time problem, *Set-Valued Anal.* **8**, no.1-2, pp.127-148
35. Cannarsa, P. & Frankowska, H. (1991): Some characterizations of optimal trajectories in control theory, *SIAM J. Control Optimization* **29**, no.6, pp.1322-1347
36. Cardaliaguet, P. (2001): Front propagation problems with nonlocal terms II, *J. Math. Anal. Appl.* **260**, no.2, pp.572-601
37. Cardaliaguet, P. (2000): On front propagation problems with nonlocal terms, *Adv. Differ. Equ.* **5**, no.1-3, pp.213-268
38. Cardaliaguet, P. & Pasquignon, D. (2001): On the approximation of front propagation problems with nonlocal terms, *M2AN, Math. Model. Numer. Anal.* **35**, no.3, pp.437-462
39. Caselles, V., Catté F., Coll, T. & Dibos, F. (1993): A geometric model for active contours, *Numerische Mathematik* **66**, pp.1-31
40. Castaing, C. & Valadier, M. (1977): *Convex analysis and measurable multifunctions*, Springer
41. Catté F., Lions, P.-L., Morel, J.-M. & Coll, T. (1992): Image selective smoothing and edge detection by nonlinear diffusion, *SIAM J. Numer. Anal.* **29**, no.1, pp.182-193
42. Céa, J. (1976): Une méthode numérique pour la recherche d"un domaine optimal, in: Glowinski, R. & Lions, J.L. (Eds.), *Computing methods in applied sciences and engineering. Part 1*, Springer, Lecture Notes in Economics and Mathematical Systems 134, pp.245-257
43. Chen, Y.G., Giga, Y. & Goto, Sh. (1991): Uniqueness and existence of viscosity solutions of generalized mean curvature flow equations, *J. Differential Geom.* **33**, no.3, pp.749-786
44. Chen, Y.G., Giga, Y. & Goto, Sh. (1989): Uniqueness and existence of viscosity solutions of generalized mean curvature flow equations, *Proc. Japan Acad. Ser.A Math. Sci.* **65**, no.7, pp.207-210
45. Clarke, F.H. (1996): A proximal characterization of the reachable set, *Syst. Control Lett.* **27**, pp.195-197
46. Clarke, F.H. (1983): *Optimization and Nonsmooth Analysis*, Wiley
47. Clarke, F.H., Ledyaev, Yu.S. & Stern R.J. (1997): Complements, approximations, smoothings & invariance properties, *J. Convex Anal.* **4**, no.2, pp.189-219
48. Clarke, F.H., Stern, R.J. & Wolenski, P.R. (1995): Proximal smoothness and the lower-C^2 property, *J. Convex Anal.* **2**, no.1/2, pp.117-144
49. Conway, E.D. (1967): Generalized Solutions of Linear Differential Equations with Discontinuous Coefficients and the Uniqueness Question for Multidimensional Quasilinear Conservation Laws, *J. Math. Anal. and Appl.* **18**, pp.238-251
50. Cornet, B. & Czarnecki, M.-O. (1999): Smooth normal approximations of epi-Lipschitzian subsets of $\mathbb{R}^n$, *SIAM J. Control Optim.* **37**, no.3, pp.710-730
51. Crandall, M.G., Ishii, H. & Lions, P.-L. (1992): User's guide to viscosity solutions of second order partial differential equations, *Bull. Amer. Math. Soc. (N.S.)* **27**, no.1, pp.1-67
52. Crandall, M.G. & Lions, P.-L. (1983): Viscosity solutions of Hamilton-Jacobi equations, *Trans. Amer. Math. Soc.* **277**, no.1, pp.1-42
53. Da Prato, G. & Frankowska, H. (1994): A stochastic Filippov theorem, *Stoch. Anal. Appl* **12**, no.4, pp.409-426
54. De Giorgi, E. (1994): *Barriers, boundaries, motion of manifolds*, Lectures held in Pavia
55. Delfour, M. & Zolésio, J.-P. (2001): *Shapes and Geometries: Analysis, Differential Calculus and Optimization*, SIAM, Advances in Design and Control
56. Delfour, M. & Zolésio, J.-P. (1991): Velocity method and Lagrangian formulation for the computation of the shape Hessian, *SIAM J. Control Optim.* **29**, no.6, pp.1414-1442

57. Demongeot, J., Kulesa, P. & Murray, J.D. (1997): Compact set valued flows. II: Applications in biological modelling, *C. R. Acad. Sci., Paris, Sér.* II, *Fasc. b* 324, no.2, pp.107-115
58. Demongeot, J. & Leitner, F. (1996): Compact set valued flows. I: Applications in medical imaging, *C. R. Acad. Sci., Paris, Sér.* II, *Fasc. b* 323, no.11, pp.747-754
59. Diekmann, O., Gyllenberg, M, Huang, H., Kirkilionis, M., Metz, J.A.J. & Thieme, H.R. (2001): On the formulation and analysis of general deterministic structured population models: II. Nonlinear theory, *J. Math. Biol.* **43**, pp.157-189.
60. Diekmann O. and Getto P. (2005): Boundedness, global existence and continuous dependence for nonlinear dynamical systems describing physiologically structured populations. *J. Differ. Equations*, **215**, pp.268-319.
61. Diestel, J. (1977): Remarks on weak compactness in $L^1(\mu,X)$, *Glasgow Math. J.* **18**, pp.87-91
62. Diestel, J., Ruess, W.M. & Schachermeyer, W. (1993): Weak compactness in $L^1(\mu,X)$, *Proc. Amer. Math. Soc.* **118**, no.2, pp.443-453
63. Diestel, J. & Uhl, J. (1977): *Vector measures*, Math. Surveys vol. 15, Amer. Math. Soc.
64. DiPerna, R.J. and Lions, P.L. (1989): Ordinary differential equations, transport theory and Sobolev spaces, *Inventiones Mathematicae*, **98**, pp.511-547
65. Donchev, T. (2002): Properties of the reachable set of control systems, *Systems Control Lett.* **46**, no.5, pp.379-386
66. Donchev, T. (2002): Properties of one-sided Lipschitz multivalued maps, *Nonlinear Analysis*, **49**, no.1, pp.13-20
67. Donchev, T. (1997): Lower semicontinuous differential inclusions. One-sided Lipschitz approach, *Colloq. Math.* **74**, no.2, pp.177-184
68. Donchev, T. & Dontchev, A.L. (2008): Extensions of Clarke's proximal characterization for reachable mappings of differential inclusions, *J. Math. Anal. Appl.* **348**, no.1, pp.454-460
69. Donchev, T. & Farkhi, E. (1998): Stability and Euler approximations of one-sided Lipschitz differential inclusions, *SIAM J. Control Optimization* **36**, no.2, pp.780-796
70. Dong, G.C. (1991): *Nonlinear Partial Differential Equations of Second Order*, AMS Translations of Mathematical Monographs 95
71. Doyen, L. (1995): Mutational equations for shapes and vision-based control, *J. Math. Imaging Vis.* **5**, no.2, pp.99-109
72. Doyen, L. (1994): Inverse function theorems and shape optimization, *SIAM J. Control Optimization* **32**, no.6, pp.1621-1642
73. Doyen, L. (1994): Shape Lyapunov functions and stabilization of reachable tubes of control problems. *J. Math. Anal. Appl.* **184**, no.2, pp.222-228
74. Doyen, L. (1993): Filippov and invariance theorems for mutational inclusions of tubes, *Set-Valued Anal.* **1**, no.3, pp.289-303
75. Effros, E.G. & Kazdan, J.L. (1971): On the Dirichlet problem for the heat equation, *Indiana Univ. Math. J.* **20**, no.8, pp.683-693
76. Engel, K.-J. & Nagel, R. (2000): *One–Parameter Semigroups of Linear Evolution Equations*, Springer
77. Evans, L.C. & Gariepy, R.F. (1992): *Measure theory and fine properties of functions*, CRC Press
78. Federer, H. (1969): *Geometric Measure Theory*, Springer
79. Federer, H. (1959): Curvature measures, *Trans. Am. Math. Soc.* **93**, pp.418–491
80. Frankowska, H. (2002): Value function in optimal control, in: Agrachev, A. A. (ed.), *Mathematical control theory*, ICTP Lect. Notes. 8, pp.515-653
81. Frankowska, H. (1989): Optimal trajectories associated with a solution of the contingent Hamilton-Jacobi equation, *Appl. Math. Optimization* **19**, no.3, pp.291-311
82. Frankowska, H. (1989): Contingent cones to reachable sets of control systems, *SIAM J. Control Optimization* **27**, no.1, pp.170-198
83. Frankowska, H., Plaskacz, S. & Rzeżuchowski, T. (1995): Measurable viability theorems and the Hamilton-Jacobi-Bellman equation, *J. Differ. Equations* **116**, no.2, pp.265-305
84. Fremlin, D.H. (1980/81): Measurable functions and almost continuous functions. *Manuscripta Math.* **33**, no.3-4, pp.387-405

85. Friedman, A. (1975): *Stochastic differential equations and applications*, Vol.1, Academic Press
86. Gautier, S. & Pichard, K. (2003): Viability results for mutational equations with delay, *Numer. Funct. Anal. Optimiz.* **24**, no.3 & 4, pp.273-284
87. Gilbarg, D. & Trudinger, N.S. (1983): Elliptic partial differential equations of second order, second edition, Springer
88. Green, J.W. & Valentine, F.A. (1960/61): On the Arzela-Ascoli theorem, *Math. Mag.* **34**, pp.199-202
89. Gorre, A. (1997): Evolutions of tubes under operability constraints. *J. Math. Anal. Appl.* **216**, no.1, pp.1-22
90. Gorre, A. (1996): Evolution de Tubes Opérables gouvernée par des équations mutationnelles, Thèse de l'Université de Paris-Dauphine
91. Gwiazda, P., Lorenz, Th. & Marciniak-Czochra, A. (2010): A nonlinear structured population model: Lipschitz continuity of measure-valued solutions with respect to model ingredients, to appear in *Journal of Differential Equations*
92. Hartman, Ph. (1964): *Ordinary Differential Equations*, Wiley
93. Hirsch, M.W., Smale, St. & Devaney R.L. (2004): *Differential Equations, Dynamical Systems, and an Introduction to Chaos* (second edition), Elsevier
94. Iannelli, M. (1995): *Mathematical Theory of Age-Structured Population Dynamics*, Applied Math. Monographs, CNR
95. Jähne, B. (2005): *Digital image processing* (sixth edition), Springer
96. Jarnik, J. & Kurzweil, J. (1977): On conditions on right hand sides of differential relations, *Ċasopis Pěst. Mat.* **102**, pp.334-349
97. Karatzas, I. & Shreve, St.E. (1991): *Brownian motion and stochastic calculus* (second edition), Springer
98. Kass, M., Witkin, A. & Terzopoulos, D. (1987): Snakes, active contour models, *ICCV* 1987, pp.259-268
99. Kelley, J.L, (1955): *General topology*, Van Nostrand, reprinted by Springer
100. Kisielewicz, M. (2001): Weak compactness of solution sets to stochastic differential inclusions with convex right-hand sides, *Topol. Methods Nonlinear Anal.* **18**, no.1, pp.149-169
101. Kisielewicz, M. (1997): Set-valued stochastic integrals and stochastic inclusions, *Stochastic Anal. Appl.* **15**, no.5, pp.783-800
102. Kisielewicz, M. (1992): Weak compactness in spaces $C(S,X)$, in: *Information theory, statistical decision functions, random processes*, Trans. 11th Prague Conf., Prague/Czech. 1990, Vol. B, pp.101-106 (1992)
103. Kisielewicz, M. (1982): Multivalued differential equations in separable Banach spaces, *J. Optimization Theory Appl.* **37**, pp.231-249
104. Kisielewicz, M., Michta, M. & Motyl, J. (2003): Set valued approach to stochastic control I & II, *Dynam. Systems Appl.* **12**, no.3-4, pp.405-466
105. Kloeden, P.E. & Lorenz, Th. (2010): Stochastic differential equations with nonlocal sample dependence, to appear in *Stoch. Anal. Appl.*
106. Kloeden, P.E. & Platen, E. (1992): *Numerical solution of stochastic differential equations*, Springer
107. Kloeden, P.E., Real, J. & Sun, Ch. (2009): Pullback attractors for a semilinear heat equation on time-varying domains, *J. Differ. Equations*, in press
108. Kloeden, P.E., Sadovsky, D.N. & Vasilyeva, I.E. (2002): Quasi-flows and equations with nonlinear differentials, *Nonlinear Anal.* **51**, no.7, pp.1143-1158
109. Kuratowski, K. & Ryll-Nardzewski, Cz. (1965): A general theorem on selectors, *Bull. Acad. Pol. Sci., Sér. Sci. Math. Astron. Phys.* **13**, pp.397-403
110. Kurzhanskij, A.B. & Filippova, T.F. (1989): On the set-valued calculus in problems of viability and control for dynamic processes: The evolution equation, *Ann. Inst. Henri Poincaré, Anal. Non Linéaire* **6**, Suppl., pp.339-363
111. Ladyzhenskaya, O.A. (1985): *The boundary value problems of mathematical physics*, Springer

112. Ladyzhenskaya, O.A., Solonnikov, V.A. & Ural'tseva, N.N. (1968): *Linear and quasi-linear equations of parabolic type*, Translations of Mathematical Monographs 23, American Mathematical Society (AMS)
113. Lakshmikantham, V. & Nieto, J.J. (2003): Differential equations in metric spaces: An introduction and an application to fuzzy differential equations, *Dyn. Contin. Discrete Impuls. Syst. Ser. A Math. Anal.* **10**, no.6, pp.991-1000
114. Lieberman, G. (1996): *Second order parabolic differential equations*, World Scientific Publishing
115. Lieberman, G. (1989): Intermediate Schauder theory for second order parabolic equations. III. The tusk conditions, *Appl. Anal.* **33**, no.1-2, pp.25-43
116. Límaco, J., Medeiros, L.A. & Zuazua, E. (2002): Existence, uniqueness and controllability for parabolic equations in non-cylindrical domains, *Mat. Contemp.* **23**, pp.49-70
117. Lorenz, Th. (2009): *Mutational Analysis. A joint framework for dynamical systems in and beyond vector spaces*, Habilitationsschrift (i.e. thesis for a postdoctoral lecture qualification), Ruprecht–Karls–University of Heidelberg
118. Lorenz, Th. (2009): Mutational inclusions: Differential inclusions in metric spaces, to appear in *Discrete Contin. Dyn. Syst. Ser. B*
119. Lorenz, Th. (2009): Nonlocal stochastic differential equations: Existence and uniqueness of solutions, to appear in *Bol. Soc. Esp. Mat. Apl.*
120. Lorenz, Th. (2008): Generalizing mutational equations for uniqueness of some nonlocal 1st-order geometric evolutions, *Set-Valued Anal.* **16**, pp.1-50
121. Lorenz, Th. (2008): Shape evolutions under state constraints: A viability theorem, *J. Math. Anal. Appl.* **340**, pp.1204-1225
122. Lorenz, Th. (2008): A viability theorem for morphological inclusions, *SIAM J. Control Optim.* **47**, no.3, pp.1591-1614
123. Lorenz, Th. (2008): Epi-Lipschitzian reachable sets of differential inclusions, *System Control Letters* **57**, pp.703-707
124. Lorenz, Th. (2008): Evolution equations in ostensible metric spaces: First-order evolutions of nonsmooth sets with nonlocal terms, *Discuss. Math., Differ. Incl. Control Optim.* **28**, pp.15-73
125. Lorenz, Th. (2008): Control problems for nonlocal set evolutions, submitted Preprint available at `http://www.ub.uni-heidelberg.de/archiv/8022`
126. Lorenz, Th. (2007): Radon measures solving the Cauchy problem of the nonlinear transport equation, Preprint available at `http://www.ub.uni-heidelberg.de/archiv/7252`
127. Lorenz, Th. (2006): Reynold's transport theorem for differential inclusions, *Set-Valued Anal.* **14**, no.3, pp.209-247
128. Lorenz, Th. (2005): Boundary regularity of reachable sets of control systems, *Syst. Control Lett.* **54**, no.9, pp.919-924
129. Lorenz, Th. (2005): Quasilinear continuity equations of measures for bounded BV vector fields, Preprint available at `http://www.ub.uni-heidelberg.de/archiv/5993`
130. Lorenz, Th. (2004): *First–order geometric evolutions and semilinear evolution equations : A common mutational approach.* Doctor thesis (PhD), Ruprecht–Karls–University of Heidelberg, 2004, `http://www.ub.uni-heidelberg.de/archiv/4949`
131. Lorenz, Th. (2001): Set-valued maps for image segmentation, *Comput. Vis. Sci.* **4**, no.1, pp.41-57
132. Lumer, G. & Schnaubelt, R. (2001): Time-dependent parabolic problems on non-cylindrical domains with inhomogeneous boundary conditions. Dedicated to Ralph S. Phillips, *J. Evol. Equ.* **1**, no.3, pp.291-309
133. Lumer, G. & Schnaubelt, R. (1999): Local operator methods and time dependent parabolic equations on non-cylindrical domains, in: Demuth, M. et al. (eds.), *Evolution equations, Feshbach resonances, singular Hodge theory*, Math. Top. 16, Wiley, pp.58-130
134. Maniglia, St. (2007): Probabilistic representation and uniqueness results for measure-valued solutions of transport equations, *J. Math. Pures Appl.* (9) **87**, no.6, pp.601-626
135. Mao, Xuerong (2008): *Stochastic differential equations and applications*, second edition, Horwood Publishing Limited

136. Matheron, G. (1975): *Radom Sets and Integral Geometry*, Wiley
137. Metz J.A.J. and Diekmann O. (1986): *The Dynamics of Physiologically Structured Populations*, Springer, Lecture Notes in Biomathematics, vol. 68
138. Michael, E. (1956): Continuous selections I, *Annals of Math.* **63**, pp.361-381
139. Michta, M. (2003): On solutions to stochastic differential inclusions, *Discrete Contin. Dyn. Syst.* 2003, suppl., pp.618-622
140. Michta, M. (2002): Optimal solutions to stochastic differential inclusions, *Appl. Math. (Warsaw)* **29**, no.4, pp.387-398
141. Molchanov, I. (2005): *Theory of Random Sets*, Springer
142. Motyl, J. (2000): Viable solutions of set-valued stochastic equation, *Optimization* **48**, no.2, pp.157-176
143. Motyl, J. (1995): On the solution of stochastic differential inclusion, *J. Math. Anal. Appl.* **192**, no.1, pp.117-132
144. Murillo Hernández, J.A. (2006): Tangential regularity in the space of directional-morphological transitions, *J. Convex Anal.* **13**, no.2, pp.423-441
145. Nagumo, M. (1942): Über die Lage der Integralkurven gewöhnlicher Differentialgleichungen, *Proc. Phys.-Math. Soc. Japan*, III. *Ser.* **24**, pp.551-559
146. Nieto, J.J. & Rodríguez-López, R. (2006): Hybrid metric dynamical systems with impulses, *Nonlinear Anal.* **64**, no.2, pp.368-380
147. Øksendal B. (2003): *Stochastic differential equations. An introduction with applications* (sixth edition), Springer
148. Osher, St. & Fedkiw, R. (2003): *Level set methods and dynamic implicit surfaces*, Springer
149. Osher, St. & Sethian, J.A. (1988): Fronts propagating with curvature-dependent speed: algorithms based on Hamilton-Jacobi formulations, *J. Comput. Phys.* **79**, no.1, pp.12-49
150. Panasyuk, A.I. (1995): Quasidifferential equations in a complete metric space under conditions of the Carathéodory type. I, *Differ. Equations* **31**, no.6, pp.901-910
151. Panasyuk, A.I. (1992): Properties of solutions of a quasidifferential approximation equation and the equation of an integral funnel, *Differ. Equations* 28, no.9, pp.1259-1266
152. Panasyuk, A.I. (1990): Equations of attainable set dynamics, part 1: Integral funnel equations, *J. Optimization Theory Appl.* **64**, no.2, pp.349-366
153. Panasyuk, A.I (1985): Quasidifferential equations in metric spaces, *Differ. Equations* **21**, pp.914-921
154. Pazy, A. (1983): *Semigroups of Linear Operators and Applications to Partial Differential Equations*, Springer
155. Perona, P. & Malik, J. (1990): Scale-space and edge detection using anisotropic diffusion, *IEEE Trans. Pattern Anal. Mach. Intel.* **12**, no.7, pp.629-639
156. Perona, P. & Malik, J. (1987): Scale space and edge detection using anisotropic diffusion, in: *Proc. of the IEEE Computer Society Workshop on Computer Vision, Miami, FL*, pp.16-27
157. Perthame, B. (2007): Transport equations in biology. Birkhäuser
158. Pichard, K. & Gautier, S. (2000): Equations with delay in metric spaces: The mutational approach, *Numer. Funct. Anal. Optimiz.* **21**, no.7 & 8, pp.917-932
159. Poliquin, R.A., Rockafellar, R.T & Thibault, L. (2000): Local differentiability of distance functions, *Trans. Am. Math. Soc.* **352**, no.11, pp.5231-5249
160. Ricceri, B. & Villani, A. (1983): Separability and Scorza–Dragoni's property, *Le Matematiche*, **37**, no.1, pp.156-161
161. Rockafellar, R.T. (1979): Clarke's tangent cones and the boundaries of closed sets in $\mathbb{R}^n$, *Nonlinear Anal. Theor. Meth. Appl.* **3**, pp.145-154
162. Rockafellar, R.T. & Wets, R. (1998): *Variational Analysis*, Springer
163. Royden, H.L. (1988): Comparison theorems for the matrix Riccati equation, *Commun. Pure Appl. Math.* **41**, no.5, pp.739–746
164. Rzeżuchowski, T. (1999): Continuous parameterization of attainable sets by solutions of differential inclusions, *Set–Valued Anal.* **7**, pp.347-355
165. Rzeżuchowski, T. (1997): Boundary solutions of differential inclusions and recovering the initial data, *Set–Valued Anal.* **5**, pp.181-193

166. Rzeżuchowski, T. (1979): On the set where all the solutions satisfy a differential inclusion, in: *Qualitative theory of differential equations*, Vol. II, Szeged 1979, Colloq. Math. Soc. János Bolyai 30 (1981), pp.903-913
167. Scorza-Dragoni, G. (1948): Un teorema sulle funzioni continue rispetto ad una e misurabili rispetto ad un'altra variabile, *Rend. Semin. Mat. Univ. Padova*, **17**, pp.102-108
168. Schaefer, H.H. (1971): Topological vector spaces, third printing corrected, Springer
169. Schwartz, L. (1973): *Radon measures*, Oxford University Press
170. Searcóid, M. (2007): *Metric Spaces*, Springer
171. Serra, J. (1982): *Image Analysis and Mathematical Morphology*, Academic Press
172. Sethian, J.A. (1999): *Level Set Methods and Fast Marching Methods*, second edition, Cambridge University Press
173. Smirnov, V. I. (1953): A Course in Higher Mathematics. Vol. 4
174. Sokolowski, J. & Zolésio, J.-P. (1992): *Introduction to Shape Optimization. Shape Sensitivity Analysis*, Springer
175. Souganidis, P.E. (1997): Front propagation: theory and applications, in: Capuzzo Dolcetta, I.at al. (eds.), *Viscosity solutions and applications* Springer Lecture Notes in Math. 1660, pp.186-242
176. Stassinopoulos, G. & Vinter, R. (1979): Continuous dependence of solutions of a differential inclusion on the right hand side with applications to stability of optimal control problems, *SIAM J. Control Optim.* **17**, no.3, pp.432-449
177. Tolstonogov, A.A. (1983): Equation of the solution funnel of a differential inclusion, *Math. Notes* **32**, pp.908-914
178. Truong-Van, B. & Truong, Xuan Duc Ha (1999): Existence results for viability problem associated to nonconvex stochastic differentiable inclusions, *Stochastic Anal. Appl.* **17**, no.4, pp.667-685
179. Ülger, A. (1991): Weak compactness in $L^1(\mu,X)$, *Proc. Amer. Math. Soc.* **113**, no.1, pp.143-149
180. Vinter, R. (2000): *Optimal Control*, Birkhäuser
181. Walter, W. (2000): *Gewöhnliche Differentialgleichungen* (seventh edition), Springer
182. Webb, G.F. (1985): *Nonlinear Age-Dependent Population Dynamics*, Dekker
183. Weickert, J. (1998): *Anisotropic diffusion in image processing*, Teubner
184. Werner, D. (2002): *Funktionalanalysis* (4th edition), Springer
185. Williams, D.J. & Shah, M. (1992): A Fast Algorithm for Active Contours and Curvature Estimation, *CVGIP : Image Understanding*, **55**, no.1, pp.14-26
186. Wolenski, P.R. & Zhuang, Yu (1998): Proximal analysis and the minimal time function, *SIAM J. Control Optimization* **36**, no.3, pp.1048-1072
187. Xu, Daoyi, Yang, Zhigua & Huang, Yumei (2008): Existence-uniqueness and continuation theorems for stochastic functional differential equations, *J. Differential Equations*, **245**, pp.1681-1703
188. Yosida, K. (1978): *Functional Analysis* (fifth edition), Springer, Grundlehren der mathematischen Wissenschaften 123
189. Ziemer, W. (1989): *Weakly Differentiable Functions*, Springer
190. Zolésio, J.-P. (1979): Identification de domaine par déformations. *Thèse de doctorat d'état*, Université de Nice

Index of Notation

Index

Lecture Notes in Mathematics

For information about earlier volumes
please contact your bookseller or Springer
LNM Online archive: springerlink.com

Vol. 1809: O. Steinbach, Stability Estimates for Hybrid Coupled Domain Decomposition Methods (2003)
Vol. 1810: J. Wengenroth, Derived Functors in Functional Analysis (2003)
Vol. 1811: J. Stevens, Deformations of Singularities (2003)
Vol. 1812: L. Ambrosio, K. Deckelnick, G. Dziuk, M. Mimura, V. A. Solonnikov, H. M. Soner, Mathematical Aspects of Evolving Interfaces. Madeira, Funchal, Portugal 2000. Editors: P. Colli, J. F. Rodrigues (2003)
Vol. 1813: L. Ambrosio, L. A. Caffarelli, Y. Brenier, G. Buttazzo, C. Villani, Optimal Transportation and its Applications. Martina Franca, Italy 2001. Editors: L. A. Caffarelli, S. Salsa (2003)
Vol. 1814: P. Bank, F. Baudoin, H. Fllmer, L.C.G. Rogers, M. Soner, N. Touzi, Paris-Princeton Lectures on Mathematical Finance 2002 (2003)
Vol. 1815: A. M. Vershik (Ed.), Asymptotic Combinatorics with Applications to Mathematical Physics. St. Petersburg, Russia 2001 (2003)
Vol. 1816: S. Albeverio, W. Schachermayer, M. Talagrand, Lectures on Probability Theory and Statistics. Ecole d'Eté de Probabilités de Saint-Flour XXX-2000. Editor: P. Bernard (2003)
Vol. 1817: E. Koelink, W. Van Assche (Eds.), Orthogonal Polynomials and Special Functions. Leuven 2002 (2003)
Vol. 1818: M. Bildhauer, Convex Variational Problems with Linear, nearly Linear and/or Anisotropic Growth Conditions (2003)
Vol. 1819: D. Masser, Yu. V. Nesterenko, H. P. Schlickewei, W. M. Schmidt, M. Waldschmidt, Diophantine Approximation. Cetraro, Italy 2000. Editors: F. Amoroso, U. Zannier (2003)
Vol. 1820: F. Hiai, H. Kosaki, Means of Hilbert Space Operators (2003)
Vol. 1821: S. Teufel, Adiabatic Perturbation Theory in Quantum Dynamics (2003)
Vol. 1822: S.-N. Chow, R. Conti, R. Johnson, J. Mallet-Paret, R. Nussbaum, Dynamical Systems. Cetraro, Italy 2000. Editors: J. W. Macki, P. Zecca (2003)
Vol. 1823: A. M. Anile, W. Allegretto, C. Ringhofer, Mathematical Problems in Semiconductor Physics. Cetraro, Italy 1998. Editor: A. M. Anile (2003)
Vol. 1824: J. A. Navarro González, J. B. Sancho de Salas, $\mathscr{C}^\infty$ – Differentiable Spaces (2003)
Vol. 1825: J. H. Bramble, A. Cohen, W. Dahmen, Multiscale Problems and Methods in Numerical Simulations, Martina Franca, Italy 2001. Editor: C. Canuto (2003)
Vol. 1826: K. Dohmen, Improved Bonferroni Inequalities via Abstract Tubes. Inequalities and Identities of Inclusion-Exclusion Type. VIII, 113 p, 2003.
Vol. 1827: K. M. Pilgrim, Combinations of Complex Dynamical Systems. IX, 118 p, 2003.
Vol. 1828: D. J. Green, Grbner Bases and the Computation of Group Cohomology. XII, 138 p, 2003.
Vol. 1829: E. Altman, B. Gaujal, A. Hordijk, Discrete-Event Control of Stochastic Networks: Multimodularity and Regularity. XIV, 313 p, 2003.
Vol. 1830: M. I. Gil', Operator Functions and Localization of Spectra. XIV, 256 p, 2003.
Vol. 1831: A. Connes, J. Cuntz, E. Guentner, N. Higson, J. E. Kaminker, Noncommutative Geometry, Martina Franca, Italy 2002. Editors: S. Doplicher, L. Longo (2004)
Vol. 1832: J. Azéma, M. Émery, M. Ledoux, M. Yor (Eds.), Séminaire de Probabilités XXXVII (2003)
Vol. 1833: D.-Q. Jiang, M. Qian, M.-P. Qian, Mathematical Theory of Nonequilibrium Steady States. On the Frontier of Probability and Dynamical Systems. IX, 280 p, 2004.
Vol. 1834: Yo. Yomdin, G. Comte, Tame Geometry with Application in Smooth Analysis. VIII, 186 p, 2004.
Vol. 1835: O.T. Izhboldin, B. Kahn, N.A. Karpenko, A. Vishik, Geometric Methods in the Algebraic Theory of Quadratic Forms. Summer School, Lens, 2000. Editor: J.-P. Tignol (2004)
Vol. 1836: C. Năstăsescu, F. Van Oystaeyen, Methods of Graded Rings. XIII, 304 p, 2004.
Vol. 1837: S. Tavaré, O. Zeitouni, Lectures on Probability Theory and Statistics. Ecole d'Eté de Probabilités de Saint-Flour XXXI-2001. Editor: J. Picard (2004)
Vol. 1838: A.J. Ganesh, N.W. O'Connell, D.J. Wischik, Big Queues. XII, 254 p, 2004.
Vol. 1839: R. Gohm, Noncommutative Stationary Processes. VIII, 170 p, 2004.
Vol. 1840: B. Tsirelson, W. Werner, Lectures on Probability Theory and Statistics. Ecole d'Eté de Probabilités de Saint-Flour XXXII-2002. Editor: J. Picard (2004)
Vol. 1841: W. Reichel, Uniqueness Theorems for Variational Problems by the Method of Transformation Groups (2004)
Vol. 1842: T. Johnsen, A. L. Knutsen, K_3 Projective Models in Scrolls (2004)
Vol. 1843: B. Jefferies, Spectral Properties of Noncommuting Operators (2004)
Vol. 1844: K.F. Siburg, The Principle of Least Action in Geometry and Dynamics (2004)
Vol. 1845: Min Ho Lee, Mixed Automorphic Forms, Torus Bundles, and Jacobi Forms (2004)
Vol. 1846: H. Ammari, H. Kang, Reconstruction of Small Inhomogeneities from Boundary Measurements (2004)
Vol. 1847: T.R. Bielecki, T. Bjrk, M. Jeanblanc, M. Rutkowski, J.A. Scheinkman, W. Xiong, Paris-Princeton Lectures on Mathematical Finance 2003 (2004)
Vol. 1848: M. Abate, J. E. Fornaess, X. Huang, J. P. Rosay, A. Tumanov, Real Methods in Complex and CR Geometry, Martina Franca, Italy 2002. Editors: D. Zaitsev, G. Zampieri (2004)
Vol. 1849: Martin L. Brown, Heegner Modules and Elliptic Curves (2004)
Vol. 1850: V. D. Milman, G. Schechtman (Eds.), Geometric Aspects of Functional Analysis. Israel Seminar 2002-2003 (2004)

Vol. 1851: O. Catoni, Statistical Learning Theory and Stochastic Optimization (2004)
Vol. 1852: A.S. Kechris, B.D. Miller, Topics in Orbit Equivalence (2004)
Vol. 1853: Ch. Favre, M. Jonsson, The Valuative Tree (2004)
Vol. 1854: O. Saeki, Topology of Singular Fibers of Differential Maps (2004)
Vol. 1855: G. Da Prato, P.C. Kunstmann, I. Lasiecka, A. Lunardi, R. Schnaubelt, L. Weis, Functional Analytic Methods for Evolution Equations. Editors: M. Iannelli, R. Nagel, S. Piazzera (2004)
Vol. 1856: K. Back, T.R. Bielecki, C. Hipp, S. Peng, W. Schachermayer, Stochastic Methods in Finance, Bressanone/Brixen, Italy, 2003. Editors: M. Fritelli, W. Runggaldier (2004)
Vol. 1857: M. Émery, M. Ledoux, M. Yor (Eds.), Séminaire de Probabilités XXXVIII (2005)
Vol. 1858: A.S. Cherny, H.-J. Engelbert, Singular Stochastic Differential Equations (2005)
Vol. 1859: E. Letellier, Fourier Transforms of Invariant Functions on Finite Reductive Lie Algebras (2005)
Vol. 1860: A. Borisyuk, G.B. Ermentrout, A. Friedman, D. Terman, Tutorials in Mathematical Biosciences I. Mathematical Neurosciences (2005)
Vol. 1861: G. Benettin, J. Henrard, S. Kuksin, Hamiltonian Dynamics – Theory and Applications, Cetraro, Italy, 1999. Editor: A. Giorgilli (2005)
Vol. 1862: B. Helffer, F. Nier, Hypoelliptic Estimates and Spectral Theory for Fokker-Planck Operators and Witten Laplacians (2005)
Vol. 1863: H. Führ, Abstract Harmonic Analysis of Continuous Wavelet Transforms (2005)
Vol. 1864: K. Efstathiou, Metamorphoses of Hamiltonian Systems with Symmetries (2005)
Vol. 1865: D. Applebaum, B.V. R. Bhat, J. Kustermans, J. M. Lindsay, Quantum Independent Increment Processes I. From Classical Probability to Quantum Stochastic Calculus. Editors: M. Schürmann, U. Franz (2005)
Vol. 1866: O.E. Barndorff-Nielsen, U. Franz, R. Gohm, B. Kümmerer, S. Thorbjønsen, Quantum Independent Increment Processes II. Structure of Quantum Lévy Processes, Classical Probability, and Physics. Editors: M. Schürmann, U. Franz, (2005)
Vol. 1867: J. Sneyd (Ed.), Tutorials in Mathematical Biosciences II. Mathematical Modeling of Calcium Dynamics and Signal Transduction. (2005)
Vol. 1868: J. Jorgenson, S. Lang, $Pos_n(R)$ and Eisenstein Series. (2005)
Vol. 1869: A. Dembo, T. Funaki, Lectures on Probability Theory and Statistics. Ecole d'Eté de Probabilités de Saint-Flour XXXIII-2003. Editor: J. Picard (2005)
Vol. 1870: V.I. Gurariy, W. Lusky, Geometry of Mntz Spaces and Related Questions. (2005)
Vol. 1871: P. Constantin, G. Gallavotti, A.V. Kazhikhov, Y. Meyer, S. Ukai, Mathematical Foundation of Turbulent Viscous Flows, Martina Franca, Italy, 2003. Editors: M. Cannone, T. Miyakawa (2006)
Vol. 1872: A. Friedman (Ed.), Tutorials in Mathematical Biosciences III. Cell Cycle, Proliferation, and Cancer (2006)
Vol. 1873: R. Mansuy, M. Yor, Random Times and Enlargements of Filtrations in a Brownian Setting (2006)
Vol. 1874: M. Yor, M. Émery (Eds.), In Memoriam Paul-Andr Meyer - Sminaire de Probabilits XXXIX (2006)
Vol. 1875: J. Pitman, Combinatorial Stochastic Processes. Ecole d'Et de Probabilits de Saint-Flour XXXII-2002. Editor: J. Picard (2006)
Vol. 1876: H. Herrlich, Axiom of Choice (2006)
Vol. 1877: J. Steuding, Value Distributions of L-Functions (2007)
Vol. 1878: R. Cerf, The Wulff Crystal in Ising and Percolation Models, Ecole d'Et de Probabilités de Saint-Flour XXXIV-2004. Editor: Jean Picard (2006)
Vol. 1879: G. Slade, The Lace Expansion and its Applications, Ecole d'Et de Probabilits de Saint-Flour XXXIV-2004. Editor: Jean Picard (2006)
Vol. 1880: S. Attal, A. Joye, C.-A. Pillet, Open Quantum Systems I, The Hamiltonian Approach (2006)
Vol. 1881: S. Attal, A. Joye, C.-A. Pillet, Open Quantum Systems II, The Markovian Approach (2006)
Vol. 1882: S. Attal, A. Joye, C.-A. Pillet, Open Quantum Systems III, Recent Developments (2006)
Vol. 1883: W. Van Assche, F. Marcellàn (Eds.), Orthogonal Polynomials and Special Functions, Computation and Application (2006)
Vol. 1884: N. Hayashi, E.I. Kaikina, P.I. Naumkin, I.A. Shishmarev, Asymptotics for Dissipative Nonlinear Equations (2006)
Vol. 1885: A. Telcs, The Art of Random Walks (2006)
Vol. 1886: S. Takamura, Splitting Deformations of Degenerations of Complex Curves (2006)
Vol. 1887: K. Habermann, L. Habermann, Introduction to Symplectic Dirac Operators (2006)
Vol. 1888: J. van der Hoeven, Transseries and Real Differential Algebra (2006)
Vol. 1889: G. Osipenko, Dynamical Systems, Graphs, and Algorithms (2006)
Vol. 1890: M. Bunge, J. Funk, Singular Coverings of Toposes (2006)
Vol. 1891: J.B. Friedlander, D.R. Heath-Brown, H. Iwaniec, J. Kaczorowski, Analytic Number Theory, Cetraro, Italy, 2002. Editors: A. Perelli, C. Viola (2006)
Vol. 1892: A. Baddeley, I. Bárány, R. Schneider, W. Weil, Stochastic Geometry, Martina Franca, Italy, 2004. Editor: W. Weil (2007)
Vol. 1893: H. Hanßmann, Local and Semi-Local Bifurcations in Hamiltonian Dynamical Systems, Results and Examples (2007)
Vol. 1894: C.W. Groetsch, Stable Approximate Evaluation of Unbounded Operators (2007)
Vol. 1895: L. Molnár, Selected Preserver Problems on Algebraic Structures of Linear Operators and on Function Spaces (2007)
Vol. 1896: P. Massart, Concentration Inequalities and Model Selection, Ecole d'Été de Probabilités de Saint-Flour XXXIII-2003. Editor: J. Picard (2007)
Vol. 1897: R. Doney, Fluctuation Theory for Lévy Processes, Ecole d'Été de Probabilités de Saint-Flour XXXV-2005. Editor: J. Picard (2007)
Vol. 1898: H.R. Beyer, Beyond Partial Differential Equations, On Linear and Quasi Linear Abstract Hyperbolic Evolution Equations (2007)
Vol. 1899: Séminaire de Probabilités XL. Editors: C. Donati-Martin, M. Émery, A. Rouault, C. Stricker (2007)
Vol. 1900: E. Bolthausen, A. Bovier (Eds.), Spin Glasses (2007)
Vol. 1901: O. Wittenberg, Intersections de deux quadriques et pinceaux de courbes de genre 1, Intersections of Two Quadrics and Pencils of Curves of Genus 1 (2007)

Vol. 1902: A. Isaev, Lectures on the Automorphism Groups of Kobayashi-Hyperbolic Manifolds (2007)
Vol. 1903: G. Kresin, V. Maz'ya, Sharp Real-Part Theorems (2007)
Vol. 1904: P. Giesl, Construction of Global Lyapunov Functions Using Radial Basis Functions (2007)
Vol. 1905: C. Prévôt, M. Röckner, A Concise Course on Stochastic Partial Differential Equations (2007)
Vol. 1906: T. Schuster, The Method of Approximate Inverse: Theory and Applications (2007)
Vol. 1907: M. Rasmussen, Attractivity and Bifurcation for Nonautonomous Dynamical Systems (2007)
Vol. 1908: T.J. Lyons, M. Caruana, T. Lévy, Differential Equations Driven by Rough Paths, Ecole d'Été de Probabilités de Saint-Flour XXXIV-2004 (2007)
Vol. 1909: H. Akiyoshi, M. Sakuma, M. Wada, Y. Yamashita, Punctured Torus Groups and 2-Bridge Knot Groups (I) (2007)
Vol. 1910: V.D. Milman, G. Schechtman (Eds.), Geometric Aspects of Functional Analysis. Israel Seminar 2004-2005 (2007)
Vol. 1911: A. Bressan, D. Serre, M. Williams, K. Zumbrun, Hyperbolic Systems of Balance Laws. Cetraro, Italy 2003. Editor: P. Marcati (2007)
Vol. 1912: V. Berinde, Iterative Approximation of Fixed Points (2007)
Vol. 1913: J.E. Marsden, G. Misiołek, J.-P. Ortega, M. Perlmutter, T.S. Ratiu, Hamiltonian Reduction by Stages (2007)
Vol. 1914: G. Kutyniok, Affine Density in Wavelet Analysis (2007)
Vol. 1915: T. Bıyıkoğlu, J. Leydold, P.F. Stadler, Laplacian Eigenvectors of Graphs. Perron-Frobenius and Faber-Krahn Type Theorems (2007)
Vol. 1916: C. Villani, F. Rezakhanlou, Entropy Methods for the Boltzmann Equation. Editors: F. Golse, S. Olla (2008)
Vol. 1917: I. Veselić, Existence and Regularity Properties of the Integrated Density of States of Random Schrdinger (2008)
Vol. 1918: B. Roberts, R. Schmidt, Local Newforms for GSp(4) (2007)
Vol. 1919: R.A. Carmona, I. Ekeland, A. Kohatsu-Higa, J.-M. Lasry, P.-L. Lions, H. Pham, E. Taflin, Paris-Princeton Lectures on Mathematical Finance 2004. Editors: R.A. Carmona, E. inlar, I. Ekeland, E. Jouini, J.A. Scheinkman, N. Touzi (2007)
Vol. 1920: S.N. Evans, Probability and Real Trees. Ecole d'Été de Probabilités de Saint-Flour XXXV-2005 (2008)
Vol. 1921: J.P. Tian, Evolution Algebras and their Applications (2008)
Vol. 1922: A. Friedman (Ed.), Tutorials in Mathematical BioSciences IV. Evolution and Ecology (2008)
Vol. 1923: J.P.N. Bishwal, Parameter Estimation in Stochastic Differential Equations (2008)
Vol. 1924: M. Wilson, Littlewood-Paley Theory and Exponential-Square Integrability (2008)
Vol. 1925: M. du Sautoy, L. Woodward, Zeta Functions of Groups and Rings (2008)
Vol. 1926: L. Barreira, V. Claudia, Stability of Nonautonomous Differential Equations (2008)
Vol. 1927: L. Ambrosio, L. Caffarelli, M.G. Crandall, L.C. Evans, N. Fusco, Calculus of Variations and Non-Linear Partial Differential Equations. Cetraro, Italy 2005. Editors: B. Dacorogna, P. Marcellini (2008)
Vol. 1928: J. Jonsson, Simplicial Complexes of Graphs (2008)
Vol. 1929: Y. Mishura, Stochastic Calculus for Fractional Brownian Motion and Related Processes (2008)
Vol. 1930: J.M. Urbano, The Method of Intrinsic Scaling. A Systematic Approach to Regularity for Degenerate and Singular PDEs (2008)
Vol. 1931: M. Cowling, E. Frenkel, M. Kashiwara, A. Valette, D.A. Vogan, Jr., N.R. Wallach, Representation Theory and Complex Analysis. Venice, Italy 2004. Editors: E.C. Tarabusi, A. D'Agnolo, M. Picardello (2008)
Vol. 1932: A.A. Agrachev, A.S. Morse, E.D. Sontag, H.J. Sussmann, V.I. Utkin, Nonlinear and Optimal Control Theory. Cetraro, Italy 2004. Editors: P. Nistri, G. Stefani (2008)
Vol. 1933: M. Petkovic, Point Estimation of Root Finding Methods (2008)
Vol. 1934: C. Donati-Martin, M. Émery, A. Rouault, C. Stricker (Eds.), Séminaire de Probabilités XLI (2008)
Vol. 1935: A. Unterberger, Alternative Pseudodifferential Analysis (2008)
Vol. 1936: P. Magal, S. Ruan (Eds.), Structured Population Models in Biology and Epidemiology (2008)
Vol. 1937: G. Capriz, P. Giovine, P.M. Mariano (Eds.), Mathematical Models of Granular Matter (2008)
Vol. 1938: D. Auroux, F. Catanese, M. Manetti, P. Seidel, B. Siebert, I. Smith, G. Tian, Symplectic 4-Manifolds and Algebraic Surfaces. Cetraro, Italy 2003. Editors: F. Catanese, G. Tian (2008)
Vol. 1939: D. Boffi, F. Brezzi, L. Demkowicz, R.G. Durán, R.S. Falk, M. Fortin, Mixed Finite Elements, Compatibility Conditions, and Applications. Cetraro, Italy 2006. Editors: D. Boffi, L. Gastaldi (2008)
Vol. 1940: J. Banasiak, V. Capasso, M.A.J. Chaplain, M. Lachowicz, J. Miękisz, Multiscale Problems in the Life Sciences. From Microscopic to Macroscopic. Będlewo, Poland 2006. Editors: V. Capasso, M. Lachowicz (2008)
Vol. 1941: S.M.J. Haran, Arithmetical Investigations. Representation Theory, Orthogonal Polynomials, and Quantum Interpolations (2008)
Vol. 1942: S. Albeverio, F. Flandoli, Y.G. Sinai, SPDE in Hydrodynamic. Recent Progress and Prospects. Cetraro, Italy 2005. Editors: G. Da Prato, M. Rckner (2008)
Vol. 1943: L.L. Bonilla (Ed.), Inverse Problems and Imaging. Martina Franca, Italy 2002 (2008)
Vol. 1944: A. Di Bartolo, G. Falcone, P. Plaumann, K. Strambach, Algebraic Groups and Lie Groups with Few Factors (2008)
Vol. 1945: F. Brauer, P. van den Driessche, J. Wu (Eds.), Mathematical Epidemiology (2008)
Vol. 1946: G. Allaire, A. Arnold, P. Degond, T.Y. Hou, Quantum Transport. Modelling, Analysis and Asymptotics. Cetraro, Italy 2006. Editors: N.B. Abdallah, G. Frosali (2008)
Vol. 1947: D. Abramovich, M. Mariño, M. Thaddeus, R. Vakil, Enumerative Invariants in Algebraic Geometry and String Theory. Cetraro, Italy 2005. Editors: K. Behrend, M. Manetti (2008)
Vol. 1948: F. Cao, J-L. Lisani, J-M. Morel, P. Mus, F. Sur, A Theory of Shape Identification (2008)
Vol. 1949: H.G. Feichtinger, B. Helffer, M.P. Lamoureux, N. Lerner, J. Toft, Pseudo-Differential Operators. Quantization and Signals. Cetraro, Italy 2006. Editors: L. Rodino, M.W. Wong (2008)
Vol. 1950: M. Bramson, Stability of Queueing Networks, Ecole d'Eté de Probabilits de Saint-Flour XXXVI-2006 (2008)

Vol. 1951: A. Moltó, J. Orihuela, S. Troyanski, M. Valdivia, A Non Linear Transfer Technique for Renorming (2009)

Vol. 1952: R. Mikhailov, I.B.S. Passi, Lower Central and Dimension Series of Groups (2009)

Vol. 1953: K. Arwini, C.T.J. Dodson, Information Geometry (2008)

Vol. 1954: P. Biane, L. Bouten, F. Cipriani, N. Konno, N. Privault, Q. Xu, Quantum Potential Theory. Editors: U. Franz, M. Schuermann (2008)

Vol. 1955: M. Bernot, V. Caselles, J.-M. Morel, Optimal Transportation Networks (2008)

Vol. 1956: C.H. Chu, Matrix Convolution Operators on Groups (2008)

Vol. 1957: A. Guionnet, On Random Matrices: Macroscopic Asymptotics, Ecole d'Eté de Probabilits de Saint-Flour XXXVI-2006 (2009)

Vol. 1958: M.C. Olsson, Compactifying Moduli Spaces for Abelian Varieties (2008)

Vol. 1959: Y. Nakkajima, A. Shiho, Weight Filtrations on Log Crystalline Cohomologies of Families of Open Smooth Varieties (2008)

Vol. 1960: J. Lipman, M. Hashimoto, Foundations of Grothendieck Duality for Diagrams of Schemes (2009)

Vol. 1961: G. Buttazzo, A. Pratelli, S. Solimini, E. Stepanov, Optimal Urban Networks via Mass Transportation (2009)

Vol. 1962: R. Dalang, D. Khoshnevisan, C. Mueller, D. Nualart, Y. Xiao, A Minicourse on Stochastic Partial Differential Equations (2009)

Vol. 1963: W. Siegert, Local Lyapunov Exponents (2009)

Vol. 1964: W. Roth, Operator-valued Measures and Integrals for Cone-valued Functions and Integrals for Cone-valued Functions (2009)

Vol. 1965: C. Chidume, Geometric Properties of Banach Spaces and Nonlinear Iterations (2009)

Vol. 1966: D. Deng, Y. Han, Harmonic Analysis on Spaces of Homogeneous Type (2009)

Vol. 1967: B. Fresse, Modules over Operads and Functors (2009)

Vol. 1968: R. Weissauer, Endoscopy for GSP(4) and the Cohomology of Siegel Modular Threefolds (2009)

Vol. 1969: B. Roynette, M. Yor, Penalising Brownian Paths (2009)

Vol. 1970: M. Biskup, A. Bovier, F. den Hollander, D. Ioffe, F. Martinelli, K. Netočný, F. Toninelli, Methods of Contemporary Mathematical Statistical Physics. Editor: R. Kotecký (2009)

Vol. 1971: L. Saint-Raymond, Hydrodynamic Limits of the Boltzmann Equation (2009)

Vol. 1972: T. Mochizuki, Donaldson Type Invariants for Algebraic Surfaces (2009)

Vol. 1973: M.A. Berger, L.H. Kauffmann, B. Khesin, H.K. Moffatt, R.L. Ricca, De W. Sumners, Lectures on Topological Fluid Mechanics. Cetraro, Italy 2001. Editor: R.L. Ricca (2009)

Vol. 1974: F. den Hollander, Random Polymers: École d'Été de Probabilités de Saint-Flour XXXVII – 2007 (2009)

Vol. 1975: J.C. Rohde, Cyclic Coverings, Calabi-Yau Manifolds and Complex Multiplication (2009)

Vol. 1976: N. Ginoux, The Dirac Spectrum (2009)

Vol. 1977: M.J. Gursky, E. Lanconelli, A. Malchiodi, G. Tarantello, X.-J. Wang, P.C. Yang, Geometric Analysis and PDEs. Cetraro, Italy 2001. Editors: A. Ambrosetti, S.-Y.A. Chang, A. Malchiodi (2009)

Vol. 1978: M. Qian, J.-S. Xie, S. Zhu, Smooth Ergodic Theory for Endomorphisms (2009)

Vol. 1979: C. Donati-Martin, M. Émery, A. Rouault, C. Stricker (Eds.), Śeminaire de Probablitiés XLII (2009)

Vol. 1980: P. Graczyk, A. Stos (Eds.), Potential Analysis of Stable Processes and its Extensions (2009)

Vol. 1981: M. Chlouveraki, Blocks and Families for Cyclotomic Hecke Algebras (2009)

Vol. 1982: N. Privault, Stochastic Analysis in Discrete and Continuous Settings. With Normal Martingales (2009)

Vol. 1983: H. Ammari (Ed.), Mathematical Modeling in Biomedical Imaging I. Electrical and Ultrasound Tomographies, Anomaly Detection, and Brain Imaging (2009)

Vol. 1984: V. Caselles, P. Monasse, Geometric Description of Images as Topographic Maps (2010)

Vol. 1985: T. Linß, Layer-Adapted Meshes for Reaction-Convection-Diffusion Problems (2010)

Vol. 1986: J.-P. Antoine, C. Trapani, Partial Inner Product Spaces. Theory and Applications (2009)

Vol. 1987: J.-P. Brasselet, J. Seade, T. Suwa, Vector Fields on Singular Varieties (2010)

Vol. 1988: M. Broué, Introduction to Complex Reflection Groups and Their Braid Groups (2010)

Vol. 1989: I.M. Bomze, V. Demyanov, Nonlinear Optimization. Cetraro, Italy 2007. Editors: G. di Pillo, F. Schoen (2010)

Vol. 1990: S. Bouc, Biset Functors for Finite Groups (2010)

Vol. 1991: F. Gazzola, H.-C. Grunau, G. Sweers, Polyharmonic Boundary Value Problems (2010)

Vol. 1992: A. Parmeggiani, Spectral Theory of Non-Commutative Harmonic Oscillators: An Introduction (2010)

Vol. 1993: P. Dodos, Banach Spaces and Descriptive Set Theory: Selected Topics (2010)

Vol. 1994: A. Baricz, Generalized Bessel Functions of the First Kind (2010)

Vol. 1995: A.Y. Khapalov, Controllability of Partial Differential Equations Governed by Multiplicative Controls (2010)

Vol. 1996: T. Lorenz, Mutational Analysis. A Joint Framework for Cauchy Problems In and Beyond Vector Spaces (2010)

Recent Reprints and New Editions

Vol. 1702: J. Ma, J. Yong, Forward-Backward Stochastic Differential Equations and their Applications. 1999 – Corr. 3rd printing (2007)

Vol. 830: J.A. Green, Polynomial Representations of GL_n, with an Appendix on Schensted Correspondence and Littelmann Paths by K. Erdmann, J.A. Green and M. Schoker 1980 – 2nd corr. and augmented edition (2007)

Vol. 1693: S. Simons, From Hahn-Banach to Monotonicity (Minimax and Monotonicity 1998) – 2nd exp. edition (2008)

Vol. 470: R.E. Bowen, Equilibrium States and the Ergodic Theory of Anosov Diffeomorphisms. With a preface by D. Ruelle. Edited by J.-R. Chazottes. 1975 – 2nd rev. edition (2008)

Vol. 523: S.A. Albeverio, R.J. Høegh-Krohn, S. Mazzucchi, Mathematical Theory of Feynman Path Integral. 1976 – 2nd corr. and enlarged edition (2008)

Vol. 1764: A. Cannas da Silva, Lectures on Symplectic Geometry 2001 – Corr. 2nd printing (2008)

LECTURE NOTES IN MATHEMATICS

Springer

Edited by J.-M. Morel, F. Takens, B. Teissier, P.K. Maini

Editorial Policy (for the publication of monographs)

1. Lecture Notes aim to report new developments in all areas of mathematics and their applications - quickly, informally and at a high level. Mathematical texts analysing new developments in modelling and numerical simulation are welcome.

   Monograph manuscripts should be reasonably self-contained and rounded off. Thus they may, and often will, present not only results of the author but also related work by other people. They may be based on specialised lecture courses. Furthermore, the manuscripts should provide sufficient motivation, examples and applications. This clearly distinguishes Lecture Notes from journal articles or technical reports which normally are very concise. Articles intended for a journal but too long to be accepted by most journals, usually do not have this "lecture notes" character. For similar reasons it is unusual for doctoral theses to be accepted for the Lecture Notes series, though habilitation theses may be appropriate.

2. Manuscripts should be submitted either to Springer's mathematics editorial in Heidelberg, or to one of the series editors. In general, manuscripts will be sent out to 2 external referees for evaluation. If a decision cannot yet be reached on the basis of the first 2 reports, further referees may be contacted: The author will be informed of this. A final decision to publish can be made only on the basis of the complete manuscript, however a refereeing process leading to a preliminary decision can be based on a pre-final or incomplete manuscript. The strict minimum amount of material that will be considered should include a detailed outline describing the planned contents of each chapter, a bibliography and several sample chapters.

   Authors should be aware that incomplete or insufficiently close to final manuscripts almost always result in longer refereeing times and nevertheless unclear referees' recommendations, making further refereeing of a final draft necessary.

   Authors should also be aware that parallel submission of their manuscript to another publisher while under consideration for LNM will in general lead to immediate rejection.

3. Manuscripts should in general be submitted in English. Final manuscripts should contain at least 100 pages of mathematical text and should always include

   - a table of contents;
   - an informative introduction, with adequate motivation and perhaps some historical remarks: it should be accessible to a reader not intimately familiar with the topic treated;
   - a subject index: as a rule this is genuinely helpful for the reader.

   For evaluation purposes, manuscripts may be submitted in print or electronic form, in the latter case preferably as pdf- or zipped ps-files. Lecture Notes volumes are, as a rule, printed digitally from the authors' files. To ensure best results, authors are asked to use the LaTeX2e style files available from Springer's web-server at:

   ftp://ftp.springer.de/pub/tex/latex/svmonot1/ (for monographs).

   Additional technical instructions, if necessary, are available on request from: lnm@springer.com.

4. Careful preparation of the manuscripts will help keep production time short besides ensuring satisfactory appearance of the finished book in print and online. After acceptance of the manuscript authors will be asked to prepare the final LaTeX source files (and also the corresponding dvi-, pdf- or zipped ps-file) together with the final printout made from these files. The LaTeX source files are essential for producing the full-text online version of the book (see www.springerlink.com/content/110312 for the existing online volumes of LNM).

   The actual production of a Lecture Notes volume takes approximately 12 weeks.

5. Authors receive a total of 50 free copies of their volume, but no royalties. They are entitled to a discount of 33.3% on the price of Springer books purchased for their personal use, if ordering directly from Springer.

6. Commitment to publish is made by letter of intent rather than by signing a formal contract. Springer-Verlag secures the copyright for each volume. Authors are free to reuse material contained in their LNM volumes in later publications: a brief written (or e-mail) request for formal permission is sufficient.

Addresses:

Professor J.-M. Morel, CMLA,
École Normale Supérieure de Cachan,
61 Avenue du Président Wilson, 94235 Cachan Cedex, France
E-mail: Jean-Michel.Morel@cmla.ens-cachan.fr

Professor F. Takens, Mathematisch Instituut,
Rijksuniversiteit Groningen, Postbus 800,
9700 AV Groningen, The Netherlands
E-mail: F.Takens@math.rug.nl

Professor B. Teissier, Institut Mathématique de Jussieu,
UMR 7586 du CNRS, Équipe "Géométrie et Dynamique",
175 rue du Chevaleret
75013 Paris, France
E-mail: teissier@math.jussieu.fr

For the "Mathematical Biosciences Subseries" of LNM:

Professor P.K. Maini, Center for Mathematical Biology,
Mathematical Institute, 24-29 St Giles,
Oxford OX1 3LP, UK
E-mail: maini@maths.ox.ac.uk

Springer, Mathematics Editorial I, Tiergartenstr. 17
69121 Heidelberg, Germany,
Tel.: +49 (6221) 487-8259
Fax: +49 (6221) 4876-8259
E-mail: lnm@springer.com